AQUANANOTECHNOLOGY
GLOBAL PROSPECTS

T0239957

AQUANANOTECHNOLOGY

GLOBAL PROSPECTS

EDITED BY

DAVID E. REISNER • T. PRADEEP

CRC Press
Taylor & Francis Group
Boca Raton London New York

CRC Press is an imprint of the
Taylor & Francis Group, an **informa** business

CRC Press
Taylor & Francis Group
6000 Broken Sound Parkway NW, Suite 300
Boca Raton, FL 33487-2742

First issued in paperback 2017

ISBN-13: 978-1-4665-1224-5 (hbk)
ISBN-13: 978-1-138-07309-8 (pbk)

Library of Congress Cataloging-in-Publication Data

Aquananotechnology : global prospects / edited by David E. Reisner, T. Pradeep.
  page txts cm
  Includes bibliographical references and index.
  ISBN 978-1-4665-1224-5 (hardback)
  1. Water--Purification--Materials. 2. Nanostructured materials--Environmental aspects. 3. Nanofiltration. I. Reisner, David Evans. II. Pradeep, T.

TD477.A64 2014
628.1'64--dc23
                                     2014027815

Visit the Taylor & Francis Web site at
http://www.taylorandfrancis.com

and the CRC Press Web site at
http://www.crcpress.com

Contents

v

Foreword

From an investment standpoint, I have always had a fondness for technology plays. As the managing director of Allen & Co. for over 34 years, I can appreciate the latent opportunities in funding new emerging technologies. I am pleased to have the chance to comment on a timely book focused on nanotechnology-based solutions to the world's water problems. There is no commodity more ubiquitous than water. The assured availability of drinking water is a key to sustainability of billions of earth's inhabitants. Water innovations are receiving unrivaled attention in the investment community.

I am familiar with some of the largest water customers in the world, in the areas of thermoelectric power, irrigation, aquaculture, public supply, and industrial use. It has often been said that Coca-Cola is the largest manufacturing consumer of drinking water in the world. It has a vested interest in clean water at an economical price. A recent press release indicates that Coca-Cola has invested in a company that has contributed to this book, WaterHealth International (WHI), and will bring safe drinking water to 1 million school children, delivering 500 million L of water a year, through installed water purification systems. These systems exploit the attractive performance characteristics of high surface area nanomaterials. For its part, in a similar vein, The Water Initiative (TWI) has pioneered point-of-use residential filtration systems in Mexico, with over 65,000 installations to date, with a focus on mitigation of arsenic and fluoride.

There are a number of emerging hot topics in the water space that have been brought to my attention. Resiliency planning for floods and droughts is paramount in light of ever-increasing storm patterns in both intensity and frequency. Storm water is indeed addressed in this book. Environmental problems associated with discharge of ballast and bilge waters, containing foreign animal species that contaminate local ecosystems, are of ever-increasing concern. We are entering an era in which wastewater is viewed as an asset. The so-called "produced water" from fracking operations is a hot-button topic. It takes 5 gal of water to generate 1 gal of fuel. Recent innovations now seek to clean up produced water to a level where it can be reinjected by tailoring the composition and using the brine to plug spent wells. These are just a few topics addressed in this comprehensive book that should appeal to both technologists and investors.

Jack Schneider

Preface

This is not a book on hydropolitics. Yet I would be remiss to ignore the implications of nano-tech solutions to water security and geopolitics. And technologists also cannot escape the geo-political implications of scarce water resources. Just Google "Friends of the Earth Middle East" (FoEME) to step into the Israeli–Palestinian–Jordanian waste–sewage–drinking water quagmire knee deep. In a recent report (February 2013) issued jointly by the Center for American Progress, the Stimson Center, and the Center for Climate and Security, noted Princeton scholar (and frequent Sunday morning talk show guest) Anne-Marie Slaughter prefaced a sleek publication addressing "The Arab Spring and Climate Change." She points out that a once-in-a-century winter drought in China (2010–2011) coupled with heat waves in Australia, Canada, Russia, and the Ukraine had a dramatic impact on wheat prices in wheat-importing states, mainly the Middle East. Only about 10% of wheat production is traded across borders, so droughts in wheat-producing countries can have critical impact on wheat importers. Egypt is the world's largest wheat importer, and the impact of rising wheat prices had a dramatic impact on the cost of food, just when Hosni Mubarak was ousted. Ironically, Mubarak used to address American leaders with "after me comes the flood." Friedman reports that Karim Sadjadpour, an Iran expert, says it's no longer "Après moi, le déluge," but "After me, the drought."

Similar droughts occurred in Libya and Syria. Fred Pearce, in his ominous book *When the Rivers Run Dry* (2006), noted that Muammar al-Gaddafi spent billions (US$27 billion) on a 3500 km pipeline to draw upon ancient aquifers under the Sahara desert, which are being rapidly depleted. One might conclude that Gaddafi would have spent less money importing Evian bottled water from Lake Geneva. Albeit obvious, the heaviness of H_2O dictates its logistics and distribution. French-Tunisian film director Abdellatif Kechiche in "La Vie d'Adèle" quotes prose poet Francis Ponge—water has "one vice, gravity." Water rejects all forms to obey gravity; it "loses all dignity because of that obsession."

Kazakhstan's Aral Sea is rapidly dwindling; you have to see the photos of fishing boats and fish packing plants rusting many miles from the current shoreline. And let's not forget the 1964 water crisis at Guantanamo Bay where the Cubans shut off the water supply from the Yaterus River due to an argument over fishing rights, forcing the Navy to bring in barges of water from Ocho Rios, Jamaica, until they could relocate a desalination unit from Point Loma, California. We could go on like this for the balance of this book and would no doubt cite the famous quip, casually attributed to Mark Twain, "Whiskey is for drinking: water is for fighting over." You see, water is the new oil. Curiously, this trend is inextricably tied in to the lower costs of fuel oil due to the dramatic rise in hydraulic fracturing ("fracking") of shale oil, a technology requiring huge amounts of water. Treatment of the so-called "produced water" from fracking operations is addressed in this book.

According to Pearce, Earth contains 1.4 billion km^3 of water. Unfortunately, 97% of the world's water is seawater. Coleridge picked up on this notion in 1798 in *The Rime of the Ancient Mariner*, "Water, water, everywhere, Nor any drop to drink." It is ironic that we are witnessing coastal flooding calamities in this day and age. Wasn't it T.S. Eliot warned "Fear death by water" in *The Waste Land*? Of course, the topic of this book relates more to technology solutions applicable to mitigating dwindling sources of potable water. Back to our water ledger.

The remaining water is fresh, yet ice comprises 2/3 of it, leaving 12 million km^3, most of which resides in aquifers. Only about 200,000 km^3 is above ground (with 180,000 km^3 tied up in lakes and permafrost). Atmospheric water vapor contains about 13,000 km^3. Throw in swamps and rivers, this leaves about 1000 km^3 for living organisms. But this is a static picture—it is the water cycle that matters. The slow part of the cycle relates to ancient aquifers, modestly replenished by rain. The fast cycle comprises evaporation from the oceans and recaptured rainwater. We can use water running from the land to the sea, which is accessible, about 9000 km^3 (1400 m^3 or 50,000 ft.^{3} per capita). Some 70% of river water goes into irrigation. The rivers are running dry. There you have it. This is the backdrop to the urgent calling for new (nano) technologies to treat wastewater, now viewed as an asset, to at least offset our shrinking water availability and expanding usage.

This new volume that I dub *"Aquananotechnology"* echoes the format of my previous CRC Press projects, viz., *Bionanotechnology: Global Prospects* and *Bionanotechnology II: Global Prospects*. It is in no way meant to be methodically comprehensive; in fact, it is a random walk of sorts, a result of having engaged authors in my professional travels. That randomness notwithstanding, I made a special effort to include as many entrepreneurial companies as possible. Any noticeable absences are not necessarily for my lack of asking (again and again!). As in the previous volumes, I make no claims as to editing of any technical content—the field is just too expansive and interdisciplinary. I am relying on reputation and credibility of authorship. I have sought out chapters from the traditional geographic "hotspots" of nanotechnology, though this distinction is somewhat shifting (India and China), if not fading. I make no apology for topic overlap. Different groups apply different solutions to the same water problems. From a sourcing standpoint, readers benefit from learning of regional expertise and commercial opportunities.

I am especially grateful to my coeditor Professor Thalappil Pradeep. He presides over the Thematic Unit of Excellence (TUE) on Water Purification Using Nanotechnology at IIT Madras in Chennai. I seized the opportunity to meet him face to face late on a hot humid June night in 2012 at a Gordon Research Conference at Mount Holyoke College. He organized the First International Conference on Emerging Technologies for Clean Water (September 14–16, 2012) at IIT Madras. I challenged him to recruit contributions from South Asia (Sri Lanka as well as India), which he did admirably. I have also actively pursued authors in parts of the world that seemed to be under the radar, e.g., Cuba, Iran, Pakistan, Argentina, and Spain. In fact, nearly a dozen countries are represented and include cities as far ranging as Lahore, Rio Cuarto, Voiron, Tainan City, Ahmedabad, Malmö, Bangalore, Havana, Chennai, Doornfontein, Beijing, and Kandy.

My Cuban contributors were a result of my attending the *NANO4 International Seminar on Nanosciences and Nanotechnologies* in Havana (September 2012). I will always remember the hospitality of Oliver Pérez (Finlay Institute and a contributor to my first bionanotech volume), who treated us to a lobster (caught with his own hands) luncheon in his Havana home. Of course, having dinner at the Havana Yacht Club in the company of "Fidelito" (Fidel's son) was hard to beat! On the other side of the globe, I was able to visit a hotbed of water technology activity at the *CAS Research Center for Eco-Environmental Sciences* (RCEES) in Beijing as part of a trip to Nanjing for the *IWA Symposium on Environmental Nanotechnology* (April 2013). This trip had a special meaning for me as my great uncle, John H. Reisner, was the Dean of the College of Agriculture and Forestry at Nanking University from 1914 to 1931. In Beijing, my thanks go to Professors Guo, Jiang, and Wang and their group members for hospitality and cook's tours of their labs at RCEES. My thanks also go to Grace Wei, and her father Liling who set up our NGI Beijing Representative Office, for sponsoring the trip. I would be remiss if I did not mention my close friend Vince Caprio for the *sine qua non* networking opportunities at the *Water Innovation Alliance* (WIA) meetings in Boston (September 2011 and 2012) and New York (February 2013). Finally, I'd like to thank Deane Dray (our sons rowed together on the Dartmouth crew team) for the networking at his *Citi*

Water Investment Conference (New York, June 2012). I am indebted to Mark Lindsay and Roc Isaacs for the valuable introduction to entrepreneur/investor Jack Schneider, former president of CBS Television, the impetus behind the launch of MTV, and the Managing Director at Allen & Co. for 34 years. I had the privilege to engage Jack in conversation a number of times at the Core Club. I am honored that he agreed to endorse the book in his Foreword.

Chapter title pages contain only the principal affiliation, with a few exceptions, in the event the work was evenly distributed across separate institutions. This is purely a formatting consideration and mitigates the gallimaufry effect from too much information. Authors' given names are spelled out for the book format. I hope I have not offended any of them. Detailed addresses/affiliations are to be found in the List of Contributors. The publisher has kindly provided a color insert section. I made efforts to pick those color images that appeared to benefit the most from a color display. You may wonder how the chapters are arranged. I sorted them alphabetically by lead author. Think of this approach vis-à-vis the L.L. Bean catalog, encouraging (no, better, forcing!) the reader to flip through all the articles, not just a specific section.

Contributions span a wide range of subject areas that fall under the aquananotechnology banner, either squarely or tangentially. In the final accounting, this volume has a strong emphasis on sorption media, with broad application to a myriad of contaminants, both geogenic and anthropogenic. Remember, it is not enough for water to be potable—it must also be palatable. Other areas include membranes, desalination, sensing, engineered polymers, magnetic nanomaterials, electrospun nanofibers, photocatalysis, endocrine disruptors, Al_{13} clusters, etc. Physics-based phenomena include subcritical water and cavitation-induced sonoluminescence (based on biomimicry processes adapted from the mantis shrimp) and fog harvesting. Emerging application areas include treatment of produced water (from fracking), bilge and ballast water, storm water, and landfill leachates. Let's not forget the energy–water nexus, which according to David Zetland (*The End of Abundance*) can be eliminated by application of microeconomics and judicious pricing by the utilities.

On a recent random walk around the University of Pennsylvania campus at a NanoBCA conference, just weeks before releasing the manuscript to the publisher, I happened upon Benjamin Franklin's Way, where granite pavers inscribed with eighteenth century axioms of Penn's founder abound, including "When the well's dry, we know the worth of water" (1746). Water, it's everywhere. Deep in the bowels (p. 445) of his "maximalist" novel *Infinite Jest*, the late David Foster Wallace relates a profound little parable. I quote verbatim:

> This wise old whiskery fish swims up to three young fish and goes, 'Morning, boys, how's the water?' and swims away; and the three young fish watch him swim away and they look at each other and go 'What the fuck is water?' and swim away.

Ah, the most obvious and important realities are often the ones most difficult to talk about and see.

It is my sincere hope that this book inspires readers grounded in many disciplines to get involved in the burgeoning area of aquananotechnology, as much good innovation occurs at the crossroads of disparate disciplines. Zhiwu, a character in Hong Kong director Wong Kar-wai's best-known film, *Fallen Angels*, states, "they say women are made of water," introducing me to a well-known Chinese proverb (女人是水做的), finding its origins in Cao Xueqin's *Dream of the Red Chamber* (1791). In this spirit, I dedicate this book to all women.

David E. Reisner
The Nano Group, Inc.
Manchester, Connecticut, USA

Editors

Kaieteur Falls, world's largest (single drop, 741 ft) waterfall in Guyana ("land of many waters") rain forest. Note water bottle, world's largest land fill component.

David E. Reisner, PhD, is a well-known early pioneer and entrepreneur in the burgeoning field of nanotechnology, having cofounded in 1996 two nanotech companies in Connecticut, Inframat® and US Nanocorp®. For nearly 15 years, he was CEO of both companies since founding, which were recognized in Y2002–Y2005 for their fast revenue growth as Deloitte & Touche *Connecticut Technology Fast50 Award* recipients. In 2004, The Nano Group Inc. was formed as a parent holding company for investment. Reisner and cofounders were featured in *Forbes* magazine in 2004.

David has more than 175 publications and is an inventor on 10 issued patents. He is the editor for the Bionanotechnology sections of both the third edition of *The BioMedical Engineering Handbook* (CRC Press) and the *Handbook of Research on Biomedical Engineering Education and Advanced Bioengineering Learning* (IGI Global). He is editor of *Bionanotechnology: Global Prospects* (CRC Press) and *Bionanotechnology II: Global Prospects* (CRC Press). He has written articles on the business of nanotechnology in *Nanotechnology Law & Business* as well as the Chinese publication *Science & Culture Review.*

David served a 3-year term as a technology pioneer for the World Economic Forum and was a panelist at the 2004 Annual Meeting in Davos. David has served on the Board of the Connecticut Venture Group and was chairman of the board of the Connecticut Technology Council from 2005 to 2009. He was a NASA *NanoTech Briefs* Nano50 awardee in 2006. For his efforts in the field of medical implantable devices, he won the first annual BEACON award for Medical Technology in 2004. He is a member of the Connecticut Academy of Science and Engineering.

David is a 1978 University Honors graduate from Wesleyan University and received his PhD from Massachusetts Institute of Technology in 1983 in the field of chemical physics. An avid hiker, he summited Kilimanjaro in 1973 and is an Adirondak "46R." David was recognized for his historic preservation efforts in 1994 when he received the Volunteer Recognition Award jointly from the Connecticut Historical Commission and the Connecticut Trust for Historic Preservation. He is known nationally for his expertise in vintage Corvette restoration and documentation. David did volunteer work in Jérémie, Haiti, soon after the January 2010 earthquake.

T. Pradeep heads the Thematic Unit of Excellence (TUE) on Water Purification Using Nanotechnology at IIT Madras.

T. Pradeep, PhD, is a professor of chemistry at the Indian Institute of Technology Madras, Chennai, India. He earned his PhD from the Indian Institute of Science in 1991 and had postdoctoral training at the Lawrence Berkeley Laboratory, University of California, Berkeley, and Purdue University, West Lafayette, Indiana. He held visiting positions at many leading universities and institutes in the United States, Europe, and Asia.

Prof. Pradeep's research interests are in molecular and nanoscale materials, and he develops instrumentation for those studies. He has authored 300 scientific papers in journals and is an inventor in 50 patents or patent applications. He is involved with the development of affordable technologies for drinking water purification. He discovered that noble metal nanoparticles can remove halocarbon pesticide residues from drinking water, and that became the first nanomaterials-based technology to be commercialized in the drinking water segment. Along with his associates, he has incubated a company to develop technologies developed in his laboratory to marketable products.

He is a recipient of several coveted recognitions in India, including the Shanti Swaroop Bhatnagar Prize (most prestigious prize in India for scientists below age 45), B.M. Birla Science Prize (an important recognition in India for scientists below age 40), National Award for Nanoscience and Nanotechnology, and the India Nanotech Innovation Award. He is an adjunct professor at the Institute of Life Sciences, Ahmedabad University, and PSG Institute of Advanced Studies, Coimbatore. He is a fellow of the Indian Academy of Sciences and Indian National Academy of Engineering.

He is the author of the introductory textbook, *Nano: The Essentials* (McGraw-Hill) and is one of the authors of the monograph, *Nanofluids* (Wiley-Interscience) and an advanced textbook, *A Textbook of Nanoscience and Nanotechnology* (McGraw-Hill). *Nano: The Essentials* is used as a textbook in more than 50 universities. It has been translated to Japanese. He has edited a book and has contributed to chapters in nine books. He is on the editorial board of journals such as *Nano Reviews*, *ACS Applied Materials & Interfaces*, *Particle*, and *Surface Innovations* and is an Associate Editor of *ACS Sustainable Chemistry & Engineering*. His other interests include education, popularization of science, and development of advanced teaching aids. He has authored a few popular science books in Malayalam and is the recipient of the Kerala Sahitya Academi Award for knowledge literature for the year 2010. For more information, please see http://www.dstuns.iitm.ac.in/pradeep-research-group.php.

Contributors

Diego F. Acevedo
Chemistry Department
Universidad Nacional de Rio Cuarto
Río Cuarto, Argentina

Zaki Ahmad
Department of Chemical Engineering
COMSATS Institute of Information
    Technology (CIIT)
Lahore, Pakistan

and

42TEK S.L.
Almazora, Spain

Juan Balach
Chemistry Department
Universidad Nacional de Rio Cuarto
Río Cuarto, Argentina

Amit K. Bansiwal
CSIR–National Environmental Engineering
    Research Institute (CSIR–NEERI)
Nagpur, India

Cesar A. Barbero
Chemistry Department
Universidad Nacional de Rio Cuarto
Río Cuarto, Argentina

Guy Baret
42TEK S.L.
Voiron, France

Karen Bellman
Department of Mechanical and
    Aerospace Engineering
The Ohio State University
Columbus, Ohio, USA

Sanjay Bhatnagar
WaterHealth International Inc.
Irvine, California, USA

Megalamane Siddaramappa Bootharaju
Department of Chemistry
Indian Institute of Technology Madras
Chennai, India

Aurelio Boza
Instituto de Ciencia y Tecnología de
    Materiales
Universidad de La Habana
La Habana, Cuba

Laura Bravo-Fariñas
Instituto de Medicina Tropical
    "Pedro Kouri"
La Habana, Cuba

Mariano M. Bruno
Chemistry Department
Universidad Nacional de Rio Cuarto
Río Cuarto, Argentina

Bob Burk
NanoH$_2$O Inc.
El Segundo, California, USA

Bárbara Cedré
Instituto Finlay
La Habana, Cuba

Chih-Cheng Chao
Department of Chemical and Materials
    Engineering
Tunghai University
Taichung, Taiwan

Alok Dhawan
Institute of Life Sciences, School of Science
and Technology
Ahmedabad University
Ahmedabad, Gujarat, India

and

Nanomaterial Toxicology Group
CSIR–Indian Institute of Toxicology
Research
Lucknow, Uttar Pradesh, India

Tony F. Diego
42TEK S.L.
Almazora, Spain

Derrick Dlamini
Department of Applied Chemistry
Faculty of Science
University of Johannesburg
Johannesburg, South Africa

Ruey-An Doong
Department of Biomedical Engineering
and Environmental Sciences
National Tsing Hua University
Hsinchu, Taiwan

Paul L. Edmiston
Department of Chemistry
The College of Wooster
and
ABSMaterials Inc.
Wooster, Ohio, USA

Ralph Exton
GE Power & Water
and
Water & Process Technologies
Trevose, Pennsylvania, USA

Lisa M. Farmen
Crystal Clear Technologies Inc.
Portland, Oregon, USA

Robina Farooq
Department of Chemical Engineering
COMSATS Institute of Information
Technology (CIIT)
Lahore, Pakistan

Anabel Fernández-Abreu
Instituto de Medicina Tropical
"Pedro Kouri"
La Habana, Cuba

Eugene A. Fitzgerald
The Water Initiative
New York, New York, USA

Ashok Kumar Ghosh
Department of Environmental and
Water Management
Anugrah Narayan College
Patna, India

Brent Giles
Lux Research Inc.
New York, New York, USA

Liang-Hong Guo
Research Center for Eco-Environmental
Sciences (RCEES)
Chinese Academy of Sciences (CAS)
Beijing, China

Gordon L. Hager
Laboratory of Receptor Biology and
Gene Expression
National Cancer Institute, National
Institutes of Health
Bethesda, Maryland, USA

Jayanta Haldar
Chemical Biology and Medicinal
Chemistry Laboratory
Jawaharlal Nehru Center for Advanced
Science Research
Bengaluru, India

Curt Hallberg
Watreco AB
Malmö, Sweden

Tom Hawkins
Puralytics
Beaverton, Oregon, USA

Richard Helferich
MetaMateria Technologies LLC
Columbus, Ohio, USA

Mike Henley
Ultrapure Water Journal
Denver, Colorado, USA

Jiaul Hoque
Chemical Biology and Medicinal
    Chemistry Laboratory
Jawaharlal Nehru Center for Advanced
    Science Research
Bengaluru, India

Natasha Hussain
Department of Chemical Engineering
COMSATS Institute of Information
    Technology (CIIT)
Lahore, Pakistan

Inocente Rodríguez Iznaga
Instituto de Ciencia y Tecnología de
    Materiales
Universidad de La Habana
La Habana, Cuba

I.P.L. Jayarathna
Chemical and Environmental Systems
    Modeling Research Group
Institute of Fundamental Studies
Kandy, Sri Lanka

Guibin Jiang
State Key Laboratory of Environmental
    Chemistry and Ecotoxicology
Research Center for Eco-Environmental
    Sciences (RCEES)
Chinese Academy of Sciences (CAS)
Beijing, China

Stephen Jolly
ABSMaterials Inc.
Wooster, Ohio, USA

Satish Vasu Kailas
Department of Mechanical Engineering
Indian Institute of Science
Bangalore, India

Asad Ullah Khan
Department of Chemical Engineering
COMSATS Institute of Information
    Technology (CIIT)
Lahore, Pakistan

Ashutosh Kumar
Institute of Life Sciences, School of Science
    and Technology
Ahmedabad University
Ahmedabad, Gujarat, India

Christopher J. Kurth
NanoH$_2$O Inc.
El Segundo, California, USA

Pawan K. Labhasetwar
CSIR–National Environmental Engineering
    Research Institute (CSIR–NEERI)
Nagpur, India

Nitin K. Labhsetwar
CSIR–National Environmental Engineering
    Research Institute (CSIR–NEERI)
Nagpur, India

Jing Lan
College of Chemistry and Pharmaceutical
    Sciences
Qingdao Agricultural University
Qingdao, China

Thomas A. Langdo
The Water Initiative
New York, New York, USA

Chun-Chi Lee
Department of Biomedical Engineering
    and Environmental Sciences
National Tsing Hua University
Hsinchu, Taiwan

Rasool Lesan-Khosh
International Centre for Nanotechnology
United Nations Industrial Development
    Organization (UNIDO)
and
Departments of Materials Science and
    Engineering
Sharif University of Technology
Tehran, Iran

Guoliang Li
State Key Laboratory of Environmental
    Chemistry and Ecotoxicology
Research Center for Eco-Environmental
    Sciences (RCEES)
Chinese Academy of Sciences (CAS)
Beijing, China

Hsing-Lung Lien
Department of Civil and Environmental
    Engineering
National University of Kaohsiung
Kaohsiung, Taiwan

S. Jagtap Lunge
CSIR–National Environmental Engineering
    Research Institute (CSIR–NEERI)
Nagpur, India

Shihabudheen M. Maliyekkal
Environmental Engineering Division
School of Mechanical and Building
    Sciences
VIT University, Chennai Campus
Chennai, India

Bhekie B. Mamba
Department of Applied Chemistry
Faculty of Science
University of Johannesburg
Johannesburg, South Africa

Monto Mani
Centre for Sustainable Technologies
Indian Institute of Science
Bangalore, India

Ali Marjowy
International Centre for Nanotechnology
United Nations Industrial Development
    Organization (UNIDO)
and
Techno-Entrepreneurship Group
Entrepreneurship Faculty
University of Tehran
Tehran, Iran

María V. Martinez
Chemistry Department
Universidad Nacional de Rio Cuarto
Río Cuarto, Argentina

Kevin M. McGovern
The Water Initiative
New York, New York, USA

Sabelo D. Mhlanga
Department of Applied Chemistry
Faculty of Science
University of Johannesburg
Johannesburg, South Africa

María Cristina Miras
Chemistry Department
Universidad Nacional de Rio Cuarto
Río Cuarto, Argentina

Mahendra K. Misra
WaterHealth International Inc.
Irvine, California, USA

María A. Molina
Chemistry Department
Universidad Nacional de Rio Cuarto
Río Cuarto, Argentina

Titus A.M. Msagati
Department of Applied Chemistry
Faculty of Science
University of Johannesburg
Johannesburg, South Africa

Stephen Musyoka
Department of Applied Chemistry
Faculty of Science
University of Johannesburg
Johannesburg, South Africa

A. Sreekumaran Nair
Amrita Center for Nanosciences and
    Molecular Medicine
Amrita Institute of Medical Sciences
Kochi, Kerala, India

Shantikumar V. Nair
Amrita Center for Nanosciences and
    Molecular Medicine
Amrita Institute of Medical Sciences
Kochi, Kerala, India

J. Catherine Ngila
Department of Applied Chemistry
Faculty of Science
University of Johannesburg
Johannesburg, South Africa

Víctor Sende Odoardo
Laboratorio ENAST La Habana
Instituto Nacional de Recursos Hidráulicos
La Habana, Cuba

Mark D. Owen
Puralytics
Beaverton, Oregon, USA

Anaisa Pérez
Instituto de Ciencia y Tecnología de
    Materiales
Universidad de La Habana
La Habana, Cuba

Ligy Philip
Department of Civil Engineering
Indian Institute of Technology (IIT)
    Madras
Chennai, India

Gabriel A. Planes
Chemistry Department
Universidad Nacional de Rio Cuarto
Río Cuarto, Argentina

Thalappil Pradeep
Department of Chemistry
Indian Institute of Technology Madras
Chennai, India

Anuradha Prakash
Department of Environmental and Water
    Management
Anugrah Narayan College
Patna, India

Shaurya Prakash
Department of Mechanical and Aerospace
    Engineering
The Ohio State University
Columbus, Ohio, USA

Jiuhui Qu
Research Center for Eco-Environmental
    Sciences (RCEES)
Chinese Academy of Sciences (CAS)
Beijing, China

Sadhana S. Rayalu
CSIR–National Environmental Engineering
    Research Institute (CSIR–NEERI)
Nagpur, India

Carlos Ángel Sánchez Recio
Natural Aqua Canarias S.L.
Canary Islands, Spain

David E. Reisner
Inframat® Corp.
Manchester, Connecticut, USA

Rick Renjilian
The Water Initiative
New York, New York, USA

Rao Revur
MetaMateria Technologies LLC
Columbus, Ohio, USA

Claudia R. Rivarola
Chemistry Department
Universidad Nacional de Rio Cuarto
Río Cuarto, Argentina

Rebeca Rivero
Chemistry Department
Universidad Nacional de Rio Cuarto
Río Cuarto, Argentina

Rusbel Coneo Rodriguez
Chemistry Department
Universidad Nacional de Rio Cuarto
Río Cuarto, Argentina

Gerardo Rodríguez-Fuentes
Instituto de Ciencia y Tecnología de
    Materiales
Universidad de La Habana
La Habana, Cuba

Aniran Ruiz
Instituto de Medicina Tropical
    "Pedro Kouri"
La Habana, Cuba

Steve Safferman
Department of Biosystems and
    Agricultural Engineering
Michigan State University
East Lansing, Michigan, USA

J. Richard Schorr
MetaMateria Technologies LLC
Columbus, Ohio, USA

Suvankar Sengupta
MetaMateria Technologies LLC
Columbus, Ohio, USA

Jaganathan Senthilnathan
Department of Civil Engineering
Indian Institute of Technology (IIT)
    Madras
Chennai, India

Ahmed Shafique
Department of Chemical Engineering
COMSATS Institute of Information
    Technology (CIIT)
Lahore, Pakistan

Rishi Shanker
Institute of Life Sciences
School of Science and Technology
Ahmedabad University
Ahmedabad, Gujarat, India

Mark A. Shannon
Department of Mechanical Science and
    Engineering
University of Illinois at
    Urbana-Champaign
Urbana, Illinois, USA

Ashutosh Sharma
Department of Chemical Engineering
Indian Institute of Technology Kanpur
Kanpur, India

Seema Sharma
Department of Environmental and
    Water Management
Anugrah Narayan College
Patna, India

Stephen Spoonamore
ABSMaterials Inc.
Wooster, Ohio, USA

Theruvakkattil Sreenivasan Sreeprasad
Department of Chemical Engineering
Kansas State University
Manhattan, Kansas, USA

Diana A. Stavreva
Laboratory of Receptor Biology and
    Gene Expression
National Cancer Institute
National Institutes of Health
Bethesda, Maryland, USA

Hongxiao Tang
Research Center for Eco-Environmental
    Sciences (RCEES)
Chinese Academy of Sciences (CAS)
Beijing, China

Maria Jose Lopez Tendero
Laurentia S.L. València Parc Tecnològic
Paterna, Valencia, Spain

Divakara S.S.M. Uppu
Chemical Biology and Medicinal
 Chemistry Laboratory
Jawaharlal Nehru Center for Advanced
 Science Research
Bengaluru, India

Sajini Vadukumpully
Amrita Center for Nanosciences and
 Molecular Medicine
Amrita Institute of Medical Sciences
Kochi, Kerala, India

Lyuba Varticovski
Laboratory of Receptor Biology and
 Gene Expression
National Cancer Institute
National Institutes of Health
Bethesda, Maryland, USA

Nishith Verma
Department of Chemical Engineering
Indian Institute of Technology Kanpur
Kanpur, India

Arya Vijayanandan
Department of Civil Engineering
Indian Institute of Technology (IIT)
 Madras
Chennai, India

Meththika Vithanage
Chemical and Environmental Systems
 Modeling Research Group
Institute of Fundamental Studies
Kandy, Sri Lanka

Dongsheng Wang
Research Center for Eco-Environmental
 Sciences (RCEES)
Chinese Academy of Sciences (CAS)
Beijing, China

Satish R. Wate
CSIR–National Environmental Engineering
 Research Institute (CSIR–NEERI)
Nagpur, India

S.S.R.M.D.H.R. Wijesekara
Chemical and Environmental Systems
 Modeling Research Group
Institute of Fundamental Studies
Kandy, Sri Lanka

Shan-Chee Wu
Graduate Institute of Environmental
 Engineering
National Taiwan University
Taipei, Taiwan

Tianmin Xie
Inframat® Corp.
Manchester, Connecticut, USA

Hanbae Yang
ABSMaterials Inc.
Wooster, Ohio, USA

Hui Zhang
Research Center for Eco-Environmental
 Sciences (RCEES)
Chinese Academy of Sciences (CAS)
Beijing, China

Zongshan Zhao
State Key Laboratory of Environmental
 Chemistry and Ecotoxicology
Research Center for Eco-Environmental
 Sciences (RCEES)
Chinese Academy of Sciences (CAS)
Beijing, China

and

Key Laboratory of Marine Chemistry
 Theory and Technology
Ministry of Education
Ocean University of China
Qingdao, China

Dedication

The editors and authors dedicate Chapter 27 to an excellent scientist, mentor, and friend, Prof. Mark A. Shannon. His contributions to the science and technology for water will be a stepping-stone for future engineers and scientists to build on for many years to come.

Water, water, everywhere,

Nor any drop to drink.

Samuel Taylor Coleridge
Rime of the Ancient Mariner (1798)

1

Bimodal UV-Assisted Nano-TiO$_2$ Catalyst—Crumb Rubber Device for Treatment of Contaminated Water

Zaki Ahmad, Robina Farooq, Asad Ullah Khan, Natasha Hussain, and Ahmed Shafique

Department of Chemical Engineering, COMSATS Institute of Information Technology (CIIT), Lahore, Pakistan

CONTENTS

1.1 Introduction

The earth is bestowed with a tremendous amount of freshwater; however, because of human neglect and adverse climatic factors, dust storms, and policy failure to conserve freshwater, we are on the verge of freshwater depletion and political conflict. Nearly 780 million people lack access to drinking water, while one-third of the global population lacks sanitation [1]. In 1 day, 20 million work hours are consumed by women in India and Africa to collect water for their families. It has been estimated that water demand would increase by 40% in the next 10 years. By 2025, nearly 1.8 billion people will live in areas plagued by water scarcity with two-thirds living in water-stressed regions [2]. Freshwater availability is sobering; only 2.5% of the 70% water covering the earth's surface is available for drinking. Our living habits have also stressed the water supplies, as witnessed in the United States where 148 trillion gallons are used per year. The story of freshwater for drinking has a potentially frightening future.

The neglect to maintain water quality even in regions abundant with freshwater has further deteriorated the resources. In India, 75% of the surface water is contaminated by human and animal wastes. In Lahore, Pakistan, once known as a city of clean, healthy water, the freshwater is now infected with pathogens, resulting in deaths and disease. In Pakistan, one in four hospital beds is occupied by victims of waterborne diseases.

The greatest challenge that mankind faces today is to solve the problem of scarcity of clean water by tapping various sustainable and nonsustainable resources and to devise cost-effective technologies for water treatment to provide healthy drinking water. These include microfiltration, ultrafiltration, reverse osmosis, and nanofiltration. The materials include polyvinylidene fluoride, polysulfones, polyacrylonitriles, and polyvinyl chloride compounds. Nanomembranes are made from cellulose acetate and cellulose nitrate blends [3,4].

1.2 Some Facts and Figures Related to Pollution

It has been reported that at least 50,000 people die daily worldwide because of waterborne diseases. In South Africa, 12 million people do not have access to adequate water supply [5]. Contaminated water may contain viruses, bacteria, and pathogens that may cause diarrhea, dysentery, typhoid, and hepatitis. Millions are exposed each year to unsafe level during storms, floods, and natural and man-made disasters. Open sewers increase the risk of contamination of various pathogenic agents. The major sources of contamination include industrial waste; agriculture and domestic wastes; and discharge of waste from the industries of manufacturing, power generation, mining, textiles, and construction sites. Most of the mortality and morbidity in developing countries is associated with infection agents, causing cholera, dysentery, typhoid, paratyphoid, scabies, yaws, skin ulcers, and trachoma. Many insects breed in water and cause malaria, dengue, yellow fever, and onchocerciasis (river blindness). Several techniques have been used for water treatment since the Egyptians discovered the principles of coagulation in 1871. They applied chemical alum for suspended particle settlement. The processes of coagulation and flocculation have been employed to separate suspended solids from water. Unit operations such as coagulation, flocculation, precipitation, ion exchange, membrane separation, aeration, and adsorption have been employed to clarify water [5]. In recent years, various types of membranes have been developed. They increase production, reduce energy costs, reduce yield costs, and are emission compliant.

Nanotechnology has shown considerable potential for remediation of contaminated waste. This technology enhances reactivity because of the increasing ratio of the aggregate reactive surface area to the aggregate particle volume as the particle size decreases. A variety of materials, such as zeolites, carbon nanotubes, biopolymers, and zero valence nanoparticles, are the focus of attention of engineers and scientists [6]. Considerable improvements in the water treatment methods during the last decades have been reported [7–9].

HaloPure® (HaloSource) is an example of technical advancement in the development of an entirely new biocidal medium in the form of chlorine rechargeable polystyrene beads. Polystyrene beads are chemically modified to enable them to bind chlorine and bromine reversibly in oxidative form. Almost no free chlorine is released when the beads are placed in water [10]. A solid form is obtained that is biocidal and rechargeable on exposure to free halogens. A powerful antibacterial component is introduced that does not need recharging. Microorganisms making contact with the surface are killed or inactivated. Advanced methods have always remained in demand for the removal of persistent organic pollutants from wastewater and groundwater. Photocatalytic methods have proven very effective. These methods are based on the generation of highly reactive hydroxyl radicals (OH·) and O_2^- using ultraviolet (UV)/titanium dioxide (TiO_2) combination leading to degradation [11–13].

Coinciding with the advancements in nanotechnology applications, photocatalysis has been successfully used by using immobilized TiO_2 nanoparticles for the degradation and mineralization of water containing organic pollutants. For photocatalytic treatment of water containing organic pollutants, immobilized nano-TiO_2 catalyst has been found highly effective in reactors using UV radiation [14].

Lately, optical fiber photoreactors have been designed for uniform distribution of light and decreased internal reflectivity. Nano-TiO_2 coatings were applied on optical fibers using titanium butoxide. The size of TiO_2 films was 15–25 nm, and the thickness of film applied ranged from 60 to 600 nm. Photodecomposition of ammonia showed the superior conductivity of TiO_2-coated optical fibers. It has been stated that OH^- radicals produced by the photocatalytic effect are 1–10,000× more effective than chlorine and ozone. The production of $(OH^•)$ radicals and $O_2^•$ is a well-known phenomenon. It can be summarized as follows: TiO_2 has a moderate band gap (1.33 eV) between their valence and conduction bands. Upon illumination by a semiconductor by photons of energy equal to or greater than the band gap, e.g., photons are absorbed, which excite the valence band ("vb") electrons, to the conduction band ("cb") creating electron $\left(e_{cb}^-\right)$ hole $\left(h_{vb}^+\right)$ charge carriers. On migration to the solid surface without recombining, the electrons and holes undergo transfer processes with the adsorbates. For instance, adsorbed water reacts with holes to produce $OH^•$ radicals, which can oxidize organic compounds. In certain cases, the final product is CO_2 and H_2O.

The following reactions exemplify the steps described above:

$$TiO_2 \xrightarrow{h\nu} TiO_2^• \left(e_{cb}^- + h_{vb}^+\right)$$

$$TiO_2^• \rightarrow \left(e_{cb}^- + h_{vb}^+\right) \xrightarrow{\text{Recombination}} TiO_2 + heat$$

$$H_2O(\text{adsorbed}) \rightarrow OH_{ads}^- + H_{ads}^+$$

$$h_{vb}^+ + OH_{ads}^- \rightarrow OH^•$$

$$e_{cb}^- + O_2 \rightarrow O_2^{•-}$$

In addition to various forms of active ions such as O_2^-, OH^-, the free radical/active species $OH^•$, and $O_2^-{}^•$ are produced.

$$e_{cb}^- + O_2 \rightarrow O_2^- (\text{ads})$$

$$O_2^{2-} (\text{ads}) + H^+ \rightarrow HO_2^• (\text{ads})$$

$$h_{vb}^+ + H_2O \rightarrow OH^• (\text{ads}) + H^+$$

$$h_{vb}^+ + O_2^- (\text{ads}) \rightarrow 2O^• (\text{ads})$$

The photocatalytic technique is a versatile and efficient disinfection process. The final decomposition products of organic pollutants are water and CO_2.

In recent years, UV radiation has been initialized to produce cost-effective cheap and clean water. Commercially produced drinking water based on solar water disinfection ("SODIS") is produced by the Swiss at EAWAG (Swiss Federal Institute for Aquatic Science and Technology; www.sodis.ch). This water is made by exposing the contaminated water to solar radiation, using solar energy to destroy pathogenic microorganisms. The principle is based on the bimodal effect of UVA light ($\lambda = 320$–400 nm) and heat, i.e., the synergistic effect. Polyethylene terephthalate (PET) water bottles are exposed to sunlight to obtain a temperature of 55°C. The bottles are placed on corrugated iron sheets or aluminum sheets. The water bottles used are either transparent PET bottles or half-blackened plastic bottles. Under normal conditions, the disinfection efficiency is about 99.9%. The dosage contains energy of 555 Wh/m^2 in UV-A in the range of 350–450 nm, corresponding to 6 h of exposure in typical temperate latitude sunshine [14].

Late entrants to the water disinfection technologies are glass beads and crumb rubber technologies [15–17]. Glass beads are manufactured from 100% virgin glass. They contain 66–73% (wt%) SiO_2, 13–14.5 wt% Na_2O, 8–9.2 wt% CaO, 3.2 wt% (max) Al_2O_3, and 4.2 wt% (max) MgO. They provide very high effective pore space owing to their spherical shape. A very smooth surface leads to deceleration of lead and manganese contamination. A large pore space allows optimal regeneration. They are chemically inert and need less water (20%) for backwashing. Photospheres are TiO_2-coated hollow spheres. They are low-density composite buoyant material. The buoyancy of glass beads allows photocatalytic application in open water systems, and it is a less expensive photocatalytic technology than construction of photocatalytic reactors. The principle of photocatalysis on which they work has been described above. The buoyancy of spheres allows reuse without a significant material loss. Titanium can be coated by sol–gel techniques.

The choice of granular media is very important. Studies on clogging phenomena were performed on zeolite, river sand, glass beads, and scoria. The overall treatment efficiency shown by glass beads was 0.93 compared with 0.68 and 0.01 for scoria (volcanic rock) and river sand, respectively. The treatment performance was measured as the difference in total suspended solid concentration (TSSC) between the inflow and outflow filters and equals

$$\text{Treatment efficiency} = \frac{\text{Inflow (TSSC)} - \text{Outflow (TSSC)} \times 100}{\text{Inflow (TSSC)}}$$

It was suggested that the zeolite filters with unrestricted flow had an infiltration rate of 62,000 mm/h compared with filters with restrictive flow, which had infiltration rate of 16,500 mm/year. The low flow rate caused increased contact time within the filter media [15]. The effect of porosity of grains, reduction in size, and packing of grains on infiltration rate is not conclusively understood [16].

Crumb rubber has shown promising results in the filtration of water. Crumb rubber is a term usually applied to recycled rubber from automotive and truck scrap tires. It is in the form of small broken pieces or powder in different meshes. Microscopic studies have shown that the micro hairs that protrude from the surface attract the solids present in the influent. Junk rubber tires provide a breeding ground for vectors of diseases like dengue. The use of recycled crumb rubber is a green technology for water treatment. It has provided in-depth filtration and provides longer filtration time and high filtration rate [17]. Because of the elasticity and compression caused by the weight of crumb rubber and a high hydraulic head loss in the media, it has the smallest pore size at the bottom and the largest at the top. This provides a higher filtration rate because of the favorable porosity gradient. It provides a new filtration technology superior to dual media sand/anthracite filters. In conventional medium, the large grains settle first and the fine grains settle last, thus clogging the filters. This is contrary to the situation in crumb rubber media. A conventional ideal filter medium consists of a coarse media at the top, a medium-sized media in the middle, and a fine media at the bottom. The disadvantage is that it causes plugging of filters and reduction of filtration time. Unlike crumb rubbers, conventional filters like anthracite/sand do not exhibit an elastic behavior.

The spread of dengue in Pakistan, Vietnam, and other south Asian countries took a heavy toll of lives, and junk rubber tires were a major component in the transmission of the virus as the tires act as containers for breeding of the mosquito species *Aedes aegypti*. Three factors, namely crumb rubber's elasticity and favorable porosity gradient for infiltration, its antipathogenic properties, and the development of green technology (recycling), were the major factors that motivated the use of crumb rubber technology for filtration. To strengthen the technology, it is believed that this technology could be applied in combination with UV exposure to provide cleaner water to flood victims, where the floodwater is highly contaminated and turbid. To provide cleaner water, this technology could be combined with a photocatalytic treatment. The effect of ultrafine particles of rubber (nanorubber) on the filtration rate was also investigated; however, they adversely affected the filtration rate, unlike crumb rubber.

1.3 Experimental

Sizes of crumb rubber ranging from 0.8 to 1.5 mm, 1.5 to 2.5 mm, and 2.5 to 5 mm were selected for the filtration column. The sizes were determined by sieve analysis using ASTM method 136-01. The density of the crumb rubber is 1130 kg/ml. The design of crumb rubber reactor constructed is shown in Figure 1.1. The filtration column was constructed from polytetrafluoroethylene, to provide a reasonably strong transparent and nonsticky surface. It was 1.5 m long and 10 cm in diameter. The influent turbid water was supplied by a water pump from a reservoir. For backwashing, an air pump was connected to the bottom. The rate of filtration was controlled by a flow meter installed at the outlet pipe. The crumb rubber media was washed and dried before pouring in the column. The depth of the fine medium and coarse media was 0.2, 0.4, and 0.6 m, to provide different contact times for the media to remove suspended and total dissolved solids (TDS), after a series of experiments. Experiments were conducted at 30, 50, and 70 m^3/h m^2 (h = hours). The principle of the design was similar to that reported by Al-Anbari [16] and Xie [17].

A stainless-steel mesh dish was used as a support. For head loss measurements, ports at equal distances were drilled throughout the column. A constant water head was maintained for the influent. For head loss measurement, the difference between the water level above the filter and the water level in the glass tube was measured.

A turbidity meter was used to measure turbidity in NTU (nephelometric turbidity unit). The light scattering of particles by a focused beam is measured using a nephelometer with a detector. Head loss was measured in meters.

For biological analysis, all equipment, including autoclaves, glassware, and Petri dishes, were sterilized. The sample size was 100 ml. The grid side of the micron filter paper was placed upward. The filtering apparatus had a magnet at the bottom. The unit was placed on top of the filtration unit. A hose was connected to a vacuum pump. The samples were

FIGURE 1.1
(See color insert.) Crumb rubber reactor.

placed on top of the filtration unit after turning on the vacuum pump. The membrane filter was removed and placed on a Petri dish, and was sealed with tape. Incubation was done at 35°C for 30 h. Dark-colored (pink) areas representing the colonies were used for counting under a dissecting microscope. The density was counted as

$$\text{Coliform}/100 \text{ ml} = (\text{No. of colonies counted}) \times 100/\text{sample size (ml)}$$

A geometric mean was taken while calculating the average coliform density. A dehydrated commercial media was used.

1.4 Results and Discussion

1.4.1 Suspended Solid Concentration

After a run of 12 h at 4 gpm/ft^2 (gpm = gallons per minute), the suspended solid concentration was reduced from 10.5 to 7 mg/l, which was a significant reduction. After a sum of 16 h, the reduction was only 8.5 mg/l. In the values reported in the literature, the suspended solid content decreased to 5 mg/l from the initial concentration of 8.4 mg/l.

The turbidity values showed a decrease of 2–3 NTU, which indicates a good turbidity removal. The head loss observed was only 4.0 in, which was much lower than in conventional filtration media such as sand/anthracite and gravel. The small head loss observed with crumb rubber is a major advantage of crumb rubber filtration. The advantage of using crumb rubber is a favorable porosity gradient, i.e., finest at the top and coarse at the bottom, which results in a lower head loss and high filtration rate. Because of higher filtration rate and lesser head loss, crumb rubber filtration holds a promising potential as a filter media compared with the porosity of sand/anthracite, 40%–43% and 47%–52%, respectively. The porosity of a porous medium is the ratio of the pore volume to the total volume of a representative sample of the medium. The specific weight of crumb rubber is closer to water (1.13). The crumb rubber media can easily be fluidized, and hence a more efficient backwashing be achieved. The factors responsible for filtration include surface adsorption and coagulation.

As verified by the Kozeny and Ergun equation, the head loss depends on the flow rate, media size, porosity, sphericity, and viscosity of water. The equation is as follows:

$$h/L = K\mu(1 - \varepsilon)^2/(\rho g \varepsilon^3)\,(a/\upsilon)V$$

where

h = head loss in depth of bed L [cm]
L = depth of filter bed [cm]
g = acceleration of gravity
ε = porosity
a/υ = grain surface area per unit of grain volume (6d for spheres)
d_{eq} = grain diameter of sphere of equal volume
V = superficial velocity above the bed
μ = absolute viscosity of fluid
ρ = mass density of fluid
K = Kozeny constant (~5)

The Kozeny equation is based normally on laminar flow conditions. It is used to depict the head loss. Ergun modified the equation, taking into account laminar transitional and inertial flow. The Ergun equation is

$$h/L = 4.17\mu(1 - \varepsilon)^2\,V/(\rho g \varepsilon^3)\,(a/\upsilon)^2 + k_2\,(1 - \varepsilon)\,V^2/(g \varepsilon^3)\,(a/\upsilon)$$

The first term is similar to Kozeny's equation, and the second term takes into account the higher velocities.

Nano-rubber is a relatively new entrant to the market and mainly used as a frictional material. In the experiment, Narpow® (SINOPEC Beijing Research Institute of Chemical Industry, "BRICI") (styrene-butadiene) with a particle size of 90 nm was used as a top layer above the medium-sized crumb rubber (0.5–1.2 mm) with a porous plate as support. It was noted that the use of nanosized rubber did not assist the filtration process; however, it removed suspended solids more than did the fine crumb rubber used in the filtration column. The total suspended solids were found to be 15 mg/l only. The porosity gradient is not significantly affected. Further studies on styrene–butadiene and carboxylic–styrene–butadiene (ultrafine rubber with particle size of 80 nm) and other sizes are in progress.

After collecting water from the crumb rubber column, the TDS was reduced to 250 mg/l and the turbidity to 2.0 NTU. The total suspended solids were reduced from 90 mg/l to undetectable levels after filtration. The filtered water from the column was then poured in transparent 1.5-liter PTE bottles. The height of horizontal water level in the column was kept at 10 cm. Special holders were constructed to place these bottles. The surface of the holders was made from corrugated iron sheets. The PTE bottles after filling were aerated for a few seconds and placed horizontally on a shiny surface of corrugated iron sheets in the sun, in a specified location such that no shadows were cast on the bottles. In another set of experiments, a surface made from shiny aluminum foil was used as a base where bottles were placed, to obtain maximum heat from the sun. A temperature of 55°C was required to kill the pathogens in the water. The time to attain the required temperature decreased with constant exposure, low wind velocity, clear sky, and magnitude of heat adsorption and reflection depending on the type of the surface. It took 10 h to obtain a temperature of 55°C on the aluminum foil compared with corrugated iron sheet, which took 15 h to reach the same temperature. To keep a stock of clean water, sufficient water was stored at a time when the temperature was favorable. After attaining the maximum temperature, the water was analyzed for pathogenic *Escherichia coli*, *Salmonella typhi*, *Shigella* spp. bacteria, and *Vibrio cholerae*. Also, analytical studies were made for protozoa (*Entamoeba histolytica* and *Giardia* spp.). The densities of coliform and *Shigella* spp. (colonies/100 ml) were determined in the filtered water. The results are summarized in Table 1.1. The bacteriological analysis of water treated by crumb rubber is shown in Table 1.2.

The synergistic effect of infrared and UV radiation above 55°C completely kills the pathogens in the water obtained from crumb rubber filtration. In one particular locality near Khaffi (Saudi Arabia), a very high count of *Shigella* spp. was determined (300) in the raw water.

A hybrid technique combining crumb rubber filtration and exposure to solar radiation (UV + SODIS) results in complete reduction of pathogens. Using the above hybrid technique, the desired temperature of 60°C was obtained only in 3 h, which shows a great reduction in time to kill the pathogens. Selection of a heat-absorbent base, the direction of the bottle holders, the types of bottles, wind speed, moisture, and humidity affect the attainment of the required temperature. The blowing of wind increases the time to reach the required temperature of 55°C under the sun. The results of field studies obtained only after solar treatment showed a 24% reduction in the number of cases of diarrhea. The technique has the advantage that it removes turbidity, TDS, and solid suspended particles (SSP) to levels accepted by the Environmental Protection Agency and World Health Organization. Although opinion against the use of PTE bottles is increasing, glass bottles

TABLE 1.1

Number of Colonies for *E. coli* and *Shigella* spp.

Sample	*E. coli* Count	*Shigella* spp. Count
1	>960	>1800
2	>1000	>7960
3	>700	4000
4	1500	6000

TABLE 1.2

Bacteriological Analysis of Water Treatment by Crumb Rubber and UV Radiation

Sample	*E. coli* Count	*Shigella* spp. Count
1	0/100 ml	0/100 ml
2	3/100 ml	2/100 ml
3	0/100 ml	0/100 ml
4	0/100 ml	0/100 ml

TABLE 1.3

Physiochemical Analysis

Tests	Crumb Rubber	Crumb Rubber and UV Radiation	WHO Standard
pH	6.8	7.8	6.0
Color (CU, color units)	8	6	6.0
Odor	None	None	None
Turbidity (NTU)	3	3	6
Total suspended solids (mg/l)	None	None	None
Total dissolved solids (mg/l)	250	250	500
Total hardness (mg/l)	180	78	500
Ca^{2+} hardness (mg/l)	67	67	75
Mg^{2+} hardness (mg/l)	65	60	50
Chloride (mg/l)	170	170	200
Iron (mg/l)	6.1	0.1	0.3
Arsenic (mg/l)	None	None	None

may be the future choice if the transmission of UV radiation is assured, such as in Corning glass. UV radiation penetrates Pyrex and Corning glasses. The presence of iron oxide prevents transmission of UV radiation. The results of physiochemical analysis of the bimodal water treatment are given in Table 1.3.

The composition shown in Table 1.3 demonstrates that the quality of water obtained by combined crumb rubber and UV radiation process complies with the WHO standards. The water produced is healthy and free of pathogens.

1.4.2 Photocatalytic Treatment

In the processes mentioned above, trace metals such as mercury (Hg), chromium (Cr), lead (Pb), and several heavy metals such as cadmium (Cd) and arsenic (As) may go undetected. Also, several organic compounds such as alcohol, carboxylic acid, phenolic derivatives, or mineral acids may be encountered. Dyes may be released in water by textile industries. In addition, organochlorine pesticides such as lindane, *p,p'*-DDT (dichlorodiphenyltrichloroethane), and methoxychlor are present in the environment. In cases where true metals, heavy metals, and agricultural pesticides may be present, the bimodal treatment (crumb rubber + solar radiation) may be extended to photocatalytic treatment by using nanoparticles of TiO_2, ZnO, and Ce_2O_3. TiO_2 solution was prepared by controlled hydrolysis, using a sol–gel technique. Tetrabutyl titanate was slowly poured into 20 ml of anhydrous ethanol with stirring in a magnetic stirrer with a speed of 850 rpm. Another solution containing 1.5 ml of acetic acid, 20 ml of anhydrous ethanol, and 1.5 ml of water was prepared. The reaction mixture was heated for 24 h. On completion of hydrolysis, a transparent solution was obtained. The bottle was rolled in a way that the lower half was coated with an *n*-TiO_2 sol coating and the other remained uncoated. The bottle was heated at 105°C in an oven until a thin layer of TiO_2 dried out. In another commercial practice, bottles of coated Pyrex glass for TiO_2 coatings were used to ensure immobilization of TiO_2 nanoparticles was complete. Finally, the bottle was annealed at 250°C for 15 h. Heating at 250°C is a prerequisite for gelation and polymerization of the *n*-TiO_2 coatings.

The sol–gel reaction proceeds in two steps, hydrolysis and polycondensation, and leads to the formation of polymers with a metal-oxide-based skeleton.

Hydrolysis:

$$M(OR)_4 + H_2O \rightarrow HO\text{-}M(OR)_3 + ROH \rightarrow M(OH)_4 + 4ROH$$

Condensation:

$$(OR)_3M\text{-}OH + HO\text{-}M(OR)_3 \rightarrow (RO)_3M\text{-}O\text{-}M(OR)_3 + H_2O$$

$$(OR)_3M\text{-}OH + RO\text{-}M(OR)_3 \rightarrow (OR)_3M\text{-}O\text{-}M(OR)_3 + ROH$$

Metal alkoxides are represented by $(MOR)_z$, where M may be Si, Ti, Zr, or Al, and R is an alkyl group. The sol–gel process involves an inorganic precursor that undergoes hydrolysis and condensation reactions.

The sol–gel methods give amorphous TiO_2 particles that become crystalline on heating. Alternatively, another solution with titanium isopropoxide (iPr) as precursor was prepared by mixing titanium (IV) in tetraisopropoxide, ethanol, and acetic acid in a molar ratio of 1:100:0.05. It was hydrolyzed by glacial acetic acid. It was found that use of titanium (IV) iPr yielded a better immobilized coating. The reactions involved are as follows:

$$Ti(O\text{-}iPr)_4 + H_2O \rightarrow Ti(OH)_4 + 4PrOH$$

$$Ti(OH)_4 \rightarrow TiO_2 + 2H_2O$$

$$Ti(O\text{-}iPr)_4 + 4EtOH \rightarrow Ti(OEt)_4 + 4PrOH$$

$$Ti(O\text{-}iPr)_4 \text{ or } Ti(OEt)_4 + H_2O \rightarrow Ti(OH)_4 + 4Pr_4OH \text{ or } EtOH$$

$$Ti(OH)_4 \rightarrow TiO_2 + 2H_2O$$

$$(EtOH = \text{ethanol})$$

After coating one exposure of water to UV radiation, it was found that the required time for killing *E. coli* and disinfecting water was reduced by 20% as a result of photocatalysis due to the release of OH˙ free radicals. The following reactions occur [18]:

Excitation:

$$TiO_2 + h\nu \rightarrow h^+_{vb} + e^-_{cb} \sim fs \,(\text{femtosecond})$$

Adsorption:

$$Ti^4 + O_L^{-2} + H_2O \rightarrow O_L H^- + Ti^4 - OH^-$$

$$Ti^4 + H_2O \rightarrow Ti^4 - H_2O$$

$$Site + R_1 \rightarrow R_{1,ads}$$

$$OH˙ + Ti^4 \rightarrow Ti^4 \int OH˙$$

where $\text{Ti}^4 \int \text{OH}^\bullet$ indicates the hydroxyl radical is still associated with the Ti^4 site

Recombination:

$$h_{vb}^+ + e_{cb}^- \rightarrow \text{heat}$$

Trapping:

$$\text{Ti}^4 - \text{OH}^- + h_{vb}^+ \rightarrow \text{Ti}^4 \int \text{OH}^\bullet$$

$$\text{Ti}^4 - \text{H}_2\text{O} + h_{vb}^+ \rightarrow \text{Ti}^4 \int \text{OH}^\bullet + \text{H}^+$$

$$R_{i,ads} + h_{vb}^+ \rightarrow R_{i,ads}^+$$

where R refers to a Rideal mechanism

$$\text{Ti}^4 + e_{cb}^- \rightarrow \text{T}_i^3$$

$$\text{Ti}^3 + \text{O}_2 \rightarrow \text{T}_i^4 - \text{O}_2^{\bullet -}$$

Hydroxyl attack:

$$\text{Ti}^4 \int \text{OH}^\bullet + R_{1,ads} \rightarrow \text{Ti}^4 + R_{2,ads}$$

$$\text{OH}^\bullet + R_{1,ads} \rightarrow R_{2,ads}$$

$$\text{Ti}^4 \int \text{OH}^\bullet + R_1 \rightarrow \text{Ti}^4 + R_2$$

$$\text{OH}^\bullet + R_1 \rightarrow R_2$$

Reaction of other radicals:

$$e_{cb}^- + \text{Ti}^4 - \text{O}_2^{\bullet -} + 2(\text{H}^+) \rightarrow \text{Ti}^4(\text{H}_2\text{O}_2)$$

$$\text{Ti}^4 - \text{O}_2^{\bullet -} + (\text{H}^+) \rightarrow \text{Ti}^4(\text{HO}_2^\bullet)$$

$$(\text{H}_2\text{O}_2) + (\text{OH}^\bullet) \rightarrow (\text{HO}_2^{\bullet -}) + (\text{H}_2\text{O})$$

The free radicals of OH$^\bullet$ and superoxides $\left(O_2^{\bullet-}\right)$ earlier are highly destructive to organic matter. Research has shown that the TiO$_2$/UV combination is highly effective in water treatment. Photocatalysis can kill animal cells, and its application has been extended to cancer treatment [19]. By observing an indigo-carmine dye pill color change, the killing of bacteria was confirmed.

Despite a dramatic progress in the decontamination of water, there are still problems related to the treatment of wastewater contaminated with dyes from the textile industries, agricultural wastes, and chemical plants chlorinated by products and chemicals. The presence of trace metals such as Hg, Cr, Pb, As, and Ni, and inorganic compounds like bromates, chromates, and halides, in water may further complicate the treatment processes.

Whereas crumb rubber and the solar distillation process are highly effective in removing pathogens and impurities, the removal of contaminants originating from industries and agricultural lands need powerful oxidants for their removal. Photocatalytic treatment, which produces OH$^\bullet$ and O$_2^{\bullet-}$ free radicals, has the capability to remove these contaminants. Whereas substantial research has been conducted on photocatalytic reactors, it is limited mainly to the availability of antireflecting transparent strong, thin surfaces, light-emitting resources, arrangement of lenses, a high ratio of activated immobilized catalysis to illuminated surface, and rigid control of angle of incidence [20]. In our ongoing research program on the development of photocatalytic reactors, consideration has been given to resolve the above issue. If we could succeed in overcoming the above challenges, mainly the limitations imposed by distribution of uniform light, prolonged contact time between the catalyst and water, and providing a large surface area for contact, we could succeed in scaling up reactors for hollow tubes for industrial use [16].

The major issue is that of creation of a surface that does not reflect light. In our work on nanocoatings, we realized these shortcomings. In our work on design, which is in progress, we have developed nanostructured surface on glasses by shotless cavitation peening aiming to obtain a grain size of 10 nm on the blasted surface. Nanostructured surfaces can be obtained by cold deformation. Sand blasting with silica particles gave a grain size of 50 nm. The process of sand blasting is performed to obtain a nanostructured surface. This process can also be used to obtain small grain sizes on steel. Because of sharp high-density boundaries, the reflection would be minimized, which would allow a uniform distribution of light. We are examining the effect of grain size and light absorption and reflection. Work on the effect of nanostructured surface on light reflection on different types of glasses, such as Corning, Pyrex, Schott, and Xo glass, is showing promising results. Current work on coating of nanosilica, hydrophobic SiO$_2$ coatings with n-TiO$_2$, and mesoporous TiO$_2$ is expected to provide a complete suppression of reflection and allow the scaling up of photocatalytic reactors. The design of the photocatalytic reactor is shown in Figure 1.2. The use of mesoporous TiO$_2$ synthesis via a chemical route provides an alternative for conventional TiO$_2$ [21]. It has an optical band gap of 3.75 eV and a diameter of 200–300 nm. Hollow spheres are excellent catalyst carriers. By making the specified glass surface nanostructured and applying a coating of mesoporous TiO$_2$ and n-SiO$_2$, on hollow tubes of glass, we expect to achieve antireflecting coatings on the hollow glass surface of photocatalytic reactors and overcome the existing challenges.

This extension of our work from a crumb rubber/UV radiation bimodal system and attempts to design large-scale photocatalytic reactors show the attempt that has been made to decontaminate water from pathogens, dyes, chemicals, agriculture runoff, and pesticides.

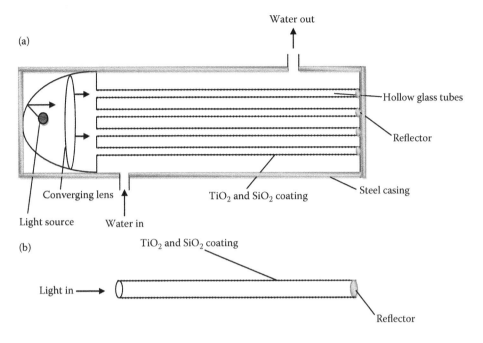

FIGURE 1.2
(a) Schematic diagram of photocatalytic reactor. (b) Schematics of one hollow tube.

1.5 Conclusion

The work conducted has shown that crumb rubber is an effective filtration medium because of its favorable porosity gradient. The bottom layer comprising coarse rubber particles does not promote clogging of the filtration column. The filtered water showed a small head loss (7 cm) at high filtration rates such as 20 m/s. The turbidity was reduced to low levels (3.0 NTU), suspended solids were not observed, and the TDS showed lower levels. Exposure of water obtained from the filtration column showed no evidence of the presence of *E. coli*, *Shigella* spp., and *V. cholerae*. The application of the bimodal technique of crumb rubber filtration and UV-supported n-TiO_2 catalysts showed a significant improvement in water quality. The physiochemical characteristics of water matched the standards invoked by WHO. The current work on the scaling up of a photocatalytic reactor is based on using hollow nanostructured tubes coated with mesoporous TiO_2 and n-SiO_2 for uniform reflection of light. It is expected that this work shall enable the scaling up of the photocatalytic reactors and allow a new generation to treat the wastewater from the textile, agricultural, and chemical industries, which is highly challenging.

Acknowledgments

The authors thank Faheemuddin Patel of KFUPM, Dhahran, Kingdom of Saudi Arabia, for his help in the work done on crumb rubber filtration, and Rizwan Ahmad of COMSATS for his contribution in designing the photocatalytic reactor. The help provided by Zahra Khan and Tayyaba Abid is appreciated.

References

1. Ross, R. World Water Day 2010, March 22, 2009, "Clean and healthy water." Pacific Institute for Studies in Development, Environment and Security, Oakland, CA (USA).
2. GAO, U.S. Water and Sanitation Aid: Millions of Beneficiaries Reported in Developing Countries, GAO-10-967, September 2010, p. 18.
3. Amjad, Z. Ed. *Reverse Osmosis: Membrane Technology, Water Chemistry, and Industrial Applications*, Van-Nostrand Reinhold, NY, 1993.
4. Pinnau, F.B.D. Formation and modification of polymeric membranes, Pinnau, F.B.D., Ed., *ACS Symposium Series*, Vol. 744, American Chemical Society, Washington, DC, 2000, pp. 1–22.
5. Available at http:/water.org./water-crisis-facts/water, 2 | 10 | 2013.
6. Shannon, M.A., Bohn, P.W., Elimelech, M., Georgiadis, J.G., Mariñas, B.J., Mayer, A.M. Science and technology for water purification in coming decades, *Nature*, 452, 301–310, 2008, doi:10.1038/nature 06599.
7. Ramandeep, S.G., Vinod, K., Ashutaosh, N., Raman, S., Vikram, B. Water pollution: Impact of pollutants and new promising techniques in purification process, *Journal of Human Ecology*, 37, 103–109, 2012.
8. Kulshrestha, S.N. A global outlook for water resources to the year 2025, *Water Resources Management*, 12, 167–184, 1998.
9. Burleigh, B., McDowell, T. Crumb rubber pre-filter for use with UV water treatment system, Multi-Disciplinary Science Design Conference, Kate Gleaon College of Engineering, Rochester Institute of Technology, Rochester, NY, Project No P 11413 (senior project), 2011.
10. Gambhir, R.S., Kapoor, V., Nirola, A., Sohl R., Bansal, V. Impact of water pollution and new promising technologies in purification process, *Journal of Human Ecology*, 37(2), 103–109, 2012.
11. McCullagh, C., Robertson, J.M.C., Bahnemann, D.W., Robertson, P.K.J. The application of TiO_2 photocatalysis for disinfection of water contaminated with pathogenic micro-organisms: A review, *Research on Chemical Intermediates*, 33(3–5), 359–375, 2007.
12. Abhang, R.M., Kumar, D., Taralkar, S.V. Design of photocatalytic reactor for degradation of phenol in wastewater, *International Journal of Chemical Engineering and Applications*, 2(5), 337–341, 2011.
13. Lo, C.-F., Wu, J.C.S. Preparation and characterization of TiO_2-coated. Optical-fiber in a photo reactor, *Journal of the Chinese Institute of Chemical Engineers*, 36(2), 119–125, 2005.
14. Mahmoodi, N.M., Arami, M., Limaee, N.Y., Gharanjing, K., Naz, F., Muhammadi, N. Nano photocatalyst using immobilized TiO_2 degradation and mineralization of water containing organic pollutants, case study of butachlor, *Materials Research Bulletin*, 42, 797–806, 2007.
15. Kandra, H.S., McCarthy, D., Deletic, A., Fletcher, T.D. Assessment of clogging phenomena in granular filter media used for stormwater treatment, NOVATECH, Conference 2010, Session 3.3, Lyon, France.
16. Al-Anbari, R.H., Wootton, K.P., Durmanic, S., Deletic, A., Fletcher, T.D. Evaluation of media for the adsorption of stormwater pollutants, 11th International Conference on Urban Drainage, Edinburgh, Scotland, UK, 2008.
17. Xie, Y.F., Killian, B.A., Gaul, A.S. Using crumb rubber as a filter media for wastewater filtration, Proceedings of the 160th American Chemical Society Rubber Division Technical Meeting, Cleveland, OH, 2001.
18. Turchi, C.S., Ollis, D.F. Photocatalytic degradation of organic water contaminates: Mechanisms involving hydroxyl radical attack, *Journal of Catalysis*, 122, 178–192, 1990.
19. Blake, D.M., Maness, P.-C., Huang, Z., Wolfrum, E.J., Huang, J. Application of the photocatalytic chemistry oftitanium dioxide to disinfection and the killingof cancer cells, *Separation and Purification Methods*, 28(1), 1–50, 1999.
20. Mukherjee, P.S., Ray, A.K. Major challenges in the design of a large-scale photocatalytic reactor for water treatment, *Chemical Engineering & Technology*, 22(3), 253–260, 1999.
21. Zhang, Y., Li, G., Wu, Y., Luo, Y., Zhang, L. The formation of mesoporous TiO_2 spheres via a facile chemical process, *Journal of Physical Chemistry B*, 109, 5478–5481, 2005.

2

Hierarchical Carbon and Hydrogels for Sensing, Remediation in Drinking Water, and Aquaculture Drug Delivery

Cesar A. Barbero, Rusbel Coneo Rodriguez, Rebeca Rivero, María V. Martinez, María A. Molina, Juan Balach, Mariano M. Bruno, Gabriel A. Planes, Diego F. Acevedo, Claudia R. Rivarola, and María Cristina Miras

Chemistry Department, Universidad Nacional de Rio Cuarto, Río Cuarto, Argentina

CONTENTS

2.1 Introduction

Nanoporous materials can be used in different aquatic nanotechnology applications. In our case, two kinds of nanoporous material are studied: nanoporous carbon and nanoporous polymeric hydrogels. Novel synthetic procedures aimed at obtaining the materials are described. The materials can be used without further modification or functionalized by incorporation of bulk conducting polymers, metallic nanoparticles, metal oxide nanoparticles, or conducting polymer nanoparticles.

2.1.1 Why Nanoporous Materials?

The development of nanotechnology could have different effects on the aquatic environment. Man-made nanoparticles could be released into the environment and harm the aquatic life [1–3]. On the other hand, nanotechnology could have different ways of improving the aquatic environment, such as by (i) detecting water contaminants; (ii) remediating contaminated water; (iii) delivering drugs, genes, or vaccines to aquaculture organisms; (iv) delivering nutraceuticals to fish; (v) controlling the growth of algae; and (vi) tagging an aquaculture organism and sensing its health [4]. While free nanoparticles could be easily released into the aquatic environment with potentially negative results [5–7], nanoporous materials and/or immobilized nanoparticles are safer since they cannot easily contaminate the aquatic environment. The large surface area of nanoporous materials gives them large adsorption capacities [8], which make them ideal materials in water remediation [9].

At the same time, catalytic activity [10], electrochemical conversion [11], and electroanalytical current signal all increase with the active surface area. Therefore, both catalytic and electrochemical devices for contaminant destruction could use nanoporous materials. The same is true for electrochemical sensors [12] and deionizers [13].

A porous material is a solid with free space (pores) that is filled by the fluid phase (vacuum, gas, pure liquid, solution) and can be filled totally or partially by another solid. To be useful in aquananotechnology, a significant portion of the pores must be open, that is, accessible to the outside aqueous solution. The International Union of Pure and Applied Chemistry (IUPAC) classifies the pores into three domains: micropores (diameter [d] < 2 nm), mesopores (2 nm < d < 50 nm), and macropores (d > 50 nm). The yardstick is the size of the molecules that can be adsorbed on the pore surface. Only small molecules (e.g., N_2) can enter micropores, while any low molecular weight molecules can enter mesopores. For molecular adsorption, macropores are less interesting because the specific surface area of a common material (e.g., carbon) only having macropores will be low (<1 m^2/g). However, large molecules (e.g., DNA) or even nanoparticles can be adsorbed inside macropores. Therefore, macropores could be used to immobilize nanoparticles in macroscopic solids to avoid release into the environment. Additionally, mass transport is usually hindered in mesopores and micropores, while mass transport in macropores is similar to solution. Therefore, waterborne species (e.g., contaminants) will have easy access to the nanoparticles inside the macroporous solid. On the other hand, it is possible to prepare materials having pores of two different kinds (e.g., macropores and mesopores). Such materials are called hierarchical and have properties related to each pore size.

2.1.2 Historical Porous Materials

Different porous solids have been used for water remediation. Zeolites are natural or synthetic microporous (subnanometric) aluminosilicate minerals that are used for ion adsorption [14]. They are used for ammonium adsorption from wastewater [15]. Ion-exchange resins are usually made of cross-linked polymers with functional groups covalently bonded to the polymer chains that interact with free ions present in the solution [16]. The porosity is given by the spaces left between linked chains (microporosity) [17], and due to the effect of porogens added during synthesis (macropores) [18]. The resins are extensively used for adsorption of deleterious ions in water treatment [19]. Activated carbon is a widely used material for water purification [20]. It is produced by carbonization of cellulosic materials [21], followed by chemical or physical activation [22]. The solid has disordered porosity (mesopores and micropores) with noncylindrical shapes whose tortuosity makes the mass transport difficult [23]. The graphene-like surface of the pores promotes the physisorption of organic substances (e.g., benzene) [24,25]. On the other hand, the surface can bear different surface groups (>C=O, –COOH, etc.) [26] that can interact with ions in water to promote chemisorption [27]. Additionally, the surface can be modified chemically to enhance adsorption [28], using organic chemistry reactions [29]. Nanoporous silica can be easily produced by sol–gel chemistry [30,31]. The pore surface is highly hydrophilic and contains mainly Si-OH groups, which could chemisorb ions from water [32]. Additionally, the well-known silane coupling chemistry [33] can be used to modify the pore surface and increase adsorption [34]. Other inorganic materials can be made into nanoporous solids, such as aluminosilicates [35,36] and metallic oxides [37]. The solids have been used to remove toxic ions from water [38]. Additionally, natural inorganic porous solids, such as diatomaceous earth, have been extensively used for water purification [39].

In the present chapter, we will discuss two types of nanoporous solids: nanoporous carbon and nanoporous hydrogels. Additionally, those porous materials could be functionalized by loading with nanoparticles made of metal, metal oxides, or conducting polymers. Therefore, we will briefly examine the literature on those materials.

2.1.2.1 Nanoporous Carbon

Besides activated carbon, it is possible to synthesize nanoporous carbon by a variety of methods [40]. Inorganic nanoporous solids (e.g., silica) can be used as a hard template by covering the pore surface with a thin layer of carbon [41]. This is achieved by the formation of an organic polymer (e.g., phenol–formaldehyde) by sequential precursor adsorption, followed by reaction [42]. The polymer is then carbonized by heating in an inert atmosphere [43]. Then, the inorganic template is removed by dissolution [44]. The inorganic template should have an interpenetrated pore topology [45]. Otherwise, only carbon nanoparticles will be obtained. Inorganic nanoparticles could also be used as hard templates of porous carbon. To do that, the nanoparticles are aggregated into opals and an organic precursor is infiltrated in the interstitial space between particles [46]. The silica nanoparticles have to be in contact with each other; otherwise, the template cannot be removed from inside the solid.

Organic species could be used as soft templates for porous carbon synthesis, among them are molecular micelles [47] and polymeric micelles [48]. We have extensively studied the synthesis of porous carbons by pyrolysis of porous polymeric (resorcinol–formaldehyde [RF]) resins [49]. The resin gels maintain nanoporosity during conventional drying through stabilization of resin nanoparticles by cationic supramolecular species. The carbon source (precursor) is subjected to a heat treatment (pyrolysis) at high temperature in the absence of oxygen to produce the carbon. Since the main research goal was to produce electrode materials for supercapacitors [50], both capacitance and response rate should be maximized. All porous carbons can be surface functionalized by linking covalent groups with the native groups present in the carbon surface [51]. Additionally, synthetic porous carbon can be produced with other elements (N, S, P) in the structure by using different precursors [52–54].

Porous materials (e.g., carbon) are usually characterized by measuring the adsorption isotherm of inert gases (e.g., N_2). By modeling the adsorption data, the surface area, pore volume, and pore size distribution can be evaluated. However, in electrochemical applications, only the surface area accessible to the electrolyte is important. It is well known that small micropores cannot be filled with electrolyte and do not contribute to the electrochemical active area but are measured by the gas adsorption isotherms [55]. Therefore, *in situ* measurements of ion adsorption phenomena such as differential capacitance [56] or probe beam deflection (PBD) [57] render more useful data of the carbon texture. PBD has shown to be very useful for the study of porous materials [58,59], including carbon aerogels [60].

2.1.2.2 Nanoporous Hydrogels

Cross-linked polymers constitute three-dimensional networks containing nanopores. The mean size of the pores is directly related to the cross-linking degree [61]. Hydrogels are cross-linked networks bearing hydrophilic groups that promote gel swelling in water [62]. The mean pore size can be calculated from the swelling capacity by using the Flory–Rehner theory [63]. Soluble species could dissolve in the water pool of the pores or interact directly with the polymer chains. Such materials have been used to remove toxic ions (cyanide, arsenic, chromate) from water [64–66]. The important parameter is the partition coefficient of the soluble species between the hydrogel and water. We have shown that loading conducting polymers [67] into the nanoporous hydrogel could significantly increase the partition of some species (e.g., tryptophan) into the material [68].

Several materials could be used to make nanocomposites with nanoporous carbon or nanoporous hydrogels, effectively changing the materials' properties.

2.1.2.3 Conducting Polymers

Conducting polymers are doped by counterions (anions or cations) behaving as polyelectrolytes that can exchange inert ions with toxic ones [69]. Additionally, since the oxidation/reduction of conducting polymers involves ion exchange, the oxidation state depends on the concentration of ions in solution [70]. Therefore, conducting polymers can be used for sensing [71] and/or removal [72] of toxic ions present in water.

2.1.2.4 Nanoparticles

Metal (e.g., gold) nanoparticles constitute one of the earliest examples of artificial nanoparticles [73] and are the most extensively studied nanotechnology objects [74]. The large surface area allows them to adsorb toxic substances present in water [75]. Metal nanoparticles have been extensively used as electrocatalysts in fuel cell research, and have potential use in electroanalysis [76] and electrochemical incineration of organic compounds [77]. PtRu catalysts are the most active for methanol oxidation [78], likely due to the formation of Ru oxides in the surface that are able to oxidize adsorbed poisons like CO. Such behavior would allow oxidation of other organic species of analytical interest that do not present reversible redox couples but have oxygenated groups or C–H bonds that can be irreversibly oxidized. Analogously, harmful organic compounds can be completely oxidized by active electrocatalysts.

Metal oxide nanoparticles (e.g., magnetite, Fe_3O_4) strongly adsorb toxic ions (e.g., arsenite) and can be used to remove them from water [79]. Conducting polymers can also be made into the form of nanoparticles [80]. The high surface area and small size makes adsorption favorable and diffusion inside the particles fast. Water purification using nanoporous materials can be made by a variety of processes: physical [81] or chemical adsorption of soluble species [82]; electrochemical [83], photochemical [84], or catalytic complete oxidation of organics [85]; electrochemical capacitive deionization [86]; absorption of soluble species in cross-linked polymers [87]; etc.

In the present chapter, we will describe different methods to synthesize nanoporous materials (carbon and hydrogels), to load them with conducting polymers or nanoparticles (metal and metal oxide), and to use them in aquatic nanotechnology.

2.2 How to Make Nanoporous Materials and Nanocomposites

2.2.1 Nanoporous Carbon

One way to form porous carbon involves producing porous precursor resins, such as RF, which are then converted into glassy carbon by carbonization. Ambient air-drying of RF gels gives nonporous resins (xerogels) by the collapse of the gel due to surface tension forces at the water/air interface during drying (Scheme 2.1). To avoid that, the gels could

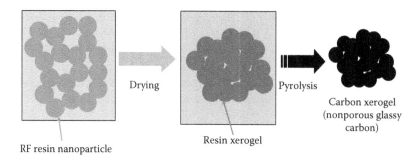

SCHEME 2.1
Formation of carbon xerogels by conventional drying of RF resin.

be dried using supercritical liquids (aerogels) [88] or low-surface-tension solvents (ambigel) [89]. In both cases, the small surface tension forces prevent the collapse of gel pores. Another way involves sublimation of the solvent from the pores by lyophilization of the gels (cryogels) [90]. All these procedures are slow and cumbersome. Therefore, the use of additives, which maintain porosity under drying in air, will be advantageous.

2.2.1.1 Use of Cationic Surfactant (CTAB) Micelles as Nanoparticle Stabilizers

Ordered mesoporous inorganic materials can be produced with a tailored pore size distribution by sol–gel techniques and using a variety of micelles of surfactant molecules as templates [91]. They are produced in various compositions, such as oxides (e.g., silica, alumina, titania, zirconia, mixed oxides), by condensation of inorganic species around the arrays of self-assembled aggregates of surfactant molecules in water [92]. The synthesis of RF organic sol-gel occurs by a condensation reaction mechanism, that is analogous to the synthesis of inorganic oxides. Accordingly, Bell and Dietz [93] reported the preparation of porous RF resins, claiming that surfactant micelles act as templates that are eliminated during carbonization. The material could be air-dried without significant contraction, avoiding the use of organic solvents or supercritical fluids [94]. We [95] have suggested that the actual mechanism involves the stabilization of resin nanoparticles (negatively charged) by cationic micelles (positively charged) (see Scheme 2.2).

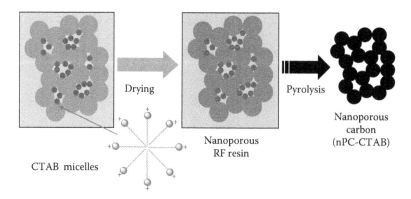

SCHEME 2.2
Formation of a nanoporous carbon by stabilization of the RF resin using cationic micelles.

FIGURE 2.1
Scanning electron micrograph of the surface of carbon produced by carbonization of a resorcinol/formaldehyde (R/F: 0.5) resin. The polymerization is catalyzed by CO_3Na_2 (catalyst/R: 0.005) in water (water/R: 10) using a stabilizer ([CTAB] = 0.077 M). The resin is carbonized at 800°C in an Ar atmosphere.

The resin could be dried in air and then carbonized to produce porous carbon (S_{sp} = 670 m²/g) (Figure 2.1). The surface does not show the large pores expected by a cylindrical micelle mechanism but individual carbon nanoparticles with interstitial nanopores.

2.2.1.2 Use of Cationic Polyelectrolyte (PDAMAC) as Nanoparticle Stabilizer

The mechanism described in Scheme 2.2 implies nanoparticle stabilization through adsorption of the cationic micelles. We thought that a cationic polymer could achieve the same effect. Tests with poly(diallyldimethylammonium chloride) (PDAMAC) reveal that large porosities could be achieved by using PDAMAC as a stabilizer (Scheme 2.3) [96].

Additionally, the pore size could be controlled by the ratio of monomer to stabilizers [97]. A fracture surface of the carbon shows the pores and the nanoparticle aggregate nature of the carbon (Figure 2.2).

2.2.1.3 Use of Fibers to Produce Hierarchical Porous Carbon

Besides molecules, larger structures can also be used to produce or maintain porosity in resins. RF monomer solution could be adsorbed onto cellulosic fiber cloths and cured into a resin. The presence of the fiber not only creates macropores (several micrometers

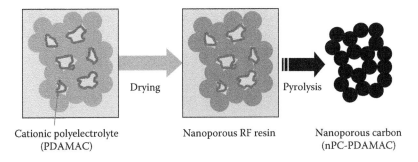

| Cationic polyelectrolyte (PDAMAC) | Drying | Nanoporous RF resin | Pyrolysis | Nanoporous carbon (nPC-PDAMAC) |

SCHEME 2.3
Formation of a nanoporous carbon by stabilization of the RF resin using a cationic polyelectrolyte (PDAMAC).

FIGURE 2.2
Scanning electron micrograph of the surface of carbon produced by carbonization of a resorcinol/formaldehyde (R/F: 0.5) resin. Polymerization is catalyzed by CO_3Na_2 (catalyst/R: 0.005) in water (water/R: 10) using a stabilizer ([PDAMAC] = 0.0013 M). The resin is carbonized at 800°C in an Ar atmosphere.

depending on the fiber size) but also stabilizes the micropores/mesopores likely by surface tension effects [98]. In that way, a spontaneous hierarchical carbon material (bearing macropores and mesopores) could be produced (Scheme 2.4).

The holes in the cloth and space between microfibrils give macropores to the material (Figure 2.3), while the stabilization of nanoparticles by the fibers gives nanopores to the material.

On the other hand, hydrophobic polymer fibers can be inserted inside the RF to produce long macropores to produce hierarchical carbon [99], using PDAMAC as a stabilizer of nanoporosity.

2.2.1.4 Use of Hard Template (Silica Nanoparticles) to Produce Hierarchical Porous Carbon

Porous carbon can be obtained using templates of nanoscale sizes [100,101]. Hierarchical porous carbon (HPC) [102–104] can be obtained using two templates in a somewhat complex process. On the other hand, we have recently shown that the volume contraction of the carbon around a remaining hard template during pyrolysis induces the formation of additional mesoporosity, and, besides the pores defined by the hard template, is able to create a hierarchical carbon material [105]. In that way, HPC can be obtained in a single pyrolysis step. Since

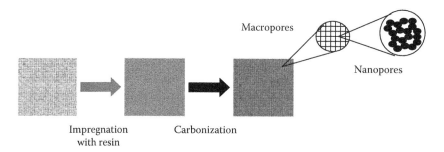

SCHEME 2.4
Synthesis of a hierarchical porous carbon (HPC-f) by carbonization of a cellulose cloth impregnated with RF resin.

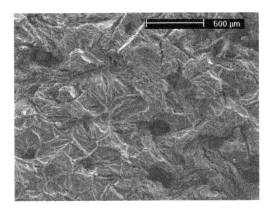

FIGURE 2.3
Scanning electron micrograph of a hierarchical carbon produced by carbonization RF resin adsorbed on a cellulose open cloth.

precursor materials are organic polymers containing hydrogen and oxygen besides carbon (approx. formula $C_{22}H_{22}O_{10}$), the pyrolysis to give pure carbon involves mass loss and volume contraction. Soft templates like polymers and surfactants are spontaneously eliminated during burning of the carbon material; therefore, posttreatments are not necessary. On the other hand, hard templates like silica and metal oxides survive to high temperatures and must be removed after pyrolysis by chemical etching. In this process, the space initially occupied by the templates is transformed into the pores in the resulting carbon materials. The whole process results in a reverse copy of the template [106]. To produce carbon with an open structure, we used silica nanoparticles as a hard template. The volume contraction during the pyrolysis of RF resin is easily observable by comparison between the piece size before and after the thermal decomposition of the precursor. However, when the carbon sample contains a rigid template, the observed contraction in the macroscale is very low owing to the presence of a noncompressible skeleton. In our case, this rigid support is composed by an array of nanoscale SiO_2 nanoparticles in close contact between them. The scanning electron microscopy (SEM) image of these SiO_2 nanoparticles reveals the existence of almost monodisperse spheres. As shown in Scheme 2.3, the hard template can be eliminated before or after pyrolysis. We have shown [19] that elimination before pyrolysis produces a carbon with one main pore size (1PSC), which is determined by the template nanoparticle size but takes into account the 20%–30% contraction of the material upon pyrolysis. However, the presence of the silica nanoparticles seems to stabilize the microporosity/mesoporosity since relatively large surface areas are measured (~150 m^2/g). It should be noted that a porous matrix with holes of ~400 nm will have ~6 m^2/g of specific surface area.

On the other hand, if the hard template is left during pyrolysis, the contraction of the solid material has to take place in the interstitial spaces between particles. In that way, the solid becomes microfractured and the carbon has at least two pore sizes (2PSC), one directly related to the hard template size (e.g., 400 nm) and another much smaller in the order of mesopores (<50 nm). Additionally, the carbon presents the sintered beads of carbon linked into a matrix through necks, which has been proposed as the building block of the RF resin. Those beads, after precipitation, form the condensed monoliths [107,108] (Scheme 2.5).

In the flat focused ion beam (FIB) cut, it is possible to observe the open three-dimensional nature of the carbon (Figure 2.4). The surface area is of ~650 m^2/g, revealing that microfractures contribute greatly to the surface area.

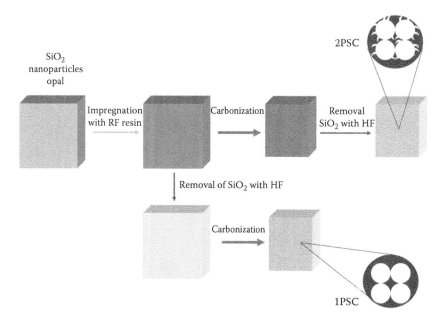

SCHEME 2.5
Formation of nanoporous carbon by hard templating of RF resin using SiO$_2$ nanoparticles.

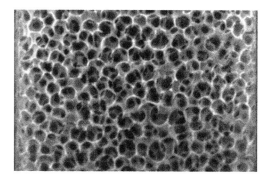

FIGURE 2.4
Scanning electron micrograph of a porous carbon (1PSC) synthesized using an opal of SiO$_2$ nanoparticles (400-nm diameter) as hard template. Cut is made using FIB of Ga ions. Lateral dimension is 5 microns.

2.2.1.5 Self-Assembly of Porous Carbon Microparticles to Produce Hierarchical Carbon Layers

Porous carbon microparticles can be fabricated by grinding the solid and sieving the particles [109] (Figure 2.5a), or by synthesizing inside water pools of microemulsion [110] (Figure 2.5b). Carbon nanoparticles can be fabricated by controlling the nucleation of the resin nanoparticles in solution [111] (Figure 2.5c).

The porous particles can be assembled by sequential dipping of a surface functionalized with charged groups (e.g., −SO$_3^-$ groups linked to the glass through silane chemistry [112]) in dispersions of charged particles and polyelectrolytes of opposite charge [113]. The technique is usually known as layer by layer (LbL) and was pioneered by G. Decher [114]. However, in our case, optical microscopy of the surface shows an increasing two-dimensional coverage of the surface, instead of an LbL growth (Figure 2.6).

FIGURE 2.5
Scanning electron micrographs of porous microparticles produced by grinding and sieving (a), synthesis inside an inverse miniemulsion using PDAMAC as nanoparticle stabilizer (b), and controlled nucleation and growth (c).

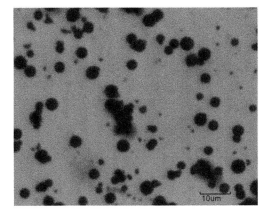

FIGURE 2.6
Optical micrograph of a glass surface covered with carbon microparticles by the step-by-step process.

Therefore, we called the process step by step (SbS) instead of LbL. While the thickness of the layer cannot be controlled by the SbS process, the relative amount of covered area can be finely controlled. In that way, the amount of material is controlled.

2.2.2 Nanoporous Hydrogels

A cross-linked polymer is made of linear polymer chains linked by short chains that act as a cross-link. Since the secondary structure is an extended three-dimensional network, all cross-linked polymers are nanoporous [115]. However, to be useful in aquananotechnology, the nanoporosity should be completely wetted by water. Hydrogels are cross-linked polymers where the polymer chains interact strongly with water (through hydrogen bonding, ion–dipole interaction, etc.). Therefore, the polymer swells in presence of water to an expanded state.

Cross-linked polyacrylamides are widely used hydrogels for different applications, including water remediation [116]. One advantage of polyacrylamides (and related polymers) is the possibility to produce different materials for copolymerization of monomeric acrylamides bearing different functional groups [117]. We have synthesized copolymer polyacrylamides including sulfonic acid ($-SO_3^-$) groups, acrylic acid [118], and N-isopropyl groups. The materials can be synthesized as bulk materials or in the form of microparticles, nanoparticles, or thin films [119]. Both microparticles and nanoparticles can be produced by similar methods to carbon materials. Microparticles (Figure 2.7a) are synthesized by radical polymerization of vinylic monomers (e.g., acrylamide) inside the water pools of an inverse water-in-oil miniemulsion. Nanoparticles (Figure 2.7b) can be made by controlled nucleation of the radical polymerization of soluble acrylamides.

The advantage of having at least one dimension of the solid in the microscale–nanoscale range being small is to achieve fast mass transport. For a diffusion process obeying Fick's law [120], the time (t) necessary to penetrate a material is given by

$$t = \frac{x^2}{2D} \tag{2.1}$$

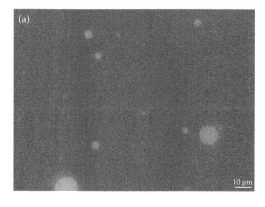
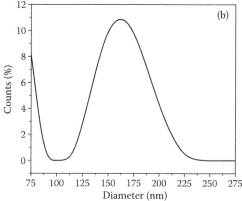

FIGURE 2.7
(a) Fluorescence microscopy of poly(acrylamide) hydrogel microparticles synthesized by radical polymerization inside water pools of inverse miniemulsion. The gels are dyed with a fluorescent dye. (b) Dynamic light scattering measurement of hydrogel nanoparticles made by controlled nucleation.

where x is the length of the path of the mobile species in the material and D is the diffusion coefficient. The diffusion coefficient of anions in hydrogels used for water remediation is on the order of 10^{-5} cm^2/s [121]. Therefore, the time for mass transport in different materials will depend on the thickness of the porous layer.

As can be seen in Table 2.1, a cube (each side = 1 cm) of nanoporous hydrogel will take 500 s (8.3 min) to be loaded with a soluble species by diffusion. On the other hand, a microparticle (d = 25 μm) will take a negligible time (<1 s) to be loaded. However, nanoparticles and microparticles are less useful because the handling of dispersions is complex and some particles could be leaked to the environment. To avoid that, the SbS assembly could be used (Figure 2.8) to produce surfaces containing small particles. It should be borne in mind that the surface does not need to be planar. If nanoparticles need to be assembled, a solid with pores of micron diameter can be used as surface.

An alternative way to produce fast transport materials without particles involves fabrication of a macroporous hydrogel where the polymer walls around macropores have small dimensions (e.g., 10 μm). The macroporous hydrogel was made using microscale templates such as *in situ* produced ice crystals (Scheme 2.6a) [122] or *in situ* produced gas bubbles (Scheme 2.6b).

The SEM micrographs of the gels, taken in low vacuum conditions (Figure 2.9), show long macropores produced by the templating effect of the ice crystals (Figure 2.9a) or

TABLE 2.1

Time Response for Water Swelling of Different Hydrogel Structures

Form of Hydrogel	Characteristic Size (μm)	Transport Time (s)
Nanoporous monolith	1000	500
Microparticle	25	0.310
Nanoparticle	0.2	0.00002
Thin film	1.5	0.001125
Macroporous monolith	10	0.05

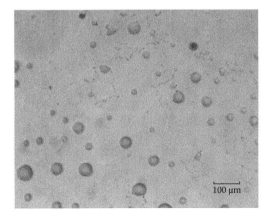

100 μm

FIGURE 2.8
Optical micrograph of hydrogel microparticles (made as described in Figure 2.6) self-assembled (SbS) onto a glass surface. Microparticles are dyed with methylene blue.

(a) Cryogelation

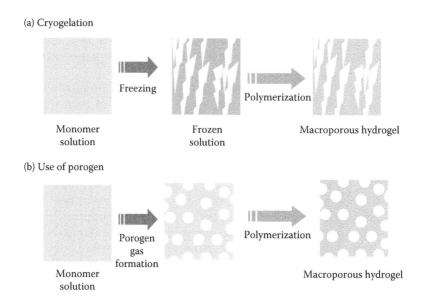

(b) Use of porogen

SCHEME 2.6
Macropore formation mechanism by templating with ice crystals in cryogelation (a) or by *in situ* gas formation (b).

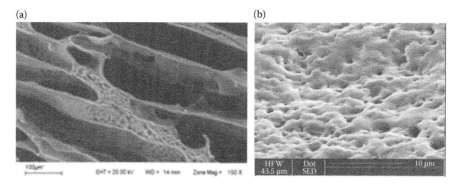

FIGURE 2.9
Scanning electron micrographs of macroporous hydrogels made by cryogelation (a) and using oxalate ion (which produces CO_2 gas by oxidation inside the gel) as porogen (b).

small spherical pores produced by the formation of gas bubbles inside the polymerization solution (Figure 2.9b).

Besides functionalization with nanoparticles, nanocomposites could be produced by *in situ* formation of another polymer inside the hydrogel. We produced different hydrogels (PAAm, PNIPAM, PAMPS) containing conducting polymers (PANI, PPy, PNMANI) by oxidative polymerization of the conductive polymer's monomer inside the hydrogel.

2.2.3 Nanocomposites

2.2.3.1 Conductive Polymer Filling the Nanopores of a Hydrogel

Conducting polymers could be used to remove pollutants by metal ion reduction [123], or nucleophilic addition of the pollutant to the polymer [124]. A simple way to produce

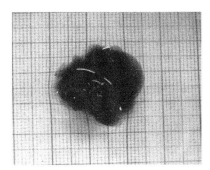

FIGURE 2.10
PNIPAM nanoporous hydrogel filled with PANI by *in situ* polymerization.

monolithic pieces of a nanocomposite containing a conducting polymer is simply *in situ* polymerization of an appropriate monomer (e.g., aniline) loaded inside the nanoporous hydrogel [67] (Figure 2.10).

2.2.3.2 Magnetite Nanoparticles Adsorbed in a Macroporous Hydrogel

Magnetite (Fe_3O_4) nanoparticles have been extensively used to adsorb toxic ions such as arsenite [125]. However, free nanoparticles in natural waters could have deleterious effects on aquatic life. A possible way to use magnetite nanoparticles without releasing them into the environment consists of adsorbing them into macroporous hydrogels (Figure 2.11).

While magnetite nanoparticles can be formed *in situ* inside nanoporous hydrogels, preformed particles do not enter a non-macroporous hydrogel likely owing to steric constraints.

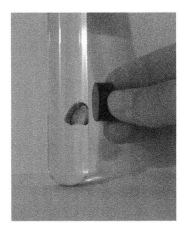

FIGURE 2.11
Magnetite nanoparticles adsorbed on the inner surface of a macroporous hydrogel. Note that the whole hydrogel mass is lifted using a permanent magnet because of the presence of nanoparticles.

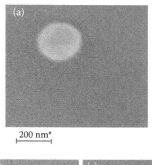

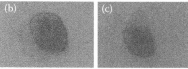

FIGURE 2.12
(a) FE-SEM of a cryofractured macroporous hydrogel showing a PANI nanosphere adsorbed on the surface of the pore. (b) Photograph of a macroporous PNIPAM hydrogel functionalized with PANI nanofibers, before thermal deswelling. (c) Photograph of a macroporous PNIPAM hydrogel functionalized with PANI nanofibers, after thermal deswelling.

2.2.3.3 Polyaniline Nanoparticles Adsorbed Inside a Macroporous Hydrogel

Conducting polymer nanoparticles could be used to remove pollutants through a specific mechanism of metal ion reduction [126] or nucleophilic addition of the pollutant (e.g., thiol) to the polymer [127]. However, free conducting polymer nanoparticles could end up in natural waters and have toxic effects on aquatic life. Indeed, we have shown that PANI nanofibers have teratogenic effects on frog larvae [128]. Therefore, a mechanism to use conducting polymer nanoparticles without releasing them to the environment is desirable. We have loaded PANI nanospheres and nanofibers into macroporous hydrogels [122]. The nanoparticles are irreversibly adsorbed on the macropores (Figure 2.12a) and maintained even under swelling/deswelling processes (Figure 2.12b,c). Therefore, the nanocomposite could be used to remove toxic substances from water.

Since PANI nanoparticles are ~200 nm in size, preformed nanoparticles do not enter a non-macroporous hydrogel owing to steric constraints.

2.3 Which Techniques Can Be Used to Measure the Relevant Properties?

Different techniques, including novel ones, have to be used to measure the properties of nanoporous materials relevant to aquatic nanotechnology applications.

2.3.1 Surface Area and Porosity of Porous Carbon

The nanoporosity of porous carbon can be measured indirectly by gas (e.g., N_2) adsorption isotherms. Figure 2.13 shows the experimental isotherm of a porous carbon and the

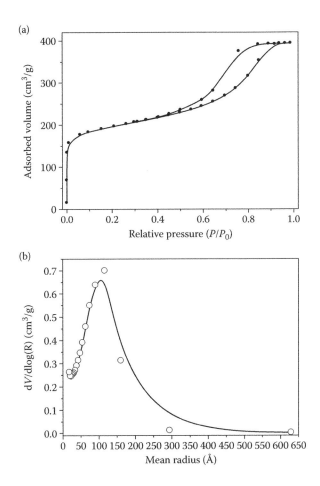

FIGURE 2.13
Nitrogen adsorption isotherm measurements of a porous carbon (HPC-f): (a) experimental data; (b) calculation of the pore size distribution using the BJH method.

corresponding pore size distribution, calculated by the Barrett–Joyner–Halenda (BJH) method. It should be borne in mind that the data are based on a model and not in the direct observation of the porosity. There are techniques, such as small angle x-ray scattering, that can directly measure the nanoporosity of ordered porous solids where the solid part has high x-ray contrast (e.g., SiO_2) [129].

2.3.2 Direct Measurement of Carbon Porosity

The porosity of carbon was measured directly using FE-SEM nanotomography. In this technique, thin (10 nm) slices of the porous solids are sequentially removed using an FIB of Ga ions and the exposed surface examined using FE-SEM (2–3 nm resolution). The surface images are collated into a three-dimensional picture of the porous solid. Figure 2.14 shows a typical nanotomography image of a porous carbon. The macroporosity (>50 nm) and some mesoporosity (>2 nm) are accessible to the technique; however, the microporosity cannot be measured.

FIGURE 2.14
FE-SEM nanotomography of a porous carbon (PC-CTAB) (dimensions: 500 × 500 × 700 nm). Gray areas represent the compact carbon domains.

2.3.3 Electrochemical Properties of Nanoporous Carbon

Applying a potential to a carbon electrode in an electrochemical cell induces the following processes: (i) charging of the electrochemical double layer; (ii) electron transfer to a surface and/or solution species through a Faradaic reaction. By measuring the current–potential relationship, it is possible to evaluate both effects.

2.3.3.1 Double-Layer Capacitance

The first process is related with the differential capacitance of the electrode, defined by

$$Cd = \frac{dQ}{dE} \qquad (2.2)$$

where Cd is the differential capacitance (F), Q the electrical charge (C), A the surface area, and E (V) the electrode potential. Since porous carbons are conductors with large specific surfaces ($S_{sp} > 100$ m^2/g), they can have large capacitances due to the relationship

$$Cd = Cd_a S = Cd_a m S_{sp} \qquad (2.3)$$

where Cd_a is the areal capacitance (F/cm^2), S the surface area (cm^2), S_{sp} the specific surface (m^2/g), and m the material mass.

As an example, an electrode of geometrical area 0.071 cm^2 (3-mm-diameter disc) covered with 1 mg of a carbon with $S_{sp} = 500$ m^2/g, which has an areal capacitance (Cd_a) of 20 μF/cm^2, will have a capacitance of 1 F.

The capacitance can be measured by cyclic voltammetry (CV) or by electrochemical impedance spectroscopy (EIS). CV applies a potential scan to the working electrode, measuring the current. This measurement gives an accurate idea of the accessible surface for any electrochemical process in aqueous media. The differential capacitance is calculated as

$$Cd = \frac{dQ}{dE} = \frac{dQ}{dt}\frac{dt}{dE} = iv \qquad (2.4)$$

where t is the time, i the current, and v the scan rate.

Using the current in amperes and the scan rate in volts per second, the differential capacitance is obtained in farads. Assuming that the capacitance surface density is constant, the capacitance gives a comparative idea of the extension of the electroactive area (Figure 2.15).

EIS uses a small potential amplitude assuring that the measurement is made on a linear system [130]. A typical Nyquist plot of a porous carbon (PC-CTAB) electrode is shown in Figure 2.16.

The response at high frequencies shows a semicircle while the response at low frequencies shows a line nearly parallel to the y-axis, suggesting that in this region the system behaves as a simple double-layer capacitance. The small deviation at low frequency shows

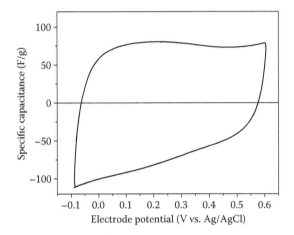

FIGURE 2.15
Cyclic voltammogram of a porous carbon (PC-PDAMAC) in 1 M H_2SO_4. Scan rate = 50 mV/s.

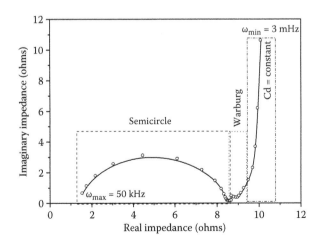

FIGURE 2.16
Impedance dependence with frequency. Nyquist plot of a porous carbon between 3 mHz and 50 kHz at 0.275 V_{SCE} in 1.0 M H_2SO_4.

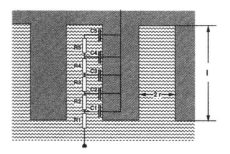

SCHEME 2.7
Transmission line model of the resistance and capacitance distribution in a porous carbon.

the influence of mass transport, related to the microporous structure of the interface between electrode and electrolyte [131].

The impedance data in the intermediate frequency region can be fitted with a straight line at 45°. In that region (Warburg zone), electron and ion transport are controlled by a process that scales with $\sqrt{t}$. Since in porous carbons, there is no diffusion-controlled mass transport of electroactive species, the process involves slow transport of double-layer ions inside the pores or the compliance to a transmission line model (TLM, Scheme 2.7) [132].

When the pore length is significantly larger than the pore diameter, the capacitance of each section of the pore sees a different potential owing to the internal resistance of the pore. Therefore, the electrical response of the pore could be modeled to a sum of RC elements. The mathematical form of the TLM shows a dependence of the signal with $\sqrt{t}$.

2.3.4 Faradaic Reactions

The other possible source of charge is the occurrence of Faradaic reactions:

$$Red = Ox + ne^- \qquad (I)$$

An oxidized species (e.g., Fe^{+3}) is reduced at the electrode to give the reduced species (e.g., Fe^{2+}). A reversible reaction means that the soluble species could be oxidized and then reduced to the original species. In that reaction, the magnitude of the current will depend on how fast the mass transport is to the electrode. Such reactions are of little interest in the detection or elimination of noxious compounds present in water.

Of greater interest are irreversible reactions where the reactant (e.g., a toxin) is irreversibly converted into another species (less toxic substance), such as

$$CN^- + OH^- + H_2O \rightarrow CO_2 + NH_3 + 2\,e^- \qquad (II)$$

Note that such a reaction does not occur in a single electrochemical step but in an initial electrooxidation step, followed by several chemical and electrochemical steps. The current–potential relationship could be very complex. However, for our purposes, the electrochemical current (i) for the oxidation of a species (Red) in the absence of the product can be expressed in a kinetic form (Butler–Volmer equation) [133]:

$$i = FAk^0 C_{Red}(0,t) e^{\frac{-\alpha nF}{RT}(E-E_{eq})} \tag{2.5}$$

where A is the active area of the electrode, k^0 the standard rate constant, F the Faraday constant, C_{Red} the concentration, α the electron transfer coefficient, E the potential, and E_{eq} the equilibrium potential.

Therefore, the current (for a given concentration) is proportional to the active area and the standard rate constant. Porous electrodes have a large surface area, increasing the current. The presence of electrocatalysts (e.g., Pt) makes the rate constant (k^0) larger, therefore increasing the current. In electrochemical sensing of toxic solutes, a large current means better sensitivity. In electrochemical incineration of organics, a larger current means faster elimination of the toxin through larger conversion of molecules per unit time.

2.3.5 Ion Transport in Porous Carbon

When an electrical conductor is placed in an electrolyte solution, an electrical double layer appears on the surface that is composed of a first layer, which comprises ions adsorbed directly onto the object due to chemical interactions, and a second layer composed of ions attracted to the surface charge by the coulomb force, electrically screening the first layer. The second layer is loosely associated with the object because it is made of free ions that move in the fluid under the influence of electric attraction (holding to the surface) and thermal motion (moving out of the surface). When the electrical charge of the conductor is changed, the ions of opposite charge move from the solution toward the electrode surface to compensate for the electronic charge. The potential controlled loading or unloading of two porous electrodes with ions is the basis for the electrochemical deionization (EDI) method of water desalination [134].

Besides the thermodynamic constraints, the electrolyte ions have to reach the electrode surface to be retained inside the porous material. However, the pores in carbon can be long, narrow, and tortuous, making the ion movement slow. Therefore, the ion transport is relevant to the applications. To study ion transport, we used PBD, an optical technique that is able to monitor ion fluxes *in situ* [135]. Figure 2.17 shows two types of behavior. In a monolithic porous carbon (Figure 2.17a), the pores are long (>100 μm) and the ions are loading through the whole measurement time span (>25 min). On the other hand, in an HPC, the macropores cut the solid matrix and makes short (<20 μm) pores. The ion loading occurs in a time span (<20 s) negligible with the measurement.

2.3.6 Swelling Degree of Hydrogels

Dry hydrogels immersed in water increase their volume (up to 250×) by water loading inside the solid material. Such a process is relevant for water remediation since, unlike rigid porous materials, it renders unnecessary the flow of water through the solid because spontaneous swelling loads the solution into the hydrogel. On the other hand, the pure water retained inside the swollen hydrogel is not available for use. A way to overcome that

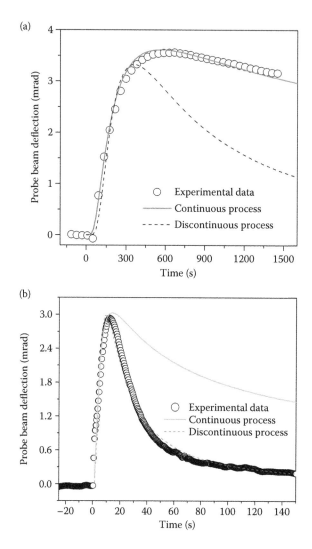

FIGURE 2.17
Probe beam deflection measurement is made after a potential pulse is applied to (a) monolithic porous electrode (NpC-CTAB); (b) hierarchical porous electrode (HPC-f). Full lines represent theoretical calculations for each type of ion transport process.

is simply the use of thermosensitive hydrogels [136], which shrink upon heating, releasing a large part (albeit not all) of the swelling water, making it available to the user.

The swelling can be easily measured by weighing the dry and swollen hydrogel and calculating the amount of water that is loaded upon swelling. Figure 2.18 shows the swelling curve of a typical hydrogel.

As it can be seen, the degree of swelling could reach several thousand percent. This means that a significant volume of external solution is taken spontaneously into the solid gel, allowing water remediation without using an external flux.

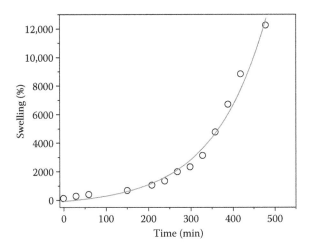

FIGURE 2.18
Swelling dynamics of a PNIPAm-based hydrogel.

2.3.7 Solute Partition in Hydrogels

Nanoporous hydrogels have polymer chains immersed in a water-rich environment. However, while the nanopores are filled with aqueous solution, the polymer regions are more hydrophobic in nature [118]. Therefore, the partition of water-soluble organic compounds between the gel and the external solution will be influenced by internal hydrophobic effects. Therefore, the gel can be considered as an immiscible solvent that could extract organic solutes from aqueous solution [137].

Therefore, the partition coefficient between the gel and solution can be defined as

$$P = \frac{m_h}{m_s} \tag{2.6}$$

where m_h is the molality in the hydrogel (moles of substance in 1 kg of dry hydrogel) and m_s is the molality of the solution (moles of substance in 1 kg of solvent).

The partition coefficient is affected by the chemical composition of the hydrogel and the conducting polymer present in a nanocomposite [68] (Table 2.2). As it can be seen, large values (>1000) of partition coefficients for some molecules (e.g., tryptophan) are

TABLE 2.2

Partition Coefficients (*P*) of Different Compounds between Swollen Hydrogel Matrix and Water at 20°C and pH 7

Hydrogels	Riboflavin	Methyl Orange	Propranolol	Tryptophan
PNIPAM/PANI	32	321	96	756
PNIPAM/PNMANI	29	4	302	1002
PNIPAM-co-2% AMPS/PANI	29	793	252	752
PNIPAM-co-2% AMPS/PNMANI	34	18	438	1211

TABLE 2.3

Diffusion Coefficients ($D \times 10^5$ cm^2/s) Measured during Drug Release from Hydrogels and Nanocomposites at 20°C

Hydrogel	Riboflavin	Methyl Orange	Propranolol	Tryptophan
PNIPAM/PANI	1.12	0.21	0.72	1.12
PNIPAM/PMANI	0.61	0.32	–	9.52
PNIPAM-co-2% AMPS/PANI	–	–	1.24	0.16
PNIPAM-co-2% AMPS/PMANI	0.61	0.21	0.02	1.12

observed while at the same time the partition of other molecules (e.g., methyl orange) is small (<50). Therefore, it is possible to selectively remove some solutes in the presence of a large concentration of others. This is important when a contaminant has to be removed from a nutritive solution or toxic solutes (ammonia, nitrite) have to be removed from aquarium water without removing nutrients. Both processes are relevant in aquaculture. Moreover, these gels can be used to selectively remove toxic solutes to improve the sensitivity of analytical techniques, in a special case of solid-phase preconcentration [138].

2.3.8 Diffusion Coefficient of Solutes in Hydrogels

To load toxic solutes from water, not only the partition coefficient should be high but also the mass transport rate should be fast enough to establish the equilibrium in a reasonable time. In the absence of electric fields, there is no migration. Since in solids convection is impossible, the only driving force is the concentration gradient that gives rise to diffusion. The diffusion coefficients of different molecules in hydrogel-based nanocomposites show values quite close to those in water; however, there is some influence in the interaction between the solute and the matrix (Table 2.3).

Unfortunately, the same kind of interactions that increase the partition coefficient seems to decrease the diffusion coefficient.

2.4 How to Use Nanoporous Materials for Aquananotechnology

Different applications of nanoporous materials and nanocomposites in aquananotechnology will be described.

2.4.1 Removal of Toxic Ions by EDI

Organic dyes are widely used in textile and food industries. About 15% of the total world production of dyes is lost during the dyeing process and is released in the textile effluents [139]. The release of those colored wastewaters in the ecosystem is a dramatic source of unsightly pollution, eutrophication, and perturbations in the aquatic life. Different technological systems for the removal of dyes have been developed, such as adsorption [140], biological [141], chemical [142], or photochemical degradation [143]. An alternative way to remove ions from solution involves EDI [144] (Scheme 2.8).

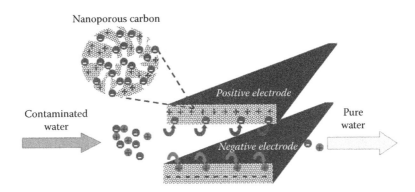

SCHEME 2.8
Description of a process for electrochemical deionization (EDI) using a nanoporous carbon electrode.

The electronic charge related with ion adsorption appears as the so-called double-layer capacitance. The amount of material adsorbed on the surface and capacitance is directly proportional to the active surface area.

$$\Gamma_{sp} = \frac{C_{sp}}{F} \Delta V S_{sp} \tag{2.7}$$

where Γ_{sp} is the specific coverage (mol/g), C_{sp} the specific capacitance (F/g), F the Faraday constant, ΔV the change of potential, and S_{sp} the specific surface (m²/g).

The accepted value C_{sp} for carbon is 20 μF/cm² [145]. For a porous carbon ($S_{sp} = 500$ m²/g) and a voltage change of 1 V, the amount of ions calculated by Equation 2.7 is ~1 mmol/g. Since the concentration of dyes in water is quite low (<0.1 mmol/l) [146], 1 g of carbon could decontaminate 10 liters of water. An advantage of capacitive deionization over simple adsorption is the fact that removal of the electrical charge results in release of the ions. Therefore, the carbon electrode could be used to decontaminate water; the pure water flushed out and replaced by waste solution where the contaminants are released. In that way, the carbon is reused without the need for regeneration.

Figure 2.19 shows the insertion (at 0.0 V vs. SCE) and release (at 0.5 V) of methylene blue dye. The cation (see chemical formula inside the graph) is inserted at 0.0 V and released at 0.5 V. The concentration of the dye was measured *in situ* by spectrophotometry at 430 nm.

2.4.2 Electroanalytical Sensing of Arsenic Ions with Nanocomposites

Redox metal oxides are widely used in electrocatalysis [147,148]. Therefore, they have also been used in electroanalysis of toxic ions [149]. Cobalt oxide nanoparticles could be easily deposited electrochemically onto flat GC electrodes [150]. They have been used for determination of hydrogen peroxide [151] and arsenic ions [152]. An important natural water contaminant present as different ionic species is arsenic [153,154]. HPC (1PSC), modified with *in situ* produced cobalt oxide nanoparticles, was used to detect arsenic ions in neutral solution (Figure 2.20). The cyclic voltammogram (CV) measured in the presence of As ions (full line) shows a clear oxidation peak that is not present in the CV measured in the absence of As ions (dashed line). The peak is likely due to the oxidation of As(III) to As(V) species, catalyzed by cobalt oxide:

$$2\,Co(OH)_2 \rightarrow 2\,CoOOH + 2\,e^- \tag{III}$$

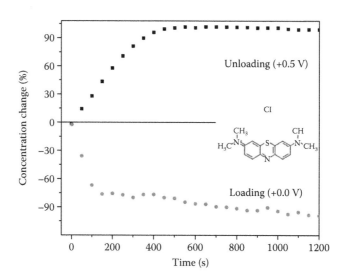

FIGURE 2.19
Time profile of methylene blue concentration (monitored by spectophotometry at 430 nm) during loading (0.0 V vs. SCE) and unloading (0.9 V vs. SCE) of dye inside porous carbon (NpC-CTAB).

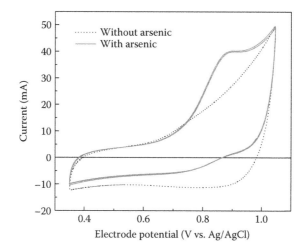

FIGURE 2.20
Cyclic voltammetry of a porous carbon electrode (2PSC) modified with CoOx nanoparticles in the absence (dashed line) and presence (full line) of AsO_3^{-3} (187.5 ppm) ions. Electrolyte = 0.1 M phosphate buffer (pH = 7). Scan rate = 5 mV/s.

$$2\ CoOOH + AsO_3^{-3} + 2\ H_2O \rightarrow 2\ Co(OH)_2 + AsO_4^{-3} + 2\ OH^- \qquad (IV)$$

It is likely that reaction IV occurs with the anion adsorbed in the nanoparticle surface. The small (<20 nm) metal oxide nanoparticles have a large surface area where such adsorption can occur. As it can be seen, low levels (<200 ppm) of arsenic can be easily detected using CV on modified 2PSC. Iron oxides can also be used to detect arsenic [155].

2.4.3 Electrochemical Incineration of Water-Soluble Organic Compounds

Organic compounds can be disposed by oxidation with air at high temperatures (incineration) [156]. However, contaminants present at low concentration in water have to be concentrated or removed from water before incineration. Electrochemical oxidation is able to oxidize organics at ambient temperature in water, providing that some amount of salts is present to give enough electrical conductivity [157]. This is usually the case with wastewater. One of the most effective catalysts for electrooxidation of organics is Pt, which is able to break the C—H bond present in most organic compounds [158]. Pure metal is quite expensive precluding its use in pure form [159]. Additionally, the electrochemical catalysis depends directly on the active surface area of the catalyst. Pt nanoparticles have large surface areas, allowing the use of small amounts of metal. However, use in electrodes requires that they be supported on a conductive material. Nanoporous carbons have both large surface area and good electrical conductivity. Therefore, they could be used to support the nanoparticles and conduct the oxidation current. However, Pt is poisoned by species present in the solution or produced during oxidation such as CO [160]. The addition of other metals (e.g., Ru) allows oxidation of the poisons on the Pt surfaces, maintaining the electrocatalyst activity [161]. The additional metal can be deposited together with Pt if the nanoparticle formation is made *in situ* by chemical reduction.

Previously, our group has investigated the electrooxidation of methanol at mesoporous Pt and Pt–Ru electrodes [162]. We have deposited PtRu nanoparticles inside a nanoporous hierarchical carbon (1PSC) by reduction of metal chlorides ($PtCl_6^{-2}$, $RuCl_3$) with formic acid. Small (2–4 nm) PtRu nanoparticles are evenly distributed inside the carbon (Figure 2.21) [105].

The electrodes are tested for the electrooxidation of methanol:

$$CH_3OH + H_2O = CO_2 + 6\ H^+ + 6\ e^- \qquad (V)$$

While the main purpose is the electrochemical destruction of the organics, it should be borne in mind that the oxidation half reaction has to be coupled with a reduction half reaction, such as

$$1.5\ O_2 + 6\ H^+ + 6\ e^- = 3\ H_2O \qquad (VI)$$

If the oxygen is provided by air, the whole electrochemical device operates as a fuel cell, producing electrical energy instead of consuming it. This is quite interesting in a waste management system. However, it requires an efficient oxygen reduction electrode.

The response of the electrode to a potential step from 0.05 V_{RHE} (a potential where methanol oxidation is negligible) to 0.55 V_{RHE} (a potential where methanol is completely oxidized) can be observed in Figure 2.22. The current density reaches 220 μA/cm^2 after 600 s of polarization. This value is higher than that measured for commercial catalysts consisting of PtRu nanoparticles supported on conventional carbon. Assuming that three adsorption sites are required for each methanol molecule to be oxidized, the measured current value implies that the whole catalyst surface is being renewed every 6 s. As a criterion of metal utilization, the current can also be expressed in terms of mass activity. With this purpose, the obtained current is divided by the total mass of the metal present at the electrode. This value is shown in the right axis of Figure 2.22. The obtained value (black triangle) after 600 s of polarization (120 Ag^{-1} at 0.55 V_{RHE}) reveals that the catalysts consist of well-dispersed, small PtRu nanoparticles and a low degree of agglomeration. The gray circle shows the

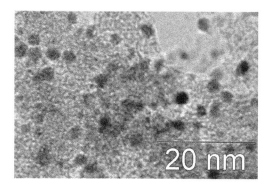

FIGURE 2.21
High-resolution transmission electron microscopy of PtRu nanoparticles deposited inside 2PSC carbon. Black spots are the high-contrast (large atomic number) metal, while the amorphous structure is carbon.

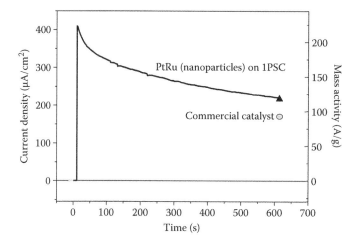

FIGURE 2.22
Current transients for methanol electrooxidation on PtRu/1PSC at 0.55 V_{RHE}, 1 M CH_3OH + 1 M H_2SO_4. $T = 60°C$. Gray circle, mass activity for a commercial catalyst (PtRu on E-TEK®) in the same conditions.

value obtained for a commercial PtRu on C in a similar condition of analysis. As it can be seen, the mass current density is clearly larger. While methanol is toxic, it is less likely to be found in wastewater contaminated with organics. However, PtRu catalysts have shown to be able to oxidize other organic compounds that can be present in wastewater [163].

2.4.4 Absorption of Water Organic Contaminants into Hydrogels and Nanocomposites

It has been shown that the partition coefficient of different solutes depends on the chemical structure of the hydrogel monomer unit, the chemical structure of the solute, and the presence of a loaded conducting polymer. The polymer chains are more hydrophobic than the water filling the pores or the external solution. Therefore, organic contaminants will preferentially partition inside the hydrogels. Such properties can be used with profit to remove toxic organics in the presence of other solutes. In Figure 2.23, it is shown how a dye

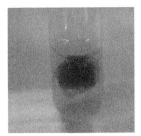

FIGURE 2.23
Photograph of a PNIPAM hydrogel immersed in a dilute solution of a dye (black eriochrome T). Note the higher coloration of the gel compared with the surrounding solution due to the preferential partition of the dye into the gel.

(black eriochrome T) is preferentially absorbed from aqueous solution. The property can be used to remove contaminant dyes from wastewater.

2.4.5 Adsorbing Nanoparticles Fixed Inside Hydrogels

Metal oxides easily adsorb arsenic anions [164]. Accordingly, magnetic nanoparticles have been extensively used to adsorb arsenic from contaminated drinking water [165]. The nanoparticles with arsenic ions adsorbed in the surface can be separated by magnetic means. However, the risk of nanoparticle leakage into the drinking water exists. It is possible to synthesize magnetite (Fe_3O_4) nanoparticles (MagNP) by a well-known chemical method [166]. These nanoparticles could be used to detect and/or remove arsenite ions from water. We have loaded magnetite (Fe_3O_4) nanoparticles into macroporous hydrogels (Figure 2.11). The whole hydrogel can be manipulated by a magnet, and arsenic ions can be adsorbed from the solution without the risk of nanoparticle leakage.

2.4.6 Toxic Metal Ion Absorption into Hydrogels Functionalized with Hydrophobically Retained Chelating Agents

Metal ions can be highly toxic to aquatic life [167]. While surface adsorption on porous solids is a highly effective process for metal ion removal from water, the amount removed per unit weight is relatively low [168]. Liquid/liquid extraction into water-immiscible organic solvents is more effective because of the use of bulk absorption instead of surface adsorption; however, it is prone to contaminate the water with the organic solvent [169]. Hydrophobic organic chelating agents could be used to complex with the hydrophilic ions and promote their transfer to the hydrophobic organic solvent [170]. Fixing chelating groups inside the polymer matrix will produce the same effect. One way to incorporate the chelating groups in polymers is copolymerization [171]; however, it requires complex synthesis of new polymers. On the basis of the hydrophobic properties of polymeric hydrogels, we thought that organic chelating agents should be able to specifically complex metal ions and promote their absorption into solid hydrogels. Figure 2.24 shows PNIPAM hydrogel loaded with phenanthroline (phen) and then immersed in an iron(II)–containing solution. As it can be seen, the gel is colored by the formation of the $Fe(phen)_3^{2+}$ complex inside the gel. The metal ion could be released into a waste solution by protonation of the phenanthroline chelate at a pH well below its pKa (pKa = 4.8 [172]).

However, numerical evaluation of the partition coefficient (Table 2.4) indicates that the hydrogel retains less of the complex than the free chelating agent. It seems that the

FIGURE 2.24
Photograph of a PNIPAM hydrogel loaded with the chelating agent phenanthroline immersed in a dilute (1 mM) solution of Fe^{2+} ions. Note the lower color intensity at the top of the gel due to the slow diffusion of the ion along the gel.

TABLE 2.4

Partition Coefficient of a Chelating Compound and Its Complex with Fe^{2+} Ions

Compound	Partition Coefficient
1,10-Phenanthroline (Phen)	59.3
Fe(Phen)$_3^{2+}$	35.7[a]

[a] Calculated per molecule of phenanthroline.

hydrophilic effect of the Fe^{2+} cation affect the partition coefficient. A possible way to increase the partition coefficient for the complex ion imply the incorporation of negatively charged groups (e.g., $-SO_3^-$) inside the hydrogel, to introduce coulombic interactions, which are absent in the chelating agent. This can be easily achieved by copolymerization of NIPAM with AMPS.

2.4.7 Use of Drug Delivery from Hydrogels in Aquaculture

Modern aquaculture uses different pharmaceuticals, including anesthetic and chemotherapeutic agents [173]. The preferred way to deliver drugs to aquaculture organisms is oral or dermal intake because other ways, such as injection or topical application, are impractical [174]. However, the drug has to survive the enzymes in the gut and/or be delivered through membranes [175]. Moreover, targeted nanodelivery vehicles can avoid leakage of drugs into the environment. Nanoporous materials could be used to deliver drugs in water [170]. A water-soluble drug absorbed inside a polymer hydrogel or adsorbed on the surface of a nanoporous carbon will be released into the solution by diffusion. While concentration gradient controlled release is useful for maintaining a constant level of drug in the organism over time [176,177], there are alternative delivery mechanisms such as temperature- or pH-driven release [178–180]. Figure 2.25 shows the concentration–time profile of a dye $\left(Ru(bpy)_3^{2+} \right)$ released from a PNIPAM hydrogel into the surrounding solution.

While spontaneous release of a drug allows one to maintain a nearly constant level of the substance in water, unlike sudden dissolution of a solid or liquid sample, a more interesting approach involves the control of the drug release using external actions. Figure 2.26 shows the change of concentration of a probe dye in water due to the volume collapse of the

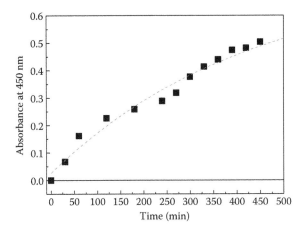

FIGURE 2.25
Release kinetics of a colored inorganic complex $\left(Ru(bpy)_3^{2+}\right)$ from a PNIPAM hydrogel.

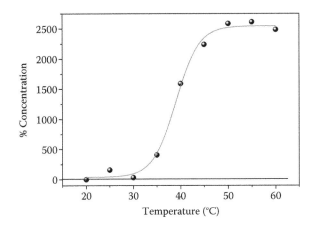

FIGURE 2.26
Change of concentration of a probe dye (Direct Red 75) in the surrounding solution of a thermosensitive PNIPAM hydrogel.

thermosensitive hydrogel driven by a temperature change. The transition temperature (T_{tr}) can be changed by introducing hydrophilic (e.g., $-SO_3^-$, increased T_{tr}) or hydrophobic (e.g., $-CH(CH_3)_2$, decreased T_{tr}) groups in the polymer chains [181]. Therefore, the release trigger temperature can be tailored to some temperature relevant to the aquatic organism (e.g., fish).

Since it was shown that it is possible to fabricate hydrogel microparticles, the delivery material could be mixed with food to be taken by aquaculture organisms.

2.5 What Future Actions Can Be Taken to Develop Aquananotechnology?

On the basis of the porous materials and nanoparticles already developed, it is possible to plan future actions to further develop aquananotechnology.

2.5.1 Forward Osmosis Using Nanocomposites Based on Hydrogels

Forward osmosis is a technique to purify water that uses a semipermeable membrane to effect separation of water from dissolved solutes. The driving force for this separation is an osmotic pressure gradient that induces a net flow of water through a semipermeable membrane into the draw solution, thus effectively separating the water in the feed from its solutes. While a high concentration of a solute will produce such osmotic gradient, the osmotic solute remains in the final solution. An alternative is to use a highly hydrophilic solid (e.g., hydrogel) to create the gradient and retain the pure water. Then, the water can be extracted from the gel using pressure, temperature, or other stimuli [182]. The large swelling capacity of nanocomposites based on polyacrylamide hydrogels and conducting polymers [65] makes them ideal candidates for forward osmosis. Moreover, the water can be released by triggering the gel collapse using microwaves [183]. Mechanically stable hydrogels with large swelling degree (Figure 2.10) will be especially useful for this purpose.

2.5.2 Drug Liberation from Hydrogel Nanoparticles

Like monolithic solids, hydrogel nanoparticles can be used to carry and release drugs for aquaculture organisms. While this approach will ensure the internalization of the pharmaceutical drug into the organisms, the possibility of leakage of nanoparticles loaded with drugs that could harm the environment increases. Therefore, some ways to control the nanoparticle transport in the aquaculture media should be devised. One way involves loading bigger (>100 nm) hydrogel nanoparticles with small (<10 nm) magnetite nanoparticles. In that way, as it is shown in Figure 2.18, the gel can be manipulated with a magnet, avoiding the release of nanoparticles into the environment.

2.5.3 Solid-Phase Extraction for Contaminant Analysis

Contaminants in drinking water are usually present in low concentration [184]. Therefore, to successfully analyze real samples, preconcentration is necessary. Solid-phase extraction is a technique widely used to analyze organics in water, using porous hydrophobic solids [185]. Nanoporous hydrogels can be used to extract less hydrophobic organics from water. Additionally, hydrogels loaded with chelating agents can be used to selectively extract metals from water for further analysis. Nanoporous carbon surfaces can be modified with functional groups that promote the chemisorption of specific analytes.

2.5.4 Detection of Harmful Waterborne Microorganisms through Fast DNA Sensing

Water is a source of harmful pathogen microorganisms affecting humans and aquatic organisms [186]. However, only some strains of bacteria are actually harmful. The classic methods of detection, involving seeding and cultivation in nutritive media, are quite slow. An alternative involves the detection of short DNA fragments only present in dangerous organisms. We have collaborated with the group of M.-C. Pham (ITODYS, U. Paris VII, France) in the development of electrochemical oligonucleotide (ODN) sensors using electroactive polymers functionalized with synthetic ODN [187]. It was also shown that porous electrodes made of multiwall carbon nanotubes glued with Nafion® could improve the sensitivity of the sensors through an increase of the effective surface area [188]. HPC (e.g., HPC-f or 2SPC) electrodes can be used as platforms for DNA sensors, allowing high sensitivity (owing to the large surface area) and easy access of the ODN (owing to the presence of macropores).

2.6 Conclusions

Nanoporous carbon can be synthesized by pyrolysis of nanoporous resin. The resin can be made from RF gels whose natural porosity is maintained during drying by stabilization of the gel nanoparticles using micelles of cationic surfactants (e.g., CTAB) or cationic polyelectrolytes (e.g., PDAMAC). Microscale-sized cellulose fibers seem also to be able to stabilize the nanoparticles, producing a carbon having both macropores (from the space between fibers) and nanopores (from the stabilized nanoparticle structure). Therefore, a hierarchical carbon is obtained.

Using silicon dioxide nanoparticles as hard template for the resin, it is possible to obtain carbon materials having mainly one or two pore sizes. Two pore sizes are produced when the hard template is removed after pyrolysis because the contraction of the solid during carbonization (owing to the loss of H and O atoms from the resin) around the hard nanoparticles creates microfractures in the material. On the other hand, if the silica is removed before pyrolysis, only one main pore size, related with the size of the silica nanoparticles, is obtained.

The carbons can be functionalized by chemical or electrochemical formation of metal (e.g., Pt–Ru) or metal oxide (e.g., CoOx) nanoparticles inside the pores. Nanoporous hydrogels can be obtained by vinylic polymerization of acrylamides in solution using bifunctional acrylamides as cross-linking agents. The hydrophobic/hydrophilic properties of the polymer matrix can be tuned by incorporation of hydrophobic (e.g., isopropyl) or hydrophilic (e.g., sulfonate) groups in the polymer chains by copolymerization. The gels can be produced as bulk monoliths, microparticles (by polymerization inside an inverse water in oil miniemulsion), thin films (by grafting the polymer chains to a surface), or nanoparticles (by controlling the nucleation and growth of the gel aggregates). Additionally, macroporous hydrogels can be produced by templating the polymerization with ice crystals (cryogelation) or porogens (e.g., gas bubbles). Hydrogels with small solid walls (e.g., nanoparticles or macroporous solids) should show fast mass transport because diffusion time depends on the square of the diffusion path length.

The hydrogels can be modified by different methods. The hydrophobic nature of the polymer part of the hydrogels allows one to retain specifically organic chelating agents. Additionally, conducting polymers can be polymerized inside the nanopores of the hydrogels. In that way, the partition properties of the hydrogels are altered because the water pool inside the gel is filled with the conducting polymer, which is relatively hydrophobic.

Moreover, macroporous hydrogels can be modified by adsorption (on the macropores surface) of nanoparticles, made of conducting polymers or metal oxides. In that way, magnetic or conductive gels are produced. Properties relevant to the applications are measured. The porosity (>10 nm) of carbons is directly measured using FIB nanotomography. The thermodynamics and kinetics of double-layer charging is studied using CV and EIS. The ion transport inside the pores is evaluated using PBD techniques. In that way, fast (discontinuous) and slow (continuous) ion transport in carbon electrodes are detected. It is shown that bulk porous carbons show slow ion transport while HPC shows fast transport. Solute partition and diffusion inside hydrogels seems strongly influenced by the interaction between molecules and the polymer matrix.

Different applications of the materials are developed:

- The removal of contaminant ions (e.g., methylene blue dye) is achieved by EDI, applying a potential to porous carbon electrodes inside an aqueous solution.
- Arsenic ions are electroanalytically measured using nanocomposites made of cobalt oxide or magnetite nanoparticles synthesized inside HPC.
- Water-soluble organic compounds are removed by complete electrochemical oxidation on nanocomposite electrodes made of HPC modified with Pt/Ru nanoparticles.
- The removal of organic soluble contaminants is achieved by absorption into hydrogels and hydrogel-based nanocomposites based on the hydrophobic nature of the polymer chains, enhanced by water pool filling with conducting polymers.
- Magnetite (Fe_3O_4) nanoparticles are shown to be absorbed inside macroporous hydrogels, avoiding their leakage to the environment. The nanoparticles can be used to remove arsenic ions from water by specific chemisorption.
- Toxic metal ions can be specifically absorbed into hydrogels that are functionalized with hydrophobically retained chelating agents.
- Model drugs to be delivered to aquaculture organisms are released spontaneously or by thermal activation.

Future uses of the developed materials in aquananotechnology were outlined. Among them are forward osmosis using nanocomposites based on hydrogels, drug liberation from hydrogel nanoparticles, solid-phase extraction for contaminant analysis, and detection of harmful waterborne microorganisms through fast DNA sensing.

A social outreach program called "nanotechnology for everyone," which tries to disseminate the knowledge about nanotechnology among wide audiences in developing countries, is described.

In summary, nanoporous materials and related nanocomposites seem to be useful tools for the development of applications in aquananotechnology.

Glossary

1PSC: hierarchical porous carbon made using a silica nanoparticle template that is removed after resin curing but before pyrolysis

2PSC: hierarchical porous carbon made using a silica nanoparticle template that is removed after pyrolysis

Ag/AgCl: reference electrode ($E_r = 0.195$ V vs. RHE)

AMPS: 2-acrylamidepropansulfonic acid; PAMPS monomer

CTAB: cetyltrimethylammonium chloride

CV: cyclic voltammetry; electrochemical technique where the electrode potential is scanned at a constant rate and the current is measured

DNA: deoxyribonucleic acid

EDI: electrochemical deionization; method to remove unwanted ions from solution by charging the double layer of porous electrodes inside the ion/s solution

EIS: electrochemical impedance spectroscopy; electrochemical technique to measure impedance

FE-SEM: focused ion beam–scanning electron microscopy; a tomographic technique to study nanoporous solids

HPC-f: hierarchical porous carbon produced by absorption of the precursors on a cellulose-based fiber cloth

IUPAC: International Union of Pure and Applied Chemistry

Macropores: according to IUPAC, pores with diameter >50 nm

Mesopores: according to IUPAC, pores with diameter between 2 and 50 nm

Micropores: according to IUPAC, pores with diameter <2 nm

NIPAM: *N*-isopropylacrylamide; PNIPAM monomer

NpC-CTAB: nanoporous carbon produced by carbonization of a resorcinol/formaldehyde resin synthesized in the presence of CTAB

NpC-PDAMAC: nanoporous carbon produced by carbonization of a resorcinol/formaldehyde resin synthesized in the presence of PDAMAC

ODN: oligonucleotide; short DNA fragment carrying genetic fingerprinting pattern

PAA: polyacrylic acid; hydrophilic polymer–bearing –COOH groups

PAAm: polyacrylamide; hydrophilic polymer

PAMPS: poly(2-acrylamidepropansulfonic acid); polyacrylamide bearing sulfonic acid groups

PANI: polyaniline; conducting polymer relatively hydrophilic and pH sensitive

PBD: probe beam deflection; optical technique where a laser beam is shined parallel to the electrode surface and the deviation is measured, which is related to instantaneous concentration gradients

PDAMAC: poly(diallyldimethylammonium chloride)

pKa: equal to $-\log(K_a)$, where K_a is the acid constant that measures the acid strength of a molecule or functional group

PNIPAM: poly(*N*-isopropylacrylamide); thermosensitive polymer

PNMANI: poly(*N*-methylaniline); conducting polymer, relatively hydrophobic

PPy: polypyrrole; hydrophobic conducting polymer, pH insensitive

RF: resorcinol–formaldehyde cross-linked resin

RHE: reversible hydrogen electrode; reference electrode (E_r = 0.0 V in the standard hydrogen scale)

SCE: saturated calomel electrode; reference electrode (E_r = 0.245 V vs. RHE)

References

1. S.K. Brara, M. Verma, R.D. Tyagi, R.Y. Surampalli, *Waste Manage.*, 30 (2010) 504.
2. J. Fabrega, S.N. Luoma, C.R. Tyler, T.S. Galloway, J.R. Lead, *Environ. Int.*, 37 (2011) 517.
3. T.M. Scown, R. van Aerle, C.R. Tyler, *Crit. Rev. Toxicol.*, 40 (2010) 653.
4. M.A. Rather, R. Sharma, M. Aklakur, S. Ahmad, N. Kumar, M. Khan, V.L. Ramya, *Fish. Aquacult. J.*, 2011 (2011) FAJ-16 1.
5. M.S. Hull, A.J. Kennedy, J.A. Steevens, A.J. Bednar, C.A. Weiss, P.J. Vikesland, *Environ. Sci. Technol.*, 43 (2009) 4169.
6. T.M. Benn, P. Westerhoff, *Environ. Sci. Technol.*, 42 (2008) 4133.
7. K. Van Hoecke, J.T.K. Quik, J. Mankiewicz-Boczek, K.A.C. De Schamphelaere, *Environ. Sci. Technol.*, 43 (2009) 4537.

8. M. Jaroniec, M. Kruk, J.P. Olivier, *Langmuir*, 15 (1999) 5410.
9. N. Zabukovec Logar, V. Kaučič, *Acta Chim. Slov.*, 53 (2006) 117.
10. P.M. Forster, A.K. Cheetham, *Top. Catal.*, 24 (2003) 79.
11. Y.-G. Guo, J.-S. Hu, L.-J. Wan, *Adv. Mater.*, 20 (2008) 2878.
12. S. Park, H. Chan Kim, T. Dong Chung, *Analyst*, 137 (2012) 3891.
13. H.J. Oh, J.H. Lee, H.J. Ahn, Y. Jeong, Y.J. Kim, C.S. Chi, *Thin Solid Films*, 515 (2006) 220.
14. S.I. Zones, M.E. Davis, *Curr. Opin. Solid State Mater. Sci.*, 1 (1996) 107.
15. T. Wen, X. Zhang, H.Q. Zhang, J.D. Liu, *Water Sci. Technol.*, 61 (2010) 1941.
16. J. Korkisch, *Handbook of Ion Exchange Resins: Their Application to Inorganic Analytical Chemistry*, Volume 6, CRC Press, Boca Raton, Florida, USA (1989).
17. S.D. Alexandratos, *Ind. Eng. Chem. Res.*, 48 (2009) 388.
18. I.M. Abrams, *React. Funct. Polym.*, 35 (1997) 7.
19. L. Dunn, M. Abouelezz, L. Cummings, M. Navvab, C. Ordunez, C.J. Siebert, K.W. Talmadge, *J. Chromatogr.*, 548 (1991) 165.
20. S. Babel, T.A. Kurniawan, *J. Hazard. Mater.*, 97 (2003) 219.
21. O. Ioannidou, A. Zabaniotou, *Renew. Sustain. Energy Rev.*, 11 (2007) 1966.
22. H. Marsh, F. Rodríguez Reinoso, *Activated Carbon*, Elsevier Science, Amsterdam, Netherlands (2005).
23. C. Pelekani, V.L. Snoeyink, *Water Res.*, 33 (1999) 1209–1219.
24. D. Das, V. Gaur, N. Verma, *Carbon*, 42 (2004) 2949.
25. P. Le Cloirec, C. Faur, *Interface Sci. Technol.*, 7 (2006) 375.
26. F. Rodriguez-Reinoso, M. Molina-Sabio, M.A. Munecas, *J. Phys. Chem.*, 96 (1992) 2707.
27. N.J. Welham, V. Berbenni, P.G. Chapman, *Carbon*, 40 (2002) 2307.
28. W. Yantasee, Y. Lin, G.E. Fryxell, K.L. Alford, B.J. Busche, C.D. Johnson, *Ind. Eng. Chem. Res.*, 43 (2004) 2759.
29. Z. Wang, M.A. Fierke, *Adv. Mater.*, 21 (2009) 265.
30. Y. Xiao, J. Sheny, Z. Xie, B. Zhou, G. Wu, *J. Mater. Sci. Technol.*, 23 (2007) 1.
31. T. Yokoi, H. Yoshitake, T. Tatsumi, *J. Mater. Chem.*, 14 (2004) 951.
32. H.I., J.H. Kim, J.M., S. Kim, J.N. Park, J.S. Hwang, J.W. Yeon, Y.J., *J. Nanosci. Nanotechnol.*, 101 (2010) 217.
33. E.P. Plueddemann, *Silane Coupling Agents*, Springer, Berlin, Germany (1982).
34. W. Yantasee, R., D., Rutledge, W. Chouyyok, V. Sukwarotwat, G., C.L. Warner, M.G. Warner, G.E. Fryxell, R.J. Wiacek, C. Timchalk, R.S., *ACS Appl. Mater. Interfaces*, 2 (2010) 2749.
35. C.T. Kresge, M.E. Leonowicz, W.J. Roth, J.C. Vartuli, J.S. Beck, *Nature*, 359 (1992) 710.
36. K.S. Smirnov, D Bougeard, *Chemical Physics*, 292 (2010) 53.
37. G.J. de A.A. Soler-Illia, C. Sanchez, B. Lebeau, J. Patarin, Chemical strategies to design textured materials: From microporous and mesoporous oxides to nanonetworks and hierarchical structures, *Chem. Rev.*, 102 (2002) 4093–4138.
38. B. Nagappa, G.T. Chandrappa, *Microporous Mesoporous Mater.*, 106 (2007) 212.
39. A.B. Cummins, L.B. Miller, *Ind. Eng. Chem.*, 26 (1934) 688.
40. C. Liang, Z. Li, S. Dai, *Angew. Chem. Int. Ed.*, 47 (2008) 3696.
41. M.-C. Chao, C.-H. Chang, H.-P. Lin, C.-Y. Tang, C.-Y. Lin, *J. Mater. Sci.*, 44 (2009) 6453.
42. H. Li, H. Xi, S. Zhu, Z. Wen, R. Wang, *Microporous Mesoporous Mater.*, 96 (2006) 357.
43. P. Yadav, S. Warule, J. Jog, S. Ogale, *Solid State Commun.*, 152 (2012) 209.
44. A.-H. Lu, F. Schüth, *Adv. Mater.*, 18 (2006) 1793.
45. J. Wu, *Cryst. Res. Technol.*, 44 (2009) 221.
46. J. Ren, J. Ding, K.-Y. Chan, H. Wang, *Chem. Mater.*, 19 (2007) 2786.
47. I. Matos, S. Fernandes, L. Guerreiro, S. Barata, A.M. Ramos, J. Vital, I.M. Fonseca, *Microporous Mesoporous Mater.*, 92 (2006) 38.
48. J. Jang, J. Bae, *Chem. Commun.*, (2005) 1200.
49. J. Balach, M.M. Bruno, N.G. Cotella, D.F. Acevedo, C.A. Barbero, *J. Power Sources*, 199 (2012) 386.
50. G. Wang, L. Zhang, J. Zhang, *Chem. Soc. Rev.*, 41 (2012) 797.
51. D. Yu, Z. Wang, N.S. Ergang, A. Stein, *Stud. Surf. Sci. Catal.*, 165 (2007) 365.

52. B. Xu, S. Hou, G. Cao, F. Wu, Y. Yang, *J. Mater. Chem.*, 22 (2012) 19088.
53. F. Böttger-Hiller, A. Mehner, S. Anders, L. Kroll, G. Cox, F. Simon, S. Spange, *Chem. Commun.*, 48 (2012) 10568.
54. N.D. Lysenko, P.S. Yaremov, M.V. Ovcharova, V.G. Ilyin, *J. Mater. Sci.*, 47 (2012) 3089.
55. E. Frackowiak, F. Beguin, *Carbon*, 39 (2001) 937.
56. B.E. Conway, *Electrochemical Supercapacitors: Scientific Fundamentals and Technological Applications*, Springer, Berlin (1999).
57. G. Lang, C.A. Barbero, *Laser Techniques for the Study of Electrode Processes*, Springer, Berlin (2012).
58. C.A. Barbero, *Phys. Chem. Chem. Phys.*, 7 (2005) 1885.
59. G. García, M.M. Bruno, G.A. Planes, J.L. Rodriguez, C. Barbero, E. Pastor, *Phys. Chem. Chem. Phys.*, 10 (2008) 6677.
60. G.A. Planes, M.C. Miras, C.A. Barbero, *Chem. Commun.*, (2005) 2146.
61. L. Ferreira, M.M. Figueiredo, M.H. Gil, M.A. Ramos, *J. Biomed. Mater. Res. B Appl. Biomater.*, 77 (2006) 55.
62. J.-Y. Sun, X. Zhao, W.R.K. Illeperuma, O. Chaudhuri, K. Hwan Oh, D.J. Mooney, J.J. Vlassak, Z. Suo, *Nature*, 489 (2012) 133.
63. P.J. Flory, J. Rehner, *J. Chem. Phys.*, 11 (1943) 512.
64. N. Sahiner, O. Ozay, N. Aktas, *Water, Air, Soil Pollut.*, 224 (2012) 1393.
65. M.A. Barakat, N. Sahiner, *J. Environ. Manage.*, 88 (2008) 955.
66. I.M.C. Lo, K. Yin, S.C.N. Tang, *J. Environ. Sci.*, 23 (2011) 1004.
67. M.A. Molina, C.R. Rivarola, C.A. Barbero, *Eur. Polym. J.*, 47 (2011) 1977.
68. M.A. Molina, C.R. Rivarola, C.A. Barbero, *Polymer*, 53 (2012) 445.
69. G. Inzelt, *Conducting Polymers: A New Era in Electrochemistry*, Springer, Berlin, Germany (2008).
70. C. Barbero, M.C. Miras, B. Schnyder, O. Hass, R. Kötz, *J. Mater. Chem.*, 4 (1994) 1775.
71. A. Mohammad-Khah, R. Ansari, A.F. Delavar, Z. Mosayebzadeh, *Bull. Korean Chem. Soc.*, 33 (2012) 1247.
72. M.R. Nabid, R. Sedghi, R. Sharifi, H. Abdi Oskooie, M.M. Heravi, *Iran. Polym. J.*, 22 (2013) 85.
73. J.J. Berzelius, *Annal. Phys.*, 98 (1831) 306.
74. D.L. Fedlheim, C.A. Foss, *Metal Nanoparticles: Synthesis, Characterization, and Applications*, CRC Press (2001).
75. W. Yantasee, C.L. Warner, T. Sangvanich, R.S. Addleman, T.G. Carter, R.J. Wiacek, G.E. Fryxell, C. Timchalk, M.G. Warner, *Environ. Sci. Technol.*, 41 (2007) 5114.
76. C.M. Welch, R.G. Compton, *Anal. Bioanal. Chem.*, 384 (2006) 601.
77. X. Zhu, S. Shi, J. Wei, F. Lv, H. Zhao, J. Kong, Q. He, J. Ni, *Environ. Sci. Technol.*, 41 (2007) 6541.
78. A.S. Aricò, S. Srinivasan, V. Antonucci, *Fuel Cells*, 1 (2001) 133.
79. S.R. Chowdhury, E.K. Yanful, *Water Environ. J.*, 25 (2011) 429.
80. J. Stejskal, I. Sapurina, *Pure Appl. Chem.*, 77 (2005) 815.
81. L. Li, Y. Wang, M.E. Vigild, S. Ndoni, *Langmuir*, 26 (2010) 13457.
82. G.U. Sumanasekera, G. Chen, K. Takai, J. Joly, N. Kobayashi, T. Enoki, P.C. Eklund, *J. Phys. Condens. Matter*, 25 (2010) 334.
83. P. Cañizares, J. Lobato, R. Paz, M.A. Rodrigo, C. Sáez, *Water Res.*, 39 (2005) 2687.
84. O. Zerbinati, S. Pittavino, *Environ. Sci. Pollut. Res. Int.*, 10 (2003) 395.
85. S.T. Christoskova, M.M. Stoyanova, *Water Res.*, 36 (2002) 2297.
86. M.A. Anderson, A.L. Cudero, J. Palma, *Electrochim. Acta*, 55 (2010) 3845–3856.
87. O. Ozay, S. Ekici, Y. Baran, N. Aktas, N. Sahiner, *Water Res.*, 43 (2009) 4403.
88. R.W. Pekala, *J. Mater. Sci.*, 24 (1989) 3221.
89. S.T. Mayer, J.L. Kaschmitter, R.W. Pekala, Method of low pressure and/or evaporative drying of aerogel. US Patent 5420168 (1995).
90. T. Yamamoto, T. Sugimoto, T. Suzuki, S.R. Mukai, H. Tamon. *Carbon*, 40 (2002) 1345.
91. L.-Z. Wang, J. Yu, J.-L. Shi, D-S. Yan, *J. Mater. Sci. Letters*, 18 (1999) 1171.
92. J. Patarin, B. Lebeau, R. Zana, *J. Colloid Interface Sci.*, 7 (2002) 107.
93. W. Bell, S. Dietz, Mesoporous carbon and polymers. US Patent 6297293 (2001).
94. K.T. Lee, S.M. Oh, *Chem. Commun.*, (2002) 2722.

95. M.M. Bruno, N.G. Cotella, M.C. Miras, T. Koch, S. Seidler, C. Barbero, *Colloids Surf. A: Phys. Eng.*, 358 (2010) 13.
96. M.M. Bruno, N.G. Cotella, M.C. Miras, C.A. Barbero, *Colloids Surf. A: Phys. Eng.*, 362 (2010) 28.
97. J. Balach, L. Tamborini, K. Sapag, D.F. Acevedo, C.A. Barbero, *Colloids Surf. A: Phys. Eng. Aspects*, 415 (2012) 343.
98. M.M. Bruno, N.G. Cotella, M.C. Miras, C.A. Barbero, *Chem. Commun.*, (2005) 5896.
99. M. Bruno, H.R. Corti, J. Balach, N.G. Cotella, C.A. Barbero, *Funct. Mater. Lett.*, 2 (2009) 135.
100. Y. Deng, C. Liu, T. Yu, F. Liu, F. Zhang, Y. Wan, L. Zhang, C. Wang, B. Tu, P.A. Webley, H. Wang, D. Zhao, *Chem. Mater.*, 19 (2007) 3271.
101. Z. Wang, F. Li, N.S. Ergang, A. Stein, *Chem. Mater.*, 18 (2006) 5543.
102. A.F. Gross, A.P. Nowak, *Langmuir*, 26 (2010) 11378.
103. O.D. Vele, A.M. Lenhoff, *Curr. Opin. Colloid Interface Sci.*, 5 (2000) 56.
104. F. Ruo-Wen, L. Zheng-Hui, L. Ye-Ru, L. Feng, X. Fei, W. Ding-Cai, *New Carbon Mater.*, 26 (2011) 171.
105. A.M. Baena-Moncada, G.A. Planes, M.S. Moreno, C.A. Barbero, *J. Power Sources*, 42 (2013) 42.
106. B. Sakintuna, Y. Yürüm, *Ind. Eng. Chem. Res.*, 44 (2005) 2893.
107. F.J. Maldonado-Hodar, M.A. Ferro-Garcia, J. Rivera-Utrilla, C. Moreno-Castilla, *Carbon*, 37 (1999) 1199.
108. R.W. Pekala, C.T. Alviso, J.D. LeMay, *J. Non-Cryst. Solids*, 125 (1990) 67.
109. J. Balach, Ph.D. Thesis, Universidad Nacional de Rio Cuarto, Argentina (2011).
110. D.L. Elbert, *Acta Biomater.*, 7 (2011) 31.
111. J. Choma, D. Jamioła, K. Augustynek, M. Marszewski, M. Gao, M. Jaroniec, *J. Mater. Chem.*, 22 (2012) 12636.
112. A. Carré, V. Lacarrière, W. Birch, *J. Colloid Interface Sci.* 260 (2003) 49.
113. W. Yuan, Z. Lu, C. Ming Li, *J. Mater. Chem.*, 21 (2011) 5148.
114. G. Decher, *Science*, 277 (1997) 1232.
115. Z. Xie, C. Wang, K.E. deKrafft, W. Lin, *J. Am. Chem. Soc.*, 133 (2011) 2056.
116. S. Saha, P. Sarkar, *J. Hazard. Mater.*, 15 (2012) 228.
117. S.A. Ezzell, C.E. Hoyle, D. Creed, C.L. McCormick, *Macromolecules*, 25 (1992) 1887.
118. M.A. Molina, C.R. Rivarola, C.A. Barbero, *Mol. Cryst. Liq. Cryst.*, 521 (2010) 265.
119. M.A. Molina, C.R. Rivarola, M.F. Broglia, D.F. Acevedo, C.A. Barbero, *Soft Matter*, 8 (2012) 307.
120. S.H. Gehrke, E.L. Cussler, *Chem. Eng. Sci.*, 44 (1989) 559.
121. D.R. Kioussis, P. Kofinas, *Polymer*, 46 (2005) 9342–9347.
122. M.A. Molina, C.R. Rivarola, M.C. Miras, D. Lescano, C.A. Barbero, *Nanotechnology*, 22 (2011) 245504.
123. Y. Kong, J. Wei, Z. Wang, T. Sun, C. Yao, Z. Chen, *J. Appl. Polym. Sci.*, 122 (2011) 2054.
124. G.M. Morales, H.J. Salavagione, D.E. Grumelli, M.C. Miras, C.A. Barbero, *Polymer*, 47 (1999) 25.
125. H.J. Shipley, S. Yean, A.T. Kan, M.B. Tomson, *Environ. Toxicol. Chem.*, 28 (2009) 509.
126. M. Trchová, J. Stejskal, *Synth. Met.*, 160 (2010) 1479.
127. C. Barbero, G.M. Morales, D. Grumelli, G. Planes, H. Salavagione, C.R. Marengo, M.C. Miras, *Synth. Met.*, 101 (1999) 694–695.
128. E.I. Yslas, L.E. Ibarra, D.O. Peralta, C.A. Barbero, V.A. Rivarola, M.L. Bertuzzi, *Chemosphere*, 87 (2012) 1374.
129. T. Suteewong, H. Sai, J. Lee, M. Bradbury, T. Hyeon, S.M. Gruner, U. Wiesner, *J. Mater. Chem.*, 20 (2010) 7807.
130. J.R. Macdonald, *Ann. Biomed. Eng.*, 20 (1992) 289.
131. M.G. Sullivan, B. Schnyder, M. Bärtsch, D. Alliata, C. Barbero, R. Imhof, R. Kötz, *J. Electrochem. Soc.*, 147 (2000) 2636.
132. R. de Levie, *Electrochim. Acta*, 9 (1964) 1231.
133. A.J. Bard, L.R. Faulkner, *Electrochemical Methods: Fundamentals and Applications*, Wiley, New York (2000).
134. A.M. Johnson, J. Newman, *J. Electrochem. Soc.*, 118 (1971) 510.
135. C. Barbero, M. C. Miras, O. Haas, R. Kötz, *J. Electrochem. Soc.*, 138 (1991) 669.
136. H.G. Schild, *Prog. Polym. Sci.*, 17 (1992) 163.

137. A.V. Juarez, L.M. Yudi, C.A. Igarzabal, M.C. Strumia, *Electrochim. Acta*, 55 (2010) 2409.
138. F. Svec, *J. Chromatogr. B*, 841 (2006) 52.
139. H. Zollinger, *Color Chemistry. Synthesis, Properties and Applications of Organic Dyes and Pigments*, 2nd Revised Edition, VCH (1991).
140. P.B. Dejohn, R.A. Hutchins, *Tex. Chem. Color.*, 8 (1976) 69.
141. S.S. Patil, V.M. Shinde, *Environ. Sci. Technol.*, 22 (1988) 1160.
142. Y.M. Slokar, A.M. Le Marechal, *Dyes Pigments*, 37 (1998) 335.
143. F. Han, V.S. Rao, K. Madapusi, S. Dharmarajan, R.R. Naidu, *Appl. Catal. A: General*, 359 (2009) 25.
144. Y. Oren, *Desalination*, 228 (2008) 10.
145. X. Zhao, H. Tian, M. Zhu, K. Tian, J.J. Wang, F. Kang, R.A. Outlaw, *J. Power Sources*, 194 (2009) 1208.
146. E.S. Beach, R.T. Malecky, R.R. Gil, C.P. Horwitz, T.J. Collins, *Catal. Sci. Technol.*, 1 (2011) 437.
147. Z. Chen, D. Higgins, A. Yu, L. Zhang, J. Zhang, *Energy Environ. Sci.*, 4 (2011) 3167.
148. L. Trotochaud, J.K. Ranney, K.N. Williams, S.W. Boettcher, *J. Am. Chem. Soc.*, 134 (2012) 17253.
149. U. Yogeswaran, S.-M. Chen, S.-H. Li, *Electroanalysis*, 20 (2008) 2324.
150. C. Barbero, G.A. Planes, M.C. Miras, *Electrochem. Commun.*, 3 (2001) 113.
151. A. Salimi, R. Hallaj, S. Soltanian, H. Mamkhezri, *Anal. Chim. Acta*, 594 (2007) 24.
152. A. Salimi, H. Mamkhezria, R. Hallaj, S. Soltanian, *Sens. Actuators B*, 129 (2008) 246.
153. K. Mandal, K.T. Suzuki, *Talanta*, 58 (2002) 201.
154. A. Davis, D. Sherwin, R. Ditmars, K.A. Hoenke, *Environ. Sci. Technol.*, 35 (2001) 2401.
155. R. Coneo Rodriguez, A. Baena Moncada, D.F. Acevedo, G.A. Planes, M.C. Miras, C.A. Barbero, *Faraday Discuss.*, (2013) accepted for publication.
156. M.-Y. Wey, H.-C. Huang, C.-L. Yeh, J.-C. Chen, *Combust. Sci. Technol.*, 175 (2003) 1211.
157. J.-F. Zhi, H.-B. Wang, T. Nakashima, T.N. Rao, A. Fujishima, *J. Phys. Chem. B*, 107 (2003) 13389.
158. E. Antolini, *Appl. Catal. B: Environ.*, 74 (2007) 337.
159. C.E. Lee, S.H. Bergens, *J. Phys. Chem. B.*, 102 (1998) 193.
160. H.A. Gasteiger, N. Markovic, P.N. Ross Jr., E. Cairns, *J. Phys. Chem.*, 98 (1994) 617.
161. G. García, J.A. Silva-Chong, O. Guillén-Villafuerte, J.L. Rodríguez, E.R. González, E. Pastor, *Catal. Today*, 116 (2006) 415.
162. G.A. Planes, G. García, E. Pastor, *Electrochem. Commun.*, 9 (2007) 839.
163. Y. Choi, G. Wang, M.H. Nayfeh, S.-T. Yau, *Appl. Phys. Lett.*, 93 (2008) 164103.
164. X. Luo, C. Wang, S. Luo, R. Dong, X. Tu, G. Zeng, *Chem. Eng. J.*, 187 (2012) 45.
165. Y. Wang, G. Morin, G. Ona-Nguema, F. Juillot, G. Calas, G.E. Brown, *Environ. Sci. Technol.*, 45 (2011) 7258.
166. L.H. Reddy, J.L. Arias, J. Nicolas, P. Couvreur, *Chem. Rev.*, 112 (2012) 5818.
167. F. Wu, Y. Mu, H. Chang, X. Zhao, J.P. Giesy, K.B. Wu, *Environ. Sci. Technol.*, 47 (2013) 446.
168. M. Kobya, E. Demirbas, E. Senturk, M. Ince, *Biores. Technol.*, 96 (2005) 1518.
169. J. Rydberg, M. Cox, C. Musikas, G.R. Choppin, *Solvent Extraction Principles and Practice*, 2nd Edition, Marcel Dekker, New York (2004).
170. H. Watarai, H. Freiser, *J. Am. Chem. Soc.*, 105 (1983) 189.
171. Ş. Tokalıoğlu, V. Yılmaz, Ş. Kartal, A. Delibaş, C. Soykan, *J. Hazard. Mater.*, 169 (2009) 593.
172. M.T. Ramírez-Silva, M. Gómez-Hernández, M. de Lourdes Pacheco-Hernández, A. Rojas-Hernández, L. Galicia, *Spectrochim. Acta A*, 60 (2004) 781.
173. Z.J. Shao, *Adv. Drug Deliv. Rev.*, 50 (2001) 229.
174. A.E. Polk, B. Amsden, D.J. Scarratt, A. Gonzal, A.O. Okhamafe, M.F.A. Goosen, *Aquacult. Eng.*, 13 (1994) 311.
175. L.J. Schep, I.G. Tucker, G. Young, R. Ledger, A.G. Butt, *J. Controlled Release*, 59 (1999) 1.
176. S.W. Kim, Y.H. Bae, T. Okano, *Pharm. Res.*, 9 (1992) 283.
177. C.-C. Lin, A.T. Metters, *Adv. Drug Deliv. Rev.*, 58 (2006) 1379.
178. Y. Qiu, K. Park, *Adv. Drug Deliv. Rev.*, 53 (2001) 321.
179. C. Gong, T. Qi, X. Wei, Y. Qu, Q. Wu, F. Luo, Z. Qian, *Curr Med. Chem.*, 20 (2013) 79.
180. Z. Li, J. Guan, *Expert Opin. Drug Deliv.*, 8 (2011) 1007.
181. S. Sershen, J. West, *Adv. Drug Deliv. Rev.*, 54 (2002) 1225.

182. D. Li, X. Zhang, J. Yao, G.P. Simon, H. Wang, *Chem. Commun.*, (2011) 1710.
183. M.A. Molina, Ph.D. Thesis, UNRC, Argentina (2011).
184. M.A. Shannon, P.W. Bohn, M. Elimelech, J.G. Georgiadis, B.J. Marin, A.M. Mayes, *Nature*, 452 (2008) 351.
185. C.F. Poole, S.K. Poole, Principles and practice of solid-phase extraction, in *Comprehensive Sampling and Sample Preparation*, J. Pawliszyn, Ed. Academic Press, Oxford (2012) 273.
186. C.D. Vecitis, M.H. Schnoor, M.S. Rahaman, J.D. Schiffman, M. Elimelech, *Environ. Sci. Technol.*, 45 (2011) 3672.
187. S. Reisberg, D.F. Acevedo, A. Korovitch, B. Piro, V. Noel, I. Buchet, L.D. Tran, C.A. Barbero, M.C. Pham, *Talanta*, 80 (2010) 1318.
188. D.F. Acevedo, S. Reisberg, B. Piro, D.O. Peralta, M.C. Miras, M.C. Pham, C.A. Barbero, *Electrochim. Acta*, 53 (2008) 4001.

3

Use of Nanomaterials in Water Remediation by a Subcritical Water Process

Guy Baret

42TEK S.L., Voiron, France

CONTENTS

3.1 Introduction

In the field of water remediation, most of the processes focus on the wastewater from houses and industry. These two origins are the most important sources of pollution of water owing to the high quantity of water used in private houses and in the industry. Specific processes have been developed to treat thousands of cubic meters of water per day at a very low cost by using filtration, settling, and bacterial treatment. Although there are exceptions, most of the polar solvents, apolar organic solvents (light solvents or heavy oils), and organic compounds are rather efficiently removed by these processes.

However, some very stable molecules cannot be degraded by these particular treatments, even the bacterial treatment. This is mainly the case for complex organic molecules. This is, for instance, the case of molecules used in medical treatments, such as animal or human proteins, hormones, blood residues, and some drugs. Among these products, hormones and anticancer drugs have been increasing very quickly for 10 or 20 years without particular attention to the effect on the environment. These classes of molecules go through the water remediation plants and end up in the rivers and, finally, the ocean.

In the rivers, downstream to the wastewater plants where the concentration of such molecules can be high, these molecules are highly suspected to interact with the living bodies, modifying natural equilibria such as the sex distribution of aquatic species, particularly fish. The danger comes from the activity of such molecules on the operation of living bodies at very low concentration, under 1 ppb. Some of these molecules can act as endocrine disruptors and then have a very strong effect at trace amounts on the hormonal equilibrium of the living bodies. The sources of these organic molecules are numerous and probably scattered from very different applications. However, the greater part comes out of the medical centers, hospitals, and medical analysis laboratories, which have a daily role in more and more clinical treatments. As the number of these important sources is very limited, it is recommended to remove these molecules before they flow into the wastewater collection system.

Owing to the high chemical stability of some of these compounds, and the fact that they are very frequently mixed with solvents, specific degradation processes should be used. The supercritical water process allows a full degradation of organic molecules. The supercritical state is a high-pressure/high-temperature state of a liquid or a gas. It is already used, and for a long time, in the food industry with supercritical carbon dioxide (SC-CO_2) for decaffeinated coffee, for instance. CO_2 has a critical point at 31.3°C and 72.9 atm. Water has a critical point at a much higher pressure and temperature, and until now, the applications of supercritical water have been limited because of these operating conditions. However, the use of nanomaterials allows one to conduct remediation processes in supercritical or subcritical water conditions.

3.2 Water Remediation in Supercritical Conditions

3.2.1 Supercritical State

Water is commonly found under three states: solid (ice), liquid (water), and gas (water vapor). The phase diagram of water shows these different domains (Figure 3.1). There are equilibria between every couple of states. When the water vapor is compressed to high pressure (A) at constant temperature, its density increases and the gas becomes more and more dense, and liquefies above a pressure that depends on the temperature. To the

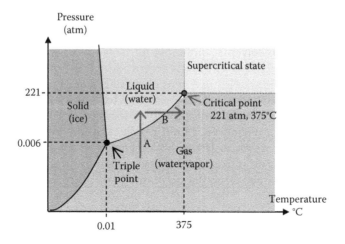

FIGURE 3.1
Phase diagram of water.

contrary, when the liquid water is heated to high temperature (B) under constant pressure, its density decreases and the liquid becomes less and less dense. Above a given temperature that depends on the pressure, the gas is the stable phase. Above a point located at 221 atm and 375°C, the liquid and the water vapor are the same phase. There is no more difference between the gas and the liquid. This point is called the critical point. The density of the fluid at the critical point is around 0.32 g/cm³.

3.2.2 Supercritical Water and Organic Compounds

Close to the critical point and above, radicals are formed by the dissociation of the H_2O molecule. The dissociation reactions depend a lot on the species present in the water. The addition of even a very low amount of an oxidizing or reducing product changes the reactions and the radicals formed. Highly oxidizing radicals are produced and then react with organic species. This is the principle of water remediation in critical water. We will see later than the use of nanocatalysts can contribute to the oxidation of organic compounds in subcritical water and that a remediation treatment can be performed in these "softer" conditions.

In these conditions, even the more stable organic molecules are oxidized to produce simple molecules such as H_2O, CO_2, N_2, and heat. In particular, all types of hormones and drugs are degraded to simple molecules in supercritical water. When the pollutant is an organic–inorganic material, the organic part is broken down to H_2O, CO_2, and N_2, and the inorganic part generally precipitate as an oxide. The presence of sulfur or phosphorous leads to the production of SO_4^{2-} and PO_4^{3-} ions.

3.2.3 Sequestration of Toxic Elements in Nanoparticles of Complex Oxides

An important parameter is drastically changing with the temperature of water: the dielectric constant. The relative dielectric constant ε_r of water decreases from room temperature to the critical point. Table 3.1 gives ε_r values for water.

As ε_r decreases, the solubility of nonpolar or slightly polar molecules increases. At the critical point, water is no longer a polar solvent and all the ionic species dissolved in water will precipitate. As the medium is highly oxidizing, oxides will be formed and complex oxides can be formed depending on the ions present in the water.

A specific interest is in the sequestration of toxic elements such as lead, cadmium, mercury, or arsenic. These elements easily form complex oxides with iron, manganese, titanium, aluminum, etc. The process to remove these elements from water is to precipitate a solution containing the toxic element and an element making the structure of the oxide. Complex oxide nanoparticles (or a solid solution of the toxic element in the oxide structure) are formed in the critical water conditions and can be removed by filtration.

TABLE 3.1

Values of Relative Dielectric Constant ε_r of Water

Pressure	Temperature (°C)			
	25°C, ε_r	300°C, ε_r	375°C, ε_r	400°C, ε_r
1 atm	79			
300 atm	80	22	10	6

An example is arsenic removal from water. The maximum content of arsenic in drinking water is set to 10 to 50 μg of arsenic per liter, depending on the country. Arsenic may be present under two oxidation states: As(III) and As(V), this last one having higher affinity with iron oxide or manganese oxide than As(III). In critical water, As(III) will be first oxidized to As(V). Then, mixing of the water flow containing As(V) with a supercritical water solution containing iron or manganese ions will lead to Fe–As or Mn–As nanooxides. For instance, $Mn_3(AsO_4)_2$ oxide is formed with manganese.

3.2.4 Process

The process for water remediation in the supercritical state is based on the heating of water containing organic pollutants near the critical point. The process steps are indicated in Figure 3.2.

The process is generally driven in a continuous mode, and all the steps are conducted in a tubular reactor. This allows a very small volume of hot pressurized water and reduces the risks linked to the handling of a large volume of supercritical water. The reactor can be very compact, and efficient heat exchange can be performed to decrease the energy intake.

Technically, the process can be achieved in two ways:

- First, by heating the flow of polluted water up to the critical point of water (Figure 3.3a). The process steps are as follows:

  1. Compression from the atmosphere up to the critical pressure of 221 atm

  2. Heating from room temperature to the critical temperature of 375°C

  3. Cooling down to room temperature

  4. Pressure release

  Oxidizing radicals are produced in the hotter part of the heating device. With the increase of temperature along the heating device, the concentration of radicals increase and more oxidizing radicals are formed. Then, the oxidation of the organics starts with the less stable organic molecules and goes on with the more stable

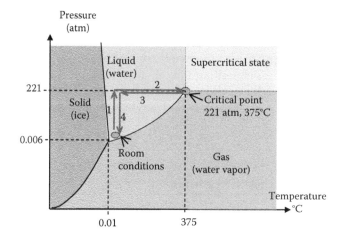

FIGURE 3.2
Process steps in the water phase diagram. 1: compression; 2: heating; 3: cooling; 4: pressure release.

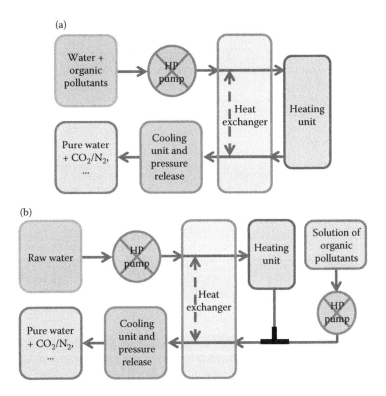

FIGURE 3.3
(a) Process steps including compression–heat exchange–heating–cooling and pressure release. (b) Process steps including compression–heat exchange–heating–mixing–cooling and pressure release.

ones. At the end of the heating step, at the critical temperature, the oxidation is generally finished.

- Second, heating a main flow of raw water to the critical point and then mixing it with the secondary flow of polluted water (Figure 3.3b). The steps are the same as above, except that cold polluted water is mixed with the supercritical water between steps 2 and 3. The reaction is very fast in the mixing device and no soak time at high temperature is required.

The choice between these two processes depends on the nature of the chemical species mixed or dissolved in the water and the concentration of the pollutants. The first process is easier to handle at large scale because there is no mixing at high temperature, and is preferred when all the pollutants are organic molecules. The decomposition of the organics leads to simple molecules such as H_2O, CO_2, N_2, and SO_4^{2-} or PO_4^{3-} ions in the presence of sulfur or phosphorous. No inorganic compound is produced, and no material deposits onto the walls of the reactor.

When the used water contains organics as well as precursors of inorganic products, there is an issue with the first process. Indeed, the inorganic products will deposit on the tube wall in the hot part of the heating unit, typically on the walls above 250°C. This is partially due to the temperature gradient in the solution from the wall to the center of the tube. Reaction happens on the hot surface, a deposit grows, and finally closes the tube. This issue can be experienced with water containing 2+ ions such as calcium or magnesium ions. It can also happen with iron-containing molecules, such as blood products. Iron oxide, Fe_2O_3, is produced, and can make a deposit on the tubing wall.

The second process, which includes a mixing at high temperature between the raw water and the flow of water containing the pollutants, avoids the formation of such a deposit on the tube wall of the heating unit. However, the problem is transferred to the mixing device, which should be specifically designed to avoid this issue.

3.2.5 Energetic Aspects

The heat quantity to bring the water up to the critical point is given by the enthalpic diagram of water (Figure 3.4).

The enthalpy at the critical point is 2095 kJ/kg water. From water at room temperature having an enthalpy of about 200 kJ/kg, the energy intake to bring 1 kg of water to the critical point is around 1900 kJ/kg, which is 0.53 kWh/kg. For equipment designed for 100 kg/h (2.4 m³/day), the energy intake will be 53 kWh/h. Hopefully, a large part of this energy can be recycled through heat exchangers between the hot water and the cold inlet. Typically, 85% of the energy can be recovered using heat exchange. Then, the energy intake is reduced to 0.08 kWh/kg, which requires a power of 8 kW for a 100 kg/h reactor. The heat exchangers have to be especially designed in order to work at high pressure and allow an efficient heat transfer.

Another way to reduce the energy intake is the addition of an organic material in water. For instance, ethanol has a combustion enthalpy of 1370 kJ/mol, which is 22,800 kJ/kg. The addition of a small quantity of ethanol to water brings a large percentage of the necessary energy. The selection of the additional organic compound (alcohol, acetone, etc.) is important and is related to the type of pollutant to degrade. Of course, the organic pollutants contained in the water to treat are to be taken into account in the energy balance. With water containing <10 g per liter of organic molecules, the energy inlet can be reduced to nearly zero.

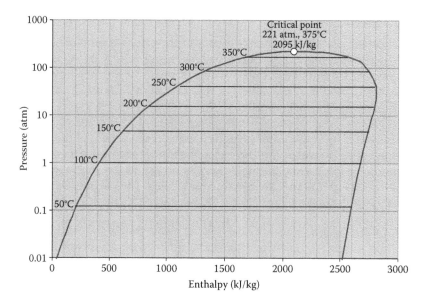

FIGURE 3.4
Enthalpic diagram of water.

3.3 Water Remediation in Subcritical Conditions

In the above description, water is heated up to the supercritical domain, above the critical point. However, depending on the stability and the content of organic pollutants, the process can be operated in the subcritical domain. There are a few advantages to work under the critical point, at lower pressure and temperature, in the liquid phase. First, the cost of equipment is reduced because all the tubing and fittings can be of lower grade and the pumping unit has a reduced output pressure. Second, the energy intake can be decreased by 20%–40%. Third, the operation of the reactor is safer at a lower pressure.

For example, a working point at 100 atm and 300°C will degrade most of the organic molecules. The enthalpy at the point 100 atm, 300°C is 1350 kJ/kg of water. From water at room temperature having an enthalpy of about 200 kJ/kg, the energy intake to bring 1 kg of water to these working conditions will be 1150 kJ/kg, which is 0.32 kWh/kg. For an equipment designed for 100 kg/h (2.4 m³/day), a feeding power of 32 kWh/h is necessary. With an energy recovery from heat exchanger of 85%, the feeding power can be as low as 4.8 kWh/h. Depending on the organic content of water, negligible energy is necessary to operate this process in a continuous mode.

The oxidizing efficiency of the subcritical process can be improved by different solutions. First, the addition of a soak time of a few seconds at the maximum temperature enhances the degradation of the most stable molecules. Such a soak time can easily be obtained by adding some tubing at the highest temperature, after the heating unit.

A second solution is in the addition of an oxidizing component in the water. For instance, addition of hydrogen peroxide (H_2O_2) or manganese permanganate ($KMnO_4$) is very efficient to attenuate the working conditions of the process. This is also an option to add compressed air in the liquid after the compression step. Compressed air can be added either before the heating, or at the maximum temperature point. This last solution avoids issues with a two-phase medium in the heat exchanger and the heating unit, which would reduce the heat exchange coefficient at the tubing internal wall.

3.3.1 Use of a Nanocatalyst

A last way consists in the action of a nanocatalyst in the subcritical water. One could disperse a nanocatalyst in the feeding water to improve the formation of oxidizing species under the critical point of water. However, this solution would use a large mass of nanocatalyst that would require a specific and expensive operation to be recovered. A better and cheaper way is to produce the nanocatalyst on the wall of the tube reactor in a first processing step, and then to process the polluted water through this prepared tubing. The nanocatalyst can be made of TiO_2 or CeO_2 pure or doped nanoparticles, which are produced as a coating onto the tube's internal surface. The catalyst activity can be increased with larger wall area, which can be done by inserting a highly porous media in the tubing at the highest temperature point. This porous media will receive the nanocatalyst coating.

3.4 Conclusion

Nanoparticles can be used in water remediation processes either to remove some specific metals from water in supercritical or subcritical water conditions, or to oxidize organic compounds in subcritical water conditions. Toxic elements can be trapped in complex oxides formed in the critical water process and recovered in a filtration step. Organic products, including very stable organic molecules such as animal or human proteins, hormones, and some drugs used in medical treatments, can be degraded in a supercritical water process to produce simple molecules such as H_2O, CO_2, and N_2. However, the process can be undertaken in subcritical water conditions, with lower energy intake, if the formation of highly oxidizing radicals and the degradation reactions are facilitated by the presence of a nanocatalyst formed on the reactor walls.

4

Reduction of Priority Pollutants by Nanoscale Zerovalent Iron in Subsurface Environments

Chun-Chi Lee,[1] **Hsing-Lung Lien,**[2] **Shan-Chee Wu,**[3]
Ruey-An Doong,[1] **and Chih-Cheng Chao**[4]

[1]*Department of Biomedical Engineering and Environmental Sciences,
National Tsing Hua University, Hsinchu, Taiwan*

[2]*Department of Civil and Environmental Engineering,
National University of Kaohsiung, Kaohsiung, Taiwan*

[3]*Graduate Institute of Environmental Engineering, National Taiwan University, Taipei, Taiwan*

[4]*Department of Chemical and Materials Engineering, Tunghai University, Taichung, Taiwan*

CONTENTS

4.1 Introduction

The development of remediation technology of soil and groundwater contaminated with priority pollutants, including chlorinated compounds, inorganic ions, and heavy metals, has recently received much attention [1–6]. However, the contamination of soil and groundwater is enormous and difficult to clean up because of the inherent complexity of the soil and groundwater properties [7]. Several technologies, such as permeable reactive barriers (PRBs), *in situ* bioremediation, and air sparging have been applied for the remediation of contaminated soils and groundwater. Laboratory-scale and field studies have demonstrated that PRBs packed with zerovalent metals (ZVMs) such as Fe, Cu, Al, Si, Zn, and Mg have been shown to be a promising strategy for the removal of organic and inorganic pollutants in contaminated subsurface environments [8–10]. Table 4.1 shows the reduction of priority pollutants by ZVMs under anaerobic conditions. Many halogenated hydrocarbons, including carbon tetrachloride (CT), tetrachloroethylene (PCE), and trichloroethylene (TCE), can be reductively degraded by ZVMs [9,11–13]. Of various materials used, zerovalent iron (ZVI) and bimetallic iron systems are the most often used materials because of their suitable redox potential for the reduction of halogenated compounds, no obvious toxic effect, and abundance in the earth's crust.

ZVI is a moderate reducing agent with relatively low redox potential ($V = -0.44$ V vs. SHE at 25°C). ZVI can react with dissolved oxygen (DO) and/or water to form ferric oxides. In the early 1990s, Gillham and coworkers found that ZVI has an excellent ability in dechlorination of chlorinated hydrocarbons under anaerobic conditions [14,15]. It is quickly recognized that the reduction of organic compounds by metal iron is a well-known classic electrochemical/corrosion process to both organic chemists and corrosion scientists. Much of the emphasis on the use of ZVI for the decomposition of priority pollutants came from the work of Matheson and Tratynek [16], who proposed the major pathways for reductive dechlorination in an anoxic Fe^0–H_2O system. On the basis of the relation of redox potential

TABLE 4.1

Reduction of Priority Pollutants by ZVM under Anaerobic Conditions

Metal	pH	Target Compound	k_{obs} (h⁻¹)	Products	References
Fe	5–8	TCE	0.165	*cis*-DCE, 1,1-DCE, VC	[53–56]
	Neutral pH	PCE	0.025	TCE, *cis*-DCE, 1,1-DCE	[57–59]
		CT	0.19–0.41	CF	[60]
	5.5	Nitrate	0.059–0.34	Ammonia	[61–63]
	7	Pentachlorophenol	0.0039	Tetrachlorophenol	[64]
Si	8.3	PCE	0.0034	TCE	[44]
	8.3	CT	0.34	CF	[44]
Mg		4-Chlorophenol	0.23	Chlorophenol	[50]
		Endosulfan	10.5	Bicyclo(2,2,1)hepta(2,5)diene	
Zn	7.0	CT	2.29	CF	[65]
	7.0	CF	0.398	DCM	[65]
Al	4<, 11>	4-Chlorophenol	0.46	1,4-Biphenol	[47,48]

between iron and target pollutant, chlorinated hydrocarbons can be reduced by accepting electrons produced from the oxidation of ZVI:

$$Fe^0 \rightarrow Fe^{2+} + 2e^- \tag{4.1}$$

$$RX + H^+ + 2e^- \rightarrow RH + X^- \tag{4.2}$$

$$Fe^0 + RX + H^+ \rightarrow Fe^{2+} + RH + X^- \tag{4.3}$$

Recent studies on the metal iron chemistry, however, showed that ZVI has several drawbacks in application [17]. These include the formation of iron oxides on the surface, the increase in solution pH, and decline in reactivity with time [18,19]. Several attempts, including the use of hydrogen gas, noble catalysts, and nanoparticles, have been made to enhance the reduction rates of priority pollutants [20–22]. The enhancement of the reactivity and change in end-product distribution by nanoscale ZVI (nZVI) has recently attracted much attention because of their small particle size (1–100 nm), large specific surface area, and unique surface morphology [7].

The robust development of nanotechnology has triggered the use of nanomaterials to environmental applications. Most environmental applications of nanotechnology fall into the following categories: treatment/remediation, nanocatalysts, nanosensing/sensor system, nano-enable energy, and green technology/pollution prevention [7]. Although several nanosized ZVMs and nanostructured materials such as nZVI, carbon nanotube, and hierarchical titanium dioxide have been synthesized and applied to the removal of many priority pollutants such as chlorinated hydrocarbons, nitroaromatics, nitrate, and heavy metal, nZVI is becoming an increasingly popular method for the treatment of hazardous and toxic chemicals in the contaminated soil and groundwater. In this chapter, an overview of using the microscale and nanoscale ZVI system for the reduction of priority pollutants under anaerobic conditions is provided. The preparation and characterization of different kinds of nZVI are first introduced and compared. The reactivity of bare ZVI and bimetallic iron systems toward reduction of priority pollutants under various environmental conditions are summarized. In addition, the stability and mobility of nZVI in the environments is discussed.

4.2 Preparation and Characterization of nZVI

Several techniques can be employed for the preparation of nZVI. The use of physical synthetic methods, such as inert gas condensation, severe plastic deformation, high-energy ball milling, and ultrasonic shot peening, can synthesize ZVI nanoparticles with diameters of 10–30 nm [23]. However, the surface reactivity of these physically synthesized iron nanoparticles is strong and is difficult for storage and application. In addition, the surface reactivity of nZVI will gradually decrease owing to the agglomeration of nanoparticles. The chemical methods include microemulsion, chemical coprecipitation, chemical vapor condensation, pulse electrodeposition, and chemical wet reduction [23]. Chemical methods are often used for the synthesis of small quantities of nZVI. In addition, the diameter of synthesized nZVI can be adjusted to the desired particle sizes, depending on the methods and starting materials used for synthesis. Choi et al. [24] used a chemical vapor condensation method to synthesize iron nanoparticles using iron pentacarbonyl ($Fe(CO)_5$) as the precursor under a helium atmosphere. Iron pentacarbonyl would decompose to ZVI and carbon monoxide (CO) at a temperature >300°C. The typical particle sizes were found on the order of 5–13 nm. Makela et al. [25] used a liquid flame spray process for the generation of single-component Ag, Pd, and Fe nanoparticles. The average diameters of synthesized nanoparticles were in the range 10–60 nm.

Although several methods are available to chemically synthesize nZVI, particles synthesized by gas-phase reduction and aqueous-phase reduction of ferrous or ferric iron by sodium borohydride ($NaBH_4$) have been the most thoroughly examined technology [26–29]. The most often studied nanomaterial synthesized by the gas-phase reduction method is reactive nanoscale iron particles (RNIP). RNIP is a commercial product from Toda Kogyo Corp. (Onoda, Japan), which is produced by the reduction of goethite (α-FeOOH) or hematite (Fe_2O_3) particles with H_2 at high temperatures of 350°C–650°C. The particle sizes and specific surface areas are in the range 50–300 nm and 7–55 $m^2 g^{-1}$, respectively. The liquid-phase reduction of metal cations by a strong reducing agent such as $NaBH_4$ and hydrazine dihydrochloride (N_2H_4·2HCl) in aqueous or nonaqueous solution is the most widely used method in the laboratory for preparation of ZVMs [30]. For nZVI, nanoparticles can be prepared by using $NaBH_4$ as the key reductant to reduce ferric chloride ($FeCl_3$·$6H_2O$) or ferrous sulfate ($FeSO_4$·$7H_2O$) to form iron metal (Fe^{BH}) [26,31]. Zhang [31] reported that excessive borohydride is typically needed to accelerate the synthesis reaction and ensure the uniform growth of iron nanoparticles. Usually the particle sizes of the freshly prepared nZVI range between 8 and 15 nm and then aggregate to around 20–100 nm.

Several studies have compared the characteristics of the nZVI prepared using different chemical methods, especially the gas- and liquid-phase reduction methods. Nurmi et al. [32] conducted a systematic investigation for comparing the physicochemical properties and reactivity of two nZVI, RNIP, and Fe^{BH}, prepared under different conditions. Table 4.2 compares the characteristics and reactivity of iron nanoparticles prepared by hydrogen reduction in gas phase and $NaBH_4$ in aqueous solution. RNIP is a two-phase material consisting of 40–70 nm α-Fe^0 and magnetite (Fe_3O_4) as the core and shell coating, respectively,

TABLE 4.2

Comparison of Physical and Chemical Properties of Fe^{BH} and RNIP

Parameters	Fe^{BH}	RNIP
Primary particle size (nm)	20–80	40–70
BET surface area ($m^2 g^{-1}$)	33.5–36.5	23–39
Major phase (core layer)	Fe^0	α-Fe^0
Shell coating	Magnetite (Fe_3O_4)	Goethite (α-FeOOH), wustite (FeO)
Surface composition[a]	Fe: 50.9%, O: 44.2%, B: 0.0%, Na: 3.0%, S: 1.9%	Fe: 20.0%, O: 49.1%, B: 16.0%, Na: 14.5%, S: 10.5%
Initial Fe^0 content (wt%)	97 ± 8%	26.9 ± 0.3
Crystallinity	Highly disordered	Crystalline
Use of hydrogen	Yes	No
End product	Mostly saturated	Unsaturated

Source: Nurmi JT et al., *Environ Sci Technol*, 39, 1221, 2005; Liu YQ et al., *Environ Sci Technol*, 39, 1338, 2005; Liu YQ, Lowry GV, *Environ Sci Technol*, 40, 6085, 2006.

[a] Determined by XPS.

while Fe[BH] material is composed of <1.5 nm crystals that are aggregated into approximately spherical 20–80 nm nanoparticles [32,33]. Although RNIP and Fe[BH] have similar primary particle sizes and specific surface areas, the elemental compositions of nanoparticle surfaces are different. X-ray photoelectron spectroscopy (XPS) showed that RNIP surfaces are mainly made up of Fe and O with small amounts of S, Na, and Ca. On the contrary, the surface of Fe[BH] contains significantly less Fe and S and more B, presumably owing to the use of $NaBH_4$ as the reductant [32]. In addition, the Fe[BH] has a higher initial Fe^0 content (97 ± 8%) than RNIP (26.9 ± 0.3%) [33]. Another important difference between Fe[BH] and RNIP is that Fe[BH] can activate and use hydrogen gas, generated from the reduction of water, for hydrodechlorination of chlorinated hydrocarbons with the production of ethane as the major end product, while RNIP cannot use hydrogen gas for dechlorination and yields primarily unsaturated products [33,34].

4.3 Core–Shell Structure of Iron Nanoparticles

Usually the freshly synthesized iron nanoparticles ignite spontaneously upon exposure to air when the particle size is too small or undergo a rapid exothermic reaction with oxygen under ambient condition [35]. This oxidation process would generate a core–shell structure with iron oxide surrounding the iron nanoparticles [33]. In aqueous solution, nZVI would react with water and trace amounts of oxygen in solution to produce a thin layer of ferrous hydroxide, and subsequently transforms to ferric oxides, resulting in the formation of core–shell iron nanoparticles. The diameters of the freshly prepared nZVI particles are, in general, >8 nm and the shell thicknesses of the freshly prepared nZVI are in the range 2–5 nm [36]. The formation of a shell layer is important for iron nanoparticles to avoid further oxidation and maintain the reactivity. Muftikian et al. [37] used XPS to identify the composition of the oxide layer of palladized ZVI and found that the reactive bimetallic Pd/Fe surfaces were formed by the stepwise reduction of Pd(IV) in solution to Pd(II), which replaced protons on the hydroxylated iron oxide surface and formed Pd(II)—O—Fe bonds. In addition, the oxide shell is an important composition of stability to protect the reactivity of the inner ZVI and the ability to transport the electron and mass across the shell layer.

The composition of ZVI influences the reactivity of nZVI toward reduction of priority pollutants. Liu et al. [33] have systematically investigated the reaction rate, pathways, and efficiency of TCE by Fe[BH] and RNIP under anaerobic conditions. The measured surface area normalized pseudo-first-order rate constant for Fe[BH] is 1.4×10^{-2} L h^{-1} m^{-2}, which is 4× higher than that for RNIP (3.1×10^{-3} L h^{-1} m^{-2}). The high reaction rate for Fe[BH] may be attributed to the oxide shell composition and the boron contents for the differences between the particle types [33]. Although the thicker and more crystalline magnetite (Fe_3O_4) shell of RNIP slows the TCE dechlorination rate and makes some nZVI inaccessible, the observed TCE dechlorination rates afforded by RNIP are likely to be rapid enough for *in situ* application. Moreover, Sohn and Fruehan [35] have demonstrated the stability and reactivity of iron nanoparticle toward nitrate in air. The freshly synthesized iron nanoparticles ignited spontaneously upon expose to air. However, an ~5 nm coating of iron oxide was formed when exposed to air slowly. The thickness of oxide shell did not increase for at least 2 months. In addition, the rate constant for nitrate reduction by nZVI decreased by a factor of ~2 after the formation of the oxide shell.

The shell layer also plays an important role in removing heavy metals. Recent studies have demonstrated that iron nanoparticles have a superior metal removal capacity than that of conventional adsorbents such as zeolites and ion exchange [23,38–40]. Shokes and Moller [41] used ZVI to remove heavy metals from acid rock drainage and found that copper (Cu) and cadmium (Cd) cemented onto the surface of the iron as ZVMs, presumably due to the positive standard potential than that of ZVI. Li and Zhang [39] used high-resolution XPS to investigate the removal mechanism for metal sequestration by nZVI. They found that dissolved metals can be removed by several mechanisms. For metal ions with standard potential very close to or more negative to that of nZVI, such as zinc (Zn(II)) and cadmium (Cd(II)), the removal mechanism is sorption/surface complex formation. On the contrary, the metal ions would be reduced to the ZVMs when the metals have greater positive standard potential than that of nZVI. Meanwhile, Ni(II) and Pb(II) can be immobilized at the nanoparticle surface by both sorption and reduction.

4.4 Effect of Environmental Parameters on the Reactivity of nZVI

Although nZVI is effective for the removal of chlorinated compounds, heavy metals, and other inorganics, the reactivity of nZVI is highly controlled by the surface characteristics and water chemistry of groundwater. The iron particle crystallizes in a body-centered cubic structure. It is estimated that about 4% of iron atoms are exposed onto the surface when the diameter of iron nanoparticles is at 50 nm. However, the ratio of iron atoms on the surface area of nanoparticles decreases to 0.001% when the particle size of the iron nanoparticles increases to 1 mm, which means that the reactivity of nZVI particles will be dramatically increased when the particle size decreases to the nanoscale. Several studies have also depicted that the reduction rate by nZVI is dependent on the specific surface area of iron. A liner relation between reduction rate and surface area of iron was observed [15,42]. Surface area-normalized pseudo-first-order rate constant (k_{SA}) for contaminant degradation between metallic nanoparticles and reducible contaminant, including chlorinated methanes, chlorinated ethanes, chlorinated aromatics, and nitrate, have been reported to be significantly influenced by the nanoparticle size, which may vary by as much as 1–2 orders of magnitude higher than those of commercial or microscale iron powders [43–48].

4.4.1 Particle Size of nZVI

The reduction efficiency and rate of the individual contaminant by nZVI at different particle sizes are varied. Liu et al. [33] reported that the k_{SA} value for TCE dechlorination by iron nanoparticles with diameters of 30–40 nm was 1.4×10^{-2} L h^{-1} m^{-2}, which is higher than that reported by Wang and Zhang [26] (3×10^{-3} L h^{-1} m^{-2}) who used 1–100 nm nZVI for TCE dechlorination. Liou et al. [47] indicate that the reactivity of iron nanoparticles with diameters of 9–10 nm for the denitrification of nitrate in aqueous solution was higher than that of iron nanoparticles with diameters of 20–60 nm. They further investigated the size effect of copper nanoparticles on the dechlorination toward CT and found that the k_{SA} of copper nanoparticles at 10 nm on resin was 110–120× higher than that of powdered copper particles, while only 10- to 20-fold increase in k_{SA} relative to that of powder copper particles were observed when the diameters of Cu nanoparticles were in the range 18–29 nm. More recently, He and Zhao [49] synthesized the monodispersed Pd/Fe bimetallic nanoparticles by applying a water-soluble starch as a stabilizer. The mean particle diameters were in the range 14.1 ± 11.7 nm. The k_{SA} value for 40 mg L^{-1} TCE dechlorination by 1 g L^{-1} starch-stabilized Pd/Fe nanoparticles was 0.67 L h^{-1} m^{-2}, which is 37× higher than that reported by Lien and Zhang [50] who used 5 g L^{-1} nonstabilized Pd/Fe for the dechlorination of 20 mg L^{-1} TCE.

4.4.2 pH Values

pH value is one of the important environmental parameters influencing the reactivity of micro- and nanosized ZVI. Several studies have demonstrated the relation between pH and the pseudo-first-order constant (k_{obs}) for dechlorination of chlorinated hydrocarbons, and found that the dechlorination efficiency and rate decreased upon increasing pH values [16,51–53]. Deng et al. [51] demonstrated that the k_{SA} values for vinyl chloride dechlorination by microscale ZVI decreased less than an order of magnitude when pH increased from 6 to 10. Chen et al. [52] found that the k_{SA} of iron corrosion linearly decreased from 0.092 to 0.018 L h^{-1} m^{-1} between pH 4.9 and 9.8, while it is significantly higher at pH 1.7 and 3.8. The dechlorination process is a consequence of direct oxidative corrosion of iron by chlorinated compounds, resulting in the production of ferrous and hydroxyl ions. At elevated pH values, the ferrous ions react with hydroxyl ions and form ferrous hydroxide and ferric oxides, and subsequently precipitate on the surface of ZVI. The precipitated iron oxides would hinder the transport of chlorinated compounds and block the reactive sites on ZVI, resulting in the decrease in reaction rate. Lin and Lo [54] indicated that the dechlorination rate of TCE by iron can be enhanced when the iron oxide shell was pretreated with HCl. Interesting, Doong et al. [9] examined the solution pH profile in zerovalent silicon (Si), nZVI, and Si/Fe systems without the addition of buffer solution, and found that the pH increased rapidly from 7.5 to 10, while the pH decreased from 8.7 to 7.8 in the first 50 h in Si solution (Figure 4.1a). The combination of Si and Fe constituted a buffer system

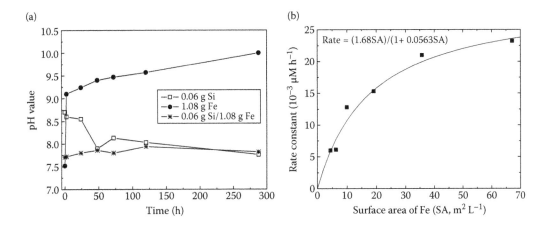

FIGURE 4.1
(a) Changes in pH values in Si, Fe, and Si/Fe systems. (b) Relation between rate constants for PCE dechlorination and surface area of Fe in the unbuffered Si/Fe–H$_2$O system containing 0.06 g Si and 0.06–2.1 g Fe. (From Doong RA et al., *Environ Sci Technol*, 37, 2575, 2003.)

and maintained the pH at a stable value (7.5–7.8), resulting in acceleration of the dechlorination rate of PCE. As shown in Figure 4.1b, the dechlorination rate of PCE was a function of Fe concentration in Si/Fe solutions without addition of buffer solution. The dechlorination rate increased linearly from 6.0×10^{-3} μM h^{-1} at 1.85 m^2 L^{-1} of Fe to 12.8×10^{-3} μM h^{-1} at 7.4 m^2 L^{-1} of Fe, and then leveled off to 23.2×10^{-3} μM h^{-1} at 64.7 m^2 L^{-1} of Fe. In addition, the final solution pH increased from 7.4 at 0.06 g of Fe to 8.1 at 2.1 g of Fe. Only 0.3 units of pH were changed as the Fe amount increased by a factor of 35, clearly showing that the combination of Si and Fe can form a natural buffering system to achieve an environmentally friendly condition near pH 7–8.

4.4.3 Transition Metal Ions

In addition to the solution pH, the impact of transition metal ions on the reductive dechlorination of chlorinated compounds and nitro-aromatics by both structural Fe(II) species and ZVI has been addressed. The deposition of small amounts of second metals such as Ni and Pd onto the iron surface has been demonstrated to effectively enhance the dechlorination efficiency and rate of chlorinated hydrocarbons [12,55,56]. The coexistence of other heavy metal ions in aqueous solution also influences the reactivity of iron nanoparticles toward contaminants. The core–shell structures of iron nanoparticles can serve as reductants as well as adsorbents to immobilize metal ions onto the surface. The adsorbed metal ions may further receive electrons produced from the core layer of ZVI. This would enhance the reduction efficiency of reducible contaminants by converting the metal ion into its low valence state form or decrease the efficiency by competing for electrons with organic contaminants. Lee and Doong [57] have shown that the k_{obs} for PCE dechlorination by zerovalent silicon in the presence of 0.1 mM Fe(II) and Ni(II) were 1.5–3.8× higher than in the absence of metal ions. Schlicker et al. [58] have shown that the existence of Cr(VI) significantly hindered the dechlorination of chloroethanes by ZVI. On the contrary, addition of catalytic ions such as Ni(II), Cu(II), Ag(I), Pd(II), and Pt(II) may enhance the reduction rate and efficiency of chloroethanes. Dries et al. [59] reported that the addition of 5–100 mg L^{-1} Ni(II) enhanced TCE reduction by 100 g L^{-1} ZVI owing to the catalytic hydrodechlorination by bimetallic Ni^0/Fe^0. However, Cr(VI) and Zn(II) lowered the TCE dechlorination rate by a factor of 2–13×. Lien et al. [60] depicted that Cu(II) enhanced the CT dechlorination by nZVI, while Pb(II) only increased the reduction rate slightly. However, Cr(VI) decreased the dechlorination rate by a factor of 2. Xie and Shang [61] investigated the impact of metal ions on the reduction of bromate by nZVI and found that the incorporation of Cu(II) led to an increase in the bromate reduction rate. Addition of Pd(VI), on the contrary, had little effect on bromate reduction by microscale ZVI.

The addition of metal ions also influences the reactivity of the Fe(II)–Fe(III) system, and could alter the long-term reactivity of ZVI in aqueous solutions. Jeong and Hayes [62] showed that the addition of transition metal ions, including Cu(II), Ni(II), Zn(II), Cd(II), and Hg(II), increased the dechlorination rates of hexachloroethane in the presence of iron-bearing minerals. The catalytic activities of Cu(II), Au(III), and Ag(I) in the reduction of chlorinated alkanes by green rust were also demonstrated [63,64]. Maithreepala and Doong [65,66] demonstrated that that addition of Ni(II), Co(II), and Zn(II) lowered the k_{obs} for CT dechlorination by structural Fe(II)–Fe(III) minerals, whereas the amendment of 0.5 mM Cu(II) into the Fe(II)–Fe(III) system significantly enhanced the efficiency and the rate of CT dechlorination. Cu(II) could be an effective metal ion to synergistically enhance the dechlorination efficiency and rate of CT by iron-bearing minerals.

4.4.4 Effect of Natural Organic Matters

The transformation process of chlorinated hydrocarbon by nZVI is a surface reaction that requires close contact of the reactive iron surface and target compounds [67–69]. Surface redox reactions comprise a series of physical and chemical processes, including (i) mass transfer of the dissolved target compounds from the aqueous phase to the iron surface, (ii) adsorption of the compounds to the iron surface, (iii) electron transfer from the iron surface to the target compounds, and (iv) desorption of the by-products from the surface [70]. Therefore, any nonreactive adsorbate that outcompetes the reactive surface sites with contaminants would result in the decrease in degradation rate [71–76].

In subsurface environments, natural organic matters (NOMs) are abundant and play important roles in both electron transfer and adsorption processes. The inhibition of the dechlorination rate of chlorinated hydrocarbons by ZVMs in the presence of organic matters was reported. Muftikian et al. [77] reported that the reactivity of Pd/Fe decreased with time because of the formation of hydroxylated iron oxide films on the surface. Tratnyek et al. [75] found that the reduction rate of TCE was inhibited by natural organic matter (NOM) in the ZVI system owing to the competitive sorption onto the surface of ZVI. The competitive reaction between organic compounds for a limited number of reactive sites on the surface of ZVI was also observed. Cho and Park [78] depicted that the rate constant for TCE reduction by ZVI decreased 1.5–5× when the solutions contained reducible co-contaminants such as CT, nitrate, and chromate. Loraine [73] also pointed out that some alcohols, such as ethanol and propanol, could inhibit the reduction of TCE by ZVI. Xie and Shang [79] depicted that the reactivity of ZVI toward bromate reduction declined by a factor of 1.3–2.0× when 5–35 mg of dissolved organic carbon (DOC) per liter of humic acid was added. However, the reduced functional groups present in humic acid would regenerate Fe(II), resulting in the maintenance of iron surface activation for bromate reduction in the long run.

Similar results were also observed by Doong and Lai [80]. In their study, humic acids were found to serve as inhibitors to compete for the reactive sites on the palladized iron with PCE. After 24 h of equilibrium of humic acid with ZVI, the adsorbed humic acids served as electron shuttles to effectively accelerate the dechlorination efficiency and rate of PCE by the palladized iron. The scanning electron microscopic (SEM) images showed that small protrusions appeared when 0.1 mM Pd(II) was amended into the solution, depicting that Pd was deposited on the surface of ZVI (Figure 4.2a). However, a mucous layer adhered onto the surface of palladized irons in the presence of humic acid, reflecting the possibility of lowering the dechlorination efficiency of Pd/Fe in the presence of NOM (Figure 4.2b). However, the quinone moiety in humic acid has different effects on PCE dechlorination by Pd/Fe. Although the addition of benzoquinone lowered the reduction rates of PCE, the values of k_{SA} for the respiked PCE after 24 h in lawsone- and hydroquinone-amended systems were 1.24× and 1.39×, respectively, higher than that in the absence of quinones (Figure 4.2c). On the contrary, a rapid and complete dechlorination of PCE by Pd/Fe was observed within 5 min in the presence of anthraquinone-2,6-disulfonate (AQDS). Respiking of 1 mg L^{-1} PCE into the batches also showed good efficiency of PCE dechlorination (Figure 4.2d). These results demonstrate that the quinone moiety in humic acid plays a pivotal role in enhancing the dechlorination efficiency of PCE by Pd/Fe.

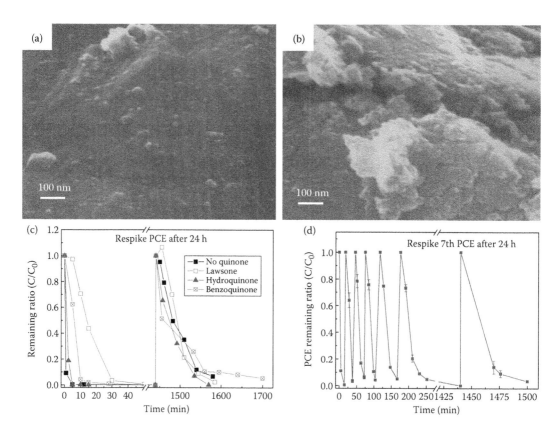

FIGURE 4.2
SEM images of Pd/Fe in the (a) absence and (b) presence of 50 mg L^{-1} humic acid after incubation for 24 h and the effect of (c) quinone compounds and (d) AQDS on the dechlorination of PCE by Pd/Fe. (Cited from Doong RA, Lai YJ, *Water Res*, 39, 2309, 2005.)

4.5 Bimetallic Iron System

4.5.1 Synthesis of Bimetallic System

Bimetallic systems, the combination of a reductant metal and a catalytic metal, have recently received much attention. The use of a bimetallic iron system can increase the decomposition efficiency and rate as well as the change in product distribution shown in Table 4.3. Mechanistic studies showed that the dechlorination pathways of chlorinated hydrocarbons by Pd/Fe or Ni/Fe changed from hydrogenolysis to hydrodechlorination, resulting in the formation of nonchlorinated hydrocarbons [57,81–83]. In addition, the introduction of a second catalytic metal could prevent toxic by-product formation by dechlorination of PCE via hydrogen reduction rather than via electron transfer, which may reduce the accumulation of hydrogen gas and enhance the reduction rate of priority pollutants.

Several methods can be used for the preparation of bimetallic nZVI systems. The physical mixing of the reductant and catalytic metals is a simple method to prepare the bimetallic iron system, while chemical combination of the reductive metal to catalytic metal can result in a dramatic increase in the degradation efficiency of priority pollutants

TABLE 4.3

Reactivity of Bimetallic Iron Nanoparticles

Bimetallic	Loading (%)	Particle Size (nm)	Target Compounds	Rate Constants		BET $(m^2\,g^{-1})$	References
				k_{obs} (h^{-1})	k_{SA} $(L\,h^{-1}\,m^{-2})$		
Ni/Fe	15.6	3–30	TCE	13.7	$(9.8 \pm 0.7) \times 10^{-2}$	59	Schrick et al. (2002)
Ni/Fe	20	<10 nm	CT	9.03	9.22×10^{-2}	39.2	Feng and Lim (2005)
			CF	0.461	4.71×10^{-3}	22–27	
Ni/Fe	20	20–40	TCE	–	3.7×10^{-2}		Tee et al. (2005)
Cu/Fe	82%	~50 μm	1,1,1-TCA	0.744 ± 0.04	–		Bransfield et al. (2006)
Pd/Fe in silica	0.1–1.0	<100 nm	1,2,4-TCB	0.4–1.86		117	Zhu et al. (2006)
Ni/Fe	10%	30 nm	PCP	2.44	–	20.9	Zhang et al. (2006)
Cu/Al	10%	–	CT	0.58	5.5×10^{-3}	5.31	Lien and Zhang (2002)
Cu/Al	10%	–	CT	0.32	3.1×10^{-3}	5.31	Lien and Zhang (2002)
Ru/Fe	1.5%	Micro	TCE	2.4 ± 0.17	–	1.66	Lin et al. (2004)
Ag/Mg	0.71%		PCP				Patel and Suresh (2006)
Pd/Fe in PVDF	1.9–2.3%	30 ± 5.7	Dichlorophenyl	1.36	6.68×10^{-2}	25	Xu and Bhattacharyya (2007)
Fe/Ni on Al$_2$O3	Fe: 10%, Ni: 90%	40	TCE	2.82	2.7×10^{-2}	20.9	Hsieh and Horng (2006)
Fe/Ni on Al$_2$O3	Fe: 10%, Ni: 90%	40	CF	0.99	9.5×10^{-3}	20.9	Hsieh and Horng (2006)
Cu/Fe	0.5–20	20–70	NO_3^{-1}	0.822	1.27×10^{-1}	16.67	Liou et al. (2005)
Ni/Fe	50	<100	As(V)	5.4	0.504	42.44	Jegadeesan et al. (2005)
Pd/Fe	10	10–20	As(V)	0.66	0.0616	42.44	Jegadeesan et al. (2005)

[47,55,57,84]. The bimetallic nZVI system can be prepared by the use of commercialized or self-synthesized nanoscale reductive metals as the reductant or adsorbent, while the catalytic metal ion serves as the precursor and adsorbate. The catalytic metal ion will then be reduced into ZVM by the addition of a reducing agent such as NaBH$_4$ or by the electrochemical reduction from the reductive metal. The structure of the chemically fabricated bimetallic materials prepared using this method is likely to be a core–shell structure in which the reductive metal acts as a core while the catalytic metal serves as a shell [29]. In addition, both the reductive and catalytic metal ions can be added in a solution and then be reduced by reducing agent simultaneously. Alloy structure is easily found in the method. Since the dechlorination process by ZVI is a surface-mediated process, the reduction efficiency of priority pollutants by a core–shell structured bimetallic system is usually better than that of the alloy bimetallic structure when the same ratio of catalytic metal to reductive metal was compared.

4.5.2 Reduction of Pollutants by Bimetallic Systems

Bimetallic systems have received significant attention because of their considerably faster reduction rates of priority pollutants. Several bimetallic particles, such as Pd/Fe, Ni/Fe, Cu/Al, and Cu/Fe, have been synthesized and applied to the reduction of a wide variety of priority pollutants, including chlorinated hydrocarbons [44,55,82,85,86], nitrobenzenes [87], pentachlorophenol [88], and anions [79,89]. Nutt et al. [81] showed that a bimetallic treatment approach involving palladium supported on gold (Pd-on-Au) increased the reaction rate in the dechlorination of TCE. A rapid and complete dechlorination of chlorinated solvents with the production of nonchlorinated hydrocarbons was also reported by using nanoscale bimetallic Pd/Fe particles [22,90,91]. In addition, bimetallic Ni/Fe nanoparticles have been found to rapidly dechlorinate chlorinated ethylenes with the formation of ethane as the main products [82,85], showing that the bimetallic system is an effective technology for accelerating the dechlorination processes and converting the chlorinated hydrocarbons to the nontoxic end products.

The deposition of a catalytic second metal such as Pd, Ni, and Cu onto the surface of a reductive metal could enhance the dechlorination efficiency and rate of chlorinated hydrocarbons and prevent the formation of toxic products by dechlorinating the chlorinated hydrocarbons via hydrogen reduction rather than through electron transfer [8,85]. In addition, the type of products obtained during the dechlorination reaction is dependent on the identity and mass loading of the second metal employed [85]. When the bimetallic system was employed to decompose the chlorinated hydrocarbons, the dechlorination efficiency could be enhanced by increasing the loading of the second metal [55,81,82]. However, higher loading of catalytic metal on the reductant metal surface inhibits the dechlorination efficiency and an optimal loading usually exists. Several plausible explanations, including the formation of a galvanic cell [85], the surface coverage of catalytic metal on the reductive metal [82], and the absorbed atomic hydrogen [86,92], have been proposed to explain this phenomenon. More recently, Parshetti and Doong [93] immobilized nZVI onto TiO_2 nanoparticles and investigated the coupled removal of TCE and 2,4-dichlorophenol (DCP) in aqueous solutions under anoxic conditions in the presence of nickel ions and UV light at 365 nm. They found that both TCE and DCP were effectively dechlorinated by Fe/TiO_2 nanocomposites, which were higher than by nZVI alone. Addition of nickel ions significantly enhanced the simultaneous photodechlorination efficiency of TCE and DCP under the illumination of UV light. The pseudo-first-order rate constants for DCP and TCE photodechlorination by Fe/TiO_2 in the presence of 20–100 µM Ni(II) was 30.4–136× and 13.2–192×, respectively, higher than those in the dark. The reaction mechanism was also proposed. As shown in Figure 4.3, the TiO_2 photocatalysts can be photoexcited by UV light to generate electron–hole pairs, while the metallization of TiO_2 with Fe prevents the recombination of holes with electrons, leading to the enhancement of the oxidizing capability of TiO_2. In addition, the Ni(II) and produced Fe(II) ions from the anaerobic corrosion of nZVI can react with photogenerated holes to form Ni(III)/Fe(III) ions and then be converted back to Ni(II)/Fe(II) ions again when reacted with electrons or hydroxyl anions, resulting in the prevention of hole–electron recombination and the increase in the total amounts of hydroxyl radicals.

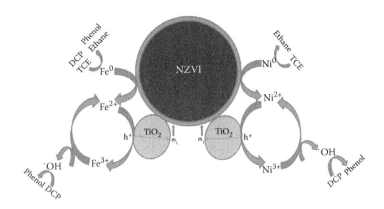

FIGURE 4.3
Enhanced photodechlorination efficiency of DCP and TCE by Fe/TiO$_2$ nanocomposites in the presence of Ni(II) ions and 365 nm UV light under anoxic conditions. (Cited from Parshetti GK, Doong RA, *Water Res*, 45, 4198, 2011.)

4.5.3 Optimal Mass Loading

Several studies have depicted the effect of additive loadings toward the dechlorination efficiency of chlorinated hydrocarbons by Fe^0 [81,82,94,95]. An optimal mass loading exists for a wide variety of bimetallic catalysts. Tee et al. [82] investigated the role of bimetallic Ni/Fe nanoparticles on the dechlorination of TCE. A higher degradation rate of TCE was observed upon increasing the Ni loading from 2 to 25 wt%, and then the k_{obs} decreased when greater Ni loading was added. Moreover, Lin et al. [55] reported that the dechlorination rate of TCE by bimetallic Ru/Fe increased as the Ru loading increased from 0.25 to 1.5 wt%. A decrease in k_{obs} was also observed when Ru loading increased to 2.0 wt%. In addition, Bransfield et al. [95] depicted that a pronounced increase in k_{obs} with increasing Cu loading was observed below 10 μmol Cu (g Fe)$^{-1}$. Lee and Doong [57] reported that an optimal mass loading of Ni at 1.47 wt% was also obtained for PCE dechlorination by zerovalent silicon. The possible explanation is that the addition of metal ions may increase the numbers of nanocrystalline Ni particles on the surface of zerovalent silicon, thus enhancing the catalytic activity for hydrodechlorination. The high loading of second dopant metal, however, leads to the aggregation of fine catalytic nanoparticles into large ones, and subsequently decreases the reaction rate of chlorinated hydrocarbons.

4.6 Reduction of Priority Pollutants by nZVI

4.6.1 Dechlorination of Chlorinated Compounds

The use of microscale and nanoscale ZVI for the reduction of organic contaminants has been widely investigated. Microscale ZVI has been demonstrated to effectively dechlorinate many priority pollutants [14–16,72,96]. Matheson and Tratnyek [16] first proposed three possible pathways for dechlorination in the Fe^0–H$_2$O system. The first pathway involves the dechlorination of chlorinated hydrocarbons by electron transfer directly from the iron surface to the adsorbed chlorinated hydrocarbons, which is believed to be the dominant pathway for dechlorination. The second pathway involves the reduction by ferrous ion, which is an immediate product of corrosion in aqueous systems. Dissolved ferrous ion is a mild reductant capable of causing dehalogenation of some chlorinated hydrocarbons. The importance of this process is the formation of a surface-bound ferrous iron, which may be attributed to the long-term stability of ZVI [65,66,97–100]. In addition, the third model for dechlorination by ZVI involves the indirect electron transfer by atomic hydrogen produced from water reduction when catalysts such as Ni, Pd, and Pt are present in solution.

Although ZVI have been successfully employed as the reactive material for treatment of soil and groundwater contaminated with chlorinated hydrocarbons, several challenges, including the production of chlorinated by-products, slow reaction rate, and inability to dechlorinate polychlorinated biphenyls (PCBs), still exist. One of the advantages for using nanoscale iron materials is that nZVI can degrade contaminants that cannot react with microscale ZVI such as PCBs and chlorinated aromatics [7,46]. Zhang [31] have demonstrated that nZVI can effectively decompose >30 organic compounds, including halogenated hydrocarbons, chlorinated aromatics, nitroaromatics, pesticides, and organic dyes. Wang and Zhang [26] used the nZVI and Pd/Fe for dechlorination of TCE and Arochlor 1254. The Pd/Fe resulted in the rapid and complete dechlorination of TCE and Arochlor 1254 after 17 h, while only <25% of PCBs with biphenyl as the final product was dechlorinated by unpalladized nZVI. The k_{SA} values for TCE and PCB dechlorination are calculated to be 10–100× higher than those of commercially available microscale ZVI particles. Lowry and Johnson [46] compared the initial rate of dechlorination of six PCBs by bare and palladized nZVI particles in a 30% ethanol/water mixture under ambient conditions. The observed k_{SA} for PCB dechlorination by nZVI and 0.05 wt% Pd/Fe were in the range $(5.5–100) \times 10^{-4}$ L yr^{-1} m^{-2} and $(3.8–17) \times 10^{-4}$ L yr^{-1} m^{-2}, respectively. Shih et al. [101] also reported that hexachlorobenzene can be decomposed by nanosized Pd/Fe bimetal. For bare nZVI, non-*ortho*-substituted congeners were found to have faster initial dechlorination rates than *ortho*-substituted congeners in the same homolog group, and chlorines in the *para* and *meta* positions were predominantly removed over chlorines in the *ortho* position, which suggests that more toxic coplanar PCB congeners are not likely to form from less toxic noncoplanar, *ortho*-substituted congeners [46].

Another advantage is that nZVI particles have a larger specific surface area and have higher surface reactivity than microscale ZVI [31,46]. However, increased surface area alone is not adequate to explain the high reactivity of nZVI. Higher density of reactive surface sites and greater intrinsic reactivity of surface sites may also be the possible reasons for the enhanced reactivity of nZVI [32]. The reduction of chlorinated hydrocarbons by iron particles has been generally believed to be a surface-mediated reaction, and Langmuir–Hinshelwood kinetics can be used to model the reaction kinetics of chlorinated hydrocarbons by nZVI [33,96]. At low concentrations of chlorinated hydrocarbons, the reaction rate can be simplified to the pseudo-first-order dechlorination reaction, while the reaction rate would change to zero-order kinetics when the surface-active sites are occupied by the overloaded substrate [42,102]. It is believed that the pseudo-first-order rate constants for dechlorination of chlorinated hydrocarbons by nZVI were higher than that by microscale ZVI. Nurmi et al. [32] depicted that the mass-normalized pseudo-first-order rate constants (k_m) for CT reduction is larger with nZVI than with microscale ZVI. However, it is unclear whether there is a nanosized effect on the k_{SA} for CT reduction. In contrast, the k_{SA} for TCE dechlorination by nZVI were found to be in the range $(2–3) \times 10^{-3}$ L h^{-1} m^{-2} [26,85], which are higher than those for normal iron filings $((3.9 \pm 3.6) \times 10^{-4}$ L h^{-1} m$^{-2})$ [42,44]. The k_{SA} values for dechlorination of CT and chloroform by nZVI were 9.22×10^{-3} and 5.13×10^{-4} L h^{-1} m^{-2}, respectively, which are 1.7–8× higher than those by microscale ZVI.

The dechlorination pathways of chlorinated hydrocarbons by ZVI are fairly well understood. The chlorinated hydrocarbons are usually reduced by ZVI and release chloride ion, whereas ZVI is oxidized and supplies electrons [33]. The proposed reaction mechanisms include β-elimination, hydrogenolysis, and hydrogenation. β-Elimination is the primary reaction pathway for the dechlorination of chlorinated ethylene, while hydrogenolysis and hydrogenation are the minor reactions [33,96]. These mechanisms can explain the formation of ethane and ethylene as the main products, with minor amounts of chlorinated intermediate and acetylene when TCE or PCE is dechlorinated by ZVI.

The β-elimination pathway involves a two-electron transfer to produce chloroacetylene from chlorinated hydrocarbons. Chloroacetylene then undergoes hydrogenolysis to form acetylene, which further undergoes a two-electron transfer to form ethane and ethene. Arnold and Roberts [96] have proposed that chlorinated ethylene could form the di-σ-bonded intermediate on the iron surface first and further reduced to less chlorinated hydrocarbons. The first step is the formation of a π-bonded surface species. In this adsorption step, the alkynes and alkenes serve as Lewis bases, while partially or fully oxidized metal ions can be represented as Lewis acids [103]. In the second step, the π-bonded intermediate forms a di-σ-bonded surface-adsorbed species, and subsequently undergoes two successive fast halide ion elimination steps to form a mono-σ-bonded vinyl surface-adsorbed species and then to acetylene.

The reaction of chlorinated hydrocarbons by nZVI is slightly different from that by microscale ZVI. The reaction mechanism and product distribution are dependent on the H_2 evolution rate and the degree of crystallinity of the nZVI. Lowry and co-workers [33,34,104] have investigated the dechlorination rate and pathway of TCE by different properties of nZVI (Fe^{BH} and RNIP). Under both iron-limited and excess iron conditions, Fe^{BH} mainly transformed TCE to ethane, C3–C6 coupling products, and some unsaturated hydrocarbons [33]. No acetylene was detected, while chlorinated by-products were detected as reactive intermediates at very low levels and disappeared quickly. For RNIP, TCE was dechlorinated primarily into acetylene and ethane under iron-limited conditions. On the contrary, ethane and ethene were produced when excess RNIP was used to dechlorinate TCE. A further study [104] has found that the highly disordered nature of nanoscale Fe^{BH} and partially oxidized Fe^{BH} provides the ability to activate and use hydrogen gas, which increases the rate and extent of hydrogenation and yields more saturated reaction products compared with crystalline Fe^{BH}, RNIP, and iron filings. However, the reaction orders for H_2 evolution and TCE dechlorination with respect to nZVI content were different, which may be attributed to the difference in the rate-controlling steps for each reaction [34]. For H_2 evolution, the formation of adsorbed H species is the rate-limiting step, while TCE reduction requires that TCE adsorbs to the iron surface, and that TCE reduction occurs via direct electron transfer or via adsorbed H species.

It has been shown that many chlorinated hydrocarbons can be effectively dechlorinated indirectly by atomic hydrogen produced from the reduction of water by the corrosion of ZVI [34,68,105]. This reaction needs a catalyst, and chlorinated hydrocarbons react with adsorbed hydrogen involving the formation of hydride complexes on the iron surface [68]. Under anaerobic conditions, hydrogen is produced after the corrosion of iron. Although the produced gases would hinder the mass transfer of the pollutants to the reactive site on the iron surface [52], the feasibility of rapid dechlorination by hydrogen is still reliable in the presence of effective catalysts such as palladium (Pd), platinum (Pt), ruthenium (Ru), copper (Cu), and nickel (Ni) [12,21,26,37,106]. As ZVI corrodes, protons from water are reduced to adsorbed H atoms and to H_2 gas at the surface of the catalytic second metal. Chlorinated hydrocarbons are then adsorbed onto the surface of bimetallic nanoparticles when the C–Cl bond is broken, and subsequently undergo the hydrodechlorination reaction by replacing the chlorine atom with hydrogen [85].

4.6.2 Reduction of Nitrate

The treatment of nitrate and nitrite contaminants in groundwater and drinking water by microscale and nanoscale ZVI has recently become one of the innovative technologies for environment remediation [35,47,89,107–110]. Usually the nitrate reduction reaction by ZVI is relatively pH sensitive, and nitrate is a well-known oxidizing inhibitor to iron corrosion owing to the formation of an overlying oxide layer [47]. The pH value controls the reduction rate and efficiency of nitrate by ZVI and affects the formation of the passive oxide layer of ZVI. Huang et al. [111] reported that microscale ZVI powder could effectively reduce nitrate only when the pH was <4, which was low enough to dissolve the passive oxide layers. It has also been demonstrated that nitrate reduction by microscale ZVI was very limited at near-neutral pH because of the formation of a black oxide film onto the surface of ZVI [112,113]. Several strategies have been employed to enhance the reduction efficiency and rate of nitrate at a near-neutral pH value. The introduction of carbon dioxide to the ZVI system for lowering solution pH was found to be effective for nitrate reduction. Li et al. [114] used a pressurized CO_2/ZVI system for the reduction of nitrate under an anoxic condition. The pressurized system has potential advantages of using less CO_2 gas and reaching equilibrium pH faster than a CO_2-bubbled system. However, the pH increased gradually with increasing oxidation of ZVI, resulting in the decrease in nitrate reduction rate.

In addition to lowering the pH value by adding CO_2 or organic buffer, the reduction efficiency and rate can be augmented by addition of ferrous (Fe^{2+}) or copper (Cu^{2+}) ions [109,112,115,116]. The major role of the surface-adsorbed Fe^{2+} is more likely to induce or accelerate phase evolutions of the iron corrosion coatings toward more active ones that could sustain higher ZVI reactivity [117]. Nitrate reduction usually consists of three stages. At the first stage, a proton directly participates in the corrosion of ZVI. The second stage is very slow because of the formation of amorphous oxides on the surface of ZVI, while the third stage is characterized by a rapid nitrate reduction concurrent with the disappearance of aqueous Fe^{2+}. Addition of Fe^{2+} would adsorb onto the surface of ZVI to form surface-bound Fe(II), resulting in the formation of structural Fe(III) and the increase in the Fe(III)/Fe(II) ratio. The transformation of amorphous iron oxides into crystalline structural magnetite triggers the rapid nitrate removal rate in the presence of Fe^{2+}.

Different from the role of Fe^{2+} in a ZVI/nitrate/water system, the addition of copper ion significantly enhances the nitrate reduction rate at near-neutral pH. Liou et al. [118] investigated the effect of three noble metals, Pd, Pt, and Cu, on the reduction efficiency and rate for nitrate [119]. They found that the nitrate reduction rate by ZVI in the presence of 0.44% Cu (atomic ratio) was three times higher than those in the presence of Pd and Pt. This may probably be due to the nitrate adsorption onto the Cu surface and rapid reduction to ammonia by a neighboring adsorbed hydrogen atom.

Another available strategy for effective reduction of nitrate at near-neutral pH is the use of nZVI. Choe et al. [120] indicated that the use of nZVI for nitrate reduction has several advantages. These include the increase in reduction rate, decrease in reductant dosage, and production of nontoxic end products. Sohn and Fruehan [35] compared the reaction kinetics of nitrate reduction by three different sizes of ZVI. Results clearly showed that both the k_m and k_{SA} values are higher with nanosized iron than with micro- and millisized iron. Liou et al. [47] used different precursor concentrations (0.01, 0.1, and 1 M $FeCl_3 \cdot 6H_2O$) for the preparation of nZVI, and found that precursor concentration controls the diameter and reactivity of nZVI. In addition, the nitrate reduction rate was significantly enhanced by nZVI without acidification and the first-order rate constants for nitrate reduction followed the order $nZVI_{0.01M} > nZVI_{0.1M} > nZVI_{1M}$. Chen et al. [121] proposed a novel technology combining electrochemical and ultrasonic methods to produce nZVI for the removal of nitrate. The nZVI with a diameter of 1–20 nm was successfully fabricated by placing platinum in the cathode and adding cetylpyridinium chloride as the dispersion agent in the solution. When using the synthesized nZVI to remove nitrate, the pseudo-first-order rate constants for nitrate removal decreased linearly upon increasing pH from 4 to 7. In addition, only around 36%–45% of nitrate was converted into ammonium in solution, while 5%–64% of nitrate was transformed to nitrogen or other nitrogen-containing gases, clearing showing that nZVI not only accelerated the nitrate reduction rate at near-neutral pH but also altered the reaction pathways for producing environmentally friendly end products.

4.6.3 Reduction of Bromate and Perchlorate

Bromate is a refractory by-product during disinfection/oxidation processes in water treatment. This compound is now classified as a possible carcinogen with the current maximum contamination level of 10 µg L^{-1} in the national primary drinking water standard of the United States. Only a few studies have applied ZVI technology to the treatment of bromate in aqueous solution. Xie and Shang [122] found that bromate can be reduced to bromide ion by microscale ZVI and the reduction reaction is a surface-mediated reaction. The precipitation of iron oxyhydroxide onto the ZVI surface resulted in the reduction of the reactivity of ZVI to bromate reduction. Several environmental parameters influence the efficiency and rate of bromate reduction by ZVI. The incorporation of copper ion led to the increase in the reduction rate of borate by ZVI, while incorporation of a palladium ion had little effect on bromate reduction [122]. The enhanced bromate reduction rate in the presence of copper is most likely the result of the newly formed active Cu(I) on the iron surface. On the contrary, the reactivity of ZVI toward bromate reduction declined by a factor of 1.3–2.0× in the presence of 5–35 mg DOC L^{-1} humic acid because of the quick complexation of humic acid with iron species [79].

Perchlorate, a commonly used agent as an energetics booster and oxidant in a variety of munitions and fireworks, has also been frequently found in groundwater. Recently, perchlorate has been recognized to be an endocrine-disrupting chemical and promulgated by the US Environmental Protection Agency in the drinking water contaminant candidate list. Several technologies, including adsorption, biological, and electrochemical processes, have been used to reduce perchlorate to chlorate or to chloride via the sequential reduction reaction [123,124]. Recently, a few studies suggested that perchlorate can also be reduced by ZVI. However, the reduction rate of perchlorate by nZVI is considerably slower than many other contaminants [125–127]. Although little degradation of perchlorate by bare ZVI was observed during the 57 days of experimental course, perchlorate was found to be partially reduced (4.1%–82%) by acid-washed ZVI after 14 days [125,128].

Several methods such as the use of nZVI and the combination of biological reduction with ZVI have been developed to enhance the removal efficiency and rate of perchlorate reduction. Cao et al. [126] have demonstrated a nearly complete reduction of perchlorate to chloride by nZVI. nZVI also can reduce chlorate, chlorite, and hypochlorite to chloride under anaerobic conditions. In addition, the activation energy of perchlorate–iron reaction was calculated to be 79.02 ± 7.75 kJ mol^{-1}. This large activation energy implies that perchlorate reduction is a kinetic-controlled reaction. Recently, the combination of microbial reaction with ZVI has been found to effectively reduce perchlorate under anaerobic conditions. Biological reduction of perchlorate by autotrophic microorganisms attached onto ZVI was investigated in both batch experiments and flow-through columns [129,130]. The hydrogen gas produced from the corrosion of ZVI could be used as an electron donor by the autotrophic microorganisms to reduce perchlorate to chloride. The degradation followed Monod kinetics with a normalized maximum utilization rate and half-velocity constants of 9200 μg (g h)$^{-1}$ and 8900 μg L^{-1}, respectively. In addition, an increase in the biomass measured by optical density at 660 nm (OD$_{660}$), from 0.025 to 0.08, led to a corresponding 4-fold increase in perchlorate reduction rate. In the presence of nitrate, however, the perchlorate reduction rate was reduced but not completely inhibited. In addition, the mass of microorganisms attached on the solid ZVI/solid in the flow-through column was found to be 3 orders of magnitude greater than that in the pore liquid, indicating that the combination of autotrophs with ZVI is a promising alternative for perchlorate reduction in the contaminated subsurface environments.

The removal efficiency of perchlorate was strongly dependent on the dosage and surface area of ZVI. Similar surface area-normalized perchlorate removal rates were calculated for the batch and long-term column studies, which means that the perchlorate removal occurs at the iron surface [125,128]. It has been found that the perchlorate is initially sorbed to the iron surface, followed by a reduction to chloride. However, the reaction mechanism for perchlorate reduction is not well established. Moore et al. [125] found that perchlorate reduction was not observed on electrolytic sources of iron or on a mixed-phase magnetite oxide, suggesting that the reactive iron phase for perchlorate reduction is neither pure ZVI nor mixed oxide alone. They proposed that a mixed valence iron hydr(oxide) coating or a sorbed Fe^{2+} surface complex represents the most likely sites for the reaction. Perchlorate is a strong oxidant with the standard redox potential of 1.389 V. Therefore, perchlorate ion is loathe to accept an additional electron because it has no low-lying unfilled electronic orbitals and the transfer of an oxygen atom is thus required for a reduction reaction [131]. Huang and Sorial [128] proposed that the perchlorate ion consists of one Cl atom in the center of a tetrahedral structure surrounded by four oxygen atoms. The outside iron atoms at the oxide film have the potential to associate with oxygen atoms in perchlorate and may eventually distort the structure that perhaps allows the loss of one oxygen atom to become chlorate. Since chlorate can be further reduced by ZVI [125], this oxygen transfer process is the rate-limiting step for perchlorate reduction.

4.6.4 Removal of Heavy Metals

In addition to the reduction of halogenated organic compounds and anions, both microscale and nanoscale ZVI have been demonstrated to be effective materials for the removal of dissolved metal ions such as Se(VI), As(V), Cr(VI), and Pb(II) [38,132,133]. Removal of divalent ions containing Cu(II), Ni(II), Cd(II), Zn(II), and Pb(II) has been well documented [10,134]. The removal of divalent metal ions in acid mine drainage by ZVI was also reported under acidic conditions [10]. In addition, the reduction of Cr(VI), Se(VI), and As(V) by ZVI has been widely examined, and it was found that these toxic metals could be reduced by ZVI to their low oxidation states and then coprecipitated with ferric oxyhydroxide on the surface of the ZVI particles [135]. Ponder et al. [38] developed supported ZVI nanoparticles (ferrogels), which are 10–30 nm in diameter, for the remediation of metal ions in aqueous solution. The ferrogels were rapidly separated and immobilized Cr(VI) and Pb(II) from aqueous solution by reducing the Cr(VI) to Cr(III) and the Pb(II) to Pb(0) while oxidizing ZVI to goethite. In addition, the rates of remediation of Cr(VI) and Pb(II) by ferrogels are up to 30× higher than those for iron filings or iron powder on an Fe molar basis. Kanel et al. [133] used nZVI for the removal of As(V). Mössbauer spectroscopy showed that the nZVI is a core–shell structure in which 19% were in zerovalent state with a coat of 81% iron oxides. A total of 25% As(V) were reduced to As(III) by nZVI after 90 days. As(V) adsorption kinetics were rapid and occurred within minutes following a pseudo-first-order expression with a k_{obs} of 0.02–0.71 min^{-1}. In addition, a bimetallic Ni/Fe system was employed for the removal of arsenate [136]. The pseudo-first-order rate constants (k_{obs}) for As(V) removal by Ni/Fe was 0.091 min^{-1}, which is 2.5× faster than that by nZVI alone (0.037 min^{-1}).

Several parameters, including temperature, initial metal concentration, pH, and inorganic anions, influence the removal efficiency and rate of metal ions by ZNI. The increase in reaction temperature from 25°C to 65°C increased the removal rates of As(VI), while competitive adsorption of phosphate and sulfate was also observed [136]. At high initial metal concentration, the removal rate shifted from pseudo-first-order rate kinetics to zero-order rate expression. The pH value of the solution is also an important factor influencing the removal capacity of Se(VI). High removal efficiency of Se(VI) by both nanosized ZVI and bimetallic Ni/Fe were observed at pH <8.0. Kumpiene et al. [134] evaluated the effects of five parameters including pH, oxidation–reduction potential (ORP), liquid-to-solid (L/S) ratio, organic matter, and microbial activity on the mobility of heavy metals in ZVI-stabilized soils. They found that the pH value was the most important factor influencing the mobility of Cr, Cu, and Zn in the ZVI-stabilized soil, while the L/S ratio and microbial activity were important factors for As stability. In addition, the studied environmental factors mostly affect the mobility of Zn, followed by Cu, As, and Cr. Also, the coexistence of metal ions influences the removal efficiency of another metal ion by ZVI. Zhang et al. [137] found that the removal of Se(VI) by ZVI appeared to be partially attributed to the reduction of Se(VI) to Se(IV) by Fe(II) oxidized from ZVI, followed by rapid adsorption of Se(IV) onto iron oxyhydroxide. The coexistence of As(V) has little influence on Se(VI) removal because As(V) was removed at much faster rates than Se(VI). However, addition of Mo(VI) significantly decreased the removal efficiency of Se(VI) by ZVI.

Several mechanisms, including adsorption, precipitation, and biologically mediated transformation, have been proposed to explain the removal of inorganic contaminants [23,132,138–140]. Direct reduction of metal ions occurring at the iron surface results in the electrochemical reduction of metal species onto the iron surface or cementation, which can be theoretically predicted by the standard reduction potentials of the metals [41]. Immobilization of the metal ions occurs by sorption to the reactive medium or precipitation from the dissolved phase. Precipitation of metals by the reduction to a less soluble form is a combination of transformation process followed by an immobilization process. Moreover, sorption is an abiotic reaction where the contaminant is attracted to the surface by hydrophobic interaction, electrostatic attraction, and/or surface complexation [17]. Usually the basis for a sorption reaction is the corrosion of ZVI by the formation of mixed-valent and crystalline ferric oxides such as green rust, lepidocrocite, and magnetite. The iron oxides on the surface of ZVI are generally covered with hydroxyl groups in aqueous solutions, and subsequently undergo the surface complexation reaction with divalent cationic metal ions [59].

$$\equiv S\text{-}OH + Me^{2+} \leftrightarrow\ \equiv S\text{-}OHMe^{+} + H^{+} \tag{4.4}$$

$$2 \equiv S\text{-}OH + Me^{2+} \leftrightarrow\ \equiv (S\text{-}OH)_2Me^{+} + 2\ H^{+} \tag{4.5}$$

$$\equiv S\text{-}OH + Me^{2+} + H_2O \leftrightarrow\ \equiv S\text{-}OHMeOH + 2\ H^{+} \tag{4.6}$$

Both sorption and precipitation processes are reversible, and therefore may require removal of the reactive materials and accumulated products, which is dependent on the stability of the immobilized compounds and the geochemistry of the groundwater. Surface precipitation and adsorption have been demonstrated to be predominant mechanisms for the removal of As(V) and As(III) by both microscale and nanoscale ZVI. The spontaneous oxidation of ZVI in aqueous solution leads to the formation of Fe(II) and ferric oxides, resulting in the uptake of As(V) on iron oxyhydroxides by a ligand-exchange mechanism that replaces the surface-bonded hydroxide ion with an irreversible and stable arsenate or inner-sphere bidentate complex [136].

4.7 Stability and Mobility of Nanoscale Iron Particles

In general, the mobility of particles in saturated aquifers is directly related to the number of particle collisions with the porous media and Brownian diffusion [7,141]. It is also commonly assumed that the nanoscale iron particles allow the particles to overcome the limitations of gravitational force and adhere to Brownian motion for particle movement and dispersion. However, the mobility of nZVI particles in saturated porous media is usually limited and the practical transport distances of nZVI are only a few centimeters or less for bare and unsupported nanoparticles, presumably attributed to the filtration from solution by attachment to aquifer materials and aggregation and sedimentation of nZVI to plug pores [142,143]. Phenrat et al. [143] depicted that the nZVI particles at 20 nm aggregated to micron-size colloids only within 10 min, subsequently assembling into fractal, chain-like clusters due to the magnetic forces between nZVI particles. At an initial nanoparticle concentration of 60 mg L^{-1}, cluster sizes increase from 20 nm to 20–70 μm within 30 min and rapidly precipitate from solution. This result also implies that the use of a stabilizer or support to modify the nZVI surface for inhibiting particle aggregation and improving mobility is needed.

Several studies have demonstrated the effectiveness of employing organic polymer or support to homogeneously disperse nZVI nanoparticles to enhance their stability and mobility, as shown in Table 4.4. Various stabilizers, including thiols, carboxylic acids, surfactants, and polymers, have been used to prevent agglomeration of nanoscale iron oxide and ZVM particles [30,144]. Sun et al. [145] found that the addition of polyvinyl alcohol–co-vinyl acetate–co-itaconic acid (PV3A) could stabilize the synthesized nZVI for >6 months. However, not all stabilizers can be applied to the nZVI because thiols and carboxylic acid can be reduced by nZVI [146]. Schrick et al. [142] used anionic support materials such as hydrophilic carbon (Fe/C) and polyacrylic acid (Fe/PAA) as stabilizers to inhibit the aggregation and reduce the sticking coefficient of nZVI. Column tests showed that nanoparticle diffusion is the dominant filtration mechanism, and the anionic surface charges of Fe/PAA and Fe/C facilitated transport through sand- and clay-rich Chagrin soil. Saleh et al. [147] employed block copolymer consisting of a hydrophobic inner shell and a hydrophilic outer shell to modify a commercial nZVI. More recently, Xu et al. [29] synthesized the bimetallic Ni/Fe and Pd/Fe nanoparticles in the PAA/polyether sulfone (PES) composite membrane for the reductive dechlorination of TCE. Cross-linked PAA/PES composite membranes containing metal ions as particle precursors were prepared by heat treatment with ethylene glycol as the cross-linker. The average particle sizes of the synthesized Ni/Fe nanoparticles were 5 ± 0.8 nm (TEM image, $n = 150$), which is a much

TABLE 4.4

Methods Used to Stabilize the Nanosized Zerovalent Irons

Materials	Description	Particle Sizes (nm)	References
Surface modification			
Carboxymethyl cellulose	The OH functional group of carboxymethyl cellulose was involved in the interaction with synthesized NZVI.	17.2–18.6	[119,129,130]
Guar gum	The mixing of RNIP with guar gum (MW 3000 kDa and viscosity 0.5 g L^{-1}) can prevent the aggregation and ensure the mobility of RNIP in the subsurface environment.	162 ± 5	[120,131]
Anionic polyelectrolytes	The adsorbed polyelectrolytes can prevent the aggregation of RNIP by the repulsion force of electrostatic double layer, osmotic, and elastic–steric.	5–40	[117,132]
Polymer modification	RNIP modified with poly(styrene sulfonate) (PSS), chitosan, polyaspartate, and PV3A can prevent the aggregation.	63–75	[113,122,123]
Supporters			
Hydrophilic carbon	The supporter-stabilized NZVI was produced by the reduction of ferrous ion adsorbed on the hydrophilic carbon.	30–100	[124]
Membrane	Polyacrylic acid, polyether sulfone, PMMA, PVDF, and nylon 66 would be utilized as the supporter to stabilize NZVI with the ethylene glycol as the cross-linker.	31–60	[126,133–135]
Magnetite	NZVI synthesized by the reduction of NaBH$_4$ was attached to the Fe$_3$O$_4$ surface due to the magnetic force. This prevents the aggregation of NZVI.	1–100	[127]
Silica	TEOS and ferrous as the precursor of silica and iron are mixed and then through the sol–gel mechanism and reduction of NaBH$_4$ the silica incorporated with NZVI was formed.	–	[128,136,137]

smaller and narrower distribution compared with those in the absence of a membrane matrix. Complete dechlorination of TCE was achieved within 1 h by Ni/Fe nanoparticles inside the PAA/PES membrane ($k_{SA} = 0.1395 \pm 0.006$ L h^{-1} m^{-2}), while the excessive agglomeration of Ni/Fe nanoparticle without the protection of membrane results in less available surface area and slow dechlorination rate ($k_{SA} = 0.0378 \pm 0.003$ L h^{-1} m^{-2}).

The stability and reactivity of bimetallic nanoparticles also can be maintained using a stabilizer. He et al. [146] developed a new strategy for stabilizing palladized iron (Pd/Fe) nanoparticles with sodium carboxylmethyl cellulose (CMC) as a stabilizer. The complexation between carboxylate groups with metals and the intermolecular hydrogen bond between CMC and the Fe particle surface were identified to be the major mechanisms for stabilizing bimetallic nanoparticles to yield stable dispersions with sizes <17.2 nm. Batch experiments showed that the k_{obs} for TCE dechlorination by CMC-stabilized Pd/Fe nanoparticles was 17× higher than that by nonstabilized counterparts. Column tests showed that the CMC-stabilized nanoparticles can be readily transported in a loamy-sand soil and then eluted nearly completely (~98%) with three bed volumes of deionized water, whereas the nonstabilized Pd/Fe nanoparticles were retained on the top of the soil column [146]. In addition, the fresh CMC-stabilized nanoparticles offer a 2× greater k_{obs} value for TCE dechlorination when compared with the starch-stabilized Pd/Fe nanoparticles [49].

Parshetti and Doong [148] immobilized bimetallic Ni/Fe in PEG/PVDF and PEG/nylon 66 membranes for dechlorination of TCE. SEM images and electron probe microanalysis (EPMA) elemental maps showed that the distribution of Fe in the nylon 66 membrane was uniform and the intensity of Ni layer was higher than that in PVDF membrane (Figure 4.4a through d). The particle sizes of bimetallic Fe/Ni in PVDF and nylon 66 membranes were 81 ± 12 and 55 ± 14 nm with the Ni layers of 12 ± 3 and 15 ± 2 nm, respectively. The efficiency and rate of TCE dechlorination increased upon increasing the mass loading of

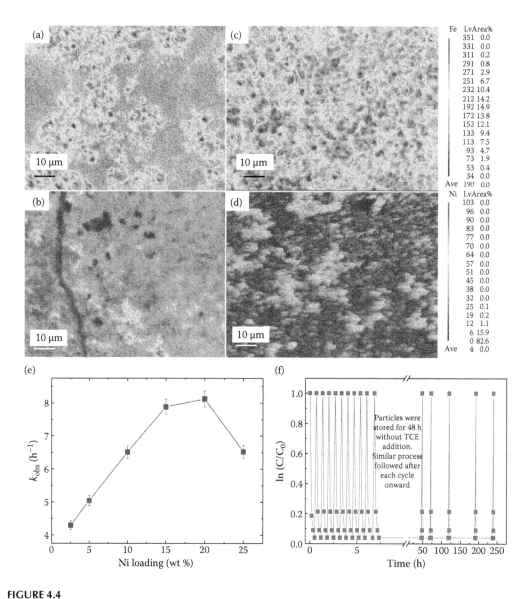

FIGURE 4.4
EPMA elemental maps of Ni and Fe species in microfiltration membranes. (a) Fe in PVDF membrane, (b) Ni in PVDF membrane, (c) Fe in nylon 66 membrane, and (d) Ni in nylon 66 membrane. (e) The pseudo-first-order rate constant for TCE dechlorination as a function of mass loading of nickel. (f) Stability and durability of Fe/Ni nanoparticles in nylon 66 membrane for TCE dechlorination. The total mass of Fe/Ni used was 7.2 mg. (Cited from Parshetti GK, Doong RA, *Water Res*, 43, 3086, 2009.)

Ni ranging between 2.5 and 20 wt%, and then decreased when the Ni loading was further increased to 25 wt% (Figure 4.4e). In addition, the stability and longevity of the immobilized Fe/Ni nanoparticles was evaluated by repeatedly injecting TCE into the solutions. A rapid and complete dechlorination of TCE by trace amounts of Fe/Ni nanoparticles was observed after 16 cycles of injection within 10 days (Figure 4.4f), indicating that the immobilization of Fe/Ni nanoparticles in the hydrophilic nylon 66 membrane can retain the longevity and high reactivity of nanoparticles toward TCE dechlorination.

4.8 Status of Field Application in the United States, Europe, and Taiwan

The application of nZVI on *in situ* groundwater remediation has been conducted in a variety of fields [149–152]. As of today, there are 58 completed or ongoing nZVI remediation projects worldwide at a pilot or field scale [152]. Among these, 36 remediation projects were conducted in the United States, including eight full-scale implementation cases. Europe is another fast-developing region where 17 sites, mainly located in the Czech Republic and Germany, have been commissioned including three full-scale deployments [153]. In Taiwan, three pilot scale tests have been implemented since 2006 (Figure 4.5) [154,155].

On the basis of the data collected from the field demonstration, there seems to be no simple guideline to ensure the success of nZVI remediation in the field. The final outcome is largely dependent on the nature of treatment such as source removal or pathway management, site hydrogeological characteristics, and the maturity of nZVI technology. In addition, the current understanding about the interaction between nZVI and the aquifer environment is still limited. The geochemical properties of the aquifer may be changed by nZVI amendment [27,153–157], while the mobility of nZVI is influenced largely by the geochemistry of aquifers. The first impact of water geochemistry on the nZVI is likely to be the particle stabilization. Bare nZVI readily forms an aggregate and becomes a larger particle (usually microscale) in the absence of suitable stabilizers [158]. Polymeric stabilization using a wide array of polymers such as PAA and CMC [159,160] is the most common way to stabilize nZVI in the aquifer. However, the concentration of stabilizer applied to an nZVI slurry drops naturally in the groundwater because of the dilution effect. The aggregation of nZVI occurs after the injection, as has been observed by SEM images of the soil sample [154].

The transport of nanoparticles in the aquifer is a complicated process involving physiochemical and biological interactions [158,161]. The nZVI transport in the aquifer is largely controlled by ionic strengths, pH and dissolved organic matter (e.g., fulvic acid and humic acid), and groundwater hydrology such as hydraulic conductivity and porosity [158]. In general, a decrease in ionic strength and an increase in dissolved organic matter concentration tend to favor the nZVI transport. The travel distance up to 3 m has been observed in the field studies within the sandy aquifer with high hydraulic conductivity [154]. Still, there are many uncertainties associated with the application of nZVI in the field, including the extent of nZVI migration laterally and vertically, the changes in reactivity due to passivation by groundwater constituents, and the change of microbial activities. To minimize this uncertainty, a detailed and accurate site assessment is required before the injection of nZVI.

A review of the recent nZVI field projects reveals large variations in key parameters [149]. For most field trials, it seems that changes in contaminant concentrations are the only information available to gauge the effectiveness of nZVI performance [149–152]. However, other factors need to be taken into account, such as the size of treatment zones and the

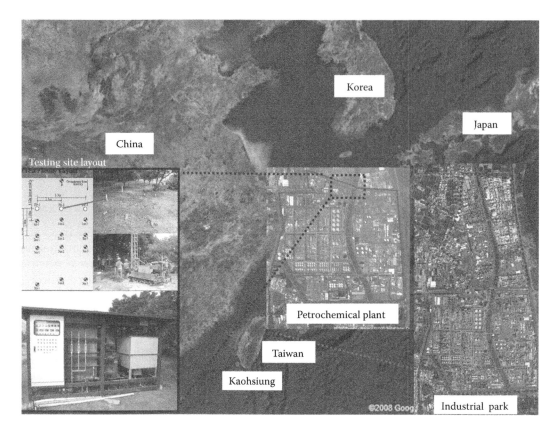

FIGURE 4.5
(See color insert.) Pilot testing site (200 m^2) of nZVI remediation conducted at a petrochemical plant in Kaohsiung, Taiwan. nZVI was synthesized on site using a semicontinuous reactor system.

operation duration, to compare field tests on a consistent basis. In other words, it may be necessary to set and apply a more generalized effectiveness indicator, which is a function of operation time, size of treated zone, and degradation efficiency (integrated over the treatment area), to enable a more meaningful comparison.

From an application point of view, there is a need to develop design parameters and guidelines for nZVI technology, so as to properly evaluate the performance and effectiveness of nZVI in the field. Through the design parameters, engineers will be able to estimate the injection amount of nZVI and the cost of treatment. Factors that may be taken into consideration as the design parameter include the dose of nZVI, contaminant removal capacity (e.g., kg contaminant (kg nZVI)$^{-1}$), and specific throughput (e.g., treated volume (kg nZVI)$^{-1}$). Currently, field data are rare and insufficient to provide a good basis for the estimation of the capacity and throughput of nZVI. Nevertheless, field experiences provide a baseline measurement of the injection dose of nZVI, although the reported results of some field cases are quite inconsistent [149,152]. For instance, a case reported high degradation efficiency (>90%) in the use of a very low nZVI dose (1.9 µg L^{-1} coated with 1% palladium) and another case determined the degradation efficiency to be about 40%–80% at a very high nZVI dose (20%) [152]. On the basis of the experience learned from the field cases in Europe, the injection dose was in the range of 1–30 g L^{-1} [153].

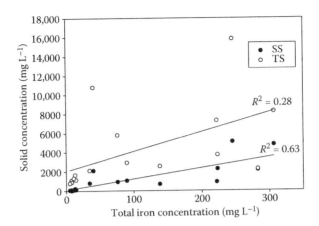

FIGURE 4.6
Correlation of total iron concentrations with total solid (TS) and suspended solid (SS) concentrations. (Cited from Wei YT et al., *J Hazard Mater*, 211, 373, 2012.)

The injection dose of nZVI is largely influenced by geochemical conditions in the groundwater. The optimum geochemical conditions of groundwater for nZVI applications, including DO, ORP, pH, and permeability, have been proposed (2–4). The nZVI in the groundwater tends to modify the geochemical conditions that facilitate the biotic and/or abiotic degradation of contaminants [27,154–156]. In general, the presence of nZVI tends to lower ORP, reduce DO, and increase pH. By monitoring the change of geochemical variables, the effective zone created by nZVI may be mapped. However, some cases reported conflicting results as no significant change of these variables after the nZVI injection was observed [157]. Other factors that may be used to map the effective zone and monitor the mobility of nZVI are total iron concentration, dissolved iron concentration, total solid, and suspended solid. Unlikely the secondary variables, e.g., ORP, pH, and DO, these factors are more directly related to nZVI if one can confirm that the solid and iron ions originated from nZVI [154]. We suggest that iron-based suspended solid may be a relatively reliable indicator in monitoring the effective zone and the mobility of nZVI according to the field study in Taiwan (Figure 4.6).

To improve the longevity, reactivity, and dosage efficiency of nZVI injection, a novel type of nZVI has recently been developed and applied in several of Taiwan's sites contaminated with chlorinated hydrocarbons [162]. This novel technology calls for the embedding of nZVI into a tunable porous supporter with high particle loading of >90%, best suitable for design and operation of the cutoff barrier of the contamination plume. Preliminary results are very encouraging, which will be published soon.

4.9 Conclusion Remarks

In this review, we have extensively reviewed and discussed the fabrication methods and physicochemical properties of nZVI. Several parameters, including particle size, pH value, metal ion concentrations, and co-contaminants, are found to significantly influence the reactivity of the nZVI toward the reduction of priority pollutants. The use of bimetallic

ZVI systems such as Ni/Fe and Pd/Fe is believed to be one of the promising technologies for remediation of chlorinated hydrocarbons and other priority pollutants, and their conversion to environmentally friendly end products. In spite of the high reactivity of nZVI for the effective reduction of many priority pollutants under anaerobic condition, the environmental application of nZVI still faces many uncertainties. The rapid aggregation of nZVI into microscale ZVI under various environmental conditions may hinder the transport of nanomaterials in soil and groundwater, which results in the change in long-term stability of nZVI during the remediation of priority pollutants. The use of adsorbed polyelectrolyte to modify the surface characteristics of nZVI or on the development of immobilization of nZVI in microfiltration membranes to minimize the aggregation and improve mobility will be an important subject requiring further research.

Close to 60 nZVI field application studies worldwide indicate that the mobility and transport of nZVI is greatly influenced by aquifer geochemistry and groundwater hydro-geology. A detailed and accurate site assessment is highly recommended to assure the cost-effectiveness of nZVI application. Polymeric stabilization remains to be the most common way to stabilize nZVI and retard its aggregation in the aquifer. Furthermore, to help better design and implement field remediation, there appears to be a need to develop a working guideline on parameters such as nZVI dose, contaminant removal capacity, and specific throughput. In addition to gauging geochemical variables such as ORP, DO, and pH, the authors suggest that iron-based suspended solids may serve as a relatively reliable indicator to monitor the effective zone and the mobility of nZVI.

Acknowledgment

The authors thank the National Science Council, Taiwan, for financial support under grant no. NSC 98-2221-E-007-030-MY3.

References

1. Yang GCC, Liu CY: Remediation of TCE contaminated soils by in situ EK-Fenton process. *J Hazard Mater* 2001, **85**(3):317–331.
2. Chang MC, Shu HY, Hsieh WP, Wang MC: Remediation of soil contaminated with pyrene using ground nanoscale zero-valent iron. *J Air Waste Manage* 2007, **57**(2):221–227.
3. Chen JL, Yang SF, Wu CC, Ton S: Effect of ammonia as a complexing agent on electrokinetic remediation of copper-contaminated soil. *Sep Purif Technol* 2011, **79**(2):157–163.
4. Yang GCC, Wu MY: Injection of nanoscale Fe3O4 slurry coupled with the electrokinetic process for remediation of NO3- in saturated soil: Remediation performance and reaction behavior. *Sep Purif Technol* 2011, **79**(2):272–277.
5. Wang JY, Huang XJ, Kao JCM, Stabnikova O: Simultaneous removal of organic contaminants and heavy metals from kaolin using an upward electrokinetic soil remediation process. *J Hazard Mater* 2007, **144**(1–2):292–299.
6. Chen ZS, Lee GJ, Liu JC: The effects of chemical remediation treatments on the extractability and speciation of cadmium and lead in contaminated soils. *Chemosphere* 2000, **41**(1–2):235–242.
7. Tratnyek PG, Johnson RL: Nanotechnologies for environmental cleanup. *Nano Today* 2006, **1**(2):44–48.

8. Farrell J, Kason M, Melitas N, Li T: Investigation of the long-term performance of zero-valent iron for reductive dechlorination of trichloroethylene. *Environ Sci Technol* 2000, **34**(3):514–521.

9. Doong RA, Chen KT, Tsai HC: Reductive dechlorination of carbon tetrachloride and tetrachloroethylene by zerovalent silicon-iron reductants. *Environ Sci Technol* 2003, **37**(11):2575–2581.

10. Wilkin RT, McNeil MS: Laboratory evaluation of zero-valent iron to treat water impacted by acid mine drainage. *Chemosphere* 2003, **53**(7):715–725.

11. Lien HL, Zhang WX: Enhanced dehalogenation of halogenated methanes by bimetallic Cu/Al. *Chemosphere* 2002, **49**(4):371–378.

12. Fennelly JP, Roberts AL: Reaction of 1,1,1-trichloroethane with zero-valent metals and bimetallic reductants. *Environ Sci Technol* 1998, **32**(13):1980–1988.

13. Agrawal A, Tratnyek PG: Reduction of nitro aromatic compounds by zero-valent iron metal. *Environ Sci Technol* 1996, **30**(1):153–160.

14. Orth WS, Gillham RW: Dechlorination of trichloroethene in aqueous solution using Fe-O. *Environ Sci Technol* 1996, **30**(1):66–71.

15. Gillham RW, Ohannesin SF: Enhanced degradation of halogenated aliphatics by zero-valent iron. *Ground Water* 1994, **32**(6):958–967.

16. Matheson LJ, Tratnyek PG: Reductive dehalogenation of chlorinated methanes by iron metal. *Environ Sci Technol* 1994, **28**(12):2045–2053.

17. Scherer MM, Richter S, Valentine RL, Alvarez PJJ: Chemistry and microbiology of permeable reactive barriers for in situ groundwater clean up. *Crit Rev Microbiol* 2000, **26**(4):221–264.

18. Mackenzie PD, Horney DP, Sivavec TM: Mineral precipitation and porosity losses in granular iron columns. *J Hazard Mater* 1999, **68**(1–2):1–17.

19. Su CM, Puls RW: Kinetics of trichloroethene reduction by zerovalent iron and tin: Pretreatment effect, apparent activation energy, and intermediate products. *Environ Sci Technol* 1999, **33**(1):163–168.

20. Lowry GV, Reinhard M: Hydrodehalogenation of 1-to 3-carbon halogenated organic compounds in water using a palladium catalyst and hydrogen gas. *Environ Sci Technol* 1999, **33**(11):1905–1910.

21. Liang LY, Korte N, Goodlaxson JD, Clausen J, Fernando Q, Muftikian R: Byproduct formation during the reduction of TCE by zero-valence iron and palladized iron. *Ground Water Monit R* 1997, **17**(1):122–127.

22. Zhang WX, Wang CB, Lien HL: Treatment of chlorinated organic contaminants with nanoscale bimetallic particles. *Catal Today* 1998, **40**(4):387–395.

23. Li XQ, Elliott DW, Zhang WX: Zero-valent iron nanoparticles for abatement of environmental pollutants: Materials and engineering aspects. *Crit Rev Solid State* 2006, **31**(4):111–122.

24. Choi CJ, Dong XL, Kim BK: Microstructure and magnetic properties of fe nanoparticles synthesized by chemical vapor condensation. *Mater Trans* 2001, **42**(10):2046–2049.

25. Makela JM, Keskinen H, Forsblom T, Keskinen J: Generation of metal and metal oxide nanoparticles by liquid flame spray process. *J Mater Sci* 2004, **39**(8):2783–2788.

26. Wang CB, Zhang WX: Synthesizing nanoscale iron particles for rapid and complete dechlorination of TCE and PCBs. *Environ Sci Technol* 1997, **31**(7):2154–2156.

27. Elliott DW, Zhang WX: Field assessment of nanoscale biometallic particles for groundwater treatment. *Environ Sci Technol* 2001, **35**(24):4922–4926.

28. Ponder SM, Darab JG, Bucher J, Caulder D, Craig I, Davis L, Edelstein N, Lukens W, Nitsche H, Rao LF *et al*: Surface chemistry and electrochemistry of supported zerovalent iron nanoparticles in the remediation of aqueous metal contaminants. *Chem Mater* 2001, **13**(2):479–486.

29. Xu J, Dozier A, Bhattacharyya D: Synthesis of nanoscale bimetallic particles in polyelectrolyte membrane matrix for reductive transformation of halogenated organic compounds. *J Nanopart Res* 2005, **7**(4–5):449–467.

30. Cushing BL, Kolesnichenko VL, O'Connor CJ: Recent advances in the liquid-phase syntheses of inorganic nanoparticles. *Chem Rev* 2004, **104**(9):3893–3946.

31. Zhang WX: Nanoscale iron particles for environmental remediation: An overview. *J Nanopart Res* 2003, **5**(3–4):323–332.

32. Nurmi JT, Tratnyek PG, Sarathy V, Baer DR, Amonette JE, Pecher K, Wang CM, Linehan JC, Matson DW, Penn RL *et al*: Characterization and properties of metallic iron nanoparticles: Spectroscopy, electrochemistry, and kinetics. *Environ Sci Technol* 2005, **39**(5):1221–1230.

33. Liu YQ, Majetich SA, Tilton RD, Sholl DS, Lowry GV: TCE dechlorination rates, pathways, and efficiency of nanoscale iron particles with different properties. *Environ Sci Technol* 2005, **39**(5):1338–1345.

34. Liu YQ, Lowry GV: Effect of particle age (Fe-o content) and solution pH on NZVI reactivity: H-2 evolution and TCE dechlorination. *Environ Sci Technol* 2006, **40**(19):6085–6090.

35. Sohn I, Fruehan RJ: The reduction of iron oxides by volatiles in a rotary hearth furnace process: Part II. The reduction of iron oxide/carbon composites. *Metall Mater Trans B* 2006, **37**(2):223–229.

36. Hwang YH, Kim DG, Shin HS: Mechanism study of nitrate reduction by nano zero valent iron. *J Hazard Mater* 2011, **185**(2–3):1513–1521.

37. Muftikian R, Fernando Q, Korte N: A method for the rapid dechlorination of low-molecular-weight chlorinated hydrocarbons in water. *Water Res* 1995, **29**(10):2434–2439.

38. Ponder SM, Darab JG, Mallouk TE: Remediation of Cr(VI) and Pb(II) aqueous solutions using supported, nanoscale zero-valent iron. *Environ Sci Technol* 2000, **34**(12):2564–2569.

39. Li XQ, Zhang WX: Sequestration of metal cations with zerovalent iron nanoparticles—A study with high resolution X-ray photoelectron spectroscopy (HR-XPS). *J Phys Chem C* 2007, **111**(19):6939–6946.

40. Cao HS, Zhang WX: Stabilization of chromium ore processing residue (COPR) with nanoscale iron particles. *J Hazard Mater* 2006, **132**(2–3):213–219.

41. Shokes TE, Moller G: Removal of dissolved heavy metals from acid rock drainage using iron metal. *Environ Sci Technol* 1999, **33**(2):282–287.

42. Johnson TL, Scherer MM, Tratnyek PG: Kinetics of halogenated organic compound degradation by iron metal. *Environ Sci Technol* 1996, **30**(8):2634–2640.

43. Lien HL, Zhang WX: Transformation of chlorinated methanes by nanoscale iron particles. *J Environ Eng-ASCE* 1999, **125**(11):1042–1047.

44. Feng J, Lim TT: Pathways and kinetics of carbon tetrachloride and chloroform reductions by nano-scale Fe and Fe/Ni particles: Comparison with commercial micro-scale Fe and Zn. *Chemosphere* 2005, **59**(9):1267–1277.

45. Li F, Vipulanandan C, Mohanty KK: Microemulsion and solution approaches to nanoparticle iron production for degradation of trichloroethylene. *Colloid Surf A* 2003, **223**(1–3):103–112.

46. Lowry GV, Johnson KM: Congener-specific dechlorination of dissolved PCBs by microscale and nanoscale zerovalent iron in a water/methanol solution. *Environ Sci Technol* 2004, **38**(19): 5208–5216.

47. Liou YH, Lo SL, Kuan WH, Lin CJ, Weng SC: Effect of precursor concentration on the characteristics of nanoscale zerovalent iron and its reactivity of nitrate. *Water Res* 2006, **40**(13):2485–2492.

48. Liou YH, Lo SL, Lin CJ: Size effect in reactivity of copper nanoparticles to carbon tetrachloride degradation. *Water Res* 2007, **41**(8):1705–1712.

49. He F, Zhao DY: Preparation and characterization of a new class of starch-stabilized bimetallic nanoparticles for degradation of chlorinated hydrocarbons in water. *Environ Sci Technol* 2005, **39**(9):3314–3320.

50. Lien HL, Zhang WX: Nanoscale iron particles for complete reduction of chlorinated ethenes. *Colloid Surf A* 2001, **191**(1–2):97–105.

51. Deng BL, Burris DR, Campbell TJ: Reduction of vinyl chloride in metallic iron-water systems. *Environ Sci Technol* 1999, **33**(15):2651–2656.

52. Chen JL, Al-Abed SR, Ryan JA, Li ZB: Effects of pH on dechlorination of trichloroethylene by zero-valent iron. *J Hazard Mater* 2001, **83**(3):243–254.

53. Shih YH, Hsu CY, Su YF: Reduction of hexachlorobenzene by nanoscale zero-valent iron: Kinetics, pH effect, and degradation mechanism. *Sep Purif Technol* 2011, **76**(3):268–274.

54. Lin CJ, Lo SL: Effects of iron surface pretreatment on sorption and reduction kinetics of trichloroethylene in a closed batch system. *Water Res* 2005, **39**(6):1037–1046.

55. Lin CJ, Lo SL, Liou YH: Dechlorination of trichloroethylene in aqueous solution by noble metal-modified iron. *J Hazard Mater* 2004, **116**(3):219–228.
56. Gui L, Gillham RW, Odziemkowski MS: Reduction of N-nitrosodimethylamine with granular iron and nickel enhanced iron. 1. Pathways and kinetics. *Environ Sci Technol* 2000, **34**(16): 3489–3494.
57. Lee CC, Doong RA: Dechlorination of tetrachloroethylene in aqueous solutions using metal-modified zerovalent silicon. *Environ Sci Technol* 2008, **42**(13):4752–4757.
58. Schlicker O, Ebert M, Fruth M, Weidner M, Wust W, Dahmke A: Degradation of TCE with iron: The role of competing chromate and nitrate reduction. *Ground Water* 2000, **38**(3):403–409.
59. Dries J, Bastiaens L, Springael D, Agathos SN, Diels L: Combined removal of chlorinated ethenes and heavy metals by zerovalent iron in batch and continuous flow column systems. *Environ Sci Technol* 2005, **39**(21):8460–8465.
60. Lien HL, Jhuo YS, Chen LH: Effect of heavy metals on dechlorination of carbon tetrachloride by iron nanoparticles. *Environ Eng Sci* 2007, **24**(1):21–30.
61. Xie L, Shang C: Chemical reduction of bromate in the presence of humic acid and ferric ions. *Abstr Pap Am Chem S* 2004, **228**:U607–U608.
62. Jeong HY, Hayes KF: Impact of transition metals on reductive dechlorination rate of hexachloroethane by mackinawite. *Environ Sci Technol* 2003, **37**(20):4650–4655.
63. O'Loughlin EJ, Burris DR: Reduction of halogenated ethanes by green rust. *Environ Toxicol Chem* 2004, **23**(1):41–48.
64. O'Loughlin EJ, Kemner KM, Burris DR: Effects of Ag-I, Au-III, and Cu-II on the reductive dechlorination of carbon tetrachloride by green rust. *Environ Sci Technol* 2003, **37**(13):2905–2912.
65. Maithreepala RA, Doong RA: Synergistic effect of copper ion on the reductive dechlorination of carbon tetrachloride by surface-bound Fe(II) associated with goethite. *Environ Sci Technol* 2004, **38**(1):260–268.
66. Maithreepala RA, Doong RA: Enhanced dechlorination of chlorinated methanes and ethenes by chloride green rust in the presence of copper(II). *Environ Sci Technol* 2005, **39**(11):4082–4090.
67. Arnold WA, Roberts AL: Kinetics of chlorinated ethylene reaction with zero-valent iron in column reactors. *Abstr Pap Am Chem S* 2000, **219**:U620–U620.
68. Li T, Farrell J: Electrochemical investigation of the rate-limiting mechanisms for trichloroethylene and carbon tetrachloride reduction at iron surfaces. *Environ Sci Technol* 2001, **35**(17):3560–3565.
69. Weber EJ: Iron-mediated reductive transformations: Investigation of reaction mechanism. *Environ Sci Technol* 1996, **30**(2):716–719.
70. Stumm W, Sulzberger B: The cycling of iron in natural environments—Considerations based on laboratory studies of heterogeneous redox processes. *Geochim Cosmochim Ac* 1992, **56**(8):3233–3257.
71. Clark CJ, Rao PSC, Annable MD: Degradation of perchloroethylene in cosolvent solutions by zero-valent iron. *J Hazard Mater* 2003, **96**(1):65–78.
72. Johnson TL, Fish W, Gorby YA, Tratnyek PG: Degradation of carbon tetrachloride by iron metal: Complexation effects on the oxide surface. *J Contam Hydrol* 1998, **29**(4):379–398.
73. Loraine GA: Effects of alcohols, anionic and nonionic surfactants on the reduction of PCE and TCE by zero-valent iron. *Water Res* 2001, **35**(6):1453–1460.
74. Alessi DS, Li ZH: Synergistic effect of cationic surfactants on perchloroethylene degradation by zero-valent iron. *Environ Sci Technol* 2001, **35**(18):3713–3717.
75. Tratnyek PG, Scherer MM, Deng BL, Hu SD: Effects of natural organic matter, anthropogenic surfactants, and model quinones on the reduction of contaminants by zero-valent iron. *Water Res* 2001, **35**(18):4435–4443.
76. Cho HH, Park JW: Reactive dechlorination of PCE using zero valent iron plus surfactants. *ACS Sym Ser* 2003, **837**:141–153.
77. Muftikian R, Nebesny K, Fernando Q, Korte N: X-ray photoelectron spectra of the palladium-iron bimetallic surface used for the rapid dechlorination of chlorinated organic environmental contaminants. *Environ Sci Technol* 1996, **30**(12):3593–3596.

78. Cho HH, Park JW: Effect of coexisting compounds on the sorption and reduction of trichloro-ethylene with iron. *Environ Toxicol Chem* 2005, **24**(1):11–16.
79. Xie L, Shang C: Role of humic acid and quinone model compounds in bromate reduction by zerovalent iron. *Environ Sci Technol* 2005, **39**(4):1092–1100.
80. Doong RA, Lai YJ: Dechlorination of tetrachloroethylene by palladized iron in the presence of humic acid. *Water Res* 2005, **39**(11):2309–2318.
81. Nutt MO, Hughes JB, Wong MS: Designing Pd-on-Au bimetallic nanoparticle catalysts for tri-chloroethene hydrodechlorination. *Environ Sci Technol* 2005, **39**(5):1346–1353.
82. Tee YH, Grulke E, Bhattacharyya D: Role of Ni/Fe nanoparticle composition on the degrada-tion of trichloroethylene from water. *Ind Eng Chem Res* 2005, **44**(18):7062–7070.
83. Lee CC, Doong RA: Concentration effect of copper loading on the reductive dechlorination of tetrachloroethylene by zerovalent silicon. *Water Sci Technol* 2010, **62**(1):28–35.
84. Lin CJ, Liou YH, Lo SL: Supported Pd/Sn bimetallic nanoparticles for reductive dechlorination of aqueous trichloroethylene. *Chemosphere* 2009, **74**(2):314–319.
85. Schrick B, Blough JL, Jones AD, Mallouk TE: Hydrodechlorination of trichloroethylene to hydrocarbons using bimetallic nickel-iron nanoparticles. *Chem Mater* 2002, **14**(12):5140–5147.
86. Cwiertny DM, Bransfield SJ, Livi KJT, Fairbrother DH, Roberts AL: Exploring the influence of granular iron additives on 1,1,1-trichloroethane reduction. *Environ Sci Technol* 2006, **40**(21): 6837–6843.
87. Xu WY, Gao TY, Fan JH: Reduction of nitrobenzene by the catalyzed Fe-Cu process. *J Hazard Mater* 2005, **123**(1–3):232–241.
88. Kim YH, Carraway ER: Dechlorination of pentachlorophenol by zero valent iron and modified zero valent irons. *Environ Sci Technol* 2000, **34**(10):2014–2017.
89. Liou YH, Lo SL, Lin CJ, Hu CY, Kuan WH, Weng SC: Methods for accelerating nitrate reduc-tion using zerovalent iron at near-neutral pH: Effects of H-2-reducing pretreatment and copper deposition. *Environ Sci Technol* 2005, **39**(24):9643–9648.
90. Lien HL, Zhang WX: Nanoscale Pd/Fe bimetallic particles: Catalytic effects of palladium on hydrodechlorination. *Appl Catal B-Environ* 2007, **77**(1–2):110–116.
91. Shih YH, Chen MY, Su YF: Pentachlorophenol reduction by Pd/Fe bimetallic nanoparticles: Effects of copper, nickel, and ferric cations. *Appl Catal B-Environ* 2011, **105**(1–2):24–29.
92. Cwiertny DM, Bransfield SJ, Roberts AL: Influence of the oxidizing species on the reactivity of iron-based bimetallic reductants. *Environ Sci Technol* 2007, **41**(10):3734–3740.
93. Parshetti GK, Doong RA: Synergistic effect of nickel ions on the coupled dechlorination of tri-chloroethylene and 2,4-dichlorophenol by Fe/TiO2 nanocomposites in the presence of UV light under anoxic conditions. *Water Res* 2011, **45**(14):4198–4210.
94. Xu J, Bhattacharyya D: Membrane-based bimetallic nanoparticles for environmental remedia-tion: Synthesis and reactive properties. *Environ Prog* 2005, **24**(4):358–366.
95. Bransfield SJ, Cwiertny DM, Roberts AL, Fairbrother DH: Influence of copper loading and sur-face coverage on the reactivity of granular iron toward 1,1,1-trichloroethane. *Environ Sci Technol* 2006, **40**(5):1485–1490.
96. Arnold WA, Roberts AL: Pathways and kinetics of chlorinated ethylene and chlorinated acety-lene reaction with Fe(O) particles. *Environ Sci Technol* 2000, **34**(9):1794–1805.
97. Doong RA, Wu SC: Reductive dechlorination of chlorinated hydrocarbons in aqueous-solutions containing ferrous and sulfide ions. *Chemosphere* 1992, **24**(8):1063–1075.
98. Kim S, Picardal FW: Enhanced anaerobic biotransformation of carbon tetrachloride in the pres-ence of reduced iron oxides. *Environ Toxicol Chem* 1999, **18**(10):2142–2150.
99. Amonette JE, Workman DJ, Kennedy DW, Fruchter JS, Gorby YA: Dechlorination of carbon tetrachloride by Fe(II) associated with goethite. *Environ Sci Technol* 2000, **34**(21):4606–4613.
100. Pecher K, Haderlein SB, Schwarzenbach RP: Reduction of polyhalogenated methanes by surface-bound Fe(II) in aqueous suspensions of iron oxides. *Environ Sci Technol* 2002, **36**(8):1734–1741.
101. Shih YH, Chen YC, Chen MY, Tai YT, Tso CP: Dechlorination of hexachlorobenzene by using nanoscale Fe and nanoscale Pd/Fe bimetallic particles. *Colloid Surf A* 2009, **332**(2–3):84–89.

102. Wust WF, Kober R, Schlicker O, Dahmke A: Combined zero- and first-order kinetic model of the degradation of TCE and cis-DCE with commercial iron. *Environ Sci Technol* 1999, **33**(23): 4304–4309.

103. Scherer MM, Balko BA, Gallagher DA, Tratnyek PG: Correlation analysis of rate constants for dechlorination by zero-valent iron. *Environ Sci Technol* 1998, **32**(19):3026–3033.

104. Liu YQ, Choi H, Dionysiou D, Lowry GV: Trichloroethene hydrodechlorination in water by highly disordered monometallic nanoiron. *Chem Mater* 2005, **17**(21):5315–5322.

105. Wang JK, Farrell J: Determining the atomic hydrogen surface coverage on iron and nickel electrodes under water treatment conditions. *J Appl Electrochem* 2006, **36**(3):369–374.

106. Cheng SF, Wu SC: The enhancement methods for the degradation of TCE by zero-valent metals. *Chemosphere* 2000, **41**(8):1263–1270.

107. Liou YH, Lo SL, Lin CJ, Kuan WH, Weng SC: Effects of iron surface pretreatment on kinetics of aqueous nitrate reduction. *J Hazard Mater* 2005, **126**(1–3):189–194.

108. Huang YH, Zhang TC: Reduction of nitrobenzene and formation of corrosion coatings in zerovalent iron systems. *Water Res* 2006, **40**(16):3075–3082.

109. Zhang TC, Huang YH: Effects of surface-bound Fe2+ on nitrate reduction and transformation of iron oxide(s) in zero-valent iron systems at near-neutral pH. *J Environ Eng-ASCE* 2006, **132**(5):527–536.

110. Liao CH, Kang SF, Hsu YW: Zero-valent iron reduction of nitrate in the presence of ultraviolet light, organic matter and hydrogen peroxide. *Water Res* 2003, **37**(17):4109–4118.

111. Huang CP, Wang HW, Chiu PC: Nitrate reduction by metallic iron. *Water Res* 1998, **32**(8):2257–2264.

112. Huang YH, Zhang TC, Shea PJ, Comfort SD: Effects of oxide coating and selected cations on nitrate reduction by iron metal. *J Environ Qual* 2003, **32**(4):1306–1315.

113. Zhang TC, Huang YH: Profiling iron corrosion coating on iron grains in a zerovalent iron system under the influence of dissolved oxygen. *Water Res* 2006, **40**(12):2311–2320.

114. Li CW, Chen YM, Yen WS: Pressurized CO2/zero valent iron system for nitrate removal. *Chemosphere* 2007, **68**(2):310–316.

115. Lin CJ, Lo SL, Liou YH: Degradation of aqueous carbon tetrachloride by nanoscale zerovalent copper on a cation resin. *Chemosphere* 2005, **59**(9):1299–1307.

116. Liu CC, Tseng DH, Wang CY: Effects of ferrous ions on the reductive dechlorination of trichloroethylene by zero-valent iron. *J Hazard Mater* 2006, **136**(3):706–713.

117. Huang YH, Zhang TC: Nitrite reduction and formation of corrosion coatings in zerovalent iron systems. *Chemosphere* 2006, **64**(6):937–943.

118. Liou YH, Lo SL, Lin CJ, Kuan WH, Weng SC: Chemical reduction of an unbuffered nitrate solution using catalyzed and uncatalyzed nanoscale iron particles. *J Hazard Mater* 2005, **127**(1–3):102–110.

119. Liou YH, Lin CJ, Weng SC, Ou HH, Lo SL: Selective decomposition of aqueous nitrate into nitrogen using iron deposited bimetals. *Environ Sci Technol* 2009, **43**(7):2482–2488.

120. Choe S, Chang YY, Hwang KY, Khim J: Kinetics of reductive denitrification by nanoscale zerovalent iron. *Chemosphere* 2000, **41**(8):1307–1311.

121. Chen SS, Hsu HD, Li CW: A new method to produce nanoscale iron for nitrate removal. *J Nanopart Res* 2004, **6**(6):639–647.

122. Xie L, Shang C: Effects of copper and palladium on the red-action of bromate by Fe(0). *Chemosphere* 2006, **64**(6):919–930.

123. Okeke BC, Giblin T, Frankenberger WT: Reduction of perchlorate and nitrate by salt tolerant bacteria. *Environ Pollut* 2002, **118**(3):357–363.

124. Coates JD, Michaelidou U, Bruce RA, O'Connor SM, Crespi JN, Achenbach LA: Ubiquity and diversity of dissimilatory (per)chlorate-reducing bacteria. *Appl Environ Microb* 1999, **65**(12):5234–5241.

125. Moore AM, De Leon CH, Young TM: Rate and extent of aqueous perchlorate removal by iron surfaces. *Environ Sci Technol* 2003, **37**(14):3189–3198.

126. Cao JS, Elliott D, Zhang WX: Perchlorate reduction by nanoscale iron particles. *J Nanopart Res* 2005, **7**(4–5):499–506.
127. Schaefer CE, Fuller ME, Condee CW, Lowey JM, Hatzinger PB: Comparison of biotic and abiotic treatment approaches for co-mingled perchlorate, nitrate, and nitramine explosives in groundwater. *J Contam Hydrol* 2007, **89**(3–4):231–250.
128. Huang H, Sorial GA: Perchlorate remediation in aquatic systems by zero valent iron. *Environ Eng Sci* 2007, **24**(7):917–926.
129. Yu XY, Amrhein C, Deshusses MA, Matsumoto MR: Perchlorate reduction by autotrophic bacteria in the presence of zero-valent iron. *Environ Sci Technol* 2006, **40**(4):1328–1334.
130. Yu XY, Amrhein C, Deshusses MA, Matsumoto MR: Perchlorate reduction by autotrophic bacteria attached to zerovalent iron in a flow-through reactor. *Environ Sci Technol* 2007, **41**(3):990–997.
131. Taube H: Observations on atom-transfer reactions. *ACS Sym Ser* 1982, **198**:151–179.
132. Blowes DW, Ptacek CJ, Benner SG, McRae CWT, Bennett TA, Puls RW: Treatment of inorganic contaminants using permeable reactive barriers. *J Contam Hydrol* 2000, **45**(1–2):123–137.
133. Kanel SR, Greneche JM, Choi H: Arsenic(V) removal kom groundwater using nano scale zerovalent iron as a colloidal reactive barrier material. *Environ Sci Technol* 2006, **40**(6):2045–2050.
134. Kumpiene J, Montesinos IC, Lagerkvist A, Maurice C: Evaluation of the critical factors controlling stability of chromium, copper, arsenic and zinc in iron-treated soil. *Chemosphere* 2007, **67**(2):410–417.
135. Manning BA, Hunt ML, Amrhein C, Yarmoff JA: Arsenic(III) and Arsenic(V) reactions with zerovalent iron corrosion products. *Environ Sci Technol* 2002, **36**(24):5455–5461.
136. Jegadeesan G, Mondal K, Lalvani SB: Arsenate remediation using nanosized modified zerovalent iron particles. *Environ Prog* 2005, **24**(3):289–296.
137. Zhang YQ, Wang JF, Amrhein C, Frankenberger WT: Removal of selenate from water by zerovalent iron. *J Environ Qual* 2005, **34**(2):487–495.
138. Waybrant KR, Blowes DW, Ptacek CJ: Selection of reactive mixtures for use in permeable reactive walls for treatment of mine drainage. *Environ Sci Technol* 1998, **32**(13):1972–1979.
139. Miehr R, Tratnyek PG, Bandstra JZ, Scherer MM, Alowitz MJ, Bylaska EJ: Diversity of contaminant reduction reactions by zerovalent iron: Role of the reductate. *Environ Sci Technol* 2004, **38**(1):139–147.
140. Li W, Wu CZ, Zhang SH, Shao K, Shi Y: Evaluation of microbial reduction of Fe(III) EDTA in a chemical absorption-biological reduction integrated NOx removal system. *Environ Sci Technol* 2007, **41**(2):639–644.
141. Tufenkji N, Elimelech M: Deviation from the classical colloid filtration theory in the presence of repulsive DLVO interactions. *Langmuir* 2004, **20**(25):10818–10828.
142. Schrick B, Hydutsky BW, Blough JL, Mallouk TE: Delivery vehicles for zerovalent metal nanoparticles in soil and groundwater. *Chem Mater* 2004, **16**(11):2187–2193.
143. Phenrat T, Saleh N, Sirk K, Tilton RD, Lowry GV: Aggregation and sedimentation of aqueous nanoscale zerovalent iron dispersions. *Environ Sci Technol* 2007, **41**(1):284–290.
144. Lin YH, Tseng HH, Wey MY, Lin MD: Characteristics, morphology, and stabilization mechanism of PAA250K-stabilized bimetal nanoparticles. *Colloid Surf A* 2009, **349**(1–3):137–144.
145. Sun YP, Li XQ, Zhang WX, Wang HP: A method for the preparation of stable dispersion of zerovalent iron nanoparticles. *Colloid Surf A* 2007, **308**(1–3):60–66.
146. He F, Zhao DY, Liu JC, Roberts CB: Stabilization of Fe-Pd nanoparticles with sodium carboxymethyl cellulose for enhanced transport and dechlorination of trichloroethylene in soil and groundwater. *Ind Eng Chem Res* 2007, **46**(1):29–34.
147. Saleh N, Sirk K, Liu YQ, Phenrat T, Dufour B, Matyjaszewski K, Tilton RD, Lowry GV: Surface modifications enhance nanoiron transport and NAPL targeting in saturated porous media. *Environ Eng Sci* 2007, **24**(1):45–57.
148. Parshetti GK, Doong RA: Dechlorination of trichloroethylene by Ni/Fe nanoparticles immobilized in PEG/PVDF and PEG/nylon 66 membranes. *Water Res* 2009, **43**(12):3086–3094.
149. Karn B, Kuiken T, Otto M: Nanotechnology and in situ remediation: A review of the benefits and potential risks. *Environ Health Persp* 2009, **117**(12):1823–1831.

150. Yan WL, Lien HL, Koel BE, Zhang WX: Iron nanoparticles for environmental clean-up: Recent developments and future outlook. *Environ Sci-Proc Imp* 2013, **15**(1):63–77.
151. Available at http://www.nanotechproject.org/inventories/remediation_map/ NM.
152. Bardos P, Elliott D, Hartog N, Henstock J, Nathanail, P: A risk/benefit approach to the application of iron nanoparticles for the remediation of contaminated sites in the environment. *Defra Research Project Final Report*, 2011.
153. Mueller NC, Braun J, Bruns J, Cernik M, Rissing P, Rickerby D, Nowack B: Application of nanoscale zero valent iron (NZVI) for groundwater remediation in Europe. *Environ Sci Pollut R* 2012, **19**(2):550–558.
154. Wei YT, Wu SC, Chou CM, Che CH, Tsai SM, Lien HL: Influence of nanoscale zero-valent iron on geochemical properties of groundwater and vinyl chloride degradation: A field case study. *Water Res* 2010, **44**(1):131–140.
155. Wei YT, Wu SC, Yang SW, Che CH, Lien HL, Huang DH: Biodegradable surfactant stabilized nanoscale zero-valent iron for in situ treatment of vinyl chloride and 1,2-dichloroethane. *J Hazard Mater* 2012, **211**:373–380.
156. He F, Zhao DY, Paul C: Field assessment of carboxymethyl cellulose stabilized iron nanoparticles for in situ destruction of chlorinated solvents in source zones. *Water Res* 2010, **44**(7):2360–2370.
157. Henn KW, Waddill DW: Utilization of nanoscale zero-valent iron for source remediation—A case study. *Remed J* 2006, 57–76.
158. Petosa AR, Jaisi DP, Quevedo IR, Elimelech M, Tufenkji N: Aggregation and deposition of engineered nanomaterials in aquatic environments: Role of physicochemical interactions. *Environ Sci Technol* 2010, **44**(17):6532–6549.
159. Jiemvarangkul P, Zhang WX, Lien HL: Enhanced transport of polyelectrolyte stabilized nanoscale zero-valent iron (nZVI) in porous media. *Chem Eng J* 2011, **170**(2–3):482–491.
160. He F, Zhao DY: Manipulating the size and dispersibility of zerovalent iron nanoparticles by use of carboxymethyl cellulose stabilizers. *Environ Sci Technol* 2007, **41**(17):6216–6221.
161. Lowry GV, Gregory KB, Apte SC, Lead JR: Transformations of nanomaterials in the environment. *Environ Sci Technol* 2012, **46**(13):6893–6899.
162. Chao CC, Sun YP: Taiwan team's proposal for collaborative research. *Presentation at EU FP7 NANOREM Collaborative Project Kick-Off Meeting*, 12 April 2013, Stuttgart.

5

Nanotoxicity: Aquatic Organisms and Ecosystems

Ashutosh Kumar,[1] **Rishi Shanker,**[1] **and Alok Dhawan**[1,2]

[1]*Institute of Life Sciences, School of Science and Technology, Ahmedabad University, Navrangpura, Ahmedabad, Gujarat, India*

[2]*Nanomaterial Toxicology Group, CSIR–Indian Institute of Toxicology Research, Mahatma Gandhi Marg, Lucknow, Uttar Pradesh, India*

CONTENTS

Nanoscience and nanotechnology have seen an exponential growth over the past decade. This is largely due to the advances in nanomaterial synthesis and imaging, or analysis tools owing to funding by national and international agencies to pursue research and innovation in this emerging area. Engineered nanoparticles (ENPs) are defined as any intentionally produced particle that (i) has a characteristic dimension between 1 and 100 nm, and (ii) possesses properties that are not shared by non-nanoscale particles with the same chemical composition [1]. The latter part of this definition arises from the fact that ENPs possess unique physicochemical properties because of their small size. Properties such as high surface area-to-volume ratio, abundant reactive sites on the surface, and a large fraction of atoms located on the exterior face, have made these novel materials the most sought after materials for consumer and industrial applications. The unique characteristics of ENPs derived from carbon (tensile strength of carbon nanotubes), metals (fluorescence properties of quantum dots; antimicrobial activity of silver nanoparticles), and metal oxides (photocatalytic properties of TiO_2) have proven to be the most commercially profitable [2,3].

Naturally occurring and incidentally generated nanoparticles are heterogeneous, whereas the ENP suspensions or powders are usually homogeneous in size, shape, and structure [4]. ENPs have found application in diverse sectors such as energy, electronics, food and agriculture, water purification, biomedical devices, imaging, biosensing and biochips, high-density data, detecting DNA sequence, environmental cleanup, household products, paints, consumer products, and sports [5]. According to the US National Nanotechnology Initiative (NNI), million ton quantities of ENPs (silica, alumina and ceria, ZnO, TiO_2, silver, carbon nanotubes, etc.) are being manufactured for various consumer products [6]. This significant increase in the ENP-containing consumer products has increased the chances of their inadvertent release in the surface and subsurface environment through landfills and other waste-disposal methods [7]. The "cradle to gate and grave" paradigm is also relevant to the release of ENPs into the environment at any stage of their life cycle, viz. production, transportation, consumer usage, and disposal.

It is also known that ENPs have excess energy at the surface, which makes particles highly reactive, mobile, thermodynamically unstable, and thus a very special class of pollutant [8]. It is also likely that some of the ENPs released into the environment induce a toxic response both in lower and higher trophic organisms of the aquatic ecosystem. Evidence is accumulating that ENPs cause toxicity to microbes, invertebrates, fishes, lower vertebrates, and others, the key components of the aquatic ecosystem. Studies have shown that ENPs adversely affect the microbes (*Escherichia coli*, *Pseudomonas aeruginosa*, and *Streptococcus aureus*) that are responsible for the maintenance of environmental health [9–13]. These observations indicate the possibility that the release of ENPs may be detrimental to biogeochemical processes in soil, such as carbon or nitrogen cycling. It is also likely that the ENPs can directly interact with the food web at different trophic levels and disturb the ecological balance. Therefore, organisms, especially those that interact strongly with their immediate environment, could be more at risk to the direct exposure of ENPs [14]. The biomagnification of ENPs across the genera is also a major concern.

Moreover, the impact of ENPs on aquatic ecosystems is governed by their distribution, which depends on various factors such as Brownian motion, inertia, gravitational influences, thermal influences, pH, and ionization [15]. The possibility of the high mobility of ENPs in water can lead to contamination of the flora and fauna. This may also result in the transfer of ENPs in the food chain, leading to the generation of nonbiodegradable pollutants [16]. Also, ENPs can affect the bioavailability of the other toxicant/pollutant by facilitating their transport [8,17]. ENPs thus elicit a negative impact on the physical, chemical, and biological strata of the aquatic ecosystem. Hence, to minimize the exposure in ENPs and thereby adverse effects on the aquatic ecosystem, it is imperative to consider the following issues: (i) bioavailability of ENPs in aquatic systems, (ii) methodological and metrological approaches for the detection and quantification of ENPs in environmental samples, and (iii) approaches and knowledge gaps in aquatic toxicity.

5.1 Bioavailability of ENPs in Aquatic Systems

The availability and uptake of ENPs to the cell is an important factor that can provide relevant information pertaining to their adverse effects on cellular systems. There have been a number of recent reports that have addressed the fate and effects of ENPs in the environment [3,18–21]. Many of these reports specifically refer to the potential impacts of ENPs on aquatic environments, since surface waters receive pollutants and contaminants, including ENPs from many sources, and act as reservoirs and channels for many environmental contaminants [3]. Understanding the potential impacts and toxicity of ENPs on the organisms within these environments is critical. At present, there is little information about the number of ENPs that enter aquatic environments and their routes of entry. The potential routes include atmospheric deposition, leaching from soil, direct input from wastewater discharges, and groundwater reservoirs [18]. Evidence for possible contamination of water sources with nanoparticles has already been reported by Mueller and Nowack [22]. They developed a model and found that the levels of silver, TiO_2 ENPs, and carbon nanotubes in freshwater were 0.3, 0.7, and 0.0005 µg/L, respectively. Kaegi et al. [23] also reported the release of TiO_2 ENPs from facade paints by natural weathering. A concentration of 13.4 µg/L TiO_2 was detected in runoff waters, and a part of this comprised TiO_2 nanoparticles. On the basis of the modeling study of ENP emissions in Switzerland, the predicted environmental concentration of TiO_2 ENPs is 16 µg/L in surface water [22]. There are a number of studies dealing with the aggregation and deposition of ENPs [24–26]. These studies have been conducted nearly without exception under controlled test conditions by adjusting the pH, ionic strength, and composition of monovalent and divalent ions, as well as the concentration of dissolved organic matter.

Consequently, there is concern about the behavior and effects of the released ENPs in the aquatic systems. This will largely depend on the hydrodynamic behavior, association with larger sediment and natural colloidal particulates, binding with organic and metal pollutants, exposure and uptake into biota, implications of ENP exposure for organism health, and ecosystem integrity [27]. The dispersion and bioavailability of the ENPs could be affected by their interaction with aquatic colloids, such as natural organic matter (NOM), humic substances, and salt ions. NOMs are generally adsorbed on the surface of the ENPs through hydrogen bonding, electrostatic interactions, and hydrophobic interactions [8]. NOMs are classified into three major classes: (i) rigid biopolymers, such as polysaccharides and peptidoglycans from phytoplankton or bacteria; (ii) fulvic compounds, mostly from terrestrial sources, originating from the decomposition products of plants; and (iii) flexible biopolymers, composed of aquagenic refractory organic matter from a recombination of microbial degradation products [28]. ENPs in aqueous ecosystems are dispersed because of the electrostatic and steric repulsion of surface charge (positive/negative) present on the particles. As the surface charges of the particle skew toward the zero value, the repulsive forces between the particles are reduced and they ultimately settle down. Because of agglomeration/aggregation, the physicochemical properties such as surface charge, size, size distribution, surface-to-volume ratio, and surface reactivity of ENPs become altered, thereby affecting their bioavailability and cellular responses. Fulvic compounds and flexible biopolymers have a tendency to modify the ENPs' surface charge, which leads to the aggregation and nonbioavailability of the particle. It has been demonstrated that the humic acid coating of hematite reversed their charge from positive to negative, leading to decreased attachment efficiencies from 1 to 0.01 mg/L to a sandy soil [29]. This leads to the increased bioavailability and decreased agglomeration of the hematite. However, the rigid biopolymers, such as polysaccharides and peptidoglycans produced by phytoplankton and bacteria, coat the ENPs and increase their mobility and bioavailability to the cell [30]. Apart from the NOMs, several other factors can also influence the aggregation and bioavailability of the ENPs, e.g., salt ions, presence of hydrophobic surfactant or polar groups on the surface of ENPs, and others. Also, the biomolecules such as proteins or polymers present in the ecosystem form a layer over the ENPs, named as "corona," which plays an important role in their biological fate. It is now well understood that corona governs the properties of the "particle-plus-corona" compound in the biological system [31,32].

Apparently, the formation of larger aggregates by high molecular weight NOM compounds favors the removal of ENPs from the sediments and is likely to decrease their bioavailability. However, solubilization by natural surfactants such as lower molecular weight NOM compounds will tend to increase the mobility and bioavailability of ENPs. Furthermore, it is now clear that ENPs can serve as transfer vectors for the pollutants in the environment. The ENPs can control the bioavailability of compounds to the cells

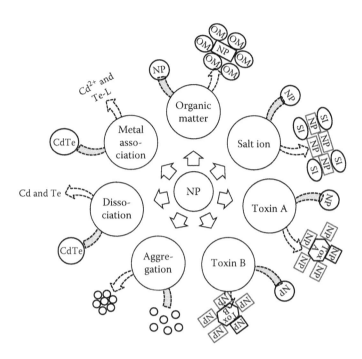

FIGURE 5.1
Factors affecting the behavior of nanoparticles and their bioavailability in aquatic ecosystems. Organic matter (OM), salt ions (SI), toxicants (ToxA and ToxB), nanoparticles (NP), cadmium telluride (CdTe), cadmium (Cd), and tellurium (Te).

depending on the properties of the pollutant. A schematic about the fate of ENPs in aquatic system and their increased/decreased bioavailability is represented in Figure 5.1.

ENPs have also been demonstrated to adsorb other pollutants on their surface owing to their high surface area-to-volume ratio and complex-forming ability. Baun et al. [33] showed in *Daphnia magna* that the toxicity of phenanthrene in the presence of C60 aggregates was significantly higher (60%) than the parent compound owing to increased bioavailability. Similarly, the bioavailability of phenanthrene to plant roots also increased upon its adsorption to alumina nanoparticles. In contrast, Knauer et al. (2007) have reported that the presence of carbon black ENPs reduced the toxicity of diuron to green algae [34]. This could be due to the adsorption of diuron on carbon black, thereby reducing its bioavailability. Also, fullerenes were found to decrease the toxicity of various chemicals to algae as a result of their decreased bioavailability [35].

5.2 Methodological and Metrological Approaches for the Detection and Quantification of ENPs in Environmental Samples

The detection of ENPs' internalization in any aquatic organism is a crucial step for understanding their behavior and toxicity. Lack of methodological and metrological approaches for the detection of ENPs in environmental samples is one of the major hurdles in mitigating their adverse health impact. The first uncertainty is in the representation of the concentration of the ENPs. Different metrics such as mass (μg/mL), number (10^n particles/mL), or surface area (μg/cm^3) have been used to express the concentration of the ENPs, which makes it difficult to correlate the existing data [36].

At present, the commonly used methods for the detection of uptake of ENPs in the cells are transmission electron microscopy (TEM), scanning electron microscopy along with backscattered electron and energy-dispersive x-ray spectroscopy (SEM + BSE + EDS), confocal and fluorescence microscopy, reflection-based imaging, and flow cytometry [37,38]. These techniques enable the tracking of ENPs in the cells as well as cellular organelles. The high resolution of TEM enables the imaging of membrane invagination, mode of ENPs' uptake, and ultrastructural changes occurring in the cells subsequent to ENP treatment. SEM, on the other hand, is used to study the morphological changes and ENP interactions with the cell. On the other hand, EDS coupled with SEM provides an additional feature to analyze the elemental composition of the specimen based on the released energy by the corresponding element [38,39]. Although these imaging techniques provide several advantages, there are certain drawbacks; for example, in TEM and SEM, the samples have to be fixed; therefore, live cell uptake cannot be monitored. It is also resource intensive, time consuming, and confined to imaging of a few cells. Furthermore, the staining process introduces electron-dense artifacts that may be construed as nanoparticles. Confocal and fluorescence microscopy, on the other hand, require that the particles be tagged with a probe or be doped with a fluorescence dye for their detection. Since the native nature of ENPs is lost, it can lead to false/incorrect interpretation of observations [39].

Flow cytometry is another technique used to assess the uptake of ENPs in the cells. This is a rapid, multiparametric, single-cell analysis with robust statistics, owing to the large number of events measured per treatment, and is cost-effective [40]. In this method, a laser beam strikes on the stream of fluid containing single-cell suspension. The light diffracted, reflected, and refracted by the cells is recorded by the photomultiplier tubes and the electronics convert these optical pulses to digital values. It is well established that the light diffracted by the cells represents the forward light scatter and is used to measure the cellular size. However, the reflected and refracted light corresponds to the side scatter, which is a combined effect of the granularity and the cellular mass of the cell. ENPs in the host cell serve as granules and reflect/refract the light based on their intrinsic property. As the ENPs enter into the cells, the side-scatter intensity of the cell increases proportionately to the concentration of the ENPs [37,41]. A fluorescent particle can give an increased signal of side scatter as well as the fluorochrome intensity in a dose-dependent manner. Most aerobic microbes divide quickly under optimal growth conditions (viz. *E. coli* in 20 min); thus, the internalization of the ENPs in the cells and their retention for several generations can easily be monitored in a short time using flow cytometry. Several studies have demonstrated the internalization of the ENPs in cell lines using flow cytometry [37,41–43]. Using the aforementioned techniques, it has been shown that ENPs can be internalized in a number of aquatic organisms (Figure 5.2).

Despite having several techniques to track the ENP uptake in different aquatic organisms, the precise mechanism of uptake is still unknown. However, it has been proposed that nonspecific diffusion, nonspecific membrane damage and specific uptake (through porins), phagocytosis, and membrane ion transporters are the possible mechanisms through which the ENPs could pass through the cell wall and membranes of different aquatic organisms [8,44].

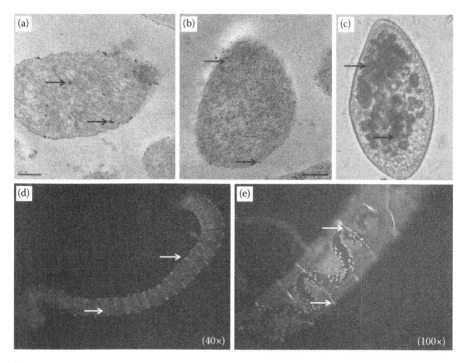

FIGURE 5.2
Internalization of the nanoparticles in different aquatic organisms: (a, b) *Escherichia coli*, (c) *Paramecium caudatum*, and (d, e) *Tubifex tubifex*.

5.3 Approaches and Knowledge Gaps in Aquatic Toxicity Studies

The frequent release and interaction of ENPs with different components of the aquatic ecosystem necessitates the development of certain strategies to test their possible hazards. It can be inferred from the current understanding of fate, behavior, and detection of different ENPs in the ecosystem that interactions of cells with ENPs are dependent on size, shape, chemical composition, surface charge, surface structure, area, solubility, and agglomeration. Hence, it is essential to study these physiochemical properties of ENPs, both in dry as well as in suspension for hazard assessment [36].

Among these physiochemical characteristics, the surface properties of the ENPs are the most important factor that governs the stability and mobility of ENPs in aqueous suspension [36]. The agglomeration tendency of the ENPs is determined by the properties such as surface area, functionalization, temperature, ionic strength, pH, concentration, size, and the solvent [45]. However, it is difficult to measure the surface properties of ENPs at the nanogram to picogram range because of the limitations of the commercially available analytical instruments. On the other hand, the concentration of ENPs in suspension is also a crucial step in designing the experiments. Since the ENPs have a tendency to agglomerate/aggregate, it results in a change in their physicochemical properties, and hence the available cellular concentration [46]. Thus, the experimental design should also consider the concentration-induced aggregation effects of the ENPs.

It may also be inferred that at lower concentrations, ENPs will tend to show less aggregation that may lead to greater uptake and response than expected at higher concentrations. However, different ways of stabilizing the ENPs (particle coating, dispersant/surfactant, sonication) mitigate agglomeration, thereby resulting in an exacerbated biological response. The stability of the ENPs inside the cell or aquatic environment as well as their interaction with cellular metabolites on the ENPs are the other key issues that need to be addressed to understand the adverse effects of ENPs. Other possible effects of ENP uptake could be the interaction with other possibly toxic substances and their mobilization and bioavailability.

The environmental fate, behavior, and bioavailability of ENPs are not well understood; therefore, their persistence and the possible interaction and impact, particularly biomagnification in food webs, is an immediate concern [7]. Hence, to evaluate and assess the effect of ENPs in an aquatic ecosystem, it is proposed that the study design should include the interaction of the ENPs at the subcellular, cellular, and organism level (Figure 5.3). A multipronged approach involving ENP characterization, uptake, biochemical studies, histopathology, genomics, proteomics, and metabolomics should be conducted to assess the impact of ENPs at subcellular and cellular levels (Figure 5.3). However, for understanding the effects at the organism level, the studies should be supplemented with life cycle and reproduction studies.

The impurities in the ENP preparation also influence the toxic outcomes; therefore, the effects of these impurities should also be considered in the study design. Elemental analysis using different techniques could help in analyzing these impurities [47]. Some of the metal oxide nanoparticles are known to release ions in the aqueous suspension, which could alter the toxicity. Hence, the quantification of soluble metal ions in the exposure medium is also a prerequisite in nanotoxicology studies [15,35,48]. Lack of reference materials, appropriate methods to monitor ENP behavior in various matrices, dose dilemma,

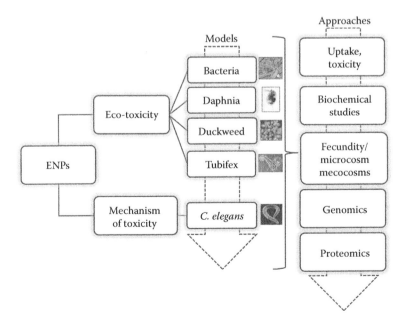

FIGURE 5.3
Aquatic toxicity of engineered nanoparticles: approaches and models.

exposure methods, and regulatory toxicology test methods are certain other hurdles that need to be addressed. Therefore, before the use of ENP-based consumer products in daily life, it is important to understand their fate in the environment to avoid imbalances in the aquatic ecosystem.

5.4 Conclusion

In summary, to unravel the potential risks of ENPs in aquatic ecosystems, it is prudent to undertake in-depth studies including environmental transport, trophic transfer, bio-concentration, and magnification and trans-strata mobility. Multidisciplinary approaches, such as particle characterization, uptake, computational modeling, and ecotoxicity models, among others, will be helpful in improving knowledge for improved study designs, for risk modeling and assessment. The lack of data on concentration and quantitative exposure of ENPs currently hampers the prediction of the environmental fate. However, our current understanding of the fate of industrial chemicals, predictive modeling, and aquatic models can provide useful insights to study design and predictive tools for rapid risk assessment of ENPs in the aquatic environment.

Acknowledgments

Funding received from the Council of Scientific and Industrial Research, New Delhi (NanoSHE; BSC-0112); the UK India Education and Research Initiative (UKIERI) standard award to Institute of Life Sciences, Ahmedabad University, India (grant no. IND/CONT/E/11-12/217); and from the Department of Biotechnology, Government of India, under the NewINDIGO Scheme for NanoLINEN project is gratefully acknowledged. Funding from the European Union Seventh Framework Programme (FP7/2007-2013) under grant agreement no. 263147 (NanoValid—Development of reference methods for hazard identification, risk assessment, and LCA of engineered nanomaterials) is also acknowledged. The financial assistance for the Centre for Nanotechnology Research and Applications (CENTRA) by the Gujarat Institute for Chemical Technology (GICT) is also acknowledged.

References

1. Bernhardt ES, Colman BP, Hochella MF, Jr., Cardinale BJ, Nisbet RM, Richardson CJ and Yin L. (2010). An ecological perspective on nanomaterial impacts in the environment. *J Environ Qual* 39: 1954–1965.
2. Wiesner MR, Lowry GV, Jones KL, Hochella MF, Jr., Digiulio RT, Casman E and Bernhardt ES. (2009). Decreasing uncertainties in assessing environmental exposure, risk, and ecological implications of nanomaterials. *Environ Sci Technol* 43: 6458–6462.

3. Wen-Che H, Westerhoff P and Posner JD. (2013). Biological accumulation of engineered nano-materials: A review of current knowledge. *Environ Sci: Processes Impacts* 15: 103–122.

4. Oberdarster G, Stone V and Donaldson K. (2007). Toxicology of nanoparticles: A historical perspective. *Nanotoxicology* 1: 2–25.

5. PEN. (2013). Project of the Emerging Nanotechnologies (PEN). Available at http://www.nanotechproject.org/inventories/consumer/browse/products/.

6. Kumari M, Khan SS, Pakrashi S, Mukherjee A and Chandrasekaran N. (2011). Cytogenetic and genotoxic effects of zinc oxide nanoparticles on root cells of Allium cepa. *J Hazard Mater* 190: 613–621.

7. Kumar A, Shanker R and Dhawan A. (2011). The need for novel approaches in ecotoxicity of engineered nanomaterials. *J Biomed Nanotechnol* 7: 79–80.

8. Navarro E, Baun A, Behra R, Hartmann NB, Filser J, Miao AJ, Quigg A, Santschi PH and Sigg L. (2008). Environmental behavior and ecotoxicity of engineered nanoparticles to algae, plants, and fungi. *Ecotoxicology* 17: 372–386.

9. Brayner R, Ferrari-Iliou R, Brivois N, Djediat S, Benedetti MF and Fievet F. (2006). Toxicological impact studies based on Escherichia coli bacteria in ultrafine ZnO nanoparticles colloidal medium. *Nano Lett* 6: 866–870.

10. Brayner R. (2008). The toxicological impact of nanoparticles. *Nano Today* 3: 48–55.

11. Wahab R, Mishra A, Yun SI, Kim YS and Shin HS. (2010). Antibacterial activity of ZnO nanoparticles prepared via non-hydrolytic solution route. *Appl Microbiol Biotechnol* 87: 1917–1925.

12. Wu B, Wang Y, Lee YH, Horst A, Wang Z, Chen DR, Sureshkumar R and Tang YJ. (2010). Comparative eco-toxicities of nano-ZnO particles under aquatic and aerosol exposure modes. *Environ Sci Technol* 44: 1484–1489.

13. Premanathan M, Karthikeyan K, Jeyasubramanian K and Manivannan G. (2011). Selective toxicity of ZnO nanoparticles toward Gram-positive bacteria and cancer cells by apoptosis through lipid peroxidation. *Nanomedicine* 7: 184–192.

14. Kumar A, Pandey AK, Shanker R and Dhawan A. (2012). Microorganism: A varsatile model for toxicity assessment of engineered nanomaterials. In *Nano-Antimicrobials: Progress and Prospects*, Edited by Nicola Cioffi, Mahendra Rai. Springer, Heidelberg, Dordrecht, London, New York: 497–524.

15. Handy RD, Owen R and Valsami-Jones E. (2008). The ecotoxicology of nanoparticles and nanomaterials: Current status, knowledge gaps, challenges, and future needs. *Ecotoxicology* 17: 315–325.

16. Mahapatra I, Clark J, Dobson PJ, Owenc R and Lead JR. (2013). Potential environmental implications of nano-enabled medical applications: Critical review. *Environ Sci: Processes Impacts* 15: 123–144.

17. Neal AL. (2008). What can be inferred from bacterium-nanoparticle interactions about the potential consequences of environmental exposure to nanoparticles? *Ecotoxicology* 17: 362–371.

18. Scown TM, van Aerle R and Tyler CR. (2010). Review: Do engineered nanoparticles pose a significant threat to the aquatic environment? *Crit Rev Toxicol* 40: 653–670.

19. Handy RD, Henry TB, Scown TM, Johnston BD and Tyler CR. (2008). Manufactured nanoparticles: Their uptake and effects on fish; a mechanistic analysis. *Ecotoxicology* 17: 396–409.

20. Sharma V, Kumar A and Dhawan A. (2012). Nanomaterials: Exposure, effects and toxicity assessment. *Proc Natl Acad Sci, India B: Biol Sci* 82: 3–11.

21. Van Hoecke K, De Schamphelaere KA, Van der Meeren P, Smagghe G and Janssen CR. (2011). Aggregation and ecotoxicity of CeO nanoparticles in synthetic and natural waters with variable pH, organic matter concentration and ionic strength. *Environ Pollut* 159: 970–976.

22. Mueller NC and Nowack B. (2008). Exposure modeling of engineered nanoparticles in the environment. *Environ Sci Technol* 42: 4447–4453.

23. Kaegi R, Ulrich A, Sinnet B, Vonbank R, Wichser A, Zuleeg S, Simmler H, Brunner S, Vonmont H and Boller MBMJ. (2008). Synthetic TiO_2 nanoparticle emission from exterior facades into the aquatic environment. *Environ Pollut* 156: 233–239.

24. Baalousha M. (2009). Aggregation and disaggreagtion of iron oxide nanoparticles: Influence of particle concentration, pH and natural organic matter. *Sci Tot Environ* 407: 2093–3101.
25. Domingos RF, Tufenkji N and Wilkinson KJ. (2009). Aggregation of titanium dioxide nanoparticles: Role of a fulvic acid. *Environ Sci Technol* 43: 1282–1286.
26. Keller AA, Wang H, Zhou D, Lenihan HS, Cherr G, Cardinale BJ, Miller R and Ji Z. (2010). Stability and aggregation of metal oxide nanoparticles in natural aqueous matrices. *Environ Sci Technol* 44: 1962–1967.
27. Moore MN. (2006). Do nanoparticles present ecotoxicological risks for the health of the aquatic environment? *Environ Int* 32: 967–976.
28. Buffle J, Wilkinson K, Stoll S, Filella M and Zhang J. (1998). A generalized description of aquatic colloidal interactions: The three-colloidal component approach. *Environ Sci Technol* 32: 2887–2899.
29. Kretzschmar R and Sticher H. (1997). Transport of humic-coated iron oxide colloids in a sandy soil: Influence of Ca^{2+} and trace metals. *Environ Sci Technol* 31: 3497–3504.
30. Kahru A, Dubourguier HC, Blinova I, Ivask A and Kasemets K. (2008). Biotests and biosensors for ecotoxicology of metal oxide nanoparticles: A minireview. *Sensors* 8: 5153–5170.
31. Lynch I and Dawson KA. (2008). Protein-nanoparticle interactions. *Nano Today* 3: 40–47.
32. Elsaesser A and Howard CV. (2011). Toxicology of nanoparticles. *Adv Drug Deliv Rev* 64: 129–137.
33. Baun A, Sorensen SN, Rasmussen RF, Hartmann NB and Koch CB. (2008). Toxicity and bioaccumulation of xenobiotic organic compounds in the presence of aqueous suspensions of aggregates of nano-C(60). *Aquat Toxicol* 86: 379–387.
34. Knauer K, Sobek A and Bucheli TD. (2007). Reduced toxicity of diuron to the freshwater green alga Pseudokirchneriella subcapitata in the presence of black carbon. *Aquat Toxicol* 83: 143–148.
35. Baun A and Hansen SF. (2008). Environmental challenges for nanomedicine. *Nanomedicine* 3: 605–608.
36. Dhawan A, Sharma V and Parmar D. (2009). Nanomaterials: A challenge for toxicologists. *Nanotoxicology* 3: 1–9.
37. Kumar A, Pandey AK, Singh SS, Shanker R and Dhawan A. (2011). A flow cytometric method to assess nanoparticle uptake in bacteria. *Cytometry A* 79A: 707–712.
38. Kumar A, Pandey AK, Singh SS, Shanker R and Dhawan A. (2011). Cellular uptake and mutagenic potential of metal oxide nanoparticles in bacterial cells. *Chemosphere* 83: 1124–1132.
39. Dhawan A and Sharma V. (2010). Toxicity assessment of nanomaterials: Methods and challanges. *Anal Bioanal Chem* 398: 589–605.
40. Shapiro GI, Edwards CD and Rollins BJ. (2000). The physiology of p16(INK4A)-mediated G1 proliferative arrest. *Cell Biochem Biophys* 33: 189–197.
41. Suzuki H, Toyooka T and Ibuki Y. (2007). Simple and easy method to evaluate uptake potential of nanoparticles in mammalian cells using a flow cytometric light scatter analysis. *Environ Sci Technol* 41: 3018–3024.
42. Shukla RK, Kumar A, Gurbani D, Pandey AK, Singh S and Dhawan A. (2013). TiO(2) nanoparticles induce oxidative DNA damage and apoptosis in human liver cells. *Nanotoxicology* 7: 48–60.
43. Shukla RK, Sharma V, Pandey AK, Singh S, Sultana S and Dhawan A. (2011). ROS-mediated genotoxicity induced by titanium dioxide nanoparticles in human epidermal cells. *Toxicol in Vitro* 25: 231–241.
44. Ghafari P, St-Denis CH, Power ME, Jin X, Tsou V, Mandal HS, Bols NC and Tang XW. (2008). Impact of carbon nanotubes on the ingestion and digestion of bacteria by ciliated protozoa. *Nat Nanotech* 3: 347–351.
45. Kahru A and Dubourguier HC. (2010). From ecotoxicology to nanoecotoxicology. *Toxicology* 269: 105–119.
46. Donaldson K and Borm P. (2004). Particle and fibre toxicology, a new journal to meet a real need. *Part Fibre Toxicol* 1: 1.
47. Nowack B. (2009). The behavior and effects of nanoparticles in the environment. *Environ Pollut* 157: 1063–1064.
48. Fairbrother A and Fairbrother JR. (2009). Are environmental regulations keeping up with innovation? A case study of the nanotechnology industry. *Ecotoxicol Environ Saf* 72: 1327–1330.

6

SonoPhotoCatalytic Cavitation (SPCC) in Water Treatment

Tony F. Diego,[1] Curt Hallberg,[2] and Maria Jose Lopez Tendero[3]

[1]42TEK S.L., Almazora, Spain

[2]Watreco AB, Malmö, Sweden

[3]Laurentia S.L., València Parc Tecnològic, Paterna, Valencia, Spain

CONTENTS

6.1 Introduction

This chapter is based on the relationship between two familiar and naturally occurring phenomena, that of sound and that of light, but that in some cases can be very closely related as shown by the extreme forms and effects produced by cavitation. Under the right circumstances, such an effect can achieve a high level of efficiency that is indeed capable of producing *sonoluminescence*, or the emission of photons by sound stimulation. A closely related field to sonoluminescence, by extension, is *sonophotocatalysis* or the capability to reproduce the photocatalytic effect upon a suitably treated surface, from the emission of photons from sonoluminescence and thus enabling a new field of study we propose to call SonoPhotoCatalytic Cavitation (SPCC). SPCC describes the combined effects of ultrasound with heterogeneous photocatalysis taking place within a cavitation, producing a resonant vortex chamber tuned to produce sonoluminescence. We draw on the parallels on how these existing technologies may be combined and applied to water remediation or potable water treatment by the advent of new materials brought about in the rapidly growing field of applied nanotechnology to produce highly efficient photocatalyst coatings combined with drag-reducing nanocoatings into both the realms of sound and light.

Cavitation and heterogeneous photocatalysis effects and processes are already being effectively used to treat water from undesired contaminants as alternative treatments. They do so without the use of additional chemical compounds and are already very effectively implemented in industrial water systems, as well as municipal potable water and wastewater treatment plants. Each of these methods in and by itself has some great and fascinating potential, but they also have some limitations on how and in what settings they may be applied effectively.

In the case of heterogeneous photocatalysis or also known as advanced oxidation process (AOP), it is the very optical properties of the water to be treated that limit its efficiency. To properly treat influent water, maximum contact surface must be achieved between the target water molecules and the photon-activated TiO_2 surface, where different highly reactive transitory species must act for mineralization of organic compounds and disinfection of water pathogens.

This can be achieved by maximizing the surface area of contact between the photocatalytic region of a reactor and the water molecules to be treated, and/or extending the residence time in the chamber. The work of maximizing the surface area can be readily achieved by using larger reactors, or by using photocatalytic mesoporous materials, greatly increasing the contact with water being treated; however, they suffer in that light cannot readily excite and reach the internal structures owing to tortuosity, and thus excite the electrons into the outer orbital to produce the hydroxyl radicals upon orbital decay.

In the case of vortex cavitation, where the contaminants may be dynamically separated from the target water by coarse centrifugal effects, sonoluminescence is rarely achieved nor detected unless specialized and sensitive equipment is used, and rarely is the case that cavitation alone has the capacity to achieve the ideal conditions of producing sonoluminescence within the target water samples.

Photocatalysis and cavitation share the potential capacity to produce highly reactive hydroxyl radicals, which are especially important with respect to not only degradation of organic compounds, but also of other impurities commonly found in water such as dissolved metal salts or dyes. In the special case of sonophotocatalytic cavitation, which is a photocatalytic reaction coupled with ultrasonic irradiation or the simultaneous irradiation of ultrasound and light with a photocatalyst, it is believed that those limitations of photocatalysis or cavitation, each in isolation, can be mitigated. It is believed that SPCC can be optimized using the current advances in sonophotocatalytic and cavitation to perform as well as either alone, and in addition, to be capable of treating water not only from these undesired contaminants, but to also provide it with an enhanced set of ancillary properties that can have profound implications in various water treatment applications and processes. Water remediation from persistent and emergent contaminants, cooling tower water systems decalcification, digester systems, emulsification and mixing, concrete hardening and curing optimization, calcium removal, ballast water discharge from biological or oil contaminants, enhancing organic reactions, mining wastewater remediation for slurry and sludge, selective contaminant isolation, and ice fabrication industries are just a few of them, which are discussed below.

To begin with, however, we will first take our initial clues from where these effects occur in nature and from the existing life forms as a guide to establish our goals. One such case is in the behavior of the mantis shrimp (aka pistol shrimp), as featured on the cover of Patek and Caldwell's book [1], which shows that these are not really "new" phenomena at all, but that have already been used for millennia by a simple class of crustaceans in a routine way for its own subsistence or defense.

While much of this work is still considered early and pioneering, the basic physical principles and observations seem to point a clear existence of these phenomena and a clear need to gain a better understanding of the mechanisms at work for them is necessary in order to be successfully implemented. Fortunately, the work of past contributions is now being pursued by researchers, and is independently being corroborated.

We begin by briefly describing the basic and unique properties of water, and move to highlighting the controlled use of cavitation applications in real-life scenarios, and then show that when implemented it can have important results applicable to energy, water, and wastewater treatment.

We then highlight some of the important achievements being carried out in the utilization of AOP and sonophotocatalysis with respect to water treatment, and indicate some overlapping applications AOP and sonophotocatalysis share, and also some novel and profound implications regarding water molecules that can be of considerable interest in the specific case of wastewater and drinking water applications. For example, the production of hydroxyl radicals is common to both cavitation and AOP.

Some of the cutting-edge research occurring at the boundary layers of AOP and some of the recent achievements in the field of sonoluminescence, or the production of electromagnetic energy from the acoustic pressures waves associated with cavitation as in sonophotocatalysis, are opening doors for these new fields of study.

The present state of the art in devices currently found in the market in which AOP is being applied, such as the Puralytics Co. Shield/600, and other devices to produce controlled cavitation, such as Watreco's IVG and VPT Vortex line of products, have shown to be very effective. Using these devices as primers, we propose a new class of products achieved by combining both of these devices jointly for a new range of devices based on SPCC.

Both cavitation and AOP produce desirable effects, coupled can avert the need for additional chemical additives to treat water contaminants, and can do so even in a passive manner without anything more than gravity and the energy from the sun as the driving forces. We propose an inline apparatus that can be retrofitted into existing water treatment systems without much additional reengineering that can both be a photocatalytic reactor as well as a sonoluminescent resonant cavitation chamber.

6.2 Water Physical–Chemical Properties

Table 6.1 gives some of the basic physical properties of pure water. Nearly every one of these properties can be affected during the treatment with the methods outlined and proposed below.

Our basic scientific knowledge about water began ca. 1800, when William Nicholson was able to achieve the first decomposition of water into hydrogen and oxygen, by using electrolysis. More than 5 years later, Joseph Louis Gay-Lussac and Alexander von Humboldt proved water to be composed of two parts hydrogen and one part oxygen, the latter giving

TABLE 1.1

Water Parameters

Molar mass	18.015 g/mol
Density max	1 g/cm^3 at 4°C
Melting point	0°C at 1 atm
Boiling point	100°C at 1 atm
Refractive index	1.333
Viscosity	0.001 Poise at 20°C
Dipole moment	1.85 D
Molecular structure	Hexagonal
Molecular shape	Bent
Angle between H atoms	104.45°
Distance between O and H atoms	0.1 nM
Dissociation constant	10–14 K_w at 25°C
Vapor pressure	2.3 kPa at 20°C
Compressibility	5.1 × 10^{-10} Pa^{-1} at 0°C at 0 Pa
Compressibility min	4.4 × 10^{-10} Pa^{-1} at 45°C at 0 Pa
Compressibility absolute min	3.9 × 10^{-10} Pa^{-1} at 0°C at 100 MPa
Bulk modulus	2.2 GPa
Heat capacity	4.2176 (J/g)·K at 100 kPa
Heat of vaporization	45.054 k J/mol at 0°C
Heat of fusion	333.55 k J/kg at 0°C
Electrical resistivity max	182 kΩ·m at 25°C
Electrical conductivity	0.055 µS/cm at 25°C
Surface tension	72.8 mN/m at 25°C

his name to the great ocean current found on the Pacific Ocean. The mass in water was thus determined to be composed of 11.1% hydrogen and 88.9% oxygen.

We now know that individual water molecules have two hydrogen atoms that are covalently bonded to a single oxygen atom and the solitary oxygen molecule attracts electrons much more strongly than hydrogen atoms, giving a net positive charge on the hydrogen atoms and a net negative charge on the oxygen atom. The presence of a charge on each of these atoms gives each water molecule a net dipole moment.

A molecule with such a charge difference is known as dipolar, and the direction of the dipole moment points from the oxygen toward the center of the hydrogens. Associated water molecules are not planar in two dimensions, but instead are three-dimensional (3D) and form specific angles, with hydrogen atoms at the vertices and oxygen in between them at its head. Since oxygen has a higher electronegativity than hydrogen, the head of the molecule with the oxygen atom has a partial negative charge with which the lone electron pairs of hydrogen tend to repel the oxygen atom. Owing to this dipole moment, the boiling point of water is above the expected result. The charge differences cause individual water molecules to be attracted to each other and to other polar molecules.

Hydrogen bonding and the resulting electrical attraction between individual water molecules make it rather difficult to separate molecules from each other, giving it a very high specific heat capacity and a high heat of vaporization. Hydrogen bonding is a relatively weak attraction compared with the covalent bonds within the water molecule itself; however, it is the reason for the high melting and high boiling point of water.

A water molecule forms a maximum of four hydrogen bonds because it can accept two and donate two hydrogen atoms. In water, the four hydrogen bonds give rise to an open structure and a 3D bonding network, resulting in the decrease of density when it is cooled below 4°C.

Water molecules experience cohesion, due to the collective action of hydrogen bonds between water molecules. However, water also has high adhesion properties because of its polar nature. On extremely smooth surfaces such as glass, the water may form a thin film because the adhesive molecular forces between glass and water molecules are stronger than the cohesive forces.

The molecules of water are not statically bound but instead are constantly moving in relation to each other, with hydrogen bonds continually breaking and recombining very quickly. However, these hydrogen bonds are sufficiently strong to create the unique properties we observe in water. However, not all hydrogen or oxygen atoms that constitute water as we know it are the same. The hydrogen isotope protium (no neutron) is the most abundantly found in natural water, accounting for >99.98% of its instances; however, water can also be found having deuterium hydrogen isotopes (3×10^{-3}%), or even tritium hydrogen isotopes, but at far lower concentrations (3×10^{-6}%). Oxygen also has three stable isotopes, with which it is found in water, the most common resulting in 99.76%; however, isotopes comprising 0.04% and 0.2% of carbon in water molecules are also found.

Pure water has a pH of 7 and is deemed neutral; however, water is also deemed amphoteric, meaning it can become an acid or a base depending on what is mixed within it. Pure water has a balanced concentration between hydroxide ions, and hydronium or hydrogen ions. However, water, being a polar liquid, can dissociate into the hydronium ion and a hydroxide ion.

Water in nature is difficult to find as pure, and to produce what is known as ultrapure water is indeed very difficult. Instead water is normally found in nature with many contaminants, and with dissolved gases such as CO_2, nitrogen, or sulfur oxides, compounds making it slightly acidic and dropping the pH value to just below 6.

Water is mostly transparent to parts of the visible spectrum, and to a part of ultraviolet light, and near red light; however, it readily absorbs most of the ultraviolet light, infrared light, and the microwave regions of the electromagnetic spectrum. The reflected part of the visible spectrum from the sun gives it a faint blue color. Interestingly, there are windows within the visible spectrum, such as 498 nm or blue-green light, that many of the biological luminescent creatures use to manifest their position or excited states.

Water and ice are poor heat conductors, and it is this specific property of water that has profound consequences for our ecosystem. Water is densest in the liquid state and at a temperature of 4°C and will sink to the bottom of deep lakes and oceans, regardless of the temperature in the atmosphere. This cold water sinking produces the downward convection of colder water, and also an expansion of water as it becomes colder than the freezing point. This denser saltwater sinks, producing convection, and the great conveyer belts of ocean currents forming to transport water away from the Arctic and Antarctic poles, leading to a global system of thermohaline circulation.

When the temperature of freshwater reaches 4°C, the layers of water near the top in contact with cold air continue to lose heat energy and their temperature falls below 4°C. On cooling below 4°C, these layers may rise up as freshwater. However, because ice is less dense than 4°C water, ice floats, and this unusual negative thermal expansion is due to intermolecular interactions within the water molecules. The intermolecular vibrations decrease, allowing the molecules to form steady hydrogen bonds with their neighbors and thereby locking into hexagonal packing.

The density of water is dependent on the dissolved salt and gas contents as well as the temperature. Ice also floats in the sea; however, the salt content of oceans lowers the freezing point by about 2°C, lowering the temperature of the density maximum of water to just above 0°C. As the surface of salt water begins to freeze at 1.9°C, the ice that forms is salt free with a density approximately equal to that of freshwater ice. Upon freezing, the density of water decreases by about 9%. This ice floats on the surface, and the salinity and density of the seawater just below it increases.

The hydrogen bonds become shorter in ice than in the liquid phase; this locking effect reduces the average coordination number of molecules as the liquid approaches nucleation. However, not all ice is less dense than the liquid form; only ordinary ice appears to be so. The melting point of ordinary hexagonal ice falls slightly under moderately high pressures. Under increasing pressure, ice undergoes a number of transitions to other allotropic forms, each of which has higher density than liquid water—ice II, ice III, HD amorphous ice, and VHD amorphous ice are examples of such. As ice transforms into its allotropes just below 210 MPa, the melting point increases, reaching 82°C at 2.216 GPa.

Water also expands significantly as the temperature increases. Water near the boiling point is about 96% as dense as water at 4°C. The melting point of ice is 0°C at standard pressure; however, pure liquid water can be supercooled well below that temperature without freezing, and it can remain in a fluid state down to −42°C.

Under special circumstances, liquid water can remain so even at 42°C below zero and ice can remain solid at temperatures of up to +82°C. Luckily this is not the case in nature's own manifestations.

Starting with the energy from the sun, and the heating of the water at the surface coupled with the density maximum of water at the liquid phase just above freezing, it is sufficient and necessary to drive the thermal cycles that drive the huge ocean currents such as the Gulf Stream in the Atlantic or the Humboldt Current, and these determine our global weather patterns, producing hurricanes, or ice ages, and the formation of new deserts where there once existed lush forests. The energy from the sun also drives the water cycle from the seas, through evaporation, to become the clouds that precipitate on land as freshwater, which simultaneously allows natural desalination.

Seawater moderates Earth's climate by buffering fluctuations in ambient temperature, allowing the oceans to absorb more heat than the atmosphere and buffering the heat resulting from global warming. The potential consequence of global warming is that the loss of ice in the poles will result in the alteration of these currents, which will have consequences on weather patterns we may just be beginning to see. In addition, the specific enthalpy of fusion of water is also very high, adding further resistance to melting of the ice on the poles, making water a good heat storage medium and heat shield.

In extreme conditions such as the abyssal ocean depths and near deep-water chimney stacks, liquid and gas phases merge into one homogeneous fluid phase exhibiting properties of both gas and liquid, or an indeterminate phase form, when the water's supercritical point is reached at a temperature of 375°C and pressure of 221 atm. The work done by Baret in producing ultrapure water uses these conditions to develop techniques in supercritical fluids that can also occur in these extreme conditions in nature.

If the vapor partial pressure is 2% of atmospheric pressure and the air is cooled from 25°C, starting at about 22°C water will start to condense, defining the dew point, and creating fog or dew. If the humidity is increased at room temperature and the temperature stays about the same, the vapor soon reaches the pressure for a phase change and then condenses out as steam. Water vapor pressure above 100% relative humidity is called supersaturated and can occur if air is rapidly cooled, for example, by rising suddenly in an updraft.

Pure water is an excellent insulator but hard to find in this condition in nature. Water in nature is almost never completely free of ions; the smallest impurity, even at parts per trillion (ppt), will make water electrically conductive. Water has a high dielectric constant. This constant shows that its ability to make electrostatic bonds with other molecules is high, meaning it can eliminate the attraction of the opposite charges of the surrounding ions.

Water is also a good solvent owing to its polarity. The ability of a substance to dissolve in water is determined by whether the substance can match the strong attractive forces that water molecules generate between other water molecules. If a substance has such properties that do not allow it to overcome these strong intermolecular forces, the molecules are "pushed away" from the water and do not readily dissolve. Contrary to the common misconception, water and hydrophobic substances do not "repel," and the hydration of a hydrophobic surface is energetically, but not entropically, favorable.

When an ionic or polar compound enters water, it is surrounded by water molecules (hydration). The relatively small size of water molecules typically allows many water molecules to surround one molecule of solute. The partially negative dipole ends of the water are attracted to positively charged components of the solute, and vice versa for the positive dipole ends. In general, ionic and polar substances such as acids, alcohols, and salts are relatively soluble in water, and nonpolar substances such as fats and oils are not. Nonpolar molecules stay together in water because it is energetically more favorable for the water molecules to hydrogen bond to each other than to engage in van der Waals interactions with nonpolar molecules.

At ambient conditions, and at standard temperature and pressure of 25°C and 1 atm (STP), water can behave more like a gel and possibly be capable of producing work, or even storing information such as at the conditions observed at the boundary interfaces identified as Exclusion Zones (EZ), as proposed by Pollack [2] based on the detailed work on photocatalytic surface boundary layers. He has been using substrates coated with TiO_2, gels, and nafion, having properties known as superhydrophilic (extreme water loving), in which water itself does not behave in anything resembling the expected way we thought it should, or with substrates on the extreme opposite end of the scale in superhydrophobic surfaces (extreme water loathing), when water can maintain perfectly its spherical shape of a drop or the least amount of volume required to hold its mass, even floating on another drop of water itself.

Pollack has noted the following changes of water properties within the EZ between the hydrophilic substrate and the water boundary layers:

- Increased density
- Stabilizing by the organized stacking of the hexagonal molecules
- Reorientation of the dipoles
- Adding charge potential between the EZ and water adjacent to it
- Increasing viscosity
- Decreasing its latent heat
- Increasing its optical index of refraction
- Allowing the water molecule to exclude solutes
- Making the water molecule become nondipolar

In biological cells and small organisms, water is in contact with membrane and protein surfaces that are hydrophilic; that is, these are surfaces that have a strong attraction to water. Langmuir observed a strong repulsive force between separate hydrophilic surfaces. To dehydrate hydrophilic surfaces is to remove the strongly held layers of water of hydration, and this requires doing substantial work against these forces, called hydration forces. These forces are very large but decrease rapidly over a range of a nanometer or less. They are important in biology, particularly when cells are dehydrated by exposure to dry atmospheres or to extracellular freezing.

We note how photosynthesis is used to produce starches and the complex molecule chains starting from basic elements in the form of minerals and nutrients and convert these into sugars and complex carbohydrates, from water and the energy provided by the sun, drawing water up against gravity to bring these nutrients from the moist soil up and through the roots and to the stems of plants and trees.

Owing to adhesion and surface tension, water exhibits capillary action whereby it rises into a narrow tube against the force of gravity. Water adheres to the inside wall of the tube and surface tension tends to straighten the surface, causing a surface rise and more water being pulled up through cohesion. The process continues as the water flows up the tube until there is enough water such that gravity balances the adhesive forces. Surface tension and capillary action are important in biology. For example, when water is carried through xylem of stems in plants, the strong intermolecular attractions (cohesion) hold the water column together and adhesive properties maintain the water attachment to the xylem and prevent tension rupture caused by transpiration pull. We note also how chemosynthesis works by extracting sulfur from deep-water chimney stacks or little volcanoes, for plant or animal life in the depths of our oceans that will never see the sun or any kind of light.

6.3 Cavitation

Viktor Schauberger, born in Austria in the 1880s, did pioneering work by systematically studying, observing, and describing nature by thinking purely about how fish swam or how birds flew with such effortless efficiency, and began applying these notions to solve technical problems needed for the industrial applications during his own time. In the specific case of cavitation, the original idea of Schauberger to work with vortices, especially in water, comes from his own thorough studies of animals and nature. Reproducing many of these resulted in various applications and inventions during his lifetime, including several types of water treatment devices. We have come a long way since Schauberger in our capacity to generate a vortex that has evolved by the availability of new precision equipment and machining methods, and with the availability of new materials, instrumentation, and fabrication processes.

To gain some understanding of how cavitation works in water, we keep in mind that water actually contains dissolved gases, and tap water for instance contains nitrogen and oxygen similarly as found in air, but in proportions that are different from that of the existing concentrations of the atmospheric air we breathe on land. By supplying mechanical and/or electromagnetic forces to matter being accelerated, such as a body or a sound wave moving through water, we can induce a set of controlled variations or pressure fluctuations events to enable cavitation to occur, or a temporary "delaminating" of a water structure or surrounding medium. This is a common occurrence in ship or airplane propellers or impellers, and can be readily noticed when accelerating too quickly makes the propeller lose "grip" with the water and air seems to get into the water, making engine revolutions per minute (rpm) spike and reducing the efficiency of the desired acceleration. Cavitation also occurs when transducers such as sonars are applied to water in amplitudes that exceed its ability to modulate at given frequencies.

Mantis shrimp (or ninja or pistol shrimp) [3] can strike its claws as fast as a speeding bullet, leaving the barrel of a gun with an acceleration of 10,400 g (102,000 m/s^2), and with top velocities of 23 m/s in nearly 1/10,000 of a second. Instantaneous forces of >1500 Nm are caused by the impact itself against the striking surface. Even more astonishing is that it does not require physical contact with the target at all, but instead when it begins to strike, it can produce a cavitational shockwave generating cavitation bubbles at the remote target, where microbubbles appear, and it is during the collapse of these bubbles that measurable forces are produced on their intended targets; the resulting shockwave alone can be enough to kill or stun much larger prey without actual contact.

Amazingly, the shockwave from the cavitation produced by the snapping gesture of the mantis shrimp can also produce very small amounts of light and very high temperatures within the collapsing bubbles, or sonoluminescence left in the wake of its strike, and although both the light and the high temperatures are short lived, they pave the way for a need to better understand the mechanisms at work in laboratory and reproducing these.

There are several ways that cavitation can manifest itself: on partially attached cavities, on travelling bubble-type cavitation, or vortex cavitation and as shear cavitation. When cavitation does finally occur, microbubbles may visibly form, and subsequently collapse. When they do collapse, these microbubbles can go on to the extreme case of emitting electromagnetic radiation, and at incredibly high temperatures, but only for time scales lasting in the range of nanoseconds or even picoseconds, making their observation quite difficult, but capable of producing catastrophic macroscale effects such as the pitting or failure of propellers on maritime vessels [4–6]. For the very brief periods of time during the minute implosions, collapsing these bubbles temperatures are capable of even melting steel [7–9].

According to NanoSpire Co., "a laser, ultrasound or other energy source is used to create small high-energy vapor bubbles through a phase transition." The collapse of cavitation bubbles in close proximity to a wall generates supersonic liquid microjets naturally pointed toward a wall or other restriction to the collapse of the bubble. NanoSpire's patented methods build on, but go beyond, the energetic process seen in nature. The size, strength, and direction of cavitation reentrant microjets can be controlled by our patented methods to a very high degree of accuracy. Cavitation microjets can travel at up to Mach 4 and are capable of drilling a hole as small as a few nanometers in a diamond. Multiple controlled bubbles are also possible, allowing lines to be machined as well as other form factors and machining applications. Cavitation allows a high aspect ratio machining to be obtained and is very repeatable" [10,11].

The development of cavitation in a liquid flow is characterized by a phase change from liquid to vapor at almost constant temperature. Water's own surface tension brings a delay in the inception of cavitation from microbubbles carried by the liquid. It is a reasonable approximation to assume that the critical pressure for the onset of cavitation is equal to the vapor pressure. In the case of vortex cavitation, and due to centripetal forces, the pressure within the vortex axis is lower than the pressure far away from it, so that a minimum pressure is expected at the vortex center axis. It helps to think of it as a centrifuge where lesser dense material is left at the center axis and denser material is thrown out to the edge of the tornado.

The dynamics of the motion of the bubble is characterized to a first approximation by the Rayleigh–Plesset equation derived from the incompressible Navier–Stokes equations and describes the motion of the radius of the bubble R as a function of time t. Here, μ is the viscosity, p the pressure, and γ the surface tension. This equation has been shown to provide reasonable estimates of the bubble dynamics under an acoustically driven field but only up to final stages before it collapses. Intricate measurements have shown that during the critical final stage of implosion, the bubble wall velocity exceeds the predicted speed of sound of the gas contained inside the bubble, and the Rayleigh–Plesset equation is no longer applicable.

$$R\ddot{R}+\frac{3}{2}\dot{R}^2 = \frac{1}{\rho}\left(p_g - P_0 - P(t) - 4\mu\frac{\dot{R}}{R} - \frac{2\gamma}{R}\right)$$

Controlled hydrodynamic cavitation is a process that seeks to optimally produce cavitation and harness the kinetic energy that is imparted to the fluid without any damage to equipment and has been demonstrated applicable to bacteria eradication, removal of dissolved gases, precipitation of certain inorganic salts, creation of hydroxyl radicals, and the formation of stable emulsions.

6.4 Advanced Oxidation Process (AOP)

The use of photocatalytic nanocoated structures such as the anatase form of titanium dioxide (TiO_2) exposed to UV light and in combination with air or water begins a reaction at the illuminated TiO_2 surface–water boundary where the water molecules start to split into H^+ and OH^- groups. When these outer electrons on TiO_2 atoms are excited by common electromagnetic radiation such as UV from the sun, it can move electrons into higher stable outer orbitals, only then to subsequently remit their energy in the form of photons again on their way back down to lower stable orbitals in a renewable and catalytically self-sustaining way that does not consume the photocatalytic surface. During this stage, photons are reemitted but at lower energy packet values, which are then accompanied by a new set of electrochemical processes producing the mysterious behaviors found in AOP, described elsewhere in this book (see Chapter 24).

The OH^- groups (hydroxyl radicals) produced are a form of very reactive free radicals that attack almost any substance but are particularly effective in reducing organic material such as the cell membranes in bacteria. In this way, it is possible to disinfect water that has been contaminated with not only bacteria but also viruses, molds and lichens, and pathogens of every type in order to achieve potable water anywhere by removing most of the contaminants commonly found in water, including dissolved metals, and metastable compounds. There are in fact at least five separate reactions: photocatalytic oxidation, photocatalytic reduction, photoadsorption, photolysis, and photodisinfection, all working together to treat contaminants in water including heavy metal ions.

Pollack, whose detailed observations describing the relationship of just how matter interacts with light, and between the substrate and the boundary interfaces in contact with the water molecule itself, is transforming the way we understand water. A new form of water called by many "the fourth state of water" is forcing us to reconsider and truly understand how water interacts and behaves with biological organisms at the smallest scales. We are only now beginning to truly understand and recognize how to harness this capacity and to implement it in ingenious manners for the use of nanotechnologies in water treatment, and certainly graphene in combination with TiO_2 will be further changing it as well.

6.5 Sonophotocatalysis

Harada [12] showed the possibility of achieving the effect of a hybrid sonochemical and photocatalytic reaction simultaneously where water was hardly decomposed to H_2 and O_2 by photocatalysis or sonolysis alone. Furthermore, in order to decompose water, he used a powdered TiO_2 photocatalyst suspended in distilled water and simultaneously irradiated the water sample by light and ultrasound. This sonophotocatalytic reaction was shown to be effective in the decomposition of water to H_2 and O_2 [13].

Separately, the work carried out by Kavitha and Palanisamy [14] showed the accelerated sonophotocatalytic degradation of tracing dye under visible light using dye sensitized TiO_2 activated by ultrasound was very effective. The effect of sonolysis, photocatalysis, and sonophotocatalysis under visible light showed the influence on the degradation rates by varying the initial substrate concentration, pH, and catalyst loading and to ascertain the synergistic effect on the degradation techniques.

Ultrasonic activation was shown to contribute to the degradation through cavitation and leading to the splitting of H_2O_2 produced by both photocatalysis and sonolysis, combined. This results in the formation of oxidative species, such as singlet oxygen O_2 and superoxide O_2^- radicals in the presence of oxygen. The increase in the amount of reactive radical species, which induce faster oxidation of the substrate and degradation of intermediates, and also the deaggregation of the photocatalyst are responsible for the synergy observed under sonification. A comparative study of photocatalysis and sonophotocatalysis using TiO_2 by Hombikat [15] using UV and ZnO was also reported.

The degradation of chitosan (polysaccharide) by means of ultrasound irradiation and its combination with heterogeneous TiO_2 was investigated [16]. Emphasis was given on the effect of additives on degradation rate constants. Ultrasound irradiation (24 kHz) was provided by a sonicator, while an ultraviolet source of 16 W was used for UV irradiation. The extent of sonolytic degradation increased with increasing ultrasound power (in the range 30–90 W), while the presence of TiO_2 in the dark generally had little effect on degradation.

On the other hand, TiO_2 sonophotocatalysis led to complete chitosan degradation in 60 min with increasing catalyst loading. TiO_2 sonophotocatalysis was always faster than the respective individual processes owing to the enhanced formation of reactive radicals as well as the possible ultrasound-induced increase of the active surface area of the catalyst. The degraded chitosan were characterized by x-ray diffraction, gel permeation chromatography, and Fourier transform infrared spectroscopy, and the average molecular weight of ultrasonicated chitosan was determined by measurements of the relative viscosity of samples. The results show that the total degree of deacetylation (DD) of chitosan did not change after degradation, and the decrease of molecular weight led to transformation of the crystal structure. A negative order for the dependence of the reaction rate on total molar concentration of chitosan solution within the degradation process was suggested [17].

6.6 Sonoluminescence

The sonoluminescence effect can occur when a sound wave of sufficient intensity induces a gaseous cavity within a liquid only to collapse quickly, and the effect was first discovered at the University of Cologne, Germany, in 1934 as a result of work on sonar technology by

H. Frenzel and H. Schultes [18] when they put an ultrasound transducer in a tank of photographic developer fluid meant to speed up the development process. However, as has often happened in science, they noticed instead tiny dots on the film after developing and realized that the bubbles in the fluid were emitting light with the ultrasound turned on.

The pioneering work by N. Marinesco and J.J. Trillat in 1933 is also credited with the early discovery of the multibubble sonoluminescence phenomenon [19].

This cavity may take the form of a preexisting bubble, or may be generated through the process of cavitation. Sonoluminescence in the laboratory can be made to be stable, so that a single bubble will expand and collapse over and over again in a periodic fashion, emitting a burst of light each time it collapses. For this to occur, a standing acoustic wave is set up within a liquid, and the bubble will sit at a pressure antinode of the standing wave. The frequencies of resonance depend on the shape and size of the container in which the bubble is contained.

The light flashes from the bubbles are extremely short, in the picosecond range with peak intensities on the order of a couple of milliwatts. While the bubbles are very small (in the range of 1 micron in diameter) for water, when they emit the light the gas contained within them is mainly atmospheric air, is crushed to nanometer ranges.

A major experimental advancement was achieved in 1989 by Felipe Gaitan and Lawrence Crum, who produced stable single-bubble sonoluminescence (SBSL) [20,21]. In SBSL, a single bubble trapped in an acoustic standing wave emits a pulse of light with each compression of the bubble within the standing wave. This technique allows a more systematic study of the phenomenon because it isolates the complex effects into one stable, predictable bubble. It was realized that the temperature inside the bubble was hot enough to melt steel [22].

Interest in sonoluminescence was again renewed when an inner temperature of such a bubble well above 1 million degrees kelvin was postulated. This temperature is thus far not conclusively proven, although recent experiments conducted by the University of Illinois at Urbana-Champaign [24] indicate temperatures quantified it to be around 20,000°K, at the bubble implosion. While the temperature of the gas vapor within the bubble may be several thousand °K and the pressure may increase to several hundred atmospheres [23].

Research has also been carried out by Dr. Klaus Fritsch of John Carroll University, University Heights, Ohio [25], and the US Navy has studied propeller-induced sonoluminescence during the Cold War at its laboratories such as that at the Naval Physics Laboratory at the University of Washington in the mid-1980s and in the work of soliton or single-photon propagation in the realm of bioluminescent fish at 498 nm.

More recently in 2002, SBSL is detailed, in a lengthy explanation, by modulating the concentrations of noble gases to enhance its effects. SBSL pulses can have very stable periods and positions. In fact, the frequency of light flashes can be more stable than the rated frequency stability of the oscillator making the sound waves driving them. However, the bubble itself undergoes significant geometric instabilities before collapse [26]. Spectral measurements have given bubble temperatures in the range from 2300°K to 5100°K, the exact temperatures depending on experimental conditions of the composition of the liquid and gas. Detection of very high bubble temperatures by spectral methods is limited because of the opacity of liquids to the very short wavelength light characteristic of very high temperatures [27].

Chemical reactions cause nitrogen and oxygen to be removed from the bubble after about 100 expansion–collapse cycles. The bubble will then begin to emit light. It would seem the phenomenon of sonoluminescence is at least roughly explained, although some details of the process remain obscure. Sonoluminescence in this respect is what is physically termed a bounded phenomenon, meaning that the sonoluminescence exists in a bounded region of parameter space for the bubble, a coupled magnetic field being one such parameter.

Noted physicist Julian Schwinger's theory of sonoluminescence [28] is based on the Casimir energy theory, which suggests that the light in sonoluminescence is generated by the vacuum within the bubble in a process similar to Hawking radiation, or the radiation generated at the event horizon of blackholes. According to this vacuum energy explanation, since the quantum theory holds that vacuum contains virtual particles, the rapidly moving interface between water and gas converts virtual photons into real photons. This is a direct relationship to the Casimir effect, and sonoluminescence may be quantum vacuum radiation [29].

6.7 Applications with Vortex Cavitation

The Watreco Company in Malmö, Sweden, and their line of products branded under the name Vortex VPT (see Figure 6.1), is one that has been successfully applying the principles of cavitation to treat water in multiple ways.

The water enters tangentially at the "fat" end and flows into the trumpet-shaped vortex chamber, generating a strong vortex inside the chamber. Normal tap water contains a lot of nanosized bubbles, and these small bubbles come in huge quantities. These small bubbles are associated with increased viscosity as they generate drag. The vortex generator generates a subpressure in the center of the vortex, as can be observed as vortices much like small tornados or hurricanes. Because of the strong pressure gradient with almost a perfect vacuum in the center, contaminants obey the Archimedes principle. As a result, less dense contaminants in water, such as the dissolved oxygen or other gases, cause the bubbles to migrate toward the lowest pressure, which happens to be in the center axis.

On the way, they expand and combine with other bubbles and they end up in the center of the vortex where we can find an elongated thin bubble of gas with very low pressures. From time to time, a part of the elongated bubble is cut off and transported away by the flow, and the water is degassed, or more correctly, "debubbled." In this way, we can achieve some very desirable effects.

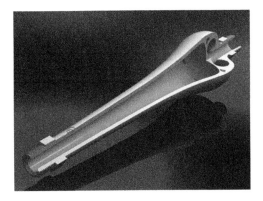

FIGURE 6.1
Cross section of the Watreco vortex generator.

6.7.1 Water Remediation

By using cavitation, solids and other substances can be separated from water molecules, as can unwanted gases as well. Many forms of mining, including coal, produce highly contaminated wastewater containing high concentrations of heavy metals. These heavy metals include aluminum (Al), cadmium (Cd), copper (Cu), iron (Fe), lead (Pb), manganese (Mn), and zinc (Zn). Results show that with addition of cavitation treatment, a noticeable improvement of removal could be achieved; >90% of the dissolved heavy metal concentrations could be removed from the residual water by "exit porting" a spatial region of the chamber. Both laboratory testing and real-life applications of the Vortex VPT indicate that it is also effective in removing various organics from influent water. Cavitation from the Vortex VPT has been shown to produce the very reactive hydroxyl radical, just like in heterogeneous photocatalysis. Adding other oxidants such as peroxide or ozone can increase its efficiencies even further [30–32].

6.7.2 Fuel Emulsification and Mixing

Fascinatingly, the Vortex VPT can also be used to achieve optimization where mixing tasks are required, such as in the manufacturing of nanomaterials, or on the oil to keep water emulsified in the mixture to produce fuel/water, fuel/ethanol, or other emulsions. For such applications, water and air can be introduced simultaneously, producing a stable emulsion even at room temperature and without emulsifiers. These applications are of special importance to the food and drug industries, and particularly the cosmetic industry, because using emulsifiers can cause undesired effects [33].

6.7.3 Cooling Water Treatment

The cooling tower applications of Watreco's Vortex VPT are currently in use worldwide for calcium carbonate removal from the influent. Water treatment results from samples containing a high concentration of $CaCO_3$ indicate that a relatively stable, low-level calcium concentration in the effluent is obtained even if the influent calcium concentration is very high. This process is ideal for the enhanced removal of calcium with the advantages of the reducing chemical usage and preventing scale formation in the wastewater treatment facility. Normally, the excess of $CaCO_3$ is usually removed through precipitation and filtration. Scale-forming calcium carbonate is removed from the cooling water through the mechanism of gas stripping of CO_2, resulting in a buffered pH of 8.8–9.0. However, with the VPT and the localized high temperatures, there is a shift in the equilibrium, calcium carbonite to aragonite, aiding in its precipitation. In addition, the treatment of cooling water without the use of any additional chemicals by the Vortex VPT includes bacterial eradication effects, providing a consistent control of bacteria, such as *Legionella*, which can develop in the warm oxygen-rich environment of recirculation cooling water.

6.7.4 Wastewater/Digester

A low-cost, environmentally friendly technology that can improve digester efficiency, increase methane production and throughput, and lower sludge generation demonstrated that VPT will result in cell lysis, or the breaking down of a cell, often by viral, enzymatic, or osmotic mechanisms that compromise its integrity, resulting in an increase in soluble Chemical Oxygen Demand (COD). This has been shown to accelerate the overall degradation reaction in the anaerobic digester, increase the decomposition of organic compounds normally found in wastewater, and increase digester gas generation [34].

6.7.5 Emerging Contaminants

Endocrine disruptors, pharmaceuticals, and other emerging compounds are now gaining interest in the treatment of both wastewater and drinking water. With the increase in soluble COD observed through cavitation, a potential result would be an increase in bioavailability of these organics. In addition, research has shown that with increased digester retention time, these compounds are removed to a greater degree. With cavitation, the removal may be similar but in a shorter amount of time. The generation of hydroxyl radicals as a result of hydrodynamic cavitation to remove trace amounts of endocrine-disrupting compounds, ECs, and microconstituents of concern has been shown to be very effective [35–37].

6.7.6 Biodiesel

A fuel alternative to conventional petroleum-based diesel engine fuel is manufactured from vegetable oils or animal fats by catalytically reacting these with a short-chain aliphatic alcohol, usually methanol or ethanol, using a process that is called either transesterification or alcoholysis. A current bottleneck is the reaction between the oil and the methanol. An improvement in efficient product conversion would be very desirable. Preliminary results show that the reaction time is reduced considerably and reaction temperatures are reduced from 140°F to 100°F. With some recent modifications, glycerin is formed almost immediately, a desired result not previously observed. The observed improvements in batch mode lend to evaluate and investigate hydrodynamic cavitation for both batch and continuous production of biodiesel on a commercial scale.

6.7.7 Ballast Water Treatment

Ballast water treatment before discharge is now becoming regulated worldwide because the discharge of ballast water from one environment into another has resulted in considerable ecological harm. The damage stems from the introduction of nonnative or exotic organisms into an environment such as the zebra mussels. Cavitation gives good results in eliminating small marine creatures in ballast water. If a certain amount of pressure and flow rate are used, the marine copepods will be crushed by the imploding vacuum bubbles. On the basis of the ability to control bacteria and other microorganism, the potential for ballast water applications is ideal.

6.7.8 Enhanced Organic Reactions

Cavitation can be effectively used for intensification of chemical reactions owing to the production of free radicals and conditions of high temperatures and pressures locally. Under the optimized conditions of sonification, the yield of sulfone was about five to six times than the conventional approach of using mechanical agitation only. In the oxidation of *p*-xylene, seven times more product could be obtained in the case of hydrodynamic cavitation than in the case of acoustic cavitation. The reaction was found to be considerably accelerated at ambient temperature by the introduction of air bubbles into water. Of course, all other gases can be introduced in this way, and they all share the same properties; the smaller the bubbles, the better transfer rate.

With the Watreco VPT, this is achieved in a similar device as in the "ice case." The only difference is that a small hose is connecting the vacuum in the center of the vortex with the air at the surface of the water, in this case, a pond. When the water is run through the vortex generator, the vacuum created literally sucks the air into the vortex cavity chamber from a suitably placed hose and further into the vortex. Here, it is ripped apart by the strong shear forces inside the vortex and at the end very fine bubbles down to nanoscale are injected into the surrounding pond water.

The result is a very efficient aerator that makes it possible to keep the oxygen levels at the top, even in harsh conditions. In this way, very little energy is needed in order to keep the fish in fish tanks alive or as in the pond case, to control algae growth by no other means than the oxygen in the air that surrounds us. Not only gas can be introduced in the bulk water, as the unit is actually a very good mixer. Various other substances can be used, even solids in powder form. However, it is preferred that the solids are dissolved or at least are in a liquid state to start, as the power has a tendency to clog the inside of an inlet tube.

6.7.9 Ice Fabrication

There are several applications where extremely small bubbles in water can either be removed or introduced depending on the actual need. One such need is to increase the flowing properties of water on artificially made ice. The idea here is to let the water flow out on the ice and into cracks and pores so the ice can be "healed" during maintenance. Today this is achieved by using hot (40°C–60°C) water that floats very smoothly; however, it also costs a lot of money because of the energy required for heating. A significant increase of pressure is required to lower the melting point of ordinary ice; for example, the pressure exerted by an ice skater on the ice reduces the melting point by approximately 0.09°C.

Using Watreco's Vortex removes the nanoscale bubbles altering the properties of the water and can achieve the same floating properties with 17°C–18°C water instead of using 40°C–60°C, saving a lot of energy in this way. Also, as the water contains fewer bubbles, the ice also does. Therefore, the ice gets stronger (fewer weak points from bubbles) but also the optical transparence gets better, making the messages from the advertisers more visible but also gives an opportunity to increase the set temperature of the ice, saving even more energy in the process.

Another use for the VPT is also in an ice arena, when the ice is built for the first time. Here, water is sprayed through a hose and a nozzle. In this case, the Watreco nozzle is shaped like a ball where the water comes out through a rather small hole. The very edge of the hole is a bit rounded and when the water runs over the rounded edge, it is exposed to Coanda forces that rip the water apart into extremely small drops, which slowly fall to the concrete piste where they freeze into thin layers of ice. In this way, the ice is "built" in thin layers, creating a very dense and strong ice. Work done with professional ice skaters by Watreco shows a hardening of the ice, improving performance for skaters, making the surface "quicker," but at the same time reducing the energy costs to make it.

6.8 SonoPhotoCatalytic Cavitation or SPCC

The combined effects of SPCC with AOP appear to offer a good solution leveraging the advantages of each individual processes to the treatment of water for the various applications discussed above. Further laboratory work is needed for SPCC, although initial indications are showing great potential for eliminating contaminants efficiently by combining both forms of irradiation (that of ultrasound and UV light) to the influent simultaneously. The common optimum operating conditions for sonochemical and photocatalytic oxidation coupled with the similarity in the mechanism of destruction, leading to a possible synergism and the possible mitigation of some of disadvantages observed for individual processing techniques due to the effects of the other technique, have prompted the development of SPCC reactors.

The present work aims at novel reactor designs for large-scale operations and further discussion about hybrid techniques with the mechanism between sonochemistry and heterogeneous photocatalysis. Different reactor configurations used for the hybrid technique need to be further analyzed, and recommendations have already been made for the design of an optimum configuration using both techniques. An optimum set of operating parameters has been based on the critical analysis of the available literature detailed in the references section. It has also been observed that sonophotocatalytic oxidation indeed results in significant enhancement of the oxidation intensity and thus the rates of degradation; however, its application on a larger scale of operation is hampered, perhaps owing to high cost and the lack of suitable design strategies associated with the sonochemical reactors. Future work is required to eliminate the drawbacks associated with the sonochemical and/or SPCC reactors.

Acknowledgments

Tony Diego gratefully acknowledges helpful discussions with Gerald Pollock at the University of Washington, Seattle.

References

1. Patek S.N. and Caldwell R.L. Extreme impact and cavitation forces of a biological hammer: Strike forces of the peacock mantis shrimp. *J. Exp. Biol.* 208(Pt 19), 3655–3664 (2005).
2. Pollack G. *The Fourth Phase of Water: Beyond Solid, Liquid, and Vapor.* Ebner and Sons Publishers (2013).
3. Lohse D., Schmitz B. and Versluis M. Snapping shrimp make flashing bubbles. *Nature* 413(6855), 477–478 (2001). doi:10.1038/35097152. PMID 11586346.
4. (a) Putterman S.J. Sonoluminescence: Sound into light. *Sci. Am.* 272, 32–37 (1995). (b) Putterman S. J. and Weninger K.R. Sonoluminescence: How bubbles turn sound into light. *Annu. Rev. Fluid Mech.* 32, 445–476 (2000).
5. (a) Brennen C.E. *Cavitation and Bubble Dynamics.* New York: Oxford University Press (1995). (b) Brennen C.E. Fission of collapsing cavitation bubbles. *J. Fluid Mechanics* 472, 153-166 (2002). doi: http://dx.doi.org/10.1017/S0022112002002288.
6. Franc J.P. and Michel J.M. Fundamentals of cavitation. In: *Fluid Mechanics and Its Applications,* vol. 76, XXII, 306 pp. New York: Springer (2004).
7. Stepanoff A.J. Cavitation properties of liquids. *J. Eng. Gas Turbines Power* 86(2), 195–199 (1964). doi:10.1115/1.3677576.
8. Fruman D.H., Reboud J.L. and Stutz B. Estimation of thermal effects in cavitation of thermosensible liquids. *Int. J. Heat Mass Transfer* 42, 3195–3204 (1999).

9. Holl J.W., Billet M.L. and Weir D.S. Thermodynamic effects on developed cavitation. *J. Fluids Eng.* 103(4), 534–542 (1981). doi: 10.1115/1.3241762.

10. Watanabe S., Hidaka T., Horiguchi H., Furukawa A. and Tsujimoto Y. Steady analysis of thermodynamic effect of partial cavitation using singularity method. *Proc. of FEDSM 2005*, Houston, TX, June 19–23 (2005).

11. Callenaere M., Franc J.P. and Michel J.M. The cavitation instability induced by the development of a re-entrant jet. *J. Fluid Mech.* 444, 223–256 (2001).

12. Harada H. Sonophotocatalytic decomposition of water using TiO_2 photocatalyst. *Ultrason. Sonochem.* 8(1), 55–58 (2001).

13. Gogate P., Rajiv K., Pandit A. and Tayal R. Cavitation: A technology on the horizon. *Curr. Sci.* 91(1), 35–46 (2006).

14. Kavitha S.K. and Palanisamy P.N. Photocatalytic and sonophotocatalytic degradation of Reactive Red 120 using dye sensitized TiO_2 under visible light. *World Acad. Sci. Eng. Technol.* 5(1), 1–6 (2011).

15. Joseph C.G., Li Puma G., Bono A. and Krishnaiah D. Sonophotocatalysis in advanced oxidation process: A short review. *Ultrason. Sonochem.* 16(5):583–589 (2009).

16. Wu Y., Huang Y., Zhou Y., Ren X. and Yang F. Degradation of chitosan by swirling cavitation. *Innovative Food Sci. Emerging Technol.* 23, 188–193 (2014).

17. Taghizadeh M.T. and Abdollahi R. Sonolytic, sonocatalytic and sonophotocatalytic degradation of chitosan in the presence of TiO_2 nanoparticles. *Ultrason. Sonochem.* 18(1), 149–57 (2011).

18. Frenzel H. and Schultes H.Z. *Phys. Chem.* B27, 421 (1934).

19. (a) Marinesco N. and Trillat J.J. *Proc. R. Acad. Sci.* 196, 858. (1933). (b) Walton A.J. and Reynolds G.T. *Adv. Phys.* 33, 595 (1984).

20. Gaitan D.F., Crum L.A., Roy R.A. and Church C.C. *J. Acoust. Soc. Am.* 91, 3166 (1992).

21. Gaitan, D. F. Transient cavitation in high-quality-factor resonators at high static pressures. *J. Acoust. Soc. Am* 127(6), 3456–3465, 20 (2010).

22. Diodati P. and Giannini G. Cavitation damage on metallic plate surfaces oscillating at 20 kHz. *Ultrason. Sonochem.* 8(1), 49–53 (2001).

23. Suslick K.S., Didenko Y., Fang M.M., Hyeon T., Kolbeck K.J., McNamara W.B. III, Mdleleni M.M. and Wong M. Acoustic cavitation and its chemical consequences. *Phil. Trans. Roy. Soc. London A*, 1999, 357, 335–353 (1999). Available at http://www.scs.illinois.edu/suslick/documents/nature.030205.pdf.

24. Flannigan D.J. and Suslick K.S. Plasma formation and temperature measurement during single-bubble cavitation. *Nature* 434, 52–55 (2005).

25. Suslick K.S. and Flint E.B. Sonoluminescence of non-aqueous liquids. *Nature* 1987, 330, 553–555. (1987).

26. Matula T.J. and Crum L.A. Evidence for gas exchange in single-bubble sonoluminescence. *Phys. Rev. Lett.* 80, 865 (1998).

27. Brenner M., Hilgenfeldt S. and Lohse D. Single bubble sonoluminescence. *Rev. Mod. Phys.* 74(2), 425–484 (2002). doi: 10.1103/RevModPhys.74.425.

28. Milton K.A. High energy theory—quantum field theory—Casimir effect. *World Scientific* (2001). Available at http://arxiv.org/pdf/hep-th/0010140.pdf.

29. Liberati S., Belgiorno F. and Visser M. Dimensional and dynamical aspects of the Casimir effect: Understanding the reality and significance of vacuum energy. Available at http://arxiv.org/abs/hep-th/0010140v1.

30. Shirsath S.R., Pinjari D.V., Gogate P.R., Sonaware S.H., Pandit A.B. Ultrasound assisted synthesis of doped TiO2 nano-particles: Characterization and comparison of effectiveness for photocatalytic oxidation of dyestuff effluent. *Ultrason. Sonochem.* 20(1), 277–286 (2013).

31. Chand R. *General Organic Acoustic, Hydrodynamic Oxidation.* University of Abertay, Scotland. Available at http://hdl.handle.net/10373/839.

32. Dynaflow Inc. J MD. Hydrodynamic Oxidation and Disinfection, CFD. Available at http://www.dynaflow-inc.com/Products/Brochures/Oxidation_Trifolded.pdf.

33. Hydro Dynamics, Inc. Georgia Hydrodynamic Fuel Technologies, Food Processing. Available at http://www.prweb.com/releases/2014/06/prweb11921687.htm.

34. Eimco Water Technologies. UT ultrasonic digester. Ovivo (Formerly Eimco Water Technologies), West Valley Salt Lake City, UT.

35. Vinu R. and Madras G. Kinetics of Sonophotocatalytic degradation of anionic dyes with Nano-TiO2. *Environ. Sci. Technol.* 43(2), 473–479 (2009). doi: 10.1021/es8025648.

36. Ragaini V., Selli E., Bianchi C.L. and Pirola C. Sono-photocatalytic degradation of 2-chlorophenol in water: Kinetic and energetic comparison with other techniques. *Ultrason. Sonochem.* 8(3), 251–258 (2001).

37. Gogate P.R. and Pandit A.B. Sonophotocatalytic reactors for wastewater treatment: A critical review. *AIChE J.* 50(5), 1051–1079 (2004). doi: 10.1002/aic.10079.

7

Bilge and Ballast Water Treatment Using Nanotechnology

Paul L. Edmiston

Department of Chemistry, The College of Wooster and ABSMaterials Inc., Wooster, Ohio, USA

CONTENTS

7.1 Introduction

A generally unrecognized, yet important, challenge in environmental protection is the disposal of wastewater generated by maritime vessels. Every ship must manage water associated with bilge pumping and ballast tank level control. Of the two types, bilge water poses the greater challenge in disposal owing to the presence of solids, colloids, and metals, in addition to free and dissolved oil. The volume of bilge water produced ranges from 0.5 to 50 m^3/day (150–65,000 gal/day) depending on the size of the vessel.[1] A study of cruise liners sailing in Alaskan waters indicated discharge rates of 25,000 gal/week.[2] Despite the large volume, bilge water only represents about 20% of the several million tons of oily wastewater discharged annually from seagoing vessels.[3] Other sources of water include engine or turbine wash water, exhaust gas scrubbing, and boiler economizer blowdowns.

The International Maritime Organization (IMO) specified in 1973 that bilge water discharged into the sea must have a residual oil content below 15 ppm.[4] (This rule is often referred to as MARPOL 73/78 as the protocol was modified and ratified in 1978.) In the subsequent 40 years, many ports have developed regulations that are more stringent in terms of water quality requirements. The ports of Long Beach and Los Angeles prohibit the discharge of bilge water and other types of engine room wastewaters within their jurisdiction.[5] Establishment of such regulations are due, in part, to the lack of shipboard treatment systems that have the ability to remove dissolved contaminants such as antifreeze, gasoline, surfactants, and cleaning agents. Moreover, while at dock, there exists the ability to pump bilge water to shore where a more extensive treatment infrastructure can be installed. It is uncertain whether stricter international regulations will be developed as maritime traffic increases and ever-larger ocean transport vessels are being built. Using statistics from the Port of Los Angeles, total tonnage to the port has increased from 26.2 million metric tons (MT) in 1973 to 158.2 million MT in 2011, a nearly 6-fold increase.[6] It may be time to reexamine wastewater discharge regulations from an international perspective to ensure the environmental quality of the oceans. However, enforcement of regulations is difficult while ships are in transit in international waters. Verification of MARPOL regulations is currently accomplished through the use of oil-in-water monitors and appropriate logbooks.

Ballast water has a different management strategy as the water is typically free of chemical contaminants. The concern is that the water is transported en masse from port to port and may contain invasive species or pathogens when discharged in a new environment.[7] It has been hypothesized that the North American invasion of freshwater zebra mussels occurred via ballast water discharged by ocean-going ships en route from the Black and/ or Caspian seas.[8,9] Treatment of ballast water generally involves either filtration or chlorination of the water before discharge to remove or kill organisms, respectively.[10] Some port facilities have ballast water exchange systems that prevent discharge of water into the environment. Overall, technologies to manage ballast water are relatively mature with a reduced need for advanced technologies outside of monitoring.

The composition of bilge water varies among vessels. Bilge fluids are becoming more complex owing to the sources of contamination from oil additives, solvents, and particles contained within the various potential input streams (Figure 7.1).[11] Certain waste streams,

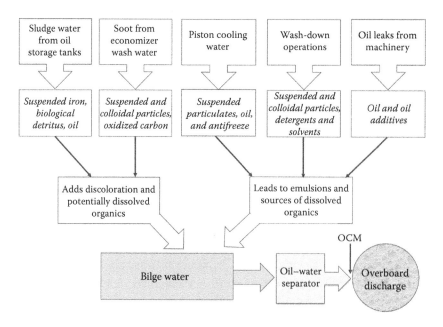

FIGURE 7.1
Diagram of the various bilge water contaminant sources. OCM, oil content monitor.

including steam condensate, boiler blowdown, and sink drainage located in various machinery spaces, can drain into the bilge. Leaks from the ship's propulsion and auxiliary systems, as well as precipitation and green water, can also enter the bilge. The presence of surface-active compounds such as surfactants and polymers leads to complex oil-in-water emulsions in bilge water that prevent physical separation methods from working effectively. Emulsion destabilization can be improved by elevating the temperature of the water, increasing the centrifugal force, or adding chemical demulsifiers. All these measures significantly increase the cost of treatment and in the case of chemical additives can lead to a new contaminant stream that enters the ocean after oil removal. Flocculation can also help to break emulsions; however, processes require addition of chemicals and increasing the pH to alkaline conditions.[12]

Current and future regulations have created a need for innovative technology to economically separate oil, especially owing to hard-to-treat emulsions found in combination with mixed contaminants characteristic of many bilge fluids. Removal of a wider variety of dispersed and dissolved chemicals is becoming more of a priority as the toxicity of some chemical additives and other surface-active compounds are much higher than that of oil. A variety of nanoengineered membranes and sorbent filters have been developed to conduct separations using molecular-level architectures to control the properties of oleophilicity and hydrophilicity. These materials are reviewed in the context of shipboard water management, including the economics of deployment in comparison with standard centrifuge-based oil–water separators.

7.2 Overview of Current Practices

Physical separation methods are most commonly used to meet MARPOL 73/78 guidelines. The primary step in treatment is the use of a bilge tank where a majority of the oil coalesces and can be removed by the phase separation into two layers (Figure 7.2). The free oil can then be pumped to a holding tank through a strainer to remove larger particles. Secondary treatment to remove dispersed oil is typically accomplished with either filtration or, more often, centrifugation to meet the 15 ppm standard. Polymer additives or dissolved iron can be added to induce coagulation or flocculation; however, such practices are

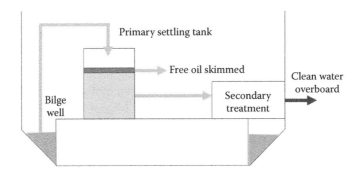

FIGURE 7.2
General treatment scheme for shipboard bilge water treatment.

not preferred by ship operators owing to cost and unavailability of consumables at foreign ports of call. Often, bilge water is heated to achieve an optimal temperature for centrifugal separation, reducing viscosity of the continuous phase and facilitating formation of larger oil droplets. A number of commercial systems are available on the market that use these various physical separation processes. Performance at <5 ppm is typically only guaranteed with use of optional final polishing filters that can be bypassed during routine operation. After separation, sludge is often dewatered and stored. In certain cases, recovered oil can be used as fuel; however, contamination may prevent usage in this manner. Oil-in-water content monitors are often employed in the discharge stream to be in compliance with IMO Resolution MEPC 107(49).

Alternative approaches to physical separation have been investigated. Electrocoagulation has been studied as a means to improve separation, especially for emulsified systems.[13] The process of electrocoagulation uses an electric current to dissolve a sacrificial iron or aluminum anode creating cations that are attracted to the fine negatively charge droplets of contaminants. Agglomeration of the droplets or particles destabilizes the suspension, leading to phase separation. Hydrolysis at the electrodes creates gas bubbles of oxygen or hydrogen that facilitate separation by flotation but also create a significant explosion hazard owing to the combustibility in enclosed environments. Laboratory-scale measurements indicated that a >95% decrease in oil and >99% decrease total suspended solids were achievable using electrocoagulation. Biodegradation has also been examined for bilge water treatment at the laboratory scale. Aerobic bioreactors were seeded with sediment from the mouth of the Cheng-Jenn River in Kaohsiung Harbor (Taiwan), which was presumed to contain bacteria necessary to degrade hydrocarbons associated with bilge water.[14] Experiments indicate that a >90% reduction in total organic carbon levels was achieved with emulsified diesel fuel being the contaminant. Practical disadvantages are the time it takes to degrade the oil (rate = 0.4 kg oil/m^3d) and the necessity to maintain the microbial population over time. An evaluation of bilge water treatment by a combination of ultrafiltration (UF) and membrane distillation (MD)[15] and reverse osmosis[16] has also been conducted. Substantial removal of organics to <5 ppm was achieved by the UF/MD method. Permeate flow varied according to oil concentration and temperature, but the system generally demonstrated ~200 kg/m^2d of permeate flux.[15] Oil–water separation methods adapted from the petroleum industry also serve as potential methods for bilge water separation.[17]

7.3 Membrane Separation

7.3.1 Nanotechnology-Based Membranes

Numerous nanotechnology-based membrane separation methods have been described for separation of oil and water that have application to bilge water treatment.[18] Separation of oily water emulsions is a relatively mature technology typically involving the use of microfiltration (MF) with pore sizes of 100–10,000 nm; however, UF, which employs smaller pore sizes (100–500 nm), is sometimes required for colloidal waste streams.[19,20] Application of nanotechnology to membrane design has allowed for improvements in the functionality of traditional levels, including higher permeability, better selectivity, and increased resistance to fouling. There are three general types of nanotechnology-based membrane materials: (i) nanostructured ceramic membranes; (ii) thin film nanocomposite organic membranes; and (iii) organic–inorganic materials. An excellent review of nano-engineered membranes was prepared by Pendergast and Hoek.[21] Ceramic membranes have the advantage of being the most mechanically stable at high pressures and able to handle mixed waste streams that contain surfactants, disinfectants, and other potential fouling agents. Most nanostructured ceramic membranes are based on zeolite materials that consist of a three-dimensional cross-linked $(Si/Al)O_4$ tetrahedral framework with interconnected cavities that create a cage-like structure, and allow for the movement of water and ions through an interconnected pore architecture.[22] The Si/Al ratio of the zeolite is the most important factor in determining the chemical stability and hydrophobic properties.[23] Increasing the relative Si content lowers surface charge and water permeability.[24] Separation is most often based on a molecular sieving mechanism making the pore size and framework density the primary factors in separation. A variety of zeolite materials have been nanoengineered to tailor the framework and the internal surfaces to promote separations. For instance, anionic sites can be introduced in the zeolite matrix to allow ion exchange type of interactions. This level of chemical and morphological control on the nanoscale allows for separations to be achieved similar to those with reverse osmosis.[25] Many nanostructured ceramic zeolite membranes have been developed that have the ability to desalinate water by promoting water permeability and rejecting ion flux.[26] Although this is a notable achievement, oil-in-water treatment relative to bilge water management does not generally require reverse osmosis levels' separation. Toward shipboard management applications, membrane permeability is more important than molecular-scale selectivity in separation. As a result, organic thin films supported on high-flux ceramic materials have been developed for oily wastewater treatment that optimizes flux and rejection of oil. Peinemann and colleagues initially demonstrated the use of coblock polymers and self-assembly to achieve highly ordered porous membranes overlaying nanocylinder-based supports.[27,28] Asatekin and Mayes used an amphiphilic copolymer comb composed of polyacrylonitrile-*graft*-poly(ethylene oxide) to modify UF membranes for improved resistance to fouling.[29] Results indicated that comb-like polymer-modified membranes allowed for ~42% chemical oxygen demand removal from refinery wastewater and could be cleaned by physical means, making them attractive for use. Faibish and Cohen reported a similar approach by covalent modification of a zirconia-based ceramic UF membrane with poly(vinylpyrrolidone).[30] Linkage of the organic polymer to the zirconium was accomplished using a silane modification of the surface. The resulting membranes were effective at separating microemulsions and demonstrated improved resistance to fouling by organics. Permeate flux of native membranes irreversibly declined upon exposure of anionic surfactants; however, polymer thin film modification prevented fouling under the same conditions.

Two Spanish groups examined the use of high-porosity ceramic membranes specifically for bilge water treatment. Benito and coworkers used a multilayer membrane with a tubular cordierite support with either α-Al_2O_3 or γ-Al_2O_3 layers deposited on the surface by dip-coating or the sol–gel process, respectively.[31] Reduction of dispersed oil levels in simulated bilge water to <15 ppm was achievable to pressures up to 0.7 MPa. García et al. tested organically modified ceramic membranes manufactured by Orelis and PCI Membrane systems in a pilot-scale system to separate hexane and water.[32] The ceramic substrate membrane was zirconia coated with polyethersulfone. The results from these tests are most insightful as they provide data for long-term use and performance. Initially, the membranes performed well with almost zero hexane in the permeate flow. However, hexane fluxes increased when the membranes were exposed to solvents, indicating partial solvation in the polymer layer may decrease separation efficiency. The results did not agree with simple models of solvent flux such as the Hagen–Poiseuille equation. Overall, the data indicate the importance of pilot-scale measurements to provide effective evaluation of technology in a real-world scale application.

Molecular-scale polymer growth and nanoscale self-assembly for structure formation show great promise in the future. Polymers are inherently "soft," meaning they can be relatively homogeneous in composition while tolerating relatively high degrees of imperfection. The soft nature also implies that the structural arrangement can be altered by changes in solutes, electrical potentials, and mechanical forces creating an opportunity for changes in functionality and self-cleaning. Additional functionality can be built by adding nanoparticles to the membrane material to either degrade contaminants or inactive bacteria. Polymer–nanocomposite membranes can be developed by incorporating either magnesium oxide (MgO)[33] or silver nanoparticles[34] as antibacterial agents, and zero-valent iron for the removal of halogenated hydrocarbons, radionuclides, and organic compounds.[35] Nanocomposite materials can also be used to reduce fouling by increasing the hydrophilicity of the membrane. Polyethersulfone UF membranes dip-coated with titania nanoparticles were created by Luo et al., which demonstrated a significant drop and contact angle and wetting performance.[36] These studies demonstrate the creative potential available in polymer–nanocomposite materials for membrane-based separations.

7.3.2 Hydrophobic Mesh Coatings

High flux and resistance to fouling are key to achieving systems that can be sufficiently robust for ship bilge water treatment. A strategy toward this end is to use large pore openings and tailor the surface chemistry and/or nanostructure to control hydrophobicity. Development of artificial superhydrophobic surfaces[37] based on "rough" nanostructured surfaces has been accomplished using a variety of approaches, including template synthesis,[38] phase separation,[39] and electrodeposition.[40] These types of approaches can be used to modify nonwoven materials for rapid oil–water separation.[41] A notable example is from the Wang group, which created smart textiles and polyurethane sponges that are able to switch between superoleophilic and superoleophobic by grafting pH-responsive block copolymers to these materials.[42] Poly(2-vinylpyridine) in conjunction with polydimethylsiloxane were used to create surfaces that have switchable oil wettability due to the protonation state of the pyridine group (Figure 7.3). At neutral pH, the pyridine groups are deprotonated and the surface is oleophilic, whereas at pH <2 the pyridines are protonated and the surface becomes hydrophilic. The resulting nonwoven material can be used for remarkably fast and switchable oil–water separations, which was demonstrated by the separation of gasoline–water mixtures (Figure 7.4). One of the most attractive features of this technology is the use of inexpensive cellulose and polypropylene textiles that show good promise for large-scale production and application to bilge water treatment and other oily–water separations.

Functionalization of Nomax® alumina fabric with hydrophilic cysteic acid surface-stabilized alumina nanoparticles were reported from the Barron group with the similar goal of creating high-flux materials for water while preventing the flux of oil through the fabric mesh.[43] Cysteic acid (Figure 7.5) functionalizes the alumina surface via the carboxylate group, and at neutral pH the molecule exists in a zwitterion form that is hydrophilic. The resulting membrane systems were capable of screening phase-separated hydrocarbons from water with high tribological endurance and stability across a pH range from 2 to 12. A similar and promising approach modifying polyester membranes and stainless-steel

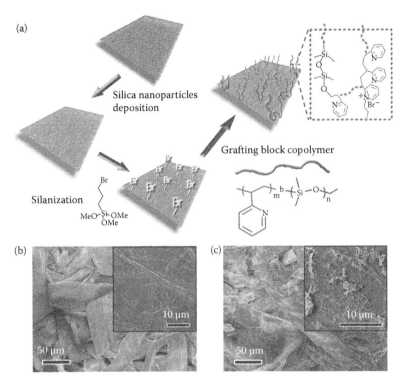

FIGURE 7.3
(See color insert.) Preparation and characterization of a surface with switchable superoleophilicity and superoleophobicity on a nonwoven textile substrate. (a) Diagram showing the preparation strategy for a surface with
switchable superoleophilicity and superoleophobicity on a nonwoven textile using a pyridine contain polymer
that changes charge state as a function of pH. (b) Scanning electron microscope (SEM) image of the raw textile. Inset: enlarged view of the surface of a single fiber. (c) SEM image of the textile after deposition of silica
nanoparticles and block copolymer grafting. Inset: enlarged view of the surface of a single fiber.

mesh was subsequently reported by Kota and coworkers, with a mixture of a hydrophilic
polymer, cross-linked polyethylene glycol diacrylate, and an oleophilic fluorinated polyhedral silsesquioxane.[44] An oil–water mixture or emulsion can be separated since the
microcrystalline regions of the polymer film reconfigure upon exposure to form a smooth,
noncrystalline surface that hydrogen bonds with the water. The water can subsequently
flow through the mesh while the oil is retained. The low cost of the mesh substrate makes
the approach commercially interesting.

Recent developments in the nanostructuring and functionalization of nonwoven surfaces indicate that such technology may be useful for bilge water treatment where the
primary goal is removal of free or dispersed oil. Several other hydrophobic mesh materials have been reported for the separation of oil and water.[45,46] Removal of dissolved compounds has not been demonstrated with mesh systems; however, there is no regulatory
necessity to remove dissolved organics. Further studies will need to encompass separation
of emulsions. If successful, mesh-based systems may be very attractive for use in bilge
water treatment as the energy requirements would be much less in the physical separation
by high-flux membranes as compared with centrifugation or UF. Overall, the area will be
interesting to follow in the coming years.

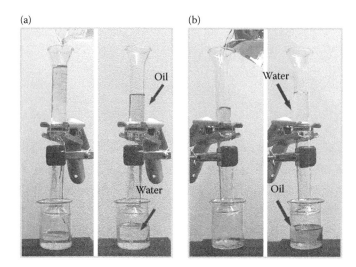

FIGURE 7.4
Controllable oil–water separation using the polymer functionalized textile. (a) The functionalized textile was fixed between two glass tubes as the separation membrane. A mixture of gasoline and water was poured into the upper glass tube. The gasoline selectively passed through the textile, whereas the water remained in the upper glass tube (right). (b) The functionalized textile was first wetted with acidic water (pH 2.0) before the water–oil separation process. Water selectively passed through the textile, whereas gasoline remained in the upper glass tube (right).

FIGURE 7.5
Structure of cysteic acid in the zwitterionic form used to increase hydrophilicity of membrane surfaces.

7.3.3 Nanofiber-Based Membranes

An alternative use of nanotechnology in membrane science is to create the membrane out of nanoengineered materials. The most common approach is to use electrospun nanofibers to create nonwoven materials that can significantly enhance separation efficiency by creating mats with very high surface-to-volume ratios (10–500 m^2/g). The fully accessible interconnected nanoporous structure allows for high flux with pore sizes that allow for MF and UF membrane applications. A unique aspect of the electrospinning technique is the ability to control the surface functionality by the choice of polymer blend or incorporation of nanofillers. An excellent review of the electrospun nanofiber membranes for environmental applications was done by Yoon et al.[47] In terms of application of bilge water treatment, coalescence of dispersed oil by fibrous membranes and beds has been established for several decades.[48–50] One of the first reported use of nanofiber composite filters for oil-in-water separations was by Shin and Chase who tested polyacrylonitrile and poly(meta-phenylene isophthalamide) electrospun membranes supported on fiberglass.[51] Composite filters showed improved separation efficiency compared with glass fiber membranes, but exhibited higher pressure drops. A similar approach was investigated by Wang and coworkers using a three-layer membrane system composed of a conventional nonwoven microfibrous support overlaid with an electrospun nanofiber midlayer and a hydrophilic nanocomposite top layer.[52] The nanofibers composing the midlayer were electrospun poly(vinyl alcohol) cross-linked with glutaraldehyde to increase structural integrity. The top layer consisted of either a polyether-*b*-polyamide copolymer or poly(vinyl alcohol) hydrogel embedded with surface-oxidized multiwall carbon nanotubes. These membranes showed high flux (330 L/m²h) and achieved 99.8% rejection of emulsified oil. Tests using simulated bilge water were done using soybean oil emulsified with a nonionic surfactant (Dow Corning 193 fluid). The nanocomposite membranes exhibited flux rates significantly higher than commercially available Pebax® 1074 (polyethyleneoxide-*b*-polyamide 12 copolymer)-coated membranes. Efficiency was >99.5% rejection of emulsified oil. Additional designs using cellulose also showed good effectiveness with simulated bilge water.[53]

Continued work on optimizing the flux, separation efficiency, and mechanical stability of nanofiber-based filters should further improve performance. In many respects, these membranes will compete with mesh designs. One of the questions concerning electrospun fibers is the throughput rate at which membranes can be achieved. Manufacturing large mats of electrospun membranes currently takes substantial amounts of time. Material costs are relatively low, so if fabrication rates can be improved there may be utility of nanofibrous UF membranes in large-scale applications such as bilge water management.

7.4 Adsorbents

An alternative approach to physical separation by centrifugation or membrane-based systems is the use of high-capacity sorbents to remove organics from water streams. Such processes differ from flocculation in that dissolved chemical additives are not added when using sorbents. Chemical flocculent residues may be part of the discharged water, which may have limitations depending on discharge regulations. Given the high levels of oil in bilge water, adsorbents are currently economical for only secondary treatment or polish steps and would be used after skimming or even centrifugation to remove hard-to-treat emulsions. One of the advantages of sorbents is the ability to remove dissolved oil or other organics, a process that can significantly improve water quality. Activated carbon is the most widely used sorbent material. A disadvantage of sorbents is the need to store the media pre- and post-use, as well as having logistics for disposal and resupply at port. In many ports of call, proper disposal facilities may be lacking. Use of water treatment media requires that a sorbent have high capacity, long working lifetime, and the potential for regeneration onboard. Nanotechnology advances have led to new materials that have functional capabilities beyond traditional media such as activated carbon for use in bilge water treatment.

7.4.1 Surface-Modified Mineral Particles

Selective absorption of oil from water has been accomplished using superhydrophobic calcium carbonate powders by Arbatan and coworkers.[54] Extraordinary levels of hydrophobicity can be achieved through a combination of surfaces that are nonpolar (low surface energy) and have microscale and nanoscale surface roughness. Typically, a material is deemed superhydrophobic when the contact angle (i.e., the angle between the edge of a drop of water and the surface) is >150° and contact angle hysteresis is <5°. Precipitated calcium carbonate powder commercially available as Precarb 100 is inherently needle-shaped but hydrophilic. To create an oleophilic surface, the powder is modified with steric acid, which absorbs to the calcium carbonate surface via carboxyl group complexing calcium ions and orienting the alkyl chain away from the surface. Steric acid modification and calcium stearate precipitation also generates a hierarchically rough surface morphology leading to superhydrophobicity (Figure 7.6). The resulting powder when mixed with a diesel–water mixture selectively absorbs the hydrocarbon, leading to clumps of solid material that can be easily removed from the water (Figure 7.7). The two main advantages

FIGURE 7.6
SEM images of the porous structure of Precarb 100 calcium carbonate. (a, b) Before the treatment with steric acid; (c, d) after treatment with steric acid.

FIGURE 7.7
Oil–water mixture (a) before powder addition; (b) right after powder addition; and (c) after separation.

of the superhydrophobic powder are that the material is environmentally compatible and is low cost. To date, it is unknown how such methods would work on emulsified oil–water systems; however, this technology development will be interesting to follow to see if it has applications for bilge water treatment.

7.4.2 Organoclays and Modified Polymers

Surface modification of clay particles has also been investigated for oil-in-water removal.[55] In general, organoclay technology has not yet involved nanoscale engineering; however, the use of organoclay is worth noting because of the low cost of the media relative to other technologies. Organoclays have been successfully used as a coalescence and filtration media for emulsified oil. Typically, bentonite or other clays are used and made oleophilic by treatment with quaternary amines that bind to the surface by a charge–charge interaction. Organoclays remove five to seven times the amount of emulsified oil compared with activated carbon.[56] Therefore, these materials are often used as prepolish media with an estimated cost savings of about 55% for the media compared with carbon. Polymer supports have been used as an alternative to clay-based systems. For instance, isobutyl methacrylate polymers (ELVACITE® 2045) have been modified with an oily coating to create an absorbent filter media available commercially as MYCELX®. The MYCELX system has been pilot tested to treat bilge water (220 gal/min) at a shipyard-based facility.[57] According to the study, bilge water could be treated for $US0.02/gal at the dock using a completed system. Neither modified clays or modified polymers can be regenerated and thus may only practically be used at onshore treatment facilities where media can be stored and replaced. It will be interesting to see if nanotechnology approaches can be applied to organoclay materials to improve performance.

7.4.3 Nanoengineered Animated Organosilica

Organosilica materials that rapidly swell have been created using the sol–gel method[58] where alkoxysilane precursors are polymerized to form Si–O–Si linkages. The name is derived from the fact that a system starts as a solution of monomers that upon polymerization yields a gel state when sufficient cross-linking increases the viscosity. Solvents can be removed after gelation to yield hard materials and ceramics with various chemical compositions and nanoscale architectures.[59,60] The sol–gel method is versatile in that properties of the final material (e.g., surface composition, pore size, morphology, hydrophobicity) can be carefully controlled by the choice of precursor, kinetics of polymerization, and postgelation processing steps.[61]

7.4.3.1 Physical Characteristics

Nanotechnological approaches have been used to create organosilica absorbents that have high capacity for the removal of organics from water.[62] The materials are created from molecular scale self-assembly to generate nanoscale particles that spontaneously assemble into an interconnected, yet highly flexible matrix (Figure 7.8). On the macroscale, the organosilica looks like glass but swells up to eight times its mass when absorbing organic liquids, as the nanoengineered matrix expands to a relaxed state.[63] Interestingly, the swelling due to incursion of organic liquids is both extremely rapid and forceful. Swelling can produce >200 N/g of force[64] derived from tension within the dry and collapsed matrix, meaning the absorption of oil can cause the sorbent to lift 20,000× its own weight. The surface area and pore volume are relatively large: 500–1000 m^2/g and 0.6–2.5 mL/g,

FIGURE 7.8
(a) Electron micrograph of swellable organosilica showing flexible interconnected nanoparticulate network.
(b) Organosilica before and after being exposed to organic solvent to induce swelling. Gravimetric gain, 680%;
volume gain, 400%.

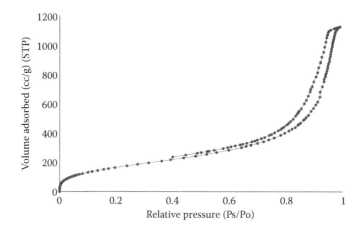

FIGURE 7.9
Nitrogen absorption–desorption isotherm for swellable organosilica in the nonswollen state.

respectively. Nitrogen porosimetry measurements indicate that the material is mesoporous (Figure 7.9). Swellable organosilica commercially available as Osorb® does not absorb water, but has high affinity for dissolved and disperse organics.

7.4.3.2 Removal of Dispersed and Dissolved Oil

The presence of surfactants in bilge water leads to colloidal systems that are difficult to treat. A highly oleophilic version of swellable organosilica, Osorb-EB, was developed that possesses mesoscale pores 50–150 nm in diameter to treat emulsified oil. The larger pore size aids in the capture of small emulsified oil droplets. Emulsified oil was prepared by adding 1000 ppm fuel oil to water containing 0.05% w/v undecyl glucoside nonionic surfactant. In batch-scale experiments, 1.5% w/v Osorb-EB was combined with the oil–water mixture and mixed by ultrasonication for 5 s. After separating the media by flotation, the oil content was measured by gas chromatography mass spectrometry (Figure 7.10) indicating that the total petroleum hydrocarbon constant, both free and dissolved, dropped from 850 to <5 ppm. Total organic

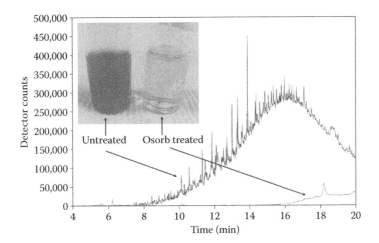

FIGURE 7.10
(See color insert.) Photographs and gas chromatograms of emulsified bilge water simulant before and after addition of 1.5% w/v Osorb-EB sorbent media. Oil content is reduced to <5 ppm.

carbon decreased from 1164 to 207 ppm, indicating that the nonionic surfactant was retained in the water although the oil was effectively stripped from the water. Currently, researchers are developing a variation of swellable organosilica that also absorbs nonionic surfactants and metal ions by altering the surface chemistry of the porous matrix. In addition, a 10 gpm pilot-scale treatment system has been built that will be used to evaluate treatment of oil–water systems. One of the key engineering steps has been accommodating the swelling.

7.4.3.3 Regeneration

Use of sorbent media on shipboard is limited by storage space of consumables and waste. As a result, physical separators such as membranes and centrifuges are attractive in limiting the physical footprint of bilge water treatment systems to make room for other equipment and cargo. Sorbent media may only find application if the materials can be regenerated for continuous reuse. Specifically, the oil must be removed from the sorbent in an efficient and autonomous manner. Currently, the only sorbent that has been demonstrated to be regenerable is swellable organosilica.[62] It was determined that the swelling is fully reversible if the absorbates are removed. The inert silica matrix allows for various processes to be used. Solvent rinsing and thermal evaporation was reported in the literature. Steam rinsing and supercritical carbon dioxide regeneration also shows promise in regenerating used swellable organosilica. Further research into sorbent technologies should be paired with developing regeneration methods if such methods are to be applied to bilge water treatment on ocean-going vessels.

7.5 Comparative Technology Status and Future of Bilge Water Management

There is very little scientific research currently being done on the treatment of bilge water despite the vast volumes that are discharged into the environment and the need for improved bilge water separations. A primary reason for this lack of interest by the scientific community is that, from a practical point of view, there is little incentive to develop new technology. Bilge water is most often discharged in international waters where enforcement and monitoring is typically nonexistent. Naval ships have sovereign immunity from environmental regulations and thus do not have to comply with MARPOL standards. Special areas with higher regulatory standards do exist in ports and environmentally sensitive bodies of water; however, shipboard bilge water storage capacity is generally sufficient to allow for temporary holding and eventual discharge once a vessel is back in the open ocean or at dock. Another difficulty for research and development activity is the lack of bilge water samples or standardized protocols to allow for measurements. To date, there has not been a comprehensive study of bilge water composition, leaving an extensive void in the general understanding of these waste streams. The constant movement of ships in the open ocean makes such studies logistically difficult. Lack of samples is compounded by the fact that there is little interest by the shipping industry to conduct these investigations.

However, there are factors that may help spur high-technology development in the area of bilge water management. The first is the growing size of the shipping industry and the number of vessels being deployed worldwide. Larger and more numerous ships lead to larger total discharges, and thus increased scrutiny of environmental impact. Cruise ship operators garner the most attention because of their greater visibility by the general public and operation in scenic regions. Fines have been assessed against cruise ship operators stemming from observed discharges. In one of the largest incidents, in 2002, Carnival Corp. pleaded guilty to six felony counts related to oily discharges in the Caribbean and seas off Florida between July 1998 and January 2001. The company paid $US 18 million in fines and agreed to establish an environmental compliance department.[65] Such incidents led to advocacy groups petitioning the US Environmental Protection Agency (EPA) Region 9 to established a No Discharge Zone off the coast of California.[66] Beginning in March 2012, all large vessels of 300 gross tons or greater are prohibited from discharging any water into navigable waters within 3 miles from shore. These relatively small, yet extreme, measures may be indicative of future regulatory steps taken by governments and international agencies. Impetus for new technology will increase if a large to mid-size navy or shipping company creates a goal of improved bilge water discharge for the purposes of environmental stewardship. The US armed forces are moving in such a direction. The National Defense Authorization Act of 1996 amended Section 312 of the Federal Water Pollution Control Act requiring the US Secretary of Defense and the Administrator of the EPA to develop uniform national discharge standards for vessels. Only Phase I of III has been completed to date. Phase I, established in 1999, identified and characterized 39 discharges (new 40 CFR part 1700), 25 needing control. The Phase I rule determines the type of vessel discharges that require control by marine pollution control devices, which include surface bilge water–oil water separator discharge. Phase II, establishing standards for discharges is still undergoing review. Phase III, implementing technology to meet standards, lies in the future. If seagoing organizations and naval forces insisted on improved bilge water management, the market would be in excess of a $US 1 billion annually offering incentives to academic and industrial researchers.

Nanotechnological developments have occurred tangentially via the goal of improved separation methods for the petroleum industry where oil–water separation is an important process. In many respects, treatment of bilge water has much in common with treatment of produced water, which may help drive innovations. Of the technologies reported within this review, physical separators represent the only technology that is widely deployed today. Phase separation methods will likely always play a key role in bilge water treatment. Recent developments using modified meshes for oil–water separation look promising, especially because of the low cost of the substrates and high flow rates. These technologies still need to be scaled and pilot tested. Of the nanotechnological approaches reported within, only swellable organosilica is commercially available. Sorbents have good potential because of their ability to remove dissolved or highly emulsified organics. However, owing to the amount of media that is required for primary treatment, it is likely that sorbent media will be used as a polishing step after physical separation. Regeneration of sorbents would be attractive for economical and logistical reasons.

As the complexity of bilge water composition increases, the need for new technology will also increase, especially if environmental protection measures are pursued. It is expected that in the future bilge water treatment will require multistep treatment systems to achieve high water quality in an economical fashion. This will require an analysis of how different technologies work in tandem, rather than separately. For instance, addition of flocculants in upstream processing may have detrimental effects on nanocomposite membrane technologies used for secondary treatment or polishing. Regardless, innovations in oil–water separation using nanotechnology are providing a promising outlook for new and effective methods of bilge water treatment.

References

1. Tomaszewska, W., Orecki, A., Karakulski, K. Treatment of bilge water using a combination of ultrafiltration and reverse osmosis. *Desalination*, 2005, 185: 203–212.
2. EPA 842-R-07-005. Cruise Ship Discharge Assessment Report, Section 4: Oily Bilge Water, December 2008. Available at http://www.epa.gov/owow/oceans/cruise_ships/disch_assess.html.
3. Ochrony, K., Morskiego, Ś., Battyckiego, M. Report of the Seminar on Reception Facilities in Ports held in Turku, Baltic Sea Environment Proceedings, No. 50, Turku, Finland, 16–19, November 1992.
4. International Maritime Organization. International Convention for the Prevention of Pollution from Ships (MARPOL) 1973–1978.
5. Port of Long Beach and Port of Los Angeles Vessel Discharge Rules and Regulations, May 2012, Section 3.3.13.
6. Available at http://www.portoflosangeles.org/maritime/tonnage.asp.
7. National Research Council. *Stemming the Tide: Controlling Introductions of Non-Indigenous Species by Ballast Water*. National Academy Press, Washington, DC, 1996.
8. Hebert, P.D.N., Muncaster, B.W., Mackie, G.L. Ecological and genetic studies on Dreissena polymorpha (Pallas): A new mollusc in the Great Lakes. *Canadian Journal of Fisheries and Aquatic Sciences*, 1989, 46: 1587–1591.
9. Duggan, I.C., Van Overdijk, C.D., Bailey, S.A., Jenkins, P.T., Limén, H., MacIsaac, H.J. Invertebrates associated with residual ballast water and sediments of cargo-carrying ships entering the Great Lakes. *Canadian Journal of Fisheries and Aquatic Sciences*, 2005, 62(11): 2463–2474.
10. Tsolaki, E., Diamadopoulos, E. Technologies for ballast water treatment: A review. *Journal of Chemical Technology and Biotechnology*, 2009, 85(1): 19–32.
11. Marine Environmental Protection Committee. Guide to diagnosing contaminants in oily bilge water to maintain, operate and troubleshoot bilge water treatments systems, July 2009. MEPC.1/Circ.677.
12. Patterson, J.W. *Industrial Wastewater Treatment Technology*, 2nd ed. Butterworths, Stoneham, MA, 1985.

13. Asselin, M., Drogui, P., Brar, S.K., Benmoussa, H., Blais, J.-F. Organics removal in oily bilge water by electrocoagulation process. *Journal of Hazardous Materials*, 2008, 151(2–3): 446–455.

14. Yang, L., Lai, C.-T., Shieh, W.K. Biodegradation of dispersed diesel fuel under high salinity conditions. *Water Research*, 2000, 34(13): 3303–3314.

15. Gryta, M., Karakulski, K., Morawski, A.W. Purification of oily wastewater by hybrid UF/MD. *Water Research*, 2001, 35(15): 3665–3669.

16. Karakulski, K., Morawski, W.A., Grzechulska, J. Purification of bilge water by hybrid ultra-filtration and photocatalytic processes. *Separation and Purification Technology*, 1998, 14(1–3): 163–173.

17. Kajitvichyanukul, P., Hung, Y.-T., Wang, L.K. Oil water separation. In *Handbook of Environmental Engineering*, Vol. 4. Eds. Wang, L.K., Hung, Y.-H., Shammas, N.K. The Humana Press, Totowa, NJ, 2002.

18. Ashaghi, K.S., Ebrahimi, M., Czermak, P. Ceramic ultra- and nanofiltration membranes for oil-field produced water treatment: A mini review. *Open Environmental Journal*, 2007, 1: 1–8.

19. Mulder, M. *Basic Principles of Membrane Technology*. Kluwer Academic Publishers, London, 1996.

20. Kesting, R.E. The four tiers of structure in integrally skinned phase inversion membranes and their relevance to the various separation regimes. *Journal of Applied Polymer Science*, 1990, 41(11–12): 2739–2752.

21. Pendergast, M., Hoek, E.M.V. A review of water treatment membrane nanotechnologies. *Energy Environmental Science*, 2011, 4(6): 1946–1971.

22. Dyer, A. *An Introduction to Zeolite Molecular Sieves*. John Wiley Sons Ltd, Chichester, UK, 1988.

23. Kumakiri, I., Yamaguchi, T., Nakao, S. Application of a Zeolite A membrane to reverse osmosis processes. *Journal of Chemical Engineering, Japan*, 2000, 33(2): 333–336.

24. Noack, M., Kölsch, P., Seefeld, V., Toussaint, P., Georgi, G., Caro, J. Influence of the Si/Al-ratio on the permeation properties of MFI-membranes. *Microporous and Mesoporous Materials*, 2005, 79(1–3): 329–337.

25. Li, L., Dong, J., Nenoff, T.M., Lee, R. Desalination by reverse osmosis using MFI zeolite membranes. *Journal of Membrane Science*, 2004, 243(1–2): 401–404.

26. Duke, M.C., O'Brien-Abraham, J., Milne, M., Zhu, B., Lin, J.Y.S., da Costa, J.C.D. Seawater desalination performance of MFI type membranes made by secondary growth. *Separation and Purification Technology*, 2009, 68(3): 343–350.

27. Jia, M.D., Peinemann, K.V., Behling, R.D. Molecular-sieving effect of the zeolite-filled silicone-rubber membranes in gas permeation. *Journal of Membrane Science*, 1991, 57(2): 289–296.

28. Peinemann, K.V., Abetz, V., Simon, P.F.W. Asymmetric superstructure formed in a block copolymer via phase separation. *Nature Materials*, 2007, 6(12): 992–996.

29. Asatekin, A., Mayes, A.M. Oil industry wastewater treatment with fouling resistant membranes containing amphiphilic comb copolymers. *Environmental Science and Technology*, 2009, 43(12): 4487–4492.

30. Faibish, R.S., Cohen, Y. Fouling and rejection behavior of ceramic and polymer-modified ceramic membranes for ultrafiltration of oil-in-water emulsions and microemulsions. *Colloids and Surfaces A: Physicochemical and Engineering Aspects*, 2001, 191(1–2): 27–40.

31. Benito, J.M., Sánchez, M.J., Pena, P., Rodríguez, M.A. Development of a new high porosity ceramic membrane for the treatment of bilge water. *Desalination*, 2007, 214(1–3): 91–101.

32. García, A., Álvarez, S., Riera, F., Álvarez, R., Coca, J. Water and hexane permeate flux through organic and ceramic membranes: Effect of pretreatment on hexane permeate flux. *Journal of Membrane Science*, 2005, 253(1–2): 139–147.

33. Stoimenov, P.K., Klinger, R.L., Marchin, G.L., Klabunde, K.J. Metal oxide nanoparticles as bactericidal agents. *Langmuir*, 2002, 18(17): 6679–6686.

34. Morones, J.R., Elechiguerra, J.L., Camacho, A., Holt, K., Kouri, J.B., Ramirez, J.T., Yacaman, M. The bactericidal effect of silver nanoparticles. *Nanotechnology*, 2005, 16(10): 2346–2353.

35. Meyer, D., Bhattacharyya, D. Impact of membrane immobilization on particle formation and trichloroethylene dechlorination for bimetallic Fe/Ni nanoparticles in cellulose acetate membranes. *Journal of Physical Chemistry B*, 2007, 111(25): 7142–7154.

36. Luo, M.L., Zhao, J.Q., Tang, W., Pu, C.S. Hydrophilic modification of poly(ether sulfone) ultra-filtration membrane surface by self-assembly of TiO$_2$ nanoparticles. *Applied Surface Science*, 2005, 249(1): 76–84.
37. Feng, X., Jiang, L. Design and creation of superwetting/antiwetting surfaces. *Advanced Materials*, 2006, 18(23): 3063–3078.
38. Feng, L., Song, Y., Zhai, J., Liu, B., Xu, J., Jiang, L., Zhu, D. Creation of a superhydrophobic surface from an amphiphilic polymer. *Angewandte Chemie*, 2003, 115(7): 824–826.
39. Han, J.T., Xu, X., Cho, K. Diverse access to artificial superhydrophobic surfaces using block copolymers. *Langmuir*, 2005, 21(15): 6662–6665.
40. Wang, S.T., Feng, L., Liu, H., Sun, T.L., Zhang, X., Jiang, L., Zhu, D.B. Manipulation of surface wettability between superhydrophobicity and superhydrophilicity on copper films. *ChemPhysChem*, 2005, 6(8): 1475–1478.
41. Wang, C., Yao, T., Wu, J., Ma, C., Fan, Z., Wang, Z., Cheng, Y., Lin, Q., Yang, B. Facile approach in fabricating superhydrophobic and superoleophilic surface for water and oil mixture separation. *ACS Applied Materials Interfaces*, 2009, 1(11): 2613–2617.
42. Zhang, L., Zhang, Z., Wang, P. Smart surfaces with switchable superoleophilicity and super-oleophobicity in aqueous media: Toward controllable oil/water separation. *NPG Asia Materials*, 2012, 4(2): e8.
43. Maguire-Boyle, S.J., Barron, A.R. A new functionalization strategy for oil/water separation membranes. *Journal of Membrane Science*, 2011, 382(1): 107–115.
44. Kota, A.K., Kwon, G., Choi, W., Mabry, J.M., Tuteja, A. Hygro-responsive membranes for effective oil–water separation. *Nature Communications*, 2012, 3: 1025.
45. Feng, L., Zhang, Z., Mai, Z., Ma, Y., Liu, B., Jiang, L., Zhu, D. A super hydrophobic and super-oleophilic coating mesh film for the separation of oil and water. *Angewandte Chemie International Edition*, 2004, 43(15): 2012–2014.
46. Xue, Z., Wang, S., Lin, L., Chen, L., Liu, M., Feng, L., Jiang, L. A novel superhydrophilic and underwater superoleophobic hydrogel-coated mesh for oil/water separation. *Advanced Materials*, 2001, 23(37): 4270–4273.
47. Yoon, K., Hsiao, B.S., Chu, B. Functional nanofibers for environmental applications. *Journal of Materials Chemistry*, 2008, 18(44): 5326–5334.
48. Daiminger, U., Nitsch, W., Plucinski, P., Hoffmann, S. Novel techniques for oil/water separation. *Journal of Membrane Science*, 1995, 99(2): 197–203.
49. Langdon, W.M., Naik, P.P., Wasan, D.T. Separation of oil dispersions from water by fibrous bed coalescence. *Environmental Science Technology*, 1972, 6(10): 905–910.
50. Davies, G.A., Jeffreys, G.V. Coalescence of droplets in packings: Factors affecting the separation of droplet dispersions. *Filtration Separation*, 1969, 6(4): 349–354.
51. Shin, C., Chase, G.G. Water-in-oil coalescence in micro-nanofiber composite filters. *AIChE Journal*, 2004, 50(2): 343–350.
52. Wang, X., Chen, X., Yoon, K., Fang, D., Hsiao, B.S., Chu, B. High flux filtration medium based on nanofibrous substrate with hydrophilic nanocomposite coating. *Environmental Science and Technology*, 2005, 39(19): 7684–7691.
53. Ma, H., Yoon, K., Rong, L., Mao, Y., Mo, Z., Fang, D., Hollider, Z., Gaiteri, J., Hsiao, B.S., Chu, B. High-flux thin-film nanofibrous composite ultrafiltration membranes containing cellulose barrier layer. *Journal of Materials Chemistry*, 2010, 20(22): 4692–4704.
54. Arbatan, T., Fang, X., Shen, W. Superhydrophobic and oleophilic calcium carbonate powder as a selective oil sorbent with potential use in oil spill clean-ups. *Chemical Engineering Journal*, 2011, 166(2): 787–791.
55. Alther, G.R. Organically modified clay removes oil from water. *Waste Management*, 1995, 15(8): 623–628.
56. Yariv, S., Cross, H. Eds. *Organo-Clay Complexes and Interactions*. CRC Press, Boca Raton, FL, 2001.
57. Alper, H. Removal of oils and organic compounds from water and air with MYCELX HRM (hydrocarbon removal matrix) technology. *Federal Facilities Environmental Journal*, 2003, 14(3): 79–101.

58. Hench, L.L., West, J.K. The sol-gel process. *Chemical Reviews*, 1990, 90(1): 33–72.
59. Mann, S., Burkett, S.L., Davis, S.A., Fowler, C.E., Mendelson, N.H., Sims, S.D., Walsh, D., Whilton, N.T. Sol-gel synthesis of organized matter. *Chemistry of Materials*, 1997, 9(11): 2300–2310.
60. Sanchez, C., Ribot, F., Lebeau, B. Molecular design of hybrid organic-inorganic nanocomposites synthesized via sol-gel chemistry. *Journal of Materials Chemistry*, 1999, 9(1): 35–44.
61. Brinker, J.C., Scherer, G.W. *Sol-Gel Science: The Physics and Chemistry of Sol-Gel Processing.* Academic Press, San Diego, CA, 1990.
62. Edmiston, P.L., Underwood, L.A. Absorption of dissolved organic species from water using organically modified silica that swells. *Separation and Purification Technology*, 2009, 66(3): 532–540.
63. Burkett, C.M., Edmiston, P.L. Highly swellable sol-gels prepared by chemical modification of silanol groups prior to drying. *Journal of Non-Crystalline Solids*, 2005, 351(40): 3174–3178.
64. Burkett, C.M., Underwood, L.A., Volzer, R.S., Baughman, J.A., Edmiston, P.L. Organic-inorganic hybrid materials that rapidly swell in non-polar liquids: Nanoscale morphology and swelling mechanism. *Chemistry of Materials*, 2008, 20(4), 1312–1321.
65. McDowell, E. For Cruise ships, a history of pollution. *New York Times*, June 16, 2002.
66. EPA-R09-OW-2010-0438, FRL-9633-9.

8

Nanoengineered Organosilica Materials for the Treatment of Produced Water

Paul L. Edmiston,[1,2] **Stephen Jolly,**[2] **and Stephen Spoonamore**[2]

[1]*Department of Chemistry, The College of Wooster, Wooster, Ohio, USA*

[2]*ABS Material Inc., Wooster, Ohio, USA*

CONTENTS

8.1 Introduction

"Produced water" is the water coextracted during the exploration and production of oil and gas from underground reservoirs. In 2007, it is estimated that 800 billion gallons of produced water was created from petroleum extraction activities worldwide.[1] A baseline portion of produced water is native to the geological formations, but is sometimes compounded by water used to stimulate the formation, which can increase the ratio of water to oil much higher than 10:1. Near the end of life of some stimulated wells, the amount of produced water can be as high as 98% of the total volume extracted.[2] On land, produced water is typically sequestered by reinjection at great depths (i.e., below the water table). Produced water generated in offshore petroleum extraction is typically treated using flotation systems or hydrocyclones to remove dispersed oils and discharged into the ocean. In the Gulf of Mexico, 91% of produced water is discharged into the sea.[3] There are a number of environmental concerns with the practice of oceanic discharge stemming from the fact that primary treatment methods do not remove certain amounts of dissolved hydrocarbons, organic chemicals, acids, salt, and metals.[4,5]

In places where water cannot be reinjected into the formation, treatment methods are required. Treatment of produced water is complicated by the number of soluble and insoluble species contained in the fluid, including natural species such as formation hydrocarbons and added chemicals such as surfactants, foam inhibitors, and scale reducers.[6] The components of produced water can be grouped into the following categories:

- Dispersed oil
- Dissolved volatile and semivolatile organics
- Organic acids
- Oilfield treatment chemicals

- Salts
- Barium
- Naturally occurring radionuclides
- Phenols

A number of treatment techniques to separate dispersed oil are currently available in oilfield management systems (flotation, skimmers, hydrocyclones, coalescers). These methods take advantage of the density difference between oil and water in a two-phase system. However, a significant amount of dissolved hydrocarbons, organic acids, and metal ions are not separated in these processes. Beyond water treatment, it may be conceivable to harvest hydrocarbons contained in produced water through the development of processes that can separate dissolved organics and metals from produced water in an economical and sustainable manner.

Development of new technology to help manage produced water streams is of great concern with petroleum operations expanding around the world in increasingly stringent regulatory environments. Toward the goal of improved treatment methods, an organosilica-based, swellable sorbent (trade name: Osorb®) has been developed that can reversibly extract dissolved organics from water and thus shows good efficacy for produced water management. The sorbent instantaneously swells up to eight times its dry mass in the presence of organic liquids, yielding substantial force during expansion. Osorb is synthesized using the process of molecular self-assembly, generating a porous material that is nanoengineered to mechanically expand when selectively absorbing organics. The pore size and hydrophobicity of Osorb is controlled in such a manner that organic solutes are selectively captured without any absorbance of water. Absorption of dissolved BTEX (benzene, toluene, ethylbenzene, and xylene), aliphatic hydrocarbons, organic acids, and various production chemicals have been measured across a range of conditions. The performance of two pilot-scale Osorb-based produced water treatment systems were characterized. The first system was a skid-mounted system that handles inputs of up to 4 gal/min (gpm). The second system was a trailer-mounted system that handles inputs of up to 60 gpm, and includes a processing unit for Osorb regeneration. The economics of using a regenerative sorbent media to treat produced water was investigated using performance metrics from bench-scale and pilot-scale tests.

8.2 Description of Swellable Organosilica Materials

Sol–gel chemical processing[7] involves the generation of inorganic networks through the formation of colloidal suspension (sol) and gelation of the sol to form a network in a continuous liquid phase (gel) (Figure 8.1). Precursors for this process are metal or metalloid elements bonded by various reactive ligands. Sol–gels are typically synthesized from alkoxysilanes, which react readily with water in the presence of an acid or base catalyst. Polymerization follows the process of hydrolysis, alcohol condensation, and water

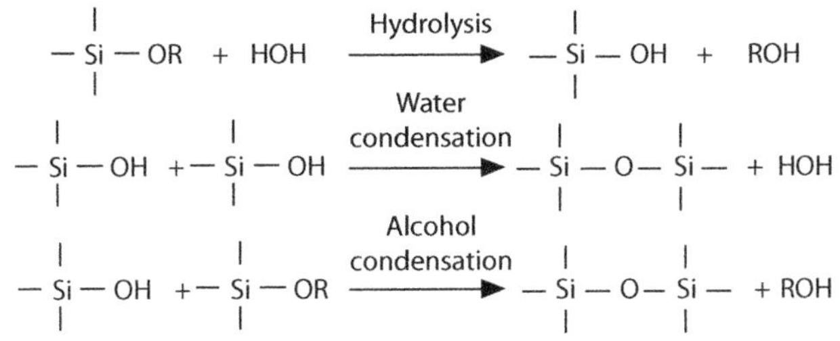

FIGURE 8.1
Sol–gel process begins with hydrolysis of alkoxysilane. Silanol can condense with another silanol or an alkoxy group to form a Si–O–Si linkage.

condensation, resulting in the three-dimensional, porous siloxane network that constitutes the gel. The solvent in the interstitial space is then removed by evaporation to create solids that have various nanoscale to microscale morphologies, possessing chemical functional groups based on the choice of precursor. A wide range of materials can be produced; from dense glasses, to highly ordered porous materials, to ultralow density aerogels.[8] The sol–gel process is a convenient method to prepare hybrid inorganic–organic materials with unique properties both in terms of chemical composition and physical microstructure.

There is a substantial variety of organically modified alkoxysilanes possessing different types of chemical groups, which facilitates a wide diversity of materials to be prepared using a common synthetic scheme. Bridged precursors are a subset of organically modified silanes where two silicon centers are tethered by an organic group $(RO)_3$-Si-Y-Si$(OR)_3$, where Y is the bridging group and -OR is the polymerizable alkoxide group. The advantage to such a structure is that polymerization can radiate from two directions and the bridge group can act as a structure directing unit.[9,10] Polycondensation of bridged precursors have been reported and subsequently used to create nanostructured materials.[11,12] Control of the sol–gel processing conditions (e.g., solvent, catalyst, aging temperature) can be used to tailor the texture of the resulting sol–gel-derived solids, which is advantageous when designing new materials.[13] Highly animated materials have been created by using molecular self-assembly as a structure-directing process to control the nanomechanical properties of polymeric organosilicate.[14] The resulting material is commercially available as Osorb.

Osorb is a nanoengineered sol–gel-derived material that rapidly swells about four to five times its dried volume and swells up to eight times its dry mass in nonpolar solvents, but not water. Before the discovery of Osorb, sol–gel-derived solids were characterized as inelastic and not able to swell after preparation. The highly animated nature of Osorb is derived from the polymerization of a bridged organosilane precursor bis(trimethoxysilylethyl) benzene (Figure 8.2) under carefully controlled conditions, including the use of THF as a

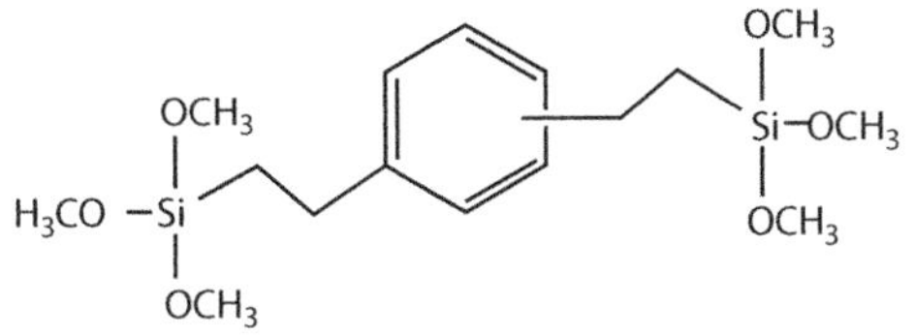

FIGURE 8.2
BTEB structure.

cosolvent and tetrabutylammonium fluoride as a catalyst. Before drying, the wet gels are rinsed to remove unreacted silane, water, and catalyst. Rinsing is followed by derivatization of the residual silanol groups using a chlorosilane reagent such as trimethylchlorosilane. Capping the silanol groups prevents cross-linking during the drying of the wet gel, which would limit the flexibility of the matrix. The identity of the silanol capping reagent has no effect on the swelling behavior.[14] However, derivatization is required to produce a material that swells. A chlorosilane possessing an alkane functional group is preferred for produced water applications as the surface becomes hydrophobic and is well suited to absorb dissolved and neat uncharged organic compounds. In addition to absorbing large amounts of nonpolar substances from the neat liquid phase, adsorption of nonpolar organic molecules is also possible from the gas phase or from aqueous solution. Because of these unique properties, there are a number of potential applications for Osorb, including the removal of dissolved organics from water.

8.2.1 Physical Characteristics

One of the most unusual aspects of Osorb is that it appears macroscopically glass-like, yet has the ability to swell up 5 mL/g upon addition of organic liquids, but does absorb water. Swelling occurs in <1 s (Figure 8.3), and there is no loss in the degree of swelling behavior when the material is cycled through many steps of absorption (swell) and evaporation (shrink). Remarkably, the swelling resulting from absorption of neat liquids is so energetic that the material expands with forces in excess of 500 N/g, which allows the material to

FIGURE 8.3
(Top) Photographs of Osorb swelling upon dropwise addition of acetone. (Middle) Scanning electron microscope images of the nanoporous organosilica matrix (left) in the dry collapsed form; (center) partially swollen state captured by swelling the matrix with a poly(2,2,3,3,4,4,4-heptafluorobutylmethacrylate) solution and allowing the solvent to dry leaving the polymer entrapped; and (right) fully expanded as captured by swollen in ethanol followed by critical point drying. (Bottom) Cartoon representing the arrangement of organosilica nanoparticles in the corresponding swollen states.

lift objects >40,000× its mass as it swells. Measurements to elucidate the structure (electron microscopy, N_2 porosity measurements, and infrared spectroscopy) indicate that the morphology of the animated matrix consists of flexibly tethered silica particles that have diameters of ~20 nm arranged in a complex and microscopically disorganized network (Figure 8.3).[15] The surface area of the dry, internally collapsed material is between 500 and 900 m²/g with internal pore volumes ranging from 0.5 to 2.8 mL/g. On the basis of the morphology of the swollen state, it has been determined that swelling results from tensile forces generated and stored by the capillary-induced collapse of the nanoporous matrix upon drying. The multitude of interparticle interactions, resulting from the high surface area, retain the compressed matrix in a state of tension. Tensile potential energy stored within bond distortions are subsequently released if interparticle noncovalent interactions are disrupted by absorbates. The high degree of interconnectivity within the pore structure allows the relaxation process to happen rapidly. The swelling process is slightly endothermic (5.2 ± 1.2 J/g), indicating that expansion is likely entropically favorable.

Swelling is completely reversible if absorbates are removed by evaporation and/or reverse mechanic pressure. Extremely high capillary forces within the nanoporous matrix upon absorbate evaporation lead to the inward collapse of the matrix, which restores the high degree of mechanical tension in the dry state after each regeneration cycle. Osorb is functionally stable up to 225°C. At higher temperatures, sintering leads to additional Si–O–Si bond formation in the collapsed network, leading to cross-linking and loss of mechanical flexibility to swell. Above 300°C, Osorb begins to thermochemically degrade. Given that the functional unit is the assembly of 20 nm nanoparticles, the material possesses all the same properties at scales from >2000 μm (bulk) down to 0.2 μm. Above the size threshold of ~2 mm, the surface area to volume ratio is low and the outside region of a particle predominantly swells, leading to potential delamination. Below 200 nm, the material is likely too small to be considered an assembly. Sizes above and below the normal values of typical sorbent media particles have limitations in terms of process systems engineering from a practical standpoint.

8.2.2 Removal of Dissolved Organics from Water

Produced water is composed of free oil and dissolved components. A focus is placed here on the removal of dissolved organics from produced water since a variety of technologies exist to physically separate free or dispersed oil in an efficient manner. It is more challenging to remove dissolved components to meet certain regulatory requirements. Toluene and benzene are notable examples of formation hydrocarbons that have a nominal solubility in water (450 and 1200 ppm, respectively). Biocides, surfactants, and methane hydrate dissolution chemicals may also be present in produced water. These additives often have relatively high solubility in water and may not be removed by physical separation or membrane technologies. Certain plays or reservoirs may have naturally occurring organics that have higher solubility in water, including organic acids (e.g., aliphatics and phenols). Polar species such as organic acids can also be more difficult to remove from produced water using standard oil–water separation methods.

One of the distinguishing characteristics of swellable organosilica is the ability to extract comparatively large amounts of dissolved organics from water. In laboratory tests (equilibrium binding studies, adsorption isotherms, breakthrough curve measurements), Osorb has been demonstrated to have the ability to absorb a wide range of dissolved organic species, including trichloroethylene, perchloroethylene, methyl *t*-butyl ether (MTBE), toluene, naphthalene, acetone, phenol, 1,4-dioxane, and 1-butanol from water (Table 8.1).[16] Partition

TABLE 8.1

Extraction of Various Compounds from Water by Osorb

Degree of Extraction	Representative Compounds
>90%	Aliphatic hydrocarbons, toluene, benzene, carbon tetrachloride, perchloroethylene, trichloroethylene, naphthalene
50%–90%	Methyl *t*-butyl ether, nonionic surfactants

coefficients for the absorption of organic species from water by Osorb range from 2.8×10^5 to 1.0×10^2. A number of aspects distinguish Osorb from traditional solid adsorbents: (i) equal to slightly higher absorbance capacity compared with activated carbon; (ii) the ability to absorb polar species such as MTBE, acetone, phenol, and 1-butanol from water; and (iii) easy absorbent bed regeneration. The magnitude of the observed partition coefficients for Osorb ($k = [X]Osorb/[X]water$) follows absorbate polarity as assessed by K_{ow} (Figure 8.4), except for the most polar species. Another exception to this trend is toluene, which has a slightly higher partition coefficient than predicted by a direct relationship upon K_{ow}. Toluene exhibited the strongest binding affinity ($k = 2.85 \times 10^5$) when the concentration of toluene was 25 ppm. At high toluene concentrations (530 ppm), Osorb absorbed 190 mg of toluene per milligram of material. On the basis of surface area measurements of the dry material using BET nitrogen absorption, it is estimated that the packed monolayer is achieved at 150 mg/mg (toluene/Osorb)[16]; thus, expansion of the matrix must play a role in the absorption under conditions of high loading. The partition coefficients for absorption of water-miscible organic solvents such as acetone and 1,4-dioxane do not follow a relationship as predicted by K_{ow} and exhibit a higher degree of absorption than expected based on polarity. Matrix expansion may also aid the binding of these polar species even though they are fully soluble in water.

A substantial amount of the data (absorption isotherms, equilibrium binding, and breakthrough curves) suggest a dynamic absorption behavior is derived from animation of the

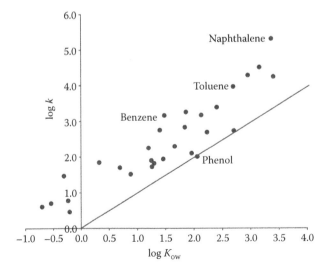

FIGURE 8.4

Plot of the partition coefficient (log k) of a compound absorbed in Osorb vs. dissolved in water vs. the compound's corresponding octanol–water partition coefficient (log K_{ow}).

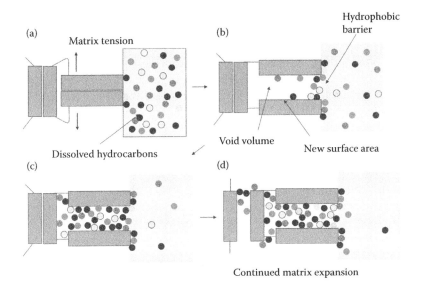

FIGURE 8.5
Proposed model for absorption of dissolved organics by swellable glass media. (a) Initial adsorption to the surface of the material. (b) Sufficient adsorption occurs to trigger matrix expansion leading to absorption across the sorbent–water boundary. (c) Pore filling leading to further percolation into the nanoporous matrix. (d) Continued matrix expansion increases available void volume.

matrix, leading to nonselective capture of organics beyond what could only be attributed to physisorption. It was observed in previous work that the force produced during swelling was not continuous with respect to swollen volume but exhibited a stepwise behavior.[16] Thus, it appears that once a particular amount of contaminant is *ad*sorbed it "unlatches" the matrix to yield void volume for subsequent *ab*sorption events (Figure 8.5). In all cases, $\log k > \log K_{ow}$; in other words, the partitioning into Osorb is greater than what would be predicted on the basis of standard liquid–liquid extraction, suggesting that there is an additional thermodynamic driving force for absorption. A likely explanation for the enhanced partition coefficients is that an increased Osorb entropy upon swelling can compensate for entropic and enthalpic barriers to absorption. What is quite unique about Osorb is that solutes condense as liquids (or solids depending on identity) in the Osorb matrix. Absorption by Osorb is fundamentally condensation in the pore structure, not adsorption through intermolecular forces of attraction between organic and the Osorb surface, although initial absorption events serve to unlock the matrix. Evidence for condensation of absorbates has been obtained using infrared spectroscopy to probe the physical state of captured solutes within the matrix.

8.2.3 Effect of Process Conditions

Variations in salt concentration, temperature, and pH are typically encountered when treating produced water streams. For instance, the total dissolved solids (TDS) in produced water can generally range from 5000 to 100,000 ppm, and in some cases can be much higher. Sodium chloride is the most abundant dissolved solid, although group I and group II metal sulfates are also relatively common. Treatment technologies need to be able to handle high variability in the amount and types of dissolved solids. Temperature of the water can also vary depending on the amount of time for a given process to complete its cycle. Produced water can be in excess of 80°C immediately upon wellhead extraction. However, temperature may drop to near freezing if water is stored above ground in cold climates before treatment. The pH of produced water is less variable and often buffered by dissolved salts (e.g., CO_3^{2-} and SO_4^{2-}). Extraction of organic acids can be highly dependent on pH, and thus understanding of how pH may affect treatment methods is critical to achieving optimal performance.

Assessment of sensitivity of water treatment with respect to process variables was conducted with Osorb across the range of values that may be anticipated with produced water. Temperature of an organic liquid has no impact on the degree Osorb swells; however, the equilibrium partitioning from a solute dissolved in water may be dependent on temperature. The amount of extraction of 200 ppm toluene by 0.2% w/v Osorb was measured at various temperatures through batch equilibrium experiments employing gas chromatography for analysis. The partition coefficient remained relatively high across all temperature ranges, declining slightly at temperatures above 60°C and below 10°C (Figure 8.6). The same trend was also observed for the partitioning of 400 ppm 1-butanol from water, indicating the phenomena is general and not dependent on polarity of the organic solute.

Whereas water temperature has little effect on extraction performance, increases in salt concentration (TDS) greatly improve partitioning of dissolved solutes. For instance, the partition coefficient for dissolved toluene (125 ppm) improves 10-fold when the TDS increases from 0 ppm to 100,000 ppm (Table 8.2). At high TDS, it is possible to extract >99% of dissolved toluene with 0.5% w/v Osorb, even at elevated temperatures. High TDS also improves the extraction of methanol from water. Elevated ionic strength does not alter the properties of the generally hydrophobic material, and the reason for the increase in extraction efficiency is more likely the well-known "salting out" effect. In short, the addition of ions tends to order the water molecules, making them participate less in solvating the dissolved organic. As a result, the organic is driven out of the aqueous phase either by

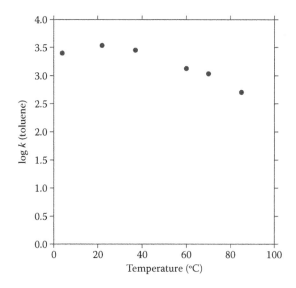

FIGURE 8.6
Dissolved toluene partition coefficient as a function of water temperature. TDS = 0 ppm; toluene concentration = 200 ppm; 0.2% w/v Osorb.

TABLE 8.2

Effect of TDS on Partition Coefficients of Toluene and Methanol

TDS, NaCl (ppm)	Partition Coefficient	
	Methanol[a]	Toluene[b]
0	1.0	3.9×10^3
5800	2.1	–
11,000	2.6	–
24,000	5.2	–
58,000	31	–
100,000	49	4.0×10^4

[a] 5000 ppm, extraction with 2.0% w/v Osorb, $T = 25°C$.
[b] 125 ppm, extraction with 0.5% w/v Osorb, $T = 70°C$.

volatilization or partitioning into a nonmiscible nonpolar solvent. The high TDS of briny produced water is ideal for the use of Osorb to remove dissolved organics.

Changes in hydrogen ion concentration do not lead to significant changes in swellable organosilica surface chemistry or an inherent ability to bind organics. The lack of sensitivity to pH is attributed to hydrophobicity and lack of acidic or basic functional groups that can be protonated or deprotonated. At pH >9.5, the organosilica matrix will begin to degrade due to the attack of hydroxide ions on the silica groups leading to dissolution of the material. The hydroxide reaction is relatively slow, allowing for applications at high pH if the contact time is limited to <8 h. The pH of the solution can have the greatest affect on the solute ionization state, which can change the affinity for absorption into the nonpolar Osorb matrix. For example, absorption of fatty acids is decreased by >60% at neutral pH compared with acidic conditions owing to the deprotonation of the carboxylic acid group above pH 4. Control of pH can thus be used to improve either the affinity or the selectivity of the extraction of organics by Osorb.

8.2.4 Variants of Osorb for Polar Water Contaminants

Typically, Osorb is synthesized using precursors and derivatizing groups that are nonpolar and lead to an oleophilic surface chemistry. For instance, surface derivatization with trimethylchlorosilane is useful in making the material hydrophobic and oleophilic. Such functionality is ideal in capturing hydrocarbons from produced water. On the other hand, there are a number of polar organic species in the produced water that need to be removed before discharge in certain regulatory environments. Some of these chemical species are naturally occurring, such as aliphatic organic acids. Phenol and alkyl phenols are common dissolved organics in steam-assisted gravity-drain water resulting from oil sand recovery operations. Finally, there are many biocides (e.g., glutaraldehyde), friction reducers (e.g., polyacrylamide), and other production chemicals that are highly water soluble and become part of some produced water streams. Removal of polar species is particularly challenging because of high water solubility and, in some cases, the ability to form azeotropes with water. These properties make the separation or extraction of polar organics thermodynamically difficult and the use of alternative methods such as biodigestion attractive. Unfortunately, the high TDS (and sometimes the presence of biocides) prohibits the use of biodigestion, limiting treatment options.

Osorb, in its standard form, has some ability to extract polar organics from water to various degrees depending the polarity of the target (Table 8.3). The affinity of more polar

TABLE 8.3

Extraction of Polar Compounds from Water by Osorb

Solute	Percent Extraction[a]
Octanoic acid	90%
p-Ethylphenol	55%
2-butoxyethanol	42%
1-Butanol	28%
1,4-Dioxane	20%
Phenol	15%
Methanol	6%
Acetone	4%

[a] Conditions: 0.5% w/v Osorb, TDS = 0 ppm, T = 25°C, [solute] = 100 ppm.

organics can be increased by changing the surface chemistry through the use of alternate derivatization routes. Derivatization using a silane with an extended polyethylenimine side chain allows for nanoparticle surfaces to become polar owing to the presence of a substantial number of -NH$_2$ groups on the polymer chain. The result is typically a ~5-fold increase in the partition coefficient for most dissolved polar organics. For example, the extraction of aliphatic organic acids from low TDS water improves significantly with polyethylenimine modification of the surface (Figure 8.7). Binding of organic acids is likely enhanced by hydrogen bond formation between the amine groups of the polyethylenimine and the organic solutes. Phenol has relatively low affinity for standard hydrophobic Osorb (k = 100) owing to the molecule's hydrogen-bonding interaction with water. Addition of hydrogen-bonding groups to the Osorb surface through the use of polyethylenimine increases the partition coefficient to k = 1400. These types of improvements demonstrate

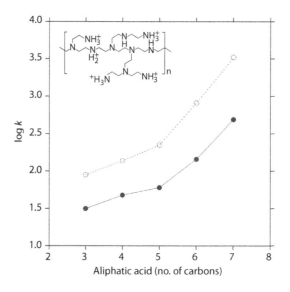

FIGURE 8.7

Comparison of partition coefficients for aliphatic organic acids (C3–C7) of Osorb (●) vs. polyethylenimine modified Osorb (○). Inset: chemical structure polyethylenimine.

that the nanostructured Osorb matrix can serve as a scaffold where either standard or novel silica modification chemistries can be employed to create a range of functionally similar materials in terms of swelling, but possessing different degrees of selectivity for organic capture. Large structural changes such as the attachment of bulky ~3000 g/mol polyethylenimine groups throughout the nanoporous matrix improves absorption of polar compounds, yet only leads to a ~40% decrease in measured swelling due to pore filling and potential cross-linking. Regeneration by solvent rinsing and subsequent drying aids to disrupt noncovalent interactions between the hydrophobic materials and the sorbates. Such methods are also successful in removing bound species from the hydrophilic polymer functional groups such as organic acids and phenols, allowing Osorb to be regenerated for additional use.

8.3 Treatment of Produced Water

Organic components in produced water are either dissolved or dispersed. Although Osorb has excellent affinity and capacity for dispersed hydrocarbons where the full extent of swelling can be utilized for capture, other physical processes (e.g., hydrocyclones) may be more economical for oil–water separation. From a cost-analysis point of view, Osorb might be useful for free organic capture when polishing or preventing of sheens in hard-to-treat waters. The distinguishing capability of Osorb is the removal of dissolved organic components in produced water. In many respects, Osorb is ideal for removal of dissolved organics from produced water as (i) most organic solutes are nonpolar hydrocarbons; (ii) the TDS levels are typically high, leading to improved performance by the salting-out effect; (iii) the nanoporous matrix has high capacity, allowing for larger volumes of water to be treated; and (iv) the media can be regenerated for economical reuse of materials. Therefore, it is hypothesized that Osorb may be best suited to be used as a polishing step in tight regulatory environments, especially if releases of BTEX is a concern. Evaluation of Osorb for produced water was conducted on both the bench scale and pilot scale to evaluate performance. Small-scale testing focused on understanding the types of produced water that could be treated with Osorb, whereas pilot-scale testing was done to determine if the processes developed at the bench scale could be incorporated into larger engineered systems.

8.3.1 Bench-Scale Testing

Bench-scale testing was completed using two types of produced water from petroleum extraction activities. Each contained relatively high (>250 ppm) amounts of dissolved organics. "Type A" produced water was obtained after significant processing to remove sediment and free oil through physical separation methods. As a result, type A water had no suspended solids and primarily contained more soluble lower molecular weight hydrocarbons including significant BTEX concentrations. "Type B" produced water was obtained much closer to the wellhead with only minor pretreatment. Type B water contained a larger fraction of higher molecular weight hydrocarbons ($>C_{12}$) in addition to a gasoline fraction. Most of the heavy oil components were in fine dispersion or adsorbed to a small amount of suspended fine particles in type B water. Both waters had nominal TDS levels around 45,000 ppm. The diversity of water types was investigated to understand the extraction capabilities of Osorb for two different classes of organic contaminants.

Type A water was well suited for Osorb treatment with >99% of the hydrocarbons removed with a 30-s application of 0.5% w/v sorbent (Figure 8.8). The rapid and complete treatment is facilitated by good mass transport dynamics of the hydrocarbons, which all exist as liquids at the experimental conditions and can readily diffuse into the expanding nanoporous matrix. It should be noted that particularly high affinity for BTEX compounds is due to π-π stacking interactions between these sorbates and the aromatic bridging group of the BTEB precursor. The aromatic bridging groups have been shown to preferentially decorate the outside surfaces of the nanoparticles that comprise the Osorb matrix. Following extraction, the light fraction of hydrocarbons could be desorbed from the Osorb at room temperature, indicating that regeneration is facile.

Type B water was more challenging to treat with Osorb in the same amount of time. The higher molecular weight species are slower to diffuse. In addition, there was likely a competition between hydrocarbons being adsorbed to suspended particles as opposed to partitioning into Osorb. Finally, the water held approximately twice the organic content of the preprocessed type A water. As a result, four times the amount of Osorb (2% w/v) was required to achieve a >99% level of organic removal within a 30 s treatment time (Figure 8.9). Although the organic content of type B water is dominated by higher molecular species, a substantial concentration of the more soluble aromatic components is also present in type B water. The amount of BTEX in treated and untreated water was analyzed separately using mass spectrometry to selectivity detect these analytes. The gas chromatography–mass spectrometry results show >99.9% of BTEX was removed concomitant with heavy petroleum removal (Figure 8.10). Co-removal of mixed contaminants is typical of Osorb, which is facilitated by the ability to swell. Specifically, absorption of one compound does not hinder the absorption of another. In contrast, absorption typically improves in a mixed-contaminant stream. Such behavior is unusual for standard adsorbents where partitioning occurs between water and active sites on a surface or finite number of molecular-scale pores. As sites saturate, the capacity of the absorbent rapidly declines. The opposite

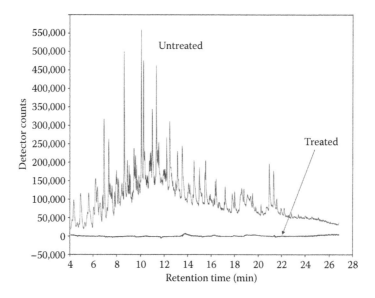

FIGURE 8.8
Gas chromatograms of type A pretreated produced water before and after treatment with 0.5% w/v Osorb. Measurements were made using the modified ISO 9377-2 method with detection by mass spectrometry.

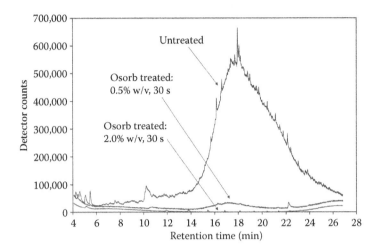

FIGURE 8.9
Gas chromatograms of type B produced water with little pretreatment before and after treatment with 0.5% w/v and 2.0% w/v Osorb. Measurements were made using the modified ISO 9377-2 method with detection by mass spectrometry.

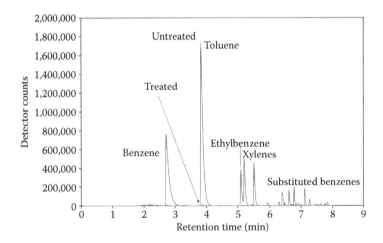

FIGURE 8.10
Gas chromatograms of type B produced water with little pretreatment before and after treatment with 0.5% w/v Osorb using selected ion monitoring (m/z = 91 amu) to detect BTEX compounds.

effect is observed in Osorb. Partitioning occurs within the matrix and swelling is induced, which leads to exposure of new surface area and pore volume as the nanoporous matrix expands. Extraction is not dictated by the amount absorbed relative to saturation, but each organic substance partitions to its own equilibrium as controlled by its polarity and thermodynamics of preference to being in a hydrophilic phase or a hydrophobic phase. The increased absorbance capacity of swelling the media in mixed systems makes the sorbent potentially useful in produced water treatment owing to the high concentration of mixed compounds in the water and the need to treat large volumes of water with as little media as possible.

8.3.2 Pilot-Scale Testing

Demonstration and analysis of Osorb treatment effectiveness at a scale relevant to actual field deployment is important in determining the practical capabilities of the technology. Many nanoengineered technologies fail to bridge the gap between laboratory and practice because of the inability to scale-up quickly or do so economically. One of the key challenges in moving Osorb to pilot-scale testing was manufacturing enough media. Manufacturing methods were thus developed to produce the flexible, nanoengineered, organosilica matrix in large reactors to facilitate a substantial amount of pilot work. Two produced water treatment systems were able to be fielded due to the success of the manufacturing development:

1. A 4 gpm skid-mounted fixed bed design, "Skid 1"
2. A 60 gpm fluidized bed design, "PWUnit#1"

The 4 gpm skid unit, Skid 1, was a straightforward system that included two bag filter vessels acting as packed columns. Skid 1 was designed to operate the bag filter vessels either in a series or in a parallel configuration. Each bag filter vessel was fitted with filled, fixed media filter bags of Osorb that could be replaced. Each filter bag contained a 4-in. layer of filter sand and 1.5 kg Osorb packed in a manner to provide room to swell, if necessary. Skid 1 was tested in conjunction with David Burnett, Director of Technology of the Global Petroleum Research Institute at Texas A&M. The Skid 1 fixed-bed treatment system was used to treat produced water at the Texas A&M testing facilities. This water was pretreated, but contained amounts of BTEX, dispersed oil, and grease. The Skid 1 fixed bed treatment system treated 100 gal of produced water at a rate of 2 gpm during testing. The samples were analyzed by an independent laboratory. The results show a substantial decrease in BTEX, oil, and grease (Table 8.4).

It was determined through testing and modeling that when moving to a produced water flow rate >10 gpm, a fluidized bed treatment method was preferential to a packed column design. A fluidized bed design (PWUnit#1) was built to operate at a 60 gpm capacity and be trailer-mounted for transportation to sites with access to produced water. Contact between the Osorb and the contaminated water occurred in a pair of 175-gallon stainless-steel tanks. Inside these tanks, a pair of eductors were used to provide sufficient turbulence and mixing for absorption (Figure 8.11). Eductors are nozzles designed to create an

TABLE 8.4

Extraction of Produced Water Hydrocarbons by 2 gpm Cartridge System

	Concentration (ppb)					Percent Reduction
Solute	Input	25 gal	50 gal	75 gal	100 gal	
Benzene	4200	114	139	194	191	95
Ethylbenzene	94	1.6	2.2	2.5	2.7	97
Naphthalene	10	5.2	5.9	7.0	7.3	27
Toluene	244	46	55	88	83	62
1,2,4-Trimethylbenzene	10	1.1	1.4	1.9	1.9	81
1,2,4-Trimethylbenzene	3.3	0.3	0.4	0.6	0.4	84
$m+p$-Xylene	39	6.3	8.6	10	11	64
o-Xylene	23	5.1	6.8	8.0	8.6	69
Oil and grease	11,500	–	–	–	0	100

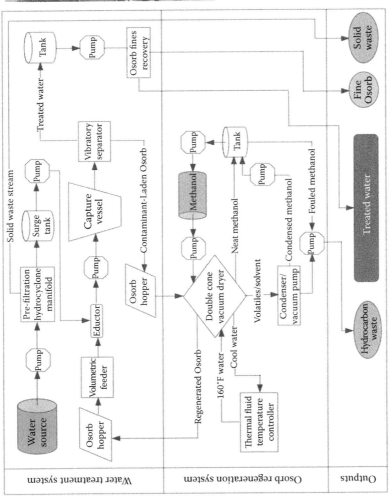

FIGURE 8.11
(See color insert.) PWUnit#1 process diagram and photograph.

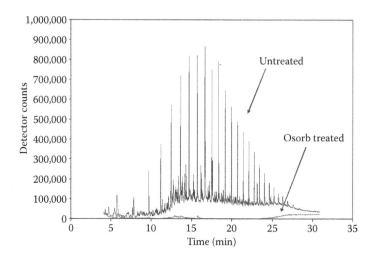

FIGURE 8.12
Gas chromatograms of Clinton formation produced water before and after treatment with PWUnit#1.

internal low pressure point when the rate of incoming water is above a certain threshold. This low pressure point then creates suction through four small openings in the back of the nozzle, allowing the nozzle to draw in the water/Osorb mixture that is present in the tank. The nozzles were able to draw in 3 gal of slurry for every 1 gal of water that is delivered from the produced water stream. The sorbent was separated from the water using a screen and sent for regeneration, which can be accomplished on site.

The 60 gpm treatment system was tested on produced water from the Clinton formation in central Ohio. The test water contained a mix of both heavy and light hydrocarbon analogs of a mixture of both type A and B produced water. A treatment of >99% was obtained in the fluidized bed design using 1% w/v Osorb addition (Figure 8.12). Regeneration was performed as part of this pilot project. These tests indicate that regeneration is the rate-limiting step in Osorb treatment, in part due to the manual labor involved in the initial design. A significant amount of investigation is being conducted to engineer an automated regeneration system.

8.4 Economics of Produced Water Treatment

The largest cost of operating an oil and gas well is the cost of managing the water. For every 1 bbl (= 42 gallons) of valuable hydrocarbons produced from a well, 2–20 bbl of produced water must be managed. Oil producers understandably view produced water from oil field operations as any other cost center. Produced water is managed in whatever manner most inexpensively removes the cost from the project. For onshore production in North America, nearly all produced water is trucked, or taken by pipeline to a disposal well. Disposal wells are primarily drilled into highly permeable deep Earth formations. In many cases, produced water is injected within the same layer producing the hydrocarbons. This process serves both the needs for disposal and the need to keep a formation stable and pressurized. The cost of disposing of water at a disposal well ranges from

$0.35 to $3.50/bbl. In many locations where the disposal wells are adjacent to hydrocarbon-producing wells, reinjection is economically viable. However, in a limited number of locations, trucking costs for transporting the produced water from the production formation to the disposal formation far exceeds the disposal well costs themselves. These costs may range as high as $23/bbl. In the Marcellus formation in Pennsylvania, the planned average cost of disposal is approximately $8.45/bbl. Even at this price point, injection disposal is generally the most attractive management practice.[17] A very small minority of land-based locations, accounting for under 2% of North American production, are obligated to recycle and process the waters. This obligation is generally in the form of regulatory requirements. Parts of Canada, and individual US states, including Alaska, Wyoming, and Colorado, have obligations for producers to recycle some portion of the produced water.

Unlike onshore production, when oil production moves offshore, especially in the EU North Sea, regions in offshore Brazil, and some other locations, operators face three challenges. The first is a lack of disposal wells; the second is a stronger regulatory obligation in many places to protect marine environments; and finally a severe limitation on footprint/weight, which makes storing even small quantities of produced water on offshore platforms physically impossible. In offshore produced water management, complex mechanical, chemical, and adsorptive systems are used to manage the tens of thousand of gallons of produced water created daily. Approximately 60%–75% of the equipment observed on a producing oil platform offshore is for the treatment of produced water. A number of well-established technologies are generally bolted together in a "treatment train" for offshore water handling. An excellent survey of the most common technologies in water treatment trains was reported by Igunnu and Chen in 2012. The primary conclusion is that there is not a single method that solves all the water-processing challenges posed by produced water management.

To provide a framework to the current situation and economics involving produced water management, four short examples are provided:

Hydraulic fracturing drill site, Central Ohio, USA, Summer 2012: Site managers reported to the authors that 50% of flow back and produced water from drilling is remixed with no treatment and reused in the drilling operation. This "on-site" reuse cost was approximately $0.50/bbl in handling and electrical pumping costs. The other 50% of the produced water was hauled offsite for disposal wells at a total cost of $4–$8/bbl. The variable to this cost is the trucking distance, as multiple disposal wells had to be used at this location. Trucking distances were 2.5–5.5 h round-trip for each trip to one of three disposal wells.

A major service company near Pittsburgh, Pennsylvania, USA, Marcellus plays, reported to the authors: Close to 100% of flow back and produced water recycling was done on site, using ozone as a biocide. The total cost was approximately $4/bbl for ozone treatment and handling. In the few situations where a full recycle for on-site reuse was not possible, costs of $7–$11/bbl were paid to dispose at disposal wells in a neighboring state. Again, the variable in disposal costs was the trucking costs.

Onshore in the Green River Basin of Wyoming, USA: A limited portion of produced water from oil and gas operations is legally obligated to be recycled and put to beneficial reuse for agricultural or wildlife habitat. Numerous technology efforts and variations on solutions have been undertaken. The authors developed technology for this market capable as one portion of a five-step treatment train. The total cost of treating the produced water from wellhead to beneficial reuse was approximately

$13/bbl total. Since most oil wells produce 5–10 bbl of water for every 1 bbl of oil, this results in a cost structure in which there is $65–$130 of water cost per $100 of oil value produced. The resulting situation is not economically viable for treating all the water and can only be applied to a very limited quantity.

Offshore EU North Sea: The authors have developed technology for use offshore, where generally a two- to four-step treatment train is bolted together. Unlike onshore treatment trains, the treated water does not have to be desalinated, as it will be injected into the already salty oceans. Systems developed by the authors have managed offshore produced water at costs between $2.80/bbl and $7.10/bbl. The variable nature in offshore treatment, unlike onshore, stems from the complexity of the chemistry to be treated and the regulatory compliance standards that the company must meet for discharge.

The economics of produced water is the difference in the value of all wells. Nearly all oil and gas wells are brought out of production, or "shut in" because of the cost of treating the waters nearly always increases over the life of the well, while the volume of hydrocarbons declines. Eventually, the valuable hydrocarbons recovered are less than the costs of operating the well, which is primarily due to the escalating costs of managing the produced water.

8.5 Summary

Produced water is likely the largest industrial waste stream in the world. The need to treat produced water is limited by the ability to deep well inject onshore and discharge to the ocean offshore. However, advanced technologies are being developed to help treat produced water where applicable. Osorb is a nanoengineered organosilica that expands in contact with organic solutes but does not absorb water. The characteristics of Osorb have been shown to be optimal for the removal of free and dissolved organics from produced water. The extraction of dissolved species is particularly unique compared with other water treatment technologies. Osorb has high affinity for dissolved BTEX and hydrocarbons. Additionally, extraction of polar organics also appears feasible, especially when altering the surface of the nanoengineered organosilica to improve the selectivity for polar compounds. Osorb has been both successfully manufactured at a pilot scale and tested using a 60 gpm system, which indicates good promise for nanoscale, engineered materials since a pathway exists for scale up. The economics of produced water treatment will favor a status quo in the near future assuming a constant regulatory environment.

Acknowledgments

The authors wish to thank the efforts of Justin Keener and Scott Buckwald of Produced Water Absorbents on the design, build out, and testing of the produced water pilot treatment systems. Funding for this work was provided by the National Science Foundation SBIR program (grant no. 1013263).

References

1. Khatib, Z.; Verbeek, P. Water to value—Produced water management for sustainable field development of mature and green fields. *J. Pet. Technol.* **2003**, *55*, 26–28.
2. Veil, J.; Puder, M.G.; Elcock, D.; Redwiek Jr., R.J. A white paper describing produced water from production of crude oil, natural gas, and coal bed methane. Available at http://www.ead.anl .gov/pub/doc/ProducedWatersWP0401.pdf (accessed November 2009).
3. Veil, J.A.; Clark, C.E. Produced water volumes and management practices in the United States, Argonne National Laboratory, September 2009. Available at http://www.ead.anl.gov/pub /doc/ANL_EVS__R09_produced_water_volume_report_2437.pdf (accessed November 2009).
4. Neff, J.M. *Bioaccumulation in Marine Organisms.* Elsevier: Oxford, 2002.
5. Reed, M.; Johnsen, S. *Produced Water 2: Environmental Issues and Mitigation Technologies.* Plenum Press: New York, 1996.
6. Stephenson, M.T. A survey of produced water studies. In *Produced Water,* Ray, J.P.; Engelhardt, F.R. (Eds.). Plenum Press: New York, 1992, pp. 1–11.
7. Brinker, J.; Scherer, G. *Sol–Gel Science.* Academic Press: San Diego, CA, 1989.
8. Corriu, R.J.P.; Mehdi, A.; Reye, C. Molecular chemistry and nanosciences: On the way to interactive materials. *J. Mater. Chem.* **2005**, *15*, 4285–4294.
9. Loy, D.A.; Shea, K.J. Bridged polysilsesquioxanes. Highly porous hybrid organic–inorganic materials. *Chem. Rev.* **1995**, *95*, 1431–1442.
10. Shea, K.J.; Moreau, J.; Loy, D.A.; Corriu, J.P.; Boury, B. Bridged polysilsesquioxanes. Molecular engineering nanostructured hybrid organic–inorganic materials. In: *Functional Hybrid Materials,* Gomex-Romero, P.; Sanchez, C. (Eds.). Wiley-VCH, Weinheim, 2004, pp. 50–85.
11. Cerveau, G.; Corriu, R.J.P.; Lepeytre, C. Organic–inorganic hybrid silica: Chemical reactivity as a tool for studying the solid arrangement as a function of molecular structure. *J. Mater. Chem.* **1995**, *5*, 793–795.
12. Corriu, R. J. P.; Mehdi, A.; Reye, C. Molecular chemistry and nanosciences: On the way to interactive materials *J. Mater. Chem.* **2005**, *15*, 4285–4294.
13. Cerveau, G.; Corriu, R.J.P.; Framery, E.; Ghosh, S.; Mutin, H.P. Hybrid materials and silica: Drastic control of surfaces and porosity of xerogels via ageing temperature, and influence of drying step on polycondensation at silicon. *J. Mater. Chem.* **2002**, *12*, 3021–3026.
14. Burkett, C.M.; Edmiston, P.L. Highly swellable sol–gels prepared by chemical modification of silanol groups prior to drying. *J. Non-Cryst. Solids* **2005**, *351*, 3174–3178.
15. Burkett, C.M.; Underwood, L.A.; Volzer, R.S.; Baughman, J.A.; Edmiston, P.L. Organic–inorganic hybrid materials that rapidly swell in non-polar liquids: Nanoscale morphology and swelling mechanism. *Chem. Mater.* **2008**, *20*, 1312–1321.
16. Edmiston, P.L.; Underwood, L.A. Absorption of dissolved organic species from water using organically modified silica that swells. *Separ. Purif. Technol.* **2009**, *66*, 532–540.
17. MWH. Produced water management. Available at http://www.spegcs.org/attachments /studygroups/1/ProducedWaterMgmt%20101.pdf.
18. Igunnu, R.T.; Chen, G.Z. Produced water treatment technologies. *Int. J. Low-Carbon Tech.* **2012**, *0*, 1–21.

9

Arsenic Removal Metrics That Commercialized the Drinking Water Market

Lisa M. Farmen

Crystal Clear Technologies Inc., Portland, Oregon, USA

CONTENTS

Water, an increasingly scarce commodity, is the world's number one environmental health determinant and the specter of even having any water at all elevates it to the number one global environmental health problem. No question that finding a clean source of drinking water on earth is no small task, and it is not getting any easier.

9.1 Background

The Safe Drinking Water Act (SDWA) of 1974 mandated that the US Environmental Protection Agency (EPA) identify and regulate drinking water contaminants that may have an adverse human health effect and that are known or anticipated to occur in public water supply systems. This law requires the EPA to determine the level of contaminants in drinking water at which no adverse health effects are likely to occur. These nonenforceable health goals, based solely on possible health risks and exposure over a lifetime with an adequate margin of safety, are called maximum contaminant level goals (MCLGs). Contaminants are any physical, chemical, biological, or radiological substances or matter in water.

The MCLG for arsenic is zero. The EPA has set this level of protection based on the best available science to prevent potential health problems. On the basis of the MCLG, the EPA has set an enforceable regulation for arsenic, called a maximum contaminant level (MCL), and is set as close to the health goals as possible, considering cost, benefits, and the ability of public water systems to detect and remove contaminants using suitable treatment technologies.[1]

In 1975, under the SDWA, the EPA established an MCL for arsenic at 0.05 mg/L. During the 1980s and early 1990s, the EPA considered changes to the MCL but did not make any. In 1996, Congress amended the SDWA and these amendments required that the EPA develop an arsenic research strategy, publish a proposal to revise the arsenic MCL by January 2000, and publish a final rule by January 2001.

On January 22, 2001, the EPA published a final Arsenic Rule in the Federal Register that revised the MCL for arsenic to 0.01 mg/L (10 μg/L). Two months later, in March 2001, the effective date of the rule was extended to provide time for the National Academy of Science to review new studies on the health effects of arsenic and for the National Drinking Water Advisory Council to review the economic issues associated with the standard. After considering the reports by the two review groups, the EPA finalized the arsenic MCL at 0.01 mg/L (10 μg/L) in January 2002.

The final rule requires all community and nontransient, noncommunity (NTNC) water systems to achieve compliance with the rule by February 2006. Adsorptive media processes are capable of achieving that level.[2] Likewise, the World Health Organization (WHO), which views arsenic as a first priority issue, set the same international arsenic removal standard, 10 μg/L, and had numerous articles written on arsenic removal.[3]

When arsenic is present above its MCL in a water supply in combination with quantities of other organic and/or inorganic contaminants, the adsorptive media process may not be the optimal method of arsenic removal. The economics of removing arsenic in addition to the volume of water to be treated and how the removed arsenic is disposed also dictate which water treatment method is selected. The EPA estimated that it would cost more than $200 million in new technology implementation to meet the new arsenic removal levels.

Arsenic contamination in drinking water has always been there as arsenic leaches from volcanic formations, runoff from orchards, and runoff from glass and electronics production and wood preservatives. How ironic that arsenic contamination in drinking water would surface when tube wells were drilled in India for the sole purpose of providing a cleaner source of drinking water. While it takes 20 years for the symptoms to develop, the long-term consumption of arsenic causes arcenosis, which builds up over time and can cause cancer and death. There are >100 million people at risk for arcenosis in Bangladesh, China (Inner Mongolia), Vietnam, Pakistan, Nepal, Myanmar, and Cambodia.[4]

Arsenic occurs in two primary forms: organic and inorganic. Organic species of arsenic are predominantly found in food stuffs. Inorganic arsenic occurs in two valence states, As^{3+} and As^{5+}. It has been well established that As^{3+} is more toxic in biological systems than As^{5+}, whereas the toxicity of organo-arsenicals is generally lower than that of inorganic arsenic species.

As^{3+} exists in five forms: $H_4AsO_3^+$, H_3AsO_3, $H_2AsO_3^-$, $HAsO_3^{2-}$, and AsO_3^{3-} with H_3AsO_3 (arsenious acid) as the primary species in natural water. As^{5+} exists in four forms in aqueous solution, depending on pH: H_3AsO_4, $H_2AsO_4^-$, $HAsO_4^{2-}$, and AsO_4^{3-}. In raw water under pH conditions of 6–9, As^{5+} exists as an anion while As^{3+} is fully protonated and exists as an uncharged molecule.[5] At the end of the day, there are two basic forms of arsenic that have to be removed to meet the 10 µg/L arsenic standard, As^{3+} and As^{5+}, with As^{3+} being more toxic.

Before new arsenic removal technologies were fully commercialized in 2006, state-of-the art technologies for arsenic removal for small community water supplies were classified into several categories as discussed below.

9.2 Conventional Arsenic Removal Technologies

Conventional water treatment technologies include precipitation, alumina, ion exchange, membrane filtration, and iron adsorption as follows:

1. Precipitate processes—coagulation/filtration, direct filtration, coagulation assisted microfiltration, enhanced coagulation, lime softening, and enhanced lime softening

2. Adsorptive process using activated alumina

3. Ion-exchange processes using anion exchange

4. Membrane filtration—reverse osmosis and electrodialysis ion exchange

5. Alternative treatment processes using granular ferric hydroxide (GFH) filtration

Evidently, these technologies were not meeting the need for small community water supplies and there are clear reasons why not:

1. Precipitate processes—Requires large treatment and retention tanks for the precipitation reaction and settling to occur, a full time on-site operator, connection to the sanitary sewer, and are still left with a hazardous waste for disposal. Most small community arsenic removal systems do not have any sanitary sewer point of connection and need a treatment system that is not active treatment but passive as there is no requirement for an operator.

2. Adsorptive process using activated alumina—This was the first adsorptive media to remove both forms of ionized arsenic but the pH has to be adjusted to pH 5.5. Incoming raw water is usually pH 6–9. Having a pH adjust system is an active treatment and requires an operator and chemical storage, pH adjustment, and monitoring of some kind.

3. Ion-exchange (IEX) resin is designed to adsorb an ion on the surface and exchange it for another ion, in this case H$^+$. The issue with IEX is that it requires regeneration at a regulated RCRA (*Resource Conservation and Recovery Act*) Part B facility to reuse the resin, and what do you do with an acidic arsenic solution? More treatment to dispose of it, and liability still has not been terminated.

4. Membrane filtration depending on the micron rating of the membrane can have arsenic permeate the membrane still requiring arsenic removal, and the membrane

reject stream is concentrated so that it can exceed discharge limits and will require a sanitary sewer connection for discharge.

5. GFH can remove arsenic and selects As^{5+} over As^{3+} and requires an oxidation step for the pretreatment to convert As^{3+} to As^{5+}. The oxidants that have proven effective are chlorine, ozone, permanganate, and solid-phase manganese dioxide media.

Since existing technology had issues with the treatment approach, waste disposal, or cost, two funding organizations stepped to the plate to empower research and development.

- The Small Business Innovation Research (or SBIR) program is a US government program, coordinated by the Small Business Administration, in which 2.5% of the total extramural research budgets of all federal agencies with extramural research budgets in excess of $100 million are reserved for contracts or grants to small businesses.[6]
- The National Academy of Engineering (NAE) announced the establishment of the Grainger Challenge Prize for Sustainability. This prize will award $1 million for a practical technology that can prevent the slow poisoning of people throughout the world as a result of arsenic contamination of drinking water.[7]

9.2.1 SBIR Program

Agencies participating in the SBIR program issue Phase I SBIR grants for research and development (R&D) to US-based small businesses for "proof of concept." Upon technical success and securing the intellectual property and proof of potential commercial success, the National Science Foundation (NSF) awarded a 2005 Phase II SBIR grant for $1M in a 2:1 matching fund only, of which the initial allocation was $500K and another $500K upon landing either an investor or customer with a $1M contribution to qualify for the remaining $500K in Phase II SBIR grant funds.

The SBIR program has expanded the grant funding to $150K for a Phase I and $1.5 M for a Phase II, which provides funding for a 2-year period to expand the results achieved in a Phase I SBIR. New R&D for arsenic removal flourished under the SBIR program and the NAE Grainger Challenge Prize in early 2000.

9.2.2 NAE Grainger Challenge[8]

The NAE Grainger Challenge prizewinners are recognized for the development, in-field verification, and dissemination of effective techniques for reducing arsenic levels in water. The systems must be affordable, reliable, easy to maintain, socially acceptable, and environmentally friendly. All of the winning systems meet or exceed the local government guidelines for arsenic removal and require no electricity. Essentially, the winning technology is a "rusty nail" (i.e., iron oxide) for arsenic removal.

The winners of the NAE Grainger Challenge were as follows:

- Abul Hussam, an associate professor in the department of chemistry and biochemistry at George Mason University, Fairfax, Virginia, will receive the Grainger Challenge Gold Award of $1 million for his SONO filter, a household water treatment system.

The Gold Award–winning SONO filter is a point-of-use method for removing arsenic from drinking water. A top bucket is filled with locally available coarse river sand and a composite iron matrix (CIM). The sand filters coarse particles and imparts mechanical stability, while the CIM removes inorganic arsenic. The water then flows into a second bucket where it again filters through coarse river sand, then wood charcoal to remove organics, and finally through fine river sand and wet brick chips to remove fine particles and stabilize water flow. The SONO filter is now manufactured and used in Bangladesh.

- Arup K. Sengupta, John E. Greenleaf, Lee M. Blaney, Owen E. Boyd, Arun K. Deb, and the nonprofit organization Water for People will share the Grainger Challenge Silver Award of $200,000 for their community water treatment system.

The system developed by the Silver Award–winning team is applied at a community's wellhead. Each arsenic removal unit serves about 300 households. Water is hand-pumped into a fixed-bed column, where it passes through activated alumina or a hybrid anion exchanger to remove the arsenic. After passing through a chamber of graded gravel to remove particulates, the water is ready to drink. This system has been used in 160 locations in West Bengal, India. The water treatment units, including the activated alumina sorbent, are being manufactured in India, and villagers are responsible for their upkeep and day-to-day operation. The active media are regenerated for reuse, and arsenic-laden sludge is contained in an environmentally safe manner with minimum leaching.

- The children's safe drinking water program at Procter & Gamble Co. (P&G), Cincinnati, Ohio, will receive the Grainger Challenge Bronze Award of $100,000 for the PUR™ Purifier of Water coagulation and flocculation water treatment system.

The PUR purifier of water technology that won the Bronze Award combines chemicals for disinfection, coagulation, and flocculation in a sachet that can treat small batches of water in the home. It is simple, portable, and treats water from any source. First, the sachet contents are stirred into a 10-liter bucket of water for 5 min. As the water rests for another 5 min, arsenic and other contaminants separate out. The water is then poured through a clean cloth to filter out the contaminants. After another 20 min, to complete the disinfection process, the water is safe to drink. As part of P&G's focal philanthropy program, the Children's Safe Water Drinking Program has worked with partners to provide 57 million sachets in more than 30 countries during the past 3 years, enough to purify more than 570 million liters of safe drinking water. Each sachet costs about 15 cents.

9.2.3 Sandia National Laboratories (SNL) Arsenic Water Technology Partnership[9] (http://www.sandia.gov/water/evaluation.htm)

With a $13M Congressional Appropriation under the Department of Energy, SNL, in Albuquerque, New Mexico, sponsored the Arsenic Water Technology Partnership with the following focus:

- Conducting research to develop innovative, arsenic removal technologies with a focus on reducing energy costs, minimizing operating costs, and minimizing quantities of waste

- Demonstrating the applicability of innovative technologies to a range of water chemistries, geographic locales, and system sizes
- Evaluating the cost-effectiveness of these technologies and providing education, training, and technology transfer assistance to user communities

The Awwa Research Foundation (AwwaRF) has managed the bench-scale research programs. WERC (Consortium for Environmental Education and Technology Development) is evaluating the economic feasibility of the technologies investigated and conducting community education/technology transfer activities.

SNL manages the pilot-scale demonstration program. During the period 2004–2006, SNL has conducted pilot treatment demonstrations at six sites. Pilot communities and technologies were matched to examine a wide range of alternative technologies and site conditions.

More than 25 arsenic removal technologies were approved for evaluation and were reviewed are described in this website. Treatment methods are categorized as adsorptive media, coagulation/filtration/membranes, or other technologies. An alphabetical index to all technologies is also provided. Other information about the technologies and related topics can be found in two summary reports and the websites for each of the Short Courses and Vendors Forums, which include links to the presentations given at these events.

9.2.4 Association of Water Treatment Professionals and Resource Center–Determined Treatment Methods

- Adsorptive media (see Table 9.1) (http://www.sandia.gov/water/docs/adsorp.doc)
- Coagulation/filtration/membranes (http://www.sandia.gov/water/docs/cfm.doc)
- Other technologies (http://www.sandia.gov/water/docs/other.doc)
- Alphabetical index (http://www.sandia.gov/water/docs/alpha.doc)

The classes of technologies tested include

1. Continuous flow systems (ion exchange, metal oxyhydroxide sorbents)
2. Batch systems (coagulation/microfiltration)
3. Reverse osmosis

This website serves as a gateway to the activities supported by SNL as part of the Arsenic Water Technology Partnership. Links are provided to

- Evaluation of innovative commercial arsenic treatment technologies (http://www.sandia.gov/water/evaluation.htm)
- Educational Short Courses and Vendors Forums held as part of the New Mexico Environmental Health Conferences in 2003–2005 (http://www.sandia.gov/water/forums.htm)
- Results of pilot demonstrations (http://www.sandia.gov/water/pubs1.htm)
- Project publications

TABLE 9.1

Adsorptive Media

Vendor	Product or Process	Website
ADA Technologies Inc.	Amended silicate	http://www.adatech.com/default.asp
AdEdge Technologies Inc.	AD33; (E-33 [granular]) E-33P (pellets)	http://www.adedgetechnologies.com
APW Inc.	Isorb, Adsorb, Hedulit, Nanolit	http://apwgroup.us
ARCTECH Inc.	HUMASORB	http://www.arctech.com/
Argonide	Alfox GR-3; Alfox18; NanoCeram	http://www.argonide.com/
Arizona State University	FeGAC	Contact: Paul Westerhoff (p.westerhoff@asu.edu)
Brimac Carbon Services	Brimac 216	http://www.brimacservices.com/
Crystal Clear Technologies	Adsorbents/ultraviolet radiation	http://www.crystalcleartechnology.us
Dow Chemical	Adsorbsia GTO	http://www.dow.com/liquidseps/prod /pt_as.htm
Eagle Picher	NXT-1; NXT-2; NXT-CF	http://www.eaglepicher.com
Engelhard	ARM 200	http://www.engelhard.com/
Graver Technologies (HydroGlobe)	MetSorb (HMRG); ActivMet; FerriMet/CF	http://www.gravertech.com/; http://www.hydroglobe.com/
Kemiron	CFH(-24) (GFH)	http://www.kemiron.com/
Kinetico Inc.	UltrAsorb-A, UltrAsorb-T, and UltrAsorb-F	http://www.kinetico.com/
Magnesium Elektron Inc./ Isolux Technologies	Isolux	http://www.zrpure.com/
MARTI [Metals & Arsenic Removal Tech Inc.] (HydroFlo)	ARTI-64	http://www.hydroflo-inc.com/; http://www.martiinc.com/
Massachusetts Institute of Technology	Kanchan Filter	http://web.mit.edu/murcott/www/arsenic; http://web.mit.edu/watsan/worldbank _summary.htm
New Mexico State University	Metal-coated aerogel	Contact: Shuguang Deng (sdeng@nmsu.edu)
Purolite	ArsenX^{np} A-530E; A-520E; A-300E; C100E	http://www.puroliteusa.com/index3.htm
ResinTech	ASM-10-HP	http://www.Resintech.com
Safe Water Technologies	Metal Ease	http://www.swtwater.com/
Sandia National Laboratories	SANS	http://www.sandia.gov/water/projects/sans .htm
SolmeteX	ArsenX^{np} NP33	http://www.solmetex.com/
University of Texas-El Paso	Fe-coated rock	Contact: Charles Turner (cturner@utep.edu)
Virotec	Bauxsol, Arsenic ProActiv, Bauxsol-GAC	http://www.Virotec.com/usa.htm

Note: The adsorptive technologies listed here are taken from the Arsenic Water Treatment Project at SNL.

9.2.5 US Environmental Protection Agency[10]

The EPA in 2006 funded arsenic removal pilot studies and lessons learned with various selected arsenic removal technologies, including costs and disposal are profiled on the EPA website (http://www.epa.gov/ordntrnt/ORD/NRMRL/wswrd/dw/arsenic/workshop/workshopagenda.htm).

9.2.5.1 *Arsenic Treatment Technology Evaluation Handbook for Small Systems*[11] *(http://water.epa.gov/drink/info/arsenic/upload/2005_11_21_arsenic_handbook_arsenic treatment-tech.pdf)*

This technical handbook is intended to help small drinking water systems make treatment decisions to comply with the revised arsenic rule. A "small" system is defined as a system serving 10,000 or fewer people. Average water demand for these size systems is normally less than 1.4 million gallons per day. Provided below is a checklist of activities that should normally take place in order to comply with the new Arsenic Rule. Many of the items on this checklist refer to a section in the *Arsenic Treatment Technology Evaluation Handbook for Small Systems* that may help in completing the activities.

9.2.5.2 *Arsenic Mitigation Checklist in the Arsenic Treatment Handbook*

1. Monitor arsenic concentration at each entry point to the distribution system (see Section 1.3.2).
2. Determine compliance status. This may require quarterly monitoring. See Section 1.3.2 for details on Arsenic Rule compliance.
3. Determine if a nontreatment mitigation strategy such as source abandonment or blending can be implemented. See Sections 2.1.1 through 2.1.3 for more detail and Decision Tree 1, Non-treatment Alternatives.
4. Measure water quality parameters. See Section 3.1.1 for more detail on water quality parameters that are used in selecting a treatment method.
   * Arsenic, total (particulate and dissolved)
   * Arsenate (As^{5+})
   * Arsenite (As^{3+})
   * Chloride
   * Fluoride
   * Iron
   * Manganese
   * Nitrate
   * Nitrite
   * Orthophosphate
   * pH
   * Silica
   * Sulfate
   * Total dissolved solids (TDS)
   * Total organic carbon

5. Determine the treatment evaluation criteria. See Section 3.1.2 for more detail on parameters that are used in selecting a treatment method.

   - Existing treatment processes
   - Target finished water arsenic concentration—technically based local limits for arsenic and TDS domestic waste discharge method
   - Land availability
   - Labor commitment
   - Acceptable percent water loss
   - Maximum source flow rate
   - Average source flow rate
   - State or primacy agency requirements that are more stringent than those of the EPA

6. Select a mitigation strategy using the decision trees provided in Section 3.2. These trees lead to the following mitigation strategies.

   - Nontreatment and treatment minimization strategies source abandonment
     - Seasonal use
     - Blending before entry to distribution system
     - Sidestream treatment
   - Enhance existing treatment processes enhanced coagulation/filtration enhanced lime softening
     - Iron/manganese filtration
   - Treatment (full stream or sidestream) ion exchange
     - Activated alumina
     - Iron-based sorbents: coagulation-assisted microfiltration; coagulation-assisted direct filtration oxidation/filtration
   - Point-of-use treatment program activated alumina
     - Iron-based sorbent
     - Reverse osmosis

7. Estimate planning-level capital and operations and maintenance costs for the mitigation strategy using the costs curves provided in Section 4. Include costs for arsenic removal and waste handling. If this planning level cost is not within a range that is financially possible, consider using different preferences in the decision trees.

8. Evaluate design considerations for the mitigation strategy. See Section 2.5 for enhancing existing treatment processes and Sections 6 through 8 for the design of new treatment processes.

9. Pilot the mitigation strategy. Although not explicitly discussed in this handbook, piloting the mitigation strategy is a normal procedure to optimize treatment variables and avoid implementing a strategy that will not work for unforeseen reasons. For many small systems, pilot testing may be performed by the vendor and result in a guarantee from the vendor that the system will perform to.

10. Develop a construction-level cost estimate and plan.

11. Implement the mitigation strategy.

12. Monitor arsenic concentration at each entry point to the distribution system to ensure that the arsenic levels are now in compliance with the Arsenic Rule assumes centralized treatment approach, not point-of-use treatment.

The *Arsenic Treatment Technology Evaluation Handbook for Small Systems* contains Table ES-1, which provides a summary of information about the different alternatives for arsenic mitigation found in this handbook. Please note that systems are not limited to using these technologies.

9.3 Success Story[12]

Titanium dioxide patent licensed from the Stevens Institute, known as MetSorb™, formed the company HydroGlobe. The award-winning patent, titled "Methods of Preparing a Surface-Activated Titanium Oxide Product and of Using Same in Water Treatment Process" is known as MetSorb, a highly effective, low-cost absorbent for reduction of arsenic and a wide variety of heavy metals from groundwater and surface water. The invention details a method for producing a surface-activated crystalline titanium oxide product having a high adsorptive capacity and high rate of adsorption with respect to dissolved contaminants. The invention further includes steps of preparing a titanium oxide precipitate from a mixture comprising a hydrolysable titanium compound.

The success part of this story is not only did this team achieve good adsorption (50,000 bed volumes [BVs]) at a relatively low cost of implementation, the Stevens Institute assisted with forming HydroGlobe and getting this patent commercially viable (http://www.stevens .edu/research/faculty_news.php?news_events_id=74).

Some of the new technologies that are in the process of commercializing are as follows:

- A nanocomposite material funded to Inframat®/MetaMateria under a US Air Force SBIR where the novel intellectual property is a nanoporous material that has an oxidizer bonded to the substrate (MnO_2–Fe_2O_3) that can convert As^{3+} to As^{5+} for full adsorption. A recent laboratory study on the oxidation of As(III) to As(V) found that solid-phase MnO_2 is preferred over chlorine, ozone, and permanganate because the latter requires stoichiometric dosing in proportion to the reducing agents, such as As^{3+}, Fe^{2+}, Mn^{2+}, and sulfide S^{2-} present in raw water. Hazardous chemicals handling and overdosing problems can be avoided when solid-phase MnO_2 media is used. The patents are US 7,655,148; US 8,216,543; China 200680038191.4; Taiwan I 359115: H. Chen, M. Wang, T.D. Xiao, D.A. Clifford.

- Crystal Clear Technologies developed a functionalized inorganic substrate that bonds both forms of arsenic and passes the Toxicity Characterization Leaching Procedure (TCLP) test. Their layering patent will allow for multiple layers of metal to be bonded on top of each other separated by a bifunctional ligand. Total arsenic loading with eight layers of metal subsequently bonded on top of one another can reach several percent.[13]

  http://groups.engr.oregonstate.edu/ewb/wp-content/uploads/crystal-clear-tech -ppt.pdf.

9.4 Metrics of Adsorptive Media Arsenic Removal Technologies

The following information is the minimum required to determine the projected cost per thousand gallons ($/kgal) of an arsenic removal treatment system:

- $Arsenic^{3+}$ influent concentration (dissolved)
- $Arsenic^{5+}$ influent concentration (dissolved)
- Total As, which is both particulate and dissolved
- Effluent As concentration allowable (10 μg/L)
- Empty bed contact time (EBCT) = time the water is in contact with the media or $(V/Q) \times 7.48$ gal/ft^3
- Flow rate in gpm
- Estimated loading of As in milligrams per gram of media (supplied by manufacturer or laboratory jar testing)

Convert the BV (50 ft^3), which equals 374 gal/BV, then take the total gallons treated; let us say 26,928,000 gallons and divide by 374 BVs, which equals 72,000 BV processed.

The EBCT determines how long of residence time is needed to remove As below the 10 μg/L level, and this number varies depending on whether the adsorptive media has the ability to remove both forms of arsenic. Recommended EBCT is 2.34–4.16 min flowing at 4.4–8.3 gal/ft^2, with a manufacturer's recommendation of 7–10 gal/ft^2.

To get cost per thousand ($/kgal), take the total treated volume/1000, which equals X, which equals cost per thousand gallons. This number should be between $0.15/kgal and $1.00/kgal for domestic applications. In Bangladesh, the treatment cost goal is under 10 paisa where 100 paisa = 1 rupee and 52 rupees = $1.00 (US).

This example above uses the Titania by HydroGlobe material, and note that this is for two media beds in parallel totaling 50 ft^3 and is capable of removing As below the MCL of 10 ppb. The ArseneX^{np} shows reaching 8 ppb with a lead-lag media bed primarily removing As^{5+}. The above arsenic removal studies and handbooks clearly show that there are numerous technologies that show efficacy for removing dissolved arsenic, and there are numerous metrics for determining which technology best fits the application.

Here is a very good "rule of thumb" for getting to the bottom line of cost for an arsenic removal system. The only components that are controlled are in bold, which are the influent and required effluent arsenic concentrations, flow rate, arsenic loading in mg/g:

Arsenic load	
Arsenic influent (total)	**0.035 mg/L**
Required arsenic effluent	**0.01 mg/L**
Total As removed	0.025 mg/L
As (+3)%	
As (+5)%	
Other ions that may load	

System flow rate	**125 gpm**
Contact time	1.6–4.34 min
Optimal contact time	3 min
Cubic feet of media	50 ft^3
EBCT in BVs	375 gal/BV
Density of media	40 lb/ft^3
Grams of media	910,428 g/50 ft^3
BV	
Total gallons to exhaustion	18,750,000 gal
Total BV to exhaustion	50,000 BV
Total As removed/gal	**0.0945 mg As/gal (jar test to confirm)**
Total As removed (mg)	1,771,875 mg As removed to exhaustion
Total As removed (g)	1772 g As removed to exhaustion
Total As removed (lb)	3.9 lb As removed to exhaustion
Total As removed/ft^3 of media	0.078 lb As removed/ft^3 media
Total As removed/g of media	0.06 mg/g As removed/g of media
Cost of treatment	
TiO$_2$ price ($/lb)	$7.00 $/lb
Density	40 lb/ft^3
Cost ($/ft^3)	$280/ft^3
Media cost ($/50 ft^3)	$14,037
k-Gallon treated	18,750 kgal treated
Cost/kgal (does not include rebed)	**0.75 $/kgal**

9.5 Arsenic in the Food Supply[14]

There is no question that water contamination (heavy metals, oxidizers such as perchlorate) goes up the food chain. Rice fields irrigated with arsenic-contaminated irrigation water allow inorganic arsenic to be adsorbed by the rice monocot plant and be disposed of in the seed of the plant, better known as rice.

Three US lawmakers introduced a bill on September 21, 2012, to limit the amount of arsenic allowed in rice and rice-based products, a legislator said. The move comes after *Consumer Reports* urged limits for arsenic in rice. Tests of more than 60 products, from Kellogg's Rice Krispies to Gerber infant cereal, showed most had some inorganic arsenic, a known carcinogen in humans.

The proposed RICE Act—Reducing Food-Based Inorganic and Organic Compounds Exposure Act—requires the Food and Drug Administration to set a maximum level of arsenic in rice and food containing rice. Democratic lawmakers Rosa DeLauro of Connecticut,

Frank Pallone of New Jersey, and Nita Lowey of New York are introducing the measure in the House of Representatives. "This is not the first time we have been alerted to the dangers of arsenic, and quite simply we must do more to ensure that our food supply is safe," DeLauro said in a statement. There are no federal standards for arsenic in most foods, including rice and rice-based products. South Korea has temporarily halted imports and domestic sales of US rice, citing concerns about possible arsenic contamination. The RICE Act could face a tough road to passage since it was introduced by Democratic lawmakers in a bitterly partisan chamber with a Republican majority. A new study from *Consumer Reports* found an increased amount of arsenic in some of the most popular brands of rice products, TODAY's national investigative correspondent Jeff Rossen reports.

The *Consumer Reports* study found higher levels of arsenic in brown rice than white rice, a result of how the two different types are processed. It also found higher levels in rice produced in southern US states than in rice from California or Asia. Earlier this year, DeLauro and Pallone introduced a bill to limit arsenic in fruit juice, after reports showed that some fruit juices contained arsenic and lead, *USA Today* reported.

9.6 Testing Protocols and Certifications for Arsenic Removal Technologies: ANSI/NSF International Testing[15]

There are so many water treatment technologies and systems to choose from; international testing standards were developed with testing protocol for treatment technologies; and component treatment systems are made from the claims each component or manufacturer is making. NSF International is a global independent public health and environmental organization that provides standards development, product certification, testing, auditing, education, and risk management services for public health and the environment. Additionally, the American National Standards Institute (ANSI) is a nonprofit organization that oversees the development of voluntary consensus standards for products, services, processes, systems, and personnel in the United States. The organization also coordinates US standards with international standards so that American products can be used worldwide.

9.7 Fate of Arsenic in Landfills[16]

In the United States, RCRA (1976) led to the establishment of federal standards for the disposal of solid waste and hazardous waste. RCRA requires that industrial wastes and other wastes must be characterized following testing protocols published by the EPA. The TCLP is one of these tests.

The Environmental Compliance Supervisor (the "gatekeeper") at a typical municipal landfill (as defined by RCRA Subtitle D) uses TCLP data to determine whether a waste may be accepted into the facility. If TCLP analytical results are below the TCLP D-list MCLs, the waste can be accepted. If they are above these levels, the waste must be taken to a hazardous waste disposal facility and the cost of disposal may increase from about $20/ton to as much as $500/ton.

Most arsenic adsorption removal technologies will pass the TCLP, which means they will be disposed in a municipal landfill. A recent dissertation by Fernando Javier Alday at the Environmental Engineering Department at the University of Arizona, titled "Iron biomineralization: implications on the fate of arsenic in landfills," discusses the fate of arsenic bound to iron-based media. His study on the leachability of primarily arsenate (As^{5+}), which is the arsenic form that iron-based media primarily adsorb, is that it can leach in a municipal landfill and would leach back into the landfill leachate and follow its fate with leachate collection system treatment or discharge, or with percolation into the local groundwater.[17]

9.8 Summary

For smaller community water treatment systems, cost per thousand gallons treated is not the only metric that needs to be evaluated in determining which technology and system gets implemented. Removing both forms of arsenic is imperative to provide clean drinking water, being able to have a system that can be remotely monitored and does not require a full time operator as well as a connection to a sanitary sewer all come into play in making the decision about which technology to implement.

A secondary consideration is where will the arsenic be disposed of and what prevents it from reentering the groundwater and having to be treated again. As cost is evaluated for treatment, what is the cost to the economy of a country for not providing clean drinking water knowing full well that people drinking arsenic-laden water are slowly being poisoned?

The United Nations Secretary General Kofi Annan asked us to face up to the threat of catastrophic world water crises, and to counter such bleak forecasts by adopting a new spirit of stewardship. To refuse this new spirit would be nothing less than a crime—and history would rightly judge us harshly for it.

We know there is arsenic in the water and we know several treatment strategies for removing it cost-effectively and even after 7 years with the new arsenic treatment rule implemented in the United States by the EPA, not all municipal drinking water systems are compliant. Without question, we all have a ways to go to remedy this problem on a global basis.

References

1. "Basic Information about Arsenic in Drinking Water" (May 21, 2012). U.S. Environmental Protection Agency. Available at http://water.epa.gov/drink/contaminants/basicinformation/arsenic.cfm.
2. Ruebel, Frederick Jr. (March 2003). "Design Manual: Removal of Arsenic from Drinking Water by Adsorptive Media" EPA/600/R-03/019. Available at http://epa.gov/nrmrl/wswrd/dw/arsenic/pubs/DesignManualRemovalofArsenicFromDrinkingWaterbyAdsorptiveMedia.pdf.
3. *WHO Guidelines for Drinking-Water Quality*, 4th edition. (2011). Available at http://www.who.int/water_sanitation_health/dwq/guidelines/en/.

4. Farmen, Lisa M. (2006). "Water, Water Everywhere & Not a Drop to Drink." *The Global Child Journal*, pages 6–9, vol 2, no. 1. Available at www.tolovechildren.org.

5. Chen, Huiming, Xie, Tianmin (2011). *Novel Multifunctional Nanocomposite Media for Removal of Arsenic and Other Contaminants from Drinking Water*, pages 21–22, Inframat® Corp.

6. "Small Business Innovation Research (SBIR) Program, Wikipedia.org" (2005). Available at http://en.wikipedia.org/wiki/SBIR, http://www.sbir.gov/.

7. "Grainger Challenge Prize for Sustainability" (February 2007). Contest/Award, National Academy of Engineering of the National Academies. Available at https://www.nae.edu/Activities/Projects/48083.aspx.

8. "National Academy of Engineering Announces Winners of the $1 Million Challenge to Provide Safe Drinking Water" (February 1, 2007). The National Academy of Engineering. Available at http://www8.nationalacademies.org/onpinews/newsitem.aspx?RecordID=02012007.

9. Siegel, Malcom, McConnell, Everett, Randy, Kirby, Carolyn (February 2006). "Arsenic Water Technology Partnership," Pilot Demonstration Project, Technology Demonstration. SAND2006-5423, Sandia National Laboratories, Albuquerque, NM. Awwa Research Foundation, WERC, Sandia Water Initiative. Available at http://www.sandia.gov/water/evaluation.htm.

10. "Agenda for Workshop on EPA's Arsenic Removal Demonstration Program: Results and Lessons Learned" (2006). U.S. Environmental Protection Agency. Available at http://www.epa.gov/ordntrnt/ORD/NRMRL/wswrd/dw/arsenic/workshop/workshopagenda.htm.

11. "Arsenic Treatment Technology Evaluation Handbook for Small Systems," (2005). U.S. Environmental Protection Agency. Washington, DC.

12. "HydroGlobe Patent Wins Thomas Alva Edison Award" (2006). Research Highlight, Stevens Institute of Technology. Available at http://www.stevens.edu/research/faculty_news.php?news_events_id=74.

13. Crystal Clear Technologies, Inc., presentation to EWB-OSU campus. (2008). Available at http://groups.engr.oregonstate.edu/ewb/wp-content/uploads/crystal-clear-tech-ppt.pdf.

14. "RICE Act: Congressmen Plan to Introduce Bill that Would Limit Arsenic Levels in Rice" (2013). HuffPost Healthy Living. Available at http://www.huffingtonpost.com/2012/09/21/rice-act-arsenic-delauro-pallone-lowey_n_1904490.html.

15. "National Sanitation Foundation, Wikipedia" (2004). Available at http://en.wikipedia.org/wiki/National_Sanitation_Foundation. "ANSI, Wikipedia," Available at http://en.wikipedia.org/wiki/ANSI.

16. "Toxicity Characterization Leaching Procedure (TCLP), Wikipedia" (1976). Available at http://en.wikipedia.org/wiki/TCLP.

17. Alday, Fernando Javier (2010) "Iron Biomineralization: Implications on the Fate of Arsenic in Landfills," Dissertation Submitted to the Dept. of Chemical and Env. Eng., Univ. of Arizona, pages 178–180.

10

Commercialization of Nano from Water Sensors to Membranes

Brent Giles

Lux Research Inc., New York, New York, USA

CONTENTS

Water technology is a difficult market for cutting-edge technologies. Water analytics, as an example, is a brutally competitive $1.5 billion business, relatively small as major markets go, and by most measures it is a very mature and consolidated market. Water membranes, surprisingly, represent a similar $1.5–1.6 billion market, once you strip away the integrated systems and treatments and focus on the membranes themselves where nanotechnologies might play. It too is highly consolidated among major players such as GE, Dow, and Hyflux. Nonmembrane treatment markets such as absorbent media are larger but dominated by low-cost materials. Next-generation technologies moving into any of these markets face a gauntlet of commodity incumbents and customers as reliably conservative as Queen Victoria.

To make a major impact, nanotechnology must find applications that allow it to exploit the superlatives we often attribute to it, i.e., dramatic miniaturization, revolutionary performance, and unique capabilities. Here we will examine one area, analytics, where this is not only possible but beginning to be realized. For contrast, we will also examine areas where, despite nano's promise, a real revolution is unlikely.

10.1 How to Win . . . Or Not

Let us start with an example of a space where highly innovative nanomaterials have been hyped and drawn the attention of major players. Consider municipal seawater reverse osmosis (RO) desalination membranes. At first glance, bringing an advanced membrane material to market should revolutionize energy requirements and vastly improve the competitiveness of desalination. Certainly membranes with extraordinary characteristics have been demonstrated in the laboratory, and these will undoubtedly find application in some industry. Thus, it is natural that advanced technology companies should at least consider this space to see what they can offer.

Recently, a major company better known in the aerospace market announced a patent on a "revolutionary" graphene oxide desalination membrane. According to published interviews, the company expects to be conducting pilot testing by 2014 or 2015. It touts the membrane as one atom thick, a thousand times stronger than steel, and vastly less resistant to the passage of water than incumbent membranes, with the implication of an unbeatable seawater desalination membrane.

None of these are the relevant metrics, however. What one really needs to know when examining a new RO membrane is the answer to questions like

- Does it resist fouling and biofouling? Ordinary membranes are often functionalized to improve this characteristic. With a single-layer nanomaterial, however, these characteristics are pretty much fixed.

- Is it robust enough to outlast the 5–7-year lifespan of current membranes? Sure, it is 1000× stronger than steel, weight for weight, but no one would dream of deploying a steel structure one atom thick. Structurally speaking, gold leaf is probably a better comparison. Adding any kind of backing will add resistance and largely negate the value of using such an exotic material. Graphene being graphene, it also is not clear that multiple layers will "play" well together, possibly limiting options to improve performance by layering.

- Does it require extensive pretreatment? A large chunk of the energy used in modern RO systems is actually in the pretreatment steps. If the graphene oxide membrane is susceptible to fouling or microtears when particles scrape it, it will need very similar pretreatment.

- Is it easy to clean? A major drawback of the current generation of polymeric membranes is that they do not withstand cleaning with chlorine. It is not clear that graphene oxide would do much better.

- Can large sheets be made to allow elements to fit the form factor of existing membranes? If so, the membranes are drop-in replacements for the large installed base of RO plants. If not, the technology is doomed to deployment at a dusty kiosk in EPCOT.

- Is it cost-competitive, at scale and right now, in a commodity market? Extreme competition between major RO membrane makers has driven margins through the floor, and a new crop of Chinese manufacturers spun up with heavy government backing are promising to make the market even less attractive. Current RO membranes cost no more than a few hundred dollars per square meter. Current graphene membranes cost tens to hundreds of thousands of dollars per square meter, pricing them completely out of the market.

By this time the thoughtful reader will be asking, yes, but what if the performance is much better than existing membranes? Would not that make it a potential winner whatever the current uncertainties?

In fact, we already know the answer to that. Even the best membranes can only go so far in reducing the energy needed to separate water from salt. The fundamental entropy term remains unchanged however the hardware is improved, and that limits energy improvements to the core process to around 40% of current best practice in RO. Building a better RO membrane is analogous to building an extremely efficient crane to lift a 100-ton weight. After the friction and motor inefficiencies are whittled away, the crane still has to lift that 100 tons. Given a modest 40% ceiling to improved performance, all of the ancillary questions we have walked through immediately come to the fore.

A more modest nano breakthrough in membranes makes the case. NanoH2O, with its zeolite nanoparticle-impregnated membranes, has worked for years to penetrate the market with what they claim is a solid improvement in performance at a slight price premium.[1] The company has raised more than $110 million from investors since 2007 without achieving profitability, this in a membrane market whose total annual value, never mind the fraction the company might reasonably address, is only 15× that raise.

If membrane performance was the key element in RO, buyers would have flocked to an improved membrane years ago. NanoH2O continues to make inroads into the market, and may one day win versus incumbents such as Dow, who according to one expert we spoke to has not significantly modified the design of their membrane for 30 years. But it is already too late for NanoH2O to make a fast or huge profit for its investors. It is operating in a commodity market and offering what amounts to incremental improvement to one component of a complex system. All RO membrane manufactures are doomed to do the same, given the fundamental physical limitations of the technology. Anyone who claims to have a "revolutionary" RO membrane either does not understand the physics or has devised what would be a real breakthrough: a cheap well-performing membrane that can actually withstand inexpensive cleaning with chlorine.

10.2 Looking for Nano Treatment Breakthroughs beyond Membranes

Nano treatment is not limited to membranes. We recently spoke to Raphael Semiat of Technion Israel Institute of Technology. Professor Semiat is bullish on nanoparticles as water treatment to remove phosphates and contaminants such as selenates, or as high surface area catalysts for advanced oxidation reactions with hydrogen peroxide. He acknowledges, however, that the particles must be effectively filtered out before release to the environment or human consumption, and that is a major sticking point before the technology can be commercialized. Cost is also a consideration: to be practical, the particles must be reusable, meaning they must be collected, reactivated, and redeployed without losing their identity as individual particles.

Such strategies cannot be dismissed out of hand, but the inherent conflict between the particles' chief advantage, high surface area and small size, and the imperative to collect and remove them, suggests a relatively narrow window for success. Particles must be able to aggregate without becoming inextricably bound, and must be free floating but easily recaptured. Any solution that meets these conflicting criteria will still be subject to significant regulatory review. Even compromises that bind particles to a substrate will face skeptical regulators who will wonder what happens as the system degrades. The large volumes of material needed for such systems will also immediately face developers with the prospect of rapidly scaling up production.

10.3 Water Quality Sensors: A Mature Industry, or One in Its Infancy?

At $1.5 billion, the sensor market is barely worth wresting from major players who rely on Edwardian chemistries and high-margin consumables for their business. The sensor market, however, is more than it seems. Every thousand dollar instrument can be a major differentiator in a power plant worth millions. Companies that would never consider playing in the traditional, commoditized market for pH and turbidity meters can find value in unique instruments, such as specialized instruments to measure antifoulants deployed in power plant cooling water.[2]

What appears to be a mature and rather unattractive industry is in fact in its infancy. Revolutionary techniques just now inching out of the laboratory and onto the market, and many more that have yet to make the leap, are making far more useful monitoring possible for both industry and municipal water.

Miniaturization and entirely new, molecular-level sensors are making this revolution possible. Growing in many cases out of the very active medical and biological research communities, next-generation sensors benefit from great progress made in, for instance, the Human Genome Project and in miniaturization of components for the telecommunications industry.

Water is a less predictable medium than medical samples, containing numerous potential interferences, and water commonly requires measurement at far lower concentrations than needed by medical devices. The water industry generally is less deep-pocketed than medicine, meaning that these higher performance systems must sell for far less than similar systems targeted to medicine. Nevertheless, technologies are making the leap.

10.4 Limits of Traditional Water Testing

To understand the nature of the revolution, it is necessary to understand the nature of existing water testing. It falls generally into three categories: laboratory tests, field tests, and online instruments.

Traditional laboratory testing requires highly trained technicians and, increasingly, very expensive and delicate equipment. Operators can choose whether they want to do tests in-house in a laboratory at great ongoing expense, or outsourced to a contract laboratory, also at great expense. Laboratory tests at this time can be arbitrarily sensitive, and this explosion in sensitivity has given rise to concerns about so-called emerging contaminants such as trace carcinogens and endocrine disruptors.

In most cases, getting results from a laboratory takes hours or days. By that time a municipality has already sent its water to the distribution system, where it has been consumed by the public, or sent their wastewater out of the plant into the environment. Most industrial processes, likewise, are well along or finished before a laboratory can return relevant results.

Many tests can be done with simple field tests. A simple pH meter requires little training, for instance, and there are simple tests for such analytes as chlorine and fluoride, which require a simple spectrometer or other analyzer costing as little as a few hundred dollars. Field tests work well in relatively clean water and for analytes in the parts-per-million (ppm) range, where extreme sensitivity is not required.

Field tests, however, have their limitations. They require an operator to be present wherever a test will be performed, or at least someone to collect and prepare a sample for testing later. Thus, at best, they are near real-time, but cannot be performed in a continuous fashion.

Online testing has existed for many years in the form of single- or dual-parameter chemical tests or relatively unreliable electrochemical tests. These tests do not require an operator except for maintenance, and can upload results automatically, often wirelessly.

A typical chemical test has a relatively large footprint, requires regular reagent replacement, and is generally suited only for locations with suitable power supplies and environmental control. The most robust of these tests measure very simple parameters such as turbidity. These are analogs of more complex parameters that would be useful to know in more detail, but most systems make do with the instrumentation available.

The industry is thus primed for a revolution that combines the best features of various types of testing, and a few characteristics that none of them share. An ideal test would

- Measure with great sensitivity, for instance parts-per-billion (ppb) levels for metals and as low as parts-per-trillion (ppt) levels for some organic chemicals
- Measure precisely the parameter needed, rather than a generic analog that offers less precise information
- Measure under difficult and dirty conditions, such as in a fast-flowing pipe or in wastewater influent
- Be automated with little or no need for maintenance
- Be capable of remote operation with little or no intervention
- Deliver real-time results to a data analysis system
- Measure multiple parameters simultaneously
- Require little or no calibration

While traditional water species such as chlorine, common metals, and fluoride occur in drinking water at ppm (1 mg/L) or tenth of a ppm range, recent research related to human and animal health has focused on species that occur in much lower concentrations. The regulations of the future for drinking and wastewater, as well as for industrial effluent, will occur in the ppb (µg/L) range or lower. Nearly every target of recent regulatory proposals has been in this range. The World Health Organization limits for arsenic, for instance, have dropped from 200 ppb to 50 ppb to 10 ppb. US and European limits have followed suit. Much lower limits exist for some organic chemicals (Figure 10.1).[3]

Contaminant	US or EU Limit, ppb (µg/L)	Contaminant Source
Antimony	5	Industry
Arsenic	10	Naturally occurring
Cadmium	5	Pipes, industry, natural sources
DBCP (1,2-Dibromo-3-chloropropane)	0.2	Agriculture
Dioxin (2,3,7,8-TCDD)	0.00003	Industry
Ethylene dibromide	0.05	Downstream oil and gas
Lindane	0.2	Agriculture
Mercury (inorganic)	2	Industry and naturally occurring
PAHs (benzo(*a*)pyrene)	0.2	Leaching from plastic infrastructure
PCBs (polychlorinated biphenyls)	0.5	Industry
Perchlorate	6	Industrial by-product
Thallium	2	Industry

FIGURE 10.1
Selected regulated chemicals for drinking water.

10.5 Metals and Inorganics: Opportunities for Next-Generation Instruments

Traditional methods struggle to measure metals at the ppb level, requiring complicated pre-concentration steps that vex even experienced laboratory operators. The US Environmental Protection Agency (EPA) lists some 24 so-called "heavy metals" of interest, and acceptable levels for discharge or drinking water for many of these tighten irregularly but inexorably throughout the developed world.[4]

The predictable if strict measurement range and the limited number of analytes make metals an ideal target for next-generation technologies, and a number are under development or currently deployed.

- Most metals require measurement at the ppb level. With the exception of traditional high-level metals, such as copper, which commonly appears at ppm concentrations, most metals and inorganics of concern hover in the range of a few ppb for most applications. This is in the range of next-generation online instruments and handheld tests.

- Metal analysis contains a limited number of analytes. A cursory glance at the periodic table shows there are only so many metals one could ever measure. Inorganic ions are likewise simple and small in number. This is a bit of an oversimplification: many metal species exist in a number of oxidation states, only one or a few of which may be of interest. Highly toxic organometallic species such as organic tin and dimethyl mercury also have the potential to broaden the list. Nevertheless, metal and inorganic analysis is relatively concrete and definable, a market with clear delineation that developers can target as a whole or in part.

- Traditional metal analyses do not translate to online instruments. Online instruments for metals are generally ion-selective electrodes, which are subject to interferences and require fairly frequent calibration, although they retain the potential for miniaturization and other improvements.[5] A number of quantitative laboratory methods for metals also exist, most notably atomic emission spectroscopy and a range of mass spectrometer–related methods.[6]

Laboratory techniques involving mass spectrometry and high-energy methods such as inductively coupled plasmas do not translate well to online instruments.

This leaves the door open to online instruments with advanced technologies that can both quantitate and speciate metals at the 1–10 ppb level. One start-up of particular note is ANDalyze, which uses short lengths of DNA to bond with and fluoresce in the presence of a metal.[7] The technique is more stable than other biologically derived reagents. The company has a field method and is developing an online system.

Based in the Netherlands, Capilix uses microchip capillary electrophoresis sensors to monitor water quality.[8] Early customers include greenhouse growers. The sensors can read a wide range of agriculturally relevant parameters, including ammonium, nitrate, phosphate, potassium, iron, and manganese. The company is also targeting the anaerobic digestor market. Metrohm acquired the company in late 2012, showing that the industry has an appetite for next-generation analytics.[9]

OndaVia, another start-up, uses enhanced Raman spectroscopy employing coated gold nanoparticles in disposable cartridges. The method could, in theory, measure a wide range of analytes, although currently the company focuses on the needs of its early customers, including key analytes such as perchlorate.[10] It is not yet clear whether similar approaches will play as online instruments, but as specialized field instruments they are already available.

Companies like Intellitect have miniaturized and combined multiple common analytes onto microchips, creating robust online instruments that can withstand difficult settings like large fast-flowing pipes.[11] Using similar technology, start-up Fluid Measurement[12] takes common parameters such as pH, dissolved oxygen, free and/or combined chlorine, turbidity and color, and miniaturizes and combines them for use in pipes and other applications. It sells the chips through integrators such as Censar Technologies.

Inorganic measurement by other next-generation techniques such as voltammetric "electronic tongues" are still in their infancy.[13] A wide variety of highly specific but somewhat unstable biologically derived (sometimes using exotic substrates such as carbon nanotubes) quantum dots or electrically conductive diamond, has begun to emerge in the space.[9–13]

10.6 Organic Species: A Moving and Expanding Target

If metal analysis is well defined with reasonable detection limits, organics analysis is the opposite. Untold billions of organic species exist. Most have unknown toxicity. Worried regulators tend to lurch from one set of contaminants to the next as research uncovers dozens of new species and then slowly begins to sort through them for those that cause the most harm.

- Organic targets are essentially infinite, favoring laboratory instruments. Only a sophisticated laboratory instrument can sort through the hundreds or thousands of organic molecules present in natural waters, which may derive from farm run-off, industrial processes, as by-products of wastewater treatment, or simply be naturally present. Even species that are naturally present can sometimes pose serious health risks. The wide variety of organics makes single-species recognition by simpler instruments less relevant.[14]

- Regulators are beginning to look for "reference" molecules. Ironically, the sheer number of organics is likely to force regulators to pick a few indicator molecules, opening the door to online instruments again. For specific applications such as food safety, i.e., measuring wash water from a specific type of vegetable, a range of sensors specific to a few known pesticides or herbicides may suffice.[15]

- Not all organics require high sensitivity. For those that do, extremely low concentrations offer both an opportunity and a challenge to next-generation instruments. Few current technologies outside the laboratory can contemplate ppt measurement. Repeated attempts to automate laboratory systems such as high-performance liquid chromatography have resulted in cumbersome, leaky white elephants. Developing such instruments for key parameters would be game changing. It would also, by its very nature, almost certainly require the precision of nanomaterials.

For organic species, even more than metals, even minute quantities of a highly bioactive species can wreak havoc on an animal or human body. For cancer-causing species, there is no safe nontoxic limit, and regulatory limits are assessed by cost/benefit analysis. Treatment itself can create by-products that are just as harmful as the species being removed.[16] Other species, such as endocrine disruptors, can be so well tailored to interfere with biological processes that even tiny quantities can have effects at the population level. This drives the relevant monitoring targets of the most harmful organic species well below the ppb level, creating a monumental challenge for laboratory instruments and a nearly insurmountable challenge for anyone trying to build an online instrument or a field test.[17]

The field is target rich. There is probably unlimited regulatory fodder among organic species. Growing pesticide use, novel chlorination and chloramination by-products, unusual by-products of industrial and municipal wastewater treatment, and waste products and by-products of industrial activity are all potential sources for highly bioactive species.[18,19] Just as the Romans never understood the impact of their lead pipes and lead pans on mental health, our chemistry-intensive society undoubtedly has created potent biological threats that will go unrecognized far into the future, while lesser threats receive attention simply because chance events or narrowly focused research reveals them.

Most current examples of commercialized technology address higher-concentration species, offering real-time solutions to what were once laboratory projects. Many also operate in unclear solutions that would stymie traditional optics techniques.

OptiEnz, founded by a Colorado State University professor, is one of the first companies to try to commercialize fiber optics fluorescence sensors using enzymes overlaid on fluorescent material. Still a very early-stage company, it has projects ranging from measuring lactose for milk processing to measuring benzene in aquifers near local shale gas fracking sites. Sensitivity ranges down to around 1 ppb; however, the technology is flexible enough to linearly measure lactose in milk at around 20%. This general technique of biological molecules on fiber optics, which has a wide history in the scientific literature, is just now edging its way into operational systems.[20]

Israeli start-up High Check Control[21] can analyze a wide range of analytes in parallel using high-resolution enhanced Raman enabled by advanced optics. With the instrument behind glass, there is little risk of fouling.

In recent years, the so-called mass spec on a chip has emerged as a possible, unproven tool for water analysis. Start-up Hydroconfidence, for instance, plans to deploy miniaturized mass spectrometers in an array around oil and gas wells to monitor aquifer quality.[22] The instruments could reassure local residents and provide ample evidence of a well's safety even in the event an aquifer was compromised by surrounding activity.

10.7 Microbes and Viruses: A Challenge Current Monitoring Does Not Address

Turbidity is a common substitute for microbial measurements in drinking water, since both microbes and the particles they travel aboard show up in these measurements. Of more specific analysis, traditional cell counting methods take at least 24–72 h to generate results. This is a very inconvenient gap: pharmaceutical processing, for instance, is generally finished before laboratory results,[23] causing entire batches of drugs to be thrown out when simple in-process corrections could save them. Likewise, natural waters testing of coastal and lake water with these delays slow down authorities' visibility on water-quality issues and can also lead to conservative behavior such as unnecessary beach closings.[24] More rapid tests such as the very common ELISA have poor sensitivity and remain relatively complex. Thus, an unfilled need exists for detectors that are very rapid. Depending on the application, operators may want to be able to identify a specific type or strain of bacteria, or may wish to broadly monitor the activity of a broad range of single-celled organisms:

- Tests must be specific, broad, and rapid. The ideal test for microbes would speciate them and determine whether cells were dead or alive, all while having a broad bandwidth to identify a range of cell species. Few if any of the technologies currently in the works can achieve this in a rapid test. Most are overly specific, to the point of missing important species, or so broad, as in the case of Dutch start-up Unisensor's cell counting technology, that they lump together living and dead cells and put huge varieties of similarly shaped cells into broad categories.[25] Such technologies may find narrow markets, but they fail to meet the industry's key needs.

- Current technologies fail to deliver. If few prospective technologies fit the need, none of the current technologies do. Cell analysis is slow and cumbersome, and inevitably requires an expert technician. Water-quality tests generally rely on a narrow subset of the species that can cause disease. Related tests are also relative failures. Wastewater's biological oxygen demand (BOD), a measure of biological activity and, indirectly, the wastewater's organic strength, takes 5 days to deliver a result wastewater operators need immediately. The key traditional substitute to BOD, chemical oxygen demand, fails to reliably track BOD's response and contains a generous dollop of mercury that requires special disposal.

  Biosensors based on living cells seem unlikely to be robust enough to replace current practice.[26] Electromechanical biosensors, similarly, face high barriers to become practical.[24] Miniaturized fluidic devices, which automate laboratory processes, may ultimately prove more promising.[27]

- Promising technologies remain in the literature. We see recent hints of next-generation technologies such as Raman scattering and high-sensitivity fluorescence, and promising technologies funded by the EPA that might transform microbe measurement.[28–30] Even ANDalyze, which we mentioned earlier, believes it could measure specific strains of living cells with some more research. It will be some time before these techniques escape the laboratory.

This is a field with more opportunity than activity, but some nanotechnology systems have been deployed for broader purposes, such as fouling.

Neosens' most innovative product is a fouling sensor, based on micro-electromechanical technology, a classic nano analytic technique. The company, acquired in 2012 by Aqualabo and now part of Orchidis Laboratoire, directly measures fouling that would otherwise be inferred indirectly by measuring levels of antifouling chemicals.

An Italian start-up, Alvim, also offers biofouling sensors; however, these measure biofouling only, using an electrical technique to measure biological activity on the sensor.

10.8 Conclusions

Judging from the amount of activity in the space, next-generation water analytics, many incorporating some form of nanotechnology, has begun transforming the space and will expand to fundamentally alter the way operators monitor water. The space is fertile ground for disruptive technology: traditional methods cannot be adapted to key applications, miniaturization opens up entirely new markets, and sensitivity must increase, in some cases, by orders of magnitude. In contrast, using nanomaterials for applications such as municipal RO membranes is a self-limiting exercise from the beginning. Passive and large-scale, such devices require refined, inexpensive large-scale manufacture from the outset, and performance is capped by basic physics laws that would not budge for new materials.

References

1. Available at www.nanoh2o.net.
2. Giles, B. *Dropwise: Leveraging the Revolutionary Value of Next-Generation Water Quality Analytics.* Lux Research, 2012. New York.
3. Available at water.epa.gov/drink/contaminants/index.cfm#List.
4. Available at www.epa.gov/reg3hwmd/bf-lr/regional/analytical/metals.htm.
5. Anastasova, S.; Radu, A.; Matzeu, G.; Zuliani, C.; Mattinen, U.; Bobacka, J.; Diamond, D. Disposable solid-contact ion-selective electrodes for environmental monitoring of lead with ppb limit-of-detection. *Electrochimica Acta* 2012, 73, 93–97.
6. Available at www.webdepot.umontreal.ca/Usagers/sauves/MonDepotPublic/CHM%203103/Contaminants%20Emergents-Water%20trends%202005.pdf.
7. Available at andalyze.com/technology.
8. Available at www.capilix.com.
9. Available at www.capilix.com/persbericht-metrohm-neemt-strategisch-belang-in-capilix-van-start-up-naar-wereldspeler.
10. Available at www.ondavia.com.
11. Available at www.intellitect-water.co.uk.
12. Available at www.fluidmeasurement.com.
13. Campos, I.; Alcaniz, M.; Aguado, D.; Barat, R.; Ferrer, J.; Gil, L.; Marrakchi, M.; Martinez-Manez, R.; Soto, J.; Vivancos, J.L. A voltammetric electronic tongue as tool for water quality monitoring in wastewater treatment plants. *Water Research* 2012, 46, 2605–2614.
14. Rather, J.A.; De Wael, K. C60-functionalized MWCNT based sensor for sensitive detection of endocrine disruptor vinclozolin in solubilized system and wastewater. *Sensors and Actuators B: Chemicals* 2012, 171–172, 907–915.
15. Irudayariaj, J. Pathogen sensors. *Sensors* 2009, 9, 8610–8612.
16. Snyder, S.A.; Westerhoff, P.; Yoon, Y.; Sedlak, D. Pharmaceuticals, personal care products, and endocrine disruptors in water: Implications for the water industry. *Environmental Engineering Science* 2003, 20, 449–469.
17. Richardson, S.D.; Ternes, T.A. Water analysis: Emerging contaminants and current issues. *Analytical Chemistry* 2005, 77, 3807–3838.
18. Available at opinionator.blogs.nytimes.com/2012/12/11/pesticides-now-more-than-ever.
19. Schriks, M.; Heringa, M.B.; van der Kooi, M.M.E.; de Voogt, P.; van Wezel, A.P. Toxicological relevance of emerging contaminants for drinking water quality. *Water Research* 2010, 44, 461–476.

20. Chiniforooshan, Y.; Ma, J.; Bock, W.J.; Hao, W.; Wang, Z.Y. Advanced fiber optic fluorescence turn-on molecular sensor for highly selective detection of copper in water. *Proceedings of SPIE* 8370, 83700A-2. 2012.
21. Available at www.high-check.com.
22. Available at hydroconfidence.com.
23. Available at www.pharmacopeia.cn/v29240/usp29nf24s0_c1231.html.
24. Xu, S. Electromechanical biosensors for pathogen detection. *Microchimica Acta* 2012, 178, 245–260.
25. Available at www.unisensor.dk.
26. Raud, M.; Tutt, M.; Jogi, E.; Kikas, T. BOD biosensors for pulp and paper industry wastewater analysis. *Environmental Science and Pollution Research* 2012, 19, 3039–3045.
27. Bridle, H.; Kersaudy-Kerhoas, M.; Miller, B.; Gavriilidou, D.; Katzer, F.; Innes, E.A.; Desmulliez, M.P.Y. Detection of cryptosporidium in miniaturized fluidic devices. *Water Research* 2012, 46, 1641–1661.
28. See, for instance, US Patent 7,880,876 B2 by Zhao et al. Lux recently spoke to the head researcher for this patent, who told us that initial efforts to measure *E. coli* in chicken and beef had proven more complicated than expected, so research had shifted to wash water from vegetables. He expected that a handheld instrument would cost around $20,000.
29. Available at www.bbc.co.uk/news/science-environment-20338540.
30. Available at cfpub.epa.gov/ncer_abstracts/index.cfm/fuseaction/display.abstractDetail/abstract/9649/report/F.

11

Nano-Photocatalytic Materials for Environmental Applications

Hui Zhang and Liang-Hong Guo

Research Center for Eco-Environmental Sciences (RCEES),
Chinese Academy of Sciences (CAS), Beijing, China

CONTENTS

11.1 Basic Principles of Photocatalysis

11.1.1 Processes of Semiconductor Photocatalysis

Photocatalysis reactions occurring on a semiconductor nanomaterial's surface have attracted intensive attention with the aim to utilize solar energy and thus address the increasing global concerns of environmental remediation and energy consumption. From the point of view of photochemistry, photocatalysis aims to enable or accelerate the specific reduction/oxidation reactions by the excited semiconductor. Typically, the electronic energy structure within a semiconductor consists of three distinguished regimes: conduction band (CB), valence band (VB), and the forbidden band (band gap, E_g). The semiconductor absorbs light and causes interband transitions if the energy of the incident photons matches or exceeds the band gap, subsequently exciting electrons from the VB into the CB in the femtosecond time scale and leaving holes in the VB. This stage is referred to as the semiconductor's "photoexcited" state. Typically, the CB electrons can act as reductants with a chemical potential of +0.5 to –1.5 V vs. the normal hydrogen electrode (NHE), while the VB holes exhibit an oxidative potential of +1.0 to +3.5 V vs. NHE.[1] The excited electrons and holes in a semiconductor migrate to the surface and can be trapped by the trapping sites there. These surface holes and electrons can oxidize and reduce surface-adsorbed species through interfacial charge transfer and surface reactions. Figure 11.1 illustrates the basic mechanism of a semiconductor photocatalytic process. During the migration process, recombination of photogenerated charge carriers may occur either in the bulk or on the surface by dissipating the energy as light or heat, thus suppressing the photocatalytic activity. It is noted that the recombination process is usually enhanced by impurities or defects in the crystal.

Compared with the conventional thermodynamics catalysis, photocatalysis can not only promote spontaneous reaction ($\Delta G < 0$), which is used to overcome the activation energy so as to accelerate the photocatalytic reaction rate, e.g., the oxidation of organic contaminants by molecular oxygen, but also nonspontaneous reactions ($\Delta G > 0$), which convert into chemical energy, e.g., photocatalytic H_2 generation and photocatalytic CO_2 conversion to hydrocarbons.

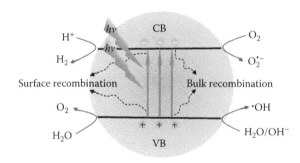

FIGURE 11.1
(See color insert.) Schematic illustration of processes involved in semiconductor photocatalysis.

11.1.2 Charge Transfer at Semiconductor Nanomaterial Interface

Compared with bulk semiconductors, nanosized semiconductors exhibit several advantages in photocatalytic reactions: (i) the magnitude of band bending, (ii) possible quantum size effect, and (iii) increased surface area. In the dark, for large particles ($r_0 \gg D$, where r_0 and D are the radius of the semiconductor particle and the depletion layer length, respectively), the major carrier (e.g., e^- in an n-type semiconductor) density at the surface is small owing to the depletion layer beneath the particle surface. In contrast, no space charge region is formed in a much smaller particle ($r_0 \ll D$). Upon light excitation, as shown in Figure 11.2, some minor carriers (e.g., h^+ in an n-type semiconductor) in large particles are transferred to the electron donor in the solution, which alleviates the positive space charge, and subsequently causes an upward band bending in the particle surface. With a much smaller particle, the photogenerated charge carriers can easily reach the surface and react with the binding species, which significantly weaken the band-bending effect.

In the case of nanoparticle systems, charge separation occurs via diffusion since the band bending is small. The average transit time (τ_D) from the interior to the surface can be expressed by Fick's diffusion model:

$$\tau_D = \frac{r_0^2}{\pi^2 D_{dc}} \qquad (11.1)$$

where D_{dc} is the diffusion coefficient. In general, τ_D is a few picoseconds. For TiO$_2$ with a radius of 6 nm, the τ_D of the electrons is 3 ps, which ensures that most charge carriers can reach the surface before recombination. It should be noted that Fick's diffusion model breaks down for particles exhibiting a quantum size effect, in which the wave function of the photogenerated charge carriers spreads over the whole semiconductor nanocluster.

Considering the reaction of photogenerated charge carriers with an acceptor molecule in solution, this semiconductor–solution interface reaction can be characterized by the rate

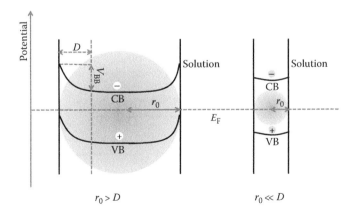

FIGURE 11.2
Schematic diagram of charge carrier transfer in large and small particles in a solution redox system. (Adapted with permission from Hoffmann, M. R., Martin, S. T., Choi, W. Y., Bahnemann, D. W., *Chem. Rev.* 95, 69. Copyright 1995, American Chemical Society.)

parameter (k_{ct}, cm s^{-1}), which is related to the observed bimolecular rate constant (k_{obs}) via the following equation[2]:

$$\frac{1}{k_{obs}} = \frac{1}{4\pi r_0^2}\left(\frac{1}{k_{ct}} + \frac{r_0}{D_{dc}}\right) \tag{11.2}$$

Equation 11.2 indicates that the rate constant for charge transfer at semiconductor–solution interface is size dependent. Larger particles usually have a faster rate than the smaller ones.

11.1.3 Reactive Oxygen Species Generated in Photocatalytic Reactions

Photocatalytic degradation of environmental organic contaminants proceeds through either direct reactions (mediated by holes and electrons) or indirect reactions (mediated by reactive oxygen species [ROS], such as $^{\bullet}OH$, $O_2^{\bullet-}$, and H_2O_2). Under light irradiation, the excited electrons and holes can directly react with surface-adsorbed organic molecules, or react with water and oxygen molecules to produce ROS, which then participate in various reactions. The redox potentials of the generated ROS are largely dependent on the intrinsic VB and CB positions of the semiconductor. Indeed, it is difficult to catalyze an oxidation (reduction) reaction by semiconductors with E_{VB} (E_{CB}) more negative (positive) than the reduction (oxidation) potential.

Hydroxyl radicals ($^{\bullet}OH$), mainly produced by oxidation of H_2O and OH^- by the holes, are recognized as one of the most oxidizing species for the photocatalytic oxidation of organic contaminants. It is noted that the surface-bound and diffusive $^{\bullet}OH$ species have different oxidation potentials, with $E_{OX}^0\left((^{\bullet}OH_{ads}) > +1.6\,V_{NHE}\right)$[3] and $E_{OX}^0\left((^{\bullet}OH_{free}) = +2.72\,V_{NHE}\right)$.[4] Electrons in the CB can be trapped by molecular oxygen to form $O_2^{\bullet-}$, which may further generate $HO_2^{\bullet}$ and H_2O_2. These ROS may complement the direct photocatalytic oxidation reactions, especially in the case where E_{VB} is insufficient for $^{\bullet}OH$ generation.

11.1.4 Nanomaterial Design for Enhanced Photocatalytic Activity

To design a high-efficiency, sunlight-induced semiconductor photocatalyst, enhancing optical absorbance and improving charge carrier separation are the two main approaches. Specially designed nanomaterials exhibit promising potentials in photocatalytic technology. Many efforts have been made to develop new photocatalyst systems to achieve these goals. Figure 11.3 shows the strategies for improving the photocatalytic activity by using

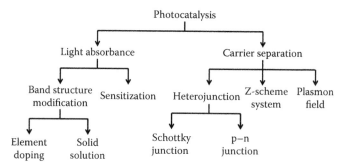

FIGURE 11.3
Schematic of strategies to improve photocatalytic activity.

nanotechnology. In the following sections, the approaches to make photocatalysts visible light active and the strategies for efficient photogenerated charge carrier separation are discussed in detail.

11.2 Electronic Band Structure Modification for Sunlight Harvesting

11.2.1 Single Semiconductor Photocatalysts

A number of semiconductors, such as TiO_2, ZnO, WO_3, Fe_2O_3, $SrTiO_3$, and CdS, can act as photoactive materials for redox/charge transfer processes owing to their electronic structure. Figure 11.4 shows the band-edge positions of several typical semiconductor photocatalysts and the electrochemical potentials required to trigger specific redox reactions.

Among these semiconductors, TiO_2 is the most widely used photocatalytic material with a band-gap energy in the 3.0–3.4 eV range. Three crystalline polymorphs of TiO_2 exist in nature: anatase, rutile, and brookite, which are composed of a TiO_6 unit and its distortions. The manner of the unit connections determines the lattice structure and electronic structure, and thus the bulk diffusion, and the redox potentials of photogenerated charge carriers. The photocatalytic activity of TiO_2 toward a specific reaction greatly depends on its physicochemical properties, e.g., crystalline structures, defects, and degree of aggregation.

The crystal defects exhibit a profound effect on the photocatalytic activity because they are intimately involved in charge-trapping and recombination processes. Experimental and theoretical results indicate that oxygen vacancies, Ti vacancies, and Ti^{n+} interstitials are the most feasible isolated or punctual defects,[6] present both on the surface and in the bulk of a semiconductor. These defects play a decisive role in the adsorption and surface reactivity of O_2, H_2O, and organic molecules. The electronic states are also associated with the defects, which cause shallow or deep donor levels that are near or below the CB, whereas the acceptor levels are located near or above the VB. This provides opportunities for tuning the energy band by doping.

A fundamentally intriguing issue in TiO_2 photocatalysis is the proposed synergistic effect between anatase and rutile crystals, which may be one of the reasons for the high

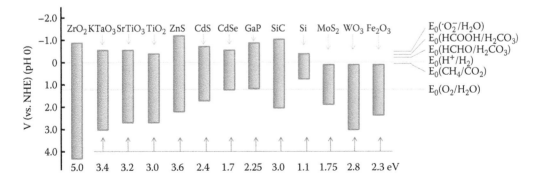

FIGURE 11.4
Band-edge positions of several semiconductor photocatalysts in contact with an aqueous electrolyte. (Adapted with permission from Navarro, R. M., Alvarez-Galván, M. C., Mano, J. A. V., Al-Zahrani, S. M., Fierro, J. L. G., *Energy Environ. Sci.*, 3, 1865. Copyright 2010, Royal Society of Chemistry.)

photocatalytic activity of Degussa P25. Electron paramagnetic resonance has provided evidences for the charge transfer across the anatase–rutile interface. However, the direction of charge transfer is still in debate. Indeed, the synergistic effect is difficult to verify because of the differences in the physicochemical properties of the pristine crystal phase as compared with the individual components in the mixture. Moreover, O_2 transfer from anatase to rutile is also proposed for the high photoactivity of P25.[7]

The surface area and aggregation state determine the adsorption behavior of TiO_2 photocatalysts toward substrate molecules, which provide the prerequisite for many photocatalytic reactions. Nanoparticulated TiO_2 in solution can aggregate to form micron-sized secondary structures, which inevitably reduce the exposed surface area to the outer environment and decrease the photocatalytic reaction activity. However, the photocatalytic activity of TiO_2 will be maintained at a high level if the aggregation is in a certain specific fashion, e.g., adjacent nanoparticles are aligned in a given crystallographic orientation, which enables a strong electronic coupling between adjacent particles, the so-called antenna effect.[8]

Semiconductor ZnO may become an alternative to TiO_2 owing to its higher carrier mobility and the more negative position of the VB, which makes the photogenerated electrons more energetic to reduce the protons, resulting in higher photocatalytic activity. In addition, the morphology of ZnO can be easily controlled by adjusting the synthetic parameters resulting in various nanostructures including nanotube, nanorod, and nanotetrapods. However, the wide-band-gap nature renders both TiO_2 and ZnO incapable of capturing the visible spectrum of the solar light. With respect to the narrow-band-gap material, WO_3, Fe_2O_3, and CdS are very attractive. However, they suffer from several disadvantages: serious photocorrosion of CdS, and the relatively positive CB of WO_3 and Fe_2O_3. To overcome these obstacles, various strategies have been made to increase the absorption of solar light.

11.2.2 Energy Band Engineering

Energy band configuration determines the absorption spectra and the redox potential of a semiconductor photocatalyst. Energy band engineering represents an effective approach to redshift the absorption edge of photocatalysts and to improve sunlight-driven photocatalytic performance. Two typical approaches have been adopted to narrow the semiconductor band gap to make photocatalysts visible light active: (i) metal and/or nonmetal doping for band-gap narrowing, and (ii) developing solid solution for continuous modulation of band structure (Figure 11.5).

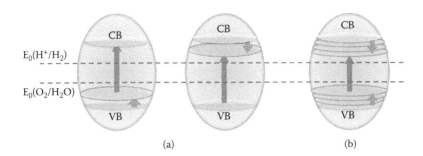

(a)   (b)

FIGURE 11.5
(See color insert.) Energy band bending strategies to narrow the band gap of semiconductor photocatalysts so as to match the solar spectrum. (a) Element doping. (b) Solid solution.

11.2.2.1 Metal Doping

Selective doping of metal ions into the crystalline matrix of semiconductors has been proven to be an effective route for improving visible light photoreactivity by creating intraband levels in the forbidden band, which can serve as either a donor level above the original VB or an acceptor level below the original CB. Thus far, metal-ion-doped TiO_2, $SrTiO_3$, and ZnS have been reported. It is reported that the photoreactivity of the metal-doped semiconductor exhibits a complex dependence on the dopant concentration, dopant energy levels within the semiconductor lattice, distribution of dopants, d-electron configuration, and other physicochemical variables.

Several studies have analyzed the doping process by transition elements: first row (V, Cr, Mn, Fe), second row (Nd, Mo), third row (W), lanthanide (Nd), and others (Ge, Sn, Pb). For the first transition row, the $3d$ states of the dopants could progressively decrease the bottom edge of the CB as the doping level increases. Both Mo and W can lower the bottom of the CB, whereas Nd can modify both the CB and VB band edges.[9] The anatase TiO_2 bandgap measurement as a function of the doping levels is shown in Figure 11.6. As described, the electronic state can be modulated by the nature and content of the metal ion in order to obtain optimum solar light absorption ability. Furthermore, doping with a transition metal ion usually induces the formation of oxygen vacancies simultaneously, which facilitate the formation of O_2^- upon chemisorptions of oxygen.

In metal-doped semiconductors, $3d$-transition element (V, Cr, Mn, Fe, Co, and Ni)-doped TiO_2 attracts considerable attention.[10] Theoretical studies reveal that the localized $3d$ states can split and mix with the CB and VB, insert an occupied level, and subsequently cause the absorption band to redshift. The extent of the redshift depends on the amount and the type of the doping metal ion. However, doping with 4 transition elements usually causes bulk defects, which can act as recombination sites. Moreover, the discrete levels formed by the localized d-states suppress the migration of carriers. To overcome these drawbacks, the formation of new VB by orbitals without $O2p$ is necessary. Orbitals of $Pb6s$ in Pb^{2+}, $Bi6s$ in Bi^{3+}, $Sn5s$ in Sn^{2+}, and $Ag4d$ in Ag^+ can form valence bands above the VB consisting of $O2p$ orbitals in metal oxide photocatalysts, such as $RbPb_2Nb_3O_{10}$ and $PbBi_2Nb_2O_9$.[11] Moreover, the degree of the contribution of these cations to the VB formation depends on the crystal

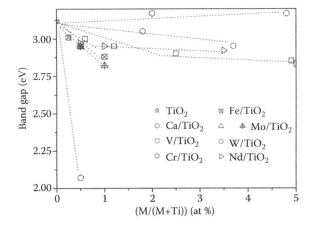

FIGURE 11.6
Effect of doping level of various metal ions on anatase TiO_2 band gap. (Adapted with permission from Kubacka A., Fernández-García M., Colón, G., *Chem. Rev.*, 112, 1555. Copyright 2012, American Chemical Society.)

structure and the doping level. A typical example is the Bi-based photocatalyst $BiVO_4$, which mainly exists in a tetragonal phase or monoclinic phase crystalline structure. With the CB composed of V3d orbitals in both cases, the top of the VB in the tetragonal $BiVO_4$ consists of only O2p orbitals, while Bi6s orbitals occupy the VB in the monoclinic case, resulting in a narrowed band gap of 2.34 eV for the monoclinic structure compared with 3.11 eV for the tetragonal structure.[12] It is noteworthy that the VB composed of Bi6s orbitals in monoclinic $BiVO_4$ can induce water oxidation to O_2 evolution by four-electron oxidation.

The doped metal ion can occupy two different positions in a semiconductor lattice: substitutional and interstitial, which is dependent on the ionic radius of the dopant compared with the matrix cation. The different positions of dopants may cause distinct effects on the electronic structure of the semiconductor. For example, in doped TiO_2 nanoparticles, Nd^{3+} and Pd^{2+} act as interstitial dopants and induce a large distribution of the potential energy, while Fe^{3+} and Pt^{2+} are in substitutional positions and have very little or no potential energy disturbance.[13]

Codoping with two suitable heteroatoms can also achieve substantial synergistic effect. For instance, codoping of Cr^{3+}/Ta^{5+}, Cr^{3+}/Sb^{5+}, Ni^{2+}/Ta^{5+}, and doping of Rh cations is effective for the sensitization of $SrTiO_3$ to visible light. Cr- or Fe-doped $La_2Ti_2O_7$ and Rh-doped $SrTiO_3$ can also function as effective visible-light-driven photocatalysts. TiO_2 codoped with Cr^{3+}/Sb^{5+}, Rh^{3+}/Sb^{5+}, and Ni^{2+}/Nb^{5+} is active for O_2 evolution.[14] In these photocatalysts, codoped metal cations can compensate the charge imbalance, and thus suppress the formation of recombination sites and maintain the visible light absorption ability.

Furthermore, metal doping can greatly change the charge carrier lifetime and then affect photoreactivity. For example, Grätzel and Howe reported that Fe^{3+}- and V^{4+}-doped TiO_2 exhibited a drastically extended lifetime of the photogenerated electron–hole pairs compared with pristine TiO_2.[15] On the other hand, doped metal ions can also act as recombination sites, especially at a high doping level. Indeed, no visible light photoreactivity is observed in Fe^{3+}-doped TiO_2 at an Fe loading of 10 wt%, while optical absorption toward visible light is considerable. Several other effects, such as surface hydrophilicity, adsorption behavior toward reactant molecules, and crystal structure transitions, would be also affected by metal doping. It is therefore necessary to control the doping process for specific photocatalytic applications.

11.2.2.2 Nonmetal Doping

Nonmetal doping is another approach used to narrow band gap and enhance the visible-light-driven photoreactivity. Compared with metal doping that is likely to form a donor level in the forbidden band, nonmetal doping usually induces the VB edge upward. In principle, there are two requirements for nonmetal doping to elevate the VB maximum of the metal oxides: (i) the nonmetal dopant should have a lower electronegativity than that of oxygen, and thus favors the newly formed VB locating at its top; (ii) the radius of the nonmetal dopant should be comparable to that of the lattice O atom to facilitate the uniform distribution of the dopant atoms within the whole matrix.

The best studied case involves doping anatase TiO_2 with N.[16] The experimental results and theoretical calculation document the presence of oxygen vacancies in both substitutional and interstitial N-doping samples without the concomitant existence of Ti^{3+} species. Essentially, N doping produces occupied states with an energy of approximately 0.0–0.2 and 0.5–0.6 eV above the VB for substitutional and interstitial doping, respectively.[17] In addition to N, C, S, B, F, I, P and N/F, N/B, N/C have all been tested as dopants of the anatase TiO_2 system. Figure 11.7 summarizes the experimentally observed electronic states for N, F, C, S and N, F codoped anatase TiO_2. For instance, C-doping at an interstitial position can decrease the TiO_2 band gap owing to the interactions present between carbon impurities and the corresponding anion

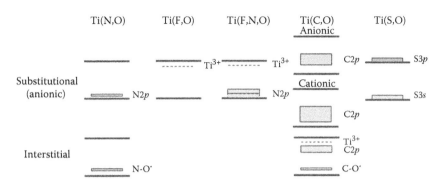

FIGURE 11.7
Electronic effect of N- and/or F-, C-, and S-doping species on the anatase electronic structure. Valence and conduction bands are represented by their top and bottom edges, respectively. (Adapted with permission from Kubacka A., Fernández-García M., Colón, G., *Chem. Rev.*, 112, 1555. Copyright 2012, American Chemical Society.)

vacancies. Halogen doping can induce in-gap states near both the VB and the CB and lead to visible light absorption up to ~800 nm. N-, F-codoped systems exhibit fully occupied N-derived gap states close to the VB as well as the additional presence of Ti^{3+} states near the CB.

Nonmetal doping extends the photosensitive region of semiconductor to visible light. However, the visible light photoreactivity is highly dependent on the chemical states of the dopant. A typical example is I-doped TiO_2. It is found that the I–O–I bond in the doped TiO_2 can induce a shoulder absorption at 550 nm, while, by combining the I–O–I and the I–O–Ti bond, the light absorption can be extended to ~800 nm due to the unoccupied I–O–I states below the CB edge and the occupied I–O–Ti states above the VB edge.[18]

An important case is that the crystalline phase structure can be changed after high-level doping with anions, such as transformation to oxynitrides, oxysulfides, and oxyhalides. The formed VB consists of N2p and S3p orbitals, in addition to O2p, resulting in significant visible light absorption. For example, compared with Ta_2O_5, oxynitride, TaON possesses a narrower band gap of 2.5 eV and exhibits higher photoreactivity for water splitting under visible light.[19]

Codoping with a cation–anion, such as Mo–C-, Ce–C-, Fe–C/N-, V–N-, In–N-, Ce–N-, Sn–N-, and W–N-doped anatase TiO_2, have been also studied.[20] One certain contribution of codoping lies in the handling of the stoichiometry, and makes it possible to control the defect formation and structural strain. However, only limited theoretical results have been presented on the synergistic effects of the codoped materials. The structural/electronic effect needs further analysis to test the potential of cation–anion codoped photomaterials compared with single-doped materials.

Regardless of metal or nonmetal doping, the localized states formed in the band gap determine the photosensitive spectral region and redox potential of the photogenerated charge carriers. However, the visible light absorption induced by doping only makes sense when the charge carriers created by visible light excitation can supply adequate redox potential to run the practical photocatalytic reactions. Thus, the challenge is to introduce suitable dopants to realize the synergistic effect.

11.2.2.3 Self-Doping

Different from impurity incorporation, self-doping (e.g., Ti^{3+} doped TiO_2, oxygen vacancies) can also provide gap tailoring. These defects can induce an additional shoulder adsorption and/or a tailor adsorption band in the visible and near-infrared region. Reduced TiO_2 (Ti^{3+} doped TiO_2) has been demonstrated to exhibit visible light photoreactivity due to the existence of a vacancy band of electronic states just below the CB. Recently, Chen et al. reported that the disorder-engineered nanophase TiO_2 exhibited substantial solar-driven photocatalytic activities, in which a lower-energy mid-gap state was derived from hybridization of the $O2p$ orbital with the $Ti3d$ orbital as the result of disorders stabilized by hydrogen.[21] Indeed, the formation of these point defects also accompanies the metal/nonmetal doping process, which needs further investigation.

11.2.2.4 Continuously Tuning Band Structure through Solid Solutions

Compared with elemental doping, forming solid solutions between wide- and narrow-band-gap semiconductors is promising for precisely controlling the electronic structures by varying the ratio of the compositions to achieve both effective visible light absorption and adequate redox potential. An important requirement for solid solutions is that the compositions should have similar crystal phase structure and lattice parameters. Three typical strategies have been explored to manipulate the band structure of the solid solution systems: continuously tuning the VB, continuously modulating the CB, and simultaneously adjusting both VB and CB.

A well-known example of VB manipulation is the $(ZnO)_x(GaN)_{1-x}$ system (Figure 11.8a). Both GaN and ZnO have wurtzite structures with a band gap over 3 eV, while the solid solution $(Ga_{1-x}Zn_x)(N_{1-x}O_x)$ exhibits visible light photoreactivity. The narrowed band gap can be explained by the p–d repulsion between $Zn3d$ and $N2p$ electrons, which shifts the VB upward without affecting CB composed of $Ga4s4p$ hybridized orbital. The quantum efficiency for water splitting at 420–450 nm reached 5% with $Rh_{2-x}Cr_xO_3$ as a cocatalyst.[22] $ZnGeN_2$–ZnO, an analog solid system, also behaves as a visible light photocatalyst for water splitting. Another example of continuous modulation of the VB is the $NaNbO_3$–$AgNbO_3$ solid solution. An enhanced visible light photoreactivity was achieved by varying the ratio of Ag/O.[23]

Continuously modulating the CB using solid solutions is also extensively explored for specific photocatalytic reactions. In the $AgGa_{1-x}In_xS_2$ solid solution, the participation of $In5s5p$ orbitals leads the CB upward and then allows for tuning the band gaps continuously. $AgGa_{0.9}In_{0.1}S_2$ exhibits the optimal photoreactivity for H_2 evolution.[25] Recently, a

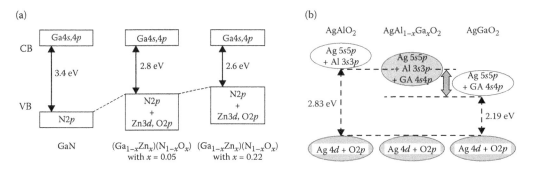

FIGURE 11.8
(a) Schematic band structures of GaN and $(Ga_{1-x}Zn_x)(N_{1-x}O_x)$ with $x = 0.05$–0.22. (b) Schematic electronic structures of $AgAlO_2$, $AgGaO_2$, and $AgAl_{1-x}Ga_xO_2$ solid solutions. (Adapted with permission from Maeda, K., Teramura, K., Takata, T., Hara, M., Saito, N., Toda, K., Inoue, Y., Kobayashi, H., Domen, K., *J. Phys. Chem. B*, 109, 20504 and Ouyang, S. X., Ye, J. H., *J. Am. Chem. Soc.*, 133, 7757, respectively. Copyright 2005 and 2011, American Chemical Society.)

series of β-AgAl$_{1-x}$Ga$_x$O$_2$ solid solutions were explored as visible-light-sensitive photocatalysts.[24] The level of CB minimum can be continuously tuned by varying the ratio of Ga/Al and thereby allows the modulation of band gaps continuously from 2.19 to 2.83 eV (Figure 11.8b). The synergistic effect between visible light absorption and redox potentials is achieved optimally in the β-AgAl$_{0.6}$Ga$_{0.4}$O$_2$ samples.

Continuous modulation of both the VB and the CB provides an effective approach to achieving both effective visible light absorption and adequate redox potential. For example, the modulation of the (AgNbO$_3$)$_{1-x}$(SrTiO$_3$)$_x$ band structure depends on the extent of both the hybridization of the Ag4d and O2p orbitals in the VB as well as the Nb4d and Ti2p orbitals in the CB, achieving the highest visible light activity with (AgNbO$_3$)$_{0.75}$(SrTiO$_3$)$_{0.25}$.[26] It is expected that the photoreactivity would be favorably promoted by changing the components of the solid solution material owing to their flexible electronic structure.

11.3 Heterogeneous Systems for Enhanced Charge Separation

Energy band engineering can extend optical absorption of photocatalysts to the visible light region, as discussed above. An alternative method to enhance photoreactivity is to improve the separation efficiency of photogenerated charge carriers and reduce bulk/surface charge recombination. Numerous strategies, such as heterojunction, plasmon–exciton coupling, and cocatalyst modification, have been explored to facilitate the separation process to increase the utilization of charge carriers and obtain high photoreactivity.

11.3.1 Metal@Semiconductor Schottky Junction

For metal@semiconductor composite material, a Schottky barrier (ϕ_{SB}) forms at the semiconductor–metal interface due to the work function difference.[27] The Schottky barrier can decrease the recombination rate of electron–hole pairs. Femtosecond diffuse reflectance spectroscopy experiments have demonstrated the effective charge separation in the Pt/TiO$_2$ composites for the enhanced photocatalytic activity.[28] Generally, a larger work function difference results in a strong Schottky barrier effect, and therefore exhibits a better activity for reduction reactions. This can explain the highest activity of Pt/TiO$_2$ among the metal–TiO$_2$ composites.

The structural geometry of the metal–semiconductor plays a critical role in the photocatalytic performance. Figure 11.9 illustrates two typical metal–semiconductor systems:

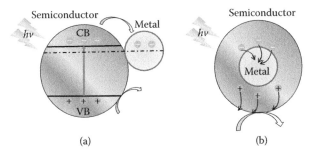

(a)   (b)

FIGURE 11.9
(See color insert.) Metal@semiconductor photocatalytic systems: (a) metal particle supported on semiconductor surface; (b) metal@semiconductor core@shell structure.

metal particle supported on semiconductors and metal@semiconductor core@shell structures. In the metal-supported semiconductor photocatalysts, the photogenerated electrons are injected into the metal nanoparticle causing reduction reactions, while the remaining photogenerated holes in the semiconductor cause oxidation reactions. For metal@semiconductor core@shell structures, photogenerated electrons will accumulate and store in the metal core, e.g., about 66 electrons in one Ag@TiO$_2$ particle under UV light irradiation.[29] These stored electrons in the metal core can transfer to the outside to react with an electron acceptor, such as O$_2$ and C$_{60}$. Although the photocatalytic reactivity is limited, core–shell structures can significantly eliminate the corrosion or dissolution of metal particles.

The loading amount, size distribution, and chemical states of metals are all crucial factors for the photoreactivity of metal@semiconductor junctions. Although impregnation and photodeposition are often adopted to deposit metal nanoparticles on semiconductors, they fail in controlling the particle size and dispersion. Experimental results demonstrated that well-dispersed metal nanoparticles provided a large metal–semiconductor interface and promoted the separation of electron–hole pairs. Moreover, noble metals may exist as several chemical states, e.g., Pt can exist as Pt^0, Pt^{II}, and Pt^{IV}, which may exhibit different activities in the photocatalytic reaction.[30] More careful and systematic studies are needed to clarify the active components of the metals in photocatalytic reactions.

11.3.2 Semiconductor/Semiconductor Heterojunction

Compared with single-phase semiconductor, multisemiconductor hybrid systems have significant advantages in promoting photocatalytic activity: (i) to extend the photoresponse region by coupling semiconductors with different band gaps; (ii) to promote electron–hole pair separation by the vectorial charge transfer in the heterojunction; and (iii) to improve the selectivity due to the reduction and oxidation reactions at different sites.

To be used under solar irradiation, a series of visible light photosensitive p–n and n–n heterojunctions[31] have been investigated: p-Si/n-TiO$_2$, p-CuMnO$_2$/n-Cu$_2$O, p-CuFeO$_2$/n-SnO$_2$, p-ZnFe$_2$O$_4$/n-SrTiO$_3$, and p-Si/n-TiO$_2$. A p-Cu$_2$O/n-WO$_3$ coupling system was developed to avoid back-reactions of photoinduced charges for photocatalytic H$_2$ production.[32] In all these p–n and n–n heterojunction photocatalysts, a much higher visible light photoreactivity was obtained owing to the effective separation of photogenerated electrons and holes compared with the single p- or n-components (Figure 11.10a). However, an obvious drawback of this mechanism is that a portion of the redox energy of the electrons/holes is released after the transfer process in the heterojunction.

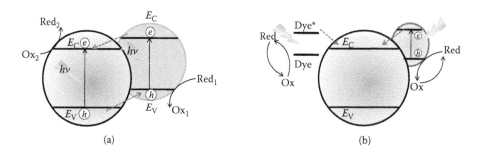

(a) (b)

FIGURE 11.10

(See color insert.) Schematic diagram illustrating the principle of charge separation between two semiconductor nanoparticles: (a) heterojunction; (b) sensitization.

The anatase-based system is one of the most broadly applied heterojunction photocatalysts. Typically, there are three different types:[33] (i) oxide–anatase system: rutile–anatase (TiO_2–TiO_2), SnO_2-, ZrO_2-, ZnO-, Bi_2O_3-, Fe_2O_3-, WO_3-, Ta_2O_5-, and Cu_2O–TiO_2; (ii) chalcogenide–anatase system: Bi_2S_3-, PbS-, CdS-, and CdSe–TiO_2; and (iii) complex oxide–anatase system: $SrTiO_3$-, $FeTiO_3$-, $ZnFeO_3$-, $BiFeO_3$-, and $LaVO_4$–TiO_2. For certain photocatalytic application, e.g., H_2 evolution from water splitting, only TiO_2 coupled with ZnO, Cu_2O, Ta_2O_5, and CdS have the potential since the e^- occupied CB is more negative than $E(H_2/H_2O) = 0$ V (vs. NHE, pH 0). Other heterojunction composites are usually used in the photocatalytic degradation of organic pollutants.

In addition to the band structure of the components, the photocatalytic performance of heterojunctions is also related to the structural geometry, the nanostructure size, and the contact interface. It usually requires a lattice match between the components to achieve a better passivation and minimize structural defects. Typically, there are "external" phase coupling or "internal" core–shell geometries for multiphase contact in the heterojunction photocatalytic system. Recently, a surface anatase–rutile heterojunction was demonstrated and exhibited unique photocatalytic activity.[34] In addition, phase contact and subsequent charge handling throughout the heterojunction interfaces are strongly size and defect dependent. For example, a size-driven p-to-n transition for Cu_2O was demonstrated as the size decreased to the nanometer region, which subsequently affected the charge carrier transfer in the Cu_2O-based heterojunction.[35]

11.3.3 Sensitization

Other than the mutually photosensitive heterojunction, coupling one photosensitive semiconductor with another nonsensitive semiconductor to fabricate a heterojunction, usually called sensitization, may also have a positive effect on the photocatalytic performance. There are two requirements for an effective sensitization process: (i) the sensitizer should have a strong absorption for visible light; (ii) the CB of the sensitizer should be higher than that of the wide-band-gap semiconductor (e.g., TiO_2) to facilitate charge transfer. Typically, dye sensitization and quantum dot (QD) sensitization are fabricated to harvest visible light and facilitate electron–hole separation for photoelectrochemical and photocatalytic applications (Figure 11.10b).

For dye sensitization, under visible light illumination, the excited dye molecule injects e^- into the CB of a wide-band-gap semiconductor to initiate the photocatalytic reaction, while the h^+ in the dyes are reduced by the sacrificial agent to regenerate the dyes and sustain the whole reaction. Synthetic organic dyes, natural pigments, and transition metal coordination compounds have been used to sensitize semiconductor photocatalysts. To facilitate electron transfer, dyes should be attached to the semiconductor surface directly by various interactions, such as covalent bands, electrostatic interactions, and hydrogen bonds.

Semiconductor nanocrystals, known as QDs, have also exhibited good performance as sensitizers in the visible light region. They have optical and electronic properties that allow them to absorb sunlight and transfer photogenerated electrons to semiconductors. Compared with dye molecules, QDs can harvest a wider range of solar spectrum and exhibit the size-dependent tunability of the band gap, which provides an opportunity to fabricate a sequential size-controlled arrangement of QDs to harvest the entire solar spectrum.[36] To favor the electron transfer, the CB difference of QD and a wide-band-gap semiconductor should be large enough to overcome the interfacial resistance. Additionally, QDs exhibit a carrier multiplication process, which can generate multiple electron–hole pairs per photon.[37] The most typical QD-sensitized photocatalyst is the CdS/TiO_2 system. Under visible light irradiation, the photoinduced electrons in CdS QDs inject into the TiO_2

CB quickly while the photoinduced holes remain in CdS. This facilitates the electron–hole separation and thus improves photoreactivity.

11.3.4 Z-Scheme Photocatalytic System

As stated above, the energy level used for photocatalytic reaction seriously limits both photocatalyst selection and visible light utilization. Semiconductors with narrow band gap can be combined together to maintain the high redox potential of the photogenerated electrons/holes and then drive specific reduction/oxidation reactions separately via a multi-photon (Z-scheme) process, which is inspired by the natural photosynthesis in green plants. In the biomimetic Z-scheme photocatalytic system, two types of semiconductor with different band structures are used to replace photosystem I (PS I) and PS II, respectively. As shown in Figure 11.11, the photogenerated electrons in PS II are transferred to the VB of PS I and then recombine with holes photogenerated at PS I, thus allowing the oxidation and reduction reaction to take place at PS I and PS II, respectively. The advantage of the Z-scheme photocatalytic system lies in extending the utilization range of solar spectra due to the reduced energy requirement to excite each photocatalyst, and keeping the photogenerated electrons/holes at a sufficient reduction/oxidation potential to drive the photocatalytic reaction, simultaneously facilitating the charge carrier separation. It is also possible to use one side of the system to drive either a reduction or an oxidation reaction.

The critical factors in a Z-scheme photocatalytic system is to find photocatalyst moieties with proper electronic structure for driving oxidation and reduction reactions separately and a reversible electron mediator acting as electron donor and acceptor in the respective half-reactions. Similar to that occurring in green plants, an indirect Z-scheme mechanism employing ionic redox couples (e.g., IO^{3-}/I^-, Fe^{3+}/Fe^{2+}, and NO_3^-/NO_2^-) as electron mediator is proposed for such a configuration, in which two isolated photocatalysts and redox mediator coexist in the reaction solution.[38] These photocatalyst systems can facilitate some difficult reactions, such as water splitting, to occur. Table 11.1 summarizes the typical Z-scheme photocatalyst systems for water splitting under visible light irradiation.

Recent developments suggest that the performance of the Z-scheme systems strongly depend on electron transfer processes between the two photocatalysts. The electron

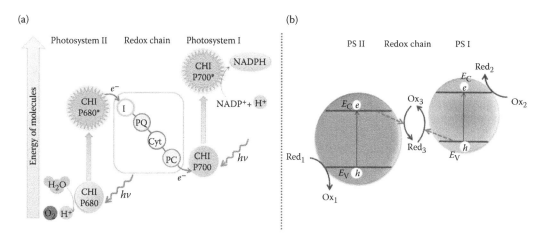

FIGURE 11.11
(See color insert.) Z-scheme mechanisms in (a) the natural photosynthesis system and (b) semiconductor photocatalyst system.

TABLE 11.1

Typical Z-Scheme Photocatalyst for Water Splitting under Visible Light Irradiation

H_2 Evolution Photocatalyst	O_2 Evolution Photocatalyst	Electron Mediator	Activity (μmol h^{-1})		Reference
			H_2	O_2	
Pt/SrTiO$_3$:Cr, Ta	Pt/WO$_3$	IO^{3-}/I$^-$	16	8	[39]
Pt/TaON	RuO$_2$/TaON	IO^{3-}/I$^-$	3	1.5	[40]
Pt/CaTaO$_2$N	Pt/WO$_3$	IO^{3-}/I$^-$	6.6	3.3	[41]
Pt/TaON	Pt/WO$_3$	IO^{3-}/I$^-$	24	12	[42]
Pt/SrTiO$_3$:Rh	BiVO$_4$	Fe^{3+}/Fe^{2+}	15	7.2	[43]
Pt/SrTiO$_3$:Rh	Bi$_2$MoO$_6$	Fe^{3+}/Fe^{2+}	19	8.9	[43]
Pt/ZrO$_2$/TaON	Pt/WO$_3$	IO^{3-}/I$^-$	52	26.6	[44]

mediator redox couple is critical to boosting effective electron relay and suppressing the backward reactions. Compared with ionic redox couples in which electron transfer process takes place between two isolated photocatalysts, the direct coupling of the components through a solid electron mediator is more favorable in retarding back reactions, which is also called a direct Z-scheme photocatalyst system. A typical example of a direct Z-scheme system is a site-selective Au@CdS/TiO$_2$ nanojunction, which exhibits a higher photocatalytic activity than single or two-component systems as a result of the vectorial electron transfer (TiO2→Au→CdS) driven by the two-step excitation process.[45] Indeed, an excellent solid-state electron mediator should possess the ability to achieve a dynamic equilibrium between electron accepting and donating processes. Recently, the reduced graphene oxide as a solid electron mediator for a Z-scheme photocatalytic water splitting system is demonstrated.[46] Another crucial factor for a Z-scheme system is the PS I/mediator/PS II contact interface, which should ensure a continuous electron flow between the photocatalysts.

In contrast to the Z-scheme process involving a redox mediator, the weaker oxidative hole and reductive electron can also be directly quenched at the solid heterojunction interface while keeping the stronger oxidative hole and reductive electron isolated on different semiconductors. This configuration appears to work without a redox mediator, such as the ZnO–CdS coupled system.[47] In relation to this, the Ru–SrTiO$_3$:Rh–BiVO$_4$ photocatalyst without a redox mediator was demonstrated to perform overall water splitting through a Z-scheme mechanism.[48] These results suggest that an intimate contact between two components is necessary in Z-scheme systems without electron mediators.

Although the catalytic activities and efficiencies of the current available Z-scheme systems are still quite low, the versatility and capability will make the Z-scheme system the focus of future research on photocatalytic applications. The challenges lie in the construction of adequate interface contacts to realize efficient electron transfer in Z-scheme systems.

11.3.5 Cocatalyst Modification

Cocatalyst coupling with semiconductors can reduce the overpotential of the photocatalytic reaction and act as a surface reduction or oxidation site, thus facilitating the reactions (Figure 11.12). To date, the most widely used cocatalysts include noble metals (e.g., Pt, Pd, Ru, Rh, Au, Ir), metal oxides (e.g., NiO, Rh$_x$Cr$_{1-x}$O$_3$), and composites (e.g., Ni/NiO, Rh/Cr$_2$O$_3$).[49] Thus, it is imperative to develop low-cost cocatalysts for replacement. Recently, a series of sulfide cocatalysts, such as MoS$_2$, WS$_2$, and PbS, have been explored. It is noted that

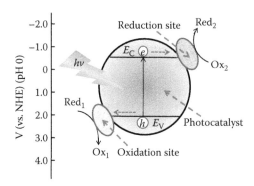

FIGURE 11.12
(See color insert.) Schematic diagram illustrating the reaction process of cocatalyst-modified semiconductor photocatalyst.

MoS$_2$ as a cocatalyst coupled with CdS exhibited a better photoreactivity than a Pt-loaded sample for H$_2$ evolution.[50]

The oxidation reaction is usually considered as the rate-determining process in the photocatalytic reaction. Coupling a suitable cocatalyst to reduce the overpotential of the oxidation reaction will greatly accelerate the overall photocatalytic process. Metal oxides, such as IrO$_2$, RuO$_2$, and cobalt phosphate, have been shown to promote oxidation reactions. Furthermore, codeposition of Pt as a reduction cocatalyst and IrO$_2$ as an oxidation cocatalyst on TiO$_2$ demonstrated additional enhancement of the reduction and oxidation processes.[51] Thus, the design and development of new types of cocatalysts play a significant role in enhancing photocatalytic efficiency.

11.3.6 Plasmon–Exciton Coupling

Surface plasmon resonance (SPR) of metallic nanostructure can be described as the resonant photon-induced collective oscillation of valance photons that subsequently generate a spatially nonhomogeneous oscillating electronic field in the neighboring region of the nanostructure.[52] By manipulating the composition, size, and morphology of metal nanoparticles, the SPR frequency of metal nanoparticles can be tuned throughout the entire solar spectrum. The investigations have demonstrated that SPR plays an important role in enhancing the rate of photocatalytic reactions of the semiconductor in plasmonic metal–semiconductor composite materials. The highest rate enhancement was observed at the wavelength corresponding to the metal SPR, which reveals that hybrid structures of plasmonic metal nanoparticles and semiconductors can induce plasmon–exciton coupling interactions, which then participate in photochemical processes. There are two nonmutually exclusive mechanisms of excited plasmonic nanostructures in the field of photochemical reactions: (i) SPR excited direct charge injection from metal to semiconductor, and (ii) plasmonic electromagnetic field induced enhancement of the photochemical reaction rates.

One of the important behaviors of plasmonic pumping in metal nanoparticles is nonradiative decay by the generation of electron–hole pairs and then transfer of the energetically excited electrons to the CB of the nearby semiconductor (Figure 11.13a). Oxidation and reduction reactions may occur on the metal and semiconductor, respectively. This is similar to the dye-sensitization process. The excellent mobility of charge carriers and high absorption cross section of a metal plasmonic nanostructure make it a promising sensitizer over the entire solar spectrum. For example, Au/TiO$_2$ shows a notable visible light photoreactivity for water

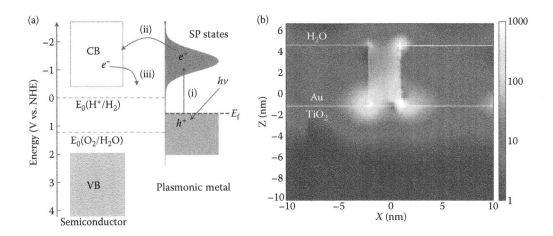

FIGURE 11.13
(See color insert.) (a) Schematic diagram of plasmon-induced charge separation and associated photochemistry at the metal–semiconductor heterojunction. (b) Optical simulations showing SPR-enhanced electric fields owing to photoexcited Au particles, permeating into a neighboring TiO_2 structure. (a: Adapted with permission from Linic, S., Christopher, P., Ingram, D.B. *Nat. Mater.*, 10, 911. Copyright 2011, Nature Publishing Group. b: Adapted with permission from Liu, Z., Hou, W., Pavaskar, P., Aykol, M., Cronin, S.B. *Nano Lett.*, 11, 1111. Copyright 2011, American Chemical Society.)

splitting due to the SPR, in which the plasmonic electrons in the metal can inject into the TiO_2 CB and then participate in the H_2 evolution reaction, leaving holes in the Au nanoparticle, which produce O_2.[53] It is important to note that only energetic electrons in metal nanoparticles can transfer to semiconductors and possess sufficient energy to execute half-reactions (e.g., H_2 evolution reaction). These results suggest that the plasmonic metal–semiconductor systems can not only enhance visible light absorption, but also greatly promote the electron–hole pair separation in the plasmonic metal nanoparticle by the electronic heterojunction.

Apart from electron transfer between the metal and semiconductor, SPR can also induce enhancement of the semiconductor photoreactivity even if the semiconductor and plasmonic metal are separated by thin nonconductive spacers. This indicates the radiative energy transfer can take place through a near-field electromagnetic mechanism. It is reported that the electron–hole formation rate in the semiconductor is significantly proportional to the electromagnetic field intensity if the semiconductor encounters the photoexcited plasmonic nanostructure. Owing to the spatial nonhomogeneity of plasmonic electromagnetic field, the highest enhancement of the SPR-induced electron–hole formation rate appears at the regions of the semiconductor closest to the plasmonic nanostructure. This mechanism is also supported by the enhancement of the photocatalytic water splitting under the visible region by coupling strongly plasmonic Au nanoparticles with TiO_2.[55] Electromagnetic simulation of the Au/TiO_2 composite film (Figure 11.13b) demonstrated that the plasmonic nanoparticles coupled the visible light effectively from the far field to the near field at the TiO_2 surface. Subsequently, most of the photogenerated charges excited by the SPR field contribute to the photocatalytic reaction. The near-field electromagnetic mechanism in the plasmonic metal–semiconductor nanostructure is further supported by the results of SPR-coupled wavelength-dependent enhancement of the semiconductor photoluminescence emission.[56]

It is important to note that the structure, size, and shape of the building blocks for the plasmonic metal–semiconductor all have significant impact on its practical performance. By tuning the relative geometric arrangement, the energy transfer mechanisms of the excited plasmonic nanostructure may be different. This requires mastering not only the controllable synthesis of metal building blocks but also the design of plasmon–exciton coupling assemblies to achieve the desired performance. For example, owing to the plasmon–exciton coupling interactions, an array of Au nanoprisms on WO_3 nanocrystalline films harvested photons in the visible to infrared region and enhanced the intrinsic absorption of WO_3.[57]

11.4 Crystal Facet Engineering for High Photocatalytic Reactivity

11.4.1 Crystal Facet Engineering

Considering the photocatalytic reactions on a semiconductor surface, the exposed crystal facets and the surface atomic configuration play a critical role in determining the surface adsorption, photocatalytic reactivity, and selectivity of the catalyst. Generally, the high-energy facets diminish quickly during the crystal growth process to minimize the total surface free energy. It is therefore necessary to develop strategies for facet-controlled growth of micro- and nanocrystallites.

Crystalline TiO_2 is chosen as the representative example to demonstrate crystal facet engineering for photocatalysts. Theoretical calculation has demonstrated that the order of the surface energy for anatase TiO_2 is 1.09 J m^{-2} for {110} > 0.90 J m^{-2} for {001} > 0.53 J m^{-2} for {100} > 0.44 J m^{-2} for {101}.[58] Although TiO_2 nanostructures with dominant {001} high-energy facets is expected to enhance surface properties, the most available anatase crystals are dominated by the less-reactive {101} facets. Therefore, controllable synthesis and assembly of anatase TiO_2 with dominant {001} facets is scientifically and technologically significant.

According to the Wulff construction, the slightly truncated octahedral bipyramid with eight {101} and two {001} facets is the common shape of the anatase crystal (Figure 11.14a,

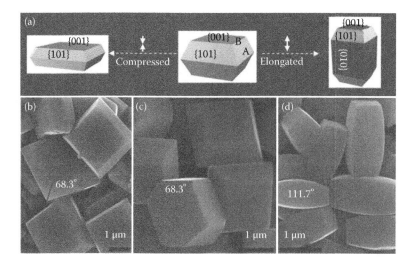

FIGURE 11.14
Morphology of anatase TiO_2 crystals. (a) Schematic of anatase TiO_2 with different percentages of {101}, {001}, and {010} facets. (b through d) SEM images of anatase crystals dominated by {001}, {101}, and {010} facets, respectively. (Adapted with permission from Pan, J., Liu, G., Lu, G. Q., Cheng, H.-M., *Angew. Chem. Int. Ed.*, 50, 2133. Copyright 2011, Wiley-VCH.)

middle). Synthesis of TiO$_2$ crystals dominated by high-energy facets needs to confine crystal growth within the kinetically controlled regime. Under liquid-phase conditions, the nucleation process can be controlled by selecting an appropriate reaction medium and surface adsorbate, which may reduce the surface energy of specific facets and thus obtain an anatase crystal dominated by high-energy surfaces. One typical example is that a fluorine ion caused an exceptional stabilization of the {001} facets of anatase TiO$_2$ crystals. Lu and coworkers first reported the synthesis of micron-sized anatase TiO$_2$ crystals with 47% {001} facets using HF as the capping agent and TiF$_4$ as the precursor.[59] This pioneering work has triggered an intensive research interest in preparing faceted anatase. Remarkably, TiO$_2$ nanosheets with exposed {001} facets up to 89% were synthesized with the involvement of a concentrated HF solution.[60]

To reduce or avoid the usage of HF, alcohols or ionic liquids are used in combination with HF of a low concentration to synthesize anatase TiO$_2$ crystals with {001} facets. Theoretical calculation indicates that, under an acidic environment, alcohols tend to dissociate to form an alkoxy group bound to coordinated unsaturated Ti^{4+} cations on {001} and {101} facets. The higher density of unsaturated Ti^{4+} cations on the {001} facet induce more obvious selective adhesion of the alkoxy group, which retards the growth of the anatase TiO$_2$ crystal along the [001] direction.[62] Another benefit of alcohols is that they provide an opportunity to control the dispersion degree and the particle size. Moreover, diethylenetriamine in isopropyl alcohol can act as a fluorine-free structure-directing agent to stabilize the high-energy {001} facets of the anatase TiO$_2$ crystals.[63]

Inspired by faceted TiO$_2$, crystal facet engineering has been extended to other semiconductor photocatalysts, including metal oxides (e.g., ZnO, SnO$_2$, Cu$_2$O, WO$_3$) and composite oxides (e.g., BiVO$_4$, Ag$_3$PO$_4$, Zn$_2$GeO$_4$) (Figure 11.15), such as ZnO nanodisks with dominant {001} facets,[64] SnO$_2$ octahedral nanocrystals with dominant {221} facets,[65] WO$_3$ octahedron enclosed with {111} facets, Ag$_3$PO$_4$ rhombic dodecahedrons with only {110} facets,[66] and cubes bounded by {100} facets. However, the capping agents adsorbed on the crystal surface need to be removed before photocatalytic applications. Recently, monoclinic BiVO$_4$ nanoplates with exposed {001} facets were synthesized by a straightforward hydrothermal route without any template or organic surfactant.[67] Nevertheless, it is still a great challenge to synthesize photocatalysts with highly reactive facets without capping agents.

11.4.2 Photoreactivity on Different Facets

11.4.2.1 Dissociative Adsorption on Crystal Facets

The adsorption of reactant molecules on a photocatalyst's surface is essential for the photocatalytic reactions to occur. Actually, the adsorption behavior of substrate molecules varies on different crystal facets. The surface interaction between the water molecule and TiO$_2$ plays an important role in the photocatalytic reaction. It is documented that water can only be molecularly adsorbed on {101} facets, while the chemically dissociated water molecules are energetically favored on {001} facets involving the bridging O-2c and Ti-5c atoms,[68] which lead to the formation of two hydroxyls terminally bounded to adjacent Ti sites and thus facilitate the formation of reactive species (Figure 11.16). Indeed, {001} facets can not only enhance photoreduction activity in water splitting but also promote photooxidation ability with respect to the decomposition of organic pollutants. It appears that {001} facets are more effective for dissociated adsorption of substrate molecules than the thermodynamically stable {101} facets.[69] This process can affect the photocatalytic reaction mechanism at the molecular level. In other words, {001} facets can mediate the dissociated adsorption of pollutant molecules by enhanced interfacial charge transfer.

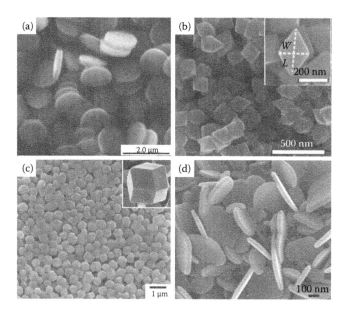

FIGURE 11.15
SEM images of (a) ZnO nanodisks, (b) SnO₂ octahedra, (c) Ag₃PO₄ rhombic dodecahedrons, and (d) monoclinic BiVO₄ nanoplates. (a: Adapted with permission from Zeng, J. H., Jin, B. B., Wang, Y. F., *Chem. Phys. Lett.*, 472, 90. Copyright 2009, Elsevier. b: Adapted with permission from Han, X. G., Jin, M. S., Xie, S. F., Kuang, Q., Jiang, Z. Y., Jiang, Y. Q., Xie, Z. X., Zheng, L. S., *Angew. Chem. Int. Ed.*, 48, 9180. Copyright 2009, Wiley-VCH. c: Adapted with permission from Bi, Y. P., Ouyang, S. X., Umezawa, N. J., Cao, Y., Ye, J. H. *J. Am. Chem. Soc.*, 133, 6490. Copyright 2011, American Chemical Society. d: Adapted with permission from Xi, G. C., Ye, J. H., *Chem. Commun.*, 46, 1893. Copyright 2010, Royal Society of Chemistry.)

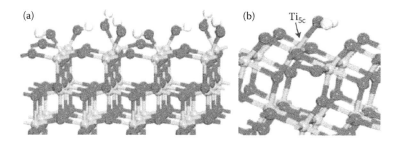

FIGURE 11.16
(See color insert.) Water adsorption behaviors on anatase (a) TiO₂ {001} vs. (b) {101} surfaces. (Adapted with permission from Selloni, A., *Nat. Mater.*, 7, 613. Copyright 2008, Nature Publishing Group.)

11.4.2.2 Reactivity on Crystal Facets

Different surface electronic structures of anatase TiO₂ {001} and {101} facets can also modify the surface-mediated photocatalytic reactions by driving a divergent diffusion of the photoexcited electrons and holes toward specific exposed crystal facets. This process results in the spatial separation of reduction and oxidation sites, subsequently reducing the recombination possibility of photogenerated charge carriers. Recently, a single-molecule imaging and kinetic analysis technology confirmed the directional flow of the photoexcited

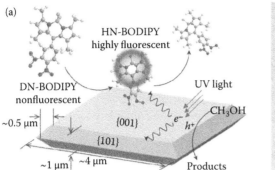
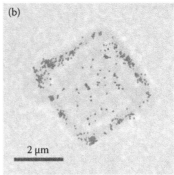

FIGURE 11.17
(See color insert.) (a) Photocatalytic generation of fluorescent HN-BODIPY from nonfluorescent DN-BODIPY over a TiO_2 crystal. (b) Transmission images of the same TiO_2 crystal immobilized on a cover glass in Ar-saturated methanol solution containing DN-BODIPY under a 488-nm laser and UV irradiation. The blue and red dots in the transmission image indicate the location of fluorescence bursts on the {001} and {101} facets of the crystal, respectively, observed during 3-min irradiation. (Adapted with permission from Tachikawa, T., Yamashita, S., Majima, T., *J. Am. Chem. Soc.*, 133, 7197. Copyright 2011, American Chemical Society.)

electrons and holes toward specific facets, which is presumably related to the facet-specific trapping sites and electronic energy levels (Figure 11.17).[71]

Determining the relationship between the photocatalytic activity and crystal facets is necessary for designing efficient photocatalysts by proper crystal facet engineering. A conventional view is that the photoreactivity is proportional to the percentage of anatase TiO_2 {001} facets. However, there are conflicting experimental findings on the optimal ratio of {001} facets to other exposed facets for photoreactivity. For instance, Pan et al. found that the {001} facet exhibited a lower reactivity than the {101} facet in photocatalytic H_2 evolution.[61] Obviously, other surface states, including surface defects, surface chemistry, and the substrate molecules, should be taken into account in explaining the function of {001} facets on the photocatalytic reaction kinetics.

11.5 Conclusions and Perspectives

The semiconductor photocatalytic technology has been an increasing topic of interest worldwide owing to its significance in environmental applications. From the semiconductor behavior upon excitation by photons to the utilization of photogenerated charge carriers, nanotechnologies including energy band engineering and heterojunction structures provide effective tools to improve photocatalytic activity. To realize a substantial breakthrough in efficiency for photocatalytic settings, a better understanding of the charge transfer processes across the surface/interface is greatly desired. Furthermore, a further insight into the interactions between environmental contaminants and semiconductor surface by *in situ* observation technologies is challenging but crucial for the optimization of the photocatalytic reactions. Environment remediation will continue to serve as an important platform to establish fundamental depth in photocatalysis, thus guiding the future design, fabrication, and modification of photocatalytic materials.

References

1. Hoffmann, M. R.; Martin, S. T.; Choi, W. Y.; Bahnemann, D. W. *Chem. Rev.* **1995**, *95*, 69.
2. Grätzel, M. *Heterogeneous Photochemical Electron Transfer*. CRC Press: Baton Rouge, FL, **1988**.
3. Tojo, S.; Tachikawa, T.; Fujitsuka, M.; Majima, T. *Chem. Phys. Lett.* **2004**, *384*, 312.
4. Schwarz, H. A.; Dodson, R. W. *J. Phys. Chem.* **1984**, *88*, 3643.
5. Navarro, R. M.; Alvarez-Galván, M. C.; Mano, J. A. V.; Al-Zahrani, S. M.; Fierro, J. L. G. *Energy Environ. Sci.* **2010**, *3*, 1865.
6. Na-Phattalung, S.; Smith, M. F.; Kim, K.; Du, M.-H.; Wei, S.-H.; Zhang, S. B.; Limpojumnong, S. *Phys. Rev. B* **2006**, *73*, 125205.
7. Cong, S.; Xu, Y. *J. Phys. Chem. C* **2011**, *115*, 21161.
8. Wang, C. Y.; Pagel, R.; Dohrmann, J. K.; Bahnemann, D. W. *C.R. Chim.* **2006**, *9*, 761.
9. Kubacka A.; Fernández-García M.; Colón, G. *Chem. Rev.* **2012**, *112*, 1555.
10. Nishikawa, T.; Shinohara, Y.; Nakajima, T.; Fujita, M.; Mishima, S. *Chem. Lett.* **1999**, *28*, 1133.
11. Kim, H. G.; Hwang, D. W.; Lee, J. S. *J. Am. Chem. Soc.* **2004**, *126*, 8912.
12. Kudo, A.; Omori, K.; Kato, H. *J. Am. Chem. Soc.* **1999**, *121*, 11459.
13. Shah, S. I.; Li, W.; Huang, C. P.; Jung, O.; Ni, C. *Proc. Natl. Acad. Sci. USA* **2002**, *99*, 6482.
14. Kitano, M.; Hara, M. *J. Mater. Chem.* **2012**, *20*, 627.
15. Grätzel, M.; Howe, R. F. *J. Phys. Chem.* **1990**, *94*, 2566.
16. Asahi, R.; Morikawa, T.; Ohwaki, T.; Aoki, K.; Taga, Y. *Science* **2001**, *293*, 269.
17. Emeline, A. V.; Kuznetsov, V. N.; Rybchuk, V. K.; Serpone, N. *Int. J. Photoenergy* **2008**, *2008*, 258394.
18. Liu, G.; Sun, C. H.; Yan, X. X.; Cheng, L.; Chen, Z. G.; Wang, X. W.; Wang, L. Z.; Smith, S. C.; Lu, G. Q.; Cheng, H. M. *J. Mater. Chem.* **2009**, *19*, 2822.
19. Chun, W. A.; Ishikawa, A.; Fujisawa, H.; Takata, T.; Kondo, J. N.; Hara, M.; Kawai, M.; Matsumoto, Y.; Domen, K. *J. Phys. Chem. B* **2003**, *107*, 1798.
20. Gai, Y.; Li, J.; Li, S. S.; Xia, J. B.; Wei, S. H. *Phys. Rev. Lett.* **2009**, *102*, 036402.
21. Chen, X.; Liu, L.; Yu, P. Y.; Mao, S. S. *Science* **2011**, *331*, 746.
22. Maeda, K.; Teramura, K.; Takata, T.; Hara, M.; Saito, N.; Toda, K.; Inoue, Y.; Kobayashi, H.; Domen, K. *J. Phys. Chem. B* **2005**, *109*, 20504.
23. Li, G. Q.; Kako, T.; Wang, D. F.; Zou, Z. G.; Ye, J. H. *J. Solid State Chem.* **2007**, *180*, 2845.
24. Ouyang, S. X.; Ye, J. H. *J. Am. Chem. Soc.* **2011**, *133*, 7757.
25. Jang, J. S.; Borse, P. H.; Lee, J. S.; Choi, S. H.; Kim, H. G. *J. Chem. Phys.* **2008**, *128*, 154717.
26. Yao, W. F.; Ye, J. H. *J. Phys. Chem. B* **2006**, *110*, 11188.
27. Zhang, Z.; Yates, J. T., Jr. *Chem. Rev.* **2012**, *112*, 5520.
28. Furube, A.; Asahi, T.; Masuhara, H.; Yamashita, H.; Anpo, M. *Chem. Phys. Lett.* **2001**, *336*, 424.
29. Hirakawa, T.; Kamat, P. V. *J. Am. Chem. Soc.* **2005**, *127*, 3928.
30. Teoh, W. Y.; Mädler, L.; Amal, R. *J. Catal.* **2007**, *251*, 271.
31. Hwang, Y. U.; Boukai, A.; Yang, P. *Nano Lett.* **2009**, *9*, 415.
32. Hu, C. C.; Nian, J. N.; Teng, H. *Sol. Energy Mater. Sol. Cells* **2008**, *92*, 1071.
33. Leung, D. Y. C.; Fu, X.; Wang, C.; Ni, M.; Leung, M. K. H.; Wang, X.; Fu, X. *ChenSunChem* **2010**, *3*, 681.
34. Zhang, J.; Xu, Q.; Feng, Z.; Li, M.; Li, C. *Angew. Chem. Int. Ed.* **2008**, *47*, 1766.
35. Scanlon, D. O.; Watson, G. W. *J. Phys. Chem. Lett.* **2010**, *1*, 2582.
36. Kamat, P. V. *J. Phys. Chem. C* **2008**, *112*, 18737.
37. Beard, M. C.; Knutsen, K. P.; Yu, P. R.; Luther, J. M.; Song, Q.; Metzger, W. K.; Ellingson, R. J.; Nozik, A. J. *Nano Lett.* **2007**, *7*, 2506.
38. Kudo, A. *MRS Bull.* **2011**, *36*, 32.
39. Abe, R.; Sayama, K.; Sugihara, H. *J. Phys. Chem. B* **2005**, *109*, 16052.
40. Higashi, M.; Abe, R.; Ishikawa, A.; Takata, T.; Ohtani, B.; Domen, K. *Chem. Lett.* **2008**, *37*, 138.
41. Higashi, M.; Abe, R.; Teramura, K.; Takata, T.; Ohtani, B.; Domen, K. *Chem. Phys. Lett.* **2008**, *452*, 120.

42. Abe, R.; Takata, T.; Sugihara, H.; Domen, K. *Chem. Commun.* **2005**, 3829.
43. Kato, H.; Hori, M.; Konta, R.; Shimodaira, Y.; Kudo, A. *Chem. Lett.* **2004**, *33*, 1348.
44. Maeda, K.; Higashi, M.; Lu, D.; Abe, Y.; Domen, K. *J. Am. Chem. Soc.* **2010**, *132*, 5858.
45. Tada, H.; Mitsui, T.; Kiyonaga, T.; Akita, T.; Tanaka, K. *Nat. Mater.* **2006**, *5*, 782.
46. Iwase, A.; Ng, Y. H.; Ishiguro, Y.; Kudo, A.; Amal, R. *J. Am. Chem. Soc.* **2011**, *133*, 11054.
47. Wang, X. W.; Liu, G.; Chen, Z. G.; Li, F.; Wang, L. Z.; Lu, G. Q.; Cheng, H. M. *Chem. Commun.* **2009**, 3452.
48. Sasaki, Y.; Nemoto, H.; Saito, K.; Kudo, A. *J. Phys. Chem. C* **2009**, *113*, 17536.
49. Osterloh, F. E. *Chem. Mater.* **2008**, *20*, 35.
50. Zong, X.; Yan, H.; Wu, G.; Ma, G.; Wen, F.; Wang, L.; Li, C. *J. Am. Chem. Soc.* **2008**, *130*, 7176.
51. Meekins, B. H.; Kamat, P. V. *J. Phys. Chem. Lett.* **2011**, *2*, 2304.
52. El-Sayed, M. A. *Acc. Chem. Res.* **2001**, *34*, 257.
53. Silva, C. G.; Juárez, R.; Marino, T.; Molinari, R.; García, H. *J. Am. Chem. Soc.* **2011**, *133*, 595.
54. Linic, S.; Christopher, P.; Ingram, D. B. *Nat. Mater.* **2011**, *10*, 911.
55. Liu, Z.; Hou, W.; Pavaskar, P.; Aykol, M.; Cronin, S. B. *Nano Lett.* **2011**, *11*, 1111.
56. Kulkarni, A. P.; Noone, K. M.; Munechika, K.; Guyer, S. R.; Ginger, D. S. *Nano Lett.* **2010**, *10*, 1501.
57. Chen, X. Q.; Li, P.; Tong, H.; Kako, T.; Ye, J. H. *Sci. Technol. Adv. Mat.* **2011**, *12*, 044604.
58. Diebold, U. *Surf. Sci. Rep.* **2003**, *48*, 53.
59. Yang, H. G.; Sun, C. H.; Qiao, S. Z.; Zou, J.; Liu, G.; Smith, S. C.; Cheng, H. M.; Lu, G. Q. *Nature* **2008**, *453*, 638.
60. Han, X. G.; Kuang, Q.; Jin, M. S.; Xie, Z. X.; Zheng, L. S. *J. Am. Chem. Soc.* **2009**, *131*, 3152.
61. Pan, J.; Liu, G.; Lu, G. Q.; Cheng, H.-M. *Angew. Chem. Int. Ed.* **2011**, *50*, 2133.
62. Yang, H. G.; Liu, G.; Qiao, S. Z.; Sun, C. H.; Jin, Y. G.; Smith, S. C.; Zou, J.; Cheng, H. M.; Lu, G. Q. *J. Am. Chem. Soc.* **2009**, *131*, 4078.
63. Chen, J. S.; Tan, Y. L.; Li, C. M.; Cheah, Y. L.; Luan, D. Y.; Madhavi, S.; Boey, F. Y. C.; Archer, L. A.; Lou, X. W. *J. Am. Chem. Soc.* **2010**, *132*, 6124.
64. Zeng, J. H.; Jin, B. B.; Wang, Y. F. *Chem. Phys. Lett.* **2009**, *472*, 90.
65. Han, X. G.; Jin, M. S.; Xie, S. F.; Kuang, Q.; Jiang, Z. Y.; Jiang, Y. Q.; Xie, Z. X.; Zheng, L. S. *Angew. Chem. Int. Ed.* **2009**, *48*, 9180.
66. Bi, Y. P.; Ouyang, S. X.; Umezawa, N. J.; Cao, Y.; Ye, J. H. *J. Am. Chem. Soc.* **2011**, *133*, 6490.
67. Xi, G. C.; Ye, J. H. *Chem. Commun.* **2010**, *46*, 1893.
68. Vittadini, A.; Selloni, A.; Rotzinger, F. P.; Grätzel, M. *Phys. Rev. Lett.* **1998**, *81*, 2954.
69. Gong, X. Q.; Selloni, A. *J. Phys. Chem. B* **2005**, *109*, 19560.
70. Selloni, A. *Nat. Mater.* **2008**, *7*, 613.
71. Tachikawa, T.; Yamashita, S.; Majima, T. *J. Am. Chem. Soc.* **2011**, *133*, 7197.

12

Engineered Polymers and Organic–Inorganic Hybrids as Antimicrobial Materials for Water Disinfection

Divakara S.S.M. Uppu, Jiaul Hoque, and Jayanta Haldar

Chemical Biology and Medicinal Chemistry Laboratory, Jawaharlal Nehru Center for Advanced Science Research, Bengaluru, India

CONTENTS

Waterborne infectious diseases continue to be the leading cause of death in many developing nations. Diarrhea, a waterborne infectious disease, kills more young children than do acquired immune immunodeficiency virus, malaria, and measles combined (WHO/UNICEF 2009).[1] Microbial contaminants include human pathogens such as *Salmonella* spp., *Shigella* spp., *Escherichia coli*, *Pseudomonas aeruginosa*, *Vibrio cholera*, rotavirus, *Cryptosporidium parvum*, *Giardia* cysts, etc. The World Health Organization (WHO, 1996)[2] recommended that any water intended for drinking should contain zero fecal and total coliform counts in any 100 mL sample. The US Environmental Protection Agency 1987 Guide Standard and

Protocol[3] for Testing Microbiological Water Purifiers says that they should show >6 $\log_{10}$, >4 $\log_{10}$, and >3 $\log_{10}$ reduction of bacteria, viruses, and protozoan parasites, respectively. Current water disinfection methods rely predominantly on chemical oxidants such as free chlorine, chloramines, and ozone, which are effective against different microbes but have disadvantages like harmful disinfection by-products (DBPs). Despite their efficiency for pathogen removal, membrane filtration and ultraviolet-based technologies have limited use as they are too expensive for use in water disinfection, mainly in developing countries.

Hence, there is a growing need for the development of economically viable and efficient antimicrobial materials for water disinfection. Inorganic nanomaterials (such as Ag, ZnO, TiO_2, and iron oxide nanoparticles) have been used as antimicrobial agents for removal of pathogens from water. To address issues such as cost, environmental toxicity, and human exposure related to the inorganic materials, development of polymers and organic–inorganic hybrids as effective materials for water disinfection is discussed in this chapter. Polymeric materials (water-insoluble quaternary ammonium polymers and *N*-halamine polymers) and organic–inorganic hybrids (polymeric composites with nanosilver and silica, organic–TiO_2 composites) can have promising potential for water disinfection. Again, an economically viable alternative is to use carbonaceous and mesoporous materials (activated carbon [AC] and clay materials and their inorganic hybrids) due to their inherent advantages like ease of availability, processability, and cost of manufacture as antimicrobial materials. Additionally, most of these materials can be used in membrane filtration technology to improve its efficiency and reduce the cost.

12.1 Polymeric Disinfectants

Polymeric disinfectants are a class of polymers having ability to inactivate or inhibit the growth of microorganisms such as bacteria, viruses, fungi, or protozoans. Antimicrobial polymers may enhance the efficiency and selectivity of currently used antimicrobial agents, while decreasing associated environmental hazards because antimicrobial polymers are nonvolatile and chemically stable. These additional benefits make these materials potential candidates for use in water purification. Water-soluble polymers, commonly known as organic polyelectrolytes, have generally been used with or without chemical coagulants to remove pathogens effectively.[4,5] Various antimicrobial polymers in the form of water-insoluble beads or coatings are also used to disinfect pathogens in water treatment processes.[6,7]

12.1.1 Water-Soluble Polymers

The main applications of water-soluble polymers (organic polyelectrolytes) in potable water production are in coagulation (the interaction of small particles to form larger particles) and flocculation (the physical process of producing interparticle contacts that lead to the formation of large particles or flocs). Conventional drinking-water treatment processes, used primarily to remove various contaminants including microorganisms from water, comprise coagulation (usually using aluminum sulfate and/or polymers), followed by flocculation, sedimentation, filtration, and disinfection.[8] Water-soluble polymers play a vital role in the conventional treatment process with or without chemical coagulants to remove pathogens. The extensive use of polymers as flocculants is due to their distinct characteristic features. Polymers are convenient to use and do not affect the pH of the medium. The addition of polymers in parts per million (ppm) amounts can drastically change the degree of aggregation of cells, and thereby are cost effective. Large tonnage use of inorganic compounds (alum and $FeCl_3$) produces large amounts of sludge—a problem absent when polymeric flocculants are used.[5] For optimal operation, the choice of an appropriate polymer is important. In addition, increasing the molecular weight of the polymer improves flocculation, probably by promoting bridge formation. Two main mechanisms of flocculation are proposed: (a) formation of macromolecular bridges between the particles, and (b) formation of the surface potential and charge reduction due to adsorption of highly charged polyelectrolytes on oppositely charged particles.[4]

These organic polymeric flocculants fall into two categories, namely natural and synthetic polymers. The polysaccharides, mainly starch and its derivatives, different types of gums, alginic acid, cellulose and its derivatives, dextran, glycogen, chitosan, etc., are among the natural polymers used in flocculation. Synthetic flocculants are broadly divided into anionic, cationic, and nonionic categories. Polyacrylamide (PAM) and poly(ethylene oxide) are nonionic. Cationic groups of polyelectrolytes are derived by introducing quaternary ammonium groups onto the polymer backbone. In addition to that, sulfonium and phosphonium groups are used to a limited extent. The most commonly used cationic polyelectrolyte is poly(diallyl dimethyl ammonium chloride). In the group of anionic polyelectrolytes, mainly two types of polymers are used; one type is polymers containing carboxyl functional groups and the other containing sulfonic acid groups. A representative of the former is poly(acrylic acid) (PAA) and its derivatives, of the latter polystyrene sulfonic acid.[9] Among polymeric flocculants, the synthetic polymers can be made by controlling the molecular weight, molecular weight distribution, and chemical structure. Thus, due to tailorability, synthetic polymers can be very efficient flocculants. Structures of some of the polymers used for removal of pathogens are shown below (Figure 12.1).

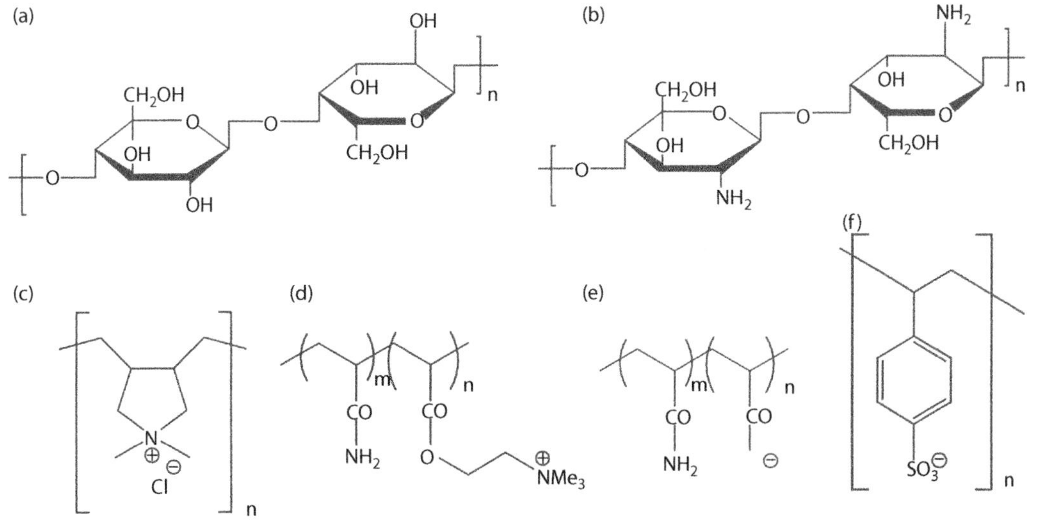

FIGURE 12.1
Representative structures of nonionic, anionic, and cationic polymers used in flocculation: (a) cellulose; (b) chitosan; (c) poly(diallyl dimethyl ammonium chloride), poly(DADMAC); (d) cationic polyacrylamide; (e) anionic polyacrylamide; (f) polystyrene sulfonate.

Several reports are available for the removal of microorganisms using both synthetic and natural polymers in the water treatment processes, some of which are discussed below. Baran examined a number of polymers for flocculating *E. coli* and found that cationic polymers are the most effective.[10] Bilanovic and Shelef studied the flocculation of microalgae with cationic polymers.[11] The authors found that increasing the ionic strength of the suspending medium leads to a decrease in the degree of flocculation.[11] Brown and Emelko found that chitosan coagulation of *C. parvum* cysts and cyst-sized polystyrene microspheres resulted in excellent turbidity and particle reductions by filtration that were comparable to those achieved when in-line filtration was preceded by alum and $FeCl_3$ coagulation during stable operation.[12] Bayat et al. found that among the different types of polymers investigated, a particular type of polyacrylamide (very high molecular weight anionic polymer) resulted in a 65.22% bacteria (*Desulfovibrio desulfuricans*) removal.[13]

12.1.2 Water-Insoluble Polymeric Matrix

Although water-soluble polymers as described in the previous section are used primarily for the removal of pathogens to a certain extent, a second level of treatment is necessary to achieve complete inactivation. The activation of pathogen is generally carried out by chlorine or ozone, which leads to toxic DBPs such as trihalomethane, halogenic acetic acids, carboxylic acids, and aldehydes. Furthermore, conventional treatment processes for water supplies are inadequate to remove viruses.[14] To overcome these challenges, antimicrobial polymers in the form of water-insoluble beads or other antimicrobial agents incorporated into the water-insoluble polymeric matrix are used. There are several reports to produce water-insoluble polymeric disinfectants by incorporating antibacterial agents into the ion exchange resins. Some of these products, especially the polyhalide–anion exchange resins, have shown promise as effective water disinfection systems.[15,16] Tyagi and Singh synthesized a hydrophilic but water-insoluble copolymer matrix to which iodine as antimicrobial agent is immobilized, and achieved a slow release of the antimicrobial agent into the medium for a prolonged duration of time.[17] The iodinated polymethyl methacrylate-*N*-vinyl-2-pyrrolidone copolymer matrix has proved to be highly effective and was found to remain effective for longer durations. The iodinated copolymer has been found to be stable, insoluble, and an active disinfectant against a variety of microbes as well as fungal species found by a zone of inhibition test. This copolymer thus holds a promising future as an efficient disinfectant for potable drinking water for prolonged duration (Figure 12.2).

Quaternary ammonium salt–based resins are also found to be effective in removal of various pathogens from water. It has been shown that cross-linked poly(*N*-benzyl-4-vinylpyridinium halide) (BVP) resin, an insoluble pyridinium-type resin, strongly removes bacteriophage T4 from aqueous solution.[18] Kawabata et al. reported the removal of various pathogenic human viruses by cross-linked BVP resin.[6] It was found that the level of infection in suspensions of enterovirus, herpes simplex virus, poliovirus, and human immunodeficiency virus was reduced 10^3–$10^5\times$ during a 2-h period and the polymers were found to be highly effective against human rotavirus, influenza virus, human adenovirus, Japanese encephalitis virus, coxsackievirus, and echovirus.

Worley et al. have shown that *N*-halamine polymers are highly antimicrobial and can be used for water purification as water-insoluble beads.[19,20] The functionalized *N*-halamine polymers are superior in overall performance (taking into account biocidal efficacy, stability at various pH and in the presence of organic receptors, rechargeability, lack of toxicity, and cost) to other biocidal polymers. The most important *N*-halamine polymers developed, taking into account their potential for economical disinfection of potable water, are the

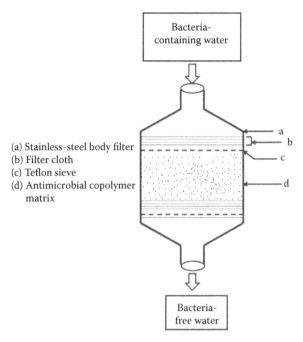

FIGURE 12.2
Schematic diagram of the refill cartridge for water purification. (Adapted from Tyagi, M. and Singh, H., *J. Appl. Polym. Sci.*, 76, 1109, 2000.)

N-halogenated poly(styrene hydantoins) (Figure 12.3a). The final polymers are amorphous solids, which are insoluble in water and can be packed into glass. It was observed that the filters made from the poly(styrene hydantoin) polymer inactivated numerous species of bacteria such as *Staphylococcus aureus*, *E. coli*, fungi, and even rotavirus within seconds of contact time in flowing water.[19,20] The mechanism of action of the *N*-halamines beads is not known. It has been proposed that a contact mechanism is involved in which, upon

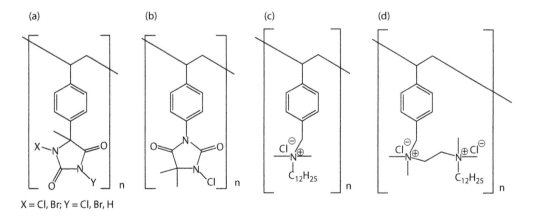

FIGURE 12.3
Structures of the biocidal polymeric beads and resins used: (a) halogenated poly(styrene hydantoin); (b) chlorinated methylated polystyrene hydantoin; (c and d) quaternary ammonium polymeric resins PQ1 and PQ2.

collision of a pathogenic cell with N-halamine beads, Cl^+ (or Br^+) is directly transferred to the cell. The X^+ penetrates the cell wall and oxidizes targets within the cell or viral particle, causing inactivation of the pathogens. It was observed that the columns did not leach out undesirable decomposition products such as total organic carbon, free and total chlorine, anions, and volatile organics such as trihalomethanes into the water.[19] Furthermore, once the halogen supply ceases, it can be regenerated on the polymers by simply exposing them to flowing aqueous free halogen (e.g., sodium hypochlorite bleach for Poly1-Cl), thus showing the applicability of these polymeric beads in water treatment process.[19]

In a report, the biocidal effects of two classes of polymers, namely water-insoluble beads of N-chlorinated polymer and derivatives of polyquats (Figure 12.3b through d) containing quaternary ammonium groups, were compared against *S. aureus* and *E. coli*. The most effective polymer, as measured by degree of inactivation in the shortest contact time of the two species of bacteria, were found to be the N-chlorinated hydantoinyl derivative of methylated polystyrene.[21]

Water-insoluble antimicrobial tablet formulations have also been made to treat waterborne pathogens. The tablets are made from gum arabic, poly(vinylalcohol), ethyl cellulose (EC), and poly(vinylpyrrolidone)-iodine. The formulation consisted of a dispersible core tablet surrounded by a hydrophilic coating of EC and poly(ethylene glycol) (PEG) mixture.[22] The tablets were found to be active against *E. coli*, *S. aureus*, *Listeria monocytogenes* Scott A, and *Salmonella typhimurium*.

12.1.3 Antimicrobial Surfaces

Although highly effective, the constant and mostly unwanted release of many biocides from antimicrobial polymeric beads as described in the previous section might cause the increased development of resistant microbes, which is a major problem of current times.[14,23] Another problem of the release-based polymeric beads includes the exhaustion of the biocide over time. The toxicity of the released biocides is another issue that needs to be addressed. Furthermore, regeneration of the resin that is used for removing microorganisms from water is extremely difficult because the capturing interaction between the resin surface and microbial cells is very strong. Thus, the resin is obviously economically unsuitable. Surface modification that effectively kills microbes without releasing the antimicrobial agent is a relatively new approach to create antimicrobial surfaces. Polymers containing quaternary ammonium groups have been successfully coated onto the surfaces such as glass, sand, and cotton, and are found to be antimicrobial (Figure 12.4).[24,25] These polymers inactivate microorganism by both electrostatic and hydrophobic interaction without releasing any active biocidal moiety into the environment.

Tiller et al. prepared an antimicrobial surface by attaching N-hexyl-poly(4-vinylpyridinium) polymer covalently. Although not tested directly for water treatment processes, these surfaces were found to inactivate various waterborne pathogens effectively.[24] It has been shown that surfaces when coated with hydrophobically modified, cationic, water-insoluble polymer N,N-dodecyl-methyl-polyethylenimines (PEIs), kill various pathogenic bacteria and viruses effectively.[25] Coating surfaces with N-alkylated PEIs, such as branched N,N-hexylmethyl-PEI via covalent attachment to glass or linear N,N-dodecylmethyl-PEI by physical deposition ("painting") onto polyethylene, enables the resultant materials to quickly and efficiently disinfect aqueous solutions of poliovirus and rotavirus.[26] The threat of poliovirus, like other enteric waterborne viruses, arises mainly from contaminated drinking sources in areas of poor sanitation. Once infected, a person can develop poliomyelitis, a debilitating disease causing muscle atrophy and paralysis. Rotavirus,

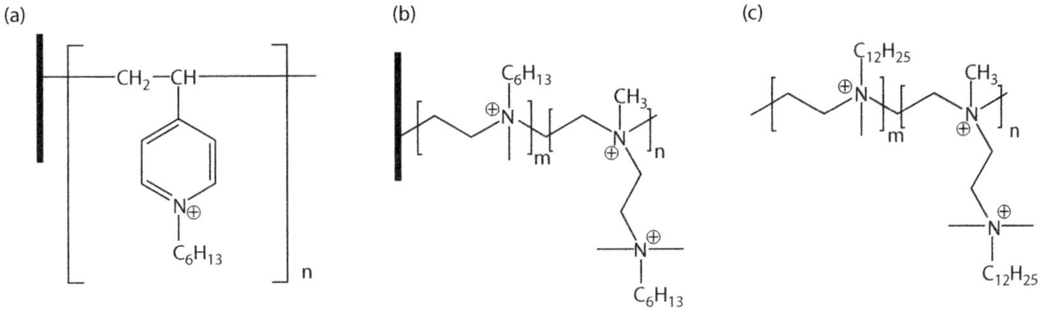

FIGURE 12.4
Schematic structural illustration of covalently immobilized (a) poly(4-vinyl-*N*-hexylpyridinium) and (b) branched *N*-hexyl,*N*-methyl-polyethyleneimine (PEI). (c) Structure of the water-insoluble *N,N*-dodecylmethyl-PEI polymer.

another waterborne human pathogen, is the chief cause of gastroenteritis and diarrhea in children.[27] These polycations being highly lethal to the various waterborne microorganisms, hold good promise to be used in a water treatment process where the surfaces of various filter media can be coated (covalently or noncovalently) to inactivate waterborne pathogens. These cationic polymers have immense potential for use in membrane-type filtration. Fabrication of these polymers into various types of membranes is possible, thereby enabling simultaneous filtration and disinfection.

12.1.4 Antimicrobial and Antifouling Polymers

Antimicrobial surfaces coated with quaternary ammonium compounds (QACs), as discussed previously, have proved to be able to efficiently kill a variety of microorganisms. A major problem with QAC surfaces is the attachment of dead microorganisms remaining on the antimicrobial coatings, which can block access to the antimicrobial functional groups for further action.[28] In addition, such polymeric antimicrobial coatings cannot fulfill the requirements of antifouling (inhibition of the attachment and subsequent formation of the biofilm onto the surface by the microorganism)—a major problem in water treatment processes, especially in the membrane-based filtration.[29] Thus, it is essential to develop materials with both antimicrobial and antifouling capabilities. PEG derivatives[30] or zwitterionic polymers[31] have been extensively used as antifouling materials to reduce microbial attachment and biofilm formation.

Although not used in water purification, these polymers can be engineered to be used in water purification to kill waterborne pathogens as well as to inhibit the biofilm formation. Recently, it has been shown that through a surface-initiated alkyne–azide click reaction, azido-terminated polymers (hyperbranched polyglycerols [HPG] or PEI or PEG) can be covalently grafted onto the propargyl bromide quaternized poly(vinylidene fluoride) (PVDF) graft poly[2-(*N,N*-dimethylamino)ethyl methacrylate] (PDMAEMA) (PVDF-*g*-PDMAEMA) membrane surfaces (Figure 12.5a). The PVDF-*g*-P[QDMAEMA-*click*-HPG] membranes exhibit good antifouling properties that are comparatively much higher than those of the PVDF-*g*-P[QDMAEMA-*click*-PEG] membranes. Furthermore, the antibacterial activity and biofilm inhibition property of the PVDF-*g*-P[QDMAEMA-*click*-QPEI] membranes were much higher than those of the PVDF-*g*-P[QDMAEMA-*click*-PEI] membranes, making the membranes potentially useful for wastewater treatment.[29]

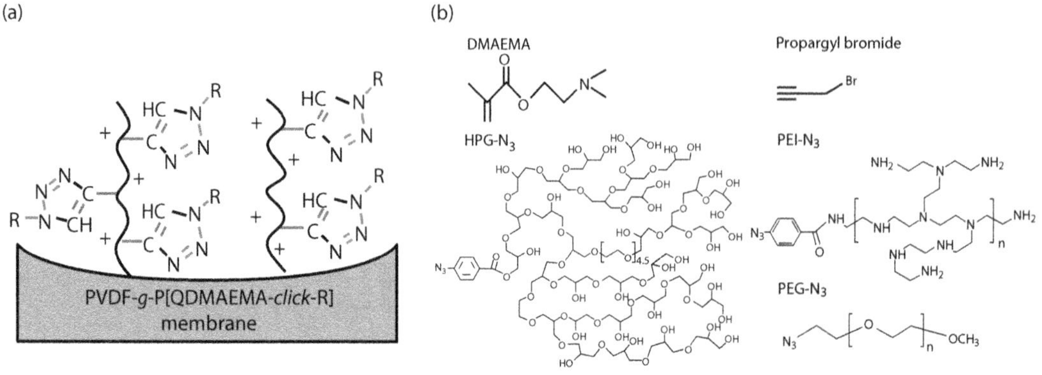

(a)

(b)

FIGURE 12.5

(a) Schematic illustration of the PVDF membrane after graft copolymerization of DMAEMA via atom transfer radical polymerization (AGET-ATRP) and surface-initiated alkyne–azide click reaction of azido-terminated hyperbranched polyglycerols (HPG-N$_3$) to form the PVDF-g-P[QDMAEMA-click-HPG]. (b) Structure of the various chemical species used in the study: PVDF = poly(vinylidene fluoride); DMAEMA = 2-(N,N-dimethylamino) ethyl methacrylate; HPG = hyperbranched polyglycerol; PEI = polyethylenimine; PEG = polyethylene glycol. (From Cai, T. et al., *Ind. Eng. Chem. Res.* 51, 15962, 2012. With permission.)

12.2 Organic–Inorganic Material Hybrids

Most of the organic–inorganic hybrids are nothing but the combination of organic materials such as polymeric and carbonaceous materials with inorganic nanomaterials. Inorganic nanomaterials such as Ag, ZnO, TiO$_2$, and iron oxide nanoparticles have been used as antimicrobial agents for removal of pathogens from water. Nanotechnology has a lot of potential for water disinfection, and recent reviews and books suggest its importance in water treatment.[32–34] To address issues such as cost, environmental toxicity, and human exposure related to the inorganic materials, organic–inorganic material hybrids such as polymer–Ag, organic–silica, and organic–TiO$_2$ have been developed as effective materials for water disinfection, particularly for developing nations.

12.2.1 Polymer–Ag Composites

Silver (Ag$^+$ or nanoparticle) has a long history of being widely used as an antimicrobial agent since ancient times.[35] It has been recently shown that silver in its nanoparticle form does not have any direct "particle-specific" antimicrobial activity.[36] It is when the Ag nanoparticle in aqueous solution undergoes oxidation and releases Ag$^+$ that it is responsible for antimicrobial activity.[36] This Ag$^+$ has a strong interaction with thiols on enzymes and bacterial cell membranes, causing denaturation and cellular damage respectively, ultimately leading to cell death. Also, it binds to bacterial cell DNA, preventing cell replication.[35–37] Thus, silver (Ag$^+$ or nanoparticle) has a large number of applications, particularly in water disinfection.[35] Interestingly, despite being effective against a broad range of microorganisms, Ag has no significant side-effects on the human body when present at sufficiently low levels.[38] According to the US Environmental Protection Agency (US EPA), the maximum contaminant level of silver ions in the drinking water must be <0.10 ppm or mg/L.[39,40]

For the controlled release of silver that can offer long-lasting antimicrobial efficacy, silver has been impregnated into various materials, including polymers, carbonaceous and

FIGURE 12.6
(a) Uncoated and (b) nAg-coated polypropylene filter. (From Heidarpour, F. et al., *Clean Technol. Environ. Policy*, 13, 499, 2011. With permission.)

mesoporous materials. Incorporation of silver into polymers has been shown to be a promising strategy for the development of materials for water disinfection. The antibacterial efficiency of cylindrical polypropylene water filters coated with a 35.0 nm layer of nano-silver particles (nAg) has been evaluated (Figure 12.6).[41] After 7-h filtration, the nano-silver-coated filters were able to remove 100% (3 $\log_{10}$ reduction) of *E. coli* contamination, whereas the uncoated filters were not able to decrease the number of bacteria. The silver leaching tests showed that the filter releases nil amounts of silver and it complies with the requirements of the US EPA. These results suggest the possibility of the use of the nano-silver-coated filters/membranes in water disinfection.[41] In another report, nAg-impregnated functionalized carbon nanotubes (multiwall carbon nanotubes [MWNTs]) polymerized with β cyclodextrin (CD) using hexamethylene diisocyanate as the linker has been reported.[42] The polymeric nanocomposites (Ag-MWNT-CD) were found to reduce bacterial cell counts in water spiked with *E. coli* to 94% within 30 min and to as low as 0 CFU/mL (colony-forming units per milliliter) within 90 min.[42]

Another strategic design of polymer–Ag nanocomposites for the development of environmentally safe water disinfection processes is using the core–shell Ag nanoparticles or nanocomposites with both antibacterial and magnetic properties. In these cases, the silver released into the system can be easily collected using a simple magnet. Alonso et al. have reported Ag@Co–NPs with a low-cost superparamagnetic Co^0–core and an antibacterial Ag–shell encapsulated on granulated cation exchange polymeric matrices for water purification applications.[43] The polymer–nanocomposites exhibited antibacterial activity and Ag and Co release from Ag@Co–NCs after 60 min of continuous operation was determined by inductively coupled plasma–mass spectrometry. All samples showed values <1.0 ppm for Ag (the limit of silver release into water as per US EPA) and 0.1 ppm for Co. These released core–shell nanoparticles can be easily collected in the treated water, and this approach has a lot of potential in real-life water purification systems, particularly for developing countries.[43]

One of the disadvantages of the polymer–Ag composites thus far developed is that the polymer used to support the inorganic material is not antibacterial. Sambhy and coworkers have developed a dual-action antibacterial composite consisting of a cationic polymer matrix and embedded silver bromide nanoparticles.[44] This composite having the antibacterial cationic polymer, poly(4-vinyl-*N*-hexylpyridinium bromide) along with the silver bromide nanoparticles made by an on-site precipitation technique has shown long-lasting antibacterial activity and inhibited the biofilm formation.[44] These dual-action polymer–Ag composites have potential for use in water disinfection. Combined with the low cost and effectiveness in its applications, the polymer–Ag technology has potential applications in water disinfection and may have large implications in developing countries.

12.2.2 Organic–Silica Composites

12.2.2.1 N-*Halamine Silica Composites*

As mentioned earlier, *N*-halamine materials are promising materials for water disinfection. *N*-chlorohydantoinyl siloxanes are also used to functionalize the surfaces of silica gel/sand particles to produce a biocidal material upon treatment with bleach solution

(Figure 12.7).[45] The functionalized silica gel was made into a cartridge filter, and the biocidal efficacy against the bacterial pathogens *S. aureus* and *E. coli* O157:H7 showed complete 6 log inactivation of the two bacterial species within 30 s of contact.[45] Moreover, the biocidal *N*-halamine moieties of the coated silica gel particles could be regenerated by simple exposure to dilute bleach. Potential uses of the biocidal silica gel includes water disinfection.

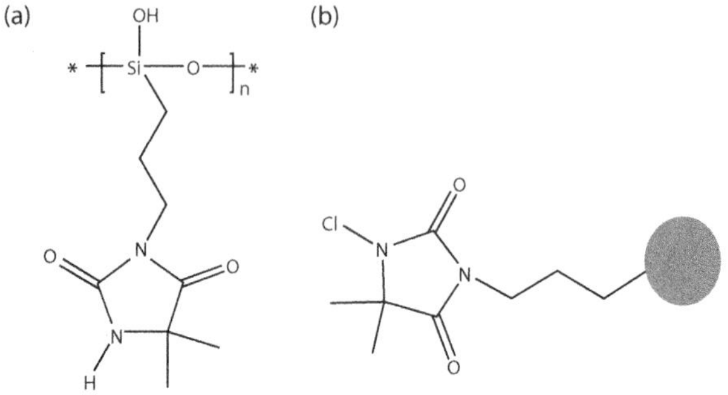

(a)　　　　　　　(b)

FIGURE 12.7
(a) Structure of *N*-Chlorohydantoinyl siloxanes employed to create antimicrobial composite. (b) Silica gel particle with covalently attached *N*-Chlorohydantoinyl siloxanes. (Adapted from Liang, J. et al., *J Appl. Polym. Sci.* 101, 3448, 2006.)

12.2.2.2 Polymer–Silica Composites

Although silver has attractive antibacterial activities, its primary shortcoming is that the particles are easily washed out since they are just impregnated into the material. This might reduce the disinfection efficiency over time and release of high amounts of silver into the water might be toxic to humans. Hayden and coworkers developed a new sand filtration water disinfection technology that relies on the antimicrobial properties of hydrophobic polycations (*N*-hexylated PEI) covalently attached to the sand's surface (Figure 12.8).[46] The applicability of the filter in water disinfection was evaluated both with water spiked with *E. coli* and with effluents from a wastewater treatment plant. In the case of water spiked with *E. coli*, >7-log reduction in bacterial count was achieved, whereas with wastewater effluent samples, the *E. coli* concentration was reduced to <2 logs. This low inactivation capability of the polymer–sand filter in case of the wastewater effluent samples compared with the aqueous sample might be due to the presence of particulate matter that diminishes the contact between the immobilized polycation and the suspended bacteria. Preliminary sand-washing methods using PBS and ethanol tested to assess potential "regeneration" approaches showed that these procedures refresh the bactericidal activity of the polymer.[46]

These polymer–sand composites and *N*-halamine hybrid composites have a lot of advantages over conventional disinfectants in terms of no harmful disinfection by-products and reduction of energy consumption. They have potential applications in water disinfection, particularly in developing countries.

12.2.3 Inorganic–Organic TiO₂ Composites

Photocatalysis applications of wide-reaching importance include degradation of environmental pollutants in aqueous contamination and wastewater treatment. One of the most popular applications of photocatalytic materials is the photodegradation of bacteria that can be used in water disinfection. The ideal antimicrobial photocatalytic materials should possess advantages such as high and lasting antibacterial efficacy, environmental safety, low toxicity, and simplicity in fabrication. TiO₂ has been the most widely researched photocatalyst, but suffers from low efficiency and a narrow light response

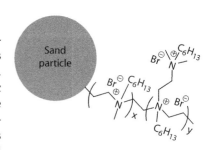

FIGURE 12.8
Covalently modified sand particle with hydrophobic polycation, *N*-hexylated polyethylenimine.

range. TiO$_2$ can only be excited under irradiation of UV light at wavelengths <380 nm, which covers only ~5% of the solar spectrum.[47] The antibacterial activity of TiO$_2$ is related to ROS production, especially hydroxyl free radicals and peroxide formed under UV-A irradiation via oxidative and reductive pathways, respectively.[48] The development of photocatalysts capable of absorbing light in the visible region of the spectrum is therefore of great interest. Metal doping has been shown to be a successful approach to enhance the photocatalytic activity of TiO$_2$ under visible light. Ag/TiO$_2$ composite materials have attracted particular attention wherein the tiny-sized Ag nanoparticles with a plasmon effect are exploited to enhance the visible light photocatalytic activity of the composites.[49–52] Recently, Ag/AgBr/TiO$_2$ that contains small particles has been prepared, and these systems are indeed more efficient than the Ag/TiO$_2$ composite materials with large TiO$_2$ particles in the photodegradation of *E. coli* under visible light.[53,54] (For more information, refer to reviews in references 55 and 56.)

A high-performance inorganic filtration membrane has been fabricated by directly growing titanate nanotube separation layers on a porous titanium membrane substrate. The resultant titanate nanotube membrane retains membrane structure, and successful separation of *E. coli* demonstrates the applicability of the titanate nanotube membrane for waterborne pathogen removal, which would be of great interest to the water purification applications.[57] Fluorescence and scanning electron microscopy (SEM) studies proved the efficiency of the membrane in pathogen removal (Figure 12.9).

Combining TiO$_2$ with carbonaceous nanomaterials is being increasingly investigated as a means to increase photocatalytic activity, and demonstrations of enhancement are plentiful. TiO$_2$-carbonaceous materials combining the photocatalytic antibacterial activity of TiO$_2$ and adsorbent properties of carbonaceous materials have a great potential for water disinfection. Graphene oxide/TiO$_2$ nanocomposites as photocatalysts for degradation of *E. coli* bacteria in an aqueous solution under solar light irradiation have been reported.[58] Photocatalytic disinfection of *E. coli* by carbon-modified TiO$_2$ photocatalysts was tested under UV and visible light irradiation. For modification purposes, five alcohols were used (methanol, ethanol, *n*-butanol, 2-butanol, and *tert*-butanol) as a source of carbon. It was found that photocatalysts with a low amount of carbon have better antibacterial ability under visible light irradiation; photocatalysts modified with methanol reduced 100% of *E. coli* after 45 min of irradiation with visible light. The photocatalysts with a higher amount of carbon had better antibacterial ability under UVA light irradiation; photocatalysts modified with 2-butanol and *n*-butanol reduced 100% of bacteria after 20 min of irradiation with UV light.[59] TiO$_2$/MWNT heterojunction arrays were synthesized and immobilized on an Si substrate as photocatalysts for inactivation of *E. coli* bacteria (Figure 12.10a). The visible-light-induced photoinactivation of

FIGURE 12.9
(See color insert.) Fluorescent microscopic images of *E. coli* feed (a) and permeate (b). (c) SEM image of retained *E. coli* on the Titanate Nanotube Membrane (TNM) after filtration (low magnification) and a high-magnification SEM image (inset). (d) Cross-section SEM image of TNM after filtration of *E. coli*. (From Zhang, H. et al. *J. Membrane Sci.*, 343, 212, 2009. With permission.)

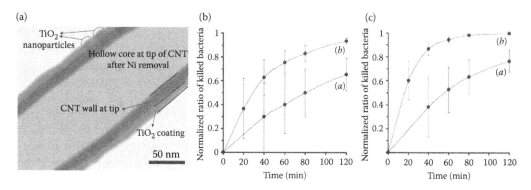

FIGURE 12.10
(a) TEM image of a TiO_2/MWNT heterojunction nanotube at tip of the CNT, after removing the Ni seed from tip of the CNT. Normalized ratio of the killed bacteria on surface of the TiO_2/MWNT samples annealed at 100°C (b) and 400°C (c) (*a*, in dark; *b*, under visible light irradiation). (From Akhavan, O. et al., *Carbon*, 47, 3280, 2009. With permission.)

the bacteria increased from MWNTs to TiO_2 to TiO_2/MWNTs, in which the bacteria could even slightly breed on the MWNTs (Figure 12.10b,c).[60] This approach can be used to make a TiO_2/MWNT composite that can be used as a filter or membrane in water disinfection. Photodegradation of bacteria with organic–TiO_2 composites have the potential for use in water disinfection if the visible light photocatalytic antimicrobial activity can be enhanced.

Overall, the polymer–Ag composites with dual mode of antimicrobial action have tremendous application wherein the polymers are also antibacterial, confer the controlled release of silver and long-lasting antimicrobial activity. Photocatalytic materials like TiO_2 also have great potential if its visible light antibacterial activity can be enhanced.

12.3 Carbonaceous and Mesoporous Materials

Advanced techniques such as membrane filtration, reverse osmosis, and ion exchange have been shown to be highly efficient in removing various types of contaminants from water. However, their high costs limit the use of membrane filtration technology in developing countries. With adsorption as one of the principal mechanisms, carbonaceous and mesoporous materials have been identified for a wide range of point-of-use (POU) and other water treatment options.

12.3.1 AC–Ag Composites

The most widely used material, AC has the best possible surface area and could be produced at low cost. A number of other forms of carbon have appeared with very large adsorption capacities. Carbonaceous materials such as AC, graphene, carbon nanofibers, fullerenes, and carbon nanotubes (single-walled carbon nanotubes [SWNTs] and MWNTs) have been used extensively in water purification and, hence, are affordable adsorbents in all commercial water technologies. AC, carbon monoliths (CMs), and AC fibers (ACFs) have been widely used in wastewater treatment to remove organic or inorganic pollutants because of their extended surface area, high adsorption amount and rate, and specific surface reactivity. The adsorption capacity of ACFs depends on many kinds of factors, such as raw materials, activation process, the nature of the pore structure, and the surface functionality. Nevertheless, problems still remain when the ACFs are used to purify drinking water because bacteria preferably adhere to solid supports made of carbon materials. It has been reported that bacteria that attach to carbon particles are highly resistant to disinfection processes owing to biofilm formation, which causes the carbon itself to become a source of bacterial contamination.[61] Water treatment systems comprising of powdered AC fail in the removal of bacterial pathogens from water. Unlike chemical contaminants, most biological adsorbates (e.g., bacteria and virus) have sizes larger than the pore size of the microporous adsorbents like AC. This size difference makes the majority of pore surface area inaccessible to the biological adsorbates, thereby limiting their removal. Hence, it would be advantageous if the AC also possessed antibacterial activity to kill airborne or waterborne bacteria.

In general reports, the antibacterial activities of ACs are available as ACs or ACFs supported with silver (AC/Ag or ACF/Ag). Silver nanoparticles deposited over AC by electrochemical deposition methods have been reported in controlling microorganisms in water. Comparison of the antibacterial activity of the Ag/C catalyst prepared by an electrochemical deposition method with that of the Ag/C catalyst prepared by a conventional impregnation technique indicates that lower amounts of the former is sufficient in controlling the microorganism, which is not the case with the latter. The main advantage of the Ag/C catalyst prepared by electrochemical deposition is that no pretreatment conditions like reduction are required for deactivation of microorganisms in water, which is not the case with the catalysts prepared by impregnation techniques.[62] Viscose-based AC fiber composite supporting silver (ACF (Ag)) containing 0.065 wt% of silver exhibited strong antibacterial activity against *E. coli* and *S. aureus*. The results of antibacterial tests for the ACF (Ag) with varying silver content showed that the ACF did not show antibacterial properties against any bacteria, but all ACF (Ag) samples exhibited strong antibacterial properties against *E. coli* and *S. aureus*. It should be mentioned that after washing for 120 h, the ACF (Ag) samples of 0.065 wt% Ag (with an initial silver content of 1.44 wt%) still exhibited strong antibacterial activity against both bacteria.[63] In the last few years, there has been growing interest in the CM as an alternative to conventional carbon materials. CMs can be produced with the desired shape and morphology with controlled composition, structure, and porosity. The potential of using a cylindrical CM containing impregnated silver with longitudinal capillary channels as a filter for water purification has been demonstrated.[64] Carbon monolith samples with a silver coating showed good antimicrobial activity against *E. coli*, *S. aureus*, and *Candida albicans*, and are therefore suitable for water purification, particularly as POU systems in developing nations.[64] Graphene is another carbon material that has been extensively used for water treatment.[65] Graphene and derivative analogs have potential applications in water disinfection owing to their inexpensive preparation from natural sources like sugar and graphene composites. They can be prepared by impregnating inorganic materials like silver that makes them antibacterial.

12.3.2 AC with Biocidal Quaternary Ammonium Groups

As shown above, in the preparation of antibacterial ACs, much effort has been devoted to the impregnation of silver in ACs. Although silver has attractive antibacterial activities, their primary shortcoming is that the particles are easily washed out since they are just deposited on the surface of the AC. Furthermore, with increasing silver content, the specific surface area of the carbon decreases greatly, resulting in reduced adsorption capability. A number

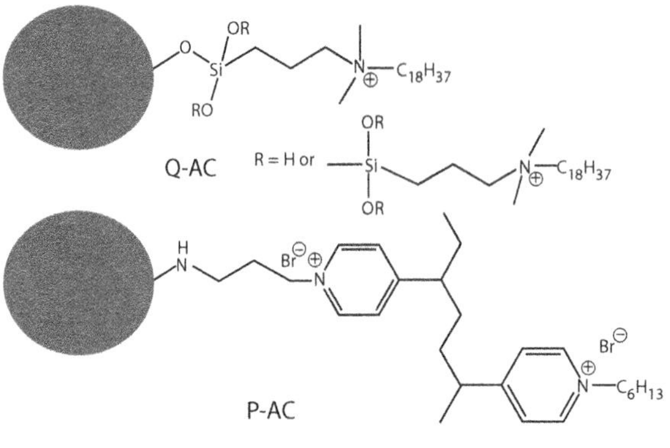

FIGURE 12.11
Schematic for the covalent functionalization of the activated carbon. (Adapted from Shi, Z. et al., *Ind. Eng. Chem. Res.*, 46, 439, 2007.)

of quaternary ammonia are known to exhibit good microbicidal properties. Various literature reports demonstrated that antibacterial/antiviral properties can be conferred on the surfaces of substrates by the covalent or noncovalent attachment of quaternary ammonium moieties.[25,66] In these approaches, the antibacterial agents will not leach out from the surface, hence providing long-term effectiveness and reducing the toxicity associated with conventional disinfectants and silver. AC was functionalized with two types of quaternary ammonium groups to achieve antibacterial properties (Figure 12.11).[61] The first type is covalent coupling of a quaternary ammonium moiety on the AC surface (Q-AC, Figure 12.11), whereas the second route used a polycation, poly-(vinyl-N-hexylpyridinium bromide) (P-AC, Figure 12.11). Both types of functionalized ACs showed highly effective antibacterial activities against *E. coli* and *S. aureus*. Furthermore, the functionalized ACs can be used in repeated antibacterial applications with little loss in efficacy.[61]

12.3.3 Carbonaceous Nanomaterials or Nanocomposites

In recent years, nanotechnology has introduced different types of nanomaterials to the water industry that can have promising outcomes. Nanosorbents such as CNTs have exceptional adsorption properties and are applied for removal of heavy metals, organics, and biological impurities. CNTs are increasingly being assessed for use in water purification owing to their high surface area, inherent antimicrobial activity, electronic properties, and ease of functionalization. Apart from their capability to act as adsorption media to remove pathogens, the inherent cytotoxic nature of pristine and/or functionalized CNTs prohibits the growth of pathogens on its surface (for an extensive review on this topic, see reference 67).

Recent work by Brady-Estévez and coworkers has shown scalable applications that use low-cost and widely available CNTs for inactivation of microbes and removal of viruses from water using a CNT-hybrid filter.[68] The filter with a PVDF-based microporous membrane having a thin layer of SWNTs removed the micron-sized bacterial cells through a sieving mechanism, whereas filtration occurred through adsorption of nanoscale viruses throughout the thickness of the CNT matrix. Their results indicated that the *E. coli* cells were completely retained by the SWNT layer owing to size exclusion, but passed readily through the base filter (PVDF). The SWNT filter described offers several potential advantages for water purification–complete bacterial retention, exceptionally high viral removal (MS2 bacteriophage, 5–7 log), and high antimicrobial activity.[68]

Electrochemical inactivation of waterborne pathogens has long been studied as an alternative to conventional water disinfection. Electrochemical MWNT filters for pathogen removal and inactivation have a lot of potential for POU drinking-water treatment (Figure 12.12a). The efficacy of an anodic MWNT microfilter has been demonstrated toward the removal and inactivation of viruses (MS2) and bacteria (*E. coli*).[69] In the absence of electrolysis, the MWNT filter is effective for complete removal of bacteria by sieving and multilog removal of viruses by depth filtration. Application of 2 and 3 V for 30 s after filtration

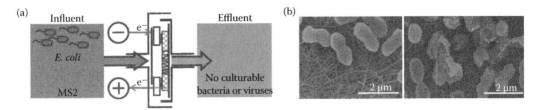

FIGURE 12.12
(a) Electrochemical MWNT filter design and (b) SEM images of *E. coli* on the MWNT filter after electrolysis. Cells exposed to electrolysis for 30 s in 10 mM NaCl at applied potentials of 0 V (left) and 3 V (right). (From Vecitis, C.D. et al., *Environ. Sci. Technol.*, 45, 3672, 2011. With permission.)

inactivated >75% of the sieved bacteria and >99.6% of the adsorbed viruses (Figure 12.12b). Electrolyte concentration and composition had no correlation to electrochemical inactivation consistent with a direct oxidation mechanism at the MWNT filter surface. Potential-dependent *E. coli* morphological changes also supported a direct oxidation mechanism.[69]

Incorporation of silver into the CNT-based electrochemical inactivation approach increases the bacterial inactivation efficiency. A multiscale device has been developed for

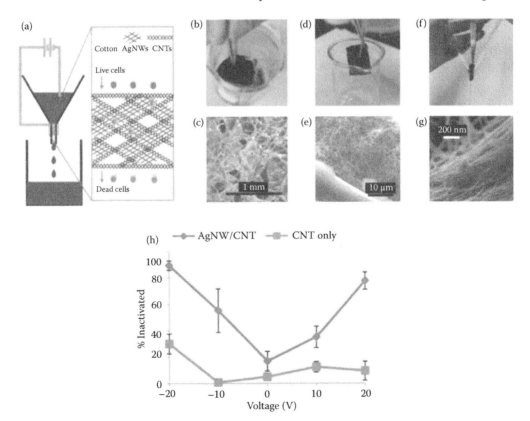

FIGURE 12.13
(See color insert.) Schematic, fabrication, and structure of cotton, AgNW/CNT device. (a) Schematic of proposed active membrane device. (b) Treatment of cotton with CNTs. (c) Treatment of device with silver nanowires (AgNWs). (d) Integration of treated cotton into funnel. (e) SEM image showing large-scale structure of cotton fibers. (f) SEM image showing AgNWs. (g) SEM image showing CNTs on cotton fibers. (h) Inactivation efficiency at five biases for AgNW/CNT cotton as well as CNT-only cotton. (From Schoen, D.T. et al., *Nano Lett.*, 10, 3628, 2010. With permission.)

the high-speed electrical sterilization of water using silver nanowires, CNTs, and cotton (Figure 12.13a through g).[70] This approach used a gravity fed device that can inactivate >98% of bacteria with only several seconds of total incubation time with a moderate bias of 20 V (Figure 12.13h). The device works by the use of an electrical mechanism rather than size exclusion, while the very high surface area of the device coupled with large electric field concentrations near the silver nanowire tips allows for effective bacterial inactivation. The CNT component of the system ensures good electrical conductivity over the entire active area of the device, so it can be placed at a controlled electric potential and used in solution as a porous electrode. The ultimate impact of this technology may be direct implementation as a cheap POU water filter for deactivating pathogens in water, or perhaps more probably as a new component to be integrated into existing filtration systems to kill microorganisms that cause biofouling in downstream filters.[70]

CNT technology has potential advantages over conventional disinfection approaches and has great application in water disinfection. Currently, the cost of CNTs stands as a major roadblock preventing them to consider for large-scale applications such as water treatment. Release of CNTs into the environment and human exposure are other major concerns at this point. Hydrothermal treatment of carbohydrates yield functionalized carbon nanomaterials that have found extensive applications in catalysis, drug delivery, etc.[71] These carbohydrate-derived carbon nanomaterials can be synthesized at very low cost and form different nanostructures such as carbon nanospheres and nanotubes/nanocables.[71] These carbon nanomaterials and their inorganic hybrids (with Ag, TiO_2, etc.) can be promising alternatives for CNTs in water disinfection owing to their inexpensive synthesis.

12.3.4 Clay–Inorganic Composites

Most of the mesoporous materials used in the water filter applications are based on ceramics, clay, sand, silicates, etc. Ceramic or sand-based materials are ideal for the development of water treatment systems for the developing countries, particularly for water disinfection. Tap water seeded with different microorganisms or untreated wastewater was passed through columns containing sand modified by the *in situ* precipitation of metallic hydroxides or unmodified sand. Columns packed with 1 kg of sand modified with a combination of ferric and aluminum hydroxide removed >99% of *E. coli, Vibrio cholerae*, poliovirus 1, and coliphage MS-2 from dechlorinated tap water.[72] This removal efficiency was consistent throughout the passage of 120 liters of water during a 30-day test. After the passage of 192 liters of tap water, these columns were still able to remove 99.9% MS-2, 80% *E. coli*, and 90% of poliovirus. Columns containing modified sand efficiently removed microorganism from water samples at the various pH and temperature (>4 log_{10} reduction at room temperature) values tested, while columns containing unmodified sand did not. In addition, these modified sand filters were able to remove coliform bacteria and coliphage from raw sewage. The modified sand seems to better remove microorganisms because of increased electrostatic interactions. Neither *E. coli* nor MS-2 was inactivated by the modified sand column effluents. The metal used for coating the sand could not be detected in the column effluents, indicating that the coatings were stable.[72]

To make these clay or silicate materials to be used for efficient water disinfection applications, they are also impregnated with antibacterial agents. In most of the cases, it is silver that is used for impregnation and silver–ceramic water filters are used extensively in water disinfection. Oyandel-Craver and Smith manufactured cylindrical colloidal-silver-impregnated ceramic filters for household (POU) water treatment and tested for performance in the laboratory with respect to flow rate and bacteria transport.[73] The filters removed between 97.8% and 100% of the applied bacteria; colloidal silver treatments improved filter performance, presumably by deactivation of bacteria. Silver concentrations in effluent filter water were initially >0.1 mg/L, but dropped below this value after 200 min of continuous operation. These results indicate that colloidal-silver-impregnated ceramic filters, which can be made using primarily local materials and labor, show promise as an effective and

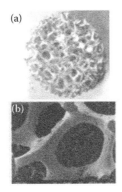
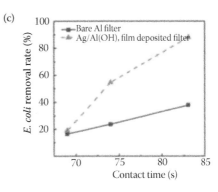

FIGURE 12.14
Photograph and SEM images of Al foam filters: (a) Al foam filter, (b) its enlarged SEM image and (c) *E. coli* removal rate versus contact time of Al alloy foam filter. (From Seo, Y.I. et al., *J. Hazard. Mater.*, 227–228, 469, 2012. With permission.)

sustainable POU water treatment technology for the world's poorest communities.[73]

A mesoporous $Ag/Al(OH)_3$ nanocomposite film on an Al foam filter capable of rapidly removing bacterial pathogens from secondary effluent has been reported.[74] The filter with $Ag/Al(OH)_3$ mesoporous nanocomposite film showed a good bacterial pathogen removal rate within a very short contact time of 83 s compared with the untreated Al foam filter (Figure 12.14). Al foam filters with deposited $Ag/Al(OH)_3$ mesoporous nanocomposite film can potentially be used as disinfection filters for tap water purification or wastewater treatment.[74]

Montmorillonite (MMT)-supported Ag/TiO_2 composite ($Ag/TiO_2/MMT$) has been prepared with Ag particles coated with TiO_2 nanoparticles well-dispersed on the surface of MMT in the composite.[75] Acting as a support for the Ag/TiO_2 composite, the MMT prevents the loss of the catalyst during recycling test. This $Ag/TiO_2/MMT$ composite exhibits high photocatalytic activity and good recycling performance in the degradation of *E. coli* under visible light. The high visible light photocatalytic activity of the $Ag/TiO_2/MMT$ composite is ascribed to the increase in surface-active centers and the localized surface plasmon effect of the Ag nanoparticles. The $Ag/TiO_2/MMT$ materials with excellent stability, recyclability, and bactericidal activities are promising photocatalysts for application in water disinfection. Overall, the carbonaceous and mesoporous materials can be the economic materials that can be used for water disinfection, particularly in developing countries.[75]

12.4 Polymer and Polymer–Nanocomposites in Membrane Filtration Technology

For the past few decades, the use of membrane filtration in drinking-water treatment, including pathogen removal, has been growing, owing to increasing drinking-water regulations and decreasing costs of purchasing and operating membrane filters.[8] A semipermeable membrane/film is used as a selective barrier to remove contaminants from water in membrane filtration. Most of the contaminants can be removed by membrane processes. The membrane processes most commonly used to remove microorganism from drinking water are microfiltration (MF), ultrafiltration (UF), nanofiltration (NF), and reverse osmosis (RO). These membranes are usually made from various polymer or polymer–inorganic composite.

12.4.1 Polymers in Membrane Filtration

Various types of polymeric materials are used to make the film-in-membrane filtration (MF, UF, NF, RO). Not all of these processes are used primarily for the removal of pathogens. For example, RO is used mainly for desalination and NF for softening and for removal of precursors of disinfectant by-products such as bromide, natural organic matter, etc. Nevertheless, in addition to their conventional applications mentioned above, these membrane filters can also be designed to remove pathogens, hence conferring them with a broader spectrum of activity. Pathogens with different sizes can be selectively removed by using polymeric membranes with specific pore size. Moreover, membranes with a particular surface charge may remove particulate or microbial contaminants of the opposite charge owing to electrostatic attraction. Membranes can also be hydrophilic (water attracting) or hydrophobic (water repelling). These terms describe how easily membranes can be wetted, as well as their ability to resist fouling to some degree. The removal of pathogens by using polymeric membranes has been reviewed in many review articles and book chapters, and is therefore not discussed in this section.[8,76]

Most MF, UF, RO, and NF membranes are made up of synthetic organic polymers. MF and UF membranes are often made from the same materials, but they are prepared under different membrane formation conditions so that different pore sizes are produced.[77] Typical MF and UF polymers include polysulfone (PSf), PVDF, poly(acrylonitrile), and poly(acrylonitrile)-poly(vinyl chloride) copolymers. Poly(ether sulfone) (PES) is also commonly used for UF membranes. MF membranes also include cellulose acetate–cellulose nitrate blends, nylons, and poly(tetrafluoroethylene).[78] RO membranes are typically either cellulose acetate or PSf coated with aromatic polyamides (PAs). NF membranes are made from cellulose acetate blends or PA composites like the RO membranes, or they could be

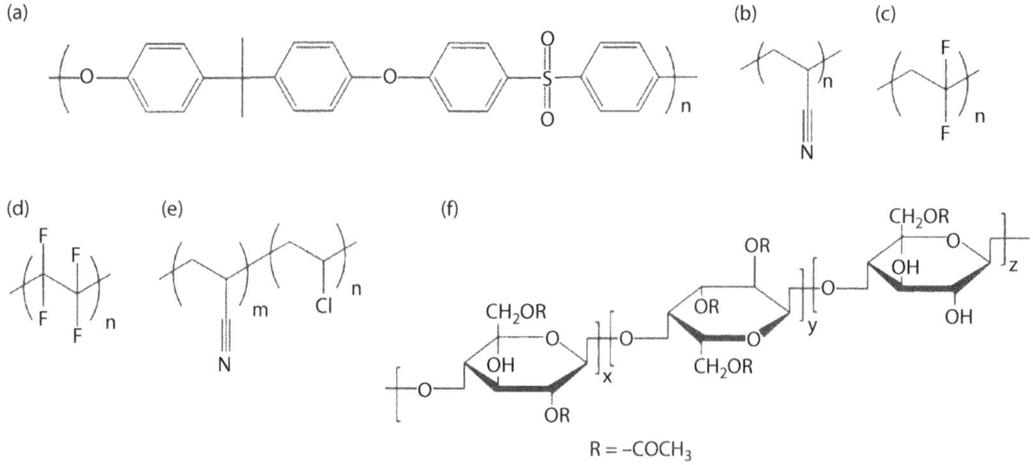

FIGURE 12.15
Structures of some representative polymers used in membrane-based water purification: (a) polysulfone (PSf); (b) polyacrylonitrile (PAN); (c) poly(vinylidene fluoride) (PVDF); (d) polytetrafluoroethylene (PTFE); (e) polyacrylonitrile-co-polyninylchloride (PAN-co-PVC); (f) cellulose acetate.

modified forms of UF membranes such as sulfonated PSf.[79] Structures of some of the polymers used in membrane filtration are shown in Figure 12.15.

12.4.2 Polymer–Inorganic Nanoparticle Composite Membrane

Although polymeric membranes remove microorganisms, these membranes cannot eliminate pathogens completely from drinking water. Furthermore, in general, these membranes are not capable of inactivating pathogens.[3,4] To meet the increasing need for potable water, the use of nanotechnology has been increasing in recent years as described earlier. Silver nanoparticles have been used along with various polymeric membranes such as polyurethane (PU), PSf, PEI, and PA membrane in water purification. In addition to nano-silver, nano-TiO_2 has also been used in the polymeric membrane to treat water and wastewater.

Chou et al. synthesized silver-loaded cellulose acetate hollow fiber membranes for water treatment. It was found that silver-loaded cellulose acetate membrane is active against *E. coli* and *S. aureus*.[80] Jain and Pradeep reported that silver nanoparticles can be coated on common PU foams by overnight exposure of the foams to the nanoparticle solutions. Repeated washing and air-drying was found to yield uniformly coated PU foam, which could be used as a drinking-water filter. When tested for antibacterial activity, PU foam coated with Ag nanoparticle was found to be completely bactericidal against *E. coli* (10^5 CFU/mL). In the zone of inhibition studies, no growth has been observed below the PU coated with nanoparticles while growth was seen in case of pure PU, which further confirms the antibacterial property of PU coated with silver nanoparticles.[81]

Nano-silver-incorporated PSf ultrafiltration membranes (nAg–PSf) (Figure 12.16a,b) exhibited antimicrobial properties and also prevented bacteria attachment to the membrane surface and reduced biofilm formation. In this study, nAg incorporated into PSf ultrafiltration membranes (nAg–PSf) showed antimicrobial properties toward a variety of bacteria, including *E. coli* and *Pseudomonas mendocina*, and the MS2 bacteriophage (Figure 12.16c).[82] Bjorge et al. reported the preparation of a polymeric nanofiber filter using PAA, PSf, PEI, and PA by an electrospinning method and their application in water treatment.[83] The results showed that the removal of pathogens was sufficient neither for culturable microorganisms (1.28–1.72) nor for coliform bacteria (1.38–1.63) to be competitive with other commercial microfiltration membranes. When tested, the antimicrobial studies showed that due to the silver nanoparticles in the functionalized membrane, a higher efficiency (log 3–log 4) could be achieved. Moreover, biofilm formation has been found to be successfully reduced in nano-silver-containing PSf membranes owing to the successive release of ionic silver over the lifetime of a membrane.

As previously discussed, titania nanoparticles are highly photoactive and exhibit antimicrobial activity under UV light.[48] Heterogeneous photocatalytic oxidation processes have been increasingly seen as an innovative and green technology for water and wastewater treatment. Molinari et al. altered commercially available porous polymeric membranes (PSf, PU, PA, PVDF) with a titania layer, by filtering a nanoparticle suspension through and applying UV/Vis irradiation and showed elevated (4-nitrophenol) photodegradation.[84] Madaeni et al. form "self-cleaning" RO membranes with the addition of titania nanoparticles.[85] Mo et al. prepared PSf-supported self-cleaning PA/titania membranes through interfacial polymerization, which contain a layer of silicon dioxide between layers of crosslinked PA and titania.[86] Flux recovery after 15 h of operation (with water cleansing and UV exposure every 3 h) is >98% for these photocatalytic membranes, significantly higher than standard water treatment membranes which correlates with the decreased membrane fouling thereby indicating the significance of the composite membrane in water purification.

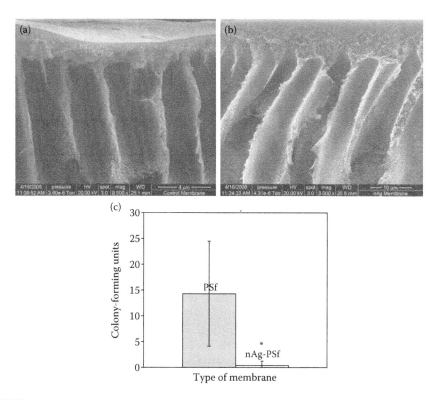

FIGURE 12.16
Cross-section of (a) PSf and (b) nAg–PSf membranes, taken with SEM. (The addition of nAg did not change the morphology of the PSf membrane.) Antibacterial properties of nAg–PSf membrane (c): impregnation of nAg (0.9 wt%) significantly decreased the number of *E. coli* grown on the membrane surface after filtration of a dilute bacteria suspension, as indicated by the number of colony-forming units (CFU) per 9.35 cm^2 membrane coupon. Error bars indicate 95% confidence intervals (n = 3). (From Zodrow, K. et al. *Water Res.*, 43, 715, 2009. With permission.)

12.4.3 Polymer–Carbonaceous Composite Membrane

Although membranes fabricated with nano-silver or nano-TiO$_2$ are found to be antimicrobial, the technology has certain limitations. Depletion and toxicity are the main problems with the silver nanoparticles and UV light is required for the titania nanoparticle to be active. Carbonaceous materials such as SWNTs and MWNTs, as previously described, are highly antimicrobial and can be fabricated into various membranes. In addition to the antimicrobial activity of CNTs, the inner cores of the nanotubes can be controlled to enable fine-tuning of the pore dimension at the nanometer scale.[87] These nanotubes have been successfully used with polymeric membranes to improve the filtration efficacy as well as removal of pathogens. Brady-Estévez et al. have successfully developed MWNT- as well as SWNT-impregnated polymeric membrane (PTFE polymer) (Figure 12.17a,b).[88] The MWNTs were deposited on a 5-µm-pore PTFE membrane (Millipore) by a sonication and filtration procedure. MS2 bacteriophage viral removal by the MWNT filter was between 1.5 and 3 log higher than that observed with a recently reported SWNT filter when examined under similar loadings (0.3 mg/cm^2) of CNTs. The greater removal of viruses by the MWNT filter is attributed to a more uniform CNT–filter matrix that allows effective removal of viruses by physicochemical (depth) filtration. It has also been found that the CNT/polymer blend membranes (MWNT and PSf) can act as nonfouling membranes where the CNT/polymer

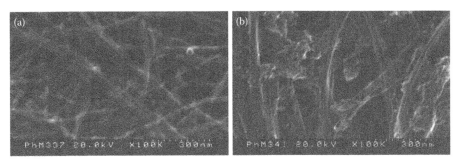

FIGURE 12.17
Field emission SEM images of CNT-hybrid filter: (a) MWNT filter under 100,000× magnification and (b) SWNT filter under 100,000× magnification. (From Brady-Estévez, A.S. et al., *Langmuir*, 26, 14975, 2010. With permission.)

blend membranes displayed 42% less flux decline and slower fouling rate than the PES membranes. It was found that the amount of foulant on bare PES membranes was 63% higher than the CNT/polymer blend membrane for a 2% MWCNT content. Thus, the CNT content of the CNT/polymer membranes was shown to mitigate the membrane fouling caused by natural water.[89]

12.4.4 Paper-Based Filtration

Although membrane filtration is highly effective, the cost of manufacture and fouling phenomenon decreases its versatile applications. Obviously, cheap POU methods to purify drinking water are necessary. Paper-based filtration methods offer the simplicity of passing the water through paper by gravity to achieve potability, while simultaneously removing contaminants. It is simple and universally accessible, usable anywhere without any special equipment, easy to transport, easy to use, inexpensive, safely disposable, possessing indefinite shelf life, and most important, extremely efficient in bacteria reduction while producing water containing safe levels of the biocide.[51] There are few reports on the development of antimicrobial paper,[90,91] some of which are discussed below and have been used for water purification.

Dankovich and Gray reported a method to deactivate pathogenic bacteria by percolation through a paper sheet containing silver nanoparticles.[90] The silver nanoparticles are deposited by the *in situ* reduction of silver nitrate on the cellulose fibers of an absorbent blotting paper sheet. The silver-nanoparticle-containing (nAg) papers, when tested, exhibited antibacterial properties toward suspensions of *E. coli* and *Enterococcus faecalis*, with log reduction values in the effluent of over log 6 and log 3, respectively. The silver loss from the nAg sheets was minimal, with values <0.1 ppm (the current US EPA and WHO limit for silver in drinking water). These results showed a promising approach toward effective emergency water treatment, which is by the percolation of bacterially contaminated water through paper embedded with silver nanoparticles.

An extremely efficient bactericidal filter paper has been developed that is capable of removing >99% of *E. coli* bacteria in a simple filtration process. The approach utilizes two active bactericidal components: a bactericidal agent, triclosan, which acts synergistically with a cationic polyelectrolyte poly(isopropanol dimethylammonium) chloride (Figure 12.18c) binder with antibacterial properties. The biocide is incorporated into the block copolymer polystyrene-block-PAA (PS-b-PAA) (Figure 12.18a) micelles attached to cellulose fibers via the cationic polyelectrolyte polyacrylamide (Figure 12.18b).[91] As the water

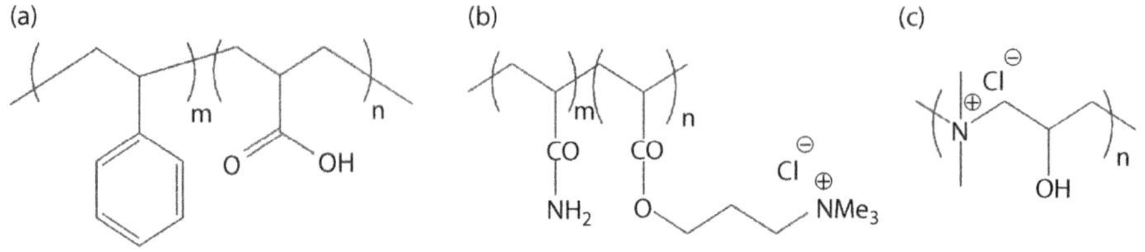

(a)   (b)   (c)

FIGURE 12.18
Structures of various polymers used to make bactericidal filter paper: (a) (polystyrene-block-polyacrylic acid (PS-b-PAA); (b) polycationic polyacrylamide; (c) poly(isopropanol dimethylammonuim) chloride (PIDMAC).

containing the bacteria is passed by gravity through the filter paper, the bactericidal agents are transferred to the bacteria through collisions with the micelles or coated fibers. A synergy between the biocide and the polyelectrolyte is responsible for the extremely high efficiency in deactivating the bacteria. This technology represents a very simple approach to provide potable water under a wide range of primitive conditions.

This chapter summarizes the use of polymeric and organic–inorganic material hybrids as efficient materials for water disinfection. Unlike inorganic disinfectants, antimicrobial polymeric materials are advantageous since they do not involve release of harmful by-products. Development of polymeric–Ag hybrids would be promising owing to the dual mode of action, wherein the inherent antimicrobial activity of the polymer coupled with controlled release of silver can confer long-lasting activity. Use of carbonaceous nanomaterials can be another viable option for developing effective materials for water disinfection if the issues such as cost (in case of CNTs) and performance can be taken care of. Most of the current water disinfection technologies, particularly membrane filtration technology, suffer from the formation of biofilms that drastically reduce their efficacy. Thus, the ultimate goal is to develop broad-spectrum (bacteria, viruses, protozoan parasites including the bacterial spores and cysts) antimicrobial water-insoluble polymeric–inorganic composites using antibacterial polymers so that they can be used in membrane filtration technology wherein the composites can be tailored to inhibit biofilm formation as well.

References

1. UNICEF and WHO. Diarrhoea: Why children are still dying and what can be done, WHO: Geneva, 2009.
2. World Health Organisation. *Guidelines for Drinking Water Quality*, vol. 2. WHO: Geneva, 1996.
3. Available at http://www.epa.gov/.
4. S. Baranya, and A. Szepesszentgyörgyi. Flocculation of cellular suspensions by polyelectrolytes. *Adv. Colloid Interface Sci.*, 111:117–129, 2004.
5. B. Boltoa, and J. Gregory. Organic polyelectrolytes in water treatment. *Water Res.*, 41:2301–2324, 2007.
6. N. Kawabata, K. Yamazak, T. Otake, I. Oishi, and Y. Minekawa. Removal of pathogenic human viruses by insoluble pyridinium-type resin. *Epidemiol. Infect.*, 105:633–642, 1990.
7. G. Sun, L. C. Allen, E. P. Luckie, W. B. Wheatley, and S. D. Worley. Disinfection of water by N-halamine biocidal polymers. *Ind. Eng. Chem. Res.*, 34:4106–4109, 1995.
8. M. W. LeChevallier, and K. K. Au. *Water Treatment and Microbial Control: A Review Document*. World Health Organization, Iwa Publishing: London, 2002.

9. W. Brostow, H. E. H. Lobland, S. Pal, and R. P. Singh. Polymeric flocculants for waste water and industrial effluent treatment. *J. Mater. Edu.*, 31:157–166, 2009.

10. A. A. Baran. Flocculation of cellular suspensions by polyelectrolytes. *Coll. Surf.*, 31:259–264, 1988.

11. D. Bilanovic, and G. Shelef. Flocculation of microalgae with cationic polymers: Effects of medium salinity. *Biomass*, 17:65–76, 1988.

12. T. J. Brown, and M. B. Emelko. Chitosan and metal salt coagulant impacts on Cryptosporidium and microsphere removal by filtration. *Water Res.*, 43:331–338, 2009.

13. O. Bayat, V. Arslan, B. Bayat, and C. Poole. Application of a flocculation ultrafiltration process for bacteria (*Desulfovibrio desulfuricans*) removal from industrial plant process water. *Biochem. Eng. J.*, 18:105–110, 2004.

14. N. Nwachcuku, and C. P. Gerba. Emerging waterborne pathogens: Can we kill them all? *Curr. Opin. Biotechnol.*, 15:175–180, 2004.

15. J. L. Lambert, and L. R. Fina. Method of disinfecting water and demand bactericide for use therein. US Patent No. 3817860, 1974.

16. J. L. Lambert, and L. R. Fina. Demand bactericide for disinfecting water and process of preparation. US Patent No. 3923665, 1975.

17. M. Tyagi, and H. Singh. Iodinated P(MMA-NVP): An efficient matrix for disinfection of water. *J. Appl. Polym. Sci.*, 76:1109–1116, 2000.

18. N. Kawabata, T. Hashizume, and T. Matsumoto. Adsorption of bacteriophage T4 by crosslinked poly(vinylpyridinium halide). *Agric. Biol. Chem.*, 50:1551–1555, 1986.

19. Y. Chen, S. D. Worley, J. Kim, C. I. Wei, T. Y. Chen, J. I. Santiago, J. F. Williams, and G. Sun. Biocidal poly(styrenehydantoin) beads for disinfection of water. *Ind. Eng. Chem. Res.*, 42:280–284, 2003.

20. V. S. Panangala, L. Liu, G. Sun, S. D. Worley, and A. Mitra. Inactivation of rotavirus by new polymeric water disinfectants. *J. Virol. Methods*, 66:263–268, 1997.

21. Y. Chen, S. D. Worley, T. S. Huang, J. Weese, J. Kim, C. I. Wei, and J. F. Williams. Biocidal polystyrene beads. III. Comparison of *N*-halamine and quat functional groups. *J. Appl. Polym. Sci.*, 92:363–367, 2004.

22. N. Mazumdar, M. L. Chikindas, and K. Uhrich. Slow release polymer–iodine tablets for disinfection of untreated surface water. *J. Appl. Polym. Sci.*, 117:329–334, 2010.

23. M. L. Cohen. Changing patterns of infectious disease. *Nature*, 406:762–767, 2000.

24. J. C. Tiller, S. B. Lee, K. Lewis, and A. M. Klibanov. Polymer surfaces derivatized with poly(vinyl-N-hexylpyridinium) kill airborne and waterborne bacteria. *Biotechnol. Bioeng.*, 79:465–471, 2002.

25. J. Haldar, D. An, L. A. de Cienfuegos, J. Chent, and A. M. Klibanov. Polymeric coatings that inactivate both influenza virus and pathogenic bacteria. *Proc. Natl. Acad. Sci. U.S.A.*, 103:17667–17671, 2006.

26. A. M. Larson, B. B. Hsu, D. Rautaray, J. Haldar, J. Chen, and A. M. Klibanov. Hydrophobic polycationic coatings disinfect poliovirus and rotavirus solutions. *Biotechnol. Bioeng.*, 108:720–723, 2011.

27. A. Bosch. Human enteric viruses in the water environment: A minireview. *Int. Microbiol.*, 1:191–196, 1998.

28. G. Cheng, H. Xue, Z. Zhang, S. Chen, and S. Jiang. A switchable biocompatible polymer surface with self-sterilizing and nonfouling capabilities. *Angew. Chem. Int. Ed.*, 47:8831–8834, 2008.

29. T. Cai, W. J. Yang, K. G. Neoh, and E. T. Kang. Poly(vinylidene fluoride) membranes with hyperbranched antifouling and antibacterial polymer brushes. *Ind. Eng. Chem. Res.*, 51:15962–15973, 2012.

30. P. Kingshott, J. Wei, D. Bagge-Ravn, N. Gadegaard, and L. Gram. Covalent attachment of poly(ethylene glycol) to surfaces, critical for reducing bacterial adhesion. *Langmuir*, 19:6912–6921, 2003.

31. G. Cheng, Z. Zhang, S. F. Chen, J. D. Bryers, and S. Y. Jiang. Inhibition of bacterial adhesion and biofilm formation on zwitterionic surfaces. *Biomaterials*, 28:4192–4199, 2007.

32. Q. Li, S. Mahendra, D. Y. Lyon, L. Brunet, M. V. Liga, D. Li, and P. J. J. Alvarez. Antimicrobial nanomaterials for water disinfection and microbial control: Potential applications and implications. *Water Res.*, 42:4591–4602, 2008.

33. N. F. Savage. *Nanotechnology Applications for Clean Water*. William Andrew Inc.: New York, 2009.

34. X. Qu, J. Brame, Q. Li, and P. J. J. Alvarez. Nanotechnology for a safe and sustainable water supply: Enabling integrated water treatment and reuse. *Acc. Chem Res.*, 46:834–843, 2013.

35. J. J. Castellano, S. M. Shafii, F. Ko, G. Donate, T. E. Wright, R. J. Mannari, W. G. Payne, D. J. Smith, and M. C. Robson. Comparative evaluation of silver-containing antimicrobial dressings and drugs. *Int. Wound J.*, 4:114–122, 2007.

36. Z. M. Xiu, Q. Zhang, H. L. Puppala, V. L. Colvin, and P. J. J. Alvarez. Negligible particle-specific antibacterial activity of silver nanoparticles. *Nano Lett.*, 12:4271–4275, 2012.

37. A. B. G. Lansdown. *Biofunctional Textiles and the Skin. Current Problems in Dermatology*, vol. 33, Hipler, U.-C., Elsner, P., Eds. Karger: Basel, Switzerland, pp. 17–34, 2006.

38. N. R. Panyala, E. M. Pena-Mendex, and J. Hovel. Silver or silver nanoparticles: A hazardous threat to the environment and human health? *J. Appl. Biomed.*, 6:117–129, 2008.

39. USEPA United State Environmental Protection Agency. Drinking water standards, 2001. Available at http://www.epa.gov/waterscience/drinkingstandards/dwstandards.pdfS.

40. USEPA United State Environmental Protection Agency. *Edition of the Drinking Water Standards and Health Advisories*. Office of Water U.S. Environmental Protection Agency: Washington, DC. Report: EPA 822-R-06-013, 2006.

41. F. Heidarpour, W. A. Wan Ab Karim Ghani, A. Fakhru'l-Razi, S. Sobri, V. Heydarpour, M. Zargar, and M. R. Mozafari. Complete removal of pathogenic bacteria from drinking water using nano silver-coated cylindrical polypropylene filters. *Clean Techn. Environ. Policy*, 13:499–507, 2011.

42. Available at http://www.wrc.org.za/Pages/DisplayItem.aspx?ItemID=8772&FromURL=%2FPages%F Default.aspx%3F.

43. A. Alonso, X. M. Berbel, N. Vigués, R. R. Rodríguez, J. Macanás, M. Muñoz, J. Mas, and D. N. Muraviev. Superparamagnetic Ag@Co-nanocomposites on granulated cation exchange polymeric matrices with enhanced antibacterial activity for the environmentally safe purification of water. *Adv. Funct. Mater.*, 23:2450–2458, 2013.

44. V. Sambhy, M. M. MacBride, B. R. Peterson, and A. Sen. Silver bromide nanoparticle/polymer composites: Dual action tunable antimicrobial materials. *J. Am. Chem. Soc.*, 128:9798–9808, 2006.

45. J. Liang, J. R. Owens, T. S. Huang, and S. D. Worley. Biocidal hydantoinylsiloxane polymers IV. *N*-halamine siloxane-functionalized silica gel. *J. Appl. Polym. Sci.*, 101:3448–3454, 2006.

46. A. O. Hayden, B. B. Hsu, A. M. Klibanov, and A. Z. Gu. An antimicrobial polycationic sand filter for water disinfection. *Water Sci. Technol.*, 63:1997–2003, 2011.

47. S. Malato, P. Fernández-Ibáñez, M. I. Maldonado, J. Blanco, and W. Gernjak. Decontamination and disinfection of water by solar photocatalysis: Recent overview and trends. *Catal. Today,* 147:1–59, 2009.

48. Y. Kikuchi, K. Sunada, T. Iyoda, K. Hashimoto, and A. Fujishima. Photocatalytic bactericidal effect of TiO_2 thin films: Dynamic view of the active oxygen species responsible for the effect. *J. Photochem. Photobiol. A. Chem.*, 106:51–56, 1997.

49. Y. Tian, and T. Tasuma. Mechanisms and applications of plasmon-induced charge separation at TiO_2 films loaded with gold nanoparticles. *J. Am. Chem. Soc.*, 127:7632–7637, 2005.

50. K. L. Kelly, and K. Yamashita. Nanostructure of silver metal produced photocatalytically in TiO_2 films and the mechanism of the resulting photochromic behavior. *J. Phys. Chem. B.*, 110:7743–7749, 2006.

51. K. Awazu, M. Fujimaki, C. Rockstuhl, J. Tominaga, H. Murakami, Y. Ohki, N. Yoshida, and T. Watanabe. A plasmonic photocatalyst consisting of silver nanoparticles embedded in titanium dioxide. *J. Am. Chem. Soc.*, 130:1676–1680, 2008.

52. T. Hirakawa, and P. V. Kamat. Charge separation and catalytic activity of Ag@TiO_2 core-shell composite clusters under UV-irradiation *J. Am. Chem. Soc.*, 127:3928–3934, 2005.

53. C. Hu, Y. Lan, J. Qu, X. Hu, and A. Wang. Ag/AgBr/TiO_2 visible light photocatalyst for destruction of azodyes and bacteria. *J. Phys. Chem. B.*, 110:4066–4072, 2006.

54. M. R. Elahifard, S. Rahimnejad, S. Haghighi, and M. R. Gholami. Apatite-coated $Ag/AgBr/TiO_2$ visible-light photocatalyst for destruction of bacteria. *J. Am. Chem. Soc.*, 129:9552–9553, 2007.

55. J. A. Byrne, P. A. Fernandez-Ibañez, P. S. M. Dunlop, D. M. A. Alrousan, and J. W. J. Hamilton. Photocatalytic enhancement for solar disinfection of water: A review. *Int. J. Photoenergy*, 2011: 12 pp., 2011.

56. M. N. Chong, B. Jin, C. W. Chow, and C. Saint. Recent developments in photocatalytic water treatment technology: A review. *Water Res.*, 44:2997–3027, 2010.

57. H. Zhang, H. Zhao, P. Liu, S. Zhang, and G. Li. Direct growth of hierarchically structured titanate nanotube filtration membrane for removal of waterborne pathogens. *J. Membrane Sci.*, 343:212–218, 2009.

58. O. Akhavan, and E. Ghaderi. Photocatalytic reduction of graphene oxide nanosheets on TiO_2 thin film for photoinactivation of bacteria in solar light irradiation. *J. Phys. Chem. C.*, 113:20214–20220, 2009.

59. M. Janus, A. Markowska-Szczupak, E. Kusaik-Nejman, and A. W. Morawski. Disinfection of *E. coli* by carbon modified TiO_2 photocatalysts. *Environ. Prot. Eng.*, 38:89–97, 2012.

60. O. Akhavan, M. Abdolahad, Y. Abdi, and S. Mohajerzadeh. Synthesis of titania/carbon nanotube heterojunction arrays for photoinactivation of *E. coli* in visible light irradiation. *Carbon*, 47:3280–3287, 2009.

61. Z. Shi, K. G. Neoh, and E. T. Kang. Antibacterial and adsorption characteristics of activated carbon functionalized with quaternary ammonium moieties. *Ind. Eng. Chem. Res.*, 46:439–445, 2007.

62. V. S. Kumar, B. M. Nagaraja, V. Shashikala, A. H. Padmasri, S. S. Madhavendra, B. D. Raju, K. S. R. Rao. Highly efficient Ag/C catalyst prepared by electro-chemical deposition method in controlling microorganisms in water. *J. Mol. Catal. A: Chem.*, 223:313–319, 2004.

63. Y. L. Wang, Y. Z. Wan, X. H. Dong, G. X. Cheng, H. M. Tao, and T. Y. Wen. Preparation and characterization of antibacterial viscose-based activated carbon fiber supporting silver. *Carbon*, 36:1567–1571, 1998.

64. M. Vukčević, A. Kalijadis, S. Dimitrijević-Branković, Z. Laušević, and M. Laušević. Surface characteristics and antibacterial activity of a silver-doped carbon monolith. *Sci. Technol. Adv. Mater.*, 9:1–7, 2008.

65. S. S. Gupta, T. S. Sreeprasad, S. M. Maliyekkal, S. K. Das, and T. Pradeep. Graphene from sugar and its application in water purification. *ACS Appl. Mater. Interfaces*, 4:4156–4163, 2012.

66. J. C. Tiller, C. J. Liao, K. Lewis, and A. M. Klibanov. Designing surfaces that kill bacteria on contact. *Proc. Natl. Acad. Sci. U.S.A.*, 98:5981–5985, 2001.

67. V. K. K. Upadhyayula, S. Deng, M. C. Mitchell, and G. B. Smith. Application of carbon nanotube technology for removal of contaminants in drinking water: A review. *Sci. Total Environ.*, 408:1–13, 2009.

68. A. S. Brady-Estévez, S. Kang, and M. Elimelech. A single-walled-carbon-nanotube filter for removal of viral and bacterial pathogens. *Small*, 4:481–484, 2008.

69. C. D. Vecitis, M. H. Schnoor, Md. S. Rahaman, J. D. Schiffman, and M. Elimelech. Electrochemical multiwalled carbon nanotube filter for viral and bacterial removal and inactivation. *Environ. Sci. Technol.*, 45:3672–3679, 2011.

70. D. T. Schoen, A. P. Schoen, L. Hu, H. S. Kim, S. C. Heilshorn, and Y. Cui. High speed water sterilization using one-dimensional nanostructures. *Nano Lett.*, 10:3628–3632, 2010.

71. D. Jagadeesan, and M. Eswaramoorthy. Functionalized carbon nanomaterials derived from carbohydrates. *Chem. Asian J.*, 5:232–243, 2010.

72. J. Lukasik, Y. F. Cheng, F. Lu, M. Tamplin, and S. R. Farrah. Removal of microorganisms from water by columns containing sand coated with ferric and aluminium hydroxides. *Wat. Res.*, 33:769–777, 1999.

73. V. A. Oyandel-Craver, and J. A. Smith. Sustainable colloidal-silver-impregnated ceramic filter for point-of-use water treatment. *Environ. Sci. Technol.*, 42:927–933, 2008.

74. Y. I. Seo, K. H. Hong, S. H. Kim, D. Chang, K. H. Lee, and Y. D. Kim. Removal of bacterial pathogen from wastewater using Al filter with Ag-containing nanocomposite film by in situ dispersion involving polyol process. *J. Hazard. Mater.*, 227–228:469–473, 2012.

75. T. S. Wu, K. X. Wang, G. D. Li, S. Y. Sun, J. Sun, and J. S. Chen. Montmorillonite supported Ag/TiO_2 nanoparticles: An efficient visible-light bacteria photodegradation material. *ACS Appl. Mater. Interfaces*, 2:544–550, 2010.

76. G. Stanfield, M. Lechevallier, and M. Snozzi. *Assessing Microbial Safety of Drinking Water Improving Approaches and Methods*, Chap. 5. IWA Publishing: London, pp. 159–178, 2003.

77. I. Pinnau, and B. D. Freeman. Formation and modification of polymeric membranes: Overview. *ACS Symposium Series*, vol. 744, American Chemical Society: Washington DC, pp. 1–22, 2000.

78. R. W. Baker. *Membrane Technology and Applications*. John Wiley & Sons: Chichester, 2004.

79. S. P. Nunes, and K. V. Peinemann. *Membrane Technology in the Chemical Industry*. Wiley-VCH: Weinheim, 2001.

80. W. L. Chou, D. G. Yu, and M. C. Yang. The preparation and characterization of silver-loading cellulose acetate hollow fiber membrane for water treatment. *Polym. Adv. Technol.*, 16:600–607, 2005.

81. P. Jain, and T. Pradeep. Potential of silver nanoparticle-coated polyurethane form as an antibacterial water filter. *Biotechnol. Bioeng.*, 90:59–63, 2005.

82. K. Zodrow, L. Brunet, S. Mahendra, D. Li, A. Zhang, Q. Li, and P. J. J. Alvarez. Polysulfone ultrafiltration membranes impregnated with silver nanoparticles show improved biofouling resistance and virus removal. *Water Res.*, 43:715–723, 2009.

83. D. Bjorge, N. Daels, S. D. Vrieze, P. Dejans, T. V. Camp, W. Audenaert, J. Hogie, P. Westbroek, K. D. Clerck, and S. W. H. V. Hulle. Performance assessment of electrospun nanofibers for filter applications. *Desalination*, 249:942–948, 2009.

84. R. Molinari, M. Mungari, E. Drioli, A. D. Paola, V. Loddo, L. Palmisano, and M. Schiavello. Study on a photocatalytic membrane reactor forwater purification. *Catal. Today*, 55:71–78, 2000.

85. S. S. Madaeni, and N. Ghaemi. Characterization of self-cleaning RO membranes coated with TiO_2 particles under UV irradiation. *J. Membr. Sci.*, 303:221–233, 2007.

86. J. Mo, S. H. Son, J. Jegal, J. Kim, and Y. H. Lee. Preparation and characterization of polyamide nanofiltration composite membranes with TiO_2 layers chemically connected to the membrane surface. *J. Appl. Polym. Sci.*, 105:1267–1274, 2007.

87. B. J. Hinds, N. Chopra, T. Rantell, R. Andrews, V. Gavalas, and L. G. Bachas. Aligned multi-walled carbon nanotube membranesc. *Science*, 303:62–65, 2004.

88. A. S. Brady-Estévez, M. H. Schnoor, C. D. Vecitis, N. B. Saleh, and M. Elimelech. Multiwalled carbon nanotube filter: Improving viral removal at low pressure. *Langmuir*, 26:14975–14982, 2010.

89. A. S. Brady-Estévez, T. H. Nguyen, L. Gutierrez, and M. Elimelech. Impact of solution chemistry on viral removal by a single-walled carbon nanotube filter. *Water Res.*, 44:3773–3780, 2010.

90. T. A. Dankovich, and D. G. Gray. Bactericidal paper impregnated with silver nanoparticles for point-of-use water treatment. *Environ. Sci. Technol.*, 45:1992–1998, 2011.

91. R. Vyhnalkova, N. Mansur-Azzam, A. Eisenberg, and T. G. M. van de Ven. Ten million fold reduction of live bacteria by bactericidal filter paper. *Adv. Funct. Mater.*, 22:4096–4100, 2012.

13

Key Water Treatment Technologies and Their Use by Several Industrial Segments and Future Potential for Nanotechnology

Mike Henley

Ultrapure Water Journal, Denver, Colorado, USA

CONTENTS

Water. Where is the market going? What opportunities lie in this industry that intimately touches our lives through drinking water in our homes, irrigation of food crops, and its use as industrial and high-purity water at manufacturing plants worldwide? And, what role will be played by nanotechnology?

This chapter will examine water treatment technologies commonly found in different segments. We will first examine some industrial segments that use water. Part of our purpose is to see the types of water treatment technologies important to different applications. After this overview, we will close with a brief look at exciting developments in the use of nanotechnology, which is the broader focus of this book. However, to understand the role nanotechnology may one day play, it is first important to have some basic background about the water treatment industry.

13.1 Major Segments

There are at least two ways to examine the water business—by the kinds of end users and by the industry's products and services.

13.1.1 End-User Categories

Water business sectors range from municipal water and wastewater all the way to high-purity water. Major industry subdivisions and examples of end-user types include the following:

- High-purity water: semiconductor, pharmaceutical, power (nuclear, sub, and supercritical boiler fossil power plants), and laboratories
- Industrial: power (lower-pressure boilers of less than 900 pounds per square inch gauge [psig]), petrochemical, oil refineries, automotive, manufacturing, and steel, among others
- Pulp and paper
- Food and beverage, bottled water
- Industrial, institutional, and large commercial cooling water
- Oil and natural gas exploration (i.e., produced water, flowback water)
- Mining
- Desalination

- Home drinking water (i.e., water softeners, and other equipment)
- Municipal drinking water
- Municipal and industrial wastewater
- Agriculture

The above list focuses on different kinds of end users for the water business companies. It should be noted that each of these categories and the subgroups have different treatment needs and water quality requirements, so there is no such thing as a "universal" water treatment method for all sectors or even within end-user groups. The treatment quality standards are often governed by government regulations, guidelines from private standard-setting organizations, or guidelines set within industries or specific industrial users. Also, local water quality conditions play a critical role.

As an example, let us briefly look at high-purity water, which is sometimes referred to as "ultrapure water" or "UPW" [1]. This kind of water may also be referred to as deionized (DI) water, where the aim is to remove contaminants dissolved or suspended, and to produce a purified water stream that is devoid of impurities that are harmful to the process using the water.

High-purity water has different meanings between the industries associated with it, and for clarity, it helps to even associate an industrial end user with the use. For instance, "semiconductor water," "solar water," "pharmaceutical water," and "laboratory water" each have unique treatment approaches and requirements. These are determined in part by the industrial application where the purified water is used, industry or government agency standards, and local water-supply conditions. Table 13.1 illustrates the major differences between producers of high-purity water. The table also highlights ways treated water is used, concerns, and guidelines or standards that are followed in the treating of high-purity water.

The high-purity water treatment segment is a small picture of the greater diversity within the water field. Thus, within the overall water business, the types of treatment technologies and the makeup of the different markets vary, and are determined partly by end-user requirements. For instance, semiconductor plants require more expensive systems than, say, a cogeneration power station with a simple boiler. Similarly, municipal drinking water technologies are not overly sophisticated. Even so, the large number of municipal water and wastewater plants makes it a very sizeable market.

13.1.2 Products and Services

Another important way to evaluate the water industry is by the types of products and services offered by companies in the business. Here is a general list of categories:

- Specialty treatment chemicals
- Treatment equipment
- Filters and membranes
- Ion-exchange (IX) resins, activated carbon, and other filter materials (i.e., greensand and multimedia materials)
- Components used in treatment systems such as pressure vessels, pumps, and other items
- Instruments and sensors
- Sludge treatment equipment

TABLE 13.1

Overview of High-Purity Water Users and their Different Concerns

Industrial End User	Types of Water	Treated Water Uses	Treated Water Concerns[a]	Primary Guidelines/ Standards[b]
Semiconductor/ electronics	Semiconductor grade	Cleaning microchips as part of the manufacturing process; immersion litho, chemical dilution	Particles, metals, and organic contaminants in water can cause device (microchip) yield loss during production, as well as failure of long-term completed chip devices	SEMI, ASTM
Solar	Photovoltaic (PV) high-purity and standard purity water	Water is used in the production of photovoltaic cells and all wet processing steps such as the rinsing of substrate wafers and panels	Particles, metals, and organic contaminants can influence product loss	SEMI, ASTM
Pharmaceuticals[c]	Bacteriostatic Water for Injection, Purified Water, Sterile Purified Water, Sterile Water for Inhalation, Sterile Water for Injection, Sterile Water for Irrigation, and Water for Injection, Water for Hemodialysis, Pure Steam, and Highly Purified Water	Container and equipment cleaning, product ingredient; intravenous fluids; some consumer products; product contact of ingredient; product contact of medical device (cleaning); reagent or solvent in drug manufacturing (but not in final product)	1. Chemical, microbial, and endotoxin contamination. 2. In some cases, specific controls are additionally needed (i.e., aluminum for Water for Hemodialysis)	USP, EP, JP, WHO
Power[d]	High-purity makeup and boiler water for subcritical and supercritical boilers	Boiler water to produce steam for turbines	Corrosion and scaling caused by water contaminants that may contribute to equipment failures and plant shutdowns	EPRI, ASME, ASTM, IAPWS

Source: Compiled by M. Henley, *Ultrapure Water Journal*, Denver, CO. Copyright 2014.

Note: ASME = American Society of Mechanical Engineers; ASTM = American Society for Testing and Materials, also known as ASTM International (ASTMI); EP = European Pharmacopeia; EPRI = Electric Power Research Institute; IAPWS = International Association for the Properties of Water and Steam; JP = Japanese Pharmacopeia; SEMI = Semiconductor Equipment and Materials International (the International Technology Roadmap for Semiconductors develops materials that serve as a basis for SEMI standards.); USP = United States Pharmacopeial Convention Inc. (USP guidelines call for using a water source that meets US EPA standards for drinking water. The US FDA uses the USP standards as a part of its inspection and validation process at pharmaceutical, biopharmaceutical, and other facilities producing products requiring pharmaceutical quality water; WHO = World Health Organization.

[a] The examples listed are commonly associated with these areas of high-purity water treatment.

[b] These are examples of the principal sources for high-purity water treatment standards and guidelines in the industries covered in this table. It should be noted that companies within an industry will develop specific standards and protocols that they follow, and that service companies serving an industry have developed guidelines that become widely recognized standards and followed in the treatment of high-purity water. Examples would include Balazs Laboratories (now known as Air Liquide-Balazs NanoAnalysis) for the semiconductor industry (*Balazs Ultrapure Water Monitoring Guidelines*), and the Babcock and Wilcox handbook, *Steam: Its Generation and Use*. In the case of semiconductor water, Deutsches Institut für Normung (DIN), ASME, and International Organization for Standardization (ISO) are viewed as reference sources for particular water treatment concerns.

[c] Besides the pharmaceuticals and biopharmaceuticals, other end users that strive for these grades of water would include medical device manufacturers, and consumer goods (e.g., cosmetics) that must meet USP quality water for their products.

[d] Power plants requiring high-purity water commonly have subcritical or supercritical boilers (3200 psig). While this is true, it should be noted that facilities with high-pressure boilers 900+ psig will also produce DI water that approaches or also meets high-purity water standards. Besides fossil-fuel power plants, nuclear power plants also require high-purity water.

- Piping and valves (within plants and for municipal infrastructure)
- Engineering firms
- Outsourcing
- Consultants
- Allied segments such as water rights acquisition, and water supply development

This second list defines business segments in which vendor companies compete. Some water business companies are found strictly in one area, while in other instances larger companies have interests in two or more of these categories.

13.2 Global Water Market

At past Water Executive Forums, B. Malarkey [2] and C. Gordon [3] have reported on market studies that place the global market in the $425 billion to $450 billion neighborhood.

More recent research from Global Water Intelligence [4] shows continued growth in the business, and places the overall water global water market at $556.8 billion for water utilities and industrial markets (excluding industrial construction). The world market is projected to reach $642.1 billion in 2018, according to Global Water Intelligence. The industrial water business (including capital and operating expenditures) was $52.8 billion in 2013.

13.2.1 Sectors

Globally and domestically, municipal water is the largest market. Other major business segments include equipment and services; engineering, procurement, and construction; pumps and valves; infrastructure; chemicals; and analytics. Of those categories, infrastructure should show future growth because of efforts by water and wastewater utilities to replace aging piping systems. Within these sectors involving water treatment, nanotechnology should find more use as this segment matures and finds greater acceptance.

Through the rest of this chapter, we will examine water treatment in these areas: power generation, pharmaceuticals, semiconductors, and desalination. Then, the chapter will conclude with a brief look at nanotechnology.

13.3 Power Industry

Electricity powers our modern world—whether it is our homes, workplaces, or the cities and towns we live in. Moreover, water plays a key role in nearly all forms of power generation—hydroelectric, cogeneration, geothermal, fossil-fuel plants, and nuclear stations. Treated water is even important in the manufacture of photovoltaic panels used in solar power. Furthermore, for all but hydroelectric facilities, it is necessary to treat nearly all the water used in power generation.

In this section of our chapter, we will examine aspects of water treatment associated with electric power generation.

13.4 Water Overview

When one examines power generation, there are several different aspects of water treatment. Generally, they include the following:

- Cooling water
- Boiler water, ranging from low-pressure boilers to supercritical boilers
- Condensate polishing
- Water used for fuel gas desulfurization (FGD)
- Drinking water for more remotely located facilities

The above list represents an industry that requires multiple water treatment technologies, ranging from the use of specialty treatment chemicals, basic filtration, and deionization treatments. In this section, we will briefly examine the needs and concerns that drive the power industry water treatment.

Please note that treated water is needed in thermoelectric plants, and also in those using gas turbines. In the latter case, such facilities often also use heat recovery steam generators. For the purposes of the remainder of this chapter section, the term "power" will refer to both types of electrical generators.

13.4.1 Cooling Water

Power plants commonly use either cooling towers or once-through systems to cool process liquids. For cooling towers, treatment approaches generally involve the use of specialty chemicals to prevent and control scaling and corrosion. Biocides are used to prevent biofouling and algae growth. While chlorine products have been a traditional biocide, plants increasingly have turned to other alternatives such as chlorine dioxide. In a cooling tower, the water is often used for several cycles before discharge.

Another approach involves once-through cooling systems that take water from the ocean or a river and then return it. This method is common at stations bordering oceans, bays, and rivers. For once-through cooling systems, one important concern revolves around protecting the water intake pipes against macroorganisms such as zebra mussels that can clog the structures. It should be noted that some once-through cooling systems are being phased out, partly because of emerging US Environmental Protection Agency (US EPA) regulations that call for alternate approaches. Environmentalists claim warmed effluent water from a cooling system can damage the surrounding ecology in an ocean or river, and that water intake systems must be designed so that they protect aquatic life from "impingement" and "entrainment" [5].

13.4.2 Boiler Water

Thermopower stations use boilers to produce steam to drive turbines to generate electricity. Boilers typically have a series of tubes that are used to produce the steam. The concern is that the tubes can fail because of corrosion, plugging, or metal weakening.

Boilers used will range from low pressure to medium pressure to high pressure. The pressure rating determines water treatment needs. For example, subcritical to supercritical boilers typically found at large utility stations require DI, high-purity water, while lower-pressure boilers commonly found at some industrial plants and cogeneration plants may only require softened water, or water treated by reverse osmosis (RO). Besides deionization technology, treatment concerns may call for the use of specialty chemicals to control scaling and corrosion. Deaeration towers, oxygen scavenger treatments, and membrane contactors are ways used to remove oxygen from the water.

13.4.3 Condensate Polishing

Condensate polishing is a treatment step in which either deep-bed IX resins, or powdered resin filter the steam condensate. The goal is to remove water contaminants (known as crud) so a power station may reuse the condensate as a feedwater, and to guarantee high-purity feedwater in once-through supercritical boilers [6].

13.4.4 FGD

FGD treatment involves using technologies such as clarification and thickening to process the wastewater from sulfur removal pollution-control systems, and to prepare the solids for disposal.

13.4.5 Drinking Water

Many power stations are in remote locations away from municipal water supplies. Consequently, one part of the facility operation is the treatment of drinking water for staff use. This is in addition to other water treatment systems.

13.5 Power Plant Water Use

Thermoelectric generation plants are a major user of water in the United States and worldwide. Estimates from the US Geological Survey (USGS) are that about 195 billion gallons of water were used daily (Bgal/d) in 2000 to produce electricity in the United States. This figure excludes hydroelectric power. Of that amount, about 59.5 Bgal/d comes from saltwater sources, and the remaining 136 Bgal/d is from freshwater. Of the total amount of water used by the power industry, 99% comes from surface water.

Although thermoelectric power stations are a major user of freshwater in the United States, irrigation is slightly higher at 137 Bgal/d. Drinking water accounts for 43 Bgal/d, while self-supplied industrial withdrawals were 20 Bgal/d. The overall amount of water use in the United States during 2000 was ~408 Bgal/d [7].

While the power industry uses large amounts of water, it should be noted that much of this water reenters the environment as either steam or treated effluent placed back into rivers or streams. Therefore, power plant water is not lost and does get reused.

13.5.1 Coming Changes

When considering the power generation sector, it is important to note that it is in the middle of the debate for how electricity will be produced in the coming decades. Historically, much of the electricity made in the United States has come from plants fueled by coal.

Other important energy sources have included natural gas, oil, and nuclear power. Now there is interest in wider use of solar and wind power. Government policies have been adopted to favor solar and wind power—thus far with mixed results. In fairness, those technologies still need to show their commercial viability outside of government subsidies.

Strong environmental regulations and federal government policies in the United States are influencing traditional fossil-fuel power plants. In the coming years, it is predicted there will be large numbers of coal-fired plant closures. In 2012, the Energy Information Administration predicted the closure of up to 175 coal-fired power plants between 2012 and 2016 because of environmental rules targeting mercury pollution and chemicals that cause acid rain [8]. The predicted closures are based on reports from power plant owners and operators, who said they expect 27 gigawatts (GW) of generating capacity from 175 coal-fired generators to shut down. It should be noted that outside of government pressure, another reason for the closure of some facilities is their age.

The 27 GW would represent 8.5% of the electricity produced by coal generators. In 2011, there were 1387 coal-fired plants in the United States, totaling nearly 318 GW of generating capability, which represents about half of the generating capacity in the United States.

13.5.2 International Business

While this section has primarily focused on the domestic power industry, the greatest water business opportunities for power generation are elsewhere in the world. R. McIlvaine [9] said that based on his research, the current top power markets are the United States, China, India, Germany, Japan, and the United Kingdom. Nations showing the greatest opportunities for future growth are China, India, Russia, Indonesia, Brazil, South Africa, Pakistan, Turkey, Poland, and Malaysia.

13.5.3 "Green" Power

As noted in Table 13.2 [10], there will be some growth in generating technologies that do not require treated water. In part, this will come because of pressure by government, and regulatory agencies. Consequently, it will have a dampening effect on the development of some conventional power plants.

TABLE 13.2

Projected US Power Capacity Additions

Energy Source	Total Additions (MW 2008–2012)
Natural gas	48,100
Coal	23,347
Petroleum	1910
Nuclear	1270
Biomass	427
Geothermal	383
Wood and wood-derived fuels	288
Wind	14,617
Solar	2395

Source: Based on *Electric Power Annual 2007*, Energy Information Administration, Washington, DC, Table 2.4, January 2009.

13.6 Power Industry Summary

The power industry is important to our economy, and a key water treatment business. As already noted, power industry water treatment involves many aspects, including the sale of new and replacement treatment equipment, condensate polishing, components (i.e., instruments), disposable items (i.e., filters, resins, etc.), specialty equipment, and mobile water treatment services. Common treatment technologies found in deionization equipment are IX, RO, and electrodeionization (EDI). For those active in nanotechnology, there certainly are areas with potential product opportunities.

The next section of this chapter will examine the pharmaceutical and biopharmaceutical industry.

13.7 Pharmaceutical Water

This section will examine pharmaceutical water treatment. For the purposes of this section, this water treatment segment includes end users who produce pharmaceutical-grade water because they are required to follow the guidelines set forth by the United States Pharmacopeial Convention Inc. (USP), or related pharmacopeia standards in other world regions (i.e., European Pharmacopeia [EP] or Japanese Pharmacopeia [JP]). Examples include pharmaceutical manufacturers, biopharmaceutical companies, some health-care device makers, kidney dialysis centers, and the makers of certain consumer products (e.g., cosmetics, and contact lens solution, among others) that require pharmaceutical-grade water.

13.7.1 Background

The pharmaceutical and health-care industry segments and their decisions on water treatment are impacted by the decisions made by outside regulatory bodies. In the case of pharmaceutical water, the US Food and Drug Administration (FDA) has oversight to inspect and validate water treatment approaches at facilities. Therefore, end-user facilities design their water systems and make their purchasing decisions to ensure they remain in compliance with the FDA and any other regulator.

In its oversight, the FDA relies on water quality standards developed through the USP, a not-for-profit organization based in Rockville, Maryland. The USP was founded in 1820 to create a system of standards for making pharmaceutical drugs in the United States. In 1848, the Drug Import Act passed by Congress recognized the USP as an official compendium of standards, a status the organization still holds today.

The USP has a water subcommittee that helps to develop and update USP guidelines related to pharmaceutical water quality, and testing approaches used to confirm water quality. These, in turn, are published in the annual *US Pharmacopeia and National Formulary* (*USP-NF*), which is a publication that provides standards for medicines, dosage forms, drug substances, excipients, medical devices, and dietary supplements [11]. The *USP-NF* contains monographs on a variety of topics, including water treatment guidelines. Different types of pharmaceutical-grade waters that are produced include Bacteriostatic Water for Injection, Purified Water, Sterile Purified Water, Sterile Water for Inhalation, Sterile Water for Injection, Sterile Water for Irrigation, and Water for Injection. The production of these waters all fall under the standards set by the USP. Some of these waters are used in medicines, while others may be used in the cleaning and preparatory steps.

The USP standards influence the pharmaceutical water business because they spell out the acceptable treatment approaches for making the different grades of water.

13.7.2 Pharmaceutical Water Technologies

The pharmaceutical and allied industries are one of the largest users of high-purity water. Common examples of products sold for pharma-grade water treatment include instruments, distillation equipment, deionization systems (frequently a combination of RO and EDI), microfiltration (MF) and ultrafiltration (UF) systems, ultraviolet (UV) light, and replacement items (i.e., membranes, filter cartridges, and other disposables).

It is common for most plants producing Purified Water to have a combination system of RO-EDI. For Water for Injection, distillation systems are commonly used. The USP guidelines specify that the source water meets drinking water standards as set forth by the EPA.

13.7.3 Pharmaceutical Industry Summary

The pharmaceutical industry is an important market for high-purity water treatment. The standards used to determine the treatment approaches come from pharmacopeias such as produced by the USP, JP, and EP. These standards, in turn, influence the type of treatment equipment selected by pharmaceutical plants to produce the different grades of water used in their processes. In addition to traditional plants producing medicines, the pharmaceutical water business extends into other types or related businesses that are also required to follow the USP guidelines and are regulated by the FDA. They include kidney dialysis, biopharmaceuticals, medical equipment and devices, and certain consumer products.

The next section of this chapter will examine semiconductor water treatment.

13.8 Semiconductor Water

The semiconductor industry is made up of companies that offer technologies and products that touch many areas of modern life. High-purity water is important to the manufacture of semiconductor wafers and flat panel display products. This water is used at different stages of manufacturing to clean devices, and its cleanliness is essential to prevent the high-purity water from being the source of product defects.

13.8.1 Industry Trends

For the water business, sales opportunities arise when new fabrication plants (fabs) are built. These facilities need treatment systems that can cost several million dollars, as well as high-purity water distribution systems. Table 13.3 [12] shows the breakdown on development trends on the number of semiconductor manufacturing, pilot, and research fabs worldwide.

TABLE 13.3

Locations of Semiconductor Fabs[a]

Country or Region	4th Qtr. 2010	4th Qtr. 2011[b]	4th Qtr. 2012[b]
America	238	238	237
China	134	153	157
Europe/Middle East	159	164	166
Japan	241	242	238
South Korea	58	59	60
Southeast Asia	46	46	48
Taiwan	101	108	111
Total	977	1010	1017

Source: C.G. Dieseldorff, "SEMI World Fab Forecast November 2011, preliminary edition," SEMI, San Jose, CA, November 2010.
[a] Includes R&D and pilot facilities.
[b] Estimated figures.

The reason to briefly examine the fab figures is twofold. One, existing fabs and fab construction provides an indication of where the industry is going. However, more important, new fab construction or planning provides an indication to those active in the semiconductor water business of where future business opportunities may lie, both for conventional water treatment technologies and products that incorporate nanotechnology.

The semiconductor water market includes a number of different aspects, as noted here:

- Pretreatment equipment
- High-purity systems (disinfection [UV, ozone]), deionization (IX, RO, and EDI), monitoring instruments (total organic carbon [TOC], pH, conductivity, and particle monitoring), high-purity distribution loops (piping, valves, and related items), heating systems for hot DI water, and final filters (MF and UF)
- Wastewater treatment to prepare semiconductor effluent for discharge or reuse
- Cooling water systems (non-high-purity water)
- Miscellaneous: design/engineering, system consulting

13.8.2 Microelectronics Industry Summary

Here are some takeaway thoughts on the semiconductor industry and water treatment:

- The semiconductor industry is a major global industry and hence, an important water treatment market.
- Future business trends indicate the greatest immediate growth will be in Asian-Pacific countries.
- High-purity water is an important ingredient to everyday fab operation. The quality standards for high-purity water imply a more sophisticated market segment.
- Opportunities exist for water technologies that can solve treatment concerns, including nanotechnology.

The next section will examine desalination.

13.9 Desalination

Thirsty. In the water-short regions of the world, desalination has become one of the best ways to provide additional water supplies—for drinking and industrial purposes. As a result, seawater desalination has become an important segment of the water business.

13.9.1 Background

Seawater generally has concentrations of dissolved salts that are from 10,000 parts per million (ppm) up to 35,000 ppm, according to the USGS [13]. The agency noted that ocean water salt content is commonly 35,000 ppm. In contrast, freshwater will have <1000 ppm in saline content.

Desalination involves the use of either thermal distillation technologies or seawater RO (SWRO) to remove salt content to make ocean or brackish water suitable for drinking or industrial use. The three primary thermal technologies are multistage flash (MSF), multi-effect distillation (MED), and vacuum compression (VC).

There are >14,000 desalination plants worldwide that can produce more than 59.9 million m^3/day of treated water, according to a report by the *International Desalination & Water Reuse Quarterly* [14]. These plants range from the size of massive desalination facilities in the Middle East and Australia with capacities in the millions of gallons per day to small facilities on islands in the Caribbean that treat a few thousand gallons per day. An example of one of the world's largest desalination plants is the Kurnell Desalination Plant near Sydney, Australia, that has a daily capacity of 66 million gal (250 ML). Desalination plants are also found on military and commercial ships.

13.9.1.1 Energy Use

One criticism of desalination is the high cost of running the pumps for SWRO, or heating the water in thermal processes. There is no question that desalination is more expensive than the treatment of freshwater resources. Briefly, here are some comparisons of the cost between membrane and thermal technologies.

N. Voutchkov [15] noted that the energy use by seawater desalination plants is between 12 and 15 kWh/1000 gal of freshwater. That compares to from 3 to 8 kWh/1000 gal for brackish RO plants. In contrast, thermal desalination can use between 25 and 40 kWh/1000 gal. Of the three principal thermal technologies, MSF uses more energy than MED plants. In turn, MED facilities consume more energy than VC plants.

Table 13.4 [16] provides a comparison of the energy use of water supply alternatives. As can be seen, desalination does require more energy than other sources. However, perhaps nanotechnology can come to play an important role in desalination, even lowering the cost.

13.9.1.2 Developing New Technologies

Another approach to save energy being pursued by some is the development of technologies that can either save energy or more efficiently remove salt. Nanotubes are one example. This approach would involve membranes made of carbon nanotubes (CNTs) and silicon. Researchers at Lawrence Livermore National Laboratory [17] found that liquids and gases will flow rapidly through the tubes. One use they found was that the nanotubes could be used for desalting or deionizing water.

TABLE 13.4

Energy Use of Various Water Supply Alternatives

Water Supply Alternative	Energy Use (kWh/kgal)
Conventional treatment of surface water	0.8–1.5
Water reclamation	2.0–4.0
Indirect potable reuse	5.0–7.5
Brackish water desalination	3.0–5.0
Pacific Ocean water desalination	10.0–14.0

Source: Based on Table 1 in N. Voutchkov, *Seawater Desalination: Current Status and Challenges*, Water Globe Consulting LLC, Stamford, CT.

Voutchkov [15] noted that nanotubes are "promising" for future use and could allow for membranes able to produce significantly more water than a conventional 8-in membrane.

13.10 Nanotechnology's Potential

Much of this chapter has examined water treatment and water use in a number of industrial-related areas. In closing, we will take a brief look at nanotechnology. Other chapters in this book will examine this subject in much greater depth.

13.10.1 Next Step

Earlier in this chapter, we have touched on different water treatment technologies. Those who have followed water treatment technologies for any time have seen changes occur among the types of technologies used. IX proved better than zeolite softening, and RO in turn has come to replace IX in a number of applications. Since the late 1980s, EDI has made steady inroads into traditional deionization applications. Similar stories are found among other types of water treatment and monitoring technologies.

Within the water business, the newest area of development involves nanotechnology. This final section will review reports from some individuals active in the development and commercialization of nanotechnology in water treatment uses. These presentations are based on presentations given at the 2009 Water Executive Forum conducted in Philadelphia.

13.10.2 Background

Broadly speaking, nanotechnology entails the manipulation of materials and objects at the 1–100 nanometer (nm) scale. The related materials may be incorporated into materials or structures that are defined in terms of their size in nm. Unlike some areas, nanotechnology includes aspects of a number of scientific fields, including chemistry, physics, biology, and material sciences, among others [18].

At present, it is unknown to what extent nanotechnology will finally influence water treatment. One development involves the use of nanomaterials in filters and membranes. The materials may be made to help a filter or membrane selectively remove specific contaminants. One example of a proposed product comes from Australia where scientists have found that nanotubes made up of boron and nitrogen atoms are able to block salt molecules [19]. If successfully commercialized, this desalination method would represent a new way to treat brackish and ocean waters. The major desalination methods currently include different types of thermal distillation, and RO.

13.11 Research Work

The following are brief summaries of research work reported on at the 2009 Water Executive Forum.

R. Sustich [20] reported on some different areas of development involving the use of nanotechnology in water. Mr. Sustich is a coauthor of *Nanotechnology Applications for Clean Water*, a book that examines uses for nanotechnology in water treatment. Other coauthors of the book published by William Andrew Publishing (January 2009) include M. Diallo, J. Duncan, N. Savage, and A. Street. The book includes work from more than 80 researchers and environmental professionals, primarily in the United States. Topics addressed include drinking water, wastewater treatment and reuse, groundwater remediation, contaminant monitoring, and social issues about the use of nanotechnology.

One example of research on the use of nanotechnologies in water treatment is the use of photocatalytic titanium nanoparticles for contaminant oxidation and disinfection. One application is to replace chlorine and other chemical disinfectants with titanium dioxide (TiO_2) in conjunction with UV light. TiO_2 improves the energy efficiency of the UV system and also has been shown to have improved disinfection results.

Another approach is the use of nanotechnology-based membrane materials. In this scenario, a thin-film composite membrane has nanoparticles dispersed on it to create a functionalized membrane for the removal of specific contaminants. One example is a thin-film membrane supported by zero-valent iron (ZVI) and bimetallic nanoparticles for the purpose of chlorine reduction.

In another instance, Sustich reported about nanoparticle pore-functionalized MF membranes that have use for dechlorination. This method involves the use of PVDF membranes modified for uniform pore size. Types of nanoparticles used can include ZVI, iron/nickel, and iron/zinc. Research has shown such membranes are capable of polychlorinated biphenyl and trichloroethylene degradation.

A fourth example cited by Sustich involved research by V.V. Tarabara, a professor at Michigan State University. Tarabara and his team have examined the use of multifunctional nanotechnology-based membrane materials that are attached to, say, ceramic membranes, resulting in "self-cleaning" catalytic surfaces. One goal is to develop membranes with additional functionalities. Figure 13.1 [19] is a series of diagrams with different nanomaterials that could be affixed to the filter surface.

13.11.1 Fullerenes

At the Water Executive Forum conference, M. Hotze [21] reported on fullerenes and their potential use in water treatment. Fullerenes are a family of carbon allotropes that are structurally similar to graphite and contain pentagonal or heptagonal rings. Other characteristics are that they are hydrophobic and insoluble in water and can be dispersed in water through various methods, of which solvent exchange and surfactant wrapping are two examples.

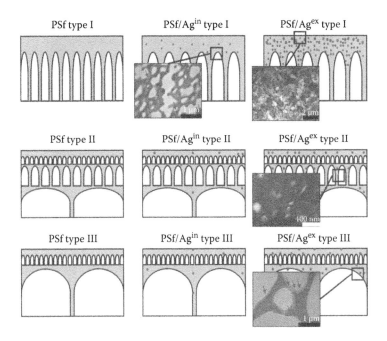

FIGURE 13.1
(See color insert.) Example of nanomaterials that could be affixed to a membrane surface. (From Figure 5.1 in Chapter 5, Multifunctional Nanomaterial-Enabled Membranes for Water Treatment, pp. 59–75, by Tarabara, V.V. Reprinted from *Nanotechnology Applications for Clean Water*, Diallo, M.; Duncan, J.; Savage, N.; Street, A.; Sustich, R., editors, Elsevier Ltd., Kidlington, UK (2009). Reprinted with permission of Elsevier.)

Outside of water treatment, some uses of fullerenes are in face creams, antiaging creams, and sporting goods such as bicycles and tennis rackets.

A potential use of fullerenes for water treatment is for membrane biofouling control. Hotze said that the fullerene coatings could have an important impact on water treatment. He noted that photosensitization chemistry could provide avenues to novel advanced oxidation processes. The antimicrobial properties of fullerenes could also be used for surface coatings on membranes. Another potential application would be for membrane biofouling control.

13.11.2 Nanocomposite RO

J. Green [22] reported about the use nanotechnology with RO membranes. He noted the invention by University of California, Los Angeles, researchers of a nanocomposite brackish RO element that has inorganic superhydrophilic nanoparticles encapsulated in a polyamide thin-film element. This membrane features high flux, fouling resistance, and high salt rejection. Figure 13.2 [22] shows the difference in fouling between commercial SWRO elements, and those combining nanocomposite materials. The standard-type SWRO is on the right side and the nanomaterial-modified SWRO is on the left side of the figure.

In the context of desalination, possible advantages of an RO membrane with nanotechnology could include the following:

- Lower energy use
- Reduce capital cost/facility footprint

FIGURE 13.2
(See color insert.) Comparison of SWRO membrane fouling. Membrane samples on the left side are from NanoH₂O elements with NP3, while those on the right are from conventional commercial SWRO. (From J. Green, "Nanocomposite Reverse Osmosis Membranes," presentation at 2009 WATER EXECUTIVE Forum, Philadelphia, March 30–31, 2009. Reprinted with author's permission.)

- More system design flexibility
- Reduce pretreatment
- Lower chemical consumption

Prototype SWRO elements can be made with the same manufacturing process using the same spiral-wound element. The nanoparticles would add a smaller percentage to the membrane cost. Green said that studies have found that nanocomposite RO membranes can be twice as productive, show fouling resistance, and have the same form factor as conventional elements.

For example, a treatment plant using the nanomaterial-modified elements can save on energy or produce more water. Alternatively, a smaller plant can be built that is 40% smaller that produces the same amount of water as a conventional RO plant. Figure 13.3 [22] illustrates how a nanomaterial-modified SWRO could improve the economics of RO systems.

13.11.3 Nano-Alumina Nonwoven Filters

F. Tepper [23] reported about nano-alumina nonwoven filters that can serve as an alternative to MF and UF membranes. One company, Argonide, has developed two filtration media. One is an electropositive media that can remove particles at high flow rates. The second is a carbon filter that has a high dynamic response and also is able to filter particles.

Argonide has patented the NanoCeram® filter media that is based on a nano-alumina fiber.

Figure 13.4 [23] illustrates the nano-alumina (AlOOH) fiber that is 2 nm in diameter. The nanofibers, which are identified by arrows, are dispersed and adhere to glass fibers and have an appearance of fuzz. In a presentation, Tepper reported that the composite fiber is formed into a nonwoven media by wet processing. The filter pore size is ~2 microns. The fibers have a charged surface and will attract particles out of the water.

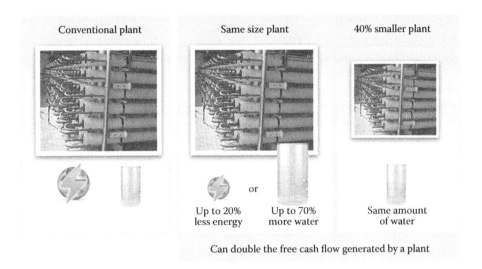

FIGURE 13.3
How a nanomaterial-modified SWRO could influence energy savings, water production, of equipment footprint size at an RO facility. (From J. Green, "Nanocomposite Reverse Osmosis Membranes," presentation at 2009 WATER EXECUTIVE Forum, Philadelphia, March 30–31, 2009. Reprinted with author's permission.)

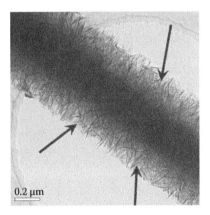

FIGURE 13.4
Example of nano alumina fiber. (From photomicrograph taken by researchers at the University of Connecticut. From F. Tepper, "Nano Alumina Non-Woven Filters: An Alternative to MF/UF Membranes," presentation at 2009 WATER EXECUTIVE Forum, Philadelphia, March 30–31, 2009. Reprinted with author's permission.)

Another approach for this technology is the use of powdered activated carbon (PAC) nanomaterial. This type of filter can be used to remove chlorine before the water moved to the RO system. Tepper noted the PAC version of the filter also could remove oil from wastewater.

One use for the nano-alumina filters is for pretreatment before an RO system as a replacement for UF membranes. The nano-alumina filters do not have a waste stream and are capable of removing water foulants down to the nanometer size. Unlike UF membranes, these filters do not have the same need for cleaning.

13.11.4 Filtration Evolution

In another presentation, R. Miller [24] said that nanotechnology provides a new option to conventional filtration technologies. Figure 13.5 [24] shows a conventional view of filtration technologies and their removal abilities—ranging from ionic contaminants (salts and metals), endotoxins, viruses, bacteria, dust and silt, cysts, and sand. Technologies and their removal abilities noted in the figure are RO (ionic and molecular), nanofiltration (ionic and molecular), UF (molecular and micro), and MF (micro and particle), and particle filtration (particles).

Miller noted that traditional filtration technologies could have limitations. For example, sieving filters that exclude contaminants because of their size may have the pitfalls of energy use, gross filtration, and waste streams. Limitations for technologies that work through chemical, adsorption, or kinetics (e.g., coalesce, IX, chemiadsorptive, and electrolytic) are residence time, chemical use, and treatment capacity.

Nanotechnology can represent the convergence of sieving and chemical–adsorption–kinetic technologies. Nanofibers used in filters offer the benefit of small diameters, large surface areas, great porosity, a variety of materials, and ease of use with different filtration media.

One example is a CNT known as Nanomesh™ developed by Seldon. The CNT technology has been shown to effectively remove TOC, *Escherichia coli*, bacteriophage MS2 virus (MS2), Taura syndrome virus (TSV), yellow head virus (YHV), white spot syndrome virus (WSSV), and infectious hypodermal and hematopoietic necrosis virus (IHHNV). For instance, at an aquaculture laboratory, the company was able to get full removal of WSSV, TSV, IHHNV, and YHV from 380 gal of seawater that had a challenge log ranging from 7.3 for TSV down to 5.5 for IHHNV. For TOC removal, the CNT nanotechnology was able to treat an influent tannic acid from 10.5 ppm down to 0.4 ppm, and an influent humic acid from 2.5 ppm down to 0.4 ppm. This performance came in treating 120 gal of water during in-house NSF P248 tests, Miller reported.

Table 13.5 [24] shows results from treating canal water from Riverside, California, with Seldon's CNT electrode technology. Contaminants removed from the water included calcium, magnesium, sodium, barium, copper, fluoride, nitrate/nitrite, and sulfate.

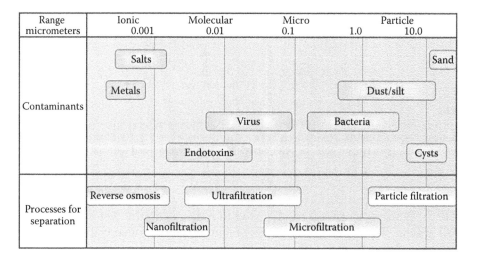

FIGURE 13.5
Filtration spectrum and current technologies. (From presentation slide from Miller, R.S. "Carbon Nanotubes: A Fundamental Shift in Filtration," presentation at 2009 WATER EXECUTIVE Forum, Philadelphia, March 30–31, 2009. Reprinted with author's permission.)

TABLE 13.5

Use of CNT Electrode Technology to Treat Canal Water in California

Contaminant	Original (mg/L)	After Treatment (mg/L)
Calcium	100	4.1
Magnesium	20	0.76
Sodium	70	10
Barium	0.12	0.016
Copper	0.009	0.001
Fluoride	0.5	<0.1
Nitrate/nitrite	13	<0.5
Sulfate	90	<1

Source: Power Point table in presentation by R.S. Miller, "Carbon Nanotubes: A Fundamental Shift in Filtration", presentation at 2009 WATER EXECUTIVE Forum, Philadelphia, March 30–31, 2009.

13.11.5 Product Development

L. Farmen [25] reported on her firm's product development strategy for its nano-coated filter media treatment product, NMX™. This product uses low-cost ligand coatings and has been shown to be capable of binding targeted metal contaminants. Proposed applications for the technology include mining and high-metal-containing wastewater, distributed purification systems, industrial and recycle process water, municipal water treatment, and home point-of-use or -entry. Expected benefits from the technology are as follows:

- Recovery and recycling of heavy metals to meet drinking water standards
- High adsorbent capacity
- In-place regeneration capability up to seven times to enhance capacity dramatically and reduce cost
- Ability to "functionalize" other substrates to enable metal removal

The company's initial focus is on removal of selenium, lead, copper, and cobalt. Work under way emphasizes development of a low-cost process for combining chitosan and selected ligands for industrial, mining waste, and battery recyclers. Development work is also examining the development of a low-cost electrodeionizer to concentrate ionic impurities in water.

13.12 Closing Thought

In this chapter, we have examined water treatment in several important areas: power generation, pharmaceuticals, semiconductors, and desalination. Nanotechnology offers great potential to improve future water treatment methods and when implemented will create new business opportunities in these areas as well as in other water treatment areas such as municipal water and wastewater, and industrial wastewater.

Author's note: This chapter is a summation of several articles by the author that were published in the *ULTRAPURE WATER® Journal* between 2009 and 2010. Additional background and information is available at www.ultrapurewater.com or www.globalwaterintel.com.

Acknowledgments

The author thanks the following individuals for their help with Table 13.1: Slava Libman, PhD, Air Liquide-Balazs NanoAnalysis; Anthony Bevilacqua, PhD, METTLER-TOLEDO Thornton; Marty Burkhart, Hi Pure Tech Inc.; William V. Collentro, Water Consulting Specialists; and Brad Buecker, Kiewit Power Engineers.

References

1. Note: This term is a registered trademark of *ULTRAPURE WATER® Journal*, an on-line technical journal that covers high-purity water treatment, which is published by Media Analytics Ltd., Oxford, U.K.
2. Malarkey, B. "Water Sector M&A Market Outlook," presentation at 2009 WATER EXECUTIVE Forum, Philadelphia, PA (March 30–31, 2009).
3. Gordon, C.R. "Future of the Water Industry," dinner presentation at 2008 WATER EXECUTIVE Forum, Philadelphia, PA (April 15, 2008).
4. Global Water Intelligence, "Global Water Market 2014: Meeting the World's Water and Wastewater Needs until 2018," Media Analytics Ltd., Oxford, U.K. (2013).
5. "Regulating Cooling Water Use at Existing Power Plants: An Overview of the Decision before EPA," Fact Sheet and Overview, Clean Air Task Force, Boston, MA (January 2004).
6. Meltzer, T.H. *High-Purity Water Preparation for the Semiconductor, Pharmaceutical, and Power Industries*, Tall Oaks Publishing Inc., Littleton, CO, p. 248 (1993).
7. Hutson, S.S.; Barber, N.L.; Kenny, J.F.; Linsey, K.S.; Lumia, D.S.; Maupin, M.A. "Estimated Use of Water in the United States in 2000," USGS Circular 1268, U.S. Geological Survey, Reston, VA. (revised issue, February 2005).
8. "Large Number of Fossil Power Plants Are Likely To Close in 2012," news brief in *ULTRAPURE WATER 29(4)*, p. 4 (July/August 2012).
9. McIlvaine, R., The McIlvaine Co., Northfield, IL, Personal Communication (December 2009).
10. *Electric Power Annual 2007*, Energy Information Administration, Washington, DC, Table 2.4 (January 2009).
11. "USP–NF—An Overview," United States Pharmacopeial Convention Inc., Rockville, MD. http://www.usp.org (accessed March 2010).
12. Dieseldorff, C.G. "SEMI World Fab Forecast November 2011, Preliminary Edition," SEMI, San Jose, CA (November 2010).
13. "Thirsty? How 'bout a Cool, Refreshing Cup of Seawater?," website article, U.S. Geological Survey, Reston, VA, http://water.usgs.gov/edu/drinkseawater.html (accessed June 2010).
14. "Total World Desalination Capacity Close to 60 million m³/d," news article at International Desalination & Water Reuse Quarterly website, http://www.desalination.biz, South Croydon, Surrey, U.K. (Nov. 8, 2009). Note: This news brief is about the 22nd Global Water Intelligence/International Desalination Association (GWI/IDA) Worldwide Desalting Plant Inventory Report.

15. Voutchkov, N. personal communication, Water Globe Consulting LLC, Stamford, CT (June 2010).
16. Voutchkov, N. "Seawater Desalination: Current Status and Challenges," Water Globe Consulting LLC, Stamford, CT.
17. "Nanotube Membranes Offer Possibility of Cheaper Desalination," press release, Lawrence Livermore National Laboratory, Livermore, CA (May 18, 2006).
18. "Evaluation of Nanotechnology for Application in Water and Wastewater Treatment and Related Aspects in South Africa," Executive Summary of Report No. KV 195/07, Foundation for Water Research, Bucks, England (August 2007).
19. Australian Broadcasting Co., "Nanotechnology Key to Faster Desalination," www.abc.net.au (accessed August 25, 2009).
20. Sustich, R. "Nanotechnology Applications for Clean Water," presentation at 2009 WATER EXECUTIVE Forum, Philadelphia, PA (March 30–31, 2009).
21. Hotze, E.M. "Fullerences and Water Treatment: Impacts, Removal, Applications," presentation at 2009 WATER EXECUTIVE Forum, Philadelphia, PA (March 30–31, 2009).
22. Green, J. "Nanocomposite Reverse Osmosis Membranes," presentation at 2009 WATER EXECUTIVE Forum, Philadelphia, PA (March 30–31, 2009).
23. Tepper, F. "Nano Alumina Non-Woven Filters: An Alternative to MF/UF Membranes," presentation at 2009 WATER EXECUTIVE Forum, Philadelphia, PA (March 30–31, 2009).
24. Miller, R.S. "Carbon Nanotubes: A Fundamental Shift in Filtration," presentation at 2009 WATER EXECUTIVE Forum, Philadelphia, PA (March 30–31, 2009).
25. Farmen, L.M. "Commercialization of Nanotechnology for Removal of Heavy Metals in Drinking Water and Wastewater," presentation at 2009 WATER EXECUTIVE Forum, Philadelphia, PA (March 30–31, 2009).

Further Readings

Air Liquide-Balazs NanoAnalysis, *Ultrapure Water Monitoring Guidelines, Revision 2.0*, Air-Liquide-Balazs NanoAnalaysis, Fremont, CA (2007).

Betz Laboratories Inc., *Betz Handbook of Industrial Water Conditioning*, Betz Laboratories Inc. (now GE Water & Process Technologies), Trevose, PA (1991).

Choules, P.; Schrotter, J-C; Leparc, J.; Gaid, K.; Lafon, D. "Improved Operation through Experience with SWRO Plants," *ULTRAPURE WATER 25(7)*, pp. 13–22 (October 2008).

Cohen, P., ed. *The ASME Handbook on Water Technology for Thermal Power Systems*, American Society of Mechanical Engineers, New York, NY (1989).

Crits, G. *Condensate Polishing—Purification Technology for Steam Power*, Tall Oaks Publishing Inc., Littleton, CO (2003).

Dietrich, J.; Voutchkov, N. "Comparison of Membrane and Granular Media Pretreatment, Toxin, and Boron Rejection Results, and Energy Recovery Update in Carlsbad, Calif.," *ULTRAPURE WATER 25(6)*, pp. 55–60 (September 2008).

Henley, M. "Pharmaceuticals Remain a Key U.S. Water Treatment Market," *ULTRAPURE WATER 23(2)*, pp. 18–21 (March 2006).

Lewis, J. "Nanotechnology May Decrease Energy Costs for Desalination," Foresight Institute, www.foresight.org (accessed June 2010).

Meltzer, T.H. *Pharmaceutical Water Systems*, Tall Oaks Publishing Inc., Littleton, CO (1997).

Stultz, S.C.; Kitto, J.B., eds., *Steam: Its Generation and Use*, 40th ed., The Babcock & Wilcox Co., Barberton, OH (1992).

14

Iron-Based Magnetic Nanomaterials in Wastewater Treatment

Zongshan Zhao,[1,2] **Jing Lan,**[3] **Guoliang Li,**[1] **and Guibin Jiang**[1]

[1]*State Key Laboratory of Environmental Chemistry and Ecotoxicology, Research Center for Eco-Environmental Sciences (RCEES), Chinese Academy of Sciences (CAS), Beijing, China*

[2]*Key Laboratory of Marine Chemistry Theory and Technology, Ministry of Education, Ocean University of China, Qingdao, China*

[3]*College of Chemistry and Pharmaceutical Sciences, Qingdao Agricultural University, Qingdao, China*

CONTENTS

14.1 Introduction

Since magnetic separation was first described for separating iron minerals by William Fullarton, it has aroused a continuously increasing worldwide concern in areas ranging from adsorption to biotechnology [1–5]. In an effort to combat the problem of water pollution, the processes based on magnetic nanomaterials (MNMs) have been considered promising options for reducing contaminants owing to their rapid, low-cost, and highly efficient advantages [6–8]. During a standard magnetic separation process, these MNMs are used for fixing specific metal species/molecules or for degrading toxic pollutants; a magnetic field is utilized as the main force for easily isolating these MNMs and the tagged contaminants from aqueous solutions [9–14]. The purification process does not generate secondary waste, and these MNMs can often be recycled.

The magnetic component of most of the MNMs used for treating wastewater is magnetite Fe_3O_4 (or its oxidation counterpart γ-Fe_2O_3). In recent years, the preparation and utilization of these iron-based MNMs with novel properties and functions have been widely studied owing to the easy synthesis, coating or modification, and the ability to control on an atomic scale [15,16]. Since the practical applications depend on the properties of these MNMs involving chemical stability, water compatibility, and the affinity to the target compounds/ions, they are often modified/functionalized with a series of media with suitable functional groups, such as silicon, phosphoric acids, carboxylic acid, and amine [15–19]. In addition, it has been proven that the size distribution, morphology, magnetic properties, and surface chemistry of MNMs depend on the preparation methods and surface coating media [20]. Therefore, a variety of synthesis approaches have been developed for preparing MNMs of high quality [16,21–24]. The synthesis approaches can be divided into three classes: (i) physical methods involving gas-phase deposition, electron beam lithography, pulsed laser ablation, laser-induced pyrolysis, powder ball billing, and aerosol, etc.; (ii) chemical methods involving coprecipitation, microemulsion, hydrothermal and electrochemical deposition, sonochemical and thermal decomposition, etc.; and (iii) biological methods mediated by fungi, bacteria, and protein, etc. Many of the prepared iron-based MNMs have presented promising potential for industrial-scale wastewater treatment at both laboratory- and field-scale tests [25,26].

In this chapter, a brief introduction on the preparation of iron-based MNMs and their applications as nanosorbents and photocatalysts for water treatment have been given. In addition, the likely fates of MNMs discharged into the environment are also discussed.

14.2 Synthesis of Iron-Based MNMs

Given their unique characteristics and great potential for water treatment on an industrial scale, many efforts have been made for synthesizing iron-based MNMs of high quality and of particular usage by developing synthesis approaches involving chemical, physical, and biological methods. Figure 14.1 presents the three most important fabrication approaches published for synthesizing superparamagnetic iron oxide nanoparticles (SPIONs), summarized by Mahmoudi et al. [27]. Consider that the applications of iron-based MNMs in water treatment greatly depend on the particle size, shape, and surface chemistry [20], as well as the degree of the structural defects or impurities present in the NMs [28]. Various preparation methods have been exploited to meet the application demand. Hydrolysis of ferrous salts in alkaline solution is the most simple and fundamental method for preparing magnetic Fe_3O_4 and γ-Fe_2O_3, the products of which are usually difficult to control in size and structure. Therefore, hydrothermal and solvothermal methods and many other chemically and physically based methods were developed for easily controlling the size distribution and tailoring the structure to extend the application spectrum of MNMs. Generally, iron-based MNMs can be divided into three categories, including simple nanocrystals, structurally functionalized MNMs, and chemically functionalized MNMs—the preparation approaches of these MNMs are introduced in detail in the following section.

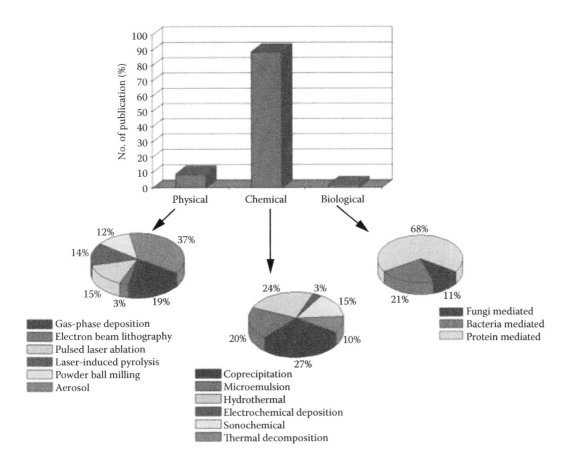

FIGURE 14.1
(See color insert.) Comparison of published work (up to date) on the synthesis of SPIONs by three different routes. (From Mahmoudi M., Sant S., Wang B. et al., *Advanced Drug Delivery Reviews*, 63, 24, 2011.)

14.2.1 Simple Nanocrystals

Generally, the magnetic component of MNMs used for water treatment is magnetite Fe_3O_4 (or its products of oxidation, $\gamma\text{-}Fe_2O_3$). Magnetite is much easier to synthesize and much cheaper than other magnetic materials composed of other metals (Co) or other oxides (of the type MFe_2O_4). Therefore, many expanding synthetic works on MNMs have evolved from around the preparation of Fe_3O_4 crystals. Fe_3O_4 has a cubic inverse spinel structure with a space grouping of *Fd-3m* [29]. The lattice constant is $a = 0.839$. In the unit cell (Figure 14.2), the oxygen ions form an *fcc* close packing, and the iron ions occupy interstitial tetrahedral sites that are occupied with Fe(II) and octahedral sites that are occupied by both Fe(II) and Fe(III), symbolized as $[Fe]_A[Fe^{2+}Fe^{3+}]_BO_4$ [30]. The high-index planes usually have higher surface energy [31] and the sequence is $\gamma(111) < \gamma(100) < \gamma(110) < \gamma(220)$ for the face-centered-cubic phase [32]. Fe_3O_4 crystals are usually overlaid with octahedral and mixed octahedral/tetrahedral layers along the <111> direction. Faster growth along the <100> direction can result in octahedral particles, while faster growth along the <111> direction can lead to cubic particles. Various morphologies of Fe_3O_4 crystals can be prepared by controlling the growth rate of different facets of the nuclei through the introduction of surfactants [33] and templates [34], or regulating the specific synthetic conditions [35].

FIGURE 14.2
(See color insert.) Crystal structure of Fe_3O_4; green atoms are Fe^{2+}, brown atoms are Fe^{3+}, white atoms are oxygen. (From Yang C., Wu J., Hou Y., *Chemical Communications*, 47, 5130, 2011.)

14.2.1.1 Spherical Nanoparticles

Since Sugimoto's group fabricated monodispersed Fe_3O_4 nanoparticles in 1980, various methods have been developed for fabricating Fe_3O_4 nanoparticles with narrow size distribution and good dispersion. In 2002, Sun's group developed a classic oleic acid (OA)–oleyl amine (OAm) system that uses both OA and OAm as surfactants for synthesizing monodispersed spherical Fe_3O_4 nanoparticles [31]. A "seed-mediated" growth method was developed to synthesize larger nanoparticles; the size of Fe_3O_4 nanoparticles (3–20 nm) can be tailored by altering the quantity of seeds. By modifying the strategy described above, Xu et al. also prepared monodispersed Fe_3O_4 nanoparticles without using polyol, providing a low-cost route [36]. In such a system, OAm was used as both a reducing agent and stabilizer, and the particle size was tunable by varying the volumetric ratio of benzyl ether and OAm.

Solvothermal methods and high-temperature liquid phase methods have also been proposed to synthesize spherical Fe_3O_4 nanoparticles. Gao's group developed a solvothermal method for synthesizing Fe_3O_4 nanoparticles with a mean diameter of 25 nm [37]. In this system, $[Zn(CO_3)_2(OH)_6]$ can accept Fe^{2+} precipitates through −OH and then prevent the agglomeration of Fe^{2+} precipitates, and sequentially superparamagnetic Fe_3O_4 nanoparticles were obtained. Park et al. synthesized monodisperse iron oxide nanoparticles with iron(III) oleates as the precursor by using a high-temperature liquid-phase method [38]. The particle size from 5 to 22 nm can be controlled by changing the decomposition temperature and ageing time. γ-Fe_2O_3 was the major phase for 5-nm iron oxide nanoparticles, while the proportion of Fe_3O_4 gradually increased with increasing particle size.

14.2.1.2 Octahedral, Dodecahedron, and Cubic Nanoparticles

It has been indicated that the shape of the particle is closely related to the crystallographic surfaces that enclose the particle [39], and the key factor for synthesizing various Fe_3O_4 nanoparticles is to tune the growth rate of specific facets. As mentioned above, the relative surface energies for Fe_3O_4 are in the order of $\gamma(111) < \gamma(100) < \gamma(110) < \gamma(220)$ owing to the distances between these faces and the coordination number with neighboring atoms [40]. Therefore, the growth rate of the (111) plane is quicker than that of other planes, and the octahedral shapes are the thermodynamically favored morphology.

Zhang et al. proposed a simple method for preparing octahedral Fe_3O_4 nanoparticles [33]. In this system, tetracosane and $Fe(OA)_3$ were used as the reaction media and precursor, respectively. OAm was used as both the surfactant and the reducing agent. The size of the obtained octahedral nanoparticles is 21 ± 2 nm. These octahedrons can also self-assemble into oriented superstructures as a result of their anisotropic shapes. Solvothermal processes have also been applied for fabricating highly crystallized Fe_3O_4 nanoparticles by Qi and coworkers [41]. In this system, the growth of the (111) facet is hindered because of the absorption of hydroxyl on the (111) facet, and thus the spherical shape of the Fe_3O_4 can be diverted to truncated octahedral particles and octahedral particles with some residual amorphous particles (Figure 14.3).

As mentioned above, since the (110) facet has the highest surface energy, magnetite nanocrystals enclosed by the (110) plane are difficult to fabricate. Nevertheless, Li's group have prepared rhombic dodecahedral (RD) Fe_3O_4 nanocrystals by using a microwave-assisted route in the presence of ionic liquid (IL) $[C_{12}Py]^+[ClO_4]^-$ [42]. ILs can be used to alter the surface condition of the Fe_3O_4 nanocrystals, and hexamethylenetetraamine/phenol adsorbed on (110) planes is favorable for the crystal growth along the [100] direction. Therefore, RD Fe_3O_4 nanocrystals enclosed by twelve (110) flakes can be obtained.

When the growth rate along the <111> direction is faster than that of the <100> direction, nanocubes can be formed for exposing distinct surfaces [43]. Yang et al. synthesized monodispersed Fe_3O_4 nanocubes with controllable sizes of 6–30 nm, by using $Fe(acac)_3$ as a precursor; benzyl ether as a solvent; and 1,2-hexadecanediol, OA, and OAm as surfactants (Figure 14.4a through c) [44]. OA controlled the size of nanoparticles owing to its carboxylic group binding selectively to crystal facets, and OAm affected the morphology owing to its comparatively weak and isotropic binding to the facets. When replacing $Fe(acac)_3$ with $Fe(OA)_3$, the stabilizer of sodium oleate (NaOL) could lead to nanocubes, potassium oleate (KOL) could generate a mixture of nanocubes with other morphologies, and dibutylammonium oleate (DBAOL) and OA could lead to spherical nanoparticles (Figure 14.4e through f) [45]. In addition, monodisperse Fe_3O_4 nanocubes can also be obtained

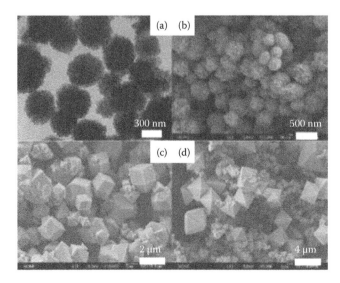

FIGURE 14.3
Morphology of Fe_3O_4 nanoparticles. Concentrations of NaOH (mol/L) are (a, b) 0, (c) 0.5, and (d) 0.7. (Modified from Qi H, Chen Q, Wang M et al., *The Journal of Physical Chemistry C*, 113, 17301, 2009.)

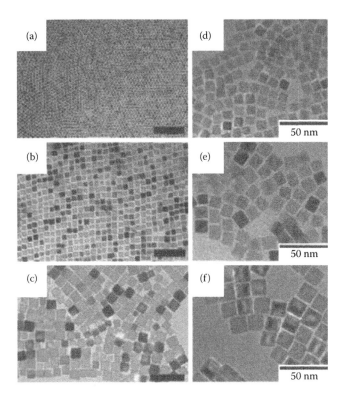

FIGURE 14.4
(a through c) Transmission electron microscopy (TEM) images of Fe$_3$O$_4$ nanocubes with various sizes: (a) 6.5 nm, (b) 15.0 nm, and (c) 30.0 nm. (d through f) TEM images of monodispersed cubic iron oxide nanoparticles. (Modified from Yang C, Wu J, Hou Y, *Chemical Communications*, 47, 5130, 2011; Yang H, Ogawa T, Hasegawa D et al., *Journal of Applied Physics*, 103, 07D526, 2008; Hwang SO, Kim CH, Myung Y et al., *The Journal of Physical Chemistry C*, 112, 13911, 2008.)

via thermal decomposition of Fe(acac)$_3$ in a mixed solution of OA and benzyl ether [41]. The anisotropic growth of nanocrystals depends on the reaction time and the monomer concentration.

14.2.2 Structurally Functionalized MNMs

14.2.2.1 One-Dimensional and Two-Dimensional Fe$_3$O$_4$ Nanomaterials

One-dimensional (1-D) nanostructures are promising building blocks owing to their specific properties such as unique electron transport behaviors. Nucleation and anisotropic growth processes are crucial for preparing 1-D Fe$_3$O$_4$ nanostructures [46]. In the nucleation process, octahedral Fe$_3$O$_4$ nanoparticles always form as a nucleus and expose the (111) facet owing to the energetic stability of the (111) facet. However, factors such as catalysts [47], substrates [48], reaction matrix [49], templates [50], and external applied magnetic field [41] will determine the growth of longitude during the elongation process.

Nanorods or nanowires are typical 1-D nanostructures. The fabrication of large-scale, single-crystalline Fe$_3$O$_4$ nanopyramid arrays by utilizing CH$_4$ and N$_2$ plasma sputtering of hematite (0001) wafers without any template or catalyst (Figure 14.5) has been reported [48]. It is believed that the energetic stability of the (111) faces determines the 1-D growth along

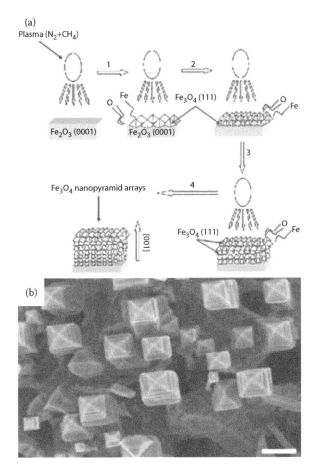

FIGURE 14.5

(a) Schematic growth mechanism of the nanopyramid arrays. (From Liu F, Cao PJ, Zhang HR et al., *Advanced Materials*, 17, 1893, 2005.) (b) Side view of the nanopyramid arrays. Scale bar: 1 μm.

the [100] direction and the incomplete pyramid-like structure resulting from the rapid sputtering velocity. Also, Fe_3O_4 nanowires can be obtained by the reduction of α-Fe_2O_3 nanowires under H_2 and Ar_2 at 400–900°C without significant change of morphology by a V–S process [49]. The oxygen vacancies in α-Fe_2O_3 nanowires played an important role on the structural transformation, maintaining the original morphology while the growth direction was converted from [110] for α-Fe_2O_3 nanowires to [110] for Fe_3O_4 nanowires. When studying the chemical vapor deposition synthesis of 1-D Fe_3O_4 nanostructures, Barth and coworkers found that the growth process was along the <110> direction and the exposed (111) facet [51]. In addition, template-assisted approaches [52,53] and hydrothermal or solvothermal processes [54] can also be used for the fabrication of 1-D Fe_3O_4 nanostructures.

2-D Fe_3O_4 NMs, such as nanorings, nanosheets, and nanoflakes, have also been prepared because of their special properties. Jia and coworkers have developed a method for preparing hematite with controllable morphology by the reduction of hematite in the presence of phosphate and sulfate [35]. The shape of the hematite nanostructure was controllable by changing the ratio of PO_4^{3-}/SO_4^{2-} because of their stronger adhesion to the (110) and (100) planes than the (001) plane [23]. Therefore, the capsule crystals preferred to grow along the [001] direction, and the following dissolution progress also takes place along the [001] direction and simultaneously nanorings can be formed. In addition, other approaches, such as a colloidal-crystal-assisted-lithography strategy [55], thermal transformation processes [23], or miniemulsion polymerization [56], can also be applied for the synthesis of Fe_3O_4 nanorings.

Fe_3O_4 nanosheets can be obtained by oxidizing Fe substrates in acidic solution in a hot plate at 70°C [57]. The fabrication of Fe_3O_4 nanoprisms has been reported by Hou's and Zhang's groups using solvothermal or hydrothermal processes [58,59]. The morphology of the Fe_3O_4 nanoprisms is controllable by varying the volume ratio between ethylene glycol (EG) and 1,3-propanediamine because of the coordination between the $-NH_2$ group and Fe^{3+} [41]. Fe_3O_4 nanoplates can be prepared by using supercritical fluid methods [60]. Lu et al. reported the fabrication of γ-Fe_2O_3 nanoplates and then converting to Fe_3O_4 nanoplates via a reduction process in the presence of poly(vinylpyrrolidone) [61]. In addition, hydrothermal or solvothermal processes have also been developed for the synthesis of Fe_3O_4 nanoprisms or nanoplates [62].

14.2.2.2 Hierarchical Superstructures

Recently, many efforts have been devoted to the self-assembly of nanoscale building blocks into 2-D and 3-D hierarchical superstructures for preventing the agglomeration and providing more tunable and unique properties [63]. In addition, Fe_3O_4 hierarchical superstructures can partially overcome the "superparamagnetic limit," which is the conflict between reducing the magnetic energy barrier and decreasing the size [50].

Quite a few self-assembled Fe_3O_4 chains have been synthesized (Figure 14.6a) on the basis of the nanoscale Kirkendall effect, in which Fe nanoparticles are first self-assembled into chainlike structures and then solid Fe spheres can be gradually oxidized into Fe_3O_4 hollow nanospheres [64–66]. The hierarchical and porous structure of Fe_3O_4 hollow submicrospheres with Fe_3O_4 nanoparticles have been fabricated by using a solvothermal method [67]. In this system, the formation of Fe_3O_4 submicrospheres composed of Fe_3O_4 nanoparticles with diameters ranging from 20 to 30 nm (Figure 14.6b) can be attributed to reduction and Ostwald ripening. Hierarchical flower-like Fe_3O_4 superstructures are another important class of hierarchical superstructures studied [63,68–70]. Zhong et al. have reported the fabrication of flower-like Fe_3O_4 superstructures using an EG-mediated self-assembly

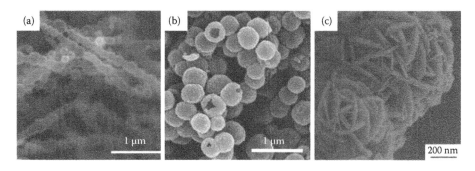

FIGURE 14.6
(a) TEM images of the chainlike arrays of Fe_3O_4 hollow nanospheres. (From Gong J, Li S, Zhang D et al., *Chemical Communications*, 46, 3514, 2010.) (b) SEM image of the hierarchical porous Fe_3O_4 hollow submicrospheres. (From Wang Y, Zhu Q, Tao L, *CrystEngComm*, 13, 4652, 2011.) (c) SEM and TEM images of the flower-like Fe_3O_4 superstructures. (From Han L, Chen Y, Wei Y, *CrystEngComm*, 14, 4692, 2012.)

process [63]. Flower-like Fe_3O_4 superstructures can also been prepared by using a hydrothermal method without any surfactant or organic solvent (Figure 14.6c) [68] or using an ultrasound-assisted hydrothermal method.

In addition, many other Fe_3O_4 hierarchical superstructures with special morphologies, such as Fe_3O_4 microspheres assembled by tetrahedral nanocrystals [71], porous hollow Fe_3O_4 beads constructed with rodlike nanoparticles [72], and nanoparticles-assembled Fe_3O_4 dendritic patterns [73] have also been fabricated.

14.2.3 Chemical-Functionalized MNMs

14.2.3.1 Monomeric Coating

Functional groups, including carboxylates, phosphates, and sulfates, are known to bind to the surface of magnetites [74]. Their introduction is favorable for both dispersibility into aqueous media and the adsorption of toxic compounds because of increasing the active polar groups.

Citric acid can be adsorbed on the surface of the magnetite nanoparticles by coordinating via one or two of the carboxylate functionalities, depending on steric necessity and the curvature of the surface [75]. At least one carboxylic acid group can be exposed to the solvent, keeping the surface negatively charged and hydrophilic. It has been indicated that carboxylates have important effects on the growth of iron oxide nanoparticles and their magnetic properties [76,77]. Increasing concentrations of citric acid can lead to significant decreases in the crystallinity of the iron oxides formed and the presence of citrate can result in changes in the surface geometry. Other coating molecules, such as gluconic acid, dimercaptosuccinic acid, and phosphorylcholine, have also been used as the coatings of iron oxide.

Alkane sulfonic and alkane phosphonic acid surfactants have been applied as efficient binding ligands on the surface of Fe_2O_3 nanoparticles for dispersion in organic solvents [78]. Two possible bonding schemes for the phosphonate ions on Fe^{3+} have been proposed by Yee et al. [79]; that is, one or two O atoms of the phosphonate group binding onto the surface. Furthermore, the phosphate ions form bidentate complexes with adjacent sites on the iron oxide surface [80].

14.2.3.2 Inorganic Materials

Many inorganic materials such as silica [81], gold [82], or gadolinium(III) [83] have been exploited to coat iron oxide nanoparticles not only to prevent the oxidation of iron oxide but also to help bind various ligands to the nanoparticle surface. Such iron-based MNMs have an inner iron oxide core and an outer metallic shell of inorganic materials.

Silica has often been exploited as a coating material for magnetic nanoparticles (MNPs) [84]. The inert silica coating on the surface of MNMs prevents their aggregation in liquid, improves their chemical stability, and helps bind various ligands. The nanoparticles are negatively charged and can be used for the removal of heavy metal cations. Three different approaches have been developed to synthesize magnetic silica nanospheres. The first method refers to the well-known Stöber process, in which silica was formed *in situ* by the hydrolysis and condensation of a sol–gel precursor, such as tetraethyl orthosilicate (TEOS) [85,86]. For example, silica colloids loaded with SPIONs can be fabricated by using this process [87]. The final size of silica colloids depends on the concentration of iron oxide nanoparticles and the type of solvent because the size of silica is closely related to the number of seeds. The second method relies on deposition of silica from silicic acid solution [88]. It has been proven that the silicic acid method is more efficient for covering a higher proportion of the magnetite surface than the TEOS method [89]. The particle size can be controlled from tens to several hundreds of nanometers by changing the ratio of SiO_2/Fe_3O_4 or repeating the coating procedure [90]. The third method is an emulsion method, in which the silica coating is confined and controlled by the introduction of micelles or inverse micelles [91,92]. In addition, the pyrolysis method has also been proposed for fabricating submicronic silica-coated magnetic-sphere aerosols [93,94].

Owing to the presence of surface silanol groups, various coupling agents can be attached to these magnetic particles by covalently bonding between silanol groups and specific ligands of coupling agents [95,96]. For example, amine groups have often been introduced on the surface of silica-coated magnetite nanoparticles by hydrolysis and condensation of an organosilane [97–99].

Gold is another important inorganic coating highly adequate to implement functionality to MNPs. To date, some protocols have been developed for fabricating MNPs coated with gold [100–103]. For example, core–shell-structured Fe/Au nanoparticles have been prepared by a reverse-micelle approach [100]. Water-soluble Au-coated magnetite nanoparticles can be prepared by the reduction of Au(III) onto the surface through iterative hydroxylamine seeding [101]. The Au shell is favorable for protecting the Fe core and providing further organic functionalization. Magnetic gold nanoshells have been developed: Fe_3O_4 nanoparticles was stabilized by oleic acid and 2-bromo-2-propionic acid, and gold seed nanoparticles can be attached to amino-modified silica particles; then, the growth of a complete gold shell ultimately leads to the formation of superparamagnetic gold nanoshells [102]. Recently, Desai et al. have also reported the fabrication of superparamagnetic FeAs@C core–shell nanoparticles with a fairly high blocking temperature (T_B) via a hot injection precipitation technique—the proposed growth mechanism is shown in Figure 14.7 [103]. The synthesis involved the use of triphenylarsine (TPA) and Fe(CO)$_5$ as the Fe and As precursor, respectively, and hexadecylamine (HDA) as a surfactant. TPA reacts with Fe(CO)$_5$ by ligand displacement at moderate temperatures (300°C). Addition of HDA assists in the

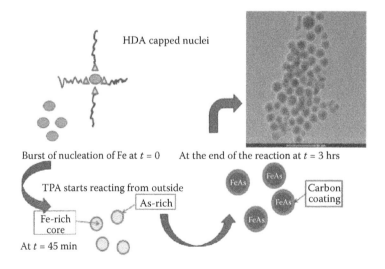

FIGURE 14.7
Proposed growth mechanism of FeAs@C nanoparticles. (From Desai P, Song K, Koza JA et al., *Chemistry of Materials*, DOI: 10.1021/cm303632c, 2013.)

formation of the nanoparticles, owing to its coordinating ability and low melting point that provides a molten flux–like condition, making this synthesis a solventless method. The decomposition of carbonaceous precursors, HDA, TPA, and $Fe(CO)_5$, leads to the formation of the carbonaceous shell coating the FeAs nanoparticles.

14.2.3.3 Polymer Stabilizers

Both *in situ* coatings and postsynthesis coatings have been developed for coating iron oxide nanoparticles [104–107]. For *in situ* coatings, nanoparticles are coated during the synthesis process. The postsynthesis coating method involves grafting the polymeric surfactants. Now, many polymers have been introduced for coating iron oxide nanoparticles, such as dextran, carboxymethylated dextran, carboxydextran, starch, arabinogalactan, glycosaminoglycan, sulfonated styrene–divinylbenzene, polyethylene glycol (PEG), polyvinyl alcohol (PVA), poloxamers, and polyoxamines.

The formation of dextran-covered magnetite was first reported by Molday and Mackenzie by the Molday coprecipitation method with *in situ* coating by dextran-40 (MW = 40,000) [106]. In this case, the dextran was functionalized to create more hydroxyl groups to allow for the binding of the amino groups. Recently, laser pyrolysis and the coprecipitation method have been proposed for fabricating dextran surface modification of pure SPIONs [107]. It is believed that the favored mechanism of adsorption of dextran on the surface of maghemite nanoparticles is the collective hydrogen bonding between dextran hydroxyl groups and the iron oxide particle surface.

PEG, a hydrophilic, water-soluble, biocompatible polymer, is another important functional medium for coating Fe_3O_4 nanoparticles [108,109]. For example, PEG-coated iron oxide nanoparticles can be fabricated via hydrolysis of $FeCl_3 \cdot 6H_2O$ in water and the subsequent treatment with PEG–poly(aspartic acid) block copolymer [110]. PVA coating onto the particle surface can prevent their agglomeration, giving rise to monodisperse particles [111,112]. The surface of Fe_3O_4 nanoparticles can be modified with PVA by precipitation of iron salts in PVA aqueous solution, and the bonding of PVA onto the surface is inevitable [113]. As is known, PVA is a unique synthetic polymer that can transform into a polymer gel [114]. Albornoz and Jacobo have reported the fabrication of an aqueous ferrofluid and the preparation of a magnetic gel with PVA and glutaraldehyde [115].

Since alginate is an electrolytic polysaccharide with many carboxyl groups, it is expected that the COO- of alginate and iron ion can interact and that the excess COO- can lead to good dispersion of the MNMs and a high adsorption effect onto heavy metals. Recently, several investigations have referred to the preparation of iron oxide nanoparticles with alginate [116]. The standard chemical synthesis consists of the following three steps: (i) gelation of alginate and ferrous ions, (ii) *in situ* precipitation of ferrous hydroxide by the alkaline treatment of alginate, and (iii) oxidation of ferrous hydroxide by using an oxidizing agent. The method is relatively complex. A new modified two-step coprecipitation method has been developed and the typical iron oxide nanoparticles are Fe_3O_4 with a core diameter of 5–10 nm.

Nowadays, the preparations of MNPs encapsulated in chitosan are also of great interest for its alkaline, nontoxic, hydrophilic, biocompatible, and biodegradable properties [117]. Such kind of MNMs can be fabricated by various approaches such as sonochemical method [118] or embedding Fe_3O_4 in chitosan [119].

Another important approach to synthesize polymeric core–shell MNPs is to use preformed synthetic polymers as a matrix for controlling the formation of magnetic cores [120,121]. For example, Underhill and Liu have offered a synthetic method for preparing an ABC triblock polymer nanosphere template for maghemite formation [122]. In this system, water-dispersible iron oxide nanoparticles with controlled sizes can be obtained by changing the triblock polymer nanosphere template. Recently, polymer gels have also been proposed for the fabrication of MNPs [123]. The nucleation and growth of iron oxide are controllable by the constrained architectures of the polymer gel [124].

The fabrication of MNPs by incorporating the iron oxide particles inside polymer particles by *in situ* polymerization has also been developed [125]. For example, Pich et al. have reported the preparation of magnetic composite particles by a two-step method [126]. In the first step, Fe_3O_4 nanoparticles are prepared, and in the second step, they are encapsulated into formed poly(styrene/acetoacetoxyethyl methacrylate) particles directly during the polymerization process. The morphology of hybrid particles can be changed by changing the monomer/iron oxide ratio.

14.2.3.4 Other Strategies

Some classic and novel adsorbents, such as activated carbon and carbon nanotubes (CNTs), have also been proposed as modified media for the fabrication of multicomponent MNMs because of their strong adsorption capacity onto toxic contaminants [127,128]. For example, iron oxide/carbon magnetic composite has been prepared by using a facile *in situ* one-pot template-free solvothermal reaction with the magnetic precursor of iron(III) in EG media and mesoporous carbon [127]. Magnetic CNTs can be fabricated by Fenton's reagent method without the addition of any cations. In this case, H_2O_2 was added slowly into an Fe^{2+} salts solution mixed with purified CNTs, and the resulting reactants were heated under a nitrogen/hydrogen flow to form magnetic CNTs, with Fe_2O_3 nanoparticles uniformly dispersed on CNTs [128].

14.3 Iron-Based MNMs in Water Treatment

Today, various techniques have been proposed for water treatment, such as adsorption, biotechnology, catalytic processes, membrane processes, ionizing radiation processes, and magnetically assistant processes [8]. When it comes to selecting the optimal method and material for practical applications, factors including quality standard, efficiency, and expenditure must be taken into consideration [129,130]; those featured with flexibility, high efficiency, reusability, environmental security and friendliness, and economic feasibility [130,131]. Since magnetism is a unique and promising physical property that independently helps in water purification, so adsorption/catalysis procedures combined with magnetic separation have been extensively proposed and developed for water treatment and environmental cleanup [8,132]. The MNMs have been proven promising on an industrial scale owing to their low cost, strong adsorption capacity, easy separation, and enhanced capacity [133–135]. It has been reported that series of organic and inorganic matters (e.g., metal ions, anti-inflammatory drugs, antibiotics, analgesics, pesticides, insecticides, dyes, surfactants, carcinogens and phenolic compounds, etc.) can be effectively removed from aqueous solutions by using MNMs. According to the removal mechanism, these MNMs can be divided into two categories, i.e., adsorbents and catalysts. In many cases, both adsorptive and catalytic processes are involved in practical water treatment.

14.3.1 Adsorption Technologies

Among water treatment technologies, adsorption is one of the most promising and frequently used techniques owing to its convenient application and superior efficiency. The features of an ideal adsorbent often embrace stronger affinity to target compounds/ions and more binding sites on its surface. Therefore, various MNMs have been synthesized and modified for water treatment by altering the particle size, shape, and surface chemistry, which have been proven efficient in removing a large number of contaminants that can be categorized into three groups, i.e., heavy metals, organic pollutants, and radioactive elements.

14.3.1.1 Sorbents for Heavy Metals

Heavy metal contamination has been of great concern because of its adverse effect on living creatures and its tendency toward bioaccumulation even at low concentrations. Therefore, removal of heavy metals from natural and industrial water is extremely urgent and has attracted considerable attention [136–140]. MNMs are one of the promising candidates for bench-scale research and field applications [141,142]. To date, large numbers of MNMs with/without modification or functionalization proposed have been referred to removal of heavy metals in different forms.

It has been reported that heavy metals in both anion and cation species can be effectively removed from aqueous solutions by using simple magnetic nanocrystals. For example, monodispersed Fe_3O_4 nanocrystals can be directly used to remove both As(III) and As(V) species from aqueous solutions via the strong and specific interactions [143,144]. Fe_3O_4 nanocrystals were also used to remove Pb(II) ions, with a maximum adsorption capacity of 36.0 mg/g, much higher than that of low-cost adsorbents [145]. γ-Fe_2O_3, as the oxidation counterpart of Fe_3O_4, was also proved to be effective in removing As(V) [146] and Cr(VI) [133,147]. The removal efficiency for those heavy metal species depends strongly on the size of the magnetic nanocrystals. The smaller the size of the MNMs, the more favorable it is for metal ions diffusing from solution onto the active sites of the sorbents, which in turn increase the removal efficiency. In a typical separation process using simple nanocrystals as sorbents, the specificity independently refers to the sorption constant K; the pH value is a key parameter on the adsorption capacity because pH controls the surface charge density of sorbents and the chemical nature of metallic ions.

For these simple nanocrystals, there are several drawbacks that are unfavorable for application in water treatment, i.e., aggregation derives from high surface energy of nanoscale dimension, easily etching nature in acid and alkaline environment, as well as the application limitation resulting from the single surface group that has poor selectivity prone to competition by coexisting interfering ions when used in adsorption [148]. Therefore, MNMs were developed by being modified with functional groups (Figure 14.8) (e.g., ligands, surfactants, polymers, silica, and carbons); designing the size and structure; and altering active functional group numbers to improve their stability, capacity, and selectivity [8,25,142,149–158], in which surface modification is a versatile technique favorable not only for protecting magnetic cores from oxidation to retain high saturation magnetism for rapid isolation but also for enhancing their water compatibility and the diffusion rate of heavy metals onto the sorbent surfaces. What is more, it is convenient to graft different functional groups to meet the purpose of improving selectivity and effectiveness in heavy metal removal.

The mechanisms of adsorption of contaminant by those functionalized MNMs mainly involve surface site binding, electrostatic interaction, and modified ligands combination

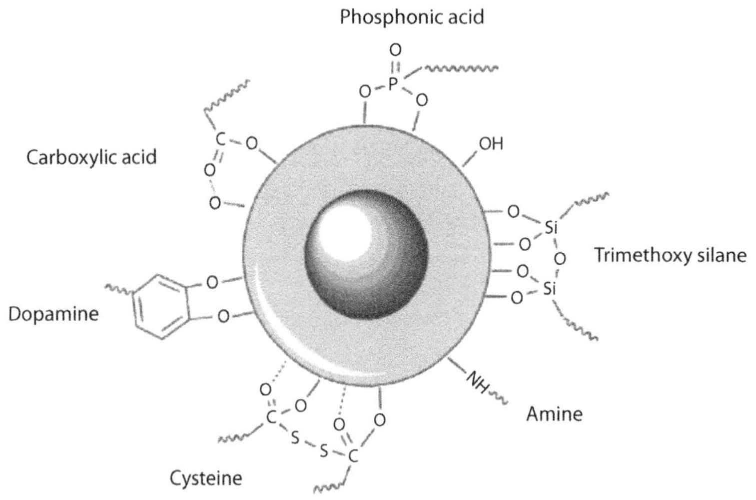

FIGURE 14.8
Common chemical moieties for the anchoring of polymers and functional groups at the surface of iron oxide MNPs. (From Dias A, Hussain A, Marcos AS et al., *Biotechnology Advances*, 29, 142, 2011.)

[154]. For example, the removal of heavy metal ions (such as Pb^{2+} and Hg^{2+}) from industrial wastewater by monodispersed Fe_3O_4–silica core–shell microspheres refers to surface site binding, i.e., the interactions between SiO_2 and heavy metal ions [141]. Such an adsorption process is reversible and the binding ions can be desorbed from SiO_2 in weak acidic water solution under ultrasound radiation. The removal of heavy metal ions (Cd^{2+}, Zn^{2+}, Pb^{2+}, and Cu^{2+}) using MNPs modified with 3-aminopropyltriethoxysilane (APS) and copolymers of acrylic acid (AA) and crotonic acid (CA) rely on the electrostatic interactions between $-COO^-$ and M^{2+} (Figure 14.9) [157]. Adsorption of heavy metals (Cu^{2+}, Cd^{2+}, Pb^{2+}, and Hg^{2+}) by Fe_3O_4 MNPs coated with humic acid also fall into this kind of mechanism, in which the electrostatic interactions between $-COO^-$ of humic acids and M^{2+} contribute partly to the adsorption [158]. To selectively and effectively remove heavy metals from aqueous solutions, mercapto and amine groups have often been proposed as important functional groups for synthesizing MNMs of high quality owing to their stronger ligand combinations on certain heavy metal ions. Thiol-containing polymer-encapsulated MNPs can effectively

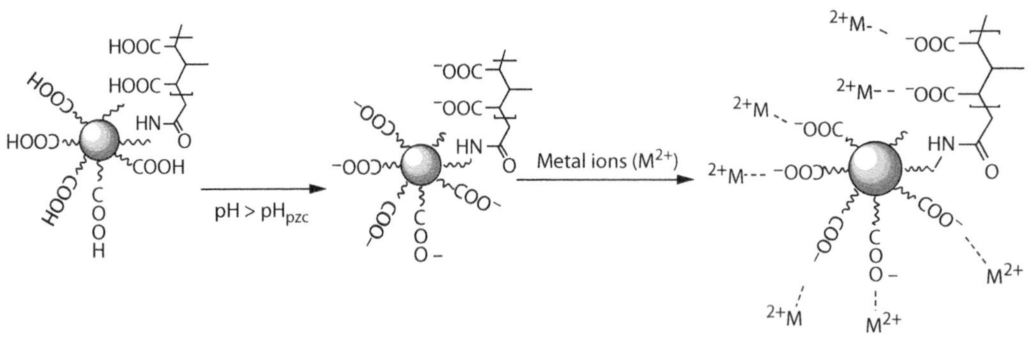

FIGURE 14.9
Schematic of the possible mechanism for adsorption of metal ions by Fe_3O_4@APS@AA-co-CA. (From Ge F, Li MM, Ye H et al., *Journal of Hazardous Materials*, 211, 366, 2012.)

adsorb Ag^+, Hg^{2+}, and Pb^{2+} from aqueous solutions [159]. Thiol-functionalized magnetic mesoporous silica was further developed to remove Pb^{2+} and Hg^{2+} from aqueous solutions, with an adsorption capacity of as high as 260 and 91.5 mg/g of sorbent, respectively [160]. It was found that the removal efficiency of Hg(II) using mercapto-functionalized nanomagnetic Fe_3O_4 polymers was highly pH dependent and related to the content of the Fe_3O_4 core [161]. An amino group was also grafted onto MNPs and proposed for removing Cu(II) and Cr(VI) from water, with an adsorption capacity of 12.43 and 11.24 mg/g, respectively [162]. Magnetic chitosan nanocomposites have been synthesized on the basis of amine-functionalized magnetite nanoparticles and applied for the removal of heavy metals [163]. Generally speaking, all those prepared nanosorbents have presented good reusability and stability in water treatment, suggesting their potential for practical applications.

However, most of the technologies on iron-based MNMs for removing heavy metals are still at a relatively early stage. More work is needed to advance knowledge of these sorbents, including the factors of practical environmental conditions and the supply of the strong magnetic field for their transfer from laboratory- to field-scale application.

14.3.1.2 Sorbents for Organic Contaminants

In the past decades, large numbers of MNMs based on magnetite (Fe_3O_4) or maghemite (γ-Fe_2O_3) have been proposed for removing organic compounds from aqueous solutions [5,142,164]. It has been reported that some MNMs with unique structures exhibit satisfying performance in removing some organic compounds existing in ionic forms. Azo dye contaminants, a kind of widespread and severe pollutant in many places worldwide, are typical target organic compounds for magnetic sorbents. In these cases, it is electrostatic attraction between the iron oxide surface and the compound species in solution that is responsible for adsorption [165]. For example, it was reported that Fe_3O_4 hollow nanospheres present an effective removal for red dye, with the maximum adsorption capacity reaching 90 mg/g [142]. 3-D flower-like iron oxide nanostructures have been proven an outstanding sorbent for removal of Orange II, with the removal capacity of 43.5 mg/g [63].

Exchange reactions are another mechanism that accounts for the adsorption of some organic contaminants. Such adsorption processes can often be divided into two steps: surface exchange reactions first take place until the surface functional sites are fully occupied, and then during the second step, the contaminants diffuse into the sorbents for further inner surface interactions [15,166,167]. Sorbents in processes are usually not single magnetic nanocrystals, but the composite magnetic sorbents are either modified with various functional groups to improve affinity toward contaminants or are constructed with unique structures to prevent aggregation and supply particular transfer paths. Usually, magnetic seeding is only used to realize the magnetic susceptibility and the concomitant comfortable separation from aqueous solutions, while functional groups play the role of selecting and adsorbing organic contaminants. To date, series of organic compounds such as dyes, pesticides, phenols, and some emerging organic pollutants, including antibiotics and endocrine-disrupting chemicals, have been proven to fall into such separation processes.

Activated carbon, one of the attractive and inexpensive materials for removing contaminants owing to its high surface area, porous structure, and easy functionalization, is one of the most frequently funtionalized media for preparing multicomponent MNMs [168,169]. Those magnetic functionalized carbon MNMs have been proven to present good performance in the removal of organic compounds such as chloroform, phenol, chlorobenzene, pesticides, and dyes [170]. For example, Fe_3O_4@C MNMs prepared using powder-activated carbon and Fe_3O_4 nanoparticles can be used to effectively remove amoxicillin

[171]. Fe_3O_4-activated carbon MNPs have been proposed for the removal of aniline, with an adsorption capacity of 90.9 mg/g at pH 6 [172]. Activated carbon/$CoFe_2O_4$ magnetic composite was developed to remove malachite green, and the adsorption capacity reached 89.29 mg/g [173]. Humic substances, dissolved organic matters in water resistant to biodegradation, known as precursors of carcinogenic trihalomethanes, can be effectively removed using magnetic mesoporous carbon, with a removal ratio up to 60% [174]. CNTs are another class of important functionalized medium for modifying magnetic components as a result of their excellent adsorption properties [175–180]. In the work of Qu et al., methylene blue and neutral red were effectively removed by multiwalled CNTs filled with Fe_2O_3 particles; the adsorption capacities were as high as 42.3 and 77.5 mg/g, respectively [181]. In addition, graphene, chitin, and even polymers such as polyaniline have also been proposed as functional media [142,182–185]. In those adsorption processes using chemically functionalized MNMs, the solution pH is usually critical for the removal of organic contaminants because it controls the chemical species of surface-active groups and the contaminants [186].

14.3.1.3 Photocatalytic Technology

Fe_2O_3, with band gap of 2.2 eV, is a promising n-type semiconductive material and a candidate catalyst for photodegradation [187]. Their inexpensive separation and low quantum yield often limit their application for photocatalysis of toxic compounds. Now, considerable efforts, including decreasing photocatalyst size to increase surface area, combining photocatalyst with some novel metal nanoparticles, and increasing hole concentration through doping, have been developed to enhance photocatalytic activity [131]. In addition, improved charge separation and inhibition of charge carrier recombination have also been taken into account for improving the overall quantum efficiency for interfacial charge transfer [188–190].

It has been reported that some species of Fe(III) oxides (i.e., α-Fe_2O_3, γ-Fe_2O_3, α-FeOOH, β-FeOOH, and γ-FeOOH) present potential for degrading organic pollutants and reducing their toxicity via the enhanced photocatalysis effect [191]. For example, electron–hole pairs can be generated through the narrow band gap of Fe_2O_3 under illumination (Equation 14.1) [192].

$$Fe_2O_3 + h\nu \rightarrow Fe_2O_3 \left(e_{cb}^- + h_{vb}^+ \right) \tag{14.1}$$

Those MNMs are illustrative to manipulate the catalytic properties of iron oxide for photocatalysis. Khedr et al. have reported that iron oxide nanoparticles synthesized by thermal evaporation and coprecipitation method can be used to photodegrade Congo red dye ($C_{32}H_{24}N_6O_6S_2$) [193]. During this process, irradiation has no pronounced effect on the catalytic decomposition capacity. Danielsen and Hayes reported the reduction of carbon tetrachloride (CCl_4) using synthetic magnetite. In this system, the primary reaction product is carbon monoxide (CO), followed by chloroform ($CHCl_3$) [194].

Although iron oxide NMs themselves have been widely used as photocatalysts, their activities frequently decline as a result of the electron–hole charge recombination on the surface at levels of nanoseconds [195]. Deposition or coating a noble metal on a magnetite (Fe_3O_4) or maghemite (γ-Fe_2O_3) surface is one of the effective approaches for solving this problem. For example, gold/iron oxide aerogels have been prepared as photocatalysts for degrading disperse Blue 79 azo dye in aqueous solution under ultraviolet light illumination

[196]. In this system, the addition of gold species caused the light absorption to shift toward the red visible region, with the band-gap energy declining. Magnetically recyclable Fe@Pt core–shell nanoparticles have been proposed as electrocatalysts for ammonia borane oxidation, which is far more active (by up to 354%) than the commercial Pt/C catalysts [197]. Different from altering band gaps of the photocatalysts, our groups have proposed a semiconductor–insulator–semiconductor structure of Fe_3O_4@TiO_2@polypyrrole for degradation of Orange II. Under radiation conditions, the photo-induced internal electric field, with electrons accumulated in the outer layer of Fe_3O_4 microspheres and correspondingly more holes concentrated in the inner layers of the PPy coating, can greatly enhance the transfer of the photo-induced electrons to the surface of the PPy coating, resulting in the enhancement of the separation of photo-induced carriers and more •OH radicals produced through e^- paths (unpublished data).

Owing to its narrow band gap, Fe_2O_3 has also been proposed as a sensitizer of photocatalysts, including TiO_2 nanoparticles [187,198]. Electrons in the valence bands of TiO_2 can be driven into Fe_2O_3 because of the formation of the built-in field in an Fe_2O_3–TiO_2 heterojunction. The increase in the electron–hole recombination time can promote photocatalytic activity of the composition [199,200].

Recently, a novel photo-Fenton-like system with the existence of iron oxides and oxalate has been proposed for water treatment. For example, the heterogeneous iron oxide–oxalate system, with iron oxides mainly acting as a photocatalyst and oxalic acid being excited to generate electron–hole pairs, has been prepared for the degradation of Orange I [201]. In this system, a strong ligand-to-metal charge transformation ability is attributed to the degradation of Orange I and the photochemical reduction of the Fe(III) complex is coupled to a Fenton reaction.

14.4 Conclusions

During the past decades, various iron-based MNMs, such as simple crystals, structure- and chemical-functionalized MNMs, with a wide range of compositions and tunable sizes, have been fabricated by developing synthesizing approaches including series of chemical, physical, and biological methods for their promising applications ranging from biotechnology to environmental remediation. In water treatment, employing iron-based MNMs to adsorb heavy metals and organic pollutants is one of the most attractive and successful applications owing to their high selectivity and effectiveness. What is more, MNMs used as photocatalysts have also presented their superiority even with a source of visible light. However, their studies/applications are still at a relatively early stage and many techniques are still at an experimental or pilot stage. More work including advancing the combination of the superior adsorption performance and magnetic properties of iron oxide to meet the actual large-scale operation is still needed. In addition, many potentially serious issues concerning their environmental fate and potential impacts on human health are inevitable because of their potential applications in many frontiers of the environment. Currently, very little information on their possible background concentrations and physical–chemical forms have been given as a result of limitations in their separation and analytical methods. Therefore, it is urgent to develop accurate and robust methodologies for determining their concentrations and forms in the environment, and then to evaluate their effects on the environment.

References

1. Shinkai M. Functional magnetic particles for medical application. *Journal of Bioscience and Bioengineering*, 2002, 94(6): 606–613.
2. Shah S A, Hashmi M U, Alam S et al. Magnetic and bioactivity evaluation of ferrimagnetic ZnFe$_2$O$_4$ containing glass ceramics for the hyperthermia treatment of cancer. *Journal of Magnetism and Magnetic Materials*, 2010, 322(3): 375–381.
3. Samoila P, Slatineanu T, Postolache P et al. The effect of chelating/combustion agent on catalytic activity and magnetic properties of Dy doped Ni-Zn ferrite. *Materials Chemistry and Physics*, 2012, 136(1): 241–246.
4. Cao Y, Bai G, Chen J et al. Preparation and characterization of magnetic microspheres for the purification of interferon α-2b. *Journal of Chromatography B*, 2006, 833(2): 236–244.
5. Zhao X, Wang J, Wu F et al. Removal of fluoride from aqueous media by Fe$_3$O$_4$@Al (OH)$_3$ magnetic nanoparticles. *Journal of Hazardous Materials*, 2010, 173(1): 102–109.
6. Nunez L, Buchholz B A, Vandegrift G F. Waste remediation using in situ magnetically assisted chemical separation. *Separation Science and Technology*, 1995, 30(7–9): 1455–1471.
7. Nunez L, Buchholz B A, Kaminski M et al. Actinide separation of high-level waste using solvent extractants on magnetic microparticles. *Separation Science and Technology*, 1996, 31(10): 1393–1407.
8. Ambashta R D, Sillanpää M. Water purification using magnetic assistance: A review. *Journal of Hazardous Materials*, 2010, 180(1): 38–49.
9. Tanaka T, Matsunaga T. Detection of HbA$_{1c}$ by boronate affinity immunoassay using bacterial magnetic particles. *Biosensors and Bioelectronics*, 2001, 16(9): 1089–1094.
10. Richardson J, Hawkins P, Luxton R. The use of coated paramagnetic particles as a physical label in a magneto-immunoassay. *Biosensors and Bioelectronics*, 2001, 16(9): 989–993.
11. Kourilov V, Steinitz M. Magnetic-bead enzyme-linked immunosorbent assay verifies adsorption of ligand and epitope accessibility. *Analytical Biochemistry*, 2002, 311(2): 166–170.
12. Okamoto Y, Kitagawa F, Otsuka K. Online concentration and affinity separation of biomolecules using multifunctional particles in capillary electrophoresis under magnetic field. *Analytical Chemistry*, 2007, 79(8): 3041–3047.
13. Hartikainen T, Nikkanen J P, Mikkonen R. Magnetic separation of industrial waste waters as an environmental application of superconductivity. *Applied Superconductivity, IEEE Transactions on*, 2005, 15(2): 2336–2339.
14. Xu J S, Zhu Y J. γ-Fe$_2$O$_3$ and Fe$_3$O$_4$ magnetic hierarchically nanostructured hollow microspheres: Preparation, formation mechanism, magnetic property, and application in water treatment. *Journal of Colloid and Interface Science*, 2012, 385: 58–65.
15. Boyer C, Whittaker M R, Bulmus V et al. The design and utility of polymer-stabilized iron-oxide nanoparticles for nanomedicine applications. *NPG Asia Materials*, 2010, 2(1): 23–30.
16. Dias A, Hussain A, Marcos A S et al. A biotechnological perspective on the application of iron oxide magnetic colloids modified with polysaccharides. *Biotechnology Advances*, 2011, 29(1): 142–155.
17. Kanel S R, Nepal D, Manning B et al. Transport of surface-modified iron nanoparticle in porous media and application to arsenic (III) remediation. *Journal of Nanoparticle Research*, 2007, 9(5): 725–735.
18. Schrick B, Hydutsky B W, Blough J L et al. Delivery vehicles for zerovalent metal nanoparticles in soil and groundwater. *Chemistry of Materials*, 2004, 16(11): 2187–2193.
19. Tiraferri A, Chen K L, Sethi R et al. Reduced aggregation and sedimentation of zero-valent iron nanoparticles in the presence of guar gum. *Journal of Colloid and Interface Science*, 2008, 324(1): 71–79.
20. Jeong U, Teng X, Wang Y et al. Superparamagnetic colloids: Controlled synthesis and niche applications. *Advanced Materials*, 2007, 19(1): 33–60.

21. Hassanjani-Roshan A, Vaezi M R, Shokuhfar A et al. Synthesis of iron oxide nanoparticles via sonochemical method and their characterization. *Particuology*, 2011, 9(1): 95–99.

22. Zhong J, Cao C. Nearly monodisperse hollow Fe_2O_3 nanoovals: Synthesis, magnetic property and applications in photocatalysis and gas sensors. *Sensors and Actuators B: Chemical*, 2010, 145(2): 651–656.

23. Fan H, Zhang T, Xu X et al. Fabrication of *n*-type Fe_2O_3 and *p*-type $LaFeO_3$ nanobelts by electrospinning and determination of gas-sensing properties. *Sensors and Actuators B: Chemical*, 2011, 153(1): 83–88.

24. Gotić M, Dražić G, Musić S. Hydrothermal synthesis of α-Fe_2O_3 nanorings with the help of divalent metal cations, Mn^{2+}, Cu^{2+}, Zn^{2+} and Ni^{2+}. *Journal of Molecular Structure*, 2011, 993(1): 167–176.

25. Girginova P I, Daniel-da-Silva A L, Lopes C B et al. Silica coated magnetite particles for magnetic removal of Hg^{2+} from water. *Journal of Colloid and Interface Science*, 2010, 345(2): 234–240.

26. White B R, Stackhouse B T, Holcombe J A. Magnetic γ-Fe_2O_3 nanoparticles coated with poly-l-cysteine for chelation of As(III), Cu(II), Cd(II), Ni(II), Pb(II) and Zn(II). *Journal of Hazardous Materials*, 2009, 161(2): 848–853.

27. Mahmoudi M, Sant S, Wang B et al. Superparamagnetic iron oxide nanoparticles (SPIONs): Development, surface modification and applications in chemotherapy. *Advanced Drug Delivery Reviews*, 2011, 63(1): 24–46.

28. Tartaj P, del Puerto Morales M, Veintemillas-Verdaguer S et al. The preparation of magnetic nanoparticles for applications in biomedicine. *Journal of Physics D: Applied Physics*, 2003, 36(13): 182–197.

29. Cornell R M, Schwertmann U. *The Iron Oxides: Structure, Properties, Reactions, Occurrences and Uses.* Wiley-Vch Verlag GmbH & Co. KGaA, Weinheim, 2003.

30. Yang C, Wu J, Hou Y. Fe_3O_4 nanostructures: Synthesis, growth mechanism, properties and applications. *Chemical Communications*, 2011, 47(18): 5130–5141.

31. Sun S, Zeng H. Size-controlled synthesis of magnetite nanoparticles. *Journal of the American Chemical Society*, 2002, 124(28): 8204–8205.

32. Wang Z L. Transmission electron microscopy of shape-controlled nanocrystals and their assemblies. *Journal of Physical Chemistry B*, 2000, 104(6): 1153–1175.

33. Zhang L, Wu J, Liao H et al. Octahedral Fe_3O_4 nanoparticles and their assembled structures. *Chemical Communications*, 2009, (29): 4378–4380.

34. Zhang D, Liu Z, Han S et al. Magnetite (Fe_3O_4) core-shell nanowires: Synthesis and magnetoresistance. *Nano Letters*, 2004, 4(11): 2151–2155.

35. Jia C J, Sun L D, Luo F et al. Large-scale synthesis of single-crystalline iron oxide magnetic nanorings. *Journal of the American Chemical Society*, 2008, 130(50): 16968–16977.

36. Xu Z, Shen C, Hou Y et al. Oleylamine as both reducing agent and stabilizer in a facile synthesis of magnetite nanoparticles. *Chemistry of Materials*, 2009, 21(9): 1778–1780.

37. Gao G, Huang P, Zhang Y et al. Gram scale synthesis of superparamagnetic Fe_3O_4 nanoparticles and fluid via a facile solvothermal route. *Crystal Engineering Communication*, 2011, 13(6): 1782–1785.

38. Park J, An K, Hwang Y et al. Ultra-large-scale syntheses of monodisperse nanocrystals. *Nature Materials*, 2004, 3(12): 891–895.

39. Burda C, Chen X, Narayanan R et al. Chemistry and properties of nanocrystals of different shapes. *Chemical Reviews-Columbus*, 2005, 105(4): 1025–1102.

40. Wiley B, Sun Y, Mayers B et al. Shape-controlled synthesis of metal nanostructures: The case of silver. *Chemistry-A European Journal*, 2005, 11(2): 454–463.

41. Qi H, Chen Q, Wang M et al. Study of self-assembly of octahedral magnetite under an external magnetic field. *Journal of Physical Chemistry C*, 2009, 113(40): 17301–17305.

42. Li X, Liu D, Song S et al. Rhombic dodecahedral Fe_3O_4: Ionic liquid-modulated and microwave-assisted synthesis and their magnetic properties. *Crystal Engineering Communication*, 2011, 13(20): 6017–6020.

43. Ding H L, Zhang Y X, Wang S et al. $Fe_3O_4@SiO_2$ core/shell nanoparticles: The silica coating regulations with a single core for different core sizes and shell thicknesses. *Chemistry of Materials*, 2012, 24(23): 4572–4580.
44. Yang H, Ogawa T, Hasegawa D et al. Synthesis and magnetic properties of monodisperse magnetite nanocubes. *Journal of Applied Physics*, 2008, 103(7): 07D526-07D526-3.
45. Hwang S O, Kim C H, Myung Y et al. Synthesis of vertically aligned manganese-doped Fe_3O_4 nanowire arrays and their excellent room-temperature gas sensing ability. *Journal of Physical Chemistry C*, 2008, 112(36): 13911–13916.
46. Liao Z M, Li Y D, Xu J et al. Spin-filter effect in magnetite nanowire. *Nano Letters*, 2006, 6(6): 1087–1091.
47. Mathur S, Barth S, Werner U et al. Chemical vapor growth of one—dimensional magnetite nanostructures. *Advanced Materials*, 2008, 20(8): 1550–1554.
48. Liu F, Cao P J, Zhang H R et al. Novel nanopyramid arrays of magnetite. *Advanced Materials*, 2005, 17(15): 1893–1897.
49. Chueh Y L, Lai M W, Liang J Q et al. Systematic study of the growth of aligned arrays of α-Fe_2O_3 and Fe_3O_4 nanowires by a vapor-solid process. *Advanced Functional Materials*, 2006, 16(17): 2243–2251.
50. Liu Z, Zhang D, Han S et al. Single crystalline magnetite nanotubes. *Journal of the American Chemical Society*, 2005, 127(1): 6–7.
51. Barth S, Estrade S, Hernandez-Ramirez F et al. Studies on surface facets and chemical composition of vapor grown one-dimensional magnetite nanostructures. *Crystal Growth and Design*, 2008, 9(2): 1077–1081.
52. Nath S, Kaittanis C, Ramachandran V et al. Synthesis, magnetic characterization, and sensing applications of novel dextran-coated iron oxide nanorods. *Chemistry of Materials*, 2009, 21(8): 1761–1767.
53. Jagminas A, Mažeika K, Reklaitis J et al. Template synthesis, characterization and transformations of iron nanowires while aging. *Materials Chemistry and Physics*, 2008, 109(1): 82–86.
54. Zhen L, He K, Xu C Y et al. Synthesis and characterization of single-crystalline $MnFe_2O_4$ nanorods via a surfactant-free hydrothermal route. *Journal of Magnetism and Magnetic Materials*, 2008, 320(21): 2672–2675.
55. Sun Z, Li Y, Zhang J et al. A universal approach to fabricate various nanoring arrays based on a colloidal-crystal-assisted-lithography strategy. *Advanced Functional Materials*, 2008, 18(24): 4036–4042.
56. Cui L, Gu H, Xu H et al. Synthesis and characterization of superparamagnetic composite nanorings. *Materials Letters*, 2006, 60(24): 2929–2932.
57. Chin K C, Chong G L, Poh C K et al. Large-scale synthesis of Fe_3O_4 nanosheets at low temperature. *Journal of Physical Chemistry C*, 2007, 111(26): 9136–9141.
58. Zeng Y, Hao R, Xing B et al. One-pot synthesis of Fe_3O_4 nanoprisms with controlled electrochemical properties. *Chemical Communications*, 2010, 46(22): 3920–3922.
59. Li X, Si Z, Lei Y et al. Direct hydrothermal synthesis of single-crystalline triangular Fe_3O_4 nanoprisms. *Crystal Engineering Communication*, 2010, 12(7): 2060–2063.
60. Li Z, Godsell J F, O'Byrne J P et al. Supercritical fluid synthesis of magnetic hexagonal nanoplatelets of magnetite. *Journal of the American Chemical Society*, 2010, 132(36): 12540–12541.
61. Lu J, Jiao X, Chen D et al. Solvothermal synthesis and characterization of Fe_3O_4 and γ-Fe_2O_3 nanoplates. *Journal of Physical Chemistry C*, 2009, 113(10): 4012–4017.
62. Zhang W D, Xiao H M, Zhu L P et al. Template-free solvothermal synthesis and magnetic properties of novel single-crystalline magnetite nanoplates. *Journal of Alloys and Compounds*, 2009, 477(1): 736–738.
63. Zhong L S, Hu J S, Liang H P et al. Self—assembled 3D flowerlike iron oxide nanostructures and their application in water treatment. *Advanced Materials*, 2006, 18(18): 2426–2431.
64. Zhang Y, Sun L, Fu Y et al. The shape anisotropy in the magnetic field-assisted self-assembly chain-like structure of magnetite. *Journal of Physical Chemistry C*, 2009, 113(19): 8152–8157.

65. Gong J, Li S, Zhang D et al. High quality self-assembly magnetite (Fe$_3$O$_4$) chain-like core-shell nanowires with luminescence synthesized by a facile one-pot hydrothermal process. *Chemical Communications*, 2010, 46(20): 3514–3516.

66. Huang J, Chen W, Zhao W et al. One-dimensional chainlike arrays of Fe$_3$O$_4$ hollow nanospheres synthesized by aging iron nanoparticles in aqueous solution. *Journal of Physical Chemistry C*, 2009, 113(28): 12067–12071.

67. Wang Y, Zhu Q, Tao L. Fabrication and growth mechanism of hierarchical porous Fe$_3$O$_4$ hollow sub-microspheres and their magnetic properties. *Crystal Engineering Communication*, 2011, 13(14): 4652–4657.

68. Han L, Chen Y, Wei Y. Hierarchical flower-like Fe$_3$O$_4$ and γ–Fe$_2$O$_3$ nanostructures: Synthesis, growth mechanism and photocatalytic properties. *Crystal Engineering Communication*, 2012, 14(14): 4692–4698.

69. Li X, Si Z, Lei Y et al. Hierarchically structured Fe$_3$O$_4$ microspheres: Morphology control and their application in wastewater treatment. *Crystal Engineering Communication*, 2011, 13(2): 642–648.

70. Zeng S, Tang K, Li T et al. Facile route for the fabrication of porous hematite nanoflowers: Its synthesis, growth mechanism, application in the lithium ion battery, and magnetic and photocatalytic properties. *Journal of Physical Chemistry C*, 2008, 112(13): 4836–4843.

71. Lv Y, Wang H, Wang X et al. Synthesis, characterization and growing mechanism of monodisperse Fe$_3$O$_4$ microspheres. *Journal of Crystal Growth*, 2009, 311(13): 3445–3450.

72. Chen Y, Xia H, Lu L et al. Synthesis of porous hollow Fe$_3$O$_4$ beads and their applications in lithium ion batteries. *Journal of Materials Chemistry*, 2012, 22(11): 5006–5012.

73. Dong W, Li X, Shang L et al. Controlled synthesis and self-assembly of dendrite patterns of Fe$_3$O$_4$ nanoparticles. *Nanotechnology*, 2009, 20(3): 035601, doi:10.1088/0957-4484/20/3/035601.

74. Sahoo Y, Pizem H, Fried T et al. Alkyl phosphonate/phosphate coating on magnetite nanoparticles: A comparison with fatty acids. *Langmuir*, 2001, 17(25): 7907–7911.

75. Sahoo Y, Goodarzi A, Swihart M T et al. Aqueous ferrofluid of magnetite nanoparticles: Fluorescence labeling and magnetophoretic control. *Journal of Physical Chemistry B*, 2005, 109(9): 3879–3885.

76. Bee A, Massart R, Neveu S. Synthesis of very fine maghemite particles. *Journal of Magnetism and Magnetic Materials*, 1995, 149(1): 6–9.

77. Liu C, Huang P M. Atomic force microscopy and surface characteristics of iron oxides formed in citrate solutions. *Soil Science Society of America Journal*, 1999, 63(1): 65–72.

78. Kreller D I, Gibson G, Novak W et al. Competitive adsorption of phosphate and carboxylate with natural organic matter on hydrous iron oxides as investigated by chemical force microscopy. *Colloids and Surfaces A: Physicochemical and Engineering Aspects*, 2003, 212(2): 249–264.

79. Yee C, Kataby G, Ulman A et al. Self-assembled monolayers of alkanesulfonic and-phosphonic acids on amorphous iron oxide nanoparticles. *Langmuir*, 1999, 15(21): 7111–7115.

80. Persson P, Nilsson N, Sjöberg S. Structure and bonding of orthophosphate ions at the iron oxide-aqueous interface. *Journal of Colloid and Interface Science*, 1996, 177(1): 263–275.

81. Zhang C, Wängler B, Morgenstern B et al. Silica-and alkoxysilane-coated ultrasmall superparamagnetic iron oxide particles: A promising tool to label cells for magnetic resonance imaging. *Langmuir*, 2007, 23(3): 1427–1434.

82 Chen M, Yamamuro S, Farrell D et al. Gold-coated iron nanoparticles for biomedical applications. *Journal Of Applied Physics*, 2003, 93(10): 7551–7553.

83 Morawski A M, Winter P M, Crowder K C et al. Targeted nanoparticles for quantitative imaging of sparse molecular epitopes with MRI. *Magnetic Resonance in Medicine*, 2004, 51(3): 480–486.

84. Woo K, Hong J, Ahn J P. Synthesis and surface modification of hydrophobic magnetite to processible magnetite@silica-propylamine. *Journal of Magnetism and Magnetic Materials*, 2005, 293(1): 177–181.

85. Lu Y, Yin Y, Mayers B T et al. Modifying the surface properties of superparamagnetic iron oxide nanoparticles through a sol-gel approach. *Nano Letters*, 2002, 2(3): 183–186.

86. Barnakov Y A, Yu M H, Rosenzweig Z. Manipulation of the magnetic properties of magnetite-silica nanocomposite materials by controlled Stöber synthesis. *Langmuir*, 2005, 21(16): 7524–7527.

87. Im S H, Herricks T, Lee Y T et al. Synthesis and characterization of monodisperse silica colloids loaded with superparamagnetic iron oxide nanoparticles. *Chemical Physics Letters*, 2005, 401(1): 19–23.

88. Butterworth M D, Bell S A, Armes S P et al. Synthesis and characterization of polypyrrole-magnetite-silica particles. *Journal of Colloid and Interface Science*, 1996, 183(1): 91–99.

89. Liu X, Xing J, Guan Y et al. Synthesis of amino-silane modified superparamagnetic silica supports and their use for protein immobilization. *Colloids and Surfaces A: Physicochemical and Engineering Aspects*, 2004, 238(1): 127–131.

90. Salazar-Alvarez G. Synthesis, characterisation and applications of iron oxide nanoparticles. PhD thesis (Salazar Alvarez 14857, KTH), Materialvetenskap, Stockholm, Sweden, 2004.

91. Tartaj P, Serna C J. Synthesis of monodisperse superparamagnetic Fe/silica nanospherical composites. *Journal of the American Chemical Society*, 2003, 125(51): 15754–15755.

92. Tartaj P, Serna C J. Microemulsion-assisted synthesis of tunable superparamagnetic composites. *Chemistry of Materials*, 2002, 14(10): 4396–4402.

93. Tartaj P, González-Carreño T, Serna C J. Synthesis of nanomagnets dispersed in colloidal silica cages with applications in chemical separation. *Langmuir*, 2002, 18(12): 4556–4558.

94. Tartaj P, González-Carreño T, Serna C J. Single-step nanoengineering of silica coated maghemite hollow spheres with tunable magnetic properties. *Advanced materials*, 2001, 13(21): 1620–1624.

95. Liu Q, Xu Z, Finch J A et al. A novel two-step silica-coating process for engineering magnetic nanocomposites. *Chemistry of Materials*, 1998, 10(12): 3936–3940.

96. Yang D, Wei K, Liu Q et al. Folic acid-functionalized magnetic $ZnFe_2O_4$ hollow microsphere core/mesoporous silica shell composite particles: Synthesis and application in drug release. *Materials Science and Engineering: C*, 2013, 33(5), 2879–2884.

97. Mornet S, Portier J, Duguet E. A method for synthesis and functionalization of ultrasmall superparamagnetic covalent carriers based on maghemite and dextran. *Journal of Magnetism and Magnetic Materials*, 2005, 293(1): 127–134.

98. del Campo A, Sen T, Lellouche J P et al. Multifunctional magnetite and silica–magnetite nanoparticles: Synthesis, surface activation and applications in life sciences. *Journal of Magnetism and Magnetic Materials*, 2005, 293(1): 33–40.

99. Yamaura M, Camilo R L, Sampaio L C et al. Preparation and characterization of (3-aminopropyl) triethoxysilane-coated magnetite nanoparticles. *Journal of Magnetism and Magnetic Materials*, 2004, 279(2): 210–217.

100. Lin J, Zhou W, Kumbhar A et al. Gold-coated iron (Fe@Au) nanoparticles: Synthesis, characterization, and magnetic field-induced self-assembly. *Journal of Solid State Chemistry*, 2001, 159(1): 26–31.

101. Lyon J L, Fleming D A, Stone M B et al. Synthesis of Fe oxide core/Au shell nanoparticles by iterative hydroxylamine seeding. *Nano Letters*, 2004, 4(4): 719–723.

102. Kim J, Park S, Lee J E et al. Designed fabrication of multifunctional magnetic gold nanoshells and their application to magnetic resonance imaging and photothermal therapy. *Angewandte Chemie*, 2006, 118(46): 7918–7922.

103. Desai P, Song K, Koza J A et al. Soft-chemical synthetic route to superparamagnetic FeAs@ C core-shell nanoparticles exhibiting high blocking temperature. *Chemistry of Materials*, 2013, 25(9): 1510–1518.

104. Berry C C, Wells S, Charles S et al. Dextran and albumin derivatised iron oxide nanoparticles: Influence on fibroblasts in vitro. *Biomaterials*, 2003, 24(25): 4551–4557.

105. Laurent S, Nicotra C, Gossuin Y et al. Influence of the length of the coating molecules on the nuclear magnetic relaxivity of superparamagnetic colloids. *Physica Status Solidi (C)*, 2004, 1(12): 3644–3650.

106. Molday R S, Mackenzie D. Immunospecific ferromagnetic iron-dextran reagents for the labeling and magnetic separation of cells. *Journal of Immunological Methods*, 1982, 52(3): 353–367.

107. Carmen Bautista M, Bomati-Miguel O, del Puerto Morales M et al. Surface characterisation of dextran-coated iron oxide nanoparticles prepared by laser pyrolysis and coprecipitation. *Journal of Magnetism and Magnetic Materials*, 2005, 293(1): 20–27.

108. Shultz M D, Calvin S, Fatouros P P et al. Enhanced ferrite nanoparticles as MRI contrast agents. *Journal of Magnetism and Magnetic Materials*, 2007, 311(1): 464–468.

109. Kohler N, Fryxell G E, Zhang M. A bifunctional poly (ethylene glycol) silane immobilized on metallic oxide-based nanoparticles for conjugation with cell targeting agents. *Journal of the American Chemical Society*, 2004, 126(23): 7206–7211.

110. Schellenberger E A, Bogdanov Jr A, Högemann D et al. Annexin V-CLIO: A nanoparticle for detecting apoptosis by MRI. *Molecular Imaging*, 2002, 1(2): 102.

111. Sairam M, Naidu B V K, Nataraj S K et al. Poly (vinyl alcohol)-iron oxide nanocomposite membranes for pervaporation dehydration of isopropanol, 1, 4-dioxane and tetrahydrofuran. *Journal of Membrane Science*, 2006, 283(1): 65–73.

112. Neuberger T, Schöpf B, Hofmann H et al. Superparamagnetic nanoparticles for biomedical applications: Possibilities and limitations of a new drug delivery system. *Journal of Magnetism and Magnetic Materials*, 2005, 293(1): 483–496.

113. Lee J, Isobe T, Senna M. Preparation of ultrafine Fe_3O_4 particles by precipitation in the presence of PVA at high pH. *Journal of Colloid and Interface Science*, 1996, 177(2): 490–494.

114. Osada Y, Gong J P. Soft and wet materials: Polymer gels. *Advanced Materials*, 1998, 10(11): 827–837.

115. Albornoz C, Jacobo S E. Preparation of a biocompatible magnetic film from an aqueous ferrofluid. *Journal of Magnetism And Magnetic Materials*, 2006, 305(1): 12–15.

116. Nishio Y, Yamada A, Ezaki K et al. Preparation and magnetometric characterization of iron oxide-containing alginate/poly (vinyl alcohol) networks. *Polymer*, 2004, 45(21): 7129–7136.

117. Zhi J, Wang Y, Lu Y et al. In situ preparation of magnetic chitosan/Fe_3O_4 composite nanoparticles in tiny pools of water-in-oil microemulsion. *Reactive and Functional Polymers*, 2006, 66(12): 1552–1558.

118. Hee Kim E, Sook Lee H, Kook Kwak B et al. Synthesis of ferrofluid with magnetic nanoparticles by sonochemical method for MRI contrast agent. *Journal of Magnetism and Magnetic Materials*, 2005, 289: 328–330.

119. Lee H S, Hee Kim E, Shao H et al. Synthesis of SPIO-chitosan microspheres for MRI-detectable embolotherapy. *Journal of Magnetism and Magnetic Materials*, 2005, 293(1): 102–105.

120. Moeser G D, Green W H, Laibinis P E et al. Structure of polymer-stabilized magnetic fluids: Small-angle neutron scattering and mean-field lattice modeling. *Langmuir*, 2004, 20(13): 5223–5234.

121. Sun S, Anders S, Hamann H F et al. Polymer mediated self-assembly of magnetic nanoparticles. *Journal of the American Chemical Society*, 2002, 124(12): 2884–2885.

122. Underhill R S, Liu G. Triblock nanospheres and their use as templates for inorganic nanoparticle preparation. *Chemistry of Materials*, 2000, 12(8): 2082–2091.

123. Zhang J, Xu S, Kumacheva E. Polymer microgels: Reactors for semiconductor, metal, and magnetic nanoparticles. *Journal of the American Chemical Society*, 2004, 126(25): 7908–7914.

124. Breulmann M, Cölfen H, Hentze H P et al. Elastic magnets: Template—controlled mineralization of iron oxide colloids in a sponge—like gel matrix. *Advanced Materials*, 1998, 10(3): 237–241.

125. Sun Y, Ding X, Zheng Z et al. Surface initiated ATRP in the synthesis of iron oxide/polystyrene core/shell nanoparticles. *European Polymer Journal*, 2007, 43(3): 762–772.

126. Pich A, Bhattacharya S, Ghosh A et al. Composite magnetic particles: 2. Encapsulation of iron oxide by surfactant-free emulsion polymerization. *Polymer*, 2005, 46(13): 4596–4603.

127. Shi S, Fan Y, Huang Y. Facile low temperature hydrothermal synthesis of magnetic mesoporous carbon nanocomposite for adsorption removal of ciprofloxacin antibiotics. *Industrial & Engineering Chemistry Research*, 2013, 52(7): 2604–2612.

128. Yu F, Chen J, Chen L et al. Magnetic carbon nanotubes synthesis by Fenton's reagent method and their potential application for removal of azo dye from aqueous solution. *Journal of Colloid and Interface Science*, 2012, 378(1): 175–183.

129. Huang D L, Zeng G M, Feng C L et al. Degradation of lead-contaminated lignocellulosic waste by Phanerochaete chrysosporium and the reduction of lead toxicity. *Environmental Science & Technology*, 2008, 42(13): 4946–4951.

130. Oller I, Malato S, Sánchez-Pérez J A. Combination of advanced oxidation processes and biological treatments for wastewater decontamination-a review. *Science of the Total Environment*, 2011, 409(20): 4141–4166.

131. Zhang L, Fang M. Nanomaterials in pollution trace detection and environmental improvement. *Nano Today*, 2010, 5(2): 128–142.

132. Mahdavian A R, Mirrahimi M A S. Efficient separation of heavy metal cations by anchoring polyacrylic acid on superparamagnetic magnetite nanoparticles through surface modification. *Chemical Engineering Journal*, 2010, 159(1): 264–271.

133. Hu J, Chen G, Lo I. Removal and recovery of Cr (VI) from wastewater by maghemite nanoparticles. *Water Research*, 2005, 39(18): 4528–4536.

134. Carabante I. *Study of Arsenate Adsorption on Iron Oxide by In Situ ATR-FTIR Spectroscopy*. Luleå University of Technology, Luleå, Sweden, 2009.

135. Fan Z, Shelton M, Singh A K et al. Multifunctional plasmonic shell–magnetic core nanoparticles for targeted diagnostics, isolation, and photothermal destruction of tumor cells. *ACS Nano*, 2012, 6(2): 1065–1073.

136. Chen A, Zeng G, Chen G et al. Simultaneous cadmium removal and 2, 4-dichlorophenol degradation from aqueous solutions by Phanerochaete chrysosporium. *Applied Microbiology and Biotechnology*, 2011, 91(3): 811–821.

137. Pang Y, Zeng G, Tang L et al. Preparation and application of stability enhanced magnetic nanoparticles for rapid removal of Cr (VI). *Chemical Engineering Journal*, 2011, 175: 222–227.

138. Plazinski W, Rudzinski W. Modeling the effect of surface heterogeneity in equilibrium of heavy metal ion biosorption by using the ion exchange model. *Environmental Science & Technology*, 2009, 43(19): 7465–7471.

139. Vilensky M Y, Berkowitz B, Warshawsky A. In situ remediation of groundwater contaminated by heavy-and transition-metal ions by selective ion-exchange methods. *Environmental Science & Technology*, 2002, 36(8): 1851–1855.

140. Çoruh S, Şenel G, Ergun O N. A comparison of the properties of natural clinoptilolites and their ion-exchange capacities for silver removal. *Journal of Hazardous Materials*, 2010, 180(1): 486–492.

141. Hu H, Wang Z, Pan L. Synthesis of monodisperse Fe_3O_4@silica core–shell microspheres and their application for removal of heavy metal ions from water. *Journal of Alloys and Compounds*, 2010, 492(1): 656–661.

142. Iram M, Guo C, Guan Y et al. Adsorption and magnetic removal of neutral red dye from aqueous solution using Fe_3O_4 hollow nanospheres. *Journal of Hazardous Materials*, 2010, 181(1): 1039–1050.

143. Yavuz C T, Mayo J T, William W Y et al. Low-field magnetic separation of monodisperse Fe3O4 nanocrystals. *Science*, 2006, 314(5801): 964–967.

144. Mayo J T, Yavuz C, Yean S et al. The effect of nanocrystalline magnetite size on arsenic removal. *Science and Technology of Advanced Materials*, 2007, 8(1): 71–75.

145. Nassar N N. Rapid removal and recovery of Pb (II) from wastewater by magnetic nanoadsorbents. *Journal of Hazardous Materials*, 2010, 184(1): 538–546.

146. Tuutijärvi T, Lu J, Sillanpää M et al. Adsorption mechanism of arsenate on crystal γ-Fe_2O_3 nanoparticles. *Journal of Environmental Engineering*, 2010, 136(9): 897–905.

147. Hu J, Lo I M, Chen G. Removal of Cr (VI) by magnetite nanoparticle. *Water Science and Technology: A journal of the International Association on Water Pollution Research*, 2004, 50(12): 139.

148. Feng Y, Gong J L, Zeng G M et al. Adsorption of Cd (II) and Zn (II) from aqueous solutions using magnetic hydroxyapatite nanoparticles as adsorbents. *Chemical Engineering Journal*, 2010, 162(2): 487–494.

149. Ashtari P, Wang K, Yang X et al. Preconcentration and separation of ultra-trace beryllium using quinalizarine-modified magnetic microparticles. *Analytica Chimica Acta*, 2009, 646(1): 123–127.

150. Huang C, Hu B. Silica-coated magnetic nanoparticles modified with γ-mercaptopropyltrim ethoxysilane for fast and selective solid phase extraction of trace amounts of Cd, Cu, Hg, and Pb in environmental and biological samples prior to their determination by inductively coupled plasma mass spectrometry. *Spectrochimica Acta Part B: Atomic Spectroscopy*, 2008, 63(3): 437–444.

151. Khajeh M, Sanchooli E. Optimization of microwave-assisted extraction procedure for zinc and iron determination in celery by Box–Behnken design. *Food Analytical Methods*, 2010, 3(2): 75–79.

152. Khajeh M. Optimization of microwave-assisted extraction procedure for zinc and copper determination in food samples by Box-Behnken design. *Journal of Food Composition and Analysis*, 2009, 22(4): 343–346.

153. Parham H, Rahbar N. Solid phase extraction–spectrophotometric determination of fluoride in water samples using magnetic iron oxide nanoparticles. *Talanta*, 2009, 80(2): 664–669.

154. Hao R, Xing R, Xu Z et al. Synthesis, functionalization, and biomedical applications of multifunctional magnetic nanoparticles. *Advanced Materials*, 2010, 22(25): 2729–2742.

155. Dawes C C, Jewess P J, Murray D A. Thiophilic paramagnetic particles as a batch separation medium for the purification of antibodies from various source materials. *Analytical Biochemistry*, 2005, 338(2): 186–191.

156. Chiang C L, Sung C S, Wu T F et al. Application of superparamagnetic nanoparticles in purification of plasmid DNA from bacterial cells. *Journal of Chromatography B*, 2005, 822(1): 54–60.

157. Ge F, Li M M, Ye H et al. Effective removal of heavy metal ions Cd^{2+}, Zn^{2+}, Pb^{2+}, Cu^{2+} from aqueous solution by polymer-modified magnetic nanoparticles. *Journal of Hazardous Materials*, 2012, 211: 366–372.

158. Liu J, Zhao Z, Jiang G. Coating Fe_3O_4 magnetic nanoparticles with humic acid for high efficient removal of heavy metals in water. *Environmental Science & Technology*, 2008, 42(18): 6949–6954.

159. Shin S, Jang J. Thiol containing polymer encapsulated magnetic nanoparticles as reusable and efficiently separable adsorbent for heavy metal ions. *Chemical Communications*, 2007, 41: 4230–4232.

160. Li G, Zhao Z, Liu J et al. Effective heavy metal removal from aqueous systems by thiol functionalized magnetic mesoporous silica. *Journal of Hazardous Materials*, 2011, 192(1): 277–283.

161. Pan S, Zhang Y, Shen H et al. An intensive study on the magnetic effect of mercapto-functionalized nano-magnetic Fe_3O_4 polymers and their adsorption mechanism for the removal of Hg (II) from aqueous solution. *Chemical Engineering Journal*, 2012, 210: 564–574.

162. Huang S H, Chen D H. Rapid removal of heavy metal cations and anions from aqueous solutions by an amino-functionalized magnetic nano-adsorbent. *Journal of Hazardous Materials*, 2009, 163(1): 174–179.

163. Liu X, Hu Q, Fang Z et al. Magnetic chitosan nanocomposites: A useful recyclable tool for heavy metal ion removal. *Langmuir*, 2008, 25(1): 3–8.

164. Park J W, Bae K H, Kim C et al. Clustered magnetite nanocrystals cross-linked with PEI for efficient siRNA delivery. *Biomacromolecules*, 2010, 12(2): 457–465.

165. Herrera F, Lopez A, Mascolo G et al. Catalytic decomposition of the reactive dye uniblue A on hematite. Modeling of the reactive surface. *Water Research*, 2001, 35(3): 750–760.

166. Ma Z Y, Guan Y P, Liu X Q et al. Preparation and characterization of micron-sized non—porous magnetic polymer microspheres with immobilized metal affinity ligands by modified suspension polymerization. *Journal of Applied Polymer Science*, 2005, 96(6): 2174–2180.

167. Hu J, Shao D, Chen C et al. Removal of 1-naphthylamine from aqueous solution by multiwall carbon nanotubes/iron oxides/cyclodextrin composite. *Journal of Hazardous Materials*, 2011, 185(1): 463–471.

168. Wang T, Liang L, Wang R et al. Magnetic mesoporous carbon for efficient removal of organic pollutants. *Adsorption*, 2012, 18(5–6): 439–444.

169. Zhu H Y, Fu Y Q, Jiang R et al. Adsorption removal of Congo red onto magnetic cellulose/Fe_3O_4/activated carbon composite: Equilibrium, kinetic and thermodynamic studies. *Chemical Engineering Journal*, 2011, 173: 494–502.

170. Bastami T R, Entezari M H. Activated carbon from carrot dross combined with magnetite nanoparticles for the efficient removal of p-nitrophenol from aqueous solution. *Chemical Engineering Journal*, 2012.

171. Kakavandi B, Esrafili A et al. Magnetic Fe_3O_4@C nanoparticles as adsorbents for removal of amoxicillin from aqueous solution. *Water Science and Technology*, 2014, 69(1): 147–155.

172. Kakavandi B, Jonidi A, Rezaei R et al. Synthesis and properties of Fe_3O_4-activated carbon magnetic nanoparticles for removal of aniline from aqueous solution: Equilibrium, kinetic and thermodynamic studies. *Iranian Journal of Environmental Health Science and Engineering*, 2013, 10(1): 19.

173. Ai L, Huang H, Chen Z et al. Activated carbon/$CoFe_2O_4$ composites: Facile synthesis, magnetic performance and their potential application for the removal of malachite green from water. *Chemical Engineering Journal*, 2010, 156(2): 243–249.

174. Kondo K, Jin T, Miura O. Removal of less biodegradable dissolved organic matters in water by superconducting magnetic separation with magnetic mesoporous carbon. *Physica C: Superconductivity*, 2010, 470(20): 1808–1811.

175. Yu F, Chen J, Chen L et al. Magnetic carbon nanotubes synthesis by Fenton's reagent method and their potential application for removal of azo dye from aqueous solution. *Journal of Colloid and Interface Science*, 2012, 378(1): 175–183.

176. Madrakian T, Afkhami A, Ahmadi M et al. Removal of some cationic dyes from aqueous solutions using magnetic-modified multi-walled carbon nanotubes. *Journal of Hazardous Materials*, 2011, 196: 109–114.

177. Yu F, Chen J, Yang M et al. A facile one-pot method for synthesis of low-cost magnetic carbon nanotubes and their applications for dye removal. *New Journal of Chemistry*, 2012, 36(10): 1940–1943.

178. Abdel Salam M, Gabal M A, Obaid A Y. Preparation and characterization of magnetic multi-walled carbon nanotubes/ferrite nanocomposite and its application for the removal of aniline from aqueous solution. *Synthetic Metals*, 2012, 161(23): 2651–2658.

179. Song X, Yang F, Wang X et al. Preparation of magnetic multi-walled carbon nanotubes and their application in active dye removal. *Micro and Nano Letters, IET*, 2011, 6(10): 827–829.

180. Shao D, Hu J, Chen C et al. Polyaniline multiwalled carbon nanotube magnetic composite prepared by plasma-induced graft technique and its application for removal of aniline and phenol. *The Journal of Physical Chemistry C*, 2010, 114(49): 21524–21530.

181. Qu S, Huang F, Yu S et al. Magnetic removal of dyes from aqueous solution using multi-walled carbon nanotubes filled with Fe_2O_3 particles. *Journal of Hazardous Materials*, 2008, 160(2): 643–647.

182. Wang C, Feng C, Gao Y et al. Preparation of a graphene-based magnetic nanocomposite for the removal of an organic dye from aqueous solution. *Chemical Engineering Journal*, 2011, 173(1): 92–97.

183. Li G, Du Y, Tao Y et al. Iron (II) cross-linked chitin-based gel beads: Preparation, magnetic property and adsorption of methyl orange. *Carbohydrate Polymers*, 2010, 82(3): 706–713.

184. Sun C, Sze R, Zhang M. Folic acid—PEG conjugated superparamagnetic nanoparticles for targeted cellular uptake and detection by MRI. *Journal of Biomedical Materials Research Part A*, 2006, 78(3): 550–557.

185. Tahmasebi E, Yamini Y, Mehdinia A et al. Polyaniline-coated Fe_3O_4 nanoparticles: An anion exchange magnetic sorbent for solid—phase extraction. *Journal of Separation Science*, 2012, 35(17): 2256–2265.

186. Pang K M, Ng S, Chung W K et al. Removal of pentachlorophenol by adsorption on magnetite-immobilized chitin. *Water, Air, and Soil Pollution*, 2007, 183(1–4): 355–365.

187. Akhavan O, Azimirad R. Photocatalytic property of Fe_2O_3 nanograin chains coated by TiO_2 nanolayer in visible light irradiation. *Applied Catalysis A: General*, 2009, 369(1): 77–82.

188. Beydoun D, Amal R, Low G et al. Role of nanoparticles in photocatalysis. *Journal of Nanoparticle Research*, 1999, 1(4): 439–458.

189. Hu X, Li G, Yu J C. Design, fabrication, and modification of nanostructured semiconductor materials for environmental and energy applications. *Langmuir*, 2009, 26(5): 3031–3039.

190. Watson S, Beydoun D, Amal R. Synthesis of a novel magnetic photocatalyst by direct deposition of nanosized TiO_2 crystals onto a magnetic core. *Journal of Photochemistry and Photobiology A: Chemistry*, 2002, 148(1): 303–313.

191. Feng W, Nansheng D, Helin H. Degradation mechanism of azo dye CI reactive red 2 by iron powder reduction and photooxidation in aqueous solutions. *Chemosphere*, 2000, 41(8): 1233–1238.

192. Bandara J, Klehm U, Kiwi J. Raschig rings-Fe_2O_3 composite photocatalyst activate in the degradation of 4-chlorophenol and Orange II under daylight irradiation. *Applied Catalysis B: Environmental*, 2007, 76(1): 73–81.

193. Khedr M H, Abdel Halim K S, Soliman N K. Synthesis and photocatalytic activity of nano-sized iron oxides. *Materials Letters*, 2009, 63(6): 598–601.

194. Danielsen K M, Hayes K F. pH dependence of carbon tetrachloride reductive dechlorination by magnetite. *Environmental Science and Technology*, 2004, 38(18): 4745–4752.

195. Rothenberger G, Moser J, Graetzel M et al. Charge carrier trapping and recombination dynamics in small semiconductor particles. *Journal of the American Chemical Society*, 1985, 107(26): 8054–8059.

196. Wang C T. Photocatalytic activity of nanoparticle gold/iron oxide aerogels for azo dye degradation. *Journal of Non-Crystalline Solids*, 2007, 353(11): 1126–1133.

197. Zhang X B, Yan J M, Han S et al. Magnetically Recyclable Fe@Pt Core-Shell Nanoparticles and Their Use as Electrocatalysts for Ammonia Borane Oxidation: The Role of Crystallinity of the Core. *Journal of the American Chemical Society*, 2009, 131(8): 2778–2779.

198. Zhang X, Lei L. Preparation of photocatalytic Fe_2O_3-TiO_2 coatings in one step by metal organic chemical vapor deposition. *Applied Surface Science*, 2008, 254(8): 2406–2412.

199. Peng L, Xie T, Lu Y et al. Synthesis, photoelectric properties and photocatalytic activity of the Fe_2O_3/TiO_2 heterogeneous photocatalysts. *Physical Chemistry Chemical Physics*, 2010, 12(28): 8033–8041.

200. Shinde S S, Bhosale C H, Rajpure K Y. Photocatalytic activity of sea water using TiO_2 catalyst under solar light. *Journal of Photochemistry and Photobiology B: Biology*, 2011, 103(2): 111–117.

201. Lei J, Liu C, Li F et al. Photodegradation of orange I in the heterogeneous iron oxide-oxalate complex system under UVA irradiation. *Journal of Hazardous Materials*, 2006, 137(2): 1016–1024.

15

Sustainable Clean Water: Closing the Cycle

Satish Vasu Kailas[1] and Monto Mani[2]
[1]*Department of Mechanical Engineering, Indian Institute of Science, Bangalore, India*
[2]*Centre for Sustainable Technologies, Indian Institute of Science, Bangalore, India*

Right from the time we wake up in the morning to the time we sleep in the evening, and even during sleep, we use much more than what we need for survival. All this is done in the name of comfort. Leaving the tap water running while brushing our teeth, taking long showers wasting water, and with water hotter than is required, are but some examples of absolute and avoidable waste. People switch on fans in winter or air conditioners in summer and use thick blankets to keep themselves warm. All of this waste can be completely avoided if we accept this as unnecessary. How to control this waste is more of a philosophical issue. However, why it is that we need to control this urge to use much more is an issue that is much more important. This chapter seeks to address this issue and on how one can make living sustainable. This chapter will also address the issue of how the approach of "closing the cycle" is the answer to sustainable living and sustainable clean water.

The sustenance of humans and their lifestyle depends on the availability of and access to resources. While the need for resources is satisfied through exploitation, and sometimes destruction of the natural environment, in the context of humans it is also accompanied with waste generation and environmental pollution. This resource exploitation and consequent waste and pollution is chiefly responsible for threatening the sustainability of both present and future generations, by disrupting nature's ability to maintain resource availability. Hence, unrestrained resource consumption threatens our own survival and sustainability.

Resources required for a community's progress include *needs*, *necessities*, and *desires* that influence the community's sustainability to varying degrees or levels (Figure 15.1). The first level, being basic or fundamental, determines the very survival and healthy existence of life, and is a prerequisite for other human activities. If air and water supplies are cut off, life as we know it will cease to exist. In fact, the cycle that sustains this planet is the conversion of oxygen to carbon dioxide and back. This cycle, shown below in Figure 15.2, is made possible by animals, which convert oxygen to carbon dioxide, and plants, which convert carbon dioxide to oxygen, with animals and plants being interlinked. This level can be termed as *basic sustainability*. The second level is essential to support modern civilization and its lifestyles, where primary education, employment, and primary health care determine the ability of the community to meet the minimum standards of life (quality of life). Although not much debate exists on issues of education, health, and individual security, a lot of debate does exist on issues such as owning individual means of transport (cars), or televisions, and the like. This level is distinct from the basic sustainability level and can be termed *essential sustainability*. The third level pertains to resource needs to satisfy a

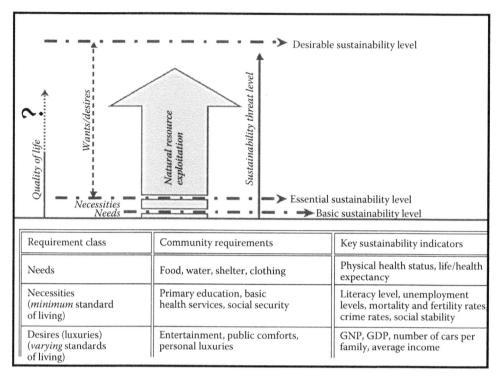

Requirement class	Community requirements	Key sustainability indicators
Needs	Food, water, shelter, clothing	Physical health status, life/health expectancy
Necessities (*minimum* standard of living)	Primary education, basic health services, social security	Literacy level, unemployment levels, mortality and fertility rates, crime rates, social stability
Desires (luxuries) (*varying* standards of living)	Entertainment, public comforts, personal luxuries	GNP, GDP, number of cars per family, average income

FIGURE 15.1
Sustainability levels. (From Mani, M. et al., *Sustainability and Human Settlements*. Sage Publications, New Delhi, Thousand Oaks, London, 2005.)

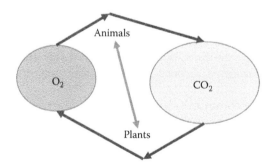

FIGURE 15.2
Simple closed cycle (illustrating that everything in the natural world is cyclic).

community's luxurious wants and desires. These can include artifacts, exotic decorations, lavish clothing including fur, etc.

Modern society should seek to achieve a healthy living environment that enables fulfillment of societal, economic, and social (including political) needs, by trying to achieve a balance between resource needs for human development and protection of environmental vitality. Sustainable development thus has a bipolar objective, the first focusing on equitable human development and the second addressing the issue of maintaining, protecting, and preserving environmental vitality, also keeping in mind the interests of future generations. Figure 15.3 illustrates a comparison between the progression underlying inappropriate development and sustainable development, as two descending and ascending systemic spirals, respectively.

Resources required to fulfill a community's luxuries are far greater than the resources required to fulfill basic needs, as Mahatma Gandhi's ubiquitous quote "Earth can provide for everyone's need, but not for everyone's greed." A distinction must also be made between luxurious living and cultural development. Cultural development implies a positive societal development, when the community has time beyond that required for basic survival, and the time is devoted toward art, architecture, literature, music, and other activities involving the betterment and refinement of human intellect and skills (Vinayak 1989). A luxurious life, on the other hand, need not be sustainable, as it does not necessarily imply a positive societal development. Crime, drugs, and violence are indications of an affluent society going wrong (Mani et al. 2005).

As in *basic sustainability*, the link between plants and animals are essential to sustain *life* on the planet (see Figure 15.2). If one looks at this interdependence, one can clearly see that everything in this planet is interlinked and everything is cyclic. There is nothing that is not a part of a cyclic process and nothing that is completely independent. The only thing that should be realized is the time scale of a cycle, from cycles lasting as low as 10^{-22} s, for a boson (God particle) (Wikipedia 2013a), to cycles that last as high as 10^{18} s, which is the time taken from the "big bang" to the "big collapse" (Wikipedia 2013b). As illustrated in Figure 15.4, the resources in nature also are part of a cycle that are expended and replenished over

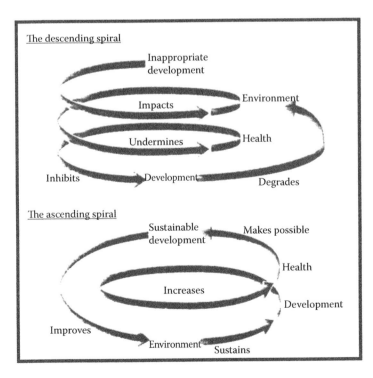

FIGURE 15.3
Comparison between inappropriate development and sustainable development. (From UNEP, *The State of the Environment: Environmental Health*. UN Environment Programme, Nairobi, 1986.)

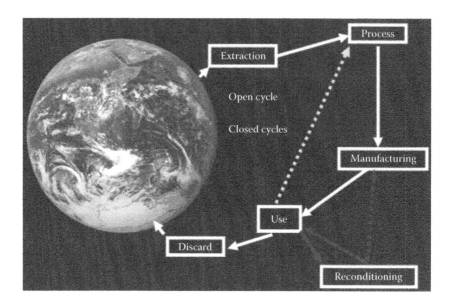

FIGURE 15.4
Material resource cycle (open loop and closed loop, where recycling is given importance, are shown).

varying time scales. Petroleum resources take millennia to get replenished, while oxygen in the air gets replenished within hours (with abundant trees).

From the earth's resources, humans extract/mine required materials, process and refine it, manufacture products, use these products, and eventually discard them. Most of what is discarded goes into landfills and also dumped into the oceans as in the Midway Islands. This cycle is open ended and termed the "open cycle," where one is not aware what happens to the material after it is discarded. As in the Midway Island, all the solid waste, plastics in particular, disintegrate and float on the ocean surface. These are consumed by birds and fledglings, which barely survive a few days.

The other cycle that is more talked about today and what is becoming a necessity are the cycles that are joined by the dashed white and dashed gray arrows (see Figure 15.4). In these cycles, the material end use is reconnected with its original extraction. This constitutes a "closed cycle." Here, the processed and manufactured product is either (i) reconditioned and reused; (ii) reconditioned, part of it remanufactured, and then reused; or (iii) the material is reprocessed, manufactured, and reused. These "closed cycles" tend to be sustainable and are becoming important because there is a limit to the availability of raw materials, a limit to the rate at which we are consuming these materials, and a limit to the carrying capacity of the waste/trash the earth can handle. However, it is not only the material usage that needs to be sustainable but also the energy used to make the product. Unfortunately, in today's industry, most of the energy comes from unsustainable sources, something that cannot continue for long.

Sustainability is the key to mankind's future. Here one should be clear of what an eco-friendly sustainable product is and what a non-eco-friendly unsustainable product is! Eco-friendly products do not cause harm to any natural cycle, for example, harvesting bamboo for construction. Non-eco-friendly products cause harm, to varying degrees, to the environment, e.g., ozone layer depletion caused by use of chlorofluorocarbons. It is important to distinguish between what is "eco-friendly" and what is "biodegradable." An eco-friendly product, such as glass, need not be biodegradable. A biodegradable product, such as genetically modified seeds, need not be eco-friendly. This distinction is crucial, as understanding of this will help in creating sustainable cycles.

The question "what makes a cycle sustainable or unsustainable" can be answered when one understands when should the cycle be "open" and when should it have to be "closed." In the cycle where we extract from nature and dispose waste into nature after use, the rate of extraction should be lower or equal to the rate at which nature can replenish the resource, for example, harvesting timber from forests. Another example is extraction of crude oil, where humans are consuming in a few hundred years a resource that has taken nature millennia to generate. At the same time, nature is unable to quickly reprocess and manage the release of huge quantities of waste exhaust CO_2 gases. Such resource use and waste generation is unsustainable and causes severe environmental disruption such as climate change. This clearly indicates that our current lifestyle and most of the products that we extensively use are unsustainable. Furthermore, in addition to the materials that go into making a product, the energy going into the manufacture should also be from sustainable sources. This energy is termed *embodied energy*. For example, the energy going into the making of photovoltaic cells (that generate energy) takes nearly a decade to be recovered through the energy generated from it. Unfortunately, the resource requirement and embodied energy of modern-day living is so intense that it is just impossible for nature to support even for a few more decades, unless our lifestyles change to a closed-loop resource cycle. Current lifestyles, as we know, simply do not fall in this category. This is indeed a challenge to scientists of today to make products that follow the closed loop and not the open loop.

The following paragraphs discuss some examples of modern-day approaches that have vitiated natural closed-loop systems resulting in vicious open systems, leading to difficult-to-manage problems and unsustainability. Furthermore, solutions have been sought that are also *open* and resulting in more complex problems. The current water crisis and severe pollution is an indicator of a closed cycle becoming open. Figure 15.5 illustrates the nature

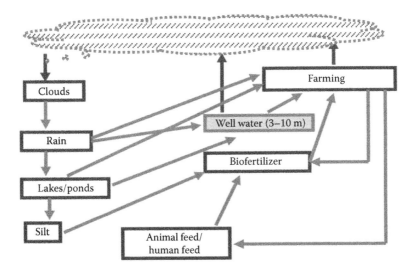

FIGURE 15.5
Closed-loop traditional farming cycle. Black arrows indicate water evaporation and condensation, which is a part of the closed natural hydrological cycle.

of traditional farming practiced for more than a few thousand years, until four decades back (1950s) in India and a hundred years in the Western industrialized countries.

The water that evaporates from various sources, including the oceans, condenses to form clouds and precipitates as rain. Rain is the essential and eternal requirement for any farming to take place. This rainwater collects in lakes and ponds and recharges the groundwater. This water was drawn (initially by sustainable means of animal and human power) during the nonrainy seasons for farming. It is important to note that the water table in most places around India was never lower than 10 m. In places around Bangalore (Karnataka, India), it was accessible within a few meters. Another point to note is the fact that the fertilizers used were biofertilizers (derived from abundant crop biodiversity) and not chemical/petroleum-derived fertilizers, simply because the latter were not available. The lake silt was also harvested annually to increase the water-storage capacity and also to fertilize the farms. This traditional sustainable close-loop system was altered to adopt the *Green Revolution*, which extensively relied on limited high-yield crops and extensive use of chemical fertilizer. Use of chemical fertilizers vitiates the soil's natural ability to generate and retain nutrients, and consequently compromised the closed-loop system (see Figure 15.6).

The use of chemical fertilizers resulted in two consequences: first, it opened up the closed-loop traditional cycle and, second, the desilting of lakes stopped, and consequently the recharging of the underground water diminished as the lakes silted up. Further, the phosphates and nitrates in chemical-based fertilizers are extensively mined and relying on petroleum derivatives (natural gas and other hydrocarbons). The required raw materials are mined from various places across the world, and these are now overexploited and running out (Gilbert 2009). They are running out simply because mining is an open-loop cycle. With their dependence on petroleum derivatives, the cost of chemical fertilizers also keeps increasing as that of the petroleum products. To insulate the farmer from steep fertilizer costs, the government provided subsidies, which today in India have reached

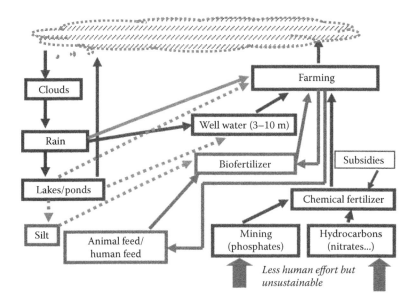

FIGURE 15.6
Chemical fertilizer–based green revolution in India (open system).

unmanageable proportions. The reliance on natural replenished fertilizer died out, resulting in an unsustainable open system as illustrated in Figure 15.7.

Today, with the lakes and ponds disappearing, the depths to which groundwater is extracted is in excess of 500 m. Groundwater recharge cannot be relied on through lake desilting alone, as the rate of groundwater withdrawal is so high that it cannot be replenished by any natural means. Furthermore, with increasing depth, the power required to draw the water increases, resulting in an ineffective use of conventional coal/petroleum-based energy. And the biggest problem that has arisen out of this deep groundwater is that it is replete with dissolved heavy metals and salts, such as sulfate, arsenic, fluoride, and iron, which result in severe soil pollution and require energy-intensive treatment. Needless to say, an open system consequently causes other closed-loop systems to open up, further exasperating the complexity of problems (Figure 15.8) facing modern society.

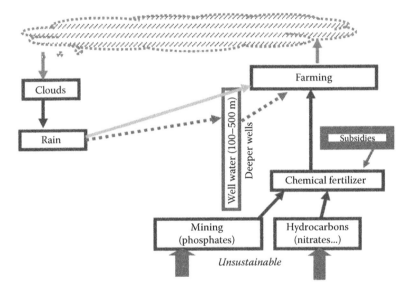

FIGURE 15.7
Farming cycle of today.

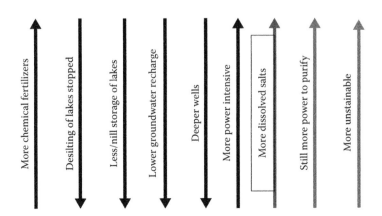

FIGURE 15.8
Effect of opening up of the cycle. It has become unsustainable.

The green revolution did, however, positively alleviate India from starvation to becoming a food surplus nation. It is the use of fertilizers and high-yield hybrid seeds that have
been the primary drivers for this turn around. Figure 15.9 is a graph that reveals the linkage
between increased fertilizer consumption and food production, for the period 1950–2000,
along with the increase in irrigated land. In other words, the increase in food production
could also be linked to the increase in irrigated land.

However, with a closer look at the trends at the onset of the green revolution (Figure 15.10),
it is evident that the increase in food production immediately followed the increase in irrigated land, while the increased adoption (and availability) of chemical fertilizer was yet to
take off. This clearly shows that the increase in food production is more closely linked to
the amount of irrigated land rather than the use of chemical fertilizer (Kailas et al. 2011).
Subsequently, owing to the forced dependence on extensive chemical fertilizer usage, the
practice of biofertilizers use declined, so much so that today modern progressive farmers
are not conversant with sustainable and natural biofertilizer use. It is also important to
note that because of a severe detrimental health impact attributed to crops grown under
heavy chemical fertilizer (and pesticide) use, worldwide there is an increased revival to go
back to traditional natural cycles that are sustainable and healthy.

One must acknowledge that increased irrigated land also dramatically accelerated the
fall in the groundwater table, which can be more attributed to the high-yielding crops
and waste rather than the effective use of water. This increase in water use for irrigation
also opened a cycle that was closed loop for millennia. All these factors led to a dip in the

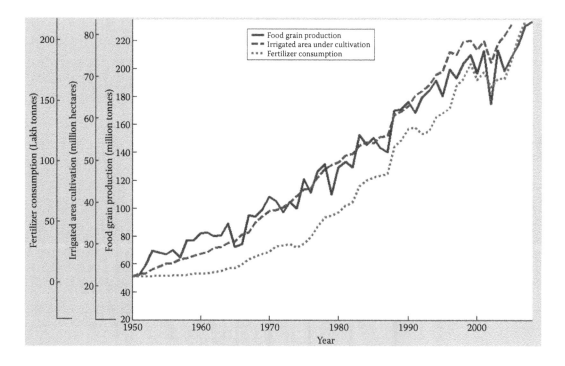

FIGURE 15.9
Graph showing increase in food production with increase in chemical fertilizer use, and the amount of irri-
gated land. (Data from RBI, *Handbook of Statistics of the Indian Economy 2008–09*. Reserve Bank of India, Alco
Corporation [for RBI], Mumbai, pp. 50–85, 2009.)

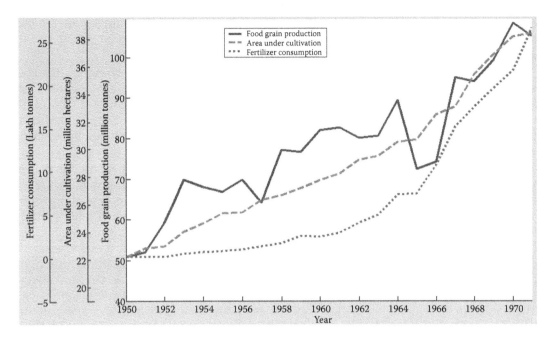

FIGURE 15.10
Closer view of the increase in food production in the early years of the green revolution. Increase in food production is more closely linked to the amount of irrigated land rather than the adoption of chemical fertilizers. (Data from RBI, *Handbook of Statistics of the Indian Economy 2008–09*. Reserve Bank of India, Alco Corporation [for RBI], Mumbai, pp. 50–85, 2009.)

groundwater level to depths that are not sustainable. We do believe that a severe water crisis is looming on the horizon.

Even eminent scientists who fostered the green revolution acknowledge its limitation and concur with the fact that traditional and organic closed-loop systems are necessary if we have to revive and sustain food production in the future. Unlike that with chemical fertilizer use, with organic farming practices, the soil nutrient increases every year with a consequent increase in its capacity to be productive (Fukuoka 1978). Furthermore, this will ensure a closed-loop regulated water use, which means that overexploitation of groundwater would be unnecessary. If one has to go back to a situation where the fresh potable groundwater is available, then one should revive the closed-loop agriculture by first desilting and recovering existing lakes and ponds.

We would like to explain with some other examples and put forward an idea on how to make the use of materials sustainable. J.C. Kumarappa (1957), hailed as the Gandhian economist, had first postulated the idea of sustainability way back in the 1950s where he defined a sustainable society as one that manages *its (economic) growth in such a way as to do no irreparable damage to its environment*. Causing no irreparable damage is in effect to operate a closed cycle. Further, Buckminister Fuller, the famous architect–engineer, refers to planet Earth as *spaceship* earth. So true is this statement that most space stations today are able to sustain themselves, simply by drawing energy from the sun, because of a closed-loop resource and nutrient cycle that has been designed and built in the spaceship.

In the preceding paragraphs, it was explained how opening the loop has led to a completely unsustainable cycle. And, more important, one must remember that the effects of opening the loop would be felt decades later. One wonders what further unpleasant effects of overexploitation of other resources are going to manifest. Unless, of course, we turn around and see things in an interconnected manner as they are in reality.

The focus is then directed to the use of modern water purification techniques like the use of nanoparticles, or nanoparticle-impregnated membranes for reverse osmosis, or nanofilms for capture of fog to condense water in arid regions, or for capture of fertilizers that enter water ways causing immense environmental problems. The simple questions that need to be asked is what materials are used to make these nanoparticles, how much energy is required to make these nanoparticles, and what happens to the nanoparticles during and after use? Let us take the example of silver nanoparticles that are used in water filters (Sumesh et al. 2011). Wet chemistry is one of the methods used for making silver nanoparticles. In this process, a silver salt is reduced by using a reducing agent along with a colloidal stabilizer. The wet process itself may not involve any energy consumption but a large amount of energy is required to make the silver salts, the reducing agent, and the colloidal stabilizer. And the purity of these salts has to be high! This adds to a higher level of energy consumption, and as mentioned earlier, most of the power produced in this world is from unsustainable sources. The other issue is the difficulty in extracting the silver nanoparticles after use. This is, without a doubt, going to be difficult and energy intensive. The extraction of silver nanoparticles would be essential as silver is something that cannot be regenerated and one has to follow the closed cycle when using silver nanoparticles. The use of nanoparticles in membranes that attract water molecules and repel dissolved salts and other contaminants is something that is gaining importance (Lind et al. 2010), and such membranes are reported to more than halve the energy consumed in reverse osmosis. The very making of these membranes involves considerable steps that consume energy during the process and for making the raw materials used for making these membranes. The reverse osmosis process itself still involves significant energy, making this process an open one and thus not sustainable. This is even without considering the disposal and loss of materials used in making these membranes that cannot be replenished. Thus, it is essential that even if we develop technologies that purify water, it is essential that they are themselves closed cycle, where the recovery or materials that cannot be replenished and use of energy sources that are sustainable are essential. This philosophy is valid for any product to make it sustainable. We would like to close the chapter by making the statements "Science without sustainability and sustainability without science are both meaningless" and "sustainability is of primary importance; efficiency and power output is of secondary importance."

References

Fukuoka, M. (1978), *One Straw Revolution*. Other India Press, Goa.

Gilbert, N. (2009), The disappearing nutrient. *Nature* 461, pp. 716–718.

Kailas, S.V., Mani, M., Dravid, Y., Umarji, V. (2011), Closing the cycle—Sustainability in natural water systems and agriculture. *Ground Report India*, special issue on *Water & Agriculture* 1(2), pp. 13–18.

Kumarappa, J.C. (1957), *Economy of Permanence*. Bhargava Bhushan Press, Varanasi.

Lind, M.L., Suk, D.E., Nguyen, T.-V., Hoek, E.M.V. (2010), Tailoring the structure of thin film nanocomposite membranes to achieve seawater RO membrane performance. *Environmental Science and Technology* 44(21), pp. 8230–8235.

Mani, M., Ganesh, L.S., Varghese, K. (2005), *Sustainability and Human Settlements*. Sage Publications, New Delhi, Thousand Oaks, London.

RBI (2009), *Handbook of Statistics of the Indian Economy 2008–2009*. Reserve Bank of India, Alco Corporation (for RBI), Mumbai, pp. 50–85.

Sumesh, E., Bootharaju, M.S., Anshup, Pradeep, T. (2011), A practical silver nanoparticle-based adsorbent for the removal of Hg2+ from water. *Journal of Hazardous Materials* 189(2011), pp. 450–457.

UNEP (1986), *The State of the Environment: Environmental Health*. UN Environment Programme, Nairobi.

Vinayak, G.K. (1989), *India and World Culture*. Sahitya Akademi Press, Delhi.

Wikipedia (2013a), Higgs Boson. Accessed May 2013. Available at http://en.wikipedia.org/wiki/Higgs_boson.

Wikipedia (2013b), Chronology of the Universe. Accessed May 2013. Available at https://en.wikipedia.org/wiki/Chronology_of_the_universe.

16

Thin Film Nanocomposite Reverse Osmosis Membranes

Christopher J. Kurth and Bob Burk

NanoH₂O Inc., El Segundo, California, USA

CONTENTS

16.1 Introduction

At the end of last year, 75 million m³ of water was being desalinated each day, about two-third of which was treated using reverse osmosis (RO) and most of the remainder using thermal approaches (primarily multistage flash and multiple-effect distillation) [1]. Hidden among that daily water production is 75,000 m³ of water being treated with a new class of RO membranes, one where improvements to the separation layer had been made by leveraging recent developments in nanotechnology to improve its baseline performance metrics (i.e., water and salt transport rate and surface properties relating to fouling). In these new membranes, the traditional separation layer, first developed in the late 1970s by North Star Research [2] and commercialized at Filmtec [3], is replaced with a blend of an interfacial formed polymer phase and nanomaterial, referred to as a thin film nanocomposite (TFN) RO membrane.

First conceived and reduced to practice at the University of California, Los Angeles, in the laboratory of Eric Hoek, the initial TFN concept involved adding zeolite nanoparticles into an interfacially formed RO membrane to give a mixed-matrix material [4]. In particular, one where the hydrophilicity of the zeolite phase could be used to minimize adhesive interactions with potential foulants, thus improving performance stability. Zeolite Linde Type A was selected from first principles; its three-dimensional (3-D) interconnected pore structure would eliminate the need to orient the nanoparticle, and its 4.9 Å pore diameter is such that water transport would be possible while hydrated ions, particularly anions, would be excluded from the pore network. In practice, incorporation of nanoparticles also lead to a reduction in membrane roughness, which would further improve performance stability, and increased water permeability. This increased water transport rate became the performance metric identified and optimized for initial commercialization. TFN technology was first released in the form of a spiral-wound *Quantum*Flux (Qfx) seawater membrane line in 2011, 4 years after the first TFN publication. This chapter will discuss academic and commercial developments related to the expanding use and potential of TFN membrane technology in salt selective membranes.

Membranes are a widely adopted technology used to separate components of a fluid stream. Depending on the identity and phase of the components to be retained and passed by the membrane, various technologies have been developed over the last century to treat liquids and gases and retain components ranging from the atomic to the macroscopic. In nearly all instances, developers of new membrane technologies are looking for ways to increase the permeability of the membrane to the components to be passed and decrease the permeability of the components to be rejected. This may be accomplished by altering the morphology or chemistry of the membrane. In the fields of gas separation and RO, different approaches were pioneered in the 1970s to alter morphology as a means of improving membrane performance in their respective fields.

16.1.1 Mixed Matrix Membrane Development

In the field of gas separation, composite barrier layers of polymer and inorganic particles were reported as early as 1973 in a fundamental work looking at the transport of gases in polymers by Paul [5]. The addition of a second phase to polymers that were known to be useful as gas separation materials was an attempt to break a flux-selectivity tradeoff curve that appeared to give an empirical maximum to a membrane's performance [6]. This approach to membrane construction has since found increasing use in gas separation [7], pervaporation [8], ion exchange [9], and as fuel cell electrolytes [10]. These relatively thick membranes were made by dispersing and then casting a polymer solution containing dispersed nanoparticles and then solidifying the membrane matrix, trapping the nanoparticles in place.

16.1.2 Thin Film Composite Membranes

In the field of desalination, the foundations of a technology shift was set in place with the discovery of monomers that, when used in an interfacial polymerization, resulted in an RO membrane suitable for use in one-pass seawater desalination. The performance of such interfacially made polyamide [2] RO membranes set in motion a multidecade shift from thermal desalination to membrane-based RO desalination. Before that point, RO membranes were made primarily from phase inversion of cellulose acetate polymers. These new membranes were made by a different process. A support membrane was prepared by casting a porous polysulfone layer above a polyester support. On top of this support layer, a water-based solution of a diamine was applied, followed by a solution of triacyl chloride in a water immiscible solvent. Immediately after contact, diamine began to diffuse into the water-immiscible solvent and an extremely fast reaction commenced. Less than a second after the second solution was applied, a thin polyamide layer had formed. This type of membrane was referred to as a thin film composite (TFC). Early variants had better rejection than the cellulosic acetate membranes, twice the permeability of water, and improved stability to high and low pH.

16.2 Academic TFN Membrane Research

Owing to the incompatibility of methods used to prepare early mixed-matrix membranes and the TFC membranes, it would be >25 years from early commercialization of TFC RO membranes and the development of mixed-matrix membrane technology until the first reported mention of a combination of these membrane methods [4]. In February of 2007 (patent priority dates suggest this work was originally conducted ~2 years earlier [11]), Byeong-Heon Jeong and Eric Hoek published results regarding TFC membranes incorporating nanoparticles in the active layer. These TFN membranes were prepared by dispersing zeolite nanoparticles in the organic solvent phase of the interfacial polymerization process. The structural differences between a TFC and TFN are illustrated in Figure 16.1.

Transmission electron microscopy (TEM) was conducted on cross sections of these membranes to verify that the nanoparticles were incorporated within the polyamide layer, and an example of these images shown in Figure 16.2.

TFN membranes prepared with increasing amounts of nanoparticles were found to have increasing permeability with similar rejection to NaCl, $MgSO_4$, and polyethylene glycol. In addition to increased permeability, membranes prepared with the zeolite particles were found to have higher hydrophilicity, a more negative surface charge, and a lower surface roughness. These characteristics suggested a potentially improved structure to resist foulant deposition.

Later, in April of 2007, Seung Yun Lee et al. reported on the incorporation of silver nanoparticles by a similar method [12]. These silver-containing TFN membranes were

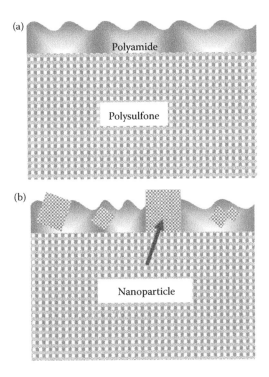

FIGURE 16.1
Schematic of (a) TFC and (b) TFN membranes.

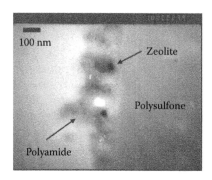

FIGURE 16.2
TEM cross section of a TFN membrane.

found to exhibit flux similar to TFC control membranes (made without silver nanoparticles), but were found to resist development of *Pseudomonas* bacteria on the surface.

An alternate approach aimed at reducing fouling was presented by Hyun Soo Lee et al. in 2008 [13]. TiO_2 nanoparticles were incorporated into an interfacially prepared RO membrane by addition to the organic solvent phase. Changes in performance were attributed to increased hydrophilicity, and the potential for these membranes to be photoactivated to degrade organic compounds was mentioned. To improve incorporation, later work by Babak Rajaeian et al. amino-silanized TiO_2 nanoparticles to allow covalent incorporation of the nanoparticles into the membrane and gave good flux and rejection [14].

Puyam Singh explored an alternate nanoparticle composition, silica (Ludox® HS-40), and used small-angle neutron scattering to explore the effect of nanoparticles on the polymer structure to help correlate performance with structure [15]. Up to about 4.5 wt% silica in the polyamide, good incorporation was observed with an alteration to the polyamide packing adjacent to the nanoparticle. Above 4.5 wt%, less effective dispersion of the nanoparticle was observed. Jadav and Singh continued the exploration of silica-based additives [16] and found the aggregate pore size of the membrane was tunable from 0.34 to 0.73 nm by adjusting the amount of Ludox HS-40 or laboratory-made silica used. Mengru Bao et al. describe the use of monodispersed spherical mesoporous nanosilica in TFN membranes and found increased hydrophilicity and permeability [17].

Mary Laura Lind and Eric Hoek, in a follow-up to the original TFN article presented in May of 2009, reported on the ability to modify the zeolite used by changing the nature of the mobile cation [18]. In this article, both Na^+ and Ag^+ were evaluated. Silver-exchanged zeolites were found to give improved performance, and were also able to resist adhesion of bacteria cells on the membrane, although antimicrobial activity was reduced relative to nonpolyamide-bound silver-containing zeolites. An article later that year also demonstrated the importance of nanoparticle zeolite size [19] with smaller zeolites that have the greatest impact on flux and rejection, while larger zeolites had the largest effect on surface properties such as charge, hydrophilicity, and roughness. In 2010, the effect of process conditions was explored [20] by investigating the effect of curing, rinsing, and posttreatments on TFN performance. This optimization work was found to give membranes having performance better than commercially available samples.

A new potential application for nanocomposites was presented by Junwoo Park et al. The subject membranes used a typical polyamide composition but used carbon nanotubes as the additive and found the resulting nanocomposite had improved resistance to chlorine [21].

Chunlong Kong et al. described an alternate approach to TFN membranes by using titanium isopropoxide, bis(triethoxysilyl)ethane, or phenyltriethoxysilane as nanoparticle precursors [22]. Phenyltriethoxysilane was found to give up to a doubling of the water transport with a negligible change of rejection.

Hang Dong et al. explored the use of surface-modified zeolite nanoparticles to improve the compatibility of the hydrophilic nanoparticles in the hydrophobic solvent as a means to decrease aggregation [23]. The resulting membranes were found to give less aggregation and slightly improved rejections relative to non-surface modified zeolite additives.

Lin Zhang et al. explored a surface modification of a different nature; they modified hydrophobic carbon nanotubes with a mixture of sulfuric and nitric acids to make them more hydrophilic and compatible with the aqueous solution [24]. The resulting membranes appeared to have carbon nanotubes spanning the membrane layer, and had more than double the permeability, but with a significant fall off in NaCl rejection.

A larger pore filler was described by Kim and Deng. They describe the preparation and use of an ordered mesoporous carbon (OMC) filler prepared around a silica template and later made hydrophilic by treatment with plasma [25]. Addition of this material resulted in a TFN membrane with increased surface hydrophilicity and pure water permeability; furthermore, bovine serum albumin adsorption was reduced with increasing amounts of OMC.

By changing the nature of the monomers to polyethyleneimine and isophthaloyl chloride, Sagar Roy et al. developed a carbon nanotube-containing TFN nanofiltration membrane [26]. Tested for organic dye rejection both in water and in methanol, the resulting membranes had nearly an order of magnitude higher permeability relative to previously reported TFC membranes. An alternate nanofiltration formulation was described by Deng Hu et al. when a piperazine trimesoyl chloride–based TFN membrane was prepared with incorporated silica nanoparticles [27]. These TFN nanofiltration membranes had approximately 20% higher permeability than TFC controls.

An alternate nanoparticle zeolite species was used by Mahdi Fathizadeh et al. when NaX was used in the fabrication process [28]. Resulting membranes had nearly twice the flux with similar salt rejections to TFC controls.

Although discussion of membranes using nanocomposite supports alone is outside the scope of this discussion, an article by Mary Pendergast et al. describes a set of membranes comprising a designed experiment looking at the effect nanocomposite supports (nTFC) and nanocomposite thin films (TFN), and combinations of both (nTFN) [29]. This nTFN approach was continued in a article by Eun-Sik Kim et al. using carbon nanotubes in the support, and silver nanoparticles in the thin film [30]. The prepared membranes showed greater resistance to *Pseudomonas aeruginosa* deposition and increased permeability. In a later article, Eun-Sik Kim et al. used carbon nanotubes both in the thin film layer for use in processing of tar sands produced water [31]. TFN membranes incorporating acid-treated multiwalled carbon nanotubes had better rejection of hydrophobic organic pollutants, higher process flux, and significantly reduced fouling.

Improved permeability and hydrophilicity was also observed by addition of alumina nanoparticles in work by Saleh and Gupta [32].

Ning Ma et al. extended the applicability of TFN membranes by exploring the performance of NaY zeolite containing TFN membranes to forward osmosis testing [33]. TFN membranes gave 80% higher permeability than baseline membranes.

Both porous and nonporous MCM-41 nanoparticles were incorporated into TFN membranes by Jun Yin et al. [34] TFN membranes were found to be more hydrophilic, more rough, more negatively charged, and with porous MCM-41 ("Mobil Composition of Matter"), of higher permeability.

To further improve compatibility, Huiqing Wu et al. introduced reactive amine groups onto mesoporous silica nanoparticles to allow covalent incorporation into a piperazine TMC-based separating layer [35]. These TFN membranes gave higher initial and sustained flux rates, as well as improved fouling stability compared with TFC controls.

In an alternate line of research, Huiqing Wu et al. form TFN membranes with the polyamide polymer phase being replaced with a polyester by the use of triethanolamine as aqueous reactant and multiwalled carbon nanotubes as the nanoparticle [36]. Further refinements in a follow-up article describe the determination of preferred surfactants to help disperse the nanotubes [37]. The resulting membranes had good flux and sulfate selectivity, which they speculate may be of interest in the chloralkali industry.

As new improvements and application areas of TFN membranes are identified and developed, the pace of research in this area is expected to rapidly increase.

16.3 Industrial TFN Development

Permeability and fouling resistance are primary drivers of cost in membrane-based water treatment systems. These performance metrics directly influence the energy intensity and capital expenditures of an RO plant, and therefore, the economics of desalination. At 70%–80% of the total expense of RO-desalinated water, energy consumption and capital expenditures are the primary reason why desalination remains expensive compared with many other freshwater sources. As a result, the commercial value of TFN technology was readily apparent and licensed shortly after discovery by Hoek et al. to NanoH2O Inc.

16.3.1 Nanocomposite Membrane Development

Commercial development of TFN technology for SWRO involved the identification of several technical failure modes specific to TFNs, and then developing methods to detect and prevent the onset of those failure modes. The importance of defect reduction can quickly be realized by recognizing the 1000-fold difference in permeability between the supporting polysulfone and the thin film layer. In an industry where 99.7% rejection of dissolved ions is expected, a parts-per-million areal defect frequency in the thin film will render the membrane unsuitable for use as the flow through defect is amplified by the locally higher defect flow rate.

Figure 16.3 illustrates one such failure mode. Scanning electron microscopy (SEM) imaging of a membrane ("B" in Table 16.1) surface reveals that nanoparticles have aggregated in the organic solution and deposited on the surface of the membrane. Identification of the deposited nanoparticles was confirmed with energy dispersive x-ray element mapping. In contrast to the nanoparticles shown in Figure 16.2, this aggregate is not contained within the barrier layer but instead sits on the surface. Although its appearance is superficially similar to that of a fouling layer, its high porosity leads to minimal resistance to flow. However, the decreased concentration of nanoparticle incorporated within the film minimizes any permeability increase as shown in Table 16.1.

A second such failure mode illustrates a secondary aggregation-based failure mode, in this case smaller aggregates that end up incorporated within the film. In Figure 16.4, a TEM image of membrane C shows smaller aggregates contained within a TFN membrane.

FIGURE 16.3
SEM image of surface-aggregated nanoparticles in membrane B.

TABLE 16.1

Membrane Performance

Membrane	Flux (gfd)	Rejection
A. Control—no nanoparticles	18.9 (±4.1)	99.66 (±0.11%)
B. Surface aggregated nanoparticles	19.5 (±0.42)	99.22 (±0.23%)
C. Smaller incorporated aggregates	25.6 (±1.3)	98.37 (±0.02%)
D. Well-dispersed nanoparticles	27.5 (±0.34)	99.37 (±0.01%)
E. Optimized nanoparticle selection	34.2 (±2.7)	99.65 (±0.01%)

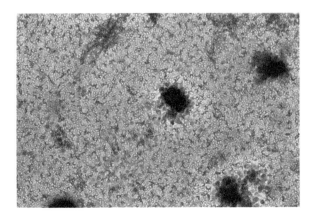

FIGURE 16.4
TEM image of nanoparticle aggregates in membrane C.

Figure 16.4 was obtained by placing the active layer against a TEM grid in a petri dish, and then adding several portions of methylene chloride to dissolve the polysulfone. Imaging with TEM through the membrane resulted in an image highlighting the location where nanoparticles were incorporated in the plane of the membrane. Particle size distribution (from light scattering) of the solutions used to prepare the membrane reveals that the nanoparticles used were present in aggregates with an average diameter just over 1 micron. The match in diameters of aggregates in solution and membrane suggests the aggregates present in solution were subsequently trapped within the forming polymer layer. Spaces between the nanoparticles in these aggregates are believed to have contributed to decreased salt rejection for this membrane (Table 16.1).

Improved solution preparation, for instance by increased sonication time and power (obtainable with a sonic probe), can be used to break up these aggregates. After appropriate preparation, the solution used to prepare the membrane in Figure 16.5 had a mean size measured by light scattering of ~100 nm, the diameter of the nanoparticles as synthesized.

The TEM image of membrane D shown in Figure 16.5 reveals a much improved distribution of nanoparticles; high areal incorporation, few aggregates, and relatively uniform surface coverage.

Once protocols were established minimizing issues with aggregates, proper selection of nanoparticle type was found to allow the production of TFN membranes having good performance (membrane E in Table 16.1). Flux is in gallons per square foot per day (gfd).

With these developments in place, scale up to a continuous manufacturing line was conducted in 2010 and the resulting membrane rolled into conventional spiral-wound modules known commercially as Qfx membranes. Membrane performance data, using industry-standard element testing, demonstrates that 20 cm (8-in) Qfx membrane modules with 37.2 m² (400 ft²) membrane area have industry-leading flux of 51.9 m³/day (13,700 gfd) with salt rejection at industry standards of 99.8%. More recent adjustment of the Qfx chemistry allowed production of Qfx membranes with salt rejection at 99.85%, significantly exceeding industry standards. When compared with membranes with 99.80% salt rejection, membranes with 99.85% rejection produce 25% better water quality. The higher rejection

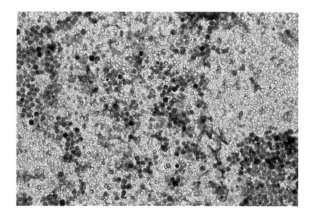

FIGURE 16.5
SEM of well-dispersed nanoparticles in membrane D.

membranes are available at lower flux 24.6 and 34 m³/day (6500 and 9000 gfd) to meet the demands of desalination plants with warm high-salinity waters and hybrid designs.

16.3.1.1 Performance Data: Las Palmas III Pilot Test Example—Increased Capacity

In most seawater RO systems, six to eight modules are loaded in series. As purified water is withdrawn from each element, the feed water becomes progressively more saline, increasing the osmotic pressure of the solution to be desalinated. At the same time, feed pressure drops slightly owing to the energy consumed promoting mixing in the feed channel. In combination, both of these lead to a decrease in net driving pressure with each subsequent element. When lower-energy systems are designed, the same increase in osmotic pressure is observed and approximately the same pressure drop is observed. With a lower feed pressure, this leads to a larger proportional difference in net driving pressure down the length of the housing, which can lead to very high flux rates for the lead element (which can exacerbate fouling). Use of an internally staged design strategy, where elements of differing permeability are used to balance the permeate flux down the length of the pressure vessel, can then be used to prevent this accelerated fouling potential. Such a design is undergoing pilot testing by EMALSA (Empresa Municipal de Aguas de Las Palmas, S.A.) in the Canary Islands.

The pilot run by EMALSA is a second-stage system used to increase recovery by desalinating the brine from an SWRO system. The pilot with TFN membranes was installed in late 2011, the pilot test utilizes commercial nanocomposite Qfx 24.6 m³/d and Qfx 34 m³/d elements in the same pressure vessel. As a result of the higher permeability, more water was permeated at the same incoming pressure. More specifically, water production increased by 50% and both water quality and energy consumption improved over the elements that were previously running in the plant. Energy savings increased by approximately 4%. Data for the first 4 months of operation is shown in Figure 16.6 and demonstrates stable performance over the test period.

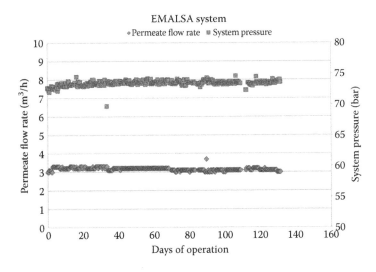

FIGURE 16.6
Stable permeate flow rate and system pressure over a 130-day period at an EMALSA pilot in the Canary Islands.

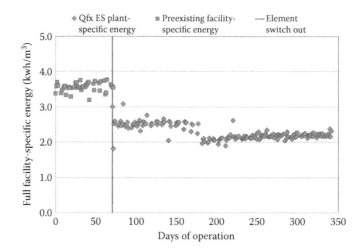

FIGURE 16.7
Change in specific energy consumption after installation of TFN membrane modules.

16.3.1.2 Performance Data: Cayman Brac—Energy Savings

The higher permeability of nanocomposite SWRO membranes in a properly designed desalination system also provides the ability to operate at lower pressures, resulting in energy savings to the plant owner when compared to other commercial membranes. At a seawater desalination plant at Cayman Brac (one of the Cayman islands) in the Caribbean, an energy saving of 28% when conventional SWRO modules was replaced with nanocomposite membranes and is visualized in Figure 16.7. A portion of this energy savings was due to recovery of flow lost over time on the existing elements from fouling; however, even when compared with the projected performance of new conventional membranes, the savings were still >16%. Stability of those data has been maintained for slightly under 1 year as of this writing.

16.4 Conclusions

In a quest to push the limits of membrane performance, researchers have reached across various fields of inquiry to introduce the TFN membrane and advance its chemistry and morphology. As a result of these efforts, TFN membranes have found new areas of application and have continued to advance performance boundaries. Now in the early stages of commercial application, end users are beginning to see the fruit of these researchers' efforts in the form of lower-cost and higher-quality water in desalination plants worldwide. As work continues, TFN membrane technology is expected to continue to improve separation options in forward and pressure-retarded osmosis, water reuse, brackish water treatment, and treatment of produced water from the oil and gas industries.

References

1. C. Gasson, and P. Allison, "Desalination Markets 2007 A Global Industry Forecast," Media Analytics Ltd, 2006.
2. J. E. Cadotte, "Interfacially synthesized reverse osmosis membrane," US Patent 42773441981.
3. R. J. Petersen, "Composite reverse osmosis and nanofiltration membranes," *Journal of Membrane Science*, vol. 83, no. 1, pp. 81–150, 1993.
4. B.-H. Jeong, E. M. V. Hoek, Y. Yan, A. Subramani, X. Huang, G. Hurwitz, A. K. Ghosh, and A. Jawor, "Interfacial polymerization of thin film nanocomposites: A new concept for reverse osmosis membranes," *Journal of Membrane Science*, vol. 294, no. 1–2, pp. 1–7, 2007.
5. D. R. Paul, and D. R. Kemp, "Containing adsorptive fillers," *Journal of Polymer Science: Polymer Symposia (Wiley)*, vol. 93, no. 41, pp. 79–93, 1973.
6. C. M. Zimmerman, A. Singh, and W. J. Koros, "Tailoring mixed matrix composite membranes for gas separations," *Journal of Membrane Science*, vol. 137, no. 1–2, pp. 145–154, 1997.
7. L. Jiang, T. S. Chung, and S. Kulprathipanja, "An investigation to revitalize the separation performance of hollow fibers with a thin mixed matrix composite skin for gas separation," *Journal of Membrane Science*, vol. 276, no. 1–2, pp. 113–125, 2006.
8. L. M. Vane, V. V. Namboodiri, and R. G. Meier, "Factors affecting alcohol–water pervaporation performance of hydrophobic zeolite–silicone rubber mixed matrix membranes," *Journal of Membrane Science*, vol. 364, no. 1–2, pp. 102–110, 2010.
9. R. Kiyono, G. Koops, M. Wessling, and H. Strathmann, "Mixed matrix microporous hollow fibers with ion-exchange functionality," *Journal of Membrane Science*, vol. 231, no. 1–2, pp. 109–115, 2004.
10. F. G. Üçtuğ, and S. M. Holmes, "Characterization and fuel cell performance analysis of polyvinylalcohol–mordenite mixed-matrix membranes for direct methanol fuel cell use," *Electrochimica Acta*, vol. 56, no. 24, pp. 8446–8456, 2011.
11. E. M. V. Hoek, Y. Yan, and B. Jeong, "Nanocomposite membranes and methods of making and using same," US Patent Appl. No. 2011/00275992011.
12. S. Y. Lee, H. J. Kim, R. Patel, S. J. Im, J. H. Kim, and B. R. Min, "Silver nanoparticles immobilized on thin film composite polyamide membrane: Characterization, nanofiltration, antifouling properties," *Polymers for Advanced Technologies*, vol. 18, no. 7, pp. 562–568, 2007.
13. H. S. Lee, S. J. Im, J. H. Kim, H. J. Kim, J. P. Kim, and B.-R. Min, "Polyamide thin-film nanofiltration membranes containing TiO_2 nanoparticles," *Desalination*, vol. 219, no. 1–3, pp. 48–56, 2008.
14. B. Rajaeian, A. Rahimpour, M. O. Tade, and S. Liu, "Fabrication and characterization of polyamide thin film nanocomposite (TFN) nanofiltration membrane impregnated with TiO_2 nanoparticles," *Desalination*, vol. 313, pp. 176–188, 2013.
15. P. S. Singh, and V. K. Aswal, "Characterization of physical structure of silica nanoparticles encapsulated in polymeric structure of polyamide films," *Journal of Colloid and Interface Science*, vol. 326, no. 1, pp. 176–185, 2008.
16. G. L. Jadav, and P. S. Singh, "Synthesis of novel silica–polyamide nanocomposite membrane with enhanced properties," *Journal of Membrane Science*, vol. 328, no. 1–2, pp. 257–267, 2009.
17. M. Bao, G. Zhu, L. Wang, M. Wang, and C. Gao, "Preparation of monodispersed spherical mesoporous nanosilica–polyamide thin film composite reverse osmosis membranes via interfacial polymerization," *Desalination*, vol. 309, pp. 261–266, 2013.
18. M. L. Lind, B. Jeong, A. Subramani, X. Huang, and E. M. V. Hoek, "Effect of mobile cation on zeolite–polyamide thin film nanocomposite membranes," *Journal of Materials Research*, vol. 24, no. 5, pp. 1624–1631, 2009.
19. M. L. Lind, A. K. Ghosh, A. Jawor, X. Huang, W. Hou, Y. Yang, and E. M. V. Hoek, "Influence of zeolite crystal size on zeolite–polyamide thin film nanocomposite membranes," *Langmuir*, vol. 25, no. 17, pp. 10139–10145, 2009.
20. M. L. Lind, D. Eumine Suk, T.-V. Nguyen, and E. M. V. Hoek, "Tailoring the structure of thin film nanocomposite membranes to achieve seawater RO membrane performance," *Environmental Science & Technology*, vol. 44, no. 21, pp. 8230–8235, 2010.

21. J. Park, W. Choi, S. H. Kim, B. H. Chun, J. Bang, and K. B. Lee, "Enhancement of chlorine resistance in carbon nanotube based nanocomposite reverse osmosis membranes," *Desalination and Water Treatment*, vol. 15, no. 1–3, pp. 198–204, 2010.

22. C. Kong, A. Koushima, T. Kamada, T. Shintani, M. Kanezashi, T. Yoshioka, and T. Tsuru, "Enhanced performance of inorganic-polyamide nanocomposite membranes prepared by metal-alkoxide-assisted interfacial polymerization," *Journal of Membrane Science*, vol. 366, no. 1–2, pp. 382–388, 2011.

23. H. Dong, X.-Y. Qu, L. Zhang, L.-H. Cheng, H. Chen, and C.-J. Gao, "Preparation and characterization of surface- modified zeolite-polyamide thin film nanocomposite membranes for desalination," *Desalination and Water Treatment*, vol. 34, pp. 6–12, 2012.

24. L. Zhang, G.-Z. Shi, S. Qiu, L.-H. Cheng, and H. Chen, "Desalination and water treatment preparation of high-flux thin film nanocomposite reverse osmosis membranes by incorporating functionalized multi-walled carbon nanotubes Preparation of high-flux thin film nanocomposite reverse osmosis membranes by incor," *Desalination and Water Treatment*, vol. 34, pp. 19–24, 2012.

25. E.-S. Kim, and B. Deng, "Fabrication of polyamide thin-film nano-composite (PA-TFN) membrane with hydrophilized ordered mesoporous carbon (H-OMC) for water purifications," *Journal of Membrane Science*, vol. 375, no. 1–2, pp. 46–54, 2011.

26. S. Roy, S. A. Ntim, S. Mitra, and K. K. Sirkar, "Facile fabrication of superior nanofiltration membranes from interfacially polymerized CNT-polymer composites," *Journal of Membrane Science*, vol. 375, no. 1–2, pp. 81–87, 2011.

27. D. Hu, Z.-L. Xu, and C. Chen, "Polypiperazine-amide nanofiltration membrane containing silica nanoparticles prepared by interfacial polymerization," *Desalination*, vol. 301, pp. 75–81, 2012.

28. M. Fathizadeh, A. Aroujalian, and A. Raisi, "Effect of added NaX nano-zeolite into polyamide as a top thin layer of membrane on water flux and salt rejection in a reverse osmosis process," *Journal of Membrane Science*, vol. 375, no. 1–2, pp. 88–95, 2011.

29. M. M. Pendergast, A. K. Ghosh, and E. M. V. Hoek, "Separation performance and interfacial properties of nanocomposite reverse osmosis membranes," *Desalination*, vol. 308, pp. 180–185, 2013.

30. E.-S. Kim, G. Hwang, M. Gamal El-Din, and Y. Liu, "Development of nanosilver and multi-walled carbon nanotubes thin-film nanocomposite membrane for enhanced water treatment," *Journal of Membrane Science*, vol. 394–395, pp. 37–48, 2012.

31. E.-S. Kim, Y. Liu, and M. Gamal El-Din, "An *in-situ* integrated system of carbon nanotubes nanocomposite membrane for oil sands process-affected water treatment," *Journal of Membrane Science*, vol. 429, pp. 418–427, 2013.

32. T. A. Saleh, and V. K. Gupta, "Synthesis and characterization of alumina nano-particles polyamide membrane with enhanced flux rejection performance," *Separation and Purification Technology*, vol. 89, pp. 245–251, 2012.

33. N. Ma, J. Wei, R. Liao, and C. Y. Tang, "Zeolite–polyamide thin film nanocomposite membranes: Towards enhanced performance for forward osmosis," *Journal of Membrane Science*, vol. 405–406, pp. 149–157, 2012.

34. J. Yin, E.-S. Kim, J. Yang, and B. Deng, "Fabrication of a novel thin-film nanocomposite (TFN) membrane containing MCM-41 silica nanoparticles (NPs) for water purification," *Journal of Membrane Science*, vol. 423–424, pp. 238–246, 2012.

35. H. Wu, B. Tang, and P. Wu, "Optimizing polyamide thin film composite membrane covalently bonded with modified mesoporous silica nanoparticles," *Journal of Membrane Science*, vol. 428, pp. 341–348, 2013.

36. H. Wu, B. Tang, and P. Wu, "MWNTs/polyester thin film nanocomposite membrane: An approach to overcome the trade-off effect between permeability and selectivity," *The Journal of Physical Chemistry C*, vol. 114, no. 39, pp. 16395–16400, 2010.

37. H. Wu, B. Tang, and P. Wu, "Optimization, characterization and nanofiltration properties test of MWNTs/polyester thin film nanocomposite membrane," *Journal of Membrane Science*, vol. 428, pp. 425–433, 2013.

17

Technological Developments in Water Defluoridation

Nitin K. Labhsetwar, S. Jagtap Lunge, Amit K. Bansiwal,
Pawan K. Labhasetwar, Sadhana S. Rayalu, and Satish R. Wate

CSIR–National Environmental Engineering Research Institute (CSIR-NEERI), Nagpur, India

CONTENTS

17.1 Introduction

Fluoride is one of the most abundant water contaminants occurring in groundwater as it poses severe problems in safe drinking water supplies. According to United Nations Educational, Scientific and Cultural Organization estimates (January 2007), >200 million people use fluoride-contaminated groundwater for drinking, with fluoride concentrations of more than the World Health Organization (WHO)–recommended value of 1.5 mg/L. The 1984 WHO guidelines mention that in areas with a warm/arid climate, the optimal fluoride concentration in drinking water should be <1 mg/L (1 ppm), while in cooler climate regions it could be acceptable up to 1.2 mg/L. This differentiation is based on the fact that perspiration occurs more in hot weather and consequently more drinking water is generally consumed. The WHO has set a guideline value (permissible upper limit) for fluoride in drinking water at 1.5 mg/L. There is minor deviation from this value for different countries.

The sources of fluoride in water are both natural and anthropogenic; however, the majority of fluoride in groundwater is of geogenic origin. Weathering of fluoride-rich minerals such as limestone, sandstone, and granite results in the release of fluoride in groundwater. The extent of fluoride contamination in groundwater depends on water geochemistry and temporal factors. The widespread occurrence of fluoride above the prescribed limits in groundwater has been reported from >30 nations around the globe, among which India, South Africa, China, Sri Lanka, Kenya, Nigeria, and Mexico are among the worst affected. In India alone, >60 million people from >17 states are considered at risk due to the consumption of fluoride-contaminated groundwater. Long-term consumption of fluoride-rich water results in a disease termed fluorosis, which can be categorized as dental or skeletal fluorosis. Besides, other clinical manifestations including gastrointestinal problems, allergies, neurological disorders anemia, and urinary tract problems due to prolonged exposure of fluoride are well documented. An estimated 25–65 million people are affected with dental, skeletal, and/or nonskeletal fluorosis in India, with a wide range of fluoride concentration of 1–48 mg/L in groundwater (Ayoob et al., 2008). As per a WHO estimate, the disability affected life years (DALY) value because of skeletal fluorosis could be as high as 17 per 1000 population in India. With depleting surface water sources and increasing water demand, the groundwater sources are under stress leading to further increase of fluoride concentration as well as higher intake.

With the presence of fluoride being a major issue of drinking water quality, extensive efforts have been made toward fluorosis mitigation, mainly focusing on defluoridation of drinking water. A large number of approaches and technological options have been reported for fluoride removal from water. However, despite intensive efforts and huge investment in the water supply sector, the objective of providing adequate potable drinking water is falling short, as at any point of time, there is considerable gap between assets created and service available, especially to the rural population in less developed countries. With exponential increase in water demand and water quality problems, as well as treatment options being region- and water input quality-sensitive, scientific efforts are still being put in for the development of improved technological options for fluoride removal. There are also challenges concerning the relatively poor translation of laboratory developments to successful technology development and its implementation in the field.

Reducing the fluorosis disease burden in a cost-effective way is the primary objective of any fluoride removal effort and, further improvements are still required to develop techno-economically feasible water purification technologies for rural applications in less developed countries. Considering the serious issues related to regeneration of adsorbents on a sustainable basis, materials with high fluoride uptake capacity are being explored. Nanomaterials, including nanocomposites, show potential properties in this regard with their high first cycle capacity. Improved water quality monitoring in the recent past has resulted not only in identification of new contaminated locations but brought out the presence of more than one pollutant in many areas. Multipollutant treatment processes with increasing number of multipollutant contaminations are, therefore, getting more attention.

In general, there has been limited work to address the minimization of waste, handling and safe disposal as well as treatment of the sludge/used materials, and management of liquid wastes generated during the regeneration and treatment processes. The concept of integrated fluorosis mitigation (IFM) and estimation of disease burden prove useful in assessing the various technological options for effective fluorosis mitigation through water treatment and other options.

The various options to tackle the high fluoride problem discussed in previous sections are preferred; however, in most of the fluoride-affected areas, alternate sources are either not available or implementation is restricted owing to techno-economic and social factors. In such cases, removal of excess fluoride to make the water potable is the only remedy. Defluoridation techniques can be implemented at a household level or community level depending on site-specific factors. The central or community-level defluoridation schemes are more suitable for urban areas where availability of skilled manpower and other resources are not limited. Such community-level defluoridation schemes can be coupled with other general treatment schemes normally available in most of the urban areas. On the other hand, household or decentralized defluoridation options are more suitable for rural areas, where population is highly scattered and other water supply infrastructure is not available. Furthermore, decentralized defluoridation can also be limited to treat the water used for drinking and cooking purposes, which results in significant cost reduction and less chemical handling and sludge generation. Several techniques are presently in practice to reduce the fluoride levels below prescribed limits, which can be categorized into four main categories, namely coagulation–precipitation, membrane processes, ion exchange, and adsorption onto various adsorbents. Brief details of these techniques along with their merits and demerits are provided in the following sections and summarized in Figure 17.1.

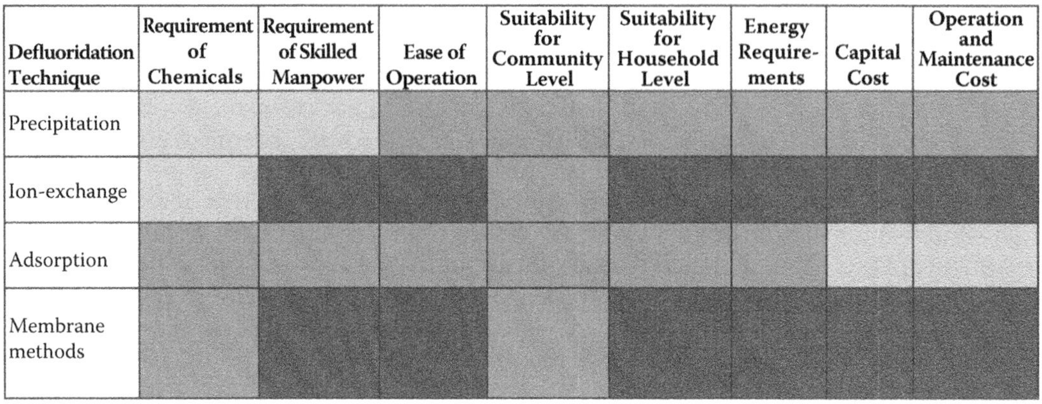

Defluoridation Technique	Requirement of Chemicals	Requirement of Skilled Manpower	Ease of Operation	Suitability for Community Level	Suitability for Household Level	Energy Require-ments	Capital Cost	Operation and Maintenance Cost
Precipitation								
Ion-exchange								
Adsorption								
Membrane methods								

Note: Light gray, advantages; gray, limitations; dark gray, can be compromised.

FIGURE 17.1
General merits and demerits of various defluoridation techniques.

17.2 Fluorosis Control Options

The ultimate objective of any defluoridation technology is to control or mitigate the health impacts of excessive fluoride intake, i.e., fluorosis. Although defluoridation of water has been considered as the most effective option for fluorosis mitigation, recent studies also point out at the importance of other parameters, including fluoride intake through food as well as the remedial potential of nutritional supplements. There have been numerous efforts to find alternate water sources; however, for obvious reasons, these options are region specific. Brief description of various options for providing water with acceptable fluoride concentrations is discussed in the following sections.

17.2.1 Alternative Water Sources

An alternate water source option becomes attractive when defluoridation technology is not effective or implemented ineffectively for some reason. A number of options are attempted and practiced in different parts of the world. Depending on hydrogeological conditions of a particular region, it could be possible to obtain a local safe water source by drawing the water from different depths. Since leaching of fluoride into groundwater is a localized phenomenon, which depends on many and complex parameters, it is sometimes possible to find a sustainable new source of water. Regular monitoring of fluoride concentration is, however, needed because mixing of water from different aquifers with different fluoride concentrations cannot be ruled out. Transporting water from a distant source has also been successfully practiced; however, the economic feasibility of this option is often difficult in less developed countries. In such cases, initial high cost is compensated in long-lasting benefits if other alternates are not effective. Mixing of high- and low-fluoride-containing waters to bring the fluoride concentration within permissible levels can be an attractive long-term solution, provided the low-fluoride water source is available within reasonable distance and that is of acceptable quality with respect to other water parameters. This has been effectively implemented in some parts.

17.2.1.1 Dual Water Sources

If there are different sources available with both high and low fluoride levels in the same region or community, stress on safe water sources in such cases can be reduced by diluting the high-fluoride water with low-fluoride water so as to bring the concentration within permissible limits. Another approach to conserve the safe water is by providing high- and low-fluoride water separately, and restricting the use of low-fluoride water only for drinking and cooking purposes. This is important considering the fact that the adverse health impact of fluoride is only through water ingestion and also that a small fraction of total water use per capita is usually used for drinking and cooking. Besides almost double cost implications, the success of both dilution and dual water supply schemes depends largely on active community awareness and participation. The use of different water sources for different purposes may also be aggravated by socioeconomic factors such as longer distances and reservations in sharing water sources with users from other neighborhoods. Awareness alone is sometimes not enough to change practices, especially because the health impacts of fluoride becomes visible only after prolonged consumption of high-fluoride water.

17.2.1.2 Rainwater Harvesting

This approach is more relevant to tropical countries like India that receive reasonable rainfall in most of the affected areas. Rainwater harvesting could be useful for fluoride control in two ways. This could be an alternate source of water, while the water collected through rainwater can be used for dilution of high-fluoride water. Harvested surface water runoff can also be effectively used to recharge high-fluoride groundwater sources. There are cost issues involved as well as implementation of such options in several habitations. Increasing water scarcity, however, makes rainwater harvesting a potential option, while advantages of water quality add to the benefit of this option if water collection and storage is done properly.

17.2.2 Artificial Recharge for Groundwater Fluoride Dilution

Artificial recharge is an *in situ* method that dilutes the concentration of groundwater fluoride in a particular aquifer. Construction of check dams in some parts of India has been demonstrated to reduce fluoride concentration in groundwater. Rainwater recharge can also be adopted using percolation tanks and recharge pits with appropriate site selections. Detailed hydrogeological investigations are required to make the artificial recharge more effective and cost-effective. Recharge using rainwater after filtration through the existing wells in the vicinity can also be cost-effective and used to improve the groundwater quality. As the users are not involved in fluoride remediation technology through artificial recharge, this presents a potential option and is being explored. An additional advantage with artificial recharge is an improved water source, which is often required in most of the areas.

17.2.3 Integrated Fluorosis Mitigation

Owing to the lowering of groundwater in many parts of India, there are significant risks of increased bacteriological and chemical contamination mainly due to arsenic, fluoride, and nitrate. Therefore, to mitigate these contaminants, there is a need for a "holistic" health management approach. This approach should begin with establishing tolerable levels of risk to human health of the specific chemicals followed by appropriate risk management strategies. In this connection, a novel concept of IFM has been developed and implemented jointly by the United Nations Children's Fund (UNICEF) and Council of Scientific and Industrial Research (CSIR)–National Environmental Engineering Institute (NEERI) to address the growing problem of fluorosis.

The scientific tool Quantitative Chemical Risk Assessment (QCRA) has been used for the first time as a diagnostic parameter before applying an appropriate fluorosis mitigation strategy. The QCRA, with data on health surveys, provides the basis for use-based separation of water sources. Groundwater consumption can also be reduced by creating alternative water resources, such as rainwater harvesting or reuse of gray water for sanitation/gardening. Development of an improved and simple process for defluoridation of drinking water at the household level could also be an effective way of defluoridation of drinking water. The IFM concept has been studied and implemented under a systematic study in the Dhar and Jhabua districts of Madhya Pradesh, India. Water quality monitoring and surveys, along with reported dental/skeletal fluorosis symptoms, have been used for identification of fluoride contamination as the chief problem in the study region. Initial interventions based on the above study have shown excellent results toward actually controlling the fluorosis in the study area. A dental fluorosis survey postintervention has shown a 47% reduction in dental fluorosis, as a result of IFM.

17.2.3.1 IFM Approach

This approach needs to begin with better understanding of health impacts of excessive fluoride intake in relation with the nutritional aspects, and establishing tolerable levels of risk to human health followed by an integrated strategy involving water management solutions, new water defluoridation techniques, as well as nutritional supplementation. It is now established beyond doubt that fluoride intake is not the only factor responsible for fluorosis, and malnutrition can play a crucial role in aggravating the problem. As a consequence, nutritional solutions can also be effective in fluorosis mitigation. Similarly, there seems to be considerable scope for improvement in existing defluoridation techniques for water, as they may not be techno-economically feasible in many rural and less developed areas.

While water can be a major source of enteric pathogens and hazardous chemicals, it is by no means the only source. In setting standards, consideration needs to be given to other sources of exposure, including food. There is limited value in establishing a strict target concentration for a chemical in drinking water, as it provides only a small proportion of total exposure (WHO, 2004c). Drinking water may be only a minor contributor to the overall intake of a particular chemical, and in some circumstances controlling the levels in drinking water, at considerable expense, may have little impact on overall exposure. Drinking water risk management strategies should therefore be considered in conjunction with other potential sources of human exposure.

Fluorosis is caused by intake of high levels of fluoride from various exposure routes, while its impact on human health also depends on various factors, including the nutritional status. Therefore, fluorosis mitigation needs a multidimensional approach, starting with proper identification of the problem and its causes. Scientific tools such as the QCRA are used for assessment of health impacts due to excessive fluoride intake in relation with the nutritional aspects, and finally establishing tolerable levels of risk to human health. Water quality monitoring and surveys along with reported dental/skeletal fluorosis symptoms lead to the identification of fluoride contamination as a problem in a region. QCRA with data on health surveys provides the basis for use-based separation of water sources. This should be followed by an IFM strategy involving various water management solutions (including rainwater harvesting, gray water reuse), improved water defluoridation techniques, and nutritional supplementation. This approach also offers a site-specific and practically more feasible solution for the effective mitigation of fluorosis.

17.2.3.1.1 IFM—A Case Study in Madhya Pradesh, India (Adopted from Integrated Fluorosis Mitigation Manual, UNICEF-CSIR-NEERI, 2007)

Madhya Pradesh is a central state in India. It has a population of >65 million with a high tribal population. More than 80% of the 48 districts of Madhya Pradesh are affected by elevated levels of fluoride in drinking water. To mitigate fluorosis, areas with both high fluoride in water and also a high prevalence of dental fluorosis were identified. Two districts (Dhar and Jhabua) were taken as implementation sites, and a dental survey of 1000 children using the dean's survey and QCRA were undertaken in 2005 (Figure 17.2). This was followed by the IFM intervention including water dilution and defluoridation, and nutritional supplementation in both schools and communities. The QCRA followed the approach outlined in the WHO Guidelines for Drinking Water Quality (vol. 3, 2004) and involved hazard identification, exposure assessment, dose response, and risk characterization. The results of the QCRA indicated that 60% of the fluoride is being consumed from food and 40% is being consumed from water.

On the basis of the QCRA, three categories of risk were determined:

1. Class A—high-risk category (DALY >100/1000 persons), where the mitigation included Information, Education and Communication (IEC), household defluoridation, nutritional supplementation, and water management

2. Class B—moderate risk category (DALY <100/1000 persons), including IEC, dilution of fluoride affected water, and nutritional supplementation

3. Class C—low-risk category (DALY <10/1000 persons); IEC and nutritional supplementation

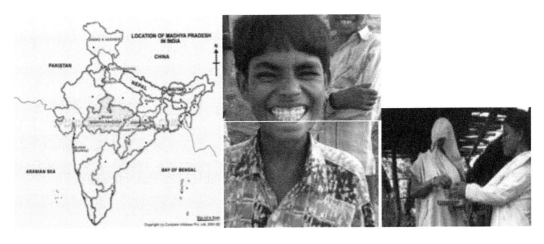

FIGURE 17.2
Study area and activities related to IFM. (Adopted from *Integrated Fluorosis Mitigation Manual*, UNICEF-CSIR-NEERI, 2007.)

Following the intervention, a further dental survey of the same 1000 children was done. Results from the dean's survey indicate a reduction in the prevalence of grade I fluorosis by 86%, grade II fluorosis by 77%, and grade III by 60%. There was a minor increase in grade II fluorosis by 21% owing to the change in the cohort age of the study group during the 2 years. Overall, the IFM resulted in a drastic reduction in the prevalence of dental fluorosis and increased the number of healthy teeth by 154%.

17.3 Technological Options for Water Defluoridation

Although the above-described options of defluoridation can be considered as effective and do not require user involvement, the feasibility of many of those options depend on hydrogeological conditions, require higher initial cost, and take a long time for planning and implementation. This led to defluoridation of drinking water as the most feasible option in most parts of the world. In the user point of view, there could be two options of defluoridation: (i) centralized or community-level water treatment, often at the source, and (ii) the treatment of water at the point of use (POU), i.e., at the household level. There are fundamental differences as well as advantages and limitations of these approaches. The centralized approach offers excellent control over defluoridation treatment and can be combined with regular drinking treatment processes, leaving no risk of quality control at the POU. Accordingly, in many developed countries, treatment at source is commonly practiced. Defluoridation treatment is carried out on a large scale under the supervision of skilled technical personnel. The same approach may not be feasible in the rural areas of less developed countries, where settlements are often scattered. Chemical treatment of the entire water demand for defluoridation process also leads to generation of large amounts of sludge/solid waste as well as wastewater. This requires safe disposal or proper treatment, further adding to the cost.

On the other hand, domestic defluoridation is attractive for the reason that only 5%–20% of the total water requirement needs to be treated, considering this small fraction is typically used for drinking and cooking purposes. This obviously makes domestic defluoridation an attractive option in developing countries where the cost of treatment is a major issue. Contradictory to this advantage, poor user awareness and implementation issues with poor control of water quality create challenges for domestic defluoridation in rural areas of many developing countries. Limitations of POU treatment are also related to the reliability of treatment units, and that all users should be encouraged to use only the treated water for drinking and cooking when untreated water is also available in the house. In general, a centralized approach is adopted also in urban areas of developing countries, and rural habitations appear to be the target areas for domestic defluoridation techniques.

Several methods for defluoridation of water have been developed; however, only a few have found widespread acceptability in terms of techno-economic feasibility. Most of these methods can be broadly categorized according to the principle of water treatment:

- Precipitation-based methods
- Membrane-based processes
- Electrochemical methods of water treatment
- Selective adsorption and ion-exchange-based methods

17.3.1 Coagulation–Precipitation-Based Defluoridation Techniques

These methods involve the addition of soluble chemicals (precipitants) to the water leading to precipitation. Fluoride is then removed either by precipitation, coprecipitation, or adsorption onto the formed precipitate. Precipitation or coprecipitation has been studied or practiced using a range of precipitants, including calcium oxide, magnesium oxide, calcium chloride, monosodium phosphate, alum, and an "alum and lime mixture."

Precipitation is recognized as one of the cheapest and well-established methods of fluoride removal from water (Ayoob et al., 2008; Steenbergen et al., 2011). The method involves stirring or mixing in one of the constituents mentioned above, allowing the precipitate time to settle; ageing; and decanting or separation. Although some of these processes are established for household and small-community-scale operation, and the chemicals are easily and often locally available, there are certain limitations associated with precipitation-based processes. Some of the limitations of precipitation-based defluoridation technologies include large quantity of sludge generation, uncontrolled pH of treated water, difficulty in establishing dose requirement, and higher dose requirements for higher F concentrations (Ayoob et al., 2008). Optimum dose of precipitant must be ensured as some of the aluminum-based precipitants could pose water quality issues with serious health impacts.

17.3.1.1 Nalgonda Technique

A comprehensive research program has been carried out (1960–1970) at the CSIR-NEERI in Nagpur, India, to develop appropriate methods for defluoridation of drinking water. As an important outcome of this program, it was concluded that the Nalgonda technique could be preferable at all levels because of the low price and ease of handling (Bulusu et al., 1979). The Nalgonda technique has been introduced in Indian villages and studied at pilot scale in, e.g., Kenya, Senegal, and Tanzania (Gitonga, 1984; Lagaude et al., 1988; Gumbo, 1987; Rao et al., 2010). The Nalgonda technology is still used in countries like Kenya, Ethiopia, and Tanzania, and the technology is modified by using local resources and knowledge. It has been applied in India at different levels. On a household scale, it is introduced in buckets or drums and at a community scale in fill-and-draw plants. For larger communities, a waterworks-like flow system is developed, where the various processes of mixing, flocculation, and sedimentation are separated in different compartments (NEERI, 1987).

The aluminum sulfate– and lime-based coagulation–flocculation sedimentation was the principle for defluoridation in the Nalgonda technique. It was developed for the low-cost use at all levels in India (household, village community, and waterworks). It is based on the combined use of alum and lime in a two-step process and has been claimed as one of the most effective techniques for fluoride removal. In the Nalgonda technique, two chemicals, alum (aluminum sulfate or calcium aluminum sulfate) and lime (calcium oxide), are added to and rapidly mixed with the fluoride-contaminated water. Induced by a subsequent gentle stirring, "cotton wool"–like floc develops (aluminum hydroxides) and is subject to removal by simple settling. The chemical reaction involving fluorides and aluminum species is complex. It is a combination of two steps involving polyhydroxy aluminum species complexation with fluorides and their adsorption on polymeric aluminum hydroxides (floc). Besides fluorides, turbidity, color, odor, pesticides, and organics are also claimed to be removed. The bacterial load is also reduced significantly. All these are achieved by the adsorption on the floc. Lime or sodium carbonate ensures the adequate alkalinity for an effective hydrolysis of aluminum salts, so that the residual aluminum does not remain in the treated water. Simultaneous disinfection is achieved with bleaching powder, which also keeps the system free from undesirable biological growths.

The method was extensively investigated under laboratory conditions and then tested in the field. Such field studies included experiments in buckets and drums (batch type) as well as in pilot plants (continuous). The technique was seen to be effective in producing clear water, with fluoride within permissible limits. The process is applicable for removal of a wide range of fluoride concentration in water. However, owing to the use of a high dose of aluminum sulfate as a coagulant, the sulfate ion concentration in treated water sometimes increases substantially.

17.3.1.2 Chemo-Defluoridation Technique

In chemo-defluoridation (Figure 17.3), the salts of calcium and phosphorous have been used to reduce the raw water fluoride concentration in the range 5–10 to <1 mg/L. The required dose of chemicals are added in the fluoride-contaminated raw water and mixed with the stick. The chemicals react with each other to form the chemical complex that sorbs fluoride and are precipitated out. After 15–20 min of mixing of the chemicals, water is allowed to flow by gravity into the sand filter at the rate of 300–400 mL/min. Filtered water with fluoride concentration <1 mg/L is collected in the third plastic container and can be used for drinking and cooking purposes. The layer of chemical complex precipitate formed on the sand filter also removes some fluoride from the water during filtration. After about 1–2 months of operation, the filter is choked by the formation of a thick layer of sludge. The Nylobolt cloth kept below the 2–5 cm of the sand layer is lifted along with the sand, and the sand is washed separately in a tub or bucket. The Nylobolt cloth is cleaned and replaced on the sand bed. Washed sand is again placed over the Nylobolt cloth. The filter is ready to use. The process is free from interference of any anion concentration in the water and does not affect the palatability of the water.

Under a project sponsored by the Rajiv Gandhi Science and Technology Commission (government of Maharashtra), 70 defluoridation plants were installed in Sakhara village in Yavatmal district in Maharashtra to test the performance of the units. In addition,

FIGURE 17.3
Chemo-defluoridation unit of CSIR-NEERI, India.

chemo-defluoridation units are also installed in Chichkavtha village of Nagpur district, and people have been using treated water for >6 months.

17.3.2 Electrocoagulation

In the electrocoagulation method, a coagulant is generated *in situ* by the electrochemical reaction (generally oxidation) of a suitable anode (commonly aluminum), which leads to flocculation. The flocculated mass is then settled and separated similar to traditional water treatment techniques. In this way, an aluminum (or iron) anode is placed in the water stream and electric current is passed that produces Al^{3+} ions, which subsequently hydrolyses to form hydroxide precipitate with fluoride. The electrocoagulation principle is quite similar to the Al coagulation–flocculation method. Instead of adding an Al salt, the Al^{3+} ions are generated by electrolytic oxidation at the anode, often offering better control. The generation of the Al^{3+} ion can be represented as Al (s) $\rightarrow Al^{3+}$ (aq) $+ 3e^-$. The settled mass is removed by any separation method like filtration. Various combinations, such as electrodialysis + ion exchange membranes, have been evaluated; however, not many are being used currently in the field (Ayoob et al., 2008). One such method developed by CSIR-NEERI India is further discussed in a later part of this chapter.

17.3.3 Contact Precipitation Method

As reported by Bailey et al. (2006) in a contact precipitation method, fluoride is removed from water through the addition of calcium- and phosphate-based compounds. The presence of a bone charcoal medium acts as a catalyst for the precipitation of fluoride in the form of CaF_2 and/or fluorapatite. Evaluation studies at a community level in Tanzania have shown results with high efficiency using field water. The method is reported to show good reliability with acceptable water quality and low-cost advantages for defluoridation of water (Fawell et al., 2006).

17.3.4 Magnesia-Based Precipitation

Magnesium oxide (MgO) can be effectively used in a precipitation–filtration type technique to remove fluoride from water. The mechanism for removal of fluoride ions is suggested to be based on chemisorption or adsorption. MgO after adding to water forms magnesium hydroxide ($Mg(OH)_2$), which can combine with fluoride to produce an insoluble magnesium fluoride (MgF_2) precipitate. Addition of MgO, however, leads to an increase in pH and other compounds are commonly added to the water to maintain the pH and promote sedimentation. The following fluoride–magnesium chemistry is reported (Ayoob et al., 2008):

$$MgO + H_2O \rightarrow Mg(OH)_2$$

$$Mg(OH)_2 + 2F^- \rightarrow MgF_2\downarrow + 2OH^-$$

The solubility product K_{sp} for MgF_2 is 7.4×10^{-11}, while the solubility for MgF_2 in pure water is 2.64×10^{-4} M.

This process is however, reported to be associated with limitations such as high initial cost, large MgO dose, alkaline pH of the treated water, as well as separation and disposal of any precipitate. The pH of the treated water was reported to be as high as

10 and above, and therefore requires acidification or carbonation. All this adds to the cost and complexity of the method, which results in difficulties in field operation. The acid requirement for neutralization could be as high as 300 mg/L expressed in terms of $CaCO_3$ per liter.

17.4 Membrane-Based Defluoridation Processes

Membrane-based processes have grown manifold during the last two decades and are finding important applications in water treatment as well as desalination. A variety of polymeric and ceramic membranes have been developed and commercially produced in recent years, leading to more interest in membrane-based water purification technologies. The specific applications of membrane fluoride removal include reverse osmosis (RO), nanofiltration (NF), electrodialysis, and Donnan dialysis. These are briefly discussed in the following sections.

17.4.1 Reverse Osmosis

Reverse osmosis is a technique in which a solvent is forced through a semipermeable membrane by applying pressure greater than the osmotic pressure of the solution. Membrane filtration processes usually remove the target particulate constituents on the basis of size exclusion; however, RO does not use the same principle. The RO process rejects the select constituents due to electrostatic repulsion at the membrane interface, chemical solubility, and diffusivity as well as straining of solutes (Boysen, 2008). The US Environmental Protection Agency lists RO as a "best available technology" for fluoride removal (Angers, 2001). Limitations of RO include a higher energy requirement associated with operation of high-pressure pumps, complexity of the process, water quality issues due to removal of all dissolved constituents in an uncontrolled separation, waste management challenge, and the loss of water rejection as well as that associated with the disposal of the high solute concentrate. However, these processes are increasingly improved and finding use for the final polishing of the treated effluent (Tetra Chemicals Europe, 2010) as well as for water treatment.

17.4.2 Nanofiltration

Nanofiltration is a low-pressure (compared with RO) membrane separation process that removes constituents on the basis of size exclusion and ion rejection principles. NF membranes have slightly larger pores than those used for RO processes. This results in a low-pressure requirement, making this a relatively low-energy process. Removal of solutes is much less complex, and outputs are faster. The selectivity of NF relative to RO is yet another distinct advantage, and much experimental and theoretical research work is being focused to obtain a better idea of the mechanism of solute retention to facilitate production and selection of targeted membranes and optimization of conditions for their most efficient use. Retention of solutes is attributed mainly to steric and charge effects. Fluoride is a small ion; however, it is more strongly hydrated than other monovalent anions owing to its high charge density. The resulting steric effect leads to fluoride being more strongly retained on NF membranes than competing monovalent anions like chloride or nitrate. This is of particular advantage in defluoridation of brackish waters with high total dissolved solids (TDS). A large percentage of feed volume, nearly 20%, is generally wasted. Elazhar et al. (2009) reported that rejection of F is quite high (~98% for an initial F concentration of 2.32 mg/L) and costs are comparable between existing NF drinking water treatment plants and an NF facility designed for selective fluoride removal.

17.4.3 Electrodialysis

Electrodialysis is similar to the RO process for separating ions, but uses an electrical gradient to separate ions through semipermeable membranes rather than using pressure. Negatively charged ions (such as F^-) migrate toward positively charged anodes, and are prevented from further migration due to a negatively charged compensating cation exchange. In this way, the main stream going through the exchange loses its contaminants. Depending on initial water quality, electrodialysis requires pretreatment to reduce fouling on the anodes. Waterworks in the United States is reported to use this technology (Amor et al., 1998).

17.4.4 Donnan Dialysis

Donnan dialysis is an irreversible ion-exchange equilibrium nonporous membrane-type separation process. Experiments have shown it to be highly efficient but expensive (Ayoob et al., 2008). Hichour et al. (1999) studied the Donnan dialysis process in a counter-current flow system in which the anion-exchange membrane was incorporated with sodium chloride and the feed used was 0.001 M NaF together with other sodium salts. Fluoride was migrated into the receiver as other ions migrated into the feed. This technique was later used for removal of fluoride from simulated solutions containing high-fluoride African groundwaters with >30 mg/L fluoride, while other ions present with the fluoride in the feed could also be brought to <1.5 mg/L (Hichour et al., 2000). Subsequently, dialysis was combined with adsorption, with the addition of aluminum oxide and zirconium oxide to the receiver to force the Donnan equilibrium in the direction of fluoride flow out of the feed. In this way, it was possible to maintain a flow of feed while leaving the receiving solution in place or renewing only in several batches. The cation composition was found to remain unchanged, whereas anions, with exception of chloride, were partially eliminated and substituted by chloride ions, resulting in a residual fluoride concentration of 1.5 mg/L (Garmes et al., 2002).

17.5 Distillation

Distillation is a physical process that can be used to remove the solid impurities from high-salinity feed water by evaporation followed by condensation. Distillation devices convert liquid water to steam and collect the condensed steam for application. This process, however, requires significant energy considering the higher latent heat of water. A domestic distillation process has been reported to remove up to 99% of fluoride from water (Brown and Aaron, 1991); however, much lower removal has also been reported. The effectiveness of distillation is not influenced by the pH of the water, while unlike adsorption processes, the distillation process does not get affected with exhausted capacity of adsorbents. Also, any problems in the distillation can be identified by the absence of water exiting the device. Energy consumption is the major challenge, and a lot of efforts have been made to design and develop energy-efficient distillation stills. The material that is removed from water gets deposited in the distillation compartment of the unit, which needs regular cleaning and is sometimes tedious. Chemicals are commonly used to effectively help remove these deposits, and the frequency of cleaning is obviously dependent on the amount and nature of the material dissolved in the input water used for distillation. Distillation can remove F and almost all other constituents in water; therefore, essential salts may need to be supplemented. Household-sized distillation systems are available, while solar stills are also being considered for cleaner energy and low cost. In this way, although distillation offers effective removal of fluoride, high-energy cost and the removal of useful water constituents are major limitations, so this is not a popular option for defluoridation.

17.6 Electrochemical Methods

Electrochemical methods have also been extensively used for removal of fluoride. This method involves electrosorption of fluoride on electrogenerated aluminum hydroxide floc, produced by the anodic dissolution of aluminum or its alloys, in an electrochemical cell (Lounici, 2004). However, the process is energy intensive, which can be overcome by using electricity generated by solar panels. Another disadvantage is that significant quantity of fluoride-rich sludge is generated, which is sometimes difficult to dispose.

17.6.1 Electrolytic Defluoridation Technique of CSIR-NEERI

A defluoridation technique based on the principle of electrolysis has been developed in NEERI, and the first plant based on this technique was installed in 2002 at Dongargaon village, Chandrapur district, Maharashtra, India. This technique requires simple equipment and is easy to operate and controlled electrically with no moving parts, thus requiring less maintenance. It can be conveniently used in rural areas where electricity is not available, as solar panels can be coupled to the unit. The technique avoids use of chemicals, and there is no problem to neutralize excess chemicals, and it produces palatable, clear, colorless, and odorless water. The process is found effective in removing excess fluoride and also bringing down the bacterial load of the raw water. The defluoridation process is based on the principle of electrolysis, using aluminum plate electrodes placed in the raw water containing excess fluoride. During the electrolysis, the anode gets ionized and fluoride is removed by complex formation, adsorption, precipitation, coagulation, and settling (Gwala et al., 2012).

An electrolytic defluoridation unit basically consists of an electrolytic reactor having an aluminum anode and cathode. When a direct current (DC) source is applied across the electrodes, the anode dissolves and hydrogen gas is released at the cathode. During the dissolution of the anode, various aqueous metallic species are produced, which depend on the solution chemistry. These metallic species act as a coagulant by combining with the fluoride ions present in the water to form large-size floc, which can be removed by sedimentation or filtration. The electrolytic defluoridation process using sacrificial aluminum electrodes has been demonstrated to be an effective process since it does not require a substantial investment, presents similar advantages as chemical coagulation, and reduces disadvantages and produces minimum sludge.

EC is an electrochemical process that comprises chemical and physical processes involving many surface and interfacial phenomena. The technology lies at the DC required for the electrolytic process, which is generated either by conversion of an alternating current electric supply by a DC conversion unit of required capacity or by a solar photovoltaic system consisting of solar panel, charge controller, and tubular battery. The solar panels as per required power (watt) and time are mounted in a southeast direction at a place where direct sunlight is received. The DC received through the solar panel is stored in the tubular battery through a charge controller and supplied to the aluminum electrodes through a DC regulator.

The main reactions involved in the EC are as follows:

$$Al \rightarrow Al^{3+} + 3e^{-} \quad \text{at the anode} \tag{17.1}$$

$$Al^{3+} + 3H_2O \rightarrow Al(OH)_3 + 3H^{+} \tag{17.2}$$

$$Al(OH)_3 + xF^{-} \rightarrow Al(OH)_{3-x}F_x + xOH^{-} \tag{17.3}$$

$$2H_2O + 2e^{-} \rightarrow H_2 + 2OH^{-} \quad \text{at the cathode} \tag{17.4}$$

On the basis of the technology, solar-power-based electrolytic defluoridation demonstration units (Figure 17.4) were installed at Dongargaon in Chandrapur district, Maharashtra, in 2002 and at Usarwara village in Durg district in Chhattisgarh state in 2010. The capacity of these units is 600 and 2000 L per batch, respectively.

Three more plants were installed at Sargapur in Seoni district, Madhya Pradesh, in 2010; at Malgaon in Balod district, Chhattisgarh, in 2011; and at Adiwasi Kanya Shiksha Parisar in Chhindwara district, Madhya Pradesh, in 2012, keeping in view the successful performance of earlier plants. These plants have a capacity to supply 2000 L per batch to a community of around 600–700 persons (Figure 17.5).

FIGURE 17.4
(See color insert.) Electro-defluoridation field units by CSIR-NEERI, Nagpur, India.

FIGURE 17.5
Electro-defluoridation units installed at different fluoride affected regions.

17.7 Adsorption- and Ion-Exchange-Based Defluoridation Methods

A wide variety of adsorbent materials have been reported in the past few years for the efficient and economical defluoridation of water. The most commonly used adsorbent for water treatment, particularly for defluoridation of water, is activated alumina (AA), which is currently used in the field for defluoridation of water in many parts of the world. The various sorbents studied and investigated for fluoride removal are reviewed in detail as follows.

17.7.1 Activated Alumina

Various alumina-based adsorbents have been reported in the past few decades for the effective removal of fluoride from drinking water, and clearly AA is the most studied adsorbent for field application. The fluoride removal efficiency of these materials mainly depends on the nature of the adsorbent. The most commonly used one for the defluoridation of drinking water is adsorption by AA, which has a high affinity for fluoride and can be regenerated with mild acid or alkali. However, fluoride removal by AA is strongly pH dependent and the presence of other co-ions also significantly interferes in the fluoride removal. The use of AA in a continuous-flow fluidized system is an economical and efficient method for defluoridating water supplies. The process could reduce the fluoride levels down to 0.1 mg/L. The operational, control, and maintenance problems, mainly clogging of bed, may be averted in this method.

Different researchers have reported different fluoride adsorption capacities for AA. Ku and Chiou et al. (2002) have reported fluoride adsorption capacity of 16.3 mg/g at pH ranging from 4.0 to 6.0. Ghorai and Pant (2005) investigated the adsorption of fluoride by AA in a fixed-bed column showing 2.41 mg/g of fluoride adsorption capacity at neutral pH. Tripathy et al. (2006) have reported the fluoride removal performance of alum-impregnated alumina significantly high at 40.38 mg/g at pH of 6.5 and fluoride concentration of 1–35 mg/L. To overcome the limitations of AA, several modified alumina have also been reported. Tripathy and Raichur (2008) have found manganese dioxide–coated AA as a promising adsorbent for enhanced defluoridation of water as compared with uncoated AA. Coated AA adsorbed more fluoride than uncoated over a wider pH range. More importantly, maximum fluoride adsorption occurred at pH ~7.0 in the case of coated alumina, which makes it a potential adsorbent for treating drinking water. Maliyekkal et al. (2008) described the improved performance of Magnesia-amended AA (MAAA) for fluoride removal. MAAA also exhibits high removal efficiency (>95%) at neutral pH. The maximum sorption capacity of fluoride deduced from the Sips equation was 10.12 mg/g. Lee et al. (2010) prepared samples of two different kinds of mesoporous alumina using aluminum tri-sec-butoxide in the presence of either cetyl-trimethyl ammonium bromide (MA-1) or stearic acid (MA-2) as a structure-directing agent, and tested for adsorptive removal of fluoride in water. MA-2 prepared using stearic acid exhibited a high adsorption capacity of 14.26 mg/g for fluoride, which is significantly higher than those of a commercial gamma alumina. More recently, Bansiwal et al. (2010) reported the fluoride removal performance of AA modified by incorporating copper oxide. They reported an interesting result that the optimal fluoride adsorption occurs at all pH ranging from 4.0 to 9.0 with 7.22 mg/g fluoride adsorption capacity. This material overcomes the drawback associated with AA showing optimal fluoride adsorption below pH 6.0 and low adsorption capacity. A high desorption efficiency of 97% was also achieved by treating a fluoride-loaded adsorbent with 4 M NaOH solution.

17.7.2 Bone Charcoal

Bone charcoal (bone char) has also been reported for the removal of fluoride from drinking water by adsorption. Bone char is obtained by heating bones to temperatures that remove organics, leaving behind hydroxyapatite (HAp), $Ca_{10}(PO_4)_6(OH)_2$ (Kaseva, 2006). Regeneration of fluoride-saturated bone char is also possible by reheating or by leaching with sodium hydroxide and, ultimately, the final product can be applied as fertilizer to household gardens.

17.7.3 Clay

A wide variety of clays have been used for removal of fluoride through adsorption or flocculation. Several clays being used for fluoride removal include ground and fired clay pot, brick chips, calcined clay, palygorskite clay, calcite, kaolin, etc. (Hamdi et al., 2009; Bårdsen et al., 1995), and removal is highly dependent on the type of clay used and its modifications. However, since clays have low fluoride removal capacity and poor hydraulic conductivity, their practical applicability for field applications is limited.

17.7.4 Ion-Exchange Resins

Ion-exchange resins, particularly strong-base-exchange resins, have also been evaluated for fluoride removal from water (Haron et al., 1995; Castel et al., 2000); however, the presence of other anions severely interferes, which results in low effective fluoride removal capacity of such resins. Several inorganic ion exchangers, e.g., metal chloride silicates, formed from barium or ferric chloride with silicic acid, also exchanged fluoride for chloride. Cation-exchange resins impregnated with alum solution have also been reported as effective defluoridation media. Polystyrene anion-exchange resins, in general, and strongly basic quaternary ammonium-type resins, in particular, are known to remove fluorides from water along with other anions.

17.7.5 Carbon

Activated carbon from coal, agricultural waste, etc., is one of the most commonly used adsorbent materials for water treatment that can effectively remove the organic as well as inorganic pollutants. In the past few decades, the fluoride removal performance of activated carbon derived from various sources has been investigated by many researchers. It has been reported that activated carbon derived from various low-cost materials exhibits good fluoride removal capacity and the optimum fluoride adsorption occurs at pH <3, suggesting that the fluoride adsorption is pH specific. Sivasamy et al. (2001) investigated fluoride removal from drinking water using different coal-based adsorbents such as lignite, fine coke, and bituminous coal as a low-cost source. They reported significantly high fluoride adsorption capacities for these coal-based adsorbents, which are 7.09, 6.90, and 7.44 mg/g for lignite, fine coke, and bituminous coal, respectively, as compared with activated carbon derived from wet coconut shell carbon (3.03 mg/g), dry coconut shell carbon (1.08 mg/g), and dry commercial activated carbon (0.19 mg/g). In the case of lignite, the higher defluoridation occurs at a pH range of 5.0–10.0. On the other hand, the higher defluoridation for fine coke and bituminous coal occurs at acidic pH <4.0. Li et al. (2001) have reported nanotubes as a support material for incorporating alumina and investigated it for fluoride removal from drinking water. Amorphous alumina supported on carbon nanotubes (CNTs) have shown the highest fluoride adsorption capacity of 28.7 mg/g, and the optimum fluoride removal occurred at pH of 5.0–9.0. Kuo et al. (2004) reported various carbonaceous materials, including six kinds of activated carbon, carbon black, four kinds of charcoals, and bone char for fluoride removal from water, and they found that the fluoride adsorption on these adsorbents depends on their specific surface area. The amount of fluoride adsorbed on the carbonaceous materials depends on the pore size distribution because the adsorption occurs in the pores, suggesting physical adsorption. Among all the reported carbonaceous materials, bituminous coal exhibits higher fluoride adsorption capacity at pH of 4.6. Parmar et al. (2006) investigated the adsorption of fluoride on corncob powder. Powdered corncob did not show remarkable adsorption but aluminum-treated corncobs exhibited good adsorption capacity of 18.9 mg/g at pH 5.0–6.5. Karthikeyan and Ilango (2007) have also prepared activated carbon from the burning and carbonization of *Morringa indica* bark for fluoride removal from drinking water. This material showed fluoride removal capacity of 0.17 mg/g at acidic pH. Daifullah et al. (2007) synthesized $KMnO_4$-modified activated carbon from the steam pyrolysis of rice husk, which is the agricultural waste, and investigated extensively for fluoride removal from water. The maximum fluoride adsorption for synthesized activated carbon occurred at a pH of 2.0, and the fluoride adsorption capacity obtained was 15.5 mg/g. Sivasankar et al. (2010) have reported fluoride removal capacities of activated tamarind fruit shell (ATFS) and ATFS modified with MnO_2, and found that ATFS and MTFS were effective at pH of 6. Alagumuthu and Rajan (2010) have reported zirconium impregnated cashew nut shell carbon for the removal of fluoride from aqueous solutions at neutral pH and low adsorbent dose. Tchomgui-Kamga et al. (2010) have synthesized charcoal adsorbents that contain dispersed aluminum and iron oxides by impregnating wood with salt solutions followed by carbonization at 500°C, 650°C, and 900°C. Substrates prepared at 650°C with aluminum and iron oxides exhibited the best efficiency with a fluoride sorption capacity of 2.31 mg/g. More than 92% removal of fluoride was achieved within 24 h from a 10 mg/L solution at neutral pH.

17.7.6 Industrial and Agricultural Waste Materials as Adsorbents

The utilization of the waste materials generated from the various industries and agricultural field as a low-cost resource is a viable option for water treatment processing. Sujana et al. (1998) have reported the use of alum sludge, which is a waste product, generated during the manufacture of alum from bauxite by the sulfuric acid process, as an adsorbent for fluoride removal from drinking water. The fluoride adsorption capacity of alum sludge was 5.35 mg/g and the maximum fluoride adsorption occurs at pH of 5.5–6.5. Cengeloglu et al. (2002) have extensively studied the potential of red mud as such and activated with acid for fluoride removal from drinking water. Red mud is a strongly alkaline waste produced during the aluminum ore processing. It was observed that red mud has an appreciable fluoride adsorption capacity, which can be further improved by treatment with acid. However, the maximum fluoride adsorption occurs at pH 5.5, and the maximum fluoride adsorption capacities were found to be 3.12 and 6.29 mg/g for as is and acid-activated mud, respectively. Mahramanlioglu et al. (2002) have reported on acid-treated spent bleach earth, a solid waste generated from the edible oil processing industry for fluoride removal from drinking water. The maximum fluoride adsorption was observed at pH 3.5, and its fluoride adsorption capacity was 7.75 mg/g.

Nigussie et al. (2007) have reported an aluminum sulfate manufacturing process waste as low-cost material generated for removal of excess fluoride from water. The material exhibits fast kinetics with adsorption capacity of 332.5 mg/g at an initial fluoride concentration of 100–550 mg/L. Tor et al. (2009) have also studied granular red mud for fluoride sorption capacity in batch and column adsorption experiments with maximum adsorption occurs at pH of 4.7 and the adsorption capacity was 0.851 mg/g. The exhausted fixed-bed column was regenerated by treating with mild alkali to desorb the fluoride and desorption efficiency of 87%–46% was achieved from the first to fourth cycle. Kemer et al. (2009) reported that precipitated waste mud (p-WM), which was obtained as a by-product from a Cu–Zn mine (Çayeli, Rize, Turkey), is a promising material to remove excess fluoride from water and wastewater. Maximum fluoride uptake was obtained at 27.2 mg/g with p-WM. The results have demonstrated that the p-WM can be used as a low-cost, highly effective, and readily available adsorbent for removal of fluoride from aqueous solutions.

17.7.7 Biopolymer-Based Adsorbents

Extensive efforts have also been made to develop defluoridation media based on biopolymers, viz. chitin and chitosan. Ma et al. (2007) have reported magnetic chitosan particles prepared by a coprecipitation method. The authors claimed excellent fluoride removal capacity of 22.49 mg/g for magnetic chitosan particles at neutral pH, and therefore, can be used as a potential adsorbent for defluoridation of water and wastewater. Regeneration studies show that loaded fluoride magnetic particles were recovered by 0.8–1.0 M NaOH solution, and the regenerated adsorbent retained the adsorption capacity up to 98%–99% of the fresh adsorbent. Kamble et al. (2007) have studied the potential of chitin, chitosan, and lanthanum (La)-incorporated chitosan as adsorbents for the removal of excess fluoride from drinking water. Lanthanum chitosan adsorbents show excellent removal of fluoride from water, which is significantly higher than unmodified biopolymers. However, it was observed that the presence of anions, viz. carbonate and bicarbonate, has a deleterious effect on the adsorption of fluoride. The adsorption capacity of 20% La-chitosan was 3.1 mg/g at pH 5 and the material can be regenerated easily.

Sahlia et al. (2007) reported removal of fluoride from water by adsorption on chitosan. Sundaram et al. (2008) synthesized a bioinorganic composite namely nano-HAp/chitosan (n-HApC) and evaluated its performance for removal of fluoride. Defluoridation capacity of the n-HApC composite was observed to be 1.560 mg/g, which is slightly higher than nano-HAp (n-HAp) having a defluoridation capacity of 1.296 mg/g. Field trials were also conducted suggesting that the n-HApC composite can be used as an effective defluoridation agent.

Viswanathan and Meenakshi (2008) also reported carboxylated chitosan beads (CCB) and lanthanum incorporated CCB (La-CCB). The adsorption capacity of CCB was observed to be 1.385 mg/g, which increased to 4.711 mg/g after La incorporation. The adsorption capacities of both CCB and La-CCB were significantly higher than fluoride adsorption capacity of raw chitosan beads (CB) (0.052 mg/g). The results of field trials indicated that the La-CCB can be effectively used for removal of fluoride. Viswanathan et al. (2009) also reported chitosan beads modified by simple protonation. Protonated chitosan beads (PCB) showed maximum defluoridation capacity of 1.664 mg/g, whereas raw chitosan beads (CB) possess only 0.052 mg/g. The removal efficiency was observed to be independent of pH; however, the presence of other coexisting anions also affects the removal. The study also established that the materials can be easily regenerated using 0.1 M HCl. The authors further extend modification of CB by multifunctional groups, namely NH_3^+ and COOH groups. The protonated and carboxylated chitosan beads (PCCB) showed a maximum defluoridation capacity of 1.8 mg/g, which was significantly higher than raw chitosan beads. Field trial results confirmed that fluoride levels below permissible limits can be achieved using PCCB, and it can also remove other ions in addition to fluoride.

Titanium-modified chitosan-based adsorbent has been explored for defluoridation of water. Titanium macrospheres (TM) prepared by a precipitation method showed an excellent fluoride removal capacity of 7.2 mg/g, which is high compared with raw and previously reported modified chitosan (Jagtap et al., 2009). The good stability and settling property of TM ensures good separation and makes it a potential material for defluoridation. Davila-Rodriguez et al. (2009) prepared a biocomposite based on chitin and a polymeric matrix, and studied the adsorption of fluoride from aqueous solutions. The biocomposite shows fluoride adsorption capacity of 0.29 mg/g at pH 5.0 and initial fluoride concentration of 15 mg/L. Sundaram et al. (2009) reported a novel nanohydroxyapatite/chitin (n-HApCh) composite for defluoridation of water with defluoridation capacity of 2.84 mg/g. The n-HApCh composite has several advantages, viz. biocompatibility, low cost, and local availability. Viswanathan et al. (2010) have prepared and investigated a new biocomposite by incorporation of zirconium (IV) tungsten–phosphate (ZrWP) into the biopolymer matrix. The adsorption capacity of fluoride from water by this ZrWP/chitosan (ZrWPCs) composite was found to be 2.025 mg/g at pH 3.0 and initial fluoride concentration of 10 mg/L. Considering the high affinity of fluoride toward alumina, Viswanathan et al. (2010) explored the possibility of preparation of alumina and biopolymeric composite. Alumina/chitosan composite was synthesized by incorporating alumina particles in the chitosan polymeric matrix, which showed maximum defluoridation capacity of 3.809 mg/g, significantly higher than both alumina and raw biopolymer. Jiménez-Reyes et al. (2010) have explored the efficacy of HAp as a potential material for the treatment of water contaminated with fluoride ions. The sorption capacity of the HAp-based adsorbent was 4.7 mg/g at pH 5–7.3 with initial fluoride concentration of 5 mg/L. Thakre et al. (2010) synthesized lanthanum-incorporated chitosan beads (LCB) using a precipitation method, and evaluated for fluoride removal from drinking water. The LCB-10 exhibited fluoride adsorption capacity of 4.7 mg/g, which is reasonably higher than the commercially used AA and significantly higher than raw chitosan. Besides having high adsorption capacity, LCB-10 also possesses relatively fast kinetics, high chemical and mechanical stability, high resistance to attrition, negligible lanthanum release, suitability for column applications, etc.

17.7.7.1 Composite-Type Adsorbent-Based Domestic Defluoridation Unit "NEERMAL" of the CSIR-NEERI

A domestic defluoridation drinking water unit named "NEERMAL" has been designed in which the developed material NEERMAL was used as defluoridation media (Figure 17.6). A pair of cartridges containing NEERMAL mixed with sand has been retrofitted in a commercial filter. NEERMAL material can be applied for defluoridation of wastewater also.

NEERMAL, a composite material based on metal oxide and carbon, nitrogen, and other elements/functional groups for removal of fluoride from water and wastewater has been developed. The composite-based adsorbent shows fluoride uptake capacity in the range of 5–14 mg/g under different conditions, with good selectivity and a wide pH range for fluoride uptake. The field trials on the prototype units are proposed for detailed evaluation with respect to its feasibility. The unit is especially designed for domestic applications in India and other developing countries. It works on gravity flow with no electricity requirement and gives fluoride levels below the prescribed limit of 1 mg/g in treated water.

17.7.7.2 Household Treatment Methods for Fluoride Removal in Rural Areas by CSIR-NEERI

The CSIR-NEERI in collaboration with UNICEF (Bhopal unit) has carried out user perception studies for assessing the acceptability of various domestic defluoridation techniques for rural applications. Some of these techniques are described here.

17.7.7.2.1 Loose Sorbent

In this method, a predefined dose of adsorbent is added to approximately 1 L of water and stirred continuously by using a stick for about 10–15 min (Figure 17.7). Stirring ensures a good contact between adsorbent and fluoride-contaminated water. After stirring, the adsorbent was separated by filtering the treated water through a piece of cotton cloth.

FIGURE 17.6
Domestic defluoridation unit prototype based on composite-type adsorbent (CSIR-NEERI, India).

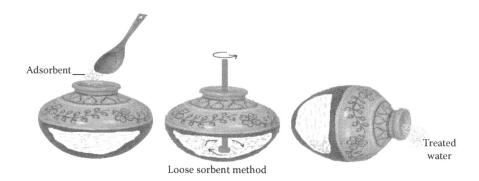

Adsorbent

Loose sorbent method

Treated water

FIGURE 17.7
Schematic of loose sorbent technique for domestic rural water defluoridation.

Treated water was then evaluated for its fluoride content and other physicochemical parameters to check its potability.

17.7.7.2.2 Sachet Technique

The method involves simple stirring of a tea bag–type sachet, containing adsorbent material for application in the field for domestic defluoridation. A prepacked sachet technique using a controlled dose for the removal of fluoride from a pot of water would be appropriate to facilitate maintaining an exact amount of adsorbent for treatment. A predefined controlled dose would facilitate the effective use of adsorbents in an uncomplicated manner by the user. For treatment of larger volumes of water, an appropriate dose of adsorbent in multiple sachets may be used. This method involves stirring of adsorbent packed sachets for 10–15 min (Figure 17.8).

17.7.7.2.3 Bamboo Column Technique

A bamboo column as shown in Figure 17.9 is closed at the bottom. The bamboo is filled with a layer of fine sand and a top layer with adsorbent. Water is fed through the top from a pot with an optimum flow rate. This treatment may be suitable for overnight treatment for a volume of water sufficient for a family. The residence time in the column was about 10 min but will depend on the input fluoride concentration.

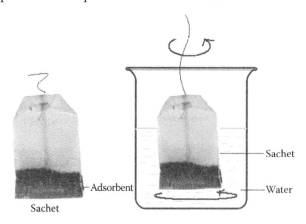

Sachet

Adsorbent

Sachet

Water

FIGURE 17.8
Schematic of Sachet-based technique for domestic rural water defluoridation.

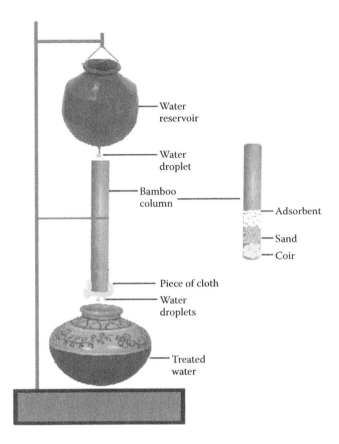

FIGURE 17.9
Schematic of bamboo-column-based technique for domestic rural water defluoridation.

17.7.8 Recent Developments in Materials for Defluoridation

Adsorption-based defluoridation processes still dominate water treatment. There have been several reports on the development of improved adsorbents in recent years. Ben et al. (2011) evaluated the cuttlefish bone as an adsorbent material (available in Tunisia) for the defluoridation of water. The results show effective sorption of fluoride on cuttlefish bone even when natural/field water was tested. The efficacy for removal of fluoride from water was found to be about 80% at pH 7.2, 1 h contact time, 15 g/L adsorbent dose, and 5 mg/L initial fluoride concentration. Despite the different anions generally present in natural waters, a fluoride concentration in agreement with the norm (<1.5 mg/L) could be achieved in a variety of natural water samples examined.

Zhang et al. (2012) reported on a new adsorbent zirconium-modified Na-attapulgite (Zr-A), synthesized by a facile method. The Zr-A adsorbent exhibited good adsorption for fluoride over a wide pH range. One of the interesting observations was its high adsorption capacity even after six reuse cycles.

Mohapatra et al. (2012) prepared a high-surface-area nanopowder of Mg-doped ferrihydrite. The results showed 0.98% Mg-doped ferrihydrite to be an effective fluoride adsorbent giving a maximum adsorption capacity of as much as 64 mg/g. The kinetics data fitted well to a pseudo-second-order kinetic model. The isothermal data followed a Langmuir model. Thermodynamic parameters confirmed the adsorption process to be spontaneous and endothermic. In a regeneration study, 89% of fluoride could be desorbed from loaded sample using 1 M NaOH.

Sivasankar et al. (2012) used *Tamarindus indica* fruit shells (TIFSs), a natural compound rich in calcium, to develop a new adsorbent. The material was impregnated with ammonium carbonate followed by carbonization, leading to ammonium carbonate activated ACA-TIFS carbon. The resulting materials and carbon arising from virgin fruit shells V-TIFS were characterized and examined as adsorbents for the removal of fluoride anions from groundwater samples. The fluoride-scavenging ability of TIFS carbons was attributed to naturally dispersed calcium compounds. X-ray diffraction showed that TIFS carbon contained a mixture of calcium oxalate and calcium carbonate. The fluoride removal of TIFS carbons was found to be 91% and 83% at a pH 7.05 for V-TIFS and ACA-TIFS carbons, respectively. The practical applicability of TIFS carbons using groundwater samples was also examined and found satisfactory. The fluoride removal was better in groundwater without hydrogen carbonate ions than those containing the ions.

Pumice stone that is functionalized by the cationic surfactant hexadecyltrimethyl ammonium (HDTMA) is used as an adsorbent for the removal of fluoride from drinking water by Asgari et al. (2012). The effects of HDTMA loading, pH (3–10), reaction time (5–60 min), and the adsorbent dosage (0.15–2.5 g/L) were investigated as a function of fluoride removal from water through different designed experiments. The results revealed that surfactant-modified pumice exhibited the optimal performance at dose 0.5 g/L, pH 6, and it was reported to remove >96% of fluoride from a solution containing 10 mg/L fluoride after 30 min of equilibration time.

17.7.8.1 Nanomaterials for Fluoride Removal

Nanoscience and nanotechnology have arisen as a promising approach in the past decade for various environmental applications. Use of nanomaterials as adsorbents for water treatment is gaining wide attention recently owing to their higher surface area. CNTs as well as other nanoforms of carbons also have appeal in water treatment. Their interesting properties owing to small size, large surface area, high mechanical strength, and remarkable electrical conductivities make them potential materials for a wide range of promising applications. Li et al. (2001) used CNTs as support for depositing Al_2O_3 and discovered the possibility of Al_2O_3/CNTs for removing fluoride from water. The adsorption isotherms inferred that the best fluoride adsorption on Al_2O_3/CNTs was observed in the pH range 5.0–9.0. The adsorption capacity for Al_2O_3/CNTs was reported to be about 13.5× higher than that of AC-300 carbon, about four times higher than that of Al_2O_3 at an equilibrium fluoride concentration of 12 mg/L. The fluoride adsorption for Al_2O_3/CNTs at pH 6.0 was reported to reach 28.7 mg/g at an equilibrium concentration of 50 mg/L. Aligned CNTs (ACNTs) were prepared by catalytic decomposition of xylene using ferrocene as a catalyst, and their performance was tested for fluoride removal from water (Li et al., 2003a). Both the surface and inner cavities of ACNTs were found to be easily available for fluoride sorption. A wide pH range of 3–9 was found optimal for fluoride removal. A high surface area of the adsorbent was reported to favor the sorption rate and adsorption capacity, while microporous diffusion in larger adsorbent particles was observed to take place slowly. Application of nanoparticles has potential for adsorption-based water treatment technologies. The maximum adsorption capacity of ACNTs was observed at pH 7.0 with 4.5 mg/g fluoride uptake at an equilibrium fluoride concentration of 15 mg/L. CNTs were modified with alumina, and the synthesized adsorbent was studied for defluoridation from water and the effects of calcination temperature, alumina loading, and pH on fluoride removal capacity were investigated (Li et al., 2003b). The optimal calcination temperature for synthesizing the adsorbents was 450°C, with the highest adsorption capacity for alumina loading of 30%. High fluoride adsorption capacity was reported for Al_2O_3/CNTs at 25°C temperature in the pH range 6.0–9.0. Fluoride adsorption was claimed to take place by ligand exchange, when the surface was positively charged and through an ion-exchange mechanism when the surface was neutral. It has also been reported that nanosized inorganic oxides were excellent reagents in various separation technologies through selective adsorption. They are known to exhibit excellent adsorption properties owing to their enhanced surface area and large interface volume, depending on the chemistry of the constituent atoms.

Maliyekkal et al. (2010) has prepared nanomagnesia (NM) by a combustion method. The combustion method was reported as advantageous from the perspectives of small size of the nanoparticle, cost-effective recovery of the material, and improvement in the fluoride adsorption capacity. The NM synthesized was crystalline with high phase purity, and the particle size varied in the range of 3–7 nm. Fluoride adsorption by NM was highly favorable and does not vary in the pH range usually encountered in groundwater. Phosphate and bicarbonate was shown to interfere with fluoride for adsorption on NM. The adsorption kinetics followed a pseudo-second-order equation, and the equilibrium data are well predicted by the Freundlich equation.

Pathak et al. (2003) optimized the synthesis of a variety of nanosized inorganic oxides through thermolysis of a polymeric-based aqueous precursor solution of the specific inorganic ions. The obtained Fe_3O_4, Al_2O_3, and ZrO_2 nanosized oxide powders were assimilated on activated charcoal through an adsorption process and used as the adsorbent for removal of trace amounts of fluoride and several other pollutants. The charcoal-embedded fine powders of the inorganic oxides were able to remove fluoride/arsenite and arsenate ions from industrial wastewater up to 0.01–0.02 mg/L levels. Wang et al. (2009) studied fluoride adsorption experiments using nanoscale aluminum oxide and hydroxide as adsorbents, owing to the high surface area of nanoparticles as compared with the conventional micron-sized adsorbents. The Langmuir fluoride adsorption capacity of n-AlOOH was found to be a moderate 3259 mg/kg. The adsorption of fluoride on n-AlOOH was observed to be strongly pH dependent. The fluoride adsorption increased with the increase in pH, reaching a maximum at pH 6.8, and then decreased with further increase in pH. The pH of the adsorbent was reported as ~7.8, which was responsible for the fluoride adsorption in acidic medium; the repulsion between the negatively charged surface and fluoride explained the low efficiency in alkaline medium. Fluoride adsorption at an initial pH >7.8 was due to van der Waal forces and not by anion exchange, which was apparent from the decrease in the final pH. At pH >7.8, n-AlOOH was observed to function as a cation exchanger as it adsorbed the sodium ions present in solution, releasing protons, resulting in decreased pH. Only the presence of sulfate and phosphate significantly was reported to be affecting fluoride uptake by n-AlOOH. Desorption studies confirmed that the fluoride could be desorbed at pH 13. Patel et al. (2009) studied the effectivity of CaO nanoparticles for fluoride removal. Colloidal particles of CaO were synthesized by a sol–gel method. The Langmuir fluoride adsorption capacity was found to be 163.3 mg/g. Fluoride adsorption onto CaO nanoparticles was inferred based on the conversion of CaO to $Ca(OH)_2$ in water, followed by fluoride adsorption through a surface chemical reaction, where hydroxide ions of calcium hydroxide were replaced by fluoride ions resulting in the formation of CaF_2. Fluoride adsorption was observed to be quite constant in a wider pH range of 2–8. A gradual decrease in fluoride uptake above pH 8 was attributed to electrostatic repulsion of fluoride to the negatively charged surface as well as to the competition by an excessive amount of hydroxyl ions. Anions such as sulfate and nitrate did not affect the adsorption of fluoride by CaO nanoparticles.

17.7.8.2 Magnetic Nanoparticles

One of the difficult challenges of nanomaterials in water treatment is related to their separation from water. Separation of the nanosized adsorbents from media is quite complicated in practice. Magnetic nanosized sorbents can, however, overcome the separation issue and seem to be promising owing to ease of their separation. Application of an external magnetic field for the recapture of the nanoparticles is a low-cost and effective option. The magnetic properties of undecorated nanoparticles could be affected by forming aggregates because of their high chemical activity, possibility of oxidation in air, loss of magnetism, dipole–dipole attraction, and agglomeration. Therefore, mounting a passive layer through postsynthetic surface modification is often practiced.

Chang et al. (2006) studied two effective types of superparamagnetic nanoscale adsorbents of bayerite/SiO_2/Fe_3O_4 via three sequential steps: (i) chemical precipitation of Fe_3O_4, (ii) coating of SiO_2 on Fe_3O_4 using an acidifying method, and (iii) further coating of bayerite ($Al(OH)_3$) on SiO_2/Fe_3O_4-adopting sol–gel (MASG) or homogeneous precipitation (MAHP) methods for the elimination of fluoride from water and comparing the adsorption potential of the prepared sorbents with that of commercial AA (CA). The Langmuir isotherm describes the equilibrium of fluoride adsorption on CA, MASG, and MAHP materials. The results showed that the fluoride adsorption capacity of CA could be improved when the initial pH value was 3.5, which was the approximate pK value of the dissociation constant of hydrofluoric acid. Among these, MASG was establish to be the most effective adsorbent with the adsorption capacity of 38 g/kg, and it competed with CA at an even higher pH value such as 6.0 vs. 3.5. Zhao et al. (2010) combined the advantages of $Al(OH)_3$ and magnetic nanoparticles to develop nanosized adsorbents with high surface area, high affinity toward fluoride, and good magnetic separation, to develop a new type of magnetic fluoride adsorbent. This nanosized adsorbent was prepared by using hydrous aluminum oxide embedded with Fe_3O_4 nanoparticles (Fe_3O_4@$Al(OH)_3$ NPs). This was applied to remove fluoride from aqueous solution. Fe_3O_4@$Al(OH)_3$ NPs exhibited effective adsorption of fluoride when the pH was between 5.0 and 7.0, which was attributed to the electrostatic attraction between the positively charged Fe_3O_4@$Al(OH)_3$ surface and fluoride. The adsorption capacity estimated by the Langmuir model was 88.48 mg/g at pH 6.5. Fluoride removal in the presence of the anion was on the order of $PO_4^{3-} < SO_4^{2-} < Br^- \approx NO_3^- \approx Cl^-$. Fluoride adsorption was claimed to be driven by both electrostatic attraction and surface complexation, which was confirmed by XPS data and experimental results.

Chai et al. (2013) synthesized a novel adsorbent of sulfate-doped Fe_3O_4/Al_2O_3 nanoparticles with magnetic separability for fluoride removal from drinking water. The fluoride adsorption isotherm was well described by the Elovich model. Fluoride adsorption capacity by a two-site Langmuir model was 70.4 mg/g at pH 7.0. This nano-adsorbent performed well over a wide pH range of 4–10, and the fluoride removal efficiencies reached up to 90% and 70% throughout the pH range of 4–10 with initial fluoride concentrations of 10 and 50 mg/L, respectively. The observed sulfate–fluoride displacement and decreased sulfur content on the adsorbent surface reveal that the anion-exchange process was an important mechanism for fluoride adsorption by the sulfate-doped Fe_3O_4/Al_2O_3 nanoparticles. With the exception of PO_4^{3-}, other coexisting anions (NO_3^-, Cl^-, and SO_4^{2-}) did not evidently inhibit fluoride removal by the nanoparticles. Findings of this study demonstrate the potential utility of the material.

Maliyekkal et al. (2010) synthesized nanoparticles of MgO by self-propagated combustion of the magnesium nitrate trapped in cellulose fibers to enhance the adsorption capacity of MgO for fluoride removal. Characterization studies revealed that the synthesized n-magnesia was crystalline with high phase purity, and the particle size varied in the range of 3–7 nm. With the help of various spectroscopic, microscopic, and macroscopic studies, the mechanism of fluoride uptake by n-magnesia was explained as fluoride removal thorough isomorphic substitution of hydroxyl groups by fluoride in a brucite lattice. This reaction was observed since both the F^- and OH^- ions are isoelectronic in nature and of similar size and with comparable ionic radii. The Langmuir maximum sorption capacity for fluoride removal was reported as high as 267.82 mg/g. Fluoride adsorption by n-magnesia was less sensitive to pH variations, and only a slight decrease in fluoride adsorption was observed at higher pH, which was due to the competition from OH^- ions. Fluoride uptake was most affected in the presence of phosphate followed by bicarbonate and nitrate. The fluoride uptake was systematically studied for n-alumina using aqueous solutions (Kumar et al., 2011). The maximum sorption capacity of n-alumina for fluoride was reported as 14.0 mg/g at 25°C, with maximum fluoride removal at pH 6.15.

17.7.8.3 Regeneration of Adsorbents

Techno-economic viability of any adsorbent significantly depends on its regeneration and reuse potential for many cycles of operation. Various regeneration media have been reported in the past few decades for the effective regeneration in fluoride adsorption studies, which obviously depend on the nature of the adsorbent used. Some of the reported regeneration methods are listed in Table 17.1.

17.7.8.4 Disposal of Adsorbent

Disposal of exhausted adsorbent is one of the important parameters that determine its overall environmental impact, and this has been a challenge in many developing countries. In any adsorption process, the exhausted material has to be either used or disposed of properly after a number of cycles of the adsorption/desorption process. Therefore, it is usually necessary to remove fluoride ions for safe disposal, which otherwise may find its way to water bodies and contaminate them. This is important for both regenerant as well as exhausted adsorbent. Safe disposal of exhausted adsorbent has been reported by a few authors. Iyenger (2005) has reported different methods for the treatment of spent media generated during the regeneration of AA and further used in the manufacturing of bricks. These methods have been discussed as follows:

- Addition of $CaCl_2$ to spent alkali regenerant to precipitate fluoride and then mixing the supernatant with an acid regenerant
- Simple mixing of spent alkali/acid regenerants
- Mixing alkali/acid regenerants and using certain additives like alum or lime to remove fluoride and to improve settling properties of the sludge

About 85% fluoride could be removed, using option 3, from a mixed alkali–acid regenerant. In this way, disposal of spent regenerants can be carried out by mixing spent alkali and acid regenerants, adjusting pH by addition of lime and settling the sludge for 24 h. The supernatant solution, with low fluoride and near-neutral pH, could then be drained off. However, the drained water could have high TDS, hardness, and sulfate. Sludge could be collected periodically and used for brick making at the village level itself. In any case, regeneration requires use of certain chemicals and minimizing its requirement would be important from a cost and environmental point of view.

TABLE 17.1

Reported Regeneration Methods of Adsorbents

Sr. No	Adsorbents	Mode of Regeneration	Regeneration Media	Efficiency (%)	Reference
1	Magnetic chitosan particle	Batch	0.8%–1% NaOH	98–99.00	Ma et al., 2007
2	Calcined MgAl–CO$_3$ layered double hydroxide (GLDH)	Column	0.1 M Na$_2$CO$_3$	1 Cycle 98.10 2 Cycle 97.70 3 Cycle 98.00 4 Cycle 97.30 5 Cycle 96.90	Fan et al., 2007
3	Laterite	Batch	0.1 M NaOH	80.40	Sarkar et al., 2007
4	Fe–Al–Ce trimetal oxide	Batch	NaOH	97.00	Mandal and Mayadevi, 2008
5	Fe(III)-loaded ligand exchange cotton cellulose	Batch	1 M NaOH	98.00	Kagne et al., 2008
6	Magnesia-amended activated alumina granules (MAAA)	Column	2% NaOH	90.00	Maliyekkal et al., 2008
7	Schwertmannite	Batch	pH 2; HCl, H$_2$SO$_4$	94.00	Maurice et al., 2008
8	Hydrous iron (III) –tin (IV) trimetal mixed oxides	Batch	0.1 M NaOH	75.00	Biswas et al., 2009
9	Zn–Al–Cl anionic clay	Column	0.01 M NaOH	1 Cycle 98.10 2 Cycle 97.70 3 Cycle 98.00 4 Cycle 97.30 5 Cycle 96.90	Mandal and Mayadevi, 2009
10	Granular ferric hydroxide	Batch	At pH 7 NaOH	65.00	Tang et al., 2009
11	Titanium macroporous (modified chitosan)	Batch	Alum at pH 12	85.00	Jagtap et al., 2009
12	Protonated chitosan beads (PCB)	Column	0.1 M NaOH 0.1 M H$_2$SO$_4$ 0.1 M HCl	85.00 89.00 94.50	Viswanathan et al., 2009a
13	Granular red mud (GRM)	Column	0.2 M NaOH	1 Cycle 87.00 2 Cycle 76.00 3 Cycle 58.00 4 Cycle 46.00	Tor et al., 2009
14	Metal ion incorporated in ion-exchange resin	Batch	0.1 M NaOH 0.1 M H$_2$SO$_4$ 0.1 M HCl	75.00 80.00 90.00	Viswanathan et al., 2009c
15	Copper oxide meso-alumina (COCA)	Batch	With 4 M NaOH	97.00	Bansiwal et al., 2010
16	Modified amberlite resin (ATU)	Batch	0.1 M NaOH	99.90	Solangi et al., 2010
17	Lanthanum hydroxide	Batch	1 M NaOH	86.30	Na and Park, 2010
18	Ceramic	Batch	0.1 M NaOH 0.1 M HCl	34.70 80.20	Chen et al., 2010

(continued)

TABLE 17.1 (Continued)

Reported Regeneration Methods of Adsorbents

Sr. No	Adsorbents	Mode of Regeneration	Regeneration Media	Efficiency (%)	Reference
19	Synthetic iron(III) –Al(III)–Cr(III) ternary mixed oxide	Batch	0.1 M NaOH 0.25 M NaOH 0.5 M NaOH 0.75 M NaOH	75.00 81.00 89.00 91.00	Biswas et al., 2010
20	Zirconium-impregnated cashew nut shell carbon	Batch	2.5% NaOH	96.20	Alagumuthu and Rajan, 2010
21	Cerium-impregnated chitosan	Batch	pH 12	93.00	Swain et al., 2010
22	Chemically modified bentonite clay	Batch	1 M NaOH	97.00	Kamble et al., 2010

Maliyekkal et al. (2008) have recoated the material with manganese oxide and reactivated the surface after a number of cycles of the adsorption/desorption process, without a considerable loss in efficiency. This recoating process is important from both an economic and disposal point of view. The recoating process reduces the amount of hazardous solid waste generated, thereby reducing the risk associated with its disposal. From the economic perspective, factors like cost of the new adsorbent and the cost associated with transportation and safe disposal of the exhausted material can be reduced. The life cycle assessment of the adsorbent or any other material can provide the more realistic overall environmental impacts of material used for defluoridation of water.

17.7.8.4.1 Domestic Defluoridation Units in Sri Lanka (from Handbook of Drinking Water Technologies by MDWS [Ministry of Drinking Water and Sanitation] and CSIR-NEERI, 2013, IGRAC [International Groundwater Resources Assessment Centre] Report 2007)

This defluoridation unit applies freshly fired brick pieces and is used in Sri Lanka for the removal of fluoride in a domestic defluoridation approach. The brick bed in the unit is placed on the top with charred coconut shells and pebbles to adjust the back pressure. Water is flown through the unit in an upflow mode. The performance of domestic units has been evaluated in rural areas of Sri Lanka as reported by Priyant and Padmassri (1997). The fluoride removal efficiency significantly depended on the quality of the freshly burnt bricks. The unit could be used for 20–40 days, with the treated water volume per day at around 8 L and input water fluoride concentration of 5 mg/L. A defluoridator made out of cement and bricks have also been recommended for low-cost field application.

17.7.8.4.2 ICOH (Inter-Country Centre for Oral Health) Domestic Defluoridator (from Handbook of Drinking Water Technologies by MDWS and CSIR-NEERI, 2013, IGRAC Report 2007)

The ICOH (at the University of Chiang Mai, Thailand) domestic defluoridator was developed in Thailand by using crushed charcoal and bone char as adsorption media. The defluoridation efficiency of this device was reported to depend on the fluoride concentration in raw water. Field trials in Thailand, Sri Lanka, and some African countries have shown interesting results for fluoride removal from water from a variety of sources.

Reports from Sri Lanka have inferred that with 300 g charcoal (mainly to remove color and odor) and 1 kg bone char, an ICOH filter can treat on an average 450 L of water containing 5 mg/L fluoride at a flow rate of 4 L/h. Regeneration of spent bone char is not recommended for these household units. Instead, it should be replaced with fresh material commercially available in local shops.

17.7.8.4.3 Domestic Defluoridation Units Developed by IIT (Indian Institute of Technology) Kanpur (Iyenger, 2005)

Extensive studies are reported from IIT Kanpur on AA-based and other adsorbents for defluoridation of water. The unit consisted of two chambers made of suitable materials such as stainless steel (SS), copolymer plastic, etc. The upper chamber was fitted with a microfilter. This has an orifice at the bottom to give a controlled flow rate of about 12 L/h for domestic application. This chamber was charged with 3 kg of AA and the depth was about 17 cm. A perforated plate made of SS was placed on the top of the AA bed to facilitate uniform distribution of raw water onto the adsorption bed. A lower chamber was provided as a treated water storage with a tap to withdraw the treated water. If desired, a lower chamber can be replaced with an earthen pot in rural areas, which not only lowers the initial cost but also keeps water cold in summer months. The filter was found to be quite user-friendly even considering the rural population. Fluoride-containing raw water is filled in the upper chamber. Water percolates through the AA adsorption bed, where fluoride is adsorbed onto the adsorbent. Treated water, collected in the lower chamber, can be withdrawn as needed.

17.7.8.4.4 Development of Hand-Pump Attached Defluoridation Units (Iyenger, 2005)

A cylindrical defluoridation unit was designed and fabricated by using mild steel sheet. The unit was designed to operate in the upflow water movement to achieve better contact. AA of grade G-87 (IPCL), with a particle size range of 0.3–0.9 mm, was filled in the unit with an approximate bed depth of 55 cm. This unit was field tested at Makkur village, Unnao district of Uttar Pradesh in India.

Experimental trials have been performed using a shallow type India Mark II hand pump ($\approx$35 ft depth). The defluoridation unit installation was executed in 1993. The hand pump discharge level was raised with an addition to its normal pedestal and by constructing an elevated platform. Users had to go up a few steps to operate the hand pump for water discharge. A bypass was also provided to draw the water directly from the hand pump for washing and bathing to make the water source suitable for this dual purpose. Raw water fluoride concentration studied was in the range of 6–7 mg/L. Regeneration of exhausted AA was done *in situ*, i.e., within the device column with a regeneration treatment time of 8–10 h. The average yield of safe water output (<1.5 mg/L fluoride) per cycle was around 25,000 L. Seventeen defluoridation cycles were reported to be required in a span of 4 years. There were no major maintenance issues observed during this period, while there was no complaint from the users regarding the operational difficulties or the palatability of treated water. However, community support during regeneration was minimal. No details about handling of spent regenerant and its disposal are available.

17.7.8.5 Green Chemical Approach

Yadav et al. (2012) explored a green chemical approach by preparation of tradition soil pots using aluminum oxalate as adsorbent for fluoride removal. The adsorbent dose was optimized in the range of 2–8 g/500 g soil, while other parameters like pH, TDS, hardness, fluoride, and aluminum parameters were also studied for defluoridation of water. The maximum fluoride removal efficiency of adsorbent was optimized. The study recommended that aluminum oxalate can be used as a potential defluoridating agent in soil pots as a green chemical approach without much environmental impacts.

17.7.8.5.1 Domestic Defluoridation Method by IISc (Indian Institute of Science) Bangalore, India (adopted from http://civil.iisc.ernet.in/~msrao/ files/applicationIIScmethod.pdf, with permission from IISc)

The IISc's method relies on a magnesium oxide–based precipitation–sedimentation–filtration technique to reduce fluoride concentrations in water to permissible levels (<1.5 ppm). It uses a domestic defluoridation unit (DDU) to implement the chemical treatment procedure. The magnesium oxide method for fluoride removal is reported to have several advantages. Magnesium oxide has limited solubility in water and therefore does not add to the TDS. All chemicals used in the process are nontoxic. The alkaline pH of the magnesium oxide–treated water is easily neutralized by addition of sodium bisulfate solution. The method does not involve any recharge process and thus avoids generation of corrosive and toxic wastes. The method is designed to treat fluoride-contaminated groundwater with varied ionic composition. Technology for reuse of sludge into stabilized mud blocks has also been developed.

A simple-to-use 15- and 100-L DDU treats fluoride-contaminated water using the IISc method. The 100-L DDU is composed of a 100-L-capacity polyvinyl chloride (PVC) drum and is equipped with a heavy-duty electrical stirrer. The 15-L DDU is composed of a 20–25-L-capacity SS drum and is equipped with a light-duty electrical stirrer or a hand-operated mechanical stirrer. The units serve as a mixing-cum-sedimentation unit. Taps are fixed above the base of the containers to drain the treated water. To trap any escaping sludge particles from the unit, a cotton cloth filter is tied to the tap of the upper unit. The cost of treating 1 L of fluoride-contaminated water by the IISc method is 17–27 paise/L (~0.4 cents) (Figure 17.10).

FIGURE 17.10
Field implementation of IISc method in Shola village, Laxmangarh district, Rajasthan, by Dr. Kushal Quango, Modi Institute of Technology, Rajasthan. (Copyright permission from IISC.)

17.7.8.5.2 TERI (The Energy and Resources Institute)-Implemented Defluoridation Unit Using Existing Nalgonda Technology (with Copyright Permission from TERI)

TERI Western Regional Centre, Goa, carried out a study to address this critical problem through community effort toward cost-effective treatment and management of fluoride contamination. The Nalgonda technique that was implemented during the project consisted of two buckets equipped with taps and a sieve on which a cotton cloth was placed. Known concentrations of alum and lime were added to the raw water bucket at the same time, and dissolved by stirring with a wooden ladle. The villagers were trained to stir fast while counting till 60 (1 min), and then slow down while counting till 300 (5 min). The floc formed was settled down for about an hour. The water was then passed through a sieve into another bucket. Both the containers were plastic buckets of 20-L capacity, supplied with covers and equipped with a tap 5 cm above the bottom to enable trapping of sludge. The treated water was then stored in the treated water bucket through the cloth, and collected for drinking or cooking.

17.7.8.5.3 Fluoride Removal by Bone Char (Handbook of Drinking Water Technologies by MDWS and CSIR-NEERI, 2013)

To produce bone char suitable as a filter material, animal bones are charred in a kiln at a defined temperature, and the oxygen content in a specific surface area of bones is increased, organic constituent are removed, and an inorganic HAp matrix remains. The charred bones are sieved and crushed to produce granular filter material. Synthetic HAp has high surface area and adsorption capacity. As it needs to be imported, it is more expensive than locally available bone char and therefore often not an option for projects in developing countries.

17.7.8.5.4 Emerging Defluoridation Technologies (Handbook of Drinking Water Technologies by MDWS and CSIR-NEERI, 2013)

17.7.8.5.4.1 Crystalactor   In the Netherlands, a new type of contact precipitator, dubbed Crystalactor, is reported to have been developed. The major part of the Crystalactor installation is the so-called pellet reactor, partially filled with a suitable seed material such as sand or minerals. The wastewater is pumped in an upward direction, maintaining the pellet bed in a fluidized state. To crystallize the target component on the pellet bed, a driving force is created by a reagent dosage and pH adjustment. By selecting the appropriate process conditions, cocrystallization of impurities is minimized and high-purity crystals are obtained. The pellets grow and move toward the reactor bottom. At regular intervals, a quantity of the largest fluidized pellets is discharged from the reactor and fresh seed material is added. After atmospheric drying, readily handled and virtually water-free pellets are obtained. A major advantage of the Crystalactor is its ability to produce high-purity, nearly dry pellets. Owing to their specific composition, the pellets can normally be recycled or reused in other plants, resulting in no residual waste for disposal. As reported, in the rare event that pellets have to be disposed of by other means, the advantage of low-volume secondary waste production still remains: water-free pellets, rather than bulky sludge. The four steps found in conventional treatment processes—coagulation, flocculation, separation, and dewatering—are combined into one as claimed in the Crystalactor. Because of the production of water-free pellets, troublesome sludge dewatering is reported to be eliminated. Furthermore, the unit is compact owing to the high surface loadings (40–120 m/h).

Cost comparisons claimed that the total treatment costs are ~25% of the costs for conventional precipitation. However, the Crystalactor could be more suitable for wastewater with high fluoride concentration (>10 mg/L). For treating drinking water, the Crystalactor may be more suitable in case of high fluoride concentrations (>10 or 20 mg/L). (Information collected from DHV B.V. Water, the Netherlands, E info-water@dhv.com).

17.7.8.5.4.2 Memstill® Technology (Source: TNO [The Netherlands Organization of Applied Scientific Research], the Netherlands)   As per the literature, the membrane distillation using the Memstill process uses a combination of flash distillation and membrane filtration (osmosis) to desalinate water. It is reported to do so at a cost that is lower than RO or distillation. The TNO has developed a membrane-based distillation concept that is claimed to radically improve the economics and ecology of existing desalination technology for seawater and brackish water. This so-called Memstill technology combines multistage flash and multi-effect distillation modes into one membrane module (Hanemaaijer et al., 2007). Cold feed water takes up heat in the condenser channel through condensation of water vapor, after which a small amount of (waste) heat is added, and flows counter-currently back via the membrane channel. Driven by the small added heat, water evaporates through the membrane, and is discharged as cold condensate. The cooled brine is disposed, or extra concentrated in a next module. The Memstill technology is claimed to produce (drinking) water at a cost well below that of existing technologies like RO and distillation. With the Memstill technology, anions like fluoride and arsenic are also removed. It is expected that the Memstill technology will be also developed for small-scale applications using solar heat (*Handbook of Drinking Water Technologies* by MDWS and CSIR-NEERI, 2013).

17.7.8.5.4.3 Water Pyramid Solution   Aqua-Aero Water Systems has developed the Water Pyramid concept for tropical, rural areas (Aqua-Aero, 2007). The Water Pyramid makes use of simple technology to process clean drinking water out of salt, brackish, or polluted water. One of the pollutants could be fluoride. Most of the energy needed to clean the water is obtained from solar radiation. The Water Pyramid with a total area of 600 m² and situated under favorable tropical conditions is claimed to produce up to 1250 L of fresh water a day. The production rate is reported to be dependent on site-specific factors such as climate and temperature, cloudiness, and wind activity. Desalination is driven by the sun, and the energy needed for pressuring the Water Pyramid is obtained using solar cells in combination with a battery back-up system. Intermittent peak demands in electricity, related to (e.g., borehole) pumping and maintenance, are covered using a small generator system (*Handbook of Drinking Water Technologies* by MDWS and CSIR-NEERI, 2013).

17.7.8.5.4.4 Solar Dew Collector System   The Solar Dew Collector system, *Solar Dew* (Solar Dew, 2007), is a new porous membrane to purify water using solar energy. The technique may be a little similar to the Water Pyramid. Water sweats through the membrane, evaporates on the membrane's surface, and increases the air humidity in the evaporation chamber. On the basis of a temperature difference, pure water condenses on the cooler surfaces of the system. The product water quality is claimed to be constant and similar to that of distilled water. The quantity depends on the intensity of the solar radiation. To avoid crystallization, the brine has to be drained periodically. The system is able to process seawater and brackish or contaminated wastewater (e.g., with heavy metals, oil residue, boron, fluoride) with an allowable pH range of 5–11 (*Handbook of Drinking Water Technologies* by MDWS and CSIR-NEERI, 2013).

17.8 Future Prospects and Outlook

Defluoridation of water is one of the most important and urgent issues facing our society today. Although many of the benefits of the existing defluoridation program are questionable, nevertheless, there are several success stories with regard to defluoridation technologies as well as fluorosis mitigation. It appears that several technological developments will ultimately lead to the adequate handling of this water-related health problem. This justifies the requirement of a variety of defluoridation solutions for widespread applications. Some of the challenges and opportunities in nanoscience and chemical engineering are expected to bring further technological advancements in defluoridation.

As discussed in this review, the surface of the materials could be modified in many ways to express specific functional groups, which promoted the defluoridation potential. In particular, biogenic and biomaterial-based composites can provide a high surface aluminol concentration, which facilitates a wide variety of surface reactions. Consequently, introduction of functional groups on the different matrices, i.e., engineering materials, by incorporation of specific functional groups for targeted applications is a developing area for water treatment. To increase the efficiency of materials, however, preservation of the materials' intrinsic activity is a necessary consideration. Therefore, more examples can be expected in the near future. Further elaboration of this field is also expected to lead to the creation of powerful strategies for nanomaterial synthesis and novel technologies that combine several methods.

The versatility of nanomaterial size, shape, and morphology introduces unique properties to nanomaterial-based water treatment systems, making it possible for researchers to develop a revolutionary class of materials and molecules that differ from traditional and conventional water treatment systems in terms of preparation, removal efficiency, and application potential. We anticipate that the development of nanomaterials-based water treatment requires input from diverse and multidisciplinary fields such as chemistry, biochemistry, physics, molecular biology, and material science.

There is a continued demand to develop robust technology to handle large volumes of water with fluctuating concentrations of coexisting anions and cations. Engineering issues particularly related to minimization of operation and maintenance requirements are of special significance for field applications. New routes to defluoridation based on constructed wetland systems are also proposed to be explored. Approaches like IFM are proving to be effective for fluorosis mitigation, which is the main objective related to defluoridation of water. This approach has clearly highlighted the importance of many other parameters for fluorosis mitigation.

17.9 Summary

Owing to continuously increasing stress on groundwater sources, fluoride will be one of the critical issues, especially for rural water management. Different approaches would be required depending on input water quality as well as other practical considerations. However, defluoridation of water is an essential approach to be practiced in future. It will be essential to adopt several defluoridation technologies to achieve the goal of attaining the stringent prescribed levels of fluoride. The tool of nanotechnology and material sciences is expected to fill crucial gaps identified in defluoridation approaches, while improving the techno-economic feasibility of already developed technologies would continue to be a challenge. There is a need to improve technologies to suit to rural applications with minimum operation and maintenance. High-capacity nanomaterials can be considered as potential options under these circumstances.

References

Alagumuthu G., Rajan M., Equilibrium and kinetics of adsorption of fluoride onto zirconium impregnated cashew nut shell carbon, *Chem. Eng. J.* 158, 451–457 (2010).

Amor Z., Malki S., Taky M., Bariou B., Mameri N., Elmidaoui A., Optimization of fluoride removal from brackish water by electrodialysis, *Desalination* 120(3), 263 (1998).

Angers J., What are the best methods for fluoride removal? *Opflow* 27(10), 6–7 (2001).

Aqua-Aero Water Systems, Water Pyramid (2007). Available at http://www.waterpyramid.nl.

Asgari G., Roshani B., Ghanizadeh G., The investigation of kinetic and isotherm of fluoride adsorption onto functionalize pumice stone, *J. Hazard. Mater.* 217, 123–132 (2012).

Ayoob S., Gupta A.K., Bhat V.T., A conceptual overview on sustainable technologies for the defluoridation of drinking water, *Crit. Rev. Environ. Sci. Technol.* 38(6), 401–470 (2008).

Bailey K., Chilton J., Dahi E., Lennon M., Jackson P., Fawell J., (eds.), *Fluoride in Drinking Water*. Geneva, Switzerland, WHO Publication (2006).

Bansiwal A., Pillewan P., Biniwale R.B., Rayalu S.S., Copper oxide incorporated mesoporous alumina for defluoridation of drinking water, *Micropor. Mesopor. Mat.* 129, 54–61 (2010).

Bårdsen A., Bjorvatn K., Fluoride sorption isotherm on fired clay. In: 1st International Workshop on Fluorosis Prevention and Defluoridation of Water. *Int. Soc. Fluoride Res.* 56–59 (1995).

Ben N.A., Walha K., Charcosset C., Ben A.R., Removal of fluoride ions using cuttle fish bones, *J. Fluor. Chem.* 132, 57–62 (2011).

Biswas K., Gupta K., Ghosh U.C., Adsorption of fluoride by hydrous iron(III) – tin(IV) bimetal mixed oxide from the aqueous solutions, *Chem. Eng. J.* 149, 196–206 (2009).

Biswas K., Gupta K., Goswami A., Ghosh U.C., Fluoride removal efficiency from aqueous solution by synthetic iron(III) – aluminum(III) – chromium(III) ternary mixed oxide, *Desalination* 255, 44–51 (2010).

Boysen R.E., Chmielewski R.C., Fouling control for polymeric membrane processes. In: Proceedings of the 2008 Integrated Petroleum Environmental Consortium at the University of Tulsa (2008).

Brown M.D., Aaron G., The effect of point of use water conditioning systems on community fluoridated water, *Pediatr. Dent.* 12, 35–38 (1991).

Bulusu K.R., Sunderasan B.B., Pathak B.N., Nawlakhe W.G., Kulkarni D.N., Thergaonkar V.P., Fluorides in water, defluoridation methods and there limitations, *J. Inst. Eng. India* 60, 1 (1979).

Castel C., Schweizer M., Simonnot M.O., Sardin M., Selective removal of fluoride ions by a two-way ion-exchange cyclic process, *Chem. Eng. Sci.* 55, 3341–3352 (2000).

Cengeloglu Y., Kir E., Ersoz M., Removal of fluoride fromaqueous solution by using red mud, *Sep. Purif. Technol.* 28, 81–86 (2002).

Chai L., Wang Y., Zhao N., Yang W., You X., Sulfate-doped Fe_3O_4/Al_2O_3 nanoparticles as a novel adsorbent for fluoride removal from drinking water, *Water Res.* 47(12), 4040–4049 (2013).

Chang C.F., Lin P.H., Höll W., Aluminum-type super paramagnetic adsorbents: Synthesis and application on fluoride removal, *Colloids Surf. A: Physicochem. Eng. Aspects* 280, 194–202 (2006).

Chen L., Wu H.X., Wang T.J., Jin Y., Zhang Y., Dou X.M., Granulation of Fe-Al- Cenano-adsorbent for fluoride removal from drinking water by spray coating on sand in a fluidized bed, *Powder Technol.* 193, 59–64 (2009).

Chen N., Zhang Z., Feng C., Li M., Zhu D., Chen R., Sugiura N., An excellent fluoride sorption behavior of ceramic adsorbent, *J. Hazard. Mater.* 183, 460–465 (2010).

Daifullah A.A.M., Yakout S.M., Elreefy S.A., Adsorption of fluoride in aqueous solutions using $KMnO_4$-modified activated carbon derived from steam pyrolysis of rice straw, *J. Hazard. Mater.* 147, 633–643 (2007).

Davila-Rodriguez J.L., Escobar-Barrios V.A., Shirai K., Rangel-Mendez J.R., Synthesis of a chitin-based biocomposite for water treatment: Optimization for fluoride removal, *J. Fluor. Chem.* 130, 718–726 (2009).

Elazhar F., Tahaikt M., Achatei A., Elmidaoui F., Taky M., El Hannouni F., Laaziz I., Jariri S., El Amrani M., Elmidaoui A., Economical evaluation of the fluoride removal by nanofiltration, *Desalination* 249(1,30), 154 (2009).

Fan J., Xu Z., Zheng S., Comment on "factors influencing the removal of fluoride from aqueous solution by calcined $Mg–Al–CO_3$ layered double hydroxides," *J. Hazard. Mater.* B139, 175–177 (2007).

Fawell J., Bailey K., Chilton J., Dahi E., Fewtrell L., Magara Y., Fluoride in Drinking Water, World Health Organization, IWA Publishing, Alliance House, London (2006).

Feenstra L., Vasak L., Yroiffioen J., Fluoride in groundwater: Overview and evaluation of removal methods, (IGRAC) International Groundwater Resources Assessment Centre Report nr. SP -1, Utrecht, September (2007).

Garmes H., Persin F., Sandeaux J., Pourcelly G., Mountadara M., Defluoridation of groundwater by a hybrid process combining adsorption and Donnan dialysis, *Desalination* 145, 287–291 (2002).

Ghorai S., Pant K.K., Equilibrium, kinetics and breakthrough studies for adsorption of fluoride on activated alumina, *Sep. Purif. Technol.* 42, 265–271 (2005).

Gitonga J.N., Partial defluoridation of borehole water. In: *Flurosis Research Strategies*, (ed. Likimani, S.), Department of Dental Surgery, University of Nairobi, Published by African Medical and Research Foundation Nairobi, (1984).

Gumbo F.J., Partial defluoridation of drinking water in Tanzania. In: Proceedings of the Second Workshop on Domestic Water Health Standards with Emphasis on Fluoride, Arusha, Tanzania (1987).

Gwala P., Andey S., Mhaisalkar V., Labhasetwar P., Pimpalkar S., Kshirsagar C., Lab scale study on electrocoagulation defluoridation process optimization along with aluminium leaching in the process and comparison with full scale plant operation, *Water Sci. Technol.* 63(12), 2788–2795 (2011).

Hamdi N., Srasra E., Retention of fluoride from industrial acidic wastewater and NaF solution by three Tunisian Clayey soils, *Fluoride* 42(1), 39–45 (2009).

Handbook on drinking water treatment technologies. Published by Government of India, MDWS and CSIR-NEERI, November 2011, Nagpur, India.

Hanemaaijer J.H., Van Medevoort J., Jansen A., Van Sonsbeek E., Hylkema H., Biemans R., Nelemans B., Stikker A., Memstill membrane distillation: A near future technology for sea water desalination. Paper presented at the International Desalination Conference, Aruba, June (2007).

Haron M.J., Wan Yunus W.M.Z., Wasay S.A., Sorption of fluoride ions from aqueous solutions by a yttrium-loaded poly(hydroxamic acid) resin, *Int. J. Environ.* 48, 245–255 (1995).

Hichour M., Persin F., Molenat J., Sandeaux J., Gavach C., Fluoride removal from diluted solutions by Donnan dialysis with anion-exchange membrane, *Desalination* 122, 53–62 (1999).

Iyenger L., *Defluoridation of Water Using Activated Alumina Technology.* Indian Institute of Technology, Kanpur for UNICEF, New Delhi, March (2005).

Jagtap S., Thakre S., Wanjari S., Kamble S., Labhsetwar N., Rayalu S.S., New modified chitosan-based adsorbent for defluoridation of water, *J. Colloid Interface Sci.* 332, 280–290 (2009).

Jagtap S., Yenkie M.K.N., Labhsetwar N., Rayalu S.S., Fluoride in drinking water and defluoridation of water, *Chem. Rev.* 112, 2454–2466 (2012).

Jiménez-Reyes M., Solache-Ríos M., Sorption behavior of fluoride ions from aqueous solutions by hydroxyapatite, *J. Hazard. Mater.* 180, 297–302 (2010).

Kagne S., Jagtap S., Dhawade P., Kamble S.P., Devotta S., Rayalu S.S., Hydrated cement: A promising adsorbent for the removal of fluoride from aqueous solution, *J. Hazard. Mater.* 154, 88–95 (2008).

Kamble S.P., Jagtap S., Labhsetwar N.K., Thakare D., Godfrey S., Devotta S., Rayalu S.S., Defluoridation of drinking water using chitin, chitosan and lanthanum-modified chitosan, *Chem. Eng. J.* 129, 173–180 (2007).

Karthikeyan G., Ilango S.S., Fluoride sorption using Morringa indica based activated carbon, *Iran. J. Environ. Health. Sci. Eng.* 4, 21–28 (2007).

Kaseva M.E., Optimization of regenerated bone char for fluoride removal in drinking water: A case study in Tanzania, *J. Water Health* 4, 139–147 (2006).

Kemer B., Ozdes D., Gundogdu A., Bulut V.N., Duran C., Soylak M., Removal of fluoride ions from aqueous solution by waste mud, *J. Hazard. Mater.* 168, 888–894 (2009).

Ku Y., Chiou H.M., The adsorption of fluoride ion from aqueous solution by activated alumina, *Water Air Soil Pollut.* 133, 349–360 (2002).

Kumar E., Bhatnagar A., Kumar U., Sillanp M., Defluoridation from aqueous solutions by nano-alumina: Characterization and sorption studies, *J. Hazard. Mater.* 186, 1042–1049 (2011).

Kuo I., Abe S., Iwasaki T., Tokimoto N., Kawasaki T., Nakamura T., Tanada S., Adsorption of fluoride ions onto carbonaceous materials, *J. Colloid Interface Sci.* 275, 35–39 (2004).

Lagaude A., Kirsche C., Travi Y., Defluoridation of ground waters in Senegal preliminary work in the case of Fatick waters, *Tech. Sci. Methods (in French)*, 83(9), 449–452 (1988).

Lee G., Chen C., Yang S.T., Ahn W.S., Enhanced adsorptive removal of fluoride using mesoporous alumina, *Micropor. Mesopor. Mater.* 127, 152–156 (2010).

Li Y.H., Wang S., Cao A., Zhao D., Zhang X., Xu C., Luan Z., Ruan D., Liang J., Wu D., Wei B., Adsorption of fluoridefrom water by amorphous alumina supported on carbon nanotubes, *Chem. Phys. Lett.* 350, 412–416 (2001).

Li Y.H., Wang S., Cao A., Zhao D., Zhang X., Xu C., Luan Z., Ruan D., Liang J., Wu D., Wei B., Removal of fluoride from water by carbon nanotube supported alumina, *Environ. Technol.* 24, 391–398 (2003b).

Li Y.H., Wang S., Zhang X., Wei J., Xu C., Luan Z., Wu D., Adsorption of fluoride from water by aligned carbon nanotubes, *Mater. Res. Bull.* 38, 469–476 (2003a).

Lounici L., Adour D., Belhocine A., Elmidaoni B., Barion N.M., Novel technique to regenerate activated alumina bed saturated by fluoride ions, *Chem. Eng. J.* 81, 153 (2001).

Lunge S., Biniwale R., Labhsetwar N., Rayalu S.S., User perception study for performance evaluation of domestic defluoridation techniques for its application in rural areas, *J. Hazard. Mater.* 191(1–3), 325–332 (2011).

Ma W., Ya F.Q., Han M., Wang R., Characteristics of equilibrium, kinetics studies for adsorption of fluoride on magnetic-chitosan particle, *J. Hazard. Mater.* 143, 296–302 (2007).

Mahramanlioglu M., Kizilcikli I., Biccer I.O., Adsorption of fluoride from aqueous solution by acid treated spent bleaching earth, *J. Fluor. Chem.* 115, 41–47 (2002).

Maliyekkal S.M., Anshup, Antony K.R., Pradeep T., High yield combustion synthesis of nano magnesia and its application for fluoride removal, *Sci. Total Environ.* 408, 2273–2282 (2010).

Maliyekkal S.M., Shukla S., Philip L., Nambi I.M., Enhanced fluoride removal from drinking water by magnesia-amended activated alumina granules, *Chem. Eng. J.* 140, 183–192 (2008).

Mandal S., Mayadevi S., Adsorption of fluoride ions by Zn–Al layered double hydroxides, *Appl. Clay Sci.* 40, 54–62 (2008).

Mandal S., Mayadevi S., Defluoridation of water using as-synthesized Zn/Al/Cl anionic clay adsorbent: Equilibrium and regeneration studies, *J. Hazard. Mater.* 167, 873–878 (2009).

Maurice E.A., Onyango S., Ochieng A., Asai S., Removal of fluoride ions from aqueous solution at low pH using schwertmannite, *J. Hazard. Mater.* 152, 571–579 (2008).

Menkouchi Sahlia M.A., Annouar S., Tahaikt M., Mountadar M., Soufianec A., Elmidaouia A., Fluoride removal for underground brackish water by adsorption on the natural chitosan and by electrodialysis, *Desalination* 212, 37–45 (2007).

Mohapatra M., Hariprasad D., Mohapatra L., Anand S., Mishra B.K. Mg-dopednanoferrihydrite-A new adsorbent for fluoride removal from aqueous solutions, *Appl. Surf. Sci.* 258, 4228–4236 (2012).

Na C.-K., Park H.-J., Defluoridation from aqueous solution by lanthanum hydroxide, *J. Hazard. Mater.* 183, 512–520 (2010).

NEERI, *Defluoridation. Technology Mission on Drinking Water in Villages and Related Water Management.* Nagpur 440020, India, National Environment Engineering Research Institute (1987).

Nigussie W., Zewge F., Chandravanshi B.S., Removal of excess fluoride from water using waste residue from alum manufacturing process, *J. Hazard. Mater.* 147, 954–963 (2007).

Parmar H.S., Patel J.B., Sudhakar P., Koshy V.J., Removal of fluoride from water with powdered corn cobs, *J. Environ. Sci. Eng.* 48, 135–138 (2006).

Patel G., Pal U., Menon S., Removal of fluoride from aqueous solution by CaO nano particles, *Sep. Sci. Technol.* 44, 2806–2826 (2009).

Pathak A., Panda A.B., Tarafdar A., Pramanik P., Synthesis of nano-sized metal oxide powders and their application in separation technology, *J. Indian Chem. Soc.* 80, 289–296 (2003).

Priyantha N., Padmashri J.P. (eds.), Procedings: Prevention of fluorosis in Sri Lanka, University of Peradeniya, Sri Lanka, 69 (1996).

Rao S.M., Mamatha P., Asha K., Soumita K., Mythri D.J., Field studies on defluoridation using magnesium oxide, *Water Manage.* 163, 147–155 (2010).

Sakhare N., Lunge S., Rayalu S.S., Bakardjiva S., Subrt J., Devotta S., Labhsetwar N., Defluoridation of water using calcium aluminate material, *Chem. Eng. J.* 203, 406–414 (2012).

Sarkar M., Banerjee A., Pramanick P.P., Sarkar A.R., Design and operation of fixed bed laterite column for the removal of fluoride from water, *Chem. Eng. J.* 131, 329–335 (2007).

Sivasamy A., Singh K.P., Mohan D., Maruthamuthu, M., Studies on defluoridation of water by coal based adsorbents, *J. Chem. Technol. Biotechnol.* 76, 717–722 (2001).

Sivasankar V., Rajkumar S., Murugesh S., Darchen A., Tamarind (Tamarindus indica) fruit shell carbon: A calcium-rich promising adsorbent for fluoride removal from groundwater, *J. Hazard. Mater.* 30(225–226), 164–172 (2012).

Sivasankar V., Ramachandramoorthy T., Chandramohan A., Fluoride removal from water using activated and MnO_2 coated Tamarind Fruit (Tamarindus indica) shell: Batch and column studies, *J. Hazard. Mater.* 177, 719–729 (2010).

Solangi I.B., Memon S., Bhanger M.I., An excellent fluoride sorption behavior of modified amberlite resin, *J. Hazard. Mater.* 176, 186–192 (2010).

Solar Dew, The Solar Dew Collector System (2007). Available at http://www.solardew.com/index2 .html.

Sujana M.G., Thakur R.S., Rao S.B., Removal of fluoride from aqueous solution by using alum sludge, *J. Colloid Interface Sci.* 206, 94–101 (1998).

Sundaram C.S., Viswanathan N., Meenakshi S., Uptake of fluoride by nano-hydroxyapatite/chitosan, a bioinorganic composite, *Bioresour. Technol.* 99, 8226–8230 (2008).

Sundaram C.S., Viswanathan N., Meenakshi S., Fluoride sorption by nano-hydroxyapatite/chitin composite, *J. Colloid Interface Sci.* 333, 58–62 (2009).

Susheela A.K., Fluorosis management programme in India, *Curr. Sci.* 77, 1250 (1999).

Swain S.K., Padhi T., Patnaik T., Patel R.K., Jha U., Dey R.K., Kinetics and thermodynamics of fluoride removal using cerium-impregnated chitosan, *J. Desalinat. Water Treat.* 13, 369–381 (2010).

Tang Y., Guan X., Wang J., Gao N., McPhail M.R., Chusuei C.C., Fluoride adsorption onto granular ferric hydroxide: Effects of ionic strength, pH, surface loading, and major co-existing anions, *J. Hazard. Mater.* 171, 774–779 (2009).

Tchomgui-Kamga E., Alonzo V., Nanseu-Njiki C.P., Audebrand N., Ngameni E., Darchen A., Preparation and characterization of charcoals that contain dispersed aluminum oxide as adsorbents for removal of fluoride from drinking water, *Carbon* 48, 333–343 (2010).

Teotia S.P.S., Teotia M., Singh R.K., Hydrogeochemical aspects of endemic skeletal fluorosis in India—An epidemiological study, *Fluoride*, 14, 69 (1981).

Tetra Chemicals Europe Treatment of Aqueous Effluents for Fluoride Removal. 2005, Tetra Technologies, Inc., Web. February 10 (2010).

Teunissen W., Bol A.A., Geus J.W., Magnetic catalyst bodies, *Catal. Today* 48, 329–336 (1999).

Thakre D., Jagtap S., Bansiwal A., Labhsetwar N., Rayalu S.S., Synthesis of La- incorporated chitosan beads for fluoride removal from water, *J. Fluor. Chem.* 131, 373–377 (2010a).

Thakre D., Rayalu S.S., Kawade R., Meshram S., Subrt J., Labhsetwar N., Magnesium incorporated bentonite clay for defluoridation of drinking water, *J. Hazard. Mater.* 180(1–3), 122–130 (2010b).

Tor A., Danaoglu N., Arslan G., Cengeloglu Y., Removal of fluoride from water by using granular red mud: Batch and column studies, *J. Hazard. Mater.* 164, 271–278 (2009).

Tripathy S.S., Bersillon J., Gopal K., Removal of fluoride from drinking water by adsorption onto alum-impregnated activated alumina, *Sep. Purif. Technol.* 50, 310–317 (2006).

Tripathy S.S., Raichur A.M., Abatement of fluoride from water using manganese dioxide-coated activated alumina, *J. Hazard. Mater.* 153, 1043–1051 (2008).

Viswanathan N., Meenakshi S., Enhanced fluoride sorption using La(III) incorporated carboxylated chitosan beads, *J. Colloid Interface Sci.* 322, 375–383 (2008).

Viswanathan N., Meenakshi S., Development of chitosan supported zirconium (IV) tungstophosphate composite for fluoride removal, *J. Hazard. Mater.* 176, 459–465 (2010a).

Viswanathan N., Meenakshi S., Enriched fluoride sorption using alumina/chitosan composite, *J. Hazard. Mater.* 178, 226–232 (2010b).

Viswanathan N., Meenakshi S., Role of metal ion incorporation in ion exchange resin on the selectivity of fluoride, *J. Hazard. Mater.* 162, 920–930 (2009).

Viswanathan N., Sundaram C.S., Meenakshi S., Removal of fluoride from aqueous solution using protonated chitosan beads, *J. Hazard. Mater.* 161, 423–430 (2009a).

Viswanathan N., Sundaram C.S., Meenakshi S., Development of multifunctional chitosan beads for fluoride removal, *J. Hazard. Mater.* 167, 325–331 (2009b).

Wang S.G., Ma Y., Shi Y.J., Gong W.X., Defluoridation performance and mechanism of nano-scale aluminum oxide hydroxide in aqueous solution, *J. Chem. Technol. Biotechnol.* 84, 1043–1050 (2009).

Yadav R.N., Yadav R., Dagar N.K., Gupta P., Singh O.P., Chandrawat M.P.S., Removal of fluoride in drinking water by green chemical approach, *J. Curr. Chem. Pharm. Sci.* 2(1), 69–75 (2012).

Yang H.H., Zhang S.Q., Chen X.L., Zhuang Z.X., Xu J.G., Wang X.R., Magnetite containing spherical silica nano particles for biocatalysis and bioseparations, *Anal. Chem.* 76, 1316–1321 (2004).

Zhang G., He Z., Xu W., A low-cost and high efficient zirconium-modified-Na -attapulgite adsorbent for fluoride removal from aqueous solutions, *Chem. Eng. J.* 183, 315–324 (2012).

Zhao X., Wang J., Wu F., Wang T., Cai Y., Shi Y., Jiang G., Removal of fluoride from aqueous media by Fe_3O_4@$Al(OH)_3$ magnetic nano particles, *J. Hazard. Mater.* 173, 102–109 (2010).

18

Nanotechnology Tunneling in the Environmental Kuznets Curve (EKC): Nano Zerovalent Iron for Underground Water Remediation

Ali Marjowy[1,2] and Rasool Lesan-Khosh[1,3]

[1]*International Centre for Nanotechnology, United Nations Industrial Development Organization (UNIDO), Tehran, Iran*

[2]*Techno-Entrepreneurship Group, Entrepreneurship Faculty, University of Tehran, Tehran, Iran*

[3]*Departments of Materials Science and Engineering, Sharif University of Technology, Tehran, Iran*

CONTENTS

18.1 Introduction

Several investigations have been conducted to disclose the relationship between environmental degradation, economic growth, technology advancement, and political and institutional factors (Kuznets, 1955; Grossman and Krueger, 1995; Jordan, 2010). For instance, according to B.R. Jordan (2010), the following equation has been introduced:

$$\text{Impact} = \text{Population} \times \text{Affluence} \times \text{Technology}$$

On the basis of this equation, the environmental impact (negative) increases proportionally with an income per capita (affluence) increment. However, there are aspects not seen in the above equation such as the fact that economic growth does not always correlate with environmental degradation. It is also possible that technological improvement reduces negative impacts from population and economic growth. Given the complexity of solving a multivariable equation between the environment and various affecting determinants, one approach is to consider the relationship between two selected variables and investigate the effect of others on the mentioned link. A useful contribution in this direction was Grossman and Krueger's (1995) work focusing on the relationship between environmental degradation and economic growth represented by income per capita (gross domestic production [GDP]) trends. The result was named an environmental Kuznets curve (EKC) after the name of Simon Kuznets (1955), who first used this approach to focus on the relationship between national income per capita and inequality (Kuznets curve [KC]). The EKC is an extension of the KC concept.

The main goal of this chapter is to characterize the role of nanotechnology in preventing or mitigating environmental degradation in developing countries. The result of the analysis will help in the development of better technology policy for those countries. The EKC has been used as a framework to pursue the aforementioned goal. We may provide a few reasons to justify using the intermediary variable of economic growth (represented by GDP) in our study. First, incorporating GDP will link our analysis to the rich context of economic growth, which is quite familiar for policy makers. As a result, effective methods in comparison, assessment, and goal setting may be adopted. Second, the results of the analysis can be quantified both in time and GDP dimensions and, third, the proposed approach can be integrated with common environmental cost–benefit analysis and broaden its scope, which leads to a better decision-making practice. On the other hand, we will try to identify the conditions and factors that may affect the elasticity of the EKC to nanotechnology advancement.

In this chapter, the special case of application of nanotechnology in water and wastewater is considered. Water pollution and scarcity are very important examples of current global environmental challenges. First, we revisit the EKC as an empirical framework for understanding economic growth–environmental relationships. Then, we will review the effect of policy tools as well as technological improvement on reshaping the EKC, i.e., tunneling of the EKC. A few important aspects of this phenomenon are discussed. The next section deals with the case of nanotechnology in water with an analysis of underground water remediation using iron nanoparticles. In the last section, an investigation of the interrelationship between technology and policy leads us to a series of integrated technology–policy guidelines focusing on international cooperation.

18.2 Nano Zerovalent Iron (nZVI) Particles in Underground Water Remediation

Groundwater is an important source of drinking water worldwide. Most groundwater is clean; however, it can become polluted as a result of human activities/economic growth. There is a range of remedial methods for groundwater treatment; however, they are quite labor intensive, marginally effective, and involve substantial expense to implement (BCC Research Report, 2010). Nanotechnology offers a number of emerging technologies that appear very promising. One of these methods is *in situ* underground remediation via iron nanoparticles. The reason for choosing this method is that it is well documented and has a strong economic justification.

nZVI is a nanoparticle proving to be useful in the environmental remediation of contaminated groundwater and soil. Because of its large surface area, nZVI is highly reactive. It has several advantages over macroscale iron for *in situ* remedial applications. The iron nanoparticles typically are mixed into sludge and injected into the ground where they form a permeable barrier that cleans groundwater. In addition, iron nanoparticles can be delivered to any depth or geological configuration. The important attribute is that they can remain catalytically active for up to 8 weeks.

In the nano-remediation of underground water, reactive nanoparticles are employed for transformation and detoxification of pollutants on site or below ground. No groundwater is pumped out for above-ground treatment, and no soil is transported to other places for treatment and disposal. In this method, the nanoparticle slurry is injected into the ground without any specific machinery. Then, the nanoparticles migrate into the ground and influence their effect on the pollutants.

18.2.1 Technology Life Cycle

Ingredients of the technology life cycle are illustrated in Figure 18.1. In this section, each component is explained for the case of the application of nZVI in underground remediation.

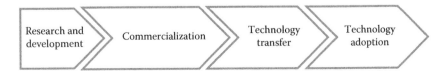

FIGURE 18.1
Technology life cycle.

18.2.1.1 Research and Development

A great deal of research has been conducted on the *in situ* underground water remediation using nanoparticles worldwide. The researchers work on the technical and economic aspects of this method, such as efficiency of the nanoparticles for removing pollutants, distance of migration into the ground, lifetime of being effective, etc. There are a lot of pilot plants implemented in the United States, Canada, Europe, and Asia (see Figure 18.2).

As an example of developing countries' contribution, a few research groups are active at Iranian universities and research centers working on this technique (in laboratory-scale projects). They are working on *in situ* removal of contaminants such as nitrates, arsenic, mercury, and oil-generated pollutants from groundwater. In addition, some groups are working on simulation of nano-iron migration into the ground in different soil conditions.

18.2.1.2 Commercialization

To commercialize the nZVI method for underground remediation, two approaches can be taken. First, have a research base infrastructure and then a successful mechanism to scale up the developed laboratory-scale to industrial-scale production. Second, the technology can be transferred from already successful developers.

As illustrated in Figure 18.2, a great deal of research and subsequent semi-industrial project plans have been implemented worldwide on the application of nano-iron in underground remediation; therefore, some nano-iron suppliers have been formed in the world. For instance, Nano Iron s.r.o. is a science and technology limited liability company in the Czech Republic, producing Fe^0 nanoparticles at the industrial scale. The company was successful in substantially decreasing the price of nZVI and has been implementing several projects in Europe.

Laboratory-scale research is also being conducted in developing countries such as IR Iran ("Iran") on the application of nZVI in water remediation. nZVI powder is also produced in semi-industrial scale in Iran by PNF Co. (Payamavaran Nanotechnology Fardanegar) with a higher price in comparison with Nano Iron in the Czech Republic, which can be mitigated by scale-up or technology transfer.

18.2.1.3 Technology Transfer and Adoption

Technology transfer is a shortcut for bringing the developing countries on track for the nZVI application for water remediation. However, even if the technology is transferred, it needs to be applied and adopted. To facilitate this process in Iran, the National Water and Wastewater Engineering Company (NWWEC), a holding company, has jurisdiction over 60 water companies in 30 provinces, all of which need to be made aware. The knowledge building can be conducted through implementing demonstration projects on using nZVI in underground remediation. However, for conducting demonstration projects, the need assessment for this technology and pollution mapping are necessary steps.

18.2.2 Product Life Cycle of nZVI

In this section, the product life cycle of iron nanoparticles in water application will be assessed (see Figure 18.3).

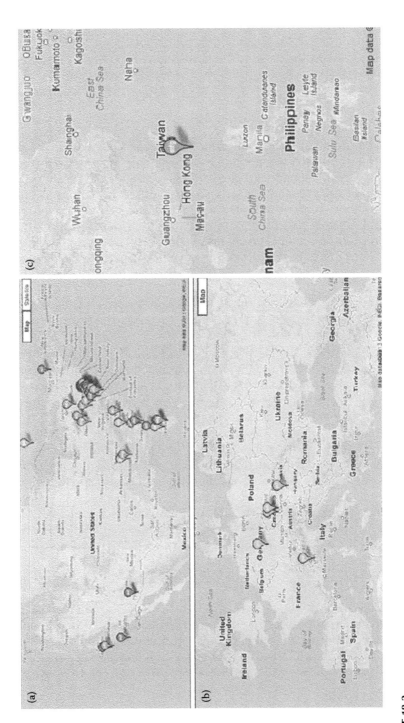

FIGURE 18.2
Map of application of nanotechnology (nZVI) for *in situ* underground water remediation in (a) United States and Canada, (b) Europe, and (c) Asia. (From www .nanotechproject.org.)

FIGURE 18.3
Product life cycle.

18.2.2.1 Raw Materials

There are two methods for nano-iron production, chemical and physical. In the case of the chemical method, the iron-based compounds have been employed to produce nanoparticles in liquid media. Then, the nanoparticles can be separated from the liquid phase via a centrifugal process or drying methods, or also can be used in nanocolloidal form. In the chemical process, there is no complicated equipment and only control of the process is vital. The chemicals used in this process, as raw materials, are not toxic and are not of any concern regarding environmental health and safety (EHS) issues. In the case of physical methods, the bulk iron is to be converted into the nanoparticulates through a physical phenomenon like electrical explosion of wires (EEW). In both cases, the raw materials are available and cheap.

18.2.2.2 Production

As mentioned before, there are two methods for nZVI particle production, chemical and physical. In the chemical method through a reduction reaction process, Fe^{2+} and Fe^{3+} are reduced into nanosize Fe^0 (zerovalent iron) particulates. The nZVI is highly active and can react easily with oxygen and consequently explode. This high reaction capability makes the nZVI very effective for removing contaminants from water.

In the physical process, an up–down approach is taken. In fact, the iron bulk material is converting into the nano-iron particles through a physical phenomenon. For instance, EEW can be applied for this purpose. In this method, through applying a high voltage and current, an explosion will occur in a narrow wire. Then, the wire is converted into the metals cluster, containing a few atoms and after agglomeration the clusters create spherical nanoparticles. As an example in developing countries, in Iran, PNF Co. has been applying this technology to supply both nanoparticles and also nanofabrication devices. The EHS issues should be considered through the whole process of nanopowder production and storage.

18.2.2.3 Use

The nZVI is to be injected into the ground and to be diffused simultaneously far away from the injection point. This process does not need any complicated devices and equipment. As a result, in the affected zone, the extracted groundwater would be clean without any contaminants. To inject the nZVI, a sludge form of iron nanocolloid is needed.

18.2.2.4 Disposal/Recycling

The injected iron nanoparticles are to stay underground and can be oxidized into iron oxide particles. Nanoparticles could be extracted with treated groundwater, which can be gathered via a magnetic technique.

18.2.3 Economics of nZVI

Table 18.1 shows the cost comparison between two conventional methods for underground remediation and nanotechnology-based nZVI. As seen, the cost for nZVI is quite lower than that of the two other conventional methods. Because of high cost and lengthy period for pump and treat remedies, *in situ* groundwater treatment is increasing.

From an economic point of view, on the basis of nZVI's relatively broad applicability, it is estimated that at least 1000 sites (about 5% of all contaminated sites) in Europe could be remediated with nZVI (Observatory Nano Focus Report, 2010). Groundwater remediation is expected to be a major use of many nanoparticles in development. In the United States, expenditures during the next 30 years for cleaning up contaminated groundwater are estimated at about $750 billion. It is projected that iron nanoparticles could save $100 billion or more in this effort. Table 18.2 summarizes the market size of iron and iron-based nanoparticles in the water treatment industry. According to the BCC Research Report (2010), several of the world's most contaminated waters are in desperate need of remediation. Nanotechnology-based remediation methods could serve as a cleanup strategy. Table 18.3 summarizes global hotspots in need of groundwater remediation. These hotspot locations can be treated via nanoparticle technologies.

TABLE 18.1

Cost Comparison between Pump and Treat, PRB, and nZVI Based for a Specific Contaminated Site in New Jersey, United States

Method	Cost, €
Pump and treat	2,760,000
PRB	1,460,000
nZVI	300,000

Source: PARS Environmental, December 2009; European Risk Observatory, EU-OSHA, Observatory Nano Focus Report, 2010. Available at http://osha.europa.eu /eu/riskobservatory.

Note: PRB, permeable reaction barriers.

TABLE 18.2

Market Size for Iron Nanoparticles in Water Treatment Industry

	2000	2005	2009	2010	2015	Compound Annual Growth Rate, %
Iron and iron-based nanoparticles	0	11	16	18	44	

Source: BCC Research Report, Nanotechnology in water treatment, 2010. Available at http://www.bccresearch.com.

TABLE 18.3

Global Hotspots for Groundwater Remediation

Hotspot Location	Source of Contamination
Dzerzinsk, Russia	Chemical and toxic byproducts from Cold War–era chemical weapons manufacturing
Linfen, Shanxi Province, China	Coal industry (arsenic, lead, etc.)
Kabwe, Zambia	Smelting and mining processes (lead, cadmium, etc.)
Chernobyl, Ukraine	Nuclear disaster
Ranipet, India	Tannery waste (hexavalent chromium and azo dyes)
Mailuu-Suu, Kyrgyzstan	Radioactive uranium mine (heavy metals and cyanides)

Source: BCC Research Report, Nanotechnology in water treatment, 2010. Available at http://www.bccresearch.com.

18.2.4 Potential Risks of Nanoparticles

To assess the potential risk of nanoparticles, we need to consider the full life cycle of the products from resources to disposal of remaining waste. In fact, the effect of exposure to nanomaterials, from handling them at water treatment plants to drinking them in treated water, is not yet known. However, there are concerns that their small size and enhanced reactivity may make them more toxic. Their small size makes their escape and diffusion into the environment quite easier.

In ongoing researches, scientists are evaluating the potential health and environmental risks of using nanotechnology in water applications. For instance, there is a special program working on EHS issues of incorporation of iron nanoparticles into groundwater in Venice (Italy) for remediation purposes. Investigations regarding the ethical, legal, and social implications of nanotechnology also are under way. Because of the generally lower scientific capacity in emerging countries, there is concern that effective regulation of any discovered risks will lag behind that of developed countries. The EHS issues should be considered at all stages of life cycle, i.e., raw materials, research and development (R&D), manufacturing, consumer use, recycling, and disposal stages, as discussed in previous sections.

In summary, it can be stated that for water treatment, nanomaterials are neither "nano-angels" nor "nano-demons." In the next section, the EKC concept is discussed.

18.3 Environmental Kuznets Curve

On the basis of the Kuznets work, Grossman and Krueger developed the EKC, which discloses the relationship between environmental degradation and economic growth (Grossman and Krueger, 1995). They used national GDP data and various indicators of local environmental conditions via panel data from GEMS (United Nations Global Environment Monitoring System). The results showed that for most indicators (pollutants in air and water), economic growth brings an initial phase of deterioration followed by a subsequent phase of improvement. The turning points, i.e., GDP peak, for different water pollutants vary; however, in most cases, they occur before a country reaches a per capita income of $8000 (see Table 18.4).

TABLE 18.4

Turning Points for Some Water Pollutants in Rivers

Pollutant	Turning Point (Peak GDP), USD (1985)	Standard Errors
Dissolved oxygen	2703	5328
BOD	7623	3307
COD	7853	2235
Nitrates	10,524	500
Lead	1887	2838
Cadmium	11,632	1096
Arsenic	4900	250
Mercury	5047	1315

Source: Grossman, G.M., Krueger, A.B., *J. Econ.*, 110, 353, 1995.

Note: BOD, biochemical oxygen demand; COD, chemical oxygen demand.

Figure 18.4 illustrates a typical EKC curve with income per capita on the X-axis and environmental degradation, i.e., pollution, on the Y-axis. Figure 18.5 illustrates the real EKC for arsenic pollution in water of a river basin.

A great deal of research and investigation have been carried out after Grossman and Krueger's work (Bruvoll and Medin, 2003; Jha and Murthy, 2003; Cole, 2004; Paudel et al., 2005; Paudel and Schafer, 2009; Jordan, 2010; Culas, 2012) to examine the accuracy of the EKC for various environmental degradation indicators such as air pollution (Bruvoll and Medin, 2003; Cole, 2004), water contaminants (Paudel et al., 2005; Paudel and Schafer, 2009), and deforestation (Culas, 2012). While much research was consistent with the EKC theory (Munasinghe and Swart, 2004; Culas, 2012; Munasinghe, 1999; Hwang, 2007) and shows a bell-shaped trend through the per capita income changes, there are some investigations that are incompatible with the EKC theory (Grossman and Krueger, 1995; Culas, 2012). Figure 18.6 shows a comparison for a number of published articles relevant to the EKC that were confirmed and rejected.

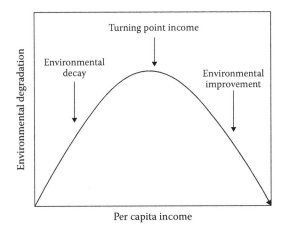

FIGURE 18.4
EKC proposed by Grossman and Krueger. (From Grossman, G.M., Krueger, A.B., *J. Econ.*, 110, 353, 1995.)

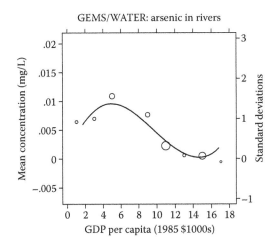

FIGURE 18.5
Relationship between per capita GDP and heavy metals contamination of rivers. (From Grossman, G.M., Krueger, A.B., *J. Econ.*, 110, 353, 1995.)

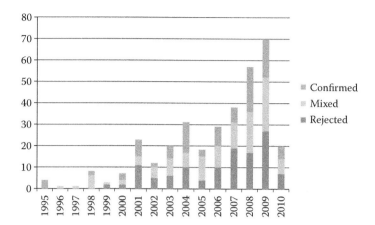

FIGURE 18.6
Confirmation of the EKC hypothesis by year. (From Jordan, B.R. *Workshop on Original Policy Research*. Georgia Tech. School of Public Policy, 2010.)

One possible interpretation of the EKC curve is that to reduce the environmental degradation impact, a certain level of economic development is needed. One explanation could be that the environment is like a luxury good (Gangadharan and Valenzuela, 2001), which at early stages of development the society and government do not care about that, but when the income reaches a certain level the society wants to pay for it. Having greater prosperity, the people demand more attention be paid to their living conditions. It is also possible that the bell-shaped curve can be caused through ceasing production of polluting products in a country and beginning to import the product from other countries, i.e., low-income or developing countries, in which they have less strict environmental regulations. In general, creating bell-shaped curves and the existence of turning points might occur through the following channels:

1. Policy change induced by the government (Culas, 2012)
2. Society awareness of environmental hazards (Culas, 2012)
3. Technology improvement (Ruttan, 1971)

In all of the above-mentioned routes, first, the government and society should be aware and care about the environmental degradation; then, when income per capita crosses a certain point, the people and government try to invest in environmental degradation mitigation. Up to now, several studies have been conducted to find the sufficient level of economic growth, i.e., income per capita, for each environmental parameter to reach the turning point (Grossman and Krueger, 1995; Culas, 2012; Munasinghe and Swart, 2004). For instance, as seen in Table 18.4, the required income per capita for a turning point for arsenic pollution in water is estimated at $4900 (Grossman and Krueger, 1995).

While EKC is a useful framework for monitoring, comparison, and analysis of environmental degradation based on income per capita, it can be misinterpreted and misused by policy makers. A possible misunderstanding is that the environment is the inevitable consequence of economic growth. The key question is that whether we need to wait for the certain economic growth to stop a specific environmental degradation? Or what if, for example, a developing low-income country does not reach the needed threshold level, and what if it reaches the threshold when it is too late for compensating the environmental damages like what is happening in deforestation?

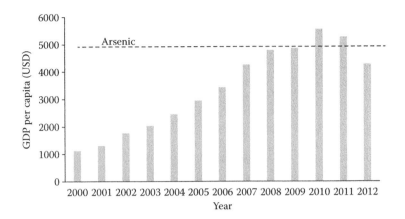

FIGURE 18.7
Trend of per capita GDP for Iran. (From World Bank. Available at http://data.worldbank.org/country/iran.)

Figure 18.7 shows how the economic boom can postpone the occurrence of the returning point in the EKC. In this figure, the trend of per capita GDP for Iran and the level of required GDP per capita for occurrence of the turning point for arsenic pollution (horizontal dotted line) are illustrated. As can be seen in the plot, it can be concluded that the time has not come for arsenic abatement for the case of Iran.

Consequently, it seems important to find a way to leapfrog the EKC and to enable low-income developing countries to avoid the need for a higher amount of per capita income for the turning point.

18.3.1 Tunneling through EKC

The fact that the EKC can be reshaped in a way that the turning point occurs at lower levels of pollution and GDP is termed "tunneling the EKC" (Culas, 2012; Munasinghe and Swart, 2004; Miah et al., 2011; Munasinghe, 1995, 1999; Rudel et al., 2005). As shown in Figure 18.8, the concept proposes that these countries can learn from developed countries' experiences to take alternative routes to leapfrog the EKC to inhibit environmental degradation damages from early stage and thus tunnel through the EKC.

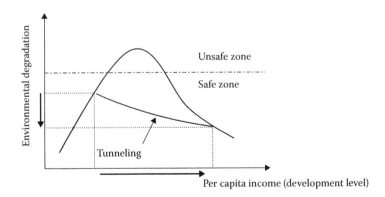

FIGURE 18.8
Tunneling through the EKC. (From Culas R.J., *Ecol. Econ.*, 79, 44, 2012.)

Many articles have been published on tunneling through the EKC (Culas, 2012; Miah et al., 2011; Munasinghe, 1995, 1999; Rudel et al., 2005). For instance, D. Miah et al. (2011) studied creating a tunnel in the EKC to lower the turning point for the case of waste emission and suspended particulates for Bangladesh in 2011, taking special policy via government. As another example, similar work has been conducted for tunneling through the EKC for deforestation of China (Culas, 2012) employing specific policy. For both cases, choosing particular policies are the key point for tunneling of the EKC. However, in general, the following routes can be taken for the leapfrogging or tunneling of the EKC:

1. Adopting particular policy via government (Culas, 2012)
2. Adopting emerging and advanced technologies (www.wikipedia.org/leapfrogging)

As mentioned before, the tunneling can also occur through the transition of pollution-generating technologies from a more developed country to a less developed one. In that case, the pollution will decline in the originating country; however, the pollution increment will transfer to the recipient. Thus, the problem is not solved and has just been transferred.

18.3.2 Policy Role

Grossman and Krueger (1995) in their article stated that "the strongest link between income and pollution in fact is via an induced policy response." Actually, these policies are, in turn, induced by societal demand. Some economists suggest that these are very important omitted variables in the EKC evaluation, and that what cleaned up the environment was not any income increment, but rather political institutions responding to public demand (Harvard Institute of Politics). In fact, there is a political mechanism underlying the EKC relationship and includes political variables in addition to income in the EKC. Political institutions have a significant effect on environmental quality for a great deal of water pollutants (Lin and Liscow, 2013). As another example, Hwang (2007) worked on the effect of regime change from communism to democracy on the environmental quality by considering water pollutants indicators such as BOD (biological oxygen demand), nitrate, and phosphorus. He studied seven European countries that moved from communism to democracy in eastern Europe. Through a more open political atmosphere and an economic growth rate increment attributed to the regime change, a better water quality was achieved (Hwang, 2007).

Culas (2012) studied adopting special policies for tunneling through the EKC for forest transition. He stated that attempts to reach an international agreement on curbing deforestation have achieved little success despite >30 years of United Nations negotiations. However, new initiatives from REDD (Reducing Emissions from Deforestation and forest Degradation) could provide financial incentives to curb deforestation. The financial incentives from REDD can make the development paths toward positive forest cover changes by shortening the forest transition periods in the countries. In summary, tunneling through the EKC is possible, by (Munasinghe and Swart, 2004):

- Adopting "win–win" policies that provide simultaneous economic, environmental, and social gains
- Using complementary measures to address harmful impacts and introducing remedies that eliminate imperfections like policy distortion, and market failures to strengthen capacity for environmental protection
- Reshaping economy-wide policies in cases where environmental and social damage was serious enough

TABLE 18.5

Water Policy in European Union and International Cooperation on Water Management

Year	Water Policy/International Commissions
1980	Groundwater protection under the Water Framework Directive
2003	New directive on groundwater protection
1976	Directive on pollution caused by certain dangerous substances discharged into the aquatic environment of the Community (also under the Water Framework Directive)
1991	Directive on Urban Waste Water Treatment
1976	Directive on Bathing Water Quality in rivers, lakes, and coastal waters
2006	New directive on bathing water quality
1998	Drinking Water Directive
1998	International Commission or the Protection of the Danube River (ICPDR; http://www.icpdr.org/pls/danubis/DANUBIS.navigator)
1996	International Commission for the Protection of the River Oder
1997	International Commission for the Protection of the River Daugava

Source: http://ec.europa.eu/environment/water.

As an example, Table 18.5 summarizes water policy in the European Union on water management during 1976–2006. These policies can be effective on the EKC turning point (http://ec.europa.eu/environment/water).

18.3.3 Technology Role

Emergence of new technologies affects the EKC turning point by two mechanisms. First, technology will provide new solutions for environment remediation and waste management. Second, technology transition in the manufacturing sector from dirty to clean can make a tunnel through the EKC. Taking emerging and advanced technologies can help leapfrog the need for a high level of income for reducing environmental degradation, but this is not an unconditional process. We will discuss these conditions in Section 18.4 with a particular focus on the need for alignment between policy tools and technology means.

According to a 2004 US EPA report, it is estimated that using traditional methods, it will take 30–35 years and cost up to $250 billion to clean up the nation's hazardous waste sites (US EPA Report, 2004). There are new emerging technologies that can result in better, cheaper, and faster site cleanup; eliminate the need for treatment and disposal of contaminated dredged soil; reduce some contaminant concentrations to near zero; and can be done *in situ*. This seems to be the only way to reduce the cost and shorten the time to reach the mentioned goal.

We continue investigating the role of technology by focusing on the effect of nanotechnology in tunneling the EKC considering the environmental medium of water in the next section.

18.3.4 Role of Nanotechnology in Tunneling the EKC: Case of Water

Among emerging technologies, nanotechnology has the fastest diffusion into the industry during the past 30 years. Nanotechnology and nanoscience involve studying and working with matter on an ultrafine scale. One nanometer is one-millionth of a millimeter, and a single human hair is around 80,000 nm in width.

Materials at the nanoscale often have different chemical and physical properties from the same material at the micro- or macroscale. For example, $n\text{-}TiO_2$ is a more effective catalyst than microscale TiO_2, and it can be used in water treatment to degrade organic pollution.

Many researchers and engineers claim that nanotechnologies offer more affordable, effective, efficient, and durable ways of alleviating water problems, i.e., removing water contaminants. Specifically, because using nanoparticles for water treatment will allow methods that are less polluting than traditional methods and require less labor, capital, land, and energy (Meridian Institute Background Paper, 2007).

According to a BCC Research Report (2010), the market for nanostructured products used in water treatment is worth an estimated $1.4 billion in 2010 and growing at a compound annual growth rate of 9.7% during the next 5 years. Currently, world-leading countries such as the United States, China, Germany, and Japan are the primary locations for nanotechnology development. However, researchers in both developed and emerging economies are active in nanotech R&D (BCC Research Report, 2010). The main way in which nanotechnology is expected to alleviate water problems is by solving technical challenges related to removing contaminants such as bacteria, viruses, heavy metals, nitrates, phosphates, and salt. Table 18.6 summarizes the market size for nanotechnology products used in the water and wastewater treatment industry.

Emerging nanotechnologies have been introduced to water and wastewater and can be categorized into three groups, including (i) nanostructures (filters and catalysts), (ii) nanoparticles, and (iii) nanodetectors/sensors.

Nanostructures contain membranes with nanopore size structure, porous materials, i.e., catalysts, filters with nanomaterial ingredients, and nanocomposites. The nanostructure filtration systems have no EHS issue and can be used safely in water treatment. In fact, the nanofilters allow water to pass through the membrane and pollutants such as bacteria, viruses, salt, heavy metals, and organic pesticides remain and are separated from the treated water. The nanocatalysts have large surface area and can be used in wastewater cleanup. The pollutants adsorbed into the catalyst body (bed) and the cleansed water can be produced. Then, the catalyst can be washed and cleaned up from the pollutants.

Nanoparticles are used with the form of dry powders or nanocolloids for water and wastewater treatment. Then, these nanoparticulates can be left in the environment or gathered via a certain technique, e.g., collecting nano-iron oxide via a magnetic field.

Nanodetectors and nanosensors can be used to characterize pollutants in water with a high level of precision. In fact, nanodetectors bring a lot of achievements, including high precision, lower cost for detection process, and fast characterization process.

In Table 18.7, a qualitative comparison for different specifications of various nanotechnologies is presented for water and wastewater applications. For instance, in the case of nanoparticles, the impact level is high, which means they can be very effective for solving water pollution; however, the impact time is long, which is mainly due to the EHS concerns. Actually, the EHS issue is a drawback for their promotion. In general, acquiring the technology of nanoparticle production is a modest challenge and can be obtained by developing countries.

TABLE 18.6

Global Market for Nanotechnology Products Used in Water Treatment ($ Million)

	2000	2005	2009	2010	2015	Compound Annual Growth Rate, %
Established products	626	863	1216	1356	2110	9.2
Emerging products	0	20	37	45	112	20.0
Total	626	883	1253	1401	2222	9.7

Source: BCC Research Report, Nanotechnology in water treatment, 2010. Available at http://www.bccresearch.com.

TABLE 18.7

Qualitative Comparison for Different Specifications of Various Nanotechnologies in Water and Wastewater Applications

Technology	Impact Level	Impact Time	Technology Level	EHS Issue
Nanoparticles	High	Long	Moderate	Under debate
Nanofilters (membrane)	High	Short	High	No problem
Nanocatalysts	High	Moderate	Moderate	No problem
Nanodetectors	High	Moderate	High	No problem

18.3.5 Practical Issues

There are various issues that need to be considered for a comprehensive evaluation of the real potential of nanotechnology for developing countries' environmental economics. We need to take into account some important aspects like level of impact, ownership of technology, timeline, and technology adoption. Shedding light on these aspects will enable us to propose some practical policy prescriptions in Section 18.4.

18.3.6 Level of Impact: Need for Full Life Cycle Economic Assessment

The key question that may arise is whether the impact of nanotechnology application is big enough to change the game in developing countries' environmental economics. A prerequisite to answer this question is undertaking a full life cycle economic assessment of the new nanotechnology methods. It will include both technology and product life cycle while considering important socioeconomic aspects. For any emerging technology, costs incurred in the R&D stage, commercialization, technology transfer, and technology adoption need to be considered.

We also need a full product life cycle assessment, which means we need to cover the product life cycle from the extraction of resources, through production, transfer and distribution, use, and recycling, up to disposal of the remaining waste. We also need to consider installation of new facilities and maintenance costs, human resource training, and safety-related costs.

Comparing conventional methods costs with new nanotechnology-based methods will provide the total savings we may expect from the application of the new technologies. Aggregation of such detailed economic calculations over a range of nanotechnology-based methods then will provide us with the overall economic impact of adopting the new nanotechnologies. The level of impact may be anticipated on the basis of the scale and number of environmental problems that will be affected by practical application of the new methods.

The important point is that even moderate reduction of costs by environmentally safe practices may change the result of an economic cost–benefit analysis. This could be particularly the case for private sector manufacture. As a result, it may enable the market mechanism in favor of environmentally friendly manufacturing methods. This is an example of how technological innovation can ease the policy instruments' tasks.

18.3.7 Ownership of Technology

Nanotechnology is an emerging area, and new products are mainly protected by domestic and foreign intellectual property rights mainly owned by large public companies in developed countries (Woodson and Harsh, 2012). This may cause a barrier for developing countries to fully benefit from the economic advantages of the emerging technologies in their environmental projects, although they may take advantage of local patent filings. This also may worsen the already present market failure in environmental economics. We will discuss more on this in Section 18.4.

18.3.8 Technology Adoption

Another important issue is whether developing countries can adopt the new nanotechnology methods in an efficient way and in widespread scale. As an example, how much would be the incurred costs and needed time for developing required human capital capabilities?

While all these may be monetized and be considered in calculations, many developing countries already realized that investment in human capital development is a long-term necessity and cannot be put in a simple cost–benefit analysis isolated from other factors. Nevertheless, for specific technologies, it is inevitable and necessary to predict and take into account the adoption stage in the economical calculations.

18.3.9 Timeline

When can we expect nanotechnology to show its impact? This is a key question regarding the subject matter of this chapter, as in many cases environmental problems may reach to an irreversible point where the damage caused to nature cannot be remediated by any means. Many nanotechnology-based promising solutions have not reached the point where they can be marketed on a wide scale. Even if they reach the market, it will take time for them to enter the mass production stage and benefit from economies of scale. This is a situation we may call "market timing failure," which means the market mechanism may provide us a solution when it is too late to be applied effectively.

18.3.10 Safety Concerns and Social Impact

As mentioned in the case study, some nanoparticles may have a side effect for the environment. The possibility of toxicity and uncontrollable dissemination through ecosystems are two major emerging concerns. It is a valid concern that while we are focusing on benefits of nanotechnology for the environment, new damages may be caused by disseminating toxic isolated nanoparticles through nature, although many applications employ agglomerated nanomaterials. These issues need to be addressed through comprehensive risk assessment as well as close monitoring and control, enforced by international and public regulatory bodies.

Figure 18.9 shows a summary of the interaction between various technology-related factors that affect the environment and its socioeconomic attributes.

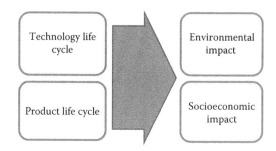

FIGURE 18.9
Analytical framework for technology assessment.

18.3.11 Tunneling through the EKC by Application of nZVI in the Water Industry

Employing nZVI particles for groundwater remediation can decrease the process cost significantly (see Table 18.5) in comparison with traditional methods. It means that the poorer and developing countries can also have cleaner groundwater (less environmental degradation) while spending less money. In other words, the EKC turning point can occur at a lower income per capita. For instance, for the case of Iran, which at the moment is not ready for a turning point for arsenic and nitrates, these pollutants can be overridden by applying nZVI for underground water remediation.

The technology of nZVI production can be transferred form developed countries having experience with this technology (see the map in Figure 18.2) to developing countries like Iran. It is noteworthy to state that countries like Iran have some laboratory-scale research experience, which can help expedite the technology diffusion and promotion.

As shown in the mentioned case study, different aspects of the technology life cycle, product life cycle, and socioeconomics effects need to be analyzed for evaluation of the effect of a new emerging nanotechnology on environment.

18.4 Policy Implications

As mentioned before, there are two main approaches for tunneling through the EKC, i.e., achieving environmental goals with less cost by appropriate policy tools and technological means. Our aim in this section is to show that they are two highly interrelated means, and should not and cannot be dealt with separately. Governing and coordinating institutes have a substantial role in realizing the vitally needed interconnections and integration. We believe that this could be a critical step in overcoming challenges of the responsible use of the technology for the benefit of the environment.

18.4.1 International Collaboration

Two proposed mechanisms for international collaboration are described below.

18.4.1.1 Collaborative Open Innovation Scheme

To address the already mentioned ownership problem and the lack of economic motivation in finding solutions to environmental problems of developing countries, an open innovation scheme may be a promising way forward. In this emerging paradigm, innovation takes place in a distributed network of individual scientists, researchers, and entrepreneurs where prohibitions on the IP rights mechanism is weaker than the centralized company based R&D framework.

There are three ways in which open innovation may help. First, it can modify the ownership pattern of the IP rights in a way that it will be owned by a network of distributed public and private institutes and companies around the world. In this way, economic benefits of the innovations will be divided more fairly among those who really need the result and have contributed to it. Second, through a properly directed open innovation scheme, R&D activities may be directed more toward solving environmental problems in developing countries. It is important to note that emerging environmental problems may arise in local areas with limited market potential, which generally may not be considered as good investment targets for big private companies. Finally, it can bring all beneficiary stakeholders together as active players and facilitate the exchange of information and experience among them.

18.4.1.2 Regulation and Governance

To address important concerns regarding safety and environmental consequences of using nanoparticles, there is a need for governing bodies' intervention both in local and international levels. International institutes that closely monitor nanotechnology development in both the developed and developing world can do the needed assessments and make required recommendations in this regard.

18.4.2 Role of International Coordinating Institutes

International coordinating institutes are the instruments to realize the mentioned goals, i.e., coordinating distributed innovative activities and facilitating regulatory and protective planning and governance. Some needed tasks are explained in more detail below.

18.4.2.1 Facilitating Knowledge and Experience Transfer

Knowledge and experience transfer between stakeholders can be accelerated via proper coordination. This is a critical point in shortening the time from laboratories to real field applications and also is a key success factor in an open innovation mechanism. This needs to be done among university researchers, industry experts, the finance sector, and industry managers as well as policy makers and decision makers.

18.4.2.2 Assisting in Provision of Realistic Technological, Socioeconomic, and Environmental Assessments

As mentioned in the previous section, the full life cycle technological, economic, societal, and environmental assessment of nanotechnology solutions is an essential requirement to enable decision makers' timely action. Assessments need to be reliable and trustworthy for the use of stakeholders. Again, international neutral institutes and organizations can contribute by managing such studies, and provide the results to decision makers in the developing world. On the other hand, in most cases, achieving reliable input data and conducting professional analysis on the data implicate close cooperation between experts and institutes from all over the world both in academic and industry sectors, which require a high level of coordination.

18.4.2.3 Facilitating Funding and Commercialization of Promising Solutions

The funding gap in bringing research results to market is a well-known phenomenon. One way to remove this gap is through mitigation of information asymmetry between technical and finance parties. In addition, stimulating cooperative joint investment plans among countries with similar environmental problems can be an effective way to accumulate the needed resources.

18.4.2.4 Coordinating and Monitoring Regulation and Ethical Aspects

Finally, the aforementioned intermediary institutes can help in achieving transparency and needed coordination in dealing with regulatory and ethical aspects.

18.5 Conclusions

Technology achievements, when aligned with policy tools, can reshape the already experienced EKC patterns in growth–environmental relationships. This is a very important opportunity that must not be overlooked by developing countries. Nanotechnology is one very promising emerging technology in this aspect. We tried to show the potential of nanotechnology in this regard by focusing on water applications and giving the example of groundwater remediation. We exercised a deeper look into the assessment of the new technology by considering technology life cycle, product life cycle, and socioeconomic aspects, all in parallel. The aforementioned typical comprehensive assessments make it clear that there are quite a number of challenges and barriers that cannot be solved in time without an effective international cooperation and coordination. Finally, a few policy recommendations were provided, and it was emphasized that international focused coordinating institutes can have a major role in implementing these policies.

References

BCC Research Report. 2010. Nanotechnology in water treatment. Available at http://www.bccrsearch .com (January 2011).

Bruvoll A. and Medin H. 2003. Factors behind the *environmental* Kuznets curve, Evidence from Norway. *Environmental and Resource Economics* **24** (1), 27–48.

Cole M. A. 2004. Trade, the pollution haven hypothesis and the environmental Kuznets curve: Examining the linkages. *Ecological Economics* **48**, 71–81.

Culas R. J. 2012. REDD and forest transition: Tunnelling through the environmental Kuznets curve. *Ecological Economics* **79**, 44–51.

European Commission. 2014. Available at http://ec.europa.eu/environment/water/index.html.

European Risk Observatory, EU-OSHA. 2010. Observatory Nano Focus Report. Available at http:// osha.europa.eu/eu.riskobservatory.

Gangadharan L. and Valenzuela M. R. 2001. The concept of joint production and *ecological economics*. *Ecological Economics* **36**, 513–531.

Grossman G. M. and Krueger A. B. 1995. Economic growth and the environment. *The Quarterly Journal of Economics* **110**, 353–377.

Harvard Institute of Politics. Available at http://www.iop.harvard.edu/programs/forum/transcripts /environment_03.13.03.pdf.

Hwang H.-C. 2007. *Water Quality and the End of Communism*, Brandes University. Available at http:// www.academia.edu/289433.

Jha R. and Murthy K. V. B. 2003. An inverse global environmental Kuznets curve. *Journal of Comparative Economics* **31** (2), 352–368.

Jordan B. R. 2010. *Workshop on Original Policy Research*. Georgia Tech. School of Public Policy. Atlanta, GA.

Kuznets S. 1955. Economic growth and income inequality. *American Economic Review* **45** (1), 1–28.

Lin C. Y. C. and Liscow Z. D. 2013. Endogeneity in the environmental Kuznets curve: An instrumental variables approach. *American Journal of Agricultural Economics* **95** (2), 268–274.

Meridian Institute Background Paper for the International Workshop on Nanotechnology, commodities and development. 2007. (May 29–31), Rio de Janeiro, Brazil.

Miah D., Masum F. H., Koike M., Akther S. and Muhammed N. 2011. A review of the environmental Kuznets curve hypothesis for deforestation policy in Bangladesh. *Environmentalist* **31**, 59–66.

Munasinghe M. 1995. Making growth more sustainable. *Ecological Economics* **15**, 121–124.

Munasinghe M. 1999. Is *environmental* degradation an inevitable consequence of *economic* growth: Tunneling through the *environmental Kuznets curve*. *Ecological Economics* **29**, 89–109.

Munasinghe M. and Swart R. (Editors) 2004. *Primer on Climate Change and Sustainable Development.* Cambridge University Press, UK.

Paudel K. P., Zapata H. and Susanto D. 2005. An empirical test of *environmental* Kuznets curve for water pollution. *Environmental and Resource Economics* **31**, 325–348.

Paudel K. P. and Schafer M. J. 2009. The environmental Kuznets curve under a new framework: The role of social capital in water pollution. *Environmental and Resource Economics* **42**, 265–278.

Rudel T. K., Coomes O. T., Moran E., Achard F., Angelsen A., Xu J. and Lambin E. 2005. Forest transitions: Towards a global understanding of land use change. *Global Environmental Change* **15**, 23–31.

Ruttan V. W. 1971. Technology and the environment. *American Journal of Agricultural Economics* **53**, 707–717.

US EPA Report, 2004.

Woodson T. and Harsh M. 2012. Pro-poor nanotechnology applications for water: Characterizing and contextualizing private sector research and development. *Nanotechnology Law and Business* **9**, 232.

World Bank. Available at http://data.worldbank.org/country/iran.

FIGURE 1.1
Crumb rubber reactor.

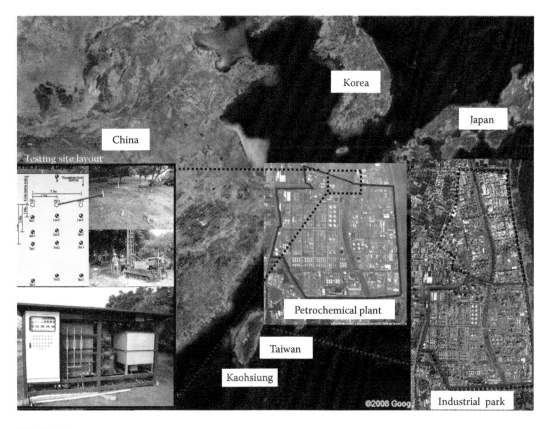

FIGURE 4.5
Pilot testing site (200 m²) of nZVI remediation conducted at a petrochemical plant in Kaohsiung, Taiwan. nZVI was synthesized on site using a semicontinuous reactor system.

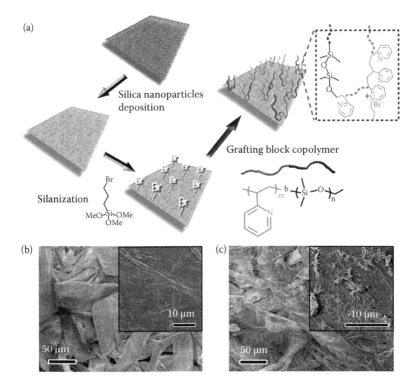

FIGURE 7.3
Preparation and characterization of a surface with switchable superoleophilicity and superoleophobicity on a nonwoven textile substrate. (a) Diagram showing the preparation strategy for a surface with switchable superoleophilicity and superoleophobicity on a nonwoven textile using a pyridine contain polymer that changes charge state as a function of pH. (b) Scanning electron microscope (SEM) image of the raw textile. Inset: enlarged view of the surface of a single fiber. (c) SEM image of the textile after deposition of silica nanoparticles and block copolymer grafting. Inset: enlarged view of the surface of a single fiber.

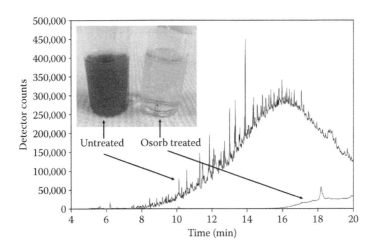

FIGURE 7.10
Photographs and gas chromatograms of emulsified bilge water simulant before and after addition of 1.5% w/v Osorb-EB sorbent media. Oil content is reduced to <5 ppm.

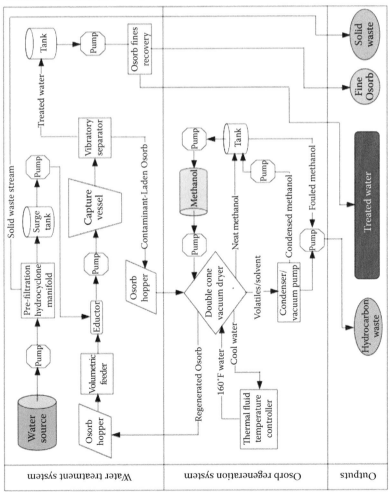

FIGURE 8.11
PWUnit#1 process diagram and photograph.

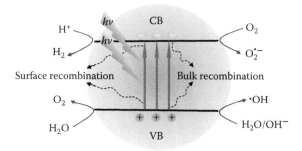

FIGURE 11.1
Schematic illustration of processes involved in semiconductor photocatalysis.

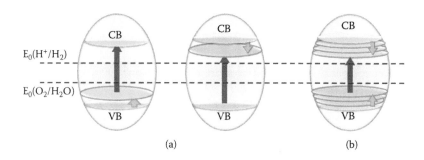

FIGURE 11.5
Energy band bending strategies to narrow the band gap of semiconductor photocatalysts so as to match the solar spectrum. (a) Element doping. (b) Solid solution.

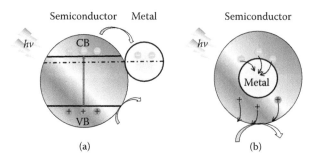

FIGURE 11.9
Metal–semiconductor photocatalytic systems: (a) metal particle supported on semiconductor surface; (b) metal–semiconductor core–shell structure.

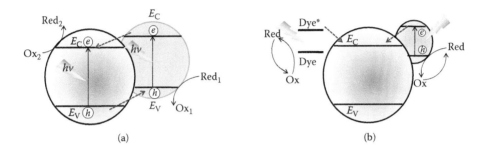

FIGURE 11.10
Schematic diagram illustrating the principle of charge separation between two semiconductor nanoparticles: (a) heterojunction; (b) sensitization.

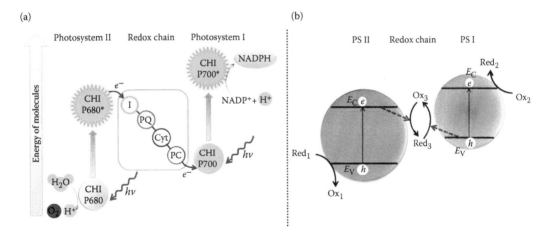

FIGURE 11.11
Z-scheme mechanisms in (a) the natural photosynthesis system and (b) semiconductor photocatalyst system.

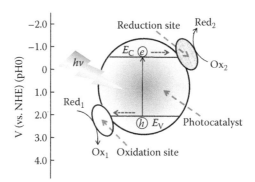

FIGURE 11.12
Schematic diagram illustrating the reaction process of cocatalyst-modified semiconductor photocatalyst.

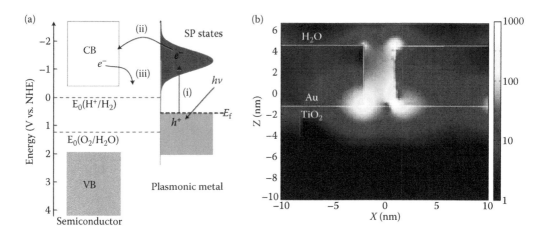

FIGURE 11.13
(a) Schematic diagram of plasmon-induced charge separation and associated photochemistry at the metal–semiconductor heterojunction. (b) Optical simulations showing SPR-enhanced electric fields owing to photoexcited Au particles, permeating into a neighboring TiO_2 structure. (a: Adapted with permission from Linic, S., Christopher, P., Ingram, D.B. *Nat. Mater.*, 10, 911. Copyright 2012, Nature Publishing Group. b: Adapted with permission from Liu, Z., Hou, W., Pavaskar, P., Aykol, M., Cronin, S.B. *Nano Lett.*, 11, 1111. Copyright 2011, American Chemical Society.)

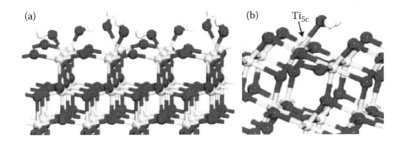

FIGURE 11.16
Water adsorption behaviors on anatase (a) TiO₂ {001} vs. (b) {101} surfaces. (Adapted with permission from Selloni, A., *Nat. Mater.*, 7, 613. Copyright 2008, Nature Publishing Group.)

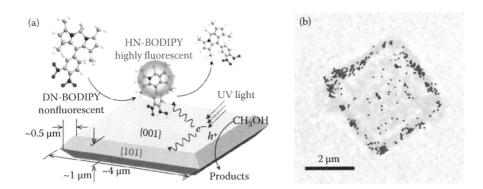

FIGURE 11.17
(a) Photocatalytic generation of fluorescent HN-BODIPY from nonfluorescent DN-BODIPY over a TiO₂ crystal. (b) Transmission images of the same TiO₂ crystal immobilized on a cover glass in Ar-saturated methanol solution containing DN-BODIPY under a 488-nm laser and UV irradiation. The blue and red dots in the transmission image indicate the location of fluorescence bursts on the {001} and {101} facets of the crystal, respectively, observed during 3-min irradiation. (Adapted with permission from Tachikawa, T., Yamashita, S., Majima, T., *J. Am. Chem. Soc.*, 133, 7197. Copyright 2011, American Chemical Society.)

FIGURE 12.9
Fluorescent microscopic images of *E. coli* feed (a) and permeate (b). (c) SEM image of retained *E. coli* on the TNM after filtration (low magnification) and a high-magnification SEM image (inset). (d) Cross-section SEM image of TNM after filtration of *E. coli*. (From Zhang, H. et al. *J. Membrane Sci.*, 343, 212, 2009. With permission.)

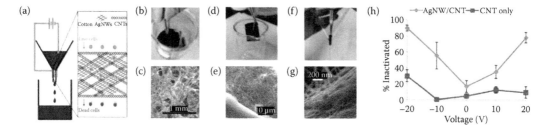

FIGURE 12.13
Schematic, fabrication, and structure of cotton, AgNW/CNT device. (a) Schematic of proposed active membrane device. (b) Treatment of cotton with CNTs. (c) Treatment of device with silver nanowires (AgNWs). (d) Integration of treated cotton into funnel. (e) SEM image showing large-scale structure of cotton fibers. (f) SEM image showing AgNWs. (g) SEM image showing CNTs on cotton fibers. (h) Inactivation efficiency at five biases for AgNW/CNT cotton as well as CNT-only cotton. (From Schoen, D.T. et al., *Nano Lett.*, 10, 3628, 2010. With permission.)

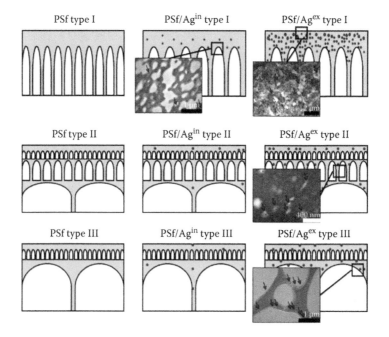

FIGURE 13.1
Example of nanomaterials that could be affixed to a membrane surface. (From Figure 5.1 in Chapter 5, Multifunctional Nanomaterial-Enabled Membranes for Water Treatment, pp. 59–75, by Tarabara, V.V. Reprinted from *Nanotechnology Applications for Clean Water*, Diallo, M.; Duncan, J.; Savage, N.; Street, A.; Sustich, R., editors, Elsevier Ltd., Kidlington, UK (2009). Reprinted with permission of Elsevier.)

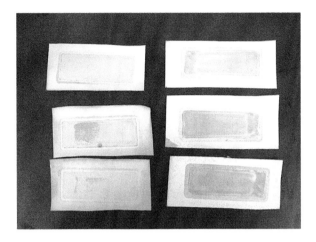

FIGURE 13.2
Comparison of SWRO membrane fouling. Membrane samples on the left side are from NanoH$_2$O elements with NP3, while those on the right are from conventional commercial SWRO. (From J. Green, "Nanocomposite Reverse Osmosis Membranes," presentation at 2009 WATER EXECUTIVE Forum, Philadelphia, March 30–31, 2009. Reprinted with author's permission.)

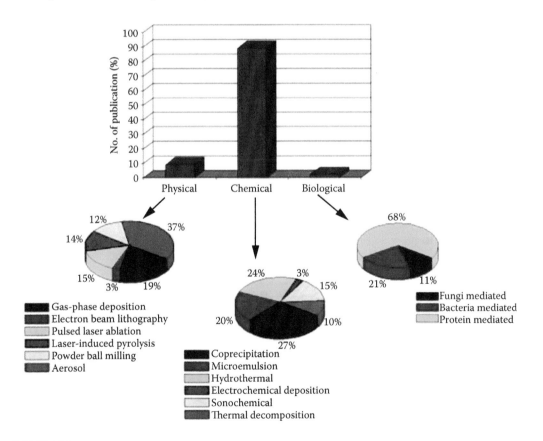

FIGURE 14.1
Comparison of published work (up to date) on the synthesis of SPIONs by three different routes. (From Mahmoudi M., Sant S., Wang B. et al., *Advanced Drug Delivery Reviews*, 63, 24, 2011.)

FIGURE 14.2
Crystal structure of Fe_3O_4; green atoms are Fe^{2+}, brown atoms are Fe^{3+}, white atoms are oxygen. (From Yang C., Wu J., Hou Y., *Chemical Communications*, 47, 5130, 2011.)

FIGURE 17.4
Electro-defluoridation field units by CSIR-NEERI, Nagpur, India.

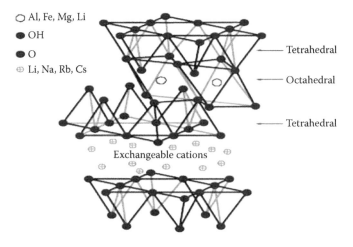

FIGURE 20.1
Layered silicate structure. (Adapted from G. Beyer, *Plast. Addit. Compound.*, 4, 22, 2002.)

FIGURE 21.3
Early gamma prototype deployed in Mexico. The technology was not yet integrated into the final product in order to test various aspects of the prototype.

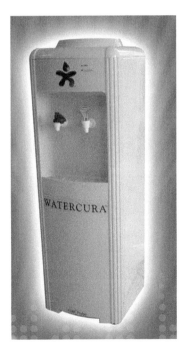

FIGURE 21.4
TWI's first product from the 3-D process using the BOP.

FIGURE 22.1
Rendering of a WHC with a cutout view showing equipment installation.

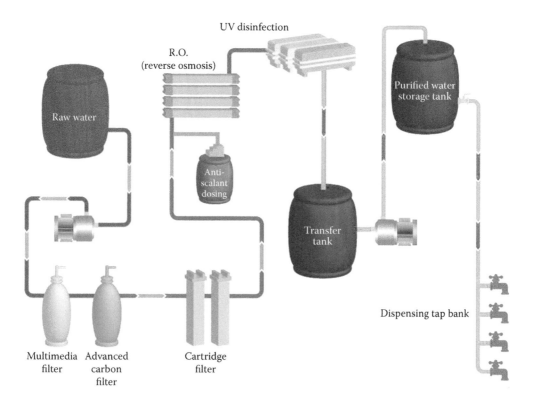

FIGURE 22.4
Schematic of WHI treatment system.

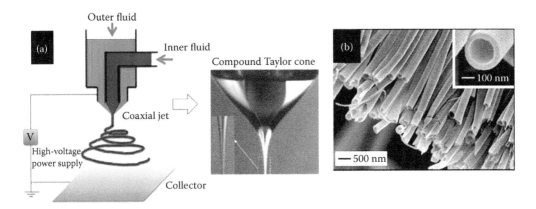

FIGURE 23.3
(a) Setup for coaxial electrospinning for the fabrication of core–shell nanofibers. (b) SEM images of an array of hollow anatase TiO_2 nanotubes. (From Feng, C. et al., *Sep. Purif. Technol.*, 102, 118, 2013.)

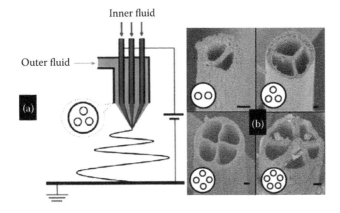

FIGURE 23.4
(a) Schematic of the multichannel coaxial electrospinning system. (b) Cross-sectional SEM images of hollow fibers with two, three, four, and five channels. Scale bars in all the images are 100 nm. (From Feng, C. et al., *Sep. Purif. Technol.*, 102, 118, 2013.)

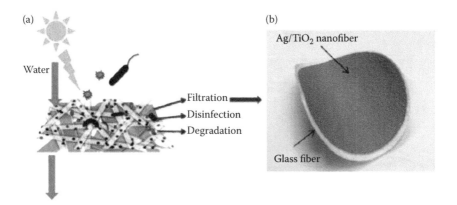

FIGURE 23.7
(a) Schematic representation of a hypothesis of a Ag/TiO$_2$ nanofiber membrane; (b) photograph of Ag/TiO$_2$ nanofiber membrane. (From Liu, L. et al., *Water Res.*, 46, 1101, 2012.)

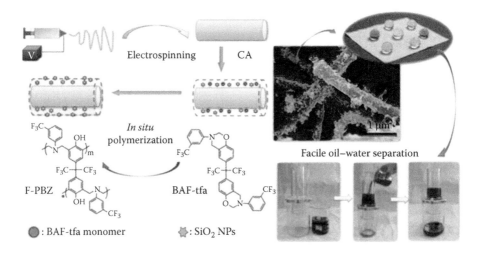

FIGURE 23.8
Schematic of electrospun CA nanofibers coated with F-PBZ and SiO$_2$ NPs and its ability to separate oil/water mixture. (From Shang, Y. et al., *Nanoscale*, 4, 7847, 2012.)

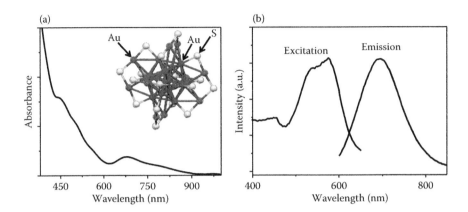

FIGURE 26.1
(a) UV/Vis optical absorption spectrum of $Au_{25}(SG)_{18}$ cluster. (b) Photoluminescence spectra of $Au_{25}(SG)_{18}$, which exhibits emission at 690 nm. (Adapted from Shibu, E.S. et al., *J. Phys. Chem C.*, 112, 12168, 2008. Copyright with permission from American Chemical Society.) Inset: crystal structure of $Au_{25}(SCH_2CH_2Ph)_{18}$. Total structure is assumed as a shell of six (–RS–Au–SR–Au–SR–) units sitting on an Au_{13} core. (Adapted from Zhu, M. et al., *J. Am. Chem. Soc.*, 130, 5883, 2008. Only Au and S atoms are shown. Copyright with permission from American Chemical Society.)

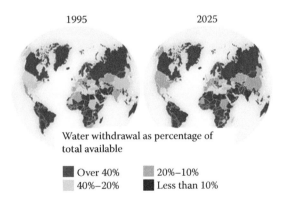

FIGURE 27.1
Water withdrawal rate as a percentage of total available water, projected till 2025. (From Service, R.F., *Science*, 313, 1088, 2006.)

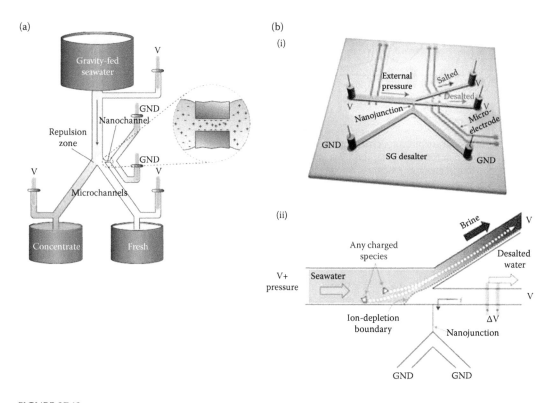

FIGURE 27.13
(a) Small-scale setup for concentration polarization desalination. (From Shannon, M.A., *Nature Nanotechnology*, 5, 248, 2010.) (b) Schematic of microchannel–nanochannel prototype tested for water desalination using concentration polarization. (From Kim, S.J., S.H. Ko, K.H. Kang, and J. Han, *Nature Nanotechnology*, 5, 297, 2010.)

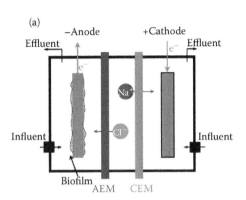
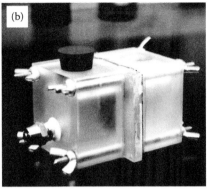

FIGURE 27.15
(a) Basic schematic representing a three-chamber cell design for coupling a microbial fuel cell with a water desalination system. (b) Microbial desalination cell prototype. (From Cao, X., X. Huang, P. Liang, K. Xiao, Y. Zhou, and B.E. Logan, *Environmental Science Technology*, 43, 7148, 2009.)

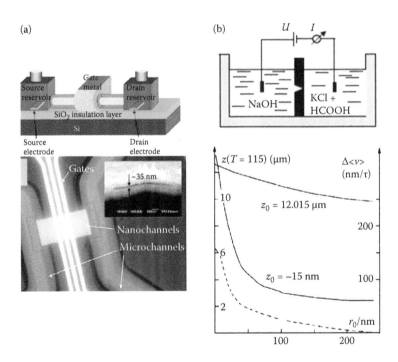

FIGURE 27.16
(a) Schematic and SEM image shows the development of a device being used as a fluidic transistor allowing electrostatic control of ions. (From Karnik, R., R. Fan, M. Yue, D. Li, P. Yang, and A. Majumdar, *Nano Letters*, 5, 943, 2005.) (b) Schematic depicts formation of a conical nanopore by electrochemical etching at potential U with current I leading to data that show pumping of the potassium ions against a concentration gradient as a function of pore diameter on the narrow side of the conical nanopore. (From Siwy, Z. and A. Fulinski, *Physical Review Letters*, 89, 19803-1, 2002.)

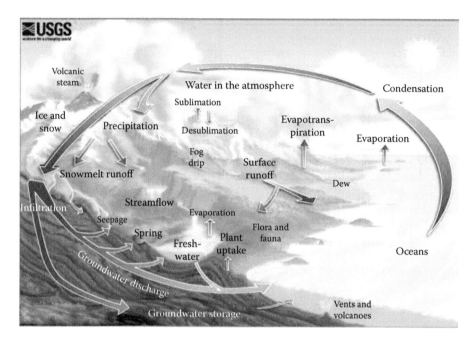

FIGURE 28.1
Natural water cycle is a balance that is dependent on many different inputs. As human activity changes these inputs, such as amount stored in groundwater or evapotranspiration from increased agricultural activity, the cycle can be disrupted or altered. (From USGS. *Summary of the Water Cycle*. [cited February 18, 2013]; Available http://ga.water.usgs.gov/edu/watercyclesummary.html, 2013.)

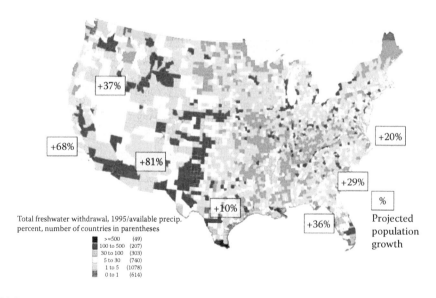

FIGURE 28.2
Projected population growth by percentage and water shortages by color. Water shortage is defined as total freshwater withdrawal divided by an area's precipitation, based on most recent available data. (From USDOE, *Energy Demands on Water Resources*. US Department of Energy, 2006.)

(a)

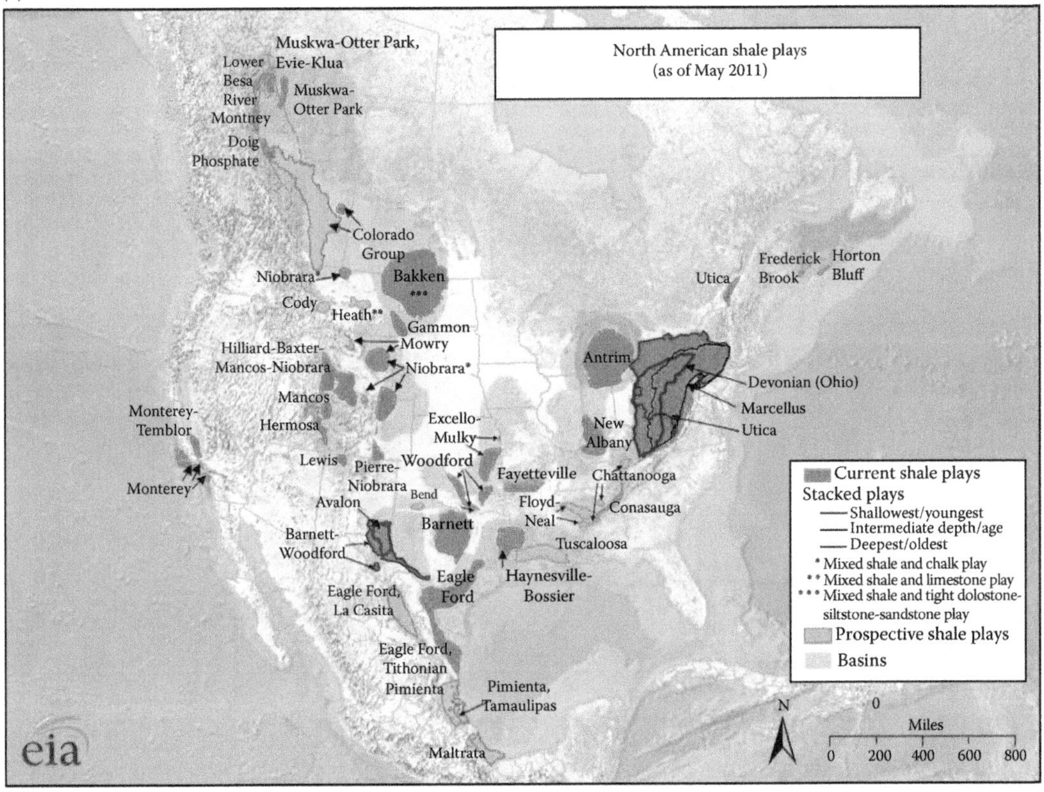

FIGURE 28.9

(a) Map of the continental United States with Mexico and Canada showing distribution of shale oil and gas reserves across North America based on the data reported by USEIA in 2011. (From USEIA, *Review of Emerging Resources: US Shale Gas and Shale Oil Plays*. US Energy Information Administration, 2011.)

(b)

Water loop—well operators have many
choices for wastewater reuse or disposal (red).

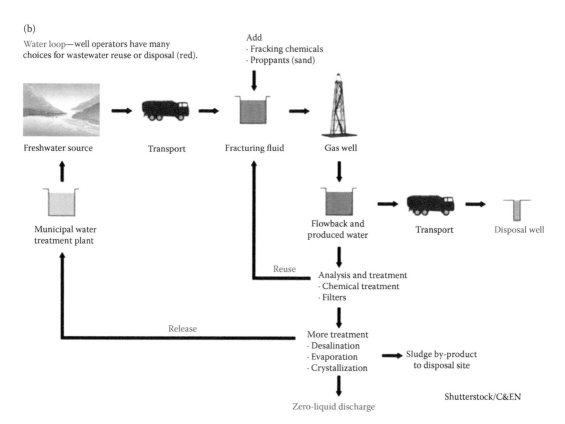

FIGURE 28.9 (Continued)

(b) Typical HF cycle of water. Initially, water, proppants, and fracking chemicals are injected into the well until high enough pressures to fracture the rock bed occur. Some of the water returns (~20%–40%) to the surface, while the rest remains in the well. Of the flowback water, reuse for further fracturing or treatment, either at a municipal water treatment plant or zero-liquid discharge facility or disposed of by deep well injection, are the pathways for managing the waste stream. (From Bomgardner, M.M., Cleaner fracking, in *Chemical and Engineering News*, 2012.)

FIGURE 29.1
AWC station with real-time weather and capture.

(a)   (b)

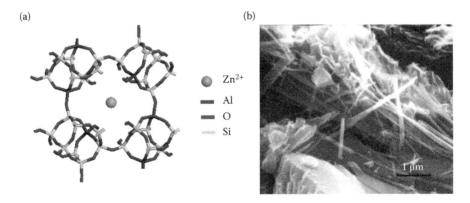

FIGURE 30.1
Representation of a Zn–clinoptilolite unit cell without water molecules (a) and clinoptilolite crystals with tabular and lath morphology (b).

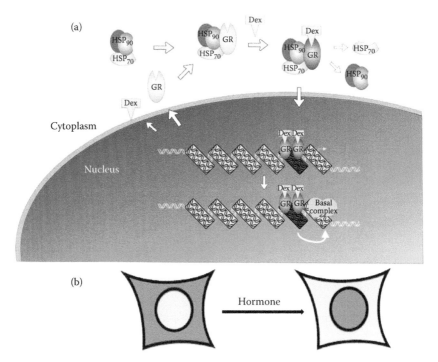

FIGURE 35.2
Glucocorticoid receptor (GR) biology. (a) At resting state, GR resides in the cytoplasm as a part of a multiprotein complex with heat-shock proteins and immunophilins. (From Pratt, W.B. and D.O. Toft, *Endocr. Rev.*, 18, 306, 1997.) Upon hormone stimulation, GR dissociates from this complex and migrates to the nucleus (cytoplasmic-to-nuclear translocation). In the nucleus, the receptor interacts with GR regulatory elements (GREs) throughout the genome, and regulates GR-mediated transcriptional responses. (b) Schematic representation of the translocation of the green fluorescent protein (GFP)-tagged GR from the cytoplasm to the nucleus in the presence of hormone or EDCs with glucocorticoid activity. This translocation is the basis of the biological screening for detection of water contaminants interacting with a specific nuclear receptor(s).

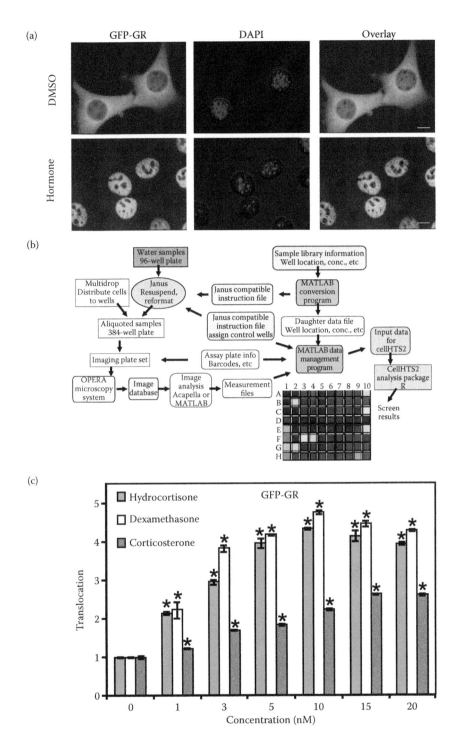

FIGURE 35.3

Application of the translocation assay for analysis of EDCs in water samples. (a) GFP-tagged glucocorticoid receptor (GFP-GR) translocates to the nucleus in mammalian cells exposed to control (DMSO, top panel) or dexamethasone (bottom panel) for 30 min. Nuclei are costained with DAPI. Scale bar, 5 μm. (b) Workflow for image-based screening of environmental contaminants with glucocorticoid activity using the Perkin Elmer OPERA image screening system. (c) GFP-GR translocates to the nucleus in a concentration-dependent manner upon treatment with known concentrations of hydrocortisone, dexamethasone, or corticosterone. An algorithm for cytoplasm and nuclear segmentation was used to determine the mean GFP-GR intensity in both compartments, and translocation was quantified as a ratio of these intensities. Each value was normalized to the control sample. Error bars represent the mean value ± s.e.m., n = 6 (P < 0.05, asterisks).

FIGURE 38.5
Al$_{13}$-PACl plant established using the electrochemical method.

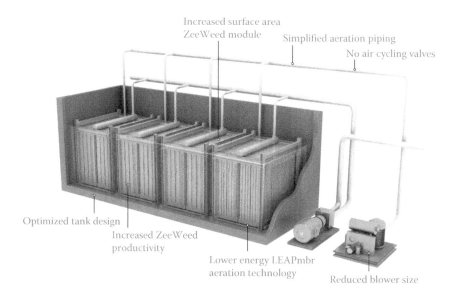

Increased surface area
ZeeWeed module

Simplified aeration piping

No air cycling valves

Optimized tank design

Increased ZeeWeed
productivity

Lower energy LEAPmbr
aeration technology

Reduced blower size

FIGURE 39.2
Building on 25 years of GE's MBR experience, LEAPmbr delivers the most advanced MBR wastewater treatment solution to date, addressing pressing water quality and operational cost issues faced by owners of municipal, industrial, and residential water/wastewater treatment facilities worldwide. The technology dramatically lowers energy costs, increases productivity, and offers flexible designs that can fit within a smaller footprint than conventional MBR technology.

19

Engineered Nanoscale Materials for Defluoridation of Groundwater

Shihabudheen M. Maliyekkal

Environmental Engineering Division, School of Mechanical and Building Sciences,
VIT University, Chennai Campus, Chennai, India

CONTENTS

19.1 Introduction to Groundwater Pollution

Water is an essential and the most critical natural resource on Earth on which all social, economic, and ecosystem functions depend. It is currently being threatened by development, overuse, and other human activities. The scarcity and quality of the freshwater are likely to be the most challenging issues to be faced by mankind in the near future. Among various available freshwater resources, groundwater often provides a water supply that is more reliable in terms of quantity and quality as compared with surface water. Groundwater drawn from deep aquifers was assumed to be free from microbial contamination and has economic and operational advantages owing to reduced treatment requirements [1]. Thus, it accounts for a significant portion of drinking water supply in many countries worldwide [2]. In India, >80% of the rural water supplies are obtained from groundwater. Over the years, a number of quantitative and qualitative changes have occurred in groundwater systems because of natural causes and human intervention. Large decreases in groundwater level and decline in water quality have become common problems in many countries worldwide. The scenario is worse in developing countries of Asia, including India and Pakistan.

Various contaminants such as fluoride [3,4]; nitrate [5,6]; pesticides [7,8]; arsenic [9,10]; heavy metals such as lead, chromium, and mercury [11,12]; endocrine-disrupting compounds [13]; nanoparticles (NPs) [14,15]; and pathogenic organisms have been found in many groundwater sources worldwide. Both geogenic and anthropogenic reasons are held responsible for the elevated concentrations of these contaminants in aquatic environments. Abuse of chemicals and drugs; improper sewerage and septic systems; indiscriminate use of pesticides [8], fertilizers, and agrochemicals [16]; urbanization [17]; and associated land use alteration with shortsighted economic benefits, storage of waste materials in excavations, such as pits or mines, acid mine drainage [18], and direct disposal of untreated or partially treated wastewater into aquatic environments [19] have caused the contamination of water systems, both surface and subsurface. Saltwater intrusion in coastal cities, overexploitation of groundwater for agricultural and drinking purposes, has also resulted in decline in groundwater quality and storage [20]. Hence, the protection of existing water resources against potential chemical and biological contamination should be carried out for a sustainable environment and economy. It is also important to remediate already polluted resources for meeting the current demands of public water supply as an interim measure.

19.2 Search for Affordable Treatment Options

The availability of safe and clean water in sufficient quantities at the point of use is a fundamental human right. We are currently facing formidable challenges in meeting rising demands of potable water. In addition to the decline in the quality of water due to pollution, erratic monsoon, unpredictable weather, draught, or flood further exerts pressure on the limited and dwindling freshwater reserves. According to estimates, 1.2 billion people worldwide do not have access to clean water [21]. Every year, more people die from the consequences of unsafe water than all forms of violence, including war and terrorism. Unfortunately, most of the affected populations are from the poorer parts of the world [22], especially rural areas where high levels of poverty prevail. Urban areas are typically better developed compared with rural areas in terms of water infrastructure. In rural areas of poor and developing countries, infrastructures are either poorly developed or nonexistent. Because of the lack of infrastructure in place, the water needs of the rural population are largely met by subsurface water. It has been established beyond doubt that water derived from the subsurface is heavily contaminated and not suitable for drinking purposes, in many locations. Fluoride is one such contaminant widely present in the groundwater sources, which needs a quick solution.

Over the years, many technologies have been developed to tackle the issue of dissolved fluoride in groundwater. However, most of the technologies did not produce the desired results in the field owing to poor efficiency or higher pricing. There is clear and urgent need for an efficient and affordable alternate treatment system for the removal of fluoride in aqueous medium. In recent years, with the advancement made in nanotechnology, nanomaterials have emerged as potential candidates for water purification. They are attractive because of their unusual properties compared with their bulk counterparts, such as larger surface area, greater density of reactive sites on the particle surfaces, short diffusion route, and higher intrinsic reactivity of the reactive surface sites [23]. They are also attractive because of the ease with which they can be anchored onto solid matrices and the ability to functionalize with different functional groups to enhance their affinity toward target molecules [11]. Because of these unique properties, it is likely that nanosized adsorbents with strong affinity to fluoride can be a useful tool in enhancing the adsorption capacity. However, the success of the process for down-to-earth applications such as water purification is largely dependent on the simplicity of large-scale material synthesis, ease of solid–liquid separation, and posttreatment handling [12].

In this chapter, the author reviews major defluoridation technologies that have been practiced in the past for water defluoridation by giving special emphasis on nanomaterials as an emerging defluoridation medium. The chapter also provides a brief description about the chemical speciation, source and occurrence of fluoride in subsurface water, and the challenges being faced in the treatment of fluoride-contaminated water.

19.3 Fluoride: A Global Groundwater Threat

Fluorine is the 13th most abundant element in the earth's crust. In nature, it occurs in various minerals such as fluorospar (CaF_2), cryolite (Na_3AlF_6), and fluorapatite ($Ca_5(PO_4)_3F$), and many others. Both natural and anthropogenic sources can contribute fluoride into the environment. However, most groundwater contaminations with fluoride are primarily due to natural reasons. In groundwater, the natural concentration of fluoride varies with the physical and chemical characteristics of the aquifer, which includes porosity, acidity of the soil and rocks, temperature, action of other chemicals, and depth of wells. In aqueous environment, it generally occurs as fluoride ion (F^-). The undissociated hydrofluoric acid (HF) and its complexes with aluminum, iron, and boron are the other most likely soluble forms of fluoride in natural water [24]. The safe drinking water limit of fluoride is <1 mg/L [4]. However, this guideline value of fluoride is not universal. The limit varies among countries and the age of persons exposed to it [25]. In India, the maximum contaminant level is restricted to 1.5 mg/L. The presence of fluoride in drinking water can be detrimental or beneficial to mankind depending on the concentration of fluoride and the duration of exposure. A small amount of fluoride in ingested water reduces the rate of occurrence of dental caries, especially in children [26]. On the contrary, excess intake of fluoride (>1 mg/L) may lead to various health problems such as osteoporosis, arthritis, brittle bones, bone cancer (osteosarcoma cancer), infertility, brain damage, Alzheimer syndrome, thyroid disorder, and DNA damage [27–30]. Fluorosis is a common symptom of high fluoride ingestion, which is manifested by mottling of teeth in mild cases and deterioration of bones and neurological damage in severe cases [31]. Excess fluoride in groundwater is reported in many countries from different continents, notably from North America, Africa, and Asia [26,32–35]. According to the latest information, fluorosis is endemic in at least 20 countries worldwide [4]. In India, it was first detected in Nellore district of Andhra Pradesh in 1937 [36] and now it is prevalent in 150 districts of 17 states [3]. Estimates show that 25 million persons in India are affected by fluorosis and approximately 66 million persons are at risk of developing fluorosis [3]. In China, cases of endemic fluorosis were reported as early as the 1930s [37]. According to a report, endemic fluorosis is prevalent in 29 provinces, municipalities, or autonomous regions in China [38]. In South Korea, excessive fluoride concentrations are frequently encountered in deep groundwater [39]. In Korea, many incidence of dental fluorosis have been reported in children who drink water with high fluoride concentrations over a long period [40]. Carrillo-Rivera et al. [41] reported that a significant amount of fluoride is found in the groundwater of San Luis Potosí, Mexico. Studies have shown that the problem of high fluoride content in groundwater is very acute in mainland Tanzania [42]. Agrawal [43] reported that in Sri Lanka, fluoride has a strong geographical control linked to climatic conditions, with high fluoride waters being restricted to the dry zone on the eastern side of the island. In some parts, wells have fluoride concentrations of >10 mg/L. Analysis of many groundwater samples collected from different parts of Ethiopia have shown excess fluoride than the guideline concentration recommended by the World Health Organization [44]. Many other cases of endemic fluorosis have been reported from various parts of the world, and many new cases are likely to be discovered in the future. The range of fluoride concentrations reported in groundwaters of developing and developed countries around the globe is shown in Figure 19.1. It is to be noted that

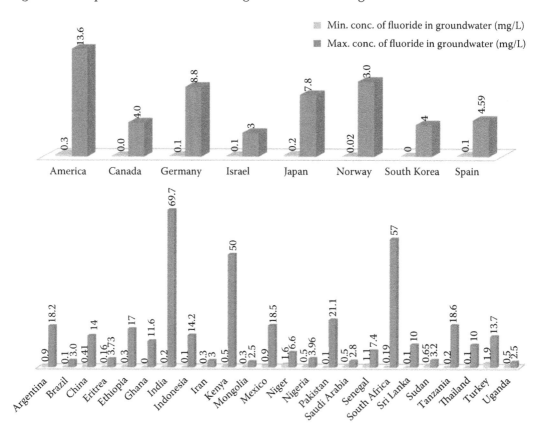

FIGURE 19.1
Range of concentrations of fluoride reported in groundwater samples around the globe. (From A.K. Susheela, *Curr. Sci.*, 77, 1250, 1999; World Health Organization (WHO), J. Fawell, K. Bailey, J. Chilton, E. Dahi, L. Fewtrell, Y. Magara. *Fluoride in Drinking Water.* ISBN: 1900222965. IWA, London, 2006; S. Naseem, T. Rafique, E. Bashir, M.I. Bhanger, A. Laghari, and T.H. Usmani, *Chemosphere*, 78, 1313, 2010; B. Salifu, K. Petrusevski, R. Ghebremichael, R. Buamah, and G. Amy, *J. Contam. Hydrol.*, 140, 34, 2012; T. Rafique, S. Naseem, T.H. Usmani, E. Bashir, F.A. Khan, and M.I. Bhanger, *J. Hazard. Mater.*, 171, 424, 2009; D.T. Jayawardana, H.M.T.G.A. Pitawala, and H. Ishiga, *J. Geochem. Explor.*, 114, 118, 2012; A. Farooqi, H. Masuda, and N. Firdous, *Environ. Pollut.*, 145, 839, 2007; X.B. Gao, F.C. Zhang, C. Wang, and Y.X. Wang, *Proc. Earth Planet. Sci.*, 7, 280, 2013.)

fluoride concentration is not uniform within the country, and the data given are based on the available information. It is also worth noting that the magnitude of the problem is high in developing countries where a high level of rural poverty prevails.

19.3.1 Fluoride-Free Water: Solution and Challenges

The first and immediate solution to this problem is to check for alternative water sources such as rainwater and surface water. The local availability of surface water in sufficient quantity and the suitability of its quality in terms of biological and chemical pollutants for drinking limits its use in many cases. Rainwater as an alternative source seems to be an ideal solution since it provides much cleaner water at a minimum cost. However, its uneven distribution is a big constraint. Thus, the option of using alternative water sources has its own limitations and may not be always viable.

Using groundwater with low fluoride contamination, from different depths around the same location, can also serve as another solution. However, the concentrations of fluoride usually vary with time in both vertical and horizontal direction. Hence, it is necessary to test the wells individually on a regular basis to judge the quality of water, which is not always possible in rural areas because of financial and technical constraints. The only viable option available in many contaminated areas is to treat the contaminated water.

Several factors must be considered while selecting the treatment of fluoride for a particular site, such as feed-flow rate, nature of the solute, solute concentration, water chemistry, and the expected removal efficiency. The socioeconomic conditions of the affected population and the type of water supply source in the affected area should also be considered while choosing the process. Another important factor to be considered is whether the affected area requires a centralized treatment system or a large number of independent treatment facilities. Management of secondary wastewater and the chemical sludge generated after the treatment are also of prime concern. Although these factors influence the selection of the treatment process, in a rural setup the cost and ease of operation should be given more weightage than others. It is also important to have public participation at the village level for the success of any technology in villages. A simple and low-cost technology may also fail in rural areas, unless the technology fits into the rural circumstances and is well accepted by the locals.

19.3.2 Overview of Defluoridation Technologies

Various defluoridation methods have been developed over the years. Processes adapted in these technologies include precipitation, ion exchange [52], sorption [53,54], and membrane processes [55] such as electrodialysis [56] and Donnan dialysis [57]. Many of the processes have been proven successful in laboratory- and pilot-scale studies, but did not produce desired results in the field. Coagulation–filtration is a well-documented process and has been traditionally used to remove many suspended and dissolved solids, including fluoride, from drinking water. The common chemicals used in this process are metal salts ($Al_2(SO_4)_3 \cdot 18H_2O$, $FeCl_3$, and $Fe_2(SO_4)_3 \cdot 7H_2O$), lime, or a combination of lime and alum. In this method, chemicals are added to contaminated water to convert dissolved fluoride into an insoluble form, which is then precipitated. Elimination of fluoride may also take place by adsorption onto insoluble flocs, which are coprecipitated. The effective operation of the coagulation process depends on the selection of a right coagulant, its optimum dose, and the pH of the system. However, the coagulation method failed in removing fluoride to a desired concentration level [58]. The Nalgonda technique (NT), a modified coagulation process, has been applied in India at different levels. NT involves the addition of lime, aluminum salt(s), and bleaching powder to raw water and proper mixing [59]. Bleaching powder is usually added to disinfect the water. This method is reported to be suitable for both community and household use, and has been successfully used at both levels [60,61]. Although this process claims high fluoride removal capacity, it has been recently reported that it removes only a small portion of fluoride (18%–33%) in the form of precipitates. It converts a greater portion of ionic fluoride (67%–82%) into soluble ion of aluminum fluoride complex [62]. Apparao and Kartikeyan [63] reported that the soluble aluminum fluoride complex is itself toxic; hence, the adoption of NT for defluoridation of water is not desirable. The large amount of fluoride-contaminated sludge produced during this process, like with other coagulation processes, is another factor that limits its use.

Various types of ion-exchange resins have been tested for the defluoridation process. It is reported that fluoride removal by anion-exchange resins is not effective because of their low selectivity toward fluoride [52]. Besides, strong base anion-exchange resins impart a taste to the treated water that may not be acceptable to the consumers. Cation-exchange resins, such as Defluoro-1 Defluoro-2, Carbon, Wasoresin-14, and polystyrene, have shown fluoride exchange capacity [64]. Many researchers have studied the fluoride exchange capacity of cation-exchange resins loaded with various metals, and the results showed that they are more preferable than anion-exchange resins [65,66]. Membrane separation is also employed for defluoridation of water. Among the different membrane techniques, reverse osmosis is more popular. These processes are relatively expensive to install and operate, especially in rural conditions. They also experience fouling, scaling and membrane degradation, which reduce the efficiency of the process [67].

Adsorption by a solid surface is considered as one of the best technologies for removing fluoride from drinking water, especially in rural areas, because of its easy handling, versatility in operation, minimal sludge production, and regeneration capability. However, the active surface area of the adsorbent, surface energy, and pH of the solution highly influence the removal efficiency. At present, several sorptive media have been reported to be successful in removing fluoride from drinking water. These include several natural, synthetic, and metal oxide–based sorbents [68]. Among the adsorbents tested, activated alumina (AA) is reported to be the most suitable one. However, adsorption by AA is a slow process and is only effective in a narrow pH range. These factors limit its use in treating fluoride-contaminated water in many cases in the field [54]. Hence, the developments of new adsorbents with tailor-made or customized physical and chemical properties are important to overcome the limitations and the use in the field.

19.4 Defluoridation of Water Using Nanoscale Materials

Since the discovery of buckyballs by Curl, Kroto, and Smalley in 1985, the field of nanotechnology has attracted overwhelming attention and has emerged rapidly [69]. Various kinds of nanosystems have been investigated for diverse applications, including medical, biotechnology, and electronics [70–72]. However, its beneficial use in purification of water has gained momentum only recently [73–75]. The advancement in nanotechnology suggests that many issues regarding water quality can be improved using nanomaterials [76]. The basic properties such as extremely small size, high surface area to volume ratio, greater density of reactive sites on the particle surfaces, and short diffusion route, and hence better kinetics, make them better candidates for purification of water including water defluoridation [11,74,77]. This section reviews some of the recent developments in defluoridation of water using nanoscale materials. Various classes of nanoscale materials considered for the discussion are as follows: (i) metals and metal oxides/hydroxide/oxyhydroxide NPs, (ii) embedded metals and metal oxides/hydroxide NPs, and (iii) carbon-based materials.

19.4.1 Metals and Metal Oxides/Hydroxides/Oxyhydroxide

Metals and their oxides/hydroxides/oxyhydroxide are important classes of materials widely used in water purification [74,77]. Recently, studies have been directed toward the controlled synthesis of these materials with respect to their size, morphology, phase, and crystallinity, expecting enhanced performance. The application of such materials in defluoridation of water is discussed in this section.

19.4.1.1 Alkaline Earth Metal–Based Oxides

The potential of magnesium oxide to scavenge fluoride from water has been known for >70 years [78,79]. Nagappa and Chandrappa [80] studied the fluoride retention potential of mesoporous MgO nanocrystals prepared through the combustion route using magnesium nitrate as the oxidizer and glycine as the fuel. NPs of size 12–23 nm prepared by this route showed a six times increase in the fluoride uptake compared with the commercial MgO and could reduce the sludge volume by 90%. However, an increase in pH was observed after treating the water using magnesium oxide. The mechanism for fluoride removal suggested is by chemisorption and the formation of MgF_2. A modified combustion route for the synthesis of nanomagnesia (NM) was proposed by Maliyekkal and coworkers [81]. The synthesis is based on the self-propagated combustion of the magnesium nitrate trapped in cellulose fibers, using urea and glycine as fuels. Various characterization studies confirmed that NM formed after combustion is crystalline and porous, with NP sizes varying from 3 to 7 nm (Figure 19.2). The proposed synthetic approach claims high product recovery, owing to reduced loss of NPs during combustion. The method also claims 20% reduction in the synthesis cost in terms of the raw materials used. The NM has been found to be superior to conventional MgO in removing fluoride from water. On the basis of the microscopic and macroscopic studies, the authors ascribed the mechanism of fluoride removal to isomorphic substitution of fluoride in brucite lattice (Figure 19.3). They did not observe evidence of MgF_2 formation under the experimental conditions (temperature $30 \pm 2°C$; fluoride concentration <50 mg/L). It is clear that MgF_2 (Ksp = 5.16×10^{-11}) has a higher-solubility product than $Mg(OH)_2$ (Ksp = 5.61×10^{-12}), and hence, $Mg(OH)_2$ will be a preferred product than MgF_2 at a lower concentration of fluoride. NPs of CaO were also investigated for the purpose [82]. The Langmuir maximum sorption capacity of the material was found to be 163.3 mg/g. The fluoride removal by the material was attributed to the formation of insoluble CaF_2. Although these compounds are efficient in removing fluoride and insensitive to normal pH variations and coexisting ions, the pH of the treated water being alkaline is a concern and needs further attention.

19.4.1.2 Alumina and Other Aluminum-Based Materials

Alumina and aluminium-based oxides and oxyhydroxides are key materials in many applications, including catalysis and molecular adsorption [83,84]. Although several adsorbents

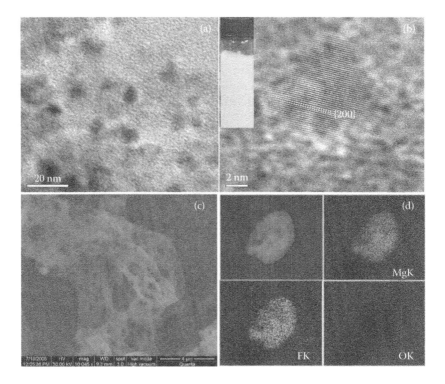

FIGURE 19.2
(a, b) High-resolution transmission electron micrographs of NM. Inset of the panel b is a photograph of the as-synthesized NM. (c) Scanning electron microscopy (SEM) micrographs of NM. (d) Elemental x-ray images of NM (MgKα, OKα, and FKα) after reacting with fluoride. (From S.M. Maliyekkal, Anshup, K.R. Antony, and T. Pradeep, *Sci. Total Environ.*, 408, 2273, 2010.)

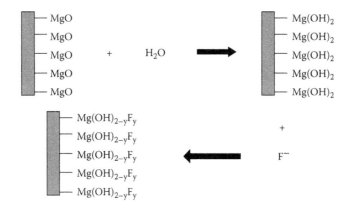

FIGURE 19.3
Scheme showing the mechanism of fluoride adsorption onto NM. (From S.M. Maliyekkal, Anshup, K.R. Antony, and T. Pradeep, *Sci. Total Environ.*, 408, 2273, 2010.)

were developed for defluoridation of water, AA is the most widely used absorbent owing to its high fluoride selectivity, relatively low cost, and ease of operation and maintenance [85,86]. Fluoride adsorption by AA is recommended as a best available technology by the US Environmental Protection Agency [87]. Extensive reports on the use of AA and its effectiveness are available in open literature [68,77]. The mechanism of F^- adsorption by AA is similar to that of a weak base ion-exchange resin. The major drawbacks of commercially available AA are its poor kinetics of adsorption and high sensitivity to pH. Fluoride adsorption capacity decreases with the increase in pH and alkalinity. The optimum pH for fluoride removal by AA is in the range of 5–6. At acidic condition (pH <6), the formation of soluble alumino–fluro complexes have been reported [88]. At pH >7, silicate and hydroxyl ions compete for the adsorption sites, reducing the F^- adsorption capacity of AA. A recent attempt by Lee et al. [89] shows that mesoporous aluminas (MA-1 and MA-2) prepared through a surfactant-assisted route is superior to commercial AA in removing fluoride. A similar study reported that the maximum fluoride uptake capacity by nanoalumina is 14.0 mg/g at 25°C, and the maximum fluoride removal occurred at pH 6.15 [90]. An excellent review on mesoporous alumina and their characteristics has also been reported [91]. Studies also show that AlOOH, a common starting precursor for alumina, is also a good adsorbent for this application. The preparation of AlOOH is relatively simple and eco-friendly, which may make the material attractive and easily acceptable by users. Wang et al. [92] studied the fluoride adsorption capacity of nanoscale AlOOH. The material showed a maximum adsorption capacity of 3.5 mg F^-/g of AlOOH, which is comparable to that of bulk AA. The adsorption of fluoride onto nano-AlOOH was strongly pH dependent and the maximum uptake was observed at pH of 6.8.

Aluminum oxide is highly soluble in water under both acidic (pH <4) and alkaline (pH >9) conditions. At pH 4, the prominent species in water is $Al(H_2O)_6^+$ ions and at alkaline pH, the dominant species is $[Al(OH)(H_2O)_5]^{2+}$. This ion can produce dimeric or polymeric complexes by further deprotonation and loss of water molecules [91]. This instability of the aluminum oxide may limit its use in the field especially when repeated reuse of the adsorbent is expected. Alumina and other aluminum-based oxides loaded with fluoride are typically regenerated by contacting with NaOH (pH >9) followed by acid wash (pH <4). The regeneration process would dissolve the materials and results in loss of the adsorbent and thus reduce the economic viability. It is also important to assure the stability of the material in water after alkali and acid wash.

19.4.1.3 Transition Metal–Based Oxides

Iron oxides exist in many forms in nature and are perhaps one of the widely studied engineered nanoscale systems for environmental remediation. Various kinds of iron NPs with diverse properties, including low toxicity, chemical inertness, biocompatibility, and superparamagnetism, have been synthesized [93–97]. Some of the unique properties of iron oxide NPs against their bulk counterparts, which make them better material for environmental remediation, are illustrated in Figure 19.4. The environmental friendliness, low cost, and the ease of synthesis are other factors that support iron NPs as an attractive choice for a mass application such as purification of water. Mohapatra et al. [98] have demonstrated the use of nanoscale goethite prepared through a new synthetic route using hydrazine sulfate as an additive. The material was investigated for its ability to remove fluoride from water. The maximum fluoride uptake was observed (59 mg F^-/g [of goethite]) between the pH 6 and 8, proving the applicability of the adsorbent for treatment of natural water systems. The same group has synthesized mixed nanoscale iron oxides comprising goethite (77%),

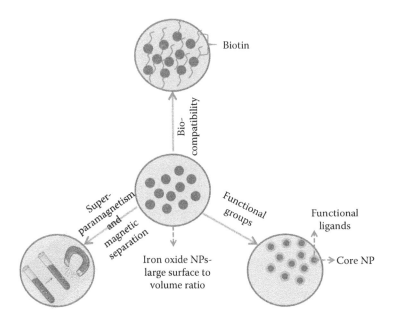

FIGURE 19.4
Some desired properties of iron oxide magnetic NPs for water purification. (From P. Xu, G.M. Zeng, D.L. Huang, C.L. Feng, S. Hu, M.H. Zhao, C. Lai, Z. Wei, C. Huang, G.X. Xie, and Z.F. Liu, *Sci. Total Environ.*, 424, 1, 2012.)

hematite (9%), and ferrihydrite (14%) through a surfactant mediated–precipitation route using cetyltrimethyl ammonium bromide. The Langmuir monolayer capacity value was estimated to be 53.19 mg/g. They found an optimum equilibrium defluoridation capacity at pH 5.75, and the capacity decreased with further increase in pH. Anion exchange (at acidic pH) and van der Waals forces of attraction (at alkaline pH) are the dominant mechanisms responsible fluoride uptake.

Zirconium oxide is another transition metal–based oxide widely studied in water purification. Recently, Cui et al. [100] have synthesized zirconium oxide NPs though a hydrothermal route and studied its potential for arsenic removal. The material showed good affinity to arsenic, both As(III) and As(V). In another effort, hydrous zirconium oxide was synthesized through simple chemical precipitation and tested for its ability to scavenge fluoride from water [101]. The material showed high affinity to fluoride, and maximum adsorption capacities of 124 and 64 mg F^-/g (of adsorbent) at pH 4 and 7, respectively, were reported. On the basis of spectroscopic studies (Raman, Fourier transform infrared [FT-IR], and ^{19}F nuclear magnetic resonance), the authors attributed the fluoride removal from water to the formation of Zr–F bonds on the hydrous zirconium oxide. They also proposed a seven-coordinate polyhedral zirconium oxyfluoride species (ZrO_2F_5) and possibly some ZrO_3F_4 formed on the adsorbent's surface by exchange reactions between surface hydroxyl groups and fluoride, based on x-ray photon spectroscopy data.

19.4.2 Embedded Metals and Metal Oxides/Hydroxides

Although metal oxides and hydroxides are efficient in scavenging fluoride from aqueous medium, most of these materials are available only as fine powders or are dispersed in aqueous suspension. Although smaller-sized particles are desirable when fast kinetics is expected for treating large quantities of water, the size below about 0.3 mm causes operational problems [25]. In adsorption-based purification of water, adsorbents in powder form, especially in nanoscale, have several practical limitations: (i) low hydraulic conductivity in packed bed, (ii) difficulty in solid–liquid separation, and (iii) risk of leaching of the NPs along with treated water [12]. Besides, the presence of coexisting ions in water can cause ion-induced aggregation of NPs and thereby reduce the reactivity. These limitations might be overcome by anchoring the NPs on suitable matrices. Anchoring can be done *in situ* or *ex situ* on natural or synthetic templates. A few examples in this direction are being reviewed.

19.4.2.1 Natural Polymers as Support Matrix

Natural cellulose fibers are gaining interest as templates because of their nanoporous surface features, low cost, and environment friendly nature. They have been used as a substrate for the *in situ* synthesis of metal and metal oxide NPs [12,102,103]. An example is shown in Figure 19.5. These kinds of composites are also interesting as adsorbent media for purification of water owing to potential synergistic properties that may arise from the combination of materials such as the inherent properties of the fibers, in particular flexibility and strength, and also the high adsorption properties of the surface-loaded NPs. However, the possible reduction in the adsorption capacity, due to blockage of adsorption sites by the interactions between the supporting media and the biopolymer, cannot be ruled out completely.

Recently, Yu et al. [104] have synthesized nanoscale hydroxyapatite supported on cellulose fibers. The nanocomposites (cellulose@hydroxyapatite) were prepared by mixing NaOH/thiourea/urea/H_2O (8:6.5:8:77.5 by weight) solution with cellulose. The composites were tested for their ability to remove fluoride from water. The adsorbent was found to be capable of removing fluoride to a level below the World Health Organization–prescribed

FIGURE 19.5
(a) Energy-dispersive x-ray spectrum of manganese oxide–impregnated cellulose fibers. (b) Field emission SEM (FESEM) micrographs of 4.64% Mn-loaded cellulose fibers. Inset of panel b shows FESEM micrographs of pristine cellulose. Photographs of manganese oxide–impregnated cellulose fibers at various manganese oxide loading: (c1) 0% Mn, (c2) 0.39% Mn, (c3) 0.66% Mn, and (c4) 3.7% Mn. (From S.M. Maliyekkal, K.P. Lisha, and T. Pradeep, *J. Hazard. Mater.*, 181, 986, 2010.)

standard (initial fluoride concentration = 10 mg/L; adsorbent dose = 3 g/L). Interestingly, the coexisting anions had no significant effect on fluoride adsorption and thus suitable for practical application. A novel Fe(III)-loaded ligand-exchange cotton cellulose (Fe(III) LECCA) was synthesized and used for fluoride removal from drinking water [105]. The fluoride uptake capacity of the material was tested as a function of pH, reaction time, temperature, co-ions, and flow rate, in batch and continuous mode. The studies indicated that the adsorbent is good in removing fluoride from water and showed a maximum equilibrium adsorption capacity of 18.55 mg/g at a temperature of 25°C. The regeneration potential of the adsorbent bed packed in column was tested in continuous mode using 1 M NaOH solution as an eluent at room temperature. The results revealed that >98% adsorbed fluoride was eluted. Interestingly, there was a slight increase in the adsorption capacity of Fe(III)LECCA with each regeneration cycle. This has been attributed to the formation of colloidal ferric hydroxide due to the reaction between loaded iron and NaOH. The adsorption mechanism was studied using FT-IR spectroscopy and by chemical analysis. A ligand-exchange mechanism (X $\equiv$ Fe(III)–L + F$^-$ $\rightarrow$ X $\equiv$ Fe(III)–F + L$^-$; X: cellulose, L: anion ligands) was proposed for the removal of fluoride in water.

   Chitosan is another interesting biopolymer extensively used in water and wastewater purification. Chitosan can be shaped to several forms, including membranes, gel beads, microspheres, films, etc. [106–108]. It is able to provide a ratio of surface area to mass that enhances the adsorption capacity and reduces the hydrodynamic limitation effects, such as column clogging and friction loss [109]. Owing to these attractive properties, many investigations have been performed using chitosan as supporting material. A detailed review on the use of chitosan and its composites as fluoride-removing media is available elsewhere [110].

19.4.2.2 Synthetic Polymers as Support Matrix

Swain et al. [111] have synthesized a novel composite (FZCA) by impregnating microparticles/NPs of a binary metal oxide (Fe(III)–Zr(IV)) in an alginate matrix, a polysaccharide composed of different proportions of β-D-mannuronic acid and α-L-guluronic acid units, linked by β-1-4 and α-1-4 bonds. The synthesized material was tested for fluoride uptake as a function of the initial concentration of fluoride, contact time, pH, and temperature. The maximum fluoride uptake was reported to be 0.981 mg/g and it was observed at pH 6. The average diameter of the composite beads was 3 mm, and the average particle size of the impregnated Fe–Zr particles varied between 70.89 and 477.7 nm. The material was regenerated using NaOH as the eluent (pH 12) and found to be effective for several cycles of adsorption. Wu et al. [112] prepared a new granular adsorbent by immobilizing a nanoscale trimetal hydroxide of Fe–Al–Ce (FAC) in porous polyvinyl alcohol (PVA) via cross-linking with boric acid. Various optimization studies were conducted to check the feasibility of using PVA as a support matrix for FAC. The effects of the concentration of PVA and FAC on the mechanical stability and fluoride adsorption capacity were also investigated. The fluoride adsorption capacity of the composite was increased with FAC concentration and decreased with PVA concentration. An FAC concentration of 12% and a PVA concentration of 7.5% were found to be the optimum values. The granules showed a fluoride adsorption capacity of 4.46 mg/g at a pH of 6.5 and an initial fluoride concentration of 19 mg/L. In another attempt, Zhao et al. [113] have used the extrusion method to granulize FAC using cross-linked PVA as the binder. The schematic diagram of the steps employed during extrusion is depicted in Figure 19.6. The stability and fluoride uptake capacity of the granules were

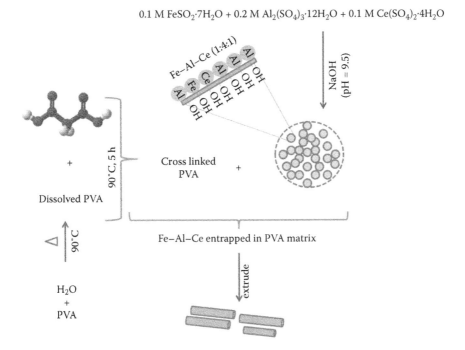

FIGURE 19.6
Schematic of steps employed during extrusion of FAC–PVA composite. (Adapted from B. Zhao, Y. Zhang, X. Doub, X. Wu, and M. Yanga, *Chem. Eng. J.*, 185, 211, 2012.)

optimized by varying the amount of FAC and the binder. The optimum conditions proposed are as follows: binder/FAC powder ratio, 1.4; average cylinder diameter of granules, 1.6 mm; drying temperature, 65°C. The granulated composite exhibited a Langmuir maximum adsorption capacity of 51.3 mg/g at pH 7.0 ± 0.2.

19.4.2.3 Inorganic Metal Oxide as Support Matrix

Chen et al. [114] developed a granulation technology for coating nanoscale FAC onto a cheap inorganic support, sand. For this, an acrylic styrene copolymer latex mixed with FAC was spray-coated onto the sand surface in a fluidized bed. Using the latex as a binder improved the stability of the coated layer. The granules had spherical shape with size varying from 2 to 3 mm. For the optimal stability and adsorption capacity, a coating amount of 27.5% was suggested. At optimum conditions, the fluoride uptake capacity of the material was found to be 2.22 mg/g (pH 7.0; initial fluoride concentration of 0.001 M). An inverse relation was observed between the coating thickness and the stability of the granules; that is, the higher the coating thickness, the lower the stability of the granules. However, fluoride uptake capacity increased with increasing coating thickness. On the basis of the FT-IR data, the reduction in adsorption capacity was attributed to the interaction between the latex surface with the active hydroxyl on the FAC adsorbent. A similar coating producer was used by the same group to support FAC onto glass beads [115]. The adsorbent was used in a packed bed for defluoridation of water. They found a direct relation between the coating temperature and the degree of cross-linking of the polymer and the associated mechanical strength of the granules. The optimal parameter conditions arrived at are as follows: coating temperature, 65°C; latex/FAC, 0.5:1; and coating amount, 27.8%. The FAC@ glass beads showed a slightly higher fluoride adsorption capacity (2.77 mg/g) compared with FAC@sand (2.22 mg/g).

19.4.3 Carbon-Based Materials

The use of carbon in water purification dates back to the Harappan civilization. Today, carbon, particularly activated carbon, has become one of the most common and trusted means for removing contaminants from water [8]. Large surface area, reactive surface functional groups, pore volume, and pore size distribution are some of the key factors that determine the success of any adsorbent in adsorption process. Over the years, various forms of carbon and their composites have been developed, expecting enhanced performance, and investigated as adsorbents for removing diverse pollutants from water [116,117]. Few recent efforts specific to defluoridation of water using carbon-based nanoscale materials are reviewed below.

Carbon nanotubes (CNTs), allotropes of carbon with a cylindrical nanostructure, have gained huge interest since their discovery. Their large surface area, small sizes, and high mechanical strength make them attractive material for water purification including defluoridation. A composite of Al_2O_3 and CNTs was prepared by depositing Al_2O_3 on CNTs, and its application for fluoride uptake from water was studied [118]. The results revealed that the adsorption process is very effective over a broad pH range (5.0–9.0) unlike AA, which is effective only at a narrow pH range of 5–6. The adsorption capacity of Al_2O_3@CNTs was found to be about 13.5 times higher than that of AC-300 carbon and 4 times higher than that of γ-Al_2O_3 at an equilibrium fluoride concentration of 12 mg/L. The mass of fluoride adsorption for Al_2O_3@CNTs at pH 6.0 reached 28.7 mg/g at an equilibrium concentration of 50 mg/L. The same group has studied the effect of calcination temperature, alumina loading, and pH on the preparation of Al_2O_3@CNTs and its ability to remove fluoride from water [119]. The optimum calcination temperature for preparing the adsorbents was observed to be 450°C. The maximum adsorption capacity was observed at an alumina loading of 30% and a pH range of 6–9. Aligned CNTs (ACNTs) were synthesized by catalytic decomposition of xylene using ferrocene as catalyst, and their performance was evaluated for fluoride removal from water [120]. Both the surface and inner cavities of ACNTs were reported to be readily accessible for fluoride adsorption. ACNTs have shown a maximum adsorption capacity of 4.5 mg/g at an equilibrium fluoride concentration of 15 mg/L. The adsorbent performed well over a broad range of pH (5–8) unlike activated carbon, which is effective only at pH 3 [121,122]. Ansari et al. [123] investigated the fluoride-scavenging capacity of multiwalled CNTs (MWCNT) from drinking water. MWCNTs showed a saturation capacity of 3.5 mg/g. Unlike ACNTs and Al_2O_3@CNTs, MWCNTs showed high sensitivity to solution pH and the optimum pH was found to be at 5 (~94% fluoride removal in 18 min). Upon increasing the pH, the adsorption capacity decreased gradually and reached 41.2% at pH 9.0. This reduction in the fluoride adsorption capacity in the alkaline pH range was ascribed to the competition for adsorption sites between hydroxyl ions and fluoride. MWCNTs exhibited a saturation capacity of 3.5 mg of fluoride per gram. The presence of coexisting anions such as chloride, nitrate, hydrogen carbonate, sulfate, and perchlorate showed a negligible effect on the adsorption of F^- onto MWCNTs.

Graphene, a one-atom-thick sheet of carbon, is a recent addition to the carbon family. Since its discovery in 2004 [124], the material has gained overwhelming attention owing to its unique physical properties, chemical properties, and low production cost compared with other graphitic forms. Compared with other allotropes of carbon, graphene has some special properties such as excellent electrical and thermal conductivity, high planar surface area, transparency to visible light, exceptional mechanical strength, and hence many possible applications. However, most of the reported works in graphene-related materials deal with catalytic or electronic processes [125]. Its possible beneficial use in water purification has attained some momentum recently [126]. The recent investigation of Li et al. [127] showed that graphene has excellent fluoride adsorption ability with a saturated monolayer adsorption capacity of 17.65 mg/g at pH 7.0 and 25°C. The study reported that adsorption is a specific process and fluoride ions are predominantly removed by the surface exchange reaction between fluoride ions in solution and hydroxyl ions on the adsorbent. The reported capacity was significantly higher than that of CNTs [120]. A composite of graphene oxide and manganese oxide (MOGO) has also been investigated for fluoride removal. The studies revealed that optimum removal of fluoride occurred in a pH range of 5.5–6.7. The composite showed maximum adsorption capacity of 11.93 mg of F^-/g of MOGO and is 8.34 times higher than that of GO [128].

19.5 Summary

Fluorosis is a chronic menace affecting a large population worldwide. Consumption of drinking water obtained from fluoride-rich ground strata is the major mechanism for fluoride intake in India and other fluoride-affected areas. An immediate and viable solution to alleviate the problem is to remove the pollutant from the contaminated groundwater resources. Many treatment methods have been developed and practiced over the years for defluoridation of water. However, many of them did not produce the desired results in the affected areas where high rural poverty prevails. Among the technologies, adsorption seems to be the most appropriate one. Surprisingly, among the numerous adsorbents that were tested for removing fluoride, AA is still the most preferred adsorbent. However, poor sorption kinetics and effectiveness in a narrow pH range, competition from other ions, poor stability, especially in acidic and alkaline pH, and energy-intensive synthesis limit the use of AA in many field situations.

Nanomaterials offer great opportunity in developing adsorbents of high capacity and faster kinetics. Many nanomaterials, especially metal oxide–based materials, showed enhanced fluoride uptake. However, most of these materials are available only as fine powders or are dispersed in aqueous suspension. Using such materials has practical limitations, such as difficulty in solid–liquid separation and low hydraulic conductivity in packed bed. To overcome these limitations, NPs were anchored on inorganic and organic matrices and tested for fluoride uptake. However, the practical utility of the nanomaterials are not well demonstrated. In short, the review indicates that although nanomaterials offer better efficacy in terms of capacity and kinetics, every technology has significant shortcomings and no single process can yet serve the purpose in all the conditions. In the future, the success of the process for down-to-earth applications such as water defluoridation is going to be largely dependent on, besides efficiency, ease of large-scale synthesis, solid–liquid separation, and posttreatment handling. It is also important to understand the local condition and the technology should be tailored according to the local needs.

References

1. Robins N.S. *Hydrology of Scotland*. HMSO, London, 1990.
2. *World Bank Groundwater in Urban Development—Assessing Management Needs and Formulating Policy Strategies*, WTP – 390. World Bank, Washington, DC, 1998.
3. Susheela A.K. Fluorosis management programme in India. *Curr. Sci.*, 77:1250–1255, 1999.
4. World Health Organization (WHO). *Guidelines for Drinking-Water Quality (Electronic Resource): Incorporating First and Second Addenda. Recommendations*, vol. 1, Third ed. Geneva, pp. 375–377, 2008. Available at http://www.epa.gov/region4/foiapgs/readingroom/hercules_inc/guidelines_for _drinking_water_quality_3v.pdf.
5. Agrawal G.D., S.K. Lunkad and T. Malkhed. Diffuse agricultural nitrate pollution of groundwaters in India. *Water Sci. Technol.*, 39(3):67–75, 1999.
6. Stuart M.E., D.C. Gooddy, J.P. Bloomfield and A.T. Williams. A review of the impact of climate change on future nitrate concentrations in groundwater of the UK. *Sci. Total Environ.*, 409(15):2859–2873, 2011.
7. Mathava K.S. and L. Philip. Adsorption, desorption characteristics of endosulfan in various Indian soils. *Chemosphere*, 62:1064–1107, 2006.
8. Maliyekkal S.M., T.S. Sreeprasad, D. Krishnan, S. Kouser, A.K. Mishra, U.V. Waghmare and T. Pradeep. Graphene: A reusable substrate for unprecedented adsorption of pesticides. *Small*, 9(2):273–283, 2013.
9. Wang S. and C.N. Mulligan. Occurrence of arsenic contamination in Canada: 3127 sources, behavior and distribution. *Sci. Total Environ.*, 366:701–721, 2006.
10. Roychowdhury T. Groundwater arsenic contamination in one of the 107 arsenic-affected blocks in West Bengal, India: Status, distribution, health effects and factors responsible for arsenic poisoning. *Int. J. Hyg. Environ. Health*, 213(6):414–427, 2010.
11. Lisha K.P., S.M. Maliyekkal and T. Pradeep. Manganese dioxide nanowhiskers: A potential adsorbent for the removal of Hg(II) from water. *Chem. Eng. J.*, 160(2):432–439, 2010.
12. Maliyekkal S.M., K.P. Lisha and T. Pradeep. A novel cellulose-manganese oxide hybrid material by in-situ soft chemical synthesis and its application for the removal of Pb(II) from water. *J. Hazard. Mater.*, 181:986–995, 2010.
13. Silva C.P., M. Otero and V. Esteves. Processes for the elimination of estrogenic steroid hormones from water: A review. *Environ. Pollut.*, 165:38–58, 2012.
14. Nowack B. and T.D. Bucheli. Occurrence, behavior and effects of nanoparticles in the environment. *Environ. Pollut.*, 150:5–22, 2007.
15. Brar S.K., M. Verma, R.D. Tyagi and R.Y. Surampalli. Engineered nanoparticles in wastewater and wastewater sludge—Evidence and impacts. *Waste Manage.*, 30:504–520, 2010.
16. Chowdary V.M., N.H. Rao and P.B.S. Sarma. Decision support framework for assessment of non-point source pollution of groundwater in large irrigation projects. *Agric. Water Manage.*, 75:194–225, 2005.
17. Schirmer M., S. Leschik and A. Musolff. Current research in urban hydrogeology—A review. *Adv. Water. Resour.*, 51:280–291, 2013.
18. Liang-Qi L., S. Ci-An, X. Xiang-Li, L. Yan-Hong and W. Fei. Acid mine drainage and heavy metal contamination in groundwater of metal sulfide mine at arid territory (BS mine, Western Australia). *Trans. Nonferrous Met. Soc. China*, 20:1488–1493, 2010.
19. Lerner D.N. Diffuse pollution of groundwater in urban areas. BHS 10th National Hydrology Symposium, Exeter, UK, 2008.
20. Collin M.L. and A.J. Melloul. Assessing groundwater vulnerability to pollution to promote sustainable urban and rural development. *J. Clean. Prod.*, 11:727–736, 2003.
21. Comprehensive Assessment of Water Management in Agriculture. *Water for Food, Water for Life: A Comprehensive Assessment of Water Management in Agriculture*. London: Earthscan, and Colombo: International Water Management Institute, 2007. Available at http://www.fao.org /nr/water/docs/summary_synthesisbook.pdf.

22. WHO-JMP-MDG Report. Meeting the MDG drinking-water and sanitation target: A midterm assessment of progress, 2004. Available at http://www.who.int/water_sanitation_health/monitoring/jmp2004/en/.

23. Tjong S.C. and H. Chen. Nanocrystalline materials and coatings. *Mat. Sci. Eng.*, R 45:1–88, 2004.

24. Pitter P. Forms of occurrence of fluorine in drinking water. *Water Res.*, 19(3):281–284, 1985.

25. Maliyekkal S.M., A. Sharma and L. Philip. Manganese oxide coated alumina: A promising sorbent for defluoridation of drinking water. *Water Res.*, 40(19):3497–3506, 2006.

26. Mahramanlioglu M., I. Kizilcikli and I.O. Bicer. Adsorption of fluoride from aqueous solution by acid treated spent bleaching earth. *J. Fluorine Chem.*, 115:41–47, 2002.

27. Sorg T.J. Treatment technology to meet the interim primary drinking water regulations for inorganics. *J. Am. Water Works Assoc.*, 70(2):105–111, 1978.

28. Tsutsui T., N. Suzuki, M. Ohmori and H. Maizumi. Cytotoxicity, chromosome aberrations and unscheduled DNA synthesis in cultured human diploid fibroblasts induced by sodium fluoride. *Mutat. Res.*, 139(4):193–198, 1984.

29. Chinoy N.J. Effects of fluoride on physiology of animals and human beings. *Indian J. Environ. Toxicol.*, 1(1):17–32, 1991.

30. Harrison P.T.C. Fluoride in water: A UK perspective. *J. Fluorine Chem.*, 126:1448–1456, 2005.

31. Fan X., D.J. Parker and M.D. Smith. Adsorption kinetics of fluoride on low cost materials. *Water Res.*, 37(20):4929–4937, 2003.

32. Lee D.R., J.M. Hargreaves, L. Badertocher, L. Rein and F. Kassir. Reverse osmosis and activated alumina water treatment plant for the California State prisons located near Blythe. *Desalination*, 103:155–161, 1995.

33. Naoki H., M. Masahiro, O. Hajime, S. Akindou, E. Takeji and K. Tooru. Measurement of fluoride ion in the river-water flowing into Lake Biwa. *Water Res.*, 30(4):865–868, 1996.

34. Gizaw B. The origin of high bicarbonate and fluoride concentration in waters of the main Ethiopian Rift Valley, East African Rift system. *J. Afr. Earth Sci.*, 22:391–402, 1996.

35. Mameri N., A.R. Yeddou, H. Lounici, D. Belhocine, H. Grib and B. Bariou. Defluorination of septentrional Sahara water of North Africa by electrocoagulation process using bipolar aluminium electrodes. *Water Res.*, 32:1604–1612, 1998.

36. Shortt W.E. Endemic fluorosis in Nellore District, South India. *Indian Med. Gaz.*, 72:396–398, 1937.

37. Lian-Fang W. and H. Jian-Zhong. Outline of control practice of endemic fluorosis in China. *Soc. Sci. Med.*, 41(8):1191–1195, 1995.

38. Zhu C., G. Bai, X. Liu and Y. Li. Screening high-fluoride and high-arsenic drinking waters and surveying endemic fluorosis and arsenism in Shaanxi province in western China. *Water Res.*, 40(16):3015–3022, 2006.

39. Kim K. and G.Y. Jeong. Factors influencing natural occurrence of fluoride-rich groundwaters: A case study in the southeastern part of the Korean Peninsula. *Chemosphere*, 58:1399–1408, 2005.

40. Choi S.H., K.H. Bae, D.H. Kim, S.M. Lee, J.Y. Kim and J.B. Kim. Prevalence of dental fluorosis at Jinyoung-up, Kimhae City, Korea. *J. Korean Acad. Dent. Health*, 28:347–361, 2004 (in Korean).

41. Carrillo-Rivera J.J., A. Cardona and W.M. Edmunds. Use of abstraction regime and knowledge of hydrogeological conditions to control high-fluoride concentration in abstracted groundwater: San Luis Potosí basin, Mexico. *J. Hydrol.*, 261(1–4):24–47, 2002.

42. Mjengera H. and G. Mkongo. Appropriate technology for use in fluoritic areas in Tanzania. 3rd Waternet/WARFSA Symposium on Water Demand Management for Sustainable Use of Water Resources, University of Dar Es Salaam, 2002.

43. Agrawal V. Groundwater quality: Focus on the fluoride problem in India. *Co Geoenviron. J.*, 10, 1997.

44. Tekle-Haimanot R., Z. Melaku, H. Kloos, C. Reimann, W. Fantaye, L. Zerihun and K. Bjorvatn. The geographic distribution of fluoride in surface and groundwater in Ethiopia with an emphasis on the Rift Valley. *Sci. Total Environ.*, 367(1):182–190, 2006.

45. World Health Organization (WHO), J. Fawell, K. Bailey, J. Chilton, E. Dahi, L. Fewtrell and Y. Magara. *Fluoride in Drinking Water*. IWA, London, 2006. ISBN: 1900222965.

46. Naseem S., T. Rafique, E. Bashir, M.I. Bhanger, A. Laghari and T.H. Usmani. Lithological influences on occurrence of high-fluoride groundwater in Nagar Parkar area, Thar Desert, Pakistan. *Chemosphere*, 78(11):1313–1321, 2010.

47. Salifu B., K. Petrusevski, R. Ghebremichael, R. Buamah and G. Amy. Multivariate statistical analysis for fluoride occurrence in groundwater in the Northern region of Ghana. *J. Contam. Hydrol.*, 140–141:34–44, 2012.

48. Rafique T., S. Naseem, T.H. Usmani, E. Bashir, F.A. Khan and M.I. Bhanger. Geochemical factors controlling the occurrence of high fluoride groundwater in the Nagar Parkar area, Sindh, Pakistan. *J. Hazard. Mater.*, 171(1–3):424–430, 2009.

49. Jayawardana D.T., H.M.T.G.A. Pitawala and H. Ishiga. Geochemical assessment of soils in districts of fluoride-rich and fluoride-poor groundwater, north-central Sri Lanka. *J. Geochem. Explor.*, 114:118–125, 2012.

50. Farooqi A., H. Masuda and N. Firdous. Toxic fluoride and arsenic contaminated groundwater in the Lahore and Kasur districts, Punjab, Pakistan and possible contaminant sources. *Environ. Pollut.*, 145(3):839–849, 2007.

51. Gao X.B., F.C. Zhang, C. Wang and Y.X. Wang. Coexistence of high fluoride fresh and saline groundwaters in the Yuncheng Basin, Northern China. *Proc. Earth Planet. Sci.*, 7:280–283, 2013.

52. Meenakshi S. and N. Viswanathan. Identification of selective ion-exchange resin for fluoride sorption. *J. Colloid Interface Sci.*, 308:438–450, 2007.

53. Ayoob S. and A.K. Gupta. Sorptive response profile of an adsorbent in the defluoridation of drinking water. *Chem. Eng. J.*, 133(1–3):273–281, 2007.

54. Maliyekkal S.M., S. Shukla, L. Philip and I.M. Nambi. Enhanced fluoride removal from drinking water by magnesia-amended activated alumina granule. *Chem. Eng. J.*, 140:183–192, 2008.

55. Sehn P. Fluoride removal with extra low energy reverse osmosis membranes: Three years of large scale field experience in Finland. *Desalination*, 223:73–84, 2008.

56. Sahli M.A.M., S. Annouar, M. Tahaikt, M. Mountadar, A. Soufiane and A. Elmidaoui. Fluoride removal for underground brackish water by adsorption on the natural chitosan and by electro-dialysis. *Desalination*, 212:37–45, 2007.

57. Hichour M., F. Persin, J. Molenat, J. Sandeaux and C. Gavach. Fluoride removal from diluted solutions by Donnan dialysis with anion-exchange membranes. *Desalination*, 122:53–62, 1999.

58. Sollo F.W. Jr., T.E. Larson and H.F. Mueller. *Fluoride Removal from Potable Water Supplies*. Completion Report No. 136 to the Office of Water Research and Technology, U. S. Department of the Interior, Washington, DC, 1978. Available at http://www.isws.illinois.edu/pubdoc/CR/ISWSCR-206.pdf.

59. Nawlakhe W.G., D.N. Kulkarni, B.N. Pathak and K.R. Bulusu. Defluoridation of water by Nalgonda Technique. *Ind. J. Environ. Health*, 17:26–65, 1975.

60. Bulusu K.R., B.B. Sundaresan, B.N. Pathak and W.G. Nawlakhe, D.N. Kulkarni, V.P. Thergaonkar. Fluorides in water, defluoridation methods and their limitations. *J. Inst. Eng.* (India), 60:1–25, 1979.

61. NEERI. *Defluoridation. Technology Mission on Drinking Water in Villages and Related Water Management*. National Environment Engineering Research Institute. Nagpur 440020, India, 1987.

62. Meenakshi S. and R.C. Maheshwari. Fluoride in drinking water and its removal. *J. Hazard. Mater.*, 137(1):456–463, 2006.

63. Apparao B.V. and G. Kartikeyan. Permissible limits of fluoride in drinking water in India in rural environment. *Indian J. Environ. Prot.*, 6(3):172–175, 1986.

64. Rao C.N.R. Fluoride and environment—A review. In: M.J. Bunch, V.M. Suresh and T.V. Kumaran, eds. Proceedings of the Third International Conference on Environment and Health, Chennai, India. Department of Geography, University of Madras and Faculty of Environmental Studies, York University, Chennai, pp. 386–399, 15–17 December, 2003.

65. Wasay S.A., M.J. Haron and S. Tokunaga. Adsorption of fluoride, phosphate, and arsnate ions on lanthanum—Impregnated alumina. *Water Environ. Res.*, 68:295–300, 1996.
66. Fang L., K.N. Ghimire, M. Kuriyama, K. Inoue and K. Makino. Removal of fluoride using lanthanum(III)-loaded adsorbents with different functional groups and polymer matrices. *J. Chem. Technol. Biotechnol.*, 78(10):1038–1047, 2003.
67. Ayoob S., A.K. Gupta and V.T.A. Bhat. A conceptual overview on sustainable technologies for defluoridation of drinking water and removal mechanisms. *Crit. Rev. Environ. Sci. Technol.*, 38:401–470, 2008.
68. Mohapatra M., S. Anand, B.K. Mishra, D.E. Giles and P. Singh. Review of fluoride removal from drinking water. *J. Environ. Manag.*, 91:67–77, 2009.
69. Curl R.F., R.E. Smalley, H.W. Kroto, S. O'Brien and J.R. Heath. How the news that we were not the first to conceive of soccer ball C-60 got to us. *J. Mol. Graphics Model*, 19(2):185–186, 2001.
70. Wang X., L. Yang, Z.G. Chen and D.M. Shin. Application of nanotechnology in cancer therapy and imaging. *CA Cancer J Clin.*, 58:97–110, 2008.
71. West J.L. and N.J. Halas. Applications of nanotechnology to biotechnology: Commentary. *Curr. Opin. Biotechnol.*, 11(2):215–217, 2000.
72. Roco M.C. The emergence and policy implications of converging new technologies integrated from the nanoscale. *J. Nanopart. Res.*, 7(2):129–143, 2005.
73. Savage N. and M.S. Diallo. Nanomaterials and water purification: Opportunities and challenges. *J. Nanopart. Res.*, 7:331–342, 2005.
74. Hristovski K., A. Baumgardner and P. Westerhoff. Selecting metal oxide nanomaterials for arsenic removal in fixed bed columns: From nanopowders to aggregated nanoparticle media. *J. Hazard. Mater.*, 147(1–2):265–274, 2007.
75. Pradeep T. and Anshup. Noble metal nanoparticles for water purification: A critical review. *Thin Solid Films*, 517(24):6441–6478, 2009.
76. Diallo M.S. and N. Savage. Nanoparticles and water quality. *J. Nanopart. Res.*, 7(4–5):325–330, 2005.
77. Bhatnagar A., E. Kumar and M. Sillanpa. Fluoride removal from water by adsorption—A review. *Chem. Eng. J.*, 171:811–840, 2011.
78. Zettlemoyer A.C., E.A. Zettlemoyer and W.C. Walker. Active magnesia. II. Adsorption of fluoride from aqueous solution. *J. Am. Chem. Soc.*, 69:1312–1315, 1947.
79. Fair G.M. and J.C. Geyer. *Water Supply and Wastewater Disposal.* John Wiley & Sons, New York, p. 632, 1954.
80. Nagappa B. and G.T. Chandrappa. Mesoporous nanocrystalline magnesium oxide for environmental, remediation. *Microporous Mesoporous Mater.*, 106(1–3):212–218, 2007.
81. Maliyekkal S.M., Anshup, K.R. Antony and T. Pradeep. High yield combustion synthesis of nanomagnesia and its application for fluoride removal. *Sci. Total Environ.*, 408:2273–2282, 2010.
82. Patel G., U. Pal and S. Menon. Removal of fluoride from aqueous solution by CaO nanoparticles. *Sep. Sci. Technol.*, 44:2806–2826, 2009.
83. Tanada S., M. Kabayama, N. Kawasaki, T. Sakiyama, T. Nakamura, M. Araki and T. Tamura. Removal of phosphate by aluminum oxide hydroxide. *J. Colloid Interface Sci.*, 257:135–140, 2003.
84. Zhang H.Y., G.B. Shan, H.Z. Liu and J.M. Xing. Preparation of (Ni/W)-γ-Al$_2$O$_3$ microspheres and their application in adsorption desulfurization for model gasoline. *Chem. Eng. Commun.*, 194:938–945, 2007.
85. Ghorai S. and K.K. Pant. Equilibrium, kinetics and breakthrough studies for adsorption of fluoride on activated alumina. *Sep. Purif. Technol.*, 42:265–271, 2005.
86. Das N., P. Pattanaik and R. Das. Defluoridation of drinking water using activated titanium rich bauxite. *J. Colloid Interface Sci.*, 292(1):1–10, 2005.
87. USEPA. Water treatment technology feasibility support document for chemical contaminants. In: Support of EPA Six-year Review of National Primary Drinking Water Regulations, EPA 815-R-03-004, 2003.
88. Hao O.J. and C.P. Huang. Adsorption characteristics of fluoride onto hydrous alumina. *J. Environ. Eng.*, 112:1054–1069, 1986.

89. Lee G., C. Chen, S.T. Yang and W.S. Ahn. Enhanced adsorptive removal of fluoride using meso-porous alumina. *Microporous Mesoporous Mater.*, 127:152–156, 2010.

90. Kumar E., A. Bhatnagar, U. Kumar and M. Sillanpää. Defluoridation from aqueous solutions by nano-alumina: Characterization and sorption studies. *J. Hazard. Mater.*, 186:1042–1049, 2011.

91. Márquez-Alvarez C., N. Žilková, J. Pérez-Pariente and J. Čejka. Synthesis, characterization and catalytic applications of organized mesoporous aluminas. *Catal. Rev.*, 50:222–286, 2008.

92. Wang S.G., Y. Ma, Y.J. Shi and W.X. Gong. Defluoridation performance and mechanism of nano-scale aluminum oxide hydroxide in aqueous solution. *J. Chem. Technol. Biotechnol.*, 84:1043–1050, 2009.

93. Hassanjani R.A., M.R. Vaezi, A. Shokuhfar and Z. Rajabali. Synthesis of iron oxide nanoparticles via sonochemical method and their characterization. *Particuology*, 9(1):95–99, 2011.

94. Huang S.H., M.H. Liao and D.H. Chen. Direct binding and characterization of lipase onto magnetic nanoparticles. *Biotechnol. Prog.*, 19(3):1095–1100, 2003.

95. Afkhami A., M. Saber-Tehrani and H. Bagheri. Modified maghemite nanoparticles as an efficient adsorbent for removing some cationic dyes from aqueous solution. *Desalination*, 263(1–3): 240–248, 2010.

96. Roco M.C. Nanotechnology: Convergence with modern biology and medicine. *Curr. Opin. Biotechnol.*, 14(3):337–346, 2003.

97. Gupta A.K. and M. Gupta. Synthesis and surface engineering of iron oxide nanoparticles for biomedical applications. *Biomaterials*, 26(18):3995–4021, 2005.

98. Mohapatra M., K. Rout, S. Gupta, P. Singh, S. Anand and B. Mishra. Facile synthesis of additive-assisted nano goethite powder and its application for fluoride remediation. *J. Nanopart. Res.*, 12:681–686, 2010.

99. Xu P., G.M. Zeng, D.L. Huang, C.L. Feng, S. Hu, M.H. Zhao, C. Lai, Z. Wei, C. Huang, G.X. Xie and Z.F. Liu. Use of iron oxide nanomaterials in wastewater treatment: A review. *Sci. Total Environ.*, 424:1–10, 2012.

100. Cui H., Q. Li, S. Gao and J.K. Shang. Strong adsorption of arsenic species by amorphous zirconium oxide nanoparticles. *J. Ind. Eng. Chem.*, 18(4):1418–1427, 2012.

101. Dou X., D. Mohan, C.U. Pittman Jr. and S. Yang. Remediating fluoride from water using hydrous zirconium oxide. *Chem. Eng. J.*, 198–199:236–245, 2012.

102. Vainio U., K. Pirkkalainen, K. Kisko, G. Goerigk, E. Kotelnikova and R. Serimaa. Copper and copper oxide nanoparticles in a cellulose support studied using anomalous small-angle X-ray scattering. *Eur. Phys. J. D*, 42:93–101, 2007.

103. Dong H. and J.P. Hinestroza. Metal nanoparticles on natural cellulose fibers: Electrostatic assembly and in situ synthesis. *ACS Appl. Mater. Interfaces*, 1(4):797–803, 2009.

104. Yu X., S. Tong, M. Ge and J. Zuo. Removal of fluoride from drinking water by cellulose@ hydroxyapatite nanocomposites. *Carbohydr. Polym.*, 92(1):269–275, 2013.

105. Zhao Y., X. Li, L. Liu and F. Chen. Fluoride removal by Fe(III)-loaded ligand exchange cotton cellulose adsorbent from drinking water. *Carbohydr. Polym.*, 72:144–150, 2008.

106. Viswanathan N., C.S. Sundaram and S. Meenakshi. Removal of fluoride from aqueous solution using protonated chitosan beads. *J. Hazard. Mater.*, 161(1):423–430, 2009.

107. Wang L.-Y., G.-H. Ma and Z.-G. Su. Preparation of uniform sized chitosan microspheres by membrane emulsification technique and application as a carrier of protein drug. *J. Controlled Release*, 106(1–2):62–75, 2005.

108. Sreeprasad T.S., S.M. Maliyekkal, D. Krishna, L. Xavier, K. Chaudhari and T. Pradeep. Transparent, luminescent, antibacterial and patternable film forming composites of graphene oxide/reduced graphene oxide. *ACS Appl. Mater. Interfaces*, 3:2643–2654, 2011.

109. Vieira R.S. and M.M. Beppu. Mercury ion recovery using natural and crosslinked chitosan membranes. *Adsorption*, 11:731–736, 2005.

110. Miretzkya P. and A.F. Cirelli. Fluoride removal from water by chitosan derivatives and composites: A review. *J. Fluorine Chem.*, 132:231–240, 2011.

111. Swain S.K., T. Patnaik, P.C. Patnaik, U. Jha and R.K. Dey. Development of new alginate entrapped Fe(III)–Zr(IV) binary mixed oxide for removal of fluoride from water bodies. *Chem. Eng. J.*, 215–216:763–771, 2013.

112. Wu H.-X., T.-J. Wang, L. Chen, Y. Jin, Y. Zhang and X.-M. Dou. Granulation of Fe–Al–Ce hydroxide nano-adsorbent by immobilization in porous polyvinyl alcohol for fluoride removal in drinking water. *Powder Technol.*, 209:92–97, 2011.

113. Zhao B., Y. Zhang, X. Doub, X. Wu and M. Yanga. Granulation of Fe–Al–Ce trimetal hydroxide as a fluoride adsorbent using the extrusion method. *Chem. Eng. J.*, 185–186:211–218, 2012.

114. Chen L., H.-X. Wu, T.-J. Wang, Y. Jin, Y. Zhang and X.-M. Dou. Granulation of Fe–Al–Ce nano-adsorbent for fluoride removal from drinking water by spray coating on sand in a fluidized bed. *Powder Technol.*, 193:59–64, 2009.

115. Chen L., T.-J. Wang, H.-X. Wu, Y. Jin, Y. Zhang and X.-M. Dou. Optimization of a Fe–Al–Ce nano-adsorbent granulation process that used spraycoating in a fluidized bed for fluoride removal from drinking water. *Powder Technol.*, 206:291–296, 2011.

116. Ruparelia J.P., S.P. Duttagupta, A.K. Chatterjee and S. Mukherji. Potential of carbon nanomaterials for removal of heavy metals from water. *Desalination*, 232:145–156, 2008.

117. Sreeprasad T.S., S.M. Maliyekkal, K.P. Lisha and T. Pradeep. Reduced graphene oxide–metal/metal oxide composites: Facile synthesis and application in water purification. *J. Hazard. Mater.*, 186:921–931, 2011.

118. Li Y.H., S. Wang, A. Cao, D. Zhao, X. Zhang, C. Xu, Z. Luan, D. Ruan, J. Liang, D. Wu and B. Wei. Adsorption of fluoride from water by amorphous alumina supported on carbon nanotubes. *Chem. Phys. Lett.*, 350:412–416, 2001.

119. Li Y.H., S. Wang, X. Zhang, J. Wei, C. Xu, Z. Luan, D. Wu and B. Wei. Removal of fluoride from water by carbon nanotube supported alumina. *Environ. Technol.*, 24:391–398, 2003.

120. Li Y.H., S. Wang, X. Zhang, J. Wei, C. Xu, Z. Luan and Z.D. Wu. Adsorption of fluoride from water by aligned carbon nanotubes. *Mater. Res. Bull.*, 38:469–476, 2003.

121. Ma Y., S.-G. Wang, M. Fan, W.-X. Gong and B.-Y. Gao. Characteristics and defluoridation performance of granular activated carbons coated with manganese oxides. *J. Hazard. Mater.*, 168:1140–1146, 2009.

122. Karthikeyan M. and K.P. Elango. Removal of fluoride from aqueous solution using graphite: A kinetic and thermodynamic study. *Indian J. Chem. Technol.*, 15:525–532, 2008.

123. Ansari M., M. Kazemipour, M. Dehghani and M. Kazemipour. The defluoridation of drinking water using multi-walled carbon nanotubes. *J. Fluorine Chem.*, 132(8):516–520, 2011.

124. Geim A.K. and K.S. Novoselov. The rise of graphene. *Nat. Mater.*, 6:183–191, 2007.

125. Sreeprasad T.S. and T. Pradeep. Graphene for environmental and biological applications. *Int. J. Mod. Phys. B.*, 26(21):1242001–1242026, 2012.

126. Wang S., H. Sun, H.M. Ang and M.O. Tadé. Adsorptive remediation of environmental pollutants using novel graphene-based nanomaterials. *Chem. Eng. J.*, 226:336–347, 2013.

127. Li Y., P. Zhang, Q. Du, X. Peng, T. Liu, Z. Wang, Y. Xia, W. Zhang, K. Wang, H. Zhu and D. Wud. Adsorption of fluoride from aqueous solution by graphene. *J. Colloid Interface Sci.*, 363:348–354, 2011.

128. Li Y., Q. Du, J. Wang, T. Liu, J. Sun, Y. Wang, Z. Wang, Y. Xia and L. Xia. Defluoridation from aqueous solution by manganese oxide coated graphene oxide. *J. Fluoride Chem.*, 148:67–73, 2013.

20

Advances in Nanostructured Polymers and Membranes for Removal of Heavy Metals in Water

**Bhekie B. Mamba, Titus A.M. Msagati, J. Catherine Ngila,
Stephen Musyoka, Derrick Dlamini, and Sabelo D. Mhlanga**
*Department of Applied Chemistry, Faculty of Science, University of
Johannesburg, Doornfontein, Johannesburg, South Africa*

CONTENTS

20.1 Introduction

The extent and dispersion of contaminants in drinking water has generated immense interest from scientists and policy makers in understanding their trend. This is due to the increased deterioration of environmental ecosystems caused by population growth, increased industrial activities, and the advent of modern technologies. The increase in human population has resulted to population density swelling in areas that are in proximity to rivers, lakes, or other sources of freshwater. Moreover, many agricultural and industrial activities are located in areas close to major water sources since water is highly needed in these sectors. The overall outcome of this trend will no doubt add massive loads of both domestic and industrial wastes to these water sources.

All these activities will certainly affect the state of water quality, resulting in waterborne diseases worldwide with the consequences being more intense in the developing world. To counter this problem, strategies to ensure that clean and safe water is available and accessible to all have been put in place, and guidelines and regulations for various contaminants (physical particulates, organic chemicals, inorganic chemicals, pathogens, etc.) have been stipulated by the relevant national and international authorities. A number of technologies have been developed for remediating contaminated water sources, and these are classified as either thermal, physico–chemical, or biological methods. These methods, however, are known to offer high efficiencies when employed to solve very specific problems of water pollution. Of late, nanotechnology has been in the forefront among other technologies that are employed for the purpose of treating water from both physiochemical and biological contaminants.

The use of different types of nanomaterials for water treatment has been reported by many researchers in various places, with some being in various stages of development. An attractive feature of these nanomaterials is that they are synthesized with specific functionalities that give them the desired peculiar properties, such as high surface area to volume ratio, which, in turn, increases the preconcentration/enrichment factors, resulting in high adsorption capacities for targeted contaminants. Some of the nanomaterials and nanoparticles that have been reported in water purification strategies include zerovalent and bimetallic iron particles, single-walled, double-walled, and multiwalled carbon nanotubes, titanium oxide, magnesium oxide, self-assembled monolayer on mesoporous supports, and nanofiltration membranes.

20.2 Polymer–Clay Nanocomposites

Polymeric composites are polymers that have been filled with synthetic or natural inorganic compounds in order to improve their chemical and physical properties or to reduce cost by acting as a diluent for the polymer.[1–3] In this context, the polymer is called a matrix. If the filler is in the nanometer range, the composite is called a nanocomposite. These nanocomposites are a new class of composites, and they possess unique properties that are typically not shared by their more conventional microscopic counterparts.[2]

Many fillers have been used in nanocomposites preparation; however, clay (hydrous layer silicates) and layered silicates have been most widely used.[4–6] Clay-based nanocomposites, which are the subject of this research, have received considerable scientific and technological attention because clay is environmentally benign, abundantly available, and there is wide knowledge available on clay-intercalation chemistry.[7]

20.2.1 Polymer–Clay Nanocomposite Formation

Any physical mixture of a polymer and clay does not necessarily form a nanocomposite with improved properties. Immiscibility leads to poor physical attraction between the polymer and the clay particles, and that consequently leads to relatively poor mechanical properties. Generally, clays are hydrophilic, and this means that it is necessary to select a hydrophilic polymer to enable miscibility when preparing nanocomposites. However, hydrophilic polymers are not compatible with all composite-fabrication procedures. Fortunately, the clay can be modified accordingly to enhance compatibility. In many cases, a surfactant is used and the resulting modified clay is called organoclay. Many compounds have been used to modify clay. For example, dimethyl sulfoxide, methanol, and octadecylamine can be used. The "popular" Cloisite® organoclays are prepared by modifying montmorillonite clay with salts of dimethyl dehydrogenated tallow quaternary ammonium.

The concept of modifying the clay takes advantage of the exchangeable ions in the gallery of the clay. The general structure of a 2:1 dioctahedral smectite clay is shown in Figure 20.1.

The structure shows that the clay possesses lithium (Li), sodium (Na), rubidium (Rb), and cesium (Cs) ions sandwiched between two layers. These ions can be replaced with surfactants under appropriate conditions to form an organoclay. The replacement of the exchange cations in the cavities of the layered clay structure by alkylammonium surfactants can render compatible the surface chemistry of the clay and a hydrophobic polymer matrix. When natural montmorillonite is used, the dispersion of the clay in the polyvinyl chloride (PVC) matrix is typically poor, and when organically modified montmorillonite is used an improved dispersion is observed.[9] Therefore, organic modification of the clay is very important and it has an effect on the nanocomposite properties. It should be mentioned that the type of surfactant, the chain length, and the packing density might play an important role[10] in determining the suitability of the surfactant to modify the clay. Replacing the exchangeable ions in the cavity of the clay does not only serve to alter the hydrophilic or polarity properties of the clay but also widens the intergallery spacing to

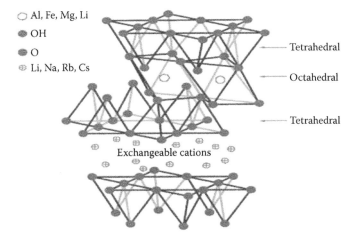

FIGURE 20.1
(See color insert.) Layered silicate structure. (Adapted from G. Beyer, *Plast. Addit. Compd.*, 4, 22, 2002.)

allow unhindered intercalation of the polymer chains in the clay. By definition, intercalation is the insertion of polymer chains between two galleries of clay. This enlarges the intergallery spacing of the clay.

The expansion of the gallery spacing by the organic modifier allows proper formation of a nanocomposite during fabrication. Three strategies are mainly employed in the preparation of nanocomposites: *in situ* polymerization,[11] the melt-blending method, and the solution-blending method.[9,12]

20.2.1.1 In Situ *Polymerization*

In the *in situ* polymerization procedure, monomers are mixed with the clay before polymerization. The main goal is to disperse the clay layers in the polymer matrix to obtain a nanocomposite with homogeneous exfoliated structures. Uniform dispersion of the clay in this method is attributed to the low monomer viscosity. According to Lingaraju et al.,[13] this procedure was used in the preparation of the "first" nanocomposite whereby researchers at the Toyota Central Research Laboratories synthesized the nylon 6–clay nanocomposite.

For most technologically essential polymers, *in situ* polymerization is limited because a suitable monomer–silicate solvent system is not always available, and it is not always compatible with current polymer-processing techniques.[14,15]

20.2.1.2 *Solution Blending*

The solution-blending technique involves the selection of a solvent that is capable of swelling the clay and dissolving the polymer matrix. A homogeneous three-component mixture of appropriate composition of the polymer, clay, and solvent is prepared through heating and mechanical and/or ultrasonic stirring. Depending on the interactions of the solvent and nanoparticles, the nanoparticle aggregates can be disintegrated in the solvent owing to the weak van der Waals forces that stack the layers together.[9] Polymer chains can then be adsorbed onto the nanoparticles. The final step of the procedure involves removal of the solvent through evaporation or by precipitation in a nonsolvent. There is entropy gain associated with the removal of the solvent molecules, and it is thought to compensate for the entropic loss of the intercalated polymer chains.[16] Therefore, it can be speculated that entropy-driven intercalation might be expected to occur even in the absence of an enthalpy gain due to favorable interactions between the macromolecules and the surface of the clay layers.[17]

This technique is used especially with water-soluble polymers, poly(acrylic acid) (PAA),[18] poly(ethylene oxide),[19] poly(ethylene vinyl alcohol), etc. The polarity of such polymers is believed to contribute an enthalpy gain helping intercalation.[20] The solution-blending procedure requires that the selected polymer be compatible with the selected solvent. An example of an organic solvent and hydrophobic polymers like polypropylene,[21] have also been given consideration. Notably, this method produces a high degree of intercalation only for certain polymer/clay/solvent systems, implying that for a given polymer one has to find the right clay, organic modifier, and solvents.

20.2.1.3 *Melt Blending*

The melt-blending procedure, also known as melt intercalation, came to prominence in the late 1990s.[22] This method involves the physical mixing of the polymer matrix and the clay in the molten state of the polymer. This process is attractive to researchers since it is the most versatile and environmentally benign among all the methods of preparing polymer–clay nanocomposites (PCNs). Nanocomposite synthesis via this method involves compounding and annealing (usually under shear) of a mixture of polymer and clay above the melting point of the polymer. During compounding and blending, the polymer melt diffuses into the cavities of the clay. This method therefore allows the processing of PCNs to be articulated directly from the precursors without using any solvent but ordinary compounding devices such as mixers and/or extruders.

Some polymers, for example, ethylene vinyl alcohol,[23] thermoplastic polyurethane,[24] and polycaprolactone (PCL),[25] are thermoplastic polymers that have been used in the study of nanocomposite synthesis by melt intercalation. Technically, the melt-blending method is much simpler and more straightforward than the methods discussed in the previous sections. In addition, although relatively new, this method has more appealing advantages that promise to greatly expand the commercial opportunities for PCN technology.[26,27] One such advantage is that this approach does not use any organic solvent, and it is compatible with existing industrial polymer extrusion and blending processes.[9] The compatibility with existing thermoplastic polymer-processing techniques minimizes capital costs, and the nonuse of organic solvents eliminates environmental concerns.[14,15]

On the negative aspects, this method forms microcomposites or tactoids at higher clay loading as a result of clay agglomeration. Furthermore, this method employs thermoplastic polymers that are generally hydrophobic, and this limits the application of the nanocomposites in water treatment. Finally, the compounding and extrusion processes are likely to amplify the range of particle shapes and sizes, principally when the clay is not uniformly exfoliated.

20.2.2 PCNs in Heavy Metal Removal from Water

PCNs have been used for heavy metal adsorption because of the suitable clay properties such as specific surface area, chemical and mechanical stability, layered structure, and high cation-exchange capacity. In addition, clays can be regarded as possessing both Brönsted and Lewis types of acidity,[2] and this acidity makes them suitable for the adsorption of heavy metals. The Brönsted acidity arises from two situations, first through the formation of H^+ ions on the surface, resulting from the dissociation of water molecules of hydrated exchangeable metal cations on the surface, as follows:

$$[M(H_2O)]_x^{n+} \rightarrow [M(OH)(H_2O)_{x-1}]^{(n-1)+} + H^+$$

Second, it can be as a result of a net negative charge on the surface due to the substitution of Si^{4+} by Al^{3+} in some of the tetrahedral positions, and the resultant charge is counterbalanced by H_3O^+ cations. On the other hand, Lewis acidity can emanate from three scenarios: (i) through dehydroxylation of some Brönsted acid sites; (ii) exposed Al^{3+} ions at the edges; and (iii) Al^{3+} arising from the breaking of Si–O–Al bonds. The resulting negative net charge is counterbalanced by exchangeable cations such as K^+, Ca^{2+}, and Mg^{2+}, adsorbed between the unit layers and around the edges, making it possible for clay to remove heavy metals from water through ion exchange, chemisorption, or physisorption.

20.2.3 Ion-Imprinted Polymers for Selective Metals Adsorption

Ion-imprinted polymers (IIPs) are chemically and physically stable materials synthesized with recognition cavities, and thus highly selective for the adsorption of targeted inorganic

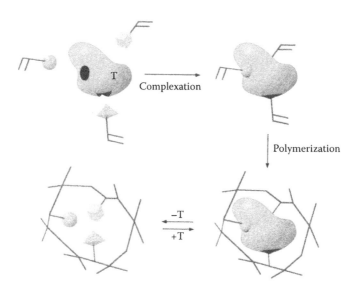

SCHEME 20.1
General steps in the synthesis of IIPs.

species such as metals in an aqueous environment. The cavities in IIPs are fabricated such that they contain adsorption sites characteristic of the target analyte in terms of shape, structure, and morphology. The procedure to synthesize IIPs includes several steps that generally involve the incorporation of a monomer, a cross-linker, an initiator, and a template.[26–28] Generally, the IIP synthesis procedure involve three main steps: (i) complexation of the template to a polymerizable ligand whereby the template interacts with the functional monomer; (ii) polymerization of the complex formed in (i), which is followed by the addition of the initiator and cross-linker that stabilizes the active sites so that the IIPs can be regenerated and reused[29–31]; finally, (iii) the third step involves the removal of the template after polymerization, and after this step the cavities or active binding sites that are morphologically similar to the target analyte will now be available in the IIP.[32,33] Scheme 20.1 shows the process of IIP synthesis.

20.2.3.1 Monomers, Cross-Linkers, and Solvents for the Imprinting Procedure

20.2.3.1.1 Monomers

The selection of appropriate monomers for IIPs considers factors such that they should have functional groups that will enhance the type of interaction with the template as well as the polymer matrix.[34] Figure 20.2 presents examples of the mostly used monomer in the synthesis of IIPs for the adsorption of various metals.

20.2.3.1.2 Cross-Linkers

In the synthetic processes for IIPs, the cross-linkers are normally in a higher proportion in the final polymer and actually the highly branched cross-linkers or those with high molecular weight are known to result into stable active sites.[35] Figure 20.3 depicts cross-linker molecules that have been reported in some applications.

Methacrylic acid
Acrylic acid
Acrylamide
2-Hydroxyethyl methacrylate
Methyl methacrylate
4-Vinylpyridine
2-Vinylpyridine

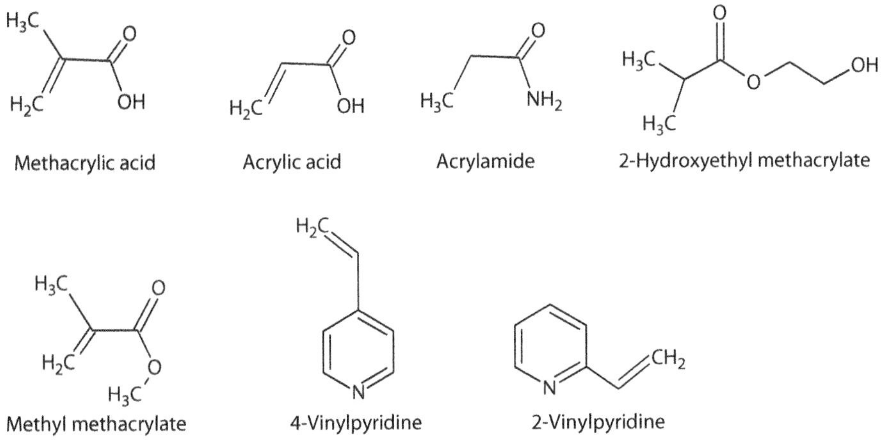

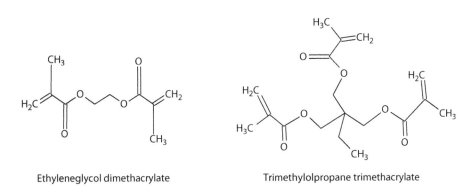

FIGURE 20.2
Some frequently used monomers for the synthesis of IIPs.

Ethyleneglycol dimethacrylate
Trimethylolpropane trimethacrylate

FIGURE 20.3
Common cross-linkers in the synthesis of IIPs.

20.2.3.1.3 Solvents (Porogens)

One attribute of a good solvent for IIPs is that it should be able to dissolve all the components of the polymerization mixture and at the same time participate in the physical makeup synthesis of the IIP.[36]

20.2.3.2 General Methods for Polymerization during IIP Synthesis

There are four polymerization methods used in the ion-imprinting technique: (i) bulk polymerization[37]; (ii) suspension polymerization[38]; (iii) precipitation polymerization[39]; and (iv) emulsion (surface) polymerization.[40] These four polymerization methods and the associated imprinting procedure may as well be categorized into two groups termed as 2D and 3D.[41] The 3D group takes place in three dimensions, implying that the cavities are distributed in the "whole" surface of the polymer. Table 20.1 gives the summary of the polymerization techniques.

TABLE 20.1

Summary of Polymerization Techniques for Synthesis of IIPs

Polymerization Technique	Size Distribution (μm)	Components
Bulk	20–50	Monomer
		Initiator
		Cross-linker
Precipitation	0.3–10	Monomer
		Solvent (optional)
Suspension	5–50	Monomer
		Continuous phase (usually water)
		Initiator (soluble in continuous phase)
		Surfactant
Emulsion	0.1–10	Monomer
		Continuous phase (usually water)
		Initiator (soluble in monomer)
		Stabilizer

20.3 Membrane Technology in Water Purification Works

Membranes are remarkable materials that form part of our daily lives. Their long history and use in biological systems have been extensively studied throughout the scientific field. The preparation of synthetic nanoscale membranes is, however, a more recent invention that has received a great audience because of its applications.[42] Membrane separation technologies are widely used in many areas of water and wastewater treatment. They can be used to produce potable water from surface water, groundwater, brackish water or seawater, or to treat industrial wastewaters before they are discharged or reused.[43] This chapter presents an overview of the various forms of nanostructured membranes and their use to date, particularly for the removal of heavy metals from water.

By definition, a membrane is a thin layer of material that is capable of separating materials as a function of their physical and chemical properties when a driving force is applied across the membrane.[43] Some key elements required for the design of membranes are as follows: it must (i) allow a high flux of water being treated, (ii) have good mechanical strength to prevent collapse of the membrane, and (iii) have good selectivity toward the desired substances.

20.3.1 Synthesis and Preparation of Membrane Materials

Numerous methods have been used for the preparation of membranes. The preparation process stands as the most significant element in membrane research since it determines the physiochemical properties of the membrane (basic structure), which govern the membrane efficiency (Figure 20.4).[44]

Depending on the preparation process used, the membranes can be classified as microfiltration (MF), ultrafiltration (UF), nanofiltration (NF), or reverse osmosis (RO) membranes. These separate molecules by size exclusion (Figure 20.5).[44]

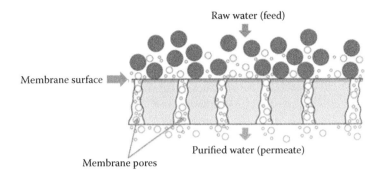

FIGURE 20.4
Basic structure of a membrane. (From http://www.hydrogroup.biz/areas-of-use/water-treatment/membrane
-filtration.html, accessed November 22, 2012.)

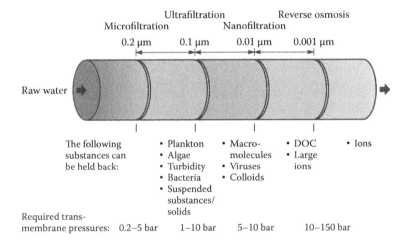

FIGURE 20.5
Classification of membranes and types of substances that can be separated by each one of them. Included are
the pressures required to drive water across each membrane type. (From http://www.hydrogroup.biz/areas-of
-use/water-treatment/membrane-filtration.html, accessed November 22, 2012.)

In most cases, it is the structure of the membrane that dictates its application. Thus, the
membranes can be classified into membrane bioreactors, low-pressure membranes, and
high-pressure membranes:

1. Membrane bioreactors: These include MF or UF membranes immersed in aeration
   tanks (vacuum system), or implemented in external pressure-driven membrane
   units, as a replacement for secondary clarifiers and tertiary polishing filters.

2. Low-pressure membranes: These include MF or UF membranes, either as a pressure system or an immersed system, providing a higher degree of suspended solids removal after secondary clarification.

3. High-pressure membranes: These include NF or RO pressure systems for treatment and production of high-quality product water suitable for indirect potable
   reuse and high-purity industrial process water.

TABLE 20.2

List of Commercial Membranes: Compositions and Applications

Membranes	Membrane Materials	Applications
Organic/polymer	Cellulose regenerated	D, UF, MF
	Cellulose nitrate	MF
	Cellulose acetate	GS, RO, D, UF, MF
	Polyamide	RO, NF, D, UF, MF
	Polysulfone	GS, UF, MF
	Poly(ether sulfone)	UF, MF
	Polycarbonate	GS, D, UF, MF
	Poly(ether imide)	UF, MF
	Poly(2,6-dimethyl-1,4-phenylene oxide)	GS
	Polyimide	GS
	Poly(vinylidene fluoride)	UF, MF
	Polytetrafluoroethylene	MF
	Polypropylene	MF
	Polyacrylonitrile	D, UF, MF
	Poly(methyl methacrylate)	D, UF
	Poly(vinyl alcohol)	PV
	Polydimethylsiloxane	PV, GS
Inorganic/ceramic	Metal (Al, Zn, Ti, Si, etc.) oxide	PV, MF
	Metal (Al, Zn, Ti, Si, etc.) nitride	PV, MF
	Metal (Al, Zn, Ti, Si, etc.) carbide	PV, MF
	Metal (Al, Zn, Ti, Si, etc.) oxide	PV, MF
	SiO_2	PV, GS

Source: Chen et al., Membrane separation: Basics and applications, In: *Handbook of Environmental Engineering, Membrane and Desalination Technologies*, Springer Science+Business Media, LLC, 2008.

Note: D, dialysis; GS, gas separation; MF, microfiltration; NF, nanofiltration; PV, pervaporation; RO, reverse osmosis; UF, ultrafiltration.

Membranes can be synthesized as *flat sheet* or *hollow fiber*. Different preparation processes are followed to prepare each type. To prepare flat-sheet membranes, the phase inversion technique is used where a polymer solution e.g., polyethersulfone, is cast on a support material (e.g., glass, polymer, metal, nonwovens) after immersing into a non-solvent (water). Hollow-fiber membranes are usually prepared by a dry–wet spinning technique based on phase inversion.[45] Other processes that have been used include track etching and film stretching.[46,47] The ideal membrane is one that is not easily susceptible to fouling.

Membranes can further be classified as organic (polymer based) or inorganic (ceramic based). Polymer-based membranes are the most widely used for water treatment purposes. Chen et al. have done an excellent review of the different types of commonly used membranes and their application (Table 20.2).[43]

20.3.2 Characterization of Membranes

The application of membrane technologies in water treatment is met with challenges relating to the fouling of the membranes. Membrane fouling refers to the attachment, accumulation, or adsorption of foulants (colloidal particles, organic matter, biological substances, inorganic salts) onto membrane surfaces and/or within the membrane, deteriorating the performance of the membranes over time. Consequently, to improve the overall performance of membranes, new components, usually nanomaterials, are added to the bulk of the membrane material by various physical or chemical methods. This usually modifies the morphology and structure of the membranes. Therefore, to gain complete understanding of the modified membranes, specific methods are used and have been categorized as follows[45]:

- Characterization of composition: Fourier transform infrared spectroscopy, nuclear magnetic resonance, x-ray photoelectron spectroscopy, differential scanning calorimetry, transmission electron microscopy (TEM).

- Characterization of morphology and structure: scanning electron microscopy, atomic force microscopy, microtomy followed by TEM.

- Characterization of performance: surface hydrophilicity/hydrophobicity by contact angle measurements, permeability and selectivity, antifouling properties.

20.3.3 Removal of Heavy Metals, Organometallics, and Metalloids Using Nanomembranes

Pollution of water by heavy metals such as mercury (Hg), cadmium (Cd), arsenic (As), chromium (Cr), thallium (Tl), lead (Pb), selenium (Se), and zinc (Zn) is a serious problem for many countries, including South Africa. The term "heavy metal" refers to any metallic chemical element that has a relatively high density and is toxic or poisonous at low concentrations. These metals are dangerous because they tend to bioaccumulate. Thus, no matter how low in concentration, they still pose serious health problems. The application of nanotechnology-enabled membranes for the removal of these metals is undergoing immense research.

The various materials and methods used to fabricate the membranes and their efficiencies can be observed in a number of studies.[48–63] A number of important points can be drawn from these studies:

- Membranes can effectively be used for the removal of heavy metals from various water sources including wastewater. Depending on the materials used, efficiencies of up to 100% can be removed.[50]

- There is an increase in research toward the incorporation of other nanomaterials/nanoparticles (e.g., TiO_2, SiO_2, carbon nanotubes, cyclodextrin) into membranes. Addition of the nanomaterials improves the antifouling properties of the membranes and sometimes the mechanical strength.[49,51–65]

- Many of the new membranes have been tested in a laboratory scale. A few studies have shown that the membranes are effective at a large scale.[50,59]

- Electrospinning and phase inversion methods are the most widely used technologies for the preparation of nanostructured membranes.

- Most of the studies have shown that the pH of the solution treated is crucial because it affects the adsorption capacity of membranes.

20.4 Nanofibers for Water Treatment Processes

Nanofiber has been broadly applied to refer to a fiber with a diameter of <1 micron. Polymeric nanofibers are rapidly finding their place in nanomaterials technology. Nanofibers are an exciting new class of material used for several value-added applications such as medical, filtration, barrier, wipes, personal care, composite, garments, insulation, and energy storage. The focus in this work is the application of nanofibers in adsorption of heavy metals from wastewater.

Electrospinning is the most popular method applied to prepare nonwoven nanofibers made of organic polymers, ceramics, and polymer/ceramic composites. It has been extensively explored as a simple and versatile method for drawing fibers from polymer solutions (or melts). The electrospinning process involves a polymeric fluid being extruded from the orifice of a needle or pipette to form a small droplet in the presence of an electric field. When the electric field is sufficiently strong, charges build up on the surface of the droplet and overcome the surface tension. A charged jet of fluid is ejected from the tip of the Taylor cone. The jet is then accelerated toward a grounded collector (Figure 20.6). The discharged polymer solution jet undergoes a rapid whipping process through which the solvent is evaporated; the liquid jet is stretched to many times its original length to produce continuous, ultrathin fibers of the polymer. In the continuous feed mode (through the use of a syringe pump), a mat of ultrathin fibers can be obtained within a relatively short period.

Biopolymers like cellulose and chitin and their derivatives are polymers that are abundant, renewable, biocompatible, and biodegradable. These factors make them ideal choices as sustainable feedstock for materials and products for use in the fields of filtration, adsorption, and preconcentration of heavy metal pollutants. However, a major challenge faces the application of these biopolymers because of their insolubility in most common solvents.[66] Biosorption processes utilizing inexpensive biomass to sequester toxic heavy metals is particularly useful for the removal of these contaminants from industrial effluents.[67] Chitin and chitosan have been used for the adsorption or the binding of heavy metals and dyes. Chitosan is a polycation polymer effective in coagulation and flocculation dewatering of activated sludge and hence used in wastewater treatment.

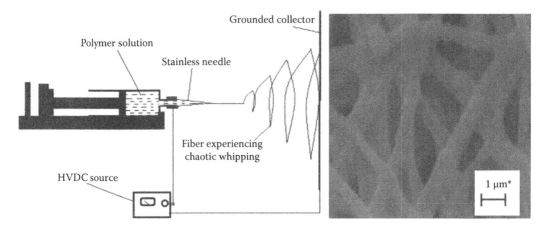

FIGURE 20.6
Schematic diagram of an electrospinning unit and SEM micrograph of electrospun mat (scale bar of 1 μm indicated).

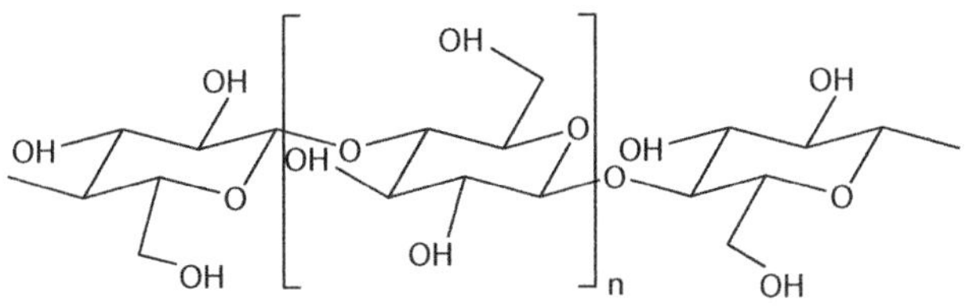

FIGURE 20.7
Chemical structure of cellulose.

20.4.1 Preparation of Cellulose and Chitosan Nanofibers

20.4.1.1 Cellulose Nanofibers

Traditionally, fibers from cellulose are formed through wet spinning of bleached wood pulp. The pulp is then dissolved in *N*-methylmorpholine *N*-oxide and passed through spinnerets to form microscale fibers. In many respects, the formation of nano- and microscale fibers from cellulose via electrospinning has mirrored the history of conventional cellulose fiber spinning.[66] This is because solubilization of cellulose without derivatization has posed challenges. Electrospinning of cellulose (Figure 20.7) has been achieved through the use of readily soluble cellulose acetate and reforming it through treatment with a strong alkaline solution. The optimized conditions for preparation of nanofibers include 16% (w/v) polymer solution, prepared in acetone/*N,N*-dimethylacetamide solvent system, a 20-gauge (0.45 mm bore diameter) stainless-steel hypodermic needle, a grounded aluminum collector at 14 cm tip-to-collector distance, and an applied high-voltage direct current current of 15 kV. A programmable syringe pump (NE-1000 single syringe pump, New Era Pump Systems Inc.) aids to pump the solution through and maintain a constant flow rate; otherwise, a glass pipette tilted at an angle can be used. Residual solvent needs to be removed before the fibers are peeled off from the foil. This is done by evaporation by heating the nanofiber mat in an oven at 200°C for 1 h.

Cellulose nanofibers are regenerated through deacetylation of the nonwoven fiber mat by soaking it in 0.3 M NaOH solution for 8 h followed by washing with double-distilled water to obtain neutral pH.

20.4.1.2 Chitin and Chitosan Nanofibers

Electrospinning of pure chitosan solutions poses many challenges[67,68] because of their intractable molecular structure. Chitosan is a positively charged polyelectrolyte at pH 2–6 due to free amino groups, which contribute to its higher solubility in comparison with chitin. However, this property makes chitosan solutions highly viscous and complicates their electrospinning.[69] Furthermore, the formation of strong hydrogen bonds in a three-dimensional network prevents the movement of polymeric chains exposed to the electrical field. Successful electrospun chitosan nanofibers from highly concentrated solutions have been reported.[70,71] Schiffman and Schauer[72] have observed that chitosan has a high solubility in polar aprotic solvents such as acetic acid and trifluoroacetic acid. Figure 20.8a and b show the chemical structures for chitin and chitosan, respectively.

Owing to the above challenges, blending of chitosan polymer with polyacrylamide (PAM) in the ratio of 7.7:2.3 (w/w) and dissolving in 60% acetic acid at 90°C yields an electrospinnable polymer solution. The optimized electrospinning parameters are as follows: flow rate, 0.60 mL/h; tip-to-collector distance, 19 cm; and applied voltage, 20 kV.

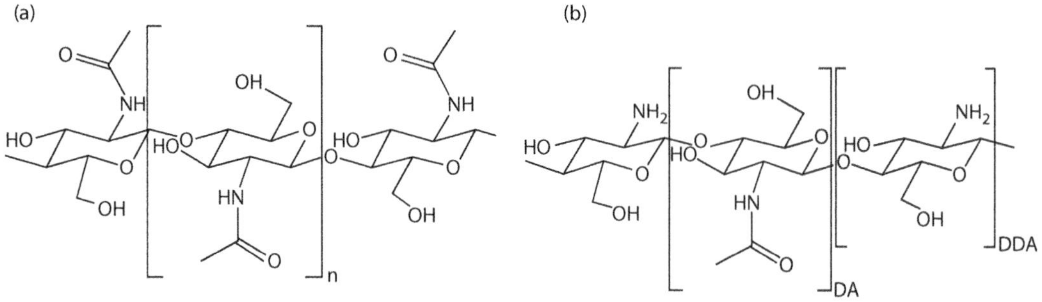

FIGURE 20.8
Structure of chitin (a) and chitosan (b).

20.4.2 Functionalization of the Nanofibers

The nanofibers can be functionalized through chemical grafting with bidentate ligands. The two that have been successfully tried are oxolane-2,5-dione and furan-2,5-dione. The cellulose nanofibers have been functionalized with oxolane-2,5-dione[73] and chitosan/PAA nanofibers have been functionalized with furan-2,5-dione. The reaction scheme for the functionalization of the nanofibers is shown in Figure 20.9.

The morphological structures of the cellulose nanofibers before and after functionalization (Figure 20.10) were obtained from SEM, and the fiber diameters were measured using Image Pro Plus 6 software. The average fiber diameter have been calculated at 385 nm.

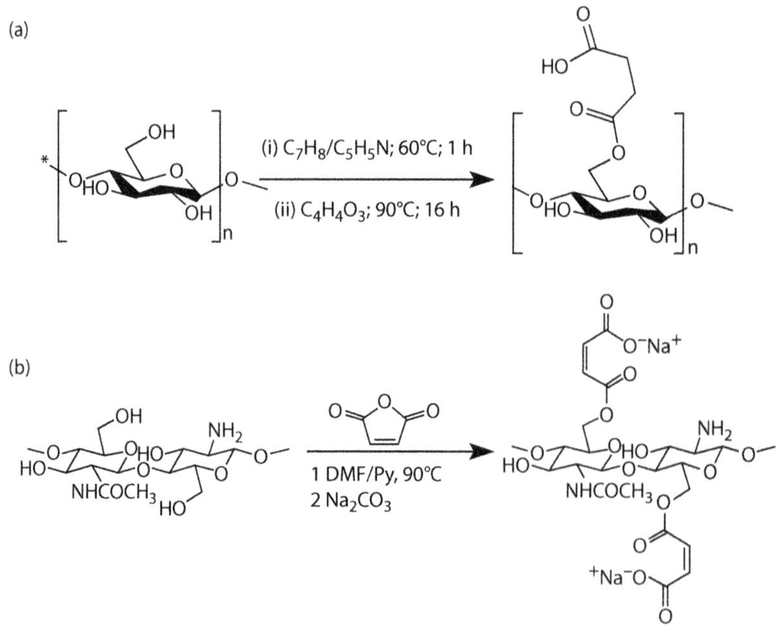

FIGURE 20.9
Functionalization scheme of nanofibers of (a) cellulose with oxolane-2,5-dione and (b) chitosan/PAM with furan-2,5-dione.

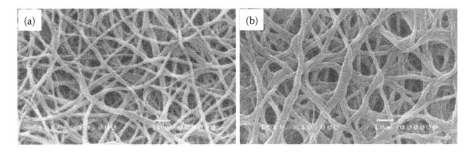

FIGURE 20.10
SEM micrographs for: cellulose-*g*-oxolane-2,5-dione nanofibers (a) and cellulose nanofibers (b).

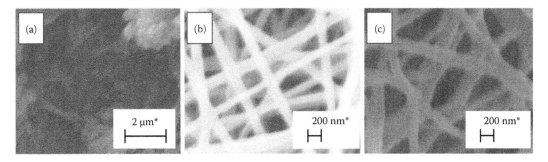

FIGURE 20.11
SEM micrographs of electrospun chitosan: (a) chitosan/PAM, (b) chitosan/PAM-*g*-furan-2,5-dione, and (c) chitosan/PAM-*g*-furan-2,5-dione nanofibers.

Those of chitosan/PAM were also obtained (Figure 20.11). The average fiber diameter was calculated at 186 nm.

20.4.3 Application of Nanofibers in Trace Metals Removal

The functionalized cellulose and chitosan nanofibers showed enhanced removal of trace metals in natural water polluted with heavy metals. The grafted functional groups of oxolane-2,5-dione and furan-2,5-dione have charged sites onto which the divalent metal cations are exchanged. Undetectable metal ions are adsorbed in these materials and when eluted with suitable acids like dilute nitric acid can be measured and their levels in the natural water determined. The desorption of the adsorbed metal ions regenerates the nanofibers allowing for continuous use up to five cycles.

20.5 Challenges

Among the properties that make clay a suitable heavy metal adsorbent is the high specific surface area. However, this property is suppressed by embedding the clay particles in a polymer despite the fact that surface area is an important parameter in adsorption technology. Furthermore, embedding the clay particles in a polymer can slow down the adsorption process because the adsorption active sites are concealed or enveloped by the polymer. This can be dominant in composites prepared by the melt-blending method, which use hydrophobic polymers. Consequently, there are long equilibration periods in adsorption experiments using composites, compared with when the filler is used independently. One of the most reported drawbacks in composites is weight increase. Weight increase tends to favor the settling of the composites in water; hence, vigorous mechanical stirring would be required. Finally, depending on the polymer used in preparing the composite, erosion of support in desorption studies can be a serious problem. Desorption of the adsorbate from an adsorbent usually occurs under harsh conditions.

The development of nanotechnology-enhanced membranes comes with challenges such as difficulty in dispersion of nanocomponents and cost related to large-scale application. Hashim et al.[64] recently reviewed the various remediation technologies for heavy-metal-contaminated groundwater. It was observed that high-pressure membrane and filtration technologies, while they have a high removal efficiency of heavy metals, their major drawback is clogging, which subsequently requires regeneration of the filter materials. However, the regeneration of the materials is not as frequent as is the case with activated carbon.[64] Thus, whether the metal removal mechanism is by filtration or adsorption, it is necessary to backwash or desorb the metal residues from the surface of the membranes. This is a challenge that cannot be avoided completely but can be made minimal. Other technical challenges may include the complexity and expense of the concentrate (residuals) disposal from high-pressure membranes. This often translates to high maintenance costs of the membranes.

20.5.1 Opportunities

Composites prepared by the melt-blending method can be molded into strips of different sizes and shapes. This makes it relatively easy to recover the composites after heavy-metal capture in water. This is crucial because to obtain clean water, the spent adsorbent should be successfully removed from the water. The size of the strip affords better handling and cleaning of the composites, compared with powder adsorbents like activated carbon and clay. Additionally, embedding the adsorbents like clay in a polymer reduces the risk of sludge formation.

On the other hand, the ability to modify the surface and structure of membranes using various chemical strategies provides opportunities for the development of nanomembranes. Other materials such as graphene sheets are now being exploited for the removal of heavy metals.[65] Membranes generally consume less energy and are suitable for large-scale use and can easily be integrated to existing plants.

20.6 Conclusions

Nanotechnology is making use of materials measuring 100 nm or less that have proven to display unique properties and characteristics that may offer solutions to water quality problems currently facing our planet. The ability of these nanomaterials and nanoparticles to alter chemical and physical properties of substances can be exploited in the processes leading to improving the quality of water significantly. In this chapter, various examples of applications have shown that nanostructured polymers and membranes that have been synthesized at the nanoscale have shown to exhibit large surface areas and pore sizes and volumes that enhanced the adsorption/removal of pollutants (physical, chemical, and microbial) from water. Further study on the use of nanostructured polymers and membranes may result in more robust solutions to the water quality and safety, thus improving the health status and economy in many societies and communities.

References

1. Trindade M.J., M.I. Dias, J. Coroado, F. Rocha. Mineralogical transformations of calcareous rich clays with firing: A comparative study between calcite and dolomite rich clays from Algarve. *Portugal. Appl. Clay Sci.*, 42: 345–355, 2009.
2. Martin R.T., S.W. Bailey, D.D. Eberl, D.S. Fanning, S. Guggenheim, H. Kodama, D.R. Pevear, J. Srodon, F.J. Wicks. Report of the Clay Minerals Society Nomenclature Committee: Revised classification of clay materials. *Clays Clay Miner.*, 39: 333–335, 1991.
3. Maes N., I. Heylen, P. Cool, E.F. Vansant. The relation between the synthesis of pillared clays and their resulting porosity. *Appl. Clay Sci.*, 12: 43–60, 1997.
4. Rousseaux D.D.J., M. Sclavons, P. Godard, J. Marchand-Brynaert. Carboxylate clays: A model study for polypropylene/clay nanocomposites. *Polym. Degrad. Stab.*, 95: 1194–1204, 2010.
5. Zulfiqar S., Z. Ahmad, M. Ishaq, M.I. Sarwer. Aromatic–aliphatic polyamide/montmorillonite clay composite materials; synthesis, nanostructure and properties. *Mater. Sci. Eng. A*, 525: 30–36, 2009.
6. Mishra S.B., A.S. Luyt. Effect of organic peroxides on the morphology, thermal and tensile properties of EVA/organoclay nanocomposites. *Express Polym. Lett.*, 2: 256–264, 2008.
7. Cho Y., S. Komarneni. Synthesis of kaolinite from micas and K-depleted micas. *Clays Clay Miner.*, 55: 565–571, 2007.
8. Beyer G. Nanocomposites: A new class of flame retardants for polymers. *Plast. Addit. Compd.*, 4: 22–27, 2002.
9. Madaleno L., J. Schjødt-Thomsen, J.C. Pinto. Morphology, thermal and mechanical properties of PVC/MMT nanocomposites prepared by solution blending and solution blending þ melt compounding. *Comp. Sci. Technol.*, 70: 804–814, 2010.
10. Zhang D., C. Zhou, C. Lin, D. Tong, W. Yu. Synthesis of clay minerals. *Appl. Clay Sci.*, 50: 1–11, 2010.
11. Zha W., C.D. Han, H.C. Moon, S.H. Han, D.H. Lee, J.K. Kim. Exfoliation of organoclay nanocomposites based on polystyrene-block-polyisoprene-block-poly(2 vinylpyridine) copolymer: Solution blending versus melt blending. *Polymer*, 51: 936–952, 2010.
12. Lee J.L., C. Zeng, X. Cia, X. Han, J. Shen, G. Xu. Polymer nanocomposite foams. *Comp. Sci. Technol.*, 65: 2344–2363, 2005.
13. Lingaraju D., K. Ramji, N.B.R.M. Rao, U.R. Lakshmi. Characterization and prediction of some engineering properties of polymer–Clay/Silica hybrid nanocomposites through ANN and regression models. *Procedia Eng.*, 10: 9–18, 2011.
14. Fornes T.D., P.J. Yoon, D.L. Hunter, H. Keskkula, D.R. Paul. Effect of organoclay structure on nylon 6 nanocomposite morphology and properties. *Polymer*, 43: 5915–5933, 2002.
15. Cho J.W., D.R. Paul. Nylon 6 nanocomposites by melt compounding. *Polymer*, 42: 1083–1094, 2001.
16. Vaia R.A., E.P. Giannelis. Lattice model of polymer melt intercalation in organically-modified layered silicates. *Macromolecules*, 30: 7990–7999, 1997.
17. Filippi S., E. Mameli, C. Marazzato, P. Magagnini. Comparison of solution-blending and melt-intercalation for the preparation of poly(ethylene-co-acrylic acid)/organoclay nanocomposites. *Eur. Polym. J.*, 43: 1645–1659, 2007.

18. Billingham J., C. Breen, J. Yarwood. Adsorption of polyamine, polyacrylic acid and polyethylene glycol on montmorillonite: An *in situ* study using ATR-FTIR. *Vibr. Spectrosc.*, 14: 19–34, 1997.

19. Malwitz M.M., S. Lin-Gibson, E.K. Hobbie, P.D. Butler, G. Schmidt. Orientation of platelets in multilayered nanocomposite polymer films. *J. Polym. Sci, Part B, Polym. Phys.*, 41: 3237–3248, 2003.

20. Qiu L., W. Chen, B. Qu. Morphology and thermal stabilization mechanism of LLDPE/MMT and LLDPE/LDH nanocomposites. *Polymer*, 47: 922–930, 2006.

21. Chiu F.-C., P.-H. Chu. Characterization of solution-mixed polypropylene/clay nanocomposites without compatibilizers. *J. Polym. Res.*, 13: 73–78, 2006.

22. Shen Z., G.P. Simon, Y.-B. Cheng. Comparison of solution intercalation and melt intercalation of polymer-clay nanocomposites. *Polymer*, 43: 4251–4260, 2002.

23. Dlamini D.S., S.B. Mishra, A.K. Mishra, B.B. Mamba. Comparative studies of the morphological and thermal properties of clay/polymer nanocomposites synthesized via melt blending and modified solution blending methods. *J. Comp. Mater.*, 45: 2211–2216, 2011.

24. Finnigan B., D. Martin, P. Halley, R. Truss, K. Campell. Morphology and properties of thermoplastic polyurethane nanocomposites incorporating hydrophilic layered silicates. *Polymer*, 45: 2249–2260, 2004.

25. Dlamini D.S., A.K. Mishra, B.B. Mamba. Adsorption behaviour of ethylene vinyl acetate and polycaprolactone-bentonite composites for Pb^{2+} Uptake. *J. Inorg. Organomet. Polym.*, 22: 342–351, 2012.

26. Mafu L.D., T.A.M. Msagati, B.B. Mamba. Ion-imprinted polymers for environmental monitoring of inorganic pollutants: Synthesis, characterization, and applications. *Environ. Sci. Pollut. Res.*, 20: 790–802, 2013

27. Mafu L.D., T.A.M. Msagati, B.B. Mamba. The enrichment and removal of Arsenic (III) from water samples using hollow fibre supported liquid membrane. *J. Phys. Chem. Earth*, 50–52: 121–126, 2012.

28. Daniel S., P.P. Rao, T.P. Rao. Investigation of different polymerization methods on the analytical performance of palladium (II) ion imprinted polymer materials. *Anal. Chim. Acta*, 536: 197–206, 2005. doi:10.1016/j.aca.2004.12.052.

29. Yan M., O Ramström (Eds). The covalent and other stoichiometric approaches. *Molecularly Imprinted Materials: Science and Technology*, pp. 59–92, 2005.

30. Zhan Y., X. Luo, S. Nie, Y. Huang, X. Tu, S. Luo. Selective separation of Cu (II) from aqueous solution with a novel Cu (II) surface magnetic ion-imprinted polymer. *Ind. Eng. Chem. Res.*, 50: 6355–6361, 2011.

31. Sellergren B., M. Lepisto, K. Mosbach. Highly enantioselective and substrate-selective polymers obtained by molecular imprinting utilizing noncovalent interactions. NMR and chromatographic studies on the nature of recognition. *J. Am. Chem. Soc.*, 110: 5853–5860, 1988.

32. Baggiani C., C. Giovannoli, L. Anfossi, C. Passini, P. Baravalle. A connection between the binding properties of imprinted and nonimprinted polymers: A change of perspective in molecular imprinting. *J. Am. Chem. Soc.*, 134: 1513–1518, 2012.

33. Villanovaa J.C.O., E. Ayresb, S.M. Carvalhoa, P.S. Patrício, F.V. Pereirad, R.L. Oréficea. Pharmaceutical acrylic beads obtained by suspension polymerization containing cellulose nano-whiskers as excipient for drug delivery. *Eur. J. Pharm. Sci.*, 42: 406–415, 2011.

34. Say R., E. Birlik, A. Ersöz, F. Yılmaz, T. Gedikbey, A. Denizli. Preconcentration of copper on ion-selective imprinted polymer microbeads. *Anal. Chim. Acta*, 480: 251–258, 2003. doi:10.1016/S0003-2670(02)01656-2.

35. Andac M., E. Özyapı, S. Senel, R. Say, A. Denizli. Ion-selective imprinted beads for aluminum removal from aqueous solutions. *Ind. Eng. Chem. Res.*, 45: 1780–1786, 2006.

36. Uezu K., H. Nakamura, J.-I. Kanno, T. Sugo, M. Goto, F. Nakashio. Metal ion-imprinted polymer prepared by the combination of surface template polymerization with postirradiation by γ-rays. *Macromolecules*, 30(96): 3888–3891, 1997.

37. Yoshikwa M. Molecularly imprinted polymeric membranes. *Bioseparation*, 10: 277–286, 2002.

38. Bi X., R.J. Lau, K.-L. Yang. Preparation of ion-imprinted silica gels functionalized with glycine, diglycine, and triglycine and their adsorption properties for copper ions. *Langmuir*, 23: 8079–8086, 2007.

39. Immanuela C.D., M.A. Pintoa, J.R. Richards, J.P. Congalidis. Population balance model versus lumped model for emulsion polymerisation: Semi-batch and continuous operation. *Chem. Eng. Res. Des.*, 86: 692–702, 2008.

40. Ito F., G. Ma, M. Nagai, S. Omi. Study on preparation of irregular shaped particle in seeded emulsion polymerization accompanied with regulated electrostatic coagulation by counter-ion species. *Colloids Surf.*, 216: 109–122, 2003.

41. Jung M.S., V.G. Gomes. Transitional emulsion polymerisation: Zero-one to pseudo-bulk. *Chem. Eng. Sci.*, 66: 4251–4260, 2011.

42. Strathmann H., L. Giorno, E. Drioli. *An Introduction to Membrane Science and Technology*. Rende: Institute on Membrane Technology, 2006.

43. Chen J.P., H. Mou, L.K. Kang, T. Matsuura, Y. Wei. Membrane separation: Basics and applications. In: L.K. Wang, J.P. Chen, Y.-T. Hung, N.K. Shammas (Eds). *Handbook of Environmental Engineering—Membrane and Desalination Technologies*. Springer Science + Business Media LLC, New York, USA, 2008.

44. Available at http://www.hydrogroup.biz/areas-of-use/water-treatment/membrane-filtration .html (accessed November 22, 2012).

45. Zhao C., J. Xue, F. Ran, S. Sun. Modification of polyethersulfone membranes—A review of methods. *Prog. Mater., Sci.*, 58: 76–150, 2013.

46. Yamazaki I.M., R. Paterson, L.P. Geraldo. A new generation of track etched membranes for microfiltration and ultrafiltration. Part I. Preparation and characterisation. *J. Membr. Sci.*, 118: 239–245, 1996.

47. Sadeghi F., A. Ajji, P.J. Carreau. Analysis of microporous membranes obtained from polypropylene films by stretching. *J. Memb. Sci.*, 292: 62–71, 2007.

48. Min M., L. Shen, G. Hong, M. Zhu, Y. Zhang, X. Wang, Y. Chen, B.S. Hsiao. Micro-nano structure poly(ethersulfones)/poly (ethyleneimine) nanofibrous affinity membranes for adsorption of anionic dyes and heavy metal ions in aqueous solution. *Chem. Eng. J.*, 197: 88–100, 2012.

49. Hadju I., M. Bodnár, Z. Csikós, S. Wei, L. Daróczi, B. Kovács, Z. Győri, J. Tamás, J. Borbély. Combined nano-membrane technology for removal of lead ions. *J. Membr. Sci.*, 409–410: 44–53, 2012.

50. Rashdi B.A.M., D.J. Johnson, N. Hilal. Removal of heavy metal ions by nanofiltration. *Desalination*, 315: 2–17, 2013.

51. Taha A., J. Qiao, F. Li, B. Zhang. Preparation and application of amino functionalized mesoporous nanofiber membrane via electrospinning for adsorption of Cr^{3+} from aqueous solution. *J. Environ. Sci.*, 24(4): 610–616, 2012.

52. Zhang X., Y. Wang, Y. You, H. Meng, J. Zhang, X. Xu. Preparation, performance and adsorption activity of TiO_2 nanoparticles entrapped PVDF hydrid membranes. *Appl. Surf. Sci.*, 263: 660–665, 2012.

53. Reiad N.A., O.E. Abdel Salam, E.F. Abadir, F.A. Harraz. Adsorptive removal of iron and manganese ions from aqueous solutions with microporous chitosan/polyethylene glycol blend membrane. *J. Environ. Sci.*, 24(8): 1425–1432, 2012.

54. Xu G.-R., J.-N. Wang, C.-J. Li. Preparation of hierarchically nanofibrous membrane and its high adaptability in hexavalent chromium removal from water. *Chem. Eng. J.*, 198–199: 310–317, 2012.

55. Daraei P., S.S. Madaeni, N. Ghaemi, E. Salehi, M.A. Khadivi, R. Moradian, B. Astinchap. Novel polyethersulfone nanocpmosite membrane prepared by PANI/Fe_3O_4 nanoparticles with enhanced performance for CU(II) removal from water. *J. Membr. Sci.*, 415–416: 250–259, 2012.

56. Saljoughi E., S.M. Mousavi. Preparation and characterization of novel polyfulfone nanofiltration membranes for removal of cadmium from contaminated water. *Sep. Purif. Technol.*, 90: 22–30, 2012.

57. Adams F.V., E.N. Nxumalo, R.W.M. Krause, E.M.V. Hoek, B.B. Mamba. Preparation and characterization of polysulfone/β-cyclodextrin polyurethane composite nanofiltration membranes. *J. Membr. Sci.*, 405–406: 291–299, 2012.

58. Salehi E., S.S. Madaeni, L. Rajabi, V. Vatanpour, A.A. Derakhshan, S. Zinadini, Sh. Ghorabi, H. Ahmadi Monfared. Novel chitosan/poly(vinyl) alcohol thin adsorptive membranes modified with amino functionalized multi-walled carbon nanotubes from Cu(II) removal from water: Preparation, characterization, adsorption kinetics and thermodynamics. *Sep. Purif. Technol.*, 89: 309–319, 2012.

59. Urgun-Demirtas M., P.L. Benda, P.S. Gillenwater, M.C. Nedri, H. Xiong, S.W. Snyder. Achieving very low mercury levels in refinery wastewater by membrane filtration. *J. Hazard. Mater.*, 215–216: 98–107, 2012.

60. Vasanth D., G. Pugazhenthi, R. Uppaluri. Biomass assisted microfiltration of chromium(VI) using Baker's yeast by ceramic membrane prepared from low cost raw materials. *Desalination*, 285: 239–244, 2012.

61. Song J., H. Oh, H. Kong, J. Jang. Polyrhodanine modified anodic aluminium oxide membrane for heavy metal ions removal. *J. Hazard. Mater.*, 187: 311–317, 2011.

62. Tian Y., M. Wu, R. Liu, Y. Li, D. Wang, J. Tan, R. Wu, Y. Huang. Electrospun membrane of cellulose acetate for heavy metal ion adsoroption in water treatment. *Carbohydr. Polym.*, 83: 743–748, 2011.

63. Ferella F., M. Prisciandro, I. De Michelis, F. Veglio. Removal of heavy metals by surfactant-enhanced ultrafiltration from waste waters. *Desalination*, 207: 125–133, 2007.

64. Hashim M.A., S. Mukhopadhay, J.N. Sahu, B. Sengupta. Remediation technologies for heavy metal contaminated ground water. *J. Environ. Manage.*, 92: 2355–2388, 2011.

65. Chang C.-F., Q.D. Truong, J.-R. Chen. Graphene sheets synthesized by ionic-liquid-assisted electrolysis for application in water purification. *Appl. Surf. Sci.*, 64: 329–334, 2013.

66. Frey M.W. Electrospinning cellulose and cellulose derivatives. *Polym. Rev.*, 48: 378–391, 2008.

67. Demirbas A. Heavy metal adsorption onto agro-based waste materials: A review. *J. Hazard. Mater.*, 157: 220–229, 2008.

68. Cai Z.-X., X.-M. Mo, K.-H. Zhang, L.-P. Fan, A.-L. Yin, C.-L. He, H.-S. Wang. Fabrication of chitosan/silk fibroin composite nanofibers for wound-dressing applications. *Int. J. Mol. Sci.*, 11: 3529–3539, 2010.

69. Pillai C.K.S., C.P. Sharma. Electrospinning of chitin and chitosan nanofibres. *Trends Biomater. Artif. Organs.*, 22: 179–201, 2009.

70. Kriegel C., K. Kit, D.J. McClements, J. Weiss. Influence of surfactant type and concentration on electrospinning of chitosan–poly(ethylene oxide) blend nanofibers. *Food Biophys.*, 4: 213–228, 2009.

71. Zhou Y., D. Yang, J. Nie. Electrospinning of chitosan/poly(vinyl alcohol)/acrylic acid aqueous solutions. *J. Appl. Polym. Sci.*, 102: 5692–5697, 2006.

72. Schiffman J.D., C.L. Schauer. Cross-linking of nanofibers. *Biomacromolecules*, 8: 594–601, 2007.

73. Musyoka S., C. Ngila, B. Moodley, A. Kindness, L. Petrik, C. Greyling. Oxolane-2,5-dione modified electrospun cellulose nanofibers for heavy metals adsorption. *J. Hazard. Mater.*, 192: 922–927, 2011.

21

Point-of-Drinking Water Purification Innovation: The Water Initiative

Eugene A. Fitzgerald, Thomas A. Langdo, Kevin M. McGovern, and Rick Renjilian
The Water Initiative, New York, New York, USA

CONTENTS

21.1 Background and Landscape

The Water Initiative (TWI) was founded in 2006 to bring new innovations in point-of-drinking (POD) water purification to the marketplace. In this chapter, we describe the TWI vision and innovation process that has led to a new approach to POD water purification.

The water industry landscape is generally composed of what TWI refers to as "big water" and "the last mile of water." Big water is composed of centralized technologies and business models, such as large regional water treatment plants. Examples would be centralized municipal water plants in developed countries, and centralized desalination plants in countries that lack sufficient freshwater sources. The last mile of water is composed of water purification systems that are closest to the end user, such as point-of-entry water purification systems (i.e., at the point-of-entry to a house or building in which all water is purified at that point) down to POD devices in which the water is purified very close to the user's lips (e.g., in the kitchen). Further examples in this category are whole-house treatment water systems in developed countries where public municipal treatment is not present, as well as consumer devices like pitcher-based products that improve the taste (palatability) of water. We call this category point of drinking or POD.

Innovation in the water purification landscape has been incremental. We define innovation as the useful embodiment of an idea in the marketplace, and so such a definition should not be confused with a new technology or other such inventions; innovation in our definition is raised on the highest pedestal above invention and science, as only an innovation has proceeded all the way to the marketplace through commercialization. Thus, despite numerous research results on forward-looking concepts, progress in big water and the last mile of water or POD has only been incremental. This incremental progress has essentially been the insertion of slightly different technologies into existing business models. The production of fundamental or transformative innovation requires the simultaneous exploration of new market applications, business models, and technology. TWI was formed with this challenge in mind: to create innovations by first establishing a company that employs a unique process to foster such innovation.

Since POD purification is currently of the most interest to TWI, we note that many new products are released every year in this sector, and most involve the same purification technology: activated carbon. Activated carbon has some purification capability (mostly in removing limited amounts of organic contaminants); however, the main feature is catalytic reduction of residual chlorine disinfectants and thus changing the taste of water that was purified in centralized water plants using batch chlorine processes. Although of value in developed countries, such products do not remove the wide range of potential contaminants in water, such as arsenic, fluoride, and pathogens. Most of these new products use new marketing or design techniques or a combination of both to garner a segment of the "taste-oriented" marketplace, particularly in developed countries. Such products also may benefit from the misconception that they actually purify water and remove additional contaminants.

If we look for existing products and technology that can actually purify water in more comprehensive ways at the POD, products that utilize reverse osmosis (RO) come to the forefront. Essentially the universal solution from a purification perspective, RO drives conventional osmosis that occurs across a membrane in the opposite direction using energy (through hydraulic pressure). Because there is a physical barrier (and not just pores), very little can pass through the barrier except water, and, therefore, almost everything is removed from the water.

However, there are many limitations to RO. While RO removes many harmful contaminants such as lead, it also removes beneficial minerals such as calcium and magnesium. The process is maintenance intensive since membranes are sensitive to fouling and scaling, frequently requiring pretreatment to improve lifetimes. In addition, they require significant energy to drive the process. Because of the previous considerations, reverse osmosis systems are the most expensive solutions. Importantly, since there is a physical barrier, purification is slow and a large volume of valuable water is rejected (80%–90% is common for home RO systems). Also note that the rejected water stream has a higher concentration of the contaminant, and depending on the situation, its runoff may need to be treated or contained to prevent ecological damage.

There are various materials processes that can add efficacy to activated carbon technology but, in general, will not produce a universal product that can compete with RO. Thus, very few products have appeared in this more advanced space, since such products typically would be more expensive than simple activated carbon solutions, and such products often only perform targeted purification of certain specific substances. Consumers in developed countries do not want to pay an increased price for partial efficacy that they do not need, and in developing countries, such solutions are risky because the consumer would first need to understand exactly what the technology problem in the water is and what limited solution such technology is trying to address. Targeted solutions are most effective when products are developed that address contaminants present in local water conditions.

Sources of new technology (not innovation) are fairly abundant but suffer from the applied context required for innovation. The wide recognition that water purification is an important worldwide problem, and a growing one, results in technology research in institutes and universities. However, since those organizations often do not participate in complete market immersion, and do not have to contend with other business models that ensure sustainability as well as innovation, the output of these endeavors are often fragments of technology that may or may not be classified as water purification innovation. Nongovernmental organizations (NGOs) have traditionally realized part of this problem and have attempted to address the problem of context by immersing themselves, generally, into a specific marketplace. However, such a context is not complete for innovation, as the organization itself does not demand a profitable business model that can result in a self-sustaining innovation. Many NGOs have realized this and attempt to partner with for-profit companies; however, extending the innovation process across organizations is difficult.

Migration of products from developed communities to developing communities does not work well since the innovations were developed in a different context. For example, RO systems are a universal technological solution but are too expensive and waste too much water for many developing regions in the world. Activated carbon systems have a lower price point and therefore could be propagated more widely; however, the technology is insufficient to meet the more complex purification demands of many other regions.

With this landscape of the relevant water industry and its technological base, we will now describe the evolution of TWI's unique innovation process and resulting penetration of previously unattainable market space.

21.2 TWI Formation and Early Innovation Iteration

In this section, we describe how TWI has purposely kept uncertainty, or options, in market application, business model, and technology open as it formulated its iterative innovation process.

The iterative innovation process requires that the technology, market application, business model, and other implementation factors interact to define the potential innovation. The need for these elements of innovation to interact is obvious upon inspection. A particular market application, for example, often requires a composite of technologies to deliver value to that market. Additionally, the technologies must fit within practical business constraints, i.e., cost, methods of production and delivery, etc. By encouraging all factors to interact instead of preselecting a technology or exact market application, the chance of converging on new innovations is much greater.

When TWI first formed in 2006, the research environment was speculating that "nanotechnology" would revolutionize many fields, including water purification. Instead of choosing a specific technology and "bringing it to market" as in a typical inefficient linear innovation model, TWI stepped back and decided to form a database of the basic materials and processes that can comprehensively remove various impurities and organisms from water, as well as dissecting current water purification technologies. This landscape of materials, processes, and devices allows TWI to rationally investigate the benefits of various technologies for specific market applications and business factors. During this process, TWI was alerted by various organizations as to what technologies were more easily applied, and which were difficult. Such information was not always accurate, and crosschecking with scientific literature and other sources was critical to understand the database objectively.

In parallel to developing an inclusive and initially unlimited set of the technology space, TWI started to investigate the various market applications for POD solutions. A new company, especially a start-up company, has few resources and investigating all market applications is a large undertaking. We therefore used the developing technology space to explore specific market application possibilities. Because of our connections that allowed us to maximize limited resources, we began to focus on Mexico. Mexico is geographically close to the United States, and both perception and initial contacts with the Mexican government confirmed that there was a significant need for POD water market applications. Although the US market was the closest market geographically, we initially speculated that taste was the major factor in most US areas, and that the Mexican consumer needed a much more comprehensive solution.

We also had to build relationships early that would help us to overcome the "foreign factor." Initially, we confronted a great deal of distrust and lack of cooperation from local officials in sharing even basic information. We were fortunate to form an excellent working relationship with Mexico's National Water Commission, CONAGUA, and its leaders. CONAGUA officials on all levels came to meetings with us and generally informed local officials that we had a beneficial purpose and mission.

TWI became aware early that the business context for technology development, especially for Mexican markets, was missing from many research programs and new products that were released. To address this issue, TWI engaged the "Base of the Pyramid" (BoP) Protocol (Prahalad and Hart, 2002). The BoP protocol was an experimental concept developed by academia and stipulates that innovation requires cocreation with the consumers in relatively economically disadvantaged communities. Thus, TWI formed teams that would coexist in Mexican communities with severe water quality problems. The BoP is a specific form of the previously described innovation process, in which the potential market applications must interact with local business factors and potential technologies to arrive at relevant innovations. A limiting factor, however, that BoP has on the innovation process is that it confines business factors (implementation) to individuals in local communities and often does not foster more universal solutions.

As TWI continued beyond this initial start-up period in which the market application was focused on water problems in Mexican communities, the technology database was expanding; the BoP was implemented for incorporation of local business factors; and we were ready to refine these areas to let them interact to converge on potential innovations.

21.3 Three D's

As TWI proceeded to refine the elements of our iterative innovation process, this process eventually evolved to a "Diagnose, Develop, Deploy" model, which we refer to as TWI's "3-D" model.

21.3.1 Market Application

TWI initially established a BoP team on the ground in Mexico. The location was determined by communication with the federal and local water institutions in Mexico. After

FIGURE 21.1
TWI began by embedding personnel into communities that required water purification for improved health.

many iterations, conversations, and visitations, TWI determined that the central valley of Mexico had severe water problems, and an archetype municipality was chosen for deploying the TWI BoP team. A local home was rented, which would act as a base for the BoP protocol and community education as well as a laboratory for field testing potential products (Figure 21.1).

The TWI team was careful to interact with the community and not introduce preconceived ideas about what was potentially in the water, what technical solutions might be available, and how such a solution could be brought to that market. After building trust, it became clear that the women in the community cared deeply about family health in general, and healthy water was an important element of this concern. Local businesses would use local water, sometimes from the tap, to create fruit-based drinks that they sell locally as well. To obtain clean water, the consumer generally walked (sometimes long distances) to local water purification dealers that appeared to use more centralized RO systems to fill large "garrafones" or plastic bottles of water. These garrafones contain 5 gallons of water and besides the cost, the people had to carry the garrafones weighing between 40 and 50 pounds inconvenient distances to their homes. Closer examination showed that often, the consumer would resort to using tap water, often filling the garrafones, particularly later in the week after payday. We also learned that the tap water "tasted funny," and the consumer did not trust the purification of the water by the local municipality. The families in these communities also purchase a lot of soda. These beverages are safe; however, the mothers are aware of the negative consequences of the high sugar content on health, especially obesity. Obesity and its consequences are not surprisingly the most prevalent cause of death.

21.3.2 Technology

In parallel to the market application refinement and the development of appropriate business models, TWI began to narrow its technology choices. After all, even though the local community had perceptions of water problems, they did not know the exact technical problem(s). Thus, cocreation of innovation without the external interjection of real technological understanding could create products that induce feelings of safety but actually could be harmful if the actual problems are not addressed.

An initial team obtained local water analyses from the municipality to understand what was technically wrong with the water. We chose Mexico with the idea that their problem was mostly pathogenic in origin, i.e., the classic "Montezuma's revenge." In several meetings with the local officials, it took some coaxing to get the water test results. We had to agree with both local and federal officials that we would keep all information confidential until we had their specific permission to disclose what the real issues were.

We were shocked to learn that arsenic was at such high levels. Thus, despite the perceived problem that pathogens were the real issue, the actual problem was levels of arsenic far above the recommended exposure of standards in Mexico (25 parts per billion [ppb] limit) and the United States (the US/WHO standard being 10 ppb). We set out to do our own analysis on local water by taking our own samples and sending the samples both to Mexican laboratories as well as laboratories in the United States. These tests eventually confirmed the high level of arsenic in the water, and also established which laboratories were most accurate and stable in their measurements.

We also visited the water treatment utility that the federal authorities had mandated a few years earlier. The municipalities used wells for the source of water, and injected chlorine into the water at the source via tanks of chlorine. Our tests confirmed that this national effort largely took care of the pathogen problem, killing most bacteria and viruses. However, the element-rich water remained, with arsenic being the most serious problem (arsenic consumption has a direct and strong correlation with increased rates of cancer, diabetes, brain function, and other health problems). As we developed the technical database, we were told by many sources that arsenic was a difficult problem to solve, and to do it economically was even more difficult. However, the database provided several possibilities, and we began to narrow down the choices for experimentation. Iterating with our developing market application knowledge, we concentrated on adsorption-based systems because of the potential for low cost and 100% water yield (all the water being purified). Furthermore, such systems are compatible with the low water pressure situations frequently encountered in Mexico, as well as rooftop cisterns that supply water during water outages (the water from the well dries up at various times) requiring people to rely on stored water.

To foster iteration and adaptation, we developed our alpha–beta–gamma prototype process. In the alpha prototype (Figure 21.2), we set up materials experiments using test columns in TWI's Cambridge, Massachusetts, laboratory. After comparison of multiple potential materials technologies for the filter media, we created more substantial test purifiers using standard water purification media bed configurations and sent them to our rented home/laboratory in Mexico (these were termed the beta prototypes). Finally, after testing for long periods in the real environment and several beta prototype iterations, we built gamma prototypes that were closer to the actual commercial form of the product.

The form of the gamma prototype (Figure 21.3) relied on the factors from understanding the market application. We learned that because of their pride in creating a healthy water environment, the Mexican people generally did not want to hide the purification system below the sink; they wanted it to be visible to visitors. They also liked to uniquely decorate their units, and wanted the unit to resemble a water cooler, as well as having hot- and cold-water dispensing capability. We thus combined these desires with the actual purification media and TWI's first product was born (TWI 1.0, Figure 21.4). It is a multistage, gravity-fed system that removes arsenic, pathogens, and improves taste in a POD device.

FIGURE 21.2
Test-column filtration experiments (alpha prototypes) used to test media in early water purification devices.

FIGURE 21.3
(See color insert.) Early gamma prototype deployed in Mexico. The technology was not yet integrated into the final product in order to test various aspects of the prototype.

FIGURE 21.4
(See color insert.) TWI's first product from the 3-D process using the BOP.

21.3.3 Business Factors

In parallel, the TWI BoP team investigated how this unit could be brought to the market-place with a profit. Profit allows self-sustaining growth of the TWI business to spread the innovation and thereby solve water purification problems on a large scale.

The total cost of the unit as conceived by iterating between the other elements of innovation resulted in a "Xerox" business model. When Xerox created the first copier, the machine was a breakthrough; however, the initial price of the machine was too great for many businesses. Thus, Xerox created a novel business model in which it rented the machines to companies, sometimes distributing the cost on a per-copy basis. TWI followed a similar model by installing the unit in a home and then charging for water use. The risks of this model were collecting continuous revenue and needing TWI to finance the units.

An additional feature of the model is a common one stipulated by the BoP protocol, in which local entrepreneurs would sell the units, a variant of the franchise model except there is no franchise fee.

21.3.4 Converging on 3-D

Once TWI established the database, and as it accumulated more knowledge about POD markets and business factors, we do not need to reinvestigate those areas. However, because different regions have different water quality problems, TWI does need to keep the options open regarding the set of technologies and what the detailed market application needs as it solves a new water problem in a different region. This realization is the key to solving water problems worldwide; it is the reason TWI is a unique worldwide pure-play

POD water purification company. Water contamination is a global issue requiring local solutions.

When solving the water purification problem in a new region, TWI must first *diagnose* the specific water problem to narrow the technology options and *develop* the most cost-effective option. Finally, it must *deploy* the solution with an appropriate business model for the given innovation.

TWI achieves scale because the specific solutions developed can be deployed in similar global regions without substantial redevelopment. For example, the central valley of Mexico has an arsenic problem because wells are drilled in element-rich earth. There are similar regions and water conditions in Argentina, China, Chile, Peru, New Hampshire, New Mexico, etc.

21.3.5 Limitations for TWI 1.0

The demand for the TWI 1.0 was ensured owing to the BoP process. Many communities began to hear of the unit, and our local entrepreneurs began to sell units. Collection efficiency surprised everyone, achieving >50%, which was better than local municipalities achieved for water fee collection. It was good, but not good enough to sustain our efforts and business model.

One problem with the Xerox model is that TWI must finance the installation of the units; break-even occurs much later as does profitability. Since TWI was a start-up company and undertook venture financing, an expensive form of capital, TWI searched for larger partners that could finance the expansion of the business. However, from microfinance organizations to development banks, no partner was willing to participate. In addition, the organic franchise model resulted in a slow rate of sales. Combined, the cash flow problem for TWI increased with time.

With this information incorporated into the innovation process, TWI realized it needed to change the implementation element of innovation, and pursued ideas to distribute the product more quickly to consumers. Thus, TWI let the technology and market application (in-home consumer) remain fixed while modifying the implementation (distribution). Sales did increase, although further analysis showed that the rate of solving home purification problems and the cash flow issue were still not optimal.

21.3.6 Lessons from BoP

TWI's innovation process also revealed important conclusions regarding the BoP protocol. The protocol is consistent with the start to a good innovation process to gain initial market application insight on the potential range of technologies required, and to determine business factors and models *related to the individual consumer*. Unfortunately, business models related to the BoP consumer all too often require long-term financing, as the consumer is cash-poor. Even though they are willing to spend a larger fraction of their income on water purification, profits can only be realized over long periods of time with tremendous commitment of financing. Thus, such models seem more appropriate for larger organizations, such as large multinational corporations, that have greater access to low-cost capital.

However, such large corporations are notoriously un-innovative and since most are public, the demand for short-term financial return outweighs the humanitarian impact of innovation. The BoP protocol may in fact be applicable to only a small set of companies since there is little intersection between innovative small organizations in which capital costs are the limiting factor, and large companies where short-term profitability and lack of innovation are the limiting factors.

In hindsight, the BoP protocol has been justified by the final result if one hypothetically multiplies a billion people by the very small amount of capital available from each consumer, resulting in a very large potential number. Such a perspective ignores the kinetic pathways required to get to that end point, in which large sums of capital must be applied for long periods. It is arguable that the financial return available from other investment opportunities substantially renders the BoP protocol a theoretical exercise as opposed to a practical one.

21.4 Scaling with TWI 2.0

Although the TWI 1.0 product was a successful innovation in that it was established in the marketplace, the growth rate was not high enough to effect the desired change and be sustainable. TWI wanted to bring pathogen- and arsenic-free water to more people more rapidly, and to do this, more iteration with the business model, technology, and market application was required to arrive at a more efficient and thus self-sustaining deployment model.

The TWI 1.0 product built brand for TWI in Mexico, especially with local communities that TWI 1.0 served, as well as with the local municipality officials and the federal authorities (CONAGUA particularly) in Mexico. Because of this brand and these relationships, TWI was encouraged to continue engaging local municipalities in alternative business plans. After all, TWI had converged on the basic set of technologies required to solve the arsenic problem, and the municipalities theoretically had the resources to deploy the solutions faster.

As market application, technology, and business factors all interact, it is not a surprise that the new market (municipal solution to the arsenic problem) changed the technology and business model. TWI needed to make a significant change to adapt to these factors. A sustainable model dictated that we generate greater cash flow allowing us to serve more people. We consciously changed our goal from a "one-to-one" to a "one-to-many" model and determined to target the municipality itself as our primary customer. This required us to rethink all the factors, as we wished to forge a path never traversed—sell a municipality on a POD model—convince them that our approach was superior to the traditional central treatment, supplemented by bottled water, approach.

We determined that an innovative model for the municipality required a device that was effective from a cost and technology standpoint, and must be durable, requiring no replacement or maintenance over extended periods.

TWI took the municipal requirements back to the laboratory and developed a new solution built around the same core 1.0 technology. New clever engineering was required, and the result is the TWI 2.0 (Figure 21.5), which is our current municipal product. The result is an ambient temperature unit that will purify the water of arsenic and pathogens for 5 years at a price point that resulted in high demand for the product. The 2.0 cleans all the water (no discharge waste) and requires no electricity for filtration, with a high 1.5 liter per minute (LPM) flow rate at low applied pressures of 10 pounds per square inch. TWI utilizes a proprietary, nanostructured, high-capacity material to effectively mitigate elevated arsenic concentrations, resulting in long lifetimes. Figure 21.6 shows a plot of arsenic

FIGURE 21.5
TWI's municipal product, the TWI 2.0 series.

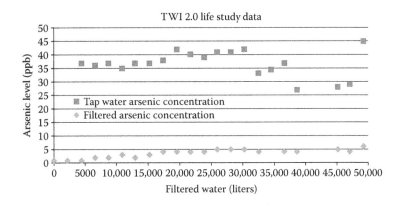

FIGURE 21.6
Plot of arsenic level in water, both unfiltered (from tap), and after passing through TWI's installed 2.0 product. The arsenic level is reduced to below the US/WHO standard of 10 ppb up to an incredible volume of 50,000 liters.

concentration in the tap water in one region in Mexico, and the arsenic concentration after TWI 2.0 as a function of the volume of water passing through the filter at a flow rate of 1.5 LPM. These levels of arsenic in the water are incredibly high, and TWI's product reduces the level to below the US/WHO limit of 10 ppb up to an incredible 50,000 liters at high flow rates for routine, everyday use.

The new business model is a direct sale to the municipality on behalf of the Mexican people. We had to make a bet with "all of our chips." We engaged in a bid process that took us over a year; we showed that our POD approach saved the municipality 70% of its costs over a 5-year period. We confronted a lot of business resistance from the bottled water owners; however, we finally prevailed when we won the first municipal contract ever in the world to install POD devices in homes. This took place in Durango, Mexico, thanks to

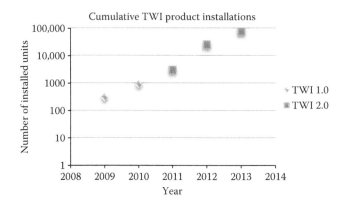

FIGURE 21.7
Cumulative number of deployed POD water purification devices by TWI (2013 estimated).

the courage of its governor (Herrera), the local water leaders, and CONAGUA. By the end of 2013, TWI estimates that it will have deployed 100,000 units, thus rapidly changing the future health of much more than a half million people.

We have currently established programs in six Mexican states with an equal number of additional states targeted over the next 12 months. We are also employing our unique iterative innovation process in other Latin countries and have just commenced in Africa.

Figure 21.7 shows the rate of deployment of TWI 1.0 and 2.0 units. Now TWI is on a foundation that allows us to continue to innovate and grow our worldwide pure-play water company.

21.5 TWI: Continuous Innovation

Since TWI has implemented a philosophy of continuous innovation from day one, and embedded that innovation process into the company operations, we have been able to utilize our 3-D process in more communities, allowing us to extend the TWI 2.0x series and develop a new TWI 3.0 product in the future.

The TWI 2.0x products include derivatives of TWI 2.0 that are a direct result of the continued 3-D process. In some regions, very high levels of arsenic are periodically diagnosed, and therefore we have developed a variant that has an even higher capacity to remove arsenic. Also, we have diagnosed regions with high levels of fluoride. Fluoride is particularly challenging because of the high concentrations requiring removal (parts per million) levels for fluoride vs. ppb levels for arsenic that challenge media adsorption capacities. Nonetheless, we have created fluoride products, utilizing proprietary techniques developed by TWI, that can last years in the field and remain economically viable for the municipality and the product user. As part of this process, TWI utilizes proprietary, nanostructured, high-capacity material to address these elevated contamination levels.

Looking forward, TWI is developing a new product series, 3.0, which combines our market application, technology, and business factor knowledge in new ways. This product will be a very portable product with filter inserts to address the particular water problems within a region. Thus, we have converged on another innovation that can span the important space between the age-old activated carbon taste product and the RO system that we mentioned at the beginning of this chapter.

Looking forward from a systematic perspective, the revolution in POD devices will eventually stimulate a rethinking of how entire water purification and distribution systems are designed. From a reliability and cost perspective, the rapid growth in developing (and developed) nations will demand a combination of central water distribution with local POD purification solutions in cities and primary reliance on POD solutions in rural areas. We hope to be at the vanguard of this paradigm shift as a global leader of decentralized water.

21.6 Conclusion

Entrenched business models and a focus on incremental technology instead of innovation processes have limited innovation in the world's third largest industry: water. TWI has embedded its unique innovation process into every aspect of its operations, resulting today in the utilization of its 3-D process (diagnose, develop, deploy) to propel true innovation and a distributive system of clean water solutions. By the end of 2013, TWI will have deployed >100,000 POD purification devices in the central valley of Mexico alone to improve taste and, importantly, remove pathogens and high levels of arsenic serving more than a half million people to live a healthier life. TWI has developed higher-capacity models of these gravity-fed and economically viable units, as well as having developed versions for removing high levels of fluoride.

This coming year, TWI will release a portable water purification system that can be tuned to the local water problem, ultimately addressing the gap in the marketplace between simple activated carbon taste filters and more expensive RO systems. We contend that just as the cell phone enabled billions to leapfrog the telecommunications equipment, disruptive POD devices will enable billions to complement and leapfrog central treatment and bottled water. We hope to serve millions of people over the next few years and to aggressively promote the paradigm shift in water innovation.

Reference

C.K. Prahalad and Stuart L. Hart, "The fortune at the bottom of the pyramid," *Strategy+Business* 26, 2002.

22

WaterHealth International: Decentralized Systems Provide Sustainable Drinking Water

Mahendra K. Misra and Sanjay Bhatnagar

WaterHealth International Inc., Irvine, California, USA

CONTENTS

22.1 Introduction

A World Health Organization (WHO)/United Nations Children's Fund report estimates that more than 800 million people in the world lack access to clean drinking water. Almost all of these people reside in developing countries. Nonavailability and inaccessibility to clean water is affecting populations in the developing world, particularly those who are very vulnerable to consuming contaminated water such as the elderly and children. Nearly 2 million children die of waterborne diseases each year—that is roughly one child every 15 s. A less vulnerable population of working adults living in regions with poor access to clean water is also affected by consuming unsafe water. This population group mostly earns daily wages, and any missed day from work due to waterborne illness affects their income. The other intangible effects of loss in quality of life as a result of having to fetch water from distant sources cannot be easily quantified. The fetching is mainly done by women in the third world, and it has been noted with amusement that this gives them cherished time out of sight of their spouses, a cultural issue observed by those drilling nearby wells that eliminate the long walk!

Up to now, the response to the water crisis has been focused solely on water accessibility, not on the critical issues of quality or sustainability, much less palatability. Government agencies and nongovernment organizations (NGOs) have been putting efforts in digging bore wells in communities to provide access to water. Generally, water from deeper bore wells do not contain pathogens and may contain higher levels of dissolved salts but are safe to drink as moderately higher levels of salts may affect the taste but not the health. Shallower wells, particularly the open ones, may contain pathogens. Over the course of time, if the deeper bore wells are not constructed or maintained properly, the water in the well can become contaminated. The contamination may occur from surface water seeping into ill-designed or ill-maintained wells, bringing with it harmful microbiological organisms from the fecal matter of human beings or livestocks or other surface contaminants such as chemicals used in agriculture. Depending on the geographical location, an additional risk with deeper wells may be the presence of other impurities such as higher levels of arsenic or fluorides that may make the water unsafe to drink. Providing access to water only solves part of the problem; it does not assure quality or sustainability. Bore wells can become contaminated or dry up in the summer when water is most needed, while trucking in water from nonreliable sources is neither safe nor economically sustainable. Community hand pumps contain movable parts that wear and break down with usage. They require ongoing service and maintenance to keep them running, and funds may not be available at all times to pay for their repair and maintenance.

Every year millions of people, mainly in developing countries, most of them children, die of diseases like cholera that are associated with inadequate water supply, sanitation, and hygiene. For example, in 1993, a mutant strain of cholera broke out in parts of India, Bangladesh, and Thailand, and thousands perished in this epidemic. One of the researchers at the Lawrence Berkeley National Laboratory (LBNL) witnessed the suffering caused by this epidemic and was determined to develop a water disinfection system suitable for this environment to address the problem. The result was a groundbreaking technology that used UV light to inactivate harmful pathogens in contaminated water and did not require much power to run it. WaterHealth licensed the ultraviolet (UV) technology from LBNL in the late 1990s, and since then has advanced the technology and developed a scalable and sustainable model for building WaterHealth Centers (WHCs) globally. The company has centers in India, Africa, and Southeast Asia, with plans to continue building facilities wherever there is a need.

To address the drinking water needs, WaterHealth believes that a safe, clean, sustainable, and affordable water supply using the locally available sources is the best solution. Using off-the-shelf proven technologies, WaterHealth has immediately deployable designs that can purify water from any available source to exceed drinking water guidelines set by the WHO. WaterHealth's strategy is to combine the use of such decentralized purification centers in partnership with local communities to create a scalable and sustainable solution. Communities with a WHC have seen a significant reduction in the number of common illnesses like cholera, dysentery, diarrhea, and other waterborne diseases. These communities not only realize immediate benefits from improved health and well-being brought by clean water, but they also benefit economically by sharing in a portion of a center's net earnings and ultimately by vesting in full ownership of that center. This unique win–win relationship is extrapolated further in subsequent sections.

In its efforts to provide purified drinking water, WaterHealth is constantly looking for advancement in purification technologies, particularly when these advances result in more effective treatment and lowering of upfront installation and ongoing operational costs. Advancement in nanotechnology as applied to water treatment promises to achieve those objectives. It will help to examine WaterHealth's business model and the economic constraints and challenging conditions it operates under to help define the framework for suitability and adaptability of emerging nanotechnologies.

22.2 WaterHealth Solution

22.2.1 WaterHealth Centers

WaterHealth firmly believes that the solution to meeting the challenge of providing safe potable water to communities is local. Transporting water from far-off places is expensive as it involves building of infrastructure and consumes significant amount of energy. Resources for building of infrastructure or energy are not readily available in developing countries. To provide clean water at an affordable price, locally available water needs to be processed and purified using state-of-the-art technologies integrated in a cost-effective way. Local water poses its own challenge with regard to contamination that may be present, and purification systems need to be designed to address the contaminant in a cost-effective manner.

A WHC is a decentralized water treatment facility that includes a water treatment system housed inside a contemporary-looking structure (see Figure 22.1). The decentralized WHCs allow underserved communities rapid access to safe water at an affordable cost, helping solve the global challenge of waterborne diseases one community at a time. The structures are modular, consisting of steel frame and synthetic panels that can be transported on a flat bed truck with ease. The water purification system comes preassembled

 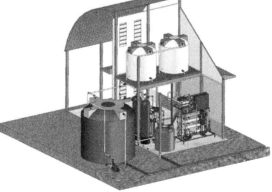

FIGURE 22.1
(See color insert.) Rendering of a WHC with a cutout view showing equipment installation.

on a skid that can also be easily transported on the same flatbed truck and requires little external pipe work. It can take less than a month to build a WHC in a community in need.

Plant capacity is sized according to the population of the community. Using the recommended 4 liters of water per person (capita) per day (LCD), and an average of five persons per household in developing countries, WHCs are sized to provide 20 liters of water per household per day. Since the focus is primarily on supply of potable water, WaterHealth International (WHI) systems are small, in contrast to water treatment systems designed to provide water to communities in developed countries. In the developed countries, all water supplied to households is treated to drinking water standards, whereas the percentage of water actually consumed for drinking constitutes only 1%–2% of the total usage. A typical point-of-entry system at a throughput of 4 gallons per minute (GPM; 15 liters per minute) for a household in a developed country can provide sufficient drinking water for a community of 5000 people. WaterHealth installations start at this throughput and typically go up to 12 GPM for a community of 15,000 people. For larger communities, a larger system can be installed, but the preference is to install more systems at several locations within the community to facilitate easy access for households. Facility wise, because WHI's systems are based on a modular design, they can be scaled to fit different needs. However, the company is predominantly deploying two versions of its community systems, one having a throughput of 4 GPM and the other 12 GPM. The standard WHC has a footprint of approximately 20 ft × 30 ft (approximately 55 m^2), plus a landscaped area for social use.

WaterHealth's business focus is on decentralized systems that often prove more efficient and more sustainable than large municipal systems among rural/peri-urban communities. Thus, the company generally installs multiple systems to serve communities of more than 10,000 people, resulting in facilities that are relatively closer to the homes of users. Decentralized community water systems are a right balance between a large metropolitan water system with piped distribution network and each household's use of a point-of-use (POU) system for treating drinking water. Large centralized treatment plants keep the cost of water treatment low, are well maintained, and adequately monitored to assure good water quality leaving the treatment facilities; however, they fail to guarantee good water quality reaching the homes because of excessive reliance on the good condition of the distribution network, the so-called last mile. Centralized treatment plants require extensive upfront capital not only to build the water treatment plant but also to build the distribution network. A common argument against the construction of a centralized treatment plant is that only 1%–2% of the water distributed to households is consumed for drinking; the rest of the water is used for landscaping or other utility purposes such as washing or bathing. All water leaving the water treatment facility has to be treated to drinking water standards. If not maintained well, the pipe network can itself become a source of contamination. Distributed water generally has chlorine added to provide residual disinfection to contain any secondary contamination arising from piped network. Added chlorine may affect the water's taste and possibly cause the formation of by-products that may be harmful to health. On the other hand, POU systems, as the name implies, ensure that drinking water is treated right before its use, and in concept, provide safe purified water at all times. The weak links in the implementation of POU systems are the users themselves. In the POU systems, there is a considerable reliance on their owners to carry out the recommended maintenance at periodic intervals. These may include change of filters, replacement of adsorbent media, or change of germicidal UV lamps. When not properly maintained, POU systems can actually worsen the water quality, examples being a build-up of a bacterial slime layer on old filters or activated carbon becoming a breeding ground for bacteria. When coupled with an ineffective UV lamp that has not been replaced, the water quality coming out of the POU unit is compromised and not safe to drink. WHCs provide the optimal balance—treating water primarily for drinking purpose as in the case of POU units while servicing, maintaining, and monitoring water quality on periodic intervals using qualified personnel as in the case of a large centralized system. Cost wise, WHCs are more effective than a centralized treatment facility or POU systems. POU systems average US$ 80 per unit; for a community with a population of 20,000 people (average of 4000 households), it will cost around US$ 320,000 to furnish each household with a POU unit. For one-third of that cost, a WHC with distribution kiosks can easily be installed serving the entire community.

22.2.2 Services Offered by WaterHealth

WaterHealth believes that the world's underserved people deserve products and services that are better than "good enough" and comparable to those available to more affluent communities. Through innovations, WaterHealth is able to offer world-class quality at costs previously thought to be unachievable. WaterHealth retains control of maintenance and operations of its centers to ensure quality. Local workers are hired from the communities it serves, and are trained to carry out routine maintenance and operations. Services offered by WaterHealth include

- Site assessment and preparation
- Conveyance of raw water from source to treatment facility
- Turnkey assembly, installation, and validation of water treatment equipment and civil works
- Building a modern, aesthetically designed, and landscaped civil structure that also serves as gathering place for the community
- Provision of specially designed water containers that minimize the potential for recontamination during customer use and storage
- Extended maintenance contracts to keep high quality and operating standards
- Recruitment, hiring, and training of local residents to operate facilities
- Overall management of WHCs
- Ongoing education programs on health and hygiene

22.2.2.1 Financing

WaterHealth works closely with the community securing the financing for WHC installations. It has a significant line of credit and loan guarantees preestablished with major banks worldwide. The first priority is for the community through its own resources, government leadership, or through private sponsors to provide the funds. To facilitate the purchase, WaterHealth provides an innovative financing program in most cases that meets qualification criteria, such as the size of the community, potential user interest, and willingness to pay. In such cases, the village leadership or other sponsors provide a down payment for the facility; a significant portion of the balance is then financed. The collection of user fees allows the repayment of financing costs over time, after which the facilities become income-generating assets for the community. The term of the financing will vary depending on the size and scope of the project, as well as the particular needs and resources of a community. WaterHealth has offered as long as 8-year financing in India.

22.2.2.2 Increase in Accessibility

Some families in a community prefer to have the water delivered to their homes for an extra amount. This opened up opportunities for local entrepreneurs to make some extra income by picking up water from WHCs and delivering them to the households. WHI has formalized this method of distributing water as it recognizes that it is an important element to conducting its business.

A 20-liter container of water is fairly heavy, and it is not easy to carry it on foot beyond a few hundred meters. In India, it is common for some people to have a mode of transport, primarily a bicycle or a motorized two-wheeler. They are able to come to the WHC from further distances to fetch water. However, overall, people who have to walk find it convenient if the distance is only a few hundred meters. It is inconvenient for households that are further away to come to the WHC. Those households that can afford the extra charges prefer to have the water delivered. WaterHealth is constantly innovating on how to provide greater accessibility to clean water to satellite communities that may be at a considerable distance from the WHCs. WaterHealth has addressed this in two ways that are proving to be extremely practical. These solutions are complementary and can be implemented as one or both.

In regions in West Africa, WaterHealth has installed what it terms as "vantage points" that serve as satellite distribution centers in neighboring communities around a primary WHC. Purified water is pumped to storage tanks installed in satellite distribution centers in neighboring communities that may be 0.5–1 km away from the primary community. In this way, convenient access to water collection points within walking distance is provided to more households. These vantage points may number from 1 to 5 depending on the size and area of the overall community.

Another way WaterHealth is increasing the access to water is by providing delivery service of prefilled containers to households that may be outside the convenient walking distance from the WHC. The delivery service provider, an independent entrepreneur, collects several containers of water from WHC and delivers them to the household for an extra amount. The delivery service provider typically uses a small truck (see Figure 22.2) that can navigate through the narrow, mostly congested, streets and can carry 60–70 standard

FIGURE 22.2
Delivery truck used by local entrepreneurs to deliver water to households.

20-liter containers. The extra amount charged by the DSP covers his/her expense to fill and transport water containers and provides an income to subsist on.

22.2.2.3 Quality

In its designs, WaterHealth uses components bought from world-class, name-brand manufacturers who pay considerable attention to the quality of their products. These components may cost more upfront but, in the long run, end up costing less over the lifetime of the equipment because they are well tested, more rugged, and tend to break down less frequently compared with their low-cost counterparts. Generally, local organizations, primarily government organizations, and benefactors that are funding the installation would like to see their money spent on buying more systems, and the lower-cost systems supplied by local integrators using non-name brand parts appear more lucrative. These organizations are either not aware of or do not pay attention to the subsequent repair and maintenance costs that can add up significantly over the lifetime of the equipment. WaterHealth has embarked on a campaign to not only educate its users but also the customers who are buying or funding the installation of WHCs.

In operating the plant, WaterHealth pays great attention to the quality of treated water, making sure that the water is safe for drinking at all times. Plants are run by highly trained operators and are serviced by skilled technicians. Periodic servicing and maintenance are carried out, and as part of this schedule, all water treatment equipment is cleaned and sanitized. WaterHealth has established its own water-testing laboratories at major service centers. Water quality is periodically monitored on site by quality technicians as well as in the laboratory.

While other players in this segment do not pay as much attention to quality, WaterHealth is able to incorporate it into its operations and service innovatively and at the same time is able to keep the cost of water affordable to the population it serves.

22.2.2.4 Education

WaterHealth has launched Dr. Water as its brand of water dispensed through WHCs. It is meant to establish a confidence with the user of guaranteed water quality. However, to promote the use of its brand, it requires proactive education among the people in the community on the benefits associated with the use of purified water, with Dr. Water being the trusted brand for purity. Education campaigns are held in several stages, and the primary objective is to communicate two key messages: (i) the health and hygiene benefits of using its safe water and (ii) the benefits of community participation in supporting the WHC in which it has an ownership interest. Educational outreach uses a multichannel approach including plant visits, workshops, seminars, signage, pamphlets, etc. WaterHealth understands that education campaigns particularly in remote areas are challenging and prefers to work with local NGOs who already have a presence in the community to spread the message on the health benefits of using clean water.

As part of its education campaign, WaterHealth tries to customize the message to the type of audience it is communicating with. For instance, when holding seminars, workshops, and plant visits for school children, the communication objective is to help children understand what may be present in contaminated water, how the contaminants cause diseases, and what they should be following as part of correct sanitary and hygienic practices. This is communicated in a simple and easy-to-understand method. Similarly, to older audiences, the same message is communicated in more detail using different means.

It always helps to enroll a trusted community figure such as a village elder or a health-care professional to reach out to the people of the community.

22.2.2.5 Employment

In addition to providing potable water, WaterHealth also generates employment for the local community. During the construction phase, local contractors and workers are generally hired to prepare the ground, drill a bore well, lay piping, and if required skills are available, to carry out other civil work to support the structure and the equipment installation. Operators for running the plant are recruited from the local community and trained. If delivering water to households is a viable business, local entrepreneurs are contracted to provide this service. In addition, there is a need for other support functions such as social education, maintenance, and revenue collection. Personnel to perform these functions are generally hired locally.

22.2.2.6 Challenges

In its efforts to provide safe drinking water, WHI is facing multiple challenges. For instance, in India, particularly in the rural areas, the concept of paying for purified water is unfamiliar. People living in these areas have access to community hand pumps, community wells, or nearby lakes and ponds where they can collect water for free. In other rural area, as part of the rural development scheme, federal or state government has installed public standpipes dispensing water that is minimally treated and disinfected. In these instances, water from a nearby river or lake is filtered through a coarse sand bed and disinfected with chlorine before being distributed to public standpipes. This water is available free of cost to the community. The problem with this water is that organics that may be present in the raw water are not removed, coarse filtration leaves finer suspended particles in the water, and chlorine addition is not controlled. Any taste and odor originating from the organics are often masked by the chlorine taste and smell. There is an added concern that chlorine may react with organic matter present to form what is termed as disinfection by-products that may be more harmful to health. People still prefer to collect water from these public taps because it is free. It will take a considerable effort on WHI's part to educate the community and help them realize the benefits of drinking clean water.

22.2.3 Partnership

On May 4, 2011, WaterHealth announced a strategic partnership with The Coca-Cola® Africa Foundation, Diageo plc (UK), and the International Finance Corporation (IFC) in forming a consortium, Safe Water for Africa (SWA), to provide sustainable access to safe drinking water in Africa. This partnership committed seed funding of US$ 6 million to provide safe drinking water to communities in Ghana, Nigeria, and Liberia, and has a goal of raising US$ 20 million to provide sustainable solutions to at least 2 million people in Africa.

The above is a good example of how WaterHealth works in close partnership with corporations who are committed to improving quality of life of people in underserved communities and realize their full potential by providing access to safe water, preventive health, education, and entrepreneurship. WaterHealth has a sustainable business model. Focusing on quality and sustainability, it installs, operates, and maintains WHCs that are small and decentralized, and charges a nominal user fee to cover its operational expenses. The WHCs provide quality drinking water to vulnerable communities for a long term of 15–20 years.

FIGURE 22.3
WHC funded by a corporate partner under construction in Ghana.

Several of the corporations and respected organizations such as the IFC have recognized the long-term sustainability of WHCs and prefer to partner with WaterHealth (Figure 22.3) as they know that their investment will create a long-lasting impact.

WaterHealth works closely with nonprofit organizations that are active in geographies where it is present. WaterHealth believes that locally active NGOs are the best partners to have because they have spent considerable effort to gain the trust of local communities, and they are intimately familiar with the local language and culture. After installing the first WHC through its subsidiary, WaterHealth Ghana, in Ghana, WaterHealth partnered with Safe Water Networks (SWN) to install five additional WHCs that were funded by SWN. Since then, as part of the SWA partnership, WaterHealth has installed close to 20 additional WHCs. One of the strengths of WaterHealth is that it believes in fostering a strong relationship with communities and works closely with them in establishing affordable usage fees for the water purification service.

22.2.4 Technology

Depending on the source of water, whether from a surface body or from a bore well, the water treatment method has to be custom designed to address the local contaminants present. Generally speaking, surface water sources have turbidity, organics, and pathogens as primary contaminants. Groundwater is generally clear, may or may not have microbiological contamination, but is generally accompanied by higher levels of dissolved solids that may include more toxic and widely regulated heavy and transition metals. Surface water may have taste, odor, and color issues emanating from the presence of organics in the water. In the groundwater, dissolved solids largely affect these attributes. WaterHealth has integrated widely used technologies in its design to treat water from both surface and ground sources.

WaterHealth has to deliberately keep its system installation and operation costs low, as it has to provide water to communities where people do not have disposable income. Unbiased by any particular technology, WHI identifies proven, cost-effective technologies and integrates them into its design in providing water at an affordable price.

Conceptually, WaterHealth design consists of six stages of water treatment. These are (i) coagulation/sedimentation, (ii) coarse filtration, (iii) organic removal, (iv) fine filtration, (v) reverse osmosis separation, and (vi) disinfection. These are shown schematically in Figure 22.4.

The first treatment stage of coagulation/sedimentation applies more to water drawn from surface bodies and is to clarify the water by adding coagulants and allowing time for the colloidal organic and inorganic particles to agglomerate and settle out of the water to form a bottom sludge layer. For the groundwater, it is just a holding tank that serves as a quiescent volume that allows larger sand particles drawn from a well to settle out. These tanks also serve the purpose of buffering the flow given that water-processing flows may not be the same as water drawn from the source. For surface water with a high level of turbidity, clarification achieved by coagulation/sedimentation can be significant with turbidity in the clarified water <5 NTU.

Coarse filtration is achieved by running the water through a bed of graded sand that removes particulates >10–20 micron in size. This step has the potential to drop the water turbidity down to <1 NTU. Filtration through a bed of activated carbon reduces levels of

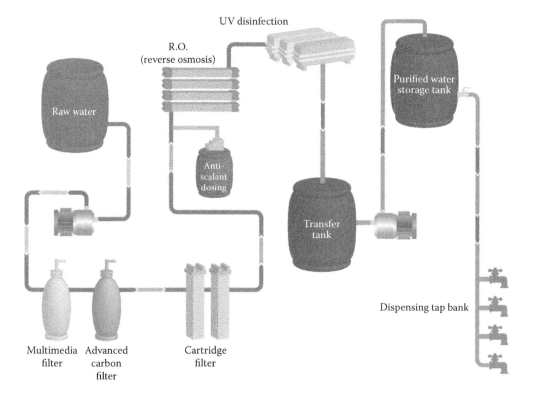

FIGURE 22.4
(See color insert.) Schematic of WHI treatment system.

dissolved organics that also addresses taste, odor, and color arising from these contaminants. Passage of water through fine 5- and 1-micron filters improves the clarity and drops the turbidity of water to <0.1 NTU.

Steps (i), (ii), and (iv) may also be replaced by treatment through an ultrafiltration (UF) membrane where reduction in turbidity to <0.1 NTU can be achieved with a single filtration step. Generally, UF membranes are tolerant to high levels of influent turbidity (200–300 NTU) and need to be maintained carefully to prolong their life. The trade-off with replacing steps (i), (ii), and (iv) with UF is generally the cost. For surface water use, a good UF design should include a controller for auto backwash as the fouling of the membrane is rapid and frequent back flow is required to dislodge the particles. The combined cost of a good-quality UF module and the controller exceeds the cost of equipment needed for steps (i), (ii), and (iv). The one-step filtration achieved by UF considerably reduces the equipment footprint and enables the system to be operated without the intervention of an operator while delivering consistent quality water downstream of the UF.

Treatment steps described thus far primarily affect the physical characteristics of the water and do not alter the water chemistry. Depending on whether higher than desired levels of dissolved solids are present in the water, a treatment step using reverse osmosis membranes is included where water is further purified to reduce the levels of dissolved salts. This step is energy intensive and is accompanied by discharge of wastewater that may contain high levels of salt. WaterHealth systems are carefully designed to balance equipment cost with energy cost and wastewater discharge.

As a final treatment step, water is disinfected using UV radiation before it is sent to overhead storage tanks. Disinfection using UV is preferred because it is cost-effective and does not involve handling or addition of chemicals such as sodium hypochlorite that are corrosive in nature and require special handling and storage. Addition of chemicals has the risk of secondary reaction with constituents in the water to produce products such as halomethanes or haloacetic acids when halogens are used as disinfectants. These by-products are carcinogenic and require that addition of halogens be tightly monitored and controlled. Disinfection by UV does not have this problem, as there is no health issue related with UV overdose of water. A disadvantage of UV disinfection is that it does not provide residual disinfection. Generally, at WaterHealth sites, any form of residual disinfection is unnecessary as the water collected by consumers is used up within a day or two. As a practice, WaterHealth offers an option to the users to purchase a 20-liter can designed to prevent secondary contamination and allow easy dispensation. The fill opening in the WHI container is narrow so that no vessel can be introduced and the container features a dispensing valve at the bottom. At places where likelihood of secondary contamination is high owing to the nature of open vessels brought by consumers, WaterHealth introduces trace amounts of chlorine to provide residual disinfection. At these sites, since the water is already disinfected, the amount of chlorine needed is extremely small. WaterHealth utilizes safer calcium hypochlorite where available and supervises the chlorine addition very closely.

In the countries where WaterHealth has operations, availability of continuous power is the biggest challenge. One way WaterHealth has overcome this challenge is to design overhead tanks where treated water is stored so that water can be dispensed to the consumers with the aid of gravity even when power is not available.

22.2.4.1 UV Waterworks™: WaterHealth's UV Disinfection Technology

WaterHealth is harnessing the strengths of UV Waterworks' design in providing safe water to communities. UV Waterworks consists of a part-cylindrical water flow channel with a UV lamp mounted over the water surface parallel to the axis of the flow channel. The UV lamp has a highly reflective cylindrical mirror mounted on top of it to reflect all UV radiation toward the water. This is illustrated in Figure 22.5. The inlet of the flow zone is designed with flow distributor and equalizer so that the flow velocities across the flow channel are very uniform. The combination of uniform flow and reflective optics imparts a very high degree of exposure to UV light. Conventional tube-in-tube UV disinfection devices commonly in use in water treatment system design provide the minimum 45 J/cm^2 required by ANSI/NSF 55. Compared with the commercial devices, UV Waterworks provides UV dosage of >120 J/cm^2.

UV Waterworks has been designed to circumvent other challenges commonly faced by conventional UV devices. The tube-in-tube design of commercial devices requires that the UV lamp be encased in a quartz cylindrical tube housed inside an external stainless-steel tube. The flow channel in these devices is provided by the annular space between the stainless-steel tube and the quartz sleeve. Occasionally, the quartz sleeve gets covered with a scale from the minerals present in water and the intensity of UV light transmitted through the quartz sleeve drops. The UV device has to be periodically dismantled to clean the quartz sleeve using chemicals and assembled again. During the assembly process, if the operator is not careful, the quartz sleeve can break easily and will then need replacement with a new sleeve.

The problems of scaling, requirement of cleaning chemicals, and the added expense of replacing the quartz sleeve have been eliminated in UV Waterworks.

UV Waterworks has been validated at independent laboratories in a number of countries. It was one of the first such devices to receive certification from the State of California as a Class A device, capable of disinfecting water that was not previously pretreated through another disinfection step. It also has been validated by laboratories in India, Mexico, the Philippines, and South Africa.

WaterHealth believes that everyone deserves both great-tasting and quality drinking water regardless of their geographical location or economic situation. As a result, WaterHealth incorporates, where relevant, reverse osmosis technology into its purification systems to improve the taste of the water, and provides WHO-quality potable water without relying on chlorine. We also use UF and arsenic, iron, manganese, and fluoride

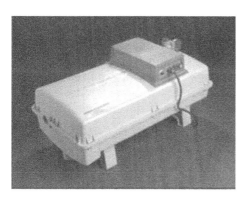
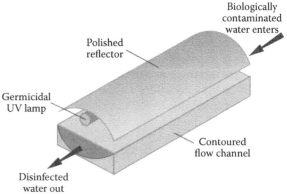

FIGURE 22.5
UV Waterworks and its working concept.

removal technologies as and when needed to provide the best-tasting (palatable) water possible.

WaterHealth's award-winning technology, UV Waterworks™ (UVW), is based on a novel approach to the use of UV light for the inactivation of microbial pathogens (Figure 22.6). WHI's technology has many advantages over preexisting technologies. First, UV treatment has the advantage that it is free of chemicals and potential carcinogenic and mutagenic disinfection by-products and environmental pollutants. Unlike conventional UV technology and other methods, UVW is designed to be highly energy efficient and work without the need for high pressure (in gravity feed mode), has superior throughput with high efficacy, and is sustainable from maintenance and cost perspectives even in remote and challenging environments with little technical infrastructure. In contrast with boiling water, it is more easily adopted and is not associated with the environmental degradation that can result from cutting down trees to provide firewood for boiling water.

In certain geographies, where likelihood of secondary contamination is present, either during water transportation or during handling, WaterHealth chlorinates the water to provide some residual disinfection. The chlorine addition, maintained at 0.2–0.5 ppm, is closely monitored and controlled to make sure it does not adversely impact the taste or generation of any harmful by-products.

Bacteria	Initial concentration (CFU/mL)	Posttreatment (CFU/mL)	Diseases
Escherichia coli	10^7	ND	Bloody diarrhea, abdominal cramps
Salmonella typhi	10^7	ND	Typhoid fever
Vibrio cholerae	10^7	ND	Cholera
Streptococcus faecalis	10^7	ND	Urinary tract infections, wound infections
Clostridium perfringens	10^7	ND	Acute food poisoning
Shigella dysenteriae	10^7	ND	Dysentery
Proteus vulgaris	10^7	ND	Urinary tract infections
Klebsiella aerogenes	10^7	ND	
Klebsiella terrigena	10^6	ND	
Enterobacter cloacae	10^7	ND	Pneumonia
Pseudomonas aeruginosa (immunotype IV)	10^7	ND	Dermatitis
Viruses	**Initial concentration (particles/mL)**	**Log reduction**	**Diseases**
Poliovirus	10^6	5.0	Polio
Rotavirus	10^6	5.0	Severe diarrhea
Protozoan cysts	**Initial concentration (organisms/L)**	**Log reduction**	**Diseases**
Cryptosporidium parvum	10^6	>5.1	Diarrhea

FIGURE 22.6
UVW effectiveness against commonly known pathogens.

22.2.4.2 Centralized Monitoring and Control

With the increasing number of installations, some of which are in remote areas that are hard to get to, WaterHealth realizes the significance of remote monitoring and control of its WHCs. It has embarked on using the available state-of-the-art technologies to manage the scale of operation and quality of service. Plants are being retrofitted or installed with sensors and inline analyzers to enable real-time parametric monitoring of the process and the water quality. Data collected are sent to a central server so that all plants can be monitored centrally, and if any problem is detected, a service crew can be dispatched in a timely manner. Taking advantage of the widespread cellular phone network in the countries WaterHealth operates in, data from WHCs are transmitted to the central server using cell phone modems. It is expected that in the long run, these remote monitoring systems will help WaterHealth to operate a global network of centers and keep the operational costs low.

22.3 Need for Low-Cost Technologies

For its raw water needs, WaterHealth draws water from whatever source is conveniently available locally. First installations of its WHCs concentrated in the coastal regions of the state of Andhra Pradesh in India and in southern regions of Ghana around the capital city of Accra. At these locations, surface water was plentiful and these sources were deemed to provide a perennial supply of water. It is well documented that with surface water bodies, the primary contaminants present in the water are (i) suspended solids affecting the clarity; (ii) organics such as tannins affecting the taste, odor, and color; and (iii) microbiological organisms that create a health hazard. Elevated levels of dissolved salts are generally not an issue for inland surface water sources. In the southern regions of Ghana, because the soil is rich in manganese ores, surface waters do get contaminated with elevated levels of iron and manganese (see Figures 22.7 and 22.8). It is interesting to note that 83% of the sites have water exceeding the allowable limit of 0.3 mg/L for iron and 16% of the sites exceed the WHO guideline of 0.4 mg/L for manganese. Levels of both iron and manganese show a seasonal variation. During wet seasons, particularly after a

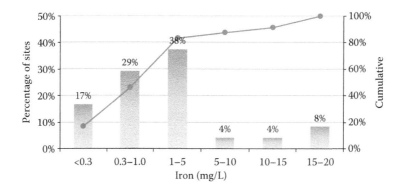

FIGURE 22.7
Distribution of iron levels in surface water sites in southern regions of Ghana.

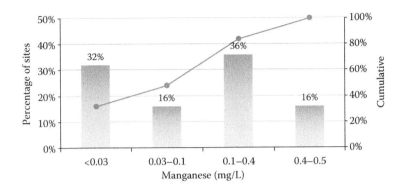

FIGURE 22.8
Distribution of manganese levels in surface water sites in southern regions of Ghana.

rain shower, runoffs bring increased amount of silt to the surface water bodies resulting in increased turbidity and levels of both iron and manganese. Water treatment systems installed at these locations need to be capable of handling the seasonal variation in iron and manganese levels, and support personnel, including operators, need to be trained to proactively identify the problem and adjust the process parameters to make sure that water dispensed from the center does not exceed the allowable limits for these contaminants. The job of monitoring and controlling the process would have been made easier if there was a low-cost technology available that was either not sensitive to wide fluctuations in levels of iron and manganese or incorporated a closed loop feedback mechanism using low-cost inline analyzers.

In the interior regions of India, groundwater provides a sustaining source of raw water. These are typically contaminated with dissolved salts, primarily bicarbonates of calcium and magnesium along with cations such as sodium and potassium and anions such as chlorides, sulfates, and nitrates. In the WaterHealth sites in India, elevated levels of fluorides are seen in some regions of Andhra Pradesh and Karnataka. In the latter state, elevated levels of nitrates, iron, and manganese and marginally elevated levels of arsenic are also common.

Generally, reverse osmosis is adapted as a universal means of treating groundwater for higher levels of dissolved salts. In WaterHealth's experience, this technique also provides a cost-effective method for reducing fluorides and marginally higher levels of arsenic instead of other adsorbent-based methods such as use of activated alumina. Compared with the use of adsorbent material that needs to be regenerated using corrosive chemicals, reverse osmosis is easier to operate and does not require frequent monitoring to look for fluoride breakthrough. Reverse osmosis is an energy-intensive process and involves discharge of significant amounts of water as a waste stream. While designing the system for achieving the right balance between cost of equipment and energy use, overall water recovery may need to be kept in the 75%–80% range, which still results in 20%–25% of the water discharged as a waste stream.

For situations where water does not have high levels of dissolved salts but has elevated levels of arsenic or fluoride, a commonly used option is to use an adsorbent such as activated alumina for arsenic or fluoride and ferric oxide–based adsorbent for arsenic. The problem with these adsorbents is that once they get depleted, they have to be sent to landfills for disposal. Even though manufacturers make all efforts to make sure that the spent adsorbents pass the regulatory leaching requirement, conditions at landfills may be such that these contaminants can leach out and find their way back into the groundwater, contaminating the local aquifers. This particularly happens for arsenic where the acidic or reducing conditions in the landfill can break apart the arsenic ferric oxide complex, releasing the toxic arsenic into the ground.

Using nanotechnology, researchers worldwide are looking for better and more cost-effective ways for treating metallic contamination and alternatives for reverse osmosis for conserving water while treating it for brackishness. Use of nanotechnology provides a means for increasing the effectiveness and loading capacity of adsorbents by depositing nanoscale particles on a substrate. These adsorbents need to be further developed to allow recovery of metallic material such as arsenic from the spent adsorbent so that they provide a more cost-effective way of producing metallic arsenic that can find uses in the semiconductor industry, in medicines for treatment of cancer, or in metallurgical applications for improving alloy properties.

22.4 Other Challenges

WaterHealth is addressing drinking water needs of communities, whether rural, peri-urban, or urban, by installing treatment centers, some or most of its installations can be in remote areas that have challenges of their own. Any new technology that is developed to address the water treatment needs of population residing in these areas also has to overcome these challenges for the technology to be practical and viable. The following sections enumerate some of these challenges.

22.4.1 Temperature, Humidity, and Dirt

To keep the energy required to a minimum, dwellings are generally not designed with any air conditioning and are generally kept open to allow circulation of air. In India, average ambient temperatures in the summer can reach over 45°C; at times during peak summer, temperatures may touch 50°C. In the arid regions and in regions where vegetation has been removed, winds can kick up a cloud of dust. During monsoon season, humidity can be over 95% accompanied by high temperatures. These environmental extremes place undue stress on electronic components. Generally, if water is drawn from the source and treated immediately, system temperatures may not reach the severe environmental temperatures due to cooler water temperatures. However, if the water is stored outside in plastic tanks that are blackened on the outside, water temperatures may be higher than the environmental temperature, and if used, can subject the piping, vessels, and treatment media to these elevated temperatures, requiring that the materials used in the water treatment system maintain their integrity and functionality at elevated temperatures and in most cases compounded by elevated pressures.

22.4.2 Accessibility

It is not given that all communities will have accessibility by wide or paved roads. Invariably, one may have to travel by dirt washboard roads that are narrow and bumpy. Water treatment equipment needs to be rugged and compact to withstand the transportation as well as allow passage through narrow crowded roads.

22.4.3 Power Availability

Power shortage is one of the biggest challenges to overcome in developing countries. Nonurban areas in almost all parts of India have regular power outages that can last for several hours. During dry seasons, particularly at the peak summer times, power outages can last for >12 h. Because of rolling brownouts, the time of a power outage and its duration can be unpredictable. WHCs generally have peak demand periods in the morning when people collect water before they rush off to work and then again in late afternoon/early evening time when children return from school or workers return from work. Typically, these are the periods when power outages occur and compromise the ability to purify water. The purification process has to be carried out around the power outages, and enough storage capacity needs to be provided to meet the demands during peak periods. This adds a layer of complexity in the plant operation and increases the cost of operation when operators have to be called in at odd hours. WaterHealth designs its systems with provision of adequate overhead storage capacity so that clean water is available during power outages and can be easily dispensed using gravity feed during these times.

It will be beneficial to have water treatment technologies that either do not require power for operation or require very little power so that it can be operated using alternate sources of energy. The purification process has to be continuous and automated so that water can be processed whenever power is available and does not require the presence of an operator.

22.5 Application of Nanotechnology

In its efforts to provide clean water at an affordable cost, WaterHealth is constantly looking for new technologies that can be incorporated into its design that help to reduce the cost of installation and operation while improving their effectiveness in treating water for a wide range of contaminants. Advances in nanotechnology in the water treatment sector are showing considerable promise. While the technologies developed need to address the challenges faced in these regions, they also need to address the three primary areas of interest:

1. Water treatment
2. Disinfection
3. Real-time remote monitoring

Generally, the regions in Asia, Africa, and Central and South America where the majority of the people not having access to clean water reside, these are the very same regions where there is a shortage of freshwater availability. Treatment methods using reverse osmosis for brackish water, for instance, particularly in small systems, waste a large fraction of water and consume a high level of energy. Developments in nanotechnology have the potential of improving these treatment methods or coming up with a new treatment method that can significantly cut down water wastage and energy consumption. Ideally, the new methods need to be such that they consume no power, other than the pumping energy, as in the case of high surface area adsorbents. If the methods do need to utilize power, they need to be low-energy-consuming technologies such as enhanced surface membrane-based systems or use of revolutionary new material such as carbon nanotubes or fullerenes, so that they can be easily integrated with alternative renewable sources of power such as solar energy or wind energy.

For disinfecting the water, chlorine is used primarily because it can be implemented relatively easily without the need for power as in the case of disinfection utilizing UV radiation, although UV disinfection is the preferred method as it does not involve addition of any chemical or adverse side effects of forming by-products that may be harmful to the health. Nanotechnology has the potential to create passive disinfection systems such as use of nanoscale silver particles deposited on a high surface area matrix that does not require any power or use of active photocatalytic material deposited on a high surface area substrate that can provide lasting disinfection effect using natural light.

Water-quality tests such as those for pathogens take several hours or days to get results back as the water sample has to be collected and microorganisms allowed to grow and multiply, before they can be detected and counted. If there is a problem with the water treatment system, there is no real-time feedback to the plant operator. By the time water test reports are back, users have already consumed the water. The response to the problem is not immediate, which can put the consumers at risk. Drinking water regulations require that bacterial counts for species such as *Escherichia coli* be undetectable in 100 mL of water. Other nonspecific bacteria counts need to be less than 100 or 1000 colony-forming units (CFU) per mL. The challenge here is that for real-time measurement, in-line tests have to be sensitive enough to detect trace amounts of bacteria and be rapid enough to detect them in a short interval of time. Here nanotechnology can be of immediate help in taking several microsamples and provide the capability of detecting bacterial proteins in a very short time.

Similarly, contaminants such as arsenic and mercury are very toxic in nature and health agencies have determined that prolonged ingestion of arsenic in drinking water containing >10 ppb can lead to cancer. Regulatory agencies are considering further lowering the maximum allowable limit to <5 ppb. The challenge here is that there is no cost-effective technique available that can consistently detect arsenic levels to 5 or 10 ppb, while at the same time be in-line and compatible with a remote monitoring system. Equipment employing adsorption or emission spectroscopy are very expensive, costing several hundred thousand dollars and are uneconomical to implement at a small community level whereas less expensive arsenic test strips that rely on conversion of arsenic in water samples to arsine gas and comparison of strip color to color standards are subjective and untenable as an in-line method compatible with remote monitoring. Here again, nanotechnology can provide the benefit of developing small, affordable sensors that will have the ability to rapidly detect trace levels of the contaminants, while integration into suitable electronics can render it compatible with remote monitoring.

To summarize, there is a need for robust cost-effective technologies for water treatment and contaminant analysis that can withstand the harsh field conditions while enabling automated operation and real-time remote monitoring.

23

Electrospun Nanofibers in Water Purification

Sajini Vadukumpully, Shantikumar V. Nair, and A. Sreekumaran Nair

Amrita Center for Nanosciences and Molecular Medicine,
Amrita Institute of Medical Sciences, Kochi, Kerala, India

CONTENTS

Nanofibers of both polymers and metal oxides have attracted increasing attention for water purification in the last 10 years because of their high surface-to-volume ratio, high porosity, and interconnected pore structures. The composition of the nanofibers can be controlled to achieve desired surface functionalities and properties. Technological advances in electrospinning facilitate fabrication of core–shell and hollow membranes, and commercial machines are available for large-scale production of nanofibers for industrial applications. This chapter summarizes recent advances in the field. The chapter begins with a modest introduction on electrospinning followed by the applications of electrospun nanofibers as antimicrobial agents and in the removal of organic contaminants, particulate materials, and heavy metals from water. The chapter concludes with the application of nanofibers in membrane distillation followed by references.

23.1 Introduction

Although the earth's crust is made up of 75% water, availability of "clean water" for drinking and other purposes is a major crisis.[1] It is estimated that >75% of the worldwide population will face fresh water shortage by 2075.[2] Given the threat to all living organisms, there is an urgent need to address this issue and to develop new technology and devices for water purification. Many technologies such as distillation, sand filtration, treatment with chemical disinfectants, reverse osmosis (RO), and membrane-based filtrations have been employed to purify water.[3,4] Among these, membrane-based filtration is a relatively new technique with certain advantages like low power consumption, scalability, nonuse of chemicals, and relatively low operational temperature.[5,6]

In general, a membrane can be defined as a semipermeable membrane that allows only certain molecules and compounds to pass through and restrict the passage of others. This is illustrated in Figure 23.1, where the fluid particles are allowed to pass through the membrane, whereas the contaminant molecules are blocked.

The separation across the membrane takes place either by a sieving mechanism or by diffusion. Larger particles are blocked on the surface due to a sieve effect, and particles smaller than the surface pores are collected by the membranes by interception or by static electrical attraction. Although advanced membranes are developed, several problems still persist such as membrane fouling and chemical stability.[7] These problems lower their operational lifetime and increase the energy demands. The filtration efficiency is highly dependent on membrane thickness, surface electronic properties, and the surface chemical functionalities. Besides the efficiency, properties like pressure drop and flux resistance are also very important to be evaluated for the filtering media.

Membrane-based filtration systems can be improved with the use of nanofibrous media. Nanofibers, due to their higher porosities and well-connected pore structures, offer higher permeability to water filtration over the conventional materials being used. Among all, electrospinning is a most successful technique to fabricate nanofibers. Electrospun nanofibers provide high filtration efficiency with a small decrease in permeability. Small pore size, high permeability, and low cost of production make them suitable to remove unwanted microsized or nanosized impurities from water.[8]

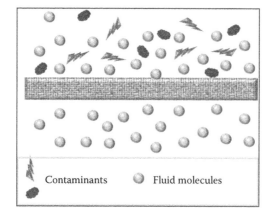

FIGURE 23.1
Schematic of a membrane filtration system.

23.2 Process of Electrospinning

Electrospinning is a well-known technique for making continuous submicron to nanosized 1-dimensional (1D) fibers.[9] It has the potential to fabricate 1D nanostructures of a wide range of material systems.[10–12] An electrospinning setup consists mainly of three components, a high voltage source, a spinneret (a syringe or capillary tube with a needle of small diameter and a grounded collecting plate, either a rotating mandrel or a metallic plate). A schematic of an electrospinning setup is shown in Figure 23.2. A high potential is applied between the needle tip and the collector. The interactions of the electrical charges in the polymeric solution with the applied field causes the polymer droplet near the needle tip to deform to a conical shape known as a Taylor cone. When the applied voltage exceeds the critical voltage at which the repulsive force overcomes the surface tension, a thin charged jet is ejected from the Taylor cone. These charged jets undergo a helical motion and are elongated continuously and during the process the solvent is thrown out. The elongated jet reaches the ground collector and solidifies as fine fibers. If the applied field is not above the critical voltage, it will cause the jet to break up into droplets. The parameters influencing the diameter and morphology of nanofibers include applied voltage, distance between the spinneret and the collector, solution feeding rate, molecular weight of the polymer, viscosity, conductivity, and surface tension of the polymeric solution. The other important factors affecting the electrospinning process are the solution temperature, humidity, and air velocity inside the electrospinning chamber. Electrospinning can be used to produce both aligned and nonaligned nanofibers.[13,14] If a rotating drum or cylinder is used as the collector, the nanofibers can be aligned in some fashion.

Electrospun nanofibers possess several attractive qualities such as high surface-to-volume ratio, high porosity, pore sizes ranging from tens of nanometers to several microns, interconnected open pore structure that allows easy accessibility of the pores to water, and high permeability of gases.[1] The composition of the nanofibers can be controlled to achieve desired surface functionalities and properties. One of the limitations associated with laboratory-scale needle-based electrospinning is the low yield of fibers. However, this has been addressed by developing multi-jet- and blowing-assisted electrospinning, which produce nanofibers on an industrial scale.[15,16] ELMARCO s.r.o. (Czech Republic) has commercialized needleless-based electrospinning machines suitable for industrial applications.[17]

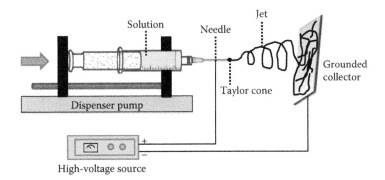

FIGURE 23.2
Schematic of an electrospinning setup.

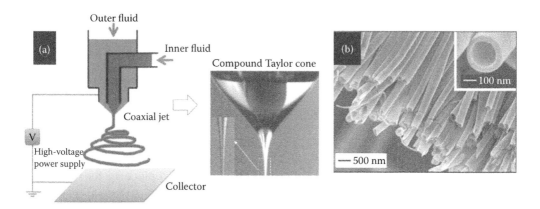

FIGURE 23.3
(See color insert.) (a) Setup for coaxial electrospinning for the fabrication of core–shell nanofibers. (b) SEM images of an array of hollow anatase TiO$_2$ nanotubes. (From Feng, C. et al., *Sep. Purif. Technol.*, 102, 118, 2013.)

23.2.1 Technological Advances in Electrospinning—Core–Shell and Hollow Fibers

Technological advances in electrospinning allow the fabrication of core–shell and hollow fibers of polymers, ceramics, and composites for water purification and other applications.[18–20] Figure 23.3a shows a schematic of the spinneret needed for coaxial electrospinning to fabricate core–shell and hollow nanofibers. The core solution is fed through the inner spinneret and shell solution through the outer one (to have core–shell geometry for fibers). For fabricating hollow fibers, the core solution through the inner spinneret is usually mineral oil. The flow rate of the solutions must be carefully controlled to get the core–sheath or hollow morphology for the fibers. Figure 23.3b shows a scanning electron microscopic (SEM) image of the hollow fibers (by removing the mineral oil at the core by treatment with hexane) fabricated by McCann et al.[21]

Yet another modification in electrospinning to have structural anisotropy for the fibers is the multifluidic compound jet electrospinning technique (Figure 23.4a). Several metallic capillaries (with gaps between them) were inserted into a plastic syringe, and two

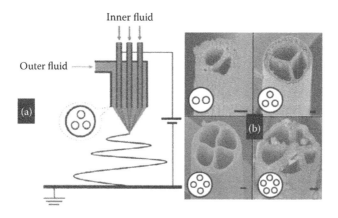

FIGURE 23.4
(See color insert.) (a) Schematic of the multichannel coaxial electrospinning system. (b) Cross-sectional SEM images of hollow fibers with two, three, four, and five channels. Scale bars in all the images are 100 nm. (From Feng, C. et al., *Sep. Purif. Technol.*, 102, 118, 2013.)

immiscible liquids were fed through the inner and outer capillaries at a specific flow rate (say mineral oil through the inner and a metal oxide precursor, titanium [IV] isopropoxide dissolved in polyvinyl pyrrolidone, through the outer). Upon electrospinning, a fiber mat was collected on the collector electrode. Figure 23.4b shows the SEM images of the two-, three-, four-, and five-channel fibers fabricated by the multichannel coaxial electrospinning. Such core–shell and hollow nanofibers may open up interesting applications for water purification processes, although experimentally these have not been tried because of technical issues.

23.3 Electrospun Nanofibers for Water Purification

The major challenges of using membranes for water filtration lie in the development of new materials with high durability, low cost, and new structures that can produce high permeation flux while maintaining a high selectivity or rejection rate.[22] In separation processes, different types of electrospun fibrous media have been used. This can be of woven or nonwoven type, depending on the conditions. Nonwoven membranes are more commonly used in filtration technology.[23] Although the random fibrous membrane forms the backbone of the filter, other components such as antimicrobial agents, particulate fillers, plasticizers, wetting agents, and softening agents can be incorporated by blending, coaxial spinning, or by coating of fibers.[24–27] Most of the filtration membranes are made of synthetic organic polymers since they are less costly and are available in large quantities. The polymers currently in use include, polysulfone, poly(acrylonitrile) (PAN), poly(vinyl alcohol) (PVA), poly(vinylidene fluoride) (PVDF), poly(ether sulfone) (PES), cellulose acetate–cellulose nitrate blends, nylons, poly(vinyl chloride) (PVC), and many others.[8] Table 23.1 summarizes the list of polymers used for making electrospun fiber mats for filtration purposes. Use of electrospun membranes for water purification can be generally classified as those used for (i) the removal of microbes, (ii) removal of organic compounds, (iii) removal of particulate contaminants, (iv) heavy metal ion removal, and (v) membrane distillation. The following sections detail the use of nanofibers for each of the applications mentioned above.

23.3.1 Removal of Microbes

For water-based filtration/purification applications, the membranes should be made hydrophilic. Hydrophilicity to the polymer fibers can be rendered by the choice of hydrophilic polymers or by suitable surface modification.[28] Surface modification techniques include plasma treatment and surface graft polymerization.[29,30] These treatments change the wetting and surface adhesion properties by altering the surface chemical composition.

Plasma treatment and surface graft polymerizations involve the treatment of the membranes with oxygen, air, or ammonia, which generates free radicals and free electrons on the surface. Surface graft polymerization is initiated by the treatment with plasma followed by UV irradiation to generate the free radicals for polymerization.[31] After plasma treatment, other functional compounds such as quaternary ammonium salts or their derivatives are covalently bound to the membranes by chemical reactions. Yao and co-workers reported the development of modified electrospun polyurethane (PU) nanofibers and tested the antimicrobial activities against *Staphylococcus aureus* and gram-negative *Escherichia coli*.[32]

TABLE 23.1

Nanofibers Processed from Polymers for Filtration Purposes

Polymer	Process	Application	Reference
PVDF	Filtration	Pretreatment of water before RO or ultrafiltration	*J. Membr. Sci.*, 2006, 281, 581–586
Polysulfone	Filtration	Prefilters for particulate removal	*J. Membr. Sci.*, 2007, 289, 210–219
Nylon 6	Filtration (to remove micron sized particles)	Water treatment	*J. Membr. Sci.*, 2006, 315, 11–19
PES	Prefiltration	Water and other liquid separation	*J. Membr. Sci.*, 2010, 365, 68–77; *J. Polym. Sci., Part B: Polym. Phys.*, 2009, 47, 2288–2300
Fluorinated copolyimide	Filtration	Water treatment	*Polym. Adv. Technol.*, 2010, 21, 861–866
Cellulose/PAN	Filtration	Bacteria and virus removal	*J. Electron. Microsc.*, 2011, 60, 201–209
PVA cross-linked with glutaraldehyde	Filtration	Oil/water emulsion filtration	*Environ. Sci. Technol.*, 2005, 39, 7684–7691
PAN/chitosan	Ultrafiltration, nanofiltration	Water filtration	*Polymer*, 2006, 47, 2434–2441
PVA/PVA hydrogel	Ultrafiltration	Oil/water emulsion	*J. Membr. Sci.*, 2006, 278, 261–268
UV cured PVA	Ultrafiltration	Oil/water emulsion	*J. Membr. Sci.*, 2009, 328, 1–5
Polyamide/PAN	Nanofiltration	Water treatment	*J. Membr. Sci.*, 2009, 326, 484–492
PVA/PAN	Ultrafiltration	Oil/water emulsion	*J. Membr. Sci.*, 2010, 356, 110–116
Interfacial polymerization using piperazine (PIP)	Ultrafiltration	Water treatment	*J. Membr. Sci.*, 2010, 365, 52–58
Polysaccharide (cellulose and chitin)	Ultrafiltration	Water purification	*Biomacromolecules*, 2011, 12, 970–976
PAN	Nanofiltration	Salt removal	*Desalination*, 2011, 279, 201–209
Polystyrene/β-cyclodextrin	Nanofiltration	Removal of organic compounds from water	*J. Membr. Sci.*, 2009, 332, 120–137
Cellulose acetate (molecularly imprinted with glutamic acid)	Adsorption	Chiral separation	*J. Membr. Sci.*, 2010, 357, 90–97
Carbonized nanofibrous (precursor PAN) membrane	Filtration	Removal of disinfection by products from water	*Sep. Purif. Technol.*, 2010, 74, 202–212
Cellulose acetate	Heavy metal ion adsorption	Water treatment	*Carbohydr. Polym.*, 2011, 83, 743–748

Source: Feng, C. et al., *Sep. Purif. Technol.*, 102, 118, 2013.

PU nanofibers were subjected to argon plasma to generate oxide and peroxide groups on the surface. It was then immersed in 4-vinylpyridine monomeric solution with exposure to UV radiation, to graft poly(4-vinylpyridine) on PU fibers. The potential of polycarbonate (PC)/chloroform solution with quaternary ammonium salt (benzyl tri-ethylammonium chloride, BTEAC) as antimicrobial nanofibrous membranes was explored by Kim et al.[33] The addition of BTEAC to PC inhibited the growth of *S. aureus, E. coli,* and *Klebsiella pneumonia.* This filtration showed 99.97% filtration efficiency of 0.3-μm particles.[33] In another piece of work, antimicrobial activity of a nylon 6 nanofiber membrane was found to increase with *N*-halamine additives namely, chlorinated 5,5-dimethylhydantoin, chlorinated 2,2,5,5-tetramethyl imidozalidin-4-one, and chlorinated 3-dodecyl-5,5-dimethylhydantoin. The growth of microbes was inhibited in the presence of the additives.[34]

Yoon et al. demonstrated a system consisting of three tiers. A "nonporous" hydrophilic coating of chitosan constitutes the first layer, an electrospun PAN nanofibrous support as the second layer, and a nonwoven polyethylene terephthalate microfibrous substrate.[35] Figure 23.5 shows the fabrication schematic of the three-tier system. PAN is used since it is resistant to most solvents, and chitosan has been used for antifouling enhancement of filtration membranes owing to its insolubility in neutral pH conditions.[36]

Other than quaternary ammonium salts, silver ions and silver compounds have long been recognized as important antimicrobial agents. Currently, silver is extensively used as silver nanoparticles (Ag NPs) due to its increased surface area.[37] Ag NPs can be incorporated onto the nanofibers by several methods. Ag NPs can be blended with polymer solution and then electrospun. This results in the embedding of Ag NPs inside the polymer matrix, which reduces the antimicrobial activity. Another method is to electrospray Ag NPs onto the fiber membrane. The most effective method to incorporate Ag NPs is by growing nanoparticles on the fiber surface. This can be achieved by mixing a silver salt with the polymer solution and then electrospinning the mixture. The composite fiber mat can be subjected to heat treatment, UV irradiation, or photoreduction for reducing silver ions to Ag NPs.[38,39] Various researchers have shown the applicability of Ag NP–incorporated polymeric nanofibers in antimicrobial studies. McCord and coworkers showed that Ag/PAN hybrid nanofibers prepared by atmospheric plasma treatment and electrospinning exhibit slow and long-lasting silver ion release, which provided robust antibacterial activity against both gram-positive *Bacillus cereus* and gram-negative *E. coli* microorganisms.[40] Likewise, silver-impregnated nanofibers have been prepared with different polymers such as PAN, PVA, gelatin, poly(lysine glycolic acid), polylactide, and nylon 6, and tested their antimicrobial activity toward different bacteria.[41–44]

Figure 23.6 shows the transmission electron microscopic (TEM) images of Ag NP prepared *in situ* on methacrylated PVA, PVA/dextran, and PVA/methyl cellulose nanofibers. It is clear that the nanoparticles are distributed uniformly over the fibers.[43] In another approach, both an antibacterial effect and dye degradation was coupled to a

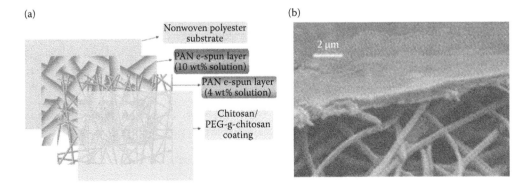

(a)

Nonwoven polyester substrate

PAN e-spun layer (10 wt% solution)

PAN e-spun layer (4 wt% solution)

Chitosan/ PEG-g-chitosan coating

(b)

2 μm

FIGURE 23.5
(a) Fabrication schematics of the electrospun scaffold with a coating layer. (b) Fractured composite membrane containing PAN nanofibrous scaffold (with 4 + 12 wt% sequential electrospinning) and chitosan coating. (From Yoon, K. et al., *Polymer*, 47, 2434, 2006.)

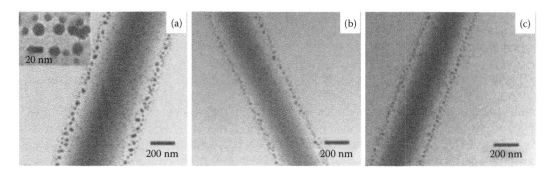

FIGURE 23.6
TEM images of PVA–Ag nanofiber (a), PVA/dextran–Ag nanofiber (b), and PVA/methylcellulose–Ag nanofiber (c). (From Mahanta, N., Valiyaveettil, S. *RSC Adv.*, 2, 11389, 2012.)

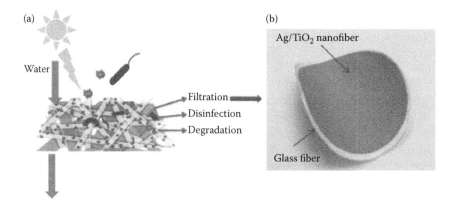

FIGURE 23.7
(See color insert.) (a) Schematic representation of a hypothesis of a Ag/TiO$_2$ nanofiber membrane; (b) photograph of Ag/TiO$_2$ nanofiber membrane. (From Liu, L. et al., *Water Res.*, 46, 1101, 2012.)

single-membrane system. Polyol synthesis was used for the deposition of Ag NPs on electrospun TiO$_2$ nanofibers (Figure 23.7). The permeate flux of the Ag/TiO$_2$ nanofiber membrane was remarkably high compared to a commercial P25 deposited membrane. The Ag/TiO$_2$ nanofiber membrane achieved 99.9% bacterial inactivation (*E. coli*) and 80.0% dye degradation under solar irradiation within 30 min.[45]

Although many research articles have been published on the antimicrobial activities of Ag-incorporated nanofibers, no significant progress has been made in testing the filtration performance and disinfectant efficiency in a flow-through system. A major leap toward this direction might replace existing conventional disinfection techniques such as chlorination and UV-assisted disinfection.

23.3.2 Removal of Organic Compounds

All the pesticides, oils, and even proteins belong to the class of organic compounds. These contaminants enter the water stream through soil erosion, surface runoff, or through leaching. Substantial research has been carried out to eliminate these organic compounds from drinking water. Electrospun nanofibrous membranes offer a promising solution for the removal of organic compounds in water. Uyar et al. studied the removal of a model organic compound, phenolphthalein, from water using beta-cyclodextrin (β-CD) incorporated polystyrene (PS) fibers.[46] A cross-linked carbon nanofiber/PVA nanofiber membrane was found to be an excellent sorbent for aromatic amines from wastewater.[47]

Oil released into water sources due to industrial discharge poses a serious threat to all the aquatic living systems. It forms a layer on the top and thus reduces the dissolved oxygen content. Shang and co-workers have been extensively working on the potential of functional electrospun nanofibers for removal of oil from water.[48]

In one of their recent publications, they have shown that a facile combination of electrospun cellulose acetate (CA) nanofibers and a novel *in situ* polymerized fluorinated polybenzoxazine (F-PBZ) functional layer that incorporated silica nanoparticles (SiO$_2$ NPs), exhibits robust oil–water separation.[48] Figure 23.8 shows the schematic of the process along with an SEM image of the prepared membrane. The same group has illustrated that micro- and nanostructures of electrospun PS fibers show much higher oil absorption capacities than the commercially available polypropylene mats (PP). This material could prove a bright candidate for oil spill cleanups (Figure 23.9).[49]

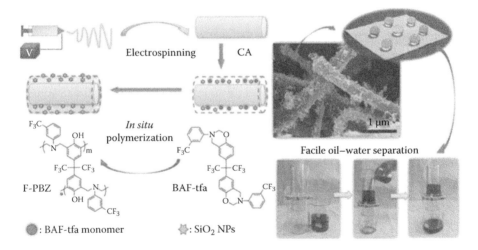

FIGURE 23.8
(See color insert.) Schematic of electrospun CA nanofibers coated with F-PBZ and SiO$_2$ NPs and its ability to separate oil–water mixture. (From Shang, Y. et al., *Nanoscale*, 4, 7847, 2012.)

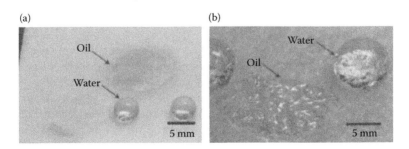

FIGURE 23.9
Water and oil droplets placed on the (a) porous PS fibrous mats and (b) commercial PP non-woven fabric. (From Lin, J. et al., *Nanoscale*, 4, 176, 2012.)

Proteins are another group of commonly found organic contaminants in water, particularly in wastewater. Specific or nonspecific protein adsorption can be achieved by suitable surface chemical functionalization of the fibers. Carboxylated electrospun carbon nanofibers were found to have better adsorption efficiency and higher permeability levels than the normal resin bed.[50] Bovine serum albumin (BSA) and lysozyme were used as the models and it was shown that surface functionalization leads to more specific binding. Sun et al. presented that poly(vinylidenefluoride-hexafluoropropylene) (PVDF-HFP) nanofiber membranes show improved hydrophilicity and protein fouling resistance via surface graft copolymerization of hydrophilic monomers.[51] In another scenario, electrospun polycaprolactone nanofibers embedded with LiCl induced a substantial increase in the protein adsorption.[52] LiCl is a kosmotrope that promotes protein salvation in aqueous solutions. Protein (BSA and protamine as models) loading on the fibers was nine times higher than that observed in the absence of the salt. Moreover, the adsorption was found to be irreversible and no protein loss was observed after repeated washings. These kinds of innovations open up ample opportunities to explore an interesting state of use of electrospun nanofibers in biomaterials-based purifiers.

23.3.3 Removal of Particulate Contaminants

The separation of micron and submicron particulate contaminants has gained importance in water purification and effluent treatment. They clog or foul the filters, thus reducing the efficiency and output. Membrane-based ultra- and microfilters can be used as a prefilter to remove these larger particles, so that the performance and lifetime of the main filter unit can be improved. Prefilters currently available include sand beds and woven and nonwoven fibrous meshes.[53] Electrospun nanofibers also find applications as prefiltration membranes. Electrospun prefilters have large surface area, hence high dirt loading capacity. In one of the investigations, a prefilter was developed using an electrospun nylon 6 fibrous membrane and polystyrene (PS) microparticles were used as a model (Figure 23.10).[54] Owing to its excellent chemical and thermal resistance and high wettability, it showed 90% separation for 0.5-μm particles of PS. In an alternative approach, electrospun nanofibrous membranes of PVDF were surface modified via graft copolymerization using methacrylic acid.[55]

It was found that only the top layer has been modified and the nanofiber morphology remained intact. Furthermore, the surface modification reduced the average membrane pore size and showed better water flux (150%–200%), higher than the commercial membrane. This proved that the nanofiber architecture is better and could result in energy-conserving membranes.

Contamination of water from engineered nanomaterials will be an emerging problem in the future due to the extensive use of nanomaterials in commercial products and their improper disposal. Engineered nanoparticles fall into the category of particulate contaminants in a 1–100 nm size range. Surface-modified electrospun PVA nanofibers were employed for the removal of nanomaterials by Mahanta and Valiyaveettil.[56] The surface hydroxyl groups of PVA nanofibers were modified using thiols and amines to improve the extraction efficiency of model nanoparticles (gold and silver). The extraction studies revealed that the amine- and thiol-modified PVA nanofibers showed ~90% extraction efficiency for Ag and Au NPs.

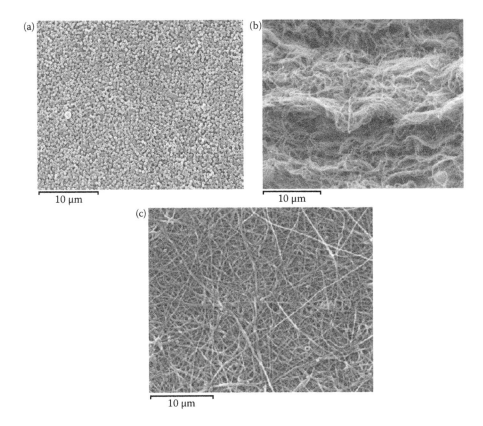

FIGURE 23.10
SEM images of 0.15 ± 0.05 mm nylon 6 membrane thicknesses after 0.5 μm PS microparticles (125 ppm) separation: (a) top surface, (b) cross section, and (c) bottom surface. (From Aussawasathien, D. et al., *J. Membr. Sci.*, 315, 11, 2008.)

23.3.4 Removal of Heavy Metal Ions

Industrial effluents release heavy metals such as lead, chromium, mercury, cadmium, copper, and arsenic, into water streams, and these cause serious problems in aquatic systems. Adsorption and filtration are the commonly used methods for removal of these contaminants. Electrospun nanofiber membranes can offer both adsorption and filtration, and have proven very effective for the removal of heavy metals from water. In recent work, Fe^{3+} ion-impregnated PVA nanofibers were used for the extraction of As(III) and As(V) compounds from aqueous media. The maximum capacity was found to be 67 mg/g and 37 mg/g for As(III) and As(V), respectively.[57] Zerovalent iron nanoparticles immobilized on multiwalled carbon nanotube (MWCNT)–reinforced poly(acrylic acid)/PVA composite fibers was also found to be an active agent for the removal of Cu(II) ions from water (Figure 23.11).[58]

The presence of MWCNTs improved the mechanical durability of the film. Cu^{2+} chemisorption occurs via chemical reduction and deposition on the Fe nanoparticle surfaces to form an Fe/Cu alloy. The uniform Fe NPs on the hybrid nanofibers offer great specific surface areas that enable very effective, high-capacity, and strong sorption of Cu(II) ions. In

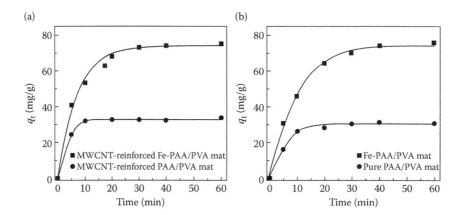

FIGURE 23.11

Removal of Cu(II) ions from solutions at different contact times by nanofibrous mats with (a) or without (b) MWCNT reinforcement ([Cu(II)] = 50 mg/L, [mat] = 0.5 mg/mL). (From Xiao, S. et al., *Colloids Surf. A*, 381, 48, 2011.)

another study, silk fibroin and a blend of silk fibroin with wool keratose was electrospun into membranes and tested for the removal of heavy metal ions (Cu^{2+} ions as the model). The adsorption capacity was significantly enhanced when nanofibers (1.65–2.88 mg/g) were used, as compared with the conventional fiber materials like wool silver (0.71 mg/g) and filter paper (0.23 mg/g). The higher adsorption capacity has been attributed to the larger surface area of the nanofiber membrane.[59] It can be concluded that a suitable choice of the polymer and surface modifier during/after the electrospinning process could achieve higher removal efficiency for heavy metal ions.

23.3.5 Membrane Distillation

Membrane distillation using hydrophobic nanofiber membranes is a convenient way for desalination.[60] This is different from conventional technologies and is an alternative to RO especially when the solute concentration is high.[60] In a thermally driven process, a nanofiber membrane with nano/submicron/micron pores acts as a separator between a warm solution and a cooler chamber that contains a liquid/gas. Being a nonisothermal process, vapor molecules migrate through the membrane pores from the warmer to the cooler side.[60] The surface of the distillation membrane should be hydrophobic to prevent the entry of water to the pores, and polymer nanofibers are appropriate for the purpose as their hydrophobicity is far more than the intrinsic hydrophobicity of the polymer. Shih[61] fabricated two composite nanofibrous membranes of PVDF and PVDF-HFP by electrospinning and used them in a direct contact membrane distillation process.

The permeate flux of the PVDF-HFP composite membrane was 4.28 kg/m² h, which was higher than the PVDF and commercial polytetrafluoroethylene membranes. Feng et al.[62] used the above PVDF–nanofiber membrane for membrane distillation, and the membrane could produce potable water (NaCl concentration <280 ppm) from saline water of NaCl concentration 6 wt% by air gap membrane distillation. This implies the membrane distillation process could eventually compete with RO and conventional distillation for the desalination purpose. Figure 23.12 shows the plot for flux versus temperature difference for 1, 3.5, and 6 wt% NaCl solutions. It is obvious that as the NaCl concentration in the

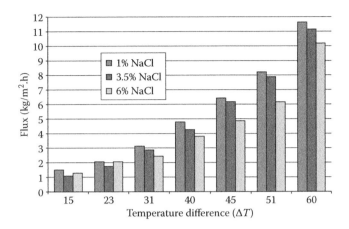

FIGURE 23.12
Flux vs. cross membrane temperature difference for different NaCl concentrations. (From Feng, C. et al., *Sep. Purif. Technol.*, 102, 118, 2013.)

feed solution increases, the flux drops slightly, indicating a small change of water vapor pressure with a change in NaCl concentration. The highest water flux achieved at the highest temperature difference (60°C) was 11–12 kg/m^2 h, which is comparable to the results obtained for other research groups.

23.4 Current Status of the Nanofiber Filtering Media

Electrospun nanofibers are widely employed in filtration processes. The US company Donaldson Co. (Minneapolis, MN)[63] produces electrospun fiber-based filter media (Ultra-Web®) for applications in industry, consumer, and defense sectors. DuPont[64] adopts an innovative new spinning process for fabricating high-efficiency, high-throughput "hybrid membrane technology (HMT) nanofiber filter media" (based on nylon 6) for superior liquid filtration systems. The company claims unique attributes such as high filtration efficiency, low filter pressure drop, high flow rate, and longer life. ELMARCO s.r.o. offers industrial nanofiber production units (using the Nanospider™ machines).[65] The company claims that low-cost nanofiber filtration media (deposited on an inexpensive substrate) not only matches but also often exceeds the performance of the costly commercial filtration media.

23.5 Future Trends and Limitations

There has been intense research in exploring the benefits of nanofibrous media in water purification in the past few decades. This trend is expected to grow as more insights about the transport and behavior of nanofibers is achieved. Currently, there are a large number of companies offering commercial nanofibers. It is evident that within a few years, nanofibrous media with unique features will be designed and created as a new-generation filtration media. However, the cost-effectiveness of these nanofibers versus the benefits must be evaluated before their use. More fundamental research on this aspect has to be performed to realize the full potential and may undoubtedly reveal more new attributes of its function as a filter in years to come.

References

1. Kaur, S.; Gopal, R.; Ng, W. J.; Ramakrishna, S.; Matsuura, T. *MRS Bull.* **2008**, 33, 21–26.
2. Veleirinho, B.; Rei, M. F.; Lopes-da-Silva, J. A. *J. Polym. Sci., Part B* **2008**, 46, 460–471.
3. Brennan, M. B. *Chem. Eng. News* **2001**, 79, 32–37.
4. Baker, R. W.; Ruthven, D. M. *Encycl. Sep. Technol.* **1997**, 2, 1212–1270.
5. Steffens, J. *Filtration* **2007**, 7, 26–28.
6. Ramakrishna, S.; Jose, R.; Archana, P. S.; Nair, A. S.; Balamurugan, R.; Venugopal, J.; Teo, W. E. *J. Mater. Sci.* **2010**, 45, 6283–6312.
7. Barhate, R. S.; Ramakrishna, S. *J. Membr. Sci.* **2007**, 296, 1–8.
8. Thavasi, V.; Singh, G.; Ramakrishna, S. *Energy Environ. Sci.* **2008**, 1, 205–221.
9. Cooley, J. F. Apparatus for electrically dispersing fluids. US Patent No: 692631, **1902**.
10. Li, D.; Wang, Y.; Xia, Y. *Nano Lett.* **2003**, 3, 1167–1171.
11. Li, D.; McCann, J. T.; Xia, Y.; Marquez, M. *J. Am. Ceram. Soc.* **2006**, 89, 1861–1869.
12. Gibson, H. S.; Gibson, P.; Wadsworth, L.; Hemphill, S.; Vontorcik, J. *Adv. Filtr. Sep. Technol.* **2002**, 15, 525–537.
13. Bazbouz, M. B.; Stylios, G. K. *J. Appl. Polym. Sci.* **2008**, 107, 3023–3032.
14. Bazbouz, M. B.; Stylios, G. K. *Eur. Polym. J.* **2007**, 44, 1–12.
15. Ding, B.; Kimura, E.; Sato, T.; Fujita, S.; Shiratori, S. *Polymer* **2004**, 45, 1895–1902.
16. Um, I. C.; Fang, D. F.; Hsiao, B. S.; Okamoto, A.; Chu, B. *Biomacromolecules* **2004**, 5, 1428–1436.
17. Available at www.elmarco.com/.
18. Li, D.; Xia, Y. *Nano Lett.* **2004**, 4, 933–938.
19. Zhang, Z.; Huang, Z.-M.; Xu, X.; Lim, C. T.; Ramakrishna, S. *Chem. Mater.* **2004**, 16, 3406–3409.
20. Yu, J. H.; Fridrikh, S. V.; Rutledge, G. C. *Adv. Mater.* **2004**, 16, 1562–1566.
21. McCann, J. T.; Li, D.; Xia, Y. *J. Mater. Chem.* **2005**, 15, 735–738.
22. Ma, H.; Burger, C.; Hsiao, B. S.; Chu, B. *J. Mater. Chem.* **2011**, 21, 7507–7510.
23. Dickenson, C. *Filters and Filtration Handbook*, 3rd Edition. Oxford, UK: Elsevier Advanced Technology, **1992**.
24. Sun, Z.; Zussman, E.; Yarin, A. L.; Wendorff, J. H.; Greiner, A. *Adv. Mater.* **2003**, 15, 1929–1936.
25. Pant, H. R.; Bajgai, M. P.; Nam, K. T.; Seo, Y. A.; Pandeya, D. R.; Hong, S. T.; Kim, H. Y. *J. Hazard. Mater.* **2011**, 185, 124–130.
26. Babel, A.; Li, D.; Xia, Y.; Jenekhe, S. A. *Macromolecules* **2005**, 38, 4705–4711.
27. Wei, M.; Lee, J.; Kang, B.; Mead, J. *Macromol. Rapid Commun.* **2005**, 26, 1127–1132.
28. Goodwin, J. *Colloids and Interfaces with Surfactants and Polymers—An Introduction*. West Sussex, England: Wiley Interscience, **2004**.
29. Ramakrishna, S.; Fujihara, K.; Teo, W. E.; Lim, T. C.; Zuwei, M. *An Introduction to Electrospinning and Nanofibers*. Hackensack, NJ: World Scientific, **2005**.
30. Zuwei, M.; Kotaki, M.; Ramakrishna, S. *J. Membr. Sci.* **2006**, 272, 179–187.
31. Yoon, K.; Hsiao, B. S.; Chu, B. *Polymer* **2009**, 50, 2893–2899.
32. Yao, C.; Li, X.; Neoh, K. G.; Shi, Z.; Kang, E. T. *J. Membr. Sci.* **2008**, 320, 259–302.
33. Kim, S. J.; Nam, Y. S.; Rhee, D. M.; Park, H. S.; Park, W. H. *Eur. Polym. J.* **2007**, 43, 3146–3152.
34. Tan, K.; Obendorf, S. K. *J. Membr. Sci.* **2007**, 305, 287–298.

35. Yoon, K.; Kim, K.; Wang, X.; Fang, D.; Hsiao, B. S.; Chu, B. *Polymer* **2006**, 47, 2434–2441.

36. Gudmund, S. B.; Thorleif, A.; Paul, S. *Chitin and Chitosan, Sources, Chemistry, Biochemistry, Physical Properties and Applications.* New York: Elsevier, **1989**.

37. Sondi, I.; Salopek-Sondi, J. B. *Colloid Interface Sci.* **2004**, 275, 177–182.

38. Hong, K. H.; Park, J. L.; Sul, I. H.; Youk, J. H.; Kang, T. J. *J. Polym. Sci. Part B* **2006**, 44, 2468–2474.

39. Lee, H. K.; Jeong, E. H.; Baek, C. K.; Youk, J. H. *Mater. Lett.* **2005**, 59, 2977–2980.

40. Shi, Q.; Vitchuli, N.; Nowak, J.; Caldwell, J. M.; Breidt, F.; Bourhamb, M.; Zhang, X.; McCord, M. *Eur. Polym. J.* **2011**, 47, 1402–1409.

41. Yu, D.-G.; Zhou, J.; Chatterton, N. P.; Li, Y.; Huang, J.; Wang, X. *Int. J. Nanomed.* **2012**, 7, 5725–5732.

42. Lee, D. Y.; Lee, K.-H.; Kim, B.-Y.; Cho, N.-I. *J. Sol-Gel Sci. Technol.* **2010**, 54, 63–68.

43. Mahanta, N.; Valiyaveettil, S. *RSC Adv.* **2012**, 2, 11389–11396.

44. Lala, N. L.; Ramaseshan, R.; Li, B.; Sundarrajan, S.; Barhate, R. S.; Liu, Y.-J.; Ramakrishna, S. *Biotechnol. Bioeng.* **2007**, 97, 1357–1365.

45. Liu, L.; Liu, Z.; Bai, H.; Sun, D. D. *Water Res.* **2012**, 46, 1101–1112.

46. Uyar, T.; Havelund, R.; Nur, Y.; Hacaloglu, J.; Besenbacher, F.; Kingshott, P. *J. Membr. Sci.* **2009**, 332, 129–137.

47. Vadukumpully, S.; Basheer, C.; Jeng, C. S.; Valiyaveettil, S. *J. Chrom. A* **2011**, 1218, 3581–3587.

48. Shang, Y.; Si, Y.; Raza, A.; Yang, L.; Mao, X.; Ding, B.; Yu, J. *Nanoscale* **2012**, 4, 7847–7854.

49. Lin, J.; Ding, B.; Yang, J.; Yu, J.; Sun, G. *Nanoscale* **2012**, 4, 176–182.

50. Schneiderman, S.; Zhang, L.; Fong, H.; Menkhaus, T. J. *J. Chrom. A* **2011**, 1218, 8989–8995.

51. Sun, F.; Li, X.; Xu, J.; Cao, P. *Chin. J. Polym. Sci.* **2010**, 28, 705–713.

52. Liu, C. X.; Zhang, S. P.; Su, Z. G.; Wang, P. *Langmuir* **2011**, 27, 760–765.

53. Warring, R. H. *Filter and Filtration Handbook.* Houston, TX: Gulf Publishing, **1982**.

54. Aussawasathien, D.; Teerawattananon, C.; Vongachariya, A. *J. Membr. Sci.* **2008**, 315, 11–19.

55. Kaur, S.; Ma, Z.; Gopal, R.; Singh, G.; Ramakrishna, S.; Matsuura, T. *Langmuir* **2007**, 23, 13085–13092.

56. Mahanta, N.; Valiyaveettil, S. *Nanoscale* **2011**, 3, 4625–4631.

57. Mahanta, N.; Valiyaveettil, S. *RSC Adv.* **2013**, 3, 2776–2783.

58. Xiao, S.; Mac, H.; Shen, M.; Wang, S.; Huang, Q.; Shia, X. *Colloids Surf. A* **2011**, 381, 48–54.

59. Ki, C. S.; Gang, E. H.; Um, I. C.; Park, Y. H. *J. Membr. Sci.* **2007**, 302, 20–26.

60. Lawson, K. W.; Lloyd, D. R. *J. Membr. Sci.* **1997**, 124, 1–25.

61. Shih, J. H. A study of composite nanofiber membrane applied in seawater desalination by membrane distillation. Master's Thesis, National Taiwan University of Science and Technology, **2011**. Available at http://pc01.lib.ntust.edu.tw/ETD-db/ETD-search/view_etd?URN=etd-0726111-153920.

62. Feng, C. Y.; Khulbe, K. C.; Tabe, S. *Desalination* **2012**, 287, 98–102.

63. Available at http://www.donaldsonfilters.com.au/.

64. Available at http://origin.dupont.com/Separation_Solutions/en_US/tech_info/hmt/hmt.html.

65. Available at http://www.elmarco.com/application-areas/liquid-filtration/.

24

Light-Activated Nanotechnology for Drinking Water Purification

Mark D. Owen and Tom Hawkins

Puralytics, Beaverton, Oregon, USA

CONTENTS

24.1 Introduction

Nearly 1 billion people lack access to any form of improved water supply within 1 km of their home [1] (Figure 24.1). The poorest 4 billion people collectively spend more than $20 billion per year on water collection and treatment. This consists of collecting water from surface sources often polluted with unknown amounts of animal waste, chemicals, heavy metals, and biological agents; partially treated piped water; shared community resources, individual water purifiers, and mobile water vendors [2].

The United Nations has asserted access to safe water as a basic human right [3]. The World Health Organization (WHO) has defined safe water as when people have access to an improved water source, when the germs are killed or removed, and when toxins are reduced to acceptable levels [4]. The United Nations (UN) Millennial 15-year goal from 2000 to 2015 was to provide access to improved water sources, and most people who remain without access are in more remote areas [5]. An improved water source can mean a

FIGURE 24.1
Polluted surface sources, long transportation distances, and inadequate storage or treatment is the water situation for billions of people.

piped source, a well and pump, a filter, or a bore hole. "Improved" does not mean the water is safe, just that access to enough water has been improved in some way. Germs make you sick tomorrow, toxins hurt you slowly and permanently, and both must be removed for water to be safe (Figure 24.2). However, the solutions provided by UN organizations and almost every other aid organization and government agency are aimed either at improving access to water or partial disinfection. Toxins are often not measured, and the solutions provided do not remove them. The majority of the world, >4 billion people by some calculations, do not have access to safe water, where toxins and germs are removed, and that number is increasing, even in places like the United States. Providing municipal treatment systems to urban communities in the world is not working, and it will not work going forward. New solutions aimed at providing safe drinking water for everyone must be developed.

A number of technologies, mostly highly refined versions of 19th and 20th century inventions, are in use to remove contaminants from drinking water, including filtration, reverse osmosis (RO), germicidal lamps, chlorination, and ozonation. However, there are >1000 new industrial contaminants introduced into the environment each year, mostly new organic chemical compounds, as shown in Figure 24.3, which were not conceived when

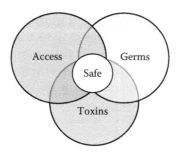

FIGURE 24.2
For water to be safe, a person must have access to enough water, it must be disinfected of germs, and detoxified of chemical and naturally occurring toxins.

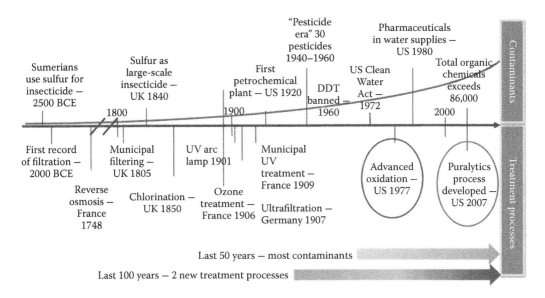

FIGURE 24.3
Current water purification status: new pollutants, old purification technologies.

these treatment technologies were invented and which cannot practically be removed by these technologies. What is needed are new treatment technologies that address 21st century contaminants, especially organic chemical compounds, with high energy and water efficiency and with simple installation and maintenance suitable for developing world deployment, providing low cost, safe drinking water.

24.2 Developing World and Crises

In the developing world, many point-of-use (POU) water purifiers for pathogenic contaminants have been developed, such as disinfecting chemicals, antimicrobial filters, and thermal or sunlight disinfection systems; however, the ability to address toxins are severely limited in these solutions. In fact, no practical point of use solutions are available that satisfy the basic requirements established by the WHO, including removal of pathogens, chemicals, and heavy metals [3].

In these guidelines, WHO establishes the need to address not only the microbiological contaminants in water that cause acute illnesses, but also the many other contaminants that lead to chronic health issues. Specifically, WHO defines safe drinking water as water that

> "does not contain any significant risk to health over a lifetime of consumption, including different sensitivities which may occur between life stages." [3]

Many common contaminants in polluted water are well known to cause cancer, defects in infants or other illnesses, as well as affecting neurological processes, even at consumption rates as low as 2 liters per day.

Water treatment is only a part of the solution. Transporting or storing water can also introduce contaminants. Additionally, water treatment solutions require skill, maintenance, and consumable supplies that might make them impractical or prohibitively expensive in actual use. In rural, remote, and crisis situations, these needs are amplified. Following a hurricane, earthquake, regional conflict, or environmental disaster, clean water becomes an immediate issue after basic triage. In each large-scale disaster—tsunamis, hurricanes/cyclones, earthquakes, etc.—water supply has become critical by day 3 of the crisis, and often remains so for months or years after the acute crisis is over.

24.3 Developed-World Water Problems

A December 2009, *New York Times* investigative series titled "Toxic Waters" began "The 35-year-old federal law regulating tap water is so out of date that the water Americans drink can pose what scientists say are serious health risks—and still be legal" [6]. Gasoline additives, pharmaceuticals, gas-fracking chemicals, industrial solvents, pesticides, and many other contaminants have been reported in numerous research papers to be present in municipal water, surface water, and groundwater in both developed and undeveloped countries. Many of these contaminants are man-made chemical compounds that require new technology to remove, including methyl *tert*-butyl ether and other petrochemical products, pharmaceuticals and personal care products, pesticides, insecticides, herbicides, cleaning solvents, textile dyes, and endocrine-disrupting compounds. Therefore, the compelling, unmet market need is for water treatment solutions that address these man-made chemical contaminants, while also treating microorganisms and other toxic compounds such as heavy metals that might be in water.

In industrial nations, point source or decentralized water purification systems are also used to further purify municipal water or groundwater to remove residual contaminants that could affect processes or products.

- Industrial users—Laboratories, food processing lines, water bottling plants, etc., purify water to remove contaminants that could destabilize production processes, to ensure consistent product quality, or to minimize risk associated with process waste.
- Institutional, commercial, and residential users of drinking water—Government facilities, schools, homes, restaurants, coffee shops, hotels, etc., purify water to ensure drinking water safety and quality.

Currently, multistage systems incorporating filters, RO, and ultraviolet (UV) sterilization processes are used for these point source solutions. These systems are almost always customized for the individual application or customer, so system integrators and value-added resellers are used to specify and install what should/could be an appliance. Current trends in contaminant monitoring, government regulations, and health awareness are expanding the need for these decentralized point source drinking water solution as shown in Figure 24.4.

Therefore, a low-cost, low-maintenance, water purification system is needed to purify water to meet the WHO Guidelines, both disinfecting and detoxifying the water, and to provide safe drinking water for anyone, anywhere in the world.

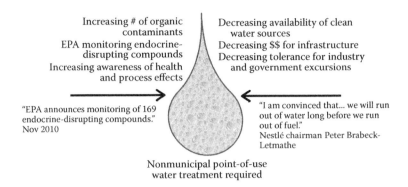

Increasing # of organic contaminants

EPA monitoring endocrine-disrupting compounds

Increasing awareness of health and process effects

"EPA announces monitoring of 169 endocrine-disrupting compounds." Nov 2010

Decreasing availability of clean water sources

Decreasing $$ for infrastructure

Decreasing tolerance for industry and government excursions

"I am convinced that... we will run out of water long before we run out of fuel." Nestlé chairman Peter Brabeck-Letmathe

Nonmunicipal point-of-use water treatment required

FIGURE 24.4
Regulations, water quality, and water demand all push toward the growing use of decentralized, point source water purification systems.

24.4 Strengths/Limitations of Current Water Treatment Technologies

Since each water treatment is constructed of multiple processes and technologies, it is important to understand the strengths and limitations of these technologies. The processes available for disinfecting and detoxifying water include

- **Chlorination:** The process of adding chlorine, a strong oxidant, to water is effective and widely used to disinfect water and to provide a residual disinfectant that disinfects pipes and containers. However, in the presence of natural organic matter in the water, undesirable by-products can be formed, and some pathogens are chlorine resistant. The taste and smell associated with this process have hindered adoption in developing countries.

- **Ozonation:** Ozone, another strong oxidant, is also used in water treatment, to provide disinfection and some chemical breakdown. The process has narrow process windows and can also produce undesirable by-products.

- **Distillation:** The process of boiling water and condensing the vapor removes a broad range of contaminants; however, this is an energy-intensive process. Some dissolved organics are transferred to the distillate, especially those with boiling points near or below that of water. Thermal distillation for desalination can be practical when using waste heat from power generation at large scales.

- **Filtration:** Including granular activated carbon, ceramic, or polymer microporous filters removes a moderately wide range of contaminants, but requires monitoring and filter replacement to assure continuous performance, and saturated filter elements require regeneration or disposal.

- **RO:** This process applies pressure to a membrane to remove ionic and high molecular weight contaminants from solution. RO effectively removes inorganic mineral salts, but typically requires significant energy and generates a waste stream that has greater volume than the purified water stream. It also fails to remove many soluble organic contaminants, including some pharmaceuticals, petroleum by-products, pesticides and herbicides, and other lower molecular weight compounds. RO is another method of desalination widely deployed.

- **UV germicidal irradiation:** UV from mercury germicidal lamps is an effective disinfectant in clear, transparent water; however, monitoring, cleaning, and prefiltration are required to assure germicidal performance.

- **Ion exchange:** This process is effective for targeting specific minerals for removal, but does not effectively remove organics, particles, pyrogens, or microorganisms, and requires frequent resin pack change or regeneration processes.

- **Continuous deionization:** This process removes only a limited number of charged organics, requires very pure feed water for efficient operation, and is commonly used only in laboratory-grade water applications.

- **UV oxidation:** This process uses a deep-UV light (185 nm) to produce ozone, hydrogen peroxide, and hydroxyl radicals, which are effective at the photodegradation and/or photolysis of organic chemicals. The ozone and hydrogen peroxide persist beyond the reactor and must be removed, and mercury lamps are very inefficient at producing this wavelength of light.

- **Advanced oxidation processes:** Use of hydroxyl radicals produced by UV activation of ozone and hydrogen peroxide is effective at oxidizing contaminants, but requires production and storage of toxic chemicals; therefore, these processes are generally impractical in smaller-scale, point source drinking water applications.

24.5 Photochemical Water Purification

Puralytics has commercialized a combination of five photochemical processes that have been shown to reduce a broad range of contaminants. These photochemical processes are driven both directly by light and indirectly through light activation of a semiconductor catalyst. These processes include

- **Photocatalytic oxidation**—an advanced oxidation process employing hydroxyl radicals produced at the surface of a photocatalyst activated by light

- **Photolysis**—the direct breaking of molecular bonds by light of appropriate wavelengths

- **Photocatalytic reduction**—reduction of a contaminant to a less toxic state at the surface of a photocatalyst

- **Photoadsorption**—the light-enhanced adsorption of contaminants to a surface

- **Photodisinfection**—using one or more bands of light to disinfect water

These new processes provide new tools to address the emerging contaminants entering our water supplies, as can be seen in Figure 24.5. In fact, these synergistic processes can improve removal of trace chemical contaminants, reduce maintenance and consumable replacement frequencies, and reduce water waste, thereby providing environmental and health benefits and reducing overall cost of ownership.

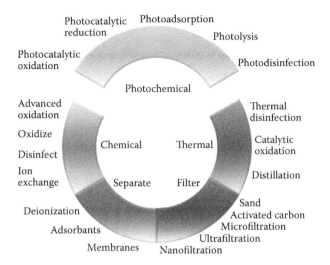

FIGURE 24.5
New photochemical purification technologies address emerging contaminants.

24.6 Puralytics Process

Puralytics has developed a unique and innovative "purification engine" for water, which is scalable and can be packaged to meet the needs of the target markets. The core technology uses light energy supplied by either semiconductor light-emitting diodes (LEDs) or sunlight to activate a nanotechnology coated fibrous mesh and thereby to enable the five simultaneous and synergistic purification processes described below.

24.6.1 Photocatalytic Oxidation

Illumination of the photocatalyst with precise wavelength photons produces highly reactive hydroxyl radicals. These break the carbon bonds in organic compounds in the water, providing destruction of the emerging contaminants, including pesticides, petrochemicals, and pharmaceuticals. Photocatalytic oxidation by a photo-activated semiconductor photocatalyst has been actively studied [7–10] as an advanced oxidation process applicable to water purification. This process offers nonselective degradation of organic contaminants in water into simpler and less toxic compounds, and ultimately into inorganic ions, CO_2, and water. Photocatalytic oxidation involves the absorption of energetic photons by the semiconductor and the subsequent production of hydroxyl radicals at the semiconductor surface. While many nanotechnology catalysts have been studied, anatase TiO_2 is a particularly effective semiconductor photocatalyst in converting light into hydroxyl radicals—a more powerful oxidizing agent than ozone and twice as powerful as chlorine—with sufficient energy to completely mineralize organic contaminants. The critical reaction pathway is

$$(TiO_2) + h\nu \rightarrow e^- + h^+ \quad [\text{electron/hole production}]$$

$$(H_2O \leftrightarrows H^+ + OH^-)_{ads} + h^+ \rightarrow H^+ + OH^\circ \quad [\text{hydroxyl radical production}]$$

$$\text{Reactant} + OH^\circ \rightarrow \text{Intermediates} \rightarrow CO_2 + H_2O + \text{minerals}$$

However, cost-effective production of sufficiently high photocatalyst surface area in contact with water, and delivery of enough energetic photons to the semiconductor to activate it, has proven difficult. Systems employing UV-activated TiO_2 slurries have been demonstrated to be effective in breaking down most organic contaminants [11,12], but require complicated, expensive systems for management of the slurry material. Puralytics uses an order-of-magnitude increase in surface area in a fixed-bed reactor with a significant improvement in mass transport over these slurry systems.

Optimized illumination sources are also needed for cost-effective water purification systems. At low UV intensities, less than ~3 mW/cm² at wavelengths below 400 nm, production of hydroxyl radicals by UV-illuminated anatase TiO_2 photocatalyst is known to be linearly proportional to the UVA intensity, while the production of hydroxyl radicals has been reported [7,8] to increase sublinearly at higher UVA intensities. Most research to date has been done with lamps illuminating a slurry. These lamps have typically been low-pressure mercury lamps emitting at 254 nm or mercury "black light" lamps emitting in the UVA band near 365–370 nm with limited optical flux and efficiency. LEDs are now able to more efficiently emit a band or bands of light that can more optimally excite photocatalytic processes, with important advantages:

- The UVA intensity can be significantly increased without exceeding the range of linear proportionality between intensity and hydroxyl radical production.
- The photocatalyst can be applied to a transparent, fixed substrate, increasing both surface area and mass transport compared with slurry systems.
- LED illumination avoids the issues associated with using lamps.

24.6.2 Photocatalytic Reduction

Free electrons produced on the illuminated photocatalyst instantly react with many positive valence compounds, including heavy metals and inorganics, reducing them to a less toxic, more elemental state. These reduced compounds demonstrate an enhance affinity for adsorption to the TiO_2 surface, where further oxidation or deposition can occur. Many inorganic compounds and heavy metals have been reported to photoreduce [8].

24.6.3 Photoadsorption

The light-activated photocatalyst strongly and irreversibly adsorbs heavy metals including mercury, lead, selenium, arsenic, permanganate, and other toxic compounds. Previous reduction reactions enhance this process. Heavy metals are permanently retained in the system, and properly managed when the catalyst is replaced. While TiO_2 is already an excellent medium for contaminant adsorption, anatase TiO_2 under exposure to UV light becomes an even more aggressive adsorber, and can also irreversibly photodeposit certain contaminants on the TiO_2 surface. Compounds involving noble metals and non-noble heavy metals with favorable redox potentials have been shown to photodegrade [13] into molecular components, photoreduce into less toxic forms, and then photodeposit onto the catalyst.

24.6.4 Photolysis

High-energy photons directly disassociate many chemical compounds, complementing and enhancing the effectiveness of the other processes. The multiple wavelengths of light used in the Puralytics process broaden the effectiveness of this process. Photolysis is the direct absorption by a contaminant molecule of photons with sufficient energy to directly dissociate chemical bonds. Shorter wavelengths are more energetic and therefore more effective over a wider range of chemical bonds. Hundreds of organic contaminants have been shown to photodegrade under UVA, UVB, and UVC light through direct photolysis.

24.6.5 Photodisinfection

The primary mechanism for sterilization of organisms is disruption of DNA molecules, thereby preventing reproduction. With multiple wavelengths, very high light intensity, and the other synergistic processes, pathogens are disinfected more effectively than standard germicidal irradiation. The combination provides improved sterilization of aggressive viruses, resistant bacteria, protozoa, and molds. UV germicidal irradiation with mercury lamps is a well-established process for sterilizing pathogens. For germicidal applications, the 250–280 nm wavelength band is effective at disrupting the DNA of microorganisms. Monochromatic radiation within this band, such as the 254 nm radiation from a low-pressure mercury lamp, sterilizes microorganisms; however, a band of wavelengths above 265 nm would be even more effective [14] and reduce dark repair of DNA [15]. Higher-pressure mercury and xenon lamps produce broadband radiation—inefficient for disinfection or for activating a semiconductor photocatalyst. Moreover, UV lamp sources are fragile, and mercury lamps in particular are environmental hazards. UV LEDs, spanning multiple wavelength bands, are effective, safe, and can uniformly illuminate a large area.

These five photochemical processes destroy a broad range of contaminants, effectively removing them from the environment, in a self-cleaning process. Since the nanomaterials are not consumed by these reactions, and only metals remain on the catalyst over time, the media does not need to be replaced until the catalyst is saturated with metals. These reactions can be enabled either through direct sunlight illumination or by using solid-state LEDs to provide the precise wavelengths of light that are needed. The combined reactions primarily produce pure H_2O, dissolved CO_2, and trace minerals as by-products.

These photochemical processes, working together within the Puralytics products, provide a disruptive new entry to the water purification market and enable new applications not currently possible.

24.7 Product Implementation

Puralytics has developed this technology into two scalable product platforms, the Shield and the SolarBag.

24.7.1 Shield

The Shield addresses industrial and commercial water treatment applications with an electrically powered, LED-activated, stand-alone purification system or system component. This unit has a small footprint and can be used in series or parallel for higher water flows or contaminant removal rates, typically for applications <20,000 gallons per day.

24.7.2 SolarBag

The SolarBag uses sunlight directly to activate the photochemical water purification processes for applications in military, disaster relief, and developing world drinking water, and for recreation and emergency preparedness. The SolarBag produces 3 liters at a time in 2–3 h of sunlight or 4–6 h on a cloudy day. It can be used up to 500× without any maintenance.

24.8 Test Results

The product technology has been tested in challenge water to exceed the *US EPA Guide Standard and Protocol for Evaluation of Microbiological Water Purifiers* [16]. Specific tests on representative contaminants were conducted using appropriate test methods at third-party test laboratories with results as shown in Figure 24.6 below.

The five photochemical processes synergistically combine to reduce or eliminate a range of contaminants as shown in Figure 24.7. More than 800 contaminants have been researched and shown to be reduced by one or more of the photochemical processes. Note that several contaminants have been reported to be reduced by two or more of these processes.

Contaminant	Compound Feed (ppm)	Product (ppm)	% Reduced	Log Reduction
Raoultella terrigena	10^6 CFU/L	ND	99.9999%	>6
Poliovirus type 1	10^6 PFU/L	ND	99.9999%	>6
Cryptosporidium parvum oocysts	2×10^8 PFU/L		99.99%	>4.1
Simian Rotavirus	10^6 PFU/L	ND	99.9999%	>6
Malathion	0.0089	<0.00006	>99%	>2.17
Pyriproxyfen	0.0071	<0.00006	>99%	>2.17
Prometon	0.0089	<0.00006	>99%	>2.07
Carbon tetrachloride	3.317	2.293	30.9%	0.160
1,2,3-Trichloropropane	2.842	0.979	65.5%	0.463
Methyl *tert*-butyl ether	2.000	0.014	99.3%	2.150
Nitrobenzene	2.626	0.025	99.0%	2.018
Trichloroethylene	2.555	0.002	99.9%	3.133
Toluene	1.694	0.001	99.9%	3.201
Caffeine	3.883	0.513	86.8%	0.879
Arsenic	0.535	0.002	99.6%	2.40
Lead	0.535	0.002	99.6%	2.40
Mercury	0.393	0.0014	99.6%	2.45
Selenium	0.617	0.028	95.5%	1.35

FIGURE 24.6
Shield system removal performance on representative contaminants as tested by Oregon Health Sciences University, Pacific Agricultural Labs, University of Arizona, and Test America.

Contaminant	Puralytics Active Purification Processes				
	Photocatalytic Oxidation	Photocatalytic Reduction	Photolysis	Photo Adsorption	Photo Disinfection
Dichromate	■	■		■	
Arsenic		■		■	
Caffeine	■		■		
Cerium		■		■	
Cobalt		■		■	
Copper		■		■	
Cryptosporidium	■				■
Estradiol	■		■		
Giardia iamblia	■				■
Heterotrophic plate count	■			■	
Lead		■		■	
Lead dioxide		■		■	
Legionella	■				■
Legionella pneumophila	■				■
Manganese oxide		■		■	
Mercury (inorganic)		■		■	
Permanganate		■		■	
Saccharomyces cerevisiae	■				■
Silver		■		■	
Staphylococcus aureus	■				■
Streptococcus faecalis	■				■
Streptococcus sobrinus	■				■
Styrene	■		■		
Sulfamethoxazole	■		■		
Total coliforms (including fecal coliform and *E. coli*)	■				■
Turbidity	■			■	
Viruses (enteric)	■				■
1, 1, 1,2-Tetrachloroethane	■				
1, 1, 1,2-Trichloroethane	■				

FIGURE 24.7
Sample of >800 contaminants that have published research showing reduction by one or more of the five photochemical processes herein reported. A complete list can be downloaded at http://www.puralytics.com.

24.9 Conclusions

A water treatment system has been developed that incorporates five photochemical processes based on light-activated nanotechnology that work synergistically together to completely destroy microorganisms and significantly reduce a broad spectrum of chemical contaminants including emerging organic chemicals of concern, as well as many inorganic chemicals and heavy metals.

References

1. "Guidelines for drinking water quality," 4th Edition, World Health Organization, 2011 (http://www.who.int/water_sanitation_health/publications/2011/dwq_chapters/en/).
2. Howard, G., and Bartram, J., "Domestic water quantity, service level, and health," World Health Organization, 2003 (http://www.who.int/water_sanitation_health/diseases/WSH03.02.pdf).
3. "Human rights and access to safe drinking water and sanitation," United Nations Human Rights Council Resolution A/HRC/RES/15/9, 2010 (http://www.right2water.eu/sites/water/files/UNHRC%20Resolution%2015-9.pdf).
4. Hammond, A., Kramer, W. J., Tran, J., Katz, R., and Walker, C., "The Next 4 Billion: Market Size and Business Strategy at the Base of the Pyramid," World Resource Institute, 2007 (http://www.wri.org/publication/next-4-billion).
5. "Drinking water equity, safety, and sustainability," JMP Thematic Report on Drinking Water, by UNICEF and the World Health Organization, March 2011.
6. Duhigg, C., "Toxic waters," *New York Times*, December 16, 2009.
7. Galvez, J. B., and Rodriquez, S. M., *Solar Detoxification*. Paris, France: UNESCO, 2003.
8. Herrmann, J.-M., "Heterogeneous photocatalysis: State of the art and present applications," *Top. Catal.* 34(1–4):49–65, 2005.
9. Bouchy, M., and Zahraa, O., "Photocatalytic reactors," *Int. J. Photoenergy* 5:191–197, 2003.
10. Hashimoto, K., Irie, H., and Fujishima, A., "TiO_2 photocatalysis: A historical overview and future prospects," *Jap. J. Appl. Phys.* 44(12):8269–8285, 2005.
11. Blake, D. M., "Bibliography of work on the heterogeneous photocatalytic removal of hazardous compounds from water and air," NREL Technical Report TP-510-31319, November 2001. Available at http://www.nrel.gov/docs/fy02osti/31319.pdf.
12. Ollis, D. F., Pelizzetti, E., and Serpone, N., "Destruction of water contaminants," *Environ. Sci. Technol.* 25(9):1523–1529, 1991.
13. Halmann, M. M., *Photodegradation of Water Pollutants*. Boca Raton, LA: CRC Press, 1996.
14. Crawford, M. H., Banas, M. A., Ross, M. P., Ruby, S. R., Nelson, J. S., Boucher, R., and Allerman, A. A., "Final LDRD Report: Ultraviolet Water Purification Systems for Rural Environments and Mobile Applications," Albuquerque, NM and Livermore, CA: Sandia National Laboratories, 2005 (www.prod.sandia.gov/cgi-bin/techlib/access-control.pl/2005/057245.pdf).
15. Zimmer, J. L., and Slawson, R. M., "Potential repair of *Escherichia coli* DNA following exposure to UV radiation from both medium- and low-pressure UV sources used in drinking water treatment," *Appl. Environ. Microbiol.* 68(7):3293–3299, 2002.
16. US EPA—Task Force Report, "Guide Standard and Protocol for Testing Microbiological Water Purifiers," United States Environmental Protection Agency, Registration Division, Office of Pesticide Programs and Criteria and Standards Division, Office of Drinking Water, Washington, DC, 1987.

25

Modified TiO$_2$-Based Photocatalytic Systems for the Removal of Emerging Contaminants from Water

Ligy Philip, Arya Vijayanandan, and Jaganathan Senthilnathan

Department of Civil Engineering, Indian Institute of Technology (IIT) Madras, Chennai, India

CONTENTS

25.1 Introduction

Scarcity of resources and pollution of existing water resources are major problems faced nowadays. Anthropogenic activities and climate change have deteriorated the quality and quantity of available water. Population growth also significantly increased the demand for water. There is a severe water crisis in sub-Saharan Africa and South Asia. According to a United Nations Children's Fund/World Health Organization report, 2012, only 63% of the world population has improved sanitation facilities and 89% of the people have access to improved drinking water resources. India also faces water scarcity. Most of the people in India lack access to treated drinking water. The problem is severe since >700 million people reside in rural areas. It is estimated that around 40 million people are affected by waterborne diseases annually.

Various pollutants are found in the water bodies because of the discharge of industrial effluents, municipal sewage, and agricultural runoff. Open defecation also contaminates the nearby water body. Pollutants present in water can be pesticides, solvents, organic matter, or pathogens. With the use of more sophisticated instruments and new technologies, more and more pollutants are detected in water bodies. These include pharmaceuticals, personal care products (PCPs), pesticides, and solvents. These compounds are categorized as "emerging contaminants" since their presence has been identified in water only recently. These compounds can have adverse effects on humans at very low concentrations, and hence it is important to remove these pollutants from water. There is a class of pollutants known as endocrine disruptors, which cause abnormalities in the endocrine system by mimicking the hormones in the body.

25.2 Emerging Contaminants

Emerging contaminants include pollutants whose presence has been identified in water recently. These compounds can cause serious health effects to the ecosystem. The removal of these compounds becomes difficult, as these are present at very low concentrations. There is no regulation in the water quality standards for these contaminants. Hence, there is no measure to monitor these compounds in the effluents of wastewater treatment plants (WWTPs). Besides, complete information about the toxicity of these compounds is also not available. These compounds are getting into the water through municipal sewage. A major part of the ingested pharmaceuticals is excreted from the body and subsequently gets discharged into the wastewater. Most of the conventional WWTPs are incapable of removing these contaminants. As a result, these compounds will be present in WWTP effluent discharged into the water bodies. Many a time, the towns/cities located in the downstream of the water bodies/rivers use this water as the water source for water supply in the cities. As a result, human beings are getting exposed to these compounds very regularly.

Emerging contaminants include endocrine-disrupting compounds (EDCs), pharmaceutically active compounds (PhACs), and PCPs. PCPs are chemicals present in cosmetics, toothpaste, sunscreens, soaps, lotions, insect repellants, etc. Parabens and phenols are mainly present in these compounds. EDCs are the compounds that are not produced by the body but mimic or act like natural hormones in the body. These include solvents, pesticides, fungicides, pharmaceuticals, and plasticizers. Depending on the effect on the body, EDCs are divided into estrogenic or antiestrogenic, androgenic or antiandrogenic, and thyroid hormones. EDCs when present in very little concentrations also cause adverse effects in the body. There have been reports indicating that EDCs in aquatic systems cause feminization of fish, presence of intersex (neither male nor female), and increased egg production. Sex hormones include estriol, estradiol, and estrone.

PhACs include pharmaceuticals consumed by human or animals and pharmaceutical metabolites. These compounds can interact with biological systems through enzyme or receptors (Comerton et al., 2009). Pharmaceuticals include analgesics, antibiotics, antihyperlipidemics, antiepileptic, and stimulants. Pharmaceuticals can be nonsteroidal anti-inflammatory drugs or analgesics, which include weak acids. These are polar in nature and contain carboxyl and hydroxyl groups. Diclofenac, ibuprofen, naproxen, paracetamol, etc., belong to this category. Among these compounds, diclofenac showed most toxicity even under concentrations <100 mg/L (Cleuvers, 2003). Blood lipid–lowering agents include gemfibrozil, clofibrate, atorvastatin, etc. They are used very widely today for controlling cholesterol and triglycerides. Levofloxacin, ampicillin, erythromycin, sulfamethoxazole, tetracycline, etc., are the antibiotics detected in water bodies. Their presence in water should be seriously considered because continuous exposure to these compounds makes the microorganisms resistant to these drugs. Jones et al. (2002) classified antibiotics as extremely toxic to microorganisms and very toxic to algae. Antiepileptic drugs affect the central nervous system by reducing neuronal activity. These compounds can be lethal to the lower-level organisms. Carbamazepine is an antiepileptic drug that has been studied extensively. This compound is known to be carcinogenic to rats (Thacker, 2005).

Emerging contaminants in water affect the ecosystem adversely. Antibiotics in water can cause changes in the microorganisms present in water. It can lead to antibiotic resistance of microorganisms. Feminization of fish was reported in Northwestern Ontario, which was the effect of EDCs such as estrone, estradiol, and nonylphenol present in the river. In a Minnesota lake, female egg-yolk protein was found in male fish. Many cases were reported regarding the reproductive abnormalities in aquatic systems due to the presence of emerging contaminants. The main challenges faced in the treatment of emerging contaminants are their high solubility, high chemical stability, and low biodegradability. These compounds will be present in very little amounts in water in the range of nanograms per liter, which makes the monitoring of these compounds difficult.

25.3 Treatment Technologies

Nowadays there is much focus on the reuse of water. Hence, high-level treatment of wastewater becomes important. If possible, wastewater should be treated to drinking water quality. In the case of industrial wastewater also, tertiary levels of treatment should be carried out to meet the discharge standards. Biological treatment can be used, but it is time consuming. Presently, activated sludge process (ASP) is the most commonly used method for wastewater treatment. It is inefficient in removing recalcitrant compounds from water. Membrane technologies such as ultrafiltration, nanofiltration, or reverse osmosis give good results but they make the system expensive and pretreatment is required to prevent clogging. Advanced oxidation processes (AOPs) are effective in degrading the compounds. Photocatalytic methods are widely used for water treatment since they produce highly oxidative species like hydroxyl and superoxide radicals. In the case of wastewater treatment, photocatalysis can be used as a pre- or posttreatment system in order to increase the efficiency of the treatment.

25.3.1 Activated Sludge Process

The ASP is the most commonly used treatment process in conventional WWTPs. The organic matter present in the wastewater gets biodegraded under aerobic conditions. This method is very efficient in removing the suspended and dissolved solids in water. The organic matter gets degraded by the microorganisms, and the flocs get settled at the bottom of the tank. ASPs are not efficient in removing most of the emerging contaminants from water.

25.3.2 Membrane Bioreactors

In membrane bioreactors (MBRs), the organic matter gets degraded first, after which the particles are removed by membrane filtration. Effluent from an MBR is superior to the effluent from an ASP. Compared with an ASP, an MBR is more efficient in removing emerging contaminants from water. In biological processes, more time is required to degrade the pollutants. The efficiency of the system can be improved by increasing the retention time, which makes the biological process time consuming in treating emerging contaminants.

25.3.3 Membrane Processes

Membrane processes include microfiltration, ultrafiltration, and nanofiltration, and reverse osmosis. The particles are removed by straining in these processes. Reverse osmosis is very efficient in removing the organic and inorganic constituents in water. The main drawback of these methods is the high cost associated with the process. The main problem faced is membrane fouling, which reduces the membrane flux.

25.3.4 Activated Carbon Adsorption

In this method, the pollutants get adsorbed in the pores present in the activated carbon. Granular activated carbon or powdered activated carbon can be used for this method. The surface area is very large for these materials, which increases the adsorption efficiency. More hydrophobic, nonpolar, and low-solubility pollutants are easily removed by this method. Removal efficiency depends on the time of contact between polluted water and activated carbon. When more organic matter is present along with the emerging contaminants, the removal efficiency of emerging contaminants decreases owing to the competition for adsorption sites.

25.3.5 Chlorination

Chlorination is a conventional oxidation process in which the organic matter, ammonia, and pathogens get removed. This process is inexpensive compared with AOPs. However, many studies have reported the formation of organochlorine compounds such as trihalomethanes and haloacetic acids, which are carcinogenic in nature. These compounds are formed as the by-products of chlorination.

25.4 Advanced Oxidation Processes

AOPs are a good alternative to existing water treatment methods. Different AOPs include photochemical and nonphotochemical processes such as Fenton, photo-Fenton, H_2O_2/UV, O_3, O_3/H_2O_2, O_3/UV, O_3/H_2O_2/UV, and photocatalysis. In these methods, highly reactive oxygen species such as hydroxyl radicals, hydrogen peroxide, and superoxide are produced. The main oxidizing species is the hydroxyl radical. It has a very high oxidizing potential. The reaction rate is very high and since these radicals are nonselective in nature, they degrade the organic pollutants as well as microorganisms present in water. AOPs are finding increased usage in water and wastewater treatment because of the nonselectivity of the oxidative species generated and the reduced time requirement for degradation. AOPs are successfully used for the treatment of emerging contaminants. Commonly used oxidizers in AOPs are hydrogen peroxide and ozone. They are capable of degrading organics in water. The intermediates formed during the reactions can be either more or less toxic than the parent compound. To reduce toxicity, these oxidizers are used in conjunction with UV radiation. AOPs can be classified into photochemical and nonphotochemical processes depending on the way hydroxyl radical is generated. In case of a photochemical process, a light source, usually UV, is used along with catalysts or oxidizers. Different AOPs are described below. Relative oxidizing powers of different oxidizing species are given in Table 25.1.

TABLE 25.1

Relative Oxidation Power of Some Oxidizing Species

S. No.	Oxidizing Species	Relative Oxidation Power
1	Chlorine	1.00
2	Hypochlorous acid	1.10
3	Permanganate	1.24
4	Hydrogen peroxide	1.31
5	Ozone	1.52
6	Atomic oxygen	1.78
7	Hydroxyl radical	2.05
8	Positively charged hole on titanium dioxide, TiO_2^+	2.35

Source: Carey, J.H., *Water Pollut. Res. J. Can.*, 27, 1, 1992.

25.4.1 Fenton's Oxidation Process

It is a widely used process in which hydrogen peroxide is the oxidizer in the presence of iron catalyst. It is simplest and economical and the reagents are easily available. This method is used for the treatment of water and wastewater and hazardous wastes. Fenton's process is known to remove many recalcitrant compounds in wastewater. In this process, hydroxyl radicals are generated by the decomposition of H_2O_2 in the presence of iron. The reactions can be represented as follows.

$$Fe^{2+} + H_2O_2 \rightarrow Fe^{3+} + OH^- + {}^{\cdot}OH$$

$$Fe^{3+} + H_2O_2 \rightarrow Fe^{2+} + HO_2^{\cdot} + H^+$$

$$Fe^{3+} + HO_2^{\cdot} \rightarrow Fe^{2+} + O_2 + H^+$$

The radicals generated oxidize the organic components in water in the presence of iron. In the case of the photo-Fenton process, the oxidation of Fe^{2+} to Fe^{3+} and the hydroxyl radical generation takes place in the presence of a light source.

There are some disadvantages in this method. Residual iron present in water, in excess amounts after the process, can be toxic to the microorganisms. This affects the performance of the subsequent biological treatment. This process is dependent on the pH of the water. Some studies reported that a pH range of 3–3.5 is optimum for the process. Hence, the pH of the wastewater should be adjusted to achieve the optimum removal. Another drawback is the sludge generation due to the precipitation of iron. As iron has coagulant properties, sludge is generated after the treatment.

25.4.2 Photolysis

This process involves the use of natural or artificial light to degrade pollutants. Photochemical reactions take place, leading to the degradation of the target compound. One example of this process is UV treatment. UV treatment is used for disinfection since it is more efficient than chlorination and it does not produce any disinfection by-products. The efficiency can be increased by using hydrogen peroxide along with UV radiation. This method depends on the absorption spectra of the target compounds selected. This method is expensive and cannot be a sustainable solution to the water and wastewater treatment.

25.4.3 Ozonation

Ozone is a strong oxidizing agent. It produces reactive oxygen species on reaction with water. The performance of ozone can be increased if a catalyst such as hydrogen peroxide or iron is used. At high pH, more hydroxyl radicals are produced during ozonation. Ozonation can be used for water and wastewater treatment. Cost is an important criterion while adopting this method.

25.4.4 Electrochemical Oxidation

In this method, contaminants are degraded by anodic oxidation. Platinum, TiO_2, PbO_2, etc., are used as anodes. Oxidants such as ozone, hydroxyl radicals, and hydrogen peroxide are formed in the bulk solution, which also helps in the degradation of the contaminant. The performance of the system depends on the electrodes and electrolytes used and the current applied. To increase conductivity, NaCl is often used as a supporting electrolyte. However, use of NaCl leads to the formation of organochlorine compounds. The introduction of dissolved iron can enhance the efficiency by reducing H_2O_2 to hydroxyl radicals, thus incorporating the benefits of Fenton's reaction. Boron-doped diamond is becoming more popular as the anode.

25.4.5 Sonolysis

This method involves the use of high-intensity ultrasound waves at high frequencies, which can produce cavitation. The cavitation bubbles act as microreactors and produce hydroxyl radicals. More volatile compounds get degraded faster since they are usually present at the liquid–gas interface due to sonochemical reactions. The presence of iron will increase the performance since H_2O_2 is produced during the sonochemical reactions.

25.4.6 Photocatalysis

Photocatalysis comes under AOPs in which oxidative species are produced under a light source. It can be divided into homogeneous and heterogeneous photocatalysis depending on the phase of the photocatalyst and the reactant. Among these, heterogeneous photocatalysis is more commonly employed.

25.4.7 Heterogeneous Photocatalysis

When the photocatalyst and the reactant are not in the same phase, the reaction is known as heterogeneous photocatalysis. Semiconductors such as TiO_2, CdS, and ZnO belong to this category. Semiconductors have a filled valence band and an empty conduction band. When a semiconductor surface is irradiated by a light source, electrons from the valence band absorb the energy of photons, get excited, and are shifted to the conduction band, leaving a hole in the valence band. Thus electron–hole formation occurs. In the absence of suitable electron or hole acceptors, the electrons and holes are recombined. The excitation of electrons occurs only if the energy of the photons (h_ν) is greater than or equal to the band-gap energy (E_G). The electrons and holes formed react with water to produce highly oxidizing species such as hydroxyl radicals and superoxide radicals. These radicals react with the pollutants or pathogens in water to degrade it. The reactions continue to produce various intermediates, and final products are formed. In most of the cases, it is completely mineralized to CO_2 and water. Among various semiconductors, TiO_2 has proven to be the most suitable one for photocatalysis applications.

25.5 TiO₂ as a Photocatalyst

TiO_2 is photocatalytically active, biologically and chemically inert, does not undergo photo-corrosion and chemical corrosion, and it is inexpensive. UV/TiO_2 has been used to remove a wide range of pollutants in the liquid and the gas phase over the past several decades. In the case of titanium dioxide, the band-gap energy is 3.2 eV and it needs UV radiation for excitation. The energy band diagram of TiO_2 is shown in Figure 25.1. The reaction mechanism of photocatalysis by semiconductors is presented in Figure 25.2.

The reactions occurring on the surface of the catalyst can be represented as follows.

$$TiO_2 + h\vartheta \rightarrow e_{cb}^- + h_{vb}^+$$

$$O_2 + e_{cb}^- \rightarrow O_2^{-\cdot}$$

$$h_{vb}^+ + H_2O \rightarrow \,^\cdot OH + H^+$$

$$^\cdot OH + \,^\cdot OH \rightarrow H_2O_2$$

$$O_2^{-\cdot} + H_2O_2 \rightarrow \,^\cdot OH + OH^- + O_2$$

$$O_2^{-\cdot} + H^+ \rightarrow \,^\cdot OOH$$

$$^\cdot OH + \text{Organic compounds or microbes} + O_2 \rightarrow CO_2 + H_2O$$

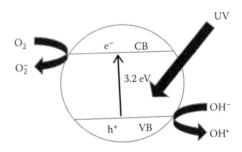

FIGURE 25.1
Energy band diagram of TiO_2.

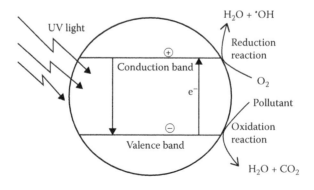

FIGURE 25.2
Reaction mechanism of photocatalysis by semiconductors.

Irradiation of an aqueous TiO_2 suspension offers an oxidation capable of mineralization of many of the recalcitrant organic substances into CO_2, H_2O, and other associated inorganic components. Numerous studies have reported that the photocatalytic activity would be affected by the crystalline structure of the TiO_2 photocatalyst. TiO_2 in the anatase form appears to be photocatalytically more active compared with the rutile form and convenient for water purification, water treatment, hazardous waste control, air purification, and water disinfection. The lower photocatalytic activity of rutile compared with that of anatase is attributed to the lower capacity of the rutile surface to adsorb molecular oxygen, leading to higher rates of e^-–h^+ pair recombination. Furthermore, the difference between rutile and anatase particles is that the main product from molecular oxygen on rutile particles is superoxide ion and it is hydrogen peroxide on anatase particles. Moreover, control of the morphology, particle size, particle size distribution, phase composition, and porosity of TiO_2 is vital to achieve optimum photocatalytic activity.

A variety of methods have been proposed for the preparation of *n*-TiO_2 such as flame synthesis, sol–gel routes, thermal hydrolysis, and reverse microemulsion method. Among these techniques, sol–gel thin film deposition offers several advantages over other techniques. The main disadvantage of the sol–gel process is agglomeration of hydrolyzed TiO_2 particles (Mohammadi et al., 2006). Many methods have been developed to disperse the aggregated TiO_2 particles such as washing with organic solvent, freeze-drying, followed by heating and adding surfactants. The most economical and efficient way to prevent the aggregation of TiO_2 is the addition of surfactants. Surfactants are well known as stabilizers, template agents, and shape directors. The highly photocatalytically active mesoporous TiO_2 with particle sizes ranging from submicron to micron were synthesized using nonionic surfactants. The preparation of surfactant-assisted *n*-TiO_2 microspheres showed better catalytic efficiency than Degussa P-25 TiO_2.

25.6 Factors Affecting the UV/TiO₂ Process

The effects of operating parameters on the photocatalytic efficiency of UV/TiO_2 are described below.

25.6.1 Substrate Concentration

The substrate concentration can influence the extent of adsorption and rate of reaction at the surface of the photocatalyst. It is an important parameter for optimization between high degradation rate and efficiency. For a steady-state balance in a photocatalytic system, the rate of mass transfer must be equal to the rate of reaction.

25.6.2 Light Intensity

Light irradiance plays a significant role in all photocatalytic reactions and determines the number of e^- and h^+ pairs created. Accordingly, an increase in the incident photon rate would result in an increase in the photocatalytic reaction rate. For a simple set of reactions including only charge–carrier generation, recombination, reduction, and oxidation, the rate of oxidation of a particular compound is proportional to the light intensity. This phenomenon indicates that high photon flux increases the probability of collision between photons and activated sites on the catalyst surface and enhances the rate of photocatalytic reaction. However, the limited surface area of catalyst particles and the diffusion rate of species involved in the reaction reduce the efficiency of photons in inducing the oxidation reaction. At higher incident light intensity, the degradation rate is proportional to the square root of the light intensity. The reason for the square root relationship has been explained by bulk recombination of e^-–h^+ pairs within the catalyst particles.

25.6.3 Photocatalyst Dosage

It is well documented that the photocatalytic reaction rate and efficiency would increase with photocatalyst dosage. The increase in the photocatalytic efficiency seems to be due to the increase in the total surface area available for the photocatalytic reaction as the dosage of the photocatalyst is increased. However, when the TiO_2 is overdosed, the number of active sites on the TiO_2 surface may become almost constant because of the decreased light penetration, increased light scattering, and the loss in surface area due to agglomeration (particle–particle interactions) at high solid concentration. In any given application, the optimum photocatalyst concentration has to be determined in order to avoid excess photocatalyst usage and to ensure total adsorption of efficient photons. The optimal photocatalyst dosage or effective optical penetration length, under given conditions, is very important in designing a slurry reactor for effective use of reactor space and photocatalyst.

25.6.4 Solution pH

The pH of an aqueous solution significantly affects all semiconductor oxides, including the surface charge on the semiconductor particles, the size of the aggregates formed, and the energy of the conduction and valence bands. At higher pH, valence band electrons become more effective and the conduction band holes become less effective. Additionally, suspended TiO_2 particles in water are known to be amphoteric. The effect of solution pH on degradation rate depending on the acidity or basic property of the substrate should also be considered while designing a photocatalytic reactor. In alkaline conditions, a high level of hydroxide ions (OH^-) induce the generation of hydroxyl free radicals ($HO^\bullet$), which come from the photooxidation of OH^- by holes forming on the TiO_2 surface. Since the hydroxyl free radical is the dominant oxidizing species in the photocatalytic process, the photodecay of organic compounds is therefore accelerated at higher solution pH.

25.6.5 Temperature

Like most photochemical reactions, a photocatalytic reaction is not significantly sensitive to minor variations in temperature. Therefore, the potentially temperature-dependent steps such as adsorption and desorption do not appear to be rate determining. However, higher temperatures may have a negative effect on the concentration of dissolved oxygen (DO) in the solution. DO concentration below a certain point may allow for e^-–h^+ recombination at the surface of the TiO_2. In this respect, the photocatalytic system is appropriate for treating contaminated water and wastewater at temperatures close to the ambient conditions.

25.6.6 Electron Scavenger

The photo-induced electrons must be removed from the TiO_2 to prevent the e^-–h^+ pair recombination and to enhance the efficiency of photocatalytic degradation. Molecular oxygen has been employed as an effective electron acceptor in most of the photocatalysis applications. Oxygen can be reduced to the superoxide, O_2^-, which may also participate in the degradation reaction of the organic molecules or be further reduced to hydrogen peroxide (H_2O_2). Owing to the electrophilic property, oxygen plays a decisive role in the mechanism of the photocatalytic degradation. Photocatalytic efficiency can be enhanced by the addition of H_2O_2. However, H_2O_2 has a dual effect on the photocatalytic degradation rate of organic substrates. Excess H_2O_2, which acts as an ·OH scavenger, produces an inhibitory effect in the photocatalytic efficiency. Electron scavenging and the consequent e^-–h^+ recombination suppression can also be achieved by the use of other inorganic oxidants. The use of an optimum concentration of inorganic peroxides has been demonstrated to enhance the degradation rate of several organic pollutants.

25.6.7 Photocatalyst Deactivation

Deactivation is an important issue for practical applications of photocatalysts. Anions (Cl^-, ClO_4^{2-}, NO_3^-, CO_3^{2-}, HCO_3^-, SO_3^{2-}, PO_4^{3-}) commonly found in neutral or polluted waters retard the oxidation rate of organic compounds either by competing for radicals or by blocking the active sites of the TiO_2 photocatalyst.

25.7 Modified TiO$_2$ for Improved Photocatalytic Activity under Visible Light

Efficient utilization of visible and other solar light is one of the major goals of modern science and engineering that will have a great impact on technological applications. Although an AOPs with TiO_2 photocatalysts has been shown to be an effective alternative in this regard, the vital snag of the TiO_2 semiconductor is that it absorbs a small portion of the solar spectrum in the UV region (band-gap energy of TiO_2 is 3.2 eV). To utilize the maximum solar energy, it is necessary to shift the absorption threshold toward the visible region. Shifting of the TiO_2 absorption into the visible light region mainly focuses on the doping with transition metals. Two general methods have been used to increase the photocatalytic activity of TiO_2 for visible light and solar light irradiation: (i) use of an organic dye as photosensitizer and (ii) doping TiO_2 with metallic and nonmetallic elements.

25.7.1 Use of Organic Dye as Photosensitizer

Use of organic dye as a photosensitizer works very well under conditions where oxygen/air is excluded and the degradation of the dye is minimized by the efficient quenching of the dye oxidation state with an appropriate electrolyte. Otherwise, the dye becomes rapidly mineralized and the photocatalytic system loses its response toward visible light in the presence of oxygen. In dye sensitization, the most relevant points are (i) the absorption spectrum of the dye in the visible region and (ii) the energy of the electron in the excited electronic state of the dye, which has to be high enough to be transferred to the semiconductor conduction band.

25.7.2 Doping with Metal Ions

Transitional metal and rare earth metal ion doping have been extensively investigated for enhancing TiO_2 photocatalytic activities. It was found that the doping of metal ions could expand the photoresponse of TiO_2 into the visible spectrum.

Doping of TiO_2 with transition metals such as iron, cobalt, nickel, manganese, chromium, vanadium, copper, molybdenum, zirconium, silver, and zinc was tried extensively. Such metal creates a "new" electron state inside the TiO_2 forbidden band, which can capture the excited electrons from the TiO_2 valance band and consequently maintain the holes. It can also allow the light absorption to be widened into the visible region to various extents depending on the type of dopant and its concentration. Therefore, photocatalysis on TiO_2 can be promoted using visible light. The photodegradation of various organic substrates depends on the nature of the dopant, concentration, and the microstructural characteristics of the catalyst. Usually there is a critical doping concentration at which any further increase in dopant concentration results in the charge carrier recombination, thus lowering the photoactivity of the prepared doped TiO_2. Excess of dopant on the particle surface of TiO_2 lessens the specific area of TiO_2 and hinders the adsorption of reactant, and thus inhibits the photocatalytic activity.

Takeuchi (2003) has investigated the electron spin resonance signals to investigate electron transfer from TiO_2 to Pt particles. It was found that Ti 3p signals increased with irradiation time and the loading of Pt reduced the amount of Ti 3p. This observation indicates the occurrence of electron transfer from TiO_2 to Pt particles. As electrons accumulate on the metal particles, their Fermi levels shift closer to the conduction band of TiO_2 (Jakob et al., 2003; Subramanian et al., 2003, 2004), resulting in more negative energy levels. Thermal instability, tendency to form charge carrier recombination centers, as well as the expensive ion implantation facilities make metal-doped TiO_2 impractical, at many times.

25.7.3 Doping with Nonmetallic Elements

TiO_2 absorption toward the visible light with nonmetallic elements such as nitrogen (N), sulfur (S), carbon (C), and phosphorus (P) has been known. Asahi et al. (2001) were the first to show an absorption increase in the visible region upon nitrogen doping. This opened the way to study TiO_2 doping with nonmetallic elements. The insertion of N or S atoms on TiO_2 produces localized states within the band gap just above the valence band. Thus, when N- or S-doped TiO_2 is exposed to visible light, electrons are promoted from these localized states to the conduction band. The substitutional (N–Ti–O) doping of N is most effective compared with other nonmetal dopants (S, P, and C) because its p states contribute to the band-gap narrowing by mixing with O 2p states. Although doping with S shows a similar band-gap narrowing, it requires larger formation energy for the substitution than that required for the substitution of N. Moreover, the ionic radius of S is larger than O; thus, it is difficult to fit it into the TiO_2 crystal.

The mechanism on enhancement of nitrogen doping is that N-doping narrowed the gap between the valence band and conduction band of N-doped TiO_2. Asahi et al. (2001) proposed that the presence of nitrogen introduced new occupied orbitals in between the valence band and the conduction band. These N-2p orbitals acted as a step up for the electrons in the O-2p orbital, which once populated had now a much smaller jump to make, to be promoted into the conduction band. Once this process occurs, electrons from the original valence band can migrate into the mid-band-gap energy level, leaving a hole in the valence band, which reacts as described before. The reaction mechanism of nonmetal-doped TiO_2 is given in Figure 25.3.

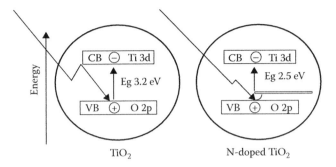

FIGURE 25.3
Reaction mechanism of N-doped TiO_2. (From Asahi, R. et al., *Science*, 293, 269, 2001.)

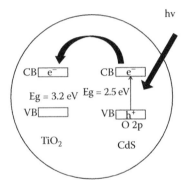

FIGURE 25.4
Coupling of TiO_2 with CdS.

25.7.4 Coupling with Semiconductors

Coupling with other semiconductors also increases photocatalytic efficiency by shifting the absorption to the visible region. In this method, a large band-gap semiconductor is coupled with a small band-gap semiconductor with a more negative conduction band level. The small band-gap semiconductor can be excited under visible light. The photogenerated electrons are transferred to the conduction band of the large band-gap semiconductor (Malato et al., 2009). The mechanism is shown in Figure 25.4.

25.7.5 Metal Ion Implantation

In this method, TiO_2 is bombarded with high-energy transition metals by applying a high voltage. This process changes the electronic structure of TiO_2 and modifies its photocatalytic activity to the visible region up to 600 nm (Malato et al., 2009). Metal ions such as Cr, Fe, Ni, and V are used for this method. Metal ion–implanted TiO_2 possesses a lot of advantages and it is known as the second-generation photocatalyst (Anpo, 2000).

25.7.6 Use of Photosensitizers

Dyes having visible light response and redox properties are used as sensitizers on the TiO_2 surface. The reactions occurring can be represented as follows (Malato et al., 2009):

$$dye + h\vartheta \rightarrow dye^*$$

$$dye^* - \xrightarrow{\ TiO_2\ } dye^+ + e^-$$

$$dye^+ + e^- \rightarrow dye$$

The dye gets activated under visible light and transfers the electrons to the semiconductor. Dyes such as thionine, toluidine blue, and methylene blue are used as photosensitizers.

25.8 Photoreactors Used in Water Treatment

Photoreactors can be divided in two groups—suspended and immobilized reactors depending on whether the catalyst used is in slurry form or in immobilized form. In case of slurry reactors, posttreatment is used to recover the catalyst. Commonly used methods are filtration, coagulation, flocculation, or centrifugation. Hence, slurry reactors become expensive. However, slurry reactors possess more photocatalytical efficiency compared with the immobilized form, due to increased surface area to volume ratio. Also, the catalyst is reused after the treatment. In the case of immobilized catalyst reactors, the catalyst is coated on the inner walls of the reactor. Selection of a suitable substrate is important for fixed-bed reactors.

The following factors should be considered while designing a photoreactor: concentration of the contaminant to be treated, flow rate of water, catalyst to be used, and electron acceptor. In case of solar photocatalysis, the collector can be a concentrating type or a non-concentrating type.

25.8.1 Practical Applications of Solar Photocatalysis

Photocatalytic treatment systems have been successfully implemented in many industries. Some of the examples are given in this section. Solar photocatalysis can remove hazardous compounds, emerging contaminants, and recalcitrant compounds from water. The reactors generally use titanium dioxide. Hence, modifications are made in the design of reactors to utilize the UV radiation in the sunlight since normal TiO_2 can be photoactivated only under UV radiation.

It has been reported that parabolic concentrating-type reactors were used for solar photocatalysis. It was used for removal of volatile organic compounds from groundwater (Mehos et al., 1992), pesticide removal (Hermann et al., 1998; Malato et al., 1998), and heavy metal removal (Prairie et al., 1992). However, Robert and Malato (2002) reported that there are some disadvantages in the parabolic concentrating-type reactors. For example, they cannot make use of the UV radiation in diffuse form. They also reported that compound parabolic collector technology (CPC) collectors show excellent performance under solar radiation.

A pilot plant, working in recycle mode, is operated in the Volkswagen AG factory in Wolfsburg (Germany) after getting promising results in laboratory-scale studies. The plant consists of 12 doubleskin sheet photoreactors. It has a total irradiated area of 27.6 m². The influent of the photoreactor is biologically pretreated. The reaction takes place for 8–11 h during daytime. A suspended catalyst slurry is used for the treatment. The catalyst particles are sedimented and reused after the reaction (Bahnemann, 2004).

Solar photocatalysis was used to treat biologically pretreated textile wastewater and was able to give a maximum degradation rate of 3 g COD/h/m² (COD: chemical oxygen demand) (Freudenhammer et al., 1997). The studies were done using a thin film fixed-bed reactor.

In Spain, under the SOLARDETOX project (Solar Detoxification Technology for the Treatment of Industrial Non-Biodegradable Persistent Chlorinated Water Contaminants), a commercial nonconcentrating solar detoxification system is used to treat water. The plant can treat 1 m³ of contaminated water. It uses a CPC with an aperture area of 100 m². In this reactor, catalyst slurry is used for treatment. After treatment, catalyst is recovered by sedimentation by adjusting the pH.

25.9 TiO₂ for Pesticide Degradation

Pesticides are widely used for the agricultural purposes in the world. These pesticides reach nearby water bodies through agricultural runoff, thereby contaminating them. Percolation of pesticides into the soil contaminates groundwater also. Many studies have revealed the presence of pesticides like DDT, lindane, endosulfan, methyl parathion, etc., in the water. Most of the pesticides belong to the category of EDCs, making pesticide-contaminated water a threat to human health. Hence, it is very important to remove these contaminants from drinking water. Among all the pesticides, organochlorine pesticides are the most difficult to degrade. Conventional methods fail to remove the pesticides completely and they are also dangerous since the intermediates of these compounds can also be toxic. Advanced oxidation is a promising method in the removal of pesticides due to its nonselectivity, high reaction rate, and the fact that it leads to the complete mineralization of the target compound.

25.9.1 Photodegradation of Methyl Parathion Using Suspended and Immobilized TiO₂ Systems under UV Light

Many researchers have studied the photocatalytic degradation of methyl parathion in aqueous TiO₂ suspensions. Pignatello and Sun (1995) have reported on the complete oxidation of methyl parathion (initial concentration 0.08 mM) in aqueous solution using the UV photoassisted Fenton (Fe^{3+} and H_2O_2) reaction. Under these conditions, methyl parathion yielded quantities corresponding to stoichiometric ratios of HNO_3, H_2SO_4, and H_3PO_4, with oxalic acid, 4-nitrophenol, and dimethylphosphoric acid being identified as intermediate species. Konstantinou et al. (2001) investigated the photocatalytic oxidation of several organophosphorus compounds like ethyl parathion, methyl parathion (1 mg/L), dichlorofenthion, and ethyl bromophos in aqueous TiO₂ (100 mg/L) suspensions. All investigated pesticides were sufficiently degraded in aqueous TiO₂ suspensions, irradiated with simulated solar light. The half-lives ranged from 10.2 to 35.5 min for the organophosphorus insecticides used for the study. The rates of catalytic disappearance depend on various parameters such as initial concentration, radiant flux, wavelength, mass, type of photocatalyst, and type of photoreactor. Sanjuan et al. (2000) described the use of 2,4,6-triphenylpyrylium ion encapsulated within Y zeolite as a photocatalyst for the photodegradation of methyl parathion (3×10^{-4} M) in aqueous suspensions. Encapsulation stabilizes the pyrylium ion leading to degradation of methyl parathion, and it was suggested that radical intermediates are involved in the degradation process.

Moctezuma et al. (2007) reported that the photocatalytic degradation of methyl parathion (initial concentration 50 mg/L) was monitored using different complementary analytical techniques to identify the organic intermediates. The results of their study clearly indicated that the first step of the photocatalytic degradation of methyl parathion was its oxidation to methyl paraoxon, and this reaction was not affected by the pH of the aqueous medium. On the other hand, the photocatalytic oxidation of methyl paraoxon was strongly influenced by the pH of the reaction solution. Hence, the pH of the solution is an important parameter for the methyl parathion degradation. Under alkaline conditions, only 60% of the total organic carbon (TOC) was converted to CO_2 after 6 h of reaction. In acidic pH, 90% of TOC was converted to CO_2 in similar condition.

Evgenidou et al. (2007) studied the photocatalytic oxidation of methyl parathion using TiO_2 and ZnO as catalysts. In the presence of TiO_2, complete disappearance of the methyl parathion was observed after 4 h, while dissolved organic carbon (DOC) was only slightly reduced in the first 2 h, indicating the formation of intermediates. At the end of treatment, >90% of the DOC was reduced. Stoichiometric release of nitrate and sulfate ions was achieved after 2 and 3 h, respectively, while phosphates were formed more slowly indicating the formation of phosphate-containing intermediates. On the other hand, in the presence of ZnO, mineralization proceeded in a slower rate since only 50% of DOC was reduced after 6 h of irradiation. Nitrate and sulfate ions contributed only 50% and 70% of the expected amount, respectively, while no formation of phosphates was observed even at the end of treatment. This indicated that TiO_2 proved to be a more efficient photocatalyst than ZnO and complete mineralization was achieved only in the presence of TiO_2. Xiaodan et al. (2006) studied the degradation of methyl parathion (50 mg/L) using anatase TiO_2, Degussa P-25, and ZnS/TiO_2, which was prepared with different ratios of ZnS and TiO_2 ($ZnS/TiO_2 = 1$, $ZnS/TiO_2 = 2$, $ZnS/TiO_2 = 3$). The combination with ZnS/TiO_2 ratio = 3 exhibited the highest photocatalytic activity among the other photocatalytic materials under visible light. The photocatalytic activity increased with increase of ZnS doping.

Kim et al. (2006) studied the photocatalytic degradation of methyl parathion using a circulating TiO_2/UV reactor. The experimental results showed that parathion was more effectively degraded in the photocatalytic condition than the photolysis or TiO_2 catalytic condition. With photocatalysis, 10 mg/L methyl parathion was completely degraded within 60 min with a TOC decrease exceeding 90%, after 150 min. Organic intermediates like 4-nitrophenol and paraoxon were also identified during the degradation of methyl parathion and these were further degraded.

25.9.2 Photodegradation of Lindane Using Suspended and Immobilized TiO_2 Systems under UV Light

Zaleska et al. (2000) studied the degradation of lindane (5 mg/L) with suspended and immobilized TiO_2 under UV light. In the presence of TiO_2 supported on glass microspheres, a rapid decrease of lindane concentration was observed at the beginning of the reaction and 30 min of irradiation eliminated 68% of lindane. A much slower concentration drop was observed during subsequent 120 min of irradiation. Lindane degradation efficiency in the $TiO_2/UV/O_2$ system using suspension anatase TiO_2 was very close to that obtained for the supported TiO_2 system. In the suspension system, 77% of lindane was eliminated after 150 min of irradiation. Hiskia et al. (1997) studied the photodegradation of lindane in aqueous solution, using near-visible and UV light (>320 nm) and in the presence of the polyoxometalate $PW_{12}O_{40}^{3-}$, it is converted to CO_2 and HCl. Initial photodecomposition took place within a few minutes, both in the presence and absence of oxygen, and the effective mineralization in the absence of oxygen suggests that OH· radicals act as the primary oxidant, in this case.

Guo et al. (2000) have reported that microporous polyoxometalates ($H_3PW_{12}O_{40}/SiO_2$ and $H_4SiW_{12}O_{40}/SiO_2$) were synthesized by encapsulating $H_3PW_{12}O_{40}$ and $H_4SiW_{12}O_{40}$ into a silica matrix via hydrolysis of tetraethyl orthosilicate. The degradation study was carried out with different concentrations of HCH (0–30 mg/L); a plot of (initial HCH degradation rate) r_o vs. Co (initial HCH concentration) exhibited a straight line; and the linear correlation coefficients (R) were 0.995 and 0.998 for $H_3PW_{12}O_{40}$ and $H_4SiW_{12}O_{40}$, respectively. This indicated that the disappearance of HCH followed a Langmuir–Hinshelwood first-order kinetic law for the initial HCH concentration chosen. Guillard et al. (1996) identified the intermediate products formed in a photocatalytically treated aqueous solution containing 1 g/dm³ of lindane. Several intermediate by-products have been identified during the degradation of lindane such as chlorocyclohexanes, chlorobenzenes, chlorophenols, chloropropanes, chloropropanones, and the pentachlorocyclohexanone isomer. The detection of heptachlorocyclohexane in the system showed that chlorine and hydrogen atoms were not only abstracted from the CHCl groups constituting lindane but also added to them.

Nienow et al. (2008) have reported that the aqueous solutions of lindane were photolyzed (k = 254 nm) under a variety of solution conditions. The initial concentrations of hydrogen peroxide (H_2O_2) and lindane varied from 0 to 20 mM and 0.21 to 0.22 µM, respectively. Lindane rapidly reacted, and the maximum reaction rate constant (9.7×10^{-3} s⁻¹) was observed at pH 7 with an initial H_2O_2 concentration of 1 mM. Thus, 90% of the lindane is destroyed in ~4 min under these conditions. In addition, within 15 min, all chlorine atoms were converted to chloride ions, indicating that chlorinated organic by-products did not accumulate. Antonaraki et al. (2010) have carried out a detailed examination of the intermediates formed during the photocatalytic degradation of lindane (initial concentration 2.4×10^{-5} M) using $PW_{12}O_{40}^{3-}$ (7×10^{-4} M) catalyst. Lindane was completely mineralized into CO_2, Cl⁻, and H_2O. The intermediates were identified using gas chromatography–mass spectrometry (GC–MS) analysis. The intermediates identified were aromatic compounds such as dichlorophenol, trichlorophenols, tetrachlorophenol, hexachlorobenzene, di- and trichloro-benzenodiol, and nonaromatic cyclic compounds such as pentachlorocyclohexene, tetrachlorocyclohexene, heptachlorocyclohexane, aliphatic compounds, tetrachloroethane, and the condensation products polychlorinated biphenyls.

Dionysiou et al. (2000) have reported on the degradation of lindane using TiO_2 immobilized on a continuous flow rotating disc and have achieved 63% lindane (initial concentration 0.016 mM) degradation. The rotating disc photoreactor operated at a hydraulic residence time of 0.25 days and at a disk angular velocity of 12 rpm. Cao et al. (2008) have reported on the degradation of lindane (0.0172 mM) using 4.2 mM persulfate $\left(S_2O_8^{2-}\right)$ and 0.108 mM ferrous (Fe^{2+}) ions in aqueous solution. Persulfate was activated by ferrous iron and produced highly potent sulfate radicals $\left(SO_4^-\right)$ with standard potential at 2.6 V. The rate of lindane oxidation was proportional to the concentration of persulfate in the aqueous solution. The final products of lindane degradation were Cl⁻ ion and CO_2, signifying complete lindane oxidation. Compared with the classical Fenton's reaction, persulfate has a relatively longer lifetime in water.

Gkika et al. (2004) have studied the photocatalytic degradation of lindane (3.7×10^{-5} M) using $PW_{12}O_{40}^{3-}$ (7×10^{-4} M) under UV light in aqueous solution. This study was carried out in the presence and absence of dioxygen. In the absence of dioxygen, neither the regeneration (reoxidation) of the catalyst nor the photodegradation of lindane took place. Under the reported experimental conditions, decomposition of lindane was achieved within 60 min. The complete photodegradation of lindane leads to its mineralization to CO_2 and Cl^- as final products.

25.9.3 Photodegradation of Dichlorvos Using Suspended and Immobilized TiO_2 Systems under UV Light

Many reports on the photocatalytic reduction of dichromate and the photocatalytic oxidation of dichlorvos are available (Zhao et al., 1995; Chen and Cheng, 1999; Khalil et al., 1998; Obee and Satyapal, 1998; Konstantinou et al., 2001; Ku and Jung, 2001; Schrank et al., 2002). Dichromate (Cr^{6+}) is a common inorganic pollutant, while dichlorvos, a widely used organophosphorous pesticide, is also an organic pollutant. Shifu and Gengyu (2005) have used dichromate and dichlorvos as inorganic and organic pollutants and TiO_2/beads as a photocatalyst. Initial concentrations of 3.8×10^{-4} M and 1.0×10^{-4} M of $Cr_2O_7^{-7}$ and dichlorvos, respectively, were used for oxidation and reduction reactions. The optimum amount of the photocatalyst used was 6.0 g/cm^3. The addition of trace amounts of Fe^{3+} or Cu^{2+} accelerated both the reactions. On the other hand, addition of Zn^{2+} and Na^+ did not influence the reactions. Acidic solution was favorable for the photocatalytic reduction of dichromate, and acidic and alkaline solutions were favorable for the photocatalytic oxidation of dichlorvos. Addition of SO_4^{2-} accelerated the photocatalytic oxidation, whereas addition of Cl^- slowed down the reaction.

Evgenidou et al. (2006) have investigated the photocatalytic degradation of two organophosphorous insecticides such as dichlorvos (initial concentration 20 mg/L) and dimethoate (initial concentration 20 mg/L) using a suspension of TiO_2 (100 mg/L) under UV light. The photocatalytic intermediates, end products, and reaction mechanisms were established in this study. The intermediate by-products such as *O,O*-dimethyl phosphonic ester, 2,2-dichlorovinyl *O*-methylphosphate, and *O,O,O*-trimethyl phosphoric ester have been identified during the degradation of dichlorvos. Harada et al. (1990) reported that the efficient degradation of dichlorvos occurred by using a Pt/TiO_2 photocatalyst under a super-high-pressure mercury lamp and produced mineralized nontoxic end products such as Cl^-, PO_4^{3-}, H^+, and CO_2.

Rahman and Muneer (2005) have studied the degradation of dichlorvos (1 and 0.5 mM) using three different photocatalysts (1 g/L), namely Degussa P-25, Hombikat UVI00 (SachtlebenChemie GmbH), and PC 500 (Millennium Inorganics). Among these three catalysts, Degussa P-25 showed highest photocatalytic activity. Oancea and Oncescu (2008) reported that the photocatalytic degradation of dichlorvos under UV irradiation was optimized with respect to the flow rate of O_2 gas. The photocatalyst concentration of dichlorvos used in the study was 1.66×10^{-4} M, and pH 4 was maintained in the system. The rate of degradation depended on the DO concentration. Decrease in reaction solution pH suggested the formation of organic acids. The presence of organic intermediates was confirmed also by TOC measurements.

25.9.4 Degradation of Dichlorvos, Methyl Parathion, and Lindane under Visible and Solar Light

Although photocatalytic degradation of pesticide using TiO_2 under UV light is possible, it may not be a practical proposition for the treatment of drinking water sources due to the high cost. Efficient utilization of visible and other solar light is one of the major goals of modern science and engineering that will have a great impact on technological applications (Ollis and Al-Ekabi, 1993). Although AOPs with TiO_2 photocatalysts have been shown to be an effective alternative in this regard, the vital snag of the TiO_2 semiconductor is that it absorbs a small portion of solar spectrum in the UV region (band-gap energy of TiO_2 is 3.2 eV). To utilize the maximum solar energy, it is necessary to shift the absorption threshold toward the visible region (Asahi et al., 2001; Chatterjee and Mahata, 2004).

Only a few studies have been reported on degradation of dichlorvos, methyl parathion, and lindane under visible and solar light. Oancea and Oncescu (2008) have studied the photocatalytic degradation of the dichlorvos (initial concentration 4.5×10^{-5} to 2×10^{-4} M) in suspended TiO_2 (6.02×10^{-5} mL) system under solar irradiation. Dichlorvos was degraded and Cl^- and PO_4^{3-} ions were formed as its mineralization end products. Chloride ion was totally released into solution, whereas PO_4^{3-} was only partially liberated and formed phosphate-containing organic compounds as stable intermediates. Vidal (1998) has studied the degradation of lindane with 0.05% suspension of TiO_2 under solar light. The initial concentration of lindane was in the range of 200–500 µg/L, as it is a representative level of pollutants in natural waters. Solar photocatalytic experiments of lindane were carried out where average radiation ranged from 700 to 800 W/m² with a maximum of 1000 W/m². Under these conditions, the lindane concentration in the solution was reduced from 200 to 500 µg/L to maximum permitted levels (0.1 µg/L) during 30 min of solar exposure. Shifuand Gengyu (2005) has studied the photocatalytic degradation of dichlorvos, monocrotophos, phorate, and parathion using TiO_2 supported on fiberglass cloth as catalyst. The results showed that 0.65×10^{-4} M/dm³ of dichlorvos, monocrotophos, phorate, and parathion could be completely degraded into PO_4^{3-} after 200 min of illumination with a 375 W medium-pressure mercury lamp.

Naman et al. (2002) studied the photodegradation of dichlorvos under UV and solar irradiation. They monitored the dechlorination of dichlorvos and found a very low efficiency of 2.9% after 7 h of exposure to sunlight. Li et al. (2005) have studied the degradation of methyl parathion with Degussa P-25 TiO_2, $H_3PW_{12}O_{40}/TiO_2$, and $H_6P_2W_{18}O_{62}/TiO_2$. The initial concentration of methyl parathion was 50 mg/L, and 0.2 g/L of each catalyst was used for their study. The photocatalytic activity of Degussa P-25 was low in this system, whereas $H_3PW_{12}O_{40}/TiO_2$ and $H_6P_2W_{18}O_{62}/TiO_2$ exhibited high visible-light photocatalytic activity for the degradation of aqueous methyl parathion. The concentration of the methyl parathion in the reaction solution decreased rapidly by using $H_3PW_{12}O_{40}/TiO_2$ or $H_6P_2W_{18}O_{62}/TiO_2$ as the visible-light-driven photocatalyst, and it only needed 40 min to completely destroy it. Some intermediates, including acetic acid and formic acid, were detected during the reaction and total disappearance of acetic acid and formic acid needed 120 and 160 min, respectively, when $H_3PW_{12}O_{40}/TiO_2$ was used as a photocatalyst.

Malato et al. (1996) have studied the degradation of technical-grade lindane (initial concentration 35.9 mg/L, TOC 9 mg/L, and chloride content 26.3 mg/L) with a chemical composition of 90% lindane, 7% calcium lignosulfite, 2.5% silicon dioxide, 0.5% sodium alkyl naphthalene sulfonate, and traces of Prussian blue dissolved directly in water under solar light. Two different types of titanium dioxide were used, Degussa P-25 and Hombikat UV-100 (200 mg/L). Two different oxidants such as peroxydisulfate (0.001 and 0.01 M) and hydrogen peroxide (0.025 M) were used in the experiment. The results showed that approximately 84% of the chloride bound to organic carbon was liberated (22 mg/L), indicating a high degree of degradation of lindane. The degradation was more efficient with sodium peroxydisulfate compared with hydrogen peroxide, although it led to an increase in salt in the treated water.

25.10 Nitrogen-Doped TiO$_2$ for Pesticide Removal

The substitution doping of nitrogen is efficient in contaminant removal since it introduces an intraband energy state. This makes the system work in the visible region. Compared with other nonmetal ions (Fe, Ag, and Cr), N-doped TiO$_2$ performed better for the removal of lindane (Figure 25.5). Complete degradation of lindane was achieved in 5.5 h. Triethylamine, when used as a precursor of nitrogen, shows more photocatalytic activity compared with other compounds such as urea, ethyl amine, and ammonium hydroxide (Senthilnathan and Philip, 2010a). The optimum removal of lindane was obtained when the molar ratio of Ti/N was 1:1.6. After N-doping, the absorption spectrum shifted to a lower energy range (400–600 nm) (Figure 25.6). The size of the particles was estimated to be 20–30 nm (Senthilnathan and Philip, 2010b).

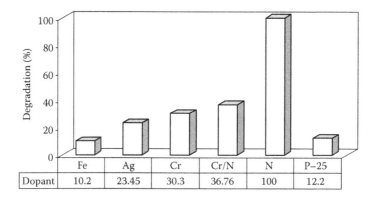

FIGURE 25.5
Comparison of lindane degradation efficiency of N-doped TiO$_2$ with metal ions–doped TiO$_2$ under visible light (concentration of lindane = 100 µg/L; doped TiO$_2$ immobilized on a Pyrex glass tube; oxygen purging rate = 300 mL/min).

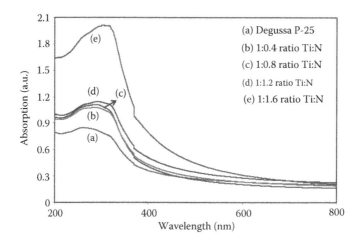

FIGURE 25.6
UV-visible absorption of Degussa P-25, anatase TiO$_2$, and N-doped TiO$_2$ (N-doped TiO$_2$ ratios = 1:0.4, 1:0.8, 1:1.2, and 1:1.6).

The activity of N-doped TiO_2 will vary when the pesticides are present individually and in a complex mixture. The three pesticides, viz., methyl parathion, dichlorvos, and lindane, were completely degraded using N-doped TiO_2 under visible light when the compounds were present individually in the water sample. Dichlorvos was degraded faster compared with other two compounds (Senthilnathan and Philip, 2010c).

The same reactions were carried out under solar light in a batch reactor, in single and mixed pesticides. The time required for degrading 33.33 µg/L of each pesticide was more compared with the time required to degrade 100 µg/L of pesticides when used individually. This clearly shows that the time required to treat mixed pesticides will be more (Senthilnathan and Philip, 2011). A thin film continuous reactor was used to study the photocatalytic activity of immobilized N-doped TiO_2 in degrading pesticides (Senthilnathan and Philip, 2012). The total volume of the reactor was 5500 mL, whereas the effective volume of the reactor was 1350 mL. Methyl parathion (250 µg/L) was used to fix the optimum flow rate. As flow rate increased, the time taken for degrading the pesticide also increased. Hence, providing more residence time can increase the degradation of pesticides. Optimum flow rate was fixed as 12.5 mL/min, and the corresponding residence time was 108 min. Different concentrations of pesticides were used for the study. No intermediate was found at the outlet of the reactor showing that the pesticides were completely degraded within the residence time (Senthilnathan and Philip, 2012). N-doped TiO_2 showed maximum efficiency under solar radiation compared with UV and visible light. In the case of mixed pesticides, also complete removal was obtained after treatment. Even after 24 h of operation, complete removal of pesticide was achieved. Proposed pathways for methyl parathion, and lindane are presented in Figures 25.7 and 25.8, respectively. The rate of degradation of

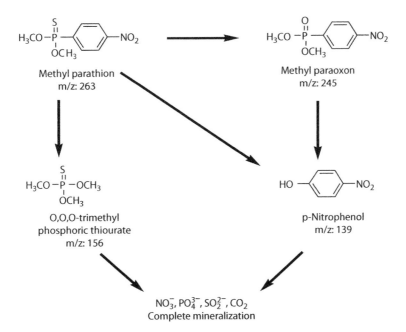

FIGURE 25.7
Possible degradation pathways of methyl parathion in immobilized TiO_2 (methyl parathion concentration = 10 mg/L; Degussa P-25 coated inner surface area = 169.56 cm²; oxygen purging rate = 300 mL/min; stirring rate = 150 rpm).

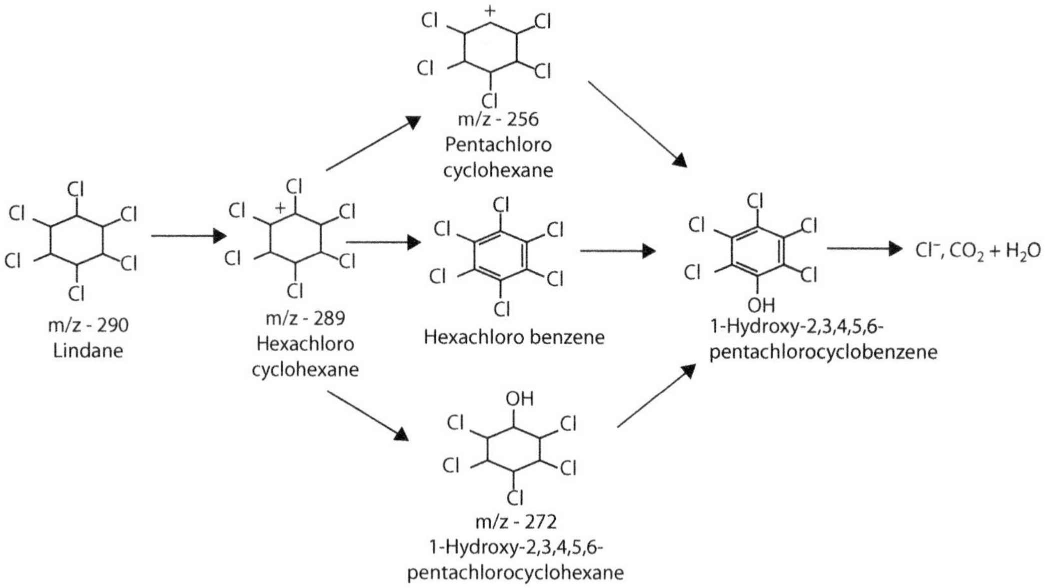

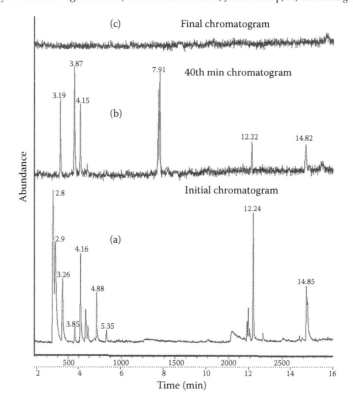

FIGURE 25.8
Proposed pathway of lindane degradation. (From Senthilnathan, J. and Philip, L., *Chem. Eng. J.*, 161, 83, 2010.)

FIGURE 25.9
GC–MS analysis of commercial-grade methyl parathion under solar radiation (sample was collected at (a) initial, (b) middle (40th min), and (c) end of the reaction for 250 µg/L of methyl parathion).

dichlorvos was higher compared with lindane and methyl parathion. No intermediate was found during the degradation of dichlorvos.

The GC–MS chromatogram of commercial-grade methyl parathion at the beginning and end of degradation is given in Figure 25.9. Similarly, GC–MS chromatograms of commercial-grade dichlorvos at the beginning and end of photocatalytic degradation are given in Figure 25.10. The intermediates formed during the degradation of methyl parathion and lindane using Degussa P-25 and N-doped TiO_2 under UV, visible, and solar radiation is given in Table 25.2.

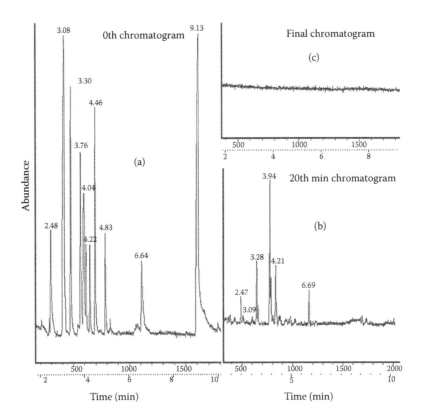

FIGURE 25.10
GC–MS analysis of commercial-grade dichlorvos under solar radiation (sample was collected at (a) initial, (b) middle (20th min), and (c) end of the reaction for 250 µg/L of dichlorvos).

TABLE 25.2

Intermediates Identified during the Degradation of Methyl Parathion, Dichlorvos, and Lindane Using UV, Visible, and Solar Radiation

Methyl Parathion	Dichlorvos	Lindane
Methyl paraxon (206)	2,2-Dichlorovinyl-*O*-methyl	Hexachlorocyclohexane (289)
O,O,O-trimethyl	phosphate (206)	Pentachlorocyclohexane (256)
phosphoricthiourate (156)	*O,O,O*-Trimethyl phosphonic	Hexachlorobenzene (284)
p-Nitrophenol (140)	ester (140)	1-Hydroxy-2,3,4,5,6-chlorocyclohexane (272)
		1-Hydroxy 2,3,4,5,6-chlorobenzene (265)
		Pentachlorocyclopentene (239)
		1,2,3,4,5-Hydroxy cyclopentene (149)
		1,2,3-Hydroxy cyclobutane (99)

25.11 TiO$_2$ for Disinfection

Studies on antimicrobial effects of titanium dioxide particles started 30 years ago. Matsunaga et al. (1985) first reported the antibacterial activity of titanium dioxide. They used a metal halide lamp for excitation of TiO$_2$ and bacterial inactivation occurred in 60–120 min. Matsunaga et al. (1988) observed that 99% of *Escherichia coli* was removed using immobilized TiO$_2$ in a continuous sterilization system. Other than *E. coli*, studies have been conducted on other organisms like poliovirus (Watts et al., 1994), cyanotoxin (Senogles et al., 2001), *Staphylococcus* (Chung et al., 2009), and cancer cells (Blake et al., 1999).

The combination of sunlight and photocatalyst can lead to a system that can degrade pathogens and organic pollutants. Many studies have reported the effectiveness of solar photocatalytic disinfection (Block et al., 1997; Blake et al., 1999; Salih, 2002; Rincón and Pulgarin, 2003). Most of the studies used suspended forms of catalyst, which, in turn, need a catalyst removal system. Less number of studies were conducted using an immobilized form of catalyst (Dunlop et al., 2002; Salih, 2002; Rincón and Pulgarin, 2003). The main concern in immobilization is regarding the stability of the catalyst. This can be overcome by using a continuous reactor with catalyst-coated glass plates or using bottles coated with catalyst on the inner wall. Better results can be obtained if reflective coatings are used (Sommer et al., 1997). This increases the intensity of solar radiation passing into water. In the normal SODIS method, the treated water should be used within 24 h because some bacteria are resistant to UV radiation and they are able to repair the damage later. The main advantage of solar photocatalysis is that there is no regrowth after treatment (Gelover et al., 2006). Since the damage to cells is irreversible, there is no chance of regrowth. Thus, it has better residual effects.

Bekbolet and Araz (1996) conducted studies on *E. coli* using a suspended form of Degussa P-25. They used cylindrical Pyrex glass vessels and a black light fluorescent lamp as the light source. Inactivation of 1000 colony-forming units (CFU)/mL *E. coli* was achieved in 60 min using 1 g/L of TiO$_2$. The reaction followed first order. Ireland et al. (1993) compared the action of Degussa on pure cultures of *E. coli* in dechlorinated tap water and on surface water samples. They reported that if a significant amount of radical scavengers are not present, rapid cell death can be achieved even in a surface water sample. They used a continuous reactor working under UV with TiO$_2$ coated on a fiberglass mesh.

Another study was conducted on deactivation of *E. coli* using Degussa P-25, a hydrothermally prepared photocatalyst (HPC) and a magnetic photocatalyst (MPC) in a batch spiral reactor (Coleman et al., 2005). They compared effectiveness of both suspended and immobilized forms of catalyst. Degussa P-25 was efficient compared with HPC and MPC. They studied the effect of buffer and catalyst loading. They reported that use of silver nanoparticles along with TiO$_2$ increased the performance of the system. Optimum catalyst loading was reported as 1 g/L.

Lonnen et al. (2005) conducted a study using the SODIS method and a solar photocatalytic method. The efficiency of both systems to degrade bacteria, fungi, and protozoa were studied. Degussa P-25 powder was coated on acetate sheets. It was inserted into Duran bottles. Simulated solar radiation was used for the studies. The bottles were irradiated for 8 h. The initial concentrations of microorganisms used were in the range of 10^4–10^6/mL.

The study reported that the batch process SODIS and solar photocatalytic methods are effective against bacterial spores and fungi. However, protozoan cysts were not degraded by both the methods. The solar photocatalytic method was 50% more efficient than the SODIS method.

Another comparative study of SODIS and solar photocatalytic disinfection was conducted by Dunlop et al. (2011). Water samples were taken in low-density polyethylene bags of 2-L capacity. Degussa P-25 was used to improve the efficiency of SODIS. Artificial sunlight was provided by xenon lamps. A 10^7 CFU/mL *E. coli* solution was taken for the studies. The effect of turbidity reduced the efficiency but complete inactivation was obtained under a longer irradiation time.

Gelover et al. (2006) studied the bacterial inactivation using 1000 CFU/mL total coliforms and 100 CFU/mL fecal coliforms. The bottles were placed in solar collectors. TiO_2 prepared by a sol–gel process was coated on Pyrex glass cylinders. It was found that addition of catalyst enhances the reaction. Bacterial regrowth was observed after the SODIS treatment.

25.11.1 Mechanism of Bacterial Degradation Using TiO_2

When TiO_2 is irradiated using photons of sufficient energy, electrons from the valence band shift to the conduction band. Thus, electron–hole pairs are generated in the semiconductor. The electron, when it reaches the surface, reduces the electron acceptor present in the solution. Similarly the photo-generated holes oxidize the electron donor present in the water. The holes react with water molecules and produce hydroxyl radicals. These radicals act as the primary oxidants in the photocatalytic system (Herrmann, 1999; Wong and Chu, 2003). Degradation depends on the hydroxyl radicals present in the system. The relation between hydroxyl radicals and *E. coli* inactivation is linear (Cho et al., 2004). If oxygen is present in the system, it is converted to superoxide radicals, thus preventing electron–hole recombination. The superoxide radical is also reactive and can oxidize cellular components. Contact between the bacteria and the catalyst particles is very important. Gogniat et al. (2006) reported that adsorption of bacterial cells to the TiO_2 particles and subsequent loss of membrane integrity are the key steps in bacterial disintegration.

Matsunaga et al. (1985) reported that inactivation of *E. coli* takes place due to the changes in coenzyme A, which inhibits the respiration of the cell. Studies have shown that damage takes place from outward to inward. Sunada et al. (2003) reported that photokilling of *E. coli* takes place in two steps. First, the outer membrane is damaged and then cytoplasmic membrane damage occurs. After that, complete disintegration takes place. Initially, the cell wall becomes permeable. Damage of cell wall layers leads to the leakage of ions from the cell. Once the cell membrane is degraded, TiO_2 particles penetrate into the interior and further damage occurs. The membrane damage occurring during photodegradation is shown in Figure 25.11. It is reported that the nucleotide of *E. coli* changes because of the leakage of ions (Chung et al., 2009). After the membrane damage, leakage of ions like K^+ occurs. Then larger molecules such as proteins come out of the cell. Thus, irreversible damage to the cell occurs. The internal components of the cell are completely degraded and complete mineralization of the cell occurs (Foster et al., 2011). The mechanism of bacterial disintegration is shown in Figure 25.12.

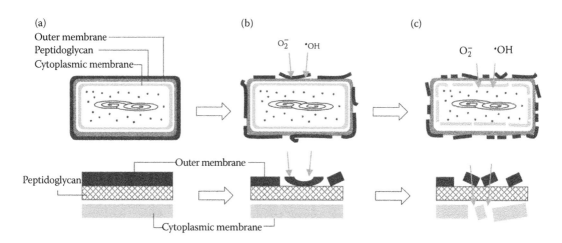

FIGURE 25.11
Photokilling of bacteria through membrane damage. (From Sunada, K. et al., *Environ. Sci. Technol.*, 37, 4785, 2003.)

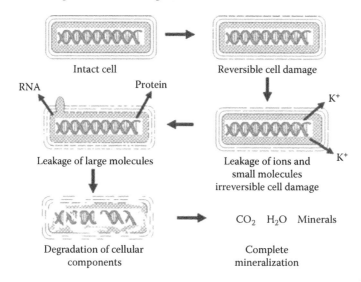

FIGURE 25.12
Mechanism of bacterial killing. (From Foster et al., *Appl. Micro. Biol.*, 90, 1847, 2011.)

25.12 Photocatalytic Degradation of *E. coli* Using Nitrogen-Doped Titanium Dioxide

Photocatalysis is an effective technology for removing pathogens from water. It has many advantages compared with the conventional disinfection methods. Chlorine is known to react with organic matter present in water and produce organochlorine products. These compounds are carcinogenic in nature. Other methods such as UV treatment and ozonation are expensive, and these methods cannot be used as sustainable water treatment methods in rural areas. Heterogeneous photocatalysis is highly efficient in removing organic compounds and microorganisms from water.

Many studies have reported the possible mechanisms of bacterial disintegration. Photocatalysis produces hydroxyl radicals, which are responsible for the disintegration of bacteria. Cho et al. (2004) reported that the relation between hydroxyl radicals and *E. coli* inactivation is linear. Apart from hydroxyl radicals, superoxide radicals also oxidize the microorganisms. The catalyst particles should be in contact with the *E. coli* because the hydroxyl radicals produced have a very short lifetime in the range of nanoseconds. The disintegration starts from the outer membrane and then the cytoplasmic membrane is degraded (Sunada et al., 2004). These ions are leaked from the cell, which causes changes in the nucleotide of *E. coli* (Chung et al., 2009). The cell organelles get exposed, and the radicals oxidize and completely mineralize the cell. Thus, the changes occurring to the cell are irreversible.

N-doped TiO$_2$ can be used for treating water using solar radiation since it gets photocatalytically activated under visible light. Antimicrobial activity of N-doped TiO$_2$ was studied using *E. coli* since it is a common indicative organism. The degradation studies were carried out in batch and continuous reactors. Comparison of the activity of Degussa P-25 and N-doped TiO$_2$ was done. Degussa P-25 showed higher photocatalytic efficiency under UV light compared with visible light. N-doped TiO$_2$ gave good results under visible light. Seventy percent of the bacteria were degraded in just 10 min. The bacterial concentration was completely removed within 1 h. The reaction followed pseudo-first-order kinetics. The activity of N-doped TiO$_2$ under UV and visible light is given in Figure 25.13.

Studies were also conducted in the immobilized form of catalyst (Arya, 2012). The catalyst was coated onto a Pyrex glass cylinder of area 169 cm^2 and was kept inside the batch reactor. The coated catalyst concentration was 0.4 mg/cm^2. There was only a marginal variation between the activity of the suspended and immobilized forms of catalyst.

Photocatalytic experiments were carried out in a thin film continuous photoreactor under sunlight with N-doped TiO$_2$. The schematic diagram of continuous reactor is given in Figure 25.14.

The flow rate of the raw water was optimized to 40 mL/min. It was found that a minimum of 30 min was needed for the efficient removal of bacteria, with an initial concentration of 1000 CFU/mL. The outlet concentration of bacteria remained the same throughout the operation of the reactor. The effect of turbidity, bicarbonate ions, and organic matter

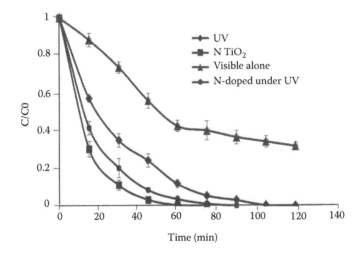

FIGURE 25.13
Effect of N-doped TiO$_2$ and light source on bactericidal inactivation.

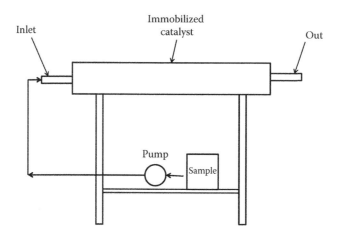

FIGURE 25.14
Schematic diagram of continuous reactor.

was studied. Large increases in turbidity reduced the efficiency since it decreases the optical penetration. The degradation rate was reduced slightly by the presence of bicarbonate ions since it is changing the pH of the solution and it acts as a radical scavenger (Coleman et al., 2005; Marugan et al., 2008). However, addition of dextrose did not show any change in the efficiency of the system.

Comparison studies of SODIS and solar photocatalytic disinfection were conducted. The bacterial solution was taken in polyethylene terephthalate (PET) bottles. The catalyst was taken in suspended and immobilized form. All the PET bottles were kept under sunlight for 6 h. There was 50% reduction of the bacteria after 15 min in the catalyst-suspended PET bottles—100% inactivation was obtained after 1 h. It took 3 h for the immobilized catalyst to completely inactivate the bacteria. In the SODIS method, complete inactivation occurred after only 6 h (Figure 25.15).

Bacterial regrowth studies were done using SODIS-treated and photocatalytically treated water samples. There was no regrowth in catalyst-treated water samples. However, bacterial colonies appeared on plates containing SODIS-treated water samples after 24 h. This is because the bacteria can repair the damage caused by the SODIS method if sufficient time is given. The change due to catalyst treatment is irreversible (Gelover et al., 2006).

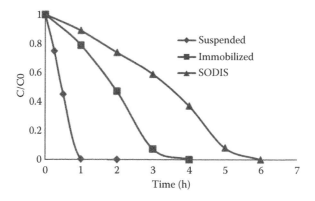

FIGURE 25.15
Kinetics of bacterial kill by SODIS and photocatalysis methods.

References

Anpo, M. (2000) Use of visible light. Second-generation titanium oxide photocatalysts prepared by the application of an advanced metal ion-implantation method. *Pure and Applied Chemistry*, 72, 1787.

Antonaraki, S., T. M. Triantis, E. Papaconstantinou, A. Hiskia (2010) Photo catalytic degradation of lindane by polyoxometalates: Intermediates and mechanistic aspects. *Catalysis Today*, 151(1–2), 119–124.

Arya, V. (2012) Development of an environmentally friendly water treatment system using N-doped TiO_2. M. Tech Thesis, Environmental and Water Resources Engineering Division, Civil Engineering, Indian Institute of Technology Madras: Chennai.

Asahi, R., T. Morikawa, T. Ohwaki, K. Aoki, Y. Taga (2001) Visible-light photocatalysis in nitrogen-doped titanium oxides. *Science*, 293, 269–271.

Bahnemann, D. (2004) Photocatalytic water treatment: Solar energy applications. *Solar Energy*, 77, 445–459.

Bekbolet, M., C. V. Araz (1996) Inactivation of *escherichia coli* by photocatalytic oxidation. *Chemosphere*, 32(5), 959–965.

Blake, D. M., P. C. Maness, Z. Huang, E. J. Wolfrum, J. Huang (1999) Application of the photocatalytic chemistry of titanium dioxide to disinfection and the killing of cancer cells. *Separation and Purification Methods*, 28(1), 1–50.

Block, S. S., V. P. Seng, D. W. Goswami (1997) Chemically enhanced sunlight for killing bacteria. *Journal of Solar Energy Engineering*, 119, 85–91.

Cao, J., W. X. Zhang, D. G. Brown, D. Sethi (2008) Oxidation of lindane with Fe^{2+} activated sodium persulfate. *Environmental Engineering Science*, 25(2), 221–228.

Carey, J. H. (1992) An introduction to AOP for destruction of organics in wastewater. *Water Pollution Research Journal Canada*, 27, 1–21.

Chatterjee, D., A. Mahata (2004) Evidence of superoxide radical formation in the photodegradation of pesticide on the dye modified TiO_2 surface using visible light. *Journal of Photochemistry and Photobiology. A: Chemical*, 165, 19–23.

Chen, S. F., X. L. Cheng (1999) Photocatalytic reduction of dichromate using TiO_2 supported on hollow glass microbeads. *Chinese Journal of Chemistry*, 17, 419–424.

Cho, M., H. Chung, W. Choi, J. Yoon (2004) Linear correlation between inactivation of *E. coli* and OH radical concentration in TiO_2 photocatalytic disinfection. *Water Research*, 38, 1069–1077.

Chung, C. J., H. I. Lin, C. M. Chou, P. Y. Hsieh, C. H. Hsiao, Z. Y. Shi, J. L. He (2009) Inactivation of *Staphylococcus* aureus and *Escherichia coli* under various light sources on photocatalytic titanium dioxide thin film. *Surface Coating Technology*, 203(8), 1081–1085.

Cleuvers, M. (2003) Aquatic ecotoxicity of pharmaceuticals including the assessment of combination effects. *Toxicology Letters*, 142, 185–194.

Coleman, H. M., C. P. Marquis, J. A. Scott, S. S. Chin, R. Amal (2005) Bactericidal effects of titanium dioxide-based photocatalysts. *Chemical Engineering Journal*, 113, 55–63.

Comerton, A. M., R. C. Andrews, D. M. Bagley (2009) Practical overview of analytical methods for endocrine-disrupting compounds, pharmaceuticals and personal care products in water and wastewater. *Philosophical Transactions of the Royal Society. Series A*, 367, 3923–3939.

Dionysiou, D. D., A. P. Khodadoust, A. M. Kern, M. T. Suidan, I. Baudin, J. M. Laîné (2000) Continuous mode photocatalytic degradation of chlorinated phenols and pesticides in water using a bench scale TiO_2 rotating disk reactor. *Applied Catalysis B: Environmental*, 24, 139–155.

Dunlop, P. S. M., J. A. Byrne, N. Manga, B. R. Eggins (2002) The photocatalytic removal of bacterial pollutants from drinking water. *Journal of Photochemistry and Photobiology A: Chemistry*, 148, 355–363.

Dunlop, P. S. M., M. Ciavola, L. Rizzo, J. A. Byrne (2011) Inactivation and injury assessment of *Escherichia coli* during solar and photocatalytic disinfection in LDPE bags. *Chemosphere*, 85(7), 1160–1166.

Evgenidou, E., I. Konstantinou, K. Fytianos, T. Albanis (2006) Study of the removal of dichlorvos and dimethoate in a titanium dioxide mediated photocatalytic process through the examination of intermediates and the reaction mechanism. *Journal of Hazardous Materials*, B137, 1056–1064.

Evgenidou, E., I. Konstantinou, K. Fytianos, I. Pouliosc, T. Albanisd (2007) Photocatalytic oxidation of methyl parathion over TiO_2 and ZnO suspensions. *Catalysis Today*, 124(3–4), 156–162.

Foster, H. A., I. B. Ditta, S. Varghese, A. Steele (2011) Photocatalytic disinfection using titanium dioxide: Spectrum and mechanism of antimicrobial activity. *Applied Microbiology and Biotechnology*, 90, 1847–1868.

Freudenhammer, H., D. Bahnemann, L. Bousselmi, S. U. Geissen, A. Ghrabi, F. Saleh, A. Salah, U. Siemon, A. Vogelpohl (1997) Detoxification and recycling of wastewater by solar-catalytic treatment. *Water Science and Technology* 35(4), 149–156.

Gelover, S., L. A. Gómez, K. Reyes, M. T. Leal (2006) A practical disinfection of water disinfection using TiO_2 films and sunlight. *Water Research*, 40, 3274–3280.

Gkika, E., P. Kormali, S. Antonaraki, D. Dimoticali, E. Papaconstantinou, A. Hiskia (2004) Polyoxometallates as effective photocatalysts in water purification from pesticides. *International Journal of Photoenergy*, 6, 227–231.

Gogniat, G., M. Thyssen, M. Denis, C. Pulgarin, S. Dukan (2006) The bactericidal effect of TiO_2 photocatalysis involves adsorption onto catalyst and the loss of membrane integrity. *FEMS Microbiology Letters*, 258(1), 18–24.

Guillard, C., P. Pichat, G. Huber, C. Hoang-Van (1996) The GC–MS analysis of organic intermediates from the TiO_2 photocatalytic treatment of water contaminated by lindane (1a,2a,3b,4a,5a,6b-hexachlorocyclohexane). *Journal of Advanced Oxidation Technology*, 1, 53–60.

Guo, Y., Y. Wang, C. Hu, Y. Wang, E. Wang, Y. Zhou, S. Feng (2000) Microporouspolyoxometalates POMs/SiO_2: Synthesis and photocatalytic degradation of aqueous organocholorine pesticides. *Chemistry of Materials*, 12, 3501–3508.

Harada, K., T. Hisanaga, K. Tanaka (1990) Photocatalytic degradation of organophosphorous insecticides in aqueous semiconductor suspensions. *Water Research*, 24, 1415–1417.

Herrmann, J. M., J. Disdier, P. Pichat, S. Malato, J. Blanco (1998) TiO_2-based solar photocatalytic detoxification of water organic pollutants case studies of 2,4-dichlorophenooxyacetic acid (2,4-D) and of benzofuran. *Applied Catalysis B: Environmental*, 17, 15–23.

Herrmann, J. M. (1999) Heterogeneous photocatalysis: Fundamentals and applications to the removal of various types of aqueous pollutants. *Catalysis Today*, 53, 115–129.

Hiskia, A., A. Mylonas, D. Tsipi, E. Papaconstantinou (1997) Photocatalytic degradation of lindane in aqueous solution. *Pesticide Science*, 50, 171–174.

Ireland, J. C., N. P. Ktosterman, E. W. Rice, R. M. Clark (1993) Inactivation of *Escherichia coli* by titanium dioxide photocatalytic oxidation. *Applied Environmental Microbiology*, 59, 1668–1670.

Jakob, M., H. Levanon, P. V. Kamat (2003) Charge distribution between UV-irradiated TiO_2 and gold nanoparticles: Determination of shift in the fermi level. *Nano Letters*, 3(3), 353–358.

Jones, O. A. H., N. Voulvoulis, J. N. Lester (2002) Aquatic environmental assessment of the top 25 English prescription pharmaceuticals. *Water Research*, 36, 5013–5022.

Khalil, L. B., W. E. Mourad, M. W. Raphael (1998) Photocatalytic reduction of environmental pollutant Cr(Vi) over some semiconductors under uv/visible light illumination. *Applied Catalysis B: Environment*, 17, 267–273.

Kim, T., J. Kim, K. Choi, K. M. Stenstrom, K. D. Zoh (2006) Degradation mechanism and the toxicity assessment in TiO_2 photocatalysis and photolysis of parathion. *Chemosphere*, 62, 926–933.

Konstantinou, J. K., T. M. Sakellarides, V. A. Sakkas, T. A. Albanis (2001) Photocatalytic degradation of selected s-Triazine herbicides and organophosphorus insecticides over aqueous TiO_2 suspensions. *Environmental Science and Technology*, 35, 398–405.

Ku, Y., I. L. Jung (2001) Photocatalytic reduction of Cr(VI) in aqueous solutions by UV irradiation with the presence of titanium dioxide. *Water Research*, 35, 135–142.

Li, L., Q. Wu, Y. Guo, C. Hu (2005) Nano size and bimodal porous polyoxotungstate anatase TiO_2 composites: Preparation and photocatalytic degradation of organophosphorus pesticide using visible-light excitation. *Microporous and Mesoporous Materials*, 87, 1–9.

Lonnen, J., S. Kilvington, S. C. Kehoe, F. Al-Touati, K. G. McGuigan (2005) Solar and photocatalytic disinfection of protozoan, fungal and bacterial microbes in drinking water. *Water Research*, 39, 877–883.

Malato, S., C. Rodriguez, J. Richter, B. Galvez, M. Vincent (1996) Photocatalytic degradation of industrial residual waters. *Solar Energy*, 56(5), 401–410.

Malato, S., J. Blanco, C. Richter, B. Braun, M. I. Maldonado (1998) Enhancement of the rate of solar photocatalytic mineralization of organic pollutants by inorganic species. *Applied Catalysis B: Environmental*, 17, 347–356.

Malato, S., P. Fernandez-Ibanez, M. I. Maldonado, J. Blanco, W. Gernjak (2009) Decontamination and disinfection of water by solar photocatalysis: Recent overview and trends. *Catalysis Today*, 147, 1–59.

Marugan, J., R. Gricken, C. Sordo, C. Cruz (2008) Kinetics of the photocatalytic disinfection of Escheria coli: Suspensions. *Applied Catalysis B: Environment*, 82, 27–36.

Matsunaga, T., R. Tomoda, T. Nakajima, H. Wake (1985) Photoelectroehemical sterilization of microbial cells by semiconductor powders. *FEMS Microbiology Letters*, 29, 211–214.

Matsunaga, T., R. Tomoda, T. Nakajima, N. Nakamura, T. Komine (1988) Continuous sterilization system that uses photosemiconductor powders. *Applied Environmental Microbiology*, 54, 1330–1333.

Mehos, M., C. Turchi, J. Pacheco, A. J. Bogel, T. Merrill, R. Stanley (1992) *Pilot-Scale Study of the Solar Detoxification of VOC Contaminated Groundwater*. NRELyTP-432-4981. National Renewable Energy Laboratory: Golden, CO.

Moctezuma, E., E. Leyva, P. Palestino, H. Lasa (2007) Photocatalytic degradation of methyl parathion: Reaction pathways and intermediate reaction products. *Journal of Photochemistry and Photobiology A: Chemistry*, 186, 71–84.

Mohammadi, M. R., M. C. Cordero-Cabrera, D. J. Fray, M. Ghorbani (2006) Preparation of high surface area titania (TiO₂) films and powders using particulate sol-gel route aided by polymeric fugitive agents. *Sensors and Actuators B: Chemical*, 120, 86–95.

Naman, S. A., Z. A. A. Khammas, F. M. Hussein (2002) Photo-oxidative degradation of insecticide dichlorvos by a combined semiconductors and organic sensitizers in aqueous media. *Journal of Photochemistry and Photobiology A: Chemistry*, 153, 229–236.

Nienow, A. M., J. C. Bezares-Cruz, I. C. Poyer, I. Hua, C. T. Jafvert (2008) Hydrogen peroxide assisted UV photodegradation of Lindane. *Chemosphere*, 72, 1700–1705.

Oancea, P., T. Oncescu (2008) The photocatalytic degradation of dichlorvos under solar irradiation. *Journal of Photochemistry and Photobiology A: Chemistry*, 199, 8–13.

Obee, T. N., S. Satyapal (1998) Photocatalytic decomposition of DMMP on titania. *Journal of Photochemistry and Photobiology A: Chemistry*, 118, 45–51.

Ollis, D. S., H. Al-Ekabi (1993) *Photocatalytic Purification and Treatment of Water and Air*. Elsevier: Amsterdam.

Pignatello, J. J., Y. Sun (1995) Complete oxidation of metolachlor and methyl parathion in water by the photo assisted fenton reaction. *Water Research*, 29(8), 1837–1844.

Prairie, M. R., J. E. Pacheco, L. R. Evans (1992) Solar detoxification of water containing chlorinated solvents and heavy metal via TiO₂ photocatalysis. *Solar Energy*, 1, 1–8.

Rahman, M. A., M. Muneer (2005) Photocatalysed degradation of two selected pesticide derivatives, dichlorvos and phosphamidon, in aqueous suspensions of titanium dioxide. *Desalination*, 181, 161–172.

Rincón, A. G., C. Pulgarin (2003) Photocatalytic inactivation of *E. coli*: Effect of (continuous intermittent) light intensity and of (suspended-fixed) TiO₂ concentration. *Applied Catalysis B: Environmental*, 44, 263–284.

Robert, D., S. Malato (2002) Solar photocatalysis: A clean process for water detoxification. *The Science of the Total Environment*, 291, 85–97.

Salih, F. M. (2002) Enhancement of solar inactivation of *Escherichia coli* by titanium dioxide photocatalytic oxidation. *Journal of Applied Microbiology*, 92, 920–926.

Sanjuan, G., M. Aguirre, H. Alvaro, H. García (2000) 2,4,6-Triphenylpyrylium ion encapsulated within Y zeolite as photocatalyst for the degradation of methyl parathion. *Water Research*, 34(1), 320–326.

Schrank, S. G., H. J. Tose, R. F. P. M. Moreira (2002) Simultaneous photocatalytic Cr(VI) reduction and dye oxidation in a TiO_2 slurry reactor. *Journal of Photochemistry and Photobiology A: Chemistry*, 147, 71–76.

Senogles, P. J., J. A. Scott, G. Shaw, H. Stratton (2001) Photocatalytic degradation of the Cyanotoxin, Cylindrospermopsin using titanium dioxide and UV irradiation. *Water Research*, 35(5), 1245–1255.

Senthilnathan, J., L. Philip (2010a) Investigation on degradation of methyl parathion using visible light in the presence of Cr+3 and N-doped TiO_2. *Advanced Materials Research*, (93–94), 280–283.

Senthilnathan, J., L. Philip (2010b) Removal of mixed pesticides from drinking water system using surfactant assisted nano TiO_2. *Water Air and Soil Pollution*, 210(1–4), 143–154.

Senthilnathan, J., L. Philip (2010c) Photocatalytic degradation of lindane under UV and visible light using N-doped TiO_2. *Chemical Engineering Journal*, 161(1–2), 83–92.

Senthilnathan, J., L. Philip (2011) Photodegradation of methyl parathion and dichorvos from drinking water using (1-1) N-doped TiO_2 under solar radiation. *Journal of Chemical Engineering*, 172, 2–3, 678–688.

Senthilnathan, J., L. Philip (2012) Elimination of pesticides and their formulation products from drinking water using thin film continuous photoreactor under solar radiation. *Solar Energy Journal*, 86(9), 2735–2745.

Shifu, C., C. Gengyu (2005) Photocatalytic degradation of organophosphorus pesticides using floating photocatalyst TiO_2:SiO_2/beads by sunlight. *Solar Energy*, 79, 1–9.

Sommer, B., A. Mariño, Y. Solarte, M. L. Salas, C. Dierolf, C. Valiente, D. Mora, R. Rechsteiner, P. Setter, W. Wirojanagud, H. Ajarmeh, A. Al-Hassan, M. Wegelin (1997) SODIS—an emerging water treatment process. *Journal of Water Supply Research Technology Aqua*, 46, 127–137.

Subramanian, V., E. Wolf, P. Kamat (2003) Green emission to probe photoinduced charge events in ZnO-Au nanoparticles, charge distribution and fermi-level equilibration. *Journal of Physical Chemistry B*, 107, 7479–7485.

Subramanian, V., E. Wolf, P. Kamat (2004) Catalysis with TiO_2/gold nano composites: Effect of metal particle size on the fermi level equilibration. *Journal American Chemical Society*, 126, 4943–4950.

Sunada, K., T. Watanabe, K. Hashimoto (2003) Bactericidal activity of copper-deposited TiO_2 thin film under weak UV light illumination. *Environmental Science Technology*, 37(20), 4785–4789.

Takeuchi, M., K. Tsujimaru, K. Sakamoto, M. Matsuoka, H. Yamashita, M. Anpo (2003) Effect of Pt loading on the photocatalytic reactivity of titanium oxide thin films prepared by ion engineering techniques. *Research on Chemical Intermediates*, 29(6), 619–629.

Thacker, P. D. (2005) Pharmaceutical data elude researchers. *Environmental Science and Technology*, 39, 193A–194A.

Vidal, A. (1998) Developments in solar photocatalysis for water purification. *Chemosphere*, 36(12), 2593–2606.

Watts, R. J., S. Kong, M. P. Orr, G. C. Miller, B. E. Henry (1994) Photocatalytic inactivation of coliform bacteria and viruses in secondary wastewater effluent. *Water Research*, 29, 95–100.

Wong, C. C., W. Chu (2003) The hydrogen peroxide-assisted photocatalytic degradation of alachlor in TiO_2 suspensions. *Environmental Science and Technology*, 37, 2310–2316.

Xiaodan, Y., W. Qingyin, J. Shicheng, G. Yihang (2006) Nanoscale ZnS/TiO_2 composites: Preparation, characterization, and visible-light photocatalytic activity. *Materials Characterization*, 57, 333–341.

Zaleska, A., J. Hupkaa, M. Wiergowski, M. Biziuk (2000) Photocatalytic degradation of lindane, p,p'-DDT and methoxychlor in an aqueous environment. *Journal of Photochemistry and Photobiology A: Chemistry*, 135(2–3), 213–220.

Zhao, M. Y., S. F. Chen, Y. W. Tao (1995) Photocatalytic degradation of organophosphorus pesticides using thin films of TiO_2. *Journal of Chemical Technology and Biotechnology*, 64, 339–344.

26

Noble Metal Nanosystems for Drinking Water Purification: From Nanoparticles to Clusters

Thalappil Pradeep and Megalamane Siddaramappa Bootharaju

Department of Chemistry, Indian Institute of Technology Madras, Chennai, India

CONTENTS

26.1 Introduction

It used to be that the quality of water was assessed on the basis of color, odor, and taste. Palatability trumped potability. People used surface water from lakes, rivers, and ponds for drinking, which was not yet harmful as contaminants derived from anthropogenic sources were not yet potent enough. Owing to the increase in population and pollution of the environment, the quantity of available surface water decreased with time and most of it succumbed to severe pollution. In several places, living beings had to depend on groundwater because of the scarcity of clean surface water. However, groundwater often contains dangerous levels of arsenic, fluoride, and other contaminants.[1]

In modern times, water sources are being progressively contaminated primarily due to industrialization and rapid increase in population. It is impossible to imagine the survival of the world without potable water. Contamination of water sources is broadly of two types, natural (geogenic) and anthropogenic. Water bodies have been polluted largely due to anthropogenic activities compared with natural sources. Toxins in water are chemical and biological. Chemical toxins include pesticides (chlorpyrifos [CP], malathion, endosulfan, atrazine, etc.), halocarbons (CCl_4, $C_6H_5CH_2Cl$, CH_2Cl_2, etc.), dyes (cationic and anionic), heavy metal ions (Hg^{2+}, Pb^{2+}, Cd^{2+}, etc.), anions (F^-, AsO_4^{3-}, AsO_3^{3-}, CN^-, NO_3^-, etc.), and organic molecules such as drugs and pharmaceuticals. Bacteria, viruses, fungi and algae are important biological pollutants. The presence of any one of these components in water makes it unsafe for drinking. Consumption of water containing the above species causes various diseases and even death. As a result, there is a great demand for technologies for clean drinking water. Important toxins of water, their permissible limits, origins, and health effects are listed in Table 26.1.

In view of the immense requirement of clean water, there is a large need of technologies for purifying water. The available methods for water purification are adsorption, disinfection, coagulation, flocculation, sedimentation, membrane filtration, and reverse osmosis. There are advantages and disadvantages associated with each method. For any method to be successful, it should be economically viable and must have reduced or no power consumption and preferably have features such as easy operation, facile synthesis of materials, recyclability, need for less manpower, and capacity to purify water in large quantities. The maximum allowed contamination levels in water are decreasing as time progresses. For example, permissible limits of arsenic and lead decreased from 200 to 10 ppb and 100 to 10 ppb, respectively (according to World Health Organization [WHO] 1990 standards). The US Environmental Protection Agency (US EPA) also decreased permissible levels of lindane, arsenic, and lead from 4.0 to 0.2 ppb, 50 to 10 ppb, and 50 to 15 ppb, respectively, during the period 1976–2001. From these data, it is clear that technologies have to be developed for sensing and removal of pollutants at ultra-low levels.[2] Technologies involving materials with high surface area and varieties of reactive adsorption sites are advantageous. This may be possible when materials are at the nanoscale (10^{-9} m).

Nanomaterials, materials with physical dimensions (diameter, thickness, grain size) <100 nm in length, are shown to be excellent candidates for the purification of water. The reasons are their large surface-to-volume ratio, unusual reactivity, as well as size-dependent optical, physical, and chemical properties. Many nanosystems have been explored by researchers in the recent past in the context of treatment of water. Some of them are nanoscale carbon (graphene, graphene oxide [GO], reduced GO [RGO], GO/RGO–metal/metal oxide composites, and carbon nanotubes), nanosized zerovalent iron, and oxides of aluminum, iron, titanium, magnesium, cerium, and manganese.[3] Among metals, noble metals are also used for decontamination of water. Use of noble metals, although not in nano form, for this application, has been documented since ancient times. People stored drinking water in containers made up of metals (copper/silver) or clay without knowing the exact reasons. Later, researchers found the inactivation of bacteria and other biological entities in these vessels. Studies showed that these metals were actually damaging the DNA of bacteria.

Noble metals are those metals with high resistance to aerial oxidation/corrosion even at high temperatures. Nobility varies from one metal to the other. Most of them are less abundant in the earth's crust, because of which, they are precious. Examples of noble

TABLE 26.1

Brief List of Pollutants, Their Origins, Health Effects, and Permissible Limits in Drinking Water, According to US EPA

S. No.	Pollutant	Source	Permissible Limit (mg/L)	Health Effects/Impact on Environment
1	Pesticides and insecticides	Pesticides used in agricultural activities, effluents, home use, cattle, and gardens	0.003 (atrazine), 0.04 (carbofuran), 0.0002 (lindane)	Problems with cardiovascular system/reproductive/blood, nervous system/liver and kidney
2	Halogenated hydrocarbons	Discharge from drug, chemical plants, and other industrial activities	0.005 (CCl_4, $C_6H_5CH_2Cl$, CH_2Cl_2)	Liver problems and increased risk of cancer
3	Mercury	Erosion of natural deposits, discharge from refineries and factories, runoff from landfills and croplands	0.002	Kidney damage, Minamata disease, and respiratory failure
4	Lead	Corrosion of household plumbing systems and erosion of natural deposits	0.015	Infants and children: delays in physical or mental development; children could show slight deficits in attention span and learning abilities. Adults: kidney problems and high blood pressure
5	Cadmium	Corrosion of galvanized pipes, erosion of natural deposits, discharge from metal refineries, runoff from waste batteries and paints	0.005	Kidney damage
6	Chromium	Discharge from steel and pulp mills; erosion of natural deposits	0.1	Allergic dermatitis
7	Copper	Corrosion of household plumbing systems and erosion of natural deposits	1.3	Short-term exposure: gastrointestinal distress. Long-term exposure: liver or kidney damage
8	Fluoride	Water additive that promotes strong teeth, erosion of natural deposits, discharge from fertilizer, and aluminum factories	4	Bone disease (pain and tenderness of the bones); children may get mottled teeth
9	Arsenic	Geological origin and electronics production wastes	0.01	High blood pressure, glucosuria, hyperpigmentation, keratoses, and cancer
10	Cyanide	Discharge from steel/metal, plastic, and fertilizer factories	0.2	Nerve damage or thyroid problems
11	Nitrate	Runoff from fertilizer use, leaking from septic tanks, sewage, and erosion of natural deposits	10	Infants (below the age of 6 months): become seriously ill and, if untreated, may die; symptoms include shortness of breath and blue baby syndrome

Source: Compiled from multiple sources on the World Wide Web.

metals are ruthenium (Ru), rhenium (Re), rhodium (Rh), copper (Cu), silver (Ag), gold (Au), palladium (Pd), platinum (Pt), osmium (Os), mercury (Hg), and iridium (Ir). In the bulk state, noble metals exhibit insolubility in water, metallic luster, malleability, electrical conductivity, and no catalytic activity. These properties change when the size of the metal is reduced. They (especially Cu, Ag, and Au) start showing dispersion in water when the size is reduced to nanoscale range. The wine red solution of Au obtained in Faraday's experiment contains finely divided metal species that absorb visible light. Gold nanoparticles (Au NPs) start exhibiting catalytic activity in many reactions such as oxidation, coupling, and reduction.[4]

Noble metal nanosystems (NMNs) can be broadly divided into four types: three-, two-, one-, and zero-dimensional (D) structures. Shapes such as stars/flowers (3D), films/plates/networks/layers (2D), fibers/tubes/rods/wires (1D), and NPs/quantum dots (QDs)/quantum clusters (QCs) (0D) are corresponding examples. QDs and QCs are structures whose sizes are typically 2–4 nm and <2 nm, respectively.[5] Optical properties, such as absorption, emission, and optical activity; physical properties; and chemical properties vary with size and shape. Silver and gold NPs protected with citrate show light yellow and wine red colors, with absorption bands at 400 and 520 nm, respectively for 10–15 nm particles. These absorption bands are due to collective oscillations of conduction band electrons when electromagnetic radiation interacts with them and are due to surface plasmon resonance (SPR). Gold nanorods, in contrast, exhibit two SPRs corresponding to longitudinal and transverse modes of plasmons. Different morphologies such as triangles, wires, and polygons exhibit multiple SPRs.

The size of QCs is in between NPs and molecules. Clusters of precise numbers of metal atoms and protecting ligands possess molecular absorption and luminescence properties (Figure 26.1). $Au_{25}(SC_2H_4Ph)_{18}$, $Au_{38}(SC_2H_4Ph)_{24}$, and $Ag_{32}(SG)_{18}$ are some of the examples of monolayer-protected noble metal clusters.[6] $Au_{25}(SC_2H_4Ph)_{18}$ is a thoroughly investigated QC along with x-ray crystal structure and its size is 1.2 nm with a HOMO-LUMO gap of 689 nm (1.8 eV). It exhibits absorption bands at 451 nm (2.75 eV) and 400 nm (3.1 eV)

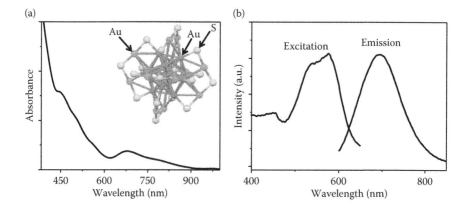

FIGURE 26.1
(See color insert.) (a) UV/Vis optical absorption spectrum of $Au_{25}(SG)_{18}$ cluster. (b) Photoluminescence spectra of $Au_{25}(SG)_{18}$, which exhibits emission at 690 nm. (Adapted from Shibu, E.S. et al., *J. Phys. Chem C.*, 112, 12168, 2008. Copyright with permission from American Chemical Society.) Inset: crystal structure of $Au_{25}(SCH_2CH_2Ph)_{18}$. Total structure is assumed as a shell of six (–RS–Au–SR–Au–SR–) units sitting on an Au_{13} core. (Adapted from Zhu, M. et al., *J. Am. Chem. Soc.*, 130, 5883, 2008. Only Au and S atoms are shown. Copyright with permission from American Chemical Society.)

due to intraband (sp→sp) and interband (d→sp) transitions, respectively, which are entirely different from the SPR peak of NPs at 520 nm. Its structure consists of an Au_{13} icosahedral core in which one Au atom is at the center. The Au_{13} core is protected by six RS–Au–SR–Au–SR units, as shown in the inset of Figure 26.1.[7] The Au_{25} cluster protected with glutathione (GSH) ligands exhibits red luminescence[8] around 700 nm (Figure 26.1b). Progress is slow in obtaining the crystal structure of silver clusters because of their poor stability, unlike in gold cluster systems. Only a few reports of crystal structure of silver clusters are available in the literature thus far. The molecular formula of one such cluster is $Ag_{14}(SC_6H_3F_2)_{12}(PPh_3)_8$ and the structure contains an Ag_6^{4+} octahedral core that is very different from analogs of gold clusters.[9] It has absorption bands at 368 and 530 nm, and it shows yellow emission both in the solid and solution states. Two other clusters with molecular formulae $Ag_{16}(DPPE)_4(SC_6H_3F_2)_{14}$ and $\{Ag_{32}(DPPE)_5(SC_6H_4CF_3)_{24}\}^{2-}$ (where DPPE is 1,2-bis(diphenylphosphino)ethane, $SC_6H_3F_2$ is 3,4-difluorothiophenol-H, and $SC_6H_4CF_3$ is 4-(trifluoromethyl)thiophenol-H) were crystallized by Yang et al.[10] The composition of these clusters reveals the presence of mixed ligands. Both the clusters exhibit molecule-like absorption spectra that contain a main peak at 485 nm. In addition to the 485-nm peak, the Ag_{32} cluster showed a shoulder at 720 nm. Both the clusters exhibit weak blue emission when excited with ultraviolet (UV) light. Photoluminescence (PL) of both the clusters was strong in CH_2Cl_2 at 440 nm when excited at 360 nm. These clusters have core–shell structures. Ag_8^{6+} and Ag_{22}^{12+} are the cores in Ag_{16} and Ag_{32}, respectively. In Ag_{16}, the Ag_8^{6+} unit was encapsulated in a complex shell of $\{Ag_8(DPPE)_4(SC_6H_3F_2)_{14}\}^{6-}$, whereas in Ag_{32}, the Ag_{22}^{12+} core was encapsulated in an $Ag_{10}(DPPE)_5(SC_6H_4CF_3)_{24}$ shell. The Ag_{22}^{12+} core was coprotected by one $\{Ag_6(DPPE)_3(SC_6H_4CF_3)_{12}\}^{6-}$, two $\{Ag_2(DPPE)(SC_6H_4CF_3)_4\}^{2-}$, and four $(SC_6H_4CF_3)^-$ units.

In this chapter, we discuss the applications of NMNs as sensors for toxins in water. It focuses mainly on the chemistry of various nanomaterials and mechanisms of interactions responsible for sensitivity. We present the sensors that are able to detect ultra-low levels of contaminants even in the presence of common interfering molecules/ions. We note that some of the applications of relevance to drinking water using NMNs have already been commercialized. Developments in this area of the past several years have been reviewed previously[2] and have also been the subject of a book chapter.[11] In the present work, we focus on the most recent literature of the past 4 years.

26.2 Sensing/Removal of Pollutants of Water Using NMNs

26.2.1 Inorganic Metal Ions

26.2.1.1 Mercury

Use of gold and silver NPs has emerged as an option for capturing various forms of mercury, at low concentrations, of significance to drinking water as these NPs possess high adsorption capacities. Adsorption capacity was found to be low when Au@citrate NPs were treated with Hg^{2+}. This is due to poor electrostatic interactions between NPs and Hg^{2+} ions. Lisha and coworkers[12] have used metal-alloying chemistry for sorption of toxic Hg^0. They found that alumina-supported Au@citrate NPs (10–20 nm in diameter) show an adsorption capacity of 4.065 g of mercury per gram of gold NPs. In this study, Hg^{2+} was reduced to Hg^0 by using dilute aqueous $NaBH_4$. Formation of an amalgam (Au_3Hg) was confirmed by x-ray diffraction (XRD) and energy-dispersive x-ray spectrometry analyses. Ojea-Jimenez et al.[13] demonstrated the use of Au@citrate NPs of 8.9 ± 1.6 nm size for sequestration of Hg^{2+} ions from Milli-Q and real waters. They have proposed that the surface of gold NPs catalyze the reduction of Hg^{2+} by citrate groups that are present on their surfaces. The formation of an Au_3Hg alloy was identified. Gold may be recovered from the alloy. Mercaptosuccinic acid (MSA)-protected Ag NPs of different sizes were studied for Hg^{2+} sorption.

Sumesh et al.[14] have synthesized Ag@MSA NPs of sizes 9 ± 2 nm and 20 ± 5 nm for which Ag/MSA mole ratios was maintained as 1:6 and 1:3, respectively (Figure 26.2a). These NPs were supported on alumina, and they were used for Hg^{2+} removal in column setups. A solution of 2 ppm Hg^{2+} was passed in separate columns containing 1:3 and 1:6 Ag@MSA materials at identical flow rates. Mercury was detected in the eluent after passing 2.0 and 5.5 L Hg^{2+} solutions in the case of 1:3 and 1:6 Ag@MSA, respectively (Figure 26.2b). Concentration of Hg^{2+} went below 85 ppb (in the case of 1:6 Ag@MSA) with an input concentration of 2 ppm. A high removal ability of 800 mg Hg/g Ag@MSA was achieved in the case of 1:6 Ag@MSA. From this, it is clear that small NPs (1:6 Ag@MSA) adsorb larger quantities of mercury owing to the presence of a larger number of functional groups per unit mass compared with NPs of a bigger size.

Gold and silver QCs were shown to have an ability to detect Hg^{2+} ions down to 2 ppb level. This concentration is the maximum contamination limit in drinking water set by the US EPA. Xie et al.[15] used red-emitting Au_{25} clusters encapsulated in a protein matrix (bovine serum albumin [BSA]) for highly selective detection of Hg^{2+}. Here, detection has been on the basis of quenching of the PL of encapsulated clusters due to a high degree of $Hg^{2+}–Au^+$ ($d^{10}–d^{10}$) interactions. The detection limit was 0.5 nM (0.1 ppb). Luminescence had partially been regained after reducing Hg^{2+} to Hg^0 by an $NaBH_4$ solution. Specific $d^{10}–d^{10}$ interactions were noticed in the $Ag^+–Cu^+$ cases also. For this, they had prepared core–shell Au@Ag clusters by reducing Ag^+ ions on as-prepared $Au_{25}BSA$ clusters. Introduction of Cu^{2+} ion quenches the luminescence of the Au@AgBSA clusters due to $Ag^+–Cu^+$ metallophilic interactions. Here, formation of Cu^+ was by the reduction of Cu^{2+} by BSA molecules.

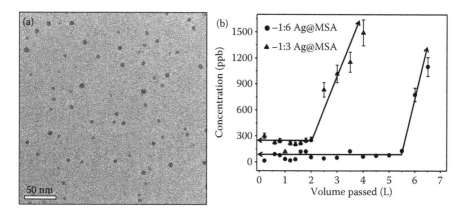

FIGURE 26.2
(a) TEM image of parent 1:6 Ag@MSA NPs. (b) Plot of the concentration of Hg^{2+} detected in the eluent as a function of the volume of Hg^{2+} solution passed through individual columns packed with 3.0 g each of 1:3 and 1:6 Ag@MSA NPs supported on alumina. Loading of 1:6 and 1:3 Ag@MSA NPs on alumina was 0.5 and 0.3 wt%, respectively. (Adapted from Sumesh, E. et al., *J. Hazard. Mater.*, 189, 450, 2011. Copyright with permission from Elsevier.)

The PL of GSH-protected silver nanoclusters was selectively quenched by Hg^{2+}. The sensitivity of these clusters for Hg^{2+} was found to be appreciable down to 0.1 nM (0.02 ppb). PL quenching was due to aggregation of clusters due to the strong interaction between Hg^{2+} and the carboxylic groups of clusters.[16]

26.2.1.2 Lead

Colorimetric, fluorescence, and electrochemical methods have been used to detect lead species in water samples. Tetra-*n*-octylammonium bromide (TOAB)-protected gold NPs were treated with different concentrations of Pb^{2+} solutions in THF.[17] It was found that the absorbance of gold particles decreased along with a red shift of the SPR peak, as Pb^{2+} concentration increased. At higher concentrations, the color of the solution turned from red to blue and finally to colorless. The red shift of the SPR peak and the blue color were attributed to aggregation of NPs and change in the dielectric constant of the particles. Pb^{2+} ions interact with Br^- ions on the surface of particles to form $PbBr_2$. During this interaction, some of the TOAB was removed, resulting in the aggregation and color change. Red emission was seen after illumination of lead ion–treated gold particle solution with UV light (300 nm), and the solution exhibits an absorption band at 300 nm due to the formation of $PbBr_2$. Wang et al.[18] reported the detection of lead in the nanomolar range using gold NPs. They propose the formation of AuPb alloy on the surface of NPs at pH 9. This alloy increases the catalytic activity of NPs in H_2O_2-mediated oxidation of Amplex UltraRed (AUR). Here, AUR is nonluminescent, whereas its oxidation product is highly luminescent (showing emission at 584 nm when excited at 540 nm). The intensity of the emission peak at 584 nm is proportional to the concentration of Pb^{2+} solution in the 0.05–5 µM range. The detection limit was 4 nM. This method was used to test the presence of Pb^{2+} in lake water and in blood samples.

Yuan et al.[19] have synthesized nonluminescent Au nanodots (Au NDs) protected with tetrakis(hydroxymethyl)phosphonium chloride (THPC) and GSH. After exchange of THPC with 11-mercaptoundecanoic acid (MUA), Au NDs exhibited a green emission due to MUA ligand-to-Au(I) metal charge transfer. These dual ligand (MUA and GSH)-stabilized Au NDs were used for selective detection of Pb^{2+} ions. After the addition of Pb^{2+} ions, green fluorescence became quenched because of the aggregation of Au NDs, which was attributed to the interaction of GSH with Pb^{2+} ions. In this method, the detection limit was 5 nM.

Yoosaf et al.[20] have synthesized gallic acid–protected gold and silver NPs that were highly stable in the pH range 4.5–5.0. The stability was attributed to strong electrostatic interactions between the NP surface and the carboxylate anion, and a high negative zeta (ζ) potential (−45 mV). Pb^{2+} ions interact with carboxylate groups of Ag/Au NPs, leading to aggregation and color change. Owing to aggregation, the SPR peak became red-shifted. The SPRs of individual particles couple in aggregates (Figure 26.3). The extent of the red shift was dependent on the concentration of lead ions. Aggregation of particles was evident from transmission electron microscopy (TEM) analyses (Figure 26.3). Interaction and shift of the SPR peak was highly specific to lead ions. Unlike Pb^{2+}, other metal ions such as Hg^{2+}, Cd^{2+}, Cu^{2+}, Zn^{2+}, and Ni^{2+} did not lead to a shift of the SPR. GSH-protected red-emitting silver NPs were found to selectively sense Pb^{2+} ions down to parts-per-quadrillion levels.[21] The reason for fluorescence quenching is believed to be the strong complexation of Pb^{2+} with GSH molecules. This led to the release of ligands into the solution resulting in the dissolution of NPs. At higher Pb^{2+} concentrations, TEM analysis showed the absence of NPs. These fluorescent Ag NPs were also tested for their ability to detect Pb species in real samples such as river water, battery, paint, and toys. The concentrations of Pb^{2+} obtained

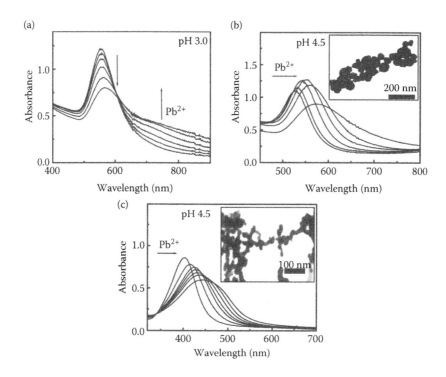

FIGURE 26.3
Absorption spectra of Au/Ag NPs in water after addition of increasing amounts (0–150 μM) of Pb^{2+} ions. (a and b) Effect of Pb^{2+} ions on Au NPs at pH 3.0 and 4.5, respectively. (c) Effect of Pb^{2+} ions on Ag NPs at pH 4.5. Insets show TEM images of Au (b) and Ag (c) NPs in the presence of 100 μM Pb^{2+} ions. (Adapted from Yoosaf, K. et al., *J. Phys. Chem. C.*, 111, 12839, 2007. Copyright with permission from American Chemical Society.)

by fluorescence quenching method were in good agreement with concentrations obtained in inductively coupled plasma–mass spectrometry data. Copper clusters encapsulated in BSA proteins were also used to selectively detect Pb^{2+} ions in ppm concentrations.[22] Luminescence of the clusters was quenched with an increase in concentration of Pb^{2+}. The reason for quenching of luminescence was obtained from dynamic light scattering (DLS) measurements. In DLS, aggregation of many clusters due to the interaction of functional groups of protein with Pb^{2+} ions was noticed.

26.2.1.3 Copper

Although copper is one of the essential elements, beyond a certain limit (>1.3 ppm in drinking water), it is toxic. Several colorimetric and fluorescence sensors have been developed in the recent past. Red luminescence of cyclodextrin (CD)-encapsulated Au_{15} clusters protected with GSH ligands was selectively quenched by Cu^{2+}.[23] The reason for quenching of luminescence was understood from x-ray photoelectron spectroscopy (XPS) analysis in which copper was found to be in the zerovalent state (Cu^0). The reduction of Cu^{2+} to Cu^0 was possible due to the enhanced reactivity at a size level of QCs. The cluster core gives electrons for the reduction of Cu^{2+}, and the reduced copper stays on the surface of reacted clusters. As a result of the change in chemical environment, fluorescence was quenched. For practical applications, the cluster was incorporated into the chitosan film and was used for copper sensing similar to a pH paper strip. Red-fluorescent gold NPs, protected with MUA, were used to selectively sense

Cu^{2+} ions down to 87 nM.[24] Detection of Cu^{2+} was on the basis of fluorescence quenching. With increase in concentration of Cu^{2+} in the range of 10^{-8}–10^{-6} M, there was a linear decrease in the fluorescence intensity. Fluorescence was recovered after the addition of a stoichiometric amount of ethylenediaminetetraacetic acid (EDTA). This indicates the blocking of the charge transfer (leading to fluorescence quenching) from the complex of S–Au to gold NPs due to the interaction of Cu^{2+} with carboxylic groups of 11-MUA. Cu^{2+} (hard acceptor) binds preferentially with carboxylate groups (hard donor). The possibility of fluorescence quenching by aggregation was ruled out as TEM analysis revealed that the particles were well dispersed. The other reason might be the paramagnetic nature of Cu^{2+} ions.

Liu et al.[25] have synthesized Ag clusters in azobenzene-modified poly(acrylic acid) (MPAA) templates. Clusters exhibit red and yellow emissions depending on the excitation wavelength, indicating the formation of clusters of different size. Fluorescence was sensitive specifically to Cu^{2+} ions and was not sensitive to a large number of other cations and anions. Fluorescence quenching was due to the binding of Cu^{2+} on the surface of clusters leading to energy transfer. Fluorescence was recovered after the addition of chelating agents such as EDTA. The energy transfer mechanism for quenching of fluorescence was confirmed by absorption spectroscopy: absorption spectra of MPAA and Cu^{2+} + MPAA were recorded in which the Cu^{2+} + MPAA complex exhibits a strong and broad absorption in the 540–800 nm range. An overlap of the emission peak of the cluster and the absorption peak of Cu^{2+} + MPAA complex indicates that energy transfer is responsible for fluorescence quenching. These clusters were integrated onto a cellulose filter paper, just by dipping. After drying, the paper strips were exposed to different concentrations of Cu^{2+} solutions and were kept under a UV lamp. There was a clear absence of emission from the high concentration Cu^{2+} solutions. These paper strips were used to detect copper ions in real water samples as well. Homocysteine and dithiothreitol (DTT)-modified Ag NPs were found to show color change from yellow to orange and then to green-brown after the addition of Cu^{2+}. This change is due to the aggregation of NPs.[26] The change of color was observed only in the case of copper, indicating high selectivity toward copper. These Ag NPs were coated on normal paper for testing copper ions. The limit of naked-eye detection of copper was 7.8 nM. It was tested in real waters such as tap and pond water.

26.2.1.4 Arsenic

Arsenic exists in groundwater in +3 and +5 oxidation states, among which As^{3+} is highly toxic. Arsenic exists mainly as anions (arsenite and arsenate). Liang et al.[27] developed a label-free colorimetric method for the detection of arsenite using G/T-rich oligonucleotides and unmodified Au NPs. Sensing was on the basis of a color change of gold NPs due to aggregation. At pH 8, arsenic exists in water as H_3AsO_3 having three hydroxyl groups that can form strong hydrogen bonds with amine groups (R_2NH, RNH_2) or carbonyl groups (C = O) of bases in G/T-rich single-stranded DNA (ssDNA). On the other hand, arsenate exists as $HAsO_4^{2-}$, which has one As=O bond that may disturb the formation of hydrogen bonds. It is well known that by the addition of common salt, Au@citrate NPs can be made to undergo aggregation. This aggregation is due to the decrease of electrostatic repulsion between negatively charged NPs. G/T-rich ssDNA has the flexible property to bind with gold NPs through N atoms of DNA. After the addition of arsenic species, arsenite forms hydrogen bonds, leading to the folding of G/T-rich ssDNA. Gold NPs were introduced into the folded DNA in which NPs were not protected. Addition of NaCl leads to aggregation of NPs resulting in a color change and appearance of a new peak around 700 nm. Aggregation of NPs was confirmed by TEM analysis. Unlike arsenic, other metal ions did not show any color change of

Au NPs. This was due to the less/no interaction of added species with G/T-rich ssDNA. Now, G/T-rich ssDNA was able to protect the Au NPs. Hence, after the addition of NaCl, there was no color change. However, there was sensitivity to Hg^{2+} due to the T–Hg–T complex, which was overcome by using EDTA. With this strategy, arsenite was detected down to 2 ppb.

Selective detection of arsenic in groundwater was done by Kalluri et al.[28] using modified gold NPs. Modification of the surface of NPs was done using thiol group (-SH)-containing molecules such as GSH, DTT, and Cys, separately. After the addition of $As^{3+/5+}$ ions into the functionalized Au NPs, there was a color change from red to blue, which was seen in the absorption spectrum as a red shift of SPR of Au NPs. The blue color was attributed to the presence of aggregated NPs and the presence was confirmed by TEM analysis. Aggregation of NPs was brought about by the attractive interaction between carboxylate groups of ligands and $As^{3+/5+}$. Other heavy metal ions did not show the color change. Selectivity toward As^{3+} was attributed to the high binding constants of As^{3+} with all the three ligands. DLS, which accurately measures the size of the particles, was used to detect low levels of arsenic, whereas the colorimetric method is applicable at high concentrations. Lower detection limits using DLS and colorimetric methods were 3 ppt and 1 ppb, respectively. Functionalized Au NPs were used to test the presence of As in Bangladeshi wells and Mississippi tap and drinking water. Bangladeshi wells were found to have >1 ppm of As, as seen by the color change of NPs.

Square-wave anodic stripping voltammetry (SWASV) is one of the highly sensitive techniques in the electrochemical analysis of trace levels of metals in different samples. Using this method, Jena et al.[29] have detected As^{3+} selectively (in presence of Cu^{2+}) with a limit of 0.02 ppb under optimized conditions. This method involves mainly two steps: (i) deposition of As^0 at an optimized potential for a particular time (−0.35 V for 100 s) and (ii) anodic stripping of deposited As^0. The anodic stripping signal was used to monitor the concentration of As^{3+} in solution. The SWASV response of the gold nanoelectrode ensembles (GNEEs) electrode at different As^{3+} concentrations (in 1 M HCl) is shown Figure 26.4a.

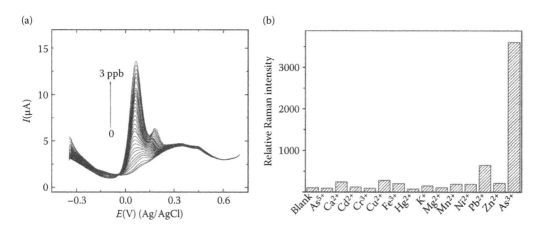

FIGURE 26.4
(a) SWASV response of the GNEE electrode toward As^{3+} at different concentrations in 1 M HCl. Each addition increased the concentration of As^{3+} by 0.1 ppb. (Adapted from Jena, B.K., and Raj, C.R., *Anal. Chem.*, 80, 4836, 2008. Copyright with permission from American Chemical Society.) (b) Metal ion–induced SERS intensity changes of a proposed SERS-based As^{3+} sensor. The concentration of each of As^{3+}, Cd^{2+}, Cu^{2+}, Cr^{3+}, Zn^{2+}, Ni^{2+}, Fe^{3+}, and As^{5+} was 1.34 µM. The concentration of other metal ions (Ca^{2+}, Hg^{2+}, K^+, Mg^{2+}, Mn^{2+}, and Pb^{2+}) was 13.4 µM. The incubation time was 2 min. (Adapted from Li, J. et al., *ACS Appl. Mater. Interfaces*, 3, 3936, 2011. Copyright with permission from American Chemical Society.)

A linear increase in current was observed up to 15 ppb. GNEEs were prepared by immobilizing colloidal gold NPs on a thiol functionalized sol–gel derived 3D silicate network preformed on a polycrystalline gold electrode.

GSH-functionalized Ag NPs based on a surface enhanced Raman spectroscopy (SERS) platform, for highly selective and sensitive detection of trace levels of As^{3+}, was developed (Figure 26.4b) by Li et al.[30] Initially, Ag NPs were functionalized with GSH and 4-mercaptopyridine (4-MPY). Here, 4-MPY was used as a Raman analyte. As^{3+} solutions of different concentrations were added into the modified Ag NPs. The color of NPs changed from yellow to brown. This color change was due to aggregation of NPs and was confirmed by microscopic analyses. When NPs undergo aggregation, the number of hotspots increases and hence Raman signals of analyte molecules become enhanced. Aggregation of Ag NPs was selective to As^{3+}; other heavy metal ions did not lead to aggregation. With this method, the detection limit for As^{3+} was 0.76 ppb. Real water samples were also analyzed using the same technique.

26.2.2 Inorganic Anions

26.2.2.1 Cyanide (CN⁻)

Au@citrate NPs have been shown to be better SERS candidates after treating with CN^- ions. Senapati et al.[31] used the SERS technique for selective and highly sensitive detection of CN^-. After the addition of CN^- ions to the Au@citrate NPs, they aggregate leading to the red shift of the SPR peak. SERS of aggregates show Raman shifts at ~300, ~370, and ~2154 cm^{-1} originating from Au–CN bending, Au–C, and C≡N stretching frequencies, respectively. This indicates that the aggregation was due to the formation of Au–C bonds. The intensity of the C≡N peak was increased with increase of CN^- ions concentration, indicating the possibility of quantification of CN^- ions. Using the SERS method, CN^- ions were detected down to 110 ppt levels. They showed the detection of CN^- ions in real samples as well. Time-dependent SERS suggested a decrease in intensity of C≡N stretching. This was attributed to the reaction of CN^- ions with gold particles leading to the formation of an $Au(CN)_2^-$ complex.

26.2.2.2 Iodide (I⁻)

$Au_{25}(SG)_{18}$ clusters were shown to selectively detect I^- ions down to the 400 nM level. After the introduction of I^- solutions of different concentrations into the Au_{25} clusters, a red shift of the emission peak was seen. Increase in the intensity of the red-shifted emission peak was noticed with an increase in I^- ion concentration. Wang et al.[32] called this as affinity-induced ratiometric and enhanced fluorescence, which is rare, whereas, in most of the cases, aggregation-induced fluorescence quenching occurs. The reason for this fluorescence enhancement was obtained by XPS analysis of I^--treated Au_{25} clusters. Quantification of elements suggests the composition to be $Au_{25}(SG)_{17.68} I_{1.09}$, indicating the partial replacement of GSH ligands by I^-. Formation of a new composition is the likely reason for the fluorescence enhancement. Another possibility is the binding of small amounts of I^- with the core of the cluster without disturbing the ligands.

Zhang et al.[33] used citrate-capped core–shell Cu@Au NPs for selective colorimetric detection of I^- ions. After the addition of I^- ions into the purple core–shell Cu@Au NPs, there was a color change from purple to red. Using the color change, they could detect 0.76 ppm of I^- within 20 min by observation with the naked eye. The color change was attributed to a transformation of the interconnected and irregularly shaped NPs to the single, separated, and nearly spherical ones, as confirmed by TEM and UV/visible (Vis) absorption spectroscopy.

26.2.2.3 Thiocyanate (SCN⁻)

Citrate-capped Au NPs were synthesized according to the literature, and Tween 20 (polysorbate surfactant) was subsequently added as a capping agent. After the addition of SCN⁻, some of the citrate ions on the surface of the Au NPs were replaced owing to the higher affinity of gold for SCN⁻ ions.[34] As a result, Tween 20 molecules, which are adsorbed on the surfaces of Au NPs, were separated and an aggregation of Au NPs occurred. Aggregation of particles was accompanied by a visible color change from red to blue within 5 min. The sensing of SCN⁻ could easily be confirmed by a UV/Vis spectrophotometer. The effects of relevant experimental parameters including concentration of Tween 20, pH of solution, incubation temperature, and time, were evaluated to optimize the method. Under the optimized conditions, this method yields excellent sensitivity, down to 11.6 ppb and selectivity toward SCN⁻. This method was also applied successfully to test the presence of SCN⁻ in saliva of smokers and real water samples.

26.2.2.4 Sulfide (S^{2-})

Au@Ag core–shell nanocubes were used for sensing of S^{2-} ions based on the change in optical properties. When S^{2-} ions (from Na_2S) were added to Au@Ag core–shell nanocubes, there was a change in the position of the SPR peak and the color of the solution.[35] Detection limits of S^{2-} using this strategy was 10 ppb by UV/Vis spectroscopy and 200 ppb by the naked eye. The SPR was tuned in the 500–750 nm window. A possible reason for the SPR tuning with concentration of S^{2-} could be the increase in thickness of the Ag_2S layer formed on the Ag shell. There were slight distortions of nanocubes after reaction with S^{2-}, which may also account for this tuning. Owing to the change in refractive index of the new composition (Ag_2S), color would be changed. The reaction can be represented as

$$4Ag + 2S^{2-} + O_2 + 2H_2O \rightarrow 2Ag_2S + 4OH^-$$

26.2.3 Organic Contaminants

26.2.3.1 Pesticides

Endosulfan, CP, and malathion are important pesticides being used in developing countries. Use of noble metal NPs for the removal of pesticides from water has been demonstrated for the first time by Nair et al.[36] Gold and silver NPs, protected with trisodium citrate, have been utilized for this purpose. Au@citrate NPs interact with endosulfan molecules leading to a color change (from wine red to blue) and a red shift of the SPR peak. SPR was sensitive down to the 2 ppm range. It was found that Au particles showed faster interaction compared with Ag particles. Au and Ag@citrate NPs were also used for the removal of CP and malathion.[37] After the introduction of CP and malathion to Ag and Au@citrate NPs, there were color changes and red shifts of SPR peaks due to the adsorption of pesticide molecules on the surface of NPs. In this method, the limit of detection was only moderate: a few ppm, typically the limit of solubilities. Ag and Au@citrate NPs were supported on neutral alumina. These materials were packed in columns and were used for the complete removal of pesticides from water passing through the columns. On this basis, Pradeep and colleagues[38–40] have demonstrated a nanotechnology-based water purifier for the first time in the world. Detection limits of CP and malathion were enhanced to parts-per-billion levels by the addition of Na_2SO_4 to the Au@citrate NPs.[41] Addition of Na_2SO_4 to the as-prepared Au particles brought change in color from wine red to purple along with small decrease in SPR peak intensity and a slight red shift. To the Au NPs + Na_2SO_4 mixture, different concentrations of CP solutions were added. The purple color of the mixture has changed at 25 ppb CP to a blue color in which the SPR peak intensity decreased and a new peak emerged around 680 nm. The intensity of the 680 nm peak further red-shifted with the increase of CP concentration. Red shift and blue color were attributed to aggregation of NPs due to adsorption of CP on the NP surface. Similar observations were made with malathion. Other salts such as NaCl, K_2SO_4, and $(NH_4)_2SO_4$ were also tested to identify the detection limit. It was found that in the presence of Na_2SO_4, the detection limit was quite low. With this method, CP and malathion were detected down to 10.8 and 40.3 ppb, respectively.

It is very important to understand the nature of interactions between the surface of NPs and the pesticide molecules. The most important interaction is adsorption. The surface chemistry of NPs and pesticides is yet to be understood. There is a great concern about the impact of pesticide molecules and their degradation products/metabolites on human health. Some molecules, such as CP, show degradation of their chemical structure even in the absence of irradiation of light such as UV (which is generally used for degradation of organic molecules). CP undergoes hydrolysis in the presence of silver NPs to give 3,5,6-trichloro-2-pyridinol (TCP), which is less toxic compared with CP and diethyl thiophosphate.[42] Aggregation of NPs was confirmed by TEM analysis and was reflected in the red shift of the SPR peak (Figure 26.5a). A new peak at 320 nm in the absorption spectrum of the reaction solution due to TCP was noticed. Formation of TCP was confirmed by electrospray ionization mass spectrometry (ESI MS). The molecular ion peak of CP at m/z (mass-to-charge ratio) 350 was completely absent after the reaction (Figure 26.5b). A new

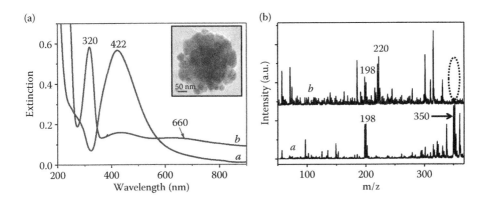

FIGURE 26.5
(a) UV/Vis absorption spectra of Ag@citrate NPs treated with 1 and 50 ppm of CP (traces *a* and *b*, respectively) for 24 h. Inset: TEM image of Ag NPs treated with CP. (b) ESI MS of parent CP and reaction product of CP + Ag NPs (traces *a* and *b*, respectively) in positive mode. Absence of the molecular ion peak of CP at m/z 350 in trace *b* is marked. (Adapted from Bootharaju, M.S., and Pradeep, T., *Langmuir*, 28, 2671, 2012. Copyright with permission from American Chemical Society.)

peak at m/z 220 in the reaction mixture was due to the TCP after addition of one sodium ion per molecule, confirming the degradation of CP to TCP. It was proposed that the degradation of CP proceeds through the formation of an AgNP–S surface complex, the presence of which was confirmed by Raman spectroscopy. In this complex, the P–O bond cleaved to give a stable aromatic species, TCP. The rate of degradation of CP increased with an increase in temperature and pH. There is a large need for the development of new methods that are simple and cost-effective for the detection of pesticides in water and other parts of the environment such as food, soil, etc. Nanomaterials-based techniques are promising for sensing pesticide molecules, which explore the changes in unusual properties of NPs, such as optical absorption, PL, and surface enhancement in Raman signals. One of the important and commonly explored properties is SPR. Metal NPs (especially Ag and Au) are well known to enhance the Raman signals of analyte molecules at their hotspots due to electromagnetic field and chemical enhancement mechanisms. The shell-thickness-dependent Raman enhancement of thiocarbamate and organophosphorous compounds by Au@Ag core–shell nanostructures has been demonstrated.[43] The core–shell NPs are better candidates compared with monometallic (Ag/Au) NPs for SERS. With the increase of the silver shell thickness to 7 nm, there was a wide range and strong SPR in the 320–560 nm range. This results in the strong enhancement in Raman signals of pesticide molecules enabling the detection of pesticides to nanogram levels.

26.2.3.2 Halocarbons

Unusual reactions of Ag and Au NPs with halocarbons were found by Nair and Pradeep.[44] Here, halocarbons such as CCl_4, $CHCl_3$, $C_6H_5CH_2Cl$, $CHBr_3$, and CH_2Cl_2 react with noble metal (Ag and Au) NPs to form corresponding metal halides and amorphous carbon. Mineralization of halocarbons by bulk silver or gold was impossible, whereas, at the nanoscale, this was possible under suitable experimental conditions. In the reaction, 2.5 mL of 1:1 (v/v) Ag@citrate NPs (in water) and isopropyl alcohol (IPA) were treated with 50 μL of CCl_4. Here, IPA was used to achieve good mixing of reactants. Absorption spectra of the reaction mixture were recorded at different intervals after mixing. The intensity of the SPR peak decreased with time accompanied by a slow disappearance of color, indicating the participation of NPs in the reaction. After 12 h of reaction, the light yellow color of the Ag NPs solution turned colorless; a gray precipitate was formed at the bottom of the reaction vessel. There was no SPR peak at this stage, indicating the completion of reaction. The precipitate was found to be AgCl by XRD. The same precipitate was analyzed with Raman spectroscopy before and after washing with ammonia. The presence of characteristic Raman features of amorphous carbon confirms the carbonaceous material. Washing the precipitate was performed to remove AgCl as $[Ag(NH_3)_2]^+Cl^-$ soluble complex. A similar reaction was also noticed with other halocarbons but at different reaction rates. In the case of Au@citrate NPs, the SPR peak was red-shifted after the introduction of benzyl chloride, indicating the aggregation of NPs along with a change in color. Reaction rate was slow (48 h) compared with that of Ag NPs. The reaction product was $AuCl_3$. The rate of reaction of halocarbons varies with the size of the particles.

Monolayer-protected noble metal QCs (especially clusters of silver), which are smaller in size (in-between molecules and NPs), show efficient reactivity with halocarbons compared with NPs.[45] An absorption spectrum and PL spectra of an $Ag_9(MSA)_7$ cluster are shown in Figure 26.6a. Reaction of CCl_4 with the $Ag_9(MSA)_7$ cluster was complete in 1.5 h. The precipitate formed in the reaction was confirmed to be AgCl (Figure 26.6b) using XRD. The mechanism of mineralization of halocarbons by noble metal NPs/clusters has

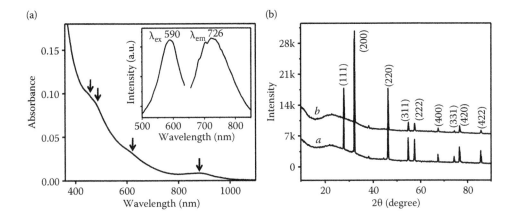

FIGURE 26.6
(a) UV/Vis absorption spectrum of Ag₉(MSA)₇ QCs. Inset is the PL spectra of Ag₉(MSA)₇ cluster emitting at 726 nm when excited at 590 nm. (b) XRD patterns of as-prepared AgCl (trace *a*) and reaction product (trace *b*) of cluster and CCl₄ in the presence of IPA. (Adapted from Bootharaju, M.S. et al., *J. Mater. Chem. A.*, 1, 611, 2013. Copyright with permission from The Royal Society of Chemistry.)

been understood to some extent. It was found that the solvent used (2-propanol) played an important role in the degradation of halocarbon. The electrons released in oxidation of IPA in the presence of metal clusters/NPs were abstracted by chlorocarbon to yield chloride ions. Decrease of pH was noticed (from ~6.0 to ~1.5) due to the formation of some acidic species. The protection monolayer on the cluster core, replaced by chloride ions, led to destabilization of clusters. As a consequence of these events, silver from cluster/NPs reacted with chloride ions to form more stable AgCl crystals.

Trichloroethene (TCE) is one of the commonly found organic compounds in groundwater and is carcinogenic. Advanced materials such as Pd-on-Au bimetallic NPs were used as catalysts for TCE hydrodechlorination (HDC) in the presence of hydrogen.[46] The reaction was done at room temperature in water and the product was ethane. Bimetallic NPs were shown to be more efficient catalysts than monometallic Pd NPs and Pd black (bulk). NPs were immobilized on silica, alumina, and magnesia to produce active oxide-supported catalysts. For Pd NPs, the higher activity was attributed to the availability of a greater amount of coordinatively unsaturated Pd atoms located on surface defects. In the case of bimetallic systems, Au NPs may be promoting the Pd activity through electronic or geometric effects, or through direct participation in the HDC reaction mechanism. The TCE HDC reaction was modeled as a Langmuir–Hinshelwood mechanism involving competitive chemisorption of dihydrogen and TCE for all three NP compositions (Pd/Au NPs with 30% and 60% Pd surface coverages, and pure Pd NPs). The HDC reaction was first-order for Pd/Au NPs and of non-first-order for Pd NPs. These differences were attributed to the differences in sorption affinities for reactant molecules. Recently, Pretzer et al.[47] found the variation of the HDC reaction rate with size of the Au particles and coverage of Pd. Maximum activity was observed for 7 nm Au particles with 60%–70% Pd coverage.

26.2.3.3 Other Organic Compounds

There are many organic contaminants, such as toluene- and sulfur-containing odorous molecules (e.g., thiophenol, thioether, thioanisole, and sulfides), present in water that cause its pollution. The main sources of these contaminants are wastewater from crude oil and petroleum refineries, waste released from manufacturing sites, pesticide and mosquito larvicide, and pharmaceutical intermediates. Scott et al.[48] developed an Au NPs/PDMS (polydimethylsiloxane) composite for the removal of the above contaminants from water. The principle of contaminant removal was based on the ability of PDMS to undergo large expansion in nonpolar solvents. The composite was prepared as films and foams among which foams exhibit large absorption of contaminants. This was attributed to the presence of a large density of micropores. Incorporation of Au NPs into a PDMS matrix led to the increase of surface area. Gupta and Kulkarni[49] have used similar composite foams for the removal of solvents such as BTEX (benzene, toluene, ethylbenzene, and xylene) and oil spills. They also showed the decolorization of garlic extract by this composite, which demonstrates promise as a food packaging material. Regeneration of composite was done by simple heat treatment and it works for several cycles. Other organic compounds, such as phenols and anilines, are being used mainly as disinfectants, antiseptics, and intermediates of dyes in organic synthesis. The "Integrated wastewater discharge standard" of the Ministry of Environmental Protection of China has set the maximum levels of aniline and phenolic compounds in discharged wastewater as <1.0 and 0.3 mg/L, respectively. Chen et al.[50] used β-CD-modified Ag NPs for sensitive and selective recognition of isomers of substituted anilines (o-, m-, p-phenylenediamine) and phenols (pyrocatechin, hydroquinone, and resorcinol). There were color changes when pyrocatechin, resorcinol, or hydroquinone was added to the β-CD-modified AgNP solutions, from apricot to yellow, dark brown, and red, respectively. These color changes were due to the different degrees of aggregation of Ag NPs, which, in turn, was due to the varying association constants between an aromatic isomer and β-CD. Binding strengths of β-CD with the above phenolic compounds decrease as follows: resorcinol > hydroquinone > pyrocatechin > phenol. The main reason for the above order is match (which follows the above order) of size of the guest, phenol isomer, and the pore volume of the host, β-CD. In the case of phenylenediamine (o, m, and p) isomers, more obvious color changes were seen. The color changes were due to the varying association constants with the guest and the hosts along with coulombic interactions between amino group and Ag NPs.

26.2.4 Biological Contamination/Sensing

Silver and copper were found to exhibit antimicrobial properties. As discussed in Section 26.1, in ancient times, people stored drinking water in utensils made of copper/silver. Nanosilver is more efficient as an antibacterial agent compared with bulk Ag due to the large surface area. The exact mechanism of antibacterial activity of Ag NPs is not known. Researchers propose the uptake of Ag NPs by bacteria, which leads to the leaching of Ag[+] ions from particles under such conditions as found inside bacteria. As a result, there is damage to DNA/cell lyses leading to the death of the bacteria. Using fluorescence of clusters and SPR peak shifts and color changes of NPs, several biological molecules were detected. A few of these are discussed below.

Dankovich and Gray[51] demonstrated the practical use of silver NPs in an antibacterial study (Figure 26.7a). In this experiment, silver NPs were incorporated into the blotting paper sheet. Water containing pathogenic bacteria, *Escherichia coli* and *Enterococcus faecalis*, was percolated through the blotting paper, and the log reduction values in the effluent were found to be about log 6 and log 3, respectively. Bacteria were found to be dead in the collected water. The quantity of silver that was leached from the paper was estimated to be <0.1 ppm (the current US EPA and WHO limits for silver in drinking water). This particular design demonstrates the effectiveness of silver NPs for emergency water purification.

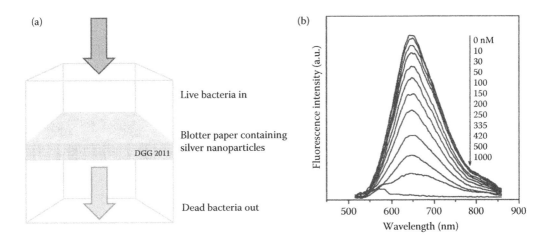

FIGURE 26.7

(a) Schematic representation of passing water containing bacteria through Ag NPs–incubated blotter paper. (Adapted from Dankovich, T.A., and Gray, D.G., *Environ. Sci. Technol.*, 45, 1992, 2011. Copyright with permission from American Chemical Society.) (b) PL spectra (λ_{ex} = 489 nm) of GSH–AgNCs (100 nM) in the presence of Cys solutions of different concentrations. (Adapted from Yuan, X. et al., *Anal. Chem.*, 85, 1913, 2013. Copyright with permission from American Chemical Society.)

Clusters are predicted to exhibit superior antimicrobial activity due to smaller size and high surface-to-volume ratio.

Yuan et al.[52] synthesized silver clusters using a cyclic reduction–decomposition method and used them for antibacterial studies with a model microbe, *Pseudomonas aeruginosa*. The microbe is known to cause urinary tract infections, pneumonia, and respiratory system infections in many burn cases and immunosuppressed AIDS patients. This particular bacterium is resistant to many conventional antibiotics, such as ampicillin, penicillin, and cephalosporin. In the experiment, red-emitting GSH-protected Ag clusters (r-Ag NCs) were used to kill the bacteria. This study showed that a very little amount of cluster (8 µg) was enough to inactivate the microbe. Growth of bacteria culture was completely stopped after treating with 500 µM of cluster (on the basis of Ag atoms). Efficiency of the r-Ag NCs for antibacterial activity was comparable to that of chloramphenicol, a common antimicrobial agent. For understanding the cause of inactivation of bacterial growth, the level of the intracellular reactive oxygen species (ROS) was measured. There was a 6-fold increase in the oxidized 2',7'-dichlorofluorescein (DCF) concentration in the r-Ag NCs-treated bacteria culture. This confirmed that the formation of ROS was due to the presence of silver clusters. The efficiency of silver in the form of clusters was higher compared with the NPs (>2 nm size) form. They also proposed that the reduction in side effects originated due to silver can be minimized (when clusters are used), as in other studies where larger-sized NPs were used.

Cysteine (Cys) is one of the essential thiol (-SH)-containing amino acids. It is recognized to be a potential neurotoxin. Deficient levels of Cys cause many health problems such as hair depigmentation, liver damage, slowed growth, skin lesions, weakness, and edema. It is important to have sensors for the selective and sensitive detection of Cys using a simple colorimetric/fluorometric method. Yuan et al.[53] have used GSH-protected Ag NCs for selective detection of Cys at <3 nM levels. Fluorescence of Ag clusters was quenched due to the interaction of Ag from clusters with the S in Cys, leading to the decomposition of clusters into thiolate–Ag(I) complexes (Figure 26.7b). Decomposition of clusters was also observed in the color change from brown to colorless along with the disappearance of characteristic features in the absorption spectrum of clusters. Fluorescence quenching was more highly selective for Cys than for the other 19 amino acids. Fluorescence of clusters was not quenched in the presence of other thiols such as GSH and proteins, e.g., BSA. Owing to the large size of these molecules, they cannot penetrate to reach the silver core through the ligand shell. In the presence of other thiols such as mercaptopropionic acid and cysteamine, fluorescence was quenched where the steric hindrance was overcome.

26.3 Devices and Market

It is important to work toward making use of materials as devices for use at community levels. Gold and silver NPs exhibit unusual properties such as optical absorption, luminescence, stability to various chemical environments, and catalytic activity. The absorption or luminescence is highly sensitive to ions/molecular interactions. Small quantities of materials are enough to cause considerably large changes in optical properties at ultra-low concentrations of analyte. This fact makes the fabrication of silver and gold nanosystems-based devices more feasible. It is also important to note that coating of nanomaterials on oxide surfaces such as alumina, silica, titania, and magnesia can be done for efficient dispersion of nanomaterials and their reuse, if the process of contaminant interaction is just physisorption. If the material involves certain chemical reactions, the reactions should be green and the products must be nontoxic. Red-emitting Au_{25} clusters, encapsulated in BSA protein, were incorporated into a nitrocellulose strip for detection of Hg^{2+}. This strip was red luminescent under a UV lamp. After dipping a piece of such a strip into various solutions of metal ions, luminescence of clusters was completely quenched selectively in the presence of Hg^{2+} ions. Similarly, β-CD-encapsulated GSH-protected Au_{15} clusters were incubated into chitosan, which was made as a film. This film was red-luminescent due to the presence of Au_{15} clusters. Those films were used to selectively detect Cu^{2+} ions down to 1 ppm, which is below the maximum contamination level (1.3 ppm), on the basis of quenching of luminescence that was seen by the naked eye.

Silver NPs, supported on neutral alumina, were used in cartridges of a water filter for the removal of pesticides, CP, and malathion. The same material was also used to decontaminate water from chlorocarbons such as CCl_4 and $C_6H_5CH_2Cl$. There are several nanosystems yet to be commercialized for water purification applications. The most important factors that are limiting commercialization of these materials are cost, reuse, and lack of detailed understanding of the effect of leachates from the materials, on human health. It is important to consider the disasters in communities where the drinking water problem is unexplainable. In such cases, materials that are cheap, highly portable, nontoxic, easy to use and distribute, and require low energy input, need to be addressed. A set-up like blotting paper containing Ag NPs for simple inactivation of bacteria is one such example.

26.4 Perspectives

Water pollution has become a major problem of the world. Water resources are contaminated severely and the complexity of the problem increases with time as the population increases steadily. Nanotechnology provides a chance to establish next-generation water-purifying systems. Permissible limits for safe drinking water norms have been decreasing with time. Novel and advanced materials are required to detect and remove contaminants at such low levels. Materials, such as graphene, possessing large surface area and high reactivity may be made as composites with NMNs (including NPs or luminescent clusters) to add more desirable properties. One of the difficulties in working with nanomaterials for water purification is the separation of materials from water. Making composite materials such as iron oxide–Ag/Au and graphene–iron oxide nanosystems definitely would enhance the separation of materials from treated water for reuse. It is important to consider the fate of nanoscale gold and silver that is entering into water during treatment. The quantity of metal ions released during treatment varies from one method to the other. Designers need to keep in mind that the designed material is applicable to remediate a large range of contaminants (instead of one or two) present in water. There must be a clear understanding of the mechanism of interactions. It is very important to focus on methods of preparation of materials. Methods involving easy handling, less time consumption for synthesis, and avoiding/minimizing special experimental conditions (such as maintaining very high and very low temperatures, expensive equipment, etc.) are always advantageous. A large number of strategies, such as microfluidics and single particle spectroscopy, may be more sensitive but have not yet entered significantly in the water segment. The strategies of sensing described above may be combined with novel materials to do simultaneous removal. In the drinking water segment, products of this kind have not yet appeared in the market place.

26.5 Summary

NMNs, including NPs and QCs, have been extensively used in sensing water contaminants such as heavy metal ions, anions, and biologically relevant molecules. All the detection methods are mainly based on changes in optical properties leading to color changes and SERS. Several molecular silver and gold clusters have been studied for metal ion detection on the basis of changes in fluorescence. The mechanisms were different for different clusters, which depend on the metal ions. In the case of NPs, detection was on the basis of aggregation of NPs leading to color changes. Importantly, several studies have extended the applications of materials in real samples as well. Most of the methods and materials were successful in detecting contaminants below the maximum contaminant levels set by the US EPA. However, the disadvantage of some of the studies is the difficulty in recycling of the materials, which is due to chemical transformation of materials and analytes. As a result, the original property of the material cannot be regained.

It is very much important to consider the reuse/recyclability when working with precious materials such as silver and gold. Silver clusters and NPs react with chlorocarbon to give silver chloride, which is one of the important ores of silver. This reaction is green as the silver chloride can again be used for extraction of silver. Bimetallic Pd-on-Au NPs were shown to be highly active catalysts for dechlorination of one of the groundwater contaminants, trichloroethane, through conversion to ethane in the presence of hydrogen. Silver NPs and clusters were utilized for antibacterial study. It was found that silver in the form of clusters was more efficient than that in the NP form.

Acknowledgments

T. Pradeep thanks the Department of Science and Technology (DST), Government of India, for constantly supporting research program on nanomaterials. M.S.B. thanks the CSIR for a senior research fellowship. The work presented here is a result of several coworkers whose names appear in the reference list.

References

1. Chouhan, S. and Flora, S.J.S., Arsenic and fluoride: Two major groundwater pollutants, *Indian J. Exp. Biol.*, 48, 666, 2010.
2. Pradeep, T. and Anshup, Noble metal nanoparticles for water purification: A critical review, *Thin Solid Films*, 517, 6441, 2009.
3. Qu, X. et al., Nanotechnology for a safe and sustainable water supply: Enabling integrated water treatment and reuse, *Acc. Chem. Res.*, 46, 834, 2013.
4. Zhang, Y. et al., Nano-gold catalysis in fine chemical synthesis, *Chem. Rev.*, 112, 2467, 2012.
5. Sajanlal, P.R. et al., Anisotropic nanomaterials: Structure, growth, assembly, and functions, *Nano Rev.*, 2, 5883, 2011.
6. Udayabhaskararao, T. and Pradeep, T., New protocols for the synthesis of stable Ag and Au nanocluster molecules, *J. Phys. Chem. Lett.*, 4, 1553, 2013.
7. Zhu, M. et al., Correlating the crystal structure of a thiol-protected Au_{25} cluster and optical properties, *J. Am. Chem. Soc.*, 130, 5883, 2008.
8. Shibu, E.S. et al., Ligand exchange of $Au_{25}SG_{18}$ leading to functionalized gold clusters: Spectroscopy, kinetics, and luminescence, *J. Phys. Chem. C.*, 112, 12168, 2008.
9. Yang, H. et al., Crystal structure of a luminescent thiolated Ag nanocluster with an octahedral Ag_6^{4+} core, *Chem. Commun.*, 49, 300, 2013.
10. Yang, H., Wang, Y. and Zheng, N., Stabilizing subnanometer Ag(0) nanoclusters by thiolate and diphosphine ligands and their crystal structures, *Nanoscale*, 5, 2674, 2013.
11. Pradeep, T. and Anshup, Detection and extraction of pesticides from drinking water using nanotechnologies. In: *Nanotechnology Applications for Clean Water*, N. Savage, M. Diallo, J. Duncan, A. Street, R. Sustich (Eds.), William Andrew Publication, New York, USA, 2009.
12. Lisha, K.P., Anshup and Pradeep, T., Towards a practical solution for removing inorganic mercury from drinking water using gold nanoparticles, *Gold Bull.*, 42, 144, 2009.
13. Ojea-Jimenez, I. et al., Citrate-coated gold nanoparticles as smart scavengers for mercury(II) removal from polluted waters, *ACS Nano*, 6, 2253, 2012.
14. Sumesh, E. et al., A practical silver nanoparticle-based adsorbent for the removal of Hg^{2+} from water, *J. Hazard. Mater.*, 189, 450, 2011.
15. Xie, J., Zheng, Y. and Ying, J.Y., Highly selective and ultrasensitive detection of Hg^{2+} based on fluorescence quenching of Au nanoclusters by Hg^{2+}–Au^+ interactions, *Chem. Commun.*, 46, 961, 2010.
16. Wang, C. et al., Fluorescent silver nanoclusters as effective probes for highly selective detection of mercury(II) at parts-per-billion levels, *Chem. Asian J.*, 7, 1652, 2012.
17. Yang, J. et al., A dual sensor of fluorescent and colorimetric for the rapid detection of lead, *Analyst*, 137, 1446, 2012.
18. Wang, C.I. et al., Catalytic gold nanoparticles for fluorescent detection of mercury(II) and lead(II) ions, *Anal. Chim. Acta*, 745, 124, 2012.
19. Yuan, Z. et al., Functionalized fluorescent gold nanodots: Synthesis and application for Pb^{2+} sensing, *Chem. Commun.*, 47, 11981, 2011.

20. Yoosaf, K. et al., In situ synthesis of metal nanoparticles and selective naked-eye detection of lead ions from aqueous media, *J. Phys. Chem. C*, 111, 12839, 2007.

21. Singh, A.K. et al., Synthesis of highly fluorescent water-soluble silver nanoparticles for selective detection of Pb(II) at the parts per quadrillion (PPQ) level, *Chem. Commun.*, 48, 9047, 2012.

22. Goswami, N. et al., Copper quantum clusters in protein matrix: Potential sensor of Pb^{2+} ion, *Anal. Chem.*, 83, 9676, 2011.

23. George, A. et al., Luminescent, freestanding composite films of Au_{15} for specific metal ion sensing, *ACS Appl. Mater. Interfaces*, 4, 639, 2012.

24. Guo, Y. et al., Stable fluorescent gold nanoparticles for detection of Cu^{2+} with good sensitivity and selectivity, *Analyst*, 137, 301, 2012.

25. Liu, X., Zonga, C. and Lu, L., Fluorescent silver nanoclusters for user-friendly detection of Cu^{2+} on a paper platform, *Analyst*, 137, 2406, 2012.

26. Ratnarathorn, N. et al., Simple silver nanoparticles colorimetric sensing for copper by paper-based devices, *Talanta*, 99, 552, 2012.

27. Liang, R.P. et al., Label-free colorimetric detection of arsenite utilizing G-/T-rich oligonucleotides and unmodified Au nanoparticles, *Chem. Eur. J.*, 19, 5029, 2013.

28. Kalluri, J.R. et al., Use of gold nanoparticles in a simple colorimetric and ultrasensitive dynamic light scattering assay: Selective detection of arsenic in groundwater, *Angew. Chem. Int. Ed.*, 48, 9668, 2009.

29. Jena, B.K. and Raj, C.R., Gold nanoelectrode ensembles for the simultaneous electrochemical detection of ultratrace arsenic, mercury, and copper, *Anal. Chem.*, 80, 4836, 2008.

30. Li, J. et al., Highly sensitive SERS detection of As^{3+} ions in aqueous media using glutathione functionalized silver nanoparticles, *ACS Appl. Mater. Interfaces*, 3, 3936, 2011.

31. Senapati, D. et al., A label-free gold-nanoparticle-based SERS assay for direct cyanide detection at the parts-per-trillion level, *Chem. Eur. J.*, 17, 8445, 2011.

32. Wang, M. et al., $Au_{25}(SG)_{18}$ as a fluorescent iodide sensor, *Nanoscale*, 4, 4087, 2012.

33. Zhang, J. et al., Colorimetric iodide recognition and sensing by citrate-stabilized core/shell Cu@Au nanoparticles, *Anal. Chem.*, 83, 3911, 2011.

34. Zhang, Z. et al., Label free colorimetric sensing of thiocyanate based on inducing aggregation of Tween 20-stabilized gold nanoparticles, *Analyst*, 137, 2682, 2012.

35. Park, G. et al., Full-color tuning of surface plasmon resonance by compositional variation of Au@Ag core–shell nanocubes with sulphides, *Langmuir*, 28, 9003, 2012.

36. Nair, A.S., Tom, R.T. and Pradeep, T., Detection and extraction of endosulfan by metal nanoparticles, *J. Environ. Monit.*, 5, 363, 2003.

37. Nair, A.S. and Pradeep, T., Extraction of chlorpyrifos and malathion from water by metal nanoparticles, *J. Nanosci. Nanotechnol.*, 7, 1871, 2007.

38. Pradeep, T. and Nair, A.S. with IIT Madras, A method of preparing purified water from water containing pesticides (chlorpyrifos and malathion), Indian Patent 200767, 2006.

39. Pradeep, T. and Nair, A.S. with IIT Madras, A device and a method for decontaminating water containing pesticides, PCT application, PCT/IN05/0002, 2005.

40. Pradeep, T. and Nair, A.S. with IIT Madras and Eureka Forbes Limited, A method to produce supported noble metal nanoparticles in commercial quantities for drinking water purification, Indian Patent Application, 2007.

41. Lisha, K.P., Anshup and Pradeep, T., Enhanced visual detection of pesticides using gold nanoparticles, *J. Envi. Sci. Health B.*, 44, 697, 2009.

42. Bootharaju, M.S. and Pradeep, T., Understanding the degradation pathway of the pesticide, chlorpyrifos by noble metal nanoparticles, *Langmuir*, 28, 2671, 2012.

43. Liu, B. et al., Shell thickness-dependent Raman enhancement for rapid identification and detection of pesticide residues at fruit peels, *Anal. Chem.*, 84, 255, 2012.

44. Nair, A.S. and Pradeep, T., Halocarbon mineralization and catalytic destruction by metal nanoparticles, *Curr. Sci.*, 84, 1560, 2003.

45. Bootharaju, M.S. et al., Atomically precise silver clusters for efficient chlorocarbon degradation, *J. Mater. Chem. A.*, 1, 611, 2013.

46. Nutt, M.O., Hughes, J.B. and Wong, M.S., Designing Pd-on-Au bimetallic nanoparticle catalysts for trichloroethene hydrodechlorination, *Environ. Sci. Technol.*, 39, 1346, 2005.
47. Pretzer, L.A. et al., Hydrodechlorination catalysis of Pd-on-Au nanoparticles varies with particle size, *J. Catal.*, 298, 206, 2013.
48. Scott, A., Gupta, R. and Kulkarni, G.U., A simple water-based synthesis of Au nanoparticle/PDMS composites for water purification and targeted drug release, *Macromol. Chem. Phys.*, 211, 1640, 2010.
49. Gupta, R. and Kulkarni, G.U., Removal of organic compounds from water by using a gold nanoparticle-poly(dimethylsiloxane) nanocomposite foam, *ChemSusChem.*, 4, 737, 2011.
50. Chen, X. et al., β-Cyclodextrin-functionalized silver nanoparticles for the naked eye detection of aromatic isomers, *ACS Nano.*, 4, 6387, 2010.
51. Dankovich, T.A. and Gray, D.G., Bactericidal paper impregnated with silver nanoparticles for point-of-use water treatment, *Environ. Sci. Technol.*, 45, 1992, 2011.
52. Yuan, X. et al., Highly luminescent silver nanoclusters with tunable emissions: Cyclic reduction–decomposition synthesis and antimicrobial properties, *NPG Asia Mater.*, 5, e39; doi:10.1038/am.2013.3, 2013.
53. Yuan, X. et al., Glutathione-protected silver nanoclusters as cysteine-selective fluorometric and colorimetric probe, *Anal. Chem.*, 85, 1913, 2013.

Further Readings

Aragay, G., Pino, F. and Merkoci, A., Nanomaterials for sensing and destroying pesticides, *Chem. Rev.*, 112, 5317, 2012.
Aragay, G., Pons, J. and Merkoci, A., Recent trends in macro-, micro-, and nanomaterial-based tools and strategies for heavy-metal detection, *Chem. Rev.*, 111, 3433, 2011.
Pradeep, T., *Nano: The Essentials*, McGraw-Hill Education (India) Pvt Limited, New Delhi, 2007.
Pradeep, T. et al., *A Textbook of Nanoscience and Nanotechnology*, McGraw-Hill Education (India) Pvt Limited, New Delhi, 2012.
Saha, K. et al., Gold nanoparticles in chemical and biological sensing, *Chem. Rev.*, 112, 2739, 2012.

27

Water Desalination: Emerging and Existing Technologies

Shaurya Prakash,[1] Mark A. Shannon,[2] and Karen Bellman[1]

[1]*Department of Mechanical and Aerospace Engineering,*
The Ohio State University, Columbus, Ohio, USA

[2]*Department of Mechanical Science and Engineering, University of*
Illinois at Urbana-Champaign, Urbana, Illinois, USA

CONTENTS

27.1 Introduction

The ability to procure or produce clean water is essential for health, energy, and security demands [1–4]. The following statistics on water quality effects emphasize the importance of advancing the science and technology to increase freshwater supplies. On the basis of World Health Organization (WHO) reports, nearly 2.4 million people die every year due to contaminated water—a child under the age of 5 dies every 20 s. The most common waterborne diseases, affecting millions of people worldwide, are malaria, cholera, and diarrhea. In addition, it has been estimated that >20 million people are affected by arsenic poisoning in the Bengal region of the Indian subcontinent (including Bangladesh). Given these brief statistics, it is not surprising that there is a renewed focus on developing new technologies for water purification [1–5], including desalination and demineralization of various source waters to enhance freshwater supplies. However, in many regions, it is not just a matter of implementing an effective water treatment system to generate high-quality potable water; the water supply itself can also be an issue. To this end, supplementing the water supply by desalination of salt water has been identified as a key step toward making progress for developing sustainable sources for freshwater [6–8]. To create effective desalination and subsequent distribution strategies, it is useful to briefly review how the water is distributed around the planet. The total available water on Earth is approximately 1.4×10^{21} liters (or about 332,500,000 mi³). Of this seemingly enormous water supply, >99% of the water is currently inaccessible to human use, and >97% of the earth's water exists in oceans, bays, seas, and saline aquifers as large reservoirs of salt water. In fact, several estimates place approximate supply of usable freshwater at ~0.7% or approximately 9.8×10^{18} liters. While this may appear to be a large number, accounting for population increase, demands on freshwater for agriculture, industry, power plant cooling; a changing climate; declining freshwater quality from worldwide contamination via industrial, municipal, and agricultural discharge; and increasing energy needs lead to a rather bleak picture for future availability of clean water for human use. In a recent editorial review [6], which reports data from the World Meteorological Organization, the combination of uneven population and water distribution is causing rapidly increasing water withdrawal rates as a fraction of total available water (Figure 27.1) and projects that by 2025 (only 11 years from now), most of the world's population will be facing serious water stresses and shortages. Although infrastructure repair, improvement, and water conservation can help relieve water stress, increasing the available freshwater supply by desalination is also an attractive option. Therefore, the ability to affordably and sustainably desalinate water can resolve many of the impending and projected water crises. In this chapter, a discussion of existing and emerging water desalination technologies is

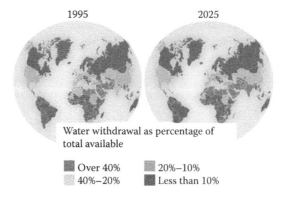

FIGURE 27.1
(See color insert.) Water withdrawal rate as a percentage of total available water, projected till 2025. (From Service, R.F., *Science*, 313, 1088, 2006.)

presented. The purpose of this chapter is to highlight the water–energy nexus, with a focus on generating freshwater from salt water and identifying emerging technologies, materials, and methods that can provide potentially more energy-efficient pathways than existing methods.

To discuss developing technologies for water purification and supply increase, it is important to review current water treatment processing as well as the energy cost of this treatment. In the United States, water is generally sourced from a surface source such as a lake or harvested from available groundwater supplies. Surface water purification is typically achieved by a series of processes, as shown schematically in Figure 27.2. Water is pumped in from the source through a screen and treated to remove odor and organisms. Next, water is transferred to the coagulation area and mixed with alum and other chemicals to help large particles like dirt form "flocs" that are heavy enough to settle out. The water is then transferred to sedimentation basins where the flocs are allowed to settle to the bottom while the clear water travels to the filtration step, where water travels through a filter of sand, gravel, and charcoal to remove smaller particles. Finally, the water is disinfected and then distributed through the system. This example is a typical treatment process; however, depending on incoming source water and regulations, additional steps may be added. For plant sizes ranging from 1 to 100 million gallons per day (MGD), the electrical energy requirement ranges from 1483 to 1407 kWh/million gallons treated. Pumping accounts for 80%–85% of these energy requirements. Groundwater treatment requires less processing than surface water; water is pumped into the plant and disinfected then distributed to the system. Energy costs for producing this water are approximately 1824 kWh/million gallons treated, almost 30% higher than surface water production due to the large energy consumption of the pumps needed to retrieve the water [9]. The total electricity demand for water purification of the public supply in the United States was 32 billion kWh in 2005 [9].

Water for desalination processing is usually obtained from either open surface sea intakes or underground beach wells [8]. Worldwide desalination capacity grew by 12.4% in 2011 to 71.9 million m³ of seawater processed per day [10]. This desalted water was supplied by 14,451 desalination plants, while another 244 were being planned or built.

Table 27.1 summarizes the 2009 desalination capacity by region [7]. Desalination plant development is expected to continue growing, with an estimated worldwide desalination capacity of 104 million m³ projected to cost $17 billion total by 2016; the majority of that investment (~$13 billion) is expected to be toward reverse osmosis (RO) plant development [8,11].

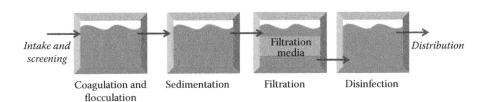

FIGURE 27.2
Schematic depicting basic water treatment of surface waters from source to distribution. Intake water is screened and coagulants are added, and flocs are allowed to form. These flocs settle in the sedimentation section. Water is subsequently filtered through gravel, sand, and charcoal filters to remove remaining particles, then disinfected and distributed.

TABLE 27.1

Worldwide Desalination Capacity

Region	Desalination Capacity (10^6 m³)	% Desalination Capacity
Middle East	31.2	52%
North America	9.6	16%
Europe	7.8	13%
Asia	7.2	12%
Africa	2.4	4%
Central America	1.8	3%
Australia	0.18	0.3%

Source: Data from Humplik, T., J. Lee, S.C. O'Hern, B.A. Fellman, M.A. Baig, S.F. Hassan, M.A. Atieh, F. Rahman, T. Laoui, R. Karnik, and E.N. Wang, *Nanotechnology*, 22, 1, 2011.

27.2 Review of Existing Water Desalination Methods

A wide variety of water desalination methods exist. In this section, a brief review of the most common methods is presented; therefore, not all techniques were included in this chapter. However, the reader is pointed to several other articles and books along with a broad variety of literature cited throughout this chapter to allow further reading given the reader's specific interests.

27.2.1 Theoretical Minimum Energy Requirement for Water Desalination

Consider an equilibrium analysis for the process of water desalination. From the second law of thermodynamics, for reversible processes the amount of energy used for such a process is independent of the method used [12]. As a consequence, given the starting and targeted salinity of the water being treated, it is possible to calculate a theoretical minimum amount of energy required for water desalination. This exercise may appear academic, but is of value since the energy–water nexus [13,14] is a major consideration toward evaluating the current state of water desalination technologies and identifying new technologies. Consider an ideal compressor for moving water vapor from a tank of seawater (typically assumed at ~35,000 ppm or mg/l salinity) to an infinitely large tank of freshwater (typically ~500 ppm salinity) [12]. Using these conditions, the minimum energy requirement for water desalination at 25°C with a recovery rate of zero (i.e., negligibly small amount of water produced from a near-infinite amount of seawater) is 2.5 kJ/l [12]. Recovery rate is defined as the ratio of freshwater produced to the inlet salt water. For a viable system, recovery ratios must be maximized in contrast to the waste or brine streams. Increase of the recovery rates to 25%, 50%, and 75% requires theoretical energy minimums of 2.9, 3,5, and 4.6 kJ/l of freshwater product, respectively [1,15]. Therefore, in principle, it should be possible to remove salt from water in an efficient manner with lower energy consumption than being achieved currently. However, several challenges exist to achieving these theoretical limits.

27.2.2 Challenges to Desalination Processes

Inherent irreversibilities present in real systems typically drive the energy requirements higher than the theoretical minima. Some of these irreversibilities relate to the presence and subsequent removal of organics and particulates and varying quality (such as pH and salinity) of source waters. Others relate to operation of mechanical and electrical equipment at varying energy efficiencies. The varying source water content can also lead to scale formation and deposition or membrane fouling in desalination plants [16]. To mitigate these problems, desalination plants often employ extensive pretreatment steps involving chemical treatment processes, including precipitation, flocculation, lime softening, ion-exchange columns, or mechanical processes such as aeration and sedimentation. For example, to minimize scaling, pretreatment of feedwater by introducing an acid followed by CO_2 degassing has shown to be an effective method for preventing alkaline scale formation [16]. Antiscalants are particularly popular because of their effectiveness at low concentrations, consequently reducing the overall chemical load. The chief chemical families from which antiscalants have been developed from are condensed polyphosphates, organophosphonates, and polyelectrolytes. Of these three classes of compounds, polyphosphates are the most economical while effectively retarding scale formation and offering corrosion protection [16]. Organophosphonates are suitable for a wider range of operating pH and temperature conditions than polyphosphates [16]. The main consequence of all pretreatment processes is that all of these involve increased energy consumption and material costs regardless of the specific desalination method employed. However, other concerns exist with the use of antiscalants. In areas where polyphosphates are used, eutrophication has been observed owing to the potential of the polyphosphate to be converted to orthophosphate, which is a major nutrient for primary producers such as algae. Phosphonates and polycarbonic acids are of some concern because of their chemical stability, giving them a long residence time in water, with the potential to disrupt natural processes due to their dispersion and complexing of magnesium and calcium ions. However, the overall toxicity of antiscalants to aquatic life is low [17].

27.2.3 Common Separation Methods

Desalination dates back to the fourth century when Greek sailors used evaporation aboard ships to produce freshwater, while membrane-based water purification including desalination became popular after World War II [18]. Approximately 86% of the available desalination production capacity employs either RO or multistage flash (MSF) distillation processes for freshwater production from either brackish water or seawater [7]. Other main technologies include multieffect distillation (MED), vapor compression (VC), and electrodialysis [19]. The primary energy requirement for MSF, MED, and VC is in the form of thermal energy, while RO requires primarily mechanical energy for pumping water, and electrodialysis requires primarily electrical energy. Other methods such as solar distillation, freezing, gas hydrate processes, membrane distillation, humidification–dehumidification processes, forward osmosis, capacitive deionization, and ion exchange are also used for desalination; however, current technology levels for these processes find limited or niche use and are not yet commercially viable on a worldwide scale for widespread implementation.

Table 27.2 summarizes energy consumption of commonly used desalination methods. Comparing energy consumption, membrane separations require the lowest energy input, by at least a factor of 6 compared with the thermal methods.

TABLE 27.2

Energy Consumption and Capacity for Various Commercially Implemented Desalination Methods

Process	MSF	MED/TVC	RO	ED[a]
Heat consumption (kJ/l)	290	145–390	–	–
Electricity consumption (kJ/l)	10.8–18	5.4–9	6.5–25.2	4.32–9
Total energy consumption (kJ/l)	300.8–308.8	150.9–399	6.5–25.3	4.32–9
Production capacity (m³/day)	<76,000	<36,000	<20,000	<19,000
Conversion to freshwater	10%–25%	23%–33%	20%–50%	80%–90%
Total capacity (m³/day)	304,000	109,091	40,000	21,111
Pretreatment required	Little	Little	Demanding	Moderate

Source: Data from Trieb, F., ed. *Concentrating Solar Power for Water Desalination.* ed. G.A.C. (DLR) and I.O.T. Thermodynamics. Federal Ministry for the Environment, Nature Conservation and Nuclear Energy: Stuttgart, Germany, 2007.

[a] Electrodialysis, for brackish water. All other technologies for seawater desalination.

Thermal processes such as MED and MSF consume between 150.9 and 399 kJ/l of energy; thus, they may not be practical for high-efficiency implementation unless coupled with a process that produces a large amount of waste heat such as a power plant. Furthermore, it should be noted that all major processes listed in Table 27.2 operate at significantly higher energy consumption than theoretical limits. Examples of combined generation plants and energy savings for the desalination process are listed in Table 27.3, with a more detailed description reported previously [8,20].

The energy consumption estimates listed in Table 27.2 often do not account for energy or material consumption for source water treatment. For example, the pretreatment requirements for RO units are very stringent to permit lower-energy operation of the desalination process and include scale control, usually managed by pH adjustment or an addition of a chemical antiscalant, thus preventing irreversible scaling damage to the membrane. Prefiltration is also employed for particulate removal, and disinfection is used to prevent biofouling when the disinfectant is compatible with the membrane material. The RO process can also require post-treatment, depending on the initial water quality to remove dissolved gases and adjust the pH and alkalinity [21]. Furthermore, most RO units operate at ~40% or lower water recovery for seawater desalination. It should be noted that scaling and corrosion are also major problems for MSF and MED desalination methods, as continued scaling decreases the efficiency of the desalination plant over time.

TABLE 27.3

Potential Energy Savings for Thermal Desalination Using Cogeneration Plants

Plant Type	Energy Savings Range
Backpressure steam turbine	5%–40%
Extraction/condensing steam turbine	0%–30%
Gas turbine/HRSG	20%–30%
Combined cycles	30%–40%

Source: Data from El-Nashar, A.M., *Desalination*, 134, 7, 2001.

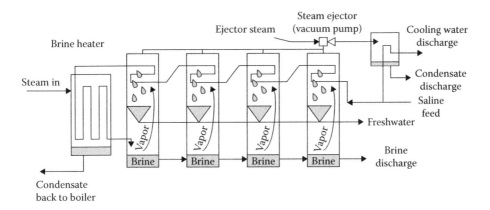

FIGURE 27.3
Block diagram schematic showing the basic elements of an MSF distillation process. (Based on Trieb, F., ed. *Concentrating Solar Power for Water Desalination*. ed. G.A.C. [DLR] and I.O.T. Thermodynamics. Federal Ministry for the Environment, Nature Conservation and Nuclear Energy: Stuttgart, Germany, 2007.)

27.2.3.1 MSF Distillation

The MSF process is well suited for highly saline or contaminated waters [22] because flashing of water vapor from the top of brine pools allows for minimal scale formation as the precipitates resulting from evaporation form in liquid rather than the critical surfaces of heat transfer [12]. The first commercial MSF plants were developed as early as the 1950s, and the method is most popular in Middle Eastern countries, particularly in Saudi Arabia, Kuwait, and the United Arab Emirates owing to MSF's ability to operate with the highly saline and particulate-laden waters of the Persian Gulf [6,23]. The desalination procedure begins by heating incoming seawater by condensing steam contained in a set of tubes running through the brine heater. The MSF process elements are depicted schematically in Figure 27.3. To reduce energy costs, MSF plants are often combined with steam cycle power plants, allowing for the utilization of the cooled steam from such production facilities [22] in combined heat and power cycles. The incoming water is often pretreated with antiscalants and heated to 90–110°C (194–230°F), with higher temperatures being avoided because of concerns of scale development [24], particularly of calcium sulfate and calcium carbonate [12]. The heated salt water is processed through a series of chambers at decreasing pressures, causing the water to "flash" or immediately boil upon entry into the chamber [24]; that is, for a given temperature, the chamber pressure is matched to the vapor pressure to induce boiling. The number of successive chambers used can be as high as 40, although many practical systems employ around 20 such distillate collection areas [22]. MSF plants are rated with performance factors such as the gained output ratio (GRO), which is the mass of desalinated water produced to mass of steam [23]. For a 20-stage MSF plant, a conventional GRO is 8 with a typical heating requirement of about 290 kJ/kg for product water [23]. Although the highest in terms of energy needed per unit product water produced, the MSF desalination process provides proven reliability as well as ability to deal with highly contaminated feedwaters.

27.2.3.2 Multi-Effect Distillation

MED processes for commercial water distillation were also introduced in the 1950s [23]. This technology borrows from the plants that were first developed to produce sugar from

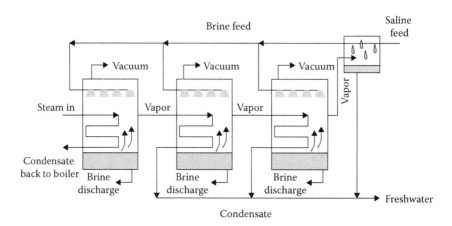

FIGURE 27.4
Block diagram schematic showing the basic elements of an MED process. (Based on Trieb, F., ed. *Concentrating Solar Power for Water Desalination.* ed. G.A.C. [DLR] and I.O.T. Thermodynamics. Federal Ministry for the Environment, Nature Conservation and Nuclear Energy: Stuttgart, Germany, 2007.)

sugar cane juice or salt through evaporation [24]. MED systems were employed for solutions at lower temperatures than MSF, and thus have found use for processes that have left-over steam heat below 100°C. Today many dedicated MED systems are used for water desalination alone. It has been estimated in the literature that the MED process offers superior thermal performance to the MSF processes; however, the scaling problems within the plants were noted to be higher [22,23]. Current systems resolve some of the scaling and corrosion issues by operating at a maximum brine temperature of approximately 70°C and, in some cases, systems employ a maximum brine temperature of 55°C allowing for utilization of low-grade waste heat [22]. The MED process begins by distributing (usually by spraying) the preheated saline feedwater onto the heat exchange surface in a thin film to encourage boiling and evaporation of water [24] through a surface area enhancement for improved heat transfer. The vapor phase of the water is then condensed in a lower-pressure chamber, much like the MSF design [22], while simultaneously creating vapor to be fed into the next chamber as shown in Figure 27.4. The MED process continues for 8–16 effects or cycles, depending on plant design [24]. It is not uncommon for MED plants to be integrated with additional heat inputs between stages, usually by thermal vapor compression (TVC) or mechanical vapor compression (MVC) [22]. Energy consumption in MED plants incorporated with the cold end of a steam cycle require about 145–390 kJ/kg in process steam withdrawn from a steam turbine and 5.4–9 kJ/l electricity for control and pumping processes [22]. Owing to the relatively low operating temperatures of the MED process, a large amount of surface area is required by the system, thus requiring large areas for MED production facilities [24]. The integration of MED plants with TVC reduces the surface area and the number of effects needed per plant capacity [24].

27.2.3.3 Vapor Compression (VC) Processes

By themselves, VC processes (both thermal and mechanical) offer simple, consistent operation that are usually deployed in small- to medium-scale desalination units. They are often used for applications such as resorts, industries, or drilling sites with a lack of direct access to freshwater supply [22]. Figure 27.5 shows a schematic representation of the unit

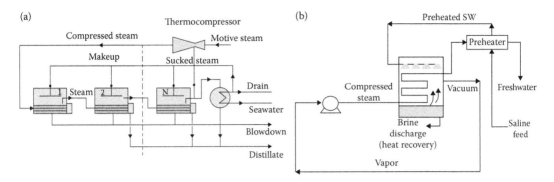

FIGURE 27.5
Schematic representations of basic unit processes involved in VC cycles for water desalination: (a) TVC system and (b) single-stage mechanical VC system. (Based on Trieb, F., ed. *Concentrating Solar Power for Water Desalination.* ed. G.A.C. [DLR] and I.O.T. Thermodynamics. Federal Ministry for the Environment, Nature Conservation and Nuclear Energy: Stuttgart, Germany, 2007.)

processes involved in VC cycles for water desalination. Mechanical vapor compression or MVC units are generally single-stage plants with a production capacity up to 3000 m^3/ day. MVC systems typically operate as single-stage units, as only the capacity and not the efficiency is increased with additional stages [23]. In the case of thermal vapor compression or TVC plants, multiple stages are used owing to increased efficiency with designed capacities up to 36,000 m^3/day [22,23]. VC systems generate the heat for evaporation by compressing feed vapor as the heat source for the heat exchanger (i.e., changes in density) as opposed to mechanically produced heat steam as in MSF and MED [24]. In TVC units, steam ejectors are used for VC whereas in MVC units, a mechanical compressor is used [22]. VC is often coupled with MED systems, which increases the efficiency of the system while raising the steam pressure requirement [22].

27.2.3.4 Reverse Osmosis

In contrast to the thermal processes discussed thus far, another class of water desalination system arises from filtration technologies relying on the use of polymeric membranes. Membrane systems [25] are becoming increasingly popular because of the low cost of polymeric membranes and relatively higher energy efficiencies. The water industry extensively uses membrane processes for pretreatment filtration depending on particle size (e.g., microfiltration, nanofiltration, or ultrafiltration), as depicted in Figure 27.6. Particles ranging in size from a few angstroms to 10 microns can be removed by filtration processes such as microfiltration, ultrafiltration, nanofiltration, electrodialysis reversal (EDR), and RO [26,27] at varying operating pressures as described in Figure 27.6a, with the length scales of relevant particles described in Figure 27.6b. Conceptual treatment of ions as hard spheres allows for filtration to work with ionic salts similar to particles if small enough pores are used in membranes. This model led to the development of RO desalination, and the first commercial RO plant began operation in 1965. In RO, a semi-permeable membrane separates the feedwater from the effluent stream. RO works by pressurizing the saline feedwater to a pressure larger than the osmotic pressure of the solution, causing the semi-permeable membranes of the system to reject most of the solute (in this case, mostly salts) while allowing solvent (freshwater) to pass through [22]. A schematic of the RO process is shown in Figure 27.7. Since osmotic pressure is determined by the salinity of the feedwater,

(a)

Separation method	Operating pressure (psi)	Particles removed
Microfiltration	10–30	Turbidity, algae, giardia, cryptosporidia, bacteria
Ultrafiltration	15–35	Macromolecules, asbestos, virus
Nanofiltration	80–150	Organic chemicals, hardness, color, radionuclides
Electrodialysis reversal	80–100	Metal ions, salt
Brackish reverse osmosis	150–300	Metal ions, salt
Seawater reverse osmosis	500–900	Metal ions, salt

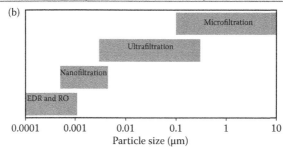

FIGURE 27.6
Membrane selectivity and operation information. (a) Operating pressure and particle removal by separation method. (b) Particle size range removed by separation process. (From Nath, K., *Membrane Separation Processes*. Prentice Hall: New Delhi, 2008; and American Membrane Technology Association. *Application of Membrane Technologies*. Available from: http://www.amtaorg.com/wp-content/uploads/1_applicationofmembranetechnologies.pdf, 2007.)

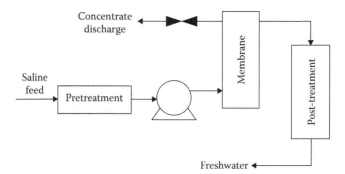

FIGURE 27.7
Simple schematic representation of the RO process showing essential unit processes. (Based on Trieb, F., ed. *Concentrating Solar Power for Water Desalination*. ed. G.A.C. [DLR] and I.O.T. Thermodynamics. Federal Ministry for the Environment, Nature Conservation and Nuclear Energy: Stuttgart, Germany, 2007.)

the pressure requirements for brackish water desalination are much lower (15–25 bar) than the pressure requirements of seawater desalination (54–80 bar) [23]. It should be noted that although the osmotic pressure of seawater is approximately 25 bar [28], higher operating pressures are required to achieve practical flows as well as balance the increasing salinity of the feedwater and overcome concentration polarization across the membranes as the desalination process progresses [1,23,29].

Concentration polarization on the external surfaces of the membrane results from the concentration imbalance created by the solution dilution on one side of the membrane and the solution concentration on the other during separation processes. It should be noted that while progress has been made in developing high-flux membranes [30,31], increasing flux often leads to increasing polarization [1]. Conceptually, it is useful to imagine the concentration polarization region as a pseudo-membrane in series with the physical membrane impeding flow and adding to the overall energy losses in the system. Furthermore, concentration polarization has been attributed toward the first step in developing permanent fouling layers on membranes [32]. More detailed information about concentration polarization is discussed later in the chapter (Section 27.2.3.6). Another challenge to the technology is the incompatibility of RO membranes with free chlorine; exposure over 1 mg/l for 200–1000 h can lead to noticeable degradation. In addition, biofilm formation can be a problem and biofouling is prevented by a constant chlorination/dechlorination process to help prevent the necessity of additional pressure [33].

Feedwater quality is of high concern in RO systems as a given membrane material operates within a fixed set of operating parameters relating to pH, organics, algae, bacteria, particulates, and other foulants found in saline water [22]. As a consequence, RO systems are typically characterized by several pretreatment stages. RO systems approach closest to the thermodynamic limits requiring approximately 6.5 kJ/l of water product for the RO process alone (see Table 27.2); however, it should be noted that the complete desalination process, including intake, pretreatment, post-treatment, and brine discharge, requires >3.6 kJ/l additional energy [5,8]. Post-treatment includes removal of boron (usually in the form of boric acid), which is of concern in many areas using RO; concentrations of boron of 0.5 mg/l is the limit set for safe drinking water. Thus, for source waters high in boron, and for desired uses of the desalinated waters to be below this limit, a second RO stage is added to take approximately 20%–30% of the product water from the first stage. The pH of the first-stage product water is raised above 10 to transform the boric acid to borate, which is more successfully rejected by the RO membranes. The two remaining product water streams are blended to bring the boron concentrations down to acceptable levels; that is, a "dilution is the solution" paradigm is implemented.

The energy consumption of RO processing has decreased from 93.6 kJ/l in 1980 to 12.2 kJ/l (average reported, not best achieved value) [11]. Owing to the increasing pressure requirements with feedwater salinity increases, 40% is a common product water recovery ratio for these systems [23], although systems often run at 50%–55% recovery generally at the expense of additional (~1.8 kJ/l) energy consumption, if run in one stage. However, two-stage RO systems with recovery ratios of approximately 60% have been used to demonstrate energy savings owing to the need to pretreat less water for a given target production (16.6 kJ/l), as well as a 33% reduction in plant size requirement and 33% reduction in the amount of brine to dispose and/or manage [34]. System size required can also be reduced by more intensive pretreatment of the feedwater.

Recent improvements in the RO process have been brought about by the use of more selective membranes and implementation of energy recovery devices that concentrate the stream exiting the pressure vessels [24]. Two classes of energy recovery devices are used for RO systems; one harvests hydraulic power of the exiting waste stream through the displacement of the energy recovery device, while the other converts the hydraulic energy to centrifugal mechanical energy, then back to hydraulic energy [11]. Each has pros and cons depending on the operating conditions of the system, but the energy transfer efficiency back to the feed stream can be as high as 95% for 36% feedwater recovery translating to an energy consumption of 7.2 kJ/l as opposed to 12.7 kJ/l without the energy recovery [35].

Therefore, the minimum energy for a given amount of product water from feedwater at specific inlet conditions is a complicated function of minimizing the material use and cost of pretreatment and post-treatment, the energy required, and the capital costs. Therefore, each location and system typically operates differently.

27.2.3.5 Electrodialysis and Electrodialysis Reversal

Electrodialysis was introduced for industrial desalination applications in the 1960s as a membrane desalination technique suited to the removal of salt ions from brackish water [36]. The conceptual idea is based on the filtration mechanism used in medical systems and the principles for artificial kidney operation. Electrodialysis systems present the first evidence of the influence of bionanotechnology on commercial water desalination processes. Currently, electrodialysis is listed as a US Environmental Protection Agency (US EPA) "best available" technology for removal of selenium, barium, nitrate and nitrite, and total dissolved solids (TDS) for water treatment plants where these concentrations are above those set by the US EPA for drinking water or for cleanup of reclaimed water for irrigation or other non-potable uses [37]. The electrodialysis process shown in Figure 27.8 removes electrically charged salt molecules of saline water by employing alternating stacks of cationic or anionic selective membranes, with ionic flux driven by an applied electric field [23]. As the cations and anions are pulled through the membranes, the salinity of the water drops at each stage, and freshwater is produced (Figure 27.8). Energy requirements of electrodialysis systems are based on the concentration of ions within the water; thus, electrodialysis can be competitive with RO processes at concentrations up to 5000 ppm [36]. A full membrane stack for this process can be composed of a bank of membranes comprising hundreds of layers of alternating anionic and cationic selective membranes [23]. One of the main challenges to electrodialysis is the high energy consumption rates needed to obtain high flux rates of desalted water. High flux demands higher flow rates, which are determined by the applied direct current (DC) at the electrodes. The power consumption

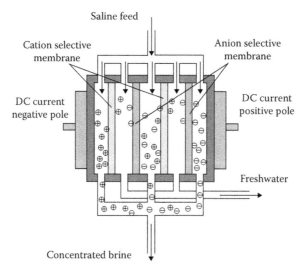

FIGURE 27.8
Schematic representation of a bank of membranes being used for electrodialysis. (Based on Trieb, F., ed. *Concentrating Solar Power for Water Desalination*. ed. G.A.C. [DLR] and I.O.T. Thermodynamics. Federal Ministry for the Environment, Nature Conservation and Nuclear Energy: Stuttgart, Germany, 2007.)

P for a given salt solution of resistance R varies with the square of the current, I, flowing through the system. To move the ions across the solution and membranes, a DC potential at the electrodes has to be added to move ions through the solution, $\eta_{solution}$, across the membrane, $\eta_{membranes}$, and overcome the overpotentials created owing to Faradaic losses at the electrodes (i.e., the transformation of electrons in the metal to ions in solution, usually via the electrolysis of water), $\eta_{overpotential}$. The power consumed can be expressed as

$$P = I \left(\eta_{solution} + \eta_{membranes} + \eta_{overpotential} \right). \tag{27.1}$$

The potentials across the solution and membranes (to the first order) are simply proportional to the current, or $\eta_{solution} = C_J I$. However, the overpotentials at the electrodes can grow exponentially with the current, such that $\eta_{overpotential} = C_F I^{(n+1)}$, where $n \geq 1$ and $C_F \gg C_J$, for concentrated salts. Therefore, the power consumption for desalination by electrodialysis can increase rapidly, such that

$$P = C_J I^2 + C_F I^{(n+2)}, \tag{27.2}$$

so that when either high fluxes are desired or concentrated solutions are being desalted, or both, the power consumption can greatly exceed both RO and thermal methods. However, at low fluxes and relatively low salt concentrations (<3000 ppm), electrodialysis can be among the most energy-efficient methods.

Other challenges in electrodialysis are similar to those with all membrane processes and relate to membrane fouling, membrane response to changing pH conditions when H^+ and OH^- ions are transported across membrane surfaces, and concentration polarization [23,38]. One key advantage of the electrodialysis process is its ability to deal with higher levels of uncharged particles in comparison with RO processes, as these particles do not travel through the membranes and are not driven to the membrane surface [23]. Electrodialysis recovery rates for brackish water range from 80% to 90%, while RO recovery rates for brackish water range from 65% to 75% [39]. Energy consumption of electrodialysis systems ranges from 4.3 to 9 kJ/l for brackish water [39]. These membranes are also tolerant to up to 0.5 ppm free chlorine dosing; can operate in a pH range 2–11; and tolerate high temperature cleaning, mechanical cleaning, and 5% hydrochloric acid cleaning [37].

Owing to the high product recovery, scaling and precipitation of salts (usually calcium carbonate, calcium sulfate, or barium sulfate is the limiting salt but this depends on the feedwater) can be a problem for electrodialysis systems [21,40]. To mitigate fouling and precipitation issues, the EDR process was developed. The EDR process periodically changes the polarity of the applied biases, allowing the system to be flushed [24]. The polarity reversal of the EDR process changes the concentrate and product water sides, making it possible to operate at a supersaturated state and increasing product recovery without chemical additives such as antiscalants [40]. Avoiding the use of antiscalants is generally preferable because feedwater must be closely monitored to achieve the correct dosing. Correct dosing is important because too little additive results in scaling, while excess doses result in membrane fouling. Hence, dosing must be changed in response to feedwater quality changes [40]. Finally, since electrodialysis and EDR processing removes charged particles, post-processing must be in place for potentially harmful electrically uncharged biological species such as viruses and bacteria.

EDR plants are an economical solution to many niche water problems. For example, the city of San Diego (California) uses reclaimed tertiary water for irrigation and industrial processing. The reclaimed water was found to be acceptable for non-potable water reuse with the exception of sodium levels and TDS (approximately 1110–1550 mg/l, where <1000 mg/l is desired), which vary based on rainfall and other factors. EDR is a good technology for delivering consistent product water quality in applications where feedwater quality is good but variable. The plant removes 55% of the salts, operates at an 85% product recovery rate, and produces 2.2 MGD of product water [41].

Projections by Siemens predict that in 2030 and beyond, Siemens will provide nearly 75% of urban water demand in Australia through continuous electrodeionization technologies reaching an estimated population of 25 million people at energy consumption levels of 1 MWh/Ml [42]. Electrodialysis- and EDR-based technologies are also being investigated and employed by GE in many of their water desalination plants. The promise of low energy usage as shown has also encouraged use of other electrical methods, such as capacitive deionization, which are being used by companies (e.g., PROINGESA) in Europe [43].

27.2.3.6 Challenges with Membrane Processes

Since RO desalination is the most commonly new installed desalination plant type worldwide, it is important to discuss some of the challenges that RO and other membrane-based technologies face. As with any technology, challenges and limitations to membrane separations must be accounted for in water treatment design and implementation. Major impedances to membranes separations include concentration polarization, fouling, and precipitation/scaling, which dictate, partially and sometimes fully, product recovery possible, maintenance protocol and procedures, and membrane lifetime. Concentration polarization occurs when, as membrane separation processes develop, concentration gradients form at the surface as the rejected feed components build up near the membrane on the reject side [44], which forms a higher concentration layer at a membrane surface as seen schematically in Figure 27.9. This, in turn, leads to an increased local osmotic pressure at the membrane surface, which causes a decrease in the effective driving pressure [45,46], leading to an eventual drop in permeate flux. As polarization persists and fouling continues to develop, particles adhered to the membrane surface can cause scaling, which occurs when solid phase precipitates out of the solution, or biofilm formation. In addition to flux impediment, biofilms can act as a source of permeate contamination. This denser, more impeding layer can be termed a cake or gel layer, as shown in Figure 27.10 [44].

Concentration polarization ($N_F < N_{Fc}$)

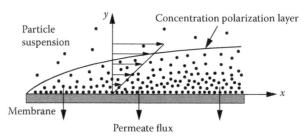

FIGURE 27.9
Concentration boundary layer formed by concentration polarization on a membrane surface. The figure shows the development of the concentration polarization boundary layer for cross-flow over the membrane, with the permeate passing through the membrane as shown. This concentration boundary layer causes flux impediment due to the increase in local concentration of species. (From Chen, J.C., Q. Li, and M. Elimelech, *Advances in Colloid and Interface Science*, 107, 83, 2004.)

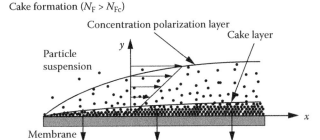

Cake formation ($N_F > N_{Fc}$)

FIGURE 27.10

Same membrane as in Figure 27.9 but figure depicts the process of membrane fouling that is initiated by concentration polarization. The cake layer is a dense particle layer adjacent to the membrane surface. This layer can irreversibly foul the membrane as the polarization layer remains over the membrane surface. (From Chen, J.C., Q. Li, and M. Elimelech, *Advances in Colloid and Interface Science*, 107, 83, 2004.)

As the cake or scale layer continues to build, it can become permanently attached to the membrane surface, causing irreversible fouling that diminishes the flux through the membrane, which cannot be recovered even after the membrane is cleaned. Cake or gel layer formation occurs at a critical pressure and solute or particles concentrate at the membrane surface [32]. Concentration polarization initiates gel layer formation; however, if the particles are very small (generally ions or non-fouling macromolecules), the polarization layer behaves as if there is an additional or virtual membrane at the physical membrane surface, as depicted schematically in Figure 27.11. Since fouling and precipitation are generally a function of concentration, the formation of an area of increased concentration next to the membrane surface is undesirable. Hence, many methods of reducing concentration polarization have been developed. Current methods of concentration polarization mitigation can be classified into three broad categories: (i) mechanical, (ii) chemical, and (iii) electrical.

Mechanical methods of polarization reduction include any method that agitates the fluid surrounding the membrane, including but not limited to physical mixing, vibration, and flow pulsing. Chemical methods include chemical surface modification of the membrane or solution to be separated. Electrical methods include applying an electrical, magnetic, or electromagnetic field on or near the membrane to increase flux by mitigating concentration

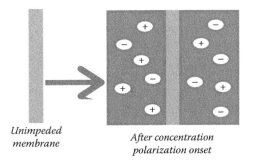

Unimpeded membrane

After concentration polarization onset

FIGURE 27.11

Schematic depicting the effect of concentration polarization. After the onset of concentration polarization, the system at hand behaves as if there were additional membranes added. Consequently, the easiest conceptual framework for visualizing concentration polarization is to recognize the presence of virtual membranes.

polarization. To understand the methods used for concentration polarization reduction, it is useful to investigate the concentration polarization modulus, c_{io}/c_{ib}, equation:

$$\frac{c_{io}}{c_{ib}} = \frac{\exp^{\frac{J_v\delta}{D_i}}}{1+E_o\left(\exp^{\frac{J_v\delta}{D_i}} - 1\right)} \tag{27.3}$$

where c_{io}/c_{ib} is the concentration ratio at the membrane surface over the concentration in the bulk, which is referred to as the polarization modulus; J_v is the volume flux through the membrane; δ is the boundary layer thickness; D_i is the diffusion coefficient of the solute in the fluid; and E_o is the membrane enrichment, or c_{ip}/c_{io} or the ratio of concentration at the membrane surface over the concentration on the permeate side. From this equation, as flux or boundary layer thickness increases, concentration polarization or c_{io}/c_{ib} also increases. In many systems, the variable that is most readily manipulated is δ, the boundary layer thickness [47]. The maximum flux through a membrane is the flux of deionized water through that membrane at a given driving force, i.e., no salt rejection causing the flux to reduce.

Mechanical methods of CP reduction tend to focus on increasing turbulence in the flow at the membrane surface in order to decrease δ, the boundary layer thickness [47]. Methods of achieving this goal include increasing the fluid velocity at the membrane surface, adding membrane spacers to disturb the flow, generating ultrasonic waves near the membrane, oscillating the membrane, pulsing the feed flow in the system, or some combination of these methods [48–54]. The disadvantage of most of these methods is that they require addition of external energy to the system. Therefore, the addition of CP mitigation methods is very dependent on membrane process and feed solution. Despite a higher energy operating budget, mechanical methods such as feed spacers remain popular as they are easy to implement in most commercial systems.

Chemical methods of polarization reduction focus on either membrane modification during membrane development, modification of the surface via surface treatment, or modification of the feed solution with the addition of a chemical such as an antiscalant. Membrane modification has been affected by fluorination of RO membranes and has been shown to increase flux by six times versus an untreated membrane for FT-30 membranes (FilmTec Corp., Minneapolis, MN) while maintaining rejection and membrane lifespan. The mechanism behind this was hypothesized to be due to a thinning of the polymer strands of the interfacial rejection layer, as shown in Figure 27.12 [55]. Membrane pretreatment has been performed [56] on the basis of membrane material and polyethylene glycol solution concentration without a decrease in membrane selectivity [56]. However, after the initial flux enhancement experienced during membrane operation, membrane performance was still hindered by concentration polarization, and as the separation continued, the flow decreased back below pretreatment conditions [56].

Electrical methods offer yet another alternative for reducing concentration polarization, including applying a DC field across or to the membrane, which has been demonstrated by many researchers [57–60]. A commercial company, Graham Tek (Singapore), utilizes both mechanical and electrical methods in their RO systems. The electrical component consists of three coils 120° out of phase energized at an alternating field of 2 kHz that are wrapped around the membrane sandwich structure. The mechanical component is an "integrated flow distributor," which creates 1–2 mm air bubbles that are allowed to flow

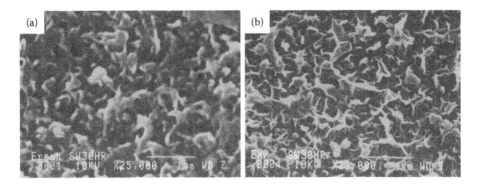

FIGURE 27.12
(a) Unexposed SW30HR membrane. (b) SW30HR membrane exposed to 15% wt. Hydrofluoric acid (HF) for 11 days. (From Mukherjee, D., A. Kulkarni, and W.N. Gill, *Journal of Membrane Science*, 97, 231, 1994.)

along the membrane surface. This system achieves sustainable fluxes of 22–27 l/m²h for 40%–50% recovery rates. From the patent report, a 380 V, 50 Hz power supply is called out to run the electric field on the coils, which is reported to run at a steady 2 A, consuming 760 W of power [61].

27.3 Emerging Water Desalination Technologies

As discussed above, a variety of methods and processes already exist to desalinate water [62–65]. However, with water, owing to high demand and increased energy consumption concerns, there has been a renewed interest in developing new technological platforms that can provide the solutions for next-generation water desalination needs [1,66]. For example, increasing energy demands have led to development of technologies for hydraulic fracturing or "fracking." The flow-back water (which can be 15%–30% of the total injection volume for a well being drilled with several million gallons needed) has significantly different salinity compared with seawater or other saline water sources. TDS can exceed 100,000 mg/l with the potential to reach nearly 300,000 mg/l. In addition, chlorides can be very high, reaching nearly 100,000 mg/l. Furthermore, flow-back water from fracking contains high particulates, including silica. Other problematic constituents include barium, calcium, strontium, and sulfates. Shallow or thin shales (e.g., the Marcellus formation in the United States) can also contain radioactive species. Therefore, desalinating and purifying flow-back water will likely require a shift in how existing technology can be used leading to use of several new or advanced methods for water treatment.

As discussed above, membrane processes are becoming increasingly popular for water desalination. The minimum thermodynamic energy needed for desalination assumes complete energy recovery, minimal polarization impedance, no membrane viscous and fouling losses, and no energy losses for membrane cleaning. Furthermore, these thermodynamic estimates assume that the source water stream comprises only of salt and water; that is, the energy required for all prefiltration steps for removal of pathogens, organic waste, colloids, etc., is not included. Two main approaches are being targeted for developing new water desalination methods. First is to improve existing technologies by integration with well-established separation or distillation technologies not previously used with water, or revisiting older methods with advanced technologies. Examples include using forward osmosis [67], humidification–dehumidification cycles [68], and trapping solar energy for thermal distillation processes [69]. The second approach relies on exploiting transport phenomena at the nanoscale and therefore employs use of nanofluidics, nanotechnology, bio-inspired methods, and nanomaterials [5,30,70–79]. Both approaches present their own sets of pros and cons, including challenges in energy consumption, management of waste streams, fabrication and cost considerations, and eventual output flux for a given water quality. Of the various methods being considered, the biotechnology- or nanotechnology-driven approaches are considered to be particularly promising as these approaches work at length scales where fundamental physical processes occur. Consequently, with recent advances in fabrication techniques, it is possible to develop systems that can manipulate these physical processes and thereby develop the next-generation water desalination systems working at near the thermodynamic energy consumption limits. Next, a discussion of some of the emerging biotechnology and nanotechnology approaches is presented.

27.3.1 Forward Osmosis

Desalination by the process of forward or direct osmosis (FO or DO) employs a highly concentrated solution (also referred to as the draw solution) to create an osmotic pressure that extracts freshwater from saline water across a semipermeable membrane [80]. Unlike RO processes, because FO uses osmotic pressure as the driving force, this method operates at a small or zero hydraulic pressure, which reduces the amount of observed fouling and allows the membrane to operate with a fairly sparse support structure [29,81]. The reliance on osmotic pressure necessitates that a major factor in the choice of the draw solution is the osmotic pressure of the draw solution [29]. The lack of hydraulic pressure means the only force the flow has to overcome is the few bars of pressure created by the physical barrier of the membrane itself [29].

Possibilities for the draw solution are numerous; for water desalination solutions of sulfur dioxide, aluminum sulfate, glucose, potassium nitrate, mixtures of glucose and fructose, and mixtures of ammonia (NH_3) and carbon dioxide (CO_2) gases have all been suggested as draw solution candidates [29]. In addition, nanoparticles such as magnetoferritin (~12 nm in diameter) have also been demonstrated [82]. However, many draw solutions add an aftertaste or an odor requiring post-treatment. Sugar (glucose and fructose primarily) has been a popular draw solution due to the minimal need for extensive post-treatment of desalinated water. The major energy requirement in FO is for treatment of the draw solution. For draw solutions of NH_3 and CO_2, which both are soluble in water, low-quality thermal energy can be used to decrease the solubility and energy needed to volatilize the NH_3 and CO_2 to remove as a gas and reuse upstream. Magnetic nanoparticles such as magnetoferritin are readily filtered out by magnetic filters as developed by industrial metal processing [82].

A major hindrance to this technology has been the development of an appropriate membrane [83]. Ideally, FO membranes should allow a high flux of water while retaining a high rejection rate of dissolved solids, demonstrate compatibility with both the feedwater and the draw solution, and handle the mechanical stresses imposed by the osmotic pressures experienced during the desalination process [83]. Currently available commercial membranes such as cellulose triacetate membranes are not compatible with many of the preferred draw solutions. Owing to the low pressure required across the membrane for FO processing, fragile membranes made from aquaporins, which are nature's perfect nanochannels and permit only water to flow through, can potentially be used for FO. Other membrane materials such as cellulose acetate, polybenzimidazole, and aromatic polyamides have also been used for FO purifications [84].

Other challenges to FO technology are overcoming the coupled effects of internal concentration polarization (ICP) and fouling. Internal concentration polarization occurs as a permeate dilutes the strength of the draw solution in the supporting layer of the membrane. This is especially detrimental to FO membranes because the osmotic pressure is the driving force for flow through the membrane [85]. One method of combating ICP is the development of the FO membrane without a fabric support layer, but instead using a supportive mesh integrated with the selective polymer layer. This solution has been shown to reduce ICP, but not eliminate the effects altogether [86]. A major advantage of FO is that many of the FO membranes used for desalination processes are compatible with chlorination; thus, less membrane maintenance and biofouling tends to occur [87]. There is the potential with FO membranes to produce a higher amount of product water from the feedwater if scaling and precipitation can be prevented. If this percentage is high enough, the salt could be left to precipitate out, thus reducing the amount of waste produced [86].

FO membrane technology has also been used to develop water purification for emergency water supply in the event of water supply disruption or natural disaster. Hydration Technology Innovations LLC (HTI, Scottsdale, AZ) commercially produces emergency pouches that produce a clean energy drink from any available water source. In addition, the first commercial FO desalination plant went into operation in 2010 in Al Khaluf, Oman. The plant design employs an FO cell to draw clean water from the feedwater and then treats the drawn solution with a RO process to produce the final product water [87]. The FO/RO cells are sized so that the waste from the RO system is recycled back as the draw solution for the FO system, thus cutting down on waste and chemicals needed for processing, which is a common practice in FO setups with final membrane processing step for product water.

27.3.2 Microfluidic and Nanofluidic Concentration Polarization for Desalination

Concentration polarization in most water desalination systems is considered to be a loss term, and immense resources and effort have been employed in better understanding and mitigating polarization effects [25,88,89]. Recent research has revealed that the imbalances created by concentration polarization (enrichment and depletion regions) can be sustained for extended periods of time when triggered across a nanochannel connecting two microchannels [90]. On the basis of this fact, researchers demonstrated the ability to desalinate water using a setup as shown in Figure 27.13, thereby exploiting polarization for separations. However, for efficient separation, pretreatment was needed for the removal of Ca^{2+} ions and physical filtration for elimination of precipitation and large debris. The membrane-less system that was tested was primarily driven by electrokinetic flows and achieved a 99% salt rejection ratio [5]. In addition to achieving seawater desalination, this system was shown to remove most solid particles, microorganisms, and biomolecules owing to their charged nature [5]. The energy consumption of this system for the actual separation process is modest due to the deflection of ions rather than physical displacement across a membrane and the low flow resistance of microchannels and energy consumption of approximately 5 Wh/l for flow rates of 0.25 µl/min [5]. Although considerable challenges lie ahead for the scaling up of a system using the concentration polarization technique, massive parallelization of the proof-of-concept device is estimated to deliver approximately 180–288 ml/min for a small-scale system [5].

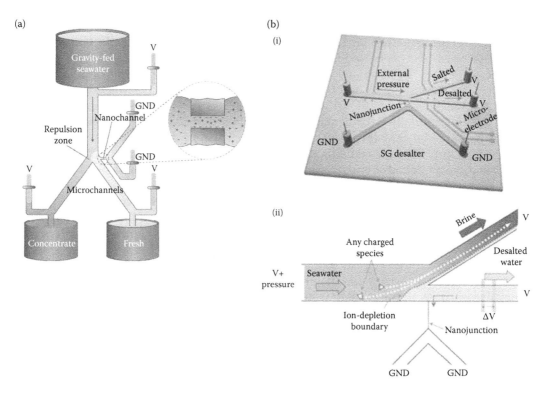

FIGURE 27.13
(See color insert.) (a) Small-scale setup for concentration polarization desalination. (From Shannon, M.A., *Nature Nanotechnology*, 5, 248, 2010.) (b) Schematic of microchannel–nanochannel prototype tested for water desalination using concentration polarization. (From Kim, S.J., S.H. Ko, K.H. Kang, and J. Han, *Nature Nanotechnology*, 5, 297, 2010.)

27.3.3 Advanced Membranes

Recently, many new classes of membranes with functional nanoscale components or those inspired by biological systems have been developed as discussed previously [91]. One kind of these advanced membranes are the fouling-resistant membranes [92,93] with integrated polymeric brushes and nanostructures that allow the same membrane to have both hydrophilic and hydrophobic regions that mitigate fouling. A second kind of advanced membrane, composed of vertically aligned carbon nanotubes (CNTs), exhibits super-high flux for water transport [30,31]. The measured fluxes were found to be higher by up to three orders of magnitude than those predicted by conventional Hagen–Poiseuille flow theory. These aligned CNT membranes were shown to have average tube diameters of 1.6 nm [30] and ultrasmooth, hydrophobic walls that do not follow the basic "no-slip" condition, thereby letting fluids slide along the wall in an essentially frictionless configuration permitting high flux through the CNTs. Water velocities at 1 bar ranging from 9.5 to 43.9 cm/s were reported, significantly exceeding the expected velocities ranging from 0.00015 to 0.00057 cm/s [31]. While water flows nearly effortlessly through the aligned CNTs, the energy cost is expended in the water molecules entering the CNTs, and in the modification of the mouths of the CNTs to reject small salt ions, which increases the energy needed. The final energy usage for these membranes is still to be determined for concentrated salt solutions with a high rejection potential.

Another type of membrane utilizing CNTs is a composite CNT membrane, which has a structure similar to traditional polymeric membranes but CNTs are mixed into a top layer such as polyamide [94]. These composite membranes have demonstrated approximately a 1.5× flux increase over traditional RO membranes and their fabrication is relatively uncomplicated [94]. In addition, functionalization of CNTs is predicted to improve the salt rejection, and although this functionalization is expected to decrease the water flux over unfunctionalized CNTs it can also permit use of larger-diameter CNTs [95]. Functionalization of CNTs has been demonstrated by chemical, biological, and "active" functionalization. Chemical functionalities such as short-chained alkanes, long-chained alkanes, long polypeptides, and highly charged dye have all been explored as gatekeepers to allow selective control over ion transport as salt ions approach the CNT entrance. For 7-nm CNTs, polypeptide functionalization gave a permeation ratio of small dye to large dye of 3.6, while the ratio in the bulk solution was 1.6. Active gatekeepers have been demonstrated with long quadra-charge dye molecules that can be controlled by an electric field [96].

A third type of membrane has also been fabricated with CNTs, which layers CNTs on existing low-pressure membranes to mitigate fouling at the membrane surface [97]. Low-pressure membranes layered with 50–80 nm CNTs were shown to have the best performance, reducing fouling onset time by almost three times that of the untreated membrane, allowing for less backwashing, chemical cleaning, or both during membrane operation [97]. However, more investigation of CNT adhesion will be needed, as some of the CNTs were removed during the backwashing cycle.

Although polymeric membranes are currently the most widely used large-scale membranes, a myriad of new materials are being developed for rapid permeation, including graphene, graphene oxide, zeolites, and boron nitride nanotubes [98–102]. These new membranes separate by molecular sieving instead of by solution/diffusion that state-of-the-art commercial polymeric membranes use [99]. Molecular dynamics simulations have shown that nanoporous graphene membranes of 1 cm^2 area could theoretically deliver a water flux of 6.7 l/day at atmospheric pressure with a 99% salt rejection in comparison to current polymeric membranes of 1 cm^2 area that deliver between 0.001 and 0.055 l/day at atmospheric pressure and similar salt rejections [99]. However, development of robust manufacturing processes still proves challenging, although 30-in sheets have been fabricated recently [7].

Researchers have developed graphene oxide membranes with thicknesses ranging from 0.1 to 10 μm, approximately 1 cm^2 in area, and as low as 4 Å in pore size. These membranes have demonstrated negligible difference between evaporation of water without a membrane and permeation through the membrane for submicron thicknesses. For the 10 μm thickness, permeation through the membrane was two times slower than unimpeded evaporation [101]. The effects of oxidation chemistry and membrane processing such as annealing on water permeation are still under investigation; however, slip flow is suggested as the transport mechanism through these membranes similar to flow through CNTs [101].

Molecular dynamics modeling has also been used to suggest that boron nitride nanotubes embedded within silicon nitride membrane would be an extremely effective material for water desalination, rejecting both cations and anions while allowing water flux at a rate of 1.6–10.7 molecules/ns [100]. In addition, this model predicted that rejection would occur even under high concentration (up to 1 M) and high pressures (approximately 612 MPa) [100]. Another simulation with boron nitride nanotubes of 0.83 nm in diameter gave a range of ~12–50 molecules/ns for a pressure range of 100–500 MPa with complete salt rejection [103]. Although many methods have been developed for boron nitride production, developing a high-purity manufacturing method with diameters on the order of those investigated by molecular dynamics has proven to be challenging [104].

Zeolite membranes have also been developed after molecular dynamics simulations suggested their desalination potential [7]. These membranes for water desalination comprised a microstructure of aluminosilicate minerals with pore sizes ranging from 3 to 8 Å. These membranes have been shown to reject 98.6% of salt with a 3.5 MPa applied pressure and 0.1 M NaCl feed solution; however, rejection drops to ~90% as the concentration of NaCl is increased to 0.3 M [7]. Another challenge to zeolite membrane technology is that salt can travel through the intercrystal defects within the zeolite structure, so further development must be done for defect-free zeolite membranes. Significant progress has been made to this effect with the development of a rapid thermal process (RTP) added to the membrane manufacture. Scanning electron microscopy (SEM) and dye testing showed significant decrease in grain boundary size or flexibility while it was observed that RTP increased separation factor to 28 over 3 for conventionally processed membranes for separation of *p*-xylene to *o*-xylene [105]. Zeolite loading in commercial thin film composite and RO membranes is also under investigation, showing increasing flux for increasing zeolite loading, but not yet reaching the flux found in advanced commercial RO membranes [7]. Molecular dynamics has also suggested another membrane type by demonstrating an induced electro-osmosis or internal recirculation between heterogeneously charged pores, which can be considered a simplistic model of the ubiquitous heterogeneous structures found in nature [77] with structure similar to aquaporins. These simulations suggest a water velocity between 0.0711 and 0.842 m/s can be achieved if this type of heterogeneous structure is achieved [75].

27.3.4 Humidification–Dehumidification Desalination

Humidification–dehumidification (HDH) desalination processes rely on distillation to achieve desalination [106]. Simple technology is required because of their ability to function at atmospheric pressure; generally, HDH units require two exchangers, an evaporator to humidify air and a condenser to collect the freshwater generated as shown in the basic system schematic in Figure 27.14. The process begins by allowing dry air to absorb vapor

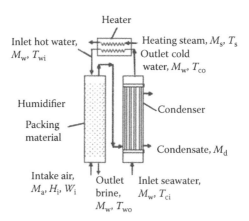

FIGURE 27.14
Schematic of conventional HDH process. (From Ettouney, H., *Desalination*, 183, 341, 2005.)

from heated feedwater in one chamber, transferring the humid air to another chamber where it comes in contact with a cool surface, thus allowing freshwater collection [63]. Energy consumption in such a system is required by the heat source as well as the pumps and blowers to move the water and vapor. Recovery rates for these systems range from 5% to 20%, and heat sources such as solar or geothermal supplies are common choices, especially since the rates of freshwater production are suited for demands of a few cubic meters per day. Temperature differences between 2°C and 5°C are required, necessitating heat consumption of approximately 20.9 kJ/m³ [106].

Researchers have also demonstrated the development of a microfluidic synthetic tree that can drive the motion of water at negative pressures. In addition, this synthetic tree structure has been shown to draw liquid from subsaturated vapor, not unlike the process of RO rejecting salt and allowing water to pass through the membrane. Consequently, the system acts as a tension-based pump that could be developed to extract or purify water from subsaturated soils [107].

27.3.5 Microbial Desalination Cells

Microbial fuel cells convert biowaste to electricity through microbial activity [108] by generating electrons available for harvesting in an external load circuit. In an innovative approach, a three-cell system was converted into a water desalination system (Figure 27.15). In a proof-of-concept study, it was shown that the three-cell design with an integrated anion-exchange membrane (AEM) and cation-exchange membrane (CEM) could generate a small amount of potential (typically less than 1 V) and desalinate water. The bacteria grow on the anode side and discharge protons into the water; however, the protons cannot pass through the AEM, so negatively charged ions from the saline water flow through the AEM to balance the positive charges produced. A similar process takes place on the cathode side, except protons are consumed, requiring positively charged ions from the saline water to cross the CEM to correct the charge imbalance. Testing of this system has been conducted with NaCl solutions at concentrations of 5, 20, and 35 g/l, which is consistent with concentrations seen from brackish water to seawater. The amount of salt removed from each concentration was at least 88 ± 2% for the 5 g/l case and up to 94 ± 3% for the 20 g/l case [108]. Comparison of the electrons harvested and NaCl removed

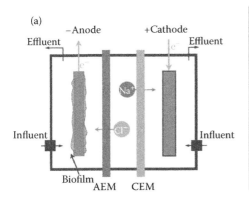
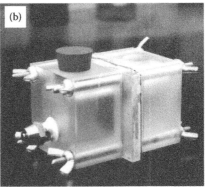

FIGURE 27.15
(See color insert.) (a) Basic schematic representing a three-chamber cell design for coupling a microbial fuel cell with a water desalination system. (b) Microbial desalination cell prototype. (From Cao, X., X. Huang, P. Liang, K. Xiao, Y. Zhou, and B.E. Logan, *Environmental Science Technology*, 43, 7148, 2009.)

revealed a charge transfer efficiency of almost 100% due to negligible effects of electrolysis (due to the low current generated) and insignificant back diffusion of ions from the electrode chambers to the desalination chamber [108]. Further work with microbial desalination cells suggested them as a pretreatment to reduce salinity before RO treatment. To this end, a microbial desalination cell was used to demonstrate salinity reduction between 43% and 63% for 5–20 g/l NaCl using an anolyte with 2 g/l acetate while generating between 295 and 424 mW/m² [109]. In another developed cell more optimized for power generation, called recirculation microbial desalination cells, reduction salinity of 20 g/l NaCl by 34% with recirculation produced a power density of 931 mW/m², while without recirculation, salinity was reduced 39% with 698 mW/m² produced power density [110].

27.3.6 Asymmetry-Driven Ion Pumps

As with nanotechnology, biology has also provided clues to developing advanced water desalination systems by following the ideas of ion transport in biological ion channels and ion pumps [111]. While existential proof for the working of these systems has been around in living organisms for a long time, implementing technologies for practical systems have continued to be a challenge due to several gaps in engineering these systems, arising from a lack of mechanistic understanding of how to manipulate ion transport in artificial systems. However, some recent studies have shown promise in advancing the

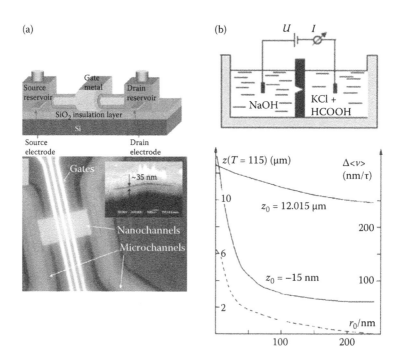

FIGURE 27.16
(See color insert.) (a) Schematic and SEM image shows the development of a device being used as a fluidic transistor allowing electrostatic control of ions. (From Karnik, R., R. Fan, M. Yue, D. Li, P. Yang, and A. Majumdar, *Nano Letters*, 5, 943, 2005.) (b) Schematic depicts formation of a conical nanopore by electrochemical etching at potential U with current I leading to data that show pumping of the potassium ions against a concentration gradient as a function of pore diameter on the narrow side of the conical nanopore. (From Siwy, Z. and A. Fulinski, *Physical Review Letters*, 89, 19803-1, 2002.)

state of knowledge toward both the science and technology of these bio-inspired systems [72,111–114]. In particular, mimicking ion-channel-gating behavior for the transport and control of ions has generated significant interest leading to the development of fluidic transistors [73] and conical nanopores for ion pumps for the potassium ion [115] (Figure 27.16). One advantage of bio-inspired systems is that the fundamental processes occur at the nanoscale and engineering these systems could allow for rapid development of bio-nanotechnology as a core area for water desalination research. Furthermore, with most biological systems operating near thermodynamic energy minima, such systems hold the promise of delivering the most energy-efficient water desalination systems ever conceived. However, significant questions remain open and would likely need extensive investigations. Molecular dynamic simulations begin to answer some of these questions and suggest that for electrokinetic flow of an electrolytic nanofilm where non-uniform charge exists (i.e., charged patches were introduced), flow of water and ions can be controlled by manipulating a combination of axial electric field and charge density of the patches. This simulation shows promise for ion control at the nanoscale [116] with potential to develop systems as these recent simulations use length scales currently accessible by fabrication technologies.

27.4 Summary and Conclusions

In this chapter, the current state of knowledge of water desalination methods was presented with a comparison between commonly used methods and opportunities for further research and growth. Throughout the chapter, the focus was on the energy–water nexus, i.e., the consumption of energy to generate freshwater from salt water. Consequently, many innovative and emerging technologies were discussed, highlighting the exciting opportunities for future development in desalination and water purification but also the need for further understanding either in basic science or engineering systems to engineer and implement these technologies in an energy-efficient manner. While biotechnology and nanotechnology provide interesting clues on developing novel systems for ion transport with applications in water desalination, many scientific and technological questions remain open.

References

1. Shannon, M.A., P.W. Bohn, M. Elimelech, J.G. Georgiadis, M.J. Marinas, and A.M. Mayes, Science and technology for water purification in the coming decades. *Nature*, 2008. **452**: pp. 301–310.
2. Torcellini, P., N. Long, and R. Judkoff, *Consumptive Water Use for U.S. Power Production*. 2003, National Renewable Energy Laboratory: Golden, CO.
3. UNICEF and WHO, *Water for Life: Making it Happen*. 2005, United Nations: Geneva.
4. Yeston, J., R. Coontz, J. Smith, and C. Ash, A thirsty world. *Science*, 2006. **313**(5790): p. 1067.
5. Kim, S.J., S.H. Ko, K.H. Kang, and J. Han, Direct seawater desalination by ion concentration polarization. *Nature Nanotechnology*, 2010. **5**: pp. 297–301.

6. Service, R.F., Desalination freshens up. *Science*, 2006. **313**: pp. 1088–1090.
7. Humplik, T., J. Lee, S.C. O'Hern, B.A. Fellman, M.A. Baig, S.F. Hassan, M.A. Atieh, F. Rahman, T. Laoui, R. Karnik, and E.N. Wang, Nanostructured materials for water desalination. *Nanotechnology*, 2011. **22**: pp. 1–19.
8. Elimelech, M., and W.A. Phillip, The future of seawater desalination: Energy, technology, and the environment. *Science*, 2011. **333**: pp. 712–717.
9. Electric Power Research Institute Inc., U.S. electricity consumption for water supply and treatment- the next half century, in *Water and Sustainability*. 2002, EPRI, Palo Alto, California.
10. Desalination and Water Reuse, The international desalination and water reuse quarterly news. *Installed desalination growth slowed in 2011–2012*, 2012.
11. Subramani, A., M. Badruzzaman, J. Oppenheimer, and J.G. Jacangelo, Energy minimization strategies and renewable energy utilization for desalination: A review. *Water Research*, 2011. **45**: pp. 1907–1920.
12. Spiegler, K.S., *Salt Water Purification*. 1962, John Wiley & Sons: New York.
13. Veerapaneni, S., B. Long, S. Freeman, and R. Bond, Reducing energy consumption for seawater desalination. *Journal of American Water Works Association*, 2007. **99**: pp. 95–106.
14. USDOE, *Energy Demands on Water Resources*. 2006, Sandia National Laboratory: Albuquerque, NM, pp. 1–80.
15. Prakash, S., K. Bellman, and M.A. Shannon, Recent advances in water desalination through biotechnology and nanotechnology, in *Bionanotechnology II: Global Prospects*, D. Reisner, ed. 2011, CRC: New York. pp. 365–382.
16. Hasson, D., and R. Semiat, Scale control in saline and wastewater desalination. *Israel Journal of Chemistry*, 2006. **46**: pp. 97–104.
17. Latteman, S., and T. Hoepner, Environmental impact and impact assessment of seawater desalination. *Desalination*, 2008. **220**: pp. 1–15.
18. Kalogirou, S.A., Seawater desalination using renewable technologies. *Progress in Energy and Combustion Science*, 2005. **31**: pp. 242–281.
19. Zhou, Y., and R.S.J. Tol, Evaluating the costs of desalination and water transport. *Water Resources Research*, 2005. **41**: pp. 1–10.
20. El-Nashar, A.M., Cogeneration for power and desalination-state of the art review. *Desalination*, 2001. **134**: pp. 7–28.
21. Crittenden, J.C., R.R. Trussell, D.W. Hand, K.J. Howe, and G. Tchobanoglous, *Water Treatment*. 2005, John Wiley & Sons: Hoboken, NJ.
22. Trieb, F., ed. *Concentrating Solar Power for Water Desalination*. ed. G.A.C. (DLR) and I.O.T. Thermodynamics. 2007, Federal Ministry for the Environment, Nature Conservation and Nuclear Energy: Stuttgart, Germany.
23. Miller, J.E., *Review of Water Resources and Desalination Technologies*. 2003, Sandia National Laboratory: Livermore, CA.
24. Buros, O.K., *The ABCs of Desalting*. 2000, International Desalination Association. http://www .idadesal.org/ABCs.pdf.
25. Lakshminarayanaiah, N., *Transport Phenomena in Membranes*. 1969, Academic Press: New York.
26. Nath, K., *Membrane Separation Processes*. 2008, Prentice Hall: New Delhi.
27. American Membrane Technology Association, Application of membrane technologies. 2007; Available from: http://www.amtaorg.com/wp-content/uploads/1_applicationofmembrane technologies.pdf.
28. Reid, C.E., Principles of reverse osmosis, in *Desalination by Reverse Osmosis*, U. Merten, ed. 1966, MIT Press: Cambridge, MA. pp. 1–14.
29. Cath, T.Y., A.E. Childress, and M. Elimelech, Forward osmosis: Principles, applications, and recent developments. *Journal of Membrane Science*, 2006. **281**: pp. 70–87.
30. Holt, J.K., H.G. Park, Y. Wang, M. Stadermann, A.B. Artyukhin, C.P. Grigoropoulos, A. Noy, and O. Bakajin, Fast mass transport through sub-2nm carbon nanotubes. *Science*, 2006. **213**: pp. 1034–1037.
31. Majumder, M., N. Chopra, R. Andrews, and B.J. Hinds, Experimental observation of enhanced liquid flow through aligned carbon nanotube membranes. *Nature*, 2005. **438**: pp. 44–45.

32. Chen, J.C., Q. Li, and M. Elimelech, In situ monitoring techniques for concentration polarization and fouling phenomena in membrane filtration. *Advances in Colloid and Interface Science,* 2004. **107**: pp. 83–108.

33. Gilbert, C., *Protecting RO Membranes from Chlorine Damage.* 2009, Dow Water Solutions. Waterworld, Pennwell Corporation: West Chester, PA.

34. Kurihara, M., H. Yamamura, T. Nakanishi, and S. Jinno, Operation and reliability of very high-recovery seawater desalination technologies by brine conversion two-stage RO desalination system. *Desalination,* 2001. **138**: pp. 191–199.

35. MacHarg, J.P., Exchanger tests verify 2.0 kWh/m³ SWRO energy use. *International Desalination and Water Reuse,* 2001. **11**(1): pp. 42–45.

36. Fritzmann, C., J. Lowenberg, T. Wintgens, and T. Melin, State-of-the-art of reverse osmosis desalination. *Desalination,* 2007. **216**: pp. 1–76.

37. US Department of the Interior Bureau of Reclamation, *Electrodialysis (ED) and Electrodialysis Reversal (EDR).* 2010.

38. Forgacs, C., N. Ishibashi, J. Leibovitz, J. Sinkovic, and K.S. Spiegler, Polarization at ion-exchange membranes in electrodialysis. *Desalination,* 1972. **10**: pp. 181–214.

39. Pilat, B., Practice of water desalination by electrodialysis. *Desalination,* 2001. **216**: pp. 385–392.

40. Allison, R.P., High water recovery with electrodialysis reversal, in *AWWA Membrane Conference.* 2003.

41. Kiernan, J., and A. von Gottberg, Selection of EDR desalting technology rather than MF/RO for the city of San Diego water reclamation project, in *American Desalting Association North American Biennial Conference and Exposition.* 1998.

42. Allison, R.P., *Electrodialysis Treatment of Surface and Waste Waters.* 2005, GE Water and Process Technologies Paper TP1032EN. GE Research: New York.

43. Innovations, *Electrochemical Capacitors for Water Desalination.* Science Daily, 2009 (Jan. 26).

44. Kujundzic, E., K. Cobry, A. Greenberg, and M. Hernandez, Use of ultrasonic sensor for characterization of membrane fouling and cleaning. *Journal of Engineered Fibers and Fabrics,* 2008. **3**: pp. 35–44.

45. Probstein, R.F., *Physiochemical Hydrodynamics: An Introduction.* 1994, John Wiley & Sons: New York.

46. Porter, M.C., Concentration polarization with membrane ultrafiltration. *Industrial & Engineering Chemistry Product Research and Development,* 1972. **11**: pp. 234–248.

47. Baker, R.W., *Membrane Technology and Application.* 2004, John Wiley & Sons: Chichester, England.

48. Schwinge, J., D. Wiley, A. Fane, and R. Guenther, Characterization of a zigzag spacer for ultrafiltration. *Journal of Membrane Science,* 2000. **172**: pp. 19–31.

49. Schwinge, J., P. Neal, D. Wiley, D. Fletcher, and A. Fane, Spiral wound modules and spacers: Review and analysis. *Journal of Membrane Science,* 2004. **242**: pp. 129–153.

50. Sablani, S.S., M.F. Goosen, R. Al-Belushi, and V. Gerardos, Influence of spacer thickness on permeate flux in spiral-wound seawater reverse osmosis systems. *Desalination,* 2002. **146**: pp. 225–230.

51. Jaffrin, M.Y., Dynamic shear-enhanced membrane filtration: A review of rotating disks, rotating membranes, and vibrating systems. *Journal of Membrane Science,* 2008. **324**: pp. 7–25.

52. Kuberkar, V., P. Czekaj, and R. Davis, Flux enhancement for membrane filtration of bacterial suspensions using high-frequency backpulsing. *Biotechnology and Bioengineering,* 1998. **60**: pp. 77–87.

53. Gomaa, H., S. Rao, and M. Taweel, Flux enhancement using oscillatory motion and turbulence promoters. *Journal of Membrane Science,* 2011. **381**: pp. 64–73.

54. Kyllonen, H., P. Pirkonen, and M. Nystrom, Membrane filtration enhanced by ultrasound: A review. *Desalination,* 2005. **181**: pp. 319–355.

55. Mukherjee, D., A. Kulkarni, and W.N. Gill, Flux enhancement of reverse osmosis membranes by chemical surface modification. *Journal of Membrane Science,* 1994. **97**: pp. 231–247.

56. Cummings, J.A., *Membrane Flux Enhancement.* 2004, US Patent Office. Patent number: US20040140259. Europe.

57. Song, W., Y. Su, X. Chen, L. Ding, and Y. Wan, Rapid concentration of protein solution by a cross-flow electro-ultrafiltration process. *Separation and Purification Technology,* 2010. **73**: pp. 310–318.

58. Huotari, H., G. Tradgardh, and I. Huisman, Crossflow membrane filtration enchanced by an external DC electric field: A review. *Transactions of IChemE*, 1999. **77**(A): pp. 461–468.
59. Huotar, H.M., I.H. Huisman, and G. Tradgardh, Electrically enhanced crossflow membrane filtration of oily waste water using the membrane as a cathode. *Journal of Membrane Science*, 1999. **156**: pp. 49–60.
60. Akay, G., and R. Wakeman, Electric field enchanced crossflow microfiltration of hydrophoically modified water soluble polymers. *Journal of Membrane Science*, 1997. **131**: pp. 229–236.
61. Winters, H., *Sustaining High System Average Flux using Electromagnetic Field (EMF) and Integrated Flow Distributor*. 2011, Graham Tek RO Research Program: Teaneck, NJ.
62. Siemens, What will the future of salt water desalination look like? 2010; Available from: http://aunz.siemens.com/PicFuture/Documents/WaterCasestudy_desalination.pdf.
63. Bourouni, K., M.T. Chaibi, and L. Tadrist, Water desalination by humidification and dehumidification of air: State of art. *Desalination*, 2001. **137**: pp. 167–176.
64. Darwish, M.A., and H. El-Dessouky, The heat recovery thermal vapour-compression desalting system: A comparison with other thermal desalination processes. *Applied Thermal Engineering*, 1996. **16**: pp. 523–527.
65. Glueckauf, E., Seawater desalination—In perspective. *Nature*, 1966. **211**: pp. 1227–1230.
66. Darwish, M.A., Desalting: Fuel energy cost in Kuwait in view of $75/barrel oil price. *Desalination*, 2007. **208**: pp. 306–320.
67. McCutcheon, J.R., and M. Elimelech, Modeling water flux on forward osmosis: Implications for improved membrane design. *AIChE Journal*, 2007. **53**: pp. 1736–1744.
68. Enezi, G.L., E. Hisham, and N. Fawzy, Low temperature humidication dehumidication desalination process. *Energy Conversion and Management*, 2006. **47**: pp. 470–484.
69. Jackson, R.D., and C.H.M. van Bavel, Solar distillation of water from soil and plant materials: A simple desert survival technique. *Science*, 1965. **149**: pp. 1377–1378.
70. Daiguji, H., P. Yang, and A. Majumdar, Ion transport in nanofluidic channels. *Nano Letters*, 2004. **4**(1): pp. 137–142.
71. Daiguji, H., Y. Oka, and K. Shirono, Nanofluidic diode and bipolar transistor. *Nano Letters*, 2005. **5**(11): pp. 2274–2280.
72. Gong, X., J. Li, H. Lu, R. Wan, J. Li, J. Hu, and H. Fang, A charge-driven molecular water pump. *Nature Nanotechnology*, 2007. **2**: pp. 709–712.
73. Karnik, R., R. Fan, M. Yue, D. Li, P. Yang, and A. Majumdar, Electrostatic control of ions and molecules in nanofluidic transistors. *Nano Letters*, 2005. **5**(5): pp. 943–948.
74. Kemery, P.J., J.K. Steehler, and P.W. Bohn, Electric field mediated transport in nanometer diameter channels. *Langmuir*, 1998. **14**(10): pp. 2884–2889.
75. Qiao, R., and N.R. Aluru, Atomistic simulation of KCl transport in charged silicon nanochannels: Interfacial effects. *Colloids and Surfaces A*, 2005. **267**: pp. 103–109.
76. Qiao, R., and N.R. Aluru, Charge inversion and flow reversal in a nanochannel electroosmotic flow. *Physical Review Letters*, 2004. **92**(19): Article no. 198301 (4 pp.).
77. Qiao, R., J.G. Georgiadis, and N.R. Aluru, Differential ion transport induced electroosmosis and internal recirculation in heterogeneous osmosis membranes. *Nano Letters*, 2006. **6**(5): pp. 995–999.
78. Qiao, R., and N.R. Aluru, Surface charge induced asymmetric electrokinetic transport in confined silicon nanochannels. *Applied Physics Letters*, 2005. **86**: pp. 143105–143107.
79. Shannon, M.A., Fresh for less. *Nature Nanotechnology*, 2010. **5**: pp. 248–250.
80. Lange, K.E., Get the salt out, in *National Geographic*. 2010, National Geographic Magazine: Washington, DC. pp. 32–35.
81. Holloway, R.W., A.E. Childress, K.E. Dennett, and T.Y. Cath, Forward osmosis for concentration of anaerobic digester concentrate. *Water Research*, 2007. **41**: pp. 4005–4014.
82. Urban, I., N. Ratcliffe, J. Duffield, G.R. Elder, and D. Patton, Functionalized paramagnetic nanoparticles for waste water treatment. *Chemical Communications*, 2010. **46**: pp. 4583–4585.
83. Yip, N.Y., A. Tiraferri, W.A. Phillip, J.D. Schiffman, and M. Elimelech, High performance thin-film composite forward osmosis membrane. *Environmental Science Technology*, 2010. **44**: pp. 3812–3818.

84. Chung, T.-S., S. Zhang, K. Wanga, J. Sua, and M. Linga, Forward osmosis processes: Yesterday, today and tomorrow. *Desalination*, 2012. **287**(78–81).

85. Tang, C.Y., Q. She, W.C.L. Lay, R. Wang, and A.G. Fane, Coupled effects of internal concentration polarization and fouling on flux behavior of forward osmosis membranes during humic acid filtration. *Journal of Membrane Science*, 2010. **354**(1–2): pp. 123–133.

86. McCutcheon, J.R., R.L. McGinnis, and M. Elimelech, A novel ammonia-carbon dioxide forward (direct) osmosis desalination process. *Desalination*, 2005. **174**: pp. 1–11.

87. Thompson, N.A., and P.G. Nicoll, Forward osmosis desalination: A commercial reality, in *IDA World Congress*. 2011, Perth, Western Australia.

88. Gray, G.T., R.L. McCutcheon, and M. Elimelech, Internal concentration polarization in forward osmosis: Role of membrane orientation. *Desalination*, 2006. **197**: pp. 1–8.

89. Pedley, T.J., Calculation of unstirred layer thickness in membrane transport experiments: A survey. *Quarterly Review of Biophysics*, 1983. **16**: pp. 115–150.

90. Pu, Q., J. Yun, H. Temkin, and S. Liu, Ion-enrichment and ion-depletion effect of nanochannel structures. *Nanoletters*, 2004. **4**(6): pp. 1099–1103.

91. Prakash, S., M.B. Karacor, and S. Banerjee, Surface modification in microsystems and nanosystems. *Surface Science Reports*, 2009. **64**: pp. 233–254.

92. Asatekin, A., A. Menniti, S. Kang, M. Elimelech, E. Morgenroth, and A.M. Mayes, Antifouling nanofiltration membranes for membrane bioreactors from self-assembling graft copolymers. *Journal of Membrane Science*, 2006. **285**: pp. 81–89.

93. Asatekin, A., S. Kang, M. Elimelech, and A.M. Mayes, Anti-fouling ultrafiltration membranes containing polyacrylonitrile-graft-poly(ethylene oxide) comb copolymer additives. *Journal of Membrane Science*, 2007. **298**: pp. 136–146.

94. Ahn, C.H., Y. Baek, C. Lee, S.O. Kim, S. Kim, S. Lee, S.-H. Kim, S.S. Bae, J. Park, and J. Yoon, Carbon nanotube-based membranes: Fabrication and application to desalination. *Journal of Industrial and Engineering Chemistry*, 2012. **18**: pp. 1551–1559.

95. Corry, B., Water and ion transport through functionalised carbon nanotubes: Implications for desalination technology. *Energy and Environmental Science*, 2011. **4**: pp. 751–759.

96. Majumder, M., A. Stinchcomb, and B.J. Hinds, Towards mimicking natural protein channels with alighned carbon nanotube membranes for active drug delivery. *Life Sciences*, 2010. **86**: pp. 563–568.

97. Ajmani, G.S., D. Goodwin, K. Marsh, D.H. Fairbrother, K.J. Schwab, J.G. Jacangelo, and H. Huang, Modification of low pressure membranes with carbon nanotube layer for fouling control. *Water Research*, 2012. **46**: pp. 5645–5654.

98. Paul, D.R., Create new types of carbon-based membranes. *Science*, 2012. **335**: pp. 413–414.

99. Wang, E.N., and R. Karnik, Graphene cleans up water. *Nature Nanotechnology*, 2012. **7**: pp. 552–554.

100. Hilder, T.A., D. Gordon, and S.-H. Chung, Salt rejection and water transport through boron nitride nanotubes. *Small*, 2009. **5**(19): pp. 2183–2190.

101. Nair, R.R., H.A. Wu, P.N. Jayaram, I.V. Grigorieva, and A.K. Geim, Unimpeded permeation of water through helium-leak-tight graphene-based membranes. *Science*, 2012. **335**: pp. 442–444.

102. Karan, S., S. Samitsu, X. Peng, K. Kurashima, and I. Ichinose, Ultrafast viscous permeation of organic solvents through diamond-like carbon nanosheets. *Science*, 2012. **2012**: p. 335.

103. Suk, M.E., A.V. Raghunathan, and N.R. Aluru, Fast reverse osmosis using boron nitride and carbon nanotubes. *Applied Physics Letters*, 2008. **92**: Article no. 133120 (3 pp.).

104. Golberg, D., Y. Bando, Y. Huang, T. Terao, M. Mitome, C. Tang, and C. Zhi, Boron nitride nanotubes and nanosheets. *ACS Nano*, 2010. **4**(6): pp. 2979–2993.

105. Choi, J., H.-K. Jeong, M.A. Snyder, J.A. Stoeger, R.I. Masel, and M. Tsapatsis, Grain boundary defect elimination in a zeolite membrane by rapid thermal processing. *Science*, 2009. **325**: pp. 590–593.

106. Ettouney, H., Design and analysis of humidification dehumidification desalination processes. *Desalination*, 2005. **183**: pp. 341–352.

107. Wheeler, T.D., and A.D. Stroock, The transpiration of water at negative pressures in a synthetic tree. *Nature*, 2008. **455**: pp. 208–212.

108. Cao, X., X. Huang, P. Liang, K. Xiao, Y. Zhou, and B.E. Logan, A new method for water desalination using microbial desalination cells. *Environmental Science Technology*, 2009. **43**(18): pp. 7148–7152.
109. Mehanna, M., T. Saito, J. Yan, M. Hickner, X. Cao, X. Huang, and B.E. Logan, Using microbial desalination cells to reduce water salinity prior to reverse osmosis. *Energy and Environmental Science*, 2010. **3**: pp. 1114–1120.
110. Qu, Y., Y. Feng, X. Wang, J. Liu, J. Lv, W. He, and B.E. Logan, Simultaneous water desalination and electricity generation in a microbial desalination cell with electrolyte recirculation for pH control. *Bioresource Technology*, 2012. **106**: pp. 89–94.
111. Hille, B., *Ionic Channels of Excitable Membranes*. 1992, Sinauer Associates Inc: Sunderland, MA.
112. Prakash, S., J. Yeom, N. Jin, I. Adesida, and M.A. Shannon, Characterization of ionic transport at the nanoscale. *Proceedings of the Institution of Mechanical Engineers, Part N: Journal of Nanosystems and Nanoengineering*, 2007. **220**(2): pp. 45–52.
113. Prakash, S., J. Lucido, J.G. Georgiadis, and M.A. Shannon, Development of a hydrogel-bridged nanofluidic system for water desalination, in *233rd National Meeting and Exposition*. 2007, American Chemical Society: Chicago, IL.
114. Hinds, B.J., A blueprint for a nanoscale pump. *Nature Nanotechnology*, 2007. **2**: pp. 673–674.
115. Siwy, Z., and A. Fulinski, Fabrication of a synthetic nanopore ion-pump. *Physical Review Letters*, 2002. **89**(19): pp. 19803-1–19803-4.
116. Zambrano, H.A., M. Pinti, A.T. Conlisk, and S. Prakash, Electrokinetic transport in a water–chloride nanofilm in contact with a silica surface with discontinuous charged patches. *Microfluidics and Nanofluidics*, 2012. **13**: pp. 735–747.

28

Challenges and Opportunities for Nanotechnology in the Energy–Water Nexus

Shaurya Prakash and Karen Bellman

Department of Mechanical and Aerospace Engineering, The Ohio State University, Columbus, Ohio, USA

CONTENTS

28.1 Introduction

Water is an essential ingredient to life. As human civilizations have evolved, the use and demands on water have grown. Modern societies are complex, interdependent networks driven by a plethora of industrial, agricultural, and residential activities involving the consumption of energy and water. For example, we use water to mine, refine, and convert energy resources while using energy to collect, treat, and distribute freshwater. Consequently, in recent years, the interdependence between generation and use of energy, and the need and use for clean water have been well documented. This interdependency or connection is referred to as the energy–water nexus. The direct science and engineering implication for the energy–water nexus is the need to solve a coupled problem rather than a single problem.

Most water withdrawals for modern societal needs occur from natural source waters. The energy and material input to treat the source water is directly related to the quality of these source waters. Figure 28.1 [1] shows the migration of water throughout an ecosystem or area as part of the natural water cycle. As human activities such as agriculture redirect precipitation to crops and away from rivers or aquifer recharge, the natural cycle is disrupted or distressed [2], and projected to result in significant challenges for biodiversity, increased rate of diseases in the ecosystems, and water shortages [3]. Therefore, to minimize higher

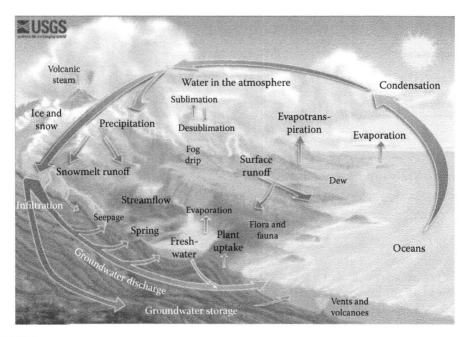

FIGURE 28.1

(See color insert.) Natural water cycle is a balance that is dependent on many different inputs. As human activity changes these inputs, such as amount stored in groundwater or evapotranspiration from increased agricultural activity, the cycle can be disrupted or altered. (From USGS. *Summary of the Water Cycle*. [cited February 18, 2013]; Available at http://ga.water.usgs.gov/edu/watercyclesummary.html, 2013.)

energy and/or material input to treating water, it is important to consider ways to mitigate the stress on the natural water cycle or to replenish the existing cycles leading to innovative science and technology development for more efficient ways of energy and water distribution and production.

Special attention to the energy–water nexus is essential because of increasing population needs driving increasing energy demands, and significantly diminished freshwater supplies [4]. For example, the US Department of Energy (USDOE) reports that freshwater withdrawals surpass precipitation in numerous regions of the United States, most significantly in the southwest, the high plains, California, and Florida, as shown in Figure 28.2 [5]. In addition, some groundwater levels in the United States have seen drops of 300–900 ft during the past 50 years because the withdrawal rate is faster than the aquifer recharge rate, causing shortages and increased salinity, as summarized in Table 28.1 [5] for a few geographic locations around the United States. For example, in Dare County, North Carolina, the population increased by 48% from 1991 to 2009, causing an increased demand on groundwater resources. Compared with the late 1980s, the salinity of the groundwater resources in the Dare County area has risen from 1000 to 2500 mg/l, requiring desalination before distribution [6].

In this chapter, the focus is on water use for energy needs, and the technology solutions available including those based on nanotechnology and/or pathways that can help minimize and eventually eliminate water use for energy production and use. Therefore, the purpose of this chapter is to provide a review of water use for a variety of common energy generation systems, identify challenges, and suggest some technology solutions with examples for the challenges toward solving the problems arising from the energy–water nexus.

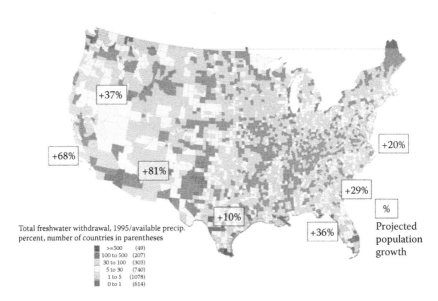

FIGURE 28.2
(See color insert.) Projected population growth by percentage and water shortages by color. Water shortage is defined as total freshwater withdrawal divided by an area's precipitation, based on most recent available data. (From USDOE, *Energy Demands on Water Resources*. US Department of Energy, 2006.)

TABLE 28.1

Representative Description of Declining Groundwater Resources

Region	Groundwater Decline
Long Island, NY	Water table declined, stream flows reduced, saltwater moving inland
West central Florida	Groundwater and surface water declining, saltwater intruding, sink holes
Baton Rouge, LA	Groundwater declining up to 200 ft
Houston, TX	Groundwater declining up to 400 ft, land subsidence up to 10 ft
Arkansas	Sparta aquifer declared "critical"
High Plains	Declines up to 100 ft, water supply (saturated thickness) reduced over half
Chicago–Milwaukee area	Groundwater serving 8.2 million people has declined as much as 900 ft
Pacific Northwest	Declines up to 100 ft
Tucson/Phoenix, AZ	Declines 300–500 ft, subsidence up to 12.5 ft
Las Vegas, NV	Declines up to 300 ft, subsidence up to 6 ft
Antelope Valley, CA	Declines >300 ft, subsidence >6 ft

Source: USDOE, *Energy Demands on Water Resources*. US Department of Energy, 2006; Vinson, D.S., H.G. Schwartz, G.S. Dwyer, and A. Vengosh, *Hydrogeology Journal*, 19, 981, 2011.

Several examples of nanotechnology that provide opportunities and potential solutions to the energy–water nexus challenges are discussed in each of the sections below. Figure 28.3 [7] shows a well-documented data plot presenting water withdrawals in the United States based on use in a variety of applications. It is evident from Figure 28.3 and the estimates provided by the USDOE and the US Department of the Interior that >40% of total water withdrawals are for thermoelectric cooling for power generation [5,8–10]. The next largest block of water withdrawal is for agricultural and livestock use at approximately 39% of all

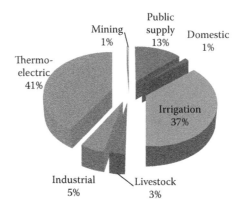

FIGURE 28.3
Water withdrawals in the United States as estimated in a 2005 report by the US DOE. Note that thermoelectric power generation, mining, and industrial use account for nearly 47% of total withdrawals. (From NETL, *Department of Energy/National Energy Technology Laboratory's Water–Energy Interface Research Program: December 2010 Update*, J.P. Ciferno, R.K. Munson, and J.T. Murphy, eds. National Energy Technology Laboratory, 2010.)

water drawn. Since water use for thermoelectric power generation is a major contributor to the energy–water nexus, a brief overview of water use, both present and projected, follows.

Recently, the USDOE has projected that thermoelectric generating capacity will increase by approximately 6% by 2035, based on the data discussed in the US Energy Information Administration's (USEIA) 2010 annual energy outlook as shown in Figure 28.4 [11,12]. Furthermore, it is estimated that water withdrawal to support electricity generation is expected to stay the same or decline slightly during the same time period. However, water consumption is expected to increase by anywhere from 14% to 26% on a national basis. The difference in water withdrawal and consumption for the estimated use should be clarified, as withdrawal (drawing water for use from a source) will likely remain unchanged because power plants retiring between 2005 and 2030 are older facilities that are more likely to employ high-withdrawal, once-through cooling. New facilities that will be built over that time period are likely to employ lower-withdrawal but high-consumption wet recirculating cooling systems. It should be noted that the projections do not account for

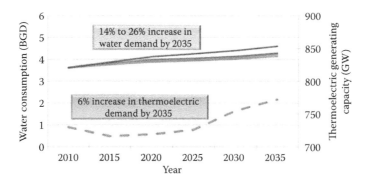

FIGURE 28.4
Plot showing water consumption in billion gallons per day in contrast to the thermoelectric power generation capacity as estimated by the US DOE with projections extended to 2035. (From NETL, *Innovations for Existing Plants: Water Energy Interface* [cited February 18, 2013]; Available at http://www.netl.doe.gov/technologies /coalpower/ewr/water/index.html#top, 2010.)

significant new technology incorporation or effects of regulations limiting carbon-based emissions. For example, current commercially available carbon capture technologies typically consume large quantities of water, and could increase water consumption by 50%–90% depending on the specific power generation platform as presented graphically in the data reported in Figure 28.5.

Following our discussion above, we note that Figure 28.3 shows water withdrawal, which represents the total water taken from source waters. For thermoelectric power generation, the most common form of water consumption is by evaporation and water consumption represents the amount post-withdrawal that is not returned to the source water. In many instances, the water may be returned to the source (usually a surface water body) but it may be at different conditions (temperature, total dissolved solids [TDS], pH, or other properties). The US Geological Survey freshwater consumption data presented in Figure 28.6 provides a comparison between water withdrawals (Figure 28.3) and water

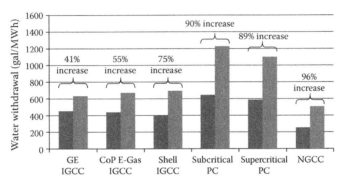

FIGURE 28.5
Water withdrawals as function of power generation method by comparing current state of the art in carbon capture technology. (From NETL, *Innovations for Existing Plants: Water Energy Interface* [cited February 18, 2013]; Available at http://www.netl.doe.gov/technologies/coalpower/ewr/water/index.html#top, 2010.)

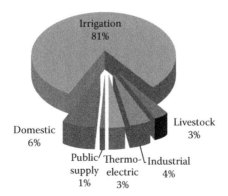

FIGURE 28.6
Consumption of water as percentage of total withdrawals. Data based on the most recent reports from the USGS (1995). Comparison with Figure 28.3 shows differences between water withdrawal and consumption. (From NETL, *Department of Energy/National Energy Technology Laboratory's Water–Energy Interface Research Program: December 2010 Update*, J.P. Ciferno, R.K. Munson, and J.T. Murphy, eds. National Energy Technology Laboratory, 2010.)

consumption. To provide context to water use, at 3% consumption, >3 billion gallons per day were consumed. The above discussion clearly illustrates that water is an essential source for energy use and generation. In the next several sections, a discussion for water use for different energy areas is presented.

28.2 Water Use in Power Generation

In a 2009 report by the US Government Accountability Office (USGAO) [13], it was shown that as of 2007, nearly 75% of electric power generation in the United States was through thermal (e.g., coal- or gas-fired power plants, nuclear, and solar–thermal, among others) methods. Water withdrawal and consumption are largely related to cooling water to operate the power plants. Specifically, as all thermal methods inherently rely on some form of a thermodynamic cycle, the heat generated needs to be dissipated at as low a temperature as possible to obtain maximum work achievable, W, which based on a simple thermodynamic analysis can be expressed as

$$W \leq \left[1 - \frac{T_C}{T_H}\right] Q \qquad (28.1)$$

where T_C is the temperature of the heat sink, T_H is the temperature of the hot reservoir, and Q is the heat generated. The use of water for cooling requirements is possible owing to the high specific heat of water, which at room temperature is 4.181 kJ/kg-K. A glance at thermodynamic tables makes this clear: the heat of vaporization of water is 2 J/mm^3 or 2 GJ/m^3; therefore, for a 1 GW power plant, 0.5 m^3 (500 liters) of water will need to be evaporated per second. It is generally known that depending on specific operating and source water conditions, including those of the surrounding environment, approximately 1%–3% of water in the cooling towers is lost to evaporation as a function of circulation rate. Of this evaporated water, 5% capture and reuse could potentially lead to savings in excess of 200 million gallons per day (MGD) across the United States.

It is commonly assumed that use of renewable sources of energy will mitigate many crises, including impending water shortages. However, the need for water withdrawal and consumption for energy use is highly dependent on the method to produce energy. For example, renewable sources such as hydroelectricity use 1.55–6.05 m^3 of water for every gigajoule of energy produced [14]. Furthermore, methods such as solar–thermal (distinct from solar-based photovoltaic approaches, see Section 28.7) are comparable to other thermal methods in water consumption. In the sections to follow, several energy generation and use methods are discussed with a brief review of water consumption for each method.

Water is used in thermoelectric power plants for a variety of activities, depending on the type of plant, including flue gas desulfurization, boiler feedwater makeup, gasification process makeup water, processing and washing of cooling systems, irrigation for wet recirculating pond systems, scrubber dilution water, and ash control, yet cooling remains the dominant use of water in thermoelectric plants [15]. Currently, there are four types of cooling schemes in these plants: once-through, wet recirculating, dry, and hybrid cooling as shown in Figure 28.7 [16]. Once-through cooling is used in many older power plants and

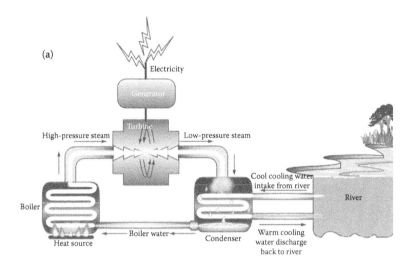

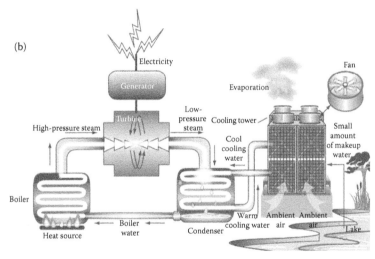

FIGURE 28.7

Cooling schemes for thermoelectric power plants. (a) Once-through cooling scheme draws water from source, which travels through a heat exchanger and then discharges the cooling water back to the source. (b) Similarly, wet recirculating cooling draws water from a source and cools power plant working fluid. The cooling water is then allowed to cool, generally in a tower or pond, and then recirculated through the heat exchanger.

works by pulling water from a source, using it to cool the system in a single loop while evaporating approximately 0.5%–1.6% of water withdrawn, typically between 100 and 400 gal/ MWh. Table 28.2 summarizes these values by plant type. The rest is discharged back to the source at approximately 3°C higher than the inlet in the United States per regulations [17]. The large amount of water needed for withdrawal to run a once-through cooled system puts the plant at risk during times of water scarcity and can worsen an existing water shortage [18]. In addition, concerns over water quality and availability, negative impacts on aquatic life, increased water return temperatures, and highly invasive zebra mussels, which foul intake structures, are also compelling economic reasons to improve thermo-electric plant cooling systems [5,15] through innovative technology approaches. For example, the Garimella group at Georgia Institute of Technology (Atlanta) has investigated the

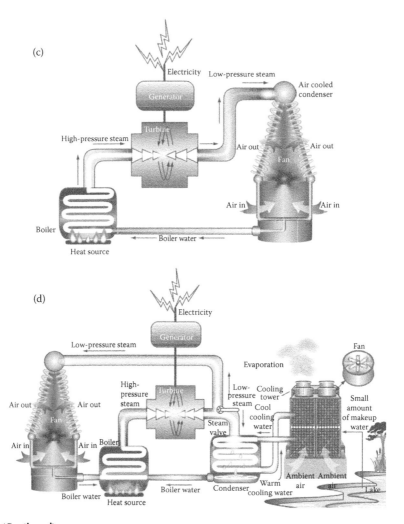

FIGURE 28.7 (Continued)
Cooling schemes for thermoelectric power plants. (c) Air cooling uses fans to blow air past heat exchanger to cool steam. (d) Hybrid cooling uses dry cooling when conditions allow efficiency to remain high, or employs water spray over the condensing tubes when conditions lead to increase in required cooling capacity. Some hybrid cooling schemes use water spray at all times. (From USGAO, *Energy–Water Nexus: Improvements to Federal Water Use Data Would Increase Understanding of Trends in Power Plant Water Use.* US Government Accountability Office, 2009.)

use of microchannel heat exchangers. In their work, distinct flow regimes within micro-tubes (dimensions in the submillimeter range) were identified as opposed to conventional heat exchangers. In a test system, with 200 MW of waste heat input, 140 MW of cooling load was obtained with nearly 90 MW cooling at 5°C with an electrical input of 23 MW [19].

Wet recirculating systems reuse cooling water and use either cooling towers or cooling ponds to allow the water temperature to decrease. These systems require <5% of the withdrawal of a once-through system, yet consume almost all of the water withdrawn [15]. The last two types of cooling schemes, dry and hybrid cooling, are considered advanced cooling methods in contrast to the widely implemented wet methods as discussed above. Dry cooling uses fans and/or blowers to create air flow past steam condenser tubes; hence,

TABLE 28.2

Summary of Water Consumption for Thermoelectric Power Plants with Once-Through and Wet Recirculating Cooling

Type of Power Generation	Cooling Type			
	Once-Through		Wet Recirculating with Cooling Tower	
	Gallons Consumed/MWh	% Withdrawal Consumed	Gallons Consumed/MWh	% Withdrawal Consumed
Fossil fuels	200–300	0.45%–1.5%	300–700	70%–100%
Natural gas	100	0.5%–1.3%	180	78%
Nuclear	400	0.67%–1.6%	400–800	31%–90%

Source: Badr, L., G. Boardman, and J. Bigger, *Journal of Energy Engineering*, 138, 246, 2012.

water withdrawal is very low. The specific heat capacity of air ranges from 1.006 kJ/kg-K for dry air to approximately 1.06 kJ/kg-K for air at 100% humidity and 30°C, and is approximately four times lower than the specific heat capacity of water, a major drawback for the system as it causes lower efficiency cooling for similar flow rates of air. This lower cooling efficiency causes decreased efficiency of net electricity production, which can be further exacerbated because of the energy needed to power fans, the higher backpressures for dry cooling, and the dependence on ambient conditions, which for a plant in a hot, dry location can reduce annual electrical production by about 2% compared with a plant in a cool, humid climate [15]. Spray cooling can be used to help alleviate the capacity issues of dry cooling during hot periods. In the spray cooling method, a spray of water is injected to the cooling stream where it evaporates and aids cooling. Although scaling and corrosion of the heat exchanger tubes can be an issue with this method, careful correlation of ambient conditions can ensure minimal corrosion and water use [20]. The spray water requirements are stringent, and therefore the water must be treated to remove contaminants likely to cause scaling or precipitation fouling. Most water treatment involves using membrane-based filters with nanoscale openings to remove most foulants. A more complete description of the membrane technologies for water treatment is available in Chapter 27.

Several approaches are being considered to reduce the water footprint for thermoelectric power generation. Prohibiting the construction of new once-through cooling for thermoelectric plants might seem counterintuitive since plants with advanced cooling actually consume more water; however, because of shortage risks discussed above, this policy could bear fruit in the long term [18] by reducing overall withdrawal. Furthermore, it has been estimated that the United States has the potential energy capacity of 3×10^6 MW of wind and solar energy, which is approximately three times the 2008 installed electricity capacity. Greater implementation of solar photovoltaics and wind turbines are of benefit due to their low water consumption as discussed in Section 28.7 [18]. Additionally, minimization of water for thermoelectric cooling cycles is under investigation through projects between the National Energy Technology Laboratory (NETL) and Sandia National Laboratory (SNL), including using reclaimed water to cool power plants, which is used in 57 power plants in the United States. Currently, most of these plants utilize water from wastewater treatment plants and 46 of them use the water for cooling towers. The volume of reclaimed water used and the decade reclaimed water began being used by number of facilities is summarized in Tables 28.3 and 28.4 [21], respectively. Production of water by trapping flue gas vapor or using waste heat to desalinate water has also been suggested as a means of water procurement for these plants [18]. Desalination at present is

TABLE 28.3

Volume of Reclaimed Water Used by Number of Power
Generation Facilities

Volume of Reclaimed Water Used (MGD)	No. of Facilities
<0.1	6
0.1–0.5	10
0.51–1	12
1.01–5	15
5.01–10	6
>10	5

Source: Veil, J.A., *Use of Reclaimed Water for Power Plant Cooling.*
Argonne National Laboratory, Environmental Science
Division, 2007.

TABLE 28.4

Use of Reclaimed Water by Decade and Number of
Power Generation Facilities

Decade Use of Reclaimed Water Began	No. of Facilities
1960–1969	2
1970–1979	3
1980–1989	8
1990–1999	18
After 2000	19

Source: Veil, J.A., *Use of Reclaimed Water for Power Plant
Cooling.* Argonne National Laboratory, Environmental
Science Division, 2007.

largely dominated by membrane removal of salts, which inherently presents a nanotechnology use in water purification [22], and may present use for advanced technologies like
humidification–dehumidification cycles combined with membrane desalination for power
plant cooling water [22].

28.3 Water Use for Fuel Production

Water has been an essential commodity in fuel production for a long time. For example,
conventional oil production for transportation needs requires approximately 1.5 gal of
water for refining every gallon of fuel produced; processing of natural gas is, by comparison, more water intensive, with the requirement for water use at approximately 4 gal of
water per gallon of natural gas. However, with diversifying fuel sources and availability
of new technologies for fuel production, a closer look is needed at water use for this essential source of energy. Table 28.5 summarizes the main methods of fuel production and
the approximate use of water for each method of fuel production. It should be noted that
each method also generates wastewater, which needs treatment requiring consumption of
energy and materials.

TABLE 28.5

Summary of Biofuel and Conventional Fuel Water Consumption

Fuel	Gallons Water Consumed per Gallon Fuel Produced
Corn ethanol with irrigation[a]	1000
Soy ethanol with irrigation	650
Cellulosic ethanol with irrigation	2–6
Conventional oil refining	1.5
Natural gas refining	4

[a] Refining without irrigation requires 4–6 gal of water per gallon fuel produced.

28.3.1 Water Use for Bio-Fuels

In a 2009 USGAO report [13], it was discussed that the production process for biofuels relies heavily on water use. As shown in Figure 28.8 [23], there are several points of water use and consumption in the biofuel life cycle. Specifically, water is used in cultivation of feedstock, fermentation, distillation, and cooling/finishing processes for biofuel generation. Depending on the specific type of biofuel produced, water needs can vary significantly. Furthermore, cultivation and irrigation of the feedstock can also be carried out through the natural water cycle by using rain-fed crops to mitigate direct use of surface or groundwater resources [13]. In this case, crops can still affect the water cycle by converting an area's precipitation to crop evapotranspiration (Figure 28.1) instead of runoff or groundwater recharge [24]. In addition, the growth of biomass for biofuels requires arable land, which has only increased globally by 9% during the last 50 years. Significant arable land development is questionable as it competes with either human interests and growing population or areas of biodiversity and natural ecosystems (such as carbon storage in the rainforest) [25].

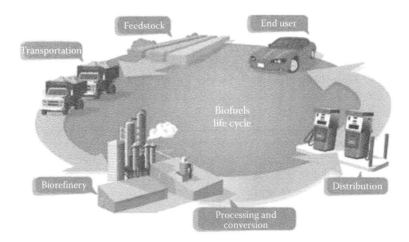

FIGURE 28.8
Essential steps for water consumption in the biofuel production cycle. The largest consumption occurs at the processing and refining stages. (Figure originally generated by USDOE and reproduced from USGAO. *Energy–Water Nexus: Coordinated Federal Approach Needed to Better Manage Energy and Water Trade-offs*. US Government Accountability Office, 2012.)

Table 28.5 shows that refining of ethanol from corn grain requires 4–6 gal of freshwater per gallon of ethanol. The requirement for irrigation per gallon increases water demand to nearly 1000 gal; the requirement for biodiesel from soy uses approximately 650 gal for cultivating the crops. By contrast, the water footprint of cellulosic ethanol is lower, ranging from 2 to 6 gal of water consumed per gallon of fuel for refining. Demands on water by liquefaction technologies are also challenging, including biomass to liquid, which consumes 2–6 gal of water for refining requirements per gallon of liquid fuel produced.

28.3.2 Shale Oil/Gas Mining and Extraction

Recent estimates by the USEIA [11] have shown the widespread availability of shale oil and gas resources across the United States (Figure 28.9a [26]), providing a broad resource for domestic hydrocarbon-based sources of energy. Advances in technologies such as hydraulic fracturing (HF) will enable enough domestic production that will likely provide a pathway to energy independence from foreign energy sources along with significant domestic economic growth. In fact, projections by the USEIA [11] show that shale gas will comprise >20% of the total US gas supply by 2020. Natural gas and oil are extracted from low-permeability coal beds and methane shale formations by coupling improved horizontal drilling methods with HF technologies. In this process, a starter hole is drilled to a known vertical depth, and then turned to drill horizontally through a target rock bed formation. The fracturing process consists of creating fissures or thin cracks with a detonation usually initiated by an electric discharge, followed by the injection of a complex mixture of water-based fluid under high pressure. The process of HF is also commonly referred to as "fracking." The cracked rock bed is an extended network of fissures that provide pathways for oil and gas to escape for modern societal energy use; Figure 28.9b [27] shows a schematic for the typical process as described here.

While the need for increased domestic US sources of energy and the implementation of available technologies is not in question, the greatest challenges lie in developing these energy sources with minimal health risks and possible ecosystem impacts associated with these extraction methods, in a resource-neutral or sustainable way. Horizontal HF techniques require large quantities of water (~3–5 million gallons per well during the drilling phase), sand, and chemicals for each job, and produce significant volumes of liquid and solid wastes that require treatment or disposal. In fact, 80%–99% of the HF fluid can be water with the remaining chemicals falling under the category of "proppants" [28,29]. The water-based slurry that is injected for HF is commonly referred to as a "frac fluid," the typical components of which are shown in Figure 28.10a. Furthermore, as there is no standard frac fluid composition, a variety of mixtures are used with varying water content due to the assorted drilling needs driven by the geology of the rock formations; however, nearly 20%–40% of the frac fluid returns and needs to be treated as flowback water; a sample composition of this flowback water is shown in Figure 28.10b. As with any large-scale industrial activity, drilling and HF also necessitate deployment of heavy equipment and new pipelines directly affecting local communities, local surface and groundwater supplies, and lands. It is therefore essential to develop these domestic energy sources in a responsible and sustainable manner with minimal impact on the ecosystem and ideally in a resource-neutral manner.

After fracking is completed, the fracking fluid is either disposed of in an underground injection well or treated, generally at a wastewater treatment plant or commercially. However, for wastewater treatment plants, the high TDS as well as heavy metal content

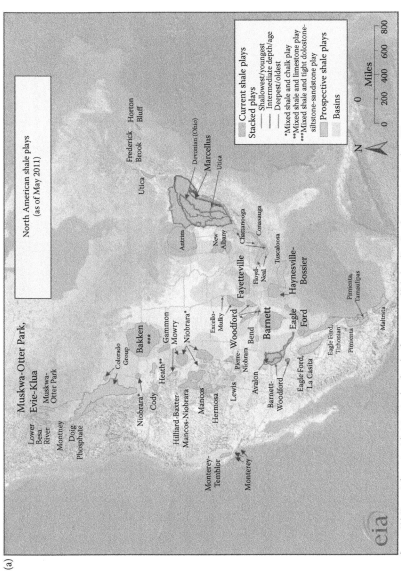

FIGURE 28.9

(See color insert.) (a) Map of the continental United States with Mexico and Canada showing distribution of shale oil and gas reserves across North America based on the data reported by USEIA in 2011. (From USEIA, *Review of Emerging Resources: US Shale Gas and Shale Oil Plays*. US Energy Information Administration, 2011.)

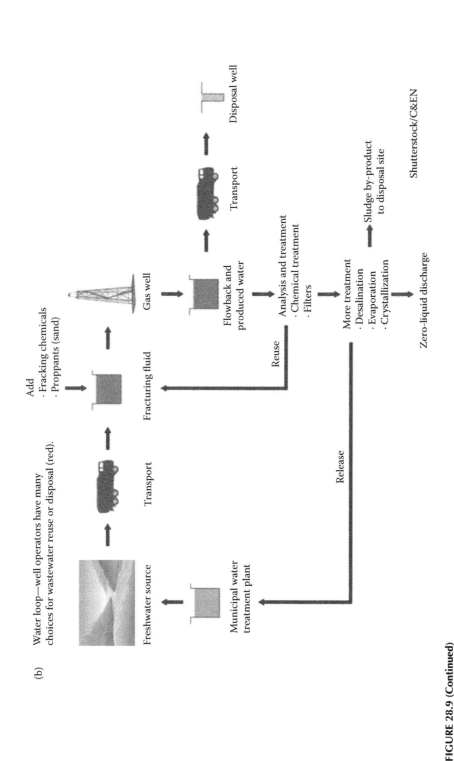

FIGURE 28.9 (Continued)

(See color insert.) (b) Typical HF cycle of water. Initially, water, proppants, and fracking chemicals are injected into the well until high enough pressures to fracture the rock bed occur. Some of the water returns (~20%–40%) to the surface, while the rest remains in the well. Of the flowback water, reuse for further fracturing or treatment, either at a municipal water treatment plant or zero-liquid discharge facility or disposed of by deep well injection, are the pathways for managing the waste stream. (From Bomgardner, M.M., Cleaner fracking, in *Chemical and Engineering News*, 2012.)

(a)		
Fracking recipe		
Example of fracturing fluid composition from a gas well in Beaver, PA		
Ingredient function	Chemical	Maximum ingredient concentration, % by mass
Carrier/base fluid	Freshwater	85.47795%
Proppant	Crystalline silica	12.66106%
Acid	Hydrochloric acid in water	1.29737%
Gelling agent	Petroleum distillate blend	0.14437%
	Polysaccharide blend	0.14437%
Cross-linker	Methanol	0.04811%
	Boric acid	0.01069%
Breaker	Sodium chloride	0.04252%
Friction reducer	Petroleum distillate, hydrotreated light	0.01499%
pH-adjusting agent	Potassium hydroxide	0.01268%
Scale inhibitor	Ethylene glycol	0.00540%
	Diethylene glycol	0.00077%
Iron control agent	Citric acid	0.00360%
Antibacterial agent	Glutaraldehyde	0.00200%
	Dimethyl benzyl ammonium chloride	0.00067%
Corrosion inhibitor	Methanol	0.00142%
	Propargyl alcohol	0.00010%

Note: Additional proprietary ingredients not listed in material safety data sheet: acid, alcohol, biocide, copolymer, disinfectant, enzyme, polymer, silica, solvent, surfactant, and weak acid. Source: FracFocus.

(b)			
Salty stuff			
The mix of fracking fluid and groundwater known as produced water contains wide variations in water chemistry			
	Shale formation		
Content (mg/l)	Barnett (TX)	Haynesville (AR, LA, TX)	Marcellus (NY, PA, WV)
TDS	40,000–185,000	40,000–205,000	45,000–185,000
Cl^-	25,000–110,000	20,000–105,000	25,000–105,000
Na^+	10,000–47,000	15,000–55,000	10,000–45,000
Ca^{2+}	2200–20,000	3100–34,000	5000–25,000
Sr^{2+}	350–3000	100–3000	500–3000
Mg^{2+}	200–3000	600–5200	500–3000
Ba^{2+}	30–500	100–2200	50–6000
Fe^{2+}/Fe^{3+}	22–100	80–350	20–200
SO_4^{2-}	15–200	100–400	10–400

TDS = total dissolved solids. Source: GE Power and Water.

FIGURE 28.10
(a) Chemical composition and (b) dissolved solids. Abundant presence of common salts makes TDS much greater than seawater. All traditional desalination methods will likely need to be updated to deal with high-TDS water as traditional membrane technologies, as discussed extensively in the main text. (From Bomgardner, M.M., Cleaner fracking, in *Chemical and Engineering News*, 2012.)

(e.g., barium or strontium) can create problems for these plants, and in Pennsylvania, many plants require that fracking wastewater not exceed 1% of the average daily value received by the plant. In addition, in May 2011, the Pennsylvania Environmental Protection Agency (EPA) requested that the fracking fluid deliveries to 15 public wastewater treatment plants be voluntarily discontinued. In 2011, regulators created average limits on TDS (500 mg/l), chloride, barium, and strontium in fracking fluid delivered to new wastewater treatment plants [30]. Commercial treatment options include crystallization (zero liquid discharge), thermal distillation/evaporation, electrodialysis, reverse osmosis (RO), ion exchange, and coagulation/flocculation and settling and/or filtration [30]. However, conventional methods for salt removal from seawater, such as RO, are inadequate to deal with frac fluids that are high in TDS including high chlorides (see Figure 28.10). It should be noted that initial water flowback from the well tends to have lower TDS as the water has been in contact with the well structure for a shorter time. Hence, as water continues to flow from the well, the salinity can increase significantly [29]. As shown in Figure 28.10, the TDS for frac fluids can reach >200,000 mg/l in contrast to 30,000–40,000 mg/l for seawater. Challenges associated with membrane fouling, concentration polarization, and degradation of RO membranes in high-chloride waters make use of most membrane technologies extremely challenging for treating frac fluids [29]. RO is generally considered uneconomical at TDS concentrations >40,000 mg/l; some success has been shown in membrane vibratory shear-enhanced processing, with lower (~18,450 mg/l) TDS waters but high (9000 mg/l) total suspended solids (TSS) [29,31]. In addition, the presence of metallic contaminants such as barium and strontium (which usually will exist in their ionic forms) exacerbates the scaling and formation of precipitated salts, adding another layer of complexity to successful treatment. Frac fluids also contain several chemicals, notably silica and a variety of gelling agents. Treatment of high-silica water remains a challenge for the water industry, and now with shale energy development this aspect is directly connected to the energy industry. Silica can form almost irreversible scale, which is a problem for water transport and piping. In water, silica comes in two forms, colloidal and reactive. Resins can be effective at removing reactive silica; however, colloidal silica contaminates the resins and causes them to be ineffective at removing the reactive silica. RO has been shown to remove approximately 80% of reactive silica and 99.8% of colloidal silica [32]. However, due to the challenges of membrane treatment, thermal distillation or crystallization has been shown to be very effective; however, these processes are energy intensive and cannot handle large volumes of water, necessitating holding tanks until the water can be treated [22]. The energy intensity of these processes are likely to be similar to other mechanical desalination processes, which require approximately 150–309 kJ/l in heat and electricity for pumping [22]. Processes such as combined ceramic microfiltration and ion exchange, membrane distillation, forward osmosis membranes, and high-efficiency RO are all being developed for the high demands of HF water treatment [33,34].

Nanotechnology also offers potential solutions to high-salinity water purification, and as the next generation of membranes continue development, treatment options for high-salinity brines such as HF flowback are likely to expand. Advanced nanostructured membranes are being constructed from a myriad of new materials, including carbon nanotube membranes, graphene, graphene oxide, zeolites, and boron nitride nanotubes [35–39]. These rapid permeation membranes separate by molecular sieving instead of by solution/diffusion that state-of-the-art commercial polymeric membranes (such as RO) use [36]. Carbon nanotube membranes have been shown to achieve water velocities approximately three times higher than predicted by Hagen–Poiseuille flow, with measured flow velocities ranging from 9.5 to 43.9 cm/s, which far exceeds the predicted 0.00015–0.00057 cm/s [40] due to slip flow within nanotubes. Nanotechnology also offers membrane-less separation devices such as those that employ concentration polarization to separate molecules and ions from water, and have been shown to remove 99% of salt when tested with seawater [41].

To reduce the costs associated with freshwater withdrawals and transportation costs, some producers have developed processes to reuse fracking fluid. Direct reuse, onsite treatment such as bag filtration, settling tanks, and mobile treatment systems have been used, while offsite treatment tends to focus on primary and secondary clarification, precipitation, and filtration. Major challenges include reducing TDS (which can reach 205,000 mg/l), calcium (which can reach 31,000 mg/l), and hardness (which can reach 55,000 mg/l as $CaCO_3$) of the flowback or produced water as they contribute to scale development in wells, which can cause reductions in well production [30]. In areas where TSS is low enough (e.g., in areas of the Marcellus shale, approximately 160 ppm) the produced water can be filtered to remove the suspended solids [42]. In order to place the TSS numbers in context, for seawater based RO plants TSS content usually varies between 1 and 15 ppm as a function of temperature (e.g., higher in summer) and geographical location. The produced water is also tested for calcium, magnesium, and other common ionic species to determine the correct amount of freshwater to blend back into the recovered water and reinject into the wells [42]. Given the lack of standards and available data for frac fluids, the approach driving the reblending for reuse of flowback water is "dilution is the solution."

28.3.3 Refining

Water resources are also consumed in the production and processing of many energy sources, including coal, natural gas, oil, uranium, and hydrogen, as summarized in Table 28.6. In the coal mining process, whether surface or underground, water is used for coal cutting, dust suppression, and coal washing. The final process is performed to remove some sulfur and increase the heat content by removing non-combustibles, and is performed on approximately 80% of the coal mined in the United States. Water consumption for mining ranges from 10 to 100 gal/ton (3.4–20.5 gal/MWh) of coal mined, while washing requires an additional 20–40 gal/ton (3.4–6.8 gal/MWh), totaling approximately 70–260 million gallons per day of water withdrawal. In addition, after water is used, treatment is required to remove coal sludge and other particulates [5]. Uranium mining and processing requires 27.3–47.8 gal/MWh, generating about 3–5 million gallons per day at a mine that has to treated for trace metals before release back into the water cycle [5]. Typical onshore oil extraction requires 5–13 gal of water per barrel of oil equivalent (boe) (2.7–7.5 gal/MWh), while enhanced recovery wells that inject water or steam to extract oil require between 81 and 14,000 gal/boe (47.8–8532 gal/MWh) [5]. In addition to water required for extraction, a typical refinery will withdraw 3–4 million gallons per day, returning only 30%–40% as wastewater while the rest is evaporated. The remaining water must be treated for residual petroleum products left from the refining process as well as increased TDS [5]. Approximately 95% of the hydrogen made in the United States is produced from steam reforming of natural gas and is used in ammonia production facilities, oil refineries, and methanol production plants. Hydrogen production by this

TABLE 28.6

Comparative Summary for Water Consumption toward Fuel Refining

Fuel	Gal Water/MWh
Coal mining and washing	6.8–27.3
Uranium mining	27.3–47.8
Onshore oil extraction	2.7–7.5
Enhanced oil recovery wells	47.8–8532.4
Hydrogen	146.8

method requires 4.9 gal of water for every kilogram of hydrogen produced (146.8 gal/MWh) [5]. In recent work [43], it has been shown that ammonia recovery can be achieved through an innovative, integrated waste management system including wastewater. If such technologies are implemented, they could significantly affect water withdrawal and consumption, perhaps leading to a change in the paradigm for wastewater being now "resource water" permitting extraction of valuable industrial chemicals and potential energy sources.

28.4 Energy Use in Wastewater Management

Effective wastewater management is essential for a number of reasons such as high biological oxygen demand, which harms aquatic systems if released untreated, odor, pathogen health hazards, heavy metal contamination, and high nitrogen and phosphorous levels, which can cause uncontrolled growth in aquatic systems. Population growth, limited land and energy resources, and increased concern about emerging contaminants such as pharmaceuticals as well as traditional pollutants such as heavy metals and pathogens are motivators for more specialized treatment [44]. Within local government entities in the United States, water and wastewater treatment is frequently one of the higher energy uses, and consumes 3%–4% the electricity used annually in the nation [45,46]. Figure 28.11 [47] displays an urban water cycle as well as energy usages of distribution and treatment. As discharge requirements become more stringent, population pressure increases, and the infrastructure ages, the energy cost of such treatment is likely to increase [45]. In addition, rising electricity rates and enhanced treatment of biosolids, such as drying or pelletizing, add additional costs to the treatment process [48]. The USEPA estimates that a 10% increase in efficiency, which can be achieved by infrastructure upgrades and implementation of energy efficient technologies, could save >10 billion kWh/year leading to $750 million per year in savings [48].

In a typical wastewater treatment plant, wastewater is treated in a series of steps, generally primary, secondary, and tertiary, as shown in Figure 28.12. The primary treatment consists of suspended solid removal by screens and sedimentation, which remove 50%–70% of the suspended solids. Secondary treatment involves coagulation and biological treatment [44]. For the solids, the final step is either further break down in a digester and disposal at a landfill, incineration, land applied, or skipping the digester completely and proceeding to one of the disposal routes stated above. The final treatment for the water usually includes disinfection and removal of residual suspended solids. Removal of excess nutrients such as phosphorus and nitrogen, which can cause eutrophication if released in excess, occurs in either the secondary or tertiary step, depending on the system, while disinfection is usually a tertiary step [49]. The water is then dispersed back into the water system by release into surface waters or percolation beds or other similar technology. The largest energy consumers in the wastewater treatment process are typically pumping and aeration. For a 10 MGD plant, pumping energy requirements range from approximately 0.6–1.2 million kWh/year based on total dynamic head of 30–60 ft (9.1–18.3 m), respectively, while aeration treatments require approximately 0.58–2.1 million kWh/year based on secondary treatment chosen [48].

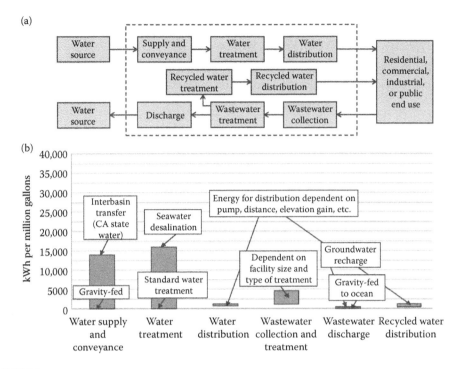

FIGURE 28.11
(a) Box-flow diagram from the work by Sanders and Webber based on a report from the California Energy Commission that shows a conceptual framework for water extraction and use in an urban environment. It should be noted that the discharge back to the source is not necessarily the same source or same quality as the initial intake. (b) Energy consumption for essential steps in the conceptual urban water cycle. It has been estimated that ~6% of US electricity consumption is for generating and conveyance of water. (From Sanders, K.T. and M.E. Webber, *Environmental Research Letters*, 7, 1, 2012.)

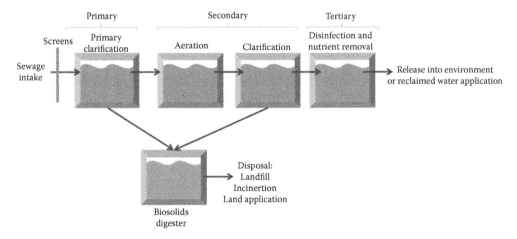

FIGURE 28.12
Typical sewage treatment process train with biosolids digester. Actual setup varies by specific wastewater treatment plant. (Based on information from Metcalf, L. and H.P. Eddy, *Wastewater Engineering: Treatment and Reuse*, G. Tchobanoglous, F.L. Burton, and H.D. Stensel, eds. Boston: McGraw Hill, 2003.)

28.5 Energy Recovery from Wastewater

Effective utilization of the energy and nutrients removed from wastewater during the treatment process is becoming more important as the need for raw materials and energy costs increase. Wastewater treatment plant biosolids have attractive energy content; the calorific value of the dry sludge is almost equal to that of brown coal [50]. In the United States, this means that the energy content in wastewater is approximately 2% of the annual electricity demand. Nitrogen and phosphorus are both valuable nutrients used in fertilizers but harmful if released into the environment untreated, as they lead to overgrowth in aquatic systems. The demand for nitrogen for fertilizer is on the order of 121 million ton per year for synthetic fertilizers made with the Haber–Bosch process, which utilizes the almost unlimited atmospheric nitrogen supply [51]. Phosphorus, however, is derived from mineral phosphate rock, the supply of which is projected to become exhausted in the next 50–100 years [51]. Ammonia is a source of nitrogen and can be harvested from wastewater. In a recent study [43], it was shown that in addition to anaerobic digestion for methane production, the additional extraction of ammonia can add an energy benefit when the system is integrated systematically with energy-efficient operation. For high total ammonia nitrogen content [TAN] of 10,000 mg/l, it was shown that with energy capture efficiency of 35% for the hydrostatic pressure and heat generated by the chemical reactions as well as utilization of the nitrogen/hydrogen product stream as the stripping gas, 26.3% more energy could be produced with the duel fuel capture (with subsequent processing of ammonia to generate hydrogen) than with methane alone. In fact, for [TAN] concentrations higher than approximately 2000 mg/l, the system received a positive benefit from methane and hydrogen production over the methane-only system. By using innovative modular system integration, it was shown [43] that it is also possible to harvest ammonia from waste streams without need for further refining to fuel and obtain a valuable industrial chemical resource.

Sewage sludge can be incinerated, pyrolyzed, gasified, or digested to produce fuel, or a combination of those methods [44]. The USEPA estimates that annual sewage sludge generation in the United States exceeds 8 million tons dry weight, with 41% of the sludge land applied, 22% incinerated, 17% landfilled, and the remaining either put through an advanced treatment process or other alternative beneficial use [52]. The USEPA has identified combined heat and power (CHP) applications involving anaerobic digesters for biogas production (mainly methane) as a reliable and beneficial addition to many wastewater treatment plants, providing the site with energy and heat independent of the electricity grid and creating higher efficiencies by avoiding energy transmission and distribution. As of 2011, CHP was feasible at 1351 wastewater treatment plants nationwide and represented an economic payback in 7 years or fewer for between 252 and 662 of these sites and an energy potential between 178 and 260 MW [53].

28.6 Energy Recovery from Water Purification Processes

In areas where excess pressure occurs in water distribution lines, which is often a result of gravity as many systems are positioned to take advantage of available hydrostatic pressure, pressure reduction valves are often used to ensure safe pressures at water taps.

Commercially available systems using water turbines can transfer the excess pressure to electricity, and current systems require a flow rate of at least 1–1.5 MGD with a pressure differential of 30–35 psi. This system delivers approximately 40 kW/gal/min [54].

Another opportunity to generate electricity while treating water can be realized through the use of multichamber microbial fuel cells to desalinate water, treat waste-water, and generate electricity. These systems work similarly to conventional microbial fuel cells; bacteria in the anode chamber is used to oxidize substrates while generating electron flow that causes reduction in the cathode chamber; the saline water in the middle is desalinated as ions are drawn out from each of the adjoining chambers [55]. Still under development, this scheme has been shown to achieve salt removals of up to 94% for 20 g/l stock solutions or create a power density of up to 931 mW/m^2 [56,57]. Nanotechnology plays a key role here in the use of high-efficiency electrode materials and development of proton exchange membranes (PEMs) for rapid transport faster than conventional systems comprising PEMs like Nafion. In a recent study [58], functionalized porous silicon membranes were used as the membrane to demonstrate robust H$_2$ fuel cell operation. Extension of such technologies to microbial fuel cells can significantly assist in increasing energy density for these systems.

One drawback of microbial fuel cell desalination is that both the anode and cathode solutions gain salinity during the process. An alternative, the microbial capacitive desalination cell, uses three-chamber microbial fuel cells relying on capacitive desalination. Here, the cell sections are separated by both ion-selective membranes and active carbon capture cloth; the first allows only the correctly charged ions to pass and the ion adsorbs to the second. These cells, with high-surface-area electrodes, can enhance desalination efficiency by 7–25× over conventional capacitive deionization methods [41,55,59].

28.7 Alternate Energy Sources

Alternate energy sources such as solar, wind, and geothermal are commonly assumed to use less water than traditional thermoelectric power generation; however, water needs are very process specific and a wide range of consumptions are realized with these power sources. Solar thermoelectric power uses parabolic troughs to concentrate and transfer heat to a fluid that creates steam in an unfired boiler. Much like fossil fuel thermoelectric power plants, water is used to cool these systems at a consumption rate of 770–920 gal/MWh. Solar power towers, which operate at higher temperatures but similar method, consume approximately 750 gal/MWh [5]. Hydropower, the largest section of renewable energy produced in the United States at approximately 7%, has a fast response time to demand, making it a valuable resource for the grid. However, the storage dams for these systems create evaporation on the order of 4500 gal/MWh from reservoirs [5].

Technologies such as solar photovoltaics and wind do not require water during operation; thus, water consumption is limited to production and maintenance such as washing. The long life cycles of these devices make water consumption over the lifetime of the device relatively small. The main drawback to these devices is the intermittent nature of the power harvested, low efficiencies, and large land requirements (72 km^2/TWh for wind and 28–64 km^2/TWh for solar; non-dual-purpose land use). Land requirements can be somewhat mitigated by installation of photovoltaics on building roofs or use of wind power sites for agriculture [60].

28.8 Summary

The inherently interconnected ability to effectively treat and distribute water as well as generate clean energy presents a complex problem. In this chapter, a mini-review of water withdrawal and consumption for various energy needs was presented. Challenges with reducing water demands were discussed with potential technology solutions. Ideally, technology development, focused on creating new solutions or fostering efficiency in production and distribution of water and energy, as well as factually driven policy development will be essential to creating viable and sustainable solutions.

References

1. USGS, *Summary of the Water Cycle*. 2013 (cited February 18, 2013); Available at http://ga.water.usgs.gov/edu/watercyclesummary.html.
2. Scanlon, B.R., I. Jolly, M. Sohpocleous, and L. Zhang, Global impacts of conversions from natural to agricultural ecosystems on water resources: Quantity versus quality. *Water Resources Research*, 2007. **43**: pp. 1–18.
3. Naiman, R.J., and M.G. Turner, A future perspective on North America's freshwater ecosystems. *Ecological Applications*, 2000. **10**: pp. 958–970.
4. Shannon, M.A., P.W. Bohn, M. Elimelech, J.G. Georgiadis, B.J. Marinas, and A.M. Mayes, Science and technology for water purification in the coming decades. *Nature*, 2008. **452**: pp. 301–310.
5. USDOE, *Energy Demands on Water Resources*. 2006, U.S. Department of Energy: Washington, DC.
6. Vinson, D.S., H.G. Schwartz, G.S. Dwyer, and A. Vengosh, Evaluating salinity sources of groundwater and implications for sustainable reverse osmosis desalination in coastal North Carolina, USA. *Hydrogeology Journal*, 2011. **19**: pp. 981–994.
7. NETL, *Department of Energy/National Energy Technology Laboratory's Water-Energy Interface Research Program: December 2010 Update*, J.P. Ciferno, R.K. Munson, and J.T. Murphy, eds. 2010, National Energy Technology Laboratory: West Virginia.
8. EPRI, US electricity consumption for water supply and treatment—The next half century, in *Water and Sustainability*. 2002, Electric Power Research Institute: Palo Alto, CA.
9. NREL, *Consumptive Water Use for U.S. Power Production*, P. Torcellini, N. Long, and R. Judkoff, eds. 2003, National Renewable Energy Laboratory: Colorado.
10. USBR, *Electrodialysis (ED) and Electrodialysis Reversal (EDR)*. 2010, US Department of the Interior Bureau of Reclamation: Washington, DC.
11. USEIA, *Annual Energy Outlook 2010 with Projections to 2035*. 2010, US Energy Information Administration: Washington, DC.
12. NETL, *Innovations for Existing Plants: Water Energy Interface*. 2010 (cited February 18, 2013); Available at http://www.netl.doe.gov/technologies/coalpower/ewr/water/index.html#top.
13. USGAO, *Energy–Water Nexus: Many Uncertainties Remain about National and Regional Effects of Increased Biofuel Production on Water Resources*. 2009, US Government Accountability Office: Washington, DC.
14. Herath, I., M. Deurer, D. Horne, R. Singh, and B. Clotheir, The water footprint of hydroelectricity: A methodological comparison from a case study in New Zealand. *Journal of Cleaner Production*, 2011. **19**: pp. 1582–1589.
15. Badr, L., G. Boardman, and J. Bigger, Review of water use in US thermoelectric power plants. *Journal of Energy Engineering*, 2012. **138**: pp. 246–257.
16. USGAO, *Energy-Water Nexus: Improvements to Federal Water Use Data Would Increase Understanding of Trends in Power Plant Water Use*. 2009, US Government Accountability Office: Washington, DC.

17. Vliet, M.T.H.V., J.R. Yearsley, F. Ludwig, S. Vögele, D.P. Lettenmaier, and P. Kabat, Vulnerability of US and European electricity supply to climate change. *Nature Climate Change*, 2012. **2**: pp. 676–681.
18. Sovacool, B.K., and K.E. Sovacool, Identifying future electricity-water tradeoffs in the United States. *Energy Policy*, 2009. **37**: pp. 2763–2773.
19. Garimella, S., A.M. Brown, and A.K. Nagavarapu, Waste heat driven absorption/vapor-compression cascade refrigeration system for megawatt scale, high-flux, low-temperature cooling. *International Journal of Refrigeration*, 2011. **34**: pp. 1776–1785.
20. EPRI, *Spray Enchancement of Air Cooled Condensers*, M. Masri, and R. Therkelsen, eds. 2003, Electric Power Research Institute.
21. Veil, J.A., *Use of Reclaimed Water for Power Plant Cooling*. 2007, Argonne National Laboratory, Environmental Science Division.
22. Prakash, S., K. Bellman, and M.A. Shannon, Recent advances in water desalination through biotechnology and nanotechnology, in *Bionanotechnology II: Global Prospects*, D. Reisner, ed. 2011, CRC: Boca Raton, FL. pp. 365–382.
23. USGAO, *Energy-Water Nexus: Coordinated Federal Approach Needed to Better Manage Energy and Water Trade-Offs*. 2012, US Government Accountability Office: Washington, DC.
24. Fingerman, K.R., G. Berndes, S. Orr, B.D. Richter, and P. Vugteveen, Impact assessment at the bioenergy-water nexus. *Biofuels Bioproducts and Biorefining*, 2011. **5**: pp. 375–386.
25. Godfray, H.C.J., J.R. Beddington, I.R. Crute, L. Haddad, D. Lawrence, J.F. Muir, J. Pretty, S. Robinson, S.M. Thomas, and C. Toulmin, Food security: The challenge of feeding 9 billion people. *Science*, 2010. **327**: pp. 812–818.
26. USEIA, *Review of Emerging Resources: US Shale Gas and Shale Oil Plays*. 2011, US Energy Information Administration: Washington, DC.
27. Bomgardner, M.M., Cleaner fracking, in *Chemical and Engineering News*. 2012.
28. USEPA, *Hydraulic Fracturing Research Study*. 2010, US Environmental Protection Agency: Washington, DC.
29. Gregory, K.B., R.D. Vidic, and D.A. Dzombak, Water management challenges associated with the production of shale gas by hydraulic fracturing. *Elements*, 2011. **7**: pp. 181–186.
30. USEPA, *Study of the Potential Impacts of Hydraulic Fracturing on Drinking Water Resources*. 2012, US Environmental Protection Agency: Washington, DC.
31. Ahmadun, R.L.-R., A. Pendashteh, L.C. Abdullah, D.R.A. Biak, S.S. Madeni, and Z.Z. Abidin, Review of technologies for oil and gas produced water treatment. *Journal of Hazardous Materials*, 2009. **170**: pp. 530–551.
32. White, M.J., J.L. Masbate, and S.G. Gare, *Reverse Osmosis Pre Treatment of High Silica Waters*. 2010, GE Power and Water: New York.
33. Brant, J.A., *Oil and Gas Produced Water Treatment Technologies*. 2011–2012, The Nexus Group, RTSEA Partnership.
34. Jiang, Q., J. Rentschler, R. Perrone, and K. Liu, Application of ceramic membrane and ion-exchange for the treatment of the flowback water from Marcellus shale gas production. *Journal of Membrane Science*, 2013. **431**: pp. 55–61.
35. Paul, D.R., Creating new types of carbon-based membranes. *Science*, 2012. **335**: pp. 413–414.
36. Wang, E.N., and R. Karnik, Graphene cleans up water. *Nature Nanotechnology*, 2012. **7**: pp. 552–554.
37. Hilder, T.A., D. Gordon, and S.-H. Chung, Salt rejection and water transport through boron nitride nanotubes. *Small*, 2009. **5**: pp. 2183–2190.
38. Nair, R.R., H.A. Wu, P.N. Jayaram, I.V. Grigorieva, and A.K. Geim, Unimpeded permeation of water through helium-leak-tight graphene-based membranes. *Science*, 2012. **335**: pp. 442–444.
39. Karan, S., S. Samitsu, X. Peng, K. Kurashima, and I. Ichinose, Ultrafast viscous permeation of organic solvents through diamond-like carbon nanosheets. *Science*, 2012. **2012**: pp. 335.
40. Majumder, M., N. Chopra, R. Andrews, and B.J. Hinds, Experimental observation of enhanced liquid flow through aligned carbon nanotube membranes. *Nature*, 2005. **438**: pp. 44–45.
41. Kim, S.J., S.H. Ko, K.H. Kang, and J. Han, Direct seawater desalination by ion concentration polarization. *Nature Nanotechnology*, 2010. **5**: pp. 297–301.

42. Mantell, M.E., *Produced water reuse and recycling challenges and opportunities across major shale plays.* Cheseapeake energy presentation, 2011.

43. Babson, D., K. Bellman, S. Prakash, and D.E. Fennell, Anaerobic digestion for methane generation and ammonia reforming for hydrogen production: A thermodynamic energy balance of a model system to demonstrate net energy feasibility. *Biomass and Bioenergy,* 2013. **56**: pp. 493–505.

44. Werther, J., and T. Ogada, Sewage sludge combustion. *Progress in Energy and Combustion Sciences,* 1999. **25**: pp. 55–116.

45. Stillwell, A.S., D.C. Hoppock, and M.E. Webber, Energy recovery from wastewater treatment plants in the United States: A case study of the energy-water nexus. *Sustainability,* 2010. **2**: pp. 945–962.

46. USGAO, *Energy-Water Nexus: Amount of Energy Needed to Supply, Use, and Treat Water is Location-Specific and Can Be Reduced by Certain Technologies and Approaches.* 2011, US Government Accountability Office: Washington, DC.

47. Sanders, K.T., and M.E. Webber, Evaluating the energy consumed for water use in the United States. *Environmental Research Letters,* 2012. **7**: pp. 1–11.

48. USEPA, *Evaluation of Energy Conservation Measures for Wastewater Treatment Facilities.* 2010, US Environmental Protection Agency: Washington, DC.

49. Tchobanoglous G., F.L. Burton, and H.D. Stensel, eds. *Wastewater Engineering: Treatment and Reuse,* 2003, McGraw Hill: Boston, MA.

50. Fytili, D., and A. Zabaniotou, Utilization of sewage sludge in EU application of old and new methods—A review. *Renewable and Sustainable Energy Reviews,* 2008. **12**: pp. 116–140.

51. Verstraete, W., and S.E. Vlaeminck, ZeroWasteWater: Short-cycling of wastewater resources for sustainable cities of the future. *International Journal of Sustainable Development and World Ecology,* 2011. **18**: pp. 253–264.

52. USEPA, *Emerging Technologies for Biosolids Management.* 2006, US Environmental Protection Agency: Washington, DC.

53. USEPA, *Opportunities for Combined Heat and Power at Wastewater Treatment Facilities: Market Analysis and Lessons from the Field.* 2011, US Environmental Protection Agency: Washington, DC.

54. Rentricity, *City of Keene Water Treatment Facility.* Rentricity: New Hampshire.

55. Forrestal, C., P. Xu, and Z. Ren, Sustainable desalination using a microbial capacitive desalination cell. *Energy and Environmental Science,* 2012. **5**: pp. 7161–7167.

56. Cao, X., X. Huang, P. Liang, K. Xiao, Y. Zhou, and B.E. Logan, A new method for water desalination using microbial desalination cells. *Environmental Science Technology,* 2009. **43**: pp. 7148–7152.

57. Qu, Y., Y. Feng, X. Wang, J. Liu, J. Lv, W. He, and B.E. Logan, Simultaneous water desalination and electricity generation in a microbial desalination cell with electrolyte recirculation for pH control. *Bioresource Technology,* 2012. **106**: pp. 89–94.

58. Moghaddam, S., E. Pengwang, Y.-B. Jiang, A.R. Garci, D.J. Burnett, C.J. Brinker, R.I. Masel, and M.A. Shannon, An inorganic-organic proton exchange membrane for fuel cells with a controlled nanoscale pore structure. *Nature Nanotechnology,* 2010. **5**: pp. 230–236.

59. Shannon, M.A., Water desalination: Fresh for less. *Nature Nanotechnology,* 2010. **5**: pp. 248–250.

60. Evans, A., V. Strezov, and T.J. Evans, Assessment of sustainability indicators for renewable energy technologies. *Renewable and Sustainable Energy Reviews,* 2009. **13**: pp. 1082–1088.

29

Nanotechnology in Passive Atmospheric Water Capture

Carlos Ángel Sánchez Recio,[1] Zaki Ahmad,[2,3] and Tony F. Diego[3]

[1]*Natural Aqua Canarias, Vistabella (La Laguna), Canary Islands, Spain*

[2]*Department of Chemical Engineering, COMSATS Institute of Information Technology (CIIT), Lahore, Pakistan*

[3]*42TEK S.L., Almazora, Spain*

CONTENTS

29.1 Introduction

The amount of renewable water in the earth's atmosphere is estimated to be roughly 12,500–12,900 km^3, or about 0.001% of the total water on Earth, consisting of a volume of approximately 1385 million km^3. It is further estimated that 2%–3% of the total water used today as drinking water is actually derived from seawater produced by large desalination plants and may cost slightly less than \$1/m^3, or about \$4/m^3 for smaller units. Bottled water can be 1000× more expensive. At these current costs of producing desalinated water through reverse osmosis, the atmospheric water capture (AWC) technique is still viewed as the most cost-effective method for many cities and townships where water availability is continuing to be a constraint. With the improvements outlined in this chapter, the case for the implementation of *hydrofarms* is clear, perhaps by more than two or three orders of magnitude.

The field of AWC for the purpose of human and animal consumption and for agricultural irrigation use is now set to be revolutionized, and is fueled by the rapidly increasing demands for water. It leverages the latest advancements and developments in the fields of material sciences and nanotechnology, more specifically the applications of superhydrophilic wonder materials such as TiO$_2$, which have the additional properties of acting as a photocatalytic decontaminant when properly applied to water treatment. These, combined with the opposite range of materials such as SiO$_2$ with superhydrophobic properties, can further optimize the capturing yields significantly.

This chapter describes the various technologies involved in AWC and focuses on the pursuit of a specialized mesh with a suitable outer layer structure that has been furnished with techniques and coatings containing both superhydrophilic and superhydrophobic capabilities, in order to achieve approximate optimization of AWC as close as possible to 100% efficient. The recent advances in nanotechnology have shown the way to create a hierarchical superhydrophobic nanostructured surface achieving water contact angles (WCA) >90° (a standardized method of determining the degree of hydrophobicity). These techniques hold great promise for areas where water is scarce, and can be furnished readily, including in those areas with a high water demand and even in those that are limited by the scarcity of clouds, or in areas where relatively low fog density preclude their use. This optimization increases the current footprint of areas that can now be made to capture atmospheric water, not only to be suitable for irrigation or cattle grazing but also for actually sustaining entire communities even in urban areas close to coastal regions and deserts.

By utilizing both superhydrophilic and superhydrophobic surfaces on the capturing substrates, a renewed interest in providing for areas of water scarcity has been achieved. Water from fog is currently harnessed at heights of at least 400 m above sea level. Below this altitude, traces of salt may still be present in the captured water.

AWC has gone beyond our own planet, as new AWC viability projects are currently being conducted by the National Aeronautics and Space Administration, including planned AWC systems such as that in the planned Mars Base Station, which intend to rely on zeolites for water capture [1].

Several types of commercial AWC systems exist today, and we differentiate between the two main ones: active and passive systems. Active systems, including state-of-the-art atmospheric water generator (AWG), are essentially dehumidifiers requiring some form of electrical power to drive a compressor used to extract moisture from the air and convert it to pure water. These are compact and can be extremely efficient. The need for electricity from the utilities networks has been mitigated by the application of efficient solar cells or wind turbines, allowing these facilities to function just about anywhere. The costs ultimately depend on the price of electricity needed to run them, including infrastructure maintenance and repair of such active systems requiring storage of energy such as batteries, which themselves are being rapidly improved by nanomaterials such as graphene.

On the other hand, passive atmospheric fog collector (PAFC) technology relies instead on wind power from well-known and established ocean weather patterns to harness water from the atmosphere without the use of additional power sources. These units began as specialized forms of fog fences and are used in coastal areas where inland winds bring fog, at high altitude areas from 400 to 1200 m, since below these altitudes traces of salt may render the captured water unusable.

Ultimately, these are meant to be solutions as low-cost units for water-scarce nations but are efficient enough to sustain reliable sources of water even on highly populated or developed megacities such as Lima, Peru. According to a 1995 study by the Ottawa International Development Research Center, the following countries are excellent sites for the implementation of this technology: Nepal, Eritrea, Yemen, Oman, and Kenya, including the entire Atlantic coast of Africa: Angola, Senegal, Congo and DR Congo, Namibia, South Africa, and Cape Verde; the entire east coast of the South Pacific, including Mexico, Chile, Bolivia, Peru, Ecuador, and Colombia; and in Asia: China, Pakistan, Mongolia, India, Bangladesh, and Sri Lanka. Gulf countries such as Oman, Saudi Arabia, and Yemen can readily afford this technology, and are eager to find economic and reliable alternatives to desalinization plants.

PAFCs can be small units for a single residence 2–5 m², or can be very large arrays with hundreds of square meters of surface area suitable for entire communities. For example, a typical unit of 1600 m² of mesh produces an average of 12,000 liters (12 m³) of water a day or roughly 7.5 L/m²/day, enough to sustain communities of 10,000 inhabitants or more. Structures of PAFCs suitably treated have been able to achieve maximum outputs in ideal conditions of 42 L/m²/day, and nearly a 6-fold increase. The variables affecting the efficiency of harvesting are increased with the larger fog droplet size, during higher wind speeds, and with narrower collection fibers/mesh widths. An analysis of wind speed, wind direction, temperature, relative humidity, dew point, barometric pressure, solar radiation, precipitation, and such atmospheric measurements will ultimately determine the feasibility of PAFC use by accurately predicting the resulting potential amount of water captured.

Today's most widely used material for atmospheric capture is polypropylene raffia, also called "raschel," which is a material classified under a class of mosquito netting fabric types (manufactured using Raschel knitting machines). This raschel can be coated with different nanomaterial surfaces by using sol–gel techniques, which can either be furnished by air gun, or controlled by immersion and subsequent forced drying.

The polypropylene itself is ultraviolet protected to avoid breakdown, allowing further efficiency increases with smaller mesh sizes and fiber widths. The application of nanotechnology-coated screens for these new systems by using treated mesh showed marked improvements in yields by nearly 30% in comparison to untreated screens [2].

With the results obtained by other researchers and the availability of optimal weather condition sites in the field, and with the combination of variable climate hosted in wind tunnel chambers found in specialized laboratories, substantial progress toward optimization of yields has been achieved. Applied research characterized by dedicated AWC harvesting methods has provided surprisingly rewarding results [3].

This research is not only being applied to raschel but also to a host of other substrates that are better able to withstand harsh environmental conditions such as wind or snow, in order to collect mist or fog and any form of water in the atmosphere, including haze, drizzle rain, dew and even snow itself.

The careful consideration of the design of the materials used and the precise placement of these fog fences, combined with their architecture, depend on the local weather conditions. Maximizing the vertical capture as well as the horizontal capture is a key consideration taken before their implementation (Figure 29.1).

Atmospheric water capture (AWC), as a general term for this technology, was first coined and adopted to distinguish it from the existing classifications for other types of water collection methods, including surface, underground, continental, etc.

By limiting the field to the water present in the atmosphere but has not yet precipitated to the ground, great strides have been made and are now set for an important evolution.

Despite ongoing interest by dedicated researchers, the new and rapid advancements in sail designs, including wind-driven turbines, and the latest developments in computer modeling methods of the atmospheric conditions or fluid dynamics coupled with and optimized by material science in nanostructured coatings, improving such systems has yet to be widely implemented. The purpose of this chapter is to place this new technology into the hands of society, an especially important issue to those areas that suffer most from the lack of geographic and economical incentives.

This refocused effort on materials and designs of structure has created a stronger case of economics for atmospheric water harvesting. While we are still a considerable way from applying the results developed to date, we have, in fact, opened up new possibilities on obtaining the most water yields from the atmosphere with architectures that can be economically built in order to provide water where it would not be economically feasible in

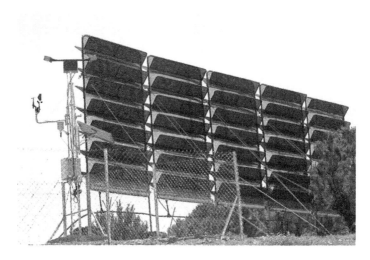

FIGURE 29.1
(See color insert.) AWC station with real-time weather and capture.

any other way. A world of design ideas and tested experimental materials and methods are now available to all.

Several projects are currently under way worldwide to optimize the methods and techniques, including two noteworthy ones. In the University of Washington (Seattle, WA), the SQWater project ("SQ" refers to squatter communities, as the original focus was on slums; of course, "SQWater" is a homonym) by Benjamin Spencer and Susan Bolton (cfpub.epa.gov /ncer_abstracts/index.cfm/fuseaction/display.abstractDetail/abstract/9866/report/0), is taking place, where a pioneering acoustic fog precipitation as a means of increasing fog collection yields will be explored in conjunction with the Universidad Nacional Mayor de San Marcos (UNMSM, Lima) and a northern urban area in Lima (cfpub.epa.gov/ncer_abstracts/index .cfm/fuseaction/display.abstractDetail/abstract/9866/report/0).

Our project is described in this chapter and led by the authors (Recio and Ahmad), both experts in water capture systems with considerable background experience, and working with the 42TEK Company, and in a joint project study in conjunction with AITEX (Instituto Tecnológico Textil), AIDICO (Instituto Tecnológico de la Construcción), and AIMME (Instituto Tecnológico Metalmecánico), three of the finest cutting-edge technical institutes dedicated to nanotechnology applications in the textile, metal, and construction industries, respectively, in Europe.

The results of these new PAFCs contemplate heterogeneous photocatalysis [4] for decontamination, renewed sail designs, new stronger substrate materials replacing nylon raschel, and will include systems capable of providing pure water by the suitable application of nanofiltration products used on these suitably coated adaptive substrates, reducing the 400 m barrier currently limiting their suitable deployments [5–7].

29.2 History of PAFCs

It is not well known when it was first discovered that it was possible to extract water from fog or sea mist even in extreme conditions such as those found in the arid Atacama Desert,

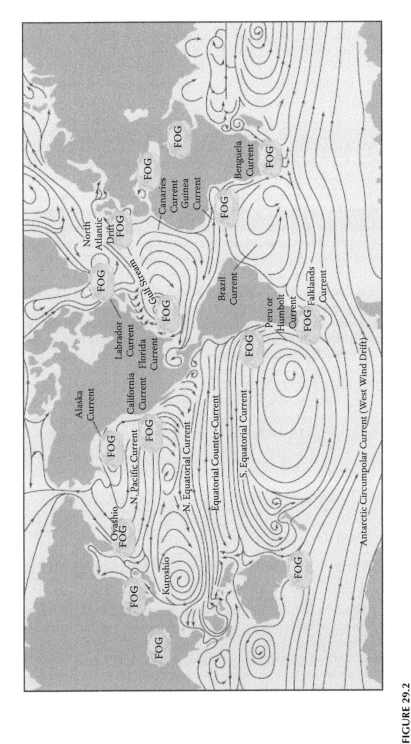

FIGURE 29.2
World map showing the main continents and oceans, and the most common locations favoring fog occurrence.

the arid Middle Eastern Desert, or the Gobi deserts; however, this technology could well have been born more than 2000 years ago. Harnessing fog from PAFCs has been historically one of the most ancient methods utilized by humanity and societies in arid areas. It is a technique that has been improved over the ages and developed by many researchers and cited throughout the years, all of whom have experimented with various types of screens and architectures, and have achieved varying results [8]. It has been a success as a low-tech solution by combining some form of mesh material strung tightly on poles, and supported by gutters to collect droplets that are fed into pipes and stored in tanks, and can be readily thought of as a fog fence. Today, it continues to be largely unknown by the general populace.

Parallel early developments in the Middle East and in South America occurred roughly simultaneously. South American legend has it that the pre-Columbian-culture people from the area of what is now known as Chile, Peru, and Bolivia, who tended to their llama herds, were among the earliest pioneers [8]. These people essentially discovered a way to use the blankets made from the wool from their very own flocks by placing these in an upright and perpendicular position facing the wind direction. Probably while seeking shelter, they found that the mist through their garments would soak them completely. It was not long before they found that the water condensed upon their blankets could be used to capture dew, and when the droplets reached a sufficient size, these would trickle down to the ground. To harness this water, they would simply place a channel and a receiver beneath them. They were thus able to store enough water to meet their own and their flock's needs, in the *altiplano* (high plains) regions, in two of the driest deserts on the planet where precipitation seldom, if ever, occurs. The name, in Spanish, for this technique of capturing water from mist is the aboriginal word *camanchaca*, a word that is also used for a very common type of fog on the Pacific coasts of South America today. The fog banks come in regularly from the Pacific Ocean breezes from the Humboldt Current and climb up the pre-Andean foothills and slopes, traveling many miles into the South American continent.

Similarly and roughly about the same time in the Middle East, air wells from caves were well known to have been tapped for harvesting of water. In the Middle Ages, dewponds in Europe were also used to collect water, especially in times of plagues or droughts (Figure 29.2).

29.3 Background and Technical Developments

Nature itself has provided us with its own multistructured and multifunctional integrated fog collection systems, as seen in a variety of plants such as the cactus which depend entirely on its own capability for fog collection. The gradient of the Laplace pressure, surface energy, and multifunction integration assist the cactus with fog collection [9–13]. Another case in point from nature is that of the Namib beetle (*Stenocara gracilipes*), which is capable of leveraging and enhancing its ability to catch water. When a moist sea breeze makes contact with the hydrophobic region of its shell, a small chemical imbalance occurs, which leads to the formation of droplets that, upon reaching sufficient size, are sent through the guttulae that extend to the mouth of the insect where they are later absorbed.

These first successful and humble attempts in capturing water from dew caught the attention of several researchers, including Boussingault in 1844, who is credited with first experimenting with various types of mist collection equipment and contrasting them. Then others, including Carlos Espinosa, German Saa, and Humberto Fuenzalida in 1959, began to do the same [14]. Notable experiments were also done by Andres Acosta Baladón [15], during 1960–1964 in the Canary and Cape Verde Islands, and more recently the Canadian–Chilean joint team of Pilar Cereceda and Robert Schemenauer, which led to the installation of a permanent station at Chungungo, Chile [16–21]. Among the many researchers who have cared about harnessing this source of drinking include a night watchman, Daniel Beysens, who carried out experiments in Croatia and Morocco, Alain Gioda of France, and Luis Santana Perez with deployments in India of a bold new system for agriculture based on these methods [22].

In 1998, the author (Recio) began work on a new project in Tenerife focused on the new design of architectural structures with cutting-edge materials but low-cost equipment [21]. The existing configuration for large fog collectors (LFCs) was unable to avoid its broad side from being a mere vertical mesh barrier against strong wind forces, often suffering damage and is incapable of doing much about the vertical precipitation or capture. He was able to improve the earlier designs because although they were already efficient in low-wind, arid areas, the same design would not operate in the Canary Islands or anywhere where strong winds could destroy them, a serious consideration to be reconciled [23–25].

Through a careful analysis of existing LFCs, he proposed, designed, and successfully constructed a new set of PAFCs based on a system design that worked well on arid, mild conditions as the starting point, and found a suitable design formula that reduces the negative impact of the high wind velocities and at the same time avoids phenomena such as vertical runoff and evaporation effects [26,27]. Early in 2000, Recio set about and developed the commercially available Aquair Optimizer® fog capturing systems through the Natural Aqua S.L. company, for the sole purpose of continuing the research and development, and maintenance and upkeep of the DYSDERA (Design and Monitoring Stations Water Gathering) project. The DYSDERA project was the first fully scaled and monitored network of PAFCs, and its work was directly focused toward the optimization of a matrix network capture and the storage and distribution of atmospheric water effectively used for all sorts of activities in rural areas, from independent supply to reforestation and fire fighting. On the basis of the original idea of using the standard fog collector (SFC), consisting of a few racks of 1 m² dressed with polypropylene mesh prepared by Marzol (Canary Islands), a jointly developed research program is under way at multiple microclimate points on Tenerife Island [28,29].

In 2006, the project (DYSDERA) received funding from the European Union (EU) and had the first TV station–controlled AWC capable of sending real-time data during the 24 h/365 days of the climate variables, which included optical penetration instrumentation and precision water flow meters being capable of establishing yield as a function of water content in fog, thus putting an end to inaccurate measurements and shattering previous capture efforts by obtaining peak yields of 42 L/m²/day.

The existing fog collectors needed to be upgraded with new shapes, fabrics, and framework types by incorporating the principles of lightness, transformability, portability, and polyvalence. Some of this progress is typified by newer and lightweight, polyvalent (having different forms) and modular space frames, fully wrapped with a light hydrophobic mesh that can collect fog and also acts as a shading/coating device and a soil humidifier for greenery and potential inhabitation. These can be easily installed in either flattened or uneven ground [30]. Conventional fog harvesting mechanisms today are effectively a pseudo-replica of two-dimensional surface phenomena in terms of water droplet and plant interaction. Again resorting to nature itself, it has been found that certain plants such as the *cacoutus fallax* have a unique hierarchical three-dimensional (3-D) arrangement formed by the leaves and fine hairs covering them, which are responsible for the retention of water droplets on foliage. These fine hairs literally wrap around the water droplet in a 3-D fashion. This 3-D structure is more similar to nanostructured hydrophobic surfaces. Parallel to these efforts, and under the direction of the author (Ahmad) [31], who was inspired by the nanostructure of the Namib beetle, research was initiated by mimicking the basic structure of an insect's shell. While incorporating in the design and testing a new mesh impregnated with hydrophilic and hydrophobic nanoparticles on opposing faces of the fog fence, a successful project has ensued, and has shown to be a feasible method of optimizing capture in an area of high concentration of mists on the west coast of Saudi Arabia [32–35].

The existing PAFCs worldwide would need to be upgraded with new shapes, fabrics, and framework types by incorporating the principles of lightness, transformability, portability, and polyvalence [36] for them to be efficient. Some progress is typified by a lightweight, polyvalent, and modular space frame fully wrapped with a light hydrophobic mesh that can collect fog and also acts as a shading/coating device and a soil humidifier for greenery and potential inhabitation. It can be easily adapted on flat or uneven grounds [37].

29.4 Materials and Methods

Careful consideration has been given to a new design that would capture other atmospheric water resources (haze, mist, drizzle, rain, ice, and snow) also present in those latitudes but practically nonexistent in arid areas. The art of the design of equipment begins with applicability to the various conditions as needed with the equipment used in dry (yet not arid) areas that were beginning to have problems with the quantity and quality of the drinking water available.

The concentration of efforts on the target goals returns us to the original LFCs. In the year 2000, a new design range of equipment available for AWC was devised, and later in 2001 circular and polyhedral configurations for slow mists or watertight channeling were introduced. The stations were installed for demonstration purposes by different agencies (Figure 29.3).

A brief analysis detected immediately that a rural environment where no water was available but with values that ranged from 1400 to 2000 microsiemens was the ideal condition for the deployment of these new PAFCs. Under such conditions, the project was able to produce water not only for consumption for drinking or for domestic use but also for sustaining cattle and irrigation. The water made available fulfilled the consumption regulation of untreated groundwater dictated by the EU and World Health Organization. Prediction of the amount of water is a heterogeneous natural phenomenon both in the composition of each face in a given frontal section. Its composition over a period of time was measured, and its relationship to yields was established.

It has become clear that not just any SFC fence section of 1 m² could be extrapolated to the rest of the square footage of the fog fence, and that the mass of water found in the fog bank itself differed from the rest of the collection sites and with the passage of time. A station that can be monitored 24 h/day, and if possible with real-time data, can detect changing weather conditions constantly and allow conclusions devoid of serious errors, which is necessary to identify how much water is actually in the mist and how much reaches the

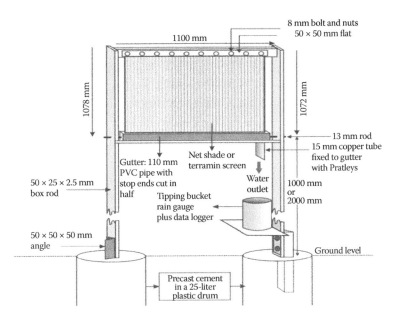

FIGURE 29.3
Model of fog collector at Cape Columbine.

fog-capturing equipment at each section. It was essential to approach the project with the knowledge of the density from different areas of the cloud; the data are necessary to proceed with further investigation of the equipment and its spatial distribution in designing a capture station (AWC). Today's Doppler radar is capable of providing such information for many areas. The approach taken to measure data bearing on the density of the fog present was previously made by direct visual observation and was enhanced by newer remote sensing methods. The second data logger counter was equipped with highly sensitive flow detectors capable of working at predetermined pressures and emitting pulses counted with every 0.5 liters of water that passed through it. The matrix network of AWC reported their data to the weather station, sent via a GPRS (precursor to 4G) wireless signal to a server located in the Station Control Research Center (Tenerife base station). This station allows watching all capturing deployments in a timely manner and analyzing the data collected from the comfort of an office desk, with the assurance that the measurements are reliable and that the measurements of the captured water are not affected by any evaporation effects. Thus, for the first time in many years, we can complete a database populated with real and meaningful data about what happens with particular fog banks or rain, and the percentage of what is captured. The observations provided a data series, now available and offered to investigators to try to make a comparative analysis between varying situations, conditions, and ensuing results. These inquiries have set clear lines for the analysis, including a methodological plan that achieved milestones for the research group and meaningful consequences that clearly identified what variables affect harvesting regimes and which have not, at least to a first approximation, which are of importance to future research. It has gone from no reliable data to a large volume of data present.

The data analysis began at once to detect seasonal conditions on AWC that, to date, has not been previously detected, including overloads such as overflowing collection systems, discrete behaviors when rainfall exceeds certain values, designs of new anchoring systems developed to sustain failure against strong winds, decantation, and inefficiencies against large volumes, overall equipment design, efficient data previously available, and clear optimization.

AWC continues to be improved by way of equipment designs becoming more efficient from the point of view of industrial and materials engineering. The first steps in applying nanotechnology advancements in AWC began 2 years ago. At present, they are being tested in various applications in Spanish laboratories to ensure the holding of a proprietary primer (serving as a bridge between the base weave and active ingredients) and to ensure that no sloughing or possible toxicity issues arise from their use. However, these steps are merely intermediary, with the final results being sought for these meshes, and the integration of nanoparticles in the raw base material of the substrates used for each type of systems is done to optimize time and costs of manufacturing.

It is also a known phenomenon that water droplets are absorbed within the mesh as in the case of most hydrophilic applications, whereas on hydrophobic surfaces droplets are formed bigger in size and roll through without being absorbed. Pollock reported that there is some evidence to show that incoming water slides over a very thin layer of greater-density water, the thickness of which is on the nanoscale. Pollock has also reported that the hydrophobic interaction extends tens of nanometers. Superhydrophobic forces can provide extra supporting force for floating objects immersed under water. A superhydrophobic surface can provide 0.4 mN force/cm^2, more than a flat surface can [37]. The vertical component of a superhydrophobic surface is increased as the air passage in the center of the surface provides extra buoyancy force. What happens at the water hydrophobic interface is not conclusively known. It is believed that water flows through small pores in hydrophobic membranes. A tiny cushion of low-density water between the hydrophobic surface and bulk water seems to exist. A dynamic interface is formed where water touches and leaves the surface repeatedly [38]. As suggested, a dynamic mechanism of droplet formation at an atomic level has not yet been fully conceived by atomic force microscopy and refractometry analysis. This would help the generation of new designs of mesh surfaces.

Metallic meshes produce greater output than a polypropylene mesh, and seem to be a better choice when considering the longevity of a PAFC's useful lifetime. There is no evidence to suggest that aluminum or corrugated iron, as was used in Oman, would be better than nets of polypropylene in terms of yield. Metals and nylons are both being studied and contrasted concerning their use as collection devices [39].

In recent work by the author (Ahmad), nanostructured hydrophobic surfaces were created in an attempt to mimic the Namib beetle's back for microcondensation of fog droplets. Stainless mesh panels were cleaned with HCl and acetone followed by washing with distilled water and drying. They were subjected to ultra shot peening at 20 kHz, the shot diameter being 1 mm. They were annealed at 350°C and cooled in air. The grain size was decreased to 25 nm. An emulsion of 20% polytetrafluoroethylene, 10% polyvinyl alcohol, and 10% sodium dodecylbenzene sulfonate in deionized water was produced. It was homogenized in an ultrasonicator at a speed of 3000 rpm. The emulsion was sprayed on a mesh panel with compressed air. The panels were subjected to laser etching immersed in an ethanol solution of fluoro-alkylsilane for 4 h. A solution of polyacrylic–acid–water–propanol was applied on the panels. The panels were transferred to an oven and dried, followed by rinsing and cleaning. The WCA was measured and was ensured to be not less than 150°. The panels were fixed on frames and tied to galvanized iron posts by cables. Drainage was through a galvanized iron pipe (15 cm). It was covered with meshed covers to prevent contamination. The reservoir was made of clay and straws (using a torrefaction process), resting in a final layer of concrete to keep the water cool. *Aloe asperifolia* was planted in the vicinity of a fog trap or PAFC. The system was connected to a data logger to record the temperature, wind velocity, humidity, and amount of trapped water. The surface of the collectors mimics the back of a Namib beetle because of the creation of hydrophobic bumps and hydrophobic troughs with epicuticular wax crystals and nanopores. Preliminary laboratory trials showed a yield 10× higher than that of meshed polypropylene. Further work is in progress [40]. The process has been recently repeated in the Dhofar region in Oman. The Dhofar region includes the southwest mountains and experiences rich and thick fog. The frequency of these fog banks is >25% per year.

29.5 Conclusions

Fog collection technology has made rapid strides from the original collection of fog and dew from sheepskins to today's nanostructured superhydrophobic/hydrophilic surfaces. The improvements in design have dramatically increased the yield from a meager 4 L/m²/day to >40 L/m²/day, and work is in progress to achieve even greater yields. The improvements in design have been incorporated not only taking into consideration substrates such as meshed polypropylene, steel, or both, or other fabrics as well, which when combined with superhydrophobic/hydrophilic coatings can withstand the effects of high winds and storms (including sand storms), increase the purity of drinking water, provide shading, and maximize the configuration of nets to provide a sustainability model and cost-effectiveness. In recent years, Doppler effect radars have been fixed to provide information about fog density and measurements of water captured via wireless data networks. Atomic force microscopy and refractometry have been used to explore the dynamics of hydrophobic surfaces in an attempt to explore the mechanism of water capture from fog formations. Polypropylene meshed nets treated with nanoparticles have been deployed. Galvanized mesh steel panels containing nanostructured substrates with microscaled mounds and troughs and nanopores are analyzed to convert the fog droplets and transform them to viable sources of drinking water, by mimicking the Namib beetle's exoskeleton to form a readily available source of water in previously inhospitable regions. With the current global water crisis and the rising demand from regions hit by shortage of drinking water, stretching from South Africa, Australia, to South America, as well as the Middle East, India, and Pakistan, each drop of water needs to be harnessed from the atmosphere and new techniques need to be devised to obtain a maximum yield. The author (Diego, 42TEK) is currently working on a project with the aforementioned institutes in Valencia, Spain, and in conjunction with Profs. Carlos Ángel Sánchez Recio and Zaki Ahmad, where a newer redesign from the original resembles the sails on sailboats, with lighter and more resistant materials capable of wind vaning to avoid excessive loads, and prepared for rapid deployments. The expertise derived from Ahmad's work in nanocoatings for optimal water is being incorporated into worldwide technology with great ramifications.

References

1. HEDS-UP, 1998. Mars Exploration Forum, January 1, pp. 171–194, Report Number LPI – Contrib. - 955 19980101.
2. Ahmad, Z., Ahmed, I., Patel, F., 2010. Fog collection by mimicking nature, *Journal of Biomimetics, Bomaterials and Tissue Engineering*, 8:35–43.

3. Sánchez Recio, C.Á., 2009. *Hidrología de Conservación de Aguas. Captación de precipitaciones horizontales y escorrentías en zonas secas*, Universidad de Valladolid, Instituto Interuniversitario de Estudios de Iberoamérica y Portugal, Valladolid.
4. Cavelier, J., Solis, D., Jaramillo, y M.A., 1996. Fog interception in mountain forests across the Central Cordillera of Panamá. *Journal of Tropical Ecology*, 12:357–369.
5. Barasoin, J.A., 1943. *El Mar de Nubes en Tenerife*, Serv. Met. Nacional, Madrid.
6. Sánchez Recio, C.Á., 2008. *Diseño y Seguimiento de Estaciones de Captura de Agua Atmosférica*, Fundación Global Nature, Drago, Santa Cruz de Tenerife.
7. Arvidsson, I., 1957. Plants as dew collectors, I.A.S.H., Publ. 43, Proc. of the Symposium on Dew, pp. 481–501.
8. Went, F.W., 1955. Fog, mist, dew, and other sources of water. En: *Yearbook of Agriculture*, US Dept. of Agriculture, Washington DC, pp. 103–109.
9. Shuttleworth, W.J., 1977. The exchange of wind-driven fog and mist between vegetation and the atmosphere. *Boundary-Layer Meteorology*, 12:463–489.
10. Stone, E.C., Went, F.W., Young, C.L., 1950. Water absorption from the atmosphere by plants growing in dry sol, *Science*, III:546–548.
11. Sudzuki, F., 1969. Utilización de la Humedad Ambiental por Prosopus tamarugo, *Phil., Boletin Técnico* 30:35–50. Univ. De Chile (*Fao. Agr. Est. Exp. Agr.*, 3:23).
12. Gioda, A., Maley, J., Espejo Guasp, R., Acosta Baladón, A.N., 1993. The fountain tree of canary island and other vegetal fog collection, Int. Symposium on Tropical Montane Cloud Forests, San Juan, Puerto Rico.
13. Muñoz Espinosa, H.R., 1967. *Captación de Agua de las Nieblas Costeras, en el Norte de Chile*, Curso Int. de Hid. Gral. y Aplicada, Escuela de Hidrologia, Madrid.
14. Acosta Baladón, A.N., 1994. Las Precipitaciones Ocultas y sus Aplicaciones en Zonas Andas, Conferencia Eso. Superior Ing. Montes, Ciudad Universitaria, Madrid. Ravena.
15. Cereceda Troncoso, P., Schemenauer, R.S., Carbajal Rojas, N., 1988. *Factores Topográficos que determinan la Distribución de las Neblinas Costeras en El Tofo, Región de Coquimbo*, Décimo Congreso Nao. de Geografía y Primera Jornada de Cartografía Temática, Pontificia Universidad Católica, Chile.
16. Cereceda, P., Schemenauer, R., Suit, y M., 1993. Producción de agua de nieblas costeras en Pera. Alisios, pág. 63–74.
17. Schemenauer, R., Suit, y M., 1993. *Producción de agua de nieblas costeras en Perá*.
18. Chaptal, L., 1932. *La Lutte Contre la Séchérésse*, La Captation de la Vapeur d'Eau Atmosphénique, La Nature, 2893, Paris.
19. Schemenauer, R.S., Osses, P., Leibbrand, M., 2004. Fog collection evaluation and operational projects in the Hajja Governorate, Yemen, Proceedings of the 3rd International Conference on Fog, Fog Collection and Dew, Cape Town, South Africa, p. 38.
20. Schemenauer, R.S., Cereceda, P., 1994. Fog collection's role in water planning for developing countries. *Natural Resources Forum*, 18:91–100. United Nations, New York.
21. Gioda, A., Blot, J., Espejo Guasp, R., Maley, J., 1993. *Les Foréts du Brouillard dans un Environnement Anide*, Résumés des Journées du Programme Environneiuent du CNRS et ORSTOM, Lyon.
22. Ceballos, L., Ortuño, y F., 1952. El bosque y el agua en Canarias, *Montes*, Madrid, 48:418–423.
23. C.M.Z de Santa Cruz de Tenerife, 1992. *Captación del agua de niebla en Tenerife*, Con colaboración del Departamento de Geografia de la ULL. Ministerio de Medio Ambiente.
24. Dorta, P., 1996. Las inversiones térmicas en Canarias, *Investigaciones Geográficas*, 15:109–204.
25. Castillo, F.E., 1965. *Evapotranspiraciones Potenciales y Balances de Agua en España*, Ministerio de Agricultura, Madrid.
26. Kenfoot, O., 1968. Mist precipitation on vegetation, *Forestry Abstract*, 29:8–20.
27. Marzol, M.V., 2001. Fog: Drinking water for rural zones, 2nd International Conference on Fog and Fog Collection, St. John's, Canada, pp. 247–250.
28. Marzol, M.V., 2002. Un sisteme de captation passive de 1 'eau du brouillard. Application et resultats obtenus aux flex Canaries (1992–2001), *Publications de L'Association Internationale de Climatologie (A.I.C.)*, 14:87–86.

29. Collection and sustainable Architecture in Atacama Coast, 5th International Conference on Fog Collection and Dew, Munster, Germany.
30. Andrews, H.G., Eccles, E.A., Schofield, W.C., Badyal, J.P.S., 2011. Three- dimensional Hierarchical structure for fog harvesting, *Langmuir*, 27:3798–3902.
31. Schernenauer, R.S., Cereceda, P., 1989. *An Investigation of the Feasibility of Monsoon Fog and Dnizzle Collection in the Dhofar Region of Southern Ornan*, Project OMA/89/005, Environment 1 Atmosphenic Environment Serv., Ottawa.
32. Ingraham, N.L., Matthews, y R.A., 1988. Fog drip as a source of groundwater recharge in northern Kenya. *Water Resources*, 24(8):1406–1410.
33. Crabbe, M., 1970. *Frequence et Intensité de la Rosée en Afnique Centrale*, Aoademie Royal des Sciences d'Outre-Mer, Bulletin des Séances 3, BruxellesDe Fina, A. Bruxelles.
34. Olivier, J., 2002. Fog water harvesting along the west coast of South Africa, Feasibilty study, *Water, South Africa*, 28(4):349–351.
35. Sung, C., 2010. Fog collection and sustainable architecture in Atacama Coast, 5th International Conference on Fog Collection and Dew, Munster, Germany, July 25–30, Fogden, pp. 2010–2089.
36. Muñoz Carpena, R., Fernández Galván, D., González Tamargo, G., Harris, y P., 1995. Diseño de una estación micrometeorológica automática de bajo coste para el cálculo de la evapotranspiración de referencia. Riegos y Drenajes XXI/N0 88. Articulo presentado en las XIII Jornadas Técnicas sobre Riegos. Puerto de la Cruz (Tenerife).
37. Bulletin de l'Observatoire du Puy de Dome, 1961. Compte Rendu du Meeting de Comparation des Capteurs de Gouttelettes.
38. R., Sudzuki, O., Pollastri, A., 1989. Coastal fog and its relation to groundwater in the IV region of Northern Chile, *Comisión Chilena de Energia Nuclear*, 79:83–91.
39. Aguas del Garoé, 1999. Captura de agua atmosférica.
40. Bouloo, J., 1993. *De la toile d'araignée. au piége á brouillard*, La Houille Blanohe 5, Paris.

Further Reading

Mallika Naguran environmental writer and founder of Gaia Discovery discussion on AWC. Available at http://www.gaiadiscovery.com.

Websites

1. Available at http://sqwater.be.washington.edu/wp/.
2. Available at http://ga.water.usgs.gov/edu/watercycleatmosphere.html.
3. Available at http://www.nebi.n/m.nih.gov/pubmed/23212396.html.
4. Available at http://cury.eas.gatech.edu/courses/ency/chapter/Ency_Atm/Fogppt.html.
5. Available at http://climatetechwiki.org/content/fog-harvesting.

30

Evaluation of a Zinc Clinoptilolite (ZZ®) for Drinking Water Treatment

Gerardo Rodríguez-Fuentes,[1] Inocente Rodríguez Iznaga,[1] Aurelio Boza,[1]
Anaisa Pérez,[1] Bárbara Cedré,[2] Laura Bravo-Fariñas,[3] Aniran Ruiz,[3]
Anabel Fernández-Abreu,[3] and Víctor Sende Odoardo[4]

[1]*Instituto de Ciencia y Tecnología de Materiales, Universidad de La Habana, La Habana, Cuba*

[2]*Instituto Finlay, La Habana, Cuba*

[3]*Instituto de Medicina Tropical "Pedro Kouri," La Habana, Cuba*

[4]*Laboratorio ENAST La Habana, Instituto Nacional de Recursos Hidráulicos, La Habana, Cuba*

CONTENTS

30.1 Introduction

The World Health Organization (WHO) *Guidelines for Drinking Water Quality* provide in its preface: *"Access to safe drinking water is essential to health, a basic human right and a component of effective policy for health protection."* The quality of drinking water is not a simple issue but one with a major influence in a myriad of problems. *"The great majority of evident water-related health problems are the result of microbial (bacteriological, viral, protozoan or other biological) contamination. Nevertheless, an appreciable number of serious health concerns may occur as a result of the chemical contamination of drinking water."*

The protection of drinking water is based on the establishment of many barriers from the source to the consumer, aimed at preventing or reducing contamination to harmless levels for human consumption. This protection includes the selection of adequate water treatment procedures as well as the best possible management and operation of distribution systems. The final manipulation of drinking water at home before human consumption is also relevant to its preservation from contamination. The storage of drinking water in poorly or seldom cleaned cisterns and tanks is the main source of microbiological contamination at homes. The most recurrent solution for the final treatment of drinking water is the home filtration system.

A great variety of more or less efficient systems have been created to remove contaminants. They are all based on the physical and chemical properties of one or various materials; thus, their useful life is limited to the effectiveness of such materials. Not one water purification system exists that can remove *every* water contaminant [1,2], although the water treatment plans that combine different processes to remove impurities and achieve microbiological purification can be rather efficient. However, the combination of all the processes involved in a water treatment plant in a compact and easy-working domestic system has until now eluded scientists.

Disinfection is of unquestionable importance in the supply of safe drinking water. The destruction of microbial pathogens is essential and usually involves the use of reactive chemical agents such as chlorine. In the past 10 years, several articles have reported the potential of silver and zinc oxide nanoparticles in drinking water treatment to eliminate mostly microbiological contamination [3]. KDF filters release small quantities of Cu and Zn [4]. Some articles have related the use of Ag/Zn zeolite for the same purpose [5].

In 1990, a new material was obtained at the Zeolites Engineering Laboratory of the University of Havana. This material with microbicidal and ion-exchange properties was given the name "ZZ®"—zinc in zeolite or a Zn^{2+} form of the purified natural clinoptilolite NZ [ZZ®: Registro de Marca No. 130531, Resolución: 4876/99, OCPI (Oficina Cubana de la Propiedad Industrial)]. ZZ was designed and developed as a controlled releaser of zinc ions to enhance their microbicidal effect against bacteria, yeast, and protozoans. The Zn ions are exchanged in the crystalline nanostructure of the clinoptilolite zeolite without modification of the unit cell. The lattice parameters of the clinoptilolite are as follows: $a =$ 17.71, $b = 17.84$, $c = 7.41$, and $\beta = 116.30°$, space group $C2/m$. This zeolite has three channels interconnected {channel A [001] 10 members 3.1 × 7.5 Å, channel B 8 members 3.6 × 4.6 Å, and channel C [100] 8 members 2.8 × 4.7 Å}. The dimensions of the channels vary owing to the considerable flexibility of the framework. The representation of a unit cell of the Zn–clinoptilolite is shown in Figure 30.1. The clinoptilolite crystals have the tabular and lath morphology observed in the micrograph of Figure 30.1.

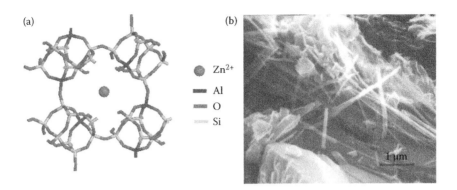

FIGURE 30.1
(See color insert.) Representation of a Zn–clinoptilolite unit cell without water molecules (a) and clinoptilolite crystals with tabular and lath morphology (b).

Many drugs based on zinc compounds for different therapeutic treatments are described in pharmacopoeia but their effectiveness depends on their capacity to release Zn^{2+} from the compound used as active ingredient. A zeolite may act as a controlled releaser of Zn^{2+} ions; however, two issues must be addressed: (i) the toxicity of the zeolite and (ii) the ability of the zeolite to release the amount of Zn^{2+} required after its incorporation to the drug without creating a health hazard.

Our early results in the formulation of the antidiarrheal drug Enterex [6], based on purified natural clinoptilolite NZ, answered the toxicological *incognita* about the zeolitic raw material. Still, the new material Zn^{2+}–NZ, named ZZ [7], was submitted to toxicological testing.

The second issue required an extensive ion-exchange study of Zn^{2+} in this zeolite considering

1. The introduction of zinc in clinoptilolite by the exchange of the natural extra-framework cations (Ca^{2+}, K^+, Mg^{2+}, and Na^+) invokes a rather difficult reaction because the selectivity of clinoptilolite for Zn^{2+} is lower with respect to the selectivity for its natural cations. Then, the chemical equilibrium of the exchange reactions must be shifted toward the introduction of Zn^{2+} in zeolite.

2. The controlled release of Zn^{2+} from the clinoptilolite should also take place through ion exchange of the ions present in drinking water or biological media at room or human body temperature (37°C), respectively. The final zinc content in water or biological media must be lower than the toxic level. The recommended zinc content for drinking water is 5 ppm because of taste (palatability), while in other biological media it depends on the human organ or animal species.

ZZ was first certified as a water purifier by the Registro Sanitario of Cuba in 1999, after several years of study [8]. Thirteen years later, other studies have been completed that show the microbicidal properties of the product and its value as a basic element of the water purification systems.

ZZ is a carrier that releases zinc into water in a slow and calibrated manner, within the limits recommended by the WHO. Zinc, which acts like a microbicide on the pathogenic microorganisms, is also a fundamental element for human and animal life. Various humanitarian projects in South American countries are intended to provide a supplement of zinc salts to combat epidemics and infections; however, the unstable supply of zinc reduces the effectiveness of these campaigns [9,10]. The continuous and measured combination of water with a zinc supplement only takes place in the ZZ purification systems.

The ZZ water purification systems based on the physical, chemical, and microbicidal properties of ZZ have been designed for use right before water consumption by humans or animals as they remove pathogenic microorganisms, clean turbid water, soften hard water, and add zinc—a crucial element to life. They are simple and reliable, and can be operated at no risk. Taking into account the habits of the different population groups, the authors have basically conceived them for use with previously treated water destined for human and animal consumption that may have contaminated along the distribution network, and for underground or elevated water tanks. They can also be used with non-treated surface water from rivers, lakes, and water reservoirs, and with water from wells or springs.

30.1.1 Why ZZ and Not a Salt or Zinc Compound?

Salt and zinc compounds are objected because of the anion or the part of the compound that is not zinc. In general, that part can have other effects or somehow remain linked to zinc, limiting its efficacy. It can also alter the taste of water, making it unpleasant (not palatable). ZZ only contributes the zinc ions—as the zeolitic material is not soluble in water—while absorbing the excess of calcium and magnesium (hardness) as well as the contaminants. ZZ can be used inside the purification systems for a more efficient action or added directly to water. A necessary condition is to ensure adequate contact with the water to be purified and then wait the short time required for the zinc to remove the pathogenic microorganisms.

Zinc is one those elements well distributed in the human body, just like carbon, hydrogen, oxygen, nitrogen, and phosphorous found in organs and biological functions. It is well documented that zinc is involved in about 200 biochemical processes. The quest for the biological role of this element leads us to the very source of man's life, his growth, and reproduction. Zn is responsible for DNA synthesis at the cell's nucleus, hence the accelerated cell division associated with zinc deficiency. These cells can be spermatozoans and the cells involved in the immunological and healing system.

The organs of the human reproductive system are rich in Zn with a maximum concentration in the seminal fluids; thus, a zinc deficiency can cause malformations in the fetus and sterility in adults. The mother's milk is the richest source of zinc. High concentrations of zinc can be found in the retina; therefore, night vision impairment and low adaptation to darkness have been found to be related to zinc deficiency. The production of insulin in the pancreas and its functioning also depend on zinc, and the immunologic system relies heavily on a good supply of zinc. This element blocks the points of attack to the free radicals that can cause death or changes in the cells leading to cancer.

The effects of zinc deficiency on health are as follows:

- *Severe deficiency:* Pustular-bullous dermatitis, chronic and severe diarrhea, alopecia, slow growth and development, slow sexual maturity, swollen liver or spleen, immune system depression, mental disturbance, recurrent infections, anorexia, night blindness, impaired healing, alteration of the sense of taste, and behavioral disorder.

- *Moderate deficiency:* Slow growth, hypogonadism, skin alterations, poor appetite, night blindness, slow healing of wounds, dysfunction of cell-mediated immunity, and abnormal neuro-sensorial changes.
- *Light deficiency:* Neuro-sensorial changes, oligospermia, hyperamonemia, reduction of the timuline serum activity, the production of interleukin-2 and the natural killer cell activity, alterations in the T-cell subpopulation, and damage of the neuropsychological functions.

In the past 10 years, there has been an increase in the volume of resources allocated to the study of new materials capable of acting as zinc carriers and suppliers, as well as the action mechanisms of this element in the biochemical processes, and also the zinc products that prevent the entry to the cells of such viruses as human immunodeficiency virus (HIV), rotavirus, and poliovirus. Several government institutions and private foundations in developing nations are working on projects to add zinc to the diet of segments of their population, basically children and elders, to counteract malnutrition, diarrhea, and infectious disease.

Humans should have a regular intake of zinc; however, in developing nations, extreme zinc deficiency prevails associated to eating habits and famines. The recommended dosage for adults is 15 mg zinc/day, contained in beef, viscera, fish, and cheese; fruits and vegetables are low in zinc. In general, human beings do not consume enough zinc, while physical activity, alcohol consumption, smoking, and contraceptives tend to increase deficiency. The recommended daily dosage of zinc is 11 mg for adults (70 kg of weight) and from 11 to 13 mg for pregnant or lactating women, while the therapeutic dose for patients with zinc deficiency is 220 mg of $ZnSO_4$ (USP quality) three times per day.

In summary, zinc is

- A micronutrient essential to humans and with a major role in reproduction, growth and development, cell metabolism, gene expression, immune response, and neurological function
- The catalytic cofactor of >300 enzymes that additionally stabilizes a great variety of protein domains
- An element found in every organ of the human body—tissues, fluids, and secretions—in quantities of 1.5–2.5 g

Zinc reacts to pathogenic microorganisms by crossing the membranes of bacteria and other microorganisms, interacting with functional groups of enzymes. Zinc ions replace the metals acting as cofactors and modify their performance inhibiting the growth of microorganisms and killing them; these are known as bacteriostatic and bactericidal effects, respectively. This microbicidal effect has been confirmed in protozoans and their cysts, and in fungi. Likewise, zinc interacts with the DNA, which accounts for its antiviral effect.

Various studies have shown that zinc is effective against a wide variety of microorganisms, especially those pathogens living in water that cause the gastrointestinal, pulmonary, and skin infections that most commonly affect humans. That is why an increasing amount of pharmaceuticals are formulated using zinc salts and compounds to combat illnesses. Other chemical elements have a similar or higher microbicidal effect than zinc. Such is the case of silver, which is more effective but highly toxic and less relevant to life.

Also, there is copper, an element with a proven bactericidal effect against a number of microorganisms and also an important element to life. Iodine, which is also used for chemical disinfection of water, cannot eliminate some pathogenic parasites; moreover, some people have adverse reactions to iodine intake. However, zinc combines both properties and is more important to life. That is why it was chosen in formulating ZZ and for use in water purification systems.

30.1.2 What Are the Bases of the ZZ Project?

ZZ is a new material with microbicidal and ion-exchange properties that can be used to purify water for human consumption, and as a viable alternative to such traditional procedures as water boiling and disinfection with chemical substances and physical methods. ZZ makes possible the design of flexible domestic and collective water purification systems, which combine the ZZ properties, the simplicity of the systems, and their low cost with the people's water storage and handling habits.

The objectives of a ZZ Project are

- Effective and efficient water purification for direct human consumption
- A broad distribution of the water purification systems to all segments of the population in a given country
- Distribution of the water purification systems to low-income segments of the population lacking the necessary infrastructure for water treatment and distribution
- The use of ZZ in population segments with eating habits limiting the incorporation of zinc to their nutrition
- The use of ZZ as an active complementary therapy for population segments affected by infectious diseases demanding zinc

This chapter offers the highlights of the ZZ water purification systems that can be developed in the conditions prevailing in Cuba and other countries that need to secure the quality of the water supplied to the population affected by natural disasters or war conflicts. These systems share similarities with other domestic water purification systems and procedures but offer the additional advantage of zinc supplementation.

30.2 Materials and Methods

A natural clinoptilolite (IUPAC code HEU) from Tasajeras, Cuba, a deposit that meets Cuban standard NC 625: 2008-Annex A [11], of a natural zeolite for human and animal health and nutrition, was used as the raw material to obtain purified natural clinoptilolite NZ [8]. The particle size of NZ chosen was 1–3 mm. The Zn^{2+}–NZ form was obtained in a pilot plant with a capacity production of 1 ton/day through the hydrothermal ion-exchange modification of NZ with 0.5 M dissolution of $ZnSO_4 \cdot 7H_2O$ USP, at boiling temperature during 10 h following the procedure described [12]. The pH, density, sterility, and humidity of Zn^{2+}–NZ were adjusted considering the sanitary properties required for products used in water purification treatments recommended by the WHO *Guidelines for Drinking Water*

Quality [13] and Cuban standard NC 827: 2010 [14]. The resulting product is denominated ZZ. The characterization of ZZ was conducted by chemical analysis and x-ray diffraction; microbiological tests were also carried out to determine (i) a possible contamination and (ii) the bactericidal effect.

As part of the project, domestic and collective water delivery systems have also been designed. The former are 1.5-L pitchers and flasks with a funnel in the upper part containing a ZZ cartridge or a load of the product. The collective systems are more complex since they are connected to the water supply pipes and their capacity can be of either 2 or 4 kg of ZZ. The ZZ systems have been certified by the Pedro Kouri Tropical Medicine Institute, the Finlay Institute, the National Water Resources Institute, the Food and Pharmacy Institute, and the Havana University Science and Technology of Materials Institute. These are all prestigious Cuban scientific institutions recognized for their world-class research. In 1999, ZZ was certified by the Registro Sanitario of Cuba [8].

The chemical and microbiological studies of ZZ were conducted in two phases: (i) *in vitro* and (ii) using a pitcher as domestic purification system.

30.2.1 Chemical Studies

The performance of the exchange of ZZ and the main cations present in drinking water was studied in kinetic and static conditions. Three samples of ZZ (20 g) were placed in three glass columns, 10.7 cm length and 2.5 cm diameter, creating a 6-cm-thick bed. The drinking water was moved through the columns at flow rates of 100, 25, and 15 mL/min during 10 h. Static experiments were designed to evaluate the behavior of the chemical composition of drinking water when in contact with ZZ for a long time. Ten experiments with different ZZ/water ratios from 1:1 to 1:10 were conducted for 48 h. The chemical composition of drinking water was determined every hour.

An experiment was designed to approach the use of ZZ in a domestic water treatment system. In a pitcher, a column was assembled with 140 g of ZZ, whose particle size was 1–1.6 mm, and placed in ion-exchange columns of 5-cm in diameter creating an 8.5-cm bed. The pitcher's volume was 2.2 L and the drinking water was moved through the column at a flow rate of 0.2 mL/min but the column remained inside the drinking water, allowing it contact with ZZ. The experiment was conducted for 45 days. Zinc content was determined every day; however, during the first 8 days, it was monitored hourly.

The content of the cations involved in the ion-exchange process was determined at different times taking drinking water samples and submitting them to chemical analysis of Ca, Mg, Na, K, and Zn, and the heavy metals Cd, Cr, Mn, Ni, and Pb, using a PYE UNICAM SP9 atomic absorption spectrophotometer.

30.2.2 Microbicidal Effect

The bactericidal effect of ZZ on bacteria that can be live in drinking water was studied using strains of reference. The minimum bactericidal dose of 10% of ZZ (10 mg of ZZ/100 mL of culture medium) was established in a study using 33 strains of gram-positive and gram-negative bacteria, and yeast strains as presented in Table 30.1.

A new series of microbiological tests was conducted using the bacterial strains that can contaminate drinking water. For each bacterial strain, a preculture in Luria–Bertani (LB) media was prepared and then incubated for 18 h at 37°C. It was checked for purity by Gram stain and 1 mL of this preculture was inoculated in a 250 mL Erlenmeyer with

TABLE 30.1

Bacterial Strains and Yeasts Removed by ZZ

Bacterial Strains and Yeasts		
Bacillus subtilis ATCC 6633	*Streptococcus cloacae* ATCC 23355	*Bacillus cereus* BSG 001
Pseudomonas aeruginosa 2327	*Salmonella typhimurium* ATCC 14628	*Streptococcus freundii* BSG 013
Pseudomonas aeruginosa ATCC 27853	*Salmonella anatum* BSG 012	*Serratia marcescens* ATCC 8100
Klebsiella pneumoniae BSG 028	*Salmonella typhimurium* BSG 030	*Arizona* ATCC 67344
Klebsiella pneumoniae BSG 036	*Proteus mirabilis* BSG 013	*Citrobacter freundii* BSG 032
Vibrio cholerae No. 01	*Proteus vulgaris* BSG 015	*Providencia* ISA 21
Staphylococcus aureus ATCC 25923	*Proteus vulgaris* ATCC 13315	*Klebsiella pneumoniae* BSG 003
Staphylococcus epidermidis BSG 021	*Proteus rettgeri* ATCC 1407	*Klebsiella pneumoniae* ISA
Shigella sonnei ATCC 75931	*Escherichia coli* 44	*Candida albicans* BSG 002
Shigella flexneri ATCC 12027	*Escherichia coli* ATCC 25922	*Candida albicans* BSG 003
Shigella flexneri BSG 009	*Escherichia coli* ATCC 25932	*Candida albicans* BSG 007

50 mL of LB media and 10% ZZ. Additionally, a control Erlenmeyer was inoculated that contained neither of these products. After introducing ZZ, the test proceeded in two ways: stirring and not stirring or static. The bacterial strains used in the test were selected considering their virulence and survival span in drinking water, as presented in Table 30.2.

The test conducted to evaluate the effect of ZZ on *Giardia lamblia* cysts was performed according to the following procedure:

- Extraction of *G. lamblia* cysts from infected human feces
- Cyst count using Neubauer chamber
- Production of the infective dose (20,000 cysts/mL)
- Inoculation of the Erlenmeyers containing drinking water control sample and those with ZZ 10%
- Cyst count after 30 min, 1 h, and 3 h

TABLE 30.2

Bacterial Strains Used to Test the Bactericidal Effect of ZZ in Drinking Water

Bacteria	Strain	Characteristics
Vibrio cholerae	C7258	Serotype Ogawa, epidemic
	C6706	Serotype Inaba, epidemic
Pseudomonas aeruginosa	O11	Serotype O11, reference
	5FQ	Serotype O9, isolated from cystic fibrosis patients
Leptospira interrogans	M20	Serovar Copenhageni serogroup icterohaemorrhagiae, reference
Shigella sonnei		
Salmonella typhi		
Aeromonas hydrophila	ATCC 7614	
Plesiomonas shigelloides	CNCTC 5132	
Escherichia coli O	149 K88 Lt(+)	
Escherichia coli O	101 K99 St(+)	
Escherichia coli	ATTC 25922	

30.3 Results and Discussion

30.3.1 Production and Utilization of ZZ

The production of ZZ requires three raw materials: natural zeolite, zinc sulfate (pharmaceutical quality USP), and water; energy is also necessary to raise the temperature of the reactor where the zeolite is processed and heat up the oven where ZZ will be dried and sterilized before packaging. The production of ZZ does not require special facilities but those of any plant where pharmaceutical raw materials are processed. Such facilities are furnished with reactors, tanks for dissolution, centrifuge, drying ovens, dust filters, packaging machines, and a quality control system. Figure 30.2 shows a diagram of a basic plant for ZZ production.

The dust filter collects ZZ powder—accounting for 20% of the production—cleaning the particles and delivering a product that can be used in the formulation of antiseptic drugs as vaginal tablets and cream for infections, dermal cream to heal burns, scratch, pressure ulcer, and wounds. This ZZ powder can also be used in the production of building materials with microbicidal activity to prevent the growth of bacteria and fungus, such as concrete and mortar for sanitary rooms, kitchens, restaurants, etc. The waste solution treatment basically consists in the separation of sludge from zinc sulfate dissolution (Figure 30.3). The sludge is a mixture of calcium sulfate and ZZ powder that can be used as plaster with microbicidal action for sanitary facilities like surgery rooms, intensive care units, and other hospital rooms.

The quality control of the process is based on the chemical analysis of the raw materials, zinc sulfate solutions, and ZZ. A chemical analysis to control the solution concentration is conducted using a volumetric method, while the raw material and ZZ are submitted to

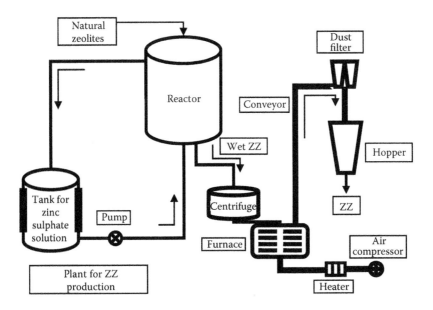

FIGURE 30.2
Diagram of the technological equipment required for ZZ production.

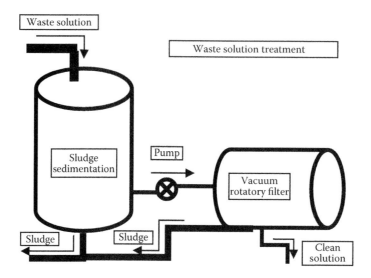

FIGURE 30.3
Waste solution treatment.

an acid digestion to allow the use of atomic absorption spectrometry. The mineral phase analysis is conducted using powder x-ray diffractometry; the patterns of NZ and ZZ are presented in Figure 30.4, showing that during the technological process of NZ to obtain ZZ, the crystal structure of the clinoptilolite remains unchanged. The mineral phase composition of NZ also remains stable: clinoptilolite–heulandite (80%), mordenita (5%), and

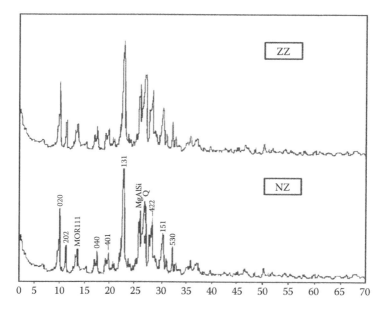

FIGURE 30.4
Diffraction patterns of NZ and ZZ showing similarity because the mineral phase composition of NZ was not modified during ZZ production.

quartz + feldspar + montmorillonite + iron oxides (15%). The unit cell parameters of clinoptilolite were not modified because the Zn^{2+} ions exchanged mainly with the Ca^{2+} ions from their cation position.

30.3.2 Chemical Studies

The chemical analysis of NZ and ZZ, described in Table 30.3, showed that zinc exchanges calcium at large (1.70%) in clinoptilolite even when the selectivity of this zeolite is lower for zinc ions. Sodium was the second cation exchanged in NZ by Zn^{2+}; however, the quantity (0.3%) is lower as compared with calcium. The ion exchange of potassium and magnesium was very small (0.1%). The hydrothermal conditions established for the ion-exchange reaction increased the zinc incorporation to the clinoptilolite structure.

Figure 30.5 shows the performance of the different cations in the ion-exchange process when ZZ was placed in a column as part of a 10.7-cm-long bed of 2.5 cm in diameter, and drinking water was moved through at a flow rate of 15 mL/min. An increase in the sodium content of water was observed because calcium ions exchanged the sodium ions of the clinoptilolite not previously removed by the zinc ions during the ZZ production. Magnesium and potassium content in the drinking water did not change, thus confirming that calcium ions only remove zinc and sodium cations [15]. These results also indicate that the counter exchange of Zn^{2+} ions was not totally verified at room temperature, saving zinc ions for further exchange.

To establish the best relation between drinking water volume and the ZZ mass, a simple experiment was conducted. Different drinking water volumes were exposed to a ZZ mass but without stirring. The time of exposure was 1, 2, 3, and 4 h, and the ratio of water volume with the ZZ mass was established from 1:1 to 10:1. The zinc content in the water was

TABLE 30.3

Chemical Composition of ZZ as Compared with NZ and Cuban Standard NC 625:2008

Elements	NZ (%)	ZZ (%)	NC 625 (%)
SiO_2	65.00	65.12	64.00–66.00
Al_2O_3	10.96	11.01	10.00–12.00
Fe_2O_3	1.70	1.70	1.5–2.20
FeO	0.47	0.45	0.20–0.5
MgO	0.40	0.36	0.30–0.90
CaO	4.27	2.57	2.50–6.00
Na_2O	1.10	0.87	1.00–2.00
K_2O	1.30	1.28	1.00–2.00
P_2O_5	0.05	0.05	0.05–0.07
H_2O	12.00	11.00	10.00–14.00
Zn	–	1.7	–
Toxic Elements			
F	1 ppm	1 ppm	<10 ppm
Pb	2 ppm	2 ppm	<10 ppm
As	0.1 ppm	0.1 ppm	<3 ppm
Cd	0.5 ppm	0.5 ppm	<2 ppm
Hg	0.2 ppm	0.2 ppm	<5 ppm

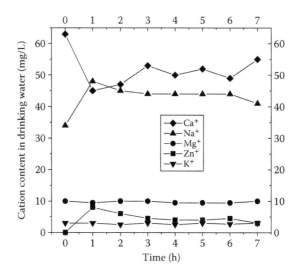

FIGURE 30.5
Performance of the different cations in the ion-exchange process with ZZ.

determined at the end of the experiment. The results (Figure 30.6) showed that zinc content increased from a ratio of 1:1 to 2:1; however, the longer the exposure, the lower the zinc content in the drinking water. The influence related to the length of the exposure indicates that the initial ion exchange of Zn^{2+} in clinoptilolite by Ca^{2+} ions was reversed. Thus, the equilibrium of clinoptilolite (ZZ) with Zn^{2+} as an extra-framework cation is better than with Ca^{2+} ions.

The amount of Zn released by the clinoptilolite structure to the drinking water was <5 mg/L, which meets the WHO *Guidelines for Drinking Water Quality* [13]. This zinc content is enough for a bactericidal effect on pathogenic bacteria, as will be shown later on in this chapter.

The kinetics of the exchange reaction of zinc ions in ZZ by calcium ions present in drinking water was studied at different water flow rates in the ion-exchange columns.

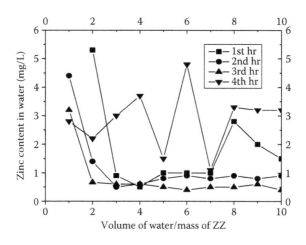

FIGURE 30.6
Zn^{2+} content in drinking water as a function of volume of water/mass of ZZ.

The elution rate of zinc ions (dQ/dt) is shown in Table 30.4. As expected, the release of zinc increased with the flow of water.

An experiment was conducted using a pitcher with a column of 8.5-cm length and 4.0-cm diameter containing 140 g of ZZ, and a flow of drinking water of 0.2 L/min, for 45 days, with daily monitoring of the volume of drinking water treated to analyze the performance of the ZZ system. Figure 30.7 describes the kinetic curves of Zn elution from ZZ for different days, showing almost identical behavior and quantity of zinc released to drinking water that remained below 2 mg/L.

The kinetic exchange parameters—elution rate (dQ/dt), internal effective diffusion (D), and internal diffusion rate constant (B)—determined during the study (Table 30.5) confirmed that their values increased indicating a better performance of ZZ. See that the three parameters improved after 8 days of continuous process of drinking water treatment.

TABLE 30.4

Elution Rate of Zn^{2+} Ions as a Function of the Flow of Drinking Water

Flow of Drinking Water	$dQ/dt \times 10^{-3}$ (meq g^{-1} seg^{-1})
15 mL/min (1)	5.26
25 mL/min	11.34
100 mL/min	28.40

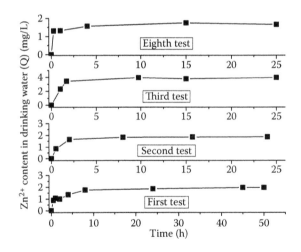

FIGURE 30.7
Kinetic curves of Zn^{2+} ions in drinking water for different days.

TABLE 30.5

Kinetic Exchange Parameters of Zn Released from ZZ

Day	dQ/dt (10^{-5} meq s^{-1})	D (10^{-11} m^2 s^{-1})	B (10^{-4} s^{-1})
1	0.68 ± 0.03	1.42 ± 0.07	0.83 ± 0.04
2	1.79 ± 0.08	2.85 ± 0.14	1.66 ± 0.08
3	3.94 ± 0.19	3.33 ± 0.16	1.94 ± 0.09
8	6.78 ± 0.33	16.33 ± 0.81	9.53 ± 0.47

Note: dQ/dt is the elution rate of Zn^{2+}, D is the internal effective diffusion coefficient, and B is the internal diffusion rate constant.

30.3.3 How Many Water Purification Systems Have Been Designed?

Both domestic and collective systems have been designed based on ion-exchange columns (Figure 30.8). The former are 2-L pitchers and flasks with a funnel in the upper part containing a ZZ cartridge or a load of the product creating a column 10 cm in height and 5 cm in diameter. The collective systems are more complex for they are connected to the water-supply pipes and their capacity can be of either 2 or 4 kg of ZZ in one or two columns.

The ZZ water purification systems (Figure 30.8) were evaluated in accordance with existing water quality parameters for human consumption [16]. The results indicate that turbidity in well water was reduced to 3 units when treated in a ZZ pitcher (Figure 30.9). In addition to its ion-exchange properties enabling the release of zinc ions through the exchange of calcium ions in the water, ZZ has another important property: the adsorption that reduces excess hardness in water, this being a major cause of sediments in the human urinary tract. Actually, water hardness is reduced by 60% (Figure 30.10).

The presence of such heavy metals as Cd, Cr, Mn, Ni, and Pb remains at lower levels than the values established for water intended for human consumption (Table 30.6). The

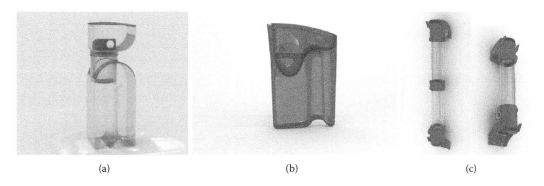

(a)   (b)   (c)

FIGURE 30.8
ZZ water purification systems: (a) flask and funnel (2 L); (b) pitcher (1.5 L); (c) collective systems.

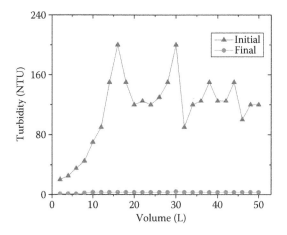

FIGURE 30.9
Water turbidity after treatment with the ZZ system.

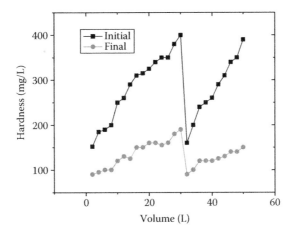

FIGURE 30.10
Water hardness after treatment with the ZZ system.

TABLE 30.6

Metal Concentration in Drinking Water Treated with ZZ

Sample/Volume Water	C(Cd) (mg/L)	C(Cr) (mg/L)	C(Mn) (mg/L)	C(Ni) (mg/L)	C(Pb) (mg/L)
I control	<Ld	<Ld	<Ld	0.03	0.02
II control	<Ld	<Ld	<Ld	0.02	<Ld
III control	<Ld	<Ld	<Ld	<Ld	<Ld
IV control	<Ld	<Ld	<Ld	0.02	<Ld
V control	<Ld	<Ld	<Ld	0.02	<Ld
2	<Ld	<Ld	0.02	0.02	<Ld
6	<Ld	<Ld	<Ld	0.03	0.03
10	<Ld	<Ld	<Ld	0.02	0.02
16	<Ld	<Ld	<Ld	0.03	<Ld
20	<Ld	<Ld	<Ld	0.02	<Ld
24	<Ld	<Ld	<Ld	0.03	<Ld
30	<Ld	<Ld	<Ld	0.02	0.02
40	<Ld	<Ld	<Ld	<Ld	<Ld
46	<Ld	<Ld	<Ld	0.02	<Ld
50	<Ld	<Ld	<Ld	0.02	0.02
60	<Ld	<Ld	<Ld	<Ld	0.03
80	<Ld	<Ld	<Ld	<Ld	0.02
100	0.002	<Ld	<Ld	0.02	0.02
125	<Ld	<Ld	<Ld	0.03	0.02
150	<Ld	<Ld	<Ld	0.02	<Ld
175	<Ld	<Ld	<Ld	0.02	0.02
200	<Ld	<Ld	<Ld	0.03	0.02
Ld (mg/L)	0.002	0.01	0.01	0.02	0.02
AMC (mg/L)	0.005	0.05	0.1	0.02	0.05

Note: Ld, experimental detection limit; AMC, admissible maximum concentration (Cuban standard NC 827: 2010).

water purified with ZZ meets the requirements set by the WHO guidelines and the Cuban standard NC 827: 2010 [13,14].

A 100 g load of ZZ allows the purification of 150 L of water. Throughout the process, the zinc content in the treated water does not increase above 3 ppm (mg/L) as observed in Figure 30.11, i.e., well under the Cuban 5 ppm standard, which is the value recommended by the WHO [10].

The study of the pitcher with ZZ cartridge (140 g) using home tap water reproduced the exchange of Zn^{2+} by Ca^{2+} as observed in Figure 30.12. The calcium content was reduced by 25% as compared with the initial value in tap water.

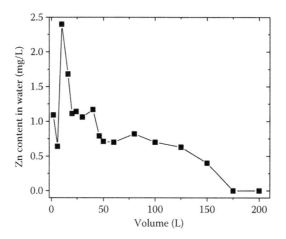

FIGURE 30.11
Zn content in drinking water treated with the ZZ purification system.

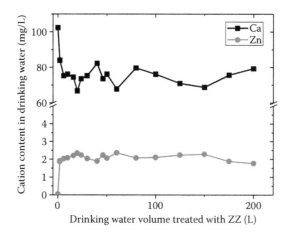

FIGURE 30.12
Zinc and calcium content in tap water treated in the ZZ pitcher.

30.3.4 Microbicidal Effect

The minimum inhibitory dosage of 10% of ZZ was established through a study that used gram-positive and gram-negative bacterial strains and yeasts. It has been determined that ZZ has a bactericidal effect on a great number of microorganisms. A specific study made possible to ascertain the microbicidal effect of ZZ on the pathogenic strains of *Vibrio cholerae* 01 and *Corynebacterium diphtheriae* [17]. The study showed that in the case of some strains, even <10% of ZZ was sufficient. Thus, the use of ZZ in the purification of water for human consumption can prevent the infections caused by these two microorganisms as well as the others showed in the Table 30.1.

Aeromona was established as a genus in 1986. The pathogenic strains common in humans are *A. hydrophila*, *A. caviae*, and *A. veronii bv. sobria*. They live in waters of low saline content and in such food as meat, milk, ice cream, and well water. The infective dose is 10^6 colony-forming units (CFU)/mL [18]. These bacteria produce gastroenteritis from diarrhea to dysentery. Table 30.7 describes the results of the test that showed the bactericidal effect of ZZ on *A. hydrophila* ATCC 7614. See that after 30 min, the bacteria were eliminated.

The infective dose of *Plesiomonas shigelloides* to produce diarrhea is approximately 10^6 CFU/mL. The time of incubation is from 1 to 14 days, and it can produce systemic infections in neonates and patients affected by hematological conditions, HIV and other illnesses [19,20]. Table 30.8 shows the bactericidal effect of ZZ on these bacteria in drinking water. The effect was observed after 30 min.

TABLE 30.7

Bactericidal Effect of ZZ on *Aeromona hydrophila* ATCC 7614

Time	CFU			
	ZZ Stirred	Control	ZZ Static	Control
0	2.6×10^9	2.6×10^9	2.6×10^9	2.6×10^9
15 min	4.6×10^3	1×10^8	4.6×10^6	1×10^8
30 min	–	1×10^7	–	1×10^7
1 h	–	1.6×10^7	–	6.5×10^7
2 h	–	6×10^7	–	3×10^7
3 h	–	1×10^7	–	6.2×10^7

TABLE 30.8

Bactericidal Effect of ZZ on *Plesiomonas shigelloides* CNCTC 5132

Time	CFU			
	ZZ Stirred	Control	ZZ Static	Control
0	8×10^8	8×10^8	8×10^8	8×10^8
15 min	3×10^5	4.5×10^7	1.6×10^5	1.6×10^7
30 min	–	7.6×10^6	–	3.3×10^6
1 h	–	5.6×10^6	–	6.3×10^6
2 h	–	3.5×10^6	–	3.1×10^6
3 h	–	3.3×10^6	–	4.2×10^6

Escherichia coli is one of the most commonly studied bacterial strains not only for its pathogenic capability but also because it is the main cause of diarrhea in poor countries, leading to the death of a million children every year [21]. The prevalence of these bacteria is lower in teenagers and adults; however, the intake of food with a high content of this microorganism (10^8–10^{10} CFU) can produce gastroenteritis with liquid diarrhea in these age groups. Table 30.9 shows the bactericidal effect of ZZ on three strains of *E. coli*.

Among microorganisms, *V. cholerae* is of the greatest importance because of the damage it causes when ingested in contaminated drinking water and food. *Vibrio cholerae* is the causal agent of a cholera epidemic [22]. This acute diarrheal illness leads to a rapid loss of body fluids and electrolyte balance, and eventually to death in 24 h as a consequence of its powerful enterotoxins. Thus, the effect of ZZ on serotypes Inaba and Ogawa was evaluated. As observed in Table 30.10, the two serotypes were eliminated after 30 min of exposure to ZZ in stirred conditions, although in static serotype Inaba required 2 h of exposure.

Shigella sonnei was studied using two different doses (10^5 and 10^2 CFU) considering that the infective dosage of these bacteria is low; that is, few cells can make a person ill depending of the strain virulence, but it generally takes from 10 to 100 cells [23]. The time of inactivation of both inocula was the same between 2 and 3 h for both conditions of the test culture, stirred and static (Table 30.11). The same result was obtained for *Salmonella typhi* using the infective dose of 10^8 CFU; in both conditions, the total inhibition of the bacteria was completed by 3 h.

Pseudomonas aeruginosa is considered an opportunist bacterium quite abundant in nature. These bacteria can infect patients with immunological deficiency, cancer, or burns,

TABLE 30.9

Bactericidal Effect of ZZ on *Escherichia coli* O:149 K88 Lt(+), *Escherichia coli* O:101 K99 St(+), and *Escherichia coli* ATTC 25922

	CFU			
Time	ZZ Stirred	Control	ZZ Static	Control
0	3×10^9	3×10^9	3×10^9	3×10^9
15 min	1.6×10^3	1×10^8	6×10^6	1×10^8
30 min	–	3×10^7	–	2.3×10^7
1 h	–	6×10^7	–	6.5×10^7
2 h	–	2.2×10^7	–	3×10^7
3 h	–	5.2×10^7	–	6.2×10^7

TABLE 30.10

Bactericidal Effect of ZZ on *Vibrio cholerae* Serotypes Ogawa and Inaba

	Vibrio cholerae C7258 (CFU)				*Vibrio cholerae* C6706 (CFU)			
Time	ZZ Stirred	Control	ZZ Static	Control	ZZ Stirred	Control	ZZ Static	Control
0	7×10^9	7×10^9	7×10^9	7×10^9	7×10^9	7×10^9	7×10^9	7×10^9
15 min	1×10^4	8×10^9	1×10^5	7.5×10^9	6.5×10^9	6×10^9	3×10^9	6×10^9
30 min	–	7.5×10^9	–	7×10^9	1×10^1	5.8×10^9	2.8×10^9	6×10^9
1 h	–	8×10^9	–	5×10^9	–	6×10^9	1×10^2	5.8×10^9
2 h	–	8×10^9	–	6×10^9	–	6.5×10^9	–	5×10^9
3 h	–	6×10^9	–	4×10^9	–	6×10^9	–	4.8×10^9

TABLE 30.11

Bactericidal Effect of ZZ on *Shigella sonnei* and *Salmonella typhi*

	Shigella sonnei (CFU)		*Salmonella typhi* (CFU)	
Time	ZZ Stirred	ZZ Static	ZZ Stirred	ZZ Static
0	1×10^5	1×10^5	1×10^5	1×10^5
15 min	1×10^5	1×10^5	1×10^5	1×10^5
30 min	1.5×10^4	1×10^5	1.5×10^4	1×10^5
1 h	1.2×10^3	1×10^5	1.2×10^3	1×10^5
2 h	1.5×10	–	1.5×10	–
3 h	–	–	–	–

or those treated in intensive care units [24]. *Pseudomonas aeruginosa* is resistant to most disinfectants and can survive in hostile environments producing several sepses in patients submitted to hemodialysis. Considering the relevance of this bacterium, the study was conducted using a reference strain and one isolated from a fibrocystic patient (5FQ). For both strains, the bactericidal effect was observed between 2 and 3 h (Table 30.12).

Leptospira interrogans is the causal agent of leptospirosis. The illness can evolve toward an epidemic transmitted by direct exposure to infected animal feces or any contact with wet soils and infected waters. Leptospirosis affects both humans and animals. This deadly disease, produced by spirochete bacteria of the *Leptospira* genus, affects the liver, kidneys, and/or nervous system [25]. *Leptospira* does not produce enteric sickness but it is transmitted by mucosal or damaged skin contact with water.

Considering the slow growth of the bacteria, the determination of viability was conducted using a microscope in dark field mode and the results were presented in qualitative form (Table 30.13). The use of ZZ in stirred conditions showed a reduction of the number

TABLE 30.12

Bactericidal Effect of ZZ on *Pseudomonas aeruginosa* Strains

	Pseudomonas aeruginosa 5F0Q (CFU)		*Pseudomonas aeruginosa* 011 (CFU)	
Time	ZZ Stirred	ZZ Static	ZZ Stirred	ZZ Static
0	3×10^9	1×10^9	5×10^8	5×10^8
15 min	2×10^9	1.8×10^9	1×10^4	5.2×10^8
30 min	1×10^4	1×10^6	1×10^4	5.8×10^8
1 h	1×10^4	1.2×10^5	1×10^4	4×10^8
2 h	1×10^2	–	–	4×10^8
3 h	–	–	–	–

TABLE 30.13

Bactericidal Effect of ZZ on *Leptospira interrogans* M20

	Cell Viability					
	0	15 min	30 min	1 h	2 h	3 h
Stirred	XXXX	XXXX	X	–	–	–
Control	XXXX	XXXX	XXXX	XXXX	XXXX	XXXX
Static	XXXX	XXXX	XXXX	X	–	–
Control	XXXX	XXXX	XXXX	XXXX	XXXX	XXXX

of cells from the first 30 min and a complete mortality after 1 h. In static condition, 2 h was required for a 100% mortality.

The results described above established the bactericidal effect of ZZ on each bacterial strain. The next step was to study the bactericidal effect using a designed purification system such as a pitcher containing a cartridge with a column of 140 g of ZZ and 2 L of water filtration capacity (Figure 30.8b). The results were similar for every bacterial strain: the time required to effectively eliminate the bacteria with ZZ was 30 min to 3 h.

The next experiment used the pitcher with ZZ and 2 L of water contaminated with a pool of bacteria—*V. cholerae, P. aeruginosa, S. typhi*, and *S. sonnei*—with a high number of cells (10^7–10^8 CFU), evaluated at two different temperatures: at 4°C, usual temperature of water consumed in tropical and subtropical countries, and at room temperature, 25°C. The full bactericidal effect was observed after 3 h at room temperature. The experiment was reproduced with another pool of bacteria—*P. shigelloides, A. hydrophila*, and *E. coli* (10^6 CFU)—with identical results after 3 h at room temperature.

Giardiasis is an infection of the small intestine produced by the protozoan *G. lamblia* (*Giardia intestinalis, Giardia duodenalis*). This parasitosis occurs worldwide, with a major prevalence in tropical and subtropical regions [25,26]. The cyst is the infection form of this protozoan released in feces, but it also infects humans through contaminated food and water. The cysts survive in the environment, although they can be destroyed by drying and heat; however, they can survive in cold water for 16 days and resist chlorination in the municipal water treatment plants. When the cysts are released in feces, they can live in wet soil and water for a few months. Mindful of this, we tested the effect of ZZ in the viability of *G. lamblia* cysts in drinking water.

The results in Table 30.14 show a 100% elimination of *G. lamblia* cysts counted after 30 min, an effect that remained after 3 h when the drinking water and ZZ were stirred. On the other hand, although with water and ZZ in static a significant reduction of cysts was also observed, a good exposure to Zn^{2+} ions is required to achieve a full microbicidal effect.

The studies of the microbicidal effect of ZZ show that this material renders absolutely inactive the following microorganisms ordinarily transmitted in drinking water: *A. hydrophila* ATCC 7614, *P. shigelloides* CNCTC 5132, *E. coli* O:149 K88 Lt(+), *E. coli* O:101 K99 St(+), *E. coli* ATTC 25922, *V. cholerae* C7258 (Ogawa), *V. cholerae* C6706 (Inaba), *P. aeruginosa* 5FQ, *P. aeruginosa* O11, *L. interrogans* M20, *S. sonnei, S. typhi*, and *G. lamblia* cysts.

It has been established that to ensure proper disinfection, water with added ZZ should be left in static for at least 30 min at room temperature before drinking it or putting it in the refrigerator. The time should be 3 h when the contamination of microorganisms is >10^5 CFU [27–30].

TABLE 30.14

ZZ Effect on *Giardia lamblia* Cysts in Drinking Water

	Time/Cysts/mL (%)			
Sample	0	30 min	1 h	3 h
Control stirred	20,000	15,000 (25%)	12,500 (37.5%)	12,500 (37.5%)
ZZ stirred	20,000	0 (100%)	0 (100%)	0 (100)
Control static	20,000	15,000 (25%)	15,000 (25%)	7500 (62.5%)
ZZ static	20,000	7500 (62.5%)	5000 (75%)	5000 (75%)

30.3.5 What Is the Useful Life of a ZZ Water Purification System?

A ZZ water purification system can last up to 10 years if it does not sustain physical damages. The only part of the system that needs replacing is the ZZ load or cartridge once its useful life has expired.

30.3.6 How Much ZZ Is Required to Purify the Drinking Water of One Person?

A person's daily water intake should be 2 L, while 100 g of ZZ can purify 150 L of water with maximum quality; therefore, one person could use 100 g of ZZ for 75 days. That is, the ZZ requirement for water purification is 500 g/person/year.

30.3.7 How Long Can the ZZ Load or Cartridge Be Used in a Domestic Purifier?

The ZZ load in a domestic purifier—the pitcher type—is of 200 g, which allows the safe purification of 300 L of water. The daily water consumption of a four-member family—approximately 10 L—would require a monthly replacement of the ZZ load, i.e., 2.5 kg annually. However, if the collective system is used containing 2 kg of ZZ, the replacement would only have to be made once a year. Table 30.15 shows the comparison in energy and cost savings of three methods of water purification: boil, bottle, and ZZ treatment.

30.3.8 What Can You Do with the Discarded Load of ZZ once the Material Has Been Replaced?

When the zinc contained in a ZZ load has been depleted, the zeolite acting as a carrier returns to its original condition, then it can be reloaded, used to improve soil in the garden or flowerpots, or simply thrown in the garbage since it is environmentally friendly.

30.3.9 Can the ZZ Production Technology Be Transferred?

The ZZ production technology has been successfully transferred. It is possible to produce ZZ anywhere with identical quality provided the raw material meets the quality standards and the technological charter are observed.

30.3.10 What Is the Production Cost of 1 Ton of ZZ and How Much Water Can Be Purified with It?

According to the current prices of raw materials and energy, the production cost of 1 ton of ZZ, compared with the cost of the electric energy required for water boiling and disinfection—based on a person's recommended daily intake of 2 L—shows that boiling water can supply 1900 persons in 1 day, while the ZZ produced with the same resources can purify the water consumed in 1 day by 750,000 persons.

30.4 Project Feasibility

A project contemplating the production of ZZ—the design of the water purification systems with ZZ according to culture and habits in a given population, and the distribution

TABLE 30.15

Comparison of Boil, Bottle, and ZZ Methods for Drinking Water Purification

	One Day				One Year					
Method	Water Volume (L)	Process Time (min)	Electric Power (kWh)	Cost (USD)	Water Volume (L)	Electric Power (kWh)	Cost (USD)	Energy Saved (kWh)	Money Saved (USD)	Time Saved (h)
One Person										
Boil	2	33.7	0.573	0.13	730	209.15	46.02	–	–	–
Bottle	2	–	–	0.75	730	–	273.75	–	–	–
ZZ (500 g)	2	4.0	5.37	0.59	730	5.37	0.59	203.78[a]	45.43[a], 273.16[b]	29[a]
Family of Four Persons										
Boil	8	142	2.5	0.55	2920	912.5	200.75	–	–	–
Bottle	8	–	–	3.00	2920	–	1095.00	–	–	–
ZZ (2 kg)	8	32	10.74	2.36	2920	10.74	2.36	901.76[a]	198.39[a], 1092.64[b]	110[a]

Note: One person drinks 2 L of water a day according to WHO.
Electric power cost in Cuba is estimated in 1 kWh = 0.22 USD.
Time of boiling water process = time to obtain 100°C + 10 min of boiling suggested by WHO.
One person needs 500 g of ZZ a year.
Bottled water price 5 L: 1.90 USD.

[a] Comparison with boiled water.
[b] Comparison with bottled water.

and maintenance of the systems—offers significant advantages with respect to the traditional systems and procedures of domestic water purification. The simplicity of the systems and the benefits of ZZ utilization speak for its implementation, for these remove the pathogenic microorganisms transmissible through water and generating epidemics in low-income populations hit by natural disasters or war conflicts, or lacking the necessary infrastructure for water purification and distribution.

The feasibility of this project is based on the amazingly simple procedure of using the new ZZ material, which does not need a substantial modification of the lifestyle of the population in terms of water purification, storage, and consumption.

The success of the project is backed by numerous studies carried out by Cuban specialists in various scientific research and productive institutions whose results have been properly checked out and certified, and can be verified by competent agencies.

The expected results of this project underline the potential to reach broad segments of the population in great need of safe drinking water and zinc supplement to help them to combat infections and epidemics.

The country and organizations financing or executing the project would be able to recover the investment basically through saving energy and resources for public health, raising the quality of water and of people's life, and reducing spending on medications for treatment of infectious diseases.

Some examples of ZZ projects include

- A project designed in Cuba with two basic objectives: water purification and energy saving by replacing the water-boiling procedure. This does not include the impact on the quality of life and health of the population receiving the benefits of the ZZ purifiers.
- The duplex filters using ZZ sold in Mexico are installed in homes, hospitals, restaurants, and companies. The results obtained have enabled promotion expansion and sale of the systems in a market where other water purification systems are available.

30.5 Conclusions

It is proven that the new ZZ material, based on the physical and chemical properties of purified natural zeolite NZ, is a nanostructured material useful for the controlled release of Zn^{2+} ions in drinking water. The zinc ions have a bactericidal effect on a wide spectrum of microorganisms suppressing their infective capacity and virulent action in humans. Furthermore, drinking water treated with ZZ exhibits a reduction of its turbidity and hardness and better quality. The content of toxic heavy metals remains at lower levels than those recommended by the *Guidelines for Drinking Water Quality*, while the presence of arsenic is reduced to lower values than those accepted by quality standards.

The water purification systems using ZZ reproduce their results, which allows for designing a wide range of systems that take ZZ as microbicidal material to address the needs of various and extensive segments of the population.

Significant energy, time, and money savings are possible with the household drinking water purification systems based on ZZ, particularly if their purchase price and operation costs are compared with the price of bottled and/or boiled water.

Another inescapable advantage of ZZ for water purification is its ease of use in cases of natural disasters or military conflicts, since just adding the product to the water, shaking it, and waiting for 30 min makes possible the massive consumption by the affected population that cannot access treated drinking water provided by the corresponding authorities.

Thus, ZZ could become the first nanostructured material with microbicidal properties extensively used by the disadvantaged world population.

Acknowledgments

The authors would like to thank their colleagues and the Cuban institutions that contributed to the studies of the ZZ® microbicide during the past two decades. We also acknowledge J. Vera García, who has significantly improved the English of the draft.

References

1. Centers for Disease Control and Prevention. Available at http://wwwnc.cdc.gov/travel/content/water-treatment.aspx.
2. Johnson, B. Available at http://www.cyber-nook.com/water/Solutions.html.
3. KDF Fluid Treatment, Inc. Available at http://kdfft.com/HowItWorks.htm.
4. Ravishankar Rai, V. and A. Jamuna Bai. Nanoparticles and their potential application as antimicrobials science against microbial pathogens: Communicating current research and technological advances. *A. Méndez-Vilas (Ed.) FORMATEX;* Microbiology Series No.3, 2011, 1: 197–209.
5. Orha, C., P. Barvinschi and G. Burtica. Structural characterization of silver modified clinoptilolite used in water treatment. *Sustain. Human. Environ.*, 2005, 1: 181–184.
6. Rodríguez-Fuentes, G., A. Iraizoz, M.A. Barrios, I. Perdomo and B. Cedré. ENTEREX: Antidiarrheic drug based on purified natural clinoptilolite. *ZEOLITES*, 1997, 19(5–6): 441–448.
7. Rodríguez-Fuentes, G. Characterization of ZZA Zn²⁺ clinoptilolite. *Stud. Surf. Sci. Catal.*, 2004, 154 C: 3052–3058.
8. Registro Sanitario of Cuba No. 009/99 - II/99, 1999.
9. Walsh C.T., H.H. Sandstead, A.S. Prasad, P.M. Newberne and P.J. Fraker. Zinc: Health effects and research priorities for the 1990s. *Environ. Health. Perspect.*, 1994, 102(2): 5–46.
10. Fischer Walker, C. and R.E. Black. Zinc and the risk for infectious disease. *Annu. Rev. Nutr.*, 2004, 24: 255–275.
11. Natural Zeolites—Requirements, NC 625: 2008, *Cuban National Bureau of Standards.* Available at www.nc.cubaindustria.cu.
12. Metodologia para la producción del microbicida ZZ, Industrial secret, IMRE-UH Property, Cuba, 1193. ZZ: Technological procedure, *IMRE-UH property.*
13. World Health Organization. *Guidelines for Drinking Water Quality*, Third Edition, 2004.
14. Drinking Water—Sanitary Requirements NC 827: 2010, *Cuban National Bureau of Standards.* Available at www.nc.cubaindustria.cu.
15. Pérez, A.I., G. Rodríguez-Fuentes and J.C. Torres. Purification of water for human consumption using natural zeolite exchanged with zinc. *Memories of 3er Workshop of Environment Cathedra*, 1997.
16. Sende, V. Report of the results of evaluation of ZZ® system for water purification. *ENAST-National Water Resources Institute Technical Report. La Habana*, 2008.

17. Cedré, B., C. Torres, G. Rodríguez-Fuentes, A. Cruz, H.M. García and L.G. García. Benefits of the Zeolitics Actives Products on Pathogenic Microorganisms. *Proc. XII Congreso Latinoamericano de Microbiología (ALAM)*, Caracas, Venezuela, 1996.

18. Galindo, C.L. and A.K. Chopra. Aeromonas and Plesiomonas species. *Book: Food Microbiology: Fundamentals and Frontier* (eds. Doyle, M.P. and Beunchal, L.R.) Washington, DC: ASM Press, 2007.

19. Mendoza, C. and P. Hernández. Incidence of Plesiomonas shigelloides in humans. *Arch. Latinoam. Nutri.*, 2001, 49(1): 67–71.

20. Obi, C.L. and P.D. Bessong. Diarrhoeagenic bacterial pathogens in HIV-positive patients with diarrhoea in rural communities of Limpopo Province, South Africa. *J. Health Popol. Nutr.*, 2002, 20(3): 230–234.

21. Prere, M.F., S.C. Bacrie and O. Baron. Bacterial aetiology of diarroea in young children: High prevalence of enterophatogenic E coli (EPEC) not belonging to the classical EPEC serogroups. *Pathol. Biol.*, 2006, 54(10): 600–602.

22. Ryan, E.T., S.B. Calderwood and F. Qadri. Live attenuated oral cholera vaccines. *Expert. Rev. Vaccines*, 2006, 5(4): 483–494.

23. Trinka, S.C., C.W. Hoge, L.L. Vanderveg, H.B. Hartman, E.V. Oaks, M.V. Venkatesan, D. Cohen and G. Robin. Vaccination against Shigellosis with Attenuated *Shigella flexneri* 2a Strain SC602. *Infect. Immun.*, 1999, 67(7): 3437–3443.

24. Rogan, M.P., C.C. Taggart, C.M. Greene, P.G. Murphy, S.J. O Neill and N.G. McElavaney. Loss of microbicidal activity and increased formation of biofilm due decreased lactoferrin activity in patients wit cystic fibrosis. *J. Infect. Dis.*, 2004, 190: 1243–1245.

25. Lashley, F.R. Emerging infectious disease: Vulnerabilities, contributing factors and approaches. *Expert. Rev. Anti Infect. Ther.*, 2004, 2(2): 299–316.

26. Garcia, L.S. and D.A. Bruckner. *Diagnostic Medical Parasitology*, 2a ed. Washington, DC: American Society for Microbiology, 1993, 31–39.

27. Cedre Marrero, B., N. Batista, L. Riverón, O.M. Martínez, C.E. Perez, R. Martínez and I. Montano. Effect of ZZ against the viability of different strain bacteria of hydric transmission. *Technical Report, Instituto Finlay, La Habana*, 2008.

28. Bravo, L. , A. Fernández, M. Ramírez, G. Martínez and A. Llop. Bactericide activity of chemical compound Zinc-Zeolite (ZZ) against the viability of microorganisms of hydric transmission. *National Laboratory of Reference EDA, Instituto de Medicina Tropical "Pedro Kourí," Technical Report, La Habana*, 2008.

29. Fernández-Abreu, A., L. Bravo-Fariñas, G. Rodríguez-Fuentes, M. Ramírez-Alvarez, A. Aguila-Sánchez, Y. Ledo-Ginarte, Y. Correa-Martínez and Y. Cruz-Infante. Evaluation of the Cuban product Zinc-Zeolite (ZZ) with strains of Aeromonas and pleisiomonas. *Lab. Ciencia.*, 2009, 4: 49–12. Available at www.labciencia.com.

30. Ruiz Espinosa, A. Results obteined of the action of ZZ® in the viability of *Giardia lamblia* cysts. Technical report. *Instituto de Medicina Tropical "Pedro Kourí," Technical Report, La Habana*, 2008.

31

Phosphorous Removal and Recovery Using Nanotechnology

J. Richard Schorr,[1] Suvankar Sengupta,[1] Rao Revur,[1] Richard Helferich,[1] and Steve Safferman[2]

[1]*MetaMateria Technologies LLC, Columbus, Ohio, USA*

[2]*Department of Biosystems and Agricultural Engineering, Michigan State University, East Lansing, Michigan, USA*

CONTENTS

31.1 Problem

Clean water is recognized as an important and growing priority and is seen as a major challenge worldwide because potable water is being used faster than it is replenished. Phosphorus in surface water is one of the most common causes of impairment. In addition to eutrophication, a more recent impact is the growth of cyanobacteria (blue green or toxic algae; Figure 31.1). The impact to ecological health and economic loss due to reduced

FIGURE 31.1
Harmful phosphor bloom.

recreational use is staggering in some locations. It is occurring in all categories of water: rivers, streams, lakes, reservoirs, estuaries, and coastal areas, often resulting in "eutrophic" or oxygen-deprived water. The amount of nutrients entering our waters has dramatically escalated during the past 50 years. A 2006 US Environmental Protection Agency report identifies approximately 207,355 miles of streams (~30%) that had "high" concentrations of phosphorus, while 108,029 miles of streams had "medium" concentrations. Harmful algae blooms nationally cause 78% of continental coastal areas to exhibit symptoms of eutrophication. Runoff from agricultural sources contributed an estimated 84% of the nonpoint source phosphate loading to surface water in the United States (Zaimes and Schultz, 2002). This results in an estimated annual $2.2 billion economic impact (Krögera et al., 2012).

Yet phosphorus is a scarce resource. Some global estimates of easy-to-obtain phosphorus reserves are just 35 years (Clabby, 2010). Morocco, Jordan, South Africa, the United States, and China account for 90% of the reserves (Zaimes and Schultz, 2002). The use of phosphorus has doubled since 1960, and this trend will increase as world food requirements are expected to double by 2030 (US Department of Agriculture). Conservation technologies must not only protect us from phosphorus but must also serve to recover this valuable resource.

Phosphorus, the 13th element, is part of all life and circulates through the environment as both inorganic and organic compounds and in soluble and particulate forms. The primary source for phosphorus is "phosphate rock," which contains ~10% phosphorus. The rock is mined and phosphorus extracted by chemical methods to form phosphoric acid, which is used for fertilizer, food, and industrial products. In agriculture, crops remove phosphorus from soils, which then need to be replenished or over time it limits crop growth. Crops are used to produce animal protein but a significant amount passes through and is found in animal waste. Municipal wastewater also contains phosphorus, albeit at tremendously lower concentrations, that originates from human excretion and common household products. International efforts exist to economically capture and recycle this phosphorus. While still in its infancy, someday one might imagine that the phosphate we consume as food will return back to food production and in the many products that require a phosphate input. Phosphorus is banned in many states for some products.

Sources of phosphate water pollution can be divided into two broad categories: point and nonpoint sources. Point sources are generally sources from industry or wastewater treatment systems that enter bodies of water from a discharge pipe or point. Nonpoint sources of water pollution result when rainfall/storm water carries or collects pollutants across large surface areas, paved or nonpaved, eventually flowing into surface water, a common occurrence from agricultural and urban storm water runoff. The concentration of phosphorus is much lower in nonpoint sources and more difficult to capture.

Phosphate ions are the common form of phosphorus found in water. This is a central phosphorus atom surrounded by four oxygen atoms and carries three negative charges in the form of electrons, so it is written as PO_4^{3-}. These negative charges have to be counterbalanced by three positive charges, such as sodium. Phosphate ions can combine with trivalent iron or aluminum cations in acidic soils to form insoluble precipitates.

Removal of phosphorus from wastewater is challenging. Unlike nitrogen, which can be microbiologically converted to nitrogen gas that escapes passively into the atmosphere, phosphorus is not volatile. Treatment options include biological or chemical/physical systems. Biological phosphorus removal entails luxury microbial uptake in an anaerobic/aerobic sequencing system. The result is excess accumulation in the biomass, beyond nutritional requirements, and removal when the excess biosolids are removed. Phosphorus can also be precipitated with an iron or aluminum salt, and compounds such as struvite can be produced by combining phosphorus with magnesium and ammonium. As with biological systems, a precipitated phosphorus complex must be physically removed from the wastewater by sedimentation or filtration.

More efficient and economical treatment systems are needed to recover phosphorus from wastewater. This is where "nanotechnology" shows great promise as an enabler for water purification to supplement traditional water purification methods. Better solutions need to be lower in overall cost, durable, and more effective for the removal of contaminants from water. As described in this chapter, nanotechnology shows exceptional performance for phosphorus removal and can provide faster removal in more compact, lower-cost systems. A novel composite is described, consisting of a highly porous, iron-based ceramic containing high concentrations of nanomaterials that are effective in sorbing phosphorus, which can be subsequently removed to allow reuse of the media and economic recovery of the phosphorus for other applications.

31.2 Phosphorous Removal Technologies

Phosphorous from domestic, industrial, and agricultural wastewater is typically removed by chemical, sorption, or biological approaches, and combinations are sometimes used.

Soluble phosphorous can be removed by precipitation with metal salts or sorption of ions onto a media. Salts of iron, aluminum, calcium, and magnesium are typically used as precipitating agents. These salts dissolve in water and chemically react to form phosphates, which can be coagulated and separated through settling and/or the use of filters. Specific, commonly used metal salts include different forms of alum (aluminum sulfate), sodium aluminate, ferric chloride, ferric sulfate, ferrous sulfate, and ferrous chloride (Bachand, 2003; Panasiuk, 2010; Camm, 2011).

Alternatively, dissolved phosphorous can be removed through "sorption" by a media, which refers to the combined processes of adsorption and precipitation of soluble phosphate. Sorption processes can be categorized into electrostatic forces, physical forces, and chemical bonding (Minton, 2002). Ion exchange is a form of electrostatic forces. As the solution passes through the media, a pollutant ion (preferred ion) may be replaced with an ion from the media (less preferred ion). The media is exhausted when all the least preferred ions are exchanged by preferred ions (Minton, 2002). During adsorption, matter (adsorbate such as molecules, ions, particles, polymers, colloids) dispersed in solution accumulates on the surface of an adsorbent (Kasprzyk-Hordern, 2004). Adsorption may occur through weaker physical sorption processes or through a stronger chemisorption where bonds are formed between the adsorbate and adsorbent. Physical forces are due to van der Waal interactions as partial charges of the adsorbate are attracted to the electrostatic charges of the adsorbent (Minton, 2002). Chemisorption is a much stronger interaction but requires an available sorption site on the adsorbent. Once chemical bonds form, desorption becomes difficult. In general, physical adsorption has rapid kinetics, followed by a slower diffusion of adsorbed matter into the matrix of the adsorbent (Kim et al., 2008).

Phosphorus-sorbing materials typically contain appreciable concentrations of aluminum (Al), iron (Fe), lanthanum (La), calcium (Ca), or magnesium (Mg). Sorption media can be naturally occurring materials, by-products of industrial processes or engineered products.

31.2.1 Chemical Precipitation

Chemical precipitation is widely used for removal of phosphorus. In wastewater treatment, the chemical may be added before primary or secondary treatment or as part of a tertiary treatment process. The effects of phosphorus and calcium concentrations, pH, temperature, and ionic strength on theoretical removal are thermodynamically modeled (Song et al., 2002); however, further studies inferred these processes to be more complex than theoretically predicted. A major concern in all precipitation processes is formation of excess sludge that needs to be removed or processed, which leads to additional costs for treatments. Sludge formation is most pronounced when lime ($CaOH_2$) is used for precipitation. Predictions concerning the best precipitation practices also vary widely in different studies (Takacs, 2006; Neethling and Gu, 2006).

The required dosage level of the precipitating chemical is related to the concentration of the soluble phosphorous. For example, if the target concentrations are >2 mg/L, a dose of 1.0 mol of aluminum or iron per mole of phosphorus is sufficient. For lower phosphorus concentrations (0.3–1.0 mg/L), the dose must be in the range of 1.2–4.0 mol of aluminum or iron per mole of phosphorus. The pH value is also an important factor for efficient removal of phosphorus, when using alum or other salts, as the solubility of their precipitates vary with pH. For alum, phosphorus removal is most efficient in a pH range 5–7 and for ferric salts it is 6.5–7.5. It should be emphasized that the impact of different competing reactions occurs and needs to be taken into account. Therefore, the amount of metal salts needed has to be determined practically in each case rather than simply basing calculations on the theoretical chemical reaction with phosphorus (Tchobanoglous and Burton, 1991). If removal involves lime, care is needed to ensure reaction between excess calcium ions and phosphate (Tchobanoglous et al., 2002). While lime can be an effective agent for phosphorus removal, its application is slightly diminished because of the high volume of produced sludge.

31.2.2 Sorption of Phosphorous

A wide variety of materials containing calcium, aluminum, iron, magnesium, and lanthanum have also been used for the sorption of phosphorus. Key media characteristics are composition, surface area, porosity, hydraulic conductivity, and cost.

Naturally occurring materials, such as various types of zeolites, bauxite, laterite, dolomite, shale, limestone, calcite, vermiculite, and iron-rich sands, have been examined. Zeolites (hydrous aluminum silicates) (Sakadevan and Bavor, 1998) can contain calcium, ferric, magnesium, and titanium oxides at much lower concentrations. Bauxite is a naturally occurring mixture of hydrous aluminum oxides and aluminum hydroxides and can also be high in ferric oxides. Laterite is low-grade bauxite (Wood and McAtamney, 1996) that is high in iron oxides and aluminum hydroxides. Shale is a fissile rock formed by the consolidation of clay, mud, or silt, and has high concentrations of iron and aluminum (Pant et al., 2001). Several industrial and wastewater by-products have also been examined as phosphorous sorption media. These include blast furnace slag, steel furnace slag, red mud, fly ash, HiClay Alumina, and aluminum- and iron-based water treatment residuals. All of these materials are either rich in aluminum, iron, and/or calcium.

Engineered media are more expensive but also much more effective. These media typically contain iron, aluminum, calcium, or lanthanum, and are engineered to produce high surface area and good hydraulic properties. Examples include activated alumina, which is characterized by high aluminum content, high specific surface area, and high macroporosity. Lightweight expanded clay aggregates contain iron. PhosLock™ is a bentonite clay treated by an ion-exchange process where lanthanum ions displace sodium ions within the clay matrix (5%). Modified diatomaceous earth utilizes lanthanum oxide and lanthanum–aluminum oxide to remove phosphorus, also used for arsenic and arsenate removal (Misra and Lenz, 2003). Other engineered media developed for arsenic removal have also been evaluated for phosphorus removal, including deposition of active nano-iron materials onto a porous polymer resin (Layne-RT) or activated carbon. Product prepared by MetaMateria uses nano-iron oxide crystals grown within a highly porous iron ceramic (Meta-PO4™). Testing shows much higher capacity than naturally occurring minerals or activated alumina (Safferman et al., 2007). A summary of performance for various media is shown in Table 31.1.

TABLE 31.1

Examples of Phosphorus Sorption Media

Sorption Media	Capacity (mg/g)	Reference
Lightweight aggregate	0.037	Zhu et al. (1997)
Steel furnace slag	0.38	Mann (1997)
Blast furnace slag	0.4	Mann (1997)
Zeolite	0.46	Drizo et al. (1999)
Bauxite	0.61	Drizo et al. (1999)
Fly ash	0.63	Mann (1997)
Shale	0.65	Drizo et al. (1999)
Limestone	0.68	Drizo et al. (1999)
Zeolite	2.15	Sakadevan and Bavor (1998)
Hematite	1.43	Sakadevan and Bavor (1998)
Goethite	16.4	Oh et al. (1999)
Alumina	17.1	Oh et al. (1999)
Allophane	51	Oh et al. (1999)
Meta-PO4 >10 mg/L	80+	Helferich (2011)
Meta-PO4 <10 mg/L	40	Helferich (2011)

31.2.3 Mechanisms for Sorption of Phosphorus

Soluble phosphate ions are removed from contaminated water by media either by bonding with the surface (for iron-, aluminum-, or lanthanum-based media) or by precipitation (for calcium- and magnesium-based media). The pH generally determines the effectiveness of the media and associated phosphorous removal mechanism. In general, alkaline environments favor phosphorus removal by calcium adsorption and precipitation, whereas low pH and acidic environments favor removal by iron and aluminum (Ugurlu and Salman, 1998).

31.2.3.1 *Aluminum- and Iron-Containing Media*

Under acidic to a neutral pH, phosphate ions are chemically adsorbed onto Fe and Al oxide surfaces through ligand exchange (Wood and McAtamney, 1996). Engineered media containing metal oxides are ideal for their high affinity toward phosphate molecules. Metal ion (cation) bonds with phosphate ion (anion), forming inner-sphere surface complexes (Figure 31.2) (Zach-Maor et al., 2011; Camm, 2011).

Several surface complexes can form with iron or other metal oxides, including mononuclear monodentate, mononuclear bidentate, and binuclear bidentate (Zach-Maor et al., 2011; Camm, 2011). Sorption capacity can be enhanced by increasing the effective surface area, which can be achieved by reducing the particle size. Sorption reactions are affected by competing species that may impede removal of pollutants. Changes in temperature can also affect sorption processes. Higher temperatures increase the vibration frequencies of sorbed molecules, making desorption more likely (Minton, 2002). The largest influence on sorption reactions is pH. In the presence of water, metal oxides are surrounded by hydroxyl groups, protons, and coordinated water molecules (Liu et al., 2001). The metal oxides are amphoteric, allowing them to act as either an acid or base. This property is influenced by the surface charge, and therefore pH, of the surrounding solution. In general, at a higher

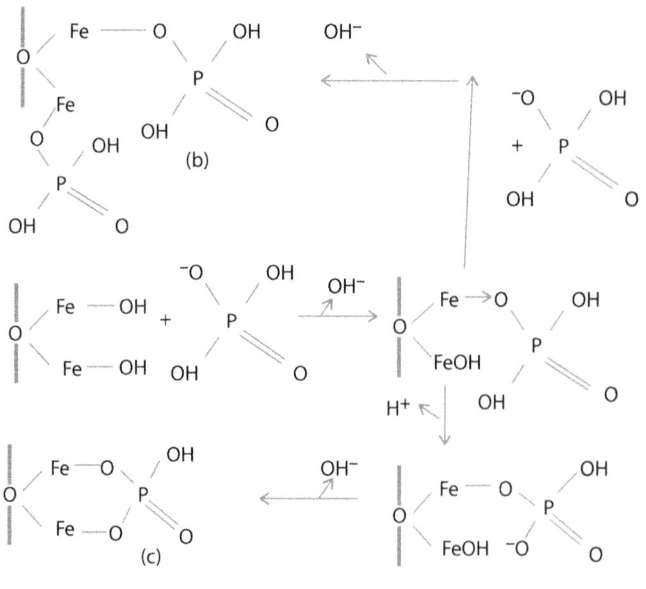

FIGURE 31.2
Types of phosphorus adsorption.

pH, surfaces are negatively charged due to the increased prevalence of hydroxide ions (OH⁻) in solution. Conversely, at a lower pH, positively charged surfaces are formed as more hydrogen ions (H⁺) are in solution (Figure 31.2). Engineered media are designed to offer high positive charge at the pH of interest to effectively bind anionic phosphates.

31.2.3.2 Calcium-Rich Media

At a higher pH, phosphate removal with a calcium-rich media occurs by precipitation of calcium phosphate (Ugurlu and Salman, 1998). Precipitates were originally amorphous calcium phosphate as shown in x-ray diffraction studies (Johansson and Gustafsson, 2000). They also found that precipitation was the primary mechanism removing phosphate from solution, while calcium and phosphorus speciation data ruled out formation of amorphous calcium phosphates. Instead, evidence indicates that precipitation to hydroxyapatite is the primary mechanism. Hydroxyapatite has low solubility and may tightly bind phosphorus. Regardless of the exact mechanism, Johansson and Gustafsson (2000) concluded that the strongest phosphorus removal is obtained in substrates from which calcium easily leaches into supersaturate solutions and for substrates that already contain seeds for hydroxyapatite or other apatite to enhance precipitation. The rate of calcium desorption may be a primary factor in sorption experiments that show there is an initial rapid removal of phosphorus followed by a lower but more sustained removal rate. The initial high removal rate is attributed to rapid desorption of calcium from the substrate leading to precipitation (Cheung and Venkitachalam, 2000). For calcium-rich substrates, the calcium content and its form are the primary factors for phosphorus sorption capacity (Brix et al., 2001; Johannson, 1999), and it was found that amorphous forms of calcium outperformed crystalline forms.

31.2.4 Biological Treatment to Remove Phosphorus

Phosphorus removal from wastewater has long been achieved through biological assimilation–incorporation as an essential element in biomass, particularly through growth of photosynthetic organisms (plants, algae, and some bacteria, such as cyanobacteria). Traditionally, this was achieved through treatment ponds containing planktonic or attached algae, rooted plants, or even floating plants (e.g., water hyacinths, duckweed). The net biomass is then removed before it decays and releases phosphorus in water (Strom, 2006c).

Enhanced biological phosphorus removal (EBPR) is a widely used method to decrease phosphorus in full-scale wastewater treatment plants (Sedlak, 1991). It is of great interest because it is possible to reach low concentrations (1.0–0.1 mg/L) and can also result in minimal sludge production and moderate operational cost (Strom, 2006a, b, c). Compared with chemical precipitation, the main advantage of EBPR is the absence of metal ions from coagulant in the sludge. While removal of biological oxygen demand (BOD), nitrogen, and phosphorus can all be achieved in a single system, it can be challenging to achieve very low concentrations of both total N and P in such systems.

The EBPR microbiology was reviewed by Mino et al. (1998), Mulkerrins et al. (2003), and Strom (2006c). In brief, phosphate-accumulating organisms (PAOs) store polyphosphate as an energy reserve in intracellular granules. Under anaerobic conditions, in the presence of fermentation products, PAOs release orthophosphate, utilizing the energy to accumulate simple organics and store them as polyhydroxyalkanoates. Under aerobic conditions, the PAOs then grow using the stored organic material to take up orthophosphate and store it as polyphosphate. Thus, PAOs, although strictly aerobic, are selected by having an upfront anaerobic zone in an activated-sludge type of biological treatment process. The PAOs are able to compete with other aerobes under these conditions because of their ability to sequester a fraction of the available organic material under the initial anaerobic conditions, while outcompeting the anaerobes because of the much higher energy yield from aerobic vs. fermentative metabolism.

The phosphate in EBPR is removed in the activated sludge, which might have 5% or more P (dry weight) as opposed to only 2%–3% in non-EBPR sludge. EBPR has been demonstrated in several systems (Tchobanoglous et al., 2002), such as the various Bardenpho processes (also remove N), including the A/O and A/A/O or A2O (removes N) processes, sequencing batch reactors, and the PhoStrip™ process (which combines EBPR with phosphate stripping and chemical removal). Simultaneous biological nutrient removal (SBNR) has also been reported in treatment systems, such as the Orbal™ oxidation ditch, not specifically designed for nutrient removal. SBNR has been examined in some detail (Littleton et al., 2000, 2002a, 2003a, b, 2007; Strom et al., 2004).

During successful operation, the EBPR is moderately priced and is also an environmentally sustainable process for phosphorus removal. On the other hand, the reliability and stability of this method can be a problematic issue. Process breakdown, performance worsening, and even failure can happen during EBPR processing (Oehmen et al., 2007). In the work by Neethling et al. (2005), factors affecting the EBPR reliability were studied. A BOD/P ratio of more than 25:1 and recycle stream control are some of the factors that help maintain reliable and high removal levels. It was also found that the lower the concentration of phosphorus achieved, the smaller the period of time it can be maintained on the given level: (i) concentration <0.1 mg/L was a quite long period (more than a month); (ii) 0.03 mg/L was a week; and (iii) <0.02 mg/L was for only a couple of days.

31.3 Nanotechnology

Nanotechnology is the manipulation of individual atoms and molecules in ways that create materials and devices having vastly different properties. "Nano" means one billionth, in this case a nanometer (nm). By convention, nanomaterials are <100 nm. These materials contain many surface atoms that are very reactive; however, a major challenge in using nanomaterials is their small size, where it can be difficult for water to easily reach all of the active surfaces and fine pores are easily plugged. These fine particles or fibers (Figure 31.3) cannot just be added to drinking water but rather must be incorporated into filtration media in ways that allow for the contaminants in water to readily come in contact with these active materials. Another problem is that water flow through fine materials is extremely slow and can require considerable pressure to obtain acceptable flow rates. Cost-effective design approaches for using nanomaterials must provide for operational efficiency.

Nanomaterials have been examined for the removal of many types of contaminants. The best approaches utilize nanomaterials in ways that allow for ease of water flow, which means attachment to a surface that water can readily pass. Because of the potential for use of nanotechnology, many organizations are working to develop filters and media that take advantage of the properties of nanomaterials for removal of contaminants from water. These range from carbon nanotubes to nanocrystals for the selective absorption of contaminants (Meridian Institute, 2006). Nanomaterials are now used in commercial products, and this can be expected to expand in the future (Weinberg et al., 2013).

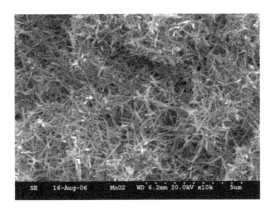

FIGURE 31.3
Example of Nano-crystals.

31.4 MetaMateria Approach to Phosphorus Removal

MetaMateria Technologies utilizes proprietary process technology to prepare nanomaterials in the form of particles, whiskers, or platelets that can take advantage of the unique properties afforded by these nanomaterials for removal of contaminants from drinking water. For phosphorus removal, nano-iron oxyhydroxide (FeOOH) crystals are grown on a novel high surface area porous ceramic structure.

31.4.1 Novel Porous Ceramic Composite

Figure 31.4 shows two examples of the highly porous ceramic composite used for phosphate removal. These contain alumino–silicate binders and filler materials, such as iron or zeolite powder. They typically have >80% open interconnected porosity, a density of 0.4–0.6 g/cm^3, and surface area of 15 m^2/g (>700,000 m^2/m^3) but this can be higher.

The novel method used to prepare the porous ceramic substrate starts with slurries of reactive materials, filler/aggregate materials, and surfactants/gas-forming agents. After mixing, liquid is poured into molds to make shapes or into a mix chamber to make aggregates. An alumino–silicate bonded matrix with interconnected porosity is formed during gas expansion of liquid (three to four times) followed by solidification due to chemical reactions. The product then goes through a curing step and other postforming procedures to prepare it for nanomodification. The phosphorous media (Meta-PO4) is prepared using

FIGURE 31.4
Foamed ceramic shape examples.

FIGURE 31.5
Water flow in a porous plate.

an iron powder as a filler, creating iron foam with a surface area of ~15 m²/g and a density of ~0.3 g/cm³.

This method of producing media develops a hierarchical structure of open interconnected porosity and provides extensive surfaces for deposition of active substances. An openness to water or gas flow exists at a low pressure drop (Figure 31.5). It also provides great flexibility in shape and size of the media for integration into systems. Composition can be adjusted to suit an application.

The high surface area of the porous structure is an ideal platform for active materials. These include (i) beneficial bacteria for bioremediation, (ii) nanomaterials to capture or breakdown contaminants, or (iii) anionic and cationic surfactants to capture contaminants such as perchlorates. Characteristics include

- *High surface area:* 100× higher than most competitive media
- *Highly porous:* ~80% with a pore hierarchy that enhances turbulent liquid flow
- *Water readily flows:* through media at low pressures
- *Available in various physical forms:* monoliths, pellets, granular
- *Composition can be adjusted:* to provide enhanced chemical reactivity
- *Fast reaction rates:* using nanomaterials
- *Cost-effective:* for most applications

31.4.2 Nanomodification

Nanomaterials are used in the Meta-PO4 media to provide large numbers of sites for sorption of phosphorus compounds. FeOOH crystals are excellent in capturing phosphate and some metal ions. Iron is known to remove phosphorus but Meta-PO4 performance is significantly enhanced by a high concentration of nano-FeOOH crystals grown on the iron foam surfaces. Surface area is >70 m²/g, and >10 wt% of nano-iron compounds are added during the modification process, which are easily accessed through interconnected porosity. Surface area and interconnecting pores are shown in Figure 31.6a and the high concentrations of 20–40 nm acicular crystals in Figure 31.6b.

During use, phosphate ions are adsorbed onto the surface of the FeOOH nanoparticles. In most cases, any FeOOH present will slow adsorption kinetics owing to diffusion resistance and, consequently, their separation from aqueous solutions after adsorption is

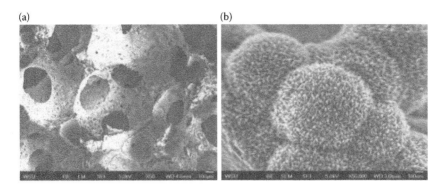

FIGURE 31.6
(a) Interconnecting pore structure; (b) surface containing 20–40 nm crystals.

more difficult. Enhanced adsorption is accomplished by reducing the size of the FeOOH using nanoparticles, which have a large surface-to-volume ratio compared with other bulk materials.

31.4.3 Media Performance

Nanomaterials significantly enhance the media to capture phosphorus. Early testing of a nanomodified granular product was done by Michigan State University (MSU) in 2007 (Safferman et al., 2007). This showed a much higher capacity and longer life than types of media, indicating the potential for nanomodified, high surface area media to remove contaminants from water. Continued development under a grant by the National Science Foundation resulted in a media with even higher capacity for phosphorus removal. Figure 31.7 shows a comparison with other materials, including a commercial iron-activated alumina. Testing with actual septic system water containing 7–10 mg/L shows that phosphorous continued to be captured for >15 months and the overall capacity (mg/g) was 5–10× higher than other commercial products.

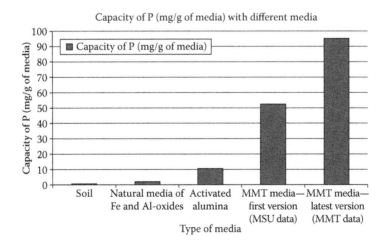

FIGURE 31.7
Capacities of natural and manufactured media for phosphorus removal.

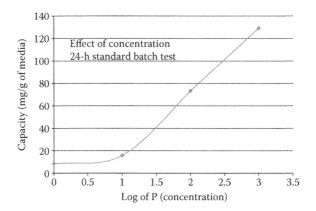

FIGURE 31.8
Effect of concentration on phosphorus removal capacity.

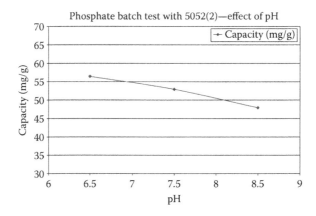

FIGURE 31.9
Effect of pH on capacities at 10 ppm of phosphorus.

Figure 31.8 shows that media capacity depends on the concentration of the phosphorus in the bulk fluid, as expected. Batch tests were conducted by mixing 100 mg of media with 100 mL of water. The desired concentration of phosphorus was added to the water, and the pH was adjusted to 6.5. After tumbling in a plastic bottle for 24 h, the media was filtered and the solution analyzed and capacity determined. Batch tests with silica and nitrate (potential competing ions) had very little effect on capacity.

The pH of water is important in sorption of phosphorus, as shown in Figure 31.9. Removal capacity is generally lowered at higher pH values (>7). Changes in capacity of one media at a 10 mg/L phosphorus concentration were evaluated. Capacity decreased modestly (~12%) as pH increased from 6.6 to 8.6.

31.4.3.1 Long-Term Testing with Actual Septic Tank Discharge

Long-term testing with actual septic tank discharge is being done using 5-cm-diameter columns containing either granular or monolith media. Figure 31.10 shows phosphorus consistently remains <1 mg/L at an empty bed contact time (EBCT) of 3 h.

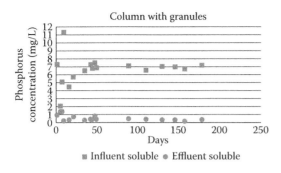
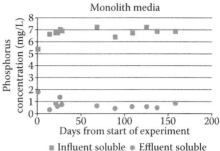

FIGURE 31.10
Longer term column testing using septic water.

31.4.3.2 Key Factors in Using Media in System

Meta-PO4 media can be made in different shapes (e.g., cubes and plates shown in Figure 31.11), and this can affect considerations in packaging media for use. Beds can contain pieces of media of different size, solid monoliths or combinations.

Time of water in contact with the phosphorus removal media is a major consideration when using any media, as well as the concentration gradient and the flow rate past the media (flux). It is desirable to have the EBCT to be 30 min or longer. For lower concentrations of phosphorus, an ECBT of 1 h or longer is preferred. Phosphorus uptake for this media appears to be following a first-order adsorption mechanism, and the EBCT depends on the diffusion of the phosphorus from the bulk liquid.

To enhance performance of the media, circulation of the water by the surface of the media shows improved removal of phosphorus. Circulation has been examined in several ways, and all improve performance. Water can be circulated through the design of water inlet, through use of airlifts, or through pumps. The ratio of media to water is important, as can be seen in Figure 31.12 by comparing ratios of 100 g of media per liter of water. Circulation helps mitigate channeling or bypass of media, which is important when using larger pieces such as the cubes (Figure 31.11). For monoliths packaged into a column, recirculation is less important. In general, a 30 min EBCT will lower phosphorus from 10 mg/L to <1 mg/L. Polishing filters can be used to lower concentration further (<0.1 mg/L). Testing shows that phosphorus can be dropped from >100 mg/L to <1 mg/L within an hour.

FIGURE 31.11
Media shapes.

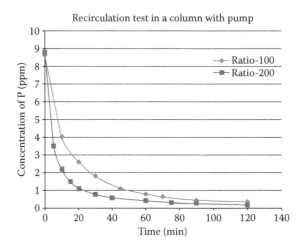

FIGURE 31.12
Effect of media: water ratio.

31.4.4 System Design Approaches

MetaMateria finds that a multichamber approach gives excellent results, especially when used with recirculation. In this approach, water flows into the first chamber where 70%–80% of the total phosphorus can be removed and then into the second chamber, which acts as a polishing filter. Once effluent from the second container approaches the desired concentration (e.g., 1 mg/L), the first chamber can be taken out of service and phosphorus removed and media regenerated for reuse. Water can now pass through the second container and into a third chamber (if available) during regeneration of media in the first chamber. For a given concentration and flow rate, the amount of media needed can be determined by desired residence time and the desired time between regenerations. This may require more than two chambers, especially at lower concentrations. The flux rate through a given area will depend on the application; however, it is expected to be 75–200 L/min for 1 m² of area, and for some applications it may be higher.

In general, criteria to be considered in designing systems utilizing Meta-PO4 media are

- Concentration of phosphorus, since it affects the average media capacity (g-P/kg-media).
- EBCT determines the time water is in contact with media.
- Initially, a 30 min ECBT is chosen, but this may be lower for some systems.
- Flux (L/min-m²) is the amount of water that flows through square meter of media.
- Total flux includes any recirculated water and is typically 75–200 L/min-m² but may be higher.
- Desired regeneration period represents time between regeneration. For lower concentrations (<10 mg/L), this may be 30 days or more. For concentrations >50 mg/L, it may be weekly.

31.4.5 Media Regeneration and Reuse

Media regeneration and reuse is important to the life cycle costs of the media, since the cost for regeneration will be significantly less than that of the initial media. After sufficient

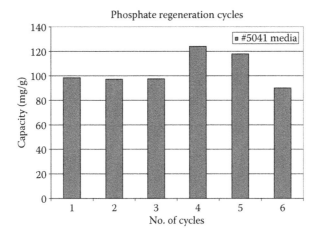

FIGURE 31.13
Capacity after regeneration.

saturation, phosphorus is chemically removed and media regenerated for reuse. Figure 31.13 shows that it is feasible to retain >90% of the original media capacity after six regenerations and >10 uses of media is expected to commonly occur; however, more verification is needed through field testing. Competing ions can reduce capacity and mechanical degradation of this very porous material with repeated uses; therefore, the lifetime is still unknown.

Regeneration can be done on site or in modules removed and regenerated off site. Economically, on-site regeneration makes the most sense when higher concentrations of phosphorus and/or larger volumes of water are treated. It is anticipated that regeneration will be done in the same module used for phosphorus capture.

The data shown in Figure 31.13 was prepared by circulating water with high concentrations of phosphorus through a bed of media for saturation, followed by removal using a mild base. Media is regenerated with a mild acid. Phosphorus saturation and extraction is considered to be one cycle. The increases in capacity shown are believed to be due to the activation of some of the iron powder used in the porous media composition, adding some additional capacity. The base iron media itself has a capacity of 15–20 mg/g.

The cost for regeneration can be 20% or less of the initial cost, which significantly reduces life-cycle costs of media. With regeneration and reuse at phosphorus concentrations >10 mg/L, it is feasible for 1 kg of media to remove in its lifetime >1 kg of phosphorus, considerably higher than other sorptive media.

In practice, the operating conditions in the field will determine the number of regeneration cycles for the media. In a two-stage operating system, the point at which the first module is regenerated will occur when the effluent reaches its desired limit. The media will not be fully saturated.

31.5 Phosphorus Recovery

Phosphorous is a valuable resource, and its recovery as a usable product is a goal that can be accomplished with use of Meta-PO4 media. Since phosphorus can be removed in a soluble form, it can be recovered through precipitation of calcium phosphate, $Ca_3(PO_4)_2$, or other compounds. Experiments show that calcium phosphate precipitates occur and can be removed by filtration or centrifugation. Analysis shows crystalline material with the expected stoichiometry (Ca:P ratio) close to 3:2. Surface area was 188 m^2/g, indicating fine particles. Impurities of silicon, iron, and aluminum (present in the original porous matrix) were low. These tests were done using standard synthetic water; however, possible contamination problems may occur under field test conditions. For many wastewater applications, it is anticipated that the recovered phosphate will be able to be used for fertilizer or other phosphoric acid applications.

Recovery of phosphorus from waste materials has been of high interest in recent years, and some 22 processes for phosphorus recovery have been identified (CEEP, 2012). There are large amounts of phosphorus available in waste streams from agriculture, sewage treatment, and many industrial water streams, especially food and pharmaceuticals. Three types of phosphorus recovery approaches have been investigated: (i) from liquid streams (precipitation or sorption/desorption), (ii) from sewage sludge, and (iii) from sludge incineration ash. One problem with sludge is that often it contains aluminum or iron metal ions from chemicals added to precipitate phosphorous. No matter which approach is used, it must be economical and many methods are not today. However, in a survey of 417 experts worldwide (CEEP, 2012), the general opinion was that economically viable processes will emerge by 2030 and half the experts believe that recovery of phosphorus from liquids is the most important route. Contaminants in wastewater and sludge do represent a serious issue.

Precipitation methods for recovery of phosphorous compounds and adsorption/desorption from liquid streams are being pursued by a number of organizations worldwide. Typically, $Ca_3(PO_4)_2$ or struvite ($MgNH_4PO_4$) compounds are precipitated on fine sand that acts as a seed for crystallization. Some 330 kg of recovered phosphorus will yield 8000 kg of struvite, which can be used as a fertilizer supplement or 1584 kg of calcium phosphate. Calcium phosphate is ideal for additional industrial processing since it is indistinguishable, in most respects, from minerals recovered from phosphate rock. Full-scale struvite recovery processes have been used for >10 years. One problem with that type of process is that it does not lower the phosphorus in water to <1 mg/L (a commonly accepted discharge limit), so additional polishing reactors are needed.

Meta-PO4 media can remove 99% of phosphorus and reach phosphorous levels <0.1 mg/L. This is much higher than most chemical precipitation processes, such as struvite crystallization, which only removes 80–90%. Once separated, soluble phosphate ions can be easily precipitated as calcium phosphate and removed by filtration or centrifugation. This provides an economic approach to both harvest and recover phosphorous, since the phosphorous is really a by-product of the removal process. Calcium phosphate can provide an economic path to production of usable phosphorus products, such as phosphoric acid. It can also be used to produce potassium phosphate compounds for use in fertilizer. The cost to reclaim the phosphorous is the cost to remove/precipitate the PO_4 ions from liquids remaining after the media is regenerated.

31.6 Economics of Phosphorus Removal

The cost to remove phosphorus will be much lower than competitive sorbents and will often compete with chemicals (iron and aluminum) used to precipitate phosphorous,

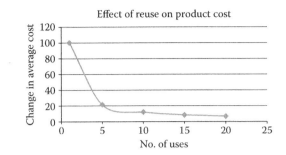

FIGURE 31.14
Effect of reuse on product cost.

which have a relatively low cost. Meta-PO4 media provides a simpler removal approach with much lower capital cost.

The cost to remove phosphorous depends on the initial cost of the media but primarily on the number of times the media can be reused and whether regeneration is done on site. Figure 31.14 illustrates how important regeneration affects average product cost, when regeneration is done on site. With five reuses, the average cost drops 80%, and 95% if media can be used 20 times.

The capital cost for a system containing Meta-PO4 media is expected to be modest, primarily consisting of containers, pumps/valves, monitoring/control, and a system for regeneration.

Media costs are usually expressed as [\$/kg-media] or [\$/kg-P removed]. Cost depends on the volume of water being processed, the number of times the media is reused, and the amount of phosphorus that must be removed. It can be as low as \$10–\$20/kg-P for larger operations and higher phosphorus concentrations (>50 mg/L). However, cost to remove phosphorous from water will be comparable and often lower than other alternatives owing to a higher capacity and the reuse characteristics of the media.

31.7 Conclusion

For phosphorous removal from water, nanomaterials provide exceptional performance compared with naturally occurring minerals and most manufactured sorbents. A high concentration of phosphorous can be adsorbed by nanoenhanced media. A key to high performance is the use of a highly porous ceramic substrate to support the active nanomaterials. The hierarchical interconnected porosity allows for easy water flow into and through the media, bringing phosphorus ions into contact with the nanomaterials. Being able to lower phosphorus to <0.01 mg/L has been shown, which is otherwise costly using chemicals. Reuse of media after phosphorous removal provides a significant economic benefit and is an alternative for many chemical precipitation applications. Phosphorous can be recovered from the regeneration liquids by precipitation as a useful commercial phosphate compound, such as calcium phosphate. Given the increasing scarcity of phosphate rock, the principal source of phosphorous, and the increasing needs of phosphorus for food and industrial production, media containing nanomaterials represents an economic and "green" long-term worldwide approach to supplying additional phosphorus.

Acknowledgment

The authors are grateful for the financial support provided by SBIR grants from the National Science Foundation and a Third Frontier grant from the Ohio Department of Development. The authors want to recognize the significant contributions made by many of the MetaMateria staff and students at Michigan State University, as well as the guidance offered by engineering and consulting firms helping to evaluate the phosphorus media.

References

Bachand, P.A.M. 2003. Potential Application of Adsorptive Media to Enhance Phosphorus.

Brix, H., Arias, C.A., and del Bubba, M. 2001. Media selection for sustainable phosphorus removal in subsurface flow constructed wetlands. *Water Science and Technology* 44(11–12): 47–54.

Camm, E. 2011. An evaluation of engineered media for phosphorus removal from greenroof stormwater runoff, MS Thesis, University of Waterloo, Waterloo, Ontario, Canada, 2011.

Centre Européen d'Estudes des Polyphosphates (CEEP). September 2012. *SCOPE Newsletter*, p. 6.

Cheung, K.C., and Venkitachalam, T.H. 2000. Improving phosphate removal of sand infiltration system using alkaline fly ash. *Chemosphere* 41: 243–249.

Clabby, C. 2010. Does peak phosphorous loom? *American Scientist* 98(4): 291.

Drizo, A., Frost, C.A., Grace, J., and Smith, K.A. 1999. Physico-chemical screening of phosphate removing substrates for use in constructed wetland systems. *Water Research* 17: 3595–3602.

Helferich, R. July 2011. Final Report to National Science Foundation, Phase I SBIR Wastewater Phosphorus Removal using Nano Enhanced Reactive Iron Media.

Johansson, L., and Gustafsson, J.P. 2000. Phosphate removal by using blast furnace slags and opoka-mechanisms. *Water Research* 34(1): 259–265.

Kasprzyk-Holdern, B. 2004. Chemistry of alumina, reactions in aqueous solution and its application in water treatment. *Advances in Colloid and Interface Science* 110: 19–48.

Kim, J., Ma, J., Howerter, K., Garofalo, H., and Sansalone, J. 2008. Interactions of phosphorus with anthropogenic and engineered particulate matter as a function of mass, number and surface area. In: W. James, K.N. Irvine, E.A. McBean, R.E. Pitt, and S.J. Wright (eds.) *Reliable Modeling of Urban Water Systems, Monograph 16*. CHI, Guelph, Ontario.

Krögera, R., Dunne, E.J., Novak, J., King, K.W., McLellan, D.E., Smith, D.R., Strock, J., Boomer, K., Tomer, M., and Noe, G.B. 2012. Downstream approaches to phosphorus management in agricultural landscapes: Regional applicability and use. *Science of the Total Environment* 442(2013): 263–274.

Littleton, H.X., Daigger, G.T., and Strom, P.F. 2007. Application of computational fluid dynamics to closed loop bioreactors: I. Characterization and simulation of fluid flow pattern and oxygen transfer. *Water Environment Research* 79(6): 600–612.

Littleton, H.X., Daigger, G.T., Strom, P.F., and Cowan, R.M. 2000. Evaluation of autotrophic denitrification and heterotrophic nitrification in simultaneous biological nutrient removal systems. *Proceedings, WEFTEC 2000, 73rd Annual Conference*, Water Environment Federation, Anaheim, CA, Session 19, 21 pp. [CD-ROM].

Littleton, H.X., Daigger, G.T., Strom, P.F., and Cowan, R.M. 2002a. Evaluation of autotrophic denitrification, heterotrophic nitrification, and PAOs in full scale simultaneous biological nutrient removal systems. *Water Science & Technology* 46(1–2): 305–312.

Littleton, H.X., Daigger, G.T., Strom, P.F., and Cowan, R.A. 2003a. Evaluation of autotrophic denitrification, heterotrophic nitrification, and biological phosphorus removal in full scale simultaneous biological nutrient removal systems. *Water Environment Research* 75: 138–150.

Littleton, H.X., Daigger, G.T., and Strom, P.F. 2003b. Summary paper: Mechanisms of simultaneous biological nutrient removal in closed loop bioreactors. *Proceedings, WEFTEC 2003, 76th Annual Conference*, Water Environment Federation, Los Angeles, CA [CD-ROM].

Liu, D., Teng, Z., Sansalone, J., and Cartledge, F.K. 2001. Surface characteristics of sorptive-filtration storm water media. II: Higher specific gravity (ρs > 1.0) oxide-coated buoyant media. *Journal of Environmental Engineering* 127(10): 879–888.

Mann, R.A. 1997. Phosphorus adsorption and desorption characteristics of constructed wetland gravels and steelworks by-products. *Australia Journal Soil Research* 35: 373–384.

Meridian Institute, Background Paper "Overview and Comparison of Conventional and Nano-Based Treatment Technologies for Water" Oct 2006 International Workshop, Chennai, India (2007).

Mino, T., Loosdrecht, M.C.M., and Heijnen J.J. 1998. Microbiology and biochemistry of the enhanced biological phosphate removal process. *Water Research* 32(11): 3193–3207.

Minton, G.R. 2002. *Stormwater Treatment: Biological, Chemical, and Engineering Principles.* Resource Planning Associates, Seattle, Washington.

Misra, M., and Lentz, P. 2003. Nano-structured lanthanum-diatomaceous Earth (DE) composites for filtration of arsenic from drinking water. *Advances in Filtration and Separation Technology* 16: 563.

Mulkerrins, D., Dobson, A.D.W., and Colleran, E. 2003. Parameters affecting biological phosphate removal from wastewaters. *Environment International* 30: 249–260.

Neethling, J.B., Bakke, B., Benisch, M., Gu, A., Stephens, H., Stensel, H.D., and Moore, R. 2005. *Factors Influencing the Reliability of Enhanced Biological Phosphorus Removal.* Water Environment Research Foundation, Alexandria.

Neethling, J.B., and Gu, A. 2006. Chemical phosphorus removal constraints—Introduction. Session P2 in WERF.

Oehmen, A., Lemos, P.C., Carvalho, G., Yuan, Z., Keller, J., Blackall, L.L., and Reis M.A.M. 2007. Advances in enhanced biological phosphorus removal: From micro to macro scale. *Water Research* 41: 2271–2300.

Oh, Y.-M., Hersterberg, D.L., and Nelson, P.V. 1999. Comparison of phosphate adsorption on clay minerals for soilless root media. *Communications in Soil Science and Plant Analysis* 30(5–6): 747–756.

Panasiuk, O. 2010. Phosphorus removal and recovery from wastewater using magnetite, Master of Science Thesis. Royal Institute of Technology, Stockholm.

Pant, H.K., Reddy, K.R., and Lemon, E. 2001. Phosphorus retention capacity of root bed media of sub-surface flow quality Trading, 91st Annual Meeting, NJWEA, Atlantic City, NJ.

Safferman, S.I., Henderson, E.M., and Helferich, R.L. 2007. Chemical Phosphorus removal from onsite generated wastewater. *Proceedings Water Environment Federation Annual Conf.*, San Diego, CA.

Sakadevan, K., and Bavor, H.J. 1998. Phosphate adsorption characteristics of soils, slags and zeolite to be used as substrates in constructed wetland systems. *Water Research* 32(4): 393–399.

Sedlak, R.I. 1991. *Phosphorus and Nitrogen Removal from Municipal Wastewater*, Second ed. Lewis Publishers, New York.

Song, Y., Hahn, H.H., and Hoffmann, E. 2002. Effects of solution conditions on the precipitation of phosphate for recovery, a thermodynamic evaluation. *Chemosphere* 48: 1029–1035.

Strom, P.F. 2006a. Introduction to phosphorus removal. Invited Presentation for Wastewater Treatment Operator's Workshop, *91st Ann. Mtg*, NJWEA, Atlantic City, NJ.

Strom, P.F. 2006b. Phosphorus removal techniques. Invited Presentation for Water Quality Trading, *91st Annual Meeting*, NJWEA, Atlantic City, NJ.

Strom, P.F. 2006c. *Technologies to Remove Phosphorus from Wastewater.* Rutgers University, New Brunswick, NJ.

Strom, P.F., Littleton, H.X., and Daigger, G.T. 2004. *Characterizing Mechanisms of Simultaneous Biological Nutrient Removal during Wastewater Treatment.* Water Environment Research Foundation, Alexandria, VA.

Takacs, I. 2006. Modeling chemical phosphorus removal processes, Session P2 in WERF.

Tchobanoglous, G., and Burto, F.L. 1991. *Wastewater Engineering, Treatment, Disposal and Reuse*, Third ed. Metcalf & Eddy, McGraw Hill International Editions, Civil Engineering Series, 733 pp.

Tchobanoglous, G., Burton, F.L., and Stensel, H.D. 2002. *Wastewater Engineering: Treatment and Reuse*, Fourth ed. Metcalf & Eddy Inc., McGraw-Hill Science Engineering, New York.

Ugurlu, A., and Salman, B. 1998. Phosphorus removal by fly ash. *Environment International* 24(8): 911–918.

Weinberg, E., Viswanathan, T., Finlay, C., and Kumar, S. 2013. Examining nanotechnology for recovery of phosphorus. *Water World Magazine*. Aug. 2011, pp. 16–20.

Wood, R.B., and McAtamney, C.F. 1996. Constructed wetlands for waste water treatment: The use of laterite in the bed medium in phosphorus and heavy metal removal. *Hydrobiologia* 340: 323–331

Zach-Maor, A., Semiat, R., and Shemer, H. 2011. Adsorption–desorption mechanism of phosphate by immobilized nano-sized magnetite layer: Interface and bulk interactions. *Journal of Colloid and Interface Science* 363: 608.

Zaimes, G.N., and Schultz, R.C. 2002. *Phosphorus in Agricultural Watershed*. Department of Forestry, Iowa State University, Ames, Iowa. Available at http://www.buffer.forestry.iastate.edu /Assets/Phosphorus_review.pdf.

Zhu, T.P.D., Jenssen, T., Mæhlum, and Krogstad, T. 1997. Phosphorus sorption and chemical characteristics of Lightweight Aggregates (LWA)—Potential filter media in treatment wetlands. *Water Science Technology* 35(5): 103–108.

32

Nanomaterials Persuasive in Long History of Pursuing Perchlorates

**Suvankar Sengupta,[1] J. Richard Schorr,[1] Rao Revur,[1]
Tianmin Xie,[2] and David E. Reisner[2]**

[1]*MetaMateria Technologies LLC, Columbus, Ohio, USA*

[2]*Inframat® Corp., Manchester, Connecticut, USA*

CONTENTS

32.1 Problem

Perchlorate is a mobile and stable ion that has been recognized as a health-threatening contaminant to groundwater, surface water, and soil. A majority of perchlorate ions result from dissolution of salts that are used in energy boosters of rockets and missiles. Perchlorate ions interfere with the iodide uptake of the thyroid gland, affecting the production of thyroid hormones. Perchlorate is an especially stable ion that is difficult to remove from water. Methods examined for the removal of perchlorate ions from contaminated soil and water sources are reviewed in this chapter, including ion exchange, bioremediation, composting, permeable reactive barrier (PRB), phytotechnology, membrane technologies, and sorption media. The use of a high surface area, nanostructured porous ceramic, was found to show much promise for the economical adsorptive capture of perchlorate, and these results are described.

Perchlorate is a stable molecule of one chlorine and four oxygen atoms that has been recognized as a health-threatening contaminant in surface water, groundwater, and soil. In March 1998, the US Environmental Protection Agency's (USEPA) Office of Water added perchlorate to the drinking water contaminant candidate list. Perchlorate ions originate as a contaminant in ground water and surface waters from the dissolution of perchlorate salts. Approximately 90% of all perchlorate salts are manufactured as ammonium perchlorate that is used as an energetic booster of solid oxidants in rockets and missiles. Ammonium perchlorate has limited shelf life, requiring periodic replacement of rocket and missile fuels. Therefore, since the 1950s, large quantities have been removed from solid rocket boosters and missiles and subsequently disposed of in Nevada, California, Utah, and other states where rockets and missiles are produced (Damian and Pontius, 1990), causing severe water contamination of perchlorate in these areas. Other uses of perchlorate compounds include pyrotechnics, matches, munitions, the chemical analytics industry, fertilizer, additives in lubricating oils, tanning, finished leather, fabric fixer, dye, electroplating, aluminum refining, rubber manufacture, paint and enamel production, as an additive in cattle feed, magnesium batteries, and as a component of automobile air bag inflators (Damian and Pontius, 1990; USEPA, 1993; Tarver et al., 1996; American Water Works Association Research Foundation [AWWARF], 1997; Urbansky, 1998; Renner, 1999; USEPA, 1998). Because of the high solubility of most perchlorate salts, the resulting perchlorate ion is extremely mobile in aqueous systems.

Although perchlorate salts are highly reactive as solids, once dissolved in water and under natural conditions, the perchlorate anion becomes relatively nonreactive and stable. The wide use of perchlorate and its high mobility in aqueous systems has caused widespread groundwater contamination of perchlorate in hundreds of locations in at least 43 states, potentially affecting the health of more than 20 million people that reside in these contaminated areas. Perchlorate contamination is of particular concern because of the persistent and toxic nature of this compound and because its stable physical and chemical properties make it challenging to treat.

Perchlorate-induced health problems are most commonly observed by their impact on or through the thyroid gland in the form of a decrease in thyroid hormone output. The thyroid gland takes up iodide ions from the bloodstream and uses the iodide to regulate metabolism along with other functions. However, in iodide uptake, the presence of ions larger than iodide, such as perchlorate, can reduce thyroid hormone production and thus disrupt metabolism. Although this property of perchlorate makes it useful as a medical treatment for Graves' disease (hyperthyroidism), it also creates a perchlorate health concern (Urbansky, 1998; EPA National Center for Environmental Assessment, 2004). Primary pathways for exposure to perchlorate in humans include ingestion of contaminated drinking water and food crops. The National Research Council reported that daily ingestion of up to 0.0007 mg of perchlorate per kilogram of body weight can occur without adversely affecting the health of the most sensitive populations.

32.2 Perchlorate Removal Technologies

Supported by the US government, several treatment technologies have been examined for perchlorate removal in contaminated water. Technologies examined for removal of perchlorate from contaminated drinking water, groundwater, and soil have included the following approaches:

- Ion exchange
- Bioreactor/bioremediation
- Composting
- Permeable reactive barrier
- Phytotechnology
- Membrane technologies (electrodialysis and reverse osmosis)
- Sorption by media such as activated carbon or other absorbents

These technology approaches are briefly summarized below.

32.2.1 Ion Exchange

Ion exchange has been used for full-scale removal of perchlorate from drinking water, groundwater, surface water, and other water discharges. During the ion-exchange process, ions held electrostatically on the surface of a solid are replaced or exchanged by ions in a solution of a similar charge. The most commonly used ion-exchange media are synthetic, strongly basic, anion-exchange resins. Ion exchange has been used at test sites to reduce perchlorate concentrations to <4 µg/L; however, it is expensive and generally too expensive for treating larger volumes of water. Water chemistry does impact the effectiveness of ion-exchange media. The media can be used both as the main filter and/or as a polishing filter for other water treatment processes such as biological treatment of perchlorate (Federal Remediation Technologies Roundtable [FRTR], 2005; Gu et al., 1999).

Monofunctional and bifunctional anion-exchange resins are commonly used for perchlorate treatment. Bifunctional resins consisting of two functional groups are used for a broader range of ionic strengths than monofunctional resins (FRTR, 2005). Some resins used for perchlorate removal include a polyvinylbenzylchloride backbone cross-linked with divinylbenzene, to form quaternary ammonium strong-base anion exchange sites (Gu et al., 1999, 2002, 2003).

Ion-exchange resin beads are usually packed into a column and contaminated water passed through the bed of resin beads. Contaminant ions are exchanged for other ions such as chlorides or hydroxides in the resin (FRTR, 2005). Ion exchange is often preceded by treatments such as filtration and oil–water separation to remove organics, suspended solids, and other contaminants that can foul the resins and reduce their effectiveness. Ion-exchange resins must be periodically regenerated to remove the adsorbed contaminants and replenish the exchanged ions (FRTR, 2005). Regeneration of a resin typically involves three steps:

1. Backwashing
2. Regeneration with a solution of ions
3. Final rinsing to remove the regenerating solution

Regeneration results in a backwash solution, a waste-regenerating solution, and a waste rinse water (EPA Office of Solid Waste and Emergency Response [OSWER], 2002). It has been reported (Gu et al., 1999) that nearly 110,000 bed volumes of water contaminated with approximately 50 µg/L perchlorate can be treated by a bifunctional resin before break-through occurs. Sodium chloride (NaCl), ammonium hydroxide (NH_4OH), ferric chloride–hydrochloric acid ($FeCl_3$–HCl), and sodium hydroxide (NaOH) are some commonly used regenerants for perchlorate-laden resins.

Factors that influence ion-exchange performance of the media include the presence of competing ions, fouling, and influent water quality. Competition for the exchange ion can reduce the effectiveness of ion exchange if ions in the resin are replaced by ions other than perchlorate—such as nitrate, sulfate, and bicarbonate—resulting in a need for more frequent bed regeneration (FRTR, 2005; Boodoo, 2003a; Gingras and Batista, 2002; Gu et al., 2002). The presence of organics, suspended solids, calcium, or iron can foul ion-exchange resins and reduce the effectiveness of the treatment system due to clogging of the resin bed (FRTR, 2005; EPA OSWER, 2002; Boodoo, 2003b; Gu et al., 2002). The presence of oxidants in the influent water has also been found to impede performance of the ion-exchange resin (FRTR, 2005).

32.2.2 Bioreactor

A bioreactor containing microorganisms capable of reducing perchlorate into a chloride ion and oxygen is frequently used for perchlorate removal from contaminated groundwater and larger amounts of surface water. This technology uses microorganisms capable of reducing perchlorate into chloride and oxygen in the presence of an electron donor in an appropriate medium to support microbial growth. Bioreactors have been used to reduce perchlorate concentrations to <4 µg/L.

Bioreactors are typically used to treat contaminated water above ground in a reactor vessel. Contaminated water comes in direct contact with microbes that selectively degrade the contaminant. Denitrification bacteria have been found capable of degrading perchlorate. The process requires an electron donor and an appropriate substrate to support bacterial growth. Perchlorate serves as the oxygen source in this process. Some commonly used electron donors are acetic acid, ethanol, methanol, and hydrogen. The addition of nutrients such as ammonia and phosphorus may be required to enhance microbial growth (Evans et al., 2002, 2003; Clark et al., 2001; USEPA, 2005).

32.2.3 Perchlorate Transformation/Biodegradation

Microbial degradation of perchlorate proceeds according to the following anaerobic reduction process:

$$ClO_4^- \rightarrow ClO_3^- \rightarrow ClO_2^- \rightarrow Cl^- + O_2$$

The rate-limiting step in this process is degradation of perchlorate to chlorate. More than 30 different strains of perchlorate-degrading microbes have been identified, with many classified in the Proteobacteria class of the bacteria kingdom. Soil and groundwater samplings have confirmed the pervasiveness of perchlorate-reducing bacteria (Polk et al., 2001; Naval Facilities Engineering Command [NAVFAC], 2000). Ongoing research suggests that perchlorate destruction involves a three-step reduction process catalyzed by two enzymes. A perchlorate reductase enzyme catalyzes reduction of perchlorate (ClO_4^-) to chlorate (ClO_3^-) and then to chlorite (ClO_2^-). A chlorite dismutase enzyme then causes a further breakdown of chlorite to chloride (Cl⁻) and oxygen (O_2) (Polk et al., 2001; Sartain and Craig, 2003; Beisel et al., 2004).

Fluidized bed reactors and packed bed reactors are two types of commercially available bioreactors. Packed or fixed bed bioreactors are made up of static sand or plastic media to support the growth of microbes. Fluidized bed bioreactors are made up of suspended sand or granular-activated carbon media to support microbial activity and growth of biomass. The activated carbon media are selected to produce a low-concentration effluent (i.e., parts per billion [ppb] levels). Fluidized systems provide larger surface area for growth of microorganisms. The fluidized bed expands with the increased growth of biofilms on the media particles. The result of this biological growth is a system capable of additional degradative performance for target contaminants in a smaller reactor volume than with a fixed bed. However, the fluidized bed reactors generally require greater pumping rates than fixed beds (Evans et al., 2002; Polk et al., 2001; Hatzinger et al., 2000; NAVFAC, 2000; Nerenberg et al., 2003).

Several factors affect the performance of a bioreactor. For instance, in the optimum range of dissolved oxygen (DO) concentration of the influent water, perchlorate reduction reaches 0.5–1.0 mg/L. When DO levels drop to <0.5 mg/L, anaerobic conditions develop that, in the presence of sulfates, result in the formation of hydrogen sulfide (USEPA, 2005). It has been reported that removal of nitrate ions from the influent water is required to achieve complete reduction of perchlorate (NAVFAC, 2000). Consistent and adequate dosage of carbon source (electron donor) and nutrients are required for growth of microorganisms on the reactor bed (FRTR, 2005; Evans et al., 2002). Furthermore, control of excessive microbial growth with a backwash strategy is essential to eliminate short circuiting and flow channeling in the bioreactor system (Evans et al., 2002; Hatzinger et al., 2000; NAVFAC, 2000; Nerenberg et al., 2003; Polk et al., 2001).

Normally, the treated effluent is suitable for discharge; however, for drinking water treatment, the effluent from bioreactors might require further treatment to remove biosolids present in the effluent (Evans et al., 2002). Fluidized bed bioreactors require a thorough mixing and upward flow of the fluid inside the reactor. One key advantage of a fluidized bed system is the availability of a large surface area for growth of biomass. However, to maintain required flow inside the reactor vessel, relatively higher pumping rates are required (USEPA, 2005). Moreover, because fixed-bed systems are more susceptible to accumulation of biosolids, they require periodic back-flushing to avoid plugging or clogging the bed (Evans et al., 2002; Hatzinger et al., 2000; Polk et al., 2001; NAVFAC, 2000; Nerenberg et al., 2003).

32.2.4 Composting

Composting has been used infrequently to treat perchlorate in contaminated soil. It is also a biological process that uses indigenous microorganisms to degrade perchlorate in the presence of appropriate soil amendments that support microbial growth. This technology has been found to reduce perchlorate concentrations in soil to as low as 0.1 mg/kg.

Under anaerobic, thermophilic conditions (54–65°C), soil contaminated with perchlorate was composted. Heat produced by microorganisms during degradation of the contaminants in the waste increases the temperature of the compost pile (FRTR, 2005; Roote, 2001). Additional information about perchlorate transformation and biodegradation, including microbial degradation pathways, is presented above, under bioreactors. Monitoring of moisture content and temperature are important for achieving maximum degradation efficiency (FRTR, 2005; Cox et al., 2000).

32.2.5 Permeable Reactive Barrier

A PRB is an *in situ* technology used to treat perchlorate-contaminated groundwater in full-scale systems. Some of the commonly used reactive materials for barriers include soybean and other edible oils, woodchips, pecan shells, cottonseed, chitin, limestone, and other composting materials. Many of these materials can provide both electron donors and the necessary nutrients for microbial growth. Soluble electron donors such as lactate, acetate, and citrate may be added to the barrier materials to further stimulate biodegradation of perchlorate.

A PRB is an *in situ* treatment zone of reactive material that degrades or immobilizes contaminants as groundwater flows through it. PRBs are installed as permanent, semipermanent, or temporary units across the flow pattern of a contaminant plume. Contaminants in groundwater that flow through a PRB are degraded chemically or biologically (FRTR, 2005). Barriers are made of reactive material that targets specific contaminants (USAFCEE, 2002; USEPA, 2005). To treat groundwater contaminated with perchlorate, the reactive barrier may be inoculated with anaerobic bacteria that can convert perchlorate into chloride and oxygen (USAFCEE, 2002). For treatment of perchlorate-contaminated groundwater, the PRB system is backfilled with reactive material that includes an electron donor to stimulate reduction of perchlorate and organic substrates to nourish the microorganisms (USAFCEE, 2002; Craig and Jacobs, 2004; Beisel et al., 2004). Proper installation of PRBs requires access to depths of the contaminated groundwater and barriers formed by trenches that surface excavation or trenching equipment may not be able to reach (FRTR, 2005). Additional maintenance may be required to unclog the barrier fouled biologically or clogged with chemical precipitates (USAFCEE, 2002; USEPA, 2005).

32.2.6 Phytotechnology

This is an emerging technology for perchlorate remediation. It involves use of plants to remove contaminants by natural processes occurring within the plant body. Selection of the best plant species is critical to achieving the treatment goals. Research is currently under way to identify the mechanism involved in perchlorate removal using this approach. A few bench-scale studies have indicated the suitability of certain plant species for remediation of perchlorate-contaminated media. Phytotechnology includes various mechanisms such as rhizosphere biodegradation, phytovolatilization, phytostabilization, and phytoextraction (FRTR, 2005). Rhizodegradation or rhizosphere degradation proceeds via activities of microorganisms present in the soil surrounding the roots. The natural substances released by plant roots provide nutrient material to the microbial population, which in turn degrade the contaminants present in soil. The mechanism of remediation of perchlorate-contaminated media by phytotechnology is not yet established. However, studies conducted at bench scale have indicated possible suitability of certain plant species for perchlorate removal (Motzer, 2001; Schnoor et al., 2004; Susarla et al., 1999).

32.2.7 Membrane Technologies

Semipermeable or permeable membrane technologies have been used for perchlorate removal. Electrodialysis and reverse osmosis are examples of membrane technologies that have been examined for perchlorate removal from groundwater, surface water, and wastewater. These are briefly discussed below.

32.2.7.1 Electrodialysis

This membrane technique uses electric current to remove perchlorate (Roquebert et al., 2000). In this technology, a current is applied to perchlorate-contaminated water as it passes through channels of alternating permeable membranes selective of anions and cations. The electric current dissociates perchlorate salts into cations and anions. Ammonium perchlorate and potassium perchlorate are two common forms of perchlorate contamination. Perchlorate ions, being negatively charged (anion), accumulate at the cationic-selective membrane and are eventually collected as concentrate or salty water. Similarly, positive ions accumulate at the anionic-selective membrane. This method produces two types of water—salty water and relatively deionized water. The deionized water is used, while the salty water is disposed of or further treated by an appropriate method before disposal (Urbansky and Schock, 1999).

The effectiveness of electrodialysis for perchlorate removal may be reduced because of fouling. Furthermore, the concentrate resulting from this method may require large quantities of water for further treatment before disposal (Urbansky and Schock, 1999), which adds costs to the application of the process.

32.2.7.2 Reverse Osmosis

Reverse osmosis is another membrane technique used for perchlorate removal (USEPA 2005b; Urbansky, 1998). Osmosis can be defined as the movement of water molecules from a region of lower solute concentration to a region of higher solute concentration through a semipermeable membrane (Urbansky and Schock, 1999). In this case, the solute is a perchlorate salt. A reverse osmosis system consists of a chamber in which perchlorate-contaminated water is placed on one side of a semipermeable membrane and freshwater is placed on the other side of the membrane. Pressure is applied at the inlet to force water molecules against the concentration gradient from the contaminated water into the freshwater section of the reverse osmosis system. This results in separation of perchlorate ions from contaminated water. Treated water can be used. The water containing perchlorate and other contaminants is further treated before disposal (USEPA, 2005).

The performance of a reverse osmosis membrane is affected by the presence of organic matter and microbes that can foul and damage the membrane. The fouling can be further enhanced by the presence of alkaline earth metals.

Reverse osmosis is normally suitable for point-of-use or small systems. Posttreatment including application of sodium chloride or sodium bicarbonate is required to make water palatable and prevent fouling of the distribution system (Urbansky, 1998).

32.2.8 Sorption Media

Various sorption media have been investigated to remove perchlorate from contaminated water. These include modified activated carbon, surface-modified zeolites, and surface-modified ceramic porous media. All these approaches rely on high surface area resulting from meso/nano porosity at the surfaces. The sorbent media are generally modified by a cationic surfactant providing a charged surface that can bond with the perchlorate ion through ion bonding.

32.2.8.1 Liquid-Phase Carbon Adsorption

Liquid-phase carbon adsorption using granular activated carbon (GAC) has been used to remove perchlorate from contaminated groundwater and surface water. In this technology, GAC is the adsorbent to remove contaminant ions from water as it passes through the GAC bed. However, GAC has a relatively small treatment capacity for perchlorate removal. Some research is being done to identify methods to improve the treatment capacity of a GAC system for perchlorate removal, including use of "tailored GAC." Tailored GAC technology was tested on a pilot-scale level. As discussed above, GAC also has been used in conjunction with bioreactors, including the role of substrate for the biodegradation processes.

Liquid-phase carbon adsorption involves use of adsorbent media (such as GAC, activated alumina, or other proprietary materials) packed into a column (FRTR, 2005; Graham et al., 2004). GAC sorbent has commonly been used to remove organic and metallic contaminants from groundwater, drinking water, and wastewater. GAC media are usually regenerated by thermal techniques to desorb and volatilize contaminants, and an off-gas unit captures the volatilized contaminants and treats it before release into the atmosphere (Graham et al., 2004). GAC media are generally considered cost-effective for water treatment when used for removal of nonpolar contaminants with low water solubility (Graham et al., 2004; FRTR, 2005). Because of the issues discussed above, activated carbon is generally considered ineffective for removal of inorganic contaminants, such as perchlorate, from water.

Carbon adsorption technology can be used in multiple beds in series to reduce the need for media regeneration. Multiple beds can also allow continuous operation because some beds can be regenerated as others continue to treat water (Graham et al., 2004). Thermal decomposition of perchlorate-contaminated GAC is a possible regeneration method for spent GAC (USEPA, 2005).

Recently, there has been discussion among experts about the types of mechanisms and effectiveness of tailored GAC for treatment of perchlorate. In addition, there have been questions raised about the potential use of tailored GAC for treatment of water contaminated with perchlorate and explosives such as royal demolition explosives, cyclotetramethylene trinitramine, and trinitrotoluene and volatile organic compounds (i.e., co-contaminated groundwater). For co-contaminated groundwater sites, practitioners have suggested the use of treatment trains consisting of standard GAC/ion-exchange resins or tailored GAC/standard GAC.

For example, a treatability study was conducted at the Massachusetts Military Reservation site investigating innovative options for *ex situ* removal of perchlorate and explosives in groundwater (Weeks et al., 2004). Life cycle cost comparisons of these different technologies and treatment trains is being examined. Contaminants in the water can reduce the effectiveness of these technologies. The performance of GAC is affected by flow rate, polarity, and water solubility as well as fouling by solid, organics, or silica (FRTR, 2005). Waste streams with high amounts of suspended solids, oil, and grease may foul the carbon. Spent carbon from the adsorption unit may require treatment before ordinary or hazardous waste disposal (FRTR, 2005; Graham et al., 2004). Contaminants with high water solubility and polarity can reduce the ability of GAC to remove contaminants from water (FRTR, 2005).

32.2.8.2 Surfactant-Modified Zeolites

Surface-modified zeolites have also been evaluated as a sorption media to remove perchlorate. Natural zeolites are hydrated alumino silicate minerals with high internal/external surface area and high cation-exchange properties. Modification of natural zeolites with quaternary amines, such as the cationic surfactant hexadecyltrimethylammonium, results in sorbent material that has anion-exchange properties. This sorbent was found to be effective in selectively removing (ClO_4^-) from water in the presence of competing ions. The sorbed (ClO_4^-) was stable against leaching by a variety of fluids over a wide pH range. The sorbent media can be regenerated by leaching with concentrated nitrate solution. The anions in the leachate can then be biodegraded. The capacity of the media was reported to about 40–47 mmol/kg (Zhang et al., 2007).

32.2.8.3 Surfactant-Modified Nanostructured Porous Media

MetaMateria Technologies has developed a nanostructured highly porous ceramic media for perchlorate removal. The approach relies on a proprietary technology to produce a high-surface-area ceramic with hierarchical porosity whose sizes range from millimeter to nanometer scale. The larger pores allow for high flow rates of water into the structure without developing back pressure. The finer meso/nano pore structure provides surfaces that can be engineered to bind targeted contaminants. The material can be economically produced in various shapes and sizes.

A novel liquid slurry method is used to prepare the porous ceramic substrate. Slurries of reactive materials are prepared containing aggregate materials and surfactants/gas-forming agents. These slurries are mixed together and the liquid is poured into molds to make desired shapes or into a mix chamber to make aggregates. What results is an alumino–silicate bonded matrix with interconnected porosity that is formed during gas expansion of liquid (about three times expansion) followed by solidification due to chemical reactions. This approach produces open porosity materials that are then taken through a curing step and other postforming procedures. This approach was used to prepare materials for the removal of perchlorate. A highly porous media, with surface area 100–300 m²/g, was prepared that contained nanopores (~10 nm) at the cell wall surface of the substrate. This provides active surface surfaces that are then modified to develop hydrophobicity and then the resulting surfaces are functionalized with ammonium cation surfactants. The surfactants are adsorbed onto the surfaces of the adsorbent through lipophilic affinity between their hydrophobic tail and porous surface. The nanostructure permits a high concentration of these surfactants. The cation heads of the surfactant then provide for adsorption of perchlorate from water through ion pair bonding.

This approach was used to successfully remove perchlorate. Test samples were prepared under different processing conditions and with varying compositions. Batch tests were conducted to evaluate the potential for each to remove perchlorate. The results are shown in Figure 32.1 for 200 ppb and Figure 32.2 for 4000 ppb concentrations of perchlorate. Testing was done using challenge water containing 50 ppb of Cl^- and SO_4^{2-} and 5 ppb of NO_3^-. Perchlorate was tested at 200 and 4000 ppb. Each media (2 g) was shaken with 100 mL of test solution for 24 h. The concentration of perchlorate was then analyzed after filtration.

These results showed that perchlorate could be completely removed by three of the surfactant-modified high-surface-area, nanostructured porous media. The other media removed 60%–90% of the perchlorate at 4000 ppb, and two of the media were found to lower the concentration to below 200 ppb.

Testing was also done with granular material in columns having a 3.8 cm diameter and a height of 13.3 cm. Effluent samples were collected periodically and analyzed for perchlorate removal. Perchlorate concentration was reduced from 4000 to 1500 ppb during the first 10 days

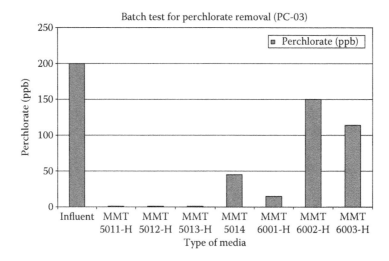

FIGURE 32.1
Batch test result for surfactant modified forous ceramic media at 200 ppb.

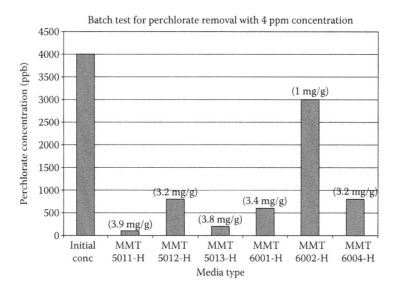

FIGURE 32.2
Batch test results for surfactant modified forous ceramic media done at 4000 ppb.

and perchlorate continued to be removed for 22 days of testing. The capacity of the media used in the column test was calculated to be 3.3 mg/g, which is slightly lower than that estimated by the batch test (3.9 mg/g). This showed that at an empty bed contact time (EBCT) of 60 min, the media does saturate and has a reasonably good capacity for removing perchlorate.

Column Test at an EBCT of 60 Min (MMT-6006)	
Total perchlorate removed (after 22 days)	83 mg
Weight of the media used in the columns	25.3 g
Capacity	3–3 mg/g

Perchlorate was lowered from 4000 to 200 ppb and then in a separate column, lowered from 200 to <2 ppb. Similar capacity was demonstrated in both batch testing and column testing. Further improvement in this media is possible, and it could be provided in different forms to meet cleanup requirements for different locations.

This limited study showed that the nanoporous structure of this ceramic media provides a base for preparation of a suitable perchlorate sorption media with good potential for commercial removal of perchlorate from water. This would be modestly priced granular product and is another example of how nanostructure facilitates the successful removal of a difficult ion to capture.

Acknowledgments

This work was funded in part by the US Air Force through a Phase II Small Business Innovation Research program (contract FA9300-04-C-0033) awarded to Inframat.

References

AWWARF. 1997. "Report on the Perchlorate Research Issue." Ontario, CA, Group: September 30–October 2, p. 4.

Beisel, T.H., Craig, M., and Perlmutter, M. 2004. "Ex-Situ Treatment of Perchlorate Contaminated Groundwater." Presented at NGWA Conference on MTBE and Perchlorate, June 3–4.

Boodoo, F. (The Purolite Co.). 2003a. Perchlorate Facts for Technology Vendors. Available at http://purolite.com/customized/uploads/pdfs/Perchlorate%20Facts%20for%20Vendors%20090203.pdf.

Boodoo, F. (The Purolite Co.). 2003b. POU/POE Removal of Perchlorate, Water Conditioning and Purification, August, p. 4. Available at http://yosemite.epa.gov/r10/CLEANUP.NSF/PH/Arkema+Technical+Documents/$FILE/Perchlorate-POUTreatment-IXResins.pdf.

Clark, R., Kavanaugh, M., McCarty, P., and Trussell, R.R. 2001. "Review of Phase 2 Treatability Study Aerojet Facility at Rancho Cordova, California—Expert Panel Final Report." July.

Cox, E., Edwards, E., Neville, S., and Girard, M. 2000. "Aerojet Bioremediation of Soil from Former Burn Area by Anaerobic Composting." Available at http://www.perchlorateinfo.com/perchloratecase-01.html.

Craig, M. (NAVFAC), and Jacobs, A. (EnSafe). 2004. "Biological PRB Used for Perchlorate Degradation in Groundwater." *Technology News Trends*, Issue 10, February, p. 2. Available at http://nepis.epa.gov/Exe/ZyNET.exe/P1000NPV.txt?ZyActionD=ZyDocument&Client=EPA&Index=2000%20Thru%202005&Docs=&Query=&Time=&EndTime=&SearchMethod=1&TocRestrict=n&Toc=&TocEntry=&QField=&QFieldYear=&QFieldMonth=&QFieldDay=&UseQField=&IntQFieldOp=0&ExtQFieldOp=0&XmlQuery=&File=D%3A%5CZYFILES%5CINDEX%20DATA%5C00THRU05%5CTXT%5C00000014%5CP1000NPV.txt&User=ANONYMOUS&Password=anonymous&SortMethod=h%7C-&MaximumDocuments=1&FuzzyDegree=0&ImageQuality=r75g8/r75g8/x150y150g16/i425&Display=p%7Cf&DefSeekPage=x&SearchBack=ZyActionL&Back=ZyActionS&BackDesc=Results%20page&MaximumPages=1&ZyEntry=1.

Damian, P., and Pontius, F.W. 1990. "From Rockets to Remediation: The Perchlorate Problem." *Environmental Protection*, 10, pp. 24–31.

EPA National Center for Environmental Assessment (NCEA). 2004. "Web Page on National Academy of Sciences Review of EPA's Draft Perchlorate Environmental Contamination: Toxicological Review and Risk Characterization." Available at http://cfpub2.epa.gov/ncea/cfm/recordisplay .cfm?deid=72117.

EPA OSWER. 2002. "Proven Alternatives for Aboveground Treatment of Arsenic in Groundwater— Engineering Forum Issue Paper." EPA-542-S-02-002, October. Available at http://www.epa .gov/tip/tsp.

Evans, P., Chu, A., Liao, S., Price, S., Moody, M., Headrick, D., Min, B., and Logan, B. 2002. "Pilot Testing of a Bioreactor for Perchlorate-Contaminated Groundwater Treatment." Presented at the Third International Conference on Remediation of Chlorinated and Recalcitrant Compounds, May 20–23.

Evans, P., Price, S., Min, B., and Logan, B. 2003. "Biotreatment and Downstream Processing of Perchlorate Contaminated Groundwater." Presented at in situ and On-Site Bioremediation— The Seventh International Symposium, June 2–5.

Federal Remediation Technologies Roundtable. 2005. "Federal Remediation Technologies Reference Guide and Screening Manual, Version 4.0." February 1. Available at http://www.frtr.gov /archives.htm.

Gingras, T.M., and Batista, J.R. 2002. "Biological Reduction of Perchlorate in Ion Exchange Regenerant Solutions Containing High Salinity and Ammonium Levels." *Journal of Environmental Monitoring*, 4, pp. 96–101.

Graham, J.R., Cannon, F.S., Parette, R., Headrick, D., and Yamamato, G. 2004. "Commercial Demonstration of the Use of Tailored Carbon for the Removal of Perchlorate Ions from Potable Water." Presented at National Groundwater Association Conference on MTBE and Perchlorate, Costa Mesa, CA, June 3–4.

Gu, B., Brown, G.M., Alexandratos, S.D., Ober, R., and Patel, V. 1999. "Selective Anion Exchange Resins for the Removal of Perchlorate ClO4—From Groundwater." ORNL/TM-13753, February.

Gu, B., Brown, G.M., and Ku, Y.-K. 2002. "Treatment of Perchlorate-Contaminated Groundwater Using Highly Selective, Regenerable Ion-Exchange Technology: A Pilot-Scale Demonstration." *Remediation*, Spring 2002, 12, pp. 51–68.

Gu, B., Ku, Y.-K., and Brown, G.M. 2003. "Bifunctional Anion Exchange Resin Pilot—Edwards AFB, CA." *Federal Facilities Environmental Journal*, 14(1), pp. 75–94.

Hatzinger, P.B., Greene, M.R., Frisch, S.A., Togna, P., Manning, J., and Guarini, W.J. (Envirogen, Inc., Lawrenceville, NJ). 2000."Biological Treatment of Perchlorate Contaminated Groundwater Using Fluidized Bed Reactors." Presented at the 2nd International Conference on Remediation of Chlorinated and Recalcitrant Compounds, Monterey, CA, May 22–25.

Motzer, W.E. 2001. "Perchlorate: Problems, Detection, and Solutions." *Environmental Forensics*, 2, pp. 301–311.

Naval Facilities Engineering Command (NAVFAC). 2000. "NASA/California Institute of Technology Jet Propulsion Laboratory, Anoxic FBR. Pasadena, CA." Available at http://www.perchlorateinfo .com/perchloratecase-40.html.

Nerenberg, R., Rittmann, B.E., Gillogly, T.E., Lehman, G.E., and Adham, S.S. 2003. "Perchlorate Reduction Using a Hollow-Fiber Membrane Biofilm Reactor: Kinetics, Microbial Ecology, and Pilot-Scale Studies." Presented at in situ and On-Site Bioremediation—The Seventh International Symposium, June 2–5.

Polk, J., Murray, C., Onewokae, C., Tolbert, D.E., Togna, A.P., Guarini, W.J., Frisch, S., and Del Vecchio, M. 2001. "Case Study of Ex-Situ Biological Treatment of Perchlorate-Contaminated Groundwater." Presented at the 4th Tri-Services Environmental Technology Symposium, June 18–20.

Renner, R. 1999. "EPA Draft Almost Doubles Safe Dose of Perchlorate in Water." *Environmental Science and Technology*, 33, pp. 110A–111A.

Roote, D. (GWRTAC). 2001. "Technology Status Report—Perchlorate Treatment Technologies, 1st Edition." May. Available at http://www.denix.osd.mil/edqw/upload/GWRTAC.PDF.

Roquebert, V., Booth, S., Cushing, R.S., Crozes, G., and Hansen, E. 2000. "Electrodialysis Reversal (EDR) and Ion Exchange as Polishing Treatment for Perchlorate Treatment." *Proceedings of the Conference on Membranes in Drinking and Industrial Water Production*, 1, pp. 481–487.

Sartain, H.S., and Craig, M. (CH2MHill). 2003. "Ex Situ Treatment of Perchlorate-Contaminated Groundwater." Presented at in situ and On-Site Bioremediation—The Seventh International Symposium, June 2–5.

Schnoor, J.L., Parkin, G.F., Just, C.L., van Aken, B., and Shourt, J.D., 2004. *Demonstration Project of Phytoremediation and Rhizodegradation of Perchlorate in Groundwater at the Longhorn Army Ammunition Plant.* The University of Iowa, Department of Civil and Environmental Engineering.

Susarla, S., Bacchus, S.T., McCutcheon, S.C., and Wolfe, N.L. 1999. "Potential Species for Phyto-remediation of Perchlorate." EPA/600/R-99/069, August.

Tarver, C.M., Urtiew, P.A., and Tao, W.C. 1996. "Shock Initiation of a Heated Ammonium Perchlorate-Based Propellant." *Combustion and Flame*, 10512, pp. 123–131.

Urbansky, E.T. 1998. "Perchlorate Chemistry: Implications for Analysis and Remediation." *Bio-remediation Journal*, 2, pp. 81–95.

Urbansky, E.T., and Schock, M.R. 1999. "Issues in Managing the Risks Associated with Perchlorate in Drinking Water." *Journal of Environmental Management*, 56, pp. 79–95. Available at http://cluin .org/download/contaminantfocus/perchlorate/urbansky1.pdf.

USAFCEE. 2002. "Perchlorate Treatment Technology Fact Sheet: Permeable Reactive Barriers." August.

USEPA. 1993. "Approaches for the Remediation of Federal Facility Sites Contaminated with Explosive or Radioactive Wastes." USEPA Office of Research and Development, Washington, DC, p. 83, EPA/625/R-93/013 with appendices.

USEPA. 1998. "Perchlorate Environmental Contamination: Toxicological Review and Risk Charac-terization Based on Emerging Information." USEPA Office of Research and Development, Washington, DC, NCEA-1-0503.

USEPA. 2005. "Perchlorate Treatment Technology Update." EPA 542-R-05-015, May. Available at http://www.epa.gov/tio/download/remed/542-r-05-015.pdf.

Weeks, K., Veenstra, S., and Hill, D. 2004. "Innovative Options for Ex-Situ Removal of Perchlorate and Explosives in Groundwater." Presented at National Defense Industry Association—30th Environmental and Energy Symposium and Exhibition, April 7.

Zhang, P., Avudzega, D.M., and Bowman, R.S. 2007. "Removal of Perchorate from Contaminated Waters using Surfactant Modified Zeolite." *Journal of Environment Quality*, 36, p. 1069.

33

Stormwater Runoff Treatment Using Bioswales Augmented with Advanced Nanoengineered Materials

Hanbae Yang and Stephen Spoonamore

ABSMaterials Inc., Wooster, Ohio, USA

CONTENTS

33.1 Introduction

Bioretention systems have the potential for managing stormwater by reducing peak runoff flow and improving surface water quality in a natural and aesthetically pleasing manner (Dietz, 2007; Davis et al., 2009). Most bioretention systems such as vegetated buffers, rain gardens, and stormwater wetlands are designed to treat runoff by employing filtration, deposition, adsorption, and infiltration through porous media (United States Environmental Protection Agency [USEPA], 2004). However, the pollutant treatment performance of these systems is highly variable and often less effective in the long term because of the limited treatment capacity of filter media (Davis et al., 2001; Hsieh and Davis, 2005; Maurakami et al., 2008; Cho et al., 2009; Yang et al., 2010). Since bioretention media compositions greatly affect pollutant removal mechanisms, development of different media compositions is critical to determining bioretention performance.

ABSMaterials Inc. developed new filter media to remediate multiple pollutants from stormwater runoff by integrating a novel absorbent nanomaterial, Osorb®, with embedded reactive metal composites in bioretention systems. Osorb is a patented, chemically inert, silica-based material that physically absorbs a wide range of organic pollutants from water (Burkett and Edmiston, 2005; Edmiston and Underwood, 2009; Edmiston, 2010). Osorb has the significant ability to absorb organic contaminants because the swelling ability of the material allows for an unmatched absorption capacity when compared with current alternate absorbents. Osorb–metal composites combine two advanced remediation materials: (i) a high-capacity organosilica sorbent, Osorb, and (ii) embedded reactive metals. The captured pollutants by Osorb, such as petroleum hydrocarbons, nitrate, biocides, endocrine-like volatiles, and chlorinated organic compounds including herbicides, will be further transformed and detoxified by reactive metals via various chemical reactions including hydrogenation, dehydrogenation, and dehalogenation (Edmiston et al., 2011). Breakdown products can be biologically mineralized in bioretention systems.

This work is part of a broader project on the development of new bioretention filter media that can be used to remediate various organic pollutants in addition to removal of nutrients by adding Osorb–metal composites in bioretention systems. The objective of the work presented here was to determine the remediative effectiveness of various Osorb–metal composites to remove a wide range of runoff pollutants. We tested five different Osorb–metal composites as an amendment to two common soil bioretention base media: (i) sand and (ii) sand–soil–compost mix among two different bioretention design configurations—(i) internal water saturated design and (ii) water unsaturated design.

This chapter is divided into two distinct sections. We will first examine the business case and market that make nanoglass materials useful in stormwater remediation. Later, we will discuss in a limited fashion the science of the Osorb nanoglass materials.

33.2 Key Market Drivers

There are three key drivers of the water–energy nexus as it affects global stormwater usage. They are as follows.

First, in 2010, 97% of the drinking water tested by the United Nations during the World Water Survey tested positive for volatile organic compounds associated with cancer, infertility, transgendering, birth defects, and numerous other health concerns. In advanced and developing nations, the problem was generally more pronounced than in underdeveloped nations. Many of these compounds, including estradiol, antidepressants, and polyaromatic hydrocarbons, increase in number and concentration in the most heavily developed nations. There are a number of technologies that are moderately effective in removing these compounds; however, they require energy-intensive inputs and often high degrees of consumables. Advanced materials capable of removing these compounds from water are in demand.

Second, today, 14% of the electricity generated globally is used to move and manage water and about one-third is used for each of three types of water: agricultural, industrial, and human direct usage. Water is second only to industrial systems in this regard and globally consumes far more energy than lighting. Many nations have created goals to specifically address reducing the electrical costs of water. Global firms, including IBM, GE, Siemens, and CH2M Hill, have created complete business units to capture what McKinsey projects to be a $3T global investment in water electrical infrastructure development by 2050.

Third, commercial chemicals, pharmaceuticals, and agrochemicals are becoming both more complex and more powerful. Many of these highly engineered chemicals elude removal by 19th and 20th century water technologies.

These three drivers confront the water treatment engineers, managers, and operators, who tend to be technology conservative and change resistant. They are, however, very aware that they lack the treatment tools needed to address increasingly complex chemistry, with lower electrical loads and costs. At the same time, public awareness and regulatory pressures are becoming less tolerant of these trace chemicals in both drinking and natural surface waters.

This awareness is high or growing for (i) municipalities, (ii) developers, (iii) universities and schools, and (iv) agricultural soil professionals. Stormwater control issues stem primarily from customers with impervious surfaces (e.g., parking lots, sports complex, high production agricultural sites, construction sites, golf courses) and insufficient management for high volumes of stormwater. High volumes of contaminated stormwater runoff are common and costly, often causing eutrophication of lakes, fishkills, erosion, sedimentation of wetlands and fish breeding sites, contamination of aquifers, and degradation of drinking water sources.

To address these issues, a growing focus is being placed on addressing both volume of stormwater flow and stormwater quality protection as far upgradient as possible. However, while a number of very good solutions are capable of high-quantity diversion of water into detention and aquifer recharge systems, there are no effective systems for long-term runoff pollutants from modern chemistry, fuels from motor vehicles, and nutrients from fertilizers. Stormwater runoff quality management is a pointed, immediate, and global issue seeking new solutions.

ABSMaterials' nanofunctional material, Osorb, which was initially developed to remove toxins at superfund sites and industrial wastewater processes, fills this need. When added to "green" stormwater systems such as enviroswales, bioswales, rain gardens, retention zones, and stormwater swales, Osorb removes and destroys nearly all persistent industrial pollutants it was designed to manage at industrial sites. Unexpectedly, the addition of Osorb to green stormwater systems also encourages more robust rhizome communities, increased genetic complexity, and a resulting biomass increase and nutrient consumption within the systems.

33.3 Functionality of Osorb Materials and Outcomes in Stormwater Systems

Affected stormwater runoff is making some waters unusable, polluting drinking water sources, surface water features, and groundwater. Adding Osorb media fill far upgradient in a system greatly improves the removal efficiency of herbicides, drugs, fuels, and other components of runoff, and provides a method for capturing high-concentration pollutants after heavy rainfall or snowmelt. Osorb media are shown to be durable and long lasting and have a high loading capacity to remove large amounts of pollutants at any given time.

Beginning in 2011, ABSMaterials began a National Science Foundation (NSF) Small Business Innovation Research (SBIR)-funded research to document how various types of Osorb material improved efficiency in removal of pollutants at three indicative stormwater locations. Two additional long-term study sites will be selected in the coming months. The research, still ongoing, is measuring the functionality, durability, and cost-effectiveness of different Osorbs with reactive metal components at each study site.

Site 1 (Figure 33.1), built in July 2011, is a twin-chamber system. Chamber 1 contains Fe–Osorb fill media. The second is a control chamber that does not contain Osorb. The unit is located at the College of Wooster in Wooster, Ohio, and drains 3.5 acres, including 80 parking spaces, a dumpster parking pad, and four buildings.

Site 2 (Figure 33.2), built in July 2012, drains ~6.5 acres of industrial site runoff, including 220 parking spaces, six loading docks, and 120,000 ft^2 of industrial rooftops. The site has been an active industrial site for >50 years with legacy chemical residues, including fuels, halons, solvents, and various chemicals associated with the site's 30-year use as a photography and microfiche processing facility.

Site 3, now in the engineering and permit phase, will drain 8.2 acres of parking roadways and rooftops at an active hospital complex.

During the first 16 months of testing at site 1 and 5 months at site 2, testing has validated substantial advantages to using nanoglass Osorbs in the runoff systems, and in most cases the field results are exceeding the results predicted from column tests conducted in a laboratory setting. The possible reasons for results are examined in the science review below; however, in summary:

1. Osorb added to stormwater systems removes at least 90%, and often 99.9%, of industrial and urban pollutants in the runoff. The concentration of volatile organic contaminants in runoff at times exceeds 400 ppm, and the Osorb nanoglass consistently reduces the concentration to <1 ppm.

2. Osorb in the stormwater system removes volatile organics directly but has the "knock-on" effect of creating extremely robust and healthy rhizome and soil

FIGURE 33.1
Tour group observes the nanoglass Osorb stormwater system at the College of Wooster. This is Study Site 1 for an NSF study of Osorb nanoglass–enabled stormwater treatment. Note: An extremely robust plant has been observed, here shown at only 12 months after initial planting. All plants are far larger and more robust than either a control bed or what would normally be expected for plants being fed waters directly from an urban parking lot and dumpster storage area.

FIGURE 33.2
At NSF Study Site 2, an Osorb stormwater system is estimated to treat approximately 6.5M gal/year of industrial stormwater runoff. Note: The foaming is associated with trace surfactants and industrial agents historically used at this site, including halons from chemical fires in 1974 and twice in the 1980s. The water has also picked up contaminants from parking lots and loading docks. This image was taken during a heavy rainfall event shortly after system was constructed in July 2012.

communities. These healthy communities result in extremely vigorous plant life and substantial and unpredicted reductions in nutrient loading associated with algae blooms and other water quality issues. This effect was not observed in the column laboratory testing conducted from 2009 to 2011.

3. Osorb nanomaterials are extremely durable and continue to perform with no degradation of performance in stormwater systems measured to date. Accelerated laboratory testing indicates 12–20+ year durability should be seen in the field.

4. These systems are passive and require no power, and they are thus extremely cost-effective. The price-per-gallon treatment of stormwater from these systems is presently $0.0008/gal to $0.0013/gal over the projected life span of the system. This is <50% of any comparable treatment system capable of this volume and quality of stormwater management.

5. Commercially interested parties, including civil engineers, Leadership in Energy and Environmental Design (LEED) architects, campus planners, and stormwater management agencies, are particularly excited to find a system this effective at treating stormwater; however, many of them simply cite the extreme healthiness of the plants in the system as the primary reason for interest. Many of them have moved away from other "green" stormwater systems because the urban runoff regularly kills off the plants in the system. Osorb captures the same pollutants and in turn chemically reacts with them, reducing them to foods accessible to the plantings, as will be discussed below.

33.4 Stormwater Industry

The runoff control industry is driven by civil engineering firms that have a very long history with the subject, and regulations set by federal, state, and local agencies. Until today,

99% of spending on stormwater management went to systems that have pipes or tunnels to move stormwater runoff away from human development and into local waterways or to central processing plants. Regulations and incentives have recently and somewhat radically changed in some locations where officials are attempting to restore natural systems, a process in the design community broadly called "biomimicry." Municipalities, large campus locations, and developers are now seeking runoff control for quality as well as quantity upgradient, but are finding the need to deal with industrial chemicals a challenge. These cultural forces and regulations can be seen as marketplace "sticks." At the same time, many architects are finding their design bonus payments attached to "carrots" in the form of LEED certifications. LEED certification incorporates green stormwater management as a critical aspect of building design.

Domestically, the US market for relieving nonpoint source pollution and stormwater volume is estimated to be $6 billion annually. The Seattle area alone has announced an initiative to build 12,000 green stormwater sites over 5 years with a goal to reduce urban runoff by 16,000,000 gallons annually. Boston, Philadelphia, and the Chesapeake Watershed Joint Management Group completed scoping documents and announced targets for this market in 2014, which is on scale with the Seattle initiative.

Internationally, especially in areas with lack of access to easy freshwater, the carrot of the marketplace is in capturing and safely harvesting urban runoff for human usage. Areas of the Mideast and West Africa have deep and acute needs for stormwater harvesting, while increasingly toxic urbanized environments and the extraction industry make the challenge of cleaning the water more difficult.

33.5 Osorb and the Need for Innovations in Stormwater Technology

Rain gardens and bioswales (landscape elements designed to remove silt and pollution from surface runoff water) are very effective at reducing runoff volumes, but have generally failed to improve runoff quality from affected urban locations. They are generally useful only as small systems capturing stormwater from single residential locations or small groups of residential properties and make up far less than 1% of the total amount spent on stormwater management in the United States.

Osorb stormwater systems incorporate blends of Osorb nanomaterials and bioretention soils to capture and reduce organic pollutants. Osorb is a highly ordered and mechanically functionalized silica material capable of hinge-like unfolding of individual and functionalized nanopores. The material has been further engineered to include embedded nanometals and micrometals inside the Osorb glass, which causes chemical species captured by the Osorb to undergo a chemical reduction (Table 33.1).

The challenge with all the species of pollutants in Table 33.1 is that they are commonly found on hard, impervious surfaces in urban or industrial settings. These chemicals are persistent, bioactive, and easily entrained in stormwater runoff. Previous systems could capture the stormwater but frequently suffered from mass plant die-offs or soil biological collapse. Often the pollutants would pass into the water table to create further problems in groundwater and well water, as well as downgradient reemergence.

Adding nanoglass allows the stormwater system to address both water volume and quality in an effective and energy-efficient manner. Furthermore, most stormwater processed

TABLE 33.1

Chemical Species Captured and/or Reduced by Osorb Materials

Product	Osorb Absorption	Osorb Reduction
Chlorinated Solvents		
Trichloroethylene (TCE)	>99%	Yes
Perchloroethylene (PCE)	>99%	Yes
Dichloroethylene (DCE)	>75%	Yes
Vinyl chloride (VC)	>75%	Yes
Trichloromethane (chloroform)	>75%	Yes
Carbon tetrachloride	>99%	Yes
Dichloromethane	>30%	No
Aromatic Compounds		
Toluene	>99%	No
Naphthalene	>99%	No
Phenol	>75%	No
Benzene	>99%	No
Ethylbenzene	>99%	No
Nitrobenzene	>99%	No
Trinitrotoluene	>99%	Yes
Chlorobenzene	>99%	Yes
Phthalate esters	>75%	No
Polychlorinated biphenyl (PCB)	>99%	Yes
Bisphenol A	>75%	No
Xylene	>99%	No
Atrazine (herbicide)	>99%	Yes
Pharmaceuticals		
Triclosan (antibacterial)	>99%	No
Fluoxetine (antidepressant)	>99%	No
Ibuprofen (anti-inflammatory)	>75%	No
Diphenhydramine (antihistamine)	>75%	No
Estradiol (birth control)	>99%	No
Imipramine (antidepressant)	>99%	No
Alcohols		
Methanol	>30%	No
Ethanol	>30%	No
Butanol	>75%	No
Hexanol	>75%	No
Solvents		
Octane	>99%	No
Hexane	>99%	No
MTBE	>75%	No
Dioxane	>75%	No
Acetone	>30%	No

Note: "Osorb reduction" refers to the chemical reduction of the compound due to nonleaching reactive metals embedded in the Osorb.

through an Osorb nanoglass system will meet or need only trivial secondary treatment to meet drinking water standards. Planning is now under way with United Nations–funded development agencies in West Africa to develop and deploy modular 8-ton units capable of treating 400 gal/day of surface water affected by industrial pollutants into potable water. Previously, no systems existed that were durable, effective, and could be operated power free for years, no less decades.

Besides the potable water solutions, discussions are now under way to offer packaged specialty soil amendments for removal of nitrates, phosphates, herbicides, fuels, pharmaceuticals, and other common runoff pollutants for first-world customers placing a premium on removing all or nearly all volatile organics from their food supply. These same volatiles are often a cause of great concern to wastewater and stormwater treatment plants that often have compliance requirements with USEPA consent decrees and/or other discharge requirements.

Testing results to date indicate that Osorb bioretention soil blends will remove >90% of stormwater volume and >95% of contaminants coming into the stormwater system. The addition of Osorb media upstream also reduces the cost for water treatment plants downstream to remove these contaminants. Twentieth century technology to remove these contaminants is focused on central wastewater treatment plants, which pay $6–12/lb for phosphate removal due to electrical energy, labor, and chemical consumables needed to treat phosphate-laden waters. The cost to remove phosphate with Osorb nanoglass is lower by an order of magnitude, at only $0.60–1.90/lb using Osorb stormwater systems upgradient. This cost savings is further compounded by the nutrients having a valuable potential impact upgradient if the plants grown with the Osorb stormwater system have commercial value.

To address this growing demand to treat these waters, substantial incentives for building upgradient and passive systems are becoming commonplace in the United States. Seattle, Portland, and Philadelphia are three cities leading a movement to manage stormwater quality and quantity upgradient in a distributed manner. Further public interest and architect/developer/contractor interest in designing to achieve LEED certifications have created a further financial and public awareness driver for green stormwater management.

LEED certification, and similar programs like STARS, Living-Machines, and BioBuilt, all place a very high value on installing smart methods for energy saving, water recycling, and stormwater savings integrated into a building or campus design. There are five categories in the LEED program where integrating Osorb nanoglass materials into a stormwater system on site can score points toward certification, with up to 13 points achieved as a result—13 LEED points for the use of any system is a relatively large number and has been driving a substantial interest in the use of the material.

Osorb nanoglass materials are enabling LEED certification points from lower operating costs, waste reduction, water quality improvement, on-site water management and reuse, native habitat restoration, technology innovation, and energy-efficient construction due to the small footprint and vastly reduced need for concrete piping. In addition, LEED-enabled tax rebates, zoning allowances, and other financial and structural incentives are benefits of integrating Osorb stormwater systems into building design. Some cities offer tax credits up to $1 million over a 10-year period (% tax credit/year depends on certification level). Los Angeles, California, requires LEED certification for all new structural buildings >50,000 ft^2. Including rain gardens that provide water reuse will help many of the companies rebuilding to attain points to become LEED certified, as well as meet requirements in cities where LEED certification is mandatory.

TABLE 33.2

Osorb Nanoengineered Glass Media vs. Alternative Technologies

Product	Remove % Nutrients (NPK)	Remove % Pharma Agents and Biocides	Remove % Hydrocarbon	$Cost/yd³	Lifetime
Osorb–soils	>80%	>90%	>95%	175–275	10+ years
Filtrexx	<10%	<25%	>80%	200–250	2–5 years
Biochar soils	<10%	<25%	<50%	25–35	5–10 years

33.6 Legacy Stormwater Technologies Replaced by Nanoengineered Glass

Currently, there are no 20th century technologies fully capable of distributed removal of volatile organics, nitrates, and phosphates. Osorb nanoglass as a soil media is truly a breakthrough media that in many cases removes 99%+ of pollutants from stormwater runoff. Osorb materials are replacing legacy systems that at best remove 50% of the pollutants. These legacy materials are, in large part, various mixtures of mulch and biochar (or granulated charcoal) soils.

A few non-nano commercial products compete with nanoglass media. One is produced by Filtrexx International. This "natural absorbent" (Filtrexx® Nutrient Control, Filtrexx® Petroleum Hydrocarbon Removal) is a soy–polymer matting that removes most petroleum products from runoff and is often specified for use on parking lot runoff next to or near sensitive natural water systems. Filtrexx's advertised retail price is approximately the same as Osorb and explicitly notes it is not as effective with biocides, pharmaceutical products, or nutrients.

Most distributed stormwater infrastructures utilize mulch and biochar soil to remove heavy oils and fuels that come with parking and roadway runoff. These systems generally demonstrate about 50% effectiveness on these species and substantial enhancement effects on plants, but almost no effect on water-soluble or molecular-level contamination. One firm, Hydro International, sells Up-Flo® Filter Mix to remove "sediments and hydrocarbons from stormwater." These sand–mulch–biochar box-filter systems require frequent replacement and do not remove pesticides, herbicides, excess pharmaceuticals, pharmaceutical agents (including endocrine disruptors), or nutrients from runoff. Up-Flo claims to remove 50% of the fuels in runoff, far short of the 90%–99% that Osorb nanomaterials remove (Table 33.2).

33.7 Estimation of Markets, Entry, and Product Reception

The customers for Osorb stormwater products have been identified as (i) professional landscape architects, LEED architects, and green developers; (ii) municipal prospects, schools, community, and organizations; and (iii) individual land owners/organic farmers with a high interest in no-biocide agents.

Professional architects and developers may require stormwater management to obtain LEED certification, especially in the circumstance where the client may want to hold their water in a cistern to reuse for landscaping. Community groups and universities are drawn toward sustainability in stormwater because of their larger campus goals and missions. These "green" flagship projects also provide an educational aspect to the community or students involved.

Green stormwater systems for single-family homes can be built for as little as a few hundred dollars, while tens of thousands of dollars are spent to address rain gardens for major parking lots, sports stadiums, or other large hard surface sites. Subsurface retention basins can cost several hundred thousand dollars. By comparison, a catchment remediation system with added Osorb would be a fraction of that cost at $25,000–55,000.

The estimated pricing for systems to be built by ABSMaterials using Osorb nanoglass to remediate pollutants in the runoff range from $4000 to $155,000 with 20%–50% of the cost of the system being the Osorb engineered soil media.

33.8 Acceptance of Engineered Materials to Assist Distributed Stormwater Systems

One may argue that the nanomaterials industry has suffered from two obstacles in the marketplace. The first is the perception that nanomaterials are solely of interest to laboratory-coat-wearing propeller heads. The second is a failure to communicate to a larger public that all chemistry functions at the molecular surface or "nanoscale." ABSMaterials has made a concerted effort to make this communication a key element of the introduction of this nanomaterial for stormwater management.

After column testing in 2010–2011 and a first fielded pilot in summer 2011, ABSMaterials employees began to contact potential customers and gauge interest—386 people from stormwater or ecological interests were contacted between November 2011 and February 2012 in Ohio and western Pennsylvania. The people ranged from civil engineers to sustainability activists. The goal of the calls was to explain the fundamentals of the materials and explore the commercial acceptability of engineered nanoglass for stormwater usage/treatments. Those who expressed interest to the point of requesting a proposal included URS Engineering, Forest City Ratner Development, Wayne County Civil Engineering, Killbuck Watershed Authority, Mahoning County Combined Sewer Authority, The University of Akron Department of Civil Engineering, Wright State University, College of Wooster, City of Euclid Ohio, Westminster College, Oberlin College, and the Nature Conservancy.

This wide-ranging and generally positive market prospecting effort was followed with efforts to convert some portion of these prospects into customers. Systems have now been built in Killbuck Watershed Authority and are in the engineering–permit phase at the College of Wooster and Westminster College—11 additional proposals for solutions are now in consideration for sites in four US states and two Canadian provinces. This conversion rate of prospects at >3% of market is a very good conversion rate for an introductory and novel product.

Beyond direct customer contacts, initial awareness relationships have already been established with a number of federal agencies, developers, and regulatory bodies that can influence what materials are specified by a building process. Soil and water conservation districts (SWCDs) in both Franklin (Columbus, OH) and Cuyahoga (Cleveland, OH) counties have expressed interest in incorporating Osorb materials in future rain garden projects. Todd Houser, stormwater manager of Cuyahoga SWCD, has become a key advocate on

ABS's product and has been pushing projects under his permit authority to closely examine the material. Likewise, the Ohio EPA SWIF program that funds stormwater improvement products has informed application parties that proposals that include Osorb-based materials to improve water quality will be favorably reviewed.

While a macro effort to introduce a product is important, it is each individual response that makes or breaks the introduction of any product. John Veney is a resident who lives next door to the twin-chamber Osorb stormwater system built at the College of Wooster in 2011. This system replaced a rather ugly ditch, which had gathered rainwater from several locations for years and was well known for overflowing into the street and being a low-grade eyesore. ABS was not aware of how bad the situation had been until Veney sent letters to both the College of Wooster and ABSMaterials expressing gratitude for the Osorb stormwater system addition to the street and for solving this long-standing problem, which included a regular flooding into his garage. He also noted the amazing native plant garden that had replaced the former mud trench and asked to be updated on how the system was working. Veney, a retiree, has since become a reference and public advocate for systems.

33.9 Introduction in Laboratory

The following section provides a brief summary of research projects on the development of new bioretention filter media to enhance pollutant removal by adding Osorb nanomaterials in bioretention systems. The research results presented here are based on work supported by the NSF SBIR program (grant no. 1113260).

33.10 Development of Osorb–Metal Composites

Five nanoscale zerovalent metals including aluminum (Al^0), iron (Fe^0), magnesium (Mg^0), nickel (Ni^0), and zinc (Zn^0) were used to create Osorb–metal composites. Osorb was used to entrap the metals to create composites. Five different Osorb–metal composites (Figure 33.3) were created by adding each metal during the synthesis of the Osorb to embed each metal within the Osorb matrix. Manufacturing was accomplished by adding 10% w/w of each metal during the manufacture of the Osorb. Each Osorb–metal composite was

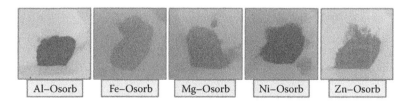

FIGURE 33.3
Five types of Osorb–metal composites: Al–Osorb, Fe–Osorb, Mg–Osorb, Ni–Osorb, and Zn–Osorb.

grinded, and three particles sizes (0.2–2.0, 0.125–0.2, and <0.125 mm) were prepared for further evaluation.

33.11 Batch- and Column-Scale Testing

All tests were carried out with simulated runoff solution with spiked concentrations of five contaminant categories (Table 33.3). Pollutant removal efficiency, hydraulic performance, and metal leaching were all studied before Fe–Osorb and Zn–Osorb were selected for field-scale testing and commercial development because of the combination of their relatively low production costs, low toxicity, long-term reactivity, and stability.

In batch tests, results showed that all five Osorb–metal composites were successful in removing a significant percentage of most tested contaminants from water. Observed results included ~99% removal of both petroleum hydrocarbons and pharmaceuticals (Table 33.4), 96% removal of pesticides (Figure 33.4), and 30%–50% removal of nutrients. No significant removal of antifreeze was observed.

Column-scale testing confirmed that the Osorb–metal composites do not negatively affect the hydraulic performance of a bioretention system, and removal results were consistent to batch-scale results with one notable exception: as the amount of Osorb–metal

TABLE 33.3

Composition of Simulated Runoff

Parameter	Pollutants	Concentration (mg/L)
Petroleum hydrocarbons	Motor oil	1000
Nutrients	Nitrate (NO_3-N)	10–20
	Phosphate (PO_4-P)	10
Herbicide	Atrazine ($C_8H_{14}ClN_5$)	0.5–1.0
Pharmaceuticals	Ethinylestradiol ($C_{20}H_{24}O_2$)	0.5–1.0
	Triclosan ($C_{12}H_7Cl_3O_2$)	0.5–1.0
Antifreeze/de-icer	Ethylene glycol ($C_2H_6O_2$)	1000

TABLE 33.4

Residual Concentrations of Runoff Pollutants after 1-Min Treatment with Osorb Composites (1% W/V)

Osorb–Metal Composites	Motor Oil (mg/L) Particle Size (mm)			Ethinylestradiol (µg/L) Particle Size (mm)			Triclosan (µg/L) Particle Size (mm)		
	>0.2	0.125–0.2	<0.125	>0.2	0.125–0.2	<0.125	>0.2	0.125–0.2	<0.125
Al–Osorb	3.8	2.5	0.6	10.3	3.1	0.1	50.1	10.3	3.1
Fe–Osorb	4.1	3.6	0.5	8.2	1.8	0.1	42.6	3.1	0.9
Mg–Osorb	3.7	2.8	0.9	7.8	2.0	0.1	48.5	8.5	4.2
Ni–Osorb	3.6	3.0	0.1	7.6	3.8	0.1	32.9	7.8	5.1
Zn–Osorb	2.8	2.0	0.5	10.0	2.4	0.1	31.4	8.8	4.5

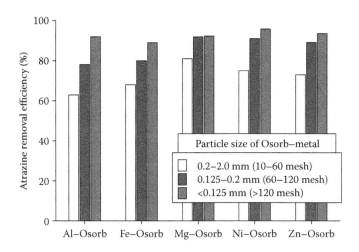

FIGURE 33.4
Removal efficiency of atrazine treated with different particle sizes of Osorb–metal composites (1% w/v) for 1 min. Initial concentration of atrazine was 1000 µg/L.

increased in both base sand and soil mix media (a mixture of 60% sand, 20% compost, and 20% topsoil), significantly higher removal efficiency of nutrients was observed, achieving 99% removal with Fe–Osorb- and Zn–Osorb-amended media (Figures 33.5 and 33.6). The results indicate that zerovalent metals in the Osorb matrix likely increased the reductive transformation and adsorption capacity for nutrient during intermittent wetting and drying conditions created in the bioretention systems.

Removal of atrazine was also significantly improved in the Osorb–metal composites amended media (Figure 33.7), in average 45% increase at the unsaturated design and 35% increase at the saturated design, indicating that the Osorb–metal composites amended media effectively capture and remove atrazine, even in the sand.

33.12 Field-Scale Testing

Two field-scale bioretention systems (rain gardens), one with standard bioretention media and one with Fe–Osorb-enhanced media, were constructed at the campus of the College of Wooster, Ohio (Figure 33.8). The systems were tested to examine the effectiveness of the Fe–Osorb-enhanced bioretention system over the standard system for runoff pollutant removal. The source of runoff was the two parking lots adjacent to the systems. The first bioretention system has a traditional underdrainage design with typical bioretention fill media that consists of 60% sand, 20% soil, and 20% compost, used as a standard (control). Fill media of the second bioretention system, however, have been mixed with 1% (w/w) of Fe–Osorb to improve treatment performance. Each bioretention system has a depth of 2.5 ft with surface area of 10 ft by 15 ft (150 ft²) to handle a 2.54 cm/h (1 in/h) rainfall.

In 2011, field-scale experiments were conducted under natural and simulated runoff events. The data show that (i) Fe–Osorb-enhanced bioretention system improved the removal efficiency of fertilizer runoff (i.e., nitrate and phosphate) by >40% compared with

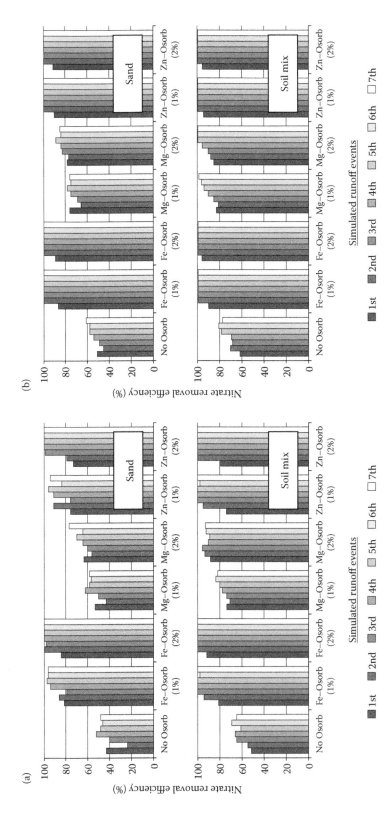

FIGURE 33.5
Removal efficiency of nitrate via unsaturated (a) and saturated (b) bioretention systems among Fe–Osorb, Mg–Osorb, and Zn–Osorb with three staggered concentrations (0%, 1%, and 2%) in two bioretention base media (sand and soil mix) from seven sequential runoff events. Loading for each event was 10 mg of nitrate-N in 1 L (10 ppm).

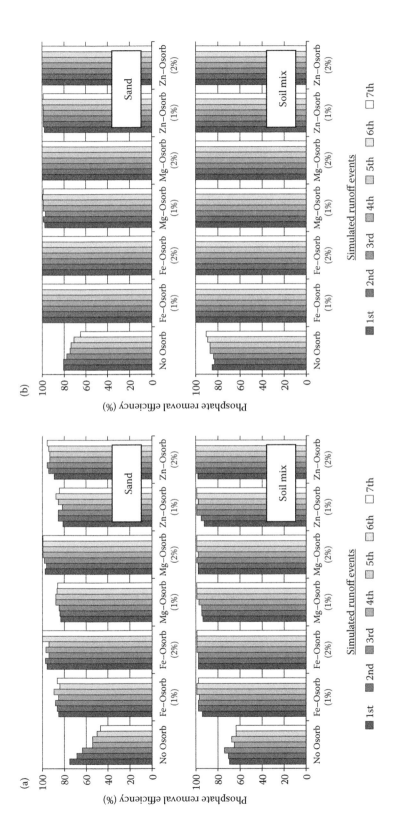

FIGURE 33.6

Removal efficiency of phosphate via unsaturated (a) and saturated (b) bioretention systems among three different Osorb–metal composites amendments (Fe–Osorb, Mg–Osorb, and Zn–Osorb) with three different amounts (0%, 1%, and 2%) in two bioretention base media (sand and soil mix) from seven sequential runoff events. Loading for each event was 10 mg of phosphate-P in 1 L (10 ppm).

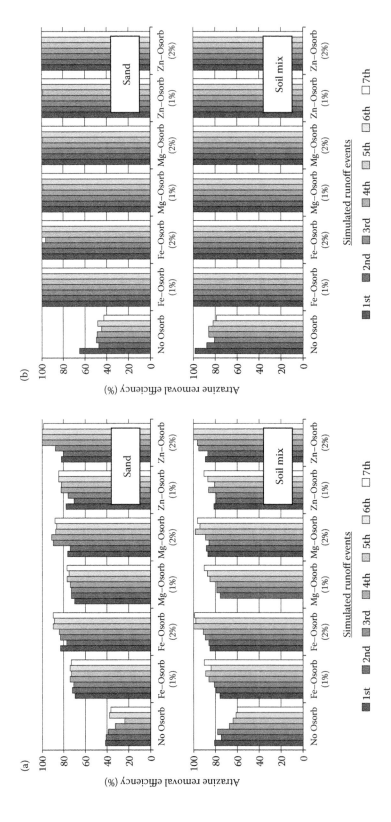

FIGURE 33.7
Removal efficiency of atrazine via unsaturated (a) and saturated (b) bioretention systems among three different Osorb–metal composites amendments (Fe–Osorb, Mg–Osorb, and Zn–Osorb) with three different amounts (0%, 1%, and 2%) in two bioretention base media (sand and soil mix) from seven sequential runoff events. Loading for each event was 0.5 mg of atrazine in 1 L (500 ppb).

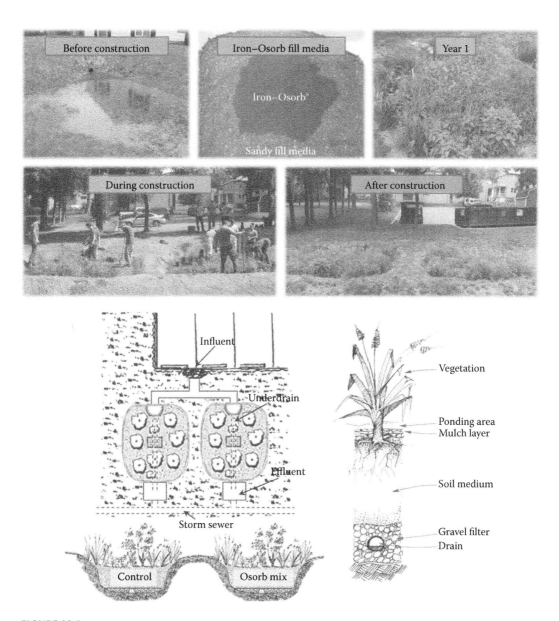

FIGURE 33.8
Site, plan, and cross-section views of field-scale bioretention systems (rain gardens) installed at the campus of the College of Wooster, Ohio. One standard model and one version enhanced with Fe–Osorb.

the standard bioretention system (Figure 33.9); (ii) a significantly lower concentration of leaching nutrients was also observed in the Fe–Osorb-enhanced bioretention system compared with the standard system (Figure 33.9); (iii) both bioretention systems significantly reduced runoff volume (~95%) and peak flow (~97%) during natural runoff events; and (iv) Fe–Osorb amendment did not alter the hydrodynamics of runoff in the fill media while maintaining a high infiltration rate. Further field tests will be continued to evaluate both hydraulic and treatment performance of the standard and Fe–Osorb-enhanced bioretention systems with a wide range of other runoff pollutants.

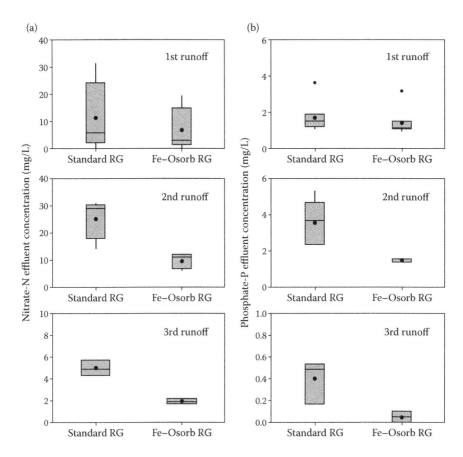

FIGURE 33.9
Nitrate (a) and phosphate (b) concentrations in effluent of standard and Fe–Osorb-enhanced bioretention systems (rain garden, RG) during three simulated runoff events. The first two runoff events (3.0-in total rainfall) were applied to each bioretention system using tap water 2 days apart with spiked concentrations of nitrate (34 g for each) and phosphate (34 g for each). No pollutants were applied for the third simulated runoff to evaluate potential leaching of nutrients.

33.13 Bacterial Growth Testing

The effects of Osorb on bacterial growth were evaluated in the presence of a biocide, triclosan. Ampicillin-resistant DH5 (alpha) *Escherichia coli* grown in Luria–Bertani (LB) media were used and tested at three different conditions: (i) control LB media, (ii) LB media + 100 ppb of triclosan, and (iii) LB media + 100 ppb of triclosan + 3 g (0.3% w/v) of Osorb, which was presterilized using ethanol. Each experiment was performed by first growing a starter culture of bacteria: 1.0 mL of the starter culture was added to 1.0 L of sterile LB media. Bacteria were grown in ampicillin (1 µg/mL) to ensure selection of only the experimental strain. Bacterial cell density was measured over time using the OD_{600} (optical density at 600 nm) method.

The results show that the bacteria grew best with Osorb even with triclosan added (Figure 33.10). Without Osorb, all the bacteria died (no growth) in the presence of 100 ppb triclosan

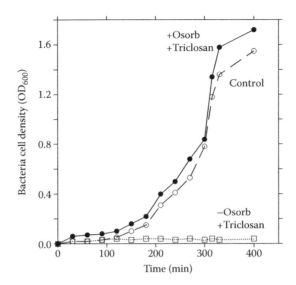

FIGURE 33.10
Bacterial cell (ampicillin-resistant DH5 [alpha] *E. coli*) growth over time at three different conditions: control LB media; LB media + 100 ppb of triclosan; and LB media + 100 ppb of triclosan + 3 g (0.3% w/v) of Osorb.

(Figure 33.10). The results indicate that Osorb protects bacteria by capturing and inactivating biocides from the media. Since the roles of microbial community in bioretention systems are substantial for pollutant removal, the results of bacterial growth suggest that the Osorb–metal composites have the potential to protect and facilitate bacterial community from toxic biocides and improve overall soil health and performance of bioretention systems.

33.14 Conclusions

New Osorb–metal composites have been initially batch, column, and field scales for enhanced remediation of runoff pollutants in bioretention systems with good, consistent, and commercially viable results. The materials are highly effective at removing multiple runoff pollutants, including nutrients (i.e., nitrate and phosphate), herbicide (i.e., atrazine), and pharmaceuticals (i.e., ethinylestradiol and triclosan).

One of the key factors to successful remediation of runoff pollutants in bioretention systems is maintaining consistent hydraulic and pollutant removal performance in the long term due to the excessive runoff volume and pollutant loads over time. Results obtained from column- and field-scale experiments have shown that the use of Osorb–metal composites in bioretention systems has the potential to simultaneously capture and remove multiple runoff pollutants such as motor oil, nitrate, phosphate, atrazine, and estradiol through enhanced physicochemical absorption and reductive transformation with excellent hydraulic performance. The ongoing effectiveness of Osorb–metal composites in field-scale bioretention systems is currently being evaluated in the long term. Future work is also focused on modification of Osorb–metal composites to maximize longevity of remediation capacity over time.

References

Burkett, C.M., Edmiston, P.L. Highly swellable sol-gels prepared by chemical modification of silanol groups prior to drying. *Journal of Non-Crystalline Solids*, 2005, 351: 3174–3178.

Cho, K.W., Song, K.G., Cho, J.W., Kim, T.G., Ahn, K.H. Removal of nitrogen by a layered soil infiltration system during intermittent storm events. *Chemosphere*, 2009, 76: 690–696.

Davis, A.P., Shokouhian, M., Sharma, H., Minami, C. Laboratory study of biological retention for urban stormwater management. *Water Environment Research*, 2001, 73: 5–14.

Davis, A.P., Hunt, W.F., Traver, R.G., Clar, M. Bioretention technology: Overview of current practice and future needs. *Journal of Environmental Engineering*, 2009, 135: 109–117.

Dietz, M.E. Low impact development practices: A review of current research and recommendations for future directions. *Water, Air, and Soil Pollution*, 2007, 186: 351–363.

Edmiston, P.L., Underwood, L.A. Absorption of dissolved organic species from water using organically modified silica that swells. *Separation Purification Technology*, 2009, 66: 532–540.

Edmiston, P.L. Swellable sol-gels, methods of making, and use thereof. US Patent 7790830 B2, US Department of Commerce, 2010.

Edmiston, P.L., Osborne, C., Reinbold, K.P., Pickett, D.C., Underwood, L.A. Pilot scale testing composite swellable organosilica nanoscale zero-valent iron—Iron-Osorb®—for in situ remediation of trichloroethylene. *Remediation Journal*, 2011, 22: 105–123.

Hsieh, C., Davis, A.P. Multiple-event study of bioretention for treatment of urban storm water runoff. *Water Science and Technology*, 2005, 51: 177–181.

Maurakami, M., Sato, N., Anegawa, A., Nakada, N., Harada, A., Komatsu, T., Takada, H., Tanaka, H., Ono, Y., Furumai, H. Multiple evaluations of the removal of pollutants in road runoff by soil infiltration. *Water Research*, 2008, 42: 2745–2755.

United States Environmental Protection Agency (USEPA). *Stormwater Best Management Practice Design Guide: Volume 1 General Considerations*. EPA/600/R-04/121. United State Environmental Protection Agency, Washington, DC, 2004.

Yang, H., McCoy, E.L., Grewal, P.S., Dick, W.A. Dissolved nutrient and atrazine removal by column-scale monophasic and biphasic rain garden model systems. *Chemosphere*, 2010, 80: 929–934.

34

Graphene: Applications in Environmental Remediation and Sensing

Theruvakkattil Sreenivasan Sreeprasad

Department of Chemical Engineering, Kansas State University, Manhattan, Kansas, USA

CONTENTS

Graphene, a single atom-thick two-dimensional (2-D) sheet of hexagonal carbon matrix,[1] is the newest member of the nanocarbon family and can be considered as the building block for other carbon-based nanomaterials, such as fullerenes and carbon nanotubes (CNTs). Graphenes possess a number of special properties that give them precedence over other carbon allotropes. For example, the band structure of graphene is unique and it behaves as a semimetal with zero band gap because the conduction band and the valence band come in contact with each other at two points (K and K') in the Brillouin zone.[2–4] Graphene also has several other features such as highest room temperature mobility,[5–7] high quantum capacitance,[8] exceptional electrical (~2000 S/cm) and thermal (5300 W/mK) conductivities,[9] transparency to visible light,[10] and exceptional mechanical strength (Young's modulus,

~1100 GPa).[11,12] Hence, graphene has found application possibilities in diverse fields, including electronics, catalysis, fuel cells, photovoltaics, biology (including targeted delivery), etc. The difficulty for bulk synthesis of graphene was a limiting factor of it being utilized in several fields. Introduction of synthetic approaches such as chemical vapor deposition (CVD),[13] chemical methods, and self-assembly processes have solved this problem, opening the door for graphene to venture into new turf.

Clean water is an essential commodity for all life forms. Cleaning up polluted water and monitoring the level of contamination are two important processes to ensure the well-being of society. A great deal of research is going on in this area. Nanomaterials, in general, with a greater number of reaction sites per unit area and exceptional surface space, are known to exhibit enhanced efficacy for removing various contaminants from air and water compared with their bulk counterparts.[14] They have proven to be useful in pollutant-sensing strategies as well.[14] Nanocarbons (e.g., CNTs and carbon onions) have also shown enhanced capacity over bulk carbon. However, until recently, graphene or graphenic materials found little application in this area. The perfect planar 2-D structures (ensuring high direct contact) and huge surface area (~2630 m^2/g)[11] of graphene makes it an ideal candidate for decontamination applications.

High mobility, quantum capacitance, and ability to conduct electrons with minimal scattering makes graphene an attractive candidate for sensing-related applications as well. The abundance of functional groups that enable easy functionalization combined with high surface area makes chemically synthesized graphene, usually referred to as graphene oxide (GO) and reduced GO (RGO) or chemically converted graphene, an exceptional prospect for environmental applications. Graphene, alone or as a hybrid in combination with other constituents, can be efficiently used for the degradation or removal of contaminants. Specific functionalization of graphene can lead to targeted sensing as well. In this chapter, a few interesting examples of graphene being used as a substrate for contaminant removal and sensing are discussed. This chapter discusses mainly the recent developments. However, a few highly important breakthroughs of the past are also noted.

34.1 Graphene for Environmental Remediation

Diverse forms of carbon such as charcoal, activated carbon (AC), and graphite have been used for water purification since ancient times. Carbon-based nanomaterials such as CNTs (alone or in combination with other materials as composites) have also found tremendous utility in this field. Very recently, graphenic materials (graphene, its chemical analogs GO and RGO) have also shown exceptional promise in this area. Pristine graphene or as composite with various other materials have been used for this purpose. Various strategies have been employed for this: (i) adsorption, (ii) capacitive deionization, (iii) photocatalytic degradation, (iv) other catalytic degradation, (v) removal using graphene-based membranes, etc. Another important aspect about graphene, its antibacterial activity, which helps graphene fight against microbial contamination, is also mentioned in the section.

34.1.1 Graphene as an Adsorbent

Adsorption, being the most effective strategy to remove contaminants from dilute solutions, is widely applied for water purification. Bulk carbon (including AC, graphite, etc.)

and nanocarbon forms such as CNTs and carbon onions, in native form and as composites, have been used for adsorption-based water remediation.[15–17] Recently, graphenic materials (and their composites) are finding huge application possibilities in this area, and studies indicated toward their high adsorption efficiency; a summary is presented in this section.

Adsorption-based removal of thiophene by single-walled CNTs and graphene was theoretically investigated, using periodic boundary conditions, van der Waals density functional, and local density approximation, by Denis and Iribarne.[18] The study provided some insights into the orientation of adsorbed molecules as well. Detailed investigation about the adsorption of various organic pollutants such as 1,2,4-trichlorobenzene (TCB), 2,4,6-trichlorophenol (TCP), 2-naphthol and naphthalene (NAPH) on graphene and GO was undertaken by Pei and coworkers.[19] A batch equilibration method and micro-Fourier transform infrared spectroscopy were employed in the study. Nonlinear isotherms for the adsorption of all four species indicated that in addition to hydrophobic interactions, various other specific interactions are involved in the adsorption process. Under alkaline pH, 2-naphthol had higher adsorption capacity on graphene compared with acidic pH. Higher π-electron density of anionic 2-naphthol than that of neutral 2-naphthol at alkaline pH, facilitating the π–π interaction with graphene was reported to be the reason for the enhancement. In the case of GO, adsorption capacity increased in the order NAPH < TCB < TCP < 2-naphthol. The adsorption mechanism was found to be different in the case of GO and graphene. Adsorption on graphene was mainly through π–π interactions. On the contrary, TCP and 2-naphthol can form H-bonding between hydroxyl groups of TCP and 2-naphthol and O-containing functional groups on GO leading to high adsorption. The removal efficiency of graphene toward various organic chemicals such as acrylonitrile, p-toluenesulfonic acid (p-TA), 1-naphthalenesulfonic acid (1-NA), and methyl blue (MB) was examined using RGO as the model graphenic material.[20] The study indicated that species with more benzene rings will be adsorbed faster with maximum adsorption capacity pointing toward the π–π stacking–based adsorption. The observed maximum adsorption capacities of p-TA, 1-NA, and MB were ~1.43, ~1.46, and ~1.52 g/g respectively. Removal of aromatic pollutants by sulfonated graphene was investigated by Zhao et al.,[21] taking naphthalene and 1-naphthol as model pollutants where the kinetics and thermodynamics of the adsorption were also investigated. Graphene demonstrated high adsorption capacity of about 2.3–2.4 mmol/g for naphthalene and 1-naphthol.[22] The study indicated that the adsorption involves stacking on the surface of graphene nanosheets (NS) with low activation energy and thermodynamically the process was found to be spontaneous and endothermic. Xu et al.[23] investigated the amputation of bisphenol A from aqueous solution using graphene. A maximum adsorption capacity of 182 mg/g was reported, which was explained on the basis of hydrogen bonding as well as π–π interactions. Graphene was found to be a highly efficient adsorbent for the removal of various pesticides from water. Graphene was found to adsorb material more than its own self-weight in certain cases.

An unprecedented adsorption of various pesticides such as chlorpyrifos, endosulfan, and malathion with maximum adsorption capacities of ~1200, 1100, and 800 mg/g, respectively, was reported by Maliyekkal et al.[24] The adsorbent was found to be highly reusable, and theoretical investigations pointed toward the adsorption being mediated by water molecules. Figure 34.1a shows the transmission electron microscopic (TEM) image of the RGO sample used in the study. The adsorption efficiency of RGO and GO (Figure 34.1b) and the adsorption energy calculated using DFT calculations (Figure 34.1c) are also given.

The utility of GO and RGO for the removal of anionic and cationic dyes such as MB, methyl violet, rhodamine B (RhB), and orange G from aqueous solutions was reported by Ramesha et al.[25] The presence of a large variety of negatively charged functionalities

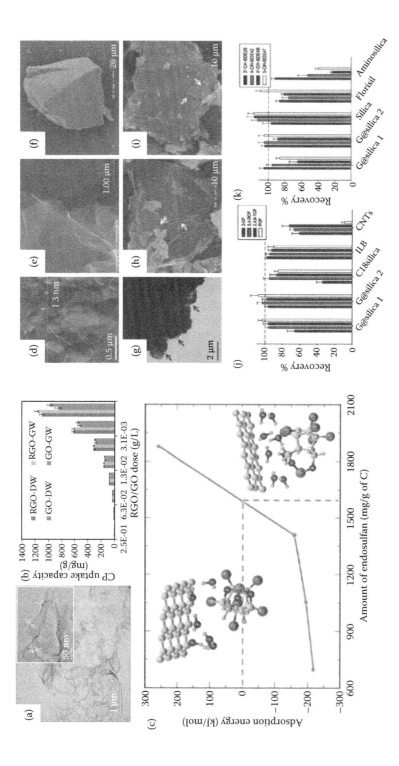

FIGURE 34.1

(a) High-resolution TEM image of RGO. Inset shows RGO with the wrinkles marked by arrows. (b) Adsorption of chlorpyrifos (CP) as a function of RGO and GO dose. (c) Energy of adsorption of endosulfan on graphene in the presence of water molecules as a function of coverage. (Adapted from Maliyekkal, S.M. et al., *Small*, Early view 2012. With permission.) (d) Atomic force microscopy and (e) SEM images of GO. (f) SEM image of bare silica. (g) TEM image of a GO@silica. (h) SEM images of GO@silica and (i) RGO@silica. (j) Comparison of the performance of graphene@silica with other adsorbents for the reversed-phase SPE of chlorophenols and (k) GO@silica with other adsorbents for the normal-phase SPE. (Adapted from Liu, Q. et al., *Angew. Chem. Int. Ed.*, 50, 5913, 2011. With permission.)

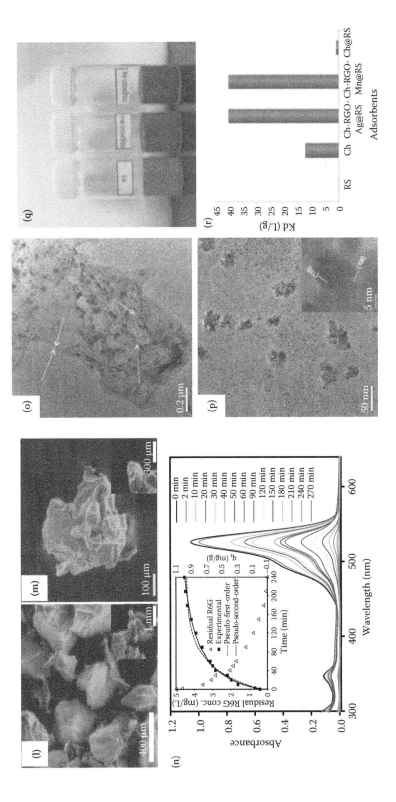

FIGURE 34.1 (Continued)

(l and m) SEM images of GSC. (n) UV/visible data showing time-dependent removal of R6G. Inset: removal of R6G as function of time (primary axis). The pseudo-second-order model fits are shown in secondary axis. (Adapted from Sreeprasad, T.S. et al., *J. Hazard. Mater.*, 246–247, 213, 2013. With permission.) (o and p) TEM images of RGO–MnO₂ (0.05 mM). (q) Photograph of different graphene composites immobilized on sand. (r) Comparison of adsorption capacities of immobilized composites. (Adapted from Sreeprasad, T.S. et al., *J. Hazard. Mater.* 186, 921, 2011. With permission.)

in GO resulted in the high adsorption efficacy of GO toward cationic dyes. Anionic dyes, on the other hand, showed negligible adsorption on GO. However, RGO was found to be highly suitable for the adsorption of anionic dyes, again pointing toward the influence of the nature of functionalities present on graphene on the adsorption of various contaminants on graphene. Yang et al.[26] and Zhang et al.[27] also reported the removal of MB by GO with high adsorption efficacy. A comparative study between the MB adsorption capacities of three different carbonaceous materials, AC, GO, and multiwalled CNTs was reported recently by Li et al.[28] The studies indicated that compared with CNTs, graphenic materials are better candidates for this application. The mechanism of adsorption was explained on the basis of the π–π electron donor–acceptor interaction and electrostatic attraction.

Graphene has been used as an adsorbent for inorganic anions and cations as well. GO was found to be an excellent substrate for the adsorption of Cu^{2+} from water.[29] Aggregation of GO induced by Cu^{2+} can lead to adsorption-aided removal of Cu with high efficiency. Compared with AC, GO exhibited 10× enhanced adsorption capacity. Deng et al.[30] reported the adaptability of functionalized graphene for the adsorption of Pb(II) and Cd(II). The as-prepared graphene sheets had 30 wt% PF_6^- (since potassium hexafluorophosphate solution was used as the electrolyte for the synthesis), and showed adsorption capacities in the order of 406.6 mg/g (pH 5.1) for Pb(II) and 73.42 mg/g (pH 6.2) for Cd(II). Graphitic oxide is reported to be a highly efficient adsorbent for arsenic and sodium with maximum adsorption capacities for arsenate, arsenite, and sodium being 142, 139, and 122 mg/g, respectively.[31] Graphene sheets synthesized by ionic liquid–assisted electrolysis were used for the adsorption of Fe^{2+} from drinking water recently.[32] Compared with GO, graphene synthesized via this method showed a 6-fold increase in adsorption capacity (maximum adsorption capacity of 299.3 mg/g). A recent report by Zhao et al.[33] suggested that GO can be an efficient adsorbent for the preconcentration of U(VI) ions from large volumes of aqueous solutions. The study proposed that the process is dominated by inner-sphere surface complexation and not by outer-sphere surface complexation or ion exchange where the oxygen-containing functional groups on the surfaces of the GO played an important role. This process also was found to be endothermic and spontaneous with a maximum capacity of 97.5 mg/g. A spontaneous, endothermic adsorption process of phosphate onto graphene was studied by Vasudevan and Lakshmi.[34] A maximum adsorption capacity of 89.37 mg/g at an initial phosphate concentration of 100 mg/L was reported in this study. Li et al.[35] demonstrated the utility of graphene for fluoride removal from drinking water. The process was found to be spontaneous endothermic in nature with a maximum adsorption capacity of 17.65 mg/g.

Assembling graphenic materials into 3-D forms has also been found to be advantageous for the decontamination of wastewater. Zhao et al.[36] recently developed a 3-D structure made up of graphenic material (through a hydrothermal process involving thiourea) called graphene sponge (GS) having tunable pore structure and surface properties that can be used as efficient, low-cost, robust, and reusable adsorbent for contaminants from water. High adsorption capacities against a variety of contaminants, including dyes, oils, and other organic solvents, were demonstrated by the GS media. For MB and diesel oil, GS can have an adsorption capacity of 184 and 129 mg/g, respectively. It was also proposed that the dye adsorption performance of GS strongly depends on its surface charge concentration and specific surface area. However, the oil adsorption capacity depends solely on the specific surface area. Bi et al.[37] also developed a similar material and used it as a sorbent for oils and other organic solvents. High adsorption efficacy was demonstrated by this material for petroleum products, fats, and toxic solvents such as toluene and chloroform (up to 86× its own weight). Dong et al.[38] reported a two-step CVD process to prepare a 3-D monolithic hybrid of graphene and CNTs for the selective removal of oils and organic solvents from the surface of water. A versatile strategy to fabricate graphene-based macroscale hydrogels and aerogels via a metal ion–induced self-assembly process was reported recently.[39] The adsorption capacity increased about 31% in the presence of 5 wt% GO in the composite. However, the maximum capacity was reported to be only 99 mg/g. A reduction of GO sheets by Fe^{2+} leading to the simultaneous formation of α-FeOOH nanorods and magnetic Fe_3O_4 nanoparticle (NP)-embedded hydrogels and aerogels was reported recently. These materials demonstrated exceptional adsorption capacities for oil and metal ions such as Cr(VI) and Pb (II).[40] The use of functionalized graphene coupled with polypyrrole (graphene–Ppy composite) as an electrically switched ion exchanger for perchlorate removal from wastewater was reported by Zhang et al.[41] Compared with Ppy films, the 3-D nanostructured graphene–Ppy nanocomposite displayed significantly improved uptake capacity for ClO^{4-}.

Nanoadsorbents anchored on substrates have the added advantage that they can be easily removed from the reaction mixture after the adsorption process. This simplifies the posttreatment handling. Recently, Liu et al.[42] devised a method to synthesize GO/RGO-immobilized silica composite through covalent binding of GO and used these composite material for solid-phase extraction (SPE). The carboxy groups on the surface of GO/RGO were covalently anchored to the amino groups of an amino-terminated silica adsorbent. Taking chlorophenols as the model system, the utility of the material in SPE was demonstrated. Figure 34.1d through k show the scanning electron microscopy (SEM) images and the efficiency of the adsorption of chlorophenols on the prepared material. Gao et al.[43] also anchored GO on a silica surface. Thiol-modified GO was anchored onto silica and was termed as "super sand" because of the 5-fold increase in adsorption capacity of the composite for removing heavy metal and dyes compared with pure sand. Generally top–down approaches (or most chemical synthesis processes) for synthesis of graphene utilize graphite as the precursor. However, recent studies indicated that graphene could be prepared from other resources as well. An *in situ* strategy to synthesize immobilized graphenic material on sand, termed graphene–sand composite (GSC) (Figure 34.1l and m), from high molecular petroleum fractions was reported recently.[44] The synthesized material was found to be highly efficient in removing pesticides and dyes from water (Figure 34.1n). The composite was able to decolorize Coca cola by removing the colored fraction from the mixture, pointing to the high activity of the material for water remediation applications. Compared with the reported adsorption efficacy of super sand, the prepared composite exhibited 12× enhanced efficacy. A similar strategy was reported for creating graphene from sugar and its immobilization on sand.[45] This composite also exhibited high capacity for removing dyes and pesticides from water. Liu et al.[46] reported a similar silica-immobilized graphene adsorbent for the removal of pesticides. The composite as expected demonstrated high adsorption capacity compared with common adsorbents such as graphite carbons, AC, pure graphene, C18 silica, and silica for the removal of 11 pesticides tested. The adsorption mechanism was also proposed, and the electron-donating abilities of the S, P, and N atoms and the strong π-bonding network of the benzene rings was understood to be the reason for the high adsorption efficacy. Synthesis of various metal/metal oxide graphene composites at room temperature through a versatile *in situ* strategy where the formed composite can be bound on solid substrates was reported by Sreeprasad et al.[47] Here, the composite is formed via a redox process without the use of any hazardous chemical reducing agents. A green strategy to anchor these composites on solid substrates such as silica or sand particles was also demonstrated. Chitosan, an abundant eco-friendly biopolymer, was used for this and the immobilization facilitated easy posttreatment handling. The immobilized composites were used for the removal of heavy metals from drinking water and compared with some common adsorbents, the composite showed highly enhanced removal efficiency. This composite was also tested in real water and found to be highly suitable for field application with appropriate modifications (Figure 34.1o through r).

Sometimes, the anchoring of other materials on graphene can result in the composites having special attributes that can improve their utility or can simplify the application process. For example, by making a graphene-based composite involving a magnetic material can result in a highly adsorbent magnetic composite. This can help in the postadsorption treatment where the adsorbent can be magnetically removed easily, simplifying the post-treatment process. Various graphene-based magnetic composites have been reported for adsorption-based remediation of polluted water. Chandra et al.[48] did the pioneering work in this direction where a water-dispersible magnetite–RGO composite was prepared and used for arsenic removal. Compared with bare magnetite particles, the composite showed high binding capacity for As(III) and As(V). Also, as proposed earlier, the composite being superparamagnetic can be separated by an external magnetic field after the removal process. A graphene-based multifunctional iron oxide sheet consisting of needle-like iron oxide NPs grown on the surface of the graphene sheets with tunable properties was reported by Koo et al.[49] recently. A paper-like material was fabricated using the composite here, which exhibited maximum adsorption capacities of 218 and 190 mg/g for As(V) and Cr(VI), respectively. A highly efficient composite for the removal of arsenate from drinking water was developed by Zhang et al.[50] This composite consisted of GO and ferric hydroxide where GO was cross linked with ferric hydroxide. At arsenate concentration of 51.14 ppm, 95% removal with an absorption capacity of 23.78 mg arsenate/g was reported over a wide range of pH from 4 to 9. Sheng et al.[51] investigated the As(V) adsorption by graphene–Fe_3O_4 (MGO) composite. They observed that the adsorption of As(V) on MGO decreased with increasing pH due to the electrostatic interaction, and the adsorption is greatly affected by the presence of coexisting ions. The presence of anions had an inhibiting effect on As(V) adsorption (more efficient at low pH), and the presence of coexisting cations had an enhancing effect (more efficient at high pH).[51]

Composite-comprising graphene sheets decorated with *Cyanobacterium metallothionein* (SmtA, a cysteine-rich metal-binding protein) was used for the ultrahigh selective adsorption of cadmium in the presence of other anionic species by Yang et al.[52] The hybrid was prepared by assembling this SmtA–GO composite on a cytopore, which showed an enrichment factor of 14.6. This hybrid was also used for selective detection of cadmium with a detection limit of 1.2 ng/L. A polymer containing graphene-based composite (polypyrrole–RGO) was used for the highly selective adsorption of Hg^{2+} recently.[53] A composite with 15 wt% GO loading showed highly selective absorption of Hg^{2+} with an enhanced adsorption efficacy up to 980 mg/g and an extremely high desorption efficiency up to 92.3%, indicating the reusability of the composite. Bhunia et al.[54] recently devised a high-temperature strategy to prepare an iron–iron oxide graphene matrix for the removal of heavy metals from water. Heating iron oxide NPs in the presence of graphene in an H_2/Ar atmosphere at elevated temperatures resulted in the formation of a highly porous matrix of iron–iron oxide on the graphene surface. A magnetic composite of graphene and magnetic cyclodextrins was reported by Fan et al.[55] for the removal of chromium from wastewater. A maximum adsorption capacity of 120 mg/g was reported in this study. Zhang et al.[56] demonstrated that modification of graphene–Fe_3O_4 composite with polyacrylic acid can make the hybrid composite a highly recyclable adsorbent for different heavy metal ions such as Cu^{2+}, Cd^{2+}, and Pb^{2+}. An interesting example reported by Zhao et al.[57] demonstrated the selective adsorption of heavy metal ions on graphene–Ppy composite leading to its selective electrochemical detection.

Magnetic graphene composites have been used for removal of organic dyes as well. Wang et al.[58] proposed the removal of fuchsine from aqueous solution using $RGO–Fe_3O_4$ composites prepared by chemical coprecipitation of Fe^{2+} and Fe^{3+} in an alkaline solution of RGO. Adsorption kinetics, capacity of the adsorbent, and the effect of the adsorbent dosage and solution pH on the removal efficiency of fuchsine were also investigated. A novel bioadsorbent composed of magnetic chitosan and GO was proposed by Fan et al.[59] for MB removal via a spontaneous and exothermic process. The substrate was found to be reusable with >90% adsorption efficacy even after four cycles of operation. A mesoporous $RGO–Mg(OH)_2$ composite prepared through a chemical deposition method was used for the removal of MB by Li et al.[60] A highly efficient magnetic graphene composite prepared via a solvothermal process for the removal of MB was reported by Ai and coworkers[61] as well. A hybrid graphene–Fe_3O_4@carbon (GFC) heterostructure nanomaterial was prepared by Fan et al.[62] recently. As-synthesized GFC showed good adsorption efficiency for MB from water in near-neutral as well as in acidic environments (about 86% and 77%, respectively).[62] A $Cu_2O–RGO$ composite (CGC) was also prepared by the same group through a similar method for the removal of dye from water, which can be used in supercapacitors.[63] The high adsorption capacity of CGC for RhB and MB was used to fabricate a reactive filtration film using this composite and was applied to remove dye from wastewater. Sun et al.[64] also used a similar strategy to fabricate $RGO–Fe_3O_4$ and reported an excellent removal efficiency for RhB (91%) and malachite green (>94%) from real water samples (industrial wastewater and lake water). An excellent adsorbent for the removal of organic pollutants from water was synthesized by coating RGO with ZnO NPs (RGO@ZnO).[65] The composite showed enhanced RhB adsorption capacity and an improved photocatalytic activity (described in Section 34.1.4) for degrading RhB compared with pure ZnO NPs. The adsorbent was found to be highly reusable as well. A graphene–Fe_3O_4 composite was used for the adsorption of aniline and *p*-chloroaniline as well.[66] A nickel–RGO composite for the removal of organic dye from water was reported recently.[67] Sui et al.[68] reported the fabrication of graphene–CNT aerosols with high adsorption efficiency for the removal organic dyes and for enrichment of heavy metal ions. Using different dye molecules such as MB, fuchsine, RhB, and acid fuchsine, as well as heavy metals such as Pb^{2+}, Hg^{2+}, Ag^+, and Cu^{2+}, the utility of the composite in water remediation was demonstrated.

34.1.2 Graphene-Based Membranes

Membranes and/or filters are important parts of the water purification industry. The possibility of creating subnanometer pores in a controllable fashion on graphene by methods such as electron beam irradiation,[69] ion bombardment,[70] or by doping[71] has been explored both theoretically and experimentally. Molecular dynamics studies about the transport of ions through 0.5-nm pores in graphene terminated with nitrogen or hydrogen was carried out by Sint et al.[72] A difference in permeation was observed between N-terminated pores and H-terminated pores. The former allow the passage of metal ions such as Li^+, Na^+, and K^+, while the latter allowed Cl^- and Br^- to pass through, but not F^-. The strongly bound hydration shell in the case of smaller ions was the reason for the low passage rates observed. Water transport through ultrathin graphene and the rate of transport across 0.75–2.75-nm-diameter pores in graphene membranes was probed by Suk and Aluru.[73] The transport values were compared with 2–10-nm-long CNTs with similar diameters. A higher flux (about two times) through graphene compared with CNTs was reported in the study.

Nanoporous materials in general are advantageous compared with traditional reverse osmosis (RO) membranes. In an RO membrane, water passage is slow and driven by diffusion processes. In nanoporous membranes, water flow is fast through the well-defined nanosized channels. Moreover, the pore size can be tuned to filter out different-sized contaminants. In most nanoporous membranes, the dimensions of the pores are comparable to the Debye screening length for electrostatic interactions and smaller than the mean free path between molecular collisions. Hence, the pores can use charge or hydrophobicity to screen ions or other molecular solutes from contaminated water. Nanocarbon materials including CNTs have found applicability in constructing nanoporous membranes.[74,75] However, low salt rejection rates and the difficulty of producing highly aligned and high-density CNT arrays limit their use. Recently, Cohen-Tanugi and Grossman,[76] using classical molecular dynamics, suggested that nanoscale pores in single-layer graphene could be used for desalination of salt water. Desalination efficiency of membrane constructed by graphene as a function of pore size, chemical functionalization, and applied pressure was investigated. Their study pointed out that the presence of hydrophilic hydroxyl groups on graphene surface can double the water flux. The study concluded that the water permeability of graphene is several orders of magnitude higher than conventional RO membranes. Zhu et al.[77] recently reported a direct production strategy for graphene-based free-standing titanate and TiO_2 nanofiber composite membranes with selective permeability for water while blocking bacteria and organic dyes. Membrane was tested against MB, methyl orange (MO), RhB, and *Bacillus coli* for their selective permeability.

Forward osmosis (FO) desalination is a comparatively new technology being proclaimed a less energy-consuming strategy for generating freshwater from seawater and other water sources. FO-based desalination works in a different way to RO-based technologies. Here, saline water is permeated through a semipermeable membrane by a draw agent having a higher osmotic pressure than the feed solution. Later, by application of suitable stimuli (depending on the nature of the material used as draw agent), pure water can be drawn out. Since the natural osmotic flow is used for the permeation process, the process is highly energy efficient and is attracting a great deal of attention. Internal concentration polarization caused by the differing solute concentrations at the transverse boundaries of the supporting layer in the membrane leading to the reduction in water flux is a big problem associated with the FO process. Recently, a graphene-based composite was used to dramatically increase water flux in the FO process.[78] Here, RGO-based hydrogels were used as draw agents. Different hydrogels with varying concentrations of RGO (0.3–3.0 wt%) were prepared using two polymers, namely poly(sodium acrylate) and poly(sodium acrylate)-poly(*N*-isopropylacrylamide). Significant enhancements in water flux were achieved for hydrogels with small amounts of RGO. Incorporation of small amounts of RGO increased the swelling ratios and softness of the hydrogels. It was proposed that addition of RGO also improved interparticle and particle–membrane contact. These factors led to the dramatic improvement of water fluxes. Moreover, the incorporation of RGO (having a high light-absorbing property) facilitated heat-induced (from the adsorbed solar light) dewetting, to draw pure water from the composite hydrogels. For the recovery process, the presence of 1.2% of RGO was found to be the optimum dose.[78] A strategy to fabricate inorganic nanofibrous membranes using GO as the cross-linker was reported by Zhang et al.[79] In this study, GO was sulfonated first to anchor -SO_3H groups on the GO surface. This sulfonated GO was treated with K-OMS-2 nanofibers to form a hierarchical membrane-like structure. The coordination interaction between the sulfonic acid group and carboxylic acid groups on GO and Mn center of the K-OMS-2 nanowire resulted in cross-linking between the two. This material was assembled into a membrane via a flow-directed assembly by filtration. This fabricated microfiltration membrane exhibited excellent rejection capacity for particle sizes >0.2 μm.

34.1.3 Capacitive Deionization

Capacitive deionization (CDI) or electrosorption using porous electrodes is another convenient way to remove various ions from aqueous solution. Carbon-based materials such as AC, carbon fibers, carbon aerogels, and CNTs have been employed as electrosorptive electrodes because of their good conductivity, high surface area, and suitable pore size distribution. Recent investigation points to the utility of graphene in this area. Li and coworkers[80] made the first attempt in this direction where they used chemically synthesized graphene as electrosorptive electrodes for CDI. The experiments were conducted in NaCl solutions at low voltage (~2 V), and the electrosorption performance was evaluated (Figure 34.2a through c). A high electrosorption capacity of 1.85 mg/g was exhibited by graphene via a physisorption-driven electrosorption. Removal of ferric ion by CDI process

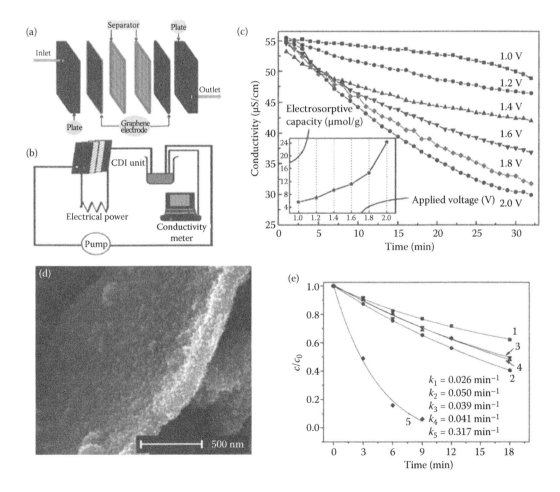

FIGURE 34.2

(a) Schematic of electrosorptive unit and (b) cell batch-mode experiment. The CDI unit consisted of graphite plate, GNF film, and separator. (c) The electrosorption of Na^+ onto the GNF electrode at different bias potentials. (Adapted from Li, H. et al., *Environ. Sci. Technol.*, 44, 8692, 2010. With permission.) (d) SEM image of heat-treated GO/TiO_2 composites. (e) Photocatalytic degradation of 10 mg/L MO at pH 4.0 by (1) neat TiO_2, (2) mixture of neat TiO_2 and GO, (3) P25, (4) mixture of P25 with GO, and (5) heat-treated composite of GO/TiO_2. (Adapted from Liang, Y. et al., *Nano Res.*, 3, 701, 2010. With permission.)

using graphene nanoflakes (GNFs) as electrodes is also reported.[81] A maximum equilibrium electrosorption capacity of 0.88 mg/g (higher than AC) was attained at a flow rate and electrical voltage of 25 mL/min and 2.0 V, respectively. The electrosorption capacities of cations on the GNFs followed the order $Fe^{3+} > Ca^{2+} > Mg^{2+} > Na^+$. The same group also evaluated the electrosorptive performance of GNF electrodes with different bias potentials, flow rates, and ionic strengths,[82] and conducted a comparative study of electrosorptive capacities of single-walled CNTs and graphene.[83]

Graphene-based composites have also been tried to get better performance from graphene as the electrode material for CDI. Wang et al.[84] recently tried a graphene–resol nanocomposite (RGO–RF) as the electrode for the removal of ferric iron. GO was aggregated due to the addition of resol, and upon calcination the structure collapsed with high pore size. A high electrosorptive capacity of 3.47 mg/g was demonstrated by RGO–RF at optimum conditions. A graphene–CNT composite also was reported to show enhanced CDI performance.[85] The insertion of CNTs helped avoid the aggregation of graphene, thereby increasing the conductivity in the vertical direction resulting in increased efficiency. Similar composites of RGO with AC[86] and mesoporous carbon (MC)[87] have also been reported. RGO–AC composite with 20% RGO content showed enhanced capacity compared with pure AC for the capacitive removal of salt ions from brackish water.[86] The graphene–MC electrode showed an enhanced adsorption capacity (731 mg/g) compared with MC alone (590 mg/g).[87] Wang et al.[88] reported a novel pyridine-based thermal strategy for preparing graphite oxide, where pyridine was used as the intercalating agent and dispersant, which used the resultant material for a CDI application. These studies indicated the utility of graphenic materials in CDI-based water remediation.

34.1.4 Photocatalytic Removal

Pioneering work in photocatalysis was done during the seventies, and since then the interest in this area has grown. A large number of catalysts have been reported, and several reviews are available on this topic.[89–91] Water remediation by nanomaterials-based photocatalysts have also been a hot area of research.[91] Here a material (the photocatalyst) catalyzes the degradation of a contaminant via light-induced reaction. Recently, graphene-based composites were also used for this purpose. Numerous reports are available in the literature where different combinations of graphenic materials and TiO_2 (one of the most efficient photocatalyst) were used for the decontamination of water. Photocatalytic degradation of MB using a composite prepared by self-assembling TiO_2 nanorods on large-area GO sheets at a water/toluene interface under ultraviolet (UV) light irradiation was reported recently.[92] The effective charge antirecombination and the effective absorption of MB on GO was proposed to be the reason for the enhanced activity. They also observed that the degradation rate of MB in the second cycle is faster than that in the first cycle. This might be due to the reduction of GO under UV light irradiation, leading to more efficient charge antirecombination by RGO. Nguyen-Phan et al.[93] probed the role of GO/RGO content on the photocatalytic efficiency of graphene-based photocatalysts by taking TiO_2–GO (prepared through a simple colloidal blending method) as the model system. They found that compared with pure TiO_2, the composites have superior adsorption and photocatalysis performance under both UV and visible radiation. It was also observed that increasing the GO content to 10 wt% resulted in the increased removal efficiency and the photodegradation rate of MB. This was explained to be due to the synergy effects, including the increase in specific surface area with GO amount as well as the formation of both π–π conjugates between dye molecules and aromatic rings on GO/RGO. The ionic interactions between MB and oxygen-containing functional groups at the edges or on the surfaces of GO sheets are also thought to increase the efficiency. It was also proposed in the study that GO is performing as the adsorbent, electron acceptor, and photosensitizer to enhance the dye photodecomposition. Chen et al.[94] reported a strategy to prepare GO–TiO_2 composites from $TiCl_3$ and GO as reactants. Visible light of wavelength >510 nm can excite the composite to induce degradation of MO. A special variety of chemically bound TiO_2 (P25) was anchored on the graphene surface by a hydrothermal process by Zhang et al.,[95] and the composite was used for the photocatalytic degradation of dyes. Liang et al.[96] prepared GO–TiO_2 composites through the direct growth of TiO_2 NPs on GO sheets. A 3-fold increase in photocatalytic degradation efficiency over conventional TiO_2 such as P25 was exhibited by the composite for the degradation of 10 mg/L MO at pH 4.0 (Figure 34.2d and e). Dong et al.[97] recently studied the shape dependence of photocatalytic activity of TiO_2–graphene composites where they anchored both spherical and rod-shaped TiO_2 NPs on graphene and compared the photodegradation of MO. They found that graphene composites having TiO_2 rods are more active than P25- and spherical NP–anchored graphene composites. A highly photoactive graphene-wrapped amorphous TiO_2 was prepared by Lee et al.[98] recently through hydrothermal process where GO reduction, TiO_2 crystallization, and formation of GO-wrapped TiO_2 NPs occur in a single step. The prepared hybrid exhibited excellent photocatalytic properties under visible light for the degradation of MB, much higher than that of bare anatase TiO_2 NPs, graphene-TiO_2 NPs (prepared by another two-step hydrothermal process), and P25 powder. Zhang et al.[99] recently suggested that the efficiency of graphene–TiO_2 composite for photocatalysis can be improved by decreasing the defects on the graphene and increasing the interfacial contact. Liu et al.[100] suggested that the use of TiO_2 NPs with specific facet-exposed structure will be beneficial to get better photocatalytic efficiency.[101] In their study, {001} facet-exposed TiO_2/graphene composite showed better photocatalyst properties than P25 and other normal TiO_2/graphene composites. This was explained to be due to the formation of a Ti–O–C bond and the formation of nanoscale Schottky interfaces at the contacts between TiO_2 and graphene. Also, it was found that positively charged dye molecules are preferentially adsorbed onto the composites due to the photogenerated charge gathered on graphene. A ternary composite containing graphene, TiO_2, and Fe_3O_4 (GTF) is also reported to have enhanced photocatalytic ability.[102] The inclusion of Fe_3O_4 helped in the easy pot-treatment removal of the catalyst. Through an *in situ* deposition strategy, TiO_2 NPs were deposited on GO by Jiang et al.[103] The composite showed excellent utility in the photocatalytic removal of pollutants at optimum conditions of solution pH, postcalcination, and at a definite GO content. The maximum photooxidative degradation rate of MO and the photoreductive conversion rate of Cr(VI) over the composites were as high as 7.4× and 5.4× than that over P25. A sonochemically prepared TiO_2–graphene photocatalyst is reported for the photocatalytic degradation of MB.[104] It was found that the photocatalytic activity of composites with only 25 wt% TiO_2 is better than that of commercial TiO_2 (100%). A TiO_2–graphene composite (prepared by a hydrothermal reaction between GO and TiO_2 in an ethanol–water mixture) for the gas-phase photodegradation of volatile organic solvent was reported by Zhang et al.[105] The photocatalytic activity and stability of the composite for the gas-phase degradation of benzene was found to be higher compared with bare TiO_2. Jiang and coworkers reported the synthesis of graphene–TiO_2 composite by *in situ* growth of TiO_2 in the interlayer of expanded graphite under solvothermal conditions.[106] Using phenol as the model system, the study demonstrated the enhanced photocatalytic performance of the composite and the enhanced activity was explained based on the increased charge separation, improved light absorbance and light absorption width, and high adsorptivity for pollutants in the composite.

Other semiconductor composites of graphene have also been found to be highly applicable for the photocatalytic decontamination of polluted water. Li and Cao[107] recently reported a ZnO–graphene composite and their efficient photocatalyst activity for the degradation and filtered removal of RhB. A composite containing ZnO, graphene, and CNT is also reported.[108] This study suggested that for this composite, the enhanced photocatalytic activity strongly depends on the presence of CNT, owing to the increased light absorption and the reduced charge recombination due to CNTs. A hybrid structure comprising ZnO@ZnS hollow dumbbells and graphene was fabricated through a polymer-assisted hydrothermal reaction and sulfurization treatment by Yu et al.[109] Owing to the hybrid structure and the presence of a new transfer pathway of electrons from ZnS to graphene, the composites exhibited superior photocatalytic activities to ZnO dumbbells and ZnO@ZnS hollow dumbbells. A facile in situ reduction strategy of GO and $ZnWO_4$ in water to form an efficient photocatalytic composite $ZnWO_4$–graphene hybrid (GAW-X) was reported by Bai et al.[110] An enhanced efficiency (~7.1× and 2.3× efficiency compared with pure $ZnWO_4$ in the presence of visible and UV light irradiation) was demonstrated by the GAW-X for the photodegradation of MO. However, visible and UV light–initiated photocatalytic processes followed different mechanisms. Visible light–induced photocatalytic activity, due to the formation of $^{\cdot}OH$ and $O_2^{\cdot-}$ via photosensitization of graphene in $ZnWO_4$–graphene. UV light induced enhanced photocatalytic activity of the composite, due to the high separation efficiency of photoinduced electron–hole pairs resulting from the promotion of an HOMO orbit of graphene in the composite.[110] Liu et al.[111] fabricated a graphene–CdS composite via a solvothermal method having significantly enhanced photocatalytic degradation efficiency toward RhB compared with bare graphene and CdS nanorods. Ye et al.[112] compared the efficiency of graphene–CdS and CNT–CdS composites for the degradation of organic dyes and found that graphene-based composites are better candidates for the application. Khan et al.[113] reported a ternary composite containing graphene, CdS, and ZnO/Al_2O_3 with superior photocatalytic activity. This study also attributed the enhancement to the enhanced surface area and effective separation of photo-induced charge carriers by the presence of GO in the composite. Different graphene-based composites containing WO_3 nanorods,[114] $BiVO_4$,[115] $BiOBr$,[116,117] $Bi_2O_2CO_3$,[118] $BiFeO_3$,[119] Fe(III),[120] $NiFe_2O_4$,[121] Ag_3PO_4,[122,123] and Ag/AgX (X = Br, Cl),[124] $InNbO_4$,[125] have been reported to show good photocatalytic degradation capacity against various pollutants. A graphene-enwrapped plasmonic Ag/AgX (X = Br, Cl) nanocomposite photocatalyst at the water–oil interface was reported by Zhu et al.[126] The photodegredation capacity of the catalyst was tested against MO under visible light irradiation. The formation of the composite led to an increase in adsorptive capacity due to the presence of GO in the composite. The smaller size of Ag/AgX NPs facilitated charge transfer, and also suppressed recombination of electron–hole pairs in Ag/AgX/GO, leading to the enhanced photocatalytic activity of the composite. Xiong et al.[127] reported the visible light–induced photocatalytic degradation of dyes over a graphene–gold NP (RGO–Au) hybrid. Spontaneous reduction of $HAuCl_4$ by RGO resulted in the anchoring of Au NPs on the graphene substrate. RhB was used as the model system to evaluate the photoactivity of the RGO–Au composite. A mechanism was proposed for the degradation as well. First, the dye gets excited to dye*, followed by an electron transfer from dye* to graphene. Later, the electron is moved to a Au NP where it gets trapped by O_2 to produce various reactive oxygen species (ROS). This ROS degrades the dye. Various graphene-based semiconductor nanocomposites have been reported to have high photocatalytic activity.

34.1.5 Graphene-Based Catalyst for Other Catalytic Degradation Methods

Graphenic materials have been used for other catalytic processes to remove contaminants as well. Sun et al.[128] demonstrated that RGO having an I_D/I_G >1.4 can activate peroxymonosulfate to produce active sulfate radicals such as $SO_4^{-\cdot}$, which, in turn, can decompose various aqueous contaminants owing to the powerful oxidizing character of the radicals. The study proved that compared with other allotropes of carbon such as AC, graphite powder, GO, and multiwalled CNTs, RGO has better activity. It was also reported that RGO-based catalysis is better compared with transition metal oxide–based catalysis for the degradation of phenol, 2,4-dichlorophenol and MB, in water. A graphene–MnOOH composite prepared by a solvothermal process involving the dissolution–crystallization and oriented attachment of MnOOH on graphene exhibited unusual catalytic performance for the thermal decomposition of ammonium perchlorate.[129] The concerted effect of graphene and MnOOH is reported to be the reason for the enhanced performance. A graphene–horseradish peroxidase composite for enzyme-catalyzed degradation of phenolic compounds was reported by Zhang et al.[130] An ethylenediamine–RGO (ED–RGO) composite for the indirect reduction of Cr(VI) to less toxic Cr(III) and subsequent removal was reported by Ma et al.[131] The removal was explained via a three-step mechanism. In the first step, Cr(VI) binds to the composite through an electrostatic interaction between the negatively charged Cr(VI) species $\left(HCrO_4^-\right)$ and the protonated amine groups on ED. Later in the second step, π electrons on the six-membered carbon ring of RGO reduced Cr(VI) to Cr(III). This Cr(III) will be liberated into the solution and will attach onto the ionized carboxylic groups on the RGO in the third step to complete the removal process. Electro-enzymatic degradation of carbofuran on a ternary GO–Fe_3O_4–hemoglobin hybrid structure was reported by Zhu et al.[132] The composite can be easily separated magnetically after the remediation process. Graphene–CdS composites were used to sonocatalytically degrade various azo dyes from water in the absence of light.[133] Shi et al.[134] reported a cobalt oxide (Co_3O_4) supported on graphene that can be used as catalyst for the sulfate radical–based oxidative removal of orange II from water.

34.1.6 Antibacterial Properties of Graphene

Drinking water contamination due to the presence of microbes is a recurring problem. Not only water bodies but also the filters used for purification purpose are also susceptible to microbial attacks. The formation of biofilms on the filter surface due to bacterial growth can impart unwanted tastes and odors to the purified water.[135] After some time, this biofilm can also lead to premature clogging of filters. Graphene can also be a solution for this problem. The antibacterial activity of GO and RGO was first investigated by Hu et al.[136] They found that graphenic materials could effectively inhibit the growth of *Escherichia coli* bacteria while showing minimal cytotoxicity toward human cells. This is an added advantage since other carbon allotropes such as CNTs are known to have cytotoxic effects. Krishnamoorthy et al.[137] studied the mechanism behind the antibacterial activity of GO/ RGO. The study was conducted on four different species of pathogenic bacteria. The study indicated that the production of ROS by graphene leads to an increase of intracellular ROS levels of the cells, making them susceptible to oxidative stress. Such oxidative stress can induce damage to cellular components, including DNA, lipids, and proteins.[138,139] Specific studies have indicated that oxidation of fatty acids by ROS can generate lipid peroxides and can subsequently stimulate a chain reaction, leading to the disintegration of the cell membrane followed by cell death.

Various graphene-based composites have been reported to have high antibacterial activity. A highly water-soluble composite composed of brilliant blue (BB, a common colorant), RGO, and tetradecyltriphenylphosphonium bromide having high antibacterial activity was reported by Cai et al.[140] The composite was active against both gram-positive and gram-negative bacteria (Figure 34.3a through h). Silver NP–graphene hybrids with enhanced antibacterial activity are also reported.[141–145] An interesting strategy to create a highly adhesive antibacterial material, inspired from mussels and their ability to attach on to diverse surfaces, was reported by Zhang et al.[146] They synthesized a composite consisting of Ag NPs, polydopamine, and GO sheets. This hybrid material exhibited strong antibacterial properties to both gram-negative and gram-positive bacteria due to the synergistic effect of GO and Ag NPs.[146] Sreeprasad et al.[147] leveraged the rich abundance of functional groups on the GO surface to anchor an antibacterial biopolymer chitosan and an antibacterial iron-binding protein (lactoferrin) and fabricated a luminescent multifunctional composite. The synergetic effect of different materials incorporated resulted in a highly antibacterial composite. The composite showed a tendency to form self-standing films that can aid in coating this composite onto suitable substrates (Figure 34.3i and j).

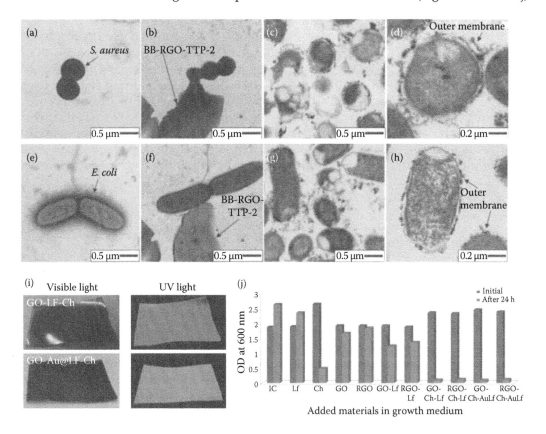

FIGURE 34.3
TEM illustrating the antibacterial activity of brilliant blue/RGO/quaternary phosphonium salt composite (BB–RGO–TTP) on *S. aureus* cell (a through d) and *E. coli* cells (e through h). The damage imparted to the outer membrane is visible in (b) and (h). (Adapted from Cai, X. et al., *Langmuir*, 27, 7828, 2011. With permission.) (i) Photographs of the composite films under visible and UV light. (j) Comparison of antibacterial activity of different materials tested in the study. (Adapted from Sreeprasad, T.S. et al., *ACS Appl. Mater. Interfaces*, 3, 2643, 2011. With permission.)

Hence, graphene adsorbents and membranes have this added advantage that there will be only minimal microbial contamination.

Photo-inactivation of bacteria (*E. coli*) by a GO–TiO$_2$ film under solar illumination was reported by Akhavan and Ghaderi.[148] The composite exhibited about 7.5× increased antibacterial activity compared with native TiO$_2$. The same group have reported an interesting process where interactions of GO with *E. coli* living in a mixed-acid fermentation environment in an anaerobic condition result in the formation of highly antibacterial graphenic material.[149] Reduction of GO to RGO occurs owing to the metabolic activity of the surviving bacteria through their glycolysis process. After the reduction process, the RGO sheets exhibited an inhibition for proliferation of the bacteria on their surfaces and the already proliferated bacteria became detached from the surface of these sheets, pointing to the antibacterial activity. Some et al.[150] devised a graphene-poly(L-lysine) composite that promotes the growth of human cell culture and exhibits high antibacterial activity. Antibacterial composites of graphene in combination with ZnO,[151] polyvinyl-*N*-carbazole,[152] lanthanum(III),[153] and chlorophenyl pendant[154] have also been reported. Another different effort to construct antibacterial graphene-based composite was reported by Wang et al.[155] where they prepared GO–benzylpenicillin (BP) anion intercalated Mg–Al layered double hydroxide (GO–BP–LDH) hybrid films. The hybrid films were fabricated via a simple solvent evaporation process. The films can release BP ions, and the release can be tuned by adjusting the composition of the film, and hence the strategy can be used for drug release applications as well. The antibacterial activity of the film was due to the presence of GO as well as the release of BP.

34.2 Graphene-Based Contaminant Sensing Strategies

The last section illustrated graphene-based strategies employed to remove contaminants from the water stream. Detecting these contaminants is as important as it is to remove these toxic materials from drinking water. However, detecting strategies should be able to detect contaminants at ultralow concentrations, since their permissible limits are very low. The maximum permissible limits of most of these contaminants are in parts per billion (ppb) or parts per trillion (ppt). Hence, smart materials have to be used to sense these toxins in a fast, sensitive, selective, reliable, and reproducible manner. Various NPs with their diverse physicochemical properties are being employed for this purpose. This section outlines the various approaches undertaken to construct graphene-based contaminant sensors.

34.2.1 Graphene-Based FET Sensors

Graphene electrically behaves as a semimetal with zero band gap. It has the maximum room temperature mobility reported. For sensors, this can be helpful for rapid signal transmission. Graphene has many other interesting properties such as ballistic transport of its charge carriers with very low scattering at room temperature,[1] a tunable band gap,[156–159] quantum interference,[160] and a large quantum capacitance.[8] Hence, a small perturbation in its electrical characteristics can lead to rapid, amplified signal response with minimal background. Therefore, graphene has been widely used for constructing various sensors based on sensitive changes of their electrical properties. Graphene-based field-effect transistors (FETs) are one of the most utilized classes of devices for sensor applications. Schedin et al.[161] did pioneering work on graphene-based sensors. In this study, single gas molecules were detected from the step-like changes in resistance caused by the change in local carrier concentration of graphene due to the interaction of gas molecules. An FET device for the detection of hydrogen gas using RGO is also reported.[162] For this, holes were introduced to RGO (with different edge-to-plane ratios) via electrochemical deposition of metal NPs. A change in the electrical conductance of the composite was seen as different concentrations of H_2 gas was introduced into the FET. Lu et al.[163] proposed an RGO-based NH_3 sensor. They found that the sensing behavior is highly dependent on the gate voltage (Vg). This was explained on the basis of the ambipolar transport of RGO and the Vg-induced change in the graphene work function. The coulomb interaction between NH_3 and the FET can also lead to such a behavior.

A large number of graphene-based FETs for the sensitive detection of chemical and biological entities are reported. Zhang and coworkers[164] developed a mercury detection device by self-assembling 1-octadecanethiol monolayers on graphene and constructing FET using this alkanethiol-modified graphene. The detection limit of the device was reported to be 10 ppm. A graphene-based FET biodevice with single bacterium detection capability was proposed by Mohanty and Berry.[165] They proposed that the attachment of a single-bacterium generated ~1400 charge carriers in a p-type chemically modified graphene resulting in a huge increase in current leading to the detection (Figure 34.4a and b). This FET was also used as a label-free DNA sensor. Huang and coworkers[166] also constructed a bacterial sensor from CVD graphene through noncovalent interaction. In addition, the sensor was used to evaluate glucose-induced metabolic activities of the bound *E. coli* bacteria in real time as well. Ohno et al.[167] recently fabricated a graphene-based FET from single-layer graphene obtained from micromechanical cleavage for chemical and biological sensing applications. The FET was able to detect the changes in solution pH with a lowest detection limit (signal-to-noise ratio = 3) of 0.025. Using the same setup, they were able to detect different charge types of biomolecules owing to their isoelectric point. Choi et al.[168] recently developed an electrical sensor for DNA. Highly water-soluble graphene sheets were synthesized via a microwave-assisted sulfonation strategy and the interactions of the functionalized graphene with DNA induced sensitive electrical changes leading to the label-free detection of DNA. Zhang et al.[169] recently reported the fabrication of a GO–polypyrene (PPr) composite film prepared by the electrochemical codeposition of GO and PPr using propylene carbonate as the organic electrolyte. They used this film as a chemoresistor-type volatile organic vapor sensor, and a normalized sensitivity of 9.87×10^{-4} ppm^{-1} was reported for toluene in this study. The mechanism of the response was also proposed. The response of the sensor to different contaminants indicated that contaminants with aromatic π electrons (such as toluene) and lone pairs (like chloroform) would show more response. This indicated that the dipolar electrostatic forces are important in the sensing process. In addition, a very low response was observed in the case of nonpolar hexane, pointing to the importance of molecular polarizability in controlling the sensitivity. Hence, the following mechanism was proposed. The absorption of polarizable analyte on PPr chains under the applied bias generated an induced dipole moment leading to the change in configuration of conjugated PPr chains into a more stretched planar structure. This resulted in enhanced electronic coupling between the monomer units of the PPr chains, increasing the overall conductivity of the polymer phase. It was also proposed that the large pyrene rings of the PPr chains allow enhanced adsorption toluene through π–π interactions, leading to higher sensitivity.[169] Myung et al.[170] recently devised a novel label-free polypeptide-based graphene–NP hybrid biosensor to detect the enzymatic activity and ultralow concentrations of specific enzymes. Change in electrical hysteresis of the hybrid by enzyme interaction was used as signal for detection.

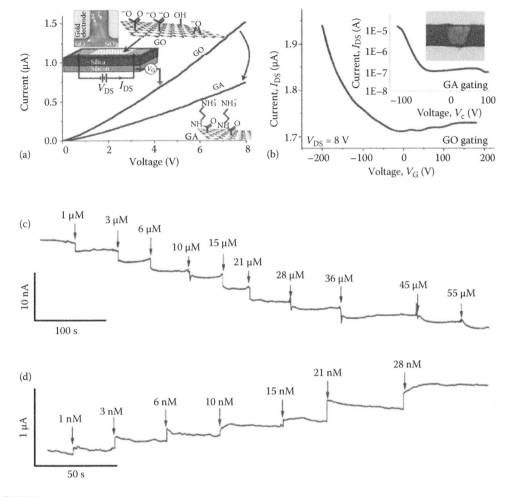

FIGURE 34.4

(a) Current-voltage behavior of the GO and graphene-amine (GA) devices with GA device showing lower conductivity than the parent GO. The insets show the schematics of GO/GA devices and their chemical structure. (b) Electrical gating studies of GO and amine–GO showing the p-type semiconductor behavior. (From Mohanty, N. and Berry, V., *Nano Lett.*, 8, 4469, 2008. With permission.) Typical real-time recording of I_{DS} from the patterned RGO FETs with the addition of Ca^{2+} ions (c) and Hg^{2+} ions (d). (Adapted from Sudibya, H. G. et al., *ACS Nano*, 5, 1990, 2011. With permission.)

Sudibya et al.[171] used chemically functionalized RGO micropatterns to construct FETs for the specific detection of heavy metal ions (Figure 34.4d and e). First, metallothionein type II (MT-II), which can bind to heavy metal ions with high affinity and selectivity, was anchored on to patterned RGO channels using specific linkers. The sensor could selectively detect ultralow concentrations of Hg^{2+} (detection limit 1 nM or 0.2 ppb) in the presence of other ions such as Mg^{2+}, Ca^{2+}, or K^+. Similarly, using calmodulin (Ca^{2+} binding protein) instead of MT-II, selective detection of Ca^{2+} was possible. It was reported that the conformational change of the functionalized protein upon metal binding is the cause for the electrical changes. The negatively charged protein came closer to the graphene channel owing to the conformational changes resulting in a stronger gating effect.[171]

34.2.2 Electrochemical Sensors

Electrochemical sensors are one of the largest classes of chemical sensors where a relationship between electricity and chemistry is used for sensing. In general, this category of sensors can be defined as devices that extract information about a sample from measurement of parameters such as potential difference, current, resistance/conductance. On the basis of this, electrochemical sensors can be categorized as voltammetric or amperometric sensors (where current in amperes is measured), potentiometric sensors (where a difference of two potentials in volts is measured), and chemiresistors or conductometric sensors (if resistance in ohms or conductance is measured). Graphene has found huge potential in this category of sensors as well. A short summary is presented in the following section.

34.2.2.1 Graphene-Based Electrodes for Voltammetric and Amperometric Sensors

In this category of electroanalytical sensors, analyte species that can become oxidized or reduced are detected from the sudden changes in current. In voltammetric sensors, current is measured by varying the potential. In amperometric sensors, a given potential is applied between two electrodes placed inside the solution containing the analyte, and the change in current is measured as the analyte is oxidized at the anode or reduced at the cathode. Graphene composites have found application as electrodes for both voltammetric and amperometric sensors. A strategy for the simultaneous voltammetric detection of catechol (CC) and hydroquinone (HQ) using a graphene-modified electrode was reported recently.[172] A graphene-modified glassy carbon electrode (GR/GCE) exhibited a well-defined peak and a significant increase of current, clearly demonstrating the utility of graphene as an efficient promoter to enhance the kinetics of the electrochemical process of catechol and hydroquinone. Yuan et al.[173] also reported graphenic MC electrodes for the simultaneous detection of HQ and CC. The detection limit for HQ and CC was reported to be 3.7×10^{-7} and 3.1×10^{-7} M in this study. A glassy carbon electrode modified with a β-CD–RGO composite was used for the determination of nitrophenol isomers. Detection limits of 0.05, 0.02, and 0.1 mg/dm^3 were reported for *para*-, *ortho*-, and *meta*-nitrophenols, respectively.[174] Electropolymerized graphene–nafion film-modified GCE was used for the detection of nitroaniline isomers recently.[175] Three nitroaniline isomers (2-nitroaniline, 3-nitroaniline, and 4-nitroaniline) were detected with a detection limit of 0.022 μg/mL using a simple pulse voltammetry technique. A novel Ni/Al layered double-hydroxide decorated graphene NS hybrid (LDHs–GNs) prepared on a cathodic substrate was used as an SPE phase for stripping voltammetric detection of organophosphate pesticides (OPs) such as MP recently.[176] The detection limit for MP in aqueous solutions was reported to be 0.6 ng/mL. A zirconia NP–graphene hybrid, prepared by a one-step co-electrodeposition approach, anchored on a cathodic substrate, was recently used for the sensitive square-wave voltammetry detection of OPs taking MP as the model pollutant.[177] A detection limit of 0.6 ng/mL was achieved in this study. An enzymeless OP sensor using Au NP-decorated GNs modified glass carbon electrode (GCE) is also reported.[178]

Liu et al.[179] recently developed a novel amperometric biosensor by anchoring acetylcholinesterase (AChE) on a 3-carboxyphenylboronic–RGO–Au NP hybrid–modified electrode for the detection of OPs and carbamate pesticides. Different pesticides such as chlorpyrifos, malathion, carbofuran, and isoprocarb showed a detection limit of 0.1, 0.5, 0.05, and 0.5 ppb respectively. Recently, Choi et al.[180] reported a freestanding flexible conductive RGO–nafion (RGON) hybrid film having superior synergistic electrochemical characteristics (such as high conductivity [1176 S/m], facile electron transfer, and low interfacial resistance). These hybrid films were used as electrochemical biosensing platforms for organophosphate detection (Figure 34.5a through c), with a sensitivity of 10.7 nA/M, detection limit of 1.37×10^{-7} M, and response time of <3 s. The device was found to be reusable for >100 cycles as well. Amino-functionalized exfoliated graphite nanoplatelet–modified electrodes were used for the determination of Pb on vegetables using differential pulse voltammetry.[181] A detection limit of 0.001 µg/L was attained in this approach. An amperometric biosensor based on AChE immobilized on CdS-decorated graphene (CdS–graphene) nanocomposite was reported by Wang et al.[182] In this study, a rapid inhibition time (2 min) was obtained due to the integration of the CdS–graphene nanocomposite. A voltammetric strategy to selectively detect Hg^{2+} from water in the presence of various other ions, using graphene-based hybrid electrode, was reported by Gong et al.[183] First, graphene was mixed with chitosan to make a homogenous dispersion and drop-casted onto the electrode surface followed by deposition of Au NPs. It was noted that the composite film greatly facilitates electron transfer, resulting in a remarkably improved sensitivity and selectivity. The detection limit of this method was found to be 6 ppt. Highly toxic Cd^{2+} was also detected using a similar method.[184] The device consisted of nafion–graphene nanocomposite film and a differential pulse anodic stripping voltameter. Cui et al.[185] fabricated a composite comprising graphene, Au NPs, and ionic liquids and modified GCE with the composite. This modified electrode was employed for the detection of paraquat, a herbicide widely used in agriculture. A detection limit of 7.3×10^{-10} M was reported. Wu et al.[186] also used a reduced β-cyclodextrin–graphene (β-CD–graphene) hybrid as a sorbent for the preconcentration and electrochemical sensing of MP.

Graphene-based materials have also been used for cyclic voltammetry–based ultrasensitive detection strategies. An electrochemical sensor using graphene–CNT–Pt NP hybrid for the determination of bisphenol A with a detection limit of 4.2×10^{-8} M is reported.[187] Graphene–CD hybrid (CD–GNs) were used as an enhanced material for ultrasensitive detection of carbendazim by cyclic voltammetry recently.[188] An increase in peak currents was observed when GNs modified GCE and the CD–GNs/GCE was used, indicating that the nanocomposite film not only shows the excellent electrical properties of GNs but also exhibits high supramolecular recognition capability of CDs. The detection limit of carbendazim was reported to be 2 nM with a signal-to-noise ratio of 3. Good recoveries (98.9% and 104.5%) were also observed.

34.2.2.2 Potentiometric Sensors

In this category of sensors, the signal is measured as the voltage difference (potential) between the working electrode and the reference electrode and the concentration of the analyte modulates the potential of the working electrode. Yuan et al.[189] recently reported a GO-based electrode for the potentiometric detection of Cu^{2+}. GO sheets grafted with 2-amino-5-mercapto-1,3,4-thiiodiazole were used as a neutral carrier in this process. A detection limit of 4.0×10^{-8} M, applicability over a wide pH range (3.0–7.0), and fast response time (15 s) were reported. A noncovalent graphene-based composite was also used for the potentiometric sensing of Zn^{2+} ions.[190] Using the π–π interaction between GO and a Zn^{2+} ions complexing ligand 1-(2-pyridylazo)-2-naphthol, a hybrid sensing membrane for the potentiometric determination of zinc ions was fabricated. A highly selective and sensitive determination of Zn^{2+} was possible using this strategy. An interesting electrochemical sensor for the selective and sensitive detection of bacteria using graphene was reported by Wan et al.[191] In this method, the nanomaterial-promoted reduction of silver ions resulting in the signal amplification of GO coupled device was used for the sensing.

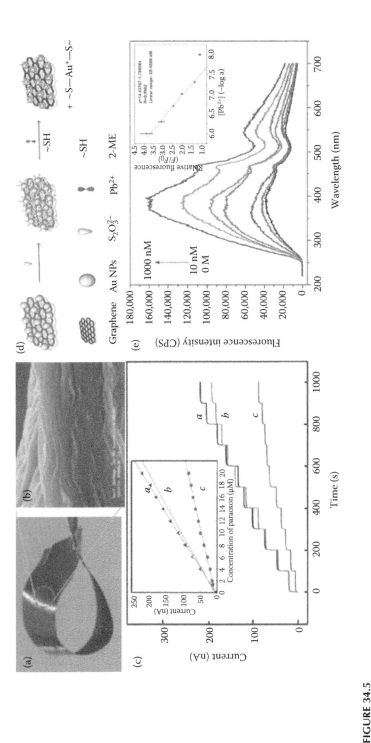

FIGURE 34.5

(a) Photograph of RGON film. (b) SEM image of the assembled film. (c) Amperometric responses with respect to 2 μM increments of paraoxon in N$_2$-saturated PBS. Applied potential, +0.85 V: (*a*) OPH-RGON, (*b*) OPH-f-RGON, and (*c*) OPH-RGO films. Inset shows the calibration curves of the current signals with respect to the concentration of paraoxon. (From Choi, B.G., *ACS Nano*, 4, 2910, 2010. With permission.) (d) Schematic representation of the sensing mechanism for the detection of Pb^{2+} ions based on leaching of Au NPs on the surface of graphene. (e) Fluorescence spectra of graphene–Au nanocomposites (NCs) in 5 mM glycine solution upon the addition of increasing concentrations of Pb^{2+} ions (0 M–1000 nM) under optimal conditions. The inset is the calibration curve of the graphene–Au NCs for detection of Pb^{2+} ions. (Adapted from Fu, X. et al., *ACS Appl. Mater. Interfaces*, 4, 1080, 2012. With permission.)

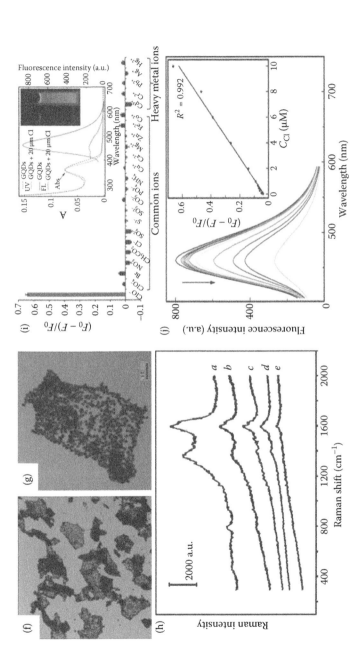

FIGURE 34.5 (Continued)

(f and g) TEM images of the GO/PDDA/Ag NPs at different magnifications. (h) SERS spectra of 9 nM folic acid obtained in the GO/PDDA/Ag NP solutions with (*a*) 0.02 mg/mL, (*b*) 0.01 mg/mL, (*c*) 0.004 mg/mL, (*d*) 0.0025 mg/mL, and (*e*) 0.002 mg/mL graphitic carbon. (Adapted from Ren, W. et al., *ACS Nano*, 5, 6425, 2012. With permission.) (i) Plot illustrating the selectivity of the GQD-based sensor for chlorine over other ions in pH 8 solution. Inset shows the UV and fluorescence (FL) spectrum of 0.14 mg/mL GQD solution in the absence and presence of 20 μM free chlorine over other ions in pH 8 solution. Inset shows the FL photos of 0.7 mg/mL GQD solution in the absence (right) and presence (left) of 100 μM free chlorine illuminated by an UV beam of 365 nm. (j) FL response of 0.14 mg/mL GQDs upon addition of various concentrations of free chlorine in a pH 8 solution (from top: 0, 0.05, 0.1, 0.3, 0.5, 0.7, 1, 2, 4, 6, 8, and 10 μM). Inset: Stern–Volmer plot of FL quenching of the GQDs by free chlorine. (Adapted from Dong, Y. et al., *Anal. Chem.*, 84, 8378, 2012. With permission.)

Similarly, electrochemical sensors for trinitrotoluene (TNT) and dinitrotoluene (DNT) (electrochemically reduced GO[192] and ionic liquid–graphene hybrid),[193] MP (graphene–chitosan),[194,195] hydrazine (AuCu NP–ionic liquid–graphene composite film, flower-like Co NP–graphene hybrid, and MnO_2–graphene composite),[196–198] dihydroxybenzene isomers (thermally reduced GO),[199,200] catechol (CuO–RGO),[201] nitrophenol (GO-coated electrode),[202] dopamine (graphene–chitosan composite[203] and single-stranded DNA-mediated immobilization of graphene on a gold electrode[204]), herbicide (paraquat, Cu_2O/polyvinyl pyrrolidone–graphene),[205] H_2O_2,[206–214] heavy metal ions (AlOOH–RGO composite),[215] kojic acid (using GSC modified with RGO),[216] and chlorophenols (Pd–graphene composite)[217] are reported where a superior sensitivity was obtained owing to the synergetic effects of graphene-based composites.

34.2.3 Colorimetric Sensors Based on Luminescence

These sensors work on the basis of color changes obtained upon interaction with targeted molecules. The color changes can be a change in visible color or a change in fluorescence. The color can be due to catalytic chemiluminescence or electrochemiluminescence (ECL) as well. A large number of graphene-based colorimetric sensors have been reported. A few recent representative examples are discussed below.

Fluorescence is very sensitive to the environment and hence the changes in fluorescence of various graphene-based composites upon interaction with contaminants have been used for constructing sensors for a large variety of contaminants. A large number of examples exist where graphene-based biosensors are used for the detection of biomolecules such as DNA,[218–220] thrombin,[221] caspase-3,[222] and bovine serum albumin.[223] Most of these strategies utilized fluorescence resonance energy transfer (FRET)-based fluorescence "turn-on" or "turn-off" leading to sensing. A similar strategy can also be used for the detection of various other molecules. A glucose sensor based on FRET from upconverting phosphors to GO was devised by Zhang et al.[224] Graphene-based fluorescence sensors for metal ions are also reported. Huang et al.[225] reported a label-free, sensitive, and selective detection strategy for Hg^{2+} using an RGO–organic dye nanoswitch. Here, RGO acts as a nanoquencher and highly selective sorbent increasing the sensitivity and selectivity. Acridine orange (AO) dye was immobilized on RGO through π–π stacking interactions, resulting in the quenching of the fluorescence of AO due to long-range resonance energy transfer. Introduction of Hg^{2+} result in the detachment of dye from RGO and the restoration of fluorescence. This was used as a turn-on sensor for Hg^{2+} detection. The detection limit for Hg^{2+} was found to be about 2.8 nM. A graphene composite with DNA duplexes of poly(dT) has also been used for the detection of Hg^{2+}.[226] An amplified fluorescence "turn-on" biosensor for detection of Pb^{2+} using graphene–DNAzyme was also reported recently.[227] The detection limit of the devise was found to be 300 pM. Fu et al.[228] introduced a GO–Au NP hybrid as fluorescent turn-on sensor for Pb^{2+}. Here, GO was used as the fluorescence source and Au NPs as the quencher. Addition of Pb^{2+} resulted in the leaching out of Au NPs and a consequent restoration of fluorescence (Figure 34.5d and e). A graphene-based strategy for the specific detection of Ag^+ utilizing the specific binding of aptamers toward target molecules (here Ag^+) was reported by Wen et al.[229] Conformational change of the aptamer (fluorescein amidite) induced by the attachment of Ag^+ leads to the recovery of fluorescence. Kundu et al.[230] reported a sensor for nitroaromatic molecules fabricated using GO–methyl cellulose hybrid. The autoluminescence of GO, which is improved by the formation of a hybrid with methyl cellulose, becomes quenched upon the addition of nitroaromatics. Here, the nitro groups create electron-deficient centers and aid in electron transfer between the hybrid and the nitroaromatic, leading to the quenching. A competitive adsorption-based method for the sensitive detection of synthetic organic dyes on RGO was also reported. The higher adsorption capacity of a dye compared with the one already anchored onto RGO was used to detect the former.[231] Wang et al.[232] proposed a graphene-conjugated oligomer hybrid probe for the detection of lectin and *E. coli*. Lee et al.[233] reported a metal–organic framework/azobenzoic acid–functionalized GO composite hydrogel (MOF/A-GO), which in the presence of Zn^{2+} can function as a chemosensor for the detection of TNT. A label-free surface plasmon resonance–based sensing strategy for the biological warfare agent *Salmonella typhi*, based on electrochemically synthesized GNS, is also available in the literature.[234]

Catalytic chemiluminescence and ECL are other phenomena used for selective sensing. Graphene–Al_2O_3 composites were fabricated through a facile one-step process involving supercritical CO_2 ($SCCO_2$), which displayed high catalytic chemiluminescence sensitivity and high selectivity to ethanol gas.[235] An appreciable response was obtained for a low concentration of 9.6 mg/mL ethanol at 200°C, indicating the promise of the nanocomposite. Amplified ECL-based sensing of acetylcholine using an RGO–quantum dots (QDs) composite was reported by Deng et al.[236] The ECL emission of QDs coated on a GO-modified electrode will be quenched by the structural defects of GO. When the double bonds are getting restored by the reduction, RGO–QDs exhibit ECL emission. ECL biosensors for choline and acetylcholine were fabricated by covalently cross-linking choline oxidase (ChO) or ChO–acetylcholinesterase on the RGO–QDs modified electrode. Linear response ranges and detection limits of 10–210 and 8.8 µM for choline, and 10–250 and 4.7 µM for acetylcholine, respectively, were obtained using these electrodes.

34.2.4 Graphene as SPE Material in Chromatographic Sensors

This section illustrates the use of graphene and its composites as the adsorbent for the SPE of pollutants and subsequent chromatography or mass spectroscopy (MS)-based sensing/detection strategy. Zhang and coworkers reported a graphene-coated solid-phase micro-extraction (SPME) fiber where GO was covalently bonded to the fused-silica substrate using 3-aminopropyltriethoxysilane as a cross-linking agent and subsequently reduced by hydrazine to make a graphene coating. Coupled with gas chromatography (GC)–MS, they used this composite structure as a high-performance SPME for polycyclic aromatic hydrocarbons (PAHs) leading to sensitive detection of PAH with good precision (<11%), low detection limits (1.52–2.72 ng/L), and wide linearity (5–500 ng/L) under the optimized conditions.[237] GO-bonded fused-silica fiber was also used as the SPME for the GC-based detection of PAH by Xu et al.[238] Use of graphene to create a high-efficiency adsorbent coating on fibers for the SPME of four triazine herbicides (atrazine, prometon, ametryn, and prometryn) from water is also reported.[239] Coupled with high-performance liquid chromatography–diode array detection (HPLC–DAD), these graphene-coated fibers were used to determine the presence of the above-mentioned four triazine herbicides. A detection limit of 0.05–0.2 ng/mL was reported in the process. Liu et al.[240] illustrated the advantages of using graphene as an adsorbent for SPE, taking eight different chlorophenols as model analytes. Luo et al.[241] recently proposed the utility of a substrate-less graphene fiber prepared by a hydrothermal process as an SPME sorbent. This was demonstrated using five different organochlorine pesticides as the model systems. The results pointed to the superiority of graphene fibers over commercial fibers with higher extraction efficiencies, higher thermal stability (up to 310°C), better reproducibility, and longer service life (more than 180× reuse). Zou et al.[242] reported the use of a polypyrrole–graphene composite-coated hybrid fiber for the SPME of phenols.

Magnetic graphene composites are also used as SPMEs. A graphene-based magnetic nanocomposite was also used for the preconcentration of the five carbamate pesticides, including metolcarb, carbofuran, pirimicarb, isoprocarb, and diethofencarb from water samples before feeding this to HPLC–DAD for their detection.[243] Here, the high adsorption capacity of the composite and the magnetic properties of the NPs, which aid in the phase separation of the adsorbent from the sample solution, were utilized, which helped save extra time required in traditional SPE. In a similar method, graphene–Fe_3O_4 magnetic NP hybrids (graphene–Fe_3O_4 magnetic NPs) were used as the SPME for the detection of triazine herbicides such as atrazine, prometon, propazine, and prometryn in environmental water samples with the aid of HPLC–DAD.[244] Extraction of neonicotinoid using magnetic graphene NPs as adsorbent was reported by Wang and coworkers.[245] In this study, graphene–Fe_3O_4 NPs were used to extract thiamethoxam, imidacloprid, acetamiprid, and thiacloprid from water samples. Using HPLC, accurate determination of analytes at spiking levels of 0.5 and 5 ng/mL was done. Graphene-based magnetic nanocomposite (graphene–Fe_3O_4) was recently used as an effective adsorbent for the preconcentration of various triazole fungicides such as myclobutanil, tebuconazole, and hexaconazole from environmental water samples before feeding it to HPLC for their detection.[246]

Shi et al.[247] used a graphene–magnetite composite for enrichment and detection of small molecules where the composite was used as a novel matrix for matrix-assisted laser desorption/ionization–time-of-flight–MS. A cyclodextrin-functionalized GO–Fe_3O_4 nanocomposite was used as a tunable stationary phase in open-tubular capillary electrochromatography by Liang et al.[248] The composite showed excellent wettability, enhanced stability against high ionic strength, suppressed electro-osmotic mobility, and less non-specific adsorption toward analytes compared with the native polydimethylsiloxane microchip. Graphene-coated stainless-steel fibers were also used for microwave-assisted headspace SPME of organochlorine pesticides in water samples recently.[249]

34.2.5 Surface-Enhanced Raman Spectroscopy-Based Sensors

Graphene-based sensors utilizing surface-enhanced Raman spectroscopy (SERS)-based sensing/detection protocols are another category of highly explored sensors. The Raman signal enhancement obtained by using the graphenic composite is used for detecting the analyte. This can occur either by enhancing the signal intensity or by removing the luminescent background of the spectrum to give a better signal. An ultrasensitive detection strategy based on the SERS property of graphene-based materials for the detection of aromatic molecules was devised by Liu et al.[250] recently. Graphene films having anchored Ag NPs were fabricated. In the composite, GO acts as the matrix that catches the aromatic molecule, and Ag NPs aid in the localized surface plasmon resonance-based SERS property. The utility of the above hybrid system was illustrated using positively charged crystal violet, negatively charged amaranth (a cosmopolitan genus of herbs), and neutral phosphorus triphenyl as model molecules. Ren et al.[251] used a similar graphene–Ag NP composite for the detection of folic acid. Ag NPs were anchored onto GO sheets via a self-assembly method with the aid of poly(diallyldimethyl ammonium chloride) (PDDA) as the functional macromolecule (Figure 34.5f through h). The hybrid showed excellent SERS properties and was used for the ultralow concentration detection (detection limit of 9 nM) of folic acid. A biosensor based on the SERS properties of a chalcogenide prism-anchored graphene multilayer is also reported.[252] The incorporation of graphene in the biosensor significantly increased the sensitivity, and the detection accuracy also increased by >100% because of high index chalcogenide glass as compared with silica glass.

34.3 New Directions

New graphene-based structures such as graphene QDs (GQDs) and graphene nanoribbons (GNRs) and carbon NPs have emerged recently with several possibilities. Sensors based on these materials are one of the newest classes of sensors. The availability of a band gap in these structures can help in constructing ultrafast transistors and electrical sensors. Representative examples of some new directions where novel graphenic structures are being used for pollutant sensing are noted here. Valentini et al.[253] recently used modified screen-printed electrodes composed of oxidized GNRs for the selective electrochemical detection of several molecules, including potassium ferricyanide, catechol, hexaammineruthenium(III) chloride, sodium hexachloroiridate(III) hydrate, dopamine, epinephrine, L-tyrosine, 3,4-dihydroxyphenylacetic acid, ascorbic acid, uric acid, 4-aceta midophenol(acetaminophen), NADH, H_2O_2, caffeic acid, guanine, and serotonin (5-HT) hydrochloride. Hydrothermally synthesized fluorescent carbon NPs were found to be a good sensor for mercury ions.[254] A carbon QD–polyamine composite was used as a Cu ion sensor as well.[255] A detection limit of 6 nm was attained in this strategy. Recently, Dong and coworkers[256] used GQDs for the detection of free chlorine in drinking water. The destruction of the passivated surface of GQDs through oxidation by free chlorine, resulting in sudden quenching of their fluorescence, led to the sensing (Figure 34.5i and j). Goh and Pumera[257] recently developed a graphene-based electrochemical sensor for detection of 2,4,6-TNT in seawater. The sensor was constructed using single-, few-, and multilayer GNRs and graphite microparticles and the performance was compared. A detection limit of 1 μg/mL was obtained in this approach. Graphdiyne (GD) is a carbon allotrope strikingly similar to graphene but contains two diacetylenic linkages between repeating carbon hexagon units. This is considered as the most stable of the various man-made diacetylenic carbon allotropes. A GD-based photocatalyst containing P25 has also been reported to have high efficiency for photocatalysis.[258] Compared with simple P25, the composite GD–P25 showed enhanced photocatalytic activity for degrading MB under UV irradiation. A study conducted by Akahvan and Ghaderi[259] recently reported the antibacterial activity of GO nanowell structures on both gram-positive and gram-negative bacteria. Nanowell structures were fabricated by electrophoretic deposition of Mg^{2+}–GO. The cell membrane damage of the bacteria caused by contact of the bacteria with the sharp edges of the nanowalls was suggested to be the mechanism of the bacterial inactivation. Gram-negative *E. coli* bacteria with an outer membrane were more resistant to this inactivation, whereas gram-positive *Staphylococcus aureus*, which does not have an outer membrane, was highly susceptible. Also, RGO nanowalls were more toxic to the bacteria than the unreduced GO nanowalls, owing to the better charge transfer between the bacteria and the more sharpened edges.

34.4 Conclusions and Future Prospects

Mankind has drawn immense benefits from the advancement in science and technology. However, on the other side of the spectrum, we are faced with the challenge of cleaning up the after-products of industrial and agricultural activities, an offshoot of this development. The large-scale use of different contaminants across the world has polluted many water bodies. Graphene, with the perfect 2-D structure, spectacular properties that can be modulated, and a surface that can be modified, is an attractive candidate for novel applications in environmental science, especially in the areas of contaminant sensing and remediation. The last few years have seen an explosion in activities that are dedicated to utilizing graphene in environmental science, and this chapter has illustrated the potential of graphene or graphene-based materials for these applications. Wastewater remediation is one area in which graphenic material has shown immense promise. The large surface area and the surface structure make them an attractive candidate for various approaches. Chemically synthesized GO/RGO is believed to be the first graphenic material to find direct application in this area. Diverse approaches including adsorption, photocatalytic and other catalytic degradation, CDI, and membrane separation–based removal have been employed for this cause. The high utility of graphene-based materials in all these strategies points toward the spectacular versatility of graphene. Hence, over the next few years, water treatment could emerge as one of the chief areas of application for chemically modified graphene. The presence of a variety of functional groups, which can be easily leveraged for functionalizing graphene with specific molecules, can lead to novel targeted sensors. The chapter illustrated how graphene in combination with different materials can be used for FET-based, electrochemical, FRET-based, or SERS-based sensing applications. The mechanism and the part played by graphene to enhance the activity are also described.

However, challenges still exist in this area. For example, although pristine graphene (graphene with pure sp^2 hybridized carbon atoms) has tremendous capability for FRET-based applications, it is highly hydrophobic, which limits the application. Similarly, GO is highly hydrophilic but its electrical conductivity is comparatively very low. Likewise, some challenges exist in the bulk production of soluble, well-defined graphene or graphene derivatives. Cytotoxicity, the cellular uptake mechanism, and the intracellular metabolic pathway of graphene and its derivates are not known in detail. Several opportunities also exist in this direction. For example, one important and immediate challenge will be to use graphene quantum structures, especially in sensing. As discussed in a previous section, the presence of a band gap in these structures can give an additional tool for creating sensors. GQDs are known to be luminescent, and this can lead to luminescence-based sensing applications through appropriate modifications. It has to be reiterated that most of the sensing applications are done with chemically derived graphene (GO/RGO) since functionalization of pristine graphene with specific molecules are not feasible. Most of the properties (especially electrical properties) of graphene are greatly diminished in GO/RGO. The conductivity, mobility, etc., are all very low in GO/RGO compared with pristine graphene. Noncovalent binding of target molecules on pristine graphene is a viable option but the weak binding is not stable. Thus, research is dedicated to formulate strategies that can anchor functional groups on graphene without disturbing its sp^2 hybridization. If this is realized, ultrafast sensors with ultralow detection capabilities can be realized. Research in this direction is still in its infancy. Another possibility is to utilize graphene bilayers and trilayers for sensing applications. The area of 2-D materials is ever growing and in the future, other 2-D materials such as MoS_2, $NbSe_2$, WS_2, and BN are expected to venture into this area. Hence, the possibilities are limitless and opportunities are waiting to be explored.

References

1. Novoselov, K.S., A.K. Geim, S.V. Morozov, D. Jiang, Y. Zhang, S.V. Dubonos, I.V. Grigorieva, and A.A. Firsov. Electric field effect in atomically thin carbon films. *Science*, 306:666–669, 2004.
2. Novoselov, K.S., E. McCann, S.V. Morozov, V.I. Fal'ko, M.I. Katsnelson, U. Zeitler, D. Jiang, F. Schedin, and A.K. Geim. Unconventional quantum Hall effect and Berry's phase of 2 pi in bilayer graphene. *Nat. Phys.*, 2:177–180, 2006.
3. Novoselov, K.S., A.K. Geim, S.V. Morozov, D. Jiang, M.I. Katsnelson, I.V. Grigorieva, S.V. Dubonos, and A.A. Firsov. Two-dimensional gas of massless Dirac fermions in graphene. *Nature*, 438:197–200, 2005.
4. Novoselov, K.S., D. Jiang, F. Schedin, T.J. Booth, V.V. Khotkevich, S.V. Morozov, and A.K. Geim. Two-dimensional atomic crystals. *P. Natl. Acad. Sci.*, 102:10451–10453, 2005.
5. Gass, M.H., U. Bangert, A.L. Bleloch, P. Wang, R.R. Nair, and A.K. Geim. Free-standing graphene at atomic resolution. *Nat. Nanotechnol.*, 3:676–681, 2008.
6. Chen, J.H., C. Jang, S.D. Xiao, M. Ishigami, and M.S. Fuhrer. Intrinsic and extrinsic performance limits of graphene devices on SiO_2. *Nat. Nanotechnol.*, 3:206–209, 2008.
7. Du, X., I. Skachko, A. Barker, and E.Y. Andrei. Approaching ballistic transport in suspended graphene. *Nat. Nanotechnol.*, 3:491–495, 2008.
8. Xia, J., F. Chen, J. Li, and N. Tao. Measurement of the quantum capacitance of graphene. *Nat. Nanotechnol.*, 4:505–509, 2009.
9. Balandin, A.A., S. Ghosh, W. Bao, I. Calizo, D. Teweldebrhan, F. Miao, and C.N. Lau. Superior thermal conduct ivity of single-layer graphene. *Nano Lett.*, 8:902–907, 2008.
10. Nair, R.R., P. Blake, A.N. Grigorenko, K.S. Novoselov, T.J. Booth, T. Stauber, N.M.R. Peres, and A.K. Geim. Fine structure constant defines visual transparency of graphene. *Science*, 320:1308, 2008.
11. Sreeprasad, T.S., and T. Pradeep. Graphene for environmental and biological applications. *Int. J. Mod. Phys. B.*, 26:1242001, 2012.
12. Sreeprasad, T.S., and V. Berry. How do the electrical properties of graphene change with its functionalization? *Small*, doi: 10.1002/smll.201202196, 9:341–350, 2013.
13. Li, X., W. Cai, J. An, S. Kim, J. Nah, D. Yang, R. Piner, A. Velamakanni, I. Jung, E. Tutuc, S.K. Banerjee, L. Colombo, and R.S. Ruoff. Large-area synthesis of high-quality and uniform graphene films on copper foils. *Science*, 324:1312–1314, 2009.
14. Pradeep, T., and Anshup. Noble metal nanoparticles for water purification: A critical review. *Thin Solid Films*, 517:6441–6478, 2009.
15. Zabihi, M., A. Ahmadpour, and A.H. Asl. Removal of mercury from water by carbonaceous sorbents derived from walnut shell. *J. Hazard. Mater.*, 167:230–236, 2009.
16. Demirbas, A . Agricultural based activated carbons for the removal of dyes from aqueous solutions: A review. *J. Hazard. Mater.*, 167:1–9, 2009.
17. Maliyekkal, S.M., K.P. Lisha, and T. Pradeep. A novel cellulose-manganese oxide hybrid material by in situ soft chemical synthesis and its application for the removal of Pb(II) from water. *J. Hazard. Mater.*, 181:986–995, 2010.
18. Denis, P.A., and F. Iribarne. Thiophene adsorption on single wall carbon nanotubes and graphene. *J. Mol. Struc.- Theochem.*, 957:114–119, 2010.
19. Pei, Z., L. Li, L. Sun, S. Zhang, X.Q. Shan, S. Yang, and B. Wen. Adsorption characteristics of 1,2,4-trichlorobenzene, 2,4,6-trichlorophenol, 2-naphthol and naphthalene on graphene and graphene oxide. *Carbon*, 51:156–163, 2013.

20. Wu, T., X. Cai, S. Tan, H. Li, J. Liu, and W. Yang. Adsorption characteristics of acrylonitrile, p-toluenesulfonic acid, 1-naphthalenesulfonic acid and methyl blue on graphene in aqueous solutions. *Chem. Eng. J.*, 173:144–149, 2011.

21. Zhao, G., L. Jiang, Y. He, J. Li, H. Dong, X. Wang, and W. Hu. Sulfonated graphene for persistent aromatic pollutant management. *Adv. Mater.*, 23:3959–3963, 2011.

22. Zhao, G., J. Li, and X. Wang. Kinetic and thermodynamic study of 1-naphthol adsorption from aqueous solution to sulfonated graphene nanosheets. *Chem. Eng. J.*, 173:185–190, 2011.

23. Xu, J., L. Wang, and Y. Zhu. Decontamination of bisphenol A from aqueous solution by graphene adsorption. *Langmuir*, 28:8418–8425, 2012.

24. Maliyekkal, S.M., T.S. Sreeprasad, D. Krishnan, S. Kouser, A.K. Mishra, U.V. Waghmare, and T. Pradeep. Graphene: A reusable substrate for unprecedented adsorption of pesticides. *Small*, 9:273–283, 2013.

25. Ramesha, G.K., A. Vijaya Kumara, H.B. Muralidhara, and S. Sampath. Graphene and graphene oxide as effective adsorbents toward anionic and cationic dyes. *J. Colloid Interface Sci.*, 361:270–277, 2011.

26. Yang, S.T., S. Chen, Y. Chang, A. Cao, Y. Liu, and H. Wang. Removal of methylene blue from aqueous solution by graphene oxide. *J. Colloid Interface Sci.*, 359:24–29, 2011.

27. Zhang, W., C. Zhou, W. Zhou, A. Lei, Q. Zhang, Q. Wan, and B. Zou. Fast and considerable adsorption of methylene blue dye onto graphene oxide. *Bull. Environ. Contam. Toxicol.*, 87:86–90, 2011.

28. Li, Y., Q. Du, T. Liu, X. Peng, J. Wang, J. Sun, Y. Wang, S. Wu, Z. Wang, Y. Xia, and L. Xia. Comparative study of methylene blue dye adsorption onto activated carbon, graphene oxide, and carbon nanotubes. *Chem. Eng. Res. Des.*, 91:361–368, 2013.

29. Yang, S.T., Y. Chang, H. Wang, G. Liu, S. Chen, Y. Wang, Y. Liu, and A. Cao. Folding/aggregation of graphene oxide and its application in Cu^{2+} removal. *J. Colloid Interface Sci.*, 351:122–127, 2010.

30. Deng, X., L. Lü, H. Li, and F. Luo. The adsorption properties of Pb(II) and Cd(II) on functionalized graphene prepared by electrolysis method. *J. Hazard. Mater.*, 183:923–930, 2010.

31. Mishra, A.K., and S. Ramaprabhu. Functionalized graphene sheets for arsenic removal and desalination of sea water. *Desalination*, 282:39–45, 2011.

32. Chang, C.F., Q.D. Truong, and J.R. Chen. Graphene sheets synthesized by ionic-liquid-assisted electrolysis for application in water purification. *Appl. Surf. Sci.*, 264:329–334, 2013.

33. Zhao, G., T. Wen, X. Yang, S. Yang, J. Liao, J. Hu, D. Shao, and X. Wang. Preconcentration of U(vi) ions on few-layered graphene oxide nanosheets from aqueous solutions. *Dalton Trans.*, 41:6182–6188, 2012.

34. Vasudevan, S., and J. Lakshmi. The adsorption of phosphate by graphene from aqueous solution. *RSC Adv.*, 2:5234–5242, 2012.

35. Li, Y., P. Zhang, Q. Du, X. Peng, T. Liu, Z. Wang, Y. Xia, W. Zhang, K. Wang, H. Zhu, and D. Wu. Adsorption of fluoride from aqueous solution by graphene. *J. Colloid Interface Sci.*, 363:348–354, 2011.

36. Zhao, J., W. Ren, and H.M. Cheng. Graphene sponge for efficient and repeatable adsorption and desorption of water contaminations. *J. Mater. Chem.*, 22:20197–20202, 2012.

37. Bi, H., X. Xie, K. Yin, Y. Zhou, S. Wan, L. He, F. Xu, F. Banhart, L. Sun, and R.S. Ruoff. Spongy graphene as a highly efficient and recyclable sorbent for oils and organic solvents. *Adv. Funct. Mater.*, 22:4421–4425, 2012.

38. Dong, X., J. Chen, Y. Ma, J. Wang, M.B. Chan-Park, X. Liu, L. Wang, W. Huang, and P. Chen. Superhydrophobic and superoleophilic hybrid foam of graphene and carbon nanotube for selective removal of oils or organic solvents from the surface of water. *Chem. Commun.*, 48:10660–10662, 2012.

39. He, Y.Q., N.N. Zhang, and X.D. Wang. Adsorption of graphene oxide/chitosan porous materials for metal ions. *Chinese Chem. Lett.*, 22:859–862, 2011.

40. Cong, H.P., X.C. Ren, P. Wang, and S.H. Yu. Macroscopic multifunctional graphene-based hydrogels and aerogels by a metal ion induced self-assembly process. *ACS Nano*, 6:2693–2703, 2012.

41. Zhang, S., Y. Shao, J. Liu, I.A. Aksay, and Y. Lin. Graphene-polypyrrole nanocomposite as a highly efficient and low cost electrically switched ion exchanger for removing ClO4- from wastewater. *ACS Appl. Mater. Interfaces*, 3:3633–3637, 2011.

42. Liu, Q., J. Shi, J. Sun, T. Wang, L. Zeng, and G. Jiang. Graphene and graphene oxide sheets supported on silica as versatile and high-performance adsorbents for solid-phase extraction. *Angew Chem. Int. Ed.*, 50:5913–5917, 2011.
43. Gao, W., M. Majumder, L.B. Alemany, T.N. Narayanan, M.A. Ibarra, B.K. Pradhan, and P.M. Ajayan. Engineered graphite oxide materials for application in water purification. *ACS Appl. Mater. Interfaces*, 3:1821–1826, 2011.
44. Sreeprasad, T.S., S.S. Gupta, S.M. Maliyekkal, and T. Pradeep. Immobilized graphene-based composite from asphalt: Facile synthesis and application in water purification. *J. Hazard. Mater.*, 246–247:213–220, 2013.
45. Gupta, S.S., T.S. Sreeprasad, S.M. Maliyekkal, S.K. Das, and T. Pradeep. Graphene from sugar and its application in water purification. *ACS Appl. Mater. Interfaces*, 4:4156–4163, 2012.
46. Liu, X., H. Zhang, Y. Ma, X. Wu, L. Meng, Y. Guo, Y. Liu, and G. Yu. Graphene-coated silica as a highly efficient sorbent for residual organophosphorus pesticides in water. *J. Mater. Chem. A*, 1:1875–1884, 2013.
47. Sreeprasad, T.S., S.M. Maliyekkal, K.P. Lisha, and T. Pradeep. Reduced graphene oxide metal/metal oxide composites: Facile synthesis and application in water purification. *J. Hazard. Mater.*, 186:921–931, 2011.
48. Chandra, V., J. Park, Y. Chun, J.W. Lee, I.C. Hwang, and K.S. Kim. Water-dispersible magnetite-reduced graphene oxide composites for arsenic removal. *ACS Nano*, 4:3979–3986, 2010.
49. Koo, H.Y., H.J. Lee, H.A. Go, Y.B. Lee, T.S. Bae, J.K. Kim, and W.S. Choi. Graphene-based multifunctional iron oxide nanosheets with tunable properties. *Chem. Eur. J.*, 17:1214–1219, 2011.
50. Zhang, K., V. Dwivedi, C. Chi, and J. Wu. Graphene oxide/ferric hydroxide composites for efficient arsenate removal from drinking water. *J. Hazard. Mater.*, 182:162–168, 2010.
51. Sheng, G., Y. Li, X. Yang, X. Ren, S. Yang, J. Hu, and X. Wang. Efficient removal of arsenate by versatile magnetic graphene oxide composites. *RSC Adv.*, 2:12400–12407, 2012.
52. Yang, T., L.H. Liu, J.W. Liu, M.L. Chen, and J.H. Wang. Cyanobacterium metallothionein decorated graphene oxide nanosheets for highly selective adsorption of ultra-trace cadmium. *J. Mater. Chem.*, 22:21909–21916, 2012.
53. Chandra, V., and K.S. Kim. Highly selective adsorption of Hg^{2+} by a polypyrrole-reduced graphene oxide composite. *Chem. Commun.*, 47:3942–3944, 2011.
54. Bhunia, P., G. Kim, C. Baik, and H. Lee. A strategically designed porous iron-iron oxide matrix on graphene for heavy metal adsorption. *Chem. Commun.*, 48:9888–9890, 2012.
55. Fan, L., C. Luo, M. Sun, and H. Qiu. Synthesis of graphene oxide decorated with magnetic cyclodextrin for fast chromium removal. *J. Mater. Chem.*, 22:24577–24583, 2012.
56. Zhang, W., X. Shi, Y. Zhang, W. Gu, B. Li, and Y. Xian. Synthesis of water-soluble magnetic graphene nanocomposites for recyclable removal of heavy metal ions. *J. Mater. Chem. A*, 1:1745–1753, 2013.
57. Zhao, Z.Q., X. Chen, Q. Yang, J.H. Liu, and X.J. Huang. Selective adsorption toward toxic metal ions results in selective response: Electrochemical studies on a polypyrrole/reduced graphene oxide nanocomposite. *Chem. Commun.*, 48:2180–2182, 2012.
58. Wang, C., C. Feng, Y. Gao, X. Ma, Q. Wu, and Z. Wang. Preparation of a graphene-based magnetic nanocomposite for the removal of an organic dye from aqueous solution. *Chem. Eng. J.*, 173:92–97, 2011.
59. Fan, L., C. Luo, X. Li, F. Lu, H. Qiu, and M. Sun. Fabrication of novel magnetic chitosan grafted with graphene oxide to enhance adsorption properties for methyl blue. *J. Hazard. Mater.*, 215–216:272–279, 2012.
60. Li, B., H. Cao, and G. Yin. Mg(OH)2@reduced graphene oxide composite for removal of dyes from water. *J. Mater. Chem.*, 21:13765–13768, 2011.
61. Ai, L., C. Zhang, and Z. Chen. Removal of methylene blue from aqueous solution by a solvo-thermal-synthesized graphene/magnetite composite. *J. Hazard. Mater.*, 15:1515–1524, 2011.
62. Fan, W., W. Gao, C. Zhang, W.W. Tjiu, J. Pan, and T. Liu. Hybridization of graphene sheets and carbon-coated Fe_3O_4 nanoparticles as a synergistic adsorbent of organic dyes. *J. Mater. Chem.*, 22:25108–25115, 2012.

63. Li, B., H. Cao, G. Yin, Y. Lu, and J. Yin. Cu$_2$O@reduced graphene oxide composite for removal of contaminants from water and supercapacitors. *J. Mater. Chem.*, 21:10645–10648, 2011.

64. Sun, H., L. Cao, and L. Lu. Magnetite/reduced graphene oxide nanocomposites: One step solvothermal synthesis and use as a novel platform for removal of dye pollutants. *Nano Res.*, 4:550–562, 2011.

65. Wang, J., T. Tsuzuki, B. Tang, X. Hou, L. Sun, and X. Wang. Reduced graphene oxide/ZnO composite: Reusable adsorbent for pollutant management. *ACS Appl. Mater. Interfaces*, 4:3084–3090, 2012.

66. Chang, Y.P., C.L. Ren, J.C. Qu, and X.G. Chen. Preparation and characterization of Fe$_3$O$_4$/graphene nanocomposite and investigation of its adsorption performance for aniline and p-chloroaniline. *Appl. Surf. Sci.*, 261:504–509, 2012.

67. Li, B., H. Cao, J. Yin, Y.A. Wu, and J.H. Warner. Synthesis and separation of dyes via Ni@reduced graphene oxide nanostructures. *J. Mater. Chem.*, 22:1876–1883, 2012.

68. Sui, Z., Q. Meng, X. Zhang, R. Ma, and B. Cao. Green synthesis of carbon nanotube-graphene hybrid aerogels and their use as versatile agents for water purification. *J. Mater. Chem.*, 22:8767–8771, 2012.

69. Hashimoto, A., K. Suenaga, A. Gloter, K. Urita, and S. Iijima. Direct evidence for atomic defects in graphene layers. *Nature*, 430:870–873, 2004.

70. Lucchese, M.M., F. Stavale, E.H.M. Ferreira, C. Vilani, M.V.O. Moutinho, R.B. Capaz, C.A. Achete, and A. Jorio. Quantifying ion-induced defects and Raman relaxation length in graphene. *Carbon*, 48:1592–1597, 2010.

71. Wei, D., Y. Liu, Y. Wang, H. Zhang, L. Huang, and G. Yu. Synthesis of N-doped graphene by chemical vapor deposition and its electrical properties. *Nano Lett.*, 9:1752–1758, 2009.

72. Sint, K., B. Wang, and P. Král. Selective ion passage through functionalized graphene nanopores. *J. Am. Chem. Soc.*, 130:16448–16449, 2008.

73. Suk, M.E., and N.R. Aluru. Water transport through ultrathin graphene. *J. Phys. Chem. Lett.*, 1:1590–1594, 2010.

74. Fornasiero, F., J.B. In, S. Kim, H.G. Park, Y. Wang, C.P. Grigoropoulos, A. Noy, and O. Bakajin. pH-tunable ion selectivity in carbon nanotube pores. *Langmuir*, 26:14848–14853, 2010.

75. Fornasiero, F., H.G. Park, J.K. Holt, M. Stadermann, C.P. Grigoropoulos, A. Noy, and O. Bakajin. Ion exclusion by sub-2-nm carbon nanotube pores. *P. Natl. Acad. Sci.*, 105:17250–17255, 2008.

76. Cohen-Tanugi, D., and J.C. Grossman. Water desalination across nanoporous graphene. *Nano Lett.*, 12:3602–3608, 2012.

77. Zhu, L., L. Gu, Y. Zhou, S. Cao, and X. Cao. Direct production of a free-standing titanate and titania nanofiber membrane with selective permeability and cleaning performance. *J. Mater. Chem.*, 21:12503–12510, 2011.

78. Zeng, Y., L. Qiu, K. Wang, J. Yao, D. Li, G.P. Simon, R. Wang, and H. Wang. Significantly enhanced water flux in forward osmosis desalination with polymer-graphene composite hydrogels as a draw agent. *RSC Adv.*, 3:887–894, 2013.

79. Zhang, T., J. Liu, and D.D. Sun. A novel strategy to fabricate inorganic nanofibrous membranes for water treatment: Use of functionalized graphene oxide as a cross linker. *RSC Adv.*, 2:5134–5137, 2012.

80. Li, H., T. Lu, L. Pan, Y. Zhang, and Z. Sun. Electrosorption behavior of graphene in NaCl solutions. *J. Mater. Chem.*, 19:6773–6779, 2009.

81. Li, H., L. Zou, L. Pan, and Z. Sun. Using graphene nano-flakes as electrodes to remove ferric ions by capacitive deionization. *Sep. Purif. Technol.*, 75:8–14, 2010.

82. Li, H., L. Zou, L. Pan, and Z. Sun. Novel graphene-like electrodes for capacitive deionization. *Environ. Sci. Technol.*, 44:8692–8697, 2010.

83. Li, H., L. Pan, T. Lu, Y. Zhan, C. Nie, and Z. Sun. A comparative study on electrosorptive behavior of carbon nanotubes and graphene for capacitive deionization. *J. Electroanal. Chem.*, 653:40–44, 2011.

84. Wang, Z., L. Yue, Z.T. Liu, Z.H. Liu, and Z. Hao. Functional graphene nanocomposite as an electrode for the capacitive removal of FeCl$_3$ from water. *J. Mater. Chem.*, 22:14101–14107, 2012.

85. Zhang, D., T. Yan, L. Shi, Z. Peng, X. Wen, and J. Zhang. Enhanced capacitive deionization performance of graphene/carbon nanotube composites. *J. Mater. Chem.*, 22:14696–14704, 2012.

86. Li, H., L. Pan, C. Nie, Y. Liu, and Z. Sun. Reduced graphene oxide and activated carbon composites for capacitive deionization. *J. Mater. Chem.*, 22:15556–15561, 2012.

87. Zhang, D., X. Wen, L. Shi, T. Yan, and J. Zhang. Enhanced capacitive deionization of graphene/mesoporous carbon composites. *Nanoscale*, 4:5440–5446, 2012.

88. Wang, H., D. Zhang, T. Yan, X. Wen, L. Shi, and J. Zhang. Graphene prepared *via* a novel pyridine-thermal strategy for capacitive deionization. *J. Mater. Chem.*, 22:23745–23748, 2012.

89. Hoffmann, M.R., S.T. Martin, W. Choi, and D.W. Bahnemann. Environmental applications of semiconductor photocatalysis. *Chem. Rev.*, 95:69–96, 1995.

90. Zhang, H., G. Chen, and D.W. Bahnemann. Photoelectrocatalytic materials for environmental applications. *J. Mater. Chem.*, 19:5089–5121, 2009.

91. Di Paola, A., E. García-López, G. Marcì, and L. Palmisano. A survey of photocatalytic materials for environmental remediation. *J. Hazard. Mater.*, 211–212:3–29, 2012.

92. Liu, J., H. Bai, Y. Wang, Z. Liu, X. Zhang, and D.D. Sun. Self-assembling TiO$_2$ nanorods on large graphene oxide sheets at a two-phase interface and their anti-recombination in photocatalytic applications. *Adv. Funct. Mater.*, 20:4175–4181, 2010.

93. Nguyen-Phan, T.D., V.H. Pham, E.W. Shin, H.D. Pham, S. Kim, J.S. Chung, E.J. Kim, and S.H. Hur. The role of graphene oxide content on the adsorption-enhanced photocatalysis of titanium dioxide/graphene oxide composites. *Chem. Eng. J.*, 170:226–232, 2011.

94. Chen, C., W. Cai, M. Long, B. Zhou, Y. Wu, D. Wu, and Y. Feng. Synthesis of visible-light responsive graphene oxide/TiO$_2$ composites with p/n heterojunction. *ACS Nano*, 4:6425–6432, 2010.

95. Zhang, H., X. Lv, Y. Li, Y. Wang, and J. Li. P25-graphene composite as a high performance photocatalyst. *ACS Nano*, 4:380–386, 2009.

96. Liang, Y., H. Wang, H. Sanchez Casalongue, Z. Chen, and H. Dai. TiO$_2$ nanocrystals grown on graphene as advanced photocatalytic hybrid materials. *Nano Res.*, 3:701–705, 2010.

97. Dong, P., Y. Wang, L. Guo, B. Liu, S. Xin, J. Zhang, Y. Shi, W. Zeng, and S. Yin. A facile one-step solvothermal synthesis of graphene/rod-shaped TiO$_2$ nanocomposite and its improved photocatalytic activity. *Nanoscale*, 4:4641–4649, 2012.

98. Lee, J.S., K.H. You, and C.B. Park. Highly photoactive, low bandgap TiO$_2$ nanoparticles wrapped by graphene. *Adv. Mater.*, 24:1084–1088, 2012.

99. Zhang, Y., N. Zhang, Z.R. Tang, and Y.J. Xu. Improving the photocatalytic performance of graphene-TiO$_2$ nanocomposites *via* a combined strategy of decreasing defects of graphene and increasing interfacial contact. *Phys. Chem. Chem. Phys.*, 14:9167–9175, 2012.

100. Brar, V.W., Y. Zhang, Y. Yayon, T. Ohta, J.L. McChesney, A. Bostwick, E. Rotenberg, K. Horn, and M.F. Crommie. Scanning tunneling spectroscopy of inhomogeneous electronic structure in monolayer and bilayer graphene on SiC. *Appl. Phys. Lett.*, 91:122102, 2007.

101. Liu, B., Y. Huang, Y. Wen, L. Du, W. Zeng, Y. Shi, F. Zhang, G. Zhu, X. Xu, and Y. Wang. Highly dispersive {001} facets-exposed nanocrystalline TiO$_2$ on high quality graphene as a high performance photocatalyst. *J. Mater. Chem.*, 22:7484–7491, 2012.

102. Lin, Y., Z. Geng, H. Cai, L. Ma, J. Chen, J. Zeng, N. Pan, and X. Wang. Ternary graphene-TiO$_2$-Fe$_3$O$_4$ nanocomposite as a recollectable photocatalyst with enhanced durability. *Eur. J. Inorg. Chem.*, 2012:4439–4444, 2012.

103. Jiang, G., Z. Lin, C. Chen, L. Zhu, Q. Chang, N. Wang, W. Wei, and H. Tang. TiO$_2$ nanoparticles assembled on graphene oxide nanosheets with high photocatalytic activity for removal of pollutants. *Carbon*, 49:2693–2701, 2011.

104. Guo, J., S. Zhu, Z. Chen, Y. Li, Z. Yu, Q. Liu, J. Li, C. Feng, and D. Zhang. Sonochemical synthesis of TiO$_2$ nanoparticles on graphene for use as photocatalyst. *Ultrason. Sonochem.*, 18:1082–1090, 2011.

105. Zhang, Y., Z.R. Tang, X. Fu, and Y.J. Xu. TiO$_2$-graphene nanocomposites for gas-phase photocatalytic degradation of volatile aromatic pollutant: Is TiO$_2$-graphene truly different from other TiO$_2$-carbon composite materials? *ACS Nano*, 4:7303–7314, 2010.

106. Jiang, B., C. Tian, W. Zhou, J. Wang, Y. Xie, Q. Pan, Z. Ren, Y. Dong, D. Fu, J. Han, and H. Fu. In situ growth of TiO₂ in interlayers of expanded graphite for the fabrication of TiO₂-graphene with enhanced photocatalytic activity. *Chem. Eur. J.*, 17:8379–8387, 2011.
107. Li, B., and H. Cao. ZnO@graphene composite with enhanced performance for the removal of dye from water. *J. Mater. Chem.*, 21:3346–3349, 2011.
108. Lv, T., L. Pan, X. Liu, and Z. Sun. Enhanced photocatalytic degradation of methylene blue by ZnO-reduced graphene oxide-carbon nanotube composites synthesized *via* microwave-assisted reaction. *Catal. Sci. Technol.*, 2:2297–2301, 2012.
109. Yu, X., G. Zhang, H. Cao, X. An, Y. Wang, Z. Shu, X. An, and F. Hua. ZnO@ZnS hollow dumbbells-graphene composites as high-performance photocatalysts and alcohol sensors. *New J. Chem.*, 36:2593–2598, 2012.
110. Bai, X., L. Wang, and Y. Zhu. Visible photocatalytic activity enhancement of ZnWO₄ by graphene hybridization. *ACS Catal.*, 2769–2778, 2012.
111. Liu, F., X. Shao, J. Wang, S. Yang, H. Li, X. Meng, X. Liu, and M. Wang. Solvothermal synthesis of graphene-CdS nanocomposites for highly efficient visible-light photocatalyst. *J. Alloy. Compd.*, 551:327–332, 2013.
112. Ye, A., W. Fan, Q. Zhang, W. Deng, and Y. Wang. CdS-graphene and CdS-CNT nanocomposites as visible-light photocatalysts for hydrogen evolution and organic dye degradation. *Catal. Sci. Technol.*, 2:969–978, 2012.
113. Khan, Z., T.R. Chetia, A.K. Vardhaman, D. Barpuzary, C.V. Sastri, and M. Qureshi. Visible light assisted photocatalytic hydrogen generation and organic dye degradation by CdS-metal oxide hybrids in presence of graphene oxide. *RSC Adv.*, 2:12122–12128, 2012.
114. An, X., J.C. Yu, Y. Wang, Y. Hu, X. Yu, and G. Zhang. WO₃ nanorods/graphene nanocomposites for high-efficiency visible-light-driven photocatalysis and NO₂ gas sensing. *J. Mater. Chem.*, 22:8525–8531, 2012.
115. Sun, Y., B. Qu, Q. Liu, S. Gao, Z. Yan, W. Yan, B. Pan, S. Wei, and Y. Xie. Highly efficient visible-light-driven photocatalytic activities in synthetic ordered monoclinic BiVO₄ quantum tubes-graphene nanocomposites. *Nanoscale*, 4:3761–3767, 2012.
116. Song, S., W. Gao, X. Wang, X. Li, D. Liu, Y. Xing, and H. Zhang. Microwave-assisted synthesis of BiOBr/graphene nanocomposites and their enhanced photocatalytic activity. *Dalton Trans.*, 41:10472–10476, 2012.
117. Tu, X., S. Luo, G. Chen, and J. Li. One-pot synthesis, characterization, and enhanced photocatalytic activity of a BiOBr-graphene composite. *Chem. Eur. J.*, 18:14359–14366, 2012.
118. Madhusudan, P., J. Yu, W. Wang, B. Cheng, and G. Liu. Facile synthesis of novel hierarchical graphene-Bi₂O₂CO₃ composites with enhanced photocatalytic performance under visible light. *Dalton Trans.*, 41:14345–14353, 2012.
119. Li, Z., Y. Shen, C. Yang, Y. Lei, Y. Guan, Y. Lin, D. Liu, and C.W. Nan. Significant enhancement in the visible light photocatalytic properties of BiFeO₃-graphene nanohybrids. *J. Mater. Chem. A*, 1:823–829, 2013.
120. Dong, Y., J. Li, L. Shi, J. Xu, X. Wang, Z. Guo, and W. Liu. Graphene oxide-iron complex: Synthesis, characterization and visible-light-driven photocatalysis. *J. Mater. Chem. A*, 1:644–650, 2013.
121. Fu, Y., H. Chen, X. Sun, and X. Wang. Graphene-supported nickel ferrite: A magnetically separable photocatalyst with high activity under visible light. *AIChE J.*, 58:3298–3305, 2012.
122. Liang, Q., Y. Shi, W. Ma, Z. Li, and X. Yang. Enhanced photocatalytic activity and structural stability by hybridizing Ag₃PO₄ nanospheres with graphene oxide sheets. *Phys. Chem. Chem. Phys.*, 14:15657–15665, 2012.
123. Hou, Y., F. Zuo, Q. Ma, C. Wang, L. Bartels, and P. Feng. Ag₃PO₄ oxygen evolution photocatalyst employing synergistic action of Ag/AgBr nanoparticles and graphene sheets. *J. Phys. Chem. C.*, 116:20132–20139, 2012.
124. Zhu, M., P. Chen, and M. Liu. Ag/AgBr/Graphene oxide nanocomposite synthesized *via* oil/water and water/oil microemulsions: A comparison of sunlight energized plasmonic photocatalytic activity. *Langmuir*, 28:3385–3390, 2012.

125. Zhang, X., X. Quan, S. Chen, and H. Yu. Constructing graphene/InNbO$_4$ composite with excellent adsorptivity and charge separation performance for enhanced visible-light-driven photocatalytic ability. *Appl. Catal. B-Environ.*, 105:237–242, 2011.

126. Zhu, M., P. Chen, and M. Liu. Graphene oxide enwrapped Ag/AgX (X = Br, Cl) nanocomposite as a highly efficient visible-light plasmonic photocatalyst. *ACS Nano*, 5:4529–4536, 2011.

127. Xiong, Z., L.L. Zhang, J. Ma, and X.S. Zhao. Photocatalytic degradation of dyes over graphene-gold nanocomposites under visible light irradiation. *Chem. Commun.*, 46:6099–6101, 2010.

128. Sun, H., S. Liu, G. Zhou, H.M. Ang, M.O. Tadé, and S. Wang. Reduced graphene oxide for catalytic oxidation of aqueous organic pollutants. *ACS Appl. Mater. Interfaces*, 4:5466–5471, 2012.

129. Chen, S., J. Zhu, H. Huang, G. Zeng, F. Nie, and X. Wang. Facile solvothermal synthesis of graphene-MnOOH nanocomposites. *J. Solid State Chem.*, 183:2552–2557, 2010.

130. Zhang, F., B. Zheng, J. Zhang, X. Huang, H. Liu, S. Guo, and J. Zhang. Horseradish peroxidase immobilized on graphene oxide: Physical properties and applications in phenolic compound removal. *J. Phys. Chem. C*, 114:8469–8473, 2010.

131. Ma, H.L., Y. Zhang, Q.H. Hu, D. Yan, Z.Z. Yu, and M. Zhai. Chemical reduction and removal of Cr(vi) from acidic aqueous solution by ethylenediamine-reduced graphene oxide. *J. Mater. Chem.*, 22:5914–5916, 2012.

132. Zhu, J., M. Xu, X. Meng, K. Shang, H. Fan, and S. Ai. Electro-enzymatic degradation of carbofuran with the graphene oxide-Fe$_3$O$_4$-hemoglobin composite in an electrochemical reactor. *Process. Biochem.*, 47:2480–2486, 2012.

133. Ghosh, T., K. Ullah, V. Nikam, C.Y. Park, Z.D. Meng, and W.C. Oh. The characteristic study and sonocatalytic performance of CdSe-graphene as catalyst in the degradation of azo dyes in aqueous solution under dark conditions. *Ultrason. Sonochem.*, 20:768–776, 2013.

134. Shi, P., R. Su, S. Zhu, M. Zhu, D. Li, and S. Xu. Supported cobalt oxide on graphene oxide: Highly efficient catalysts for the removal of orange II from water. *J. Hazard. Mater.*, 229–230:331–339, 2012.

135. Trogolo, J. Filter media: Bacterial growth on filters—The silver solution. *Filtr. Separat.*, 43:28–29, 2006.

136. Hu, W., C. Peng, W. Luo, M. Lv, X. Li, D. Li, Q. Huang, and C. Fan. Graphene-based antibacterial paper. *ACS Nano*, 4:4317–4323, 2010.

137. Krishnamoorthy, K., M. Veerapandian, L.H. Zhang, K. Yun, and S.J. Kim. Antibacterial efficiency of graphene nanosheets against pathogenic bacteria *via* lipid peroxidation. *J. Phys. Chem. C*, 116:17280–17287, 2012.

138. Liu, S., T.H. Zeng, M. Hofmann, E. Burcombe, J. Wei, R. Jiang, J. Kong, and Y. Chen. Antibacterial activity of graphite, graphite oxide, graphene oxide, and reduced graphene oxide: Membrane and oxidative stress. *ACS Nano*, 5:6971–6980, 2011.

139. Li, Y., Y. Liu, Y. Fu, T. Wei, L. Le Guyader, G. Gao, R.S. Liu, Y.Z. Chang, and C. Chen. The triggering of apoptosis in macrophages by pristine graphene through the MAPK and TGF-beta signaling pathways. *Biomaterials*, 33:402–411, 2012.

140. Cai, X., S. Tan, M. Lin, A. Xie, W. Mai, X. Zhang, Z. Lin, T. Wu, and Y. Liu. Synergistic antibacterial brilliant blue/reduced graphene oxide/quaternary phosphonium salt composite with excellent water solubility and specific targeting capability. *Langmuir*, 27:7828–7835, 2011.

141. Ma, J., J. Zhang, Z. Xiong, Y. Yong, and X.S. Zhao. Preparation, characterization and antibacterial properties of silver-modified graphene oxide. *J. Mater. Chem.*, 21:3350–3352, 2011.

142. Bao, Q., D. Zhang, and P. Qi. Synthesis and characterization of silver nanoparticle and graphene oxide nanosheet composites as a bactericidal agent for water disinfection. *J. Colloid Interface Sci.*, 360:463–470, 2011.

143. Liu, L., J. Liu, Y. Wang, X. Yan, and D.D. Sun. Facile synthesis of monodispersed silver nanoparticles on graphene oxide sheets with enhanced antibacterial activity. *New J. Chem.*, 35:1418–1423, 2011.

144. Zhang, D., X. Liu, and X. Wang. Green synthesis of graphene oxide sheets decorated by silver nanoprisms and their anti-bacterial properties. *J. Inorg. Biochem.*, 105:1181–1186, 2011.

145. Das, M.R., R.K. Sarma, R. Saikia, V.S. Kale, M.V. Shelke, and P. Sengupta. Synthesis of silver nanoparticles in an aqueous suspension of graphene oxide sheets and its antimicrobial activity. *Colloids Surf. B*, 83:16–22, 2011.

146. Zhang, Z., J. Zhang, B. Zhang, and J. Tang. Mussel-inspired functionalization of graphene for synthesizing Ag-polydopamine-graphene nanosheets as antibacterial materials. *Nanoscale*, 5:118–123, 2013.

147. Sreeprasad, T.S., M.S. Maliyekkal, K. Deepti, K. Chaudhari, P.L. Xavier, and T. Pradeep. Transparent, luminescent, antibacterial and patternable film forming composites of graphene oxide/reduced graphene oxide. *ACS Appl. Mater. Interfaces*, 3:2643–2654, 2011.

148. Akhavan, O., and E. Ghaderi. Photocatalytic reduction of graphene oxide nanosheets on TiO_2 thin film for photoinactivation of bacteria in solar light irradiation. *J. Phys. Chem. C*, 113:20214–20220, 2009.

149. Akhavan, O., and E. Ghaderi. Escherichia coli bacteria reduce graphene oxide to bactericidal graphene in a self-limiting manner. *Carbon*, 50:1853–1860, 2012.

150. Some, S., S.M. Ho, P. Dua, E. Hwang, Y.H. Shin, H. Yoo, J.S. Kang, D.K. Lee, and H. Lee. Dual functions of highly potent graphene derivative-poly-l-Lysine composites to inhibit bacteria and support human cells. *ACS Nano*, 6:7151–7161, 2012.

151. Kavitha, T., A.I. Gopalan, K.P. Lee, and S.Y. Park. Glucose sensing, photocatalytic and antibacterial properties of graphene-ZnO nanoparticle hybrids. *Carbon*, 50:2994–3000, 2012.

152. Mejias Carpio, I.E., C.M. Santos, X. Wei, and D.F. Rodrigues. Toxicity of a polymer-graphene oxide composite against bacterial planktonic cells, biofilms, and mammalian cells. *Nanoscale*, 4:4746–4756, 2012.

153. Wang, X., N. Zhou, J. Yuan, W. Wang, Y. Tang, C. Lu, J. Zhang, and J. Shen. Antibacterial and anticoagulation properties of carboxylated graphene oxide-lanthanum complexes. *J. Mater. Chem.*, 22:1673–1678, 2012.

154. Mondal, T., A.K. Bhowmick, and R. Krishnamoorti. Chlorophenyl pendant decorated graphene sheet as a potential antimicrobial agent: Synthesis and characterization. *J. Mater. Chem.*, 22:22481–22487, 2012.

155. Wang, Y., D. Zhang, Q. Bao, J. Wu, and Y. Wan. Controlled drug release characteristics and enhanced antibacterial effect of graphene oxide-drug intercalated layered double hydroxide hybrid films. *J. Mater. Chem.*, 22:23106–23113, 2012.

156. Zhang, Y., T.T. Tang, C. Girit, Z. Hao, M.C. Martin, A. Zettl, M.F. Crommie, Y.R. Shen, and F. Wang. Direct observation of a widely tunable bandgap in bilayer graphene. *Nature*, 459:820–823, 2009.

157. Gomez-Navarro, C., R.T. Weitz, A.M. Bittner, M. Scolari, A. Mews, M. Burghard, and K. Kern. Electronic transport properties of individual chemically reduced graphene oxide sheets. *Nano Lett.*, 7:3499–3503, 2007.

158. Barone, V.N., O. Hod, and G.E. Scuseria. Electronic structure and stability of semiconducting graphene nanoribbons. *Nano Lett.*, 6:2748–2754, 2006.

159. Mohanty, N., D. Moore, Z. Xu, T.S. Sreeprasad, A. Nagaraja, A.A. Rodriguez, and V. Berry. Nanotomy-based production of transferable and dispersible graphene nanostructures of controlled shape and size. *Nat. Commun.*, 3:844, 2012.

160. Young, A.F., and P. Kim. Quantum interference and Klein tunnelling in graphene heterojunctions. *Nat. Phys.*, 5:222–226, 2009.

161. Schedin, F., A.K. Geim, S.V. Morozov, E.W. Hill, P. Blake, M.I. Katsnelson, and K.S. Novoselov. Detection of individual gas molecules adsorbed on graphene. *Nat. Mater.*, 6:652–655, 2007.

162. Vedala, H., D.C. Sorescu, G.P. Kotchey, and A. Star. Chemical sensitivity of graphene edges decorated with metal nanoparticles. *Nano Lett.*, 11:2342–2347, 2011.

163. Lu, G., K. Yu, L.E. Ocola, and J. Chen. Ultrafast room temperature NH3 sensing with positively gated reduced graphene oxide field-effect transistors. *Chem. Commun.*, 47:7761–7763, 2011.

164. Zhang, T., Z. Cheng, Y. Wang, Z. Li, C. Wang, Y. Li, and Y. Fang. Self-assembled 1-octadecanethiol monolayers on graphene for mercury detection. *Nano Lett.*, 10:4738–4741, 2010.

165. Mohanty, N., and V. Berry. Graphene-based single-bacterium resolution biodevice and DNA transistor: Interfacing graphene derivatives with nanoscale and microscale biocomponents. *Nano Lett.*, 8:4469–4476, 2008.

166. Huang, Y., X. Dong, Y. Liu, L.J. Li, and P. Chen. Graphene-based biosensors for detection of bacteria and their metabolic activities. *J. Mater. Chem.*, 21:12358–12362, 2011.

167. Ohno, Y., K. Maehashi, and K. Matsumoto. Chemical and biological sensing applications based on graphene field-effect transistors. *Biosens. Bioelectron.*, 26:1727–1730, 2010.

168. Choi, B.G., H. Park, M.H. Yang, Y.M. Jung, S.Y. Lee, W.H. Hong, and T.J. Park. Microwave-assisted synthesis of highly water-soluble graphene towards electrical DNA sensor. *Nanoscale*, 2:2692–2697, 2010.

169. Zhang, L., C. Li, A. Liu, and G. Shi. Electrosynthesis of graphene oxide/polypyrene composite films and their applications for sensing organic vapors. *J. Mater. Chem.*, 22:8438–8443, 2012.

170. Myung, S., P.T. Yin, C. Kim, J. Park, A. Solanki, P.I. Reyes, Y. Lu, K.S. Kim, and K.B. Lee. Label-free polypeptide-based enzyme detection using a graphene-nanoparticle hybrid sensor. *Adv. Mater.*, 24:6080, 2012.

171. Sudibya, H.G., Q. He, H. Zhang, and P. Chen. Electrical detection of metal ions using field-effect transistors based on micropatterned reduced graphene oxide films. *ACS Nano*, 5:1990–1994, 2011.

172. Du, H., J. Ye, J. Zhang, X. Huang, and C. Yu. A voltammetric sensor based on graphene-modified electrode for simultaneous determination of catechol and hydroquinone. *J. Electroanal. Chem.*, 650:209–213, 2011.

173. Yuan, X., D. Yuan, F. Zeng, W. Zou, F. Tzorbatzoglou, P. Tsiakaras, and Y. Wang. Preparation of graphitic mesoporous carbon for the simultaneous detection of hydroquinone and catechol. *Appl. Catal. B-Environ.*, 129:367–374, 2013.

174. Liu, Z., X. Ma, H. Zhang, W. Lu, H. Ma, and S. Hou. Simultaneous determination of nitrophenol isomers based on β-cyclodextrin functionalized reduced graphene oxide. *Electroanalysis*, 24:1178–1185, 2012.

175. Lin, X., Y. Ni, and S. Kokot. Voltammetric analysis with the use of a novel electro-polymerised graphene-nafion film modified glassy carbon electrode: Simultaneous analysis of noxious nitroaniline isomers. *J. Hazard. Mater.*, 243:232–241, 2012.

176. Liang, H., X. Miao, and J. Gong. One-step fabrication of layered double hydroxides/graphene hybrid as solid-phase extraction for stripping voltammetric detection of methyl parathion. *Electrochem. Commun.*, 20:149–152, 2012.

177. Gong, J., X. Miao, H. Wan, and D. Song. Facile synthesis of zirconia nanoparticles-decorated graphene hybrid nanosheets for an enzymeless methyl parathion sensor. *Sens. Actuators B-Chem.*, 162:341–347, 2012.

178. Gong, J., X. Miao, T. Zhou, and L. Zhang. An enzymeless organophosphate pesticide sensor using Au nanoparticle-decorated graphene hybrid nanosheet as solid-phase extraction. *Talanta*, 85:1344–1349, 2011.

179. Liu, T., H. Su, X. Qu, P. Ju, L. Cui, and S. Ai. Acetylcholinesterase biosensor based on 3-carboxyphenylboronic acid/reduced graphene oxide-gold nanocomposites modified electrode for amperometric detection of organophosphorus and carbamate pesticides. *Sens. Actuators B-Chem.*, 160:1255–1261, 2011.

180. Choi, B.G., H. Park, T.J. Park, M.H. Yang, J.S. Kim, S.Y. Jang, N.S. Heo, S.Y. Lee, J. Kong, and W.H. Hong. Solution chemistry of self-assembled graphene nanohybrids for high-performance flexible biosensors. *ACS Nano*, 4:2910–2918, 2010.

181. Ion, I., and A.C. Ion. Differential pulse voltammetric analysis of lead in vegetables using a surface amino-functionalized exfoliated graphite nanoplatelet chemically modified electrode. *Sens. Actuators B-Chem.*, 166–167:842–847, 2012.

182. Wang, K., Q. Liu, L. Dai, J. Yan, C. Ju, B. Qiu, and X. Wu. A highly sensitive and rapid organophosphate biosensor based on enhancement of CdS-decorated graphene nanocomposite. *Anal. Chim. Acta*, 695:84–88, 2011.

183. Gong, J., T. Zhou, D. Song, and L. Zhang. Monodispersed Au nanoparticles decorated graphene as an enhanced sensing platform for ultrasensitive stripping voltammetric detection of mercury(II). *Sens. Actuators B-Chem.*, 150:491–497, 2010.
184. Li, J., S. Guo, Y. Zhai, and E. Wang. Nafion-graphene nanocomposite film as enhanced sensing platform for ultrasensitive determination of cadmium. *Electrochem. Commun.*, 11:1085–1088, 2009.
185. Cui, F., L. Chu, and X. Zhang. Nanocomposite of graphene based sensor for paraquat: Synergetic effect of nano-gold and ionic liquids on electrocatalysis. *Anal. Methods*, 4:3974–3980, 2012.
186. Wu, S., X. Lan, L. Cui, L. Zhang, S. Tao, H. Wang, M. Han, Z. Liu, and C. Meng. Application of graphene for preconcentration and highly sensitive stripping voltammetric analysis of organophosphate pesticide. *Anal. Chim. Acta*, 699:170–176, 2011.
187. Zheng, Z., Y. Du, Z. Wang, Q. Feng, and C. Wang. Pt/graphene-CNTs nanocomposite based electrochemical sensors for the determination of endocrine disruptor bisphenol A in thermal printing papers. *Analyst*, 138:693–701, 2013.
188. Guo, Y., S. Guo, J. Li, E. Wang, and S. Dong. Cyclodextrin-graphene hybrid nanosheets as enhanced sensing platform for ultrasensitive determination of carbendazim. *Talanta*, 84:60–64, 2011.
189. Yuan, X., Y. Chai, R. Yuan, Q. Zhao, and C. Yang. Functionalized graphene oxide-based carbon paste electrode for potentiometric detection of copper ion(ii). *Anal. Methods*, 4:3332–3337, 2012.
190. Jaworska, E., W. Lewandowski, J. Mieczkowski, K. Maksymiuk, and A. Michalska. Non-covalently functionalized graphene for the potentiometric sensing of zinc ions. *Analyst*, 137: 1895–1898, 2012.
191. Wan, Y., Y. Wang, J. Wu, and D. Zhang. Graphene oxide sheet-mediated silver enhancement for application to electrochemical biosensors. *Anal. Chem.*, 83:648–653, 2010.
192. Chen, T.W., Z.H. Sheng, K. Wang, F.B. Wang, and X.H. Xia. Determination of explosives using electrochemically reduced graphene. *Chem. Asian J.*, 6:1210–1216, 2011.
193. Guo, S., D. Wen, Y. Zhai, S. Dong, and E. Wang. Ionic liquid-graphene hybrid nanosheets as an enhanced material for electrochemical determination of trinitrotoluene. *Biosens. Bioelectron.*, 26:3475–3481, 2011.
194. Yang, S., S. Luo, C. Liu, and W. Wei. Direct synthesis of graphene-chitosan composite and its application as an enzymeless methyl parathion sensor. *Colloids Surf. B*, 96:75–79, 2012.
195. Zhao, L., F. Zhao, and B. Zeng. Electrochemical determination of methyl parathion using a molecularly imprinted polymer-ionic liquid-graphene composite film coated electrode. *Sens. Actuators B-Chem.*, 176:818–824, 2013.
196. Shang, L., F. Zhao, and B. Zeng. Electrocatalytic oxidation and determination of hydrazine at an AuCu nanoparticles - graphene - ionic liquid composite film coated glassy carbon electrode. *Electroanalysis*, 24:2380–2386, 2012.
197. He, Y., J. Zheng, and S. Dong. Ultrasonic-electrodeposition of hierarchical flower-like cobalt on petalage-like graphene hybrid microstructures for hydrazine sensing. *Analyst*, 137:4841–4848, 2012.
198. Lei, J., X. Lu, W. Wang, X. Bian, Y. Xue, C. Wang, and L. Li. Fabrication of MnO_2/graphene oxide composite nanosheets and their application in hydrazine detection. *RSC Adv.*, 2:2541–2544, 2012.
199. Li, S.J., C. Qian, K. Wang, B.Y. Hua, F.B. Wang, Z.H. Sheng, and X.H. Xia. Application of thermally reduced graphene oxide modified electrode in simultaneous determination of dihydroxybenzene isomers. *Sens. Actuators B-Chem.*, 174:441–448, 2012.
200. Ong, B.K., H.L. Poh, C.K. Chua, and M. Pumera. Graphenes prepared by Hummers, Staudenmaier and Hofmann methods for analysis of TNT-based nitroaromatic explosives in seawater. *Electroanalysis*, 24:2085–2093, 2012.
201. Zhao, Y., X. Song, Q. Song, and Z. Yin. A facile route to the synthesis copper oxide/reduced graphene oxide nanocomposites and electrochemical detection of catechol organic pollutant. *Cryst. Eng. Comm.*, 14:6710–6719, 2012.
202. Li, J., D. Kuang, Y. Feng, F. Zhang, Z. Xu, and M. Liu. A graphene oxide-based electrochemical sensor for sensitive determination of 4-nitrophenol. *J. Hazard. Mater.*, 201–202:250–259, 2012.

203. Liu, B., H.T. Lian, J.F. Yin, and X.Y. Sun. Dopamine molecularly imprinted electrochemical sensor based on graphene-chitosan composite. *Electrochim. Acta*, 75:108–114, 2012.

204. Wang, L., X. Qin, S. Liu, Y. Luo, A.M. Asiri, A.O. Al-Youbi, and X. Sun. Single-stranded DNA-mediated immobilization of graphene on a gold electrode for sensitive and selective determination of dopamine. *Chem. Plus. Chem.*, 77:19–22, 2012.

205. Ye, X., Y. Gu, and C. Wang. Fabrication of the Cu_2O/polyvinyl pyrrolidone-graphene modified glassy carbon-rotating disk electrode and its application for sensitive detection of herbicide paraquat. *Sens. Actuators B-Chem.*, 173:530–539, 2012.

206. Shan, C., H. Yang, D. Han, Q. Zhang, A. Ivaska, and L. Niu. Water-soluble graphene covalently functionalized by biocompatible poly-l-lysine. *Langmuir*, 25:12030–12033, 2009.

207. Liu, S., J. Tian, L. Wang, and X. Sun. A method for the production of reduced graphene oxide using benzylamine as a reducing and stabilizing agent and its subsequent decoration with Ag nanoparticles for enzymeless hydrogen peroxide detection. *Carbon*, 49:3158–3164, 2011.

208. Chen, L., X. Wang, X. Zhang, and H. Zhang. 3D porous and redox-active prussian blue-in-graphene aerogels for highly efficient electrochemical detection of H_2O_2. *J. Mater. Chem.*, 22:22090–22096, 2012.

209. Sheng, Q., M. Wang, and J. Zheng. A novel hydrogen peroxide biosensor based on enzymatically induced deposition of polyaniline on the functionalized graphene-carbon nanotube hybrid materials. *Sens. Actuators B-Chem.*, 160:1070–1077, 2011.

210. Qiu, J.D., L. Shi, R.P. Liang, G.C. Wang, and X.H. Xia. Controllable deposition of a platinum nanoparticle ensemble on a polyaniline/graphene hybrid as a novel electrode material for electrochemical sensing. *Chem. Eur. J.*, 18:7950–7959, 2012.

211. Liu, R., S. Li, X. Yu, G. Zhang, S. Zhang, J. Yao, B. Keita, L. Nadjo, and L. Zhi. Facile synthesis of Au-nanoparticle/polyoxometalate/graphene tricomponent nanohybrids: An enzyme-free electrochemical biosensor for hydrogen peroxide. *Small*, 8:1398–1406, 2012.

212. Mohammad-Rezaei, R., and H. Razmi. Reduced graphene oxide|carbon ceramic electrode modified with CdS-hemoglobin as a sensitive hydrogen peroxide biosensor. *Electroanalysis*, 24:2094–2101, 2012.

213. Zhu, M., N. Li, and J. Ye. Sensitive and selective sensing of hydrogen peroxide with iron-tetrasulfophthalocyanine-graphene-Nafion modified screen-printed electrode. *Electroanalysis*, 24:1212–1219, 2012.

214. Cao, X., Z. Zeng, W. Shi, P. Yep, Q. Yan, and H. Zhang. Three-dimensional graphene network composites for detection of hydrogen peroxide. *Small*, 9:1703–1707, 2012.

215. Gao, C., X.Y. Yu, R.X. Xu, J.H. Liu, and X.J. Huang. AlOOH-reduced graphene oxide nanocomposites: One-pot hydrothermal synthesis and their enhanced electrochemical activity for heavy metal ions. *ACS Appl. Mater. Interfaces*, 4:4672–4682, 2012.

216. Wang, Y., D. Zhang, and J. Wu. Electrocatalytic oxidation of kojic acid at a reduced graphene sheet modified glassy carbon electrode. *J. Electroanal. Chem.*, 664:111–116, 2012.

217. Shi, J.J., and J.J. Zhu. Sonoelectrochemical fabrication of Pd-graphene nanocomposite and its application in the determination of chlorophenols. *Electrochim. Acta*, 56:6008–6013, 2011.

218. Liu, M., H. Zhao, S. Chen, H. Yu, and X. Quan. Interface engineering catalytic graphene for smart colorimetric biosensing. *ACS Nano*, 6:3142–3151, 2012.

219. Jang, H., Y.K. Kim, H.M. Kwon, W.S. Yeo, D.E. Kim, and D.H. Min. A graphene-based platform for the assay of duplex-DNA unwinding by helicase. *Angew Chem. Int. Ed.*, 49:5703–5707, 2010.

220. He, S., B. Song, D. Li, C. Zhu, W. Qi, Y. Wen, L. Wang, S. Song, H. Fang, and C. Fan. A graphene nanoprobe for rapid, sensitive, and multicolor fluorescent DNA analysis. *Adv. Funct. Mater.*, 20:453–459, 2010.

221. Chang, H., L. Tang, Y. Wang, J. Jiang, and J. Li. Graphene fluorescence resonance energy transfer aptasensor for the thrombin detection. *Anal. Chem.*, 82:2341–2346, 2010.

222. Wang, H., Q. Zhang, X. Chu, T. Chen, J. Ge, and R. Yu. Graphene oxide-peptide conjugate as an intracellular protease sensor for caspase-3 activation imaging in live cells. *Angew Chem. Int. Ed.*, 50:7065–7069, 2011.

223. Xu, Y., A. Malkovskiy, and Y. Pang. A graphene binding-promoted fluorescence enhancement for bovine serum albumin recognition. *Chem. Commun.*, 47:6662–6664, 2011.
224. Zhang, C., Y. Yuan, S. Zhang, Y. Wang, and Z. Liu. Biosensing platform based on fluorescence resonance energy transfer from upconverting nanocrystals to graphene oxide. *Angew Chem. Int. Ed.*, 50:6851–6854, 2011.
225. Huang, W.T., Y. Shi, W.Y. Xie, H.Q. Luo, and N.B. Li. A reversible fluorescence nanoswitch based on bifunctional reduced graphene oxide: Use for detection of Hg2+ and molecular logic gate operation. *Chem. Commun.*, 47:7800–7802, 2011.
226. Zhang, J.R., W.T. Huang, W.Y. Xie, T. Wen, H.Q. Luo, and N.B. Li. Highly sensitive, selective, and rapid fluorescence Hg2+ sensor based on DNA duplexes of poly(dT) and graphene oxide. *Analyst*, 137:3300–3305, 2012.
227. Zhao, X.H., R.M. Kong, X.B. Zhang, H.M. Meng, W.N. Liu, W. Tan, G.L. Shen, and R.Q. Yu. Graphene-DNAzyme based biosensor for amplified fluorescence "turn-on" detection of Pb^{2+} with a high selectivity. *Anal. Chem.*, 83:5062–5066, 2011.
228. Fu, X., T. Lou, Z. Chen, M. Lin, W. Feng, and L. Chen. "Turn-on" fluorescence detection of lead ions based on accelerated leaching of gold nanoparticles on the surface of graphene. *ACS Appl. Mater. Interfaces*, 4:1080–1086, 2012.
229. Wen, Y., F. Xing, S. He, S. Song, L. Wang, Y. Long, D. Li, and C. Fan. A graphene-based fluorescent nanoprobe for silver(i) ions detection by using graphene oxide and a silver-specific oligonucleotide. *Chem. Commun.*, 46:2596–2598, 2010.
230. Kundu, A., R.K. Layek, and A.K. Nandi. Enhanced fluorescent intensity of graphene oxide-methyl cellulose hybrid in acidic medium: Sensing of nitro-aromatics. *J. Mater. Chem.*, 22:8139–8144, 2012.
231. Huang, S.T., Y. Shi, N.B. Li, and H.Q. Luo. Fast and sensitive dye-sensor based on fluorescein/reduced graphene oxide complex. *Analyst*, 137:2593–2599, 2012.
232. Wang, L., K.Y. Pu, J. Li, X. Qi, H. Li, H. Zhang, C. Fan, and B. Liu. A graphene-conjugated oligomer hybrid probe for light-up sensing of lectin and escherichia coli. *Adv. Mater.*, 23:4386–4391, 2011.
233. Lee, J.H., S. Kang, J. Jaworski, K.Y. Kwon, M.L. Seo, J.Y. Lee, and J.H. Jung. Fluorescent composite hydrogels of metal-organic frameworks and functionalized graphene oxide. *Chem. Eur. J.*, 18:765–769, 2012.
234. Singh, V.V., G. Gupta, A. Batra, A.K. Nigam, M. Boopathi, P.K. Gutch, B.K. Tripathi, A. Srivastava, M. Samuel, G.S. Agarwal, B. Singh, and R. Vijayaraghavan. Greener electrochemical synthesis of high quality graphene nanosheets directly from pencil and its SPR sensing application. *Adv. Funct. Mater.*, 22:2352–2362, 2012.
235. Jiang, Z., J. Wang, L. Meng, Y. Huang, and L. Liu. A highly efficient chemical sensor material for ethanol: Al_2O_3/Graphene nanocomposites fabricated from graphene oxide. *Chem. Commun.*, 47:6350–6352, 2011.
236. Deng, S., J. Lei, L. Cheng, Y. Zhang, and H. Ju. Amplified electrochemiluminescence of quantum dots by electrochemically reduced graphene oxide for nanobiosensing of acetylcholine. *Biosens. Bioelectron.*, 26:4552–4558, 2011.
237. Zhang, S., Z. Du, and G. Li. Layer-by-layer fabrication of chemical-bonded graphene coating for solid-phase microextraction. *Anal. Chem.*, 83:7531–7541, 2011.
238. Xu, L., J. Feng, J. Li, X. Liu, and S. Jiang. Graphene oxide bonded fused-silica fiber for solid-phase microextraction-gas chromatography of polycyclic aromatic hydrocarbons in water. *J. Sep. Sci.*, 35:93–100, 2012.
239. Wu, Q., C. Feng, G. Zhao, C. Wang, and Z. Wang. Graphene-coated fiber for solid-phase microextraction of triazine herbicides in water samples. *J. Sep. Sci.*, 35:193–199, 2012.
240. Liu, Q., J. Shi, L. Zeng, T. Wang, Y. Cai, and G. Jiang. Evaluation of graphene as an advantageous adsorbent for solid-phase extraction with chlorophenols as model analytes. *J. Chromatogr. A*, 1218:197–204, 2011.
241. Luo, Y.B., B.F. Yuan, Q.W. Yu, and Y.Q. Feng. Substrateless graphene fiber: A sorbent for solid-phase microextraction. *J. Chromatogr. A*, 1268:9–15, 2012.

242. Zou, J., X. Song, J. Ji, W. Xu, J. Chen, Y. Jiang, Y. Wang, and X. Chen. Polypyrrole/graphene composite-coated fiber for the solid-phase microextraction of phenols. *J. Sep. Sci.*, 34:2765–2772, 2011.

243. Wu, Q., G. Zhao, C. Feng, C. Wang, and Z. Wang. Preparation of a graphene-based magnetic nanocomposite for the extraction of carbamate pesticides from environmental water samples. *J. Chromatogr. A*, 1218:7936–7942, 2011.

244. Zhao, G., S. Song, C. Wang, Q. Wu, and Z. Wang. Determination of triazine herbicides in environmental water samples by high-performance liquid chromatography using graphene-coated magnetic nanoparticles as adsorbent. *Anal. Chim. Acta*, 708:155–159, 2011.

245. Wang, W., Y. Li, Q. Wu, C. Wang, X. Zang, and Z. Wang. Extraction of neonicotinoid insecticides from environmental water samples with magnetic graphene nanoparticles as adsorbent followed by determination with HPLC. *Anal. Methods*, 4:766–772, 2012.

246. Wang, W., X. Ma, Q. Wu, C. Wang, X. Zang, and Z. Wang. The use of graphene-based magnetic nanoparticles as adsorbent for the extraction of triazole fungicides from environmental water. *J. Sep. Sci.*, 35:2266–2272, 2012.

247. Shi, C., J. Meng, and C. Deng. Enrichment and detection of small molecules using magnetic graphene as an adsorbent and a novel matrix of MALDI-TOF-MS. *Chem. Commun.*, 48:2418–2420, 2012.

248. Liang, R.P., C.M. Liu, X.Y. Meng, J.W. Wang, and J.D. Qiu. A novel open-tubular capillary electrochromatography using β-cyclodextrin functionalized graphene oxide-magnetic nanocomposites as tunable stationary phase. *J. Chromatogr. A*, 1266:95–102, 2012.

249. Ponnusamy, V.K., and J.F. Jen. A novel graphene nanosheets coated stainless steel fiber for microwave assisted headspace solid phase microextraction of organochlorine pesticides in aqueous samples followed by gas chromatography with electron capture detection. *J. Chromatogr. A*, 1218:6861–6868, 2011.

250. Liu, X., L. Cao, W. Song, K. Ai, and L. Lu. Functionalizing metal nanostructured film with graphene oxide for ultrasensitive detection of aromatic molecules by surface-enhanced Raman spectroscopy. *ACS Appl. Mater. Interfaces*, 3:2944–2952, 2011.

251. Ren, W., Y. Fang, and E. Wang. A binary functional substrate for enrichment and ultrasensitive SERS spectroscopic detection of folic acid using graphene oxide/Ag nanoparticle hybrids. *ACS Nano*, 5:6425–6433, 2011.

252. Maharana, P.K., and R. Jha. Chalcogenide prism and graphene multilayer based surface plasmon resonance affinity biosensor for high performance. *Sens. Actuators B-Chem.*, 169:161–166, 2012.

253. Valentini, F., D. Romanazzo, M. Carbone, and G. Palleschi. Modified screen-printed electrodes based on oxidized graphene nanoribbons for the selective electrochemical detection of several molecules. *Electroanalysis*, 24:872–881, 2012.

254. Guo, Y., Z. Wang, H. Shao, and X. Jiang. Hydrothermal synthesis of highly fluorescent carbon nanoparticles from sodium citrate and their use for the detection of mercury ions. *Carbon*, 52:583–589, 2013.

255. Dong, Y., R. Wang, G. Li, C. Chen, Y. Chi, and G. Chen. Polyamine-functionalized carbon quantum dots as fluorescent probes for selective and sensitive detection of copper ions. *Anal. Chem.*, 84:6220–6224, 2012.

256. Dong, Y., G. Li, N. Zhou, R. Wang, Y. Chi, and G. Chen. Graphene quantum dot as a green and facile sensor for free chlorine in drinking water. *Anal. Chem.*, 84:8378–8382, 2012.

257. Goh, M.S., and M. Pumera. Graphene-based electrochemical sensor for detection of 2,4,6-trinitrotoluene (TNT) in seawater: The comparison of single-, few-, and multilayer graphene nanoribbons and graphite microparticles. *Anal. Bioanal. Chem.*, 399:127, 2011.

258. Wang, S., L. Yi, J.E. Halpert, X. Lai, Y. Liu, H. Cao, R. Yu, D. Wang, and Y. Li. A novel and highly efficient photocatalyst based on P25-graphdiyne nanocomposite. *Small*, 8:265–271, 2012.

259. Akhavan, O., and E. Ghaderi. Toxicity of graphene and graphene oxide nanowalls against bacteria. *ACS Nano*, 4:5731–5736, 2010.

35

Quantitative High Throughput Assay for Detection of Biologically Active Endocrine Disrupting Chemicals in Water

Diana A. Stavreva, Lyuba Varticovski, and Gordon L. Hager
Laboratory of Receptor Biology and Gene Expression, National Cancer Insitute, National Institutes of Health, Bethesda, Maryland, USA

Water sources are frequently contaminated by pollutants originating from municipal and industrial wastewater effluents, as well as runoffs from agricultural areas. Without question, the contamination of freshwater will increase as the human population continues to grow, and wastewater will become an even larger fraction of the flow. Of the vast variety of man-made chemicals worldwide that have been introduced in the environment, endocrine-disrupting chemicals (EDCs) are among the most hazardous because of their activity at very low doses (WHO, 2013). As their name suggests, they interfere with the endocrine (hormonal) system, which governs the development and function of all tissues and organs (Figure 35.1). Many of the EDCs exert their effects through direct interaction with endocrine hormone receptors such as estrogen receptors (ERs), androgen receptors (ARs), thyroid hormone receptors (TRs), and other nuclear receptors involved in metabolism and differentiation: aryl hydrocarbon (AhR), retinoid X, peroxisome proliferator-activated, liver X, and farsenoid X receptors (Swedenborg et al., 2009). The World Health Organization (WHO) defines an endocrine disruptor as an "exogenous substance or a mixture that alters function(s) of the endocrine system and consequently causes adverse health effects in an intact organism, its progeny, or (sub) populations" (WHO, 2011). The Scientific Statement of the Endocrine Society postulates that EDCs have effects on reproduction, breast development and cancer, prostate cancer, neuroendocrinology, thyroid metabolism, obesity and cardiovascular endocrinology (Diamanti-Kandarakis et al., 2009). The most sensitive window for the action of EDCs is during fetal development, which may have long-lasting consequences. Thus, the harmful effects of exposure to EDCs may not be immediately apparent and could manifest later in life, making it more difficult to discern from other causes. Many EDCs could also induce harmful traits that are carried over to future generations (called transgenerational effects). These transgenerational effects frequently have an epigenetic origin. In other words, they do not result in changes of DNA sequence but are nonetheless inheritable. Thus, contamination of water sources with EDCs threatens the integrity of aquatic ecosystems and poses a serious concern for human and animal health (Diamanti-Kandarakis et al., 2009; Deblonde et al., 2011).

Among the best-studied EDCs in water are contaminants with estrogenic activity (van der Linden et al., 2013; Leusch et al., 2006; Muller et al., 2008; Schiliro et al., 2004). Harmful effects of estrogenic water contaminants (Iwanowicz et al., 2009; Alvarez et al., 2009; Blazer et al., 2011) and synthetic progestogens (Zeilinger et al., 2009; Paulos et al., 2010) specifically on fish reproduction have been previously documented. Increased susceptibility to infections

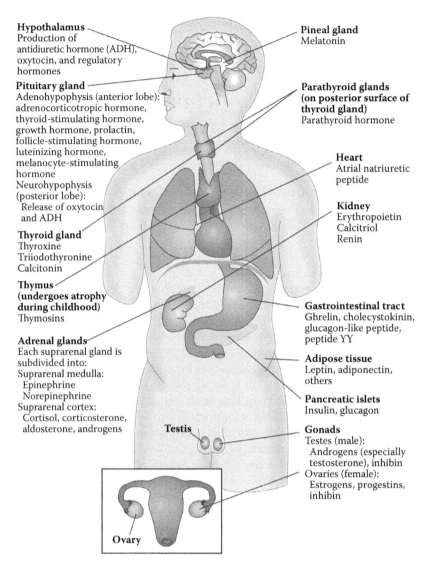

FIGURE 35.1
Schematic representation of the human endocrine (hormonal) system. (Adapted from WHO, *State of the Science of Endocrine Disrupting Chemicals*, 2012. Edited by Å. Bergman, J.J. Heindel, S. Jobling et al., IOMC, 2012. http://www.who.int/ceh/publications/endocrine/en/.)

indicative of a weakened immune system has been associated with fishkills in the Potomac river watershed (Ripley et al., 2008; Blazer et al., 2010; Ankley et al., 2009), suggesting a possible contamination with additional, yet unidentified, classes of EDCs. Contaminants interfering with the thyroid hormonal function such as triiodothyronine (T3)-like activity were also reported in effluents from wastewater treatment plants (WWTPs) in Japan (Murata and Yamauchi, 2008), and anti-T3 hormonal activity was found in WWTP effluent in Thailand (Ishihara et al., 2009). In addition, several AhR-activating contaminants have been detected in the United States and elsewhere (Long et al., 2003, 2007).

Because of increased concern about contamination of the environment with EDCs as well as the complexity of their biological effects (Diamanti-Kandarakis et al., 2009;

van der Linden et al., 2013; WHO, 2011, 2013; Ishihara et al., 2009; Murata and Yamauchi, 2008; Ma et al., 2005; Kusk et al., 2011), significant attention and investment have been devoted to their detection. A variety of methods, including laborious chemical methods of isolation and identifications by a combination of high-performance liquid chromatography (HPLC), liquid or gas chromatography (GC), and/or mass spectroscopy (MS), as well as "omic" approaches (genomics, transcriptomics, proteomics, and/or metabolomics on fish and other affected organisms), have been described (Mansilha et al., 2010; Jordan et al., 2011). However, high cost, lack of uniform quantification, and uncertainty of the biological response prevented their use for efficient detection and monitoring of EDCs in the environment, particularly in the water.

In an attempt to streamline the detection of EDCs in water sources, a number of assays have been introduced (Roy and Pereira, 2005). One of the first described assays is a yeast estrogen screen assay, which is based on ER gene–modified yeast cells (Routledge and Sumpter, 1996). These cells carry human ER and ER-responsive gene, which codes for β-galactosidase. In the presence of estrogen, the activity of this artificial gene is turned "on," producing the enzyme. β-Galactosidase in turn modifies the yellow chlorophenol red–β-galactopyranoside, present in the growth medium, to a red product that is quantified spectrophotometrically at 540 nm. Another approach has been the use of luciferase reporter assay. In this assay, the expression of luciferase gene is governed by regulatory elements responsive to a specific hormone-activated receptor, and accumulation of the protein luciferase is an indication of receptor activation. Recently, several luciferase-expressing cell lines capable of detecting EDCs that influence the function of the AR, AhR, and TR have been developed and implemented (Kusk et al., 2011). For example, AhR-activating contaminants in water have been detected by the chemical-activated luciferase gene expression (CALUX) assay (Murk et al., 1996; Long et al., 2003, 2007). TR-mediated luciferase gene expression assay was used to detect T3-like and anti-T3 activity in water (Murata and Yamauchi, 2008; Ishihara et al., 2009). The T-screen is another bioassay based on thyroid hormone–dependent cell proliferation of a rat pituitary tumor cell line (GH3) in serum-free medium (Gutleb et al., 2005). A cell line (MVLN) derivative of ER-positive MCF-7 breast cancer cells carrying ER response element (ERE)–luciferase reporter vector (Bonefeld-Jorgensen et al., 2005) has been used to test for the potency of water contamination with estrogens. Methods to detect the rate of hormone synthesis and breakdown in response to EDCs have also been described. Hilscherova et al. (2004) implemented a human adrenocortical carcinoma cell line (H295R) to measure the expression of steroidogenic genes by using real-time PCR. Another interesting approach is the use of the transgenic zebrafish (Lee et al., 2012). This model has the potential to assess the impact of environmental estrogens on a variety of body systems. The transgenic zebrafish is engineered to contain an estrogen-inducible promoter composed of multiple EREs and a Gal4ff–UAS system. An estrogenic signal is detected when the chemical–ER complex binds to the ERE, which activates Gal4ff. Subsequently, Gal4ff protein binds to UAS to induce the fluorescent reporter (green fluorescent protein, GFP). In spite of many attempts of detect EDCs in the environment, most assays are manually demanding, require considerable time for analysis and are frequently limited to detection of a single hormone.

Our laboratory has recently developed a unique high throughput assay for detection and biological testing of EDCs in mammalian cells (Stavreva et al., 2012a). We engineered mouse mammary cells that express GFP-tagged nuclear steroid receptors, such as glucocorticoid receptor (GR) (Walker et al., 1999). This assay is based on the fact that, in the absence of the hormone, the receptor resides in the cytoplasm bound to various heat shock proteins and immunophilins in a large multiprotein complex (Pratt and Toft, 1997). Upon hormone binding, GR dissociates from this complex and translocates to the cell nucleus

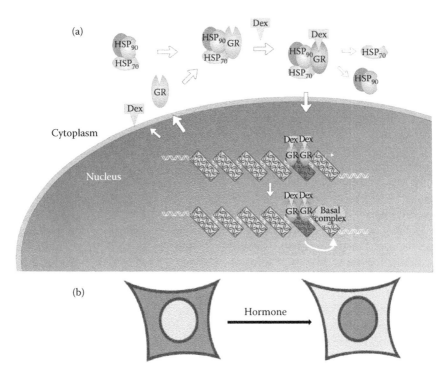

FIGURE 35.2
(See color insert.) Glucocorticoid receptor (GR) biology. (a) At resting state, GR resides in the cytoplasm as a part of a multiprotein complex with heat-shock proteins and immunophilins. (From Pratt, W.B. and D.O. Toft, *Endocr. Rev.*, 18, 306, 1997.) Upon hormone stimulation, GR dissociates from this complex and migrates to the nucleus (cytoplasmic-to-nuclear translocation). In the nucleus, the receptor interacts with GR regulatory elements (GREs) throughout the genome, and regulates GR-mediated transcriptional responses. (b) Schematic representation of the translocation of the green fluorescent protein (GFP)-tagged GR from the cytoplasm to the nucleus in the presence of hormone or EDCs with glucocorticoid activity. This translocation is the basis of the biological screening for detection of water contaminants interacting with a specific nuclear receptor(s).

(Figure 35.2a), where it interacts with GR genomic regulatory elements (GREs) to elicit hormone-specific transcription regulation (John et al., 2008).

Similarly to the GR, AR is largely cytoplasmic in the absence of its ligand, and rapidly translocates to the nucleus in response to testosterone (Klokk et al., 2007). We used cell lines engineered to express GFP-tagged GR and AR constructs under a tetracycline-repressible promoter (Walker et al., 1999; Klokk et al., 2007). In these cells, tetracycline suppresses the expression of the receptors, and removal of the drug results in protein expression. When cells are plated on coverslips overnight in a medium supplemented with hormone-free (charcoal-stripped) serum without tetracycline, cytoplasmic expression of GFP-tagged receptors can be readily observed, and translocation to the nucleus occurs in 30 min (Figure 35.2b). Upon exposure to corticosteroids (such as dexamethasone) or a vehicle control. Cells are fixed, counterstained with 4',6-diamidino-2-phenylindole (DAPI) for visualization of the nuclei, and translocation is visualized on a fluorescent microscope. Images are acquired in the green (GFP-GR and GFP-AR) and UV (DAPI) channel, and receptor translocation is scored as "negative," "partial," or "complete" (representative negative and complete translocation of the GFP-GR are shown in Figure 35.3a).

We further developed an automated, quantitative high throughput read-out of the translocation assay using a Perkin Elmer OPERA imaging system (Figure 35.3b). For these

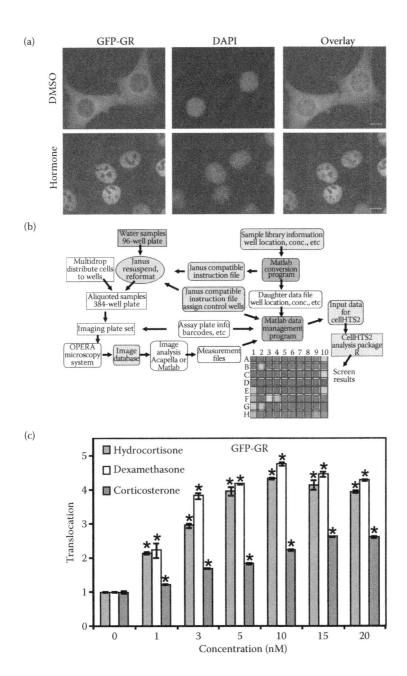

FIGURE 35.3
(See color insert.) Application of the translocation assay for analysis of EDCs in water samples. (a) GFP-tagged glucocorticoid receptor (GFP-GR) translocation to the nucleus in mammalian cells exposed to control (DMSO, top panel) or dexamethasone (bottom panel) for 30 min. Nuclei are costained with DAPI. Scale bar, 5 μm. (b) Workflow for image-based screening of environmental contaminants with glucocorticoid activity using the Perkin Elmer OPERA image screening system. (c) GFP-GR translocation to the nucleus in a concentration-dependent manner upon treatment with known concentrations of hydrocortisone, dexamethasone, or corticosterone. An algorithm for cytoplasm and nuclear segmentation was used to determine the mean GFP-GR intensity in both compartments, and translocation was quantified as a ratio of these intensities. Each value was normalized to the control sample. Error bars represent the mean value ± s.e.m., n = 6 ($P < 0.05$, asterisks).

experiments, cells were plated on 96- or 384-well plates, and after 30-min treatment followed by fixation, stained with DRAQ5 (BioStatus Limited) for 15 min. Cells were either immediately imaged on the Perkin Elmer OPERA imaging system, or kept in phosphate-buffered saline at 4°C for later imaging. An algorithm was customized using the Acapella image analysis software development kit (Perkin Elmer) to automatically segment both the nucleus and cytoplasm of each cell in the digital micrographs. The algorithm also calculated the mean GFP-GR or GFP-AR intensity in both compartments, and translocation was measured as a ratio of these intensities. Each value was further normalized to the value for the control (dimethyl sulfoxide, DMSO) sample. A graph demonstrating the concentration-dependent GFP-GR translocation to the nucleus in response to synthetic glucocorticoid (dexamethasone), human glucocorticoid (hydrocortisone), and rodent glucocorticoid (corticosterone) is shown in Figure 35.3c. Confident in the sensitivity of this assay, we tested >100 water samples collected from 14 states in the United States for glucocorticoid and androgen activities. We discovered a previously unrecognized glucocorticoid activity in 27% of the samples (Figure 35.4) and androgen activity in 35% of the samples (Figure 35.5). Some of these samples were discrete "grab" water samples (circles) and the others were collected by polar organic chemical integrative samplers (POCIS) deployed in the water at the collection sites over a period of time (triangles).

Contamination of water sources with androgen/anti-androgen activity was previously demonstrated in the Netherlands (van der Linden et al., 2013), United Kingdom (Kirk et al., 2002), New Zealand (Leusch et al., 2006), Denmark (Kusk et al., 2011), and China (Chang et al., 2009a). Contamination of water sources with glucocorticoids and its negative effects on fish health were also reported (Chang et al., 2007, 2009a; Schriks et al., 2010; Kugathas and Sumpter, 2011).

The major advantages of the translocation assay over other previously described assays are that they are rapid, quantitative, inexpensive, and compatible with high throughput screening. Receptor translocation is completed in 20–30 min, and multiple cell lines or multiple color-tagged receptors in the same cell line can be tested and utilized simultaneously. In contrast to the chemical methods, this assay does not reveal the chemical structure of the active contaminants. Instead, it detects biologically active EDCs capable of interacting with a specific nuclear receptor (in this case, GR or AR). To identify a specific or multiple chemical structures responsible for these effects, we tested one sample (SS97) positive in the translocation assay by chemical detection methods to perform "forensic chemistry." Sample SS97 tested positive in the GFP-GR translocation assay (Figure 35.6a) and was also capable of inducing transcriptional response from a GR-regulated gene, *Per1*.

To identify the active glucocorticoid contaminant(s), we analyzed SS97 fractions of the HPLC retention on a C18 column (Chang et al., 2009a) for GFP-GR translocation (data not shown). Positive fractions were further analyzed by ultra-performance liquid chromatography/MS and GC/MS (Mansilha et al., 2010). Resulting mass spectra were searched extensively in the National Institute of Standards and Technology/Environmental Protection Agency/National Institutes of Health Mass Spectral Library, and in the Wiley Mass Spectra Database of Androgens, Estrogens, and other Steroids 2010. In spite of this detailed analysis, we found no evidence of any known glucocorticoids in this sample. However, comparison of the mass spectra of chromatographic peaks 1–3 (Figure 35.6c) with standard spectra from the Atomic Emission Spectroscopy (AES) 2010 database suggested similarities to known androstane-class compounds that are potentially capable of activating AR. One of these compounds, androst-4-en-3,6-dione (peak 2), was synthesized (Hunter and Priest, 2006) and its translocation activity tested in cell lines expressing GFP-GR or GFP-AR. We confirmed that androst-4-en-3,6-dione induced GFP-AR (Figure 35.6d) but not GFP-GR translocation (data not shown).

Some of the EDCs are resistant to a complete biodegradation and persist in the effluents of the WWTPs (Chang et al., 2009b). Thus, current wastewater treatment methods

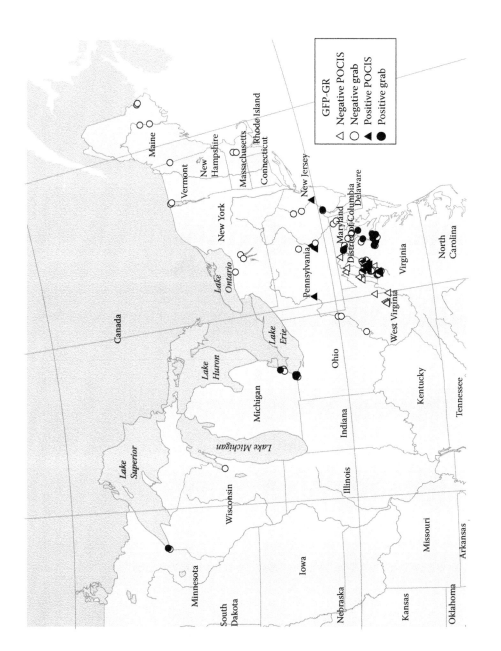

FIGURE 35.4
Geographic locations of the collection sites and their contamination with glucocorticoid activity. Open symbols mark negative samples. Samples positive for gluco-corticoid activity are marked with black. Triangles indicate grab samples, while the circles indicate the use of POCIS membranes. (For complete sample description—collection method as well as the time of collection and translocation activity—see Stavreva, D.A. et al., *Sci Rep.*, 2, 1, 2012a.)

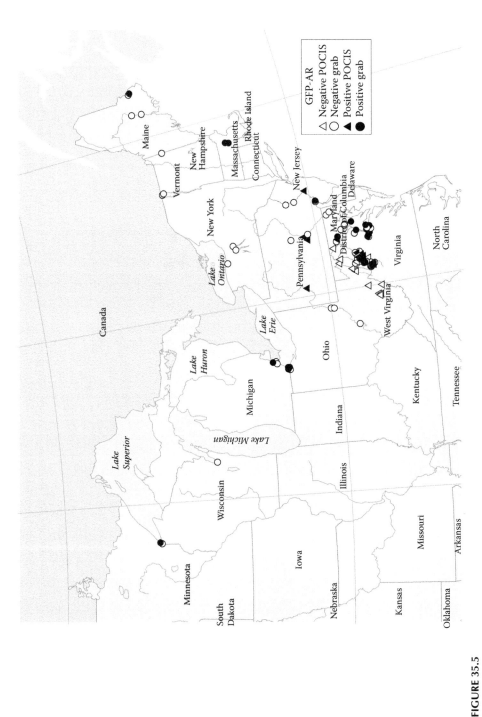

FIGURE 35.5
Geographic locations of the collection sites and their contamination with androgen activity. Open symbols mark negative samples. Samples positive for androgen activity are marked with black. Triangles indicate grab samples, while the circles indicate the use of POCIS membranes. (For complete sample description—collection method as well as the time of collection and translocation activity—see Stavreva, D.A. et al., *Sci Rep.*, 2, 1, 2012a.)

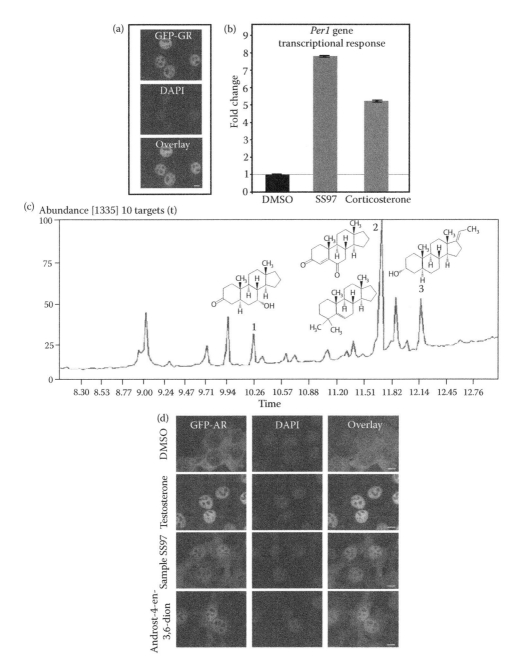

FIGURE 35.6

Analysis of water sample SS97 for glucocorticoid and androgen contamination. (a) Sample SS97 collected by a POCIS induces complete GFP-GR translocation in 30 min. Scale bar, 5 μm. (b) Transcriptional activation of the GR-regulated *Per1* gene by sample SS97 (concentrated 100×) is compared with transactivation induced by corticosterone. Data are normalized to the solvent (DMSO). Error bars represent the mean ± s.e.m., n = 3. (c) GC/MS total ion chromatogram of one of the HPLC fractions of sample SS97 revealed the presence of a complex mixture of volatile hydrocarbons, as indicated by the peaks. Database searching of the extracted MS spectra corresponding to peaks 1–3 showed structural similarity to known androstane-type steroids. (d) Representative images of GFP-AR nuclear translocation in response to 100 nM of testosterone, androst-4-ene-3,6-dione, and sample SS97 (100× concentrated). Scale bar, 5 μm.

are not efficient enough to prevent contamination of environmental surface waters with EDCs (Kusk et al., 2011). In addition, many EDCs are metabolized, producing derivatives that are not present in the currently existing libraries and thus cannot be identified by chemical methods. Complex mixtures with endocrine-altering properties in the water further complicate the process of identifying active constituents. Our results unambiguously demonstrate that the translocation-based assay can detect biologically active EDCs that are not identifiable by chemical methods. The results of our "forensic chemistry" analysis also indicate that the translocation-based assay is superior to the currently used chemical methods, as it successfully detected not only androgenic but also glucocorticoid activity in the water, the latter undetectable by the currently known chemical methods.

Finally, should we be concerned by the presence of these two classes of EDCs in the water sources? It is true that none of the samples were taken from drinking water reservoirs, and most were highly concentrated. However, when several different concentrations of two randomly selected samples were tested, we discovered that both samples were positive for glucocorticoid and androgen activity at 10× concentrations (10× higher than the original water sample). In addition, one of the samples was also positive for glucocorticoid activity at 1× concentration, suggesting that the level of glucocorticoids at that location is sufficiently high to elicit biological responses.

Naturally occurring glucocorticoids acting through glucocorticoid and mineralocorticoids receptors (GR and MR, respectively) are released in mammalian organisms in a complex circadian and ultradian manner (Lightman, 2006; Stavreva et al., 2009, 2012b). Excess exposure to glucocorticoid is associated with immune suppression and a variety of other deleterious side effects (Schacke et al., 2002). Exposures to stress hormones, including glucocorticoids, during the prenatal period have programming effects on the hypothalamic–pituitary–adrenal axis, brain neurotransmitter systems, and cognitive abilities of the offspring (Darnaudery and Maccari, 2008). Negative effects on the immune function on the offspring upon prenatal exposure to endogenous or synthetic glucocorticoids were also reported (Merlot et al., 2008). Moreover, disruption of the AR signaling in males by environmental chemicals is well documented (Luccio-Camelo and Prins, 2011). Studies using guinea pigs and monkeys demonstrate that prenatal exposure to androgens is associated with irreversible changes in sexual and social behaviors of the female progenies later in life (Luccio-Camelo and Prins, 2011; Wallen and Hassett, 2009). Thus, contamination of water sources with both glucocorticoid and androgenic activity could have deleterious short-term effects, as well as long-term consequences.

In conclusion, implementation of this novel approach for detection and monitoring of biologically active EDCs in water (Stavreva et al., 2012a) may facilitate both the development and testing of appropriate filtration methods for efficient removal of EDCs from water sources, as well as the establishment of regulatory guidelines. This approach could be readily extended to other nuclear receptors and applied to detection of various classes of EDCs in the environment.

References

Alvarez, D.A., W.L. Cranor, S.D. Perkins, V.L. Schroeder, L.R. Iwanowicz, R.C. Clark, C.P. Guy, A.E. Pinkney, V.S. Blazer, and J.E. Mullican. 2009. Reproductive health of bass in the Potomac, U.S.A., drainage: Part 2. Seasonal occurrence of persistent and emerging organic contaminants. *Environ. Toxicol. Chem.* 28:1084–1095.

Ankley, G.T., D.C. Bencic, M.S. Breen, T.W. Collette, R.B. Conolly, N.D. Denslow, S.W. Edwards, D.R. Ekman, N. Garcia-Reyero, K.M. Jensen, J.M. Lazorchak, D. Martinovic, D.H. Miller, E.J. Perkins, E.F. Orlando, D.L. Villeneuve, R.L. Wang, and K.H. Watanabe. 2009. Endocrine disrupting chemicals in fish: Developing exposure indicators and predictive models of effects based on mechanism of action. *Aquat. Toxicol.* 92:168–178.

Blazer, V.S., L.R. Iwanowicz, H. Henderson, P.M. Mazik, J.A. Jenkins, D.A. Alvarez, and J.A. Young. 2011. Reproductive endocrine disruption in smallmouth bass (Micropterus dolomieu) in the Potomac River basin: Spatial and temporal comparisons of biological effects. *Environ. Monit. Assess.* 184(7):4309–4334.

Blazer, V.S., L.R. Iwanowicz, C.E. Starliper, D.D. Iwanowicz, P. Barbash, J.D. Hedrick, S.J. Reeser, J.E. Mullican, S.D. Zaugg, M.R. Burkhardt, and J. Kelble. 2010. Mortality of centrarchid fishes in the Potomac drainage: Survey results and overview of potential contributing factors. *J. Aquat. Anim. Health* 22:190–218.

Bonefeld-Jorgensen, E.C., H.T. Grunfeld, and I.M. Gjermandsen. 2005. Effect of pesticides on estrogen receptor transactivation in vitro: A comparison of stable transfected MVLN and transient transfected MCF-7 cells. *Mol. Cell. Endocrinol.* 244:20–30.

Chang, H., J. Hu, and B. Shao. 2007. Occurrence of natural and synthetic glucocorticoids in sewage treatment plants and receiving river waters. *Environ. Sci. Technol.* 41:3462–3468.

Chang, H., Y. Wan, and J. Hu. 2009a. Determination and source apportionment of five classes of steroid hormones in urban rivers. *Environ. Sci. Technol.* 43:7691–7698.

Chang, H.S., K.H. Choo, B. Lee, and S.J. Choi. 2009b. The methods of identification, analysis, and removal of endocrine disrupting compounds (EDCs) in water. *J. Hazard. Mater.* 172:1–12.

Darnaudery, M., and S. Maccari. 2008. Epigenetic programming of the stress response in male and female rats by prenatal restraint stress. *Brain Res. Rev.* 57:571–585.

Deblonde, T., C. Cossu-Leguille, and P. Hartemann. 2011. Emerging pollutants in wastewater: A review of the literature. *Int. J. Hyg. Environ. Health* 214:442–448.

Diamanti-Kandarakis, E., J.P. Bourguignon, L.C. Giudice, R. Hauser, G.S. Prins, A.M. Soto, R.T. Zoeller, and A.C. Gore. 2009. Endocrine-disrupting chemicals: An Endocrine Society scientific statement. *Endocr. Rev.* 30:293–342.

Gutleb, A.C., I.A. Meerts, J.H. Bergsma, M. Schriks, and A.J. Murk. 2005. T-Screen as a tool to identify thyroid hormone receptor active compounds. *Environ. Toxicol. Pharmacol.* 19:231–238.

Hilscherova, K., P.D. Jones, T. Gracia, J.L. Newsted, X. Zhang, J.T. Sanderson, R.M. Yu, R.S. Wu, and J.P. Giesy. 2004. Assessment of the effects of chemicals on the expression of ten steroidogenic genes in the H295R cell line using real-time PCR. *Toxicol. Sci.* 81:78–89.

Hunter, A.C., and S.M. Priest. 2006. An efficient one-pot synthesis generating 4-ene-3,6-dione functionalised steroids from steroidal 5-en-3beta-ols using a modified Jones oxidation methodology. *Steroids* 71:30–33.

Ishihara, A., F.B. Rahman, L. Leelawatwattana, P. Prapunpoj, and K. Yamauchi. 2009. In vitro thyroid hormone-disrupting activity in effluents and surface waters in Thailand. *Environ. Toxicol. Chem.* 28:586–594.

Iwanowicz, L.R., V.S. Blazer, C.P. Guy, A.E. Pinkney, J.E. Mullican, and D.A. Alvarez. 2009. Reproductive health of bass in the Potomac, U.S.A., drainage: Part 1. Exploring the effects of proximity to wastewater treatment plant discharge. *Environ. Toxicol. Chem.* 28:1072–1083.

John, S., P.J. Sabo, T.A. Johnson, M.H. Sung, S.C. Biddie, S.L. Lightman, T.C. Voss, S.R. Davis, P.S. Meltzer, J.A. Stamatoyannopoulos, and G.L. Hager. 2008. Interaction of the glucocorticoid receptor with the global chromatin landscape. *Mol. Cell.* 29:611–624.

Jordan, J., A. Zare, L.J. Jackson, H.R. Habibi, and A.M. Weljie. 2011. Environmental contaminant mixtures at ambient concentrations invoke a metabolic stress response in goldfish not predicted from exposure to individual compounds alone. *J. Proteome. Res.* 11(2):1133–1143.

Kirk, L.A., C.R. Tyler, C.M. Lye, and J.P. Sumpter. 2002. Changes in estrogenic and androgenic activities at different stages of treatment in wastewater treatment works. *Environ. Toxicol. Chem.* 21:972–979.

Klokk, T.I., P. Kurys, C. Elbi, A.K. Nagaich, A. Hendarwanto, T. Slagsvold, C.Y. Chang, G.L. Hager, and F. Saatcioglu. 2007. Ligand-specific dynamics of the androgen receptor at its response element in living cells. *Mol. Cell. Biol.* 27:1823–1843.

Kugathas, S., and J.P. Sumpter. 2011. Synthetic glucocorticoids in the environment: First results on their potential impacts on fish. *Environ. Sci. Technol.* 45:2377–2383.

Kusk, K.O., T. Kruger, M. Long, C. Taxvig, A.E. Lykkesfeldt, H. Frederiksen, A.M. Andersson, H.R. Andersen, K.M. Hansen, C. Nellemann, and E.C. Bonefeld-Jorgensen. 2011. Endocrine potency of wastewater: Contents of endocrine disrupting chemicals and effects measured by in vivo and in vitro assays. *Environ. Toxicol. Chem.* 30:413–426.

Lee, O., A. Takesono, M. Tada, C.R. Tyler, and T. Kudoh. 2012. Biosensor zebrafish provide new insights into potential health effects of environmental estrogens. *Environ. Health Perspect.* 120:990–996.

Leusch, F.D., H.F. Chapman, M.R. van den Heuvel, B.L. Tan, S.R. Gooneratne, and L.A. Tremblay. 2006. Bioassay-derived androgenic and estrogenic activity in municipal sewage in Australia and New Zealand. *Ecotoxicol. Environ. Saf.* 65:403–411.

Lightman, S.L. 2006. Patterns of exposure to glucocorticoid receptor ligand. *Biochem. Soc. Trans.* 34:1117–1118.

Long, M., P. Laier, A.M. Vinggaard, H.R. Andersen, J. Lynggaard, and E.C. Bonefeld-Jorgensen. 2003. Effects of currently used pesticides in the AhR-CALUX assay: Comparison between the human TV101L and the rat H4IIE cell line. *Toxicology* 194:77–93.

Long, M., A. Stronati, D. Bizzaro, T. Kruger, G.C. Manicardi, P.S. Hjelmborg, M. Spano, A. Giwercman, G. Toft, J.P. Bonde, and E.C. Bonefeld-Jorgensen. 2007. Relation between serum xenobiotic-induced receptor activities and sperm DNA damage and sperm apoptotic markers in European and Inuit populations. *Reproduction* 133:517–530.

Luccio-Camelo, D.C., and G.S. Prins. 2011. Disruption of androgen receptor signaling in males by environmental chemicals. *J. Steroid Biochem. Mol. Biol.* 127:74–82.

Ma, M., J. Li, and Z. Wang. 2005. Assessing the detoxication efficiencies of wastewater treatment processes using a battery of bioassays/biomarkers. *Arch. Environ. Contam. Toxicol.* 49:480–487.

Mansilha, C., A. Melo, H. Rebelo, I.M. Ferreira, O. Pinho, V. Domingues, C. Pinho, and P. Gameiro. 2010. Quantification of endocrine disruptors and pesticides in water by gas chromatography-tandem mass spectrometry. Method validation using weighted linear regression schemes. *J. Chromatogr. A* 1217:6681–6691.

Merlot, E., D. Couret, and W. Otten. 2008. Prenatal stress, fetal imprinting and immunity. *Brain Behav. Immun.* 22:42–51.

Muller, M., F. Rabenoelina, P. Balaguer, D. Patureau, K. Lemenach, H. Budzinski, D. Barcelo, M.L. de Alda, M. Kuster, J.P. Delgenes, and G. Hernandez-Raquet. 2008. Chemical and biological analysis of endocrine-disrupting hormones and estrogenic activity in an advanced sewage treatment plant. *Environ. Toxicol. Chem.* 27:1649–1658.

Murata, T., and K. Yamauchi. 2008. 3,3′,5-Triiodo-L-thyronine-like activity in effluents from domestic sewage treatment plants detected by in vitro and in vivo bioassays. *Toxicol. Appl. Pharmacol.* 226:309–317.

Murk, A.J., J. Legler, M.S. Denison, J.P. Giesy, G.C. van de, and A. Brouwer. 1996. Chemical-activated luciferase gene expression (CALUX): A novel in vitro bioassay for Ah receptor active compounds in sediments and pore water. *Fundam. Appl. Toxicol.* 33:149–160.

Paulos, P., T.J. Runnalls, G. Nallani, P.T. La, A.P. Scott, J.P. Sumpter, and D.B. Huggett. 2010. Reproductive responses in fathead minnow and Japanese medaka following exposure to a synthetic progestin, Norethindrone. *Aquat. Toxicol.* 99:262.

Pratt, W.B., and D.O. Toft. 1997. Steroid receptor interactions with heat shock protein and immunophilin chaperones. *Endocr. Rev.* 18:306–360.

Ripley, J., L. Iwanowicz, V. Blazer, and C. Foran. 2008. Utilization of protein expression profiles as indicators of environmental impairment of smallmouth bass (Micropterus dolomieu) from the Shenandoah River, Virginia, USA. *Environ. Toxicol. Chem.* 27:1756–1767.

Routledge, E.J., and J.P. Sumpter. 1996. Estrogenic activity of surfactants and some of their degradation products assessed using a recombinant yeast screen. *Environ. Toxicol. Chem.* 15:241–248.

Roy, P., and B.M. Pereira. 2005. A treatise on hazards of endocrine disruptors and tool to evaluate them. *Indian J. Exp. Biol.* 43:975–992.

Schacke, H., W.D. Docke, and K. Asadullah. 2002. Mechanisms involved in the side effects of glucocorticoids. *Pharmacol. Ther.* 96:23–43.

Schiliro, T., C. Pignata, E. Fea, and G. Gilli. 2004. Toxicity and estrogenic activity of a wastewater treatment plant in Northern Italy. *Arch. Environ. Contam. Toxicol.* 47:456–462.

Schriks, M., J.A. van Leerdam, S.C. van der Linden, B. van der Burg, A.P. van Wezel, and P. de Voogt. 2010. High-resolution mass spectrometric identification and quantification of glucocorticoid compounds in various wastewaters in the Netherlands. *Environ. Sci. Technol.* 44:4766–4774.

Stavreva, D.A., A.A. George, P. Klausmeyer, L. Varticovski, D. Sack, T.C. Voss, R.L. Schiltz, V.S. Blazer, L.R. Iwanowicz, and G.L. Hager. 2012a. Prevalent gluccorticoid and androgen activity in U.S. water sources. *Sci. Rep.* 2:1–8.

Stavreva, D.A., L. Varticovski, and G.L. Hager. 2012b. Complex dynamics of transcription regulation. *Biochim. Biophys. Acta* 1819:657–666.

Stavreva, D.A., M. Wiench, S. John, B.L. Conway-Campbell, M.A. McKenna, J.R. Pooley, T.A. Johnson, T.C. Voss, S.L. Lightman, and G.L. Hager. 2009. Ultradian hormone stimulation induces glucocorticoid receptor-mediated pulses of gene transcription. *Nat. Cell Biol.* 11:1093–1102.

Swedenborg, E., J. Ruegg, S. Makela, and I. Pongratz. 2009. Endocrine disruptive chemicals: Mechanisms of action and involvement in metabolic disorders. *J. Mol. Endocrinol.* 43:1–10.

van der Linden, S.C., M.B. Heringa, H.Y. Man, E. Sonneveld, L.M. Puijker, A. Brouwer, and B.B. van der. 2013. Detection of multiple hormonal activities in wastewater effluents and surface water, using a panel of steroid receptor CALUX bioassays. *Environ. Sci. Technol.* 32:548–561.

Walker, D., H. Htun, and G.L. Hager. 1999. Using inducible vectors to study intracellular trafficking of GFP-tagged steroid/nuclear receptors in living cells. *Methods (Companion to Methods in Enzymology)* 19:386–393.

Wallen, K., and J.M. Hassett. 2009. Sexual differentiation of behaviour in monkeys: Role of prenatal hormones. *J. Neuroendocrinol.* 21:421–426.

WHO. 2011. Global assessment of the state-of-the-science of endocrine disruptors. WHO/PCS/EDC/02. 2.

WHO. 2013. State of the science of endocrine disrupting chemicals 2012: Summary for decision-makers. WHO/HSE/PHE/IHE/2013. 1.

Zeilinger, J., T. Steger-Hartmann, E. Maser, S. Goller, R. Vonk, and R. Lange. 2009. Effects of synthetic gestagens on fish reproduction. *Environ. Toxicol. Chem.* 28:2663–2670.

36

Novel Carbon-Based Nanoadsorbents for Water Purification

Nishith Verma and Ashutosh Sharma

Department of Chemical Engineering, Indian Institute of Technology Kanpur, Kanpur, India

CONTENTS

With its many interesting physicochemical properties, carbon continues to attract the attention of scientists and engineers alike. Consequently, it is the most cited element in the scientific literature. If you search for "carbon" on the webpage of any scientific journal that publishes reports on materials, the chance that the search will hit the page limit is significantly high. Therefore, in the context of wastewater treatment, it is not surprising that although the utility of carbon in its allotropic form of charcoal was recognized many years ago, dating back to the prehistoric era, it remains the core material of modern water filters. In recent times, with nanotechnology applied to the many facets of environmental remediation, either in process intensification or product development, several groups are on the forefront of developing novel carbon-based nanoadsorbents for water purification.

This chapter presents an overview of such carbon-based materials developed as adsorbents, the methods of their preparation, and their ability to remove dissolved solutes in wastewater, primarily arsenic and fluoride, two common contaminants present in groundwater. This chapter also discusses the surface characterization of the adsorbents using several state-of-the-art analytical instruments, including atomic force microscopy, scanning electron microscopy (SEM), Fourier transform infrared spectroscopy, and Brunauer–Emmett–Teller (BET) surface area analyzer and pore size distribution (PSD) measurements, as well as the relevance of these characterization tests to the physicochemical properties of the prepared materials.

36.1 Water and Water Pollutants

Water is indispensable for life. In fact, it can be considered as *the molecule of life*. The daily requirement of water may vary from one person to another; however, there is a certain minimum daily uptake of pure water that is essential for sustaining the natural growth of human life [1]. Therefore, providing pure and fresh water, free of toxic solutes, is indispensable for healthy living, and providing safe drinking water to every resident is the top priority of any government.

Groundwater often has high amounts of inorganic constituents because of the dissolution of rock-forming minerals and the release of industrial effluents. Fluoride and arsenic ions are the most commonly occurring inorganic contaminants in groundwater. The presence of these ions in drinking water, in excess of permissible limits, results in undesirable health effects. Unfortunately, a significant increase in the levels of arsenic and fluoride in surface water has been reported in several regions in India and its neighboring countries during the past decade [2]. The primary reason for this increase is the rapid decline in the water table with increasing use of surface water for agricultural purposes and for urban water supplies. The risks to human health have increased with the increased concentrations of these solutes in potable water.

Fluoride is considered to be a reactive element. It may combine with other elements by forming covalent and ionic bonds. It is mainly found in alkaline rocks, alkaline soils, and industrial effluents, such as pharmaceuticals, semiconductor waste, cosmetics, coal power plants, and fertilizer manufacturing plants. According to the World Health Organization (WHO) norms, the upper limit for the fluoride concentration in drinking water is 1.5 mg/l. Small quantities of fluoride can protect against dental cavities and the weakening of bones; however, large amounts can impair health by causing dental fluorosis and damage to the bones (skeletal fluorosis). As many as 200 million people worldwide are affected by fluoride contamination of drinking water [3].

The contamination of groundwater by arsenic also affects vast regions in India, Bangladesh, China, Mexico, the United States, Cambodia, and Argentina. The maximum arsenic level allowed in drinking water has been set at 10 µg/l (WHO, 2001) since 1993, whereas the permitted limit for industrial effluents is <0.2 mg/l. Drinking arsenic-rich water over a long period results in various adverse health effects, including skin problems; cancers of the bladder, kidney, and lung; disease in the blood vessels of the legs and feet; and possibly diabetes, high blood pressure, and reproductive disorders. A study in 2007 found that >137 million people in >70 countries are most likely affected by arsenic-contaminated drinking water. Wastewater effluents from a few metallurgical industries, such as gold, silver, copper and zinc, electronics, and pharmaceuticals, may contain >1 mg/l of arsenic [2,4].

36.2 Wastewater Treatment and Adsorption

Various methods are in place for the treatment of wastewater containing inorganic contaminants, including fluoride and arsenic. The methods are based on the principles of precipitation and coagulation, chemical oxidation, sedimentation, filtration, adsorption, osmosis, ion exchange, etc. Currently, adsorption is being used extensively for the removal of inorganic pollutants from aqueous solutions. There are some strategic advantages in treating wastewater by adsorption. For example, the amount of water required for the treatment is much less compared with other technologies. Therefore, the technology is suitable in regions where water is scarce, especially in developing countries such as India. Adsorption is also relatively less energy intensive. Therefore, countries with an underdeveloped power infrastructure can afford such technology. The most important feature of adsorption is that, in general, the method does not involve a pretreatment or posttreatment step. Therefore, the operation is simple and requires fewer chemicals than other methods. However, the viability of such a technique is greatly dependent on the development of suitable adsorbents.

There are many adsorbents currently in use. Activated alumina and alumina-supported metal oxides are the most common adsorbents studied for defluoridation applications. In recent years, low-cost adsorbents, such as calcite, fluorspar, quartz, fly ash, red mud, bentonite, and the oxides of a few metals, have been investigated for their fluoride removal capabilities [5–10]. Zeolites have also gained significant interest, primarily because of their ion-exchange capacity and large surface area. The application of zeolites as adsorbents is attributed to the latter property. Furthermore, large deposits of naturally available zeolites in many countries, including Greece, the United Kingdom, Italy, Mexico, Iran, and Jordan, have rendered this material extremely cost-effective by suitable surface modifications [11].

The most widely studied adsorption media for arsenic removal include iron hydroxides and iron oxides, such as amorphous hydrous ferric oxide, ferrihydrite, and goethite; activated alumina; nanoactive alumina; silica and phyllosilicates; cellulose sponges; sand; and zeolite [12]. Studies have revealed that iron(III) has a high affinity toward inorganic arsenic species and is selective in the sorption process because arsenic forms stronger surface complexes or the complexes migrate into the iron structure. Consequently, concerted efforts have been made, especially in the last decade, to develop Fe-impregnated adsorbents to increase their adsorption capacity for arsenic [13–16]. Carbon and alumina are the two most common materials studied as a substrate for iron, primarily because of their porous structure, which provides a large active surface area.

36.3 Carbon-Based Adsorbents

Owing to their versatility and ready availability, activated carbons (ACs) have been widely used as adsorbents for wastewater treatment. The relatively superior adsorption capacity of AC is a product of its high surface area, the controlled pore structure, and the surface chemistry. The latter is determined by the presence of heteroatoms, such as oxygen, nitrogen, hydrogen, phosphorus, and sulfur. The amount and chemical forms of these heteroatoms depend on the precursor of the carbon, the history of its preparation, and the surface treatment conditions. ACs have been prepared from a variety of raw materials, including paper mill sludge [17], coffee grounds [18], particle board waste [19], rattan sawdust [20], corncobs [21], lignin [22], apricot stone shells [23], activated sludge [24], and coconut shells [25].

AC in its granular form, also known as granular AC (GAC), has been tested for defluoridation, although without much success [26,27]. The application of modified GAC to remove dissolved solutes has drawn interest [28–32]. Although modified GACs have shown considerable potential in removing organic pollutants, such as phenolic compounds, and, to some extent, the arsenic and manganese in wastewater, only partial success has been reported for defluoridation. In one such study, the adsorptive removal of fluoride has been carried out using carbon derived from zirconium ion–impregnated coconut fiber (ZICFC) [33]. In another study, defluoridation of water has been performed using aluminum (Al)-impregnated AC [34].

As mentioned earlier, the affinity of iron (Fe(III)) toward inorganic arsenic has been well established by many researchers. Consequently, several studies have been carried out, mostly in the last decade, to develop Fe-based adsorbents to increase their adsorption capacity for arsenic. Conceptually, impregnating iron/iron oxide onto AC produces an adsorbent with a large surface area and increased affinity for arsenic.

36.4 Novel Carbon Nanoadsorbents

The use of GAC or powdered AC (PAC) in water purification has been in commercial practice for a long time. In the last decade, the fabric form of carbon, namely, AC fiber (ACF), has been developed and effectively used for several adsorption applications, including air and water pollution control, because of its relatively large BET surface area [35,36]. ACFs are primarily microporous materials possessing a large surface area (~1200–1800 m²/g), which is a key attribute for an adsorbent. In ACFs, micropores, responsible for adsorption, are directly connected to the external surface with a small diffusion length because of the narrow diameter of the fiber (usually 1–10 μm). Therefore, the mass transfer diffusion resistance to the solute is negligible, and the removal rate is adsorption controlled. Figure 36.1 schematically describes the PSD in a carbon fiber. ACFs also exhibit catalytic activities attributed to different surface functional groups, such as hydroxyl, carboxylic, and quinone. Depending on the type of adsorbate (basic or acidic, anionic or cationic) to be removed, the surface of an ACF may be functionalized with suitable reagents. ACFs can also be impregnated with metal catalysts in certain applications.

With the recent success of nanotechnology in the development of several nanomaterials, nanostructured carbons, such as carbon nanotubes and carbon nanofibers (CNFs), have been developed and applied as adsorbents in several environmental remediation

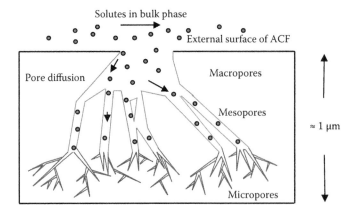

FIGURE 36.1
Schematic representation of PSD in an ACF. (Reprinted from *Activated Carbon Surfaces in Environmental Remediation*, 1st ed., T.J. Bandosz, Copyright 2006, with permission from Elsevier.)

applications. In this chapter, we describe two types of such carbon-based nanoadsorbents: a hierarchical web of ACFs/CNFs and carbon micro-/nanoparticles, which have shown tremendous potential as adsorbents for fluorides and arsenic.

A hierarchal web of carbon micro-/nanofibers is prepared using micron-sized ACFs as a substrate. CNFs are grown on ACFs by catalytic chemical vapor deposition (CVD). The use of ACFs as a substrate to grow CNFs obviates the need for removing the substrate in a postsynthesis step, and a hierarchal carbon web of microfibers and nanofibers prepared in this manner can be directly used in end applications. Figure 36.2 describes the tip-growth mechanism for growing a CNF on an ACF by CVD.

Briefly, a web of ACFs/CNFs is prepared by impregnation of ACFs with a suitable transition metal (Ni/Fe/Cu/Zn/Ag) catalyst, calcinations at high temperature, reduction by hydrogen to convert metal oxides to their metallic state, growing of CNFs by CVD, and subsequent sonication of the prepared web to dislodge the metal nanoparticles from the tips [38–42]. The final step opens up the tip and allows the solute to diffuse into the pores and to be adsorbed on the interior surface of the material. Most important, a hierarchal web prepared in this manner may be reimpregnated with another type of metal catalyst (different from the parent catalyst used for growing the CNF), which may be required for certain end applications. Thus, there are two categories of distinct conditioning when preparing a web of carbon micro-/nanofibers for different end applications. In one category, the metal catalyst, for example Fe, has two functions: (i) to grow the hierarchal structure and (ii) to remove contaminant solutes, such as As, in the wastewater [41]. In the other category, the parent catalyst, for example Ni, must be removed after growing the CNFs, and the web must be reprocessed by impregnating it with another type of metal, such as Al [40]. In the latter category, the sequence of steps for the incorporation of metals within the micro-/nanopores of the carbon web is qualitatively similar to that for the parent catalyst. However, the operating conditions are distinctively different. In either of the two categories, the hierarchal web may also be functionalized with surface functional groups for enhancing its adsorptive/catalytic efficiency. For example, basic groups, such as hydroxyls or ketones, may be incorporated to remove the solutes, which are acidic, or vice versa.

For the synthesis of carbon micro-/nanoparticles, the polymeric precursors are first synthesized as beads, using suspension polymerization, and then carbonized and activated by physical activation using steam or CO_2 to develop micropores and mesopores in the beads. Iron is incorporated into the polymeric beads in an intermediate step during

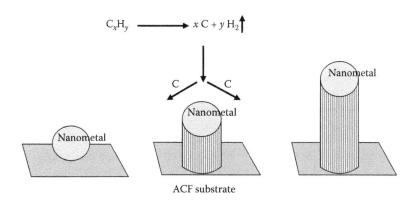

FIGURE 36.2
Growing CNF on ACF by CVD using hydrocarbon source.

polymerization [43]. In other words, AC is used as a substrate to *in situ* deposit Fe into its micropores and mesopores. Fe-doped AC beads may then be milled to produce micro-sized and nanosized particles that show increased reactivity. It has been shown that the Fe-doped ACs prepared this way are more effective at removing dissolved arsenic from water than those synthesized via the traditional route of preparing Fe-impregnated AC. Alternatively, the polymeric beads may be milled to produce the nanoparticles and then carbonized and activated to produce carbon-based nanoadsorbents.

36.5 Surface Morphology of a Hierarchical Web of ACF/ CNF and Micro-/Nanocarbon Particles

In this section, we present and discuss representative SEM images of these materials at different stages of the preparation. As observed in the image of the as-received (untreated) ACF substrate, shown in Figure 36.3a, the surface of the carbon fibers is smooth, with the diameter in the range of 10–20 μm. The SEM image in Figure 36.3b shows the homogeneous dispersion of nickel nitrate on the surface of the ACF substrate following the impregnation and before the calcination step. Figure 36.3c presents an SEM image of the ACF surface after the reduction step. Fine nanoparticles are uniformly dispersed on the surface of the fiber. No signs of any damage, including deformation and loss of integrity of the fiber, were observed in the samples. CNFs were grown uniformly and densely on the ACF substrate, as shown in Figures 36.3d and e corresponding to 400,000× and 100,000× magnifications, respectively. The diameters of most of the CNFs are approximately 30–40 nm, with a length of up to several microns. Bright, spherical nanoparticles can be observed at the tip of the CNFs, with the size approximately the same as the diameter of the CNF (see Figure 36.3d). However, in some cases, the diameters of the CNFs may not be the same as that of the Ni particles because the Ni metal particles may undergo agglomeration, sintering, and fragmentation during the growth of the CNFs. As seen in the SEM images, the nanofibers are not straight but have a crooked morphology with a three-dimensional network structure. The SEM image shown in Figure 36.3f shows the morphology of a CNF after sonication, which essentially removed most of the Ni catalysts from the CNF. The density distribution and average diameter of the CNFs remained almost unchanged after sonication. In Section 36.4, it was mentioned that in the preparations of the Al-based CNFs required for defluoridation applications, the parent catalyst, Ni, was removed after growing the CNF and the web was reprocessed by impregnating it with Al.

Figure 36.4a shows the SEM images of Al–CNF samples prepared after sonication of Ni–CNF samples and reprocessing, viz. impregnation, calcinations, and reduction. As observed, there is a distinct change in the morphology of CNFs after loading with Al. The fibers were observed to be interwoven because of the agglomeration. The energy-dispersive x-ray (EDX) spectra shown in Figure 36.4b revealed the presence of 5%–7% (w/w) of Al with 0.4%–1.2% (w/w) of Ni.

Figure 36.5 shows SEM images of Fe–CNF adsorbents. These adsorbents were synthesized in a one-step method for the removal of As from wastewater, in which Fe had dual roles: (i) to grow the CNFs and (ii) to remove As. Figure 36.5a shows the uniform and well-dispersed growth of the CNFs on the ACF substrate. The shining iron particles on the tip of the nanofiber are evidence of the tip growth of the CNF. The diameter of the nanofiber

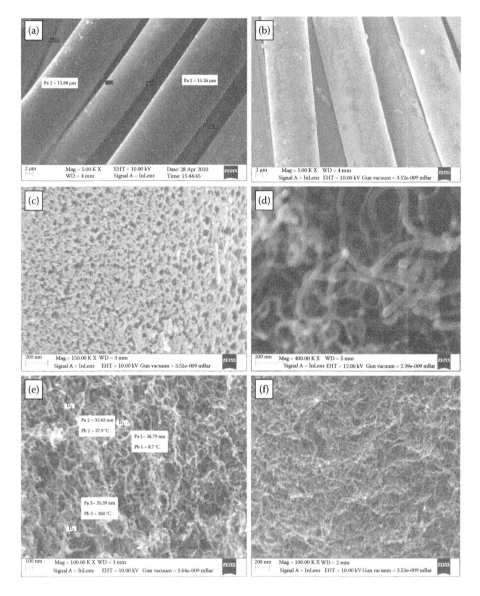

FIGURE 36.3
SEM images of ACF (a) as-received, (b) impregnated with nickel nitrate, (c) post Ni reduction, (d and e) CNF grown on ACF, and (f) after ultrasonication. (From A. Gupta et al., *I&EC Res.*, 48, 9697, 2009.)

is between 10 and 20 nm. After the growth of the CNFs, the samples are sonicated. The purpose of sonication is to remove the iron particles from the tips of the grown CNFs so that the interior surface is available for adsorption. Figure 36.5b shows an SEM image of the CNFs after sonication. As shown, there are fewer iron particles at the tips of the fibers. The surface morphology of the CNFs is altered after adsorption. Figure 36.5c presents an SEM image of the adsorbent (CNFs) after treating arsenic-laden wastewater. The EDX analysis shown in Figure 36.5d confirmed the presence of arsenic and Fe on the prepared adsorbents.

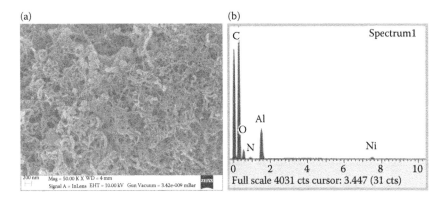

FIGURE 36.4
(a) SEM image of post-Al-reduction sample and (b) EDX spectra of image in (a). (From A. Gupta et al., *I&EC Res.*, 48, 9697, 2009.)

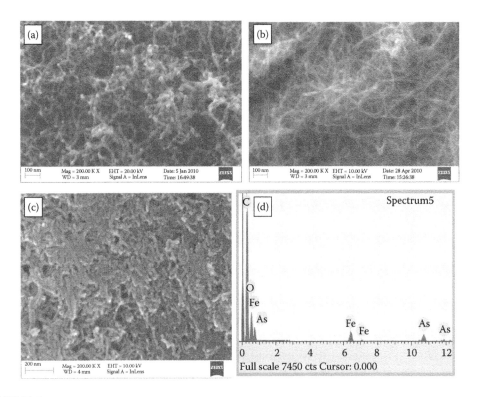

FIGURE 36.5
SEM images of Fe–CNF (a) before and (b) after sonication and (c) postadsorption with As(V), and (d) EDX spectra of the image in (c). (From A. Gupta et al., *I&EC Res.*, 49, 7074, 2010.)

We will now discuss the SEM images of the other adsorbent, namely, micro-/nano-carbon beads. Figure 36.6 shows representative images and the EDX spectra of different specimens of Fe-doped polymeric beads and carbonized/activated Fe beads (before and after adsorption with arsenic). Figure 36.6a shows that the Fe-doped beads were approximately spherical (average diameter ~0.6 mm), with a smooth surface. The SEM image

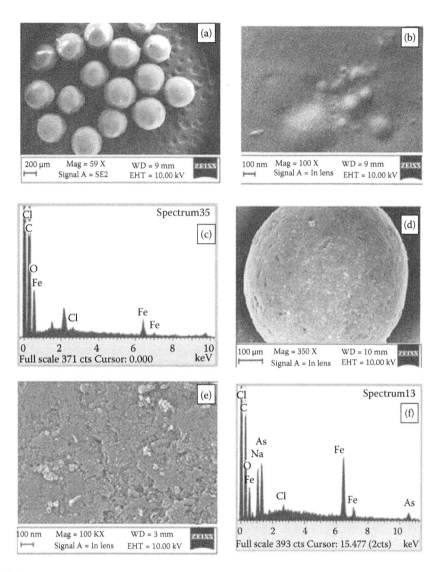

FIGURE 36.6
(a) SEM image of Fe-doped phenolic spherical beads; (b) SEM image of the surface of polymeric beads dispersed with heterogeneous phase; (c) EDX spectra of Fe/As on the surface; (d) SEM image of Fe-doped activated beads; (e) SEM image of the surface of arsenic-treated Fe-doped activated beads; and (f) EDX spectra of Fe/As on the surface. (From A.K. Sharma et al., *Chem. Eng. Sci.*, 65, 3591, 2010.)

(Figure 36.6b) at large magnification shows the absence of pores on the external surface of the beads. The SEM image also shows the dispersion of a heterogeneous phase on the external surface of beads. Porosity develops in the beads after carbonization and subsequent activation. Figure 36.6c is the representative EDX spectrum. The concentrations of Fe obtained from the spectra ranged between 0.6% and 2.5% (w/w). Figure 36.6d shows an SEM image of the activated beads. A distinct change in the surface morphology because of the activation is evident from the images. The surface morphology also changed following adsorption with arsenic, as observed in the corresponding SEM image (Figure 36.6e) of the

FIGURE 36.7
SEM images of activated Fe-doped carbon nanoparticles (a) before and (b) after As adsorption. (From A.K. Sharma et al., *Chem. Eng. Sci.*, 65, 3591, 2010.)

activated specimen. Figure 36.6f is the representative EDX spectra after arsenic adsorption. The EDX spectra confirmed the presence of Fe and As in the specimen.

The SEM image shown in Figure 36.7a corresponds to the specimen of the carbon nanoparticles (~100 nm) produced after the milling of the phenolic spherical beads, followed by carbonization and activation. The surface of the particles is rough and porous. Figure 36.7b is an SEM image of the sample adsorbed with arsenic.

36.6 Surface Area and PSD

In general, adsorbents should have relatively large internal surface areas. The solutes adsorb onto the surface by physisorption and/or chemisorption. The adsorbents with a large surface area provide a large number of active sites to which the solute can adsorb. Such materials also act as an efficient support to the metal catalysts required for certain end applications. Furthermore, adsorbents should have the required PSD of micropores, mesopores, and macropores, depending on the size of the target adsorbate molecules.

The total surface area of the above-discussed carbon-based nanoadsorbents was determined from the multipoint adsorption/desorption isotherms. The isotherms were measured from the pore volume of the adsorbents using nitrogen as the adsorbate molecule at 77 K. The well-known BET equation was applied in its linearized form in the region of the relative saturation pressure range between 0.05 and 0.35 to calculate the specific surface area. The BET plot is usually linear in this range. Moreover, most of the micropores are filled in this range. The total pore volume was measured from the amount of N_2 adsorbed at the relative pressure close to unity (0.9994). The PSD was determined by the Barret–Joyner–Halenda method.

The surface characteristics data from the ACF, the ACF containing Ni in its reduced state, and ACF/CNF samples are reported in Table 36.1. The ACF substrate exhibited a very high surface area, indicating high porosity. After impregnation with the metal salt, the surface area decreased as the salt blocked some of the pores. Interestingly, the CNF samples

TABLE 36.1

Surface Characterization of Different Samples

Sample	BET (m²/g)	Micropore Volume (cm³/g)	Mesopore Volume (cm³/g)	Macropore Volume (cm³/g)
ACF	1375	0.620	0.052	0.037
Metal–ACF	1260	0.466	0.231	0.021
CNF	970	0.356	0.163	0.019

Source: B. Mekala et al., *I&EC Res.*, 50, 13092, 2011.

TABLE 36.2

Surface Characterization Data of Adsorbents

Sample	Average Size	BET Surface Area (m²/g)	Total Pore Volume	Mesopore Volume (cm³/g)	Micropore Volume (cm³/g)
PBFe	0.6 mm	0.14	0.001	0	0
PBFe–Act	0.6 mm	231	0.142	0.0102	0.128
PBFe–Act–BM	100 nm	288	0.198	0.0146	0.172
PBFe–BM–Act	100 nm	781	0.468	0.0318	0.426

Source: A.K. Sharma et al., *Chem. Eng. Sci.*, 65, 3591, 2010.
Note: PBFe represents the phenolic beads, PBFe–Act represents the activated beads, PBFe–Act–BM represents nanoparticles prepared by first carbonization/activation then ball milling, and PBFe–BM–Act represents the nanoparticles prepared by ball milling first and then activation.

showed less BET area than the ACF samples. In principle, the nanofibers must have a large surface area because the N_2 used as the probe molecule for the BET measurement is unable to penetrate the nanopores of nanofibers. It has been suggested that CO_2 should be used to measure the surface area of nanopores [44].

Table 36.2 summarizes the BET surface area, total pore volume, and PSD of the carbon-based micro-/nanoadsorbents. The data show that the porosity of the material depends on the sequence of activation and milling of the polymeric beads. Milling before activation is found to be more effective at creating a large BET area than activation followed by the milling of the beads. The BET area of the nanoparticles was ~780 m²/g in the former route to synthesis, more than double of that (~300 m²/g) obtained in the latter route. The pore volume was increased from ~0.1 to 0.4 cm³/g under the identical conditions. Table 36.2 also describes the percentage of the volume represented by the different pore sizes (macro-, meso-, and micro-) obtained in the samples. As shown, the pores in the activated samples are mostly in the micropore range, contributing 90% of the total pore volume.

36.7 Nongraphitic ACF

ACF and CNF are nongraphitic and amorphous materials. Raman analysis was performed to characterize the ACF and CNF studied herein at various stages of preparation in order to verify that the materials are amorphous or graphitic. The laser Raman spectra of the samples were taken using a confocal Raman instrument (Alpha model; Witec, Germany). The data were collected using an Ar ion laser with a 514-nm wavelength as the excitation source and a charged coupled device as the detector in the range of 800–2000 cm⁻¹ at room temperature in air. Figure 36.8 describes the Raman intensities of ACF, Ni–ACF, and CNF grown on ACF. Table 36.3 presents the corresponding data extracted from the Raman graphs.

Typically, there are two peaks observed for carbon in the Raman spectra. The Raman peak in the range of 1250–1450 cm⁻¹ corresponds to the D-band and is attributed to the disordered components of carbon, whereas the peak in the range of 1550–1600 cm⁻¹ corresponds to the G-band and is attributed to the ordered component (graphite). The I_D/I_G ratio reflects the extent of disorder in the carbons. In addition, the smaller the ratio, the more highly ordered the graphitic structure. The I_D/I_G ratio of the ACF sample was determined

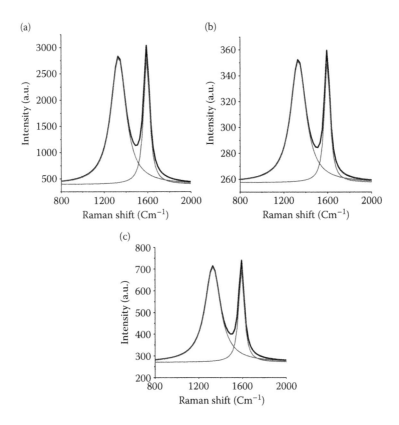

FIGURE 36.8
Raman spectra of (a) ACF, (b) Ni–ACF, and (c) CNF grown on ACF. (From A. Gupta et al., *I&EC Res.*, 48, 9697, 2009.)

TABLE 36.3

Laser Raman Spectra Parameters

		Peak Shift (cm⁻¹)		Area		
S. No.	Sample	D-Band	G-Band	D-Band	G-Band	I_D/I_G
1	ACF	1331	1591	626,583	245,667	2.55
2	Ni–ACF	1333	1596	24,826	8833	2.81
3	CNF	1328	1591	117,779	40,635	2.89

Source: A. Gupta et al., *I&EC Res.*, 48, 9697, 2009.

to be 2.55, indicating the presence of a relatively large amount of disordered graphite components in ACF. This disorder increases during nickel impregnation. As also observed, CNF grown on ACF retains short-range graphitic characteristics with an I_D/I_G ratio of 2.89.

36.8 ACF/CNF as Adsorbents

The adsorption capacity of an adsorbent may be tested under batch and/or flow conditions. In a typical batch experiment, a fixed amount of adsorbent is mixed with a fixed volume of water containing an initially predetermined concentration of the solute and is allowed to come to equilibrium at a constant temperature. The concentration of the solute in the adsorbent phase is theoretically determined by the mass balance, $q = \dfrac{V(C_0 - C)}{w}$, where C_0 and C are the aqueous-phase solute concentrations before and after equilibrium is attained. V is the volume of the solution in contact with the adsorbent of weight w. Figure 36.9 presents the equilibrium concentrations of the fluoride ions in the solid phase as a function of the aqueous-phase concentration for five types of samples: the ACF substrate, Al–ACF (ACF impregnated with Al), CNF, and Al–CNF with and without sonication (acid treatment). The lines regressed through the data are essentially the equilibrium isotherms of fluoride ions vs. Al-impregnated carbon fibers at 35°C. As observed, the solid-phase concentrations are between 0.10 and 15 mg/g, corresponding to the liquid-phase concentrations between 0.1 and 50 mg/l. For each sample, the data are best explained by the Freundlich isotherm, $q = K \times C^n$, where q is the amount of adsorbed fluoride ions in milligrams per gram of ACF, K is the Freundlich constant, and n is the power of isotherm. The values of the constant n less than unity indicate a favorable adsorption for the systems reported here.

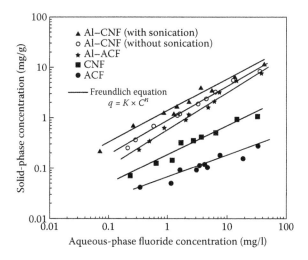

FIGURE 36.9
Comparative performance of Al–CNF (with and without sonication), Al–ACF, CNF, and ACF for fluoride removal. Solid lines show Freundlich isotherms, $q = K \times C^n$; where $K = 1.30, 0.85, 0.59, 0.17, 0.06$ and $n = 0.61, 0.69, 0.75, 0.59, 0.41$ for Al–CNF (with and without sonication), Al–ACF, CNF, and ACF, respectively. (From A. Gupta et al., *I&EC Res.*, 48, 9697, 2009.)

From the data shown in Figure 36.9, it may be concluded that the equilibrium loading of fluoride is larger on Al–CNF than on Al–ACF. The equilibrium loading of fluoride is also larger (approximately 10×) on Al–CNF than CNF (the sonicated samples without metals). In addition, sonication enhanced the equilibrium capacity of the prepared nanofibers because of the dislodging of Ni catalysts from the tips and other surfaces of the grown CNF, thereby creating additional sites for the incorporation of Al in the subsequent impregnation step.

It is important to compare the loading (mg/g) of fluoride ions on Al–CNF to those reported in the literature for fluoride ions on different adsorbents such as AC granules, activated alumina, and natural clays. From Figure 36.9, it may be noted that the loading of fluoride ions on Al–CNF is 0.25–17 mg/g corresponding to aqueous-phase fluoride concentrations between 0.06 and 50 ppm. These values compare well with 0.58 mg/g on activated alumina corresponding to 1 ppm of fluoride in water, 1–5 mg/g on the ZICFC corresponding to 20–100 ppm of fluoride in water, and 2 mg/g on clays and 2.7 mg/g on activated titanium-rich bauxite, both corresponding to 10 ppm of fluoride ions in the solution. In another study, magnetic chitosan particles saw loading of fluoride ions at 16.5 mg/g, corresponding to an aqueous-phase concentration of 100 ppm [40]. From the data presented in this section, it may be concluded that surface loading (equilibrium concentration) of fluoride ions on Al–CNF is larger than those reported in the literature, either on carbon or alumina.

A *breakthrough curve* is essentially the time history of the effluent concentration from a packed bed subject to a step change in the inlet concentration, which is influenced by the transport processes in the voids between the fibers and within the pores of the fibers. In general, an adsorbate of relatively small size and large BET area exhibits a breakthrough curve suppressed for a longer duration. Consequently, such a breakthrough characteristic is reflective of a good adsorbate used under dynamic conditions. In such a case, the inter-particle and intraparticle diffusion resistances are usually small, and the process is likely to be controlled by the adsorption/desorption kinetic rate. On the other hand, an immediate breakthrough in the packed bed means large interparticle and/or intraparticle diffusion resistance, and consequently small uptake of the solute by the adsorbent.

The total uptake of the solute during the adsorption time is determined from the breakthrough curve (i.e., hatched area marked in Figure 36.10 for the uppermost curve) by calculating the integral along the curve between the two time limits: the incipience of the adsorption test and the instance when the bed is saturated. Mathematically, the uptake of the solute (mg) is determined as follows:

$$\text{Uptake (mg)} = Q\left(C_{in}T - \int_0^T C_{exit}\, dt \right) \tag{36.1}$$

where T is the total time of adsorption until the bed is saturated with the solute in the influent stream. It may be mentioned that the specific uptake (mg/g of the adsorbent) must be approximately equal to the equilibrium loading of the solute obtained under the batch conditions, indicating the complete utilization of the adsorbent under flow conditions. In the typical experiments performed to study breakthrough characteristics using the perforated tubular reactor wrapped with fresh (unsaturated) carbon fibers, the aqueous solution of fluoride at a predetermined concentration is fed at a constant flow rate to the reactor and the concentrations at the outlet of the tube are measured. Figure 36.10 describes the breakthrough curves obtained for three different inlet concentration levels of fluoride in the aqueous solution. It is important to mention that the fluoride uptakes calculated from the respective breakthrough curves shown in the figure matched approximately with

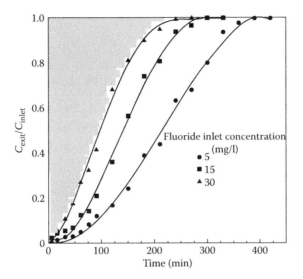

FIGURE 36.10
Breakthrough data for varying inlet fluoride concentration (W_{Al-CNF} = 3 g, Q = 0.01 l/min, T = 303 K). (From A. Gupta et al., *I&EC Res.*, 48, 9697, 2009.)

the equilibrium ion loading shown in Figure 36.9 for all three concentrations, within the experimental and calculation errors, which suggests that the Al–CNF-based adsorbents are suitable for developing water purifiers in defluoridation applications.

Fe–CNFs have also exhibited significant adsorption capacity for As(V). Figure 36.11 describes the adsorption isotherms for As(V) on different ACF/CNF-based adsorbents. From the isotherms shown in the figure, the equilibrium loading of arsenic may be observed to be much higher on Fe–CNF (micro-/nanoweb) compared with the other

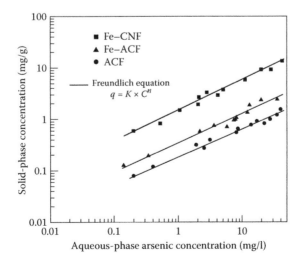

FIGURE 36.11
Comparative performance of Fe–CNF, Fe–ACF, and ACF for arsenic removal. Solid lines show Freundlich isotherms, $q = K \times C^n$, where K = 1.5, 0.36, 0.2, and n = 0.59, 0.56, 0.52 for Fe–CNF, Fe–ACF, and ACF, respectively. (From A. Gupta et al., *I&EC Res.*, 49, 7074, 2010.)

adsorbents, Fe–ACF and ACF. The loading of arsenic on Fe–CNF is approximately 10× and 8× larger than ACF and Fe–ACF, respectively.

Generally, carbon–iron-based adsorbents have exhibited large arsenic loading capacities compared with other adsorbents. From the equilibrium isotherms presented in Figure 36.11, one may note that the loading of arsenic ions on Fe–CNF is 0.6–14 mg/g, corresponding to aqueous-phase arsenic concentrations between 0.2 and 40 ppm at pH 6.5. These values are comparable to 10 mg/g on $FePO_4$ (amorphous and crystalline) corresponding to 0.5–100 mg/l of arsenic at pH 6.0–6.7, 4.5 mg/g on the GAC-based iron containing adsorbent at pH 5.0, 6.5 mg/g obtained for the cationic surfactant–modified PAC corresponding to 10 ppm of arsenic in water, and 1–4 mg/g on laterite and iron-modified AC corresponding to 0–4 ppm of arsenic in water at pH 8.18. In most of the other studies reported in the literature, arsenic loading is lower than that reported here. Some examples include arsenic loading of 2.4 mg/g obtained for coconut shell carbon, corresponding to 0–200 mg/l As concentration at pH 5; 0.02 mg/g loading obtained on Fe^{3+}-impregnated carbon at 3.2 ppm arsenic concentration and pH 7.0; 1.92 mg/g loading for GAC–Fe–H_2O_2 at 0–30 mg/l and pH 4.2; and 0.036 mg/g loading for AC impregnated with Fe^{2+} and Fe^{3+} at 0.3 ppm and pH 6.5–8.5. There are very few studies with higher arsenic loading, such as 51.3 mg/g at pH 6.0 and 43.6 mg/g at pH 8.0 for iron-modified AC, corresponding to 20 mg/l of aqueous-phase concentration; 37.46 mg/g at pH 5.0 for FeO/AC, corresponding to 37.46 mg/l of arsenic in water; and 26 mg/g at pH 7.6–8.0 for hydrous FeO–GAC, corresponding to 0.3 ppm of arsenic in water [41]. In general, variation in the arsenic loading on the adsorbents is because of different types of carbon, surface functional groups, textural properties, the amount of iron loaded, and operating conditions, including pH.

Similar to the case of the adsorbents developed for fluorides, Fe–CNF was also tested under flow conditions to ascertain its arsenic uptake (mg/g) compared with the equilibrium loadings under batch conditions, and determine its potential applications as a material for water purifiers. Figure 36.12 describes the breakthrough data for different inlet

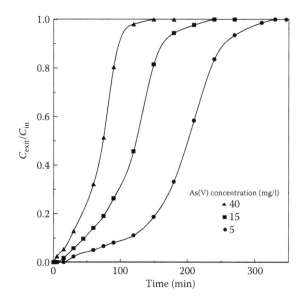

FIGURE 36.12
Breakthrough data for varying inlet As(V) concentration ($W_{Fe–CNF} = 3$ g, $Q = 0.01$ l/min, pH 6.5, $T = 303$ K). (From A. Gupta et al., *I&EC Res.*, 49, 7074, 2010.)

arsenic concentrations under constant solution flow rate and amount of adsorbent. For each case, the solute uptake determined by the area above the breakthrough curve was found to be within 10% of corresponding equilibrium loading, suggesting the almost complete utilization of the adsorbent under flow conditions.

36.9 Carbon Micro-/Nanoparticles as Adsorbents for Arsenic

It was mentioned in Section 36.4 that the novelty of such carbon micro-/nanoparticles is the *in situ* incorporation of Fe in their polymeric matrix during the polymerization stage. Furthermore, the carbon beads (~0.8 mm) produced after carbonization and activation of the micron-sized beads contained uniform and high loading (~2.5% w/w) of Fe particles. The carbon nanoparticles (~100 nm) produced after ball milling of the polymeric beads, followed by carbonization and activation, retained the Fe–nanoparticles and were found to have even higher adsorption capacity than the micron-sized particles. Another salient feature of these adsorbents is that the maximum removal of As(III) or As(V) occurs at pH between 6.5 and 7.5. Moreover, the variation in pH of the solution increased marginally (not more than 20%) during adsorption. Considering that pH of ground or potable water is commonly in the vicinity of 7 (usually between 6 and 7.5), the maximum adsorption of arsenic at pH ~7 and marginal variation in pH observed during adsorption suggest that pretreatment or posttreatment of water may not be required using these adsorbents.

Figure 36.13 presents the equilibrium concentration of arsenic ions (As(III) and As(V)) in the solid phase as a function of the aqueous-phase concentration for four types of samples: Fe-doped phenolic beads (PBFe), carbonized/activated beads (PBFe–Act), and Fe-incorporated carbon nanoparticles produced before and after the carbonization step (PBFe–Act–BM and PBFe–BM–Act, respectively). The lines regressed through the data are essentially the equilibrium isotherms at 30°C over the aqueous-phase concentration range of 0.1–10 mg/l or ppm. The saturation capacity of PBFe–BM–Act for As(III) is approximately 13 mg/g, while for As(V) it is 5 mg/g. Revisiting Section 36.8, these values are considerably

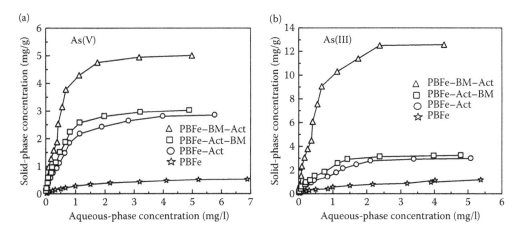

FIGURE 36.13
Adsorption equilibrium isotherms for various Fe-impregnated adsorbents ($T = 303$ K, pH 6.5 for As(V) (a) and pH 7 for As(III) (b)). (From A.K. Sharma et al., *Chem. Eng. Sci.*, 65, 3591, 2010.)

higher than the reported literature data in most cases and comparable in a few cases, for both As(III) and As(V). As observed in Figure 36.13, adsorption of arsenic ions by PBFe–BM–Act is significantly larger than that by the other adsorbents. Additionally, the adsorption of PBFe–Act–BM is shown to be approximately 10%–15% higher than that of PBFe–Act, implying the milling of activated beads results in a small increase of the adsorption capacity. These results are also consistent with the BET surface area and pore volumes of the various samples presented in Table 36.2.

Similar to the column studies carried out for the nanofiber-based adsorbents, the performance of the adsorbents was evaluated by examining the breakthrough response. Figure 36.14 describes the breakthrough curves obtained for three inlet concentrations of As(III) and As(V) in the column packed with PBFe–Act beads. The breakthrough time may be defined as the time at which the concentration of the effluent stream increases to 1% of the inlet concentration. For clarity, the enlarged section of the figure corresponding to the breakthrough times for three concentrations (1, 2, and 3 ppm of arsenic) is now produced as

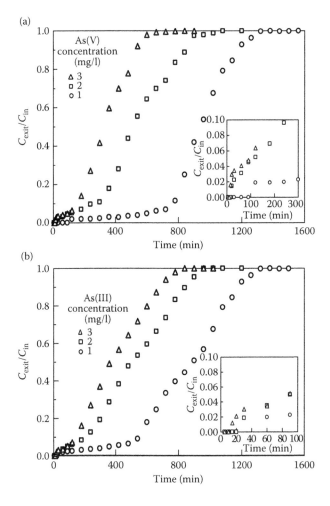

FIGURE 36.14
Breakthrough data for varying inlet arsenic concentration (amount of PBFe–Act = 1 g, Q = 0.003 l/min, pH 7 for As(III) (b) and pH 6.5 for As(V) (a), T = 303 K). (From A.K. Sharma et al., *Chem. Eng. Sci.*, 65, 3591, 2010.)

an inset in Figure 36.14. As seen in the figure, the adsorption times are approximately the same for As(III) and As(V) at the same inlet concentrations. Furthermore, the solute uptakes calculated from the breakthrough curves are found to be consistent with the batch equilibrium data, reconfirming the potential application of such materials for water purifiers.

References

1. Available at http://en.wikipedia.org/wiki/Water#Health_and_pollution.
2. A. Mukherjee, M.K. Sengupta, M.A. Hossain, S. Ahamed, B. Das, B. Nayak, D. Lodh, M.M. Rahman, D. Chakraborti. Arsenic contamination in groundwater: A global perspective with emphasis on the Asian scenario. *J. Health Popul. Nutr.*, 24(2):142–163, 2006.
3. WHO. Guidelines for drinking-water quality first addendum to third edition. Available at http://www.who.int/water_sanitation_health/dwq/gdwq0506.pdf (accessed March 15, 2008).
4. P.L. Smedley, D.G. Kinniburgh. Sources and behaviour of arsenic in natural water. In *United Nations Synthesis Report on Arsenic in Drinking Water*, Panamerican Health Organization, WA, 2005.
5. V. Gopal, K.P. Elango. Equilibrium, kinetic and thermodynamics studies of adsorption of fluoride into plaster of Paris. *J. Hazard. Mater.*, 141(1):98–105, 2007.
6. A.K. Chaturvedi, K.P. Yadava, K.C. Pathak, V.N. Singh. Defluoridation of water by adsorption on fly ash. *Water, Air Soil Pollut.*, 49:51–61, 1990.
7. Y. Cengeloglu, E. Kir, M. Ersoz. Removal of fluoride from aqueous solution by using red mud. *Sep. Purif. Technol.*, 28:81–86, 2002.
8. M. Srimurali, A. Pragathi, J. Karthikeyan. A study on removal of fluoride from drinking water by adsorption onto low-cost materials. *Environ. Pollut.*, 99:285–289, 1998.
9. S.M. Maliyekkal, A.K. Sharma, L. Philip. Manganese-oxide-coated alumina: A promising sorbent for defluoridation of water. *Water Res.*, 40(19):3497–3506, 2006.
10. N. Das, P. Pattanaik, R. Das. Defluoridation of drinking water using activated titanium rich bauxite. *J. Colloid Interface Sci.*, 292(1):1–10, 2005.
11. A. Măicăneanu, C. Indolean, M. Burcă, M. Stanca, H. Bedelean, M. Cornelia. Organics removal from aqueous solutions using suspended and immobilized Romanian bentonites. *Stud. Univ. Babes-Bolyai, Seria Chem.*, 56(1):81–93, 2011.
12. D. Mohan, C.U. Pittman Jr. Arsenic removal from water/wastewater using adsorbents—A critical review. *J. Hazard. Mater.*, 142:1–53, 2007.
13. J.A. Munoz, A. Gonzalo, M. Valiente. Arsenic adsorption by Fe(III)-loaded open-celled cellulose sponge. Thermodynamic and selectivity aspects. *Environ. Sci. Technol.*, 36(15):3405–3411, 2002.
14. E.A. Deliyanni, D.N. Bakoyannakis, A.I. Zouboulis, K.A. Matis. Sorption of As(V) ions by akaganeite-type nanocrystals. *Chemosphere*, 50:155–163, 2003.
15. W. Chen, R. Parette, J. Zoua, F.S. Cannon, B.A. Dempsey. Arsenic removal by iron-modified activated carbon. *Water Res.*, 41(9):1851–1858, 2007.
16. G. Muñiz, V. Fierro, A. Celzard, G. Furdin, G. Gonzalez-Sánchez, M.L. Ballinas. Synthesis, characterization and performance in arsenic removal of iron-doped activated carbons prepared by impregnation with Fe(III) and Fe(II). *J. Hazard. Mater.*, 165:893–899, 2009.
17. N. Khalili, J. Vyas, W. Weangkaew, S. Westfall, S. Parulekar, R. Sherwood. Synthesis and characterization of activated carbon and bioactive adsorbent produced from paper mill sludge. *Sep. Purif. Technol.*, 26:295–304, 2002.
18. A. Namane, A. Mekarzia, K. Benrachedi, N. Belhaneche, A. Hellal. Determination of the adsorption capacity of activated carbon made from coffee grounds by chemical activation with $ZnCl_2$ and H_3PO_4. *J. Hazard. Mater.*, 119(1–3):189–194, 2005.

19. P. Girods, A. Dufour, V. Fierro, Y. Rogaume, C. Rogaume, A. Zoulalian, A. Celzar. Activated carbons prepared from wood particleboard wastes: Characterization and phenol adsorption capacities. *J. Hazard. Mater.*, 166(1):491–501, 2009.

20. B. Hamed, A. Rahman. Removal of phenol from aqueous solutions by adsorption onto activated carbon prepared from biomass material. *J. Hazard. Mater.*, 160:576–581, 2008.

21. A. Nasser, S. Hendawy, B. Girgis. Adsorption characteristics of activated carbons obtained from corncobs. *Colloids Surf. A: Physicochem. Eng. Asp.*, 180:209–221, 2001.

22. E. Serrano, T. Cordero, R. Mirasol, L. Cotoruelo. Removal of water pollutants with activated carbons prepared from H3PO4 activation of lignin from kraft black liquors. *Water Res.*, 38:3043–3050, 2004.

23. A. Daifullah, B. Girgis. Removal of some substituted phenols by activated carbon obtained from agricultural waste. *Water Res.*, 32:1169–1177, 1998.

24. M. Martin, E. Serra, A. Ros, M. Balaguer, M. Rigola. Carbonaceous adsorbents from sewage sludge, and their application in a combined activated sludge-powdered activated carbon (AS-PAC) treatment. *Carbon*, 42:1383–1388, 2004.

25. M. Radhika, K. Palanivelu. Adsorptive removal of chloro phenols from aqueous solution by low cost adsorbent-kinetics and isotherm analysis. *J. Hazard. Mater.*, 138:116–124, 2006.

26. V.K. Gupta, I. Ali, V.K. Saini. Defluoridation of wastewaters using waste carbon slurry. *Water Res.*, 41:3307–3312, 2007.

27. X. Fan, D.J. Parker, M.D. Smith. Adsorption kinetics of fluoride on low cost materials. *Water Res.*, 37(20):4929–4937, 2003.

28. V. Srivastava, I.D. Mall, I.M. Mishra. Adsorption of toxic metal ions onto activated carbon study of sorption behaviour through characterization and kinetics. *Chem. Eng. Process.*, 47(8):1269–1280, 2008.

29. W. Cheng, S.A. Dastgheib, T. Karanfil. Adsorption of dissolved natural organic matter by modified activated carbons. *Water Res.*, 39(11):2281–2290, 2005.

30. H.H. Huang, M.C. Lu, J.N. Chen, C.T. Lee. Catalytic decomposition of hydrogen peroxide and 4-chlorophenol in the presence of modified activated carbons. *Chemosphere*, 51(9):935–943, 2003.

31. M. Barkat, D. Nibou, S. Chegrouche, A. Mellah. Kinetics and thermodynamics studies of chromium(VI) ions adsorption onto activated carbon from aqueous solutions. *Chem. Eng. Process.: Process Intensif.*, 48(1):38–47, 2009.

32. P. Mondal, C.B. Majumder, B. Mohanty. Effects of adsorbent dose, its particle size and initial arsenic concentration on the removal of arsenic, iron and manganese from simulated ground water by Fe3+ impregnated activated carbon. *J. Hazard. Mater.*, 150(3):695–702, 2008.

33. R.S. Sathish, N.S.R. Raju, G.S. Raju, G.N. Rao, K.A. Kumar, C. Janardhana. Equilibrium and kinetic studies for fluoride adsorption from water on zirconium impregnated coconut shell carbon. *Sep. Sci. Technol.*, 42(4):769–788, 2007.

34. R.L. Ramos, J. Ovalle-Turrubiarters, M.A. Sanchez-Castillo. Adsorption of fluoride from aqueous solution on aluminum-impregnated carbon. *Carbon*, 37(4):609–617, 1999.

35. S.M. Manocha. Porous carbons. *Sadhana*, 28(1–2):335–348, 2003.

36. C.L. Mangun, M.A. Daley, R.D. Braatz, J. Economy. Effect of pore size on adsorption of hydrocarbons in phenolic-based activated carbin fibres. *Carbon*, 36(1–2):123–131, 1998.

37. T.J. Bandosz. *Activated Carbon Surfaces in Environmental Remediation*, 1st Edition. Elsevier Ltd., NY, 2006.

38. N. Verma, A. Sharma. Process for synthesis of sonicated hierarchal web of carbon micronano-fiber and applications thereof. Indian Patent Application No. 1157/DEL/2009, 2009.

39. R. Singhal, A. Sharma, N. Verma. Micro/nano hierarchal web of activated carbon fibers for catalytic gas adsorption and reaction. *I&EC Res.*, 47(10):3700–3707, 2008.

40. A. Gupta, D. Deva, A. Sharma, N. Verma. Adsorptive removal of fluoride by micro-nano hierarchal web of activated carbon fibers. *I&EC Res.*, 48(21):9697–9707, 2009.

41. A. Gupta, D. Deva, A. Sharma, N. Verma. Fe-grown carbon nanofibers for the removal of As (V) in wastewater. *I&EC Res.*, 49(15):7074–7084, 2010.

42. A. Chakraborty, D. Deva, A. Sharma, N. Verma. Adsorbents based on carbon microfibers and carbon nanofibers for the removal of phenol and lead from water. *J. Colloid Interface Sci.*, 359(1):228–239, 2011.

43. A.K. Sharma, N. Verma, A. Sharma, D. Deva, N. Sankararamakrishnan. Iron doped phenolic resin based activated carbon micro and nanoparticles by milling: Synthesis, characterization and application in arsenic removal. *Chem. Eng. Sci.*, 65(11):3591–3601, 2010.

44. D. Lozano-Castello, D. Cazorla-Amoros, A. Linares-Solano. Usefulness of CO_2 adsorption at 273 K for the characterization of porous carbons. *Carbon*, 42:1231–1236, 2004.

45. B. Mekala, S. Mandal, G.N. Mathur, A. Sharma, N. Verma. Modification of activated carbon fiber by metal dispersion and surface functionalization for the removal of 2-chloroethanol. *I&EC Res.*, 50(23):13092–13104, 2011.

37

Engineered Nanomaterials for Landfill Leachate Treatment in the Humid Tropics: The Sri Lankan Perspective

Meththika Vithanage,[1] **S.S.R.M.D.H.R. Wijesekara,**[1] **I.P.L. Jayarathna,**[1] **Anuradha Prakash,**[2] **Seema Sharma,**[2] **and Ashok Kumar Ghosh**[2]

[1]*Chemical and Environmental Systems Modeling Research Group, Institute of Fundamental Studies, Kandy, Sri Lanka*
[2]*Department of Environmental and Water Management, Anugrah Narayan College, Patna, India*

CONTENTS

37.1 Introduction

37.1.1 Municipal Solid Waste Dumping

Municipal solid waste (MSW) management is a major environmental problem in many countries worldwide [1]. Globally, hundreds of megatons of waste is generated each year (i.e., MT/year $>10^8$) and MSW is a dominant stream that contributes a significant fraction to the total content of waste [2]. Most countries worldwide practice different waste management methods; several are with successful stories, and some are poorly designed or poorly managed with demerits [3]. Developed countries use advanced MSW management methods such as waste incineration and bioelectricity for generation of electricity instead of conventional waste management methods [4,5]. The management of waste has become a critical concern for most of the developing countries because of the absence of appropriate waste management methods [6]. However, only a few local bodies and some institutes in developing countries are practicing composting, anaerobic digestion, and valuable materials recovery using MSW as solutions instead of open dumping and burning of waste [7,8]. On the other hand, population increments with rapid urbanization, industrialization, and modernization as way of life are generating more waste content that is complex in nature. Mainly, MSW is the major fraction for the total waste content that comprises biodegradable material, plastics, metals, construction and demolition waste, electronic and hazardous waste, etc. However, it is difficult to categorize the different types of waste because of their complex characteristics and disposal procedures.

Most of the Asian countries practice open dumping and nonengineered disposal of waste since it is easy and cheap [1]. Sri Lanka's commercial city, Colombo, in the western province itself has about 60 locations as open dumps that are currently being maintained by the local authorities, well an example of this situation [9]. Furthermore, recent studies have revealed that MSW of about 3700 tons is being generated each day in Sri Lanka [10].

One of the most critical aspects associated with open dump sites is the formation of leachate, which is well understood as a typical contaminant containing a large amount of pollutants [11]. The higher biodegradable fraction or organic content that is typically disposed in large content in Sri Lankan dump sites may raise many issues due to the formation and direct discharge of highly polluted leachate into the environment [12]. Frequently, this complex leachate is released to the nearest water body without any treatment mechanisms. This can lead to much adverse impact on the local soil and water sources [13]. Unfortunately, most of the open dump sites are located close to the water bodies, which even magnify the adverse impacts with an accelerated rate. In addition, the landfill leachate is shown to be highly toxic to higher plants, algae, invertebrates, and fish [14]. Therefore, economically viable, environmentally friendly, and socially acceptable solutions are needed and commissioned for the responsible authorities immediately to overcome landfill leachate–related issues to ensure environmental sustainability.

37.1.2 Characteristics of Landfill Leachate

Many countries have identified landfill leachate as a typical contaminant for surface water, groundwater, and soil fertility due to its toxicity [15]. High concentrations of Cd(II), Hg(II), Ni(II), Mn(II), Cu(II), Zn(II), and Pb(II) have been reported associated with leachate, and these metals have been identified as enhancing their transportation with dissolved organic carbon derivatives by anaerobic degradation of organic compounds present in the leachate such as humic, fulvic, and hydrophilic compounds [11,16,17]. Furthermore, inorganic ions

TABLE 37.1

Composition of Acetogenic and Methanogenic Landfill Leachates

	Values from Literature			
	Acetogenic Leachate		Methanogenic Leachate	
Parameter[a]	Al-Wabel [1]	Robinson [16]	Hunce [18]	Robinson [16]
pH	5.9–6.3	5.5–7.0	7.9	7.5–8.5
Conductivity	6.3–42.5	7–30	37.2	<1–<1
BOD	–	4000–30,000	4250	<500–1000
COD	13,900–22,350	10,000–50,000	8038	2000–6000
Alkalinity	–	2000–10,000	13,200	10,000–30,000
Ammonium–nitrogen	–	750–2000	1430	1500–3000
Nitrate–nitrogen	–	<1–<1	–	<0.1–<0.1
Phosphate	–	5–20	22.8	1000–3000
Chloride		1000–2000	7000	2000–4000
Zinc	0.108–0.226	5–20	1.767	<0.01–0.05
Cadmium	<0.002	<0.1–<0.2	<0.005	<0.02–0.1
Nickel	0.384–0.718	<0.1–<1	0.597	<0.05–0.1
Chromium	0.21–0.336	1–<0.5	0.354	0.02–5
Copper	0.124–0.246	<0.1–0.1	0.145	<0.3–2
Lead	<0.04	<0.1–<0.5	<0.05	<0.05–0.2

[a] All values are in mg/l except pH and EC (mS/cm).

(such as NO_3^-, NO_2^-, NH_4^+, SO_4^{3-}, PO_4^{3-}, and Cl^-) and xenobiotic organic compounds (such as phenols, halogenated hydrocarbons, chlorinated aliphatics, and aromatic hydrocarbons) are reported in the landfill leachate, which may cause serious biological effects [11,19]. Mainly, landfill leachate can be classified as acetogenic (acetate production by anaerobic bacteria) and methanogenic (methane production by anaerobic bacteria) leachate according to their characteristics (Table 37.1).

The recorded chemical oxygen demand (COD), biological oxygen demand (BOD), and other solids, such as total suspended solids, volatile suspended solids, and total solids for acetogenic leachate, are considerably high compared with methanogenic leachate [16]. However, landfill leachate characteristics can be different from temperate to tropical countries owing to the climatic differences and variation in consumer patterns [6]. In addition, solid waste composition and age of the landfill significantly influence the determination of the composition of landfill leachate [6]. Although many characterization studies and extensive review of the composition of landfill leachates are available, these are restricted to the developed nations. Only a few studies are reported on the landfill leachate, its transport, and its fate in the tropics, especially from the Asian region [19,20,21]. Therefore, characterization of leachate plays a significant role on deciding appropriate treatment practices.

37.1.3 Common Treatment Practices for Landfill Leachate

Leachate treatment technologies fall into two basic types, biological and physical/chemical. In larger systems and depending on the treatment goals, integrated systems that combine the two are often used. The most common biological treatment is activated sludge, which is a suspended-growth process that uses aerobic microorganisms to biodegrade organic contaminants in leachate. With conventional activated sludge treatment, the leachate is aerated in an open tank with diffusers or mechanical aerators. Air stripping, adsorption, and membrane filtration are the major physical leachate treatment methods; coagulation–flocculation, chemical precipitation, and chemical and electrochemical oxidation methods are the common chemical methods used for landfill leachate treatment [22–27]. Li et al. [28] reported that ammonium removal can be achieved by chemical precipitation. In the landfill leachate treatment, a 66% COD and 50% ammonia removal were obtained by nanofiltration [27]. Evaporation and reverse osmosis have been used for the treatment of industrial landfill leachate [29]. Furthermore, combined processes have been successfully applied together, with coagulation–flocculation + biological treatment [30]; photochemical oxidation + activated sludge; Fe(III) chloride coagulation + photooxidation; and ozonation + adsorption [31–33]. Several researchers have investigated the efficiency of ozonation for treating landfill leachate [26,34–36]. Activated carbon adsorption systems have also been used in the treatment of landfill leachates for removal of dissolved organics; however, most of these techniques are generally considered expensive treatment options and often must be combined with other treatment technologies to achieve desired results.

37.1.4 Nanomaterials in Leachate Treatment

Most commonly encountered engineered nanoparticles (ENPs) fall into one of the following categories: carbonaceous nanoparticles, metal oxides, quantum dots, zerovalent metals, and nanopolymers, with new products gradually being added to the list. Presently, ENPs are widely used in many applications, including wastewater and polluted plumes treatments and hazardous waste treatment, owing to the unique sorptive and reactive properties that contribute for efficient treatment or degradation of contaminants [37]. Mainly, these nanomaterials possess unique atomic structure and distinctive chemical (such as behavior of excellent electron donor) and mechanical or physical properties (such as large surface-to-volume ratio involved in surprising surface and quantum effects or sorptive properties) [37]. This may be due to the small particle size (<100 nm) that falls in the transitional zone between individual atoms or molecules and the corresponding bulk material, which can modify the physicochemical properties of the material (e.g., performing exceptional feats of conductivity, reactivity, and optical sensitivity). For instance, nanoscale particulates such as nano-zerovalent iron (NZVI), iron oxide (γ-Fe_2O_3, Fe_2O_3), silver, and gibbsite particulates are used for wastewater treatment purposes [38–42]. Among these materials, NZVI has been tested widely for a wide range of pollutant plumes and contaminated sites.

Nanosized particles that can degrade a wide range of hazardous pollutants are successfully used for environmental remediation [43]. To date, large-scale treatment plants incorporated with zerovalent iron (ZVI) (i.e., iron shavings) can be applied to industrial wastewater treatment [44]. However, comparative studies have shown that nanoparticulate (<100 nm) zerovalent iron (NZVI) degrades contaminants more effectively than ZVI with micron-sized particles [45]. Different approaches have been used to bring ZVI in contact with contamination. These include construction of permeable reactive barriers (PRBs), into which contaminants flow, and *in situ* injection, where ZVI is delivered actively to the plume. The *in situ* injection of NZVI has shown significant efficiency with low operating costs compared with conventional ZVI PRBs [37,46]. Figure 37.1 shows a typical PRB system that is used for removing contaminants from groundwater.

These nanoparticles immobilize heavy metals such as Cr(VI) and As(V), chlorinated hydrocarbons such as trichloroethene and pesticides (DDT, chlorpyrifos, lindane, etc.), and nutrients (nitrates, phosphates, etc.) [38,47–49]. Although significant numbers of publications in the field of pollution prevention using nanomaterials are reported, the use of nanomaterials in landfill leachate treatment is still lacking. A few studies have demonstrated the use of NZVI

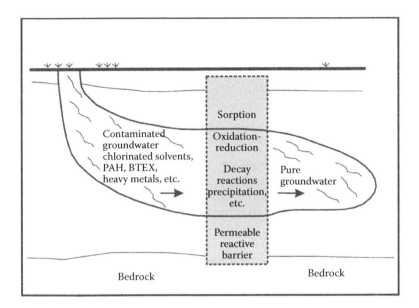

FIGURE 37.1
Schematic diagram of a PRB. Source of a contaminant plume flows through PRB; sorption, oxidation–reduction, and precipitation occurs while treating contaminant plume.

in leachate treatment, but restricted to developed and temperate countries [46]. In addition to NZVI, iron oxide nanoparticles (IONPs), silver nanoparticle (AgNPs), and gibbsite nanoparticles (GNPs) have been applied for removal of heavy metals, organic dyes, and wastewater treatment [40,46,50,51]. Therefore, a case study is described here, focusing on synthesizing and characterizing nanoparticles (NZVI, iron oxide, gibbsite, silver) to investigate their efficiency for the treatment of landfill leachate in the tropics, with a focus on Sri Lanka.

37.2 Materials and Methods

37.2.1 Synthesis of Nanoparticles

37.2.1.1 Synthesizing NZVI

The NZVI was prepared by modifying a borohydride reduction method, dropping an aqueous borohydride solution into an aqueous ferrous solution [52]. We used $FeSO_4 \cdot 7H_2O$ because of its high reactivity with borohydride, less oxidation after synthesis, and requirement for significantly less amount of borohydride; hence, the cost involved is low for large-scale applications [53].

In brief, the starch-stabilized NZVI (hereafter referred as S-NZVI) was synthesized by adding 100 ml of 0.5 M sodium borohydride solution into a flask containing 100 ml of 0.14 M $FeSO_4 \cdot 7H_2O$ solution with 5% (w/w) starch. The dropwise borohydride addition was approximately completed in 2 h to control the particle size. An additional 10 ml sodium borohydride was added to the solution, and the mixture was stirred under nitrogen gas bubbling for a further 15 min to complete the reaction. The precipitated NZVI particles were separated by centrifugation (Beckman GP Centrifuge) at 3000 rpm for 10 min, and washed by resuspending the material in absolute ethanol. This procedure was repeated three times. After the final centrifugation, the particles were dried in a freeze dryer for ~45 min and the prepared particles were stored at <4°C. The ferrous iron was reduced to the zerovalent iron by addition of borohydride according to the following reaction:

$$2Fe^{2+}_{(aq)} + BH^-_{4(aq)} + 3H_2O_{(l)} \rightarrow 2Fe^0_{(s)} + H_2BO^-_{3(aq)} + 4H^+_{(aq)} + 2H_{2(g)} \tag{37.1}$$

37.2.1.2 Synthesis of Iron Oxide Nanoparticles

IONPs were synthesized by a modified coprecipitation method [54]. In this method, ferromagnetic iron oxide nanoparticles were synthesized by using coprecipitation of ferrous Fe(II) and ferric Fe(III) ions in an alkaline medium. First, 3.25 g (0.020 mol) $FeCl_3$ and 2.00 g (0.010 mol) $FeCl_2 \cdot 4H_2O$ powder was dissolved in 60 ml aqueous acid, obtained by combining 50 ml deionized water with 10 ml of 1 mol/dm^3 HCl solution. The molar ratio of Fe(II)/Fe(III) solution was 0.5. Then, the resulting solution of Fe(II)/Fe(III) was added dropwise into 100 ml of 1 mol/dm^3 NaOH solution under vigorous stirring. After all the Fe(II)/Fe(III) solution was added, the reaction mixture was stirred for a further 30 min to prevent any coagulation of particles. Next, the obtained colloidal solution was centrifuged at 3000 rpm (Beckman GP centrifuge) at ambient conditions for 15 min, and the precipitate was washed several times with deionized water to remove any remaining Cl^- and Na^+ ions in the precipitate. Finally, the precipitate was dried at 60°C in an oven. The obtained iron oxide particles were ground thoroughly, and the obtained iron oxide powder was used for other experiments [55]. Synthesized particles were characterized by the x-ray diffraction (XRD), transmission electron microscopy (TEM), and Brunauer–Emmett–Teller (BET) techniques.

37.2.1.3 Synthesis of Gibbsite Nanoparticles

Gibbsite nanoparticles were synthesized by a previously described method [56]. An amorphous aluminum hydroxide suspension was produced by a titration of 1 mol/dm^3 $AlCl_3$ with 6 mol/dm^3 NaOH until pH reached ~4.6. The suspension was stirred throughout the process. Particular attention was paid to minimize CO_2 contamination of the solution by continuous nitrogen purging (99.99%). At the end of the titration, the suspension was sealed and stirred continuously at least for 6 h. It was transferred into dialysis membranes and was subjected to dialysis against double-distilled water. After 2 days, the excess liquid phase was separated from aluminum hydroxide gel and the dialysis was continued at 50°C for 60 days. The dialysis was performed using prewashed and hydrated cellulose dialysis membranes (Sigma-Aldrich, D-9402). The open end of each bag was sealed with dialysis clips (Sigma-Aldrich, Australia). The dialysis water was refreshed daily and maintained at a constant temperature throughout this process. Synthesis particles were characterized by using XRD, TEM, and BET techniques.

37.2.1.4 Synthesis of Ag Nanoparticles from Bacteria

The pure cultures of *Bacillus cereus*, *Bacillus subtilis*, *Bacillus koreensis*, and *Bacillus megaterium* strains were prepared and they were inoculated in MGYP nutrient media (glucose 1%, malt extract 1%, yeast extract 0.3%, peptone 0.5%) [57]. After inoculation of the microbes in media, silver $AgNO_3$ was added in the concentration of 100 mg/l in different sets of flasks. The concentration of silver salt was gradually increased from 100 to 20,000 mg/l to check the tolerance of microbes at higher concentration of metal salts. Temperature and rotations were maintained at 37°C and 100 rpm, respectively.

The viability of bacteria was checked with increasing concentrations of $AgNO_3$ in the culture media, and it was observed that these strains could sustain the concentrations of $AgNO_3$ up to 20,000 mg/l. Absorbance of AgNPs was taken on a UV-Vis spectrophotometer at regular intervals from 1 to 72 h. After 72 h of incubation and on a shaker, the biomass was centrifuged at 10,000 rpm on a REMI Deep Freeze Shaker for 15 min. The supernatant was filtered out and the filtrate was washed three to five times with distilled water and was then dried in hot air oven at 200°C.

37.2.2 Chemical Characteristics of Landfill Leachate

The landfill leachate used in the experiment was collected from the Gohagoda open dump site, situated in the World Heritage city, Kandy. The collected leachate samples were characterized for pH, temperature, conductivity, alkalinity, total dissolved solids, and dissolved oxygen at field conditions, and immediately transferred into the laboratory under 4°C conditions for other experiments such as analysis of nutrients, COD, Cl^-, solids, total organic carbon, dissolved organic carbon, and metals. The methods of analysis are given in Table 37.2.

TABLE 37.2

Summary of Major Chemical Constituents of Leachate from Gohagoda Open Landfill, Kandy; Methods of Analysis and Comparison with Literature

Constituents	Method/Reference	Gohagoda Leachates	Acetogenic Leachates	Methanogenic Leachates
pH	ROSS sure-flow combination epoxy body electrode	8.0–8.60	5.5–7	7.5–8.5
EC	Conductivity meter (Orion 5-Star series)	8.96–29.6	7–30	<1
BOD_5	Winkler method [58]	21.6–1458	4000–30,000	<500–1000
COD	Spectrophotometer (HACH DRB 200)	70–69,700	10,000–50,000	2000–6000
TOC	TOC analyzer (Analytikjena Multi N/C 2100)	3514.9	3000–20,000	<10–300
Alkalinity	Titrimetric method [58]	6200–19,320	2000–10,000	10,000–30,000
Ammonia–nitrogen	Iron-selective electrode	6.12–576.8	750–2000	1500–3000
Nitrate–nitrogen	Cadmium reduction method	0.4–172.2	<1	<1
Nitrite–nitrogen	Diazotization method	0.04–112.7	< 0.1	<0.1
Phosphate	Ascorbic acid method	7.65–172.2	5–20	1000–3000
Zinc	Atomic adsorption spectrophotometer (GBC 933 Australia)	0.2–1.15	5–20	<0.01–0.05
Cadmium		0.004–0.062	<0.1–<0.2	<0.02–0.01
Nickel		0.133–0.532	<0.1–<1	<0.05–0.1
Chromium		0.021–0.323	0.1–<0.5	0.02–5
Copper		0.048–0.257	<0.1–0.1	<0.3–2
Lead		0.015–0.416	<0.1–<0.5	<0.05–0.2
Iron		1.49–15.25	100–800	0.1–0.5
Manganese		0.155–1.203	10–50	5–50

Source: H. Robinson, *Commun. Waste Resour. Manage.* 8, 19, 2007.
Note: All values are in mg/l except pH and EC (mS/cm).

37.2.2.1 Treatment of Landfill Leachate

Batch kinetic experiments were carried out to understand the removal behavior of nutrients from the landfill leachate from Gohagoda landfill. In brief, 1 g/l of the synthesized S-NZVI, IONP, GNP, and AgNP were added to 50 ml of leachate. Leachate without any nanomaterial was used as control sample and treated in the same conditions to see any changes in the matrix. The resulting slurries were shaken at 100 rpm (EYELA B603 shaker), and samples were collected at predetermined time intervals (1, 4, 12, 24, and 120 h). After ultracentrifugation at 10,000 rpm (Model RC26PLUS, SORVALL), the supernatant composition was analyzed for nutrients removal.

37.2.2.2 Synthetic Leachate for Metal Treatment

The sampled leachate had relatively low concentrations of metals compared with those reported in soils contaminated by landfill leachate [13]. To simulate a "worst-case scenario" for metal concentrations, adsorption studies were conducted using synthetic leachate [13]. In the present study, we focused on Zn(II) and Pb(II) since these metals have been found as the most abundant metal ions in the leachate from the studied landfill [13]. The nanomaterial solutions (S-NZVI, IONP, GNP, and AgNP; 1 g/l) were purged with N_2 for 10 min to remove dissolved oxygen and an additional 1 h to equilibrate the solution and solid in a closed system to avoid oxidation.

The reactions were started by adding synthetic leachate to 1 g/l of synthesized nanomaterial suspension to a final concentration of Zn(II) and Pb(II), approximately 0.5 and 1.4 mg/l, respectively. The resulting slurries were shaken at 100 rpm, and samples were collected at predetermined time intervals (1, 4, 12, 24, and 120 h). After ultracentrifugation, the supernatant composition was separated for analysis of dissolved metal concentration.

In addition, isotherm studies were performed on Pb(II) and Zn(II) for each nanoparticulate solutions that were initially prepared, same as for the synthetic leachate experiment (1 g/l nanomaterial solutions with 10 min N_2 purging and 1 h equilibration period). The reactions were started by extracting 20 ml of the slurry per experiment and adding 40–1000 μl synthetic leachate. This resulted in initial metal concentrations ranging from 1.25 to 41.25 mg/l for Zn(II) and 0.60 to 37.35 mg/l for Pb(II). The pH was adjusted to 6.0 after addition of synthetic leachate using 0.1 M HNO_3, and the centrifuge tubes were then shaken at 100 rpm overnight. To separate solid and solution, the tubes were ultracentrifuged at 10,000 rpm at room temperature and the supernatant was transferred to a separate tube for metal analysis in isotherm studies. Control experiments were performed in the absence of nanoparticles but with otherwise identical experimental conditions. Dissolved metal concentrations were determined with atomic absorption spectroscopy (GBC 933A).

37.3 Results and Discussion

37.3.1 Physiochemical Characteristics of Leachate

The characteristics of the leachate used in this study are given in Table 37.2. The chemical constituents comparison between Gohagoda landfill leachates and other landfill leachates is summarized in Table 37.2. Low COD and BOD were observed; the basic pH values for Gohagoda leachates strongly suggest that the landfill is in the methanogenic phase [16].

However, during the leachate characterization period, we noticed a very high COD (50,000–67,900 mg/l), indicating acetogenic-phase characteristics on a few occasions [16]. It is obvious that the uncontrolled dumping of the waste can yield similar results due to an irregular degradation pattern of the waste. The manual mixing of the waste can influence reporting of high concentration of nutrients such as nitrate–nitrogen, ammonia–nitrogen, and phosphate. In addition, most of the chemical constituents tested exceeded the permissible levels of wastewater discharge in the country [59].

37.3.2 Characterization of Nanoparticles

37.3.2.1 Starch-Coated NZVI

XRD patterns of prepared S-NZVI are shown in Figure 37.2. According to Figure 37.2, the characteristic peak at 44.7° indicates the crystallization of Fe^0 for S-NZVI [52]. In addition to iron, a very faint peak observed at below 8.0° possibly suggests a trace of green rust, and broad peaks around 20° and 35° suggest the presence of some ferrihydrite as a result of oxidation. The BET surface area of the S-NZVI was recorded as 136.0 m^2/g, indicating very high surface area-to-volume ratio. The TEM images of S-NZVI are shown in Figure 37.2. Accordingly, it is clear that Fe^0 particles are present in nanoscale and appear in spheroidal shape. Furthermore, the needle- or cubic shape–like structures can be due to the presence of trace amounts of goethite and maghemite.

37.3.2.2 Iron Oxide Nanoparticles

The crystal structure of the synthesized IONPs was investigated by XRD using Cu K_α radiation. As shown in Figure 37.3, the XRD pattern of the synthesized nanoparticles was well matched with the "γ" phase of the iron oxide (γ-Fe_2O_3) crystalline [60]. There are six characteristic peaks at 2Θ = 31.7°, 36.7°, 41.1°, 53.4°, 57.0°, and 62.6°, corresponding to the Miller indices of γ-Fe_2O_3 at (220), (311), (400), (422), (511), and (440), respectively [61]. The BET surface area of the IONP was recorded as 150.0 m^2/g [62]. TEM data in Figure 37.3 indicate that the γ-Fe_2O_3 nanoparticle has a

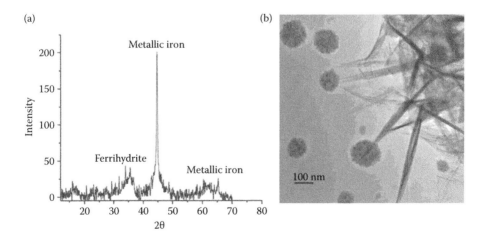

FIGURE 37.2
XRD pattern (a) and TEM image (b) for synthesized starch-stabilized NZVI. Peaks refer to ferrihydrite and metallic iron. Scale bar represents 100 nm, and aggregation of spherical shape nanoparticles is imaged.

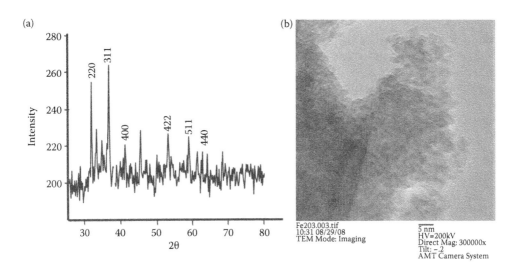

FIGURE 37.3
XRD pattern (a) and TEM image (b) for synthesized γ-Fe₂O₃ nanoparticles. Scale bar represents 5 nm, and aggregated cubic structure shape of nanoparticles is imaged.

spinel cubic structure and particle size between the 5 and 20 nm range. The calculated average particle size of the γ-Fe₂O₃ nanoparticle (14.3 nm) agreed well with TEM data.

37.3.2.3 Nanogibbsite

Figure 37.4 shows the XRD pattern of the synthesized gibbsite particles after a 60-day dialysis period. The 2Θ values of the synthesized gibbsite nanoparticles at 18.24°, 20.24°, 21.46°, 26.91°, 36.50°, 37.58°, 39.42°, 41.5°, 44.3°, 45.4°, 50.72°, 54.22°, and 63.8° are well fitted with literature data [56,63]. Synthesized gibbsite particles obtained after 60 days of dialysis were examined with TEM (Figure 37.4). The nanogibbsite crystals showed a pseudo-hexagonal morphology with an average crystallite size of 90 ± 25 nm. Gibbsite crystals exhibit plate-like crystal structures [64]. According to the TEM images, the thicknesses of

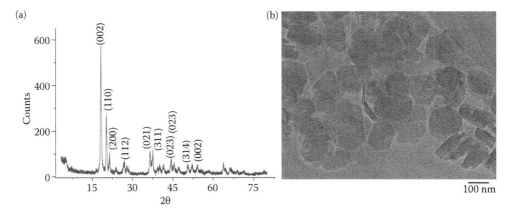

FIGURE 37.4
XRD pattern (a) and TEM image (b) for synthesized gibbsite nanoparticles. Scale bar represents 100 nm, and aggregated hexagonal structure shape of nanoparticles is imaged.

the nanocrystals vary between 25 and 60 nm. In addition, the surface area of the GNP was recorded as 56 m^2/g.

37.3.2.4 Silver Nanoparticles

Figure 37.5 shows an XRD pattern that can be indexed on the basis of the structure of silver. The XRD pattern thus obtained clearly shows [111,200,220], and [311] planes, and demonstrates that the synthesized AgNPs by the *Bacillus* were crystalline in nature. The diffraction peaks were found to be broad around their bases, indicating that the silver particles are nanoscale. The peak broadening at half-maximum intensity of the XRD lines is due to a reduction in crystallite size, flattening, and microstrains within the diffracting domains. The TEM image of a nanoparticle that was synthesized by *B. megaterium* indicated that the nanoparticles were in the size range of 10–20 nm and well crystalline [57]. In addition, the surface area of the AgNP was recorded as 25.0 m^2/g.

Figure 37.6 shows the appearance of the 1 g/l of each nanoparticle suspension at 0 day and after 30 days. According to visual observations, there is no color change after 30 days

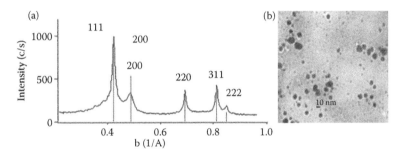

FIGURE 37.5
XRD pattern (a) and TEM image (b) for synthesized AgNPs. Scale bar represents 10 nm.

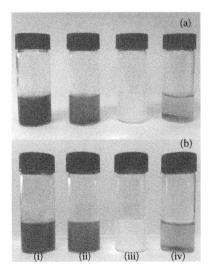

FIGURE 37.6
Photographs of changing colors in different 1 g/l nanoparticle suspension: (a) initial appearance and (b) after 1 month of (i) S-NZVI, (ii) IONP, (iii) GNP, and (iv) AgNP.

except in the S-NZVI suspension. This may be due to the oxidation of S-NZVI into iron oxide in aqueous media. Therefore, Figure 37.6 exhibits the stability of different nanomaterials in the aqueous phase.

37.3.3 Nutrient Treatment Using Different Nanoparticles

Nitrogen compounds are considered a major constituent in landfill leachate and lead to serious environmental issues. They exist as ammonium, ammonia, nitrate, and nitrites, which are main by-products of protein and urea degradation [46]. The removal densities of nitrate–N and nitrite–N with time are shown in Figure 37.7a and b. The highest removal density was obtained for S-NZVI: 1.654×10^{-5} and 1.102×10^{-7} mol/m^2 for nitrate–N and nitrite–N removal, respectively, in the first 60 min. All other tested materials showed constant nitrate–N removal. The reduction of nitrate–N was very slow with IONP and GNP. However, AgNP showed a slight removal in nitrate–N, which may be due to the slow oxidation of Ag to Ag$^+$.

The nitrogen removal process for nitrate–N and nitrite–N could mainly be associated with the oxidation–reduction reaction with S-NZVI. This mechanism can be explained as below. In acidic media, the nitrate reduction produces ammonium and ferrous ions (Reaction 37.2) [65].

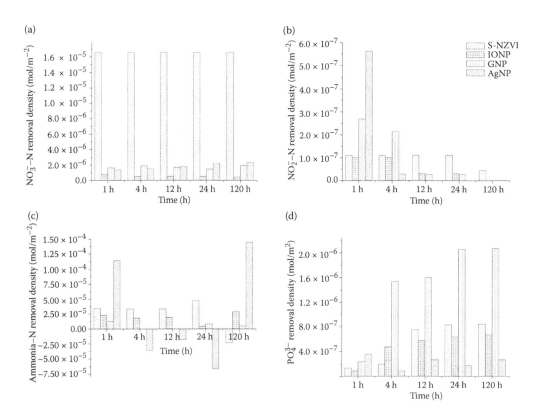

FIGURE 37.7
Batch experiment results for (a) nitrate–nitrogen $\left(NO_3^- - N\right)$, (b) nitrite–nitrogen $\left(NO_2^- - N\right)$, (c) ammonia–nitrogen $(NH_3–N)$, and (d) phosphate $\left(PO_4^{3-}\right)$ while reacting leachate with S-NZVI, IONP, GNP, and AgNP solutions (1 g/l).

The reduction of nitrate could also occur in a different pathway producing nitrite and nitrogen gas. The reactions involved are shown in Reactions 37.3 through 37.6 [65]:

$$4Fe^0_{(s)} + NO^-_{3(aq)} + 10H^+_{(aq)} \rightarrow 4Fe^{2+}_{(aq)} + NH^+_{4(aq)} + 3H_2O_{(l)} \tag{37.2}$$

$$5Fe^0_{(s)} + 2NO^-_{3(aq)} + 12H^+_{(aq)} \rightarrow 5Fe^{2+}_{(aq)} + N_{2(g)} + H_2O_{(l)} \tag{37.3}$$

$$Fe^0_{(s)} + NO^-_{3(aq)} + 2H^+_{(aq)} \rightarrow Fe^{2+}_{(aq)} + NO^-_{2(aq)} + H_2O_{(l)} \tag{37.4}$$

$$2Fe^0_{(s)} + 2H_2O_{(l)} + O_{2(aq)} \rightarrow 2Fe^{2+}_{(aq)} + 4OH^-_{(aq)} \tag{37.5}$$

$$Fe^0_{(s)} + 2H_2O_{(l)} \rightarrow Fe^{2+}_{(aq)} + H_{2(g)} + 2OH^-_{(aq)} \tag{37.6}$$

In slightly basic to neutral pH in the leachate, nitrate reduction through nitrite could be more favorable. Although the nitrite–N concentration in the tested leachate sample was very low, it increased over time with the reduction of nitrate by S-NZVI. This is in agreement with the literature [65,66].

The ammonium concentrations should also be considered for a better understanding of the process (Figure 37.7c). In anoxic conditions, ammonium can be converted to nitrogen gas, whereas nitrite is the electron acceptor [46]. However, in this experiment, we observed formation as well as decline of the ammonium concentrations. H_2 formed as in Reaction 37.6 could be used by denitrifying bacteria/denitrifiers such as *Micrococcus*, *Bacillus*, and *Pseudomonas* to convert nitrate to nitrogen gas or nitrous oxide [46].

37.3.4 Phosphate $\left(PO_4^{3-}\right)$ Removal by Nanomaterials

According to the results from the batch experiment, GNP showed the highest phosphate removal density compared with the other sorbents. Phosphate removal has reached the equilibrium within the first 12 h for S-NZVI, indicating 7.484×10^{-7} mol/m² removal density (Figure 37.7d). The second best material for phosphate removal was S-NZVI, which indicated 8.0×10^{-6} mol/m² after 120 h. This may be due to the coagulation effect of GNP. Except AgNP, other sorbent materials (IONP and NZVI) showed 70% phosphate removal density during the study period. The phosphate has a strong tendency to be adsorbed on metal surfaces, through surface OH bonds, and precipitated with iron species [67,68]. During the corrosive leaching of the Fe(III) ion, the pure metal phosphates are precipitated in the intermediate at high pH range. At lower pH, phosphate could be adsorbed onto iron hydroxides, oxyhydroxides, and mixed valence green rust. Furthermore, phosphate removal can also increase through a biological phosphate removal process since the anoxic environment is favorable for the microorganism (*Microlunatus phosphovorus*, *Lampropedia* spp., etc.) activity that contributed to degrade phosphate [69].

37.3.5 BOD Removal by Nanomaterials

Figure 37.8 shows the BOD values of control and nanomaterial-treated leachate samples with time. The control sample and nanoparticle-treated sample, except S-NZVI, showed

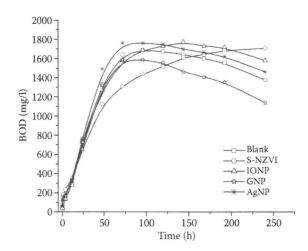

FIGURE 37.8
Batch experiment results for BOD while reacting leachate with S-NZVI, IONP, GNP, and AgNP solutions (1 g/l).

a typical curve for carbonaceous matter degradation with a suitable bacterial flora and adequate amount of nutrients present. In the first 5 days, it showed an increase in BOD values, and thereafter, BOD values decreased. This may be due to limiting of the aerobic environmental condition and it leads to decreases of aerobic microbes and macroorganisms. Compared with the rate of increase of BOD values, AgNPs showed rapid BOD increase during the first 5 days. This can be due to Ag acting as an inhibitor for bacterial activities; therefore, bacteria can respond by increasing its activity at the beginning [70]. However, S-NZVI showed the slowest BOD increasing rate. This may be due to the inhibition of bacterial activity by S-NZVI [44]. However, S-NZVI showed an increase in BOD values with time continuously. This may be due to nitrifying bacteria developing by oxidation of reduced nitrogen compounds (ammonia) to nitrate. IONPs showed a typical BOD curve compared with the other results. In the first 5 days, a rapid increase of BOD values was observed. After that, the increase in BOD values became slow, and then equilibrium was established for S-NZVI. After 5 days, other materials demonstrated a decrease in BOD values. This may be mainly due to the absence of aerobic conditions in the system.

37.3.6 Metal Ion Removal in Synthetic Leachate

Among the tested materials, S-NZVI and IONP showed considerable metal removal. S-NZVI and IONP demonstrated very high removal efficiencies of approximately 95% in the first hour of reaction time for tested metal ions (Figure 37.9). In addition, GNP showed a slow removal density compared with the other three absorbents, indicating approximately 5.843×10^{-8} and 2.784×10^{-8} mol/m^2 for Pb(II) and Zn(II) removal densities, respectively, within 72 h. This observation agreed with the previous literature [71]. However, AgNP showed very low or negligible Zn(II) removal density and for Pb(II), which seemed a gradual removal pattern, but a 2.602×10^{-8} mol/m^2 Pb(II) removal density after 72 h.

Heavy metal removal may occur through several different mechanisms such as adsorption, reduction, and precipitation. In the case of S-NZVI, the metals may have been reduced due to the high oxidation capacity of the NZVI. Also, the pH of the media plays a major role here. For instance, first, the medium pH rises when adding S-NZVI and IONP to the

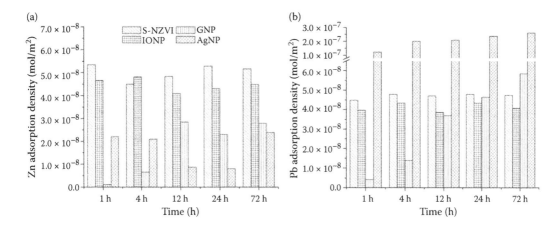

FIGURE 37.9

Batch experiment results for the synthetic leachate metals: (a) Zn^{2+} and (b) Pb^{2+} removal behaviors while reacting leachate with S-NZVI, IONP, GNP, and AgNP solutions (1 g/l).

leachate by the production of OH^- groups. An increase in OH^- concentration is more favorable for metal hydroxide precipitation. In the case of GNP, the main mechanism may have been adsorption.

37.3.7 Adsorption Isotherm Studies

Metal adsorptions were modeled with the Langmuir isotherm model given by Equation 37.7:

$$\Gamma_{ads} = \frac{k[M]\Gamma_{max}}{\left(1 + k[M]\right)} \tag{37.7}$$

where Γ_{ads} is the removal capacity (mg/g); [M], the initial solution concentration (mg/l); Γ_{max}, the maximum adsorption capacity (mg/g); and k, the equilibrium constant for the overall adsorption process. The Langmuir model assumes that all adsorption sites have equal affinity for the adsorbate and therefore only monolayer adsorption occurs.

The Langmuir isotherm modeling data are tabulated in Table 37.3. According to the adsorbed concentrations, the maximum Pb(II) adsorption density was recorded for IONP

TABLE 37.3

Langmuir Isotherm Model Data for Pb(II) and Zn(II) with S-NZVI, IONP, GNP, and AgNP at pH 6.0

Experiment		Langmuir Isotherm		
		K	Γ_{max} (mg/g)	R^2
S-NZVI	Zn(II)	18.4546	24.974	0.986
	Pb(II)	6.7561	25.847	0.973
IONP	Zn(II)	2.9629	18.106	0.952
	Pb(II)	1.8207	122.119	0.991
GNP	Pb(II)	0.2495	3.649	0.990
AgNP	Pb(II)	0.8651	5.436	0.994

at 122 mg/g, while S-NZVI showed 25.8 mg/g. The high sorption of IONP may be due to precipitation at the surface. A high but comparatively low uptake was observed for GNP and AgNP, indicating 3.65 and 5.44 mg/g, respectively, for Pb(II) treatment. The tested S-NZVI and IONP indicated 24.97 and 18.11 mg/g, respectively, for Zn(II) adsorption. In addition, no considerable Zn(II) sorption was observed for GNP and AgNP. On the basis of the results, a monolayer formation can be suggested for the metal adsorption for S-NZVI, while for Pb on IONP it may be a multilayer formation. Therefore, S-NZVI can be considered as a potential material for metal treatment and nutrient removal in landfill leachates.

37.4 Conclusions

In this study, we synthesized and characterized four different nanoparticles: S-NZVI, IONP, GNP, and AgNP. Thereafter, these materials were applied to landfill leachate from the Gohagoda open dump site in Sri Lanka, with the synthetic leachate targeting the effect on nutrients and metals removal. Characterization results confirmed the formation of expected nanoparticles, and batch experiments demonstrated that some of these materials could be used for application in landfill leachate treatments. In summary, synthesized S-NZVI showed the highest removal densities for nitrate–N (1.65×10^{-5} mol/m^2) and phosphate (1.95×10^{-7} mol/m^2). Furthermore, results from adsorption studies indicated the effective removal of Pb(II) and Zn(II) for S-NZVI and IONP, >90% within the first 60 min. Therefore, it can be concluded that the starch-coated NZVI can be effectively used to treat tropical landfill leachates. However, it is important to investigate the cost–benefit differences of using S-NZVI for such treatment, especially for developing nations. Also, further studies may be vital in enhancing the nutrient and metal treatment of other nanoparticles, IONP, GNP, and AgNP, by application of different coatings or activating them using different methods.

References

1. Al-Wabel, M.I., W.S. Al Yehya, A.S. Al-Farraj, S.E. El-Maghraby, *Journal of the Saudi Society of Agricultural Sciences.* 10 (2011) 65.
2. Tuck, C.O., E. Perez, I.T. Horvath, R.A. Sheldon, M. Poliakoff, *Science.* 337 (2012) 695.
3. Magutu, P.O., C.O. Onsongo, *Integrated Waste Management.* II (2011).
4. Ecke, H., H. Sakanakura, T. Matsuto, N. Tanaka, A. Lagerkvist, *Waste Management & Research.* 18 (2000) 41.
5. Logan, B.E., K. Rabaey, *Science.* 337 (2012) 686.
6. Trankler, J., C. Visvanathan, P. Kuruparan, O. Tubtimthai, *Waste Management.* 25 (2005) 1013.
7. Ekanayake, K.M., B. Basnayake, S.K. Gunathilake, A.S.H. Chandrasena, Proceedings of the Seventh International Summer Symposium, Japanese Society of Civil Engineers, Tokyo, Japan, July 30 (2005) 339.
8. Basnayake, B.F.A., S.A. Karunarathne, H.Y.R. Gunarathne, S. Murugathasan, APLAS International Symposium, Sapporo, Hokkaido, Japan, October 22–24 (2008).
9. SAARC SOE Report Sri Lanka, Environmental Economics & Global Affairs Division, Ministry of Environment and Natural Resources, Sri Lanka, 2002.

10. Menikpura, S.N.M., B.F.A. Basnayake, *Renewable Energy*. 34 (2009) 1587.
11. Asadi, M. The 7th International Conference on Environmental Engineering, Faculty of Environmental Engineering, Vilnius Gediminas Technical University, Vilnius, Lithuania, (2008) 484, May 22–23, 2008.
12. Ariyawansha, R.T.K., B.F.A. Basnayake, K.P.M.N. Pathirana, A.S.H. Chandrasena, 21st Annual Congress, Postgraduate Institute of Agriculture, University of Peradeniya, Peradeniya, Sri Lanka, (2009). November 19–20, 2009.
13. Wijesekara, S.S.R.M.D.H.R., R.M.A.U. Rajapaksha, I.P.L. Jayarathne, B.F.A. Basnayake, M. Vithanage, The 12th Annual Conference of Thai Society of Agricultural Engineering "International Conference on Agricultural Engineering," Chon-Chan Pattaya Resort, Chonburi, Thailand, March 31–April 1 (2011).
14. Langler, G.J. *Aquatic Toxicity and Environmental Impact of Landfill Leachate*. PhD Thesis, University of Brighton (2004).
15. Sun, B., F.J. Zhao, E. Lombi, S.P. McGrath, *Environmental Pollution*. 113 (2001) 111.
16. Robinson, H. *Communications in Waste and Resource Management*. 8 (2007) 19.
17. Christensen, J.B., D.L. Jensen, T.H. Christensen, *Water Research*. 30 (1996) 3037.
18. Mor, S., K. Ravindra, R.P. Dahiya, A. Chandra, *Environmental Monitoring and Assessment*. 118 (2006) 435.
19. Hunce, S.Y., D. Akgul, G. Demir, B. Mertoglu, *Waste Management*. 32 (2012) 1394.
20. Kale, S.S., A. Kadam, S. Kumar, N.J. Pawar, *Environmental Monitoring and Assessment*. 162 (2010) 327.
21. Vasanthi, P., S. Kaliappan, R. Srinivasaraghavan, *Environmental Monitoring and Assessment*. 143 (2008) 227.
22. Amokrane, A., C. Comel, J. Veron, *Water Research*. 31 (1997) 2775.
23. Ahn, D.H., C. Yun-Chul, C. Won-Seok, *Journal of Environmental Science and Health. A: Toxic/Hazardous Substances & Environmental Engineering*. 37 (2002) 163.
24. Chiang, L., J. Chang, C. Chung, *Environmental Engineering Science*. 18 (2001) 369.
25. Lin, S.H., C.H. Chang, *Water Research*. 34 (2000) 4243.
26. Steensen, M. *Water Science and Technology*. 35 (1997) 249.
27. Marttinen, S.K., R.H. Kettunen, K.M. Somunen, R.M. Soimasuo, J.A. Rintala, *Chemosphere*. 46 (2002) 851.
28. Li, X.Z., Q.L. Zhao, X.D. Hao, *Waste Management*. 19 (1999) 409.
29. Di Palma, L., P. Ferrantelli, C. Merli, E. Petrucci, *Waste Management*. 22 (2002) 951.
30. Kargi, F., M.Y. Pamukoglu, *Biotechnology Letters*. 25 (2003) 695.
31. Koh, I., X. Chen-Hamacher, K. Hicke, W. Thiemann, *Journal of Photochemistry and Photobiology A: Chemistry*. 162 (2004) 261.
32. Wang, Z., Z. Zhang, Y. Lin, N. Deng, T. Tao, K. Zhuo, *Journal of Hazardous Materials*. 95 (2002) 153.
33. Rivas, F.J., F. Beltra, O. Gimeno, B. Acedo, F. Carvalho, *Water Research*. 37 (2003) 4823.
34. Baig, S., I. Coulomb, P. Courant, P. Liechti, *Ozone Science and Engineering*. 21 (1999) 1.
35. Kuo, W.S. *Ozone Science and Engineering*. 21 (1999) 539.
36. Silva, A.C., M. Dezotti, G.L. Sant'Anna Jr, *Chemosphere*. 55 (2004) 207.
37. Li, X., D.W. Elliott, W. Zhang, *Critical Reviews in Solid State and Materials Sciences*. 31 (2006) 111.
38. Hwang, Y.H., D.G. Kim, H.S. Shin, *Journal of Hazardous Materials*. 185 (2011) 1513.
39. Huittinen, N., T. Rabung, J. Lützenkirchen, S.C. Mitchell, B.R. Bickmore, J. Lehtoa, H. Geckeisb, *Journal of Colloid and Interface Science*. 332 (2009) 158.
40. Matei, E., A. Predescu, E. Vasile, A. Predescu, Nanosafe2010: International Conference on Safe Production and Use of Nanomaterials (2011). Maison Minatec Congress Center, Grenoble, France. November 16–18, 2010.
41. Hua, M., S. Zhang, B. Pan, W. Zhang, L. Lv, Q. Zhang, *Journal of Hazardous Materials*. 211 (2012) 317.
42. Pan, B., H. Qiu, B. Pan, G. Nie, L. Xiao, L. Lv, W. Zhang, Q. Zhang, S. Zheng, *Water Research*. 44 (2010) 815.
43. He, F., D. Zhao, *Environmental Science and Technology*. 41 (2007) 6216.

44. Ma, L., W.X. Zhang, *Environmental Science and Technology*. 42 (2008) 5384.
45. Alidokht, L., A.R. Khataee, A. Reyhanitabar, S. Oustan, *Desalination*. 270 (2011) 105.
46. Jun, D., Z. Yongsheng, Z. Weihong, H. Mei, *Journal of Hazardous Materials*. 161 (2009) 224.
47. Shea, P.J., T.A. Machacek, S.D. Comfort, *Environmental Pollution*. 132 (2004) 183.
48. Blowes, D.W., C.J. Patacek, S.G. Benner, C.W.T. McRae, T.A. Bennett, R.W. Puls, *Journal of Contaminant Hydrology*. 45 (2000) 123.
49. He, F., D. Zhao, *Environmental Science and Technology*. 39 (2005) 3314.
50. Bootharaju, M.S., T. Pradeep, *Journal of Physical Chemistry C*. 114 (2010) 8328.
51. Rocher, V., J.M. Siaugue, V. Cabuil, A. Bee, *Water Research*. 42 (2008) 1290.
52. Alidokht, L., A.R. Khataee, A. Reyhanitabar, S. Oustan, *Desalination*. (2010).
53. Choi, I.C., A.B.M. Giasuddin, S.R. Kanel, US 2008/009054 A1 (2008).
54. Yu, S., G.M. Chow, *Journal of Materials Chemistry*. 14 (2004) 2781.
55. Jayarathne, L., W.J. Ng, A. Bandara, M. Vitanage, C.B. Dissanayake, R. Weerasooriya, *Colloids and Surfaces A: Physicochemical and Engineering Aspects*. 403 (2012) 96.
56. Kumara, C.K., W.J. Ng, A. Bandara, R. Weerasooriya, *Journal of Colloid and Interface Science*. 352 (2010) 252.
57. Prakash, A., S. Sharma, N. Ahmad, A. Ghosh, P. Sinha, *International Research Journal of Biotechnology*. 1 (2010) 071.
58. American Public Health Association, *Standard Methods for the Examination of Water and Wastewater*. New York (2005).
59. National Environmental (Protection and Quality) Regulations (2008) No. 47 of 1980, No.1 of (1A), 31A. Ministry of Environment and Natural Resources, Sri Lanka.
60. Ma, W., F.Q. Ya, M. Han, R. Wang, *Journal of Hazardous Materials*. 143 (2007) 296.
61. Zhong, L.S., J.S. Hu, H.P. Liang, A.M. Cao, W.G. Song, L.J. Wan, *Advanced Materials*. 18 (2006) 2426.
62. Jayarathne, L., W.J. Ng, A. Bandara, M. Vitanage, C.B. Dissanayake, R. Weerasooriya, *Colloids and Surfaces A: Physicochemical and Engineering Aspects*. 403 (2012) 96.
63. Liu, H., J. Hu, J. Xu, Z. Liu, J. Shu, H.K. Mao, J. Chen, *Physics and Chemistry of Minerals*. 31 (2004) 240.
64. Wang, S.L., C.T. Johnston, *American Mineralogist*. 85 (2000) 739.
65. Hsu, J.C., C.H. Lio, Y.L. Wei, *Sustainable Environment Research*. 21 (2011) 353.
66. Su, C., R.W. Puls, *Environmental Science and Technology*. 38 (2004) 2715.
67. Emmerik, T.J.V., D.E. Sandström, O.N. Antzutkin, M.J. Angove, B.B. Johnson, *Langmuir*. 23 (2007) 3205.
68. Zeng, H., B. Fisher, D.E. Giammar, *Environmental Science and Technology*. 42 (2008) 147.
69. Mino, T., M.C.M.V. Loosdrecht, J.J. Heijnen, *Water Research*. 32 (1998) 3193.
70. Perez, M.A. *The Effects of Silver Nanoparticles on Wastewater Treatment and Escherichia Coli Growth*. Florida State University (2012).
71. Mason, S.E., C.R. Iceman, K.S. Tanwar, T.P. Trainor, A.M. Chaka, *Journal of Physical Chemistry C*. 113 (2009) 2159.

38

Al$_{13}$ Cluster Nanoflocculants for Remediation of Dissolved Organic Compounds and Colloidal Particles in Water

Dongsheng Wang, Hongxiao Tang, and Jiuhui Qu

Research Center for Eco-Environmental Sciences (RCEES),
Chinese Academy of Sciences (CAS), Beijing, China

CONTENTS

38.1 Introduction

Poly-aluminum chlorides (PACls) of the form Al$_n$Cl$_{(3n-m)}$OH$_m$ are typically used as coagulants in water and wastewater treatment to remove dissolved organic compounds and suspended colloidal particles. The efficiency of the coagulation process depends largely on the highly efficient flocculant, matching reactor (flocculator), and auto-dosing control technique (as in the FRD system suggested in the early 1990s; Tang, 1998) besides the source water quality. Among the influencing factors, the deciding one is the physicochemical properties and the speciation distribution of the coagulant. Traditionally, the most widely applied coagulants are alum and ferric chloride. However, inorganic polymer flocculants (IPFs) developed based on the salts in the 1960s are typically applied instead because of their high efficiency with relatively low cost, as well as other merits such as tight floc formation, efficient sediment quality and suitability at low temperature, high natural organic matter, wider alkalinity region, etc. (Bottero et al., 1987; Hu et al., 2006; Tang and Luan, 1996; Tang, 1998; Van Benschoten and Edzwald, 1990; Wang et al., 2002; Zhao et al., 2013). Therefore, IPFs could be regarded as second-generation coagulants.

Nano-IPFs, or simplified as "nanoflocculants," are defined as inorganic compounds with particle size distribution (PSD) in the nanoscale. The nano-IPF as a new kind of species can be prepared by special physicochemical processes and under certain separation and purification. In industrial production, the nano-Al_{13} coagulant is defined as an IPF-PACl containing Al_{13} species >70% as a component. Since the particle size is quite small, i.e., between molecular and colloidal scale, nano-Al_{13} coagulants exhibit special physicochemical properties and can be tailor-made to fit the application demanded and to meet a certain PSD (the PSD effect of IPF is also addressed in this chapter). Thus, the nano-IPFs could be considered further as the third generation of coagulants. The relative theory and application will contribute certainly to the science and technology of coagulation.

38.2 Loss of Al_{13} in High-Concentration PACl

Since Al_{13} is only one of the hydrolysis products of Al(III), pH becomes a significant factor for the stability and distribution of Al_{13}. When the concentration of PACl is increased, the spontaneous hydrolysis becomes more significant. In many experiments, it is indicated that Al_{13} could not exist in high-concentration PACl solutions with >2.0 mol Al/L (Huang et al., 2006a,b; Kloprogge et al., 1992; Zhao et al., 2013). As shown in Figure 38.1, the content of Al_{13} changes obviously with pH in PACl of 0.01, 0.1, 0.5, and 2.0 mol Al/L. A pH zone can be observed at 3.5–5.0. Below pH 3.5, no Al_{13} can be observed in the PACl solution. Therefore, how to prepare a commercial PACl with a high concentration of Al_{13} is a big challenge.

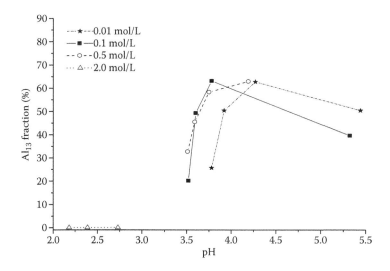

FIGURE 38.1
Change of Al_{13} and pH in PACl solution. (From Huang, L. et al., *J. Environ. Sci.* 18, 872, 2006a.)

38.3 Preparation and Stability of Pure Al$_{13}$

Through a SO_4^{2-}–Ba^{2+} method (Xu et al., 2003), a series of purified Al$_{13}$ solutions can be prepared as shown in Table 38.1. It can be seen that the samples prepared at concentrations of 0.01–2.0 mol Al/L contain mainly Al$_b$ species (based on the ferron reactivity as shown in the ferron assay, the species can be divided into three classes: the rapid reactive Al$_a$, the moderate reactive Al$_b$, and the inert species Al$_c$), i.e., >97%. No Al$_c$ fraction exists significantly as the ferron method is analytically erroneous around 1%. It seems also that the Al$_a$ fraction could be contributed from the surface Al of Al$_{13}$, i.e., as the result of rapid dissolution of the surface Al by ferron. It needs to be noted that the speciation characterization by the ferron method is operationally defined, i.e., Al$_a$ is the fraction reacted suddenly in 1 min. Some researchers defined Al$_a$ as the fraction reacted within 30 s (Bertsch, 1996). Therefore, the Al$_a$ calculated here is significantly higher. The samples obtained therefore contain mainly Al$_{13}$ species. It indicates that the Al$_{13}$ samples prepared can be diluted into various concentrations. High-concentration Al$_{13}$ solutions are relatively stable and therefore become the valid proof for a commercial preparation. Characterization using ^{27}Al-nuclear magnetic resonance (NMR) analysis exhibits only a sharp single response at 63.0 ppm (Wang et al., 2011). No peak of Al$_m$ (monomer) at 0 ppm or other species can be observed for all the samples. The peak area at 63.0 ppm increases rapidly with the increase of concentration. A very good linear equation with an R^2 of 0.9996 can be attained, i.e., $y = 4.085x + 0.0231$, where y is the peak area and x is the concentration. It indicates that the only species observed in the above samples are Al$_{13}$. The results show that the Al$_{13}$ detected by ^{27}Al-NMR has a good relation with Al$_b$ under these conditions (Bi et al., 2014).

Al$_{13}$ solutions at concentrations between 0.11 and 2.1 mol Al/L were further aged at room temperature to investigate the effect of concentration on the stability of Al$_{13}$. The results of the ferron assay are shown in Table 38.2. It seems that the Al$_{13}$ solution is in a state of pseudo-stability and tends to aggregate with aging. Concentration has a significant role for its stability at >0.5 mol Al/L. At a higher concentration of 2.1 mol Al/L, the Al$_{13}$ solution undergoes slow aggregation with aging and becomes turbid after 1 week, and only 87% of Al$_b$ remains after 1 month of aging. The 0.1 mol Al/L solution shows quite stable features and undergoes only a minor change in 1 month of aging. It has also been confirmed by NMR analysis. It indicates that the decreased Al$_{13}$ in the ^{27}Al-NMR spectrum contributes partially to those remaining in the Al$_b$ fraction during aging.

TABLE 38.1

Speciation Distribution of Purified Al$_{13}$ Solutions by Ferron Assay

Al$_{total}$ (mol Al/L)	Al$_a$%	Al$_b$%	Al$_c$%
0.01	2.35	97.44	0.20
0.055	2.20	97.77	0.03
0.11	2.31	97.33	0.36
0.42	2.06	97.38	0.56
1.06	2.02	96.80	1.18
2.11	1.91	96.59	1.60

TABLE 38.2

Speciation Analysis of Pure Al_{13} Solutions under Aging: Ferron Assay

Conc. (mol/L)	Aging	pH	Speciation $Al_a\%$	$Al_b\%$	$Al_c\%$
0.11	0	5.50	2.31	97.33	0.36
	1	5.51	1.58	97.51	0.90
	7	5.42	1.43	97.44	1.13
	30	5.27	1.21	94.81	3.98
0.42	0	5.16	2.06	97.38	0.56
	1	5.15	0.96	97.20	1.84
	7	5.00	1.80	94.71	3.49
	30	4.83	2.42	86.16	11.42
1.06	0	4.96	2.02	96.80	1.18
	1	4.93	1.08	96.07	2.85
	7	4.88	1.01	94.49	4.50
	30	4.67	0.93	85.70	13.37
2.11	0	4.81	1.91	96.59	1.50
	1	4.80	0.47	90.27	9.26
	7	4.70	0.86	86.79	12.35
	30	4.36	1.04	77.31	21.65

38.4 Particle Size Effect of Coagulation with Nano-IPF

With the purified Al_{13} available, a series of nano-IPFs can be tailor-made with different PSDs; therefore, the PSD effect of coagulation with nano-IPF can be examined (Zhao et al., 2013). Table 38.3 shows the speciation distribution of Al_{13} aggregates by the ferron assay. Figures 38.2 and 38.3 show the PSD effect of coagulation with nano-IPF at different doses and pH conditions. From Figures 38.2 and 38.3, it could be seen that the coagulation with nano-IPF exhibits a certain PSD effect in regard to the coagulant speciation and PSD features therein. With increasing dose, the nano-IPFs exhibit efficient charge-neutralization capability and turbidity removal as compared with traditional salts and low-Al_{13} PACls. This becomes more obvious in the aspects of floc formation kinetics and floc size (Figure 38.3).

TABLE 38.3

Speciation Distribution of Al_{13} Aggregates by Ferron Assay

Al_T	$Al_a\%$	$Al_b\%$	$Al_c\%$	Average PSD[a] (nm)
$B = 2.6$	2.80	97.4	−0.20	8.9
$B = 2.7$	2.80	94.3	2.90	11.9
$B = 2.8$	2.40	92.8	4.80	23.1

[a] Determined by photo correlation spectroscopy, the diameter of Al_{13} is around 2 nm; B = OH/Al ratio.

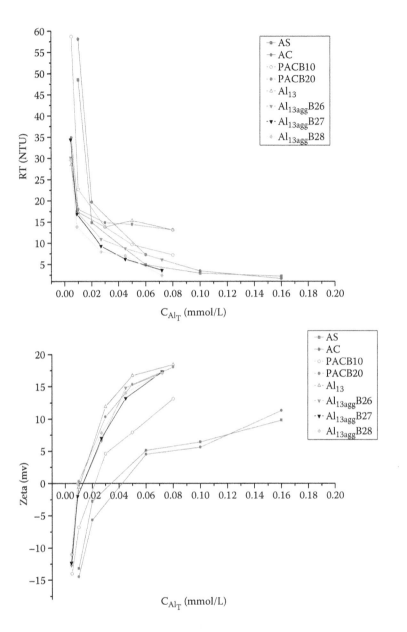

FIGURE 38.2
PSD effect of coagulation with nano-IPF.

38.5 Industrialization of Nano-IPF

As discussed above, it can be seen nano-Al$_{13}$ can be prepared in high concentration and exhibits efficient coagulation behavior. The possible industrialization for this kind of commercial product has also been demonstrated. The schematic routes are summarized in Figure 38.4.

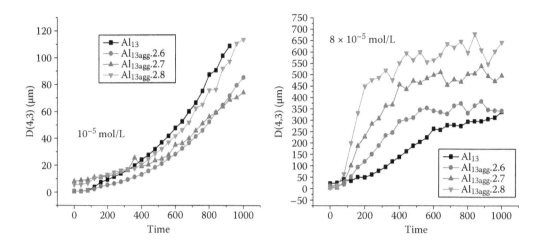

FIGURE 38.3
PSD effect of coagulation with nano-IPF: growth of floc.

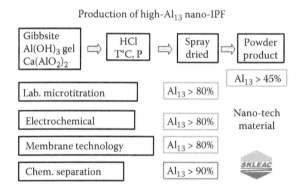

FIGURE 38.4
Schematic description of the industrialization of nano-IPF. SKLEAC, State Key Laboratory of Electroanalytical Chemistry.

The current industrial production through high-temperature/high-pressure dissolution–second alkalification process can yield only <45% Al_{13}. On the other hand, in laboratory microtitration, it is easy to reach as high as 80% Al_{13}. For an industrial-scale production, the possible ways to reach high yields include electrochemical preparation, membrane technology, and chemical separation and purification. Figure 38.5 shows a demo-plant established in Beijing, China, using the electrochemical preparation method.

The separation and purification of Al_{13} in PACl with various concentrations (0.05–0.5 mol/L) using the ultrafiltration (UF) method have also been investigated (Huang et al., 2006b). The relation between the species distribution data obtained by the ferron assay and the interception data of UF membranes with different molecular weight cutoffs (MWCOs), in which the concentration factor is neglected, is presented in Figure 38.6.

It can be clearly discovered that if the content of Al_b rises in PACl, the fraction with an MWCO >6 kilodaltons (kD) does not augment obviously, the 3–6 kD MWCO fraction

FIGURE 38.5
(See color insert.) Al$_{13}$-PACl plant established using the electrochemical method.

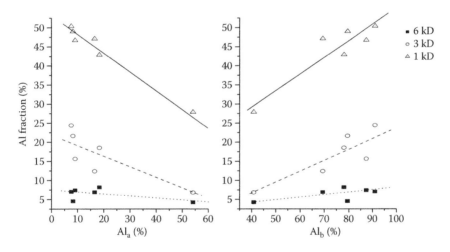

FIGURE 38.6
Relation between Al$_a$ or Al$_b$ by UF on various MWCO membranes.

increases somewhat, and the 1–3 kD MWCO fraction enhances greatly. If the content of Al$_a$ enhances, the fraction with an MWCO >1 kD decreases and that with an MWCO <1 kD increases. It also indicates that the particle size of Al$_b$ is close to the pore diameter of a 1-kD membrane. The experimental results show that the PSD of PACls varies markedly with the B value and total aluminum concentration with a 3–6 kD MWCO membrane. The interception rate differs significantly with the Al$_a$, Al$_b$, and Al$_c$ species (hydrolyzed speciation according to chemical ferron assay). By choosing the combination of 3 kD and 1 kD UF membrane, it is possible to separate and purify Al$_{13}$. The operation pressure exhibits little effect on the selective separation of Al$_{13}$ with other species, while it does decrease the separation time needed. High concentrations of Al$_{13}$ (about 0.5 mol Al/L) could be obtained with a purity of >95%.

38.6 Summary

Environmental nanotechnology will bring significant transformation in the field of environmental monitoring and treatment. Preparation and application of nanomaterials, nano-adsorbents, and sensors bring more and more attractive topics and breakthroughs in this area. The development of nano-IPFs is obviously one of the important contributions in the sequential potential changes.

The physicochemical properties and speciation distribution of coagulants are among the deciding factors in the coagulation process. IPFs developed from the traditional salts can be regarded as the second generation of coagulants. The third generation based on the nano-species is then discussed concerning aspects of possibility, stability, coagulation behavior, and mechanism, and the general production methods. However, on the basis of the rapid development of the hydrolysis chemistry of Al(III) and nanospecies, several knotty problems remain to be answered: What are the optimum coagulation species and coagulation mechanisms involved? How to design molecularly the optimum coagulation species? How to make the third-generation coagulants meet the change of water quality? How to develop the water treatment process fit for the needs of application of new coagulants, the reactors, the micro-interface process, and the process control? All these need further deep investigation and discussion. A large volume of research previously paid much attention on nano-Al_{13} species being considered as one of the optimum coagulation species owing to its special physicochemical properties and stability. However, with more recognition on the Al_6, Al_8, KEGGIN–non-KEGGIN Al_{13}, Al_{30}, and Al_{13} aggregates, answers to what is the optimum species and how to maintain its optimum performance will be reached in the near future.

Acknowledgments

The authors want to thank the students involved in this research, namely S.F. Wang, Q. Gao, G.H. Li, C.Q. Ye, L. Huang, W. Sun, X.H. Wu, Y.J. Chen, Y. Xu, C.H. Feng, M.Q. Yan, and Z. Bi. This research was supported by a 973 Nano-drinking water project under 2011CB933700; the NSF of China under 51221892, 51025830, 51078348, and 20477054; and also the 863 program under 2002AA001290.

References

Bi, Z., Chen, Y.J., Wang, S.F., Wang, D.S. (2014) Hydrolyzed Al(III)-clusters. II: Speciation transformation and stability of Al_{13} aggregates, *Colloids Surf. A*, 440, 59–62.
Bertsch, P.M. (1996) *The Environmental Chemistry of Aluminum*, CRC Press: NY; Chapter 4.
Bottero, J.Y., Axelos, M., Tchoubar, D., Cases, J.M., Fripiat, J.J., Fiessinger, F. (1987) Mechanism of formation of aluminum trihydroxide from Keggin Al_{13} polymers, *JCIS*, 117, 47.
Hu, C.Z., Liu, H.J., Qu, J.H., Wang, D.S., Ru, J. (2006) Coagulation behavior of aluminum salts in eutrophic water: Significance of Al_{13} species and pH control, *ES&T*, 40(1), 325–331.
Huang, L., Tang, H.X., Wang, D.S., Wang, S.F., Deng, Z.J. (2006a) Al(III) speciation distribution and transformation in high concentration PACl solutions, *J. Environ. Sci.*, 18(5), 872–879.

Huang, L., Wang, D.S., Tang, H.X., Wang, S.F. (2006b) Separation and purification of nano-Al$_{13}$ by UF method, *Colloids Surf. A*, 275(1–3), 200–208.

Kloprogge, J.T., Seykens, D., Jansen, J.B.H., Geus, J.W.A. (1992) ^{27}Al NMR study on the optimalization of the development of the Al$_{13}$ polymer, *J. Non-Cryst. Solids*, 142(1–2), 94–102.

Tang, H.X. (1998) The flocculation morphology of hydroxyl polyaluminum chloride, *Acta Circum. Sinica*, 18(1), 1–10 (in Chinese).

Tang, H.X., Luan, Z.K. (1996) The differences of behavior and coagulating mechanism between inorganic polymer flocculants and traditional coagulants, in: Hahn, H.H. et al. (Eds.) *Chemical Water and Wastewater Treatment (IV)*, Springer-Verlag, Berlin, 83–93.

Van Benschoten, J., Edzwald, J.K. (1990) Chemical aspects of coagulation using aluminum salts. I. Hydrolytic reactions of aluminum and polyaluminum chloride, *Water Res.*, 24(12), 1519.

Wang, D.S., Tang, H.X., Gregory, J. (2002) Relative importance of charge-neutralization and precipitation during coagulation with IPF-PACl: Effect of sulfate, *ES&T*, 36(8), 1815–1820.

Wang, D.S., Wang, S.F., Wu, X.H., Huang, C.P., Chow, C.W.K. (2011) Hydrolyzed Al(III) clusters: Speciation stability of nano-Al13, *J. Environ. Sci.*, 23(5) 705–710.

Xu, Y., Wang, D.S., Liu, H., Tang, H.X. (2003) Optimization on the separation and purification of Al$_{13}$, *Colloids Surf.*, 231, 1–9.

Zhao, Y.M., Xiao, F., Wang, D.S., Yan, M.Q., Bi, Z. (2013) Disinfection byproduct precursor removal by enhanced coagulation and their distribution in chemical fractions, *J. Environ. Sci.*, 25(11) 2207–2213.

39

GE Water Provides Portfolio of Nanotech Membrane-Based Solutions

Ralph Exton

GE Power & Water and Water & Process Technologies, Trevose, Pennsylvania, USA

CONTENTS

Water is one of the world's most valuable resources. It's the lifeblood of agriculture, essential to the generation of power, the preventer of disease, and the curator of pollution. Yet, it is under a constant threat because of climate change and resulting drought, explosive population growth, and mere waste. By 2025, an estimated two-thirds of the global population will not have access to clean water. Communities around the world simply cannot be fed or fueled without sustainable water supplies. The water crisis is very real—and the magnitude of this global challenge is growing rapidly.

Ninety-seven percent of the world's water is salt water; yet, despite continued advances in technology, desalination remains an energy-intensive, expensive process requiring costly infrastructure. Water reuse is becoming a necessity. However, traditionally separate wastewater collection and water supply systems, and aging water and wastewater infrastructure in the developed world are not designed to accommodate this pressing need. These challenges underscore the need for technological innovation to transform the way we treat, distribute, and reuse water.

For more than 100 years, General Electric (GE) has been at the forefront of technology, pushing the limits of science, and it is quickly advancing the industrial and municipal water treatment and recycling necessary to combat water scarcity challenges. Utilizing proven, reliable technologies from GE, forward-thinking industrial leaders are creating a more sustainable water supply, saving money, generating new revenue sources, and helping the environment by reclaiming and reusing wastewater. Municipalities are leveraging GE's experience and expertise to safely recover wastewater and generate new incomes by using this reclaimed water for agricultural and industrial applications. Furthermore, it also has the support of most Americans. According to a recent GE report, which surveyed 3000 consumers in the United States, China, and Singapore, >80% support "toilet-to-turf" uses—agricultural irrigation, landscaping, industrial processing and manufacturing, toilet flushing, and car washing. It is a significant finding, considering that 36 US states face water shortages in the coming year and 5.3 billion people worldwide will be vulnerable to water shortages in the next 10+ years.

The adoption of advanced technology, including nanotechnology—referencing the increasingly small pore size of membranes or filters for removal of contaminants—combined with forward-thinking government policy and education will play a major role in the evolution of water and wastewater treatment and reuse. This chapter discusses several of these advanced membrane solutions, explores specific GE innovations, and also discusses how industry collaboration and policy can play a stronger role in making water reuse a priority.

39.1 Nanofiltration

Nanofiltration (NF), the popularity of which has grown considerably during the last several years, offers opportunities to develop next-generation water supply systems. The NF membranes and their properties are quite diverse but can generally be described as having rejection characteristics that range from "loose" reverse osmosis (RO) (hyperfiltration) to "tight" ultrafiltration (UF). The uniqueness of these membranes is highlighted by their ability to selectively reject different dissolved salts, and have high rejection of low molecular weight dissolved components. NF membranes are mainly used to partially soften potable water, allowing some minerals to pass into the product water and thus increase the stability of the water and prevent it from being aggressive to distribution piping material. Additionally, NF membranes are finding increasing use for purifying industrial effluents and minimizing waste discharge (Figure 39.1).

FIGURE 39.1
GE's Nanofiltration Membranes' innovative spacer design and unique three-layer technology contribute to lower fouling and optimal cleaning properties in tough-to-treat applications.

39.2 Ultrafiltration

Many filtration technologies are both sustainable and energy efficient in removing suspended solids, turbidity, viruses, and pathogens from water. UF is a pressure-driven barrier to suspended solids, bacteria, viruses, endotoxins, and other pathogens, to produce water with very high purity and low silt density. It serves as a pretreatment for surface water, sea water, and biologically treated municipal effluent before RO and other membrane systems. GE's UF membranes are supplied with nominal pores size in the range of 20–40 nm and achieve >6-log (99.9999%) removal of *Cryptosporidium* and *Giardia lamblia*.

UF is also used in industry to separate suspended solids from solution. Industrial applications include power generation, food and beverage processing, pharmaceutical production, biotechnology, and semiconductor manufacturing (Figure 39.2).

39.2.1 Technology LEAP: GE's ZeeWeed Membrane Bioreactor

As the global membrane bioreactor (MBR) leader, with >5,000,000 m^3d (1.32 bgd) of ZeeWeed capacity, GE's MBR systems continue to combine proven UF technology with biological treatment for municipal, commercial, and industrial wastewater treatment and water reuse applications. Building on more than two decades of ZeeWeed MBR product innovation, the company recently introduced LEAPmbr to address our customers' key wastewater treatment challenges, and provide the low-energy and advanced performance solution demanded by the global wastewater treatment and reuse market—dramatically lowering energy costs, increasing productivity, and offering a flexible design that can fit within a smaller footprint than conventional MBR technology.

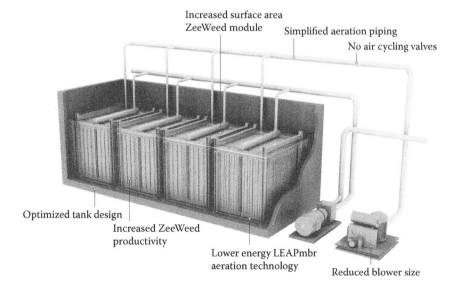

FIGURE 39.2
(See color insert.) Building on 25 years of GE's MBR experience, LEAPmbr delivers the most advanced MBR wastewater treatment solution to date, addressing pressing water quality and operational cost issues faced by owners of municipal, industrial, and residential water/wastewater treatment facilities worldwide. The technology dramatically lowers energy costs, increases productivity, and offers flexible designs that can fit within a smaller footprint than conventional MBR technology.

Innovations in the LEAPmbr product include the increased surface area of the ZeeWeed membrane module, simplified aeration piping, elimination of air cycling valves, optimization of the membrane tank design, increased productivity of the ZeeWeed modules, reduction in blower size, and introduction of LEAPmbr Aeration Technology. LEAPmbr Aeration Technology employs large bubbles, released at shorter cycle times as a single-phase flow from aerators below the modules. This produces a better scouring action, and the net result is a more efficient membrane air scour process. Extensive full-scale field validation has been performed at multiple plants to challenge the performance in a range of operating environments.

The increased productivity in LEAPmbr technology provides a reduction in membrane cassette requirements with fewer membrane modules necessary to treat a given capacity. This productivity boost comes without a compromise to quality and reliability as the ZeeWeed membrane remains at the core of this technology, continuing a legacy of being a highly reliable and high-performing membrane in the industry (Figure 39.2).

39.2.2 ZeeWeed Building Blocks

A full-scale ZeeWeed treatment facility comprises a given number of modular components: modules, cassettes, and trains. A module is the basic building block and the heart of a ZeeWeed system. Each module contains thousands of vertically strung membrane fibers that have millions of microscopic pores in each strand. Water is filtered by applying a slight vacuum to the end of each fiber, which draws the water through the tiny pores and into the fibers themselves. The pores form a physical barrier that allows clean water to pass through while blocking unwanted material such as suspended solids, bacteria, pathogens, and certain viruses.

Modules are joined together to form a cassette, which is the smallest operable unit of the filtration system (Figure 39.3). Each cassette can have a variety of module configurations depending on the amount of water that the cassette is required to treat. Multiple cassettes are joined to form what is known as a process train. The train is a production unit

FIGURE 39.3
ZeeWeed 500 membranes provide reliable wastewater treatment for industrial reuse and environmental compliance.

containing a number of cassettes immersed in a membrane tank. Multiple process trains form a ZeeWeed treatment plant.

39.2.3 Treatment Process

Feed water flows into the membrane tanks and treated water is drawn through the membranes during production by applying a vacuum to the inside of the membrane fibers. The water removed by permeation is replaced with feed water to maintain a constant level in the tank.

The particles that are rejected by the membrane pores remain in the process tank and are periodically removed by a process called a backwash. During a backwash, filtered water is reversed through the membrane fiber to dislodge any particles that may be physically lodged in the membrane fiber. Simultaneously, aeration scours any solids that are attached on the surface of the fibers.

39.2.4 Coupling ZeeWeed to Upstream Processes

ZeeWeed membrane systems can remove particles that are larger than the pores on the membrane fiber. Contaminants that exist in dissolved form, or are smaller than the pore size, can also be removed by the membranes if they are first transformed into insoluble species or larger particles. Treatment processes commonly coupled to ZeeWeed to accomplish such conversions include enhanced coagulation and oxidation. Typical examples you can achieve via this treatment process are found in Table 39.1.

TABLE 39.1

Treatment Results

Potable/Process Water	
Turbidity	<0.05 NTU
Bacteria	>4 log removal
Giardia cysts	>4 log removal
Cryptosporidium oocysts	>4 log removal
Virus rejection	>2.5 log
Total suspended solids	<1 mg/L
Total organic carbon	50%–90% removal[a]
Color	<5 PCU
Iron	<0.05 mg/L
Manganese	<0.02 mg/L
SDI	<1
Wastewater Effluent (as Part of an MBR Process)	
BOD	<2 mg/L
TSS	<2 mg/L
NH_3-N	<0.5 mg/L
TN	<3 mg/L[b]
TP	<0.05 mg/L[b]
Turbidity	<0.2 NTU
Fecal coliform	<10 CFU/100 mL
Transmissivity	>75%

[a] With coagulant addition.
[b] With appropriate design and/or chemical addition.

39.3 Hyperfiltration

RO, also known as hyperfiltration, represents the state of the art in water treatment technology. Developed in the late 1950s under US government funding as a method of desalinating sea water, today RO is one of the most convenient and thorough methods to filter water. It is used by most water bottling plants and by many industries that require ultra-refined water in manufacturing.

During the RO process, water molecules are forced through a 0.0001-micron semipermeable membrane by water pressure. Long sheets of the membrane are ingeniously sandwiched together and rolled up around a hollow central tube in a spiral fashion. This rolled-up configuration is commonly referred to as a spiral wound membrane or module. They are available in different sizes for processing different quantities of water. Typically, a module for home water treatment is as small as 2-in diameter and 10-in long, while one for industrial use may be 4-in diameter and 40-in long (Figure 39.4).

For the membrane to be usable, it must be in some type of container (membrane housing) so pressure can be maintained on its surface. It is this osmotic pressure that supplies the energy to force the water through the membrane, separating it from unwanted substances. Substances left behind are automatically diverted to a waste drain to prevent the build up in the system that occurs with conventional filtering devices. This is accomplished by using a part of the unprocessed water (feed water) to carry away the rejected substances to the drain, thus keeping the membrane clean. This is the reason why RO membranes can last so long and perform like new with minimum maintenance even after years of operation.

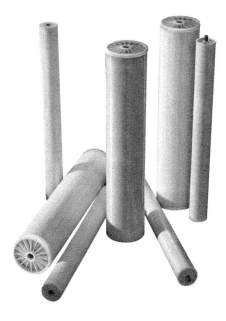

FIGURE 39.4
GE offers multiple RO configurations, including brackish, low-energy, low-fouling, high-temperature, seawater, and wastewater membranes, which offer high salt rejection and energy-savings models.

39.4 Electrodialysis Reversal and Electrodeionization

An alternative to RO, the capabilities of electrodialysis reversal (EDR)—a process commercialized by GE in the 1950s—are helping many municipal and industrial users in desalination applications where high water recovery is important for resource conservation and cost control. Both EDR and RO use semipermeable membranes to filter out dissolved ions from water. However, whereas RO uses the application of pressure to overtake osmotic pressure and force water through the membrane, EDR uses voltage potential and polarity reversal to remove unwanted constituents by pulling them through the membranes.

During the last 10–15 years, numerous advances in membrane and system technology have made EDR an especially attractive technology. Improved membrane technology now allows for one-step machine manufacture of ion-exchange membranes, reducing costs and lowering membrane resistivity. Moreover, new high-performance spacers (placed between the membranes) allow >90%–95% transport of contaminants like nitrates, arsenic, and perchlorate, speeding the process, reducing the number of membrane stacks required, and shrinking costs. Major improvements to the EDR system design have taken shape in the form of the next-generation GE EDR 2020 product line. This new design streamlines the process flow with simpler hydraulics and standardized components, substantially lowering the capital and operating costs of EDR demineralization.

In recent years, completely new EDR installations have taken place worldwide for a variety of applications. At the Ruth Fisher School in Arizona, an EDR system was installed with the objective of removing inorganic components from groundwater and reducing nitrates to meet US Environmental Protection Agency drinking water standards. The nitrate concentration in the feed is >130 mg/L; however, the EDR system produces water with extremely low total dissolved solids and nitrate concentrations. At the Bermuda Water Works, EDR is used to reduce hardness in the island's existing water supply. The brackish water lens under the island is contaminated from septic tank leach fields, making nitrate removal essential. The 600,000 gpd plant removes 86% of the nitrates while achieving 90% water recovery (Figure 39.5).

Electrodeionization (EDI) is a modification of conventional electrodialysis systems with ion-exchange resins installed in the electrodialysis stacks. The technology—similar to RO and EDR in that it uses semipermeable membranes with ion exchange media to provide a high efficiency demineralization process—is a continuous electrochemical process of water purification where ion-specific membranes, mixed bed resin and a DC voltage across them, replace the standard acid-caustic chemical regeneration process. An EDI cell, like GE's E-Cell*, consists of a series of thin chambers that alternately contain mixed bed resin for water purification and a concentrate water flow to carry away impurities.

While the fundamental concept is somewhat simple with the basic desalting unit being an electrodialysis dilute cell filled with mixed-bed ion-exchange resin, some complex chemical reactions take place within the resin-filled cell, helping to produce very high purity water. When flow enters the resin-filled diluting compartment of an EDI stack, several processes are set in motion. Strong ions are scavenged out of the feed stream by the mixed-bed resin. Under the influence of the strong DC field applied across the stack of components, charged ions are pulled off the resin and drawn toward the respective, oppositely charged electrodes, cathode or anode. As these strongly charged species, such as sodium and chloride, migrate toward the ion-exchange membrane, they are continuously removed and

FIGURE 39.5
EDR technology achieves high water recovery for water scarce areas. Because of the polarity reversal design, EDR for wastewater reuse is a self-cleaning, durable membrane system ideal for turbid wastewater. GE's wastewater EDR systems reclaim more than 20 million gallons per day (75,000 m³/day) of wastewater for other uses.

transferred into the adjacent concentrating compartments. As the strong ions are removed from the dilute process stream, the conductivity becomes quite low. This relatively pure water helps to set the stage for further chemical reactions. The electrical potential splits the water at the surface of the resin beds, producing hydrogen and hydroxyl ions, which act as continuous regenerating agents of the ion-exchange resin. These regenerated resins, in turn, act as microregions of high or low pH permitting ionization of neutral or weakly ionized aqueous species, such as carbon dioxide or silica. Once these species acquire a charge, they become subject to the influence of the strong DC field and are removed from the diluting compartment through the ion-exchange membranes. The membranes used in EDI stacks are flat sheet, homogeneous, ion-exchange membranes, which help to provide efficient ion transfer.

First introduced commercially about two decades ago, EDI is a polishing technology that requires RO as pretreatment; the combination of RO–EDI provides a continuous, chemical-free system that may achieve overall water recoveries of greater than 90%.

39.5 Role of Regulatory and Policy

Amidst the growing water scarcity and concern about the future availability and quality of water, Americans strongly support reusing water to help the United States drive economic competitiveness and protect the environment. The recent GE survey noted earlier also revealed Americans' overwhelming (>80%) belief that industry and government should play a stronger role in making water reuse a priority. However, there are major stumbling blocks to making this reality. Cost-effective technologies already exist to solve virtually all water challenges, providing more access to clean water, and enabling industries and municipalities to recycle resources. However, the focus needs to be placed on the human side of the equation. We have seen individual companies, associations, and municipalities step up; take the crisis into their own hands; and enforce high standards for water safety. However, we can no longer be independent water users.

Rather, we must continue to change today's approach to water management through collective action from organizations, governments, and individuals, in order to ensure a more prosperous and water-filled future. To shepherd in this new era of stewardship, industry and government must work closely together. Industrial and commercial manufacturers are the largest users of water worldwide. Obvious factors such as population growth, increasing regulation/industrialization, and weather patterns are cited as increasing concerns negatively affecting profitability and economic growth. These unpredictable variables make it difficult to accurately calculate and plan for profitable water risk management.

Companies are turning to inventive ideas and tools to help them assess and respond to increasing water risk globally. Programs such as the Aqueduct Alliance, a consortium of private and public sectors, nonprofits, and academia, bring influential entities together to provide "unprecedented levels of water risk information for business and government." The alliance evaluates social issues, including willingness and ability of governments to address water scarcity, leveraging new points of data for risk management.

This online tool allows companies, investors, financial intermediaries, and public sector decision makers measure, map, and respond to geographically explicit water risks and opportunities. It works by quantifying multiple drivers of water risk that can be easily and transparently aggregated and disaggregated. A stand-alone tool, Aqueduct, identifies hotspots of water risk that can constrain a company's access to water, increase its costs, and disrupt its operations. When overlaid with the locations of production facilities and/or key suppliers, these maps highlight the water risks a company is exposed to, while pinpointing where solutions are required.

Beyond Aqueduct, initiatives like WaterMatch—an innovative new, free online website dedicated to promoting beneficial reuse of municipal effluent for global industrial use—are a step forward in responsible water management, and in partnerships between industries and municipalities. In addition to putting fundamental environmental best practices into place, industrial companies can take advantage of such programs. For instance, as part of its water sustainability program, Coca Cola has developed a forum for government plants to evaluate water risk, providing reports on individual risk and educating companies on water consumption.

Governments must also take strong action to promote more water reuse and recycling. There are four well-documented, major types of policies currently in play worldwide to increase this initiative:

- **Education and outreach** efforts support recognition awards and certificate programs; information dissemination; and reporting of water consumption, discharge, and recycling data.
- **Removing barriers** by revising plumbing codes and alleviating stringent permit and inspection requirements for recycled water allows companies and communities to meet obligations that are otherwise difficult to attain.
- **Incentives** such as direct subsidies reduce government payments for the reintroduction of recovered water and provide regulatory relief for recycled water users through structured pricing mechanisms.
- **Mandates and regulations**, through the requirement of water recovery systems and recycled water for certain large-volume activities (e.g., irrigation), continue to reinforce these initiatives along with the strong need for government participation.

Government involvement through regulation and policy is necessary to solving the world's global water crisis. Coalitions that advocate for legislation focused on sustainable programs in support of water recycling are also critical. The WateReuse Association, as one example, supports the "Water Quality Protection and Job Creation Act of 2011" (H.R.3145), which aims to renew federal commitment to addressing the United States substantial needs for wastewater infrastructure by investing $13.8 billion over 5 years through the state revolving fund and other efforts to improve water quality.

Government and industry alike compete for the same finite water resources, and as this supply rapidly dwindles, working together has never been more important. Governments must promote greater water recycling through fundamental frameworks, policies, and incentives that define clearer regulations. Industries should also do more by implementing available recycling technologies—tapping into existing, underutilized water resources. Through more creative, public partnerships among industries, communities, municipalities, and government, we can protect and enhance performance and competitiveness, as well as the needs and interest of key stakeholders—the most important of which is our environment.

Future success depends on our ability to work together. New levels of efficiency must continually be enforced through education, conservation, governance, and incentive. Understanding the risks, and the opportunities, will place businesses and governments in a more competitive position to lead and succeed in a carbon- and water-constrained economy, and ultimately secure a future of water sustainability.

39.6 Nanotech Could Cut Oil Sands Carbon Emissions by 25%

Although the oil sands deposits in Canada contain as much as 173 billion barrels of economically viable oil, which is topped only by Saudi Arabia's reserves, only about 1 million barrels are produced each day. The problem is that extracting the oil can be a difficult process, and one that is more energy consuming than traditional crude oil production. With the oil industry in Canada expecting to increase production to 4 million barrels a day by 2020, there is a growing need to not only treat the massive amounts of water used in the extraction process but to also tackle the greenhouse gas emissions that are a result of the energy used while mining. Scientists at GE Global Research, the hub of technology development for all of GE's businesses, have partnered with the University of Alberta and Alberta Innovates—Technology Futures, Alberta's technology incubator, to use nanotechnology to create a new filtration system that tackles the twin problems of water treatment and carbon capture. The work has the potential to cut carbon emissions during the extraction process by 25%.

In oil sands, the oil is as thick as peanut butter and does not flow. Called "bitumen," it must be processed to take on the characteristics of light crude oil; it can either be mined at the surface, which only works for 20% of the operations in Canada, or by drilling deep underground wells. In drilling, steam or other solvents must be injected so the thick oil can be pumped to the surface. It is that process—turning the water to steam, pumping it in and out, and then upgrading the material into light crude—that uses a great deal of energy and water.

At the heart of the new nanotechnology research are naturally occurring zeolites identified by the University of Alberta. These materials are rocks with molecularly sized pores that allow small molecules to enter while excluding larger molecules. Zeolites are widely used in the chemical industry as catalysts, and this project seeks to form these materials into membranes that can be used for high-temperature gas separation and for filtration of contaminated water.

Index

Page numbers followed by f and t indicate figures and tables, respectively.

Printed and bound by CPI Group (UK) Ltd, Croydon, CR0 4YY

18/10/2024

01776254-0011